Digital Electronics with VHDL

William Kleitz

Tompkins Cortland Community College

PEARSON

Prentice Hall

Upper Saddle River, New Jersey
Columbus, Ohio

Library of Congress Cataloging-in-Publication Data
Kleitz, William.
 Digital electronics with VHDL/William Kleitz.
  p. cm.
 ISBN 0-13-110080-7
 1. Digital electronics. 2. Electronic circuit design—Data processing. 3. VHDL (Computer
hardware description language) I. Title.

 TK7868.D5K555 2004
 621.381—dc21   2003040506

Editor in Chief: Stephen Helba
Acquisitions Editor: Dennis Williams
Development Editor: Kate Linsner
Production Editor: Christine M. Buckendahl
Production Coordination: Carlisle Publishers Services
Design Coordinator: Diane Ernsberger
Cover Designer: Keith Van Norman
Cover art: Digital Vision
Production Manager: Pat Tonneman
Marketing Manager: Ben Leonard

This book was set in Times by Carlisle Communications, Ltd., and was printed and bound by Courier
Kendallville, Inc. The cover was printed by Phoenix Color Corp.

MultiSIM® is a trademark of Electronics Workbench.

Altera is a trademark and service mark of Altera Corporation in the United States and other countries. Altera products are the
intellectual property of Altera Corporation and are protected by copyright laws and one or more U.S. and foreign patents and
patent applications.

Material based on or adapted from figures and text owned by Xilinx, Inc., courtesy of Xilinx, Inc. © 1999–2001. All rights reserved.

Pearson Education Ltd.
Pearson Education Singapore Pte. Ltd.
Pearson Education Canada, Ltd.
Pearson Education—Japan

Pearson Education Australia Pty. Limited
Pearson Education North Asia Ltd.
Pearson Educación de Mexico, S. A. de C. V.
Pearson Education Malaysia Pte. Ltd.

10 9 8 7 6 5 4 3 2 1
ISBN: 0-13-110080-7

To Patty, Shirelle, and Hayley

Contents

chapter **3**

Basic Logic Gates   58

chapter **4**

Programmable Logic Devices: CPLDs and FPGAs with VHDL Design   104

chapter **5**

Boolean Algebra and Reduction Techniques   144

chapter **6**

Exclusive-OR and Exclusive-NOR Gates   222

chapter **7**

Arithmetic Operations and Circuits   244

chapter **8**

Code Converters, Multiplexers, and Demultiplexers   292

chapter **9**

Logic Families and Their Characteristics 360

chapter **10**

Flip-Flops and Registers 404

chapter **11**

Practical Considerations for Digital Design   458

chapter **12**

Counter Circuits and VHDL State Machines   504

chapter **13**

Shift Registers   592

chapter **14**

Multivibrators and the 555 Timer   644

chapter **15**

Interfacing to the Analog World   678

chapter **16**

Semiconductor, Magnetic, and Optical Memory   714

chapter **17**

Microprocessor Fundamentals   754

chapter **18**

The 8051 Microcontroller   776

Preface

Digital Electronics with VHDL provides the fundamentals of digital circuitry to students in engineering and technology curricula, supplying all of the information needed for an entry level student to be brought up to speed in this emerging field. As an instructor and author, my goals in producing this book have been to present practical examples, make it easy to read, and provide everything necessary for motivated students to teach themselves this new subject matter. I trust that readers will find these goals successfully achieved.

The digital circuits are introduced using fixed-function 7400 ICs and evolve into CPLDs (Complex Programmable Logic Devices) programmed with VHDL (VHSIC Hardware Description Language). Coverage begins with the basic logic gates used to perform arithmetic operations and proceeds up through sequential logic and memory circuits used to interface with modern PCs.

Digital electronic ICs (integrated circuits) and CPLDs are the "brains" behind common microprocessor-based systems like those cfound in automobiles, personal computers, and automated factory control systems. The most exciting recent development in this field is that students can now design, simulate, and implement their circuits using a programming language called VHDL instead of wiring individual gates and devices to achieve the required function.

Each topic in this text consistently follows a specific sequence of steps, making the transition from problem definition, to practical example, to VHDL definitions for CPLD implementation. To accomplish this, the text first introduces the theory of operation of the digital logic and then implements the design in integrated circuit form. Once the fixed-function IC logic is thoroughly explained, the next step is to implement the design as a graphic design file using CPLD software and then to implement it using the hardware descriptive language VHDL. Several examples are used to bolster students' understanding of the subject before moving on to system-level design and troubleshooting applications of the logic. Over the years, this step-by-step method has proven to be the most effective method to build the fundamental understanding of digital electronics before proceeding to implement the logic design in VHDL.

The Altera MAX+PLUS II software included on the textbook CD allows students to either graphically design their circuits by drawing the logic (using logic gates or 7400 macrofunctions) or use VHDL to define their logic. The design can then be simulated on a PC before using the same software to download the logic to a CPLD on one of the commercially available CPLD programmer boards.

Over 1000 four-color illustrations are used to exemplify the operation of complex circuit operations. Most of the illustrations contain annotations describing the inputs and outputs, and many have circuit operational notes. The VHDL program listings are enriched with many annotations, providing a means for students to teach themselves the intricacies of the language.

Each chapter begins with an outline, objectives, and introduction, and concludes with a summary, glossary, design and troubleshooting problems, schematic interpretation problems, Electronics Workbench MultiSIM® problems, CPLD problems, and answers to the review questions.

Prerequisites

Although not mandatory, it is helpful if students using this text have an understanding of, or are concurrently enrolled in, a basic electricity course. Otherwise, all of the fundamental concepts of basic electricity required to complete this text are presented in Appendix F.

Chapter Organization

Basically, the text can be divided into two halves: Chapters 1 to 8 cover basic digital logic and combinational logic, and Chapters 9 to 18 cover sequential logic and digital systems. Chapters 1 and 2 provide the procedures for converting between the various number systems and introduce the student to the electronic signals and switches used in digital circuitry. Chapter 3 covers the basic logic gates and introduces the student to timing analysis and troubleshooting techniques. Chapter 4 explains how to implement designs using CPLDs. Chapter 5 shows how several of the basic gates can be connected together to form combinational logic. Boolean algebra, De Morgan's theorem, VHDL programming, and Karnaugh mapping are used to reduce the logic to its simplest form. Chapters 6, 7, and 8 discuss combinational logic used to provide more advanced functions like parity checking, arithmetic operations, and code converting.

The second half of this book begins with a discussion of the operating characteristics and specifications of the TTL and CMOS logic families (Chapter 9). Chapter 10 introduces flip-flops and the concept of sequential timing analysis. Chapter 11 makes students aware of the practical limitations of digital ICs and some common circuits that are used in later chapters to facilitate the use of medium-scale ICs. Chapters 12 and 13 expose students to the operation and use of several common medium-scale ICs and their VHDL equivalents used to implement counter and shift register systems. Chapter 14 deals with oscillator and timing circuits built with digital ICs and with the 555 timer IC. Chapter 15 teaches the theory behind analog and digital conversion schemes and the practical implementation of ADC and DAC IC converters. Chapter 16 covers semiconductor, magnetic, and optical memory as it applies to PCs and microprocessor systems. Chapter 17 introduces microprocessor hardware and software to form a bridge between digital electronics and a follow-up course in microprocessors. Chapter 18 provides a working knowledge of one of today's most popular microcontrollers, the 8051. The book concludes with several appendices used to supplement the chapter material.

Features

Margin Annotation Icons

Several annotations are given in the page margins throughout the text. These are intended to highlight particular points that are made on the page. They can be used as the catalyst to develop a rapport between the instructor and the students and to initiate team discussions among the students. Four different icons distinguish the annotations.

Common Misconception: These annotations point out areas of digital electronics that have typically been stumbling blocks for students and need careful attention. Pointing out these potential problem areas helps students avoid making that mistake.

Team Discussion: These annotations are questions that initiate a discussion about a particular topic. The instructor can use them to develop cooperative learning by encouraging student interaction.

Helpful Hint: These annotations offer suggestions for circuit analysis and highlight critical topics presented in that area of the text. Students use these tips to gain insights regarding important concepts.

Inside Your PC: These annotations are used to illustrate practical applications of the theory in that section as it is applied inside a modern PC. This will help students understand many of the terms used to describe the features that define the capability of a PC.

Basic Problem Sets

A key part of learning any technical subject matter is for students to have practice solving problems of varying difficulty. The problems at the end of each chapter are grouped together by section number. Within each section are several basic problems designed to get students to solve a problem using the fundamental information presented in the chapter. In addition to the basic problems, there are three other problem types:

D (Design) Problems designated with the letter D ask students to modify an existing circuit or to design an original circuit to perform a specific task. This type of exercise stimulates creative thinking and instills a feeling of accomplishment on successful completion of a circuit design.

T (Troubleshooting) Problems designated with the letter T present students with a malfunctioning circuit to be diagnosed or ask for a procedure to follow to test for proper circuit operation. This develops analytical skills and prepares students for troubleshooting tasks that would typically be faced on the job.

C (Challenging) Problems designated with the letter C are the most challenging to solve. They require a thorough understanding of the material covered and go a step beyond by requiring students to develop some of their own strategies to solve a problem that is different from the examples presented in the chapter. This type of problem also expands analytical skills and develops critical thinking techniques.

MultiSIM Problems

Electronics Workbench®/MultiSIM® is a software simulation tool that is used to reinforce the theory presented in each chapter. It provides an accurate simulation of digital and analog circuit operations along with a simulation of instruments used by technicians to measure IC, component, and circuit characteristics. With this software, students have the ability to build and test most of the circuits presented in this text.

Several MultiSIM problems are included at the end of each chapter. One of the CDs included in the back of this textbook contains the textbook edition of the MultiSIM® software and all of the circuit files and instructions needed to solve each problem. There are three types of problems:

(1) *Circuit interaction problems* require students to change input values and take measurements at the outputs to verify circuit operation.

(2) *Design problems* require students to design, or modify, a circuit to perform a particular task.

(3) *Troubleshooting problems* require students to find and fix the fault that exists in the circuit that is given.

Schematic Interpretation Problems

These problems are designed to give students experience interpreting circuits and ICs in complete system schematic diagrams. Students are asked to identify certain components in the diagram, describe their operation, modify circuit elements, and design new circuit interfaces. This provides experience working with real-world, large-scale schematics like the ones that students will see on the job.

CPLD Problems

Complex Programmable Logic Device (CPLD) problems are included at the end of several chapters. Designing digital logic with CPLDs is becoming very popular in situations where high complexity and programmability are important. The CPLD problems use the Altera MAX+PLUS II software included on the CD to solve designs that were previously implemented using 7400-series ICs. Students are asked to solve the design using a graphic design approach as well as a VHDL solution. After compiling the design, students are then asked to perform a software simulation of the circuit before downloading the implementation to an actual CPLD.

VHDL Programming

The VHDL programming language has become a very important tool in the design of digital systems. Throughout the text, digital design solutions are first done with fixed-function 7400-series logic gates, and then the same solution is completed using the VHDL hardware description language. It is important for today's technician to be able to read and modify VHDL programs as well as in some cases to write original programs to implement intermediate-level digital circuits.

Microprocessor Fundamentals

The "brains" behind most high-level digital systems is the microprocessor. The basic understanding of microprocessor software and hardware is imperative for the technician to design and troubleshoot digital systems. Chapter 17 provides the fundamentals of microprocessor software and hardware. Chapter 18 covers one of today's workhorses, the 8051. Its internal architecture, hardware interfacing, and software programming are introduced and then demonstrated by solving several complete data acquisition applications.

To the Instructor

Teaching and Learning Digital Electronics: I would like to share with you some teaching strategies that I've developed over the past 25 years of teaching digital electronics. Needless to say, students have become very excited about learning digital electronics because of the increasing popularity of the digital computer and the expanding job opportunities for digital technicians and engineers. Students are also attracted to the subject area because of the availability of inexpensive digital ICs and CPLDs, which have enabled them to construct useful digital circuits in the lab or at home at a minimal cost.

Laboratory Experimentation: Giving the students the opportunity for hands-on laboratory experience is a very useful component of any digital course. An important feature of this text is that there is enough information given for any of the circuits so that they can be built and tested in the lab and that you can be certain they will give the same response as shown in the text.

Altera MAX+PLUS II Software: Altera Corporation, a leading supplier of CPLDs and FPGAs, has supplied the design, simulation, and programming software (MAX+PLUS II) on the enclosed CD. It is suggested that each school enroll in the Altera University Program at www.altera.com. Enrollment ensures that the college will be kept up to date on the latest products and software updates.

CPLD Programming Board: The final step in any CPLD design process is to implement the logic design in an actual CPLD by programming it with the supplied software. This lab experience is achieved by downloading the design created by MAX+PLUS II to a CPLD programming board containing a CPLD. The two programming boards that are recommended for this exercise are the UP2 (or UP1) University Design Laboratory Kit sold by Altera (www.altera.com) or the RSR PLDT-2 sold by Electronics Express (www.elexp.com). One of those programming boards is required for each lab station.

Student Projects: I always encourage students to build some of the fundamental building-block circuits that are presented in this text. The circuits that I recommend are the 5-V power supply in Figure 11–43, the 60-Hz pulse generator in Figure 11–44, the cross-NAND switch debouncer in Figure 11–40, and the seven-segment LED display in Figure 12–48. Having these circuits provides a starting point for students to test many of the other circuits in the text at their own pace, at home.

Team Discussions: As early as possible in the course, I take advantage of the *Team Discussion* margin annotations. These are cooperative learning exercises where the students are encouraged to form teams, discuss the problem, and present their conclusions to the class. These activities give them a sense of team cooperation and create a student network connection that will carry on throughout the rest of their studies.

Circuit Illustrations: Almost every topic in the text has an illustration associated with it. Because of the extensive art program, I normally lecture directly from illustration to illustration. To do this, I have an overhead transparency made of every figure in the text. All figures and tables in the text are also available in PowerPoint® format for instructors adopting the text.

Testing: Rather than let a long period of time elapse between tests, I try to give a half-hour quiz each week. Aside from the daily homework, this forces the students to study at least once per week. I also believe that it is appropriate to allow them to have a formula sheet for the quiz or test (along with a TTL or CMOS databook). They can write anything they want on the sheet. Making up the formula sheet is a good way for them to study and eliminates a lot of routine memorization that they would not normally have to do on the job.

The Learning Process: Students' knowledge is generally developed by learning the theory and the tools required to understand a particular topic, working through the examples provided, answering the review questions at the end of each section, and finally, solving the problems at the end of the chapter. I always encourage the students to re-work the solutions given in the examples without looking at the solutions in the book until they are done. This gives them extra practice and a secure feeling of knowing the detailed solution is right at their disposal.

Unique Learning Tools

Special features included in this textbook to enhance the learning and comprehension process are:

- Two CD-ROMs packaged with each copy of this text containing:
  1. (a) Electronics Workbench®/MultiSIM® software (textbook edition) with
     (b) MultiSIM circuit data files.
     (c) Altera MAX+PLUS II graphic and VHDL design software.
     (d) Solutions to in-text Altera CPLD examples.
  2. Texas Instruments' fixed-function data sheets.

- Annotated and completely explained VHDL solutions to common digital circuits.

- Step-by-step tutorial for using MAX+PLUS II software for design and CPLD programming.

- Over 100 MultiSIM exercises aimed at enhancing students' understanding of fundamental concepts, troubleshooting strategies, and circuit design procedures.

- Over 200 examples that are worked out step by step to clarify problems that are normally stumbling blocks.

- Over 1000 detailed illustrations with annotations that give visual explanations and serve as the basis for all discussions. Color operational notes are included on several of the illustrations to describe the operation of a particular part of the figure.

- A four-color format that provides a visual organization to the various parts of each section.

- More than 1000 problems and questions to enhance problem-solving skills. A complete range of problems, from straightforward to very challenging, is included.

- Troubleshooting applications and problems throughout the text to teach testing and debugging procedures.

- References to manufacturers' data sheets throughout the book that provide a valuable experience with real-world problem solving.

- Timing waveforms that illustrate the timing analysis techniques used in industry and to give a graphical picture of the sequential operations of digital ICs and CPLDs.

- Several tables of commercially used ICs, a source for state-of-the-art circuit design.

- Several photographs to illustrate specific devices and circuits discussed in the text.

- Performance-based objectives at the beginning of each chapter that outline the goals to be achieved.

- Review questions summarizing each section, with answers at end of each chapter, to see that each learning objective is met.

- A summary at the end of each chapter reviewing the topics covered.

- A glossary at the end of each chapter that summarizes the terminology just presented.

- A supplementary index of ICs that provides a quick way to locate a particular IC by number.

Extensive Supplements Package

In addition to the two CDs previously described which are included with this text, an extensive package of supplementary material is available to aid in the teaching and learning process.

- **Instructor's Resource Manual** (ISBN 0-13-111231-7), with solutions and answers to in-text problems as well as solutions to the laboratory manuals.

- **PowerPoint® Transparencies** (ISBN 0-13-111232-5), which include all figures and tables from the text, in addition to Lecture Notes prepared by Chris Konrad (Iowa Western Community College).

- **Two laboratory manuals,** providing hands-on laboratory experience to reinforce the material presented in the textbook:

    (1) (Altera CPLDs): *Laboratory Manual to accompany Digital Electronics With VHDL,* by Steve Waterman (DeVry University), (ISBN 0-13-113756-5).

    (2) (Fixed-Function Logic): *Laboratory Manual (A Troubleshooting Approach) to Accompany Digital Electronics,* by Michael Wiesner and Vance Venable (both of Heald College) (ISBN 0-13-092291-9).

- **PH TestGen** (ISBN 0-13-111222-8) by Les Taylor (DeVry University), Mike Durren (Lake Michigan College), and Michael Agina (Texas Southern University), with over 1000 test questions that can be used to develop customized quizzes, tests, or final exams.

- **Companion Website,** a student resource containing additional multiple-choice questions by Kenneth Lawell (Brown Institute) and Obrin Griffin (Ivy Tech State College) as well as other textbook-related links, found at http://www.prenhall.com/kleitz.

To The Student: Getting the Most from This Textbook

Digital electronics is the foundation of computers and microprocessor-based systems found in automobiles, industrial control systems, and home entertainment systems. You are beginning your study of digital electronics at a good time. Technological advances made during the past 25 years have provided us with ICs that can perform complex tasks with a minimum amount of abstract theory and complicated circuitry. Before you are finished studying this book, you'll be developing exciting designs that you've always wondered about but can now experience firsthand.

The study of digital electronics also provides the prerequisite background for your future studies in microprocessors and microcomputer interfacing. It provides the job skills for you to become a computer service technician, production test technician, or digital design technician, or to fill many other positions related to computer and microprocessor-based systems.

This book is written as a learning tool, not just as a reference. The concept and theory of each topic is presented first. Then, an explanation of its operation is given. This is followed by several worked-out examples and, in some cases, a system design application. The review questions at the end of each chapter will force you to dig back into the reading to see that you have met the learning objectives given at the beginning of the chapter. The problems at the end of each chapter will require more analytical reasoning, but the procedures for their solutions were already given to you in the examples. One good way to prepare for homework problems and tests is to cover the solutions to the examples and try to work them out yourself. If you get stuck, you've got the answer and an explanation for the answer right there.

I also suggest that you take advantage of your MultiSIM® and MAX+PLUS II software. The more practice you get, the easier the course will be. I wish you the best of luck in your studies and future employment.

Professor Bill Kleitz
Tompkins Cortland Community College

Acknowledgments

Thanks are due to the following professors for reviewing my work in the past and providing valuable suggestions.

Dale A. Amick, High Tech Institute

Scott Boldwyn, Missouri Technical School

Henry Baskerville, Heald Institute

Darrell Boucher, Jr., High Plains Institute of Technology

Steven R. Coe, DeVry University

Terry Collett, Lake Michigan College

Mike Durren, Lake Michigan College

Doug Fuller, Humber College

Julio R. Garcia, San Jose State University

Norman Grossman, DeVry University

Anthony Hearn, Community College of Philadelphia

Donald P. Hill, RETS Electronic Institute

Nazar Karzay, Ivy Tech State College

Charles L. Laye, United Electronics Institute

David Longobardi, Antelope Valley College

William Mack, Harrisburg Area Community College

Robert E. Martin, Northern Virginia Community College

Lew D. Mathias, Ivy Tech State College

Serge Mnatzakanian, Computer Learning Center

Chrys A. Panayiotou, Brevard Community College

Richard Parett, ITT Technical Institute

Bob Redler, Southeast Community College

Dr. Lee Rosenthal, Fairleigh Dickinson University

Ron Scott, Northeastern University

Edward Small, Southeast College of Technology

Ron L. Syth, ITT Technical Institute

Edward Troyan, LeHigh Carbon Community College

Vance Venable, Heald Institute of Technology

Donnie L. Williams, Murray State College

Ken Wilson, San Diego City College

I extend a special thank you to Patty Alessi, who has influenced my writing style by helping me explore new, effective teaching strategies. I am grateful to Scott Wager, Mitch Wiedemann, and Bill Sundell of Tompkins Cortland Community College; Kevin White of Bob Dean Corporation; Dick Quaif of DQ Systems; Alan Szary and Paul Constantini of Precision Filters, Inc.; and Jim Delsignore of Axiohm Corporation for their technical assistance. I am also appreciative of Interactive Image Technology, Texas Instruments, Inc., and Altera Corporation for their software products, and Philips Corporation for providing many of the data sheets used in this book. Also, thanks to my students of the past 25 years who have helped me to develop better teaching strategies and have provided suggestions for clarifying several of the explanations contained in this book, and to Dennis Williams, Kate Linsner, and Christine Buckendahl from the marketing, editorial, and production staff at Prentice Hall.

1 Number Systems and Codes

OUTLINE

OBJECTIVES

Upon completion of this chapter, you should be able to:

- Determine the weighting factor for each digit position in the decimal, binary, octal, and hexadecimal numbering systems.
- Convert any number in one of the four number systems (decimal, binary, octal, and hexadecimal) to its equivalent value in any of the remaining three numbering systems.
- Describe the format and use of binary-coded decimal (BCD) numbers.
- Determine the ASCII code for any alphanumeric data by using the ASCII code translation table.

INTRODUCTION

Digital circuitry is the foundation of digital computers and many automated control systems. In a modern home, digital circuitry controls the appliances, alarm systems, and heating systems. Under the control of digital circuitry and microprocessors, newer automobiles have added safety features, are more energy efficient, and are easier to diagnose and correct when malfunctions arise.

Other uses of digital circuitry include the areas of automated machine control, energy monitoring and control, inventory management, medical electronics, and music. For example, the numerically controlled (NC) milling machine can be programmed by a production engineer to mill a piece of stock material to prespecified dimensions with very accurate repeatability, within 0.01% accuracy. Another use is energy monitoring and control. With the high cost of energy, it is very important for large industrial and commercial users to monitor the energy flows within their buildings. Effective control of heating, ventilating, and air-conditioning can reduce energy bills significantly. More and more grocery stores are using the universal product code (UPC) to check out and total the sale of grocery orders as well as to control inventory and replenish stock automatically. The area of medical electronics uses digital thermometers, life-support systems, and monitors. We have also seen more use of digital electronics in the reproduction of music. Digital reproduction is less susceptible to electrostatic noise and therefore can reproduce music with greater fidelity.

Digital electronics evolved from the principle that transistor circuitry could easily be fabricated and designed to output one of two voltage levels based on the levels placed at its inputs. The two distinct levels (usually + 5 volts [V] and 0 V) are HIGH and LOW and can be represented by 1 and 0.

The binary numbering system is made up of only 1's and 0's and is therefore used extensively in digital electronics. The other numbering systems and codes covered in this chapter represent groups of binary digits and therefore are also widely used.

1–1   Digital Versus Analog

Digital systems operate on discrete digits that represent numbers, letters, or symbols. They deal strictly with ON and OFF states, which we can represent by 0's and 1's. *Analog* systems measure and respond to continuously varying electrical or physical magnitudes. Analog devices are integrated electronically into systems to continuously monitor and control such quantities as temperature, pressure, velocity, and position and to provide automated control based on the levels of these quantities. Figure 1–1 shows some examples of digital and analog quantities.

Review Questions *

1–1. List three examples of *analog* quantities.

1–2. Why do computer systems deal with *digital* quantities instead of *analog* quantities?

1–2   Digital Representations of Analog Quantities

Most naturally occurring physical quantities in our world are **analog** in nature. An analog signal is a continuously variable electrical or physical quantity. Think about a mercury-filled tube thermometer; as the temperature rises, the mercury expands in analog fashion

*Answers to Review Questions are found at the end of each chapter.

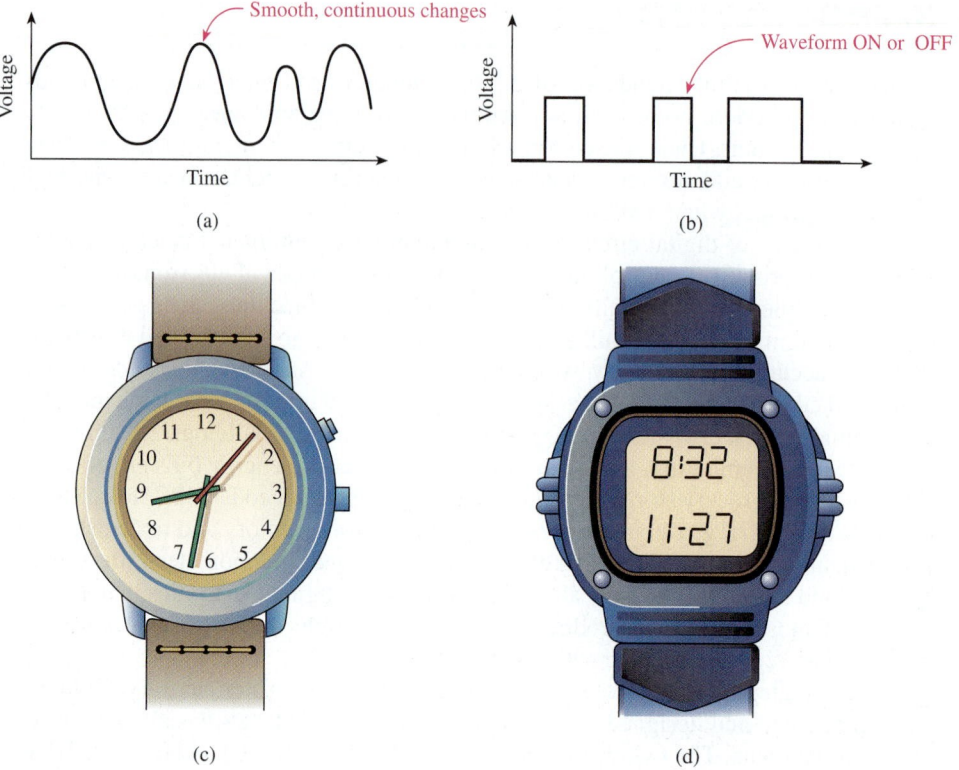

Figure 1–1 Analog versus digital: (a) analog waveform; (b) digital waveform; (c) analog watch; (d) digital watch.

and makes a smooth, continuous motion relative to a scale measured in degrees. A baseball player swings a bat in an analog motion. The velocity and force with which a musician strikes a piano key are analog in nature. Even the resulting vibration of the piano string is an analog, sinusoidal vibration.

So why do we need to use **digital** representations in a world that is naturally analog? The answer is that if we want an electronic machine to interpret, communicate, process, and store analog information, it is much easier for the machine to handle it if we first convert the information to a digital format. A digital value is represented by a combination of ON and OFF voltage levels that are written as a string of 1's and 0's.

For example, an analog thermometer that registers 72 degrees can be represented in a digital circuit as a series of ON and OFF voltage levels. (We'll learn later that the number 72 converted to digital levels is 0100 1000.) The convenient feature of using ON/OFF voltage levels is that the circuitry used to generate, manipulate, and store them is very simple. Instead of dealing with the infinite span and intervals of analog voltage levels, all we need to use is ON or OFF voltages (usually $+5$ V = ON and 0 V = OFF).

A good example of the use of a digital representation of an analog quantity is the audio recording of music. Compact disks (CDs) and digital audio tapes (DATs) are becoming commonplace and are proving to be superior means of recording and playing back music. Musical instruments and the human voice produce analog signals, and the human ear naturally responds to analog signals. So, where does the digital format fit in? Although the process requires what appears to be extra work, the recording industries convert analog signals to a digital format and then store the information on a CD or DAT. The CD or DAT player then converts the digital levels back to their corresponding analog signals before playing them back for the human ear.

To accurately represent a complex musical signal as a digital string (a series of 1's and 0's), several samples of an analog signal must be taken, as shown in

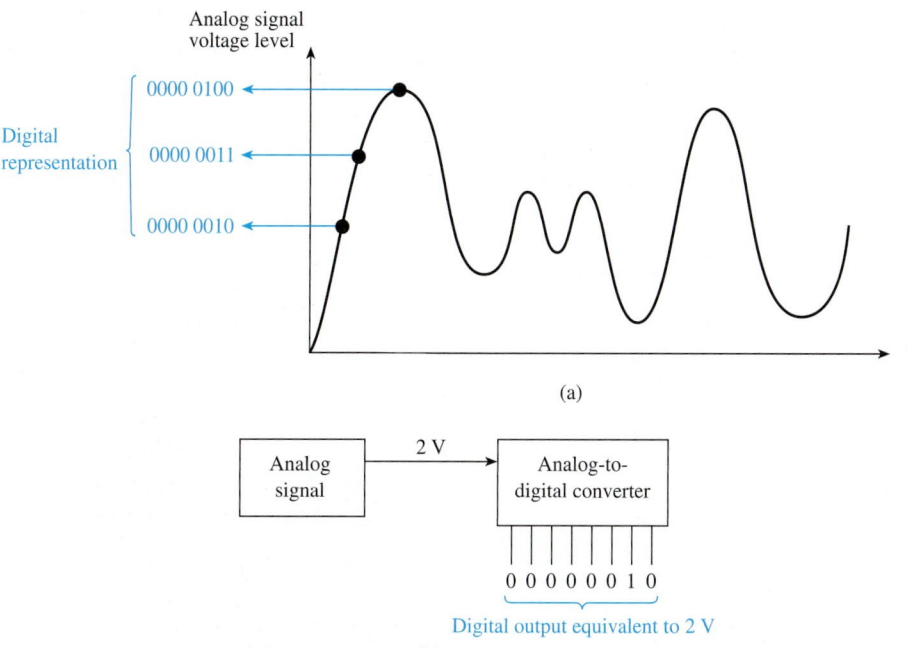

(a)

Helpful Hint

One of the more interesting uses of analog-to-digital (A-to-D) and digital-to-analog (D-to-A) conversion is in CD audio systems. Also, several A-to-D and D-to-A examples are given in Chapter 15.

Inside Your PC

A typical 4-minute song requires as many as 300 million ON/OFF digital levels (bits) to be represented accurately. To be transmitted efficiently over the Internet, data compression schemes such as the MP3 standard are employed to reduce the number of bits tenfold. (For information about specifications, visit the MP3 Web site listed in Appendix A.)

(b)

Figure 1–2   (a) Digital representation of three data points on an analog waveform; (b) converting a 2-V analog voltage into a digital output string.

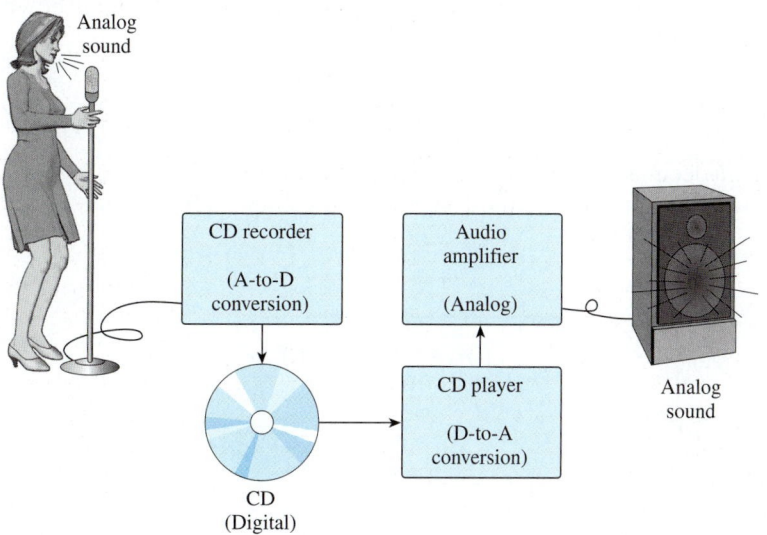

Inside Your PC

The CD player uses the optics of a laser beam to look for pits or non-pits on the CD as it spins beneath it. These pits, which were burnt into the CD by the CD recorder, represent the 1's and 0's of the digital information the player needs to recreate the original data. A CD contains up to 650 million bytes of digital 1's and 0's (1 byte = 8 bits).

Another optical storage medium is the digital versatile disk (DVD). A DVD is much denser than a CD. It can hold up to 17 billion bytes of data!

Figure 1–3   The process of converting analog sound to digital and then back to analog.

Figure 1–2(a). The first conversion illustrated is at a point on the rising portion of the analog signal. At that point, the analog voltage is 2 V. Two volts are converted to the digital string 0000 0010, as shown in Figure 1–2(b). The next conversion is taken as the analog signal in Figure 1–2(a) is still rising, and the third is taken at its highest level. This process will continue throughout the entire piece of music to be recorded. To play back the music, the process is reversed. Digital-to-analog conversions are made to recreate the original analog signal (see Figure 1–3). If a high enough number of samples are taken of the original analog signal, an almost exact reproduction of the original music can be made.

It certainly is extra work, but digital recordings have virtually eliminated problems such as electrostatic noise and the magnetic tape hiss associated with earlier methods of audio recording. These problems have been eradicated because, when imperfections are introduced to a digital signal, the slight variation in the digital level does not change an ON level to an OFF level, whereas a slight change in an analog level is easily picked up by the human ear as shown in Figure 1–4.

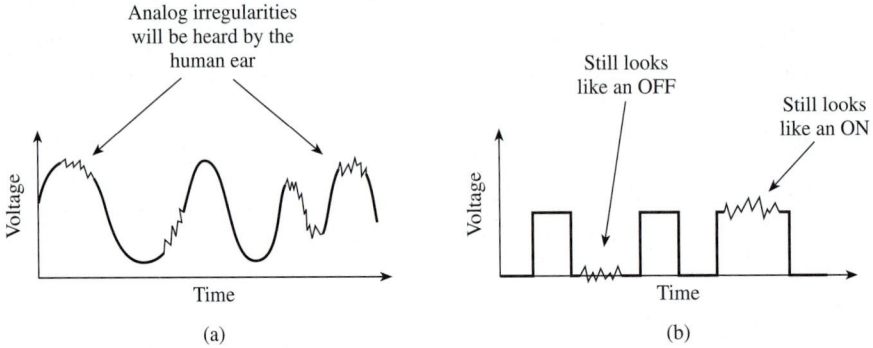

Figure 1–4 Adding unwanted electrostatic noise to: (a) an analog waveform; (b) a digital waveform.

1–3 Decimal Numbering System (Base 10)

In the **decimal** numbering system, each position contains 10 different possible digits. These digits are 0, 1, 2, 3, 4, 5, 6, 7, 8, and 9. Each position in a multidigit number will have a weighting factor based on a power of 10.

EXAMPLE 1–1

In a four-digit decimal number, the least significant position (rightmost) has a weighting factor of 10^0; the most significant position (leftmost) has a weighting factor of 10^3:

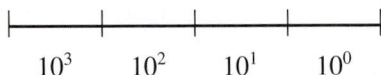

$$10^3 \quad 10^2 \quad 10^1 \quad 10^0$$

where $10^3 = 1000$
$10^2 = 100$
$10^1 = 10$
$10^0 = 1$

To evaluate the decimal number 4623, the digit in each position is multiplied by the appropriate weighting factor:

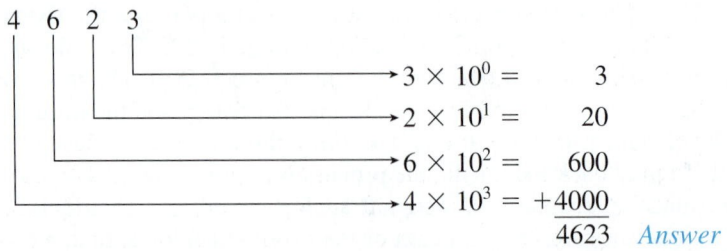

$$
\begin{aligned}
3 \times 10^0 &= 3 \\
2 \times 10^1 &= 20 \\
6 \times 10^2 &= 600 \\
4 \times 10^3 &= +4000 \\
\hline
&4623 \quad \textit{Answer}
\end{aligned}
$$

Example 1–1 illustrates the procedure used to convert from some number system to its decimal (base 10) equivalent. (In the example, we converted a base 10 number to a base 10 answer.) Now let's look at base 2 (binary), base 8 (octal), and base 16 (hexadecimal).

1–4   Binary Numbering System (Base 2)

Digital electronics use the **binary** numbering system because it uses only the digits 0 and 1, which can be represented simply in a digital system by two distinct voltage levels, such as $+5 \text{ V} = 1$ and $0 \text{ V} = 0$.

The weighting factors for binary positions are the powers of 2 shown in Table 1–1.

TABLE 1–1	Powers-of-2 Binary Weighting Factors

128	64	32	16	8	4	2	1
2^7	2^6	2^5	2^4	2^3	2^2	2^1	2^0

$$2^0 = 1$$
$$2^1 = 2$$
$$2^2 = 4$$
$$2^3 = 8$$
$$2^4 = 16$$
$$2^5 = 32$$
$$2^6 = 64$$
$$2^7 = 128$$

EXAMPLE 1–2

Convert the binary number 01010110_2 to decimal. (Notice the subscript 2 used to indicate that 01010110 is a base 2 number. A capital letter B can also be used, i.e., 01010110B.)

Solution: Multiply each binary digit by the appropriate weight factor and total the results.

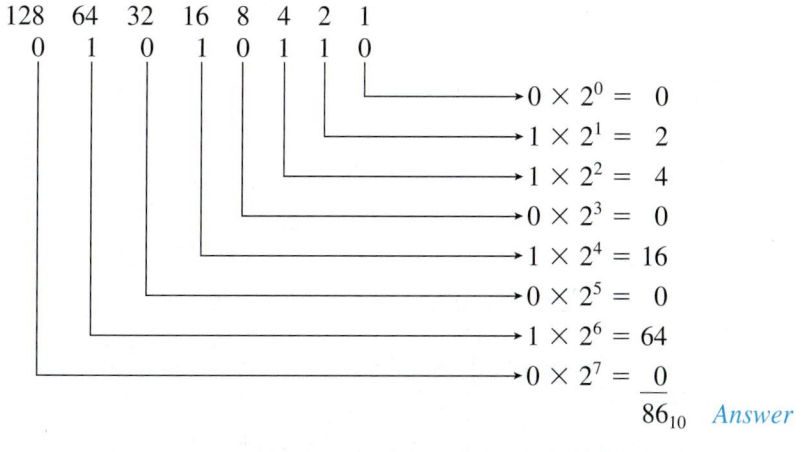

```
128  64  32  16  8  4  2  1
 0    1   0   1  0  1  1  0
```

$$0 \times 2^0 = 0$$
$$1 \times 2^1 = 2$$
$$1 \times 2^2 = 4$$
$$0 \times 2^3 = 0$$
$$1 \times 2^4 = 16$$
$$0 \times 2^5 = 0$$
$$1 \times 2^6 = 64$$
$$0 \times 2^7 = 0$$

86_{10} *Answer*

Although seldom used in digital systems, binary weighting for values less than 1 is possible (fractional binary numbers). These factors are developed by successively dividing the weighting factor by 2 for each decrease in the power of 2. This is also useful to illustrate why 2^0 is equal to 1, not zero (see Figure 1–5).

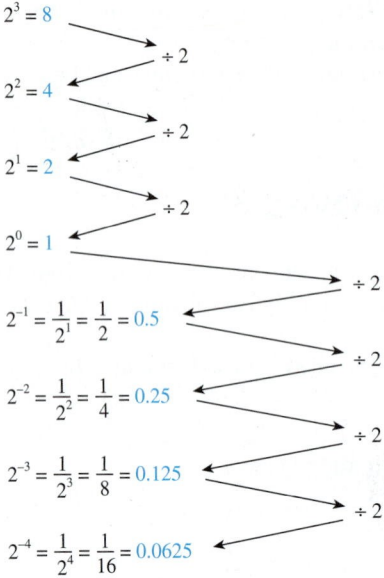

$$2^3 = 8$$

$\div 2$

$$2^2 = 4$$

$\div 2$

$$2^1 = 2$$

$\div 2$

$$2^0 = 1$$

$\div 2$

$$2^{-1} = \frac{1}{2^1} = \frac{1}{2} = 0.5$$

$\div 2$

$$2^{-2} = \frac{1}{2^2} = \frac{1}{4} = 0.25$$

$\div 2$

$$2^{-3} = \frac{1}{2^3} = \frac{1}{8} = 0.125$$

$\div 2$

$$2^{-4} = \frac{1}{2^4} = \frac{1}{16} = 0.0625$$

Figure 1–5 Successive division by 2 to develop fractional binary weighting factors and show that 2^0 is equal to 1.

EXAMPLE 1–3

Convert the fractional binary number 1011.1010_2 to decimal.

Solution: Multiply each binary digit by the appropriate weighting factor given in Figure 1–5, and total the results. (We will skip the multiplication for the binary digit 0 because it does not contribute to the total.)

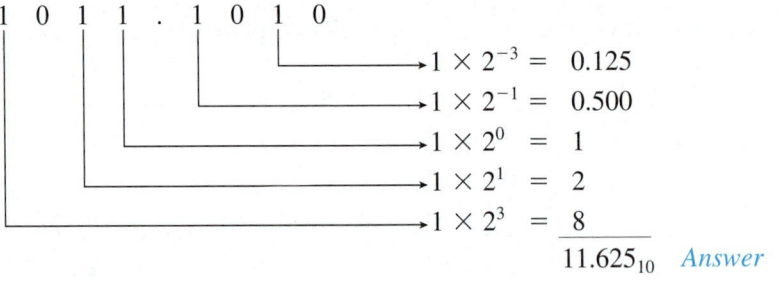

1 0 1 1 . 1 0 1 0

$1 \times 2^{-3} = 0.125$

$1 \times 2^{-1} = 0.500$

$1 \times 2^0 = 1$

$1 \times 2^1 = 2$

$1 \times 2^3 = 8$

11.625_{10} *Answer*

Review Questions

1–3. Why is the binary numbering system commonly used in digital electronics?

1–4. How are the weighting factors determined for each binary position in a base 2 number?

1–5. Convert $0110\ 1100_2$ to decimal.

1–6. Convert 1101.0110_2 to decimal.

1–5   Decimal-to-Binary Conversion

The conversion from binary to decimal is usually performed by the digital computer for ease of interpretation by the person reading the number. On the other hand, when a person enters a decimal number into a digital computer, that number must be converted to binary before it can be operated on. Let's look at *decimal-to-binary* conversion.

EXAMPLE 1–4

Convert 133_{10} to binary.

Solution: Referring to Table 1–1, we can see that the largest power of 2 that will fit into 133 is 2^7 ($2^7 = 128$), but that will still leave the value 5 ($133 - 128 = 5$) to be accounted for. Five can be taken care of by 2^2 and 2^0 ($2^2 = 4$, $2^0 = 1$). So the process looks like this:

$$
\begin{array}{r}
133 \\
-128 \rightarrow 2^7 \\
\hline
5 \\
-\;\;4 \rightarrow 2^2 \\
\hline
1 \\
-\;\;1 \rightarrow 2^0 \\
\hline
0
\end{array}
\qquad
\begin{array}{|c|c|c|c|c|c|c|c|}
\hline
1 & 0 & 0 & 0 & 0 & 1 & 0 & 1 \\
\hline
2^7 & 2^6 & 2^5 & 2^4 & 2^3 & 2^2 & 2^1 & 2^0 \\
\end{array}
$$

Answer: 10000101_2

Note: The powers of 2 needed to give the number 133 were first determined. Then all other positions were filled with zeros.

EXAMPLE 1–5

Convert 122_{10} to binary.

Solution:

$$
\begin{array}{r}
122 \\
-64 \rightarrow 2^6 \\
\hline
58 \\
-32 \rightarrow 2^5 \\
\hline
26 \\
-16 \rightarrow 2^4 \\
\hline
10 \\
-\;\;8 \rightarrow 2^3 \\
\hline
2 \\
-\;\;2 \rightarrow 2^1 \\
\hline
0
\end{array}
\qquad
\begin{array}{cccccccc}
0 & 1 & 1 & 1 & 1 & 0 & 1 & 0 \\
2^7 & 2^6 & 2^5 & 2^4 & 2^3 & 2^2 & 2^1 & 2^0 \\
\end{array}
$$

Answer: $0\,1\,1\,1\,1\,0\,1\,0_2$

Helpful Hint

This is a good time to realize that a useful way to learn new material like this is to re-solve the examples with the solutions covered up. That way, when you have a problem, you can uncover the solution and see the correct procedure.

Another method of converting decimal to binary is by *successive division*. Successive division involves dividing repeatedly by the number of the base to which you are converting. Continue the process until the answer is 0. For example, to convert 122_{10} to base 2, use the following procedure:

$$122 \div 2 = 61 \quad \text{with a remainder of } 0 \quad \text{(LSB)}$$
$$61 \div 2 = 30 \quad \text{with a remainder of } 1$$
$$30 \div 2 = 15 \quad \text{with a remainder of } 0$$
$$15 \div 2 = 7 \quad \text{with a remainder of } 1$$
$$7 \div 2 = 3 \quad \text{with a remainder of } 1$$
$$3 \div 2 = 1 \quad \text{with a remainder of } 1$$
$$1 \div 2 = 0 \quad \text{with a remainder of } 1 \quad \text{(MSB)}$$

The first remainder, 0, is the **least significant bit (LSB)** of the answer; the last remainder, 1, is the **most significant bit (MSB)** of the answer. Therefore, the answer is

$$1\ 1\ 1\ 1\ 0\ 1\ 0_2 \quad \longleftarrow \text{LSB}$$

However, because most computers or digital systems deal with groups of 4, 8, 16, or 32 **bits** (Binary digITs), we should keep all our answers in that form. Adding a leading zero to the number $1\ 1\ 1\ 1\ 0\ 1\ 0_2$ will not change its numeric value; therefore, the 8-bit answer is

$$1\ 1\ 1\ 1\ 0\ 1\ 0_2 = 0\ 1\ 1\ 1\ 1\ 0\ 1\ 0_2$$

Common Misconception

Remember not to reverse the LSB and MSB when listing the binary answer.

EXAMPLE 1–6

Convert 152_{10} to binary using successive division.

Solution:

$$152 \div 2 = 76 \quad \text{remainder } 0 \quad \text{(LSB)}$$
$$76 \div 2 = 38 \quad \text{remainder } 0$$
$$38 \div 2 = 19 \quad \text{remainder } 0$$
$$19 \div 2 = 9 \quad \text{remainder } 1$$
$$9 \div 2 = 4 \quad \text{remainder } 1$$
$$4 \div 2 = 2 \quad \text{remainder } 0$$
$$2 \div 2 = 1 \quad \text{remainder } 0$$
$$1 \div 2 = 0 \quad \text{remainder } 1 \quad \text{(MSB)}$$

Answer: $1\ 0\ 0\ 1\ 1\ 0\ 0\ 0_2$

Review Questions

1–7. Convert 43_{10} to binary.

1–8. Convert 170_{10} to binary.

1–6   Octal Numbering System (Base 8)

The **octal** numbering system is a method of grouping binary numbers in groups of three. The eight allowable digits are 0, 1, 2, 3, 4, 5, 6, and 7.

The octal numbering system is used by manufacturers of computers that utilize 3-bit codes to indicate instructions or operations to be performed. By using the octal representation instead of binary, the user can simplify the task of entering or reading computer instructions and thus save time.

In Table 1–2, we see that when the octal number exceeds 7, the least significant octal position resets to zero, and the next most significant position increases by 1.

TABLE 1–2	Octal Numbering System	
Decimal	**Binary**	**Octal**
0	000	0
1	001	1
2	010	2
3	011	3
4	100	4
5	101	5
6	110	6
7	111	7
8	1000	10
9	1001	11
10	1010	12

1–7   Octal Conversions

Converting from *binary to octal* is simply a matter of grouping the binary positions in groups of three (starting at the least significant position) and writing down the octal equivalent.

EXAMPLE 1–7

Convert $0\ 1\ 1\ 1\ 0\ 1_2$ to octal.

Solution:

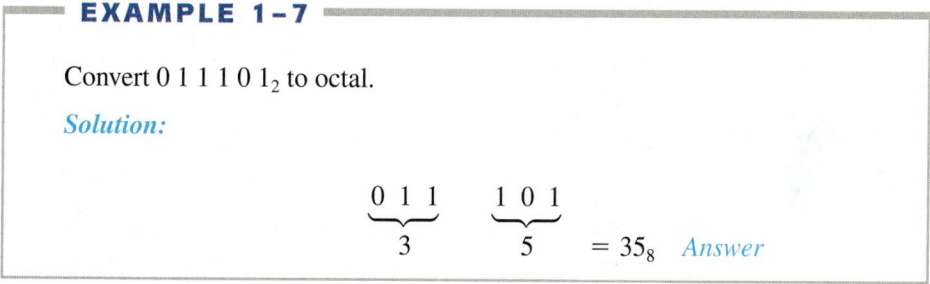

EXAMPLE 1–8

Convert $1\ 0\ 1\ 1\ 1\ 0\ 0\ 1_2$ to octal.

Solution:

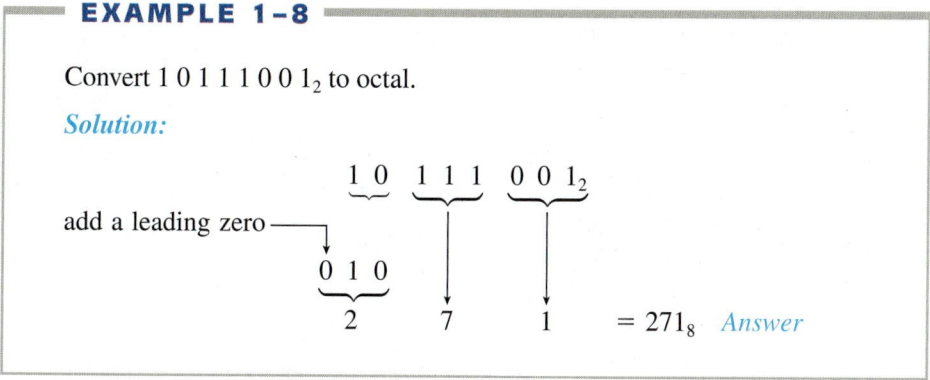

To convert *octal to binary,* you reverse the process.

EXAMPLE 1–9

Convert $6\ 2\ 4_8$ to binary.

Solution:

$$\underbrace{6}\quad\underbrace{2}\quad\underbrace{4}$$
$$1\ 1\ 0\quad 0\ 1\ 0\quad 1\ 0\ 0 = 1\ 1\ 0\ 0\ 1\ 0\ 1\ 0\ 0_2 \quad \textit{Answer}$$

To convert from *octal to decimal,* follow a process similar to that in Section 1–3 (multiply by weighting factors).

EXAMPLE 1–10

Convert $3\ 2\ 6_8$ to decimal.

Solution:

$$3 \quad 2 \quad 6$$

$$6 \times 8^0 = 6 \times\ 1 = \quad 6$$
$$2 \times 8^1 = 2 \times\ 8 = \quad 16$$
$$3 \times 8^2 = 3 \times 64 = \underline{192}$$
$$214_{10} \quad \textit{Answer}$$

To convert from *decimal to octal,* the successive-division procedure can be used.

EXAMPLE 1–11

Convert $4\ 8\ 6_{10}$ to octal.

Solution:

$$
\left.
\begin{array}{l}
486 \div 8 =\ 60 \quad \text{remainder}\quad 6\\
60 \div 8 =\quad 7 \quad \text{remainder}\quad 4\\
7 \div 8 =\quad 0 \quad \text{remainder}\quad 7
\end{array}
\right\} \quad 746_8
$$

$$486_{10} = 746_8 \quad \textit{Answer}$$

Check:

$$7 \quad 4 \quad 6$$

$$6 \times 8^0 =\quad 6$$
$$4 \times 8^1 =\quad 32$$
$$7 \times 8^2 = \underline{448}$$
$$486 \quad \sqrt{}$$

Review Questions

1–9. The only digits allowed in the octal numbering system are 0 to 8. True or false?

1–10. Convert 111011_2 to octal.

1–11. Convert 263_8 to binary.

1–12. Convert 614_8 to decimal.

1–13. Convert 90_{10} to octal.

1–8 Hexadecimal Numbering System (Base 16)

The **hexadecimal** numbering system, like the octal system, is a method of grouping bits to simplify entering and reading the instructions or data present in digital computer systems. Hexadecimal uses 4-bit groupings; therefore, instructions or data used in 8-, 16-, or 32-bit computer systems can be represented as a two-, four-, or eight-digit hexadecimal code instead of using a long string of binary digits (see Table 1–3).

TABLE 1–3	Hexadecimal Numbering System	
Decimal	**Binary**	**Hexadecimal**
0	0000	0
1	0001	1
2	0010	2
3	0011	3
4	0100	4
5	0101	5
6	0110	6
7	0111	7
8	1000	8
9	1001	9
10	1010	A
11	1011	B
12	1100	C
13	1101	D
14	1110	E
15	1111	F
16	0001 0000	1 0
17	0001 0001	1 1
18	0001 0010	1 2
19	0001 0011	1 3
20	0001 0100	1 4

Hexadecimal (hex) uses 16 different digits and is a method of grouping binary numbers in groups of four. Because hex digits must be represented by a single character, letters are chosen to represent values greater than 9. The 16 allowable hex digits are 0, 1, 2, 3, 4, 5, 6, 7, 8, 9, A, B, C, D, E, and F.

To signify a hex number, a subscript 16 or the letter H is used (that is, $A7_{16}$ or A7H). Two hex digits are used to represent 8 bits (also known as a *byte*). Four bits (one hex digit) are sometimes called a *nibble*.

1-9   Hexadecimal Conversions

To convert from *binary to hexadecimal,* group the binary number in groups of four (starting in the least significant position) and write down the equivalent hex digit.

EXAMPLE 1-12

Convert $0\ 1\ 1\ 0\ 1\ 1\ 0\ 1_2$ to hex.

Solution:

$$\underbrace{0\ 1\ 1\ 0}_{6}\quad \underbrace{1\ 1\ 0\ 1_2}_{D} \quad = 6D_{16}\quad \textit{Answer}$$

To convert *hexadecimal to binary,* use the reverse process.

EXAMPLE 1-13

Convert $A9_{16}$ to binary.

Solution:

$$\begin{array}{cc} A & 9 \\ 1010 & 1001 \end{array} = 1 0 1 0 1 0 0 1_2 \quad \textit{Answer}$$

To convert *hexadecimal to decimal,* use a process similar to that in Section 1–3.

EXAMPLE 1-14

Convert $2\ A\ 6_{16}$ to decimal.

Solution:

$$
\begin{aligned}
6 \times 16^0 &= 6 \times 1 = 6 \\
A \times 16^1 &= 10 \times 16 = 160 \\
2 \times 16^2 &= 2 \times 256 = \underline{512} \\
&\qquad\qquad\qquad 678_{10} \quad \textit{Answer}
\end{aligned}
$$

EXAMPLE 1-15

Redo Example 1–14 by converting first to binary and then to decimal.

Solution:

$$\underbrace{2}_{0010}\quad \underbrace{A}_{1010}\quad \underbrace{6}_{0110} = 2 + 4 + 32 + 128 + 512 = 678_{10} \quad \textit{Answer}$$

To convert from *decimal to hexadecimal,* use successive division. (*Note:* Successive division can always be used when converting from base 10 to any other base numbering system.)

Helpful Hint

At this point, you may be asking if you can use your hex calculator key instead of the hand procedure to perform these conversions. It is important to master these conversion procedures before depending on your calculator so that you understand the concepts involved.

EXAMPLE 1–16

Convert 151_{10} to hex.

Solution:

$$151 \div 16 = 9 \quad \text{remainder } 7 \quad \text{(LSD)}$$
$$9 \div 16 = 0 \quad \text{remainder } 9 \quad \text{(MSD)}$$
$$151_{10} = 97_{16} \quad \textit{Answer}$$

Check:

$$97_{16}$$
$$7 \times 16^0 = \quad 7$$
$$9 \times 16^1 = \underline{144}$$
$$151 \quad \sqrt{}$$

Team Discussion

Which is the largest number—142_8, 142_{10}, or 142_{16}?

EXAMPLE 1–17

Convert 498_{10} to hex.

Solution:

$$498 \div 16 = 31 \quad \text{remainder} \quad 2 \quad \text{(LSD)}$$
$$31 \div 16 = 1 \quad \text{remainder} \quad 15 \quad (= F)$$
$$1 \div 16 = 0 \quad \text{remainder} \quad 1 \quad \text{(MSD)}$$
$$498_{10} = 1\,F\,2_{16} \quad \textit{Answer}$$

Check:

$$1\,F\,2_{16} \qquad 2 \times 16^0 = 2 \times \quad 1 = \quad 2$$
$$F \times 16^1 = 15 \times \quad 16 = 240$$
$$1 \times 16^2 = \quad 1 \times 256 = \underline{256}$$
$$498 \quad \sqrt{}$$

Review Questions

1–14. Why is hexadecimal used instead of the octal numbering system when working with 8- and 16-bit digital computers?

1–15. The *successive-division* method can be used whenever converting from base 10 to any other base numbering system. True or false?

1–16. Convert $0110\ 1011_2$ to hex.

1–17. Convert $E7_{16}$ to binary.

1–18. Convert $16C_{16}$ to decimal.

1–19. Convert 300_{10} to hex.

1–10   Binary-Coded-Decimal System

The binary-coded-decimal (**BCD**) system is used to represent each of the 10 decimal digits as a 4-bit binary code. This code is useful for outputting to displays that are always numeric (0 to 9), such as those found in digital clocks or digital voltmeters.

To form a BCD number, simply convert each decimal digit to its 4-bit binary code.

EXAMPLE 1–18

Convert $4\ 9\ 6_{10}$ to BCD.

Solution:

$$\underbrace{4}_{0100}\quad \underbrace{9}_{1001}\quad \underbrace{6}_{0110} = 0100\quad 1001\quad 0110_{BCD} \quad \textit{Answer}$$

To convert *BCD to decimal,* just reverse the process.

EXAMPLE 1–19

Convert $0111\ 0101\ 1000_{BCD}$ to decimal.

Solution:

$$\underbrace{0111}_{7}\quad \underbrace{0101}_{5}\quad \underbrace{1000}_{8} = 758_{10} \quad \textit{Answer}$$

EXAMPLE 1–20

Convert $0110\ 0100\ 1011_{BCD}$ to decimal.

Solution:

$$0110\quad 0100\quad 1011$$
$$6\qquad 4\qquad *$$

———

*This conversion is impossible because 1011 is not a valid binary-coded decimal. It is not in the range 0 to 9.

1–11   Comparison of Numbering Systems

Table 1–4 compares numbers written in the five number systems commonly used in digital electronics and computer systems.

1–12   The ASCII Code

To get information into and out of a computer, we need more than just numeric representations; we also have to take care of all the letters and symbols used in day-to-day

TABLE 1–4	Comparison of Numbering Systems			
Decimal	Binary	Octal	Hexadecimal	BCD
0	0000	0	0	0000
1	0001	1	1	0001
2	0010	2	2	0010
3	0011	3	3	0011
4	0100	4	4	0100
5	0101	5	5	0101
6	0110	6	6	0110
7	0111	7	7	0111
8	1000	1 0	8	1000
9	1001	1 1	9	1001
10	1010	1 2	A	0001 0000
11	1011	1 3	B	0001 0001
12	1100	1 4	C	0001 0010
13	1101	1 5	D	0001 0011
14	1110	1 6	E	0001 0100
15	1111	1 7	F	0001 0101
16	0001 0000	2 0	1 0	0001 0110
17	0001 0001	2 1	1 1	0001 0111
18	0001 0010	2 2	1 2	0001 1000
19	0001 0011	2 3	1 3	0001 1001
20	0001 0100	2 4	1 4	0010 0000

processing. Information such as names, addresses, and item descriptions must be input and output in a readable format. But remember that a digital system can deal only with 1's and 0's. Therefore, we need a special code to represent all **alphanumeric** data (letters, symbols, and numbers).

Most industry has settled on an input/output (I/O) code called the American Standard Code for Information Interchange (ASCII). The **ASCII code** uses 7 bits to represent all the alphanumeric data used in computer I/O. Seven bits will yield 128 different code combinations, as listed in Table 1–5.

Each time a key is depressed on an ASCII keyboard, that key is converted into its ASCII code and processed by the computer. Then, before outputting the computer contents to a display terminal or printer, all information is converted from ASCII into standard English.

To use the table, place the 4-bit group in the least significant positions and the 3-bit group in the most significant positions.

EXAMPLE 1–21

100 0111 is the code for G.

3-bit group   4-bit group

EXAMPLE 1–22

Using Table 1–5, determine the ASCII code for the lowercase letter *p*.

Solution: 1110000 (*Note:* Often, a leading zero is added to form an 8-bit result, making *p* = 0111 0000.)

Team Discussion

Have you ever tried displaying non-ASCII data to your PC screen using a disk utility program? If you were to read a file created by the IRS for your tax return, which fields would be ASCII?

TABLE 1–5		American Standard Code for Information Interchange						

	MSB							
LSB	000	001	010	011	100	101	110	111
0000	NUL	DLE	SP	0	@	P		p
0001	SOH	DC_1	!	1	A	Q	a	q
0010	STX	DC_2	"	2	B	R	b	r
0011	ETX	DC_3	#	3	C	S	c	s
0100	EOT	DC_4	$	4	D	T	d	t
0101	ENQ	NAK	%	5	E	U	e	u
0110	ACK	SYN	&	6	F	V	f	v
0111	BEL	ETB	'	7	G	W	g	w
1000	BS	CAN	(	8	H	X	h	x
1001	HT	EM	)	9	I	Y	i	y
1010	LF	SUB	*	:	J	Z	j	z
1011	VT	ESC	+	;	K	[	k	{
1100	FF	FS	,	<	L	\	l	|
1101	CR	GS	–	=	M	]	m	}
1110	SO	RS	.	>	N	↑	n	~
1111	SI	US	/	?	O	—	o	DEL

Definitions of control abbreviations:

ACK	Acknowledge	GS	Group separator
BEL	Bell	HT	Horizontal tab
BS	Backspace	LF	Line feed
CAN	Cancel	NAK	Negative acknowledge
CR	Carriage return	NUL	Null
DC_1–DC_4	Direct control	RS	Record separator
DEL	Delete idle	SI	Shift in
DLE	Data link escape	SO	Shift out
EM	End of medium	SOH	Start of heading
ENQ	Enquiry	SP	Space
EOT	End of transmission	STX	Start text
ESC	Escape	SUB	Substitute
ETB	End of transmission block	SYN	Synchronous idle
ETX	End text	US	Unit separator
FF	Form feed	VT	Vertical tab
FS	Form separator		

Review Questions

1–20. How does BCD differ from the base 2 binary numbering system?

1–21. Why is ASCII code required by digital computer systems?

1–22. Convert 947_{10} to BCD.

1–23. Convert $1000\ 0110\ 0111_{BCD}$ to decimal.

1–24. Determine the ASCII code for the letter E.

1–13 Applications of the Numbering Systems

Because digital systems work mainly with 1's and 0's, we have spent considerable time working with the various number systems. Which system is used depends on how the data were developed and how they are to be used. In this section, we will work with several applications that depend on the translation and interpretation of these digital representations.

The ABC Corporation chemical processing plant uses a computer to monitor the temperature and pressure of four chemical tanks, as shown in Figure 1–6(a). Whenever a temperature or a pressure exceeds the danger limit, an internal tank sensor applies a 1 to its corresponding output to the computer. If all conditions are OK, then all outputs are 0.

Helpful Hint

This and the following five applications illustrate the answer to the common student question, "Why are we learning this stuff?"

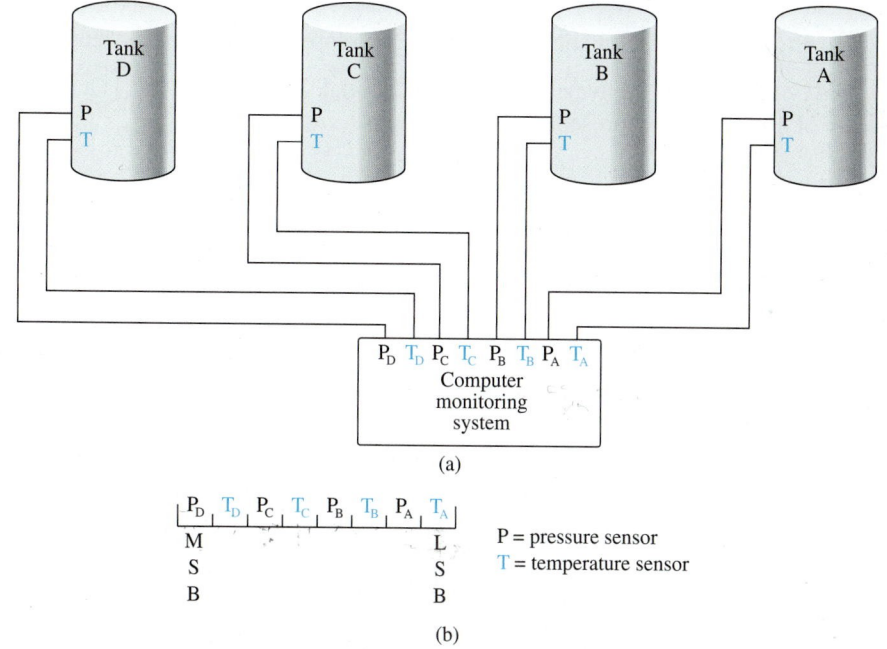

(a)

P = pressure sensor
T = temperature sensor

(b)

Figure 1–6 (a) Circuit connections for chemical temperature and pressure monitors at the ABC Corporation chemical processing plant; (b) layout of binary data read by the computer monitoring system.

(a) If the computer reads the binary string 0010 1000, what problems exist?

Solution: Entering that binary string into the chart of Figure 1–6(b) shows us that the pressure in tanks C and B is dangerously high.

(b) What problems exist if the computer is reading 55H (55 hex)?

Solution: 55H = 0101 0101, meaning that all temperatures are too high

(c) What hexadecimal number is read by the computer if the temperature and pressure in both tanks D and B are high?

Solution: CCH (1100 1100 = CCH)

(d) Tanks A and B are taken out of use, and their sensor outputs are connected to 1's. A computer programmer must write a program to ignore these new circuit conditions. The computer program must check that the value read is always less than what decimal equivalent when no problem exists?

Solution: $< 31_{10}$, because, with the 4 low-order bits HIGH, if TC goes HIGH, then the binary string will be 0001 1111, which is equal to 31_{10}.

(e) In another area of the plant, only three tanks (A, B, and C) have to be monitored. What octal number is read if tank B has a high temperature and pressure?

Solution: 14_8 ($001\ 100_2 = 14_8$)

APPLICATION 1–2

A particular brand of CD player has the capability of converting 12-bit signals from a CD into their equivalent analog values.

(a) What are the largest and smallest hex values that can be used in this CD system?

Solution: Largest: FFF_{16}; smallest: 000_{16}

(b) How many different analog values can be represented by this system?

Solution: FFF_{16} is equivalent to 4095 in decimal. Including 0, this is a total of 4096 unique representations.

APPLICATION 1–3

Typically, digital thermometers use BCD to drive their digit displays.

(a) How many BCD bits are required to drive a 3-digit thermometer display?

Solution: 12; 4 bits for each digit

(b) What 12 bits are sent to the display for a temperature of 147 degrees?

Solution: 0001 0100 0111

Common Misconception

You may have a hard time visualizing why we add or subtract 1 to determine memory locations. Answer this question: How many problems must you solve if your teacher assigns problems 5 through 10? (You would subtract 5 from 10 and then add 1.) How about if you solve 8 problems starting with 10: Would the last problem be 18 or 17?

APPLICATION 1–4

Most PC-compatible computer systems use a 20-bit address code to identify each of over 1 million memory locations.

(a) How many hex characters are required to identify the address of each memory location?

Solution: Five (Each hex digit represents 4 bits.)

(b) What is the 5-digit hex address of the 200th memory location?

Solution: 000C7H (200_{10} = C8H; but the first memory location is 00000H, so we have to subtract 1).

(c) If 50 memory locations are used for data storage starting at location 000C8H, what is the location of the last data item?

Solution: 000F9H (000C8H = 200_{10}, 200 + 50 = 250_{10}, 250 − 1 = 249_{10}, 249_{10} = F9H [We had to subtract 1 because location C8H (200_{10}) received the first data item, so we needed only 49 more memory spaces.])

APPLICATION 1–5

If the part number 651-M is stored in ASCII in a computer memory, list the binary contents of its memory locations.

Solution:

$$6 = 011\ 0110$$
$$5 = 011\ 0101$$
$$1 = 011\ 0001$$
$$\text{-} = 010\ 1101$$
$$M = 100\ 1101$$

Because most computer memory locations are formed by groups of 8 bits, let's add a zero to the leftmost position to fill each 8-bit memory location. (The leftmost position is sometimes filled by a parity bit, which is discussed in Chapter 6.)

Therefore, the serial number, if strung out in five memory locations, would look like

0011 0110   0011 0101   0011 0001   0010 1101   0100 1101

If you look at these memory locations in hexadecimal, they will read

36   35   31   2D   4D

The address settings of your PC I/O devices are given as hexadecimal numbers. They can be determined on a Windows-based machine by pressing the sequence: My Computer > Control Panel > System > Device Manager > Properties > I/O. Determine from the list on your screen what the address settings are for your keyboard, printer, and floppy disk drive.

APPLICATION 1-6

To look for an error in a BASIC program, a computer programmer uses a debugging utility to display the ASCII codes of a particular part of her program. The codes are displayed in hex as 474F5430203930. Assume that the leftmost bit of each ASCII string is padded with a 0.

(a) Translate the program segment that is displayed.

Solution: GOT0 90.

(b) If you know anything about programming in BASIC, try to determine what the error is.

Solution: Apparently a number zero was typed in the GOTO statement instead of the letter O. Change it, and the error should go away.

Summary

In this chapter, we have learned that

1. Numeric quantities occur naturally in analog form but must be converted to digital form to be used by computers or digital circuitry.

2. The binary numbering system is used in digital systems because the 1's and 0's are easily represented by ON or OFF transistors, which output 0 V for 0 and 5 V for 1.

3. Any number system can be converted to decimal by multiplying each digit by its weighting factor.

4. The weighting factor of the least significant digit in any numbering system is always 1.

5. Binary numbers can be converted to octal by forming groups of 3 bits and to hexadecimal by forming groups of 4 bits, beginning with the LSB. Each group is then converted to an octal or hex digit.

6. The successive-division procedure can be used to convert from decimal to binary, octal, or hexadecimal.

7. The binary-coded-decimal system uses groups of 4 bits to drive decimal displays such as those in a calculator.

8. ASCII is used by computers to represent all letters, numbers, and symbols in digital form.

Helpful Hint

Skimming through the glossary terms is a good way to review the chapter. You should also feel that you have a good understanding of all the topics listed in the objectives at the beginning of the chapter.

Glossary

Alphanumeric: Characters that contain alphabet letters as well as numbers and symbols.

Analog: A system that deals with continuously varying physical quantities such as voltage, temperature, pressure, or velocity. Most quantities in nature occur in analog, yielding an infinite number of different levels.

ASCII Code: American Standard Code for Information Interchange. ASCII is a 7-bit code used in digital systems to represent all letters, symbols, and numbers to be input or output to the outside world.

BCD: Binary-coded decimal. A 4-bit code used to represent the 10 decimal digits 0 to 9.

Binary: The base 2 numbering system. Binary numbers are made up of 1's and 0's, each position being equal to a different power of 2 (2^3, 2^2, 2^1, 2^0, and so on).

Bit: A single binary digit. The binary number 1101 is a 4-bit number.

Decimal: The base 10 numbering system. The 10 decimal digits are 0, 1, 2, 3, 4, 5, 6, 7, 8, and 9. Each decimal position is a different power of 10 (10^3, 10^2, 10^1, 10^0, and so on).

Digital: A system that deals with discrete digits or quantities. Digital electronics deals exclusively with 1's and 0's or ONs and OFFs. Digital codes (such as ASCII) are then used to convert the 1's and 0's to a meaningful number, letter, or symbol for some output display.

Hexadecimal: The base 16 numbering system. The 16 hexadecimal digits are 0, 1, 2, 3, 4, 5, 6, 7, 8, 9, A, B, C, D, E, and F. Each hexadecimal position represents a different power of 16 (16^3, 16^2, 16^1, 16^0, and so on).

Least Significant Bit (LSB): The bit having the least significance in a binary string. The LSB will be in the position of the lowest power of 2 within the binary number.

Most Significant Bit (MSB): The bit having the most significance in a binary string. The MSB will be in the position of the highest power of 2 within the binary number.

Octal: The base 8 numbering system. The eight octal numbers are 0, 1, 2, 3, 4, 5, 6, and 7. Each octal position represents a different power of 8 (8^3, 8^2, 8^1, 8^0, and so on).

Section 1–4

1–1. Convert the following binary numbers to decimal.

(a) 0110   **(b)** 1011   **(c)** 1001   **(d)** 0111

(e) 1100   **(f)** 0100 1011   **(g)** 0011 0111

(h) 1011 0101   **(i)** 1010 0111   **(j)** 0111 0110

Section 1–5

1–2. Convert the following decimal numbers to 8-bit binary.

(a) 186_{10} **(b)** 214_{10} **(c)** 27_{10} **(d)** 251_{10} **(e)** 146_{10}

Sections 1–6 and 1–7

1–3. Convert the following binary numbers to octal.

(a) 011001   **(b)** 11101   **(c)** 1011100

(d) 01011001   **(e)** 1101101

1–4. Convert the following octal numbers to binary.

(a) 46_8 **(b)** 74_8 **(c)** 61_8 **(d)** 32_8 **(e)** 57_8

1–5. Convert the following octal numbers to decimal.

(a) 27_8 **(b)** 37_8 **(c)** 14_8 **(d)** 72_8 **(e)** 51_8

1–6. Convert the following decimal numbers to octal.

(a) 126_{10} **(b)** 49_{10} **(c)** 87_{10} **(d)** 94_{10} **(e)** 108_{10}

Sections 1–8 and 1–9

1–7. Convert the following binary numbers to hexadecimal.

(a) 1011 1001   **(b)** 1101 1100   **(c)** 0111 0100

(d) 1111 1011   **(e)** 1100 0110

1–8. Convert the following hexadecimal numbers to binary.

(a) $C5_{16}$ **(b)** FA_{16} **(c)** $D6_{16}$ **(d)** $A94_{16}$ **(e)** 62_{16}

1–9. Convert the following hexadecimal numbers to decimal.

(a) 86_{16} **(b)** $F4_{16}$ **(c)** 92_{16} **(d)** AB_{16} **(e)** $3C5_{16}$

1–10. Convert the following decimal numbers to hexadecimal.

(a) 127_{10} **(b)** 68_{10} **(c)** 107_{10} **(d)** 61_{10} **(e)** 29_{10}

Section 1–10

1–11. Convert the following BCD numbers to decimal.

(a) $1001\ 1000_{BCD}$ **(b)** $0110\ 1001_{BCD}$ **(c)** $0111\ 0100_{BCD}$

(d) $0011\ 0110_{BCD}$ **(e)** $1000\ 0001_{BCD}$

1–12. Convert the following decimal numbers to BCD.

(a) 87_{10} **(b)** 142_{10} **(c)** 94_{10} **(d)** 61_{10} **(e)** 44_{10}

Section 1–12

1–13. Use Table 1–5 to convert the following letters, symbols, and numbers to ASCII.

(a) %

(b) $14

(c) N-6

(d) CPU

(e) Pg

1–14. Insert a zero in the MSB of your answers to Problem 1–13, and list your answers in hexadecimal.

Section 1–13

C *   **1–15.** The computer monitoring system at the ABC Corporation shown in Figure 1–6 is receiving the following warning codes. Determine the problems that exist for each code (H stands for hex).

(a) $0010\ 0001_2$ **(b)** $C0_{16}$ **(c)** 88H   **(d)** 024_8 **(e)** 48_{10}

C   **1–16.** What is the BCD representation that is sent to a 3-digit display on a voltmeter that is measuring 120 V?

1–17. A computer programmer observes the following hex string when looking at a particular section of computer memory: 736B753433.

(a) Assume that the memory contents are ASCII codes with leading zeros and translate this string into its alphanumeric equivalent.

(b) The programmer realizes that the program recognizes only capital (uppercase) letters. Convert all letters in the alphanumeric equivalent to capital letters, and determine the new hex string.

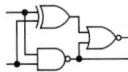

Schematic Interpretation Problems

(*Note:* Appendix G contains four schematic diagrams of actual digital systems. At the end of each chapter, you will have the opportunity to work with these diagrams to gain experience with real-world circuitry and observe the application of digital logic that was presented in the chapter.)

S *   **1–18.** Locate the HC11D0 master board schematic in Appendix G. Determine the component name and grid coordinates of the following components. (*Example:* Q3 is a 2N2907 located at A3.)

(a) U1   **(b)** U16   **(c)** Q1   **(d)** P2

S   **1–19.** Find the date and revision number for the HC11D0 master board schematic.

S   **1–20.** Find the quantity of the following devices that are used on the watchdog timer schematic.

(a) 74HC85   **(b)** 74HC08   **(c)** 74HC74   **(d)** 74HC32

---

*The letter **C** signifies problems that are more **C**hallenging and thought provoking.

*The letter **S** designates a **S**chematic interpretation problem.

Electronics Workbench®/MultiSIM® Exercises

Electronics Workbench® (EWB) is a software simulation tool that is used to reinforce the theory presented in each chapter. It provides an accurate simulation of digital and analog circuit operation along with a simulation of instruments used by a technician to measure IC, component, and circuit characteristics. With this software, you have the ability to build and test most of the circuits presented in this text.

The problems at the end of each chapter are based on circuits from the data disk provided with the textbook. The problems are based on the circuits and theory presented in the section corresponding to the file name. Before attempting any EWB problems, you must thoroughly understand the material presented in that textbook section. The problem definition for each EWB circuit is fully explained in the Description Window that appears in each EWB file.

The problems are basically of three types: (1) *circuit interaction problems* require the student to change input values and take measurements at the outputs to verify circuit operation; (2) *design problems* require the student to design, or modify, a circuit to perform a particular task; and (3) *troubleshooting problems* require the student to find and fix the fault that exists in the circuit that is given.

To get the best view of the circuits, set your computer monitor resolution to 800 × 600 or better. You will notice that the EWB problems use a slightly different notation to represent certain variables. For example, $\overline{A}$ is represented by A', C_p is represented by Cp, and 2^0 is represented by 2^0.

E1–1. (*Note:* You need to understand binary to hexadecimal conversions [Section 1–8] before attempting this exercise.) Load the circuit file for **Section 1–08.** This circuit is used to demonstrate the conversion between the binary and hexadecimal numbering systems similar to Examples 1–12 and 1–13. The Word Generator is used to drive eight binary lights and two hexadecimal displays. Read the instructions for the circuit in the *Description* window at the bottom of the screen.

(a) What 8-bit binary number will you see on the lights if you press *Step* five times? (An ON light is a 1.) Try it.

(b) How many times must you press *Step* to get the binary number 0000 1011? Try it.

(c) What hexadecimal number will you see if you press *Step* 14 times? Try it.

(d) How many times must you press *Step* to see the hexadecimal number 1b? Try it.

E1–2. (*Note:* You need to understand the operation of the chemical plant monitoring system presented in Figure 1–6 before attempting this exercise.) Load the circuit file for **Section 1–13.** Turn the power switch ON. The hex display should read 00H, which indicates that there are no high temperature or pressure levels.

(a) Read the instructions for the circuit in the *Description* window at the bottom of the screen. What would you expect the hex display to read if there is a high temperature in Tank D? To check your answer, raise the temperature in Tank D by pressing the indicated key several times. Return the temperature to a low level by holding the (Ctrl) key as you press 2 repeatedly.

(b) What would you expect the display to read if all temperatures are high? Check your answer, then return the levels to a low state.

(c) What levels are too high if the hex display reads 0CH? Check your answer by raising the levels on the appropriate tank(s). Return all levels to a low state.

(d) What levels are too high if the hex display reads AAH? Check your answer by raising the levels on the appropriate tanks(s). Return all levels to a low state.

Answers to Review Questions

1–1. Temperature, pressure, velocity, weight, sound

1–2. Because digital quantities are easier for a computer system to store and interpret

1–3. Because it uses only two digits, 0 and 1, which can be represented by using two distinct voltage levels

1–4. By powers of 2

1–5. 108_{10}

1–6. 13.375_{10}

1–7. $0010\ 1011_2$

1–8. $1010\ 1010_2$

1–9. False

1–10. 73_8

1–11. $010\ 110\ 011_2$ or $1011\ 0011_2$

1–12. 396_{10}

1–13. 132_8

1–14. Because hexadecimal uses 4-bit groupings

1–15. True

1–16. $6B_{16}$

1–17. $1110\ 0111_2$

1–18. 364_{10}

1–19. $12C_{16}$

1–20. BCD is used only to represent decimal digits 0 to 9 in 4-bit groupings.

1–21. To get alphanumeric data into and out of a computer

1–22. $1001\ 0100\ 0111_{BCD}$

1–23. 867_{10}

1–24. $0100\ 0101_{ASCII}$

2

Digital Electronic Signals and Switches

OUTLINE

OBJECTIVES

Upon completion of this chapter, you should be able to:

- Describe the parameters associated with digital voltage-versus-time waveforms.
- Convert between frequency and period for a periodic clock waveform.
- Sketch the timing waveform for any binary string in either the serial or parallel representation.
- Discuss the application of manual switches and electromechanical relays in electric circuits.
- Explain the basic characteristics of diodes and transistors when they are forward biased and reverse biased.
- Calculate the output voltage in an electric circuit containing diodes or transistors operating as digital switches.
- Perform input/output timing analysis in electric circuits containing electromechanical relays or transistors.
- Explain the operation of a common-emitter transistor circuit used as a digital inverter switch.

INTRODUCTION

As mentioned in Chapter 1, digital electronics deals with 1's and 0's. These logic states will typically be represented by a high and a low voltage level (usually $1 = 5$ V and $0 = 0$ V).

In this chapter, we see how these logic states can be represented by means of a timing diagram and how electronic switches are used to generate meaningful digital signals.

2-1 Digital Signals

A digital signal is made up of a series of 1's and 0's that represent numbers, letters, symbols, or control signals. Figure 2–1 shows the **timing diagram** of a typical digital signal. Timing diagrams are used to show the HIGH and LOW (1 and 0) levels of a digital signal as it changes relative to time. In other words, it is a plot of *voltage versus time*. The y axis of the plot displays the voltage level and the x axis, the time. Digital systems respond to the digital state (0 or 1), not the actual voltage levels. For example, if the voltage levels in Figure 2–1(a) were not exactly 0 V and $+5$ V, the digital circuitry would still interpret it as the 0 state and 1 state, and respond identically. The actual voltage level standards of the various logic families are discussed in detail in Chapter 9.

Figure 2–1(a) is a timing diagram showing the bit configuration 1 0 1 0 as it would appear on an **oscilloscope.** Notice in the figure that the LSB comes first in time. In this case, the LSB is transmitted first. The MSB could have been transmitted first as long as the system on the receiving end knows which method is used.

Figure 2–1(b) is a photograph of an oscilloscope, which is a very important test instrument for making accurate voltage-versus-time measurements.

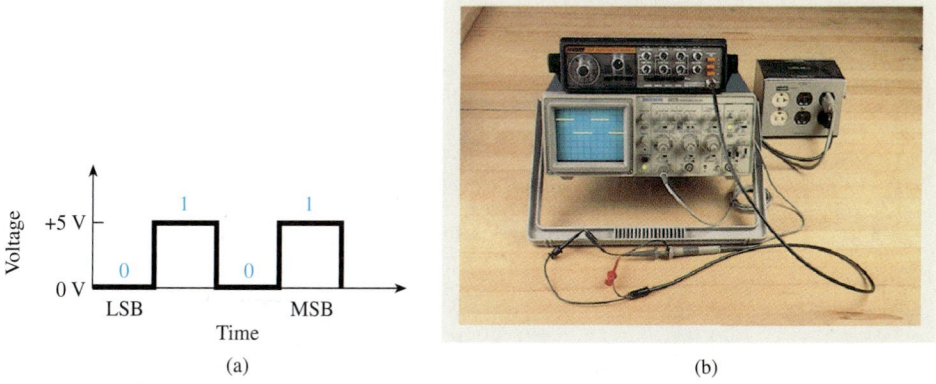

Figure 2-1   (a) Typical digital signal; (b) An oscilloscope displaying the digital waveform from a clock generator instrument.

2-2 Clock Waveform Timing

Most digital signals require precise timing. Special clock and timing circuits are used to produce clock waveforms to trigger the digital signals at precise intervals (timing circuit design is covered in Chapter 14).

Figure 2–2 shows a typical *periodic clock waveform* as it would appear on an oscilloscope displaying voltage versus time. The term *periodic* means that the waveform is repetitive, at a specific time interval, with each successive pulse identical to the previous one.

Figure 2–2 shows eight clock pulses, which we label 0, 1, 2, 3, 4, 5, 6, and 7. The **period** of the clock waveform is defined as the length of time from the falling edge of one pulse to the falling edge of the next pulse (or rising edge to rising edge) and is abbreviated t_p in Figure 2–2. The **frequency** of the clock waveform is defined as the reciprocal of the clock period. Written as a formula,

$$f = \frac{1}{t_p} \quad \text{and} \quad t_p = \frac{1}{f}$$

Figure 2–2 Periodic clock waveform as seen on an oscilloscope displaying voltage versus time.

The basic unit for frequency is *hertz* (Hz), and the basic unit for period is *seconds* (s). Frequency is often referred to as cycles per second (cps) or pulses per second (pps).

Team Discussion

An interesting exercise is to sketch the waveform from a 10-cps clock that is allowed to run for 1 s. How long did it take to complete one cycle? How did you find that time? Next, repeat for a 1-MHz clock.

EXAMPLE 2–1

What is the frequency of a clock waveform whose period is 2 microseconds (μs)?

Solution:

$$f = \frac{1}{t_p} = \frac{1}{2\ \mu s} = 0.5 \text{ megahertz } (0.5 \text{ MHz or } 500 \text{ kHz})$$

Hint: To review engineering notation, see Table 2–1.

Helpful Hint

Frequency and time calculations can often be made without a calculator if you realize some of the common reciprocal relationships (e.g., 1/milli = kilo, 1/micro = mega). When using a calculator, if the result is not a power of 3, 6, 9, or 12, then the answer must be converted to one of these common engineering prefixes using algebra or, if available, the *ENG* key on your calculator.

TABLE 2–1	Common Engineering Prefixes	
Prefix	**Abbreviation**	**Power of 10**
Giga	G	10^9
Mega	M	10^6
Kilo	k	10^3
Milli	m	10^{-3}
Micro	μ	10^{-6}
Nano	n	10^{-9}
Pico	p	10^{-12}

EXAMPLE 2–2

If the frequency of a waveform is 4.17 MHz, what is its period?

Solution:

$$t_p = \frac{1}{f} = \frac{1}{4.17 \text{ MHz}} = 0.240\ \mu s \text{ (or 240 ns)}$$

Digital communications concerns itself with the transmission of bits (1's and 0's). The rate, or frequency, at which they are transmitted is given in bits-per-second (bps). Common transmission rates for a PC connected to the Internet via a telephone line are 28.8 kilobits-per-second (28.8 kbps) and 56 kbps.

EXAMPLE 2–3

Sketch and label the *x* and *y* axis representing a 56-kbps clock waveform used for PC communication. (Assume that the voltage levels are 0.5 V and 7 V.)

Solution: The waveform is shown in Figure 2–3.

$$t_p = \frac{1}{f} = \frac{1}{56 \text{ kbps}} = 17.9 \ \mu s$$

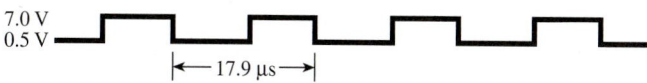

Figure 2–3   Solution to Example 2–3.

Team Discussion

For those students who have a PC: Do you know (or could you find out) at what frequency your internal microprocessor operates?

EXAMPLE 2–4

Determine the frequency of the waveform in Figure 2–4.

Solution:

$$f = \frac{1}{t_p} = \frac{1}{34.7 \ \mu s} = 28.8 \text{ kHz (or 28.8 kbps)}$$

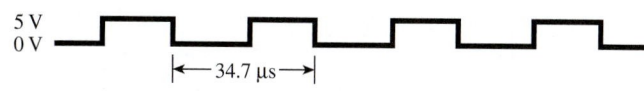

Figure 2–4   Waveform for Example 2–4.

Common Misconception

The period is labeled from rising edge to rising edge (or falling edge to falling edge) and is not just the positive pulse.

Review Questions

2–1. What are the labels on the *x* axis and *y* axis of a digital signal measured on an oscilloscope?

2–2. What is the relationship between clock frequency and clock period?

2–3. What is the time period from the rising edge of one pulse to the rising edge of the next pulse on a waveform whose frequency is 8 MHz?

2–4. What is the frequency of a periodic waveform having a period of 50 ns?

2–3   Serial Representation

Binary information to be transmitted from one location to another will be in either **serial** or **parallel** format. The serial format uses a single electrical conductor (and a common ground) for the data to travel on. The serial format is inexpensive because it only uses a single conductor and one set of input/output circuitry, but it is slow because it can only transmit 1 bit for each clock period. Communication over telephone lines (like the Internet) and computer-to-computer communication (like office networks) use serial communication (see Figure 2–5). The ports labels COM on a PC are most often used for the serial communication connection to telephone lines. A plug-in card is used in a PC to provide network serial communication (e.g., Ethernet).

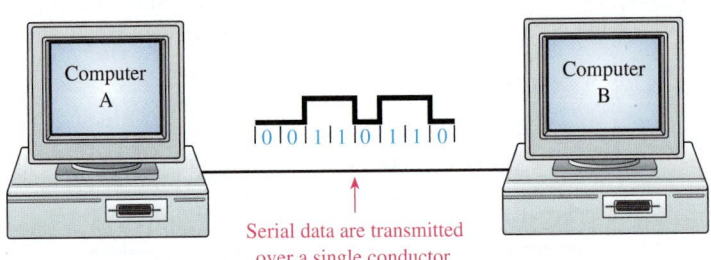

Serial data are transmitted over a single conductor.

Figure 2–5   Serial communication between computers.

Serial communication can be sped up by using extremely high-speed clock signals. Modern Internet connections and office networks communicate at speeds exceeding 1 million bps. Several standards have been developed for high-speed serial communications, the most common of which are V.90, ISDN, T1, T2, T3, Universal Serial Bus (USB), Ethernet, 10baseT, 100baseT, cable, and DSL.

Let's use Figure 2–6 to illustrate the serial representation of the binary number 0 1 1 0 1 1 0 0. The serial representation (S_o) is shown with respect to some clock waveform (C_p), and its LSB is drawn first. Each bit from the original binary number occupies a separate clock period, with the change from one bit to the next occurring at each *falling* edge of C_p (C_p is drawn just as a reference) .

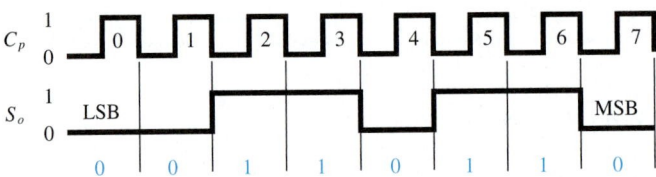

Figure 2–6   Serial representation of the binary number 0110110.

0110110 0.

2–4   Parallel Representation

The parallel format uses a separate electrical conductor for each bit to be transmitted (and a common ground). For example, if the digital system is using 8-bit numbers, eight lines are required (see Figure 2–7). This tends to be expensive, but the entire 8-bit number can be transmitted in one clock period, making it very fast.

Inside a computer, binary data are almost always transmitted on parallel channels (collectively called the *data bus*). Two parallel data techniques commonly used by computers to communicate to external devices are the Centronics printer interface (port LPT1) and the Small Computer Systems Interface (SCSI, pronounced *scuzzy*).

Figure 2–8 illustrates the same binary number that was used in Figure 2–6 (01101100), this time in the parallel representation.

If the clock period were 2 μs, it would take 2 μs × 8 periods = 16 μs to transmit the number in serial and only 2 μs × 1 period = 2 μs to transmit the same 8-bit num-

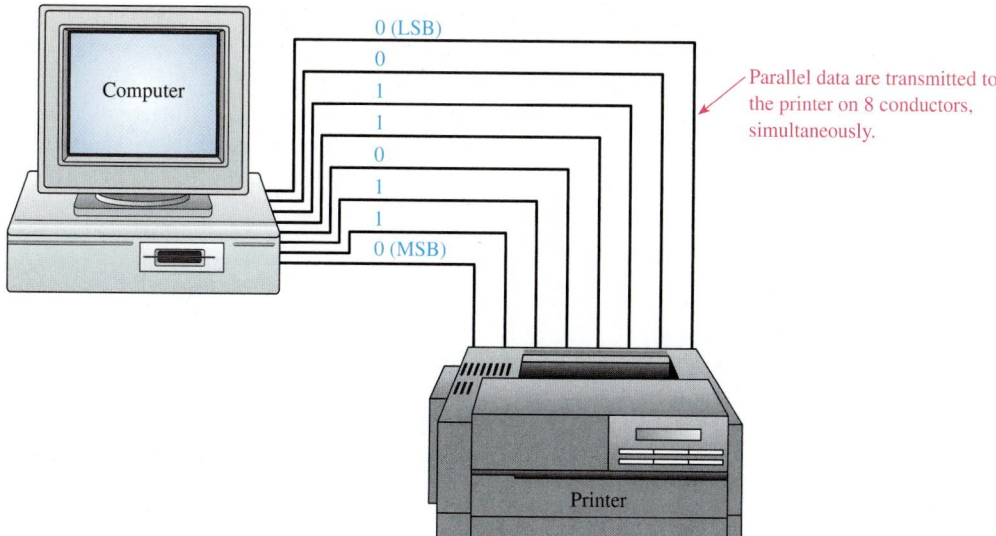

Figure 2–7 Parallel communication between a computer and a printer.

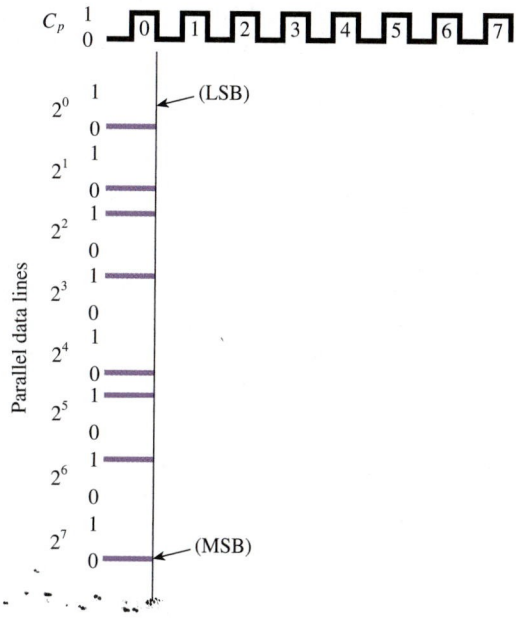

Figure 2–8 Parallel representation of the binary number 01101100.

Team Discussion

What other devices might use parallel communication? How about serial communication?

Inside Your PC

The printer port on PCs (labeled LPT1 on Windows-based machines) use an 8-bit parallel scheme (8 bits are called 1 byte). They have a transmission speed of 115 kBps. (The B signifies BYTES-per-second instead of bits). Much higher speeds are achieved using an SCSI plug-in computer board. The first SCSI standard specified an 8-bit parallel transmission. The speed was 5 MBps. Newer SCSI standards use 16-bit parallel and provide speeds up to 160 MBps! (For more information, visit the SCSI Web site listed in Appendix A.)

ber in parallel. Thus, you can see that when speed is important, parallel transmission is preferred over serial transmission.

The following examples further illustrate the use of serial and parallel representations.

EXAMPLE 2–5

Sketch the serial and parallel representations of the 4-bit number 0 1 1 1. If the clock frequency is 5 MHz, find the time to transmit using each method.

Solution: Figure 2–9 shows the representation of the 4-bit number 0 1 1 1.

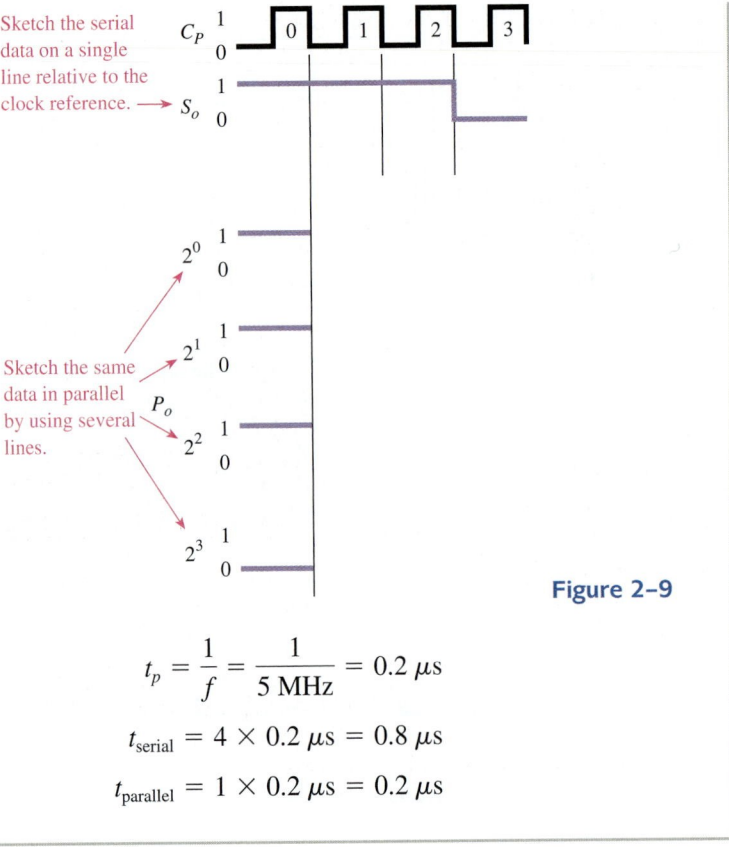

Sketch the serial data on a single line relative to the clock reference. → S_o

Sketch the same data in parallel by using several lines. P_o

Figure 2–9

$$t_p = \frac{1}{f} = \frac{1}{5 \text{ MHz}} = 0.2 \ \mu s$$

$$t_{\text{serial}} = 4 \times 0.2 \ \mu s = 0.8 \ \mu s$$

$$t_{\text{parallel}} = 1 \times 0.2 \ \mu s = 0.2 \ \mu s$$

EXAMPLE 2–6

Sketch the serial and parallel representations (least significant digit first) of the hexadecimal number 4A. (Assume a 4-bit parallel system and a clock frequency of 4 kHz.) Also, what is the state (1 or 0) of the serial line 1.2 ms into the transmission?

Solution: $4A_{16} = 0\ 1\ 0\ 0\ 1\ 0\ 1\ 0_2$.

$$t_p = \frac{1}{f} = \frac{1}{4 \text{ kHz}} = 0.25 \text{ ms}$$

Therefore, the increment of time at each falling edge increases by 0.25 ms. Because each period is 0.25 ms, 1.2 ms will occur within the 0 period of the number 4, which, on the S_o line, is a *0* **logic state** (see Figure 2–10).

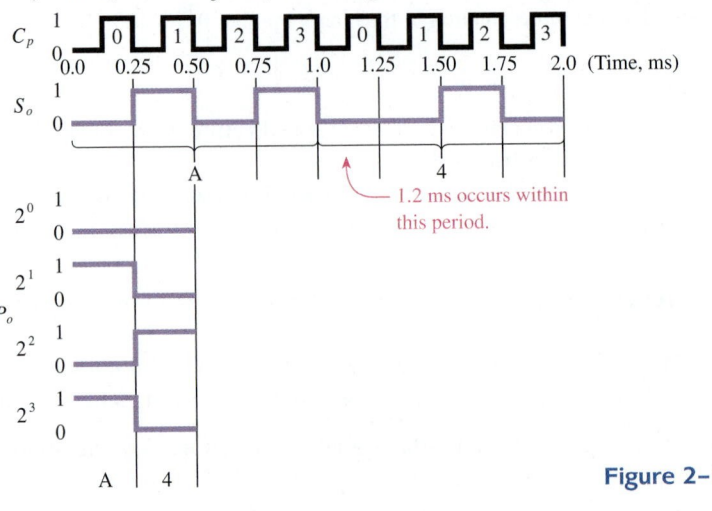

1.2 ms occurs within this period.

Figure 2–10

34

Review Questions

2–5. What advantage does parallel have over serial in the transmission of digital signals?

2–6. Which system requires more electrical conductors and circuitry, serial or parallel?

2–7. How long will it take to transmit three 8-bit binary strings in serial if the clock frequency is 5 MHz?

2–8. Repeat Questions 2–7 for an 8-bit parallel system.

2–5   Switches in Electronic Circuits*

The transitions between 0 and 1 digital levels are caused by switching from one voltage level to another (usually 0 V to +5 V). One way that switching is accomplished is to make and break a connection between two electrical conductors by way of a manual switch or an electromechanical relay. Another way to switch digital levels is by use of semiconductor devices such as **diodes** and **transistors.**

Manual switches and relays have almost *ideal* ON and OFF resistances in that when their contacts are closed (ON) the resistance (measured by an ohmmeter) is 0 ohms (Ω), and when their contacts are open (OFF), the resistance is infinite. Figure 2–11 shows the manual switch. When used in a digital circuit, a single-pole, double-throw manual switch can produce 0 and 1 states at some output terminal, as shown in Figures 2–12 and 2–13, by moving the switch (SW) to the up or down position.

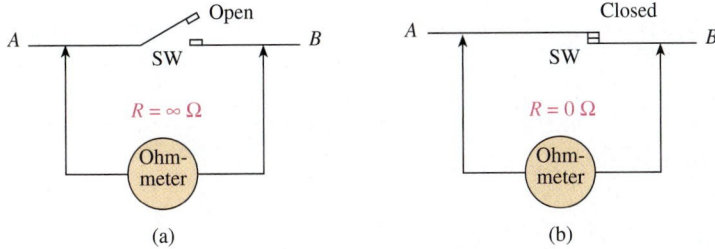

Figure 2–11   Manual switch: (a) switch open, $R = \infty$ ohms; (b) switch closed, $R = 0$ ohms.

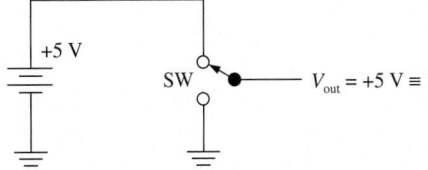

Figure 2–12   1-Level output.

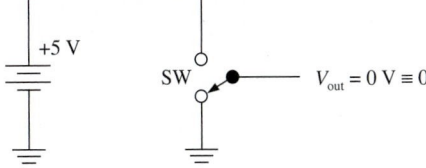

Figure 2–13   0-Level output.

---

*The fundamentals of basic electricity are provided in Appendix F. Ohm's law, simple series circuits, open circuits, and short circuits are explained to help you understand the electrical principles used in the remainder of this chapter.

2–6   A Relay as a Switch*

An electromechanical **relay** has contacts like a manual switch, but it is controlled by external voltage instead of being operated manually. Figure 2–14 shows the physical layout of an electromechanical relay. In Figure 2–14(a) the magnetic coil is energized by placing a voltage at terminals C_1–C_2; this will cause the lower contact to bend downward, opening the contact between X_1 and X_2. This relay is called *normally closed* (NC) because, at rest, the contacts are touching, or closed. In Figure 2–14(b), when the coil is energized, the upper contact will be attracted downward, making a connection between X_1 and X_2. This is called a *normally open* (NO) relay because, at rest, the contacts are not touching, they are open.

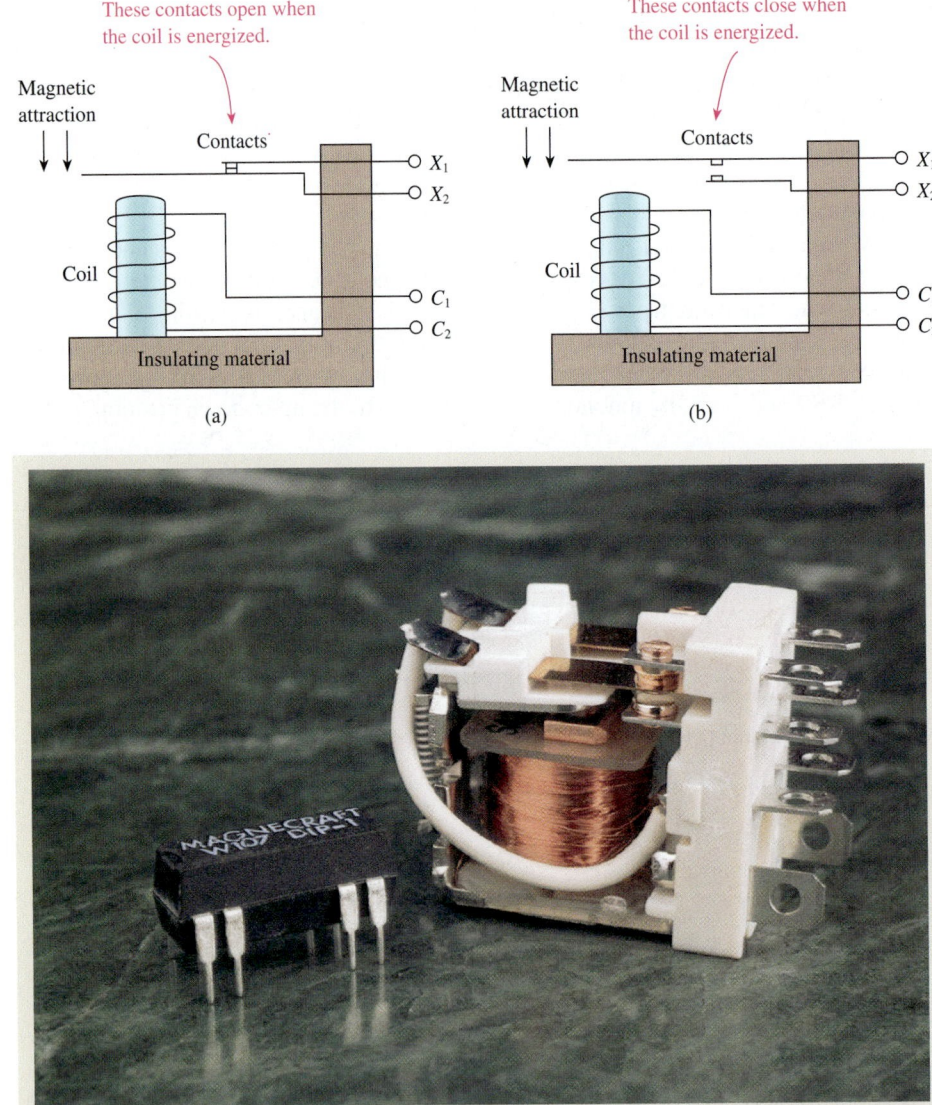

(a)

(b)

(c)

Figure 2–14   Physical representation of an electromechanical relay: (a) normally closed (NC) relay; (b) normally open (NO) relay; (c) photograph of actual relays.

---

*Systems requiring complex relay switching schemes are generally implemented using programmable logic controllers (PLC®s). PLC®s are microprocessor-based systems that are programmed to perform complex logic operations, usually to control electrical processes in manufacturing and industrial facilities. They use a programming technique called *ladder logic* to monitor and control several processes, eliminating the need for individually wired relays. PLC® is a registered trademark of Allen-Bradley Corporation.

A relay provides total isolation between the triggering source applied to C_1-C_2 and the output X_1-X_2. This total isolation is important in many digital applications, and it is a feature that certain semiconductor switches (e.g., transistors, diodes, and integrated circuits) cannot provide. Also, the contacts are normally rated for currents much higher than the current rating of semiconductor switches.

There are several disadvantages, however, of using a relay in electronic circuits. To energize the relay coil, the triggering device must supply several milliamperes, whereas a semiconductor requires only a few microamperes to operate. A relay is also much slower than a semiconductor. It will take several milliseconds to switch, compared to microseconds (or nanoseconds) for a semiconductor switch.

In Figure 2–15 a relay is used as a shorting switch in an electric circuit. The +5-V source is used to energize the coil, and the +12-V source is supplying the external electric circuit. When the switch (SW) in Figure 2–15(a) is closed, the relay coil will become energized, causing the relay contacts to open, which will make V_{out} change from 0 V to 6 V with respect to ground. The voltage-divider equation (see Appendix F) is used to calculate V_{out}:

$$V_{out} = \frac{12 \text{ V} \times 5 \text{ k}\Omega}{5 \text{ k}\Omega + 5 \text{ k}\Omega} = 6 \text{ V}$$

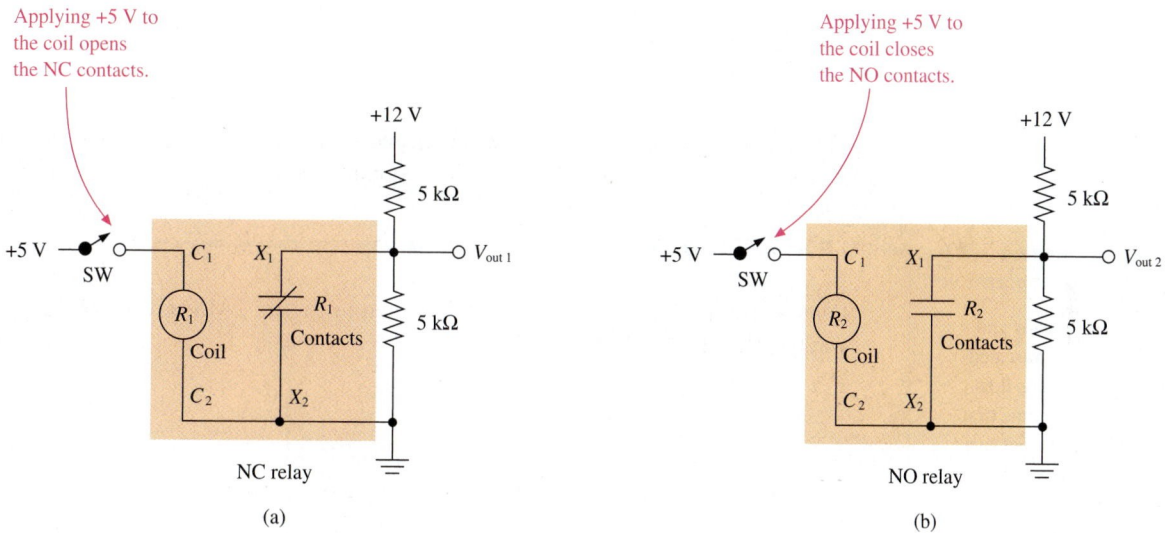

Figure 2–15 Symbolic representation of an electromechanical relay: (a) NC relay used in a circuit; (b) NO relay used in a circuit.

When the switch in Figure 2–15(b) is closed, the relay coil becomes an **energized relay coil,** causing the relay contacts to close, changing $V_{out\,2}$ from 6 V to 0 V.

Now, let's go a step further and replace the 5-V battery and switch with a clock oscillator and use a timing diagram to analyze the results. In Figure 2–16, the relay is triggered by the clock waveform, C_p. The diode D_1 is placed across the relay coil to protect it from arcing each time the coil is de-energized. Timing diagrams are very useful for comparing one waveform to another because the waveform changes states (1 or 0) relative to time. The timing diagram in Figure 2–17 shows that when the clock goes HIGH (1), the relay is energized, causing $V_{out\,3}$ to go LOW (0). When C_p goes LOW (0), the relay is de-energized, causing $V_{out\,3}$ to go to +5 V (using the voltage divider equation, $V_{out} = [10 \text{ V} \times 5 \text{ k}\Omega]/[5 \text{ k}\Omega + 5 \text{ k}\Omega] = 5 \text{ V}$).

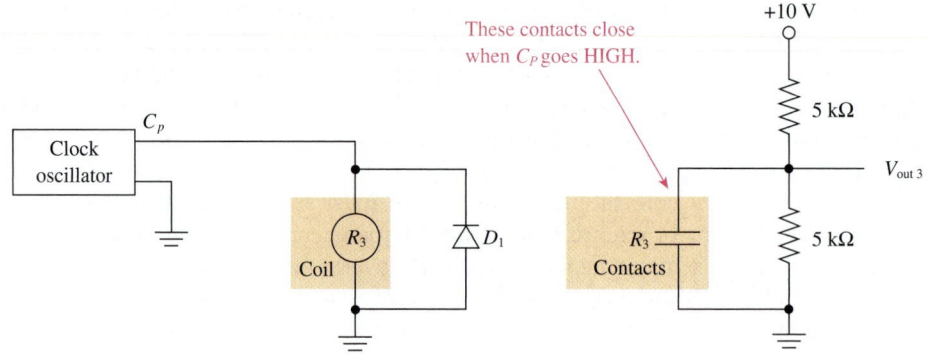

Figure 2–16 Relay used in a digital circuit.

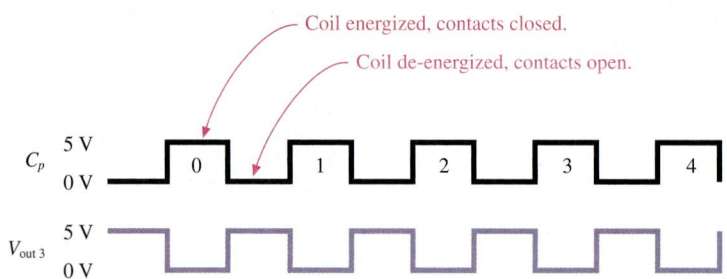

Figure 2–17 Timing diagram for Figure 2–16.

The following examples illustrate electronic switching and will help to prepare you for more complex timing analysis in subsequent chapters.

Common Misconception

The effects of opens and shorts are often miscalculated. Occasionally, it is instructive to assume that an open is equivalent to a 10-MΩ resistor and calculate the voltage across it using the voltage divider equation. Appendix F provides several examples of opens and shorts to illustrate their effect on circuits.

EXAMPLE 2–7

Draw a timing diagram for the circuit shown in Figure 2–18, given the C_p waveform in Figure 2–19.

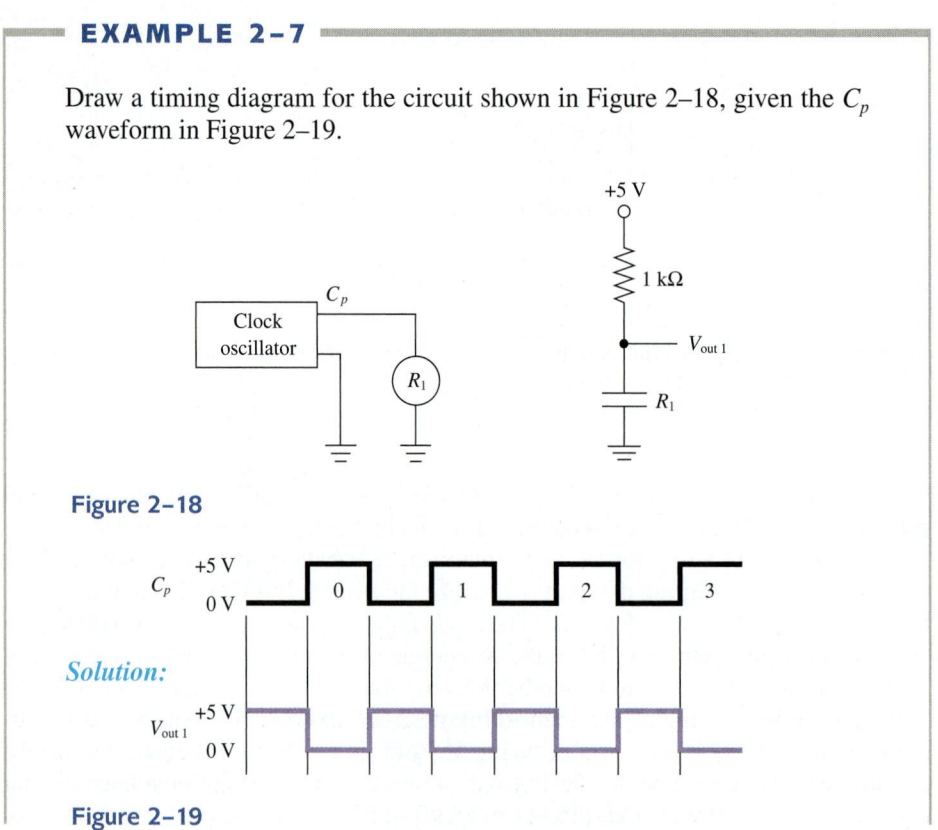

Figure 2–18

Solution:

Figure 2–19

Explanation: When C_p is LOW, the R_1 coil is de-energized, the R_1 contacts are open, $I_{1 \text{ k}\Omega} = 0$ A, $V_{\text{drop } 1 \text{ k}\Omega} = I \times R = 0$ V, and $V_{\text{out1}} = 5$ V $- 0_{V \text{ drop}} = 5$ V. When C_p is HIGH, the R_1 coil is energized, the R_1 contacts are closed, and $V_{\text{out1}} = 0$ V. (See Appendix F for a review of opens and shorts.)

EXAMPLE 2–8

Draw a timing diagram for the circuit shown in Figure 2–20, given the C_p waveform in Figure 2–21.

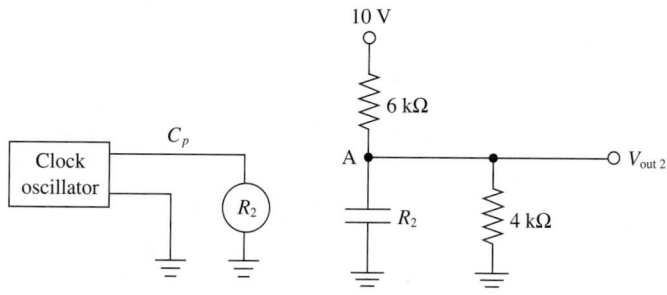

Figure 2–20

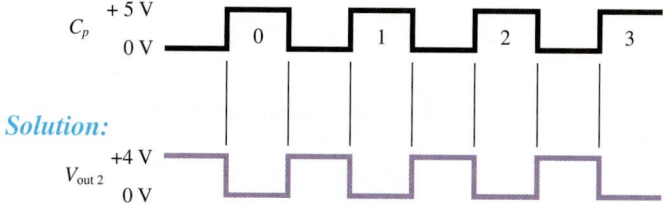

Figure 2–21

Explanation: When the R_2 contacts are closed (R_2 is energized), the voltage at point A is 0 V, making V_{out2} equal to 0 V. When the R_2 contacts are open (R_2 is de-energized), the voltage at point A is

$$V_A = \frac{10 \text{ V} \times 4 \text{ k}\Omega}{6 \text{ k}\Omega + 4 \text{ k}\Omega} = 4 \text{ V} \text{ and } V_{\text{out2}} = V_A = 4 \text{ V}.$$

Review Questions

2–9. Describe the operation of a relay coil and relay contacts.

2–10. How does a normally open relay differ from a normally closed relay?

2–7 A Diode as a Switch

Manual switches and electromechanical relays have limited application in today's digital electronic circuits. Most digital systems are based on semiconductor technology, which uses diodes and transistors. In Chapter 9, we discuss in detail the formation of digital circuits using transistors and diodes. Most electronics students should also take a separate course in electronic devices to cover the in-depth theory of the operation of diodes and transistors. However, without getting into a lot of detail, let's look at how a diode and a transistor can operate as a simple ON/OFF switch.

A diode is a semiconductor device that allows current to flow in one direction but not the other. Figure 2–22 shows a diode in both the conducting and nonconducting states. The term *forward biased* refers to a diode whose anode voltage is *more positive* than its cathode, thus allowing current flow in the direction of the arrow. (**Bias** is the voltage necessary to cause a semiconductor device to conduct or cut off current flow.) A reverse-biased diode will not allow current flow because its anode voltage is *equal to* or *more negative* than its cathode. A diode is analogous to a check valve in a water system (see Figure 2–23).

A diode is not a perfect short in the forward-biased condition, however. The voltage-versus-current curve shown in Figure 2–24 shows the characteristics of a diode. Notice in the figure that for the reverse-biased condition, as V_{rev} becomes more negative, there is still practically zero current flow.

In the forward-biased condition, as V_{forw} becomes more positive, no current flows until a 0.7-V cut-in voltage is reached.* After that point, the voltage across the diode

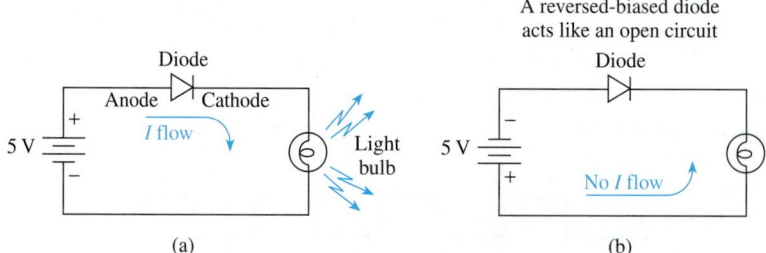

Figure 2–22   Diode in a series circuit: (a) forward biased; (b) reverse biased.

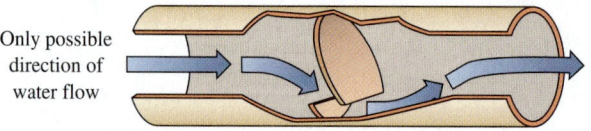

Figure 2–23   Water system check valve.

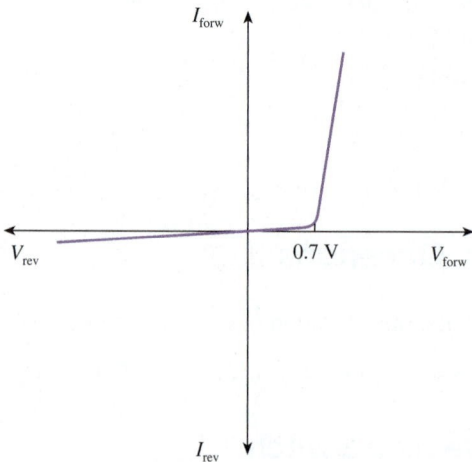

Figure 2–24   Diode voltage versus current characteristic curve.

---

*0.7 V is the typical cut-in voltage of a silicon diode, whereas 0.3 V is typical for a germanium diode. We will use the silicon diode because it is most commonly used in industrial applications.

(V_{forw}) will remain at approximately 0.7 V, and I_{forw} will flow, limited only by the external resistance of the circuit and the 0.7-V internal voltage drop.

What this means is that current will flow only if the anode is more positive than the cathode, and under those conditions, the diode acts like a short circuit except for the 0.7 V across its terminals. This fact is better illustrated in Figure 2–25.

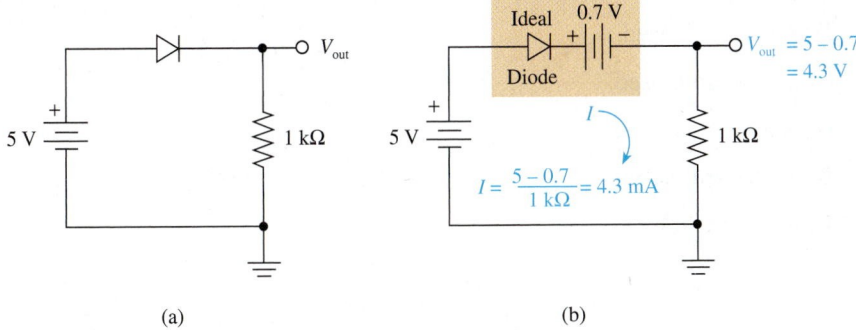

(a)   (b)

Figure 2–25   Forward-biased diode in an electric circuit: (a) original circuit; (b) equivalent circuit showing the diode voltage drop and $V_{out} = 5 - 0.7 = 4.3$ V.

The following examples and the problems at the end of the chapter demonstrate the effect that diodes have on electric circuits.

EXAMPLE 2–9

Determine if the diodes shown in Figure 2–26 are forward or reverse biased.

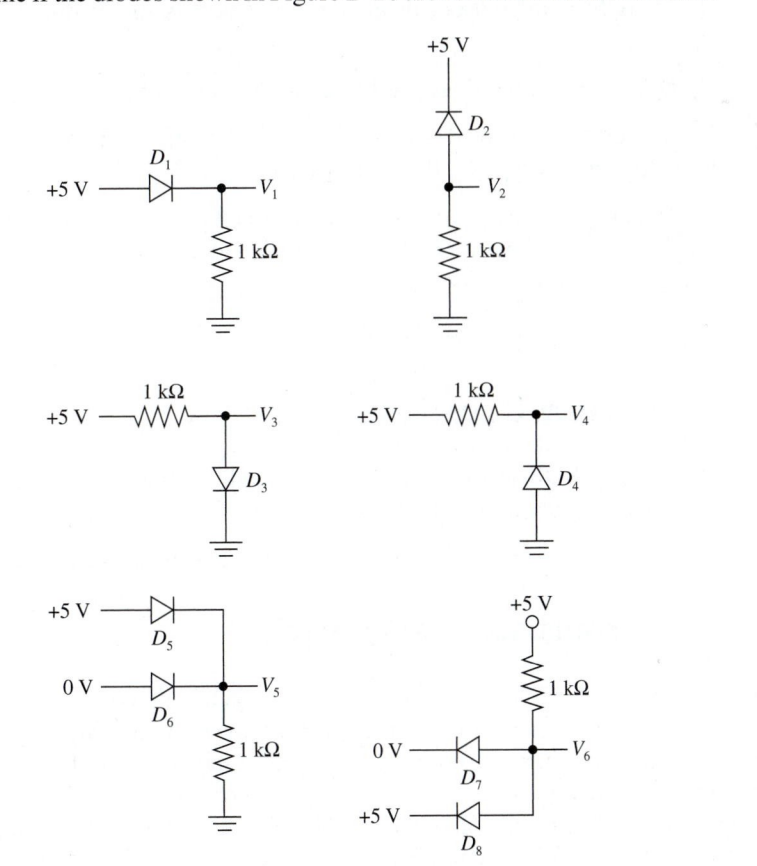

Figure 2–26

Solution: The diode is forward biased if the anode is more positive than the cathode.

D_1 is forward biased.

D_2 is reverse biased.

D_3 is forward biased.

D_4 is reverse biased.

D_5 is forward biased.

D_6 is reverse biased.

D_7 is forward biased.

D_8 is reverse biased.

EXAMPLE 2–10

Determine V_1, V_2, V_3, and V_4 (with respect to ground) for the circuits in Example 2–9.

Solution: V_1: D_1 is forward biased, dropping 0.7 V across its terminals. Therefore, $V_1 = 4.3$ V (5.0 − 0.7).

V_2: D_2 is reverse biased. No current will flow through the 1-kΩ resistor, so $V_2 = 0$ V.

V_3: D_3 is forward biased, dropping 0.7 V across its terminals, making $V_3 = 0.7$ V.

V_4: D_4 is reverse biased, acting like an open. Therefore, $V_4 = 5$ V.

V_5: Because D_6 is reverse biased (open), it has no effect on the circuit. D_5 is forward biased, dropping 0.7 V, making $V_5 = 4.3$ V.

V_6: D_8 is reverse biased (open), so it has no effect on the circuit. D_7 is forward biased, so it has +0.7 V on its anode side, which is +0.7 above the 0-V ground level, making $V_6 = +0.7$ V.

Review Questions

2–11. To forward bias a diode, the anode is made more _____ (positive, negative) than the cathode.

2–12. A forward-biased diode has how many volts across its terminals?

2–8   A Transistor as a Switch

The bipolar transistor is a very commonly used switch in digital electronic circuits. It is a three-terminal semiconductor component that allows an input signal at one of its terminals to cause the other two terminals to become a short or an open circuit. The transistor is most commonly made of silicon that has been altered into *N*-type material and *P*-type material. *N*-type silicon is made by bombarding pure silicon with atoms having structures with *one more* electron than silicon does. *P*-type silicon is made by bombarding pure silicon with atoms having structures with *one less* electron than silicon does.

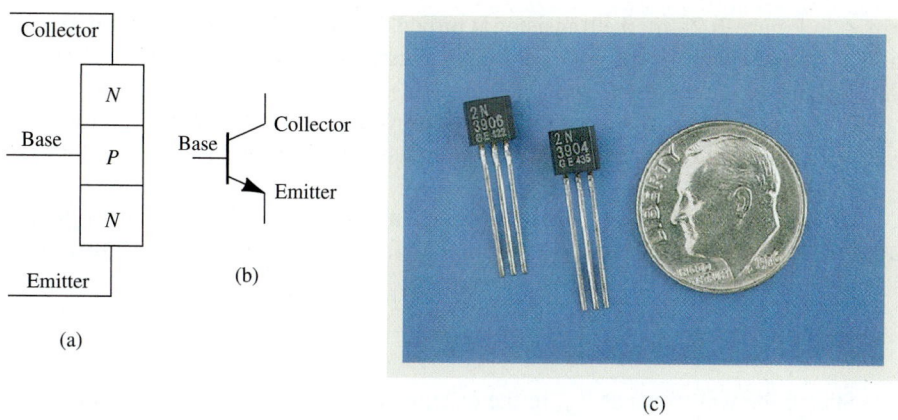

Figure 2–27 The *NPN* bipolar transistor: (a) physical layout; (b) symbol; (c) photograph.

Three distinct regions make up a bipolar transistor: *emitter, base,* and *collector.* They can be a combination of *N-P-N*-type material or *P-N-P*-type material bonded together as a three-terminal device. Figure 2–27 shows the physical layout and symbol for an *NPN* transistor. (In a *PNP* transistor, the emitter arrow points the other way.)

In an electronic circuit, the input signal (1 or 0) is usually applied to the base of the transistor, which causes the collector–emitter junction to become a short or an open circuit. The rules of transistor switching are as follows:

1. In an *NPN* transistor, applying a positive voltage from base to emitter causes the collector-to-emitter junction to short (this is called "turning the transistor ON"). Applying a negative voltage or 0 V from base to emitter causes the collector-to-emitter junction to open (this is called "turning the transistor OFF").

2. In a *PNP** transistor, applying a negative voltage from base to emitter turns it ON. Applying a positive voltage or 0 V from base to emitter turns it OFF.

Figure 2–28 shows how an *NPN* transistor functions as a switch in an electronic circuit. In the figure, resistors R_B and R_C are used to limit the base current and the collector

A positive voltage on the base of an NPN causes *C*-to-*E* to short.

Common Misconception

Students often think that the input signal to the base of a transistor must somehow be part of the output at the collector or emitter, but it is not. Once you determine if the *C*-to-*E* is a short or an open, you can ignore the base circuit altogether.

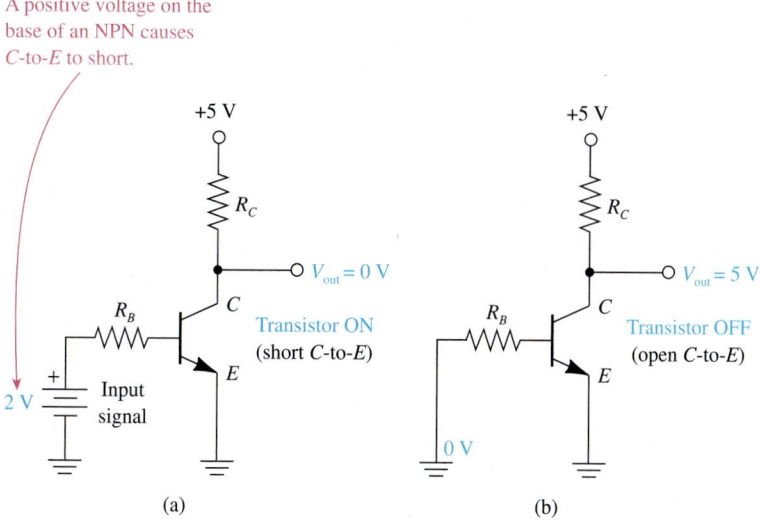

Figure 2–28 *NPN* transistor switch: (a) transistor ON; (b) transistor OFF.

PNP transistor circuits are analyzed in the same way as *NPN* circuits except that all voltage and current polarities are reversed. *NPN* circuits are much more common in industry and will be used most often in this book.

current. In Figure 2–28(a), the transistor is turned ON because the base is more positive than the emitter (input signal = +2 V). This causes the collector-to-emitter junction to short, placing ground potential at V_{out} (V_{out} = 0 V).

In Figure 2–28(b), the input signal is removed, making the base-to-emitter junction 0 V, turning the transistor OFF. With the transistor OFF, there is no current through R_C, so V_{out} = 5 V − (0 A × R_C) = 5 V.

Digital input signals are usually brought in at the base of the transistor, and the output is taken off the collector or emitter. The following examples use timing analysis to compare the input and output waveforms.

EXAMPLE 2–11

Sketch the waveform at V_{out} in the circuit shown in Figure 2–29, given the input signal C_p in Figure 2–30.

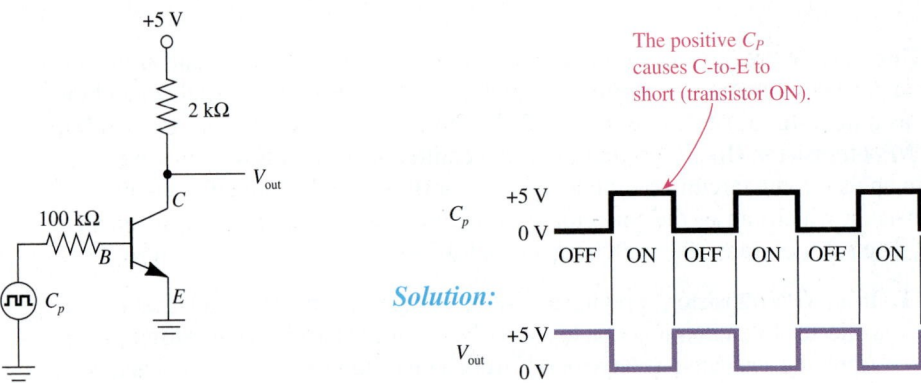

Figure 2–29   **Figure 2–30**

Explanation: When C_p = 0 V, the transistor is OFF and the equivalent circuit is as shown in Figure 2–31(a).

$$I_C = 0 \text{ A}$$

Therefore,

$$V_C = 5 \text{ V} − (0 \text{ A} \times 2 \text{ k}\Omega) = 5 \text{ V}$$

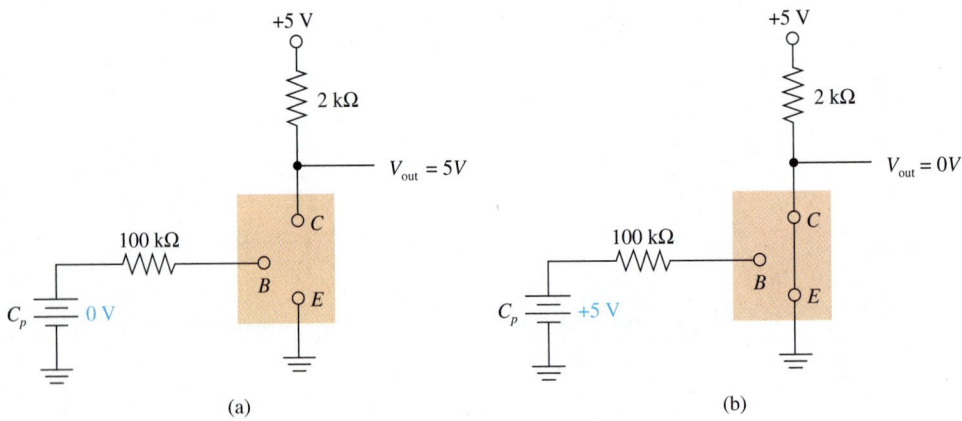

Figure 2–31   Equivalent circuits: (a) transistor OFF; (b) transistor ON.

When C_p = +5 V, the transistor is ON and the equivalent circuit is as shown in Figure 2–31(b). The collector is shorted directly to ground; therefore, V_{out} = 0 V.

EXAMPLE 2–12

Sketch the waveform at V_{out} in the circuit shown in Figure 2–32, given the input signal C_p in Figure 2–33.

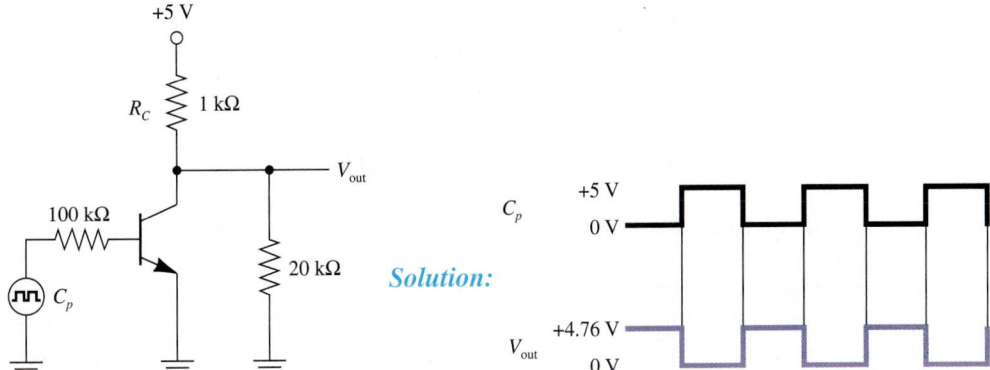

Figure 2–32

Figure 2–33

Solution:

Explanation: When $C_p = 0$ V, the transistor is OFF and the equivalent circuit is as shown in Figure 2–34(a). From the voltage-divider equation,

$$V_{out} = \frac{5\text{ V} \times 20\text{ k}\Omega}{20\text{ k}\Omega + 1\text{ k}\Omega} = 4.76\text{ V}$$

Next, when $C_p = +5$ V, the transistor is ON and the equivalent circuit is as shown in Figure 2–34(b). Now the collector is shorted to ground, making $V_{out} = 0$ V. Notice the difference in V_{out} as compared to Example 2–11, which had no load resistor connected to V_{out}.

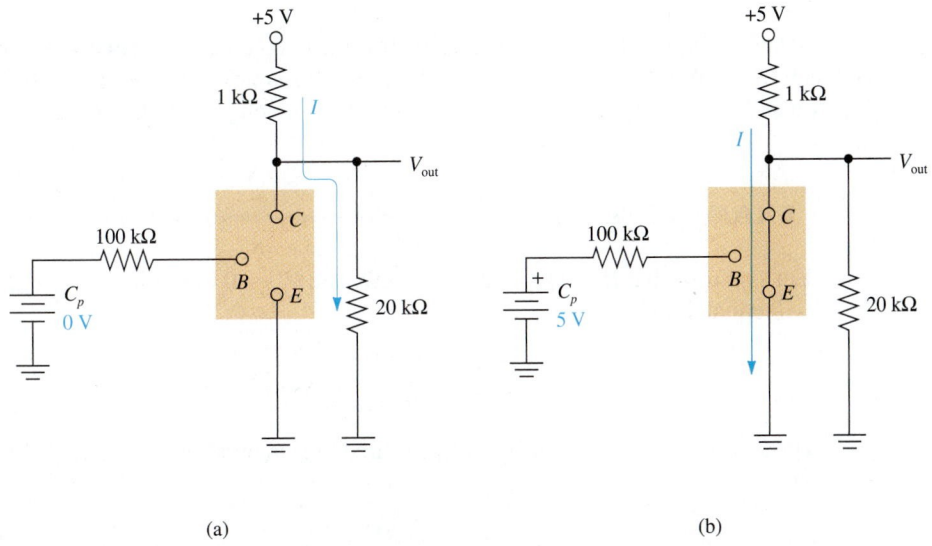

(a)

(b)

Figure 2–34 Equivalent circuits: (a) transistor OFF; (b) transistor ON.

2–13. Name the three pins on a transistor.

2–14. To turn ON an *NPN* transistor, a _____ (positive, negative) voltage is applied to the base.

2–15. When a transistor is turned ON, its collector-to-emitter becomes a _____ (short, open).

2–9   The TTL Integrated Circuit

Transistor–transistor logic (**TTL**) is one of the most widely used integrated-circuit technologies. TTL integrated circuits use a combination of several transistors, diodes, and resistors integrated together in a single package.

One basic function of a TTL integrated circuit is as a complementing switch, or **inverter.** The inverter is used to take a digital level at its input and complement it to the opposite state at its output (1 becomes 0, 0 becomes 1). Figure 2–35 shows how a common-emitter-connected transistor switch can be used to perform the same function.

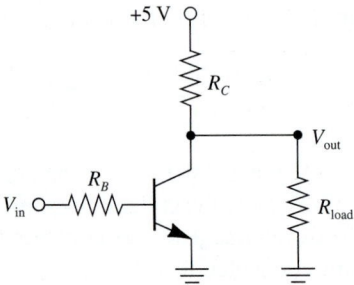

Figure 2–35   Common-emitter transistor circuit operating as an inverter.

When V_{in} equals 1 (+5 V), the transistor is turned on (called **saturation**) and V_{out} equals 0 (0 V). When V_{in} equals 0 (0 V), the transistor is turned off (called **cutoff**) and V_{out} equals 1 (approximately 5 V), assuming that R_L is much greater than R_C ($R_L >> R_C$).

EXAMPLE 2–13

Let's assume that $R_C = 1\ k\Omega$, $R_L = 10\ k\Omega$, and $V_{in} = 0$ in Figure 2–35. V_{out} will equal 4.55 V:

$$\frac{5\ V \times 10\ k\Omega}{1\ k\Omega\ +\ 10\ k\Omega} = 4.55\ V$$

But if R_L decreases to 1 kΩ by adding more loads in parallel with it, V_{out} will drop to 2.5 V:

$$\frac{5\ V \times 1\ k\Omega}{1\ k\Omega\ +\ 1\ k\Omega} = 2.5\ V$$

We can see from Example 2–13 that the 1-level output of the inverter is very dependent on the size of the load resistor (R_L), which can typically vary by a factor of 10. So right away you might say, "Let's keep R_C very small so that R_L is always much greater than R_C" ($R_L >> R_C$). Well, that's fine for the case when the transistor is cutoff ($V_{out} = 1$), but when the transistor is saturated ($V_{out} = 0$), the transistor collector current will be excessive if R_C is very small ($I_C = 5$ V/R_C; see Figure 2–36).

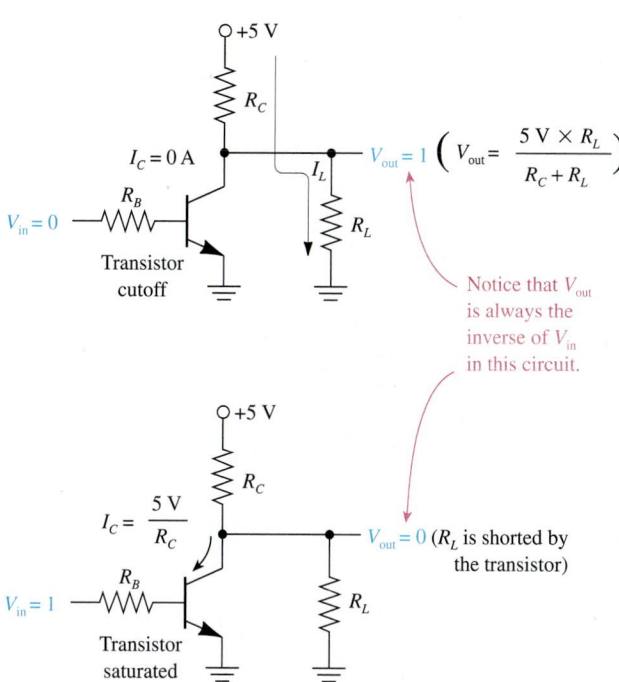

Figure 2–36 Common-emitter calculations.

Helpful Hint

If you understand the idea that V_{out} varies depending on the size of the connected load, it will help you understand why gate outputs in the upcoming chapters are not exactly 0 V and 5 V. We discuss TTL and CMOS input/output characteristics in Chapter 9.

Therefore, it seems that when the transistor is cutoff ($V_{out} = 1$), we want R_C to be small to ensure that V_{out} is close to 5 V, but when the transistor is saturated, we want R_C to be large to avoid excessive collector current.

This idea of needing a variable R_C resistance is accommodated by the TTL **integrated circuit** (Figure 2–37). It uses another transistor (Q_4) in place of R_C to act like a varying resistance. Q_4 is cutoff (acts like a high R_C) when the output transistor (Q_3) is saturated, and then Q_4 is saturated (acts like a low R_C) when Q_3 is cutoff. (In other words, when one transistor is ON, the other one is OFF.) This combination of Q_3 and Q_4 is referred to as the **totem-pole** arrangement.

Transistor Q_1 is the input transistor used to drive Q_2, which is used to control Q_3 and Q_4. Diode D_1 is used to protect Q_1 from negative voltages that might inadvertently be placed at the input. D_2 is used to ensure that when Q_3 is saturated, Q_4 will be cut off totally. V_{CC} is the abbreviation used to signify the power supply to the integrated circuit.

TTL is a very popular family of integrated circuits. It is much more widely used than RTL (resistor–transistor logic) or DTL (diode–transistor logic) circuits, which were the forerunners of TTL. Details on the operation and specifications of TTL ICs are given in Chapter 9. In that chapter, you will learn why V_{out} is not exactly 0 V and 5 V (it is more typically 0.2 V and 3.4 V).

A single TTL integrated-circuit (IC) package such as the 7404 has six complete logic circuits fabricated into a single silicon **chip,** each logic circuit being the equivalent of Figure 2–37. The 7404 has 14 metallic pins connected to the outside of a plastic case containing the silicon chip. The 14 pins, arranged 7 on a side, are aligned on 14 holes of

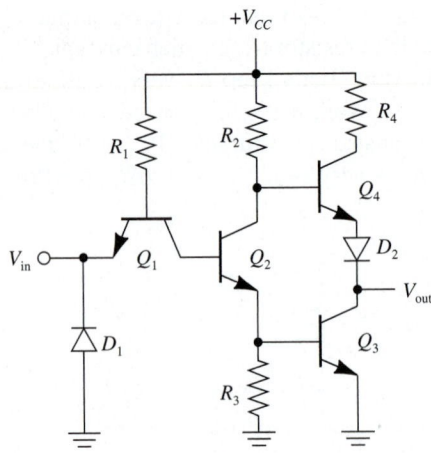

Figure 2–37 Schematic of a TTL inverter circuit.

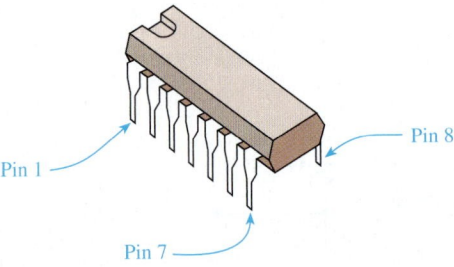

Figure 2–38 A 7404 TTL IC chip.

a printed-circuit board, where they are then soldered. The 7404 is called a 14-pin **DIP** (dual-in-line package) and costs less than 24 cents. Figure 2–38 shows a sketch of a 14-pin DIP IC. In subsequent chapters, we will see how to use ICs in actual digital circuitry.

ICs are configured as DIPs to ensure that the mechanical stress exerted on the pins when being inserted into a socket is equally distributed and that, although most of these pins serve as conductors to either the gates' inputs or outputs, some simply provide structural support and are simply anchored to the IC casing. These latter pins are denoted by the letters NC, meaning that they are *not* physically or electrically *connected* to an internal component.

The pin configuration of the 7404 is shown in Figure 2–39. The power supply connections to the IC are made to pin 14 (+5 V) and pin 7 (ground), which supplies

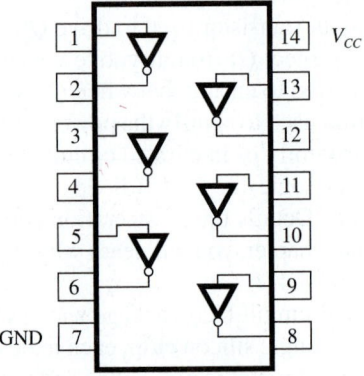

Figure 2–39 A 7404 hex inverter pin configuration.

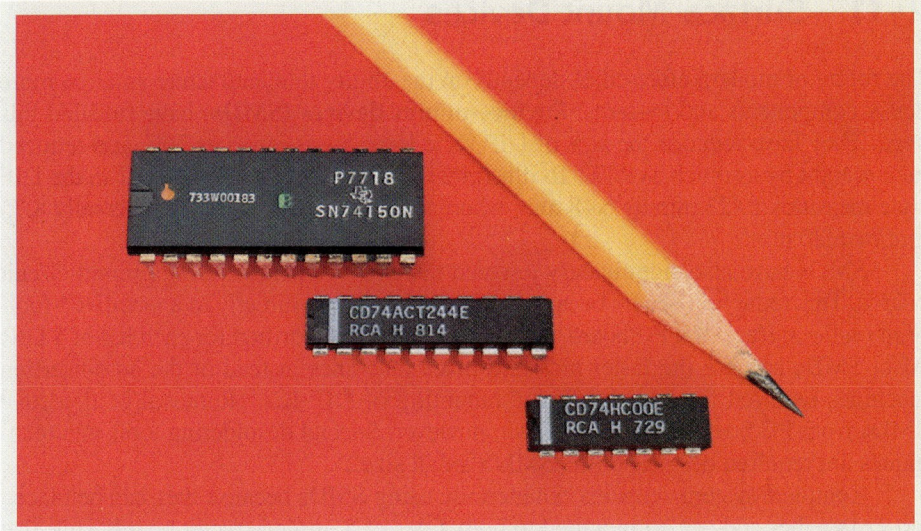

Figure 2–40   Photograph of three commonly used ICs: the 74HC00, 74ACT244, and 74150.

power to all six logic circuits. In the case of the 7404, the logic circuits are called *inverters*. The symbol for each inverter is a triangle with a circle at the output. The circle is used to indicate the inversion function. Although *never* shown in the pin configuration top view of digital ICs, each gate is electrically tied internally to both V_{CC} and ground. The entire circuit shown in Figure 2–37 is contained inside each of the six inverters.

Figure 2–40 shows three different ICs next to a pencil to give you an idea of their sizes.

2–10   The CMOS Integrated Circuit

Another common IC technology used in digital logic is the **CMOS** (complementary metal oxide semiconductor). CMOS uses a complementary pair of metal oxide semiconductor field-effect transistors (MOSFETs) instead of the bipolar transistors used in TTL chips. (Complete coverage of TTL and CMOS is given in Chapter 9.)

The major advantage of using CMOS is its low power consumption. Because of that, it is commonly used in battery-powered devices such as hand-held calculators and digital thermometers. The disadvantage of using CMOS is that generally its switching speed is slower than TTL and it is susceptible to burnout due to electrostatic charges if not handled properly. Figure 2–41 shows the pin configuration for a 4049 CMOS **hex inverter.**

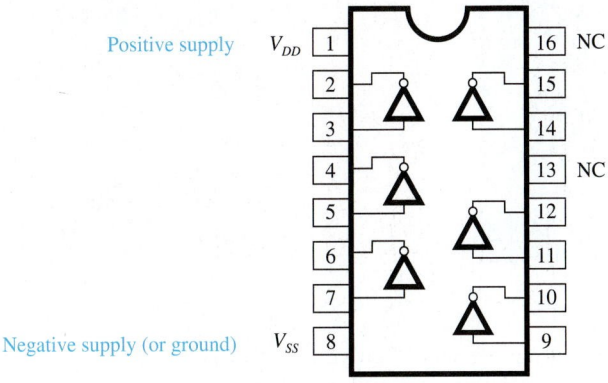

Figure 2–41   A 4049 CMOS hex inverter pin configuration.

2–11 Surface-Mount Devices

The future of modern electronics depends on the ability to manufacture smaller, more dense components and systems. **Surface-mount devices (SMDs)** have fulfilled this need. They have reduced the size of DIP-style logic by as much as 70% and reduced their weight by as much as 90%. To illustrate the size difference, a 7400 IC in the DIP style measures 19.23 mm by 6.48 mm, whereas the equivalent 7400 SMD is only 8.75 mm by 6.20 mm.

SMDs have also significantly lowered the cost of manufacturing printed-circuit boards. This reduction occurs because SMDs are soldered directly to a metalized footprint on the surface of a PC board, whereas holes must be drilled for each leg of a DIP. Also, SMDs can use the faster pick-and-place machines instead of the autoinsertion machines required for "through-hole" mounting of DIP ICs. (Removal of defective SMDs from PC boards is more difficult, however. Special desoldering tools and techniques are required because of the SMD's small size.)

Complete system densities can increase using SMDs because they can be placed closer together and can be mounted to both sides of a printed-circuit board. This also tends to decrease the capacitive and inductive problems that occur in digital systems operating at higher frequencies. (This topic is discussed further in Chapter 9.)

The most popular SMD package styles are the SO (small outline) and the PLCC (plastic leaded chip carrier) shown in Figure 2–42. The SO is a dual-in-line plastic

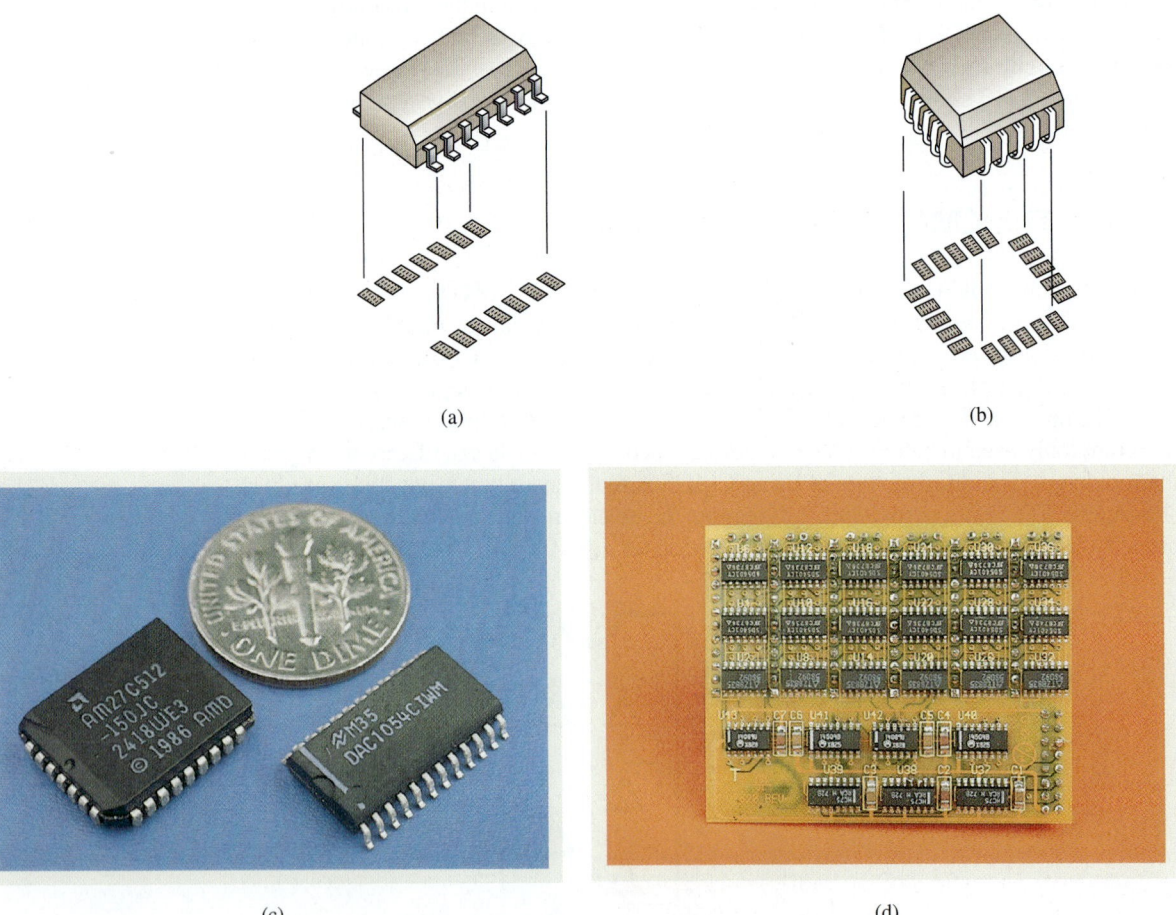

(a) (b)

(c) (d)

Figure 2–42 Typical surface-mount devices (SMDs) and their footprints: (a) small outline (SO); (b) plastic leaded chip carrier (PLCC); (c) photograph of actual SMDs; (d) photograph of SMDs mounted on a printed-circuit board.

package with leads spaced 0.050 in. apart and bent down and out in a gull-wing format. The PLCC is the most common SMD for ICs requiring a higher pin count (those having more than 28 pins). The PLCC is square, with leads on all four sides. They are bent down and under in a J-bend configuration. They, too, are soldered directly to the metalized footprint on the surface of the circuit board.

The SO package is available for the most popular lower-complexity TTL and CMOS digital logic and analog IC devices. PLCCs are available to implement more complex logic, such as microprocessors, microcontrollers, and large memories.

Review Questions

2–16. In a common-emitter transistor circuit, when V_{out} is 0, R_C should be _____ (small, large), and when V_{out} is 1, R_C should be _____ (small, large).

2–17. Which transistor in the schematic of the TTL circuit in Figure 2–37 serves as a variable R_C resistance?

Summary

In this chapter, we have learned that

1. The digital level for 1 is commonly represented by a voltage of 5 V in digital systems. A voltage of 0 V is used for the 0 level.

2. An oscilloscope can be used to observe the rapidly changing voltage-versus-time waveform in digital systems.

3. The frequency of a clock waveform is equal to the reciprocal of the waveform's period.

4. The transmission of binary data in the serial format requires only a single conductor with a ground reference. The parallel format requires several conductors but is much faster than serial.

5. Electromechanical relays are capable of forming shorts and opens in circuits requiring high current values but not high speed.

6. Diodes are used in digital circuitry whenever there is a requirement for current to flow in one direction but not the other.

7. The transistor is the basic building block of the modern digital IC. It can be switched on or off by applying the appropriate voltage at its base connection.

8. TTL and CMOS ICs are formed by integrating thousands of transistors in a single package. They are the most popular ICs used in digital circuitry today.

9. SMD-style ICs are gaining popularity over the through-hole style DIP ICs because of their smaller size and reduced manufacturing costs.

Glossary

Bias: The voltage necessary to cause a semiconductor device to conduct or cut off current flow. A device can be forward or reverse biased, depending on what action is desired.

Chip: The term given to an integrated circuit. It comes from the fact that each integrated circuit comes from a single chip of silicon crystal.

CMOS: Complementary metal oxide semiconductor. A family of integrated circuits used to perform logic functions in digital circuits. The CMOS is noted for its low power consumption but sometimes slow speed.

Cutoff: A term used in transistor switching that signifies the collector-to-emitter junction is turned off, or is not allowing current flow.

Diode: A semiconductor device used to allow current flow in one direction but not the other. As an electronic switch, it acts like a short in the forward-biased condition and like an open in the reverse-biased condition.

DIP: Dual-in-line packages. The most common pin layout for integrated circuits. The pins are aligned in two straight lines, one on each side.

Energized Relay Coil: By applying a voltage to the relay coil, a magnetic force is induced within it; this is used to attract the relay contacts away from their resting positions.

Frequency: A measure of the number of cycles or pulses occurring each second. Its unit is the hertz (Hz), and it is the reciprocal of the period.

Hex Inverter: An integrated circuit containing six inverters on a single DIP package.

Integrated Circuit: The fabrication of several semiconductor and electronic devices (transistors, diodes, and resistors) onto a single piece of silicon crystal. Integrated circuits are being used to perform the functions that once required several hundred discrete semiconductors.

Inverter: A logic circuit that changes its input into the opposite logic state at its output (0 to 1 and 1 to 0).

Logic State: A 1 or 0 digital level.

Oscilloscope: An electronic measuring device used in design and troubleshooting to display a waveform of voltage magnitude (y axis) versus time (x axis).

Parallel: A digital signal representation that uses several lines or channels to transmit binary information. The parallel lines allow for the transmission of an entire multibit number with each clock pulse.

Period: The measurement of time from the beginning of one periodic cycle or clock pulse to the beginning of the next. Its unit is the second(s), and it is the reciprocal of frequency.

Relay: An electric device containing an electromagnetic coil and normally open or normally closed contacts. It is useful because, by supplying a small triggering current to its coil, the contacts will open or close, switching a higher current on or off.

Saturation: A term used in transistor switching that signifies that the collector-to-emitter junction is turned on, or conducting current heavily.

Serial: A digital signal representation that uses one line or channel to transmit binary information. The binary logic states are transmitted 1 bit at a time, with the LSB first.

Surface-Mounted Device: A newer style of integrated circuit, soldered directly to the surface of a printed circuit board. They are much smaller and lighter than the equivalent logic constructed in the DIP through-hole-style logic.

Timing Diagram: A diagram used to display the precise relationship between two or more digital waveforms as they vary relative to time.

Totem Pole: The term used to describe the output stage of most TTL integrated circuits. The totem-pole stage consists of one transistor in series with another, configured in such a way that when one transistor is saturated, the other is cut off.

Transistor: A semiconductor device that can be used as an electronic switch in digital circuitry. By applying an appropriate voltage at the base, the collector-to-emitter junction will act like an open or a shorted switch.

TTL: Transistor–transistor logic. The most common integrated circuit used in digital electronics today. A large family of different TTL integrated circuits is used to perform all the logic functions necessary in a complete digital system.

▬▬ Problems ▬▬

Section 2–1 and 2–2

2–1. Determine the period of a clock waveform whose frequency is

(a) 2 MHz   (b) 500 kHz   (c) 4.27 MHz   (d) 17 MHz

Determine the frequency of a clock waveform whose period is

(e) 2 μs   (f) 100 μs   (g) 0.75 ms   (h) 1.5 μs

Sections 2–3 and 2–4

2–2. Sketch the serial and parallel representations (similar to Figure 2–10) of the following numbers, and calculate how long they will take to transmit (clock frequency = 2 MHz).

(a) $45B_{16}$ (b) $A3C_{16}$

2–3. (a) How long will it take to transmit the number 33_{10} in serial if the clock frequency is 3.7 MHz? (Transmit the number as an 8-bit binary number.)

(b) Is the serial line HIGH or LOW at 1.21 μs?

2–4. (a) How long will it take to transmit the three ASCII-coded characters $14 in 8-bit parallel if the clock frequency is 8 MHz?

(b) Repeat for $78.18 at 4.17 MHz.

Sections 2–5 and 2–6

2–5. Draw the timing diagram for V_{out1}, V_{out2}, and V_{out3} in Figure P2–5.

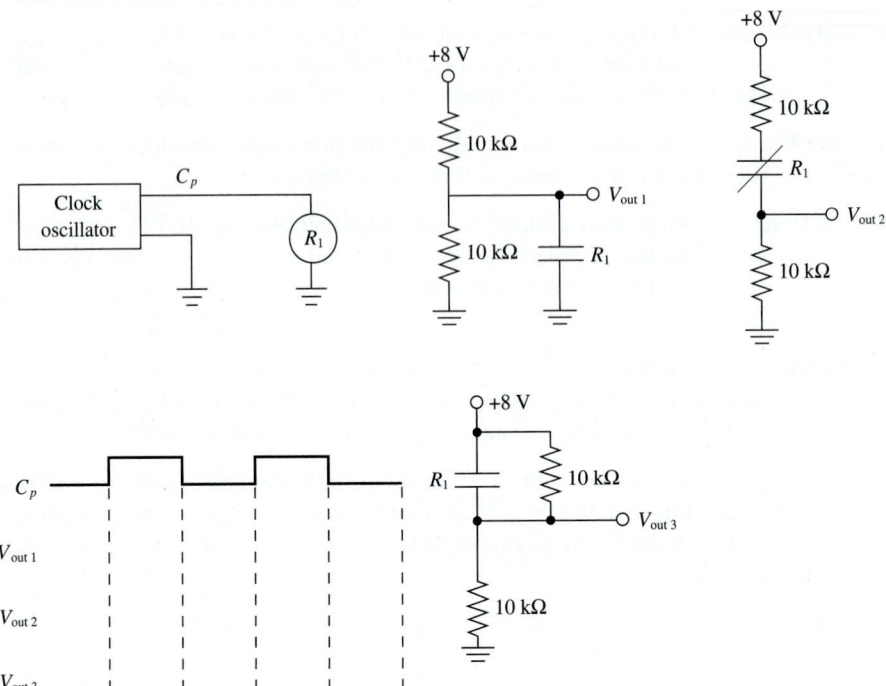

Figure P2–5

Section 2–7

2–6. Determine if the diodes in Figure P2–6 are reverse or forward biased.

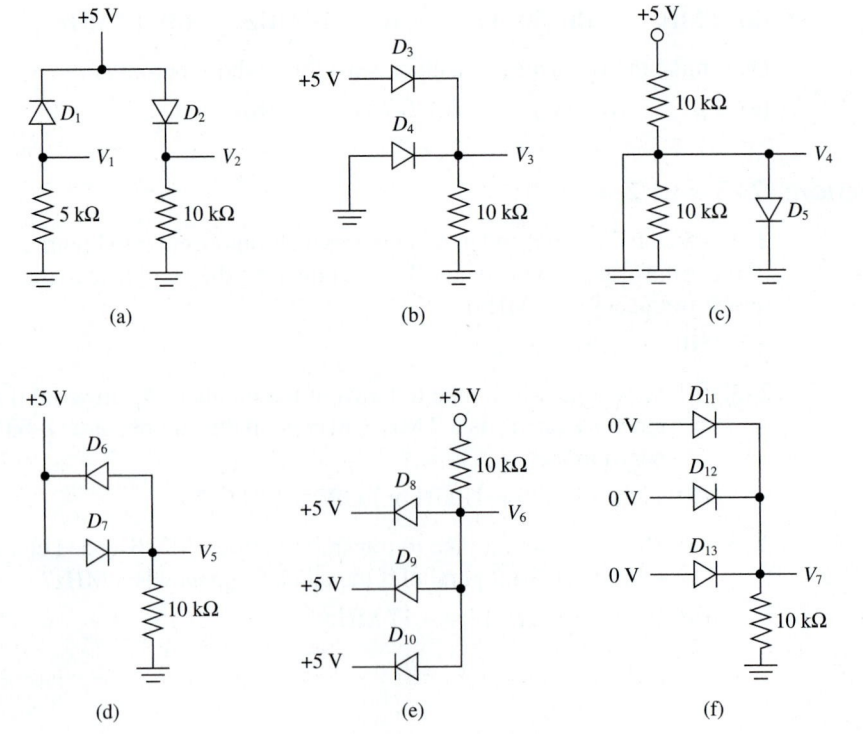

Figure P2–6

2–7. Determine V_1, V_2, V_3, V_4, V_5, V_6, and V_7 in the circuits of Figure P2–6.

2–8. In Figure P2–6, if the cathode of any one of the diodes D_8, D_9, or D_{10} is connected to 0 V instead of +5 V, what happens to V_6?

2–9. In Figure P2–6, if the anode of any of the diodes D_{11}, D_{12}, or D_{13}, is connected to +5 V instead of 0 V, what happens to V_7?

Section 2–8

2–10. Find V_{out1} and V_{out2} for the circuits of Figure P2–10.

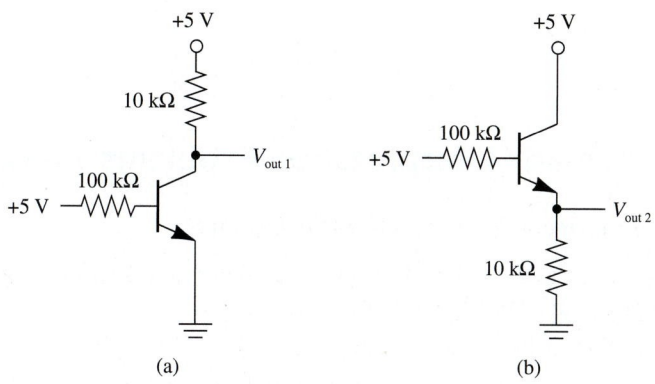

(a)   (b)

Figure P2–10

2–11. Sketch the waveforms at V_{out} in the circuit of Figure 2–32 using $R_C = 6$ kΩ.

Section 2–9

2–12. To use a common-emitter transistor circuit as an inverter, the input signal is connected to the _____ (base, collector, or emitter) and the output signal is taken from the _____ (base, collector, or emitter).

2–13. Determine V_{out} for the common-emitter transistor inverter circuit of Figure 2–35 using $V_{in} = 0$ V, $R_B = 1$ MΩ, $R_C = 330$ Ω, and $R_{load} = 1$ MΩ.

2–14. If the load resistor (R_{load}) used in Problem 2–13 is changed to 470 Ω, describe what happens to V_{out}.

2–15. In the circuit of Figure 2–35 with $V_{in} = 0$ V, V_{out} will be almost 5 V as long as R_{load} is much greater than R_C. Why not make R_C very small to ensure that the circuit will work for all values of R_{load}?

2–16. In Figure 2–35, if $R_C = 100$ Ω, find the collector current when $V_{in} = +5$ V.

2–17. Describe how the totem-pole output arrangement in a TTL circuit overcomes the problems faced when using the older common-emitter transistor inverter circuit.

2–18. Sketch the waveform at C_p and V_{out} for Figure P2–18.

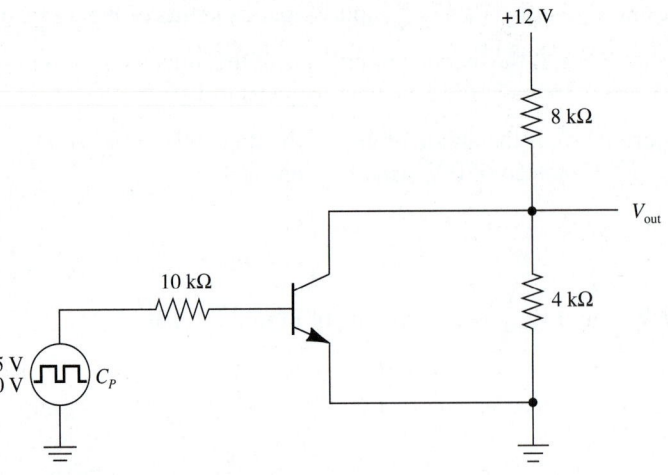

+12 V

8 kΩ

V_{out}

10 kΩ

4 kΩ

5 V
0 V C_P

Figure P2–18

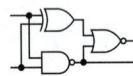

See Appendix G for the schematic diagrams.

Schematic Interpretation Problems

(S) **2–19.** Y1 in the 4096/4196 control card schematic sheet 1 is a crystal used to generate a very specific frequency.

(a) What is its rated frequency?

(b) What time period does that create?

(S) **2–20.** Repeat Problem 2–19 for the crystal X1 in the HC11D0 master board schematic.

(S) **2–21.** The circuit on the HC11D0 schematic is capable of parallel as well as serial communication via connectors P_3 and P_2. Which is parallel, and which is serial? (*Hint:* TX stands for transmit, RX stands for receive.)

(S) **2–22.** Is diode D_1 of the HC11D0 schematic forward or reverse biased? (*Hint:* $V_{CC} = 5$ V.)

(S) **2–23.** The transistor Q_1 in the HC11D0 schematic is turned ON and OFF by the level of pin 2 on U3:A. At what level must pin 2 be to turn Q_1 ON, and what will happen to the level on the line labeled RESET B when that happens?

Electronics Workbench®/MultiSIM® Exercises

E2–1. Load the circuit file for **Section 2–3.** Read the instructions in the *Description* window.

(a) Determine the three ASCII characters that are transmitted in serial.

(b) Determine the number of serial bits transmitted.

E2–2. Load the circuit file for **Section 2–4.** Read the instructions in the *Description* window.

(a) Determine the three ASCII characters that are transmitted in parallel.

(b) How many clock pulses did it take to complete the transmission?

E2–3. Load the circuit file for **Section 2–6a.** Read the instructions in the *Description* window. The normally open relay contacts are used to create a short across the lower 5-kΩ resistor when Cp goes HIGH.

(a) Measure the voltage levels of Cp and Vout3 with the oscilloscope. Note the relationship between the two waveforms.

(b) Change the upper resistor to 2kΩ and the lower resistor to 8kΩ. Predict the new voltage levels, then measure them with the oscilloscope.

(c) If the normally closed relay contacts were used, what change would you expect in the Vout3 waveform? Try it.

C   **E2–4.** Load the circuit file for **Section 2–6b.** Read the instructions in the *Description* window. The normally closed relay contacts are used to create an open between the two resistors when Cp goes HIGH.

(a) Measure the voltage levels of Cp and Vout4 with the oscilloscope. Note the relationship between the two waveforms. (The top waveform is Vout4).

(b) Change the upper resistor to 2kΩ and the lower resistor to 8kΩ. Predict the new voltage levels, then measure them with the oscilloscope.

(c) If the normally open relay contacts were used instead of the normally closed contacts, what change would you expect in the Vout4 waveform? Try it.

E2–5. Load the circuit file for **Section 2–7.** Read the instructions in the *Description* window. Before turning the power switch ON, predict the voltage V1, V2, V3 and V4.

(a) Turn the switch ON and check your answers.

(b) Reverse all six diodes, and predict what V1, V2, V3 and V4 will become. Turn the power switch ON, and check your answers.

E2–6. Load the circuit file for **Section 2–8.** Read the instructions in the *Description* window.

(a) Measure the voltage levels of Cp and Vout with the oscilloscope. Note the relationship between the two waveforms.

(b) Change the upper resistor to 2kΩ and the lower resistor to 2kΩ. Predict the new voltage levels, then measure them with the oscilloscope.

Answers to Review Questions

2–1. x axis, time; y axis, voltage

2–2. The clock frequency is the reciprocal of the clock period.

2–3. 125 ns

2–4. 20 MHz

2–5. It is faster.

2–6. Parallel

2–7. 4.80 μs

2–8. 600 ns

2–9. The relay coil is energized by placing a voltage at its terminals. The contacts will either make a connection (NO relay) or break a connection (NC relay) when the coil is energized.

2–10. An NO relay makes connection when energized. An NC relay breaks connection when energized.

2–11. Positive

2–12. Approximately 0.7 V

2–13. Emitter, base, collector

2–14. Positive

2–15. Short

2–16. Large, small

2–17. Q_4

3

Basic Logic Gates

OUTLINE

OBJECTIVES

Upon completion of this chapter, you should be able to:

- Describe the operation and use of AND gates and OR gates.
- Construct truth tables for two-, three-, and four-input AND and OR gates.
- Draw timing diagrams for AND and OR gates.
- Describe the operation, using timing analysis, of an ENABLE function.
- Sketch the external connections to integrated-circuit chips to implement AND and OR logic circuits.
- Explain how to use a logic pulser and a logic probe to troubleshoot digital integrated circuits.
- Describe the operation and use of inverter, NAND, and NOR gates.
- Construct truth tables for two-, three-, and four-input NAND and NOR gates.
- Draw timing diagrams for inverter, NAND, and NOR gates.
- Use the outputs of a Johnson shift counter to generate specialized waveforms utilizing various combinations of the five basic gates.
- Develop a comparison of the Boolean equations and truth tables for the five basic gates.

INTRODUCTION

Logic **gates** are the basic building blocks for forming digital electronic circuitry. A logic gate has one output terminal and one or more input terminals. Its output will be HIGH (1) or LOW (0) depending on the digital level(s) at the input terminal(s). Through the use of logic gates, we can design digital systems that will evaluate digital input levels and produce a specific output response based on that particular logic circuit design. The five basic logic gates are the AND, OR, NAND, NOR, and inverter.

3–1   The AND Gate

Let's start by looking at the two-input AND gate whose schematic symbol is shown in Figure 3–1. The operation of the AND gate is simple and is defined as follows: *The out-*

Figure 3–1   Two-input AND gate symbol.

put, X, is HIGH if input A AND input B are both HIGH. In other words, if $A = 1$ *AND* $B = 1$, then $X = 1$. If either A or B or both are LOW, the output will be LOW.

The best way to illustrate how the output level of a gate responds to all the possible input-level combinations is with a **truth table.** Table 3–1 is a truth table for a two-

TABLE 3–1	Truth Table for a Two-Input AND Gate	
Inputs		
A	*B*	**Output *X***
0	0	0
0	1	0
1	0	0
1	1	1

input AND gate. On the left side of the truth table, all possible input-level combinations are listed, and on the right side, the resultant output is listed.

From the truth table, we can see that the output at X is HIGH *only* when *both A AND B* are HIGH. If this AND gate is a TTL integrated circuit, HIGH means $+5$ V and LOW means 0 V (i.e., 1 is defined as $+5$ V and 0 is defined as 0 V).

One example of how an AND gate might be used is in a bank burglar alarm system. The output of the AND gate will go HIGH to turn on the alarm if the alarm activation key is in the ON position *AND* the front door is opened. This setup is illustrated in Figure 3–2.

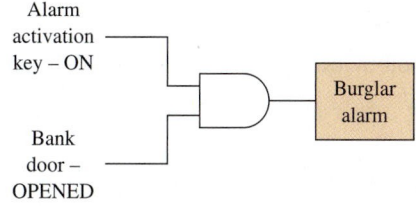

Figure 3–2   AND gate used to activate a burglar alarm.

Another way to illustrate the operation of an AND gate is by use of a series electric circuit. In Figure 3–3, using manual and transistor switches, the output at X is HIGH if *both* switches A AND B are HIGH (1).

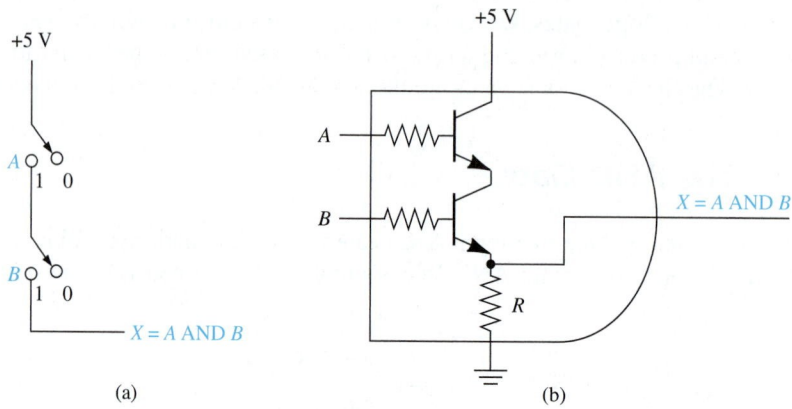

(a)   (b)

Figure 3–3   Electrical analogy for an AND gate: (a) using manual switches; (b) using transistor switches.

Figure 3–3 also shows what is known as the **Boolean equation** for the AND function, $X = A$ AND B, which can be thought of as X *equals 1 if* A AND B *both equal 1.* The Boolean equation for the AND function can more simply be written as $X = A \cdot B$ or just $X = AB$ (which is read as "X equals A AND B"). Boolean equations will be used throughout the rest of the book to depict algebraically the operation of a logic gate or a combination of logic gates.

AND gates can have more than two inputs. Figure 3–4 shows a four-input and an eight-input AND gate. The truth table for an AND gate with four inputs is shown in

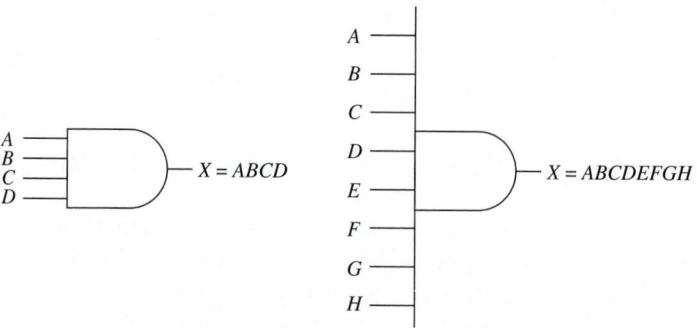

Figure 3–4   Multiple-input AND gate symbols.

Table 3–2. To determine the total number of different combinations to be listed in the truth table, use the equation

$$\text{number of combinations} = 2^N, \qquad \text{where } N = \text{number of inputs} \qquad \textbf{3–1}$$

Therefore, in the case of a four-input AND gate, the number of possible input combinations is $2^4 = 16$.

When building the truth table, be sure to list all 16 *different* combinations of input levels. One easy way to ensure that you do not inadvertently overlook a combination of these variables or duplicate a combination is to list the inputs in the order of a

TABLE 3–2	Truth Table for a Four-Input AND Gate			
A	B	C	D	X
0	0	0	0	0
0	0	0	1	0
0	0	1	0	0
0	0	1	1	0
0	1	0	0	0
0	1	0	1	0
0	1	1	0	0
0	1	1	1	0
1	0	0	0	0
1	0	0	1	0
1	0	1	0	0
1	0	1	1	0
1	1	0	0	0
1	1	0	1	0
1	1	1	0	0
1	1	1	1	1

The output at X is HIGH only if *all* inputs are HIGH.

Common Misconception

When you build a truth table, you might mistakenly omit certain input combinations if you don't set the variables up as a binary counter.

binary counter (0000, 0001, 0010, . . . ,1111). Also notice in Table 3–2 that the A column lists eight 0's, then eight 1's; the B column lists four 0's, four 1's, four 0's, four 1's; the C column lists two 0's, two 1's, two 0's, two 1's, and so on; and the D column lists one 0, one 1, one 0, one 1, and so on.

3–2   The OR Gate

The OR gate also has two or more inputs and a single output. The symbol for a two-input OR gate is shown in Figure 3–5. The operation of the two-input OR gate is defined as follows: *The output at X will be HIGH whenever input A OR input B is HIGH or both are HIGH.* As a Boolean equation, this can be written $X = A + B$ (which is read as "X equals A <u>OR</u> B"). Notice the use of the $+$ symbol to represent the OR function.

Figure 3–5   Two-input OR gate.

The truth table for a two-input OR gate is shown in Table 3–3.

TABLE 3–3	Truth Table for a Two-Input OR Gate	
Inputs		
A	B	Output X
0	0	0
0	1	1
1	0	1
1	1	1

From the truth table you can see that X is 1 whenever A OR B is 1 or if *both* A and B are 1. Using manual or transistor switches in an electric circuit, as shown in Figure 3–6, we can observe the electrical analogy to an OR gate. From the figure, we see that the output at X will be 1 if A *or* B, or *both*, are HIGH (1).

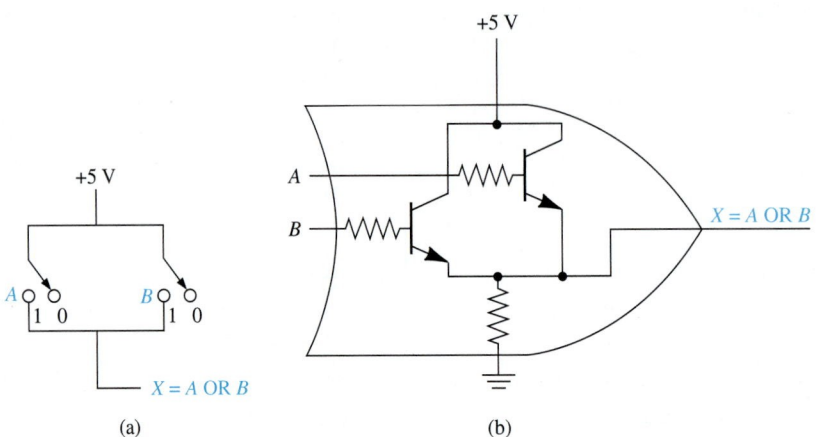

(a)   (b)

Figure 3–6 Electrical analogy for an OR gate: (a) using manual switches; (b) using transistor switches.

OR gates can also have more than two inputs. Figure 3–7 shows a three-input OR gate, and Figure 3–8 shows an eight-input OR gate. The truth table for the three-input OR gate will have eight entries ($2^3 = 8$), and the eight-input OR gate will have 256 entries ($2^8 = 256$).

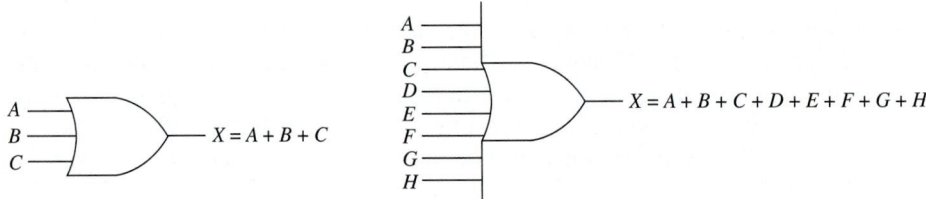

Figure 3–7 Three-input OR gate symbol.

Figure 3–8 Eight-input OR gate symbol.

Let's build a truth table for the three-input OR gate.

The truth table of Table 3–4 is built by first using Equation 3–1 to determine that there will be eight entries, then listing the eight combinations of inputs in the order of a binary counter (000 to 111), and then filling in the output column (X) by realizing that

The output at X is HIGH if *any* input is HIGH.

TABLE 3–4	Truth Table for a Three-Input OR Gate		
A	*B*	*C*	*X*
0	0	0	0
0	0	1	1
0	1	0	1
0	1	1	1
1	0	0	1
1	0	1	1
1	1	0	1
1	1	1	1

X will always be HIGH as long as at least one of the inputs is HIGH. When you look at the completed truth table, you can see that the only time the output is LOW is when *all* the inputs are LOW.

EXAMPLE 3-1

Determine the output at W, X, Y, and Z in Figure 3–9.

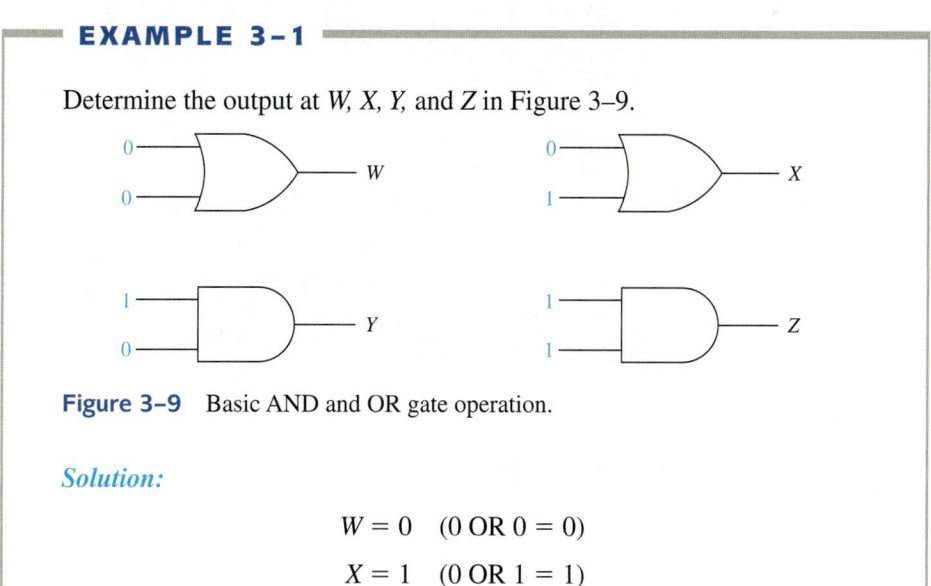

Figure 3-9   Basic AND and OR gate operation.

Solution:

$$W = 0 \quad (0 \text{ OR } 0 = 0)$$
$$X = 1 \quad (0 \text{ OR } 1 = 1)$$
$$Y = 0 \quad (1 \text{ AND } 0 = 0)$$
$$Z = 1 \quad (1 \text{ AND } 1 = 1)$$

Review Questions

3–1. All inputs to an AND gate must be HIGH for it to output a HIGH. True or false?

3–2. What is the purpose of a truth table?

3–3. What is the purpose of a Boolean equation?

3–4. What input conditions must be satisfied for the output of an OR gate to be LOW?

3–3   Timing Analysis

Another useful means of analyzing the output response of a gate to varying input-level changes is by means of a *timing diagram*. A timing diagram, as described in Chapter 2, is used to illustrate graphically how the output levels change in response to input-level changes.

The timing diagram in Figure 3–10 shows the two input waveforms (A and B) that are applied to a two-input AND gate and the X output that results from the AND operation. (For TTL and most CMOS logic gates, $1 = +5$ V and $0 = 0$ V.) As you can see, timing analysis is very useful for visually illustrating the level at the output for varying input-level changes.

Timing waveforms can be observed on an *oscilloscope* or a *logic analyzer.* A dual-trace oscilloscope can display *two* voltage-versus-time waveforms on the same x axis. That is ideal for comparing the relationship of one waveform relative to another.

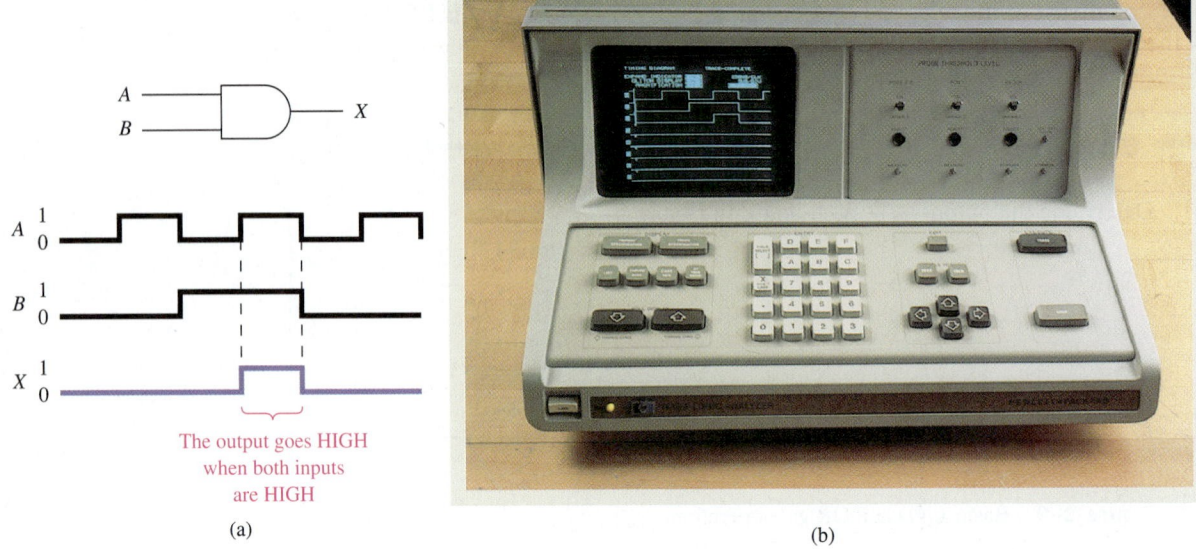

The output goes HIGH
when both inputs
are HIGH

(a)

(b)

Figure 3–10 Timing analysis of an AND gate: (a) waveform sketch; (b) actual logic analyzer display.

The other timing analysis tool is the logic analyzer. Among other things, it can display up to 16 voltage-versus-time waveforms on the same x axis (see Figure 3–10[b]). It can also display the levels of multiple digital signals in a *state table,* which lists the binary levels of all the waveforms, at predefined intervals, in binary, hexadecimal, or octal. Timing analysis of 8 or 16 channels concurrently is very important when analyzing advanced digital and microprocessor systems in which the interrelationship of several digital signals is critical for proper circuit operation.

EXAMPLE 3–2

Sketch the output waveform at X for the two-input OR gate shown in Figure 3–11, with the given A and B input waveforms in Figure 3–12.

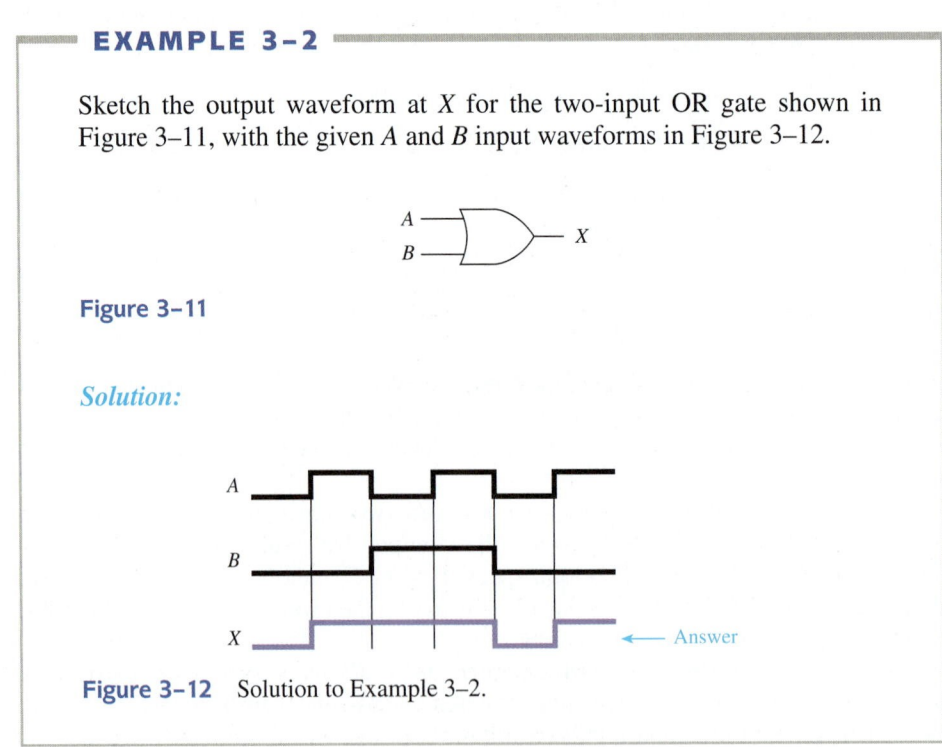

Figure 3–11

Solution:

Figure 3–12 Solution to Example 3–2.

EXAMPLE 3–3

Sketch the output waveform at *X* for the three-input AND gate shown in Figure 3–13, with the given *A, B,* and *C* input waveforms in Figure 3–14.

Figure 3–13

Solution:

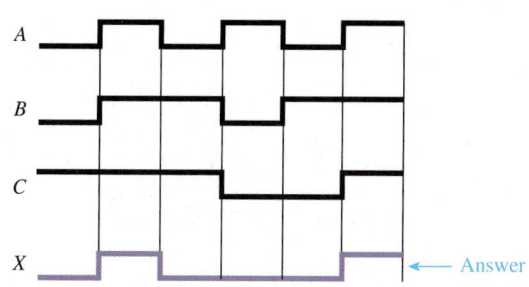

Figure 3–14   Solution to Example 3–3.

EXAMPLE 3–4

The input waveform at *A* and the output waveform at *X* are given for the AND gate in Figure 3–15. Sketch the input waveform that is required at *B* to produce the output at *X* in Figure 3–16.

Figure 3–15

Solution:

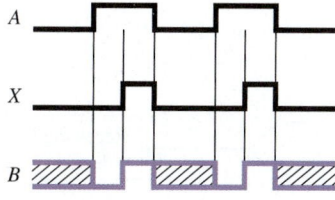

= Don't care (*B* can be HIGH or LOW to get the same output at *X*.)

Figure 3–16   Solution to Example 3–4.

3–4   Enable and Disable Functions

AND and OR gates can be used to **enable** or **disable** a waveform from being transmitted from one point to another. For example, let's say that you wanted a 1-MHz clock oscillator to transmit only four pulses to some receiving device. You would want to *enable* four clock pulses to be transmitted and then *disable* the transmission from then on.

The clock frequency of 1 MHz converts to 1 μs (1/1 MHz) for each clock period. Therefore, to transmit four clock pulses, we have to provide an *enable* signal for 4 μs. Figure 3–17 shows the circuit and waveforms to *enable* four clock pulses. For the HIGH clock pulses to get through the AND gate to point *X*, the second input to the AND gate (enable signal input) must be HIGH; otherwise, the output of the AND gate will be LOW. Therefore, when the enable signal is HIGH for 4 μs, four clock pulses pass through the AND gate. When the enable signal goes LOW, the AND gate *disables* any further clock pulses from reaching the receiving device.

An OR gate can also be used to disable a function. The difference is that the enable signal input is made HIGH to disable, and that the output of the OR gate goes HIGH when it is disabled, as shown in Figure 3–18.

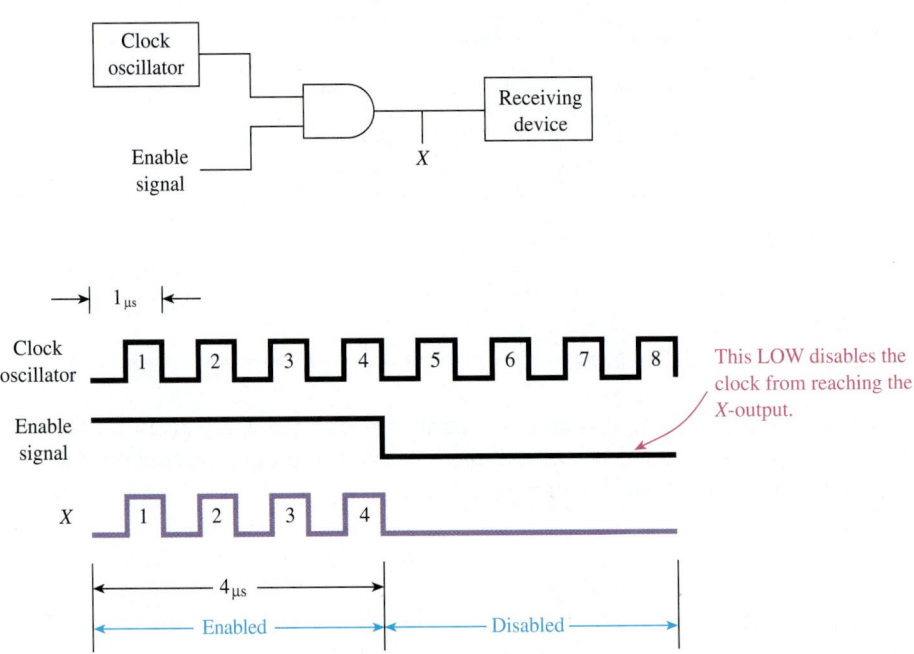

Figure 3–17 Using an AND gate to enable/disable a clock oscillator.

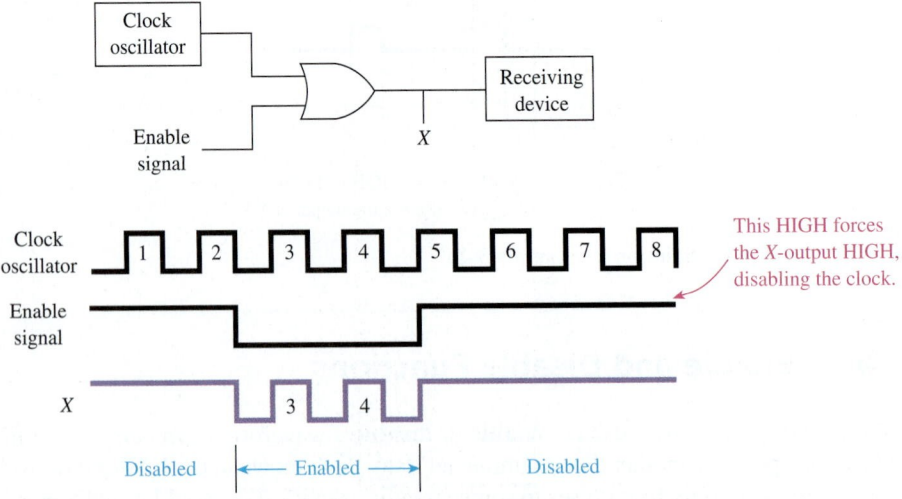

Figure 3–18 Using an OR gate to enable/disable a clock oscillator.

CHAPTER 3 | BASIC LOGIC GATES

Review Questions

3–5. Describe the purpose of a *timing diagram.*

3–6. Under what circumstances would diagonal "don't care" hash marks be used in a timing diagram?

3–7. A _____ (HIGH, LOW) level is required at the input to an AND gate to *enable* the signal at the other input to pass to the output.

3–5 Using IC Logic Gates

AND and OR gates are available as ICs. The IC pin layout, logic gate type, and technical specifications are all contained in the logic data manual supplied by the manufacturer of the IC. For example, referring to a TTL or a CMOS logic data manual, we can see that there are several AND and OR gate ICs. To list just a few:

1. The 7408 (74HC08) is a quad two-input AND gate.

2. The 7411 (74HC11) is a triple three-input AND gate.

3. The 7421 (74HC21) is a dual four-input AND gate.

4. The 7432 (74HC32) is a quad two-input OR gate.

In each case, the HC stands for high-speed CMOS. For example, the 7408 is a TTL AND gate, and the 74HC08 is the equivalent CMOS AND gate. The terms *quad* (four), *triple* (three), and *dual* (two) refer to the number of separate gates on a single IC.

Let's look in more detail at one of these ICs, the 7408 (see Figure 3–19). The 7408 is a 14-pin DIP IC. The power supply connections are made to pins 7 and 14. This supplies the operating voltage for all four AND gates on the IC. Pin 1 is identified by a small indented circle next to it or by a notch cut out between pin 1 and 14 (see Figure 3–19). Let's make the external connections to the IC to form a clock oscillator enable circuit similar to Figure 3–17.

In Figure 3–20, the first AND gate in the IC was used and the other three are ignored. The IC is powered by connecting pin 14 to the positive power supply and pin 7 to ground. The other connections are made by following the original design from Figure 3–17. The clock oscillator signal passes on to the receiving device when the switch is in the *enable* (1) position, and it stops when in the *disable* (0) position .

The pin configurations for some other logic gates are shown in Figure 3–21.

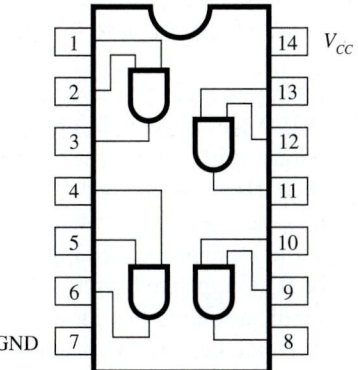

Common Misconception

Students often think that a gate output receives its HIGH or LOW voltage level from its input pin. You need to be reminded that each gate has its own totem-pole output arrangement and receives its voltage from V_{CC} or ground.

Figure 3–19 The 7408 quad two-input AND gate IC pin configuration.

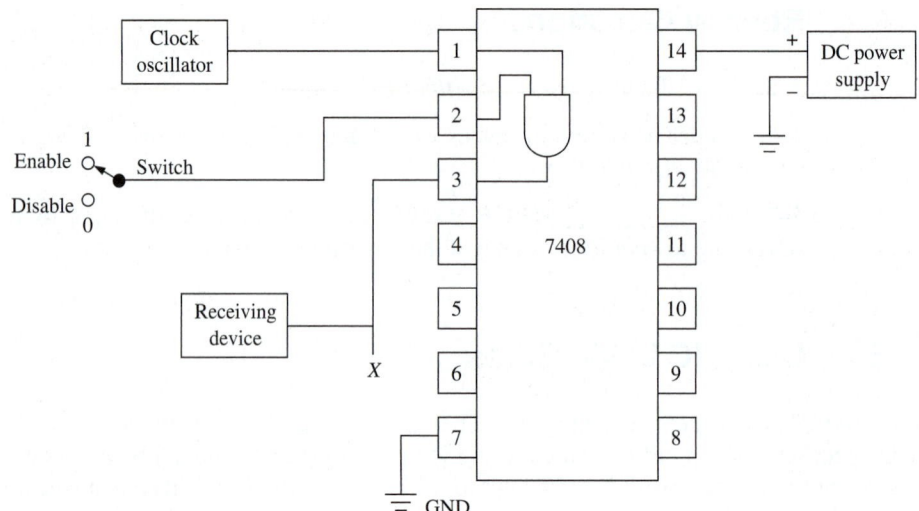

Figure 3–20   Using the 7408 TTL IC in the clock enable circuit of Figure 3–17.

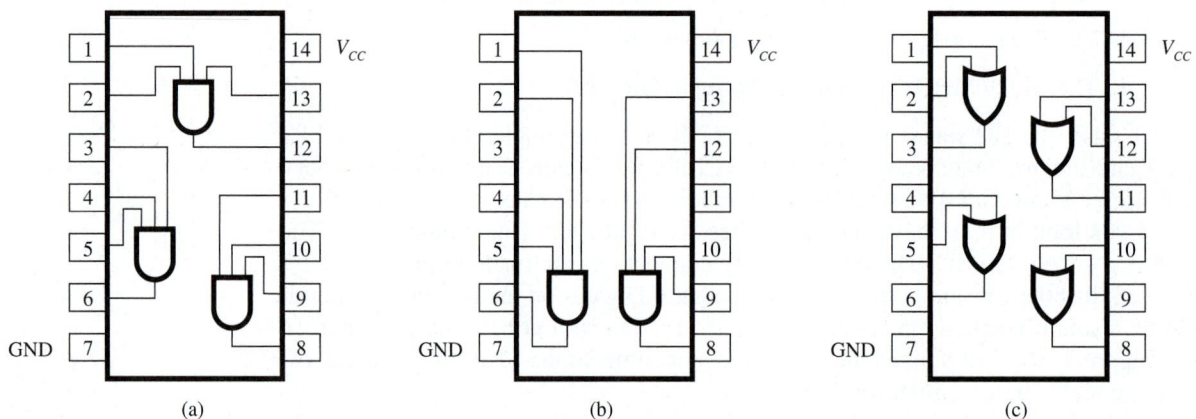

Figure 3–21   Pin configurations for other popular TTL and CMOS AND and OR gate ICs: (a) 7411 (74HC11); (b) 7421 (74HC21); (c) 7432 (74HC32).

3–6   Introduction to Troubleshooting Techniques

Like any other electronic device, ICs and digital electronic circuits can go bad. **Troubleshooting** is the term given to the procedure used to find the **fault,** or *trouble,* in the circuits.

To be a good troubleshooter, you must first *understand the theory and operation* of the circuit, devices, and ICs that are suspected to be bad. If you understand how a particular IC is *supposed* to operate, it is a simple task to put the IC through a test or to exercise its functions to see if it operates as you expect.

There are two simple tools that we will start with to test the ICs and digital circuits. They are the logic pulser and logic probe (see Figure 3–22). The **logic probe** has a metal tip that is placed on the IC pin, printed-circuit board trace, or device lead that you want to test. It also has an indicator lamp that glows, telling you the digital level at that point. If the level is HIGH (1), the lamp glows brightly. If the level is LOW (0), the

Figure 3–22 Logic pulser and logic probe.

lamp goes out. If the level is **floating** (open circuit, neither HIGH nor LOW), the lamp is dimly lit. Table 3–5 summarizes the states of the logic probe.

TABLE 3–5	Logic Probe States
Logic Level	**Indicator Lamp**
HIGH (1)	On
LOW (0)	Off
Float	Dim

The **logic pulser** is used to provide digital pulses to a circuit being tested. By applying a pulse to a circuit and simultaneously observing a logic probe, you can tell if the pulse signal is getting through the IC or device as you would expect. As you become more and more experienced at troubleshooting, you will find that most IC and device faults are due to an open or short at the input or output terminals.

Figure 3–23 shows four common problems that you will find on printed-circuit boards that will cause opens or shorts. Figure 3–23(a) shows an IC that was inserted into its socket carelessly, causing pin 14 to miss its hole and act like an open. In Figure 3–23(b), the printed-circuit board is obviously cracked, which causes an open circuit across each of the copper traces that used to cross over the crack. Poor soldering results in the *solder bridge* evident in Figure 3–23(c). In the center of this photo, you can see where too much solder was used, causing an electrical bridge between two adjacent IC pins and making them a short. Experienced troubleshooters will also visually inspect printed-circuit boards for components that may appear to be darkened from excessive heat. Notice the four transistors in the middle of Figure 3–23(d). The one on the lower left looks charred and is probably burned out, thus acting like an open.

Figure 3–23 Four common printed-circuit faults: (a) misalignment of pin 14; (b) cracked board; (c) solder bridge; (d) burned transistor.

CHAPTER 3 | BASIC LOGIC GATES

The following troubleshooting examples will illustrate some basic troubleshooting techniques using the logic probe and pulser.

EXAMPLE 3–5

The IC AND gate in Figure 3–24 is suspected of having a fault and you want to test it. What procedure should you follow?

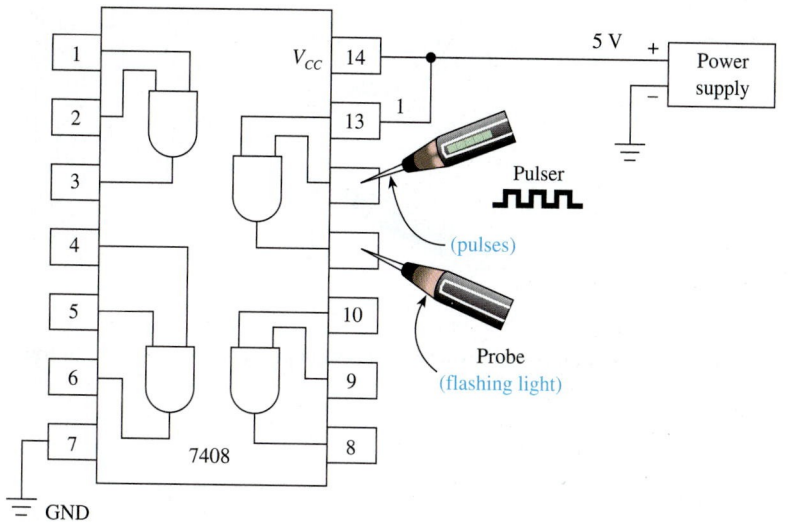

Helpful Hint

You should be aware that these troubleshooting examples assume that the IC is removed from the circuit board. In-circuit testing will often give false readings because of the external circuitry connected to the IC. In that case, the circuit schematic must be studied to determine how the other ICs may be affecting the readings.

Figure 3–24   Connections for troubleshooting one gate of a quad AND IC.

Solution: First you apply power to V_{CC} (pin 14) and GND (pin 7). Next you want to check each AND gate with the pulser/probe. Because it takes a HIGH (1) on *both* inputs to an AND gate to make the output go HIGH, if we put a HIGH (+5 V) on one input and pulse the other, we would expect to get pulses at the output of the gate. Figure 3–24 shows the connections to test one of the gates of a quad AND IC. When the pulser is put on pin 12, the light in the end of the probe flashes at the same speed as the pulser, indicating that the AND gate is passing the pulses through the gate (similar in operation to the clock enable circuit of Figure 3–17).

The next check is to reverse the connections to pins 12 and 13 and check the probe. If the probe still flashes, that gate is okay. Proceed to the other three gates and follow the same procedure. When one of the gate outputs does not flash, you have found the fault.

As mentioned earlier, *the key to troubleshooting an IC is understanding how the IC works.*

EXAMPLE 3–6

Sketch the connections for troubleshooting the first gate of a 7421 dual AND gate.

Solution: The connections are shown in Figure 3–25. The probe should be flashing if the gate is good. Check each of the four inputs with the pulser by keeping three inputs high and pulsing the fourth while you look at the probe. In any case, if the probe does not flash, you have found a bad gate.

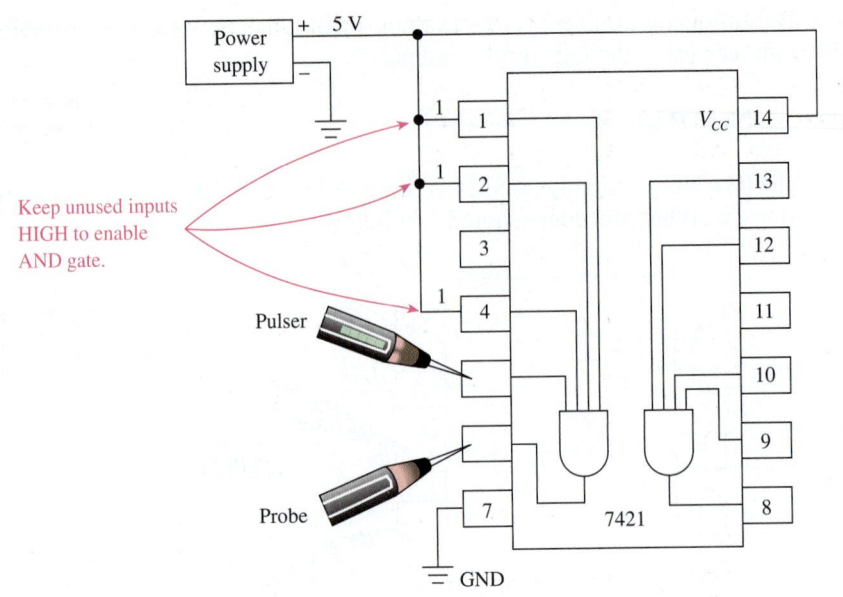

Figure 3–25 Connections for troubleshooting one gate of a 7421 dual four-input AND gate.

EXAMPLE 3-7

Sketch the connections for troubleshooting the first gate of a 7432 quad OR gate.

Solution: The connections are shown in Figure 3–26. The probe should be flashing if the gate is good. Notice that the second input to the OR gate being checked is connected to a LOW (0) instead of a HIGH. The reason for this is that the output would *always* be HIGH if one input were connected HIGH. Because one input is connected LOW instead, the output will flash together with the pulses from the logic pulser if the gate is good.

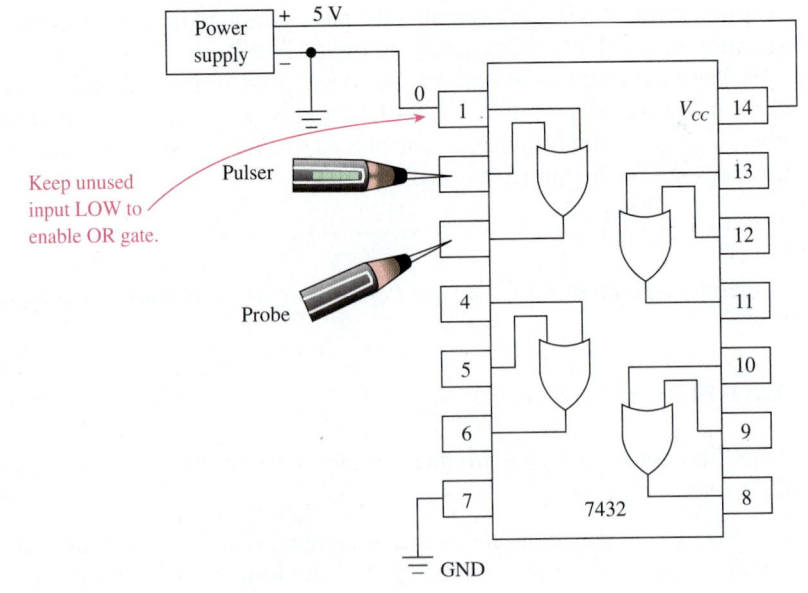

Figure 3–26 Connections for troubleshooting one OR gate of a 7432 IC.

3–8. Which pins on the 7408 AND IC are used for power supply connections, and what voltage levels are placed on those pins?

3–9. How is a *logic probe* used to troubleshoot digital ICs?

3–10. How is a *logic pulser* used to troubleshoot digital ICs?

3–7   The Inverter

The inverter is used to complement, or invert, a digital signal. It has a single input and a single output. If a HIGH level (1) comes in, it produces a LOW-level (0) output. If a LOW level (0) comes in, it produces a HIGH-level (1) output. The symbol and truth table for the inverter gate are shown in Figure 3–27. (*Note:* The circle is the part of the symbol that indicates inversion. The inversion circle will be used on other gates in upcoming sections.)

Input ——————▷○—— Output
A   X

Input A	Output X
0	1
1	0

Figure 3–27   Inverter symbol and truth table.

The operation of the inverter is very simple and can be illustrated further by studying the timing diagram of Figure 3–28. The timing diagram graphically shows us the operation of the inverter. When the input is HIGH, the output is LOW, and when the input is LOW, the output is HIGH. The output waveform is, therefore, the exact complement of the input.

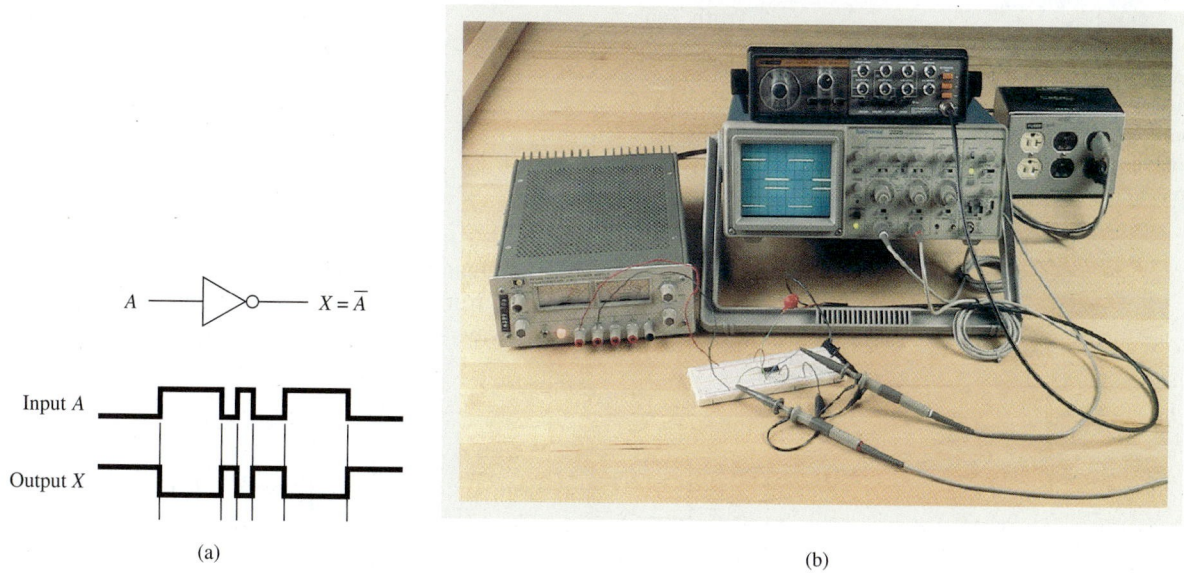

A ——▷○—— $X = \overline{A}$

Input A

Output X

(a)

(b)

Figure 3–28   Timing analysis of an inverter gate: (a) waveform sketch; (b) oscilloscope display.

The Boolean equation for an inverter is written $X = \overline{A}$ ($X = \text{NOT}\,\overline{A}$). The *bar* over the A is an **inversion bar,** used to signify the **complement.** The inverter is sometimes referred to as the NOT gate.

3–8   The NAND Gate

The operation of the NAND gate is the same as the AND gate except that its output is inverted. You can think of a NAND gate as an AND gate with an inverter at its output. The symbol for a NAND gate is made from an AND gate with the inversion circle (bubble) at its output, as shown in Figure 3–29.

Input A —— ⟫○ —— Output $X = \overline{AB}$
Input B ——

Figure 3–29   Symbol for a NAND gate.

In digital circuit diagrams, you will find the small circle used whenever complementary action (inversion) is to be indicated. The circle at the output acts just like an inverter, so a NAND gate can be drawn symbolically as an AND gate with an inverter connected to its output, as shown in Figure 3–30.

$A \overset{1}{——}$ ⟫ $\overset{1}{—}$ ▷○ $\overset{0}{—} X = \overline{AB}$
$B \underset{1}{——}$

Figure 3–30   AND–INVERT equivalent of a NAND gate with $A = 1$, $B = 1$.

The Boolean equation for the NAND gate is written $X = \overline{AB}$. The inversion bar is drawn over (A and B), meaning that the output of the NAND is the complement of (A and B) [**NOT** (A and B)]. Because we are inverting the output, the truth table outputs in Table 3–6 will be the complement of the AND gate truth table outputs. The easy way to construct the truth table is to think of how an AND gate would respond to the inputs and then invert your answer. From Table 3–6, we can see that the output is LOW when *both* inputs A and B are HIGH (just the opposite of an AND gate). Also, the output is HIGH whenever either input is LOW.

TABLE 3–6	Two-Input NAND Gate Truth Table	
A	B	X
0	0	1
0	1	1
1	0	1
1	1	0

Output is always HIGH unless both inputs are HIGH.

NAND gates can also have more than two inputs. Figure 3–31 shows three- and eight-input NAND gate symbols. The truth table for a three-input NAND gate (see Table 3–7) shows that the output is always HIGH unless *all* inputs go HIGH.

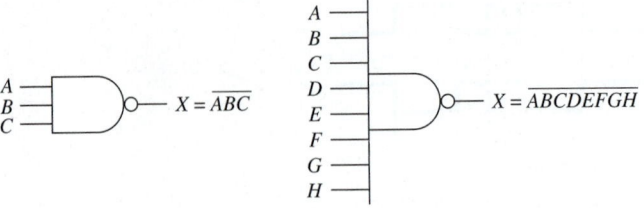

Figure 3–31   Symbols for three- and eight-input NAND gates.

TABLE 3–7	Truth Table for a Three-Input NAND Gate		
A	B	C	X
0	0	0	1
0	0	1	1
0	1	0	1
0	1	1	1
1	0	0	1
1	0	1	1
1	1	0	1
1	1	1	0

Timing analysis can also be used to illustrate the operation of NAND gates. The following examples will contribute to your understanding.

EXAMPLE 3–8

Sketch the output waveform at X for the NAND gate shown in Figure 3–32, with the given input waveforms in Figure 3–33.

Figure 3–32

Solution:

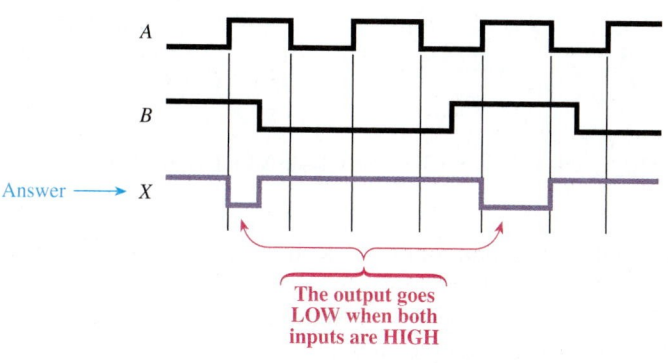

The output goes LOW when both inputs are HIGH

Figure 3–33 Timing analysis of a NAND gate.

EXAMPLE 3–9

Sketch the output waveform at X for the NAND gate shown in Figure 3–34, with the given input waveforms at A, B, and Control.

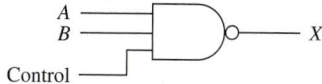

Figure 3–34

Solution: In Figure 3–35, the Control input waveform is used to *enable/disable* the NAND gate. When it is LOW, the output is stuck HIGH. When it goes HIGH, the output will respond LOW when *A* and *B* go HIGH.

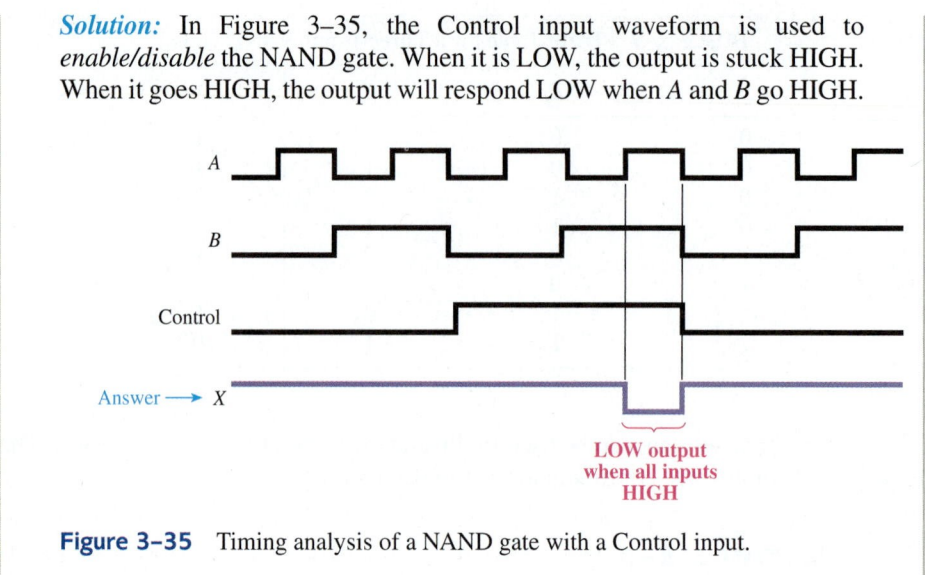

Figure 3–35   Timing analysis of a NAND gate with a Control input.

3–9   The NOR Gate

The operation of the NOR gate is the same as that of the OR gate except that its output is inverted. You can think of a NOR gate as an OR gate with an inverter at its output. The symbol for a NOR gate and its equivalent OR–INVERT symbol are shown in Figure 3–36.

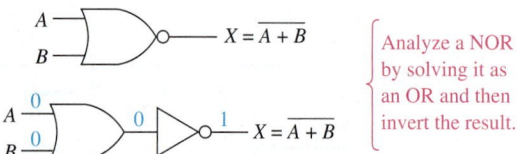

Figure 3–36   NOR gate symbol and its OR–INVERT equivalent with $A = 0$, $B = 0$.

The Boolean equation for the NOR function is $X = \overline{A + B}$. The equation is stated "*X* equals *not* (*A* or *B*)." In other words, *X* is LOW if *A* or *B* is HIGH. The truth table for a NOR gate is given in Table 3–8. Notice that the output column is the complement of the OR gate truth table output column.

TABLE 3–8	Truth Table for a NOR Gate	
A	*B*	$X = \overline{A + B}$
0	0	1
0	1	0
1	0	0
1	1	0

Output is always LOW unless both inputs are LOW.

Now let's study some timing analysis examples to get a better grasp of NOR gate operation.

EXAMPLE 3-10

Sketch the output waveform at X for the NOR gate shown in Figure 3–37, with the given input waveforms in Figure 3–38.

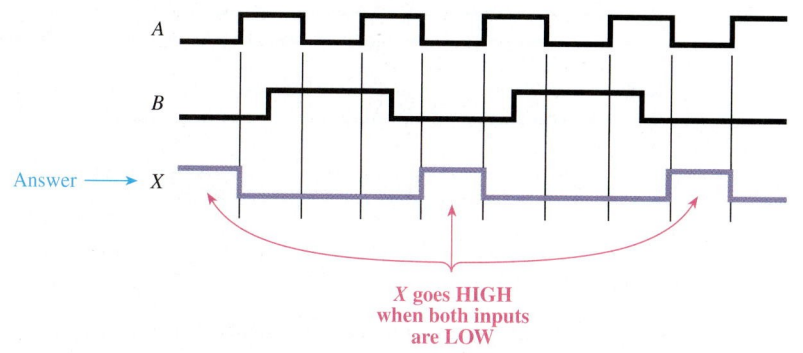

Figure 3-37

Solution:

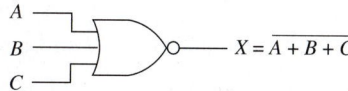

Figure 3-38   NOR gate timing analysis.

Helpful Hint

To solve a timing analysis problem, it is useful to look at the gate's truth table to see what the *unique* occurrence is for that gate. In the case of the NOR, the odd occurrence is when the output goes HIGH due to all LOW inputs.

EXAMPLE 3-11

Sketch the output waveform at X for the NOR gate shown in Figure 3–39, with the given input waveforms in Figure 3–40.

Figure 3-39

Solution:

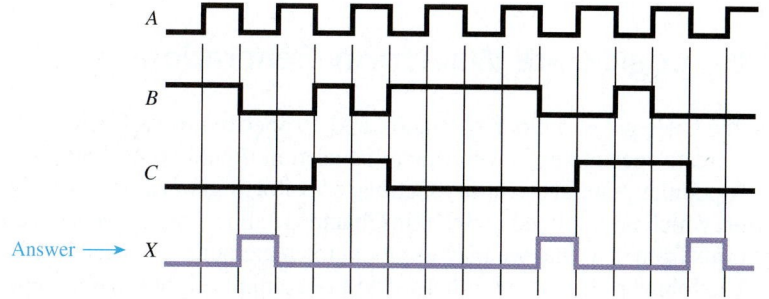

Figure 3-40   Three-input NOR gate timing analysis.

EXAMPLE 3–12

Sketch the waveform at the *B* input of the gate shown in Figure 3–41 that will produce the output waveform shown in Figure 3–42 for *X*.

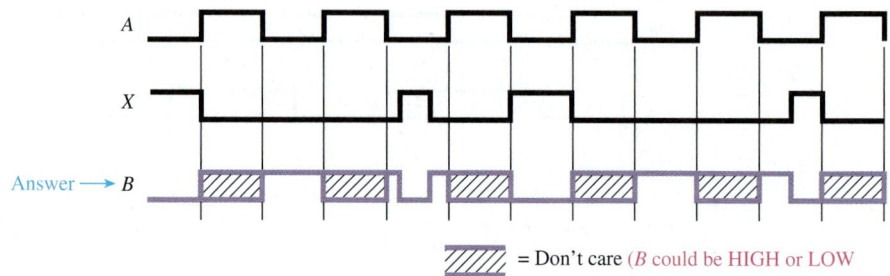

Figure 3–41

Solution:

Figure 3–42   Input waveform requirement to produce a specific output.

═══ **Review Questions** ═══

3–11. What is the purpose of an inverter in a digital circuit?

3–12. How does a NAND gate differ from an AND gate?

3–13. The output of a NAND gate is always HIGH unless *all* inputs are made _____ (HIGH, LOW).

3–14. Write the Boolean equation for a three-input NOR gate.

3–15. The output of a two-input NAND gate is _____ (HIGH, LOW) if $A = 1$, $B = 0$.

3–16. The output of a two-input NOR gate is _____ (HIGH, LOW) if $A = 0$, $B = 1$.

3–10   Logic Gate Waveform Generation

Using the basic gates, a clock oscillator, and a **repetitive waveform** generator circuit, we can create specialized waveforms to be used in digital control and sequencing circuits. A popular general-purpose repetitive **waveform generator** is the **Johnson shift counter,** which is explained in detail in Chapter 13. For now, all we need are the output waveforms from it so that we may use them to create our own specialized waveforms.

The Johnson shift counter that we will use outputs eight separate repetitive waveforms: *A, B, C, D;* and their complements, $\overline{A}, \overline{B}, \overline{C}, \overline{D}$. The input to the Johnson shift counter is a clock oscillator (C_p). Figure 3–43 shows a Johnson shift counter with its input and output waveforms.

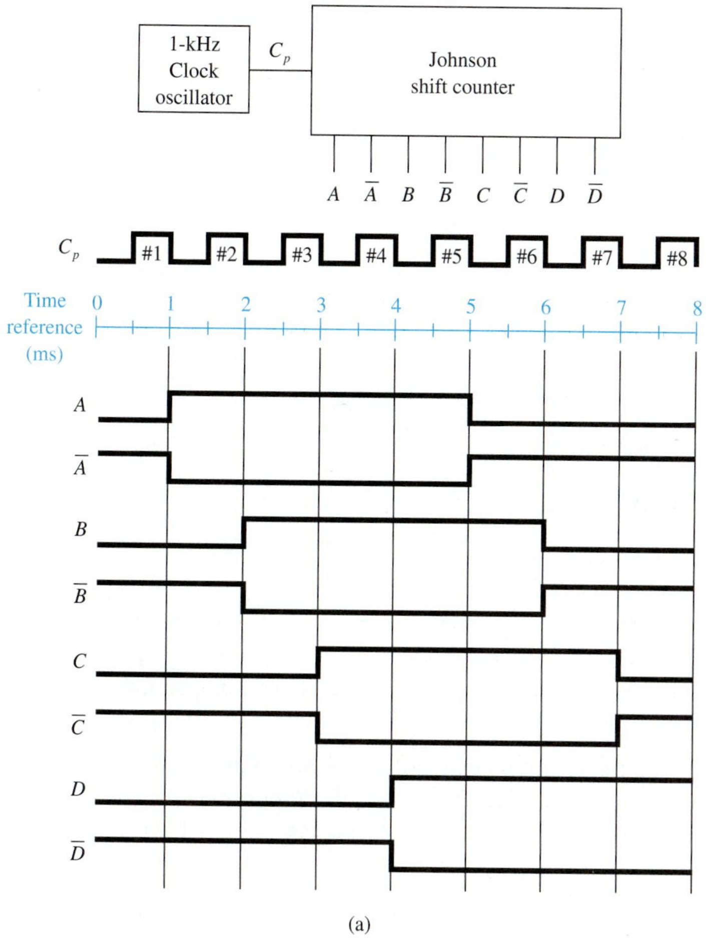

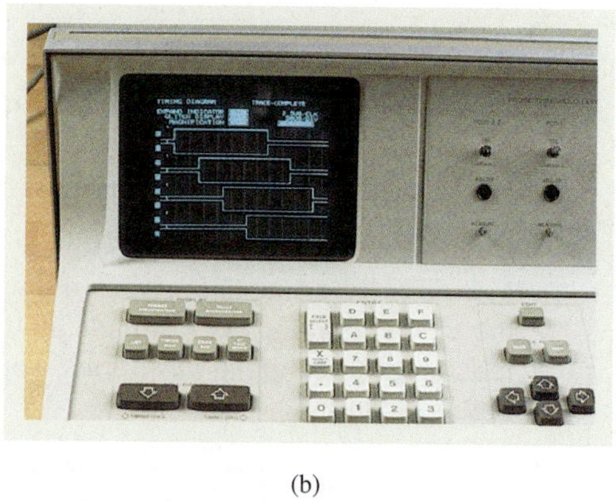

Figure 3–43 Johnson shift counter waveform generation: (a) waveform sketch; (b) logic analyzer display.

The clock oscillator produces the C_p waveform, which is input to the Johnson shift counter. The shift counter uses C_p and internal circuitry to generate the eight repetitive output waveforms shown.

Now, if one of those waveforms is exactly what you want, you are all set. But let's say we need a waveform that is HIGH for 3 ms, from 2 until 5 on the millisecond time reference scale. Looking at Figure 3–43, we can see that this waveform is not available.

Using some logic gates, however, will enable us to get any waveform that we desire. In this case, if we feed the A and B waveforms into an AND gate, we will get our HIGH level from 2 to 5, as shown in Figure 3–44.

Helpful Hint

The circuitry and operation of the Johnson shift counter are given in Chapter 13. For now, you need to know only that it is used to provide a combination of sequential waveforms that we will use to create specialized waveforms and improve our understanding of the basic gates. It is helpful if you have a photocopy of Figure 3–43(a) to work on for aligning the waveforms to solve the problems.

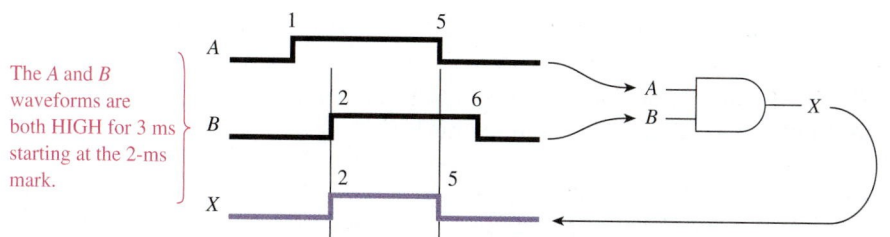

The A and B waveforms are both HIGH for 3 ms starting at the 2-ms mark.

Figure 3–44 Generating a 3-ms HIGH pulse using an AND gate and a Johnson shift counter.

Team Discussion

Could we obtain a LOW pulse from 4 to 5 instead of a HIGH by using the complemented signals of *A* and *D?*

Working through the following examples will help you to understand logic gate operation and waveform generation.

EXAMPLE 3–13

Which Johnson counter outputs will you connect to an AND gate to get a 1-ms HIGH-level output from 4 to 5 ms?

Solution: Referring to Figure 3–43, we see that the two waveforms that are *both* HIGH from 4 to 5 ms are *A* and *D;* therefore, the circuit of Figure 3–45 will give us the required output.

Figure 3–45 Solution to Example 3–13.

EXAMPLE 3–14

Which Johnson counter outputs must be connected to a three-input AND gate to enable just the C_p#4 pulse to be output?

Solution: Referring to Figure 3–43, we see that the C and $\overline{D}$ waveforms are both HIGH only during the C_p 4 *period.* To get just the C_p#4 *pulse,* you must provide C_p as the third input. Now, when you look at all three input waveforms, you see that they are all HIGH only during the C_p#4 *pulse* (see Figure 3–46).

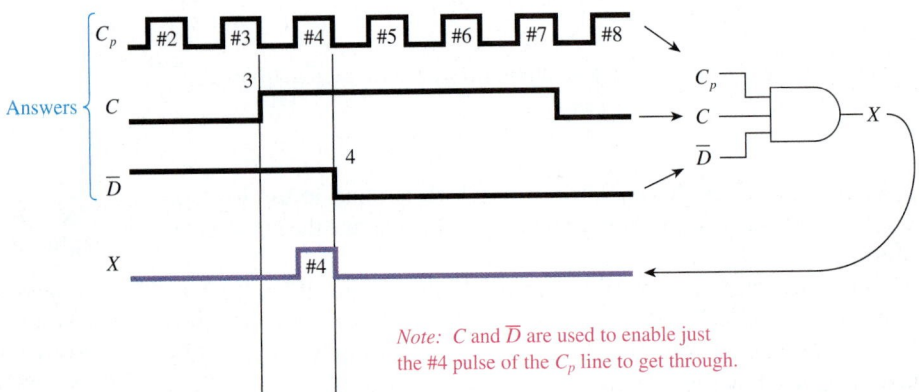

Note: C and $\overline{D}$ are used to enable just the #4 pulse of the C_p line to get through.

Figure 3–46 Solution to Example 3–14.

EXAMPLE 3–15

Sketch the output waveform that will result from inputting A, $\overline{B}$, and $\overline{C}$ into the three-input OR gate shown in Figure 3–47.

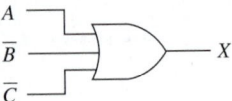

Figure 3–47

Solution: The output of an OR gate is always HIGH unless *all* inputs are LOW. Therefore, the output is always HIGH except between 5 and 6, as shown in Figure 3–48.

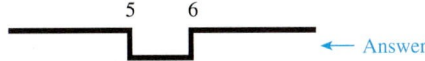

Figure 3–48   Solution to Example 3–15.

EXAMPLE 3–16

Sketch the output waveform that will result from inputting C_p, $\overline{B}$, and C into the NAND gate shown in Figure 3–49.

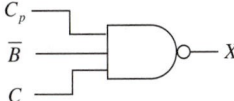

Figure 3–49

Solution: From reviewing the truth table of a NAND gate, we determine that the output is always HIGH unless *all* inputs are HIGH. Therefore, the output will always be HIGH except during pulse 7, as shown in Figure 3–50.

Figure 3–50   Solution to Example 3–16.

EXAMPLE 3–17

Sketch the output waveforms that will result from inputting A, B, and D into the NOR gate shown in Figure 3–51.

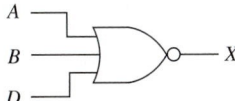

Figure 3–51

Solution: Reviewing the truth table for a NOR gate, we determine that the output is always LOW except when *all* inputs are LOW. Therefore, the output will always be LOW except from 0 to 1, as shown in Figure 3–52.

Figure 3–52   Solution to Example 3–17.

SECTION 3–10 I LOGIC GATE WAVEFORM GENERATION

EXAMPLE 3–18

Sketch the output waveforms for the gates shown in Figure 3–53. The inputs are connected to the Johnson shift counter of Figure 3–43.

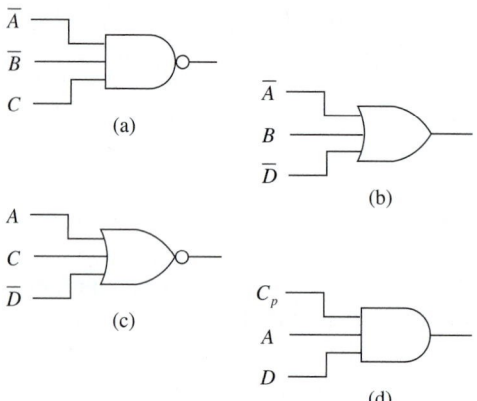

Figure 3–53

Solution: The output waveforms are shown in Figure 3–54.

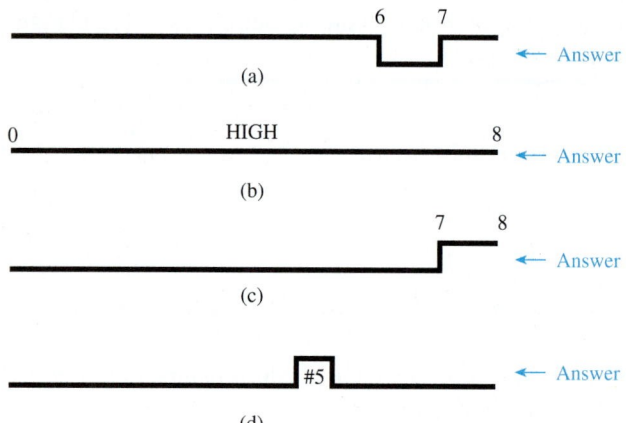

Figure 3–54   Solution to Example 3–18.

EXAMPLE 3–19

Determine which shift counter waveforms from Figure 3–43 will produce the output waveforms shown in Figure 3–55.

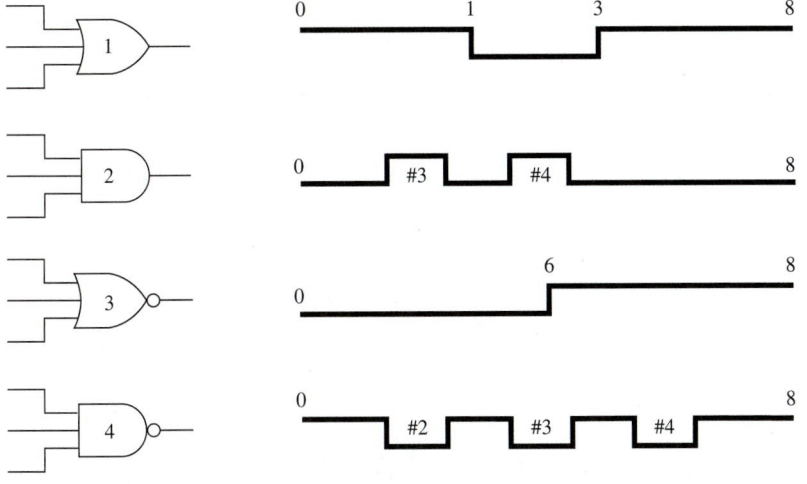

Answers: $1 = \overline{A}, C, D$ $2 = C_p, B, \overline{D}$ $3 = A, B, \overline{D}$ $4 = C_p, A, \overline{D}$

Figure 3–55   Solution to Example 3–19.

EXAMPLE 3–20

By using combinations of gates, we can obtain more specialized waveforms. Sketch the output waveforms for the circuit shown in Figure 3–56.

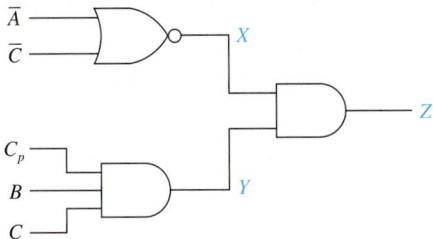

Figure 3–56

Solution: The output waveforms are shown in Figure 3–57. The second AND gate is used to AND X with Y. Therefore, Z is only HIGH in areas where X and Y are both HIGH. (*Note:* The X and Y waveforms must be aligned carefully to get the correct output at Z.)

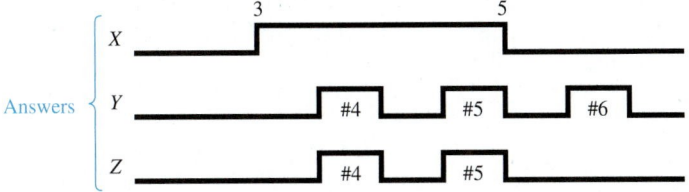

Figure 3–57   Solution to Example 3–20.

EXAMPLE 3–21

Sketch the output waveforms for the circuit shown in Figure 3–58.

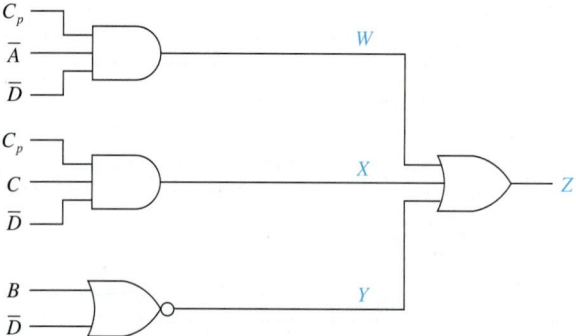

Figure 3–58

Solution: The output waveforms are shown in Figure 3–59. (*Note:* The output at Z is HIGH whenever W or X or Y are HIGH.)

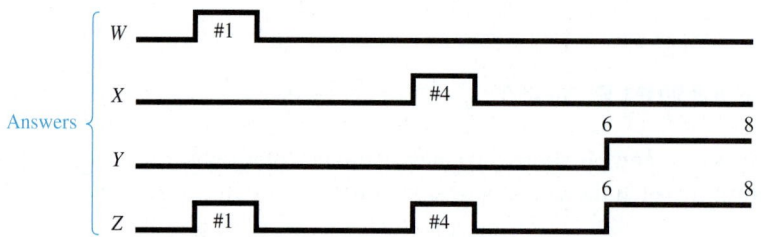

Figure 3–59   Solution to Example 3–21.

3–11   Using IC Logic Gates

All the logic gates are available in various configurations in the TTL and CMOS families. To list just a few: The 7404 TTL and the 4049 CMOS are **hex** (six) inverter ICs, the 7400 TTL and the 4011 CMOS are **quad** (four) two-input NAND ICs, and the 7402 TTL and the 4001 CMOS are quad two-input NOR ICs. Other popular NAND and NORs are available in three-, four-, and eight-input configurations. Consult a TTL or CMOS data manual for the availability and pin configuration of these ICs. The pin configurations for the hex inverter, the quad NOR, and the quad NAND are given in Figures 3–60 and 3–61. (High-speed CMOS 74HC04, 74HC00, and 74HC02 have the same pin configuration as the TTL ICs.)

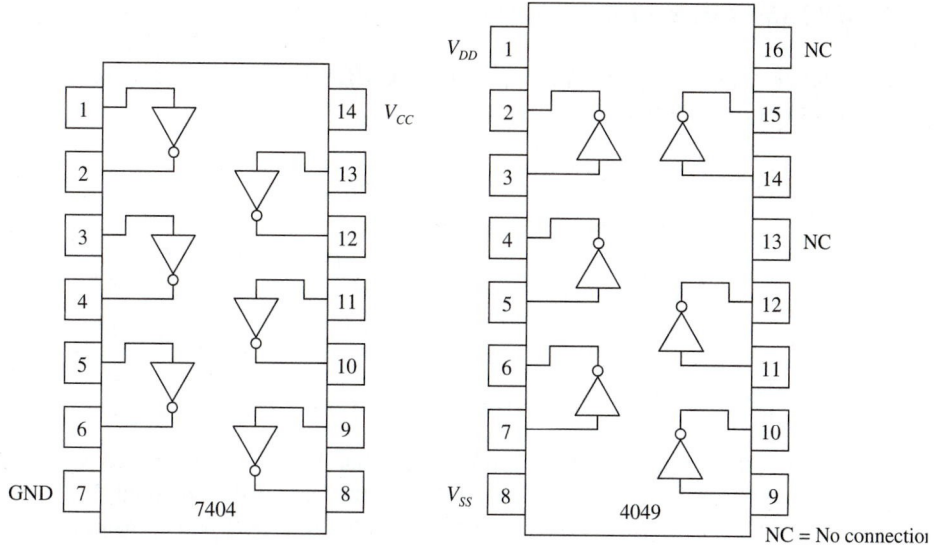

Figure 3–60 7404 TTL and 4049 CMOS inverter pin configurations.

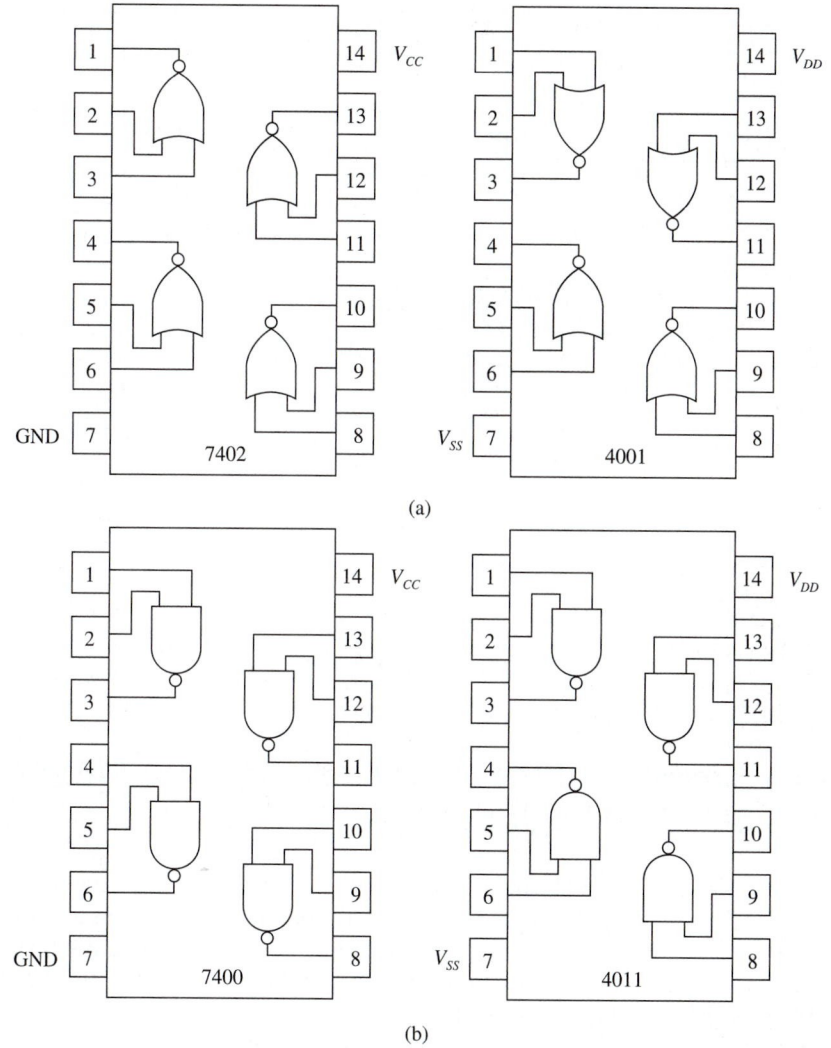

Figure 3–61 (a) 7402 TTL NOR and 4001 CMOS NOR pin configurations; (b) 7400 TTL NAND and 4011 CMOS NAND pin configurations.

EXAMPLE 3–22

Draw the external connections to a 4011 CMOS IC to form the circuit shown in Figure 3–62.

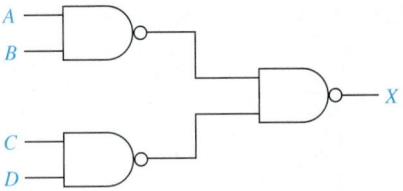

Figure 3–62

Solution: Referring to Figure 3–63, notice that V_{DD} is connected to the +5-V supply and V_{SS} to ground. According to the CMOS data manual, V_{DD} can be any positive voltage from +3 to +15 V with respect to V_{SS} (usually ground).

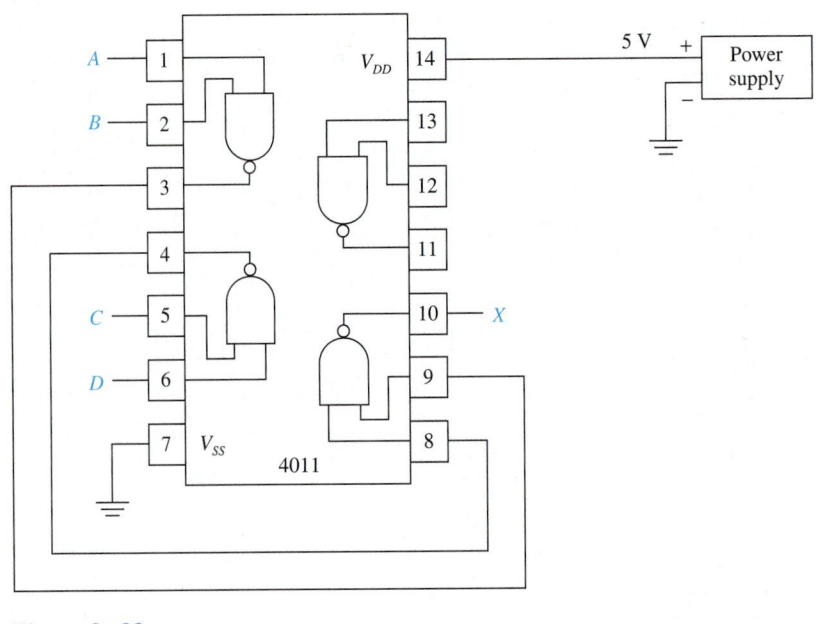

Figure 3–63

3–12 Summary of the Basic Logic Gates and IEEE/IEC Standard Logic Symbols

By now you should have a thorough understanding of the basic logic gates: inverter, AND, OR, NAND, and NOR. In Chapter 5, we will combine several gates to form complex logic functions. Figure 3–64(a) shows the author designing a logic system using a computer-aided design (CAD) system to combine the functions of the NAND, AND, and OR gates shown in Figure 3–64(b).

Because the basic logic gates are the building blocks for larger-scale ICs and digital systems, it is very important that the operation of these gates be second nature to you.

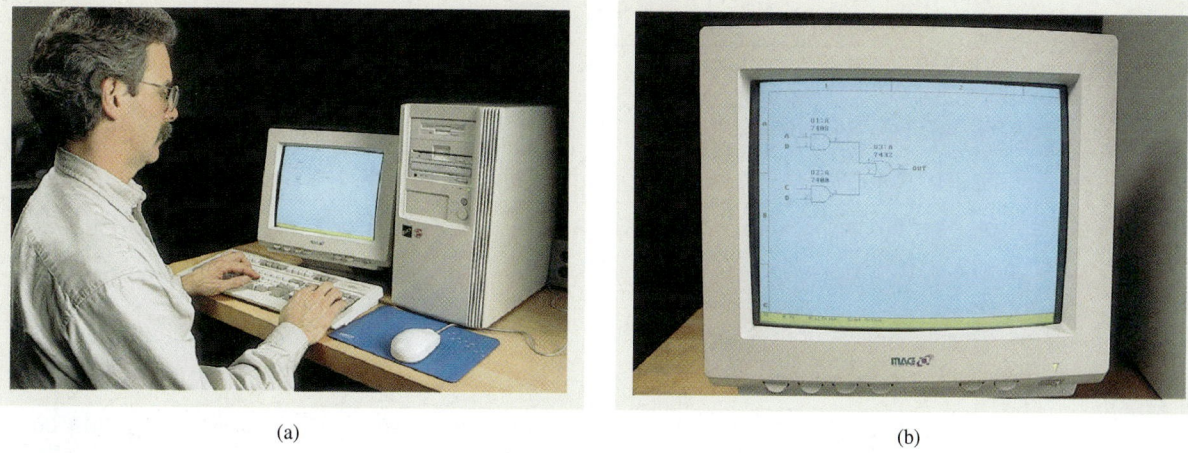

(a)   (b)

Figure 3–64 Computer-aided design (CAD): (a) the author designing with a CAD system; (b) final combinational circuit.

A summary of the basic logic gates is given in Figure 3–65. You should memorize these logic symbols, Boolean equations, and truth tables. Also, a table of the most common IC gates in the TTL and CMOS families is given in Table 3–9. You will need to refer to a TTL or CMOS data book for the pin layout and specifications.

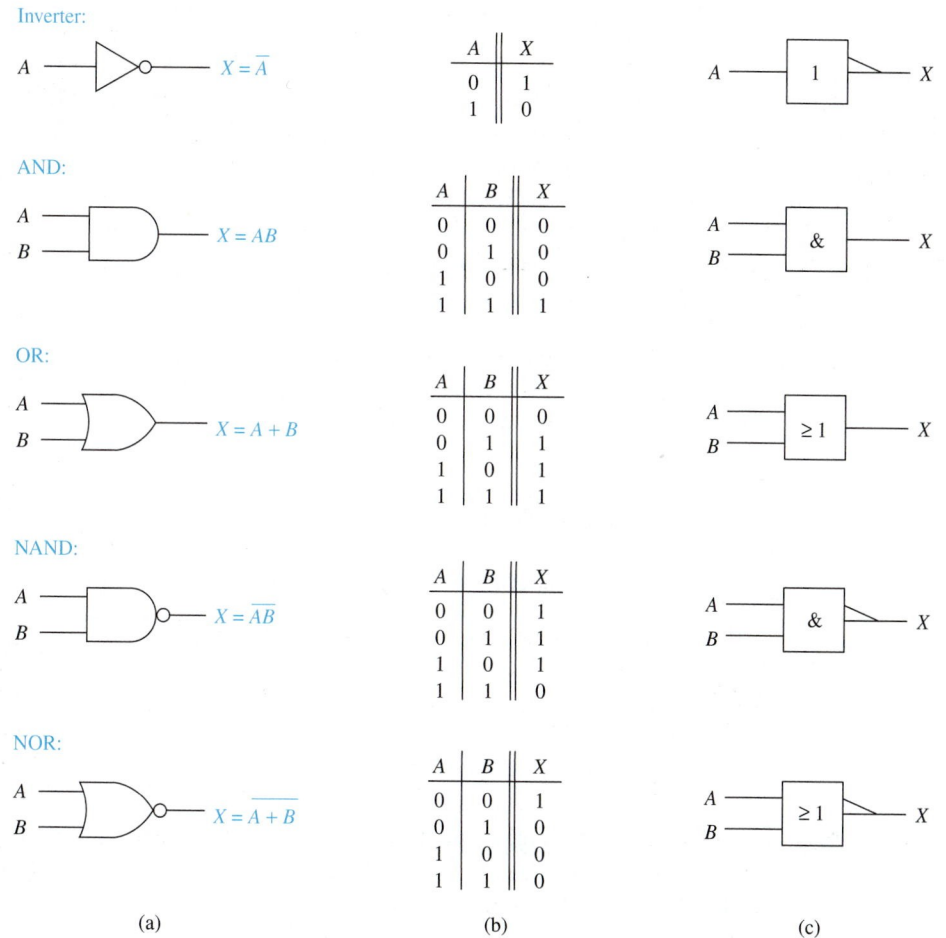

(a)   (b)   (c)

Figure 3–65 Summary of logic gates: (a) traditional logic symbols with Boolean equation; (b) truth tables; (c) IEEE/IEC standard logic symbols.

TABLE 3–9 | Common IC Gates in the TTL and CMOS Families

Gate Name	Number of Inputs per Gate	Number of Gates per Chip	Part Number Standard TTL	Standard CMOS	High-Speed CMOS
Inverter	1	6	7404	4069	74HC04
AND	2	4	7408	4081	74HC08
	3	3	7411	4073	74HC11
	4	2	7421	4082	—
OR	2	4	7432	4071	74HC32
	3	3	—	4075	74HC4075
	4	2	—	4072	—
NAND	2	4	7400	4011	74HC00
	3	3	7410	4013	74HC10
	4	2	7420	4012	74HC20
	8	1	7430	4068	—
	12	1	74134	—	—
	13	1	74133	—	—
NOR	2	4	7402	4001	74HC02
	3	3	7427	4025	74HC27
	4	2	7425	4002	74HC4002
	5	2	74260	—	—
	8	1	—	4078	—

Also, in Figure 3–65(c), we introduce the IEEE/IEC standard logic symbols. This alternate standard for logic symbols was developed in 1984. It uses a method of determining the complete logical operation of a device just by interpreting the notations on the symbol for the device. This includes the basic gates as well as the more complex digital logic functions. Unfortunately, this standard has not achieved widespread use, but you will see it used in some newer designs. Most digital IC data books will show both the traditional and the new standard logic symbols, although most circuit schematics still use the traditional logic symbols. For this reason, the summary in Figure 3–65 shows both logic symbols, but throughout the remainder of this text we will use the traditional logic symbols. (A complete description of the IEEE/IEC standard for logic symbols is provided in Appendix C.)

Review Questions

3–17. What is the function of the Johnson shift counter in this chapter?

3–18. What are the part numbers of a TTL inverter IC and a CMOS NOR IC?

3–19. What type of logic gate is contained within the 7410 IC? the 74HC27 IC?

Summary

In this chapter, we have learned that

1. The AND gate requires that all inputs are HIGH to get a HIGH output.

2. The OR gate outputs a HIGH if any of its inputs are HIGH.

3. An effective way to measure the precise timing relationships of digital waveforms is with an oscilloscope or a logic analyzer.

4. Besides providing the basic logic functions, AND and OR gates can also be used to enable or disable a signal to pass from one point to another.

5. Several ICs are available in both TTL and CMOS that provide the basic logic functions.

6. Two important troubleshooting tools are the logic pulser and the logic probe. The pulser is used to inject pulses into a circuit under test. The probe reads the level at a point in a circuit to determine if it is HIGH, LOW, or floating.

7. An inverter provides an output that is the complement of its input.

8. A NAND gate outputs a LOW when all of its inputs are HIGH.

9. A NOR gate outputs a HIGH when all of its inputs are LOW.

10. Specialized waveforms can be created by using a repetitive waveform generator and the basic gates.

11. Manufacturers' data manuals are used by the technician to find the pin configuration and operating characteristics for the ICs used in modern circuitry.

Glossary

Boolean Equation: A logic expression that illustrates the functional operation of a logic gate or combination of logic gates.

Complement: A change to the opposite digital state. A1 becomes a 0, and a 0 becomes a 1.

Disable: To disallow or deactivate a function or circuit.

Enable: To allow or activate a function or circuit.

Fault: The problem in a nonfunctioning electrical circuit. It is usually due to an open circuit, short circuit, or defective component.

Floating: A logic level in a digital circuit that is neither HIGH nor LOW. It acts like an open circuit to anything connected to it.

Gate: The basic building block of digital electronics. The basic logic gate has one or more inputs and one output, and is used to perform one of the following logic functions: AND, OR, NOR, NAND, INVERT, exclusive-OR, or exclusive-NOR.

Hex: When dealing with integrated circuits, a term specifying *six* gates on a single IC package.

Inversion Bar: A line over variables in a Boolean equation signifying that the digital state of the variables is to be complemented. For example, the output of a two-input NAND gate is written $X = \overline{AB}$.

Johnson Shift Counter: A digital circuit that produces several repetitive digital waveforms useful for specialized waveform generation.

Logic Probe: An electronic tool used in the troubleshooting procedure to indicate a HIGH, LOW, or float level at a particular point in a circuit.

Logic Pulser: An electronic tool used in the troubleshooting procedure to inject a pulse or pulses into a particular point in a circuit.

NOT: When reading a Boolean equation, the word used to signify an inversion bar. For example, the equation $X = \overline{AB}$ is read "X equals NOT AB."

Quad: When dealing with integrated circuits, the term specifying *four* gates on a single IC package.

Repetitive Waveform: A waveform that repeats itself after each cycle.

Troubleshooting: The work that is done to find the problem in a faulty electrical circuit.

Truth Table: A tabular listing that is used to illustrate all the possible combinations of digital input levels to a gate and the output that will result.

Waveform Generator: A circuit used to produce specialized digital waveforms.

Problems

Section 3–1

3–1. Build the truth table for a three-input AND gate.

3–2. Build the truth table for a four-input AND gate.

3–3. If we were to build a truth table for an eight-input AND gate, how many different combinations of inputs would we have?

3–4. Describe in words the operation of an AND gate.

Section 3–2

3–5. Describe in words the operation of an OR gate.

3–6. Write the Boolean equation for

(a) A three-input AND gate

(b) A four-input AND gate

(c) A three-input OR gate

Section 3–3

3–7. Sketch the output waveform at X for the two-input AND gates shown in Figure P3–7.

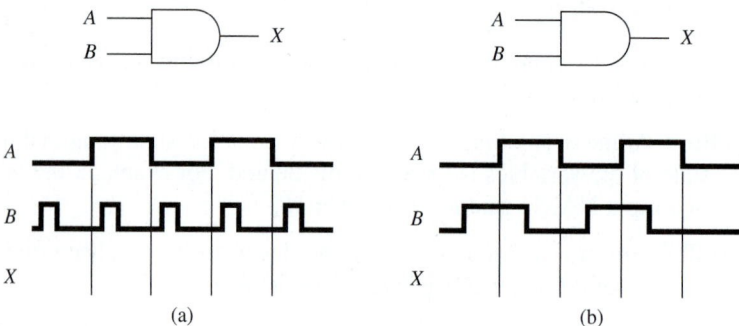

Figure P3–7

CHAPTER 3 | BASIC LOGIC GATES

3–8. Sketch the output waveform at *X* for the two-input OR gates shown in Figure P3–8.

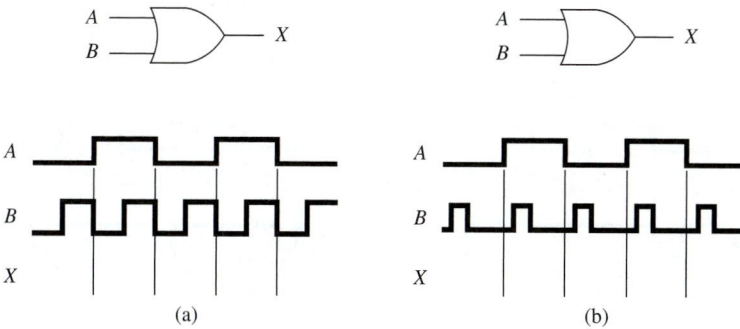

(a)

(b)

Figure P3–8

3–9. Sketch the output waveform at *X* for the three-input AND gates shown in Figure P3–9.

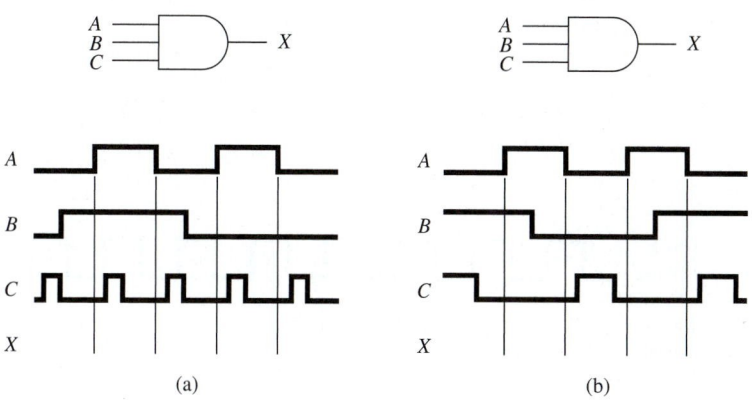

(a)

(b)

Figure P3–9

C **3–10.** The input waveform at *A* is given for the two-input AND gates shown in Figure P3–10. Sketch the input waveform at *B* that will produce the output at *X*.

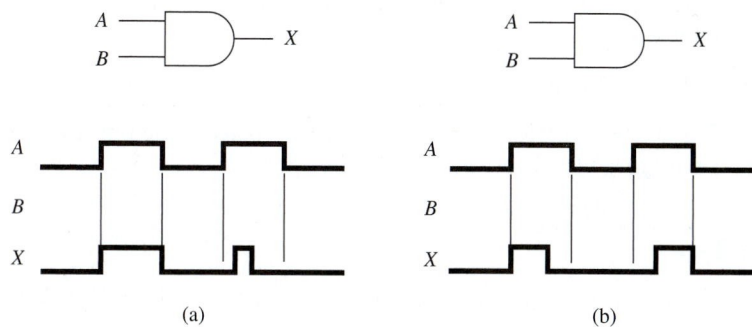

(a)

(b)

Figure P3–10

C

3–11. Repeat Problem 3–10 for the two-input OR gates shown in Figure P3–11.

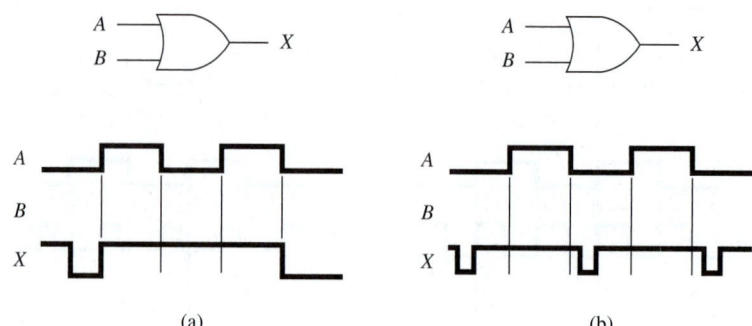

(a)   (b)

Figure P3–11

Section 3–4

3–12. Using Figure P3–12, sketch the waveform for the *enable signal* that will allow pulses 2, 3 and 6, 7 to get through to the receiving device.

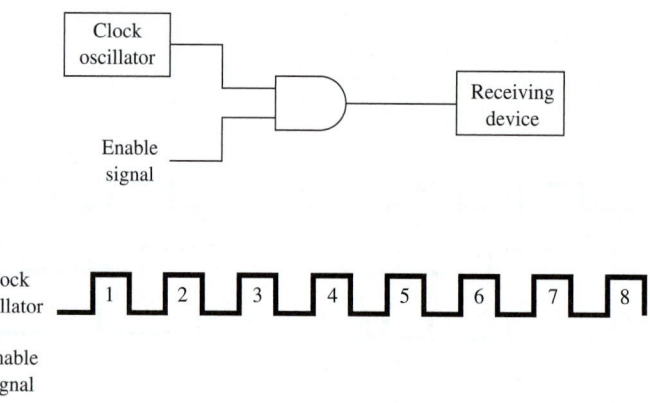

Figure P3–12

3–13. Repeat Problem 3–12, but this time sketch the waveform that will allow only the even pulses (2, 4, 6, 8) to get through.

Section 3–5

3–14. How many separate OR gates are contained within the 7432 TTL IC?

3–15. Sketch the actual pin connections to a 7432 quad two-input OR TTL IC to implement the circuit of Figure 3–18.

3–16. How many inputs are there on each AND gate of a 7421 TTL IC?

3–17. The 7421 IC is a 14-pin DIP. How many of the pins are *not* used for anything?

Section 3–6

T *   **3–18.** What are the three logic levels that can be indicated by a logic probe?

---

*The letter **T** designates a problem that involves **T**roubleshooting.

T   **3–19.** What is the function of the logic pulser?

T   **3–20.** When troubleshooting an OR gate such as the 7432, when the pulser is applied to one input, should the other input be connected HIGH or LOW? Why?

T   **3–21.** When troubleshooting an AND gate such as the 7408, when the pulser is connected to one input, should the other input be connected HIGH or LOW? Why?

C T   **3–22.** The clock enable circuit shown in Figure P3–22 is not working. The enable switch is up in the enable position. A logic probe is placed on the following pins and gets the following results. Find the cause of the problem.

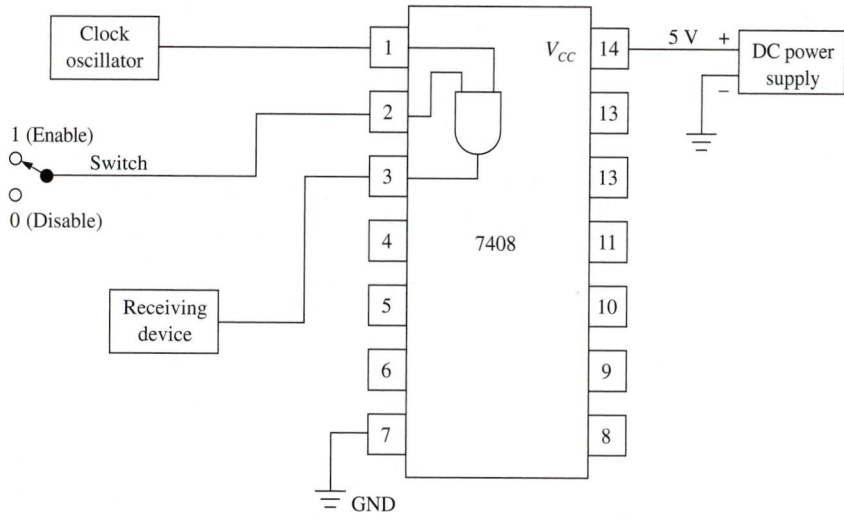

Figure P3–22

Probe on Pin	Indicator Lamp
1	Flashing
2	On
3	Off
7	Off
14	On

C T   **3–23.** Repeat Problem 3–22 for the following troubleshooting results.

Probe on Pin	Indicator Lamp
1	Flashing
2	Off
3	Off
7	Off
14	On

C T   **3–24.** Repeat Problem 3–22 for the following troubleshooting results.

Probe on Pin	Indicator Lamp
1	Flashing
2	On
3	Off
7	Dim
14	On

Section 3–7

3–25. For Figure P3–25, write the Boolean equation at X. If $A = 1$, what is X?

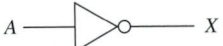

Figure P3–25

3–26. For Figure P3–26, write the Boolean equation at X and Z. If $A = 0$, what is X? What is Z?

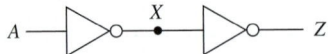

Figure P3–26

3–27. Using Figure P3–26, sketch the output waveform at X and Z if the timing waveform shown in Figure P3–27 is input at A.

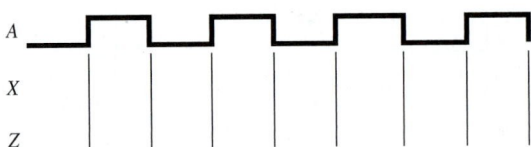

Figure P3–27

Section 3–8

3–28. For Figure P3–28, write the Boolean equation at X and Y.

Figure P3–28

3–29. Build a truth table for each gate in Figure P3–28.

3–30. Using Figure P3–28, sketch the output waveforms for X and Y, given the input waveforms shown in Figure P3–30.

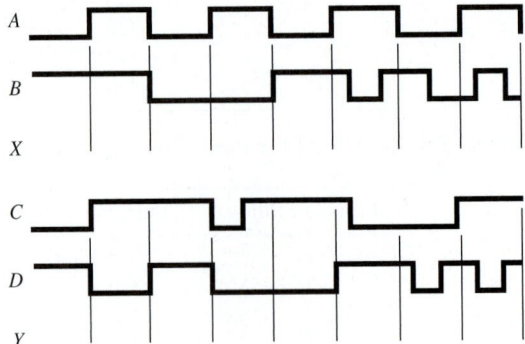

Figure P3–30

CHAPTER 3 I BASIC LOGIC GATES

3–31. Using Figure P3–31, sketch the waveforms at X and Y with the switches in the down (0) position. Repeat with the switches in the up (1) position.

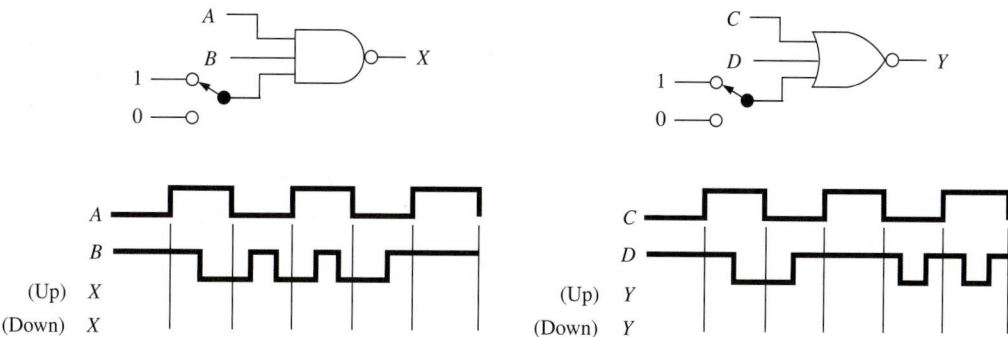

Figure P3–31

3–32. In words, what effect does the switch have on each circuit in Figure P3–31?

3–33. For Figure P3–33, write the Boolean equation at X and Y.

3–34. Make a truth table for the first NOR gate in Figure P3–33.

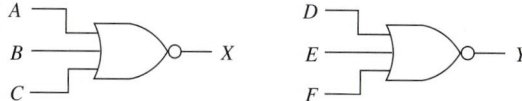

Figure P3–33

3–35. Referring to Figure P3–33, sketch the output at X and Y, given the input waveforms in Figure P3–35.

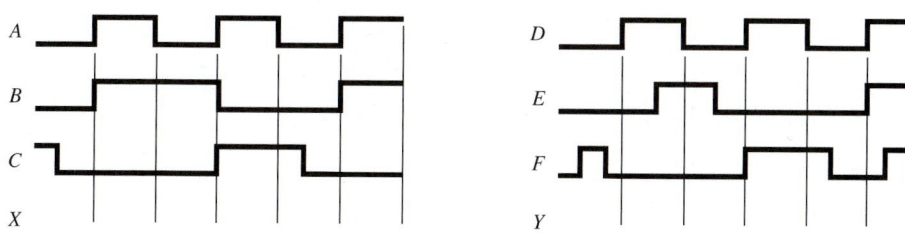

Figure P3–35

3–36. The Johnson shift counter outputs shown in Figure 3–43 are connected to the inputs of the logic gates shown in Figure P3–36. Sketch and label the output waveform at *U, V, W, X, Y,* and *Z.*

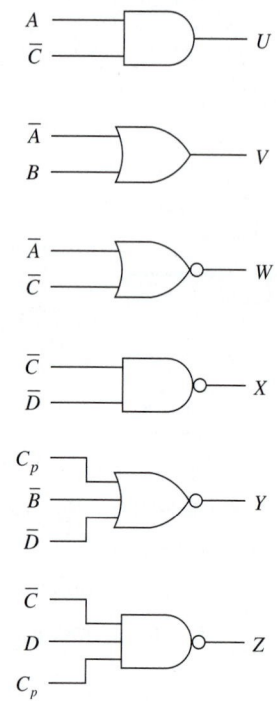

Figure P3–36

3–37. Repeat Problem 3–36 for the gates shown in Figure P3–37.

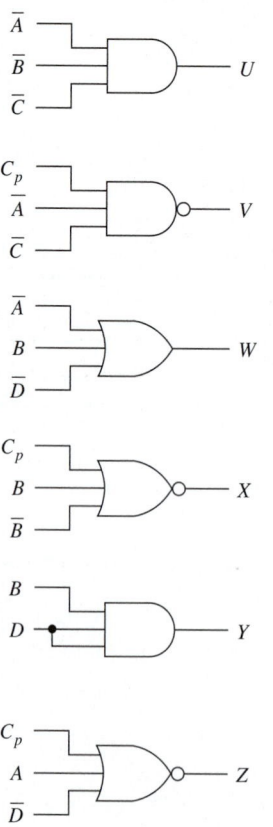

Figure P3–37

3–38. Using the Johnson shift counter outputs from Figure 3–43, label the inputs to the logic gates shown in Figure P3–38 so that they will produce the indicated output.

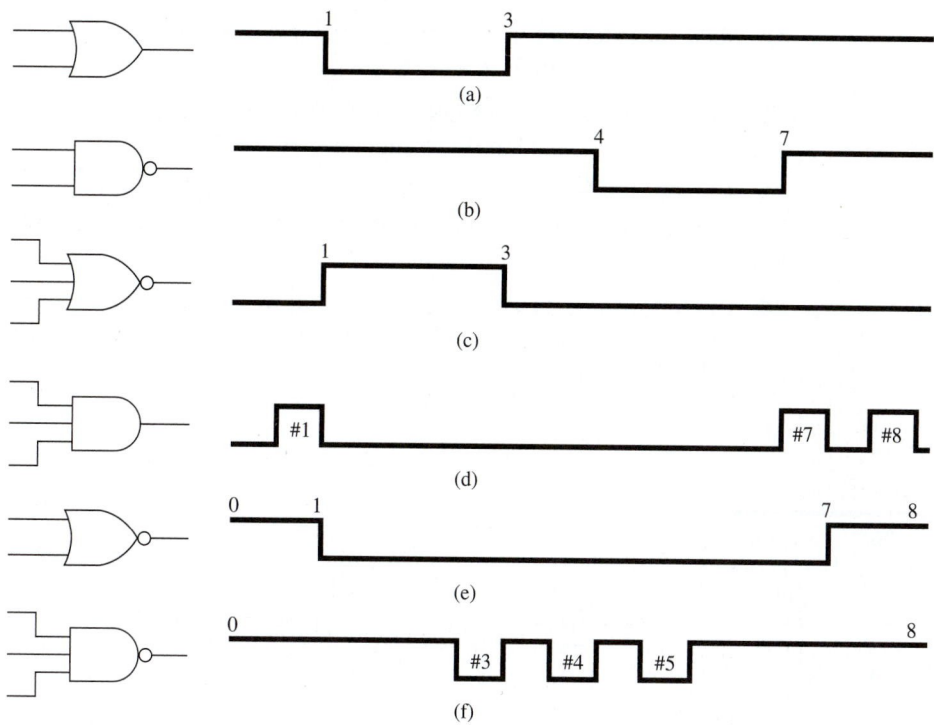

Figure P3–38

3–39. Determine which lines from the Johnson shift counter are required at the inputs of the circuits shown in Figure P3–39 to produce the waveforms at *U, V, W,* and *X.*

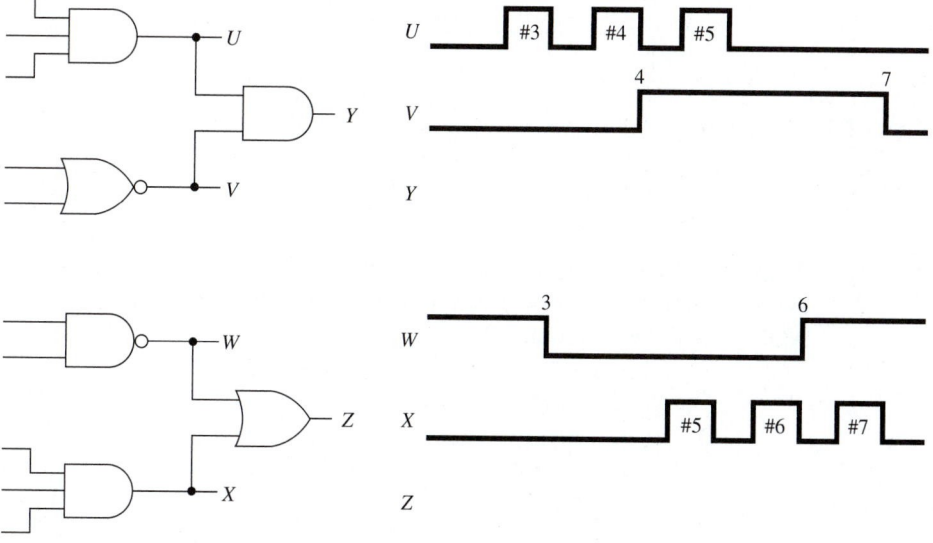

Figure P3–39

C   **3–40.** The waveforms at *U, V, W,* and *X* are given in Figure P3–39. Sketch the waveforms at *Y* and *Z*.

Section 3–11

3–41. Make the external connections to a 7404 inverter IC and a 7402 NOR IC to implement the function $X = \overline{A} + B$.

T   **3–42.** When troubleshooting a NOR gate like the 7402, with the logic pulser applied to one input, should the other input be held HIGH or LOW? Why?

T   **3–43.** When troubleshooting a NAND gate like the 7400, with the logic pulser applied to one input, should the other input be held HIGH or LOW? Why?

T   **3–44.** The following data table was built by putting a logic probe on every pin of the hex inverter shown in Figure P3–44. Are there any problems with the chip? If so, which gate(s) are bad?

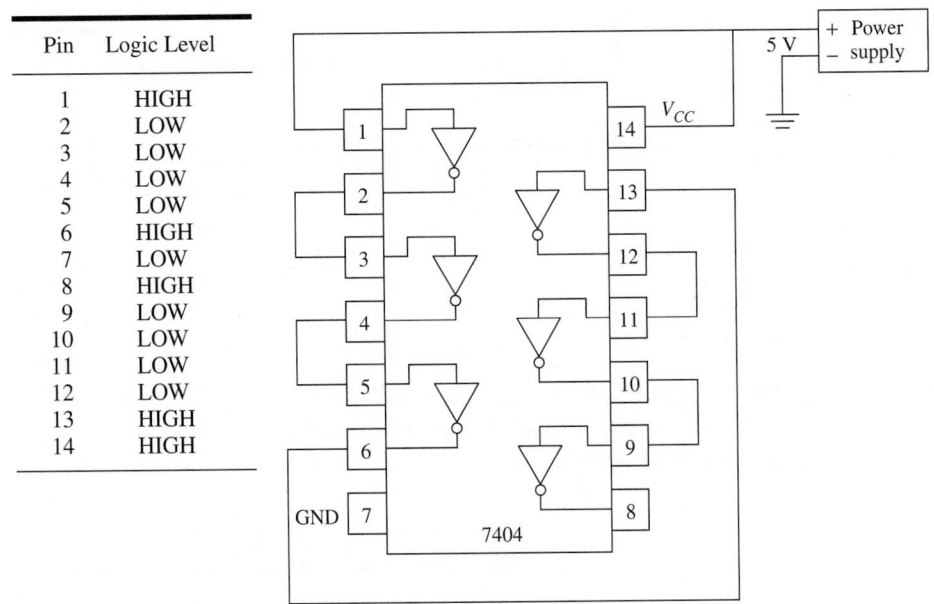

Pin	Logic Level
1	HIGH
2	LOW
3	LOW
4	LOW
5	LOW
6	HIGH
7	LOW
8	HIGH
9	LOW
10	LOW
11	LOW
12	LOW
13	HIGH
14	HIGH

Figure P3–44

C T   **3–45.** The logic probe in Figure P3–45 is always OFF (0) whether the switch is in the up or down position. Is the problem with the inverter or the NOR, or is there no problem?

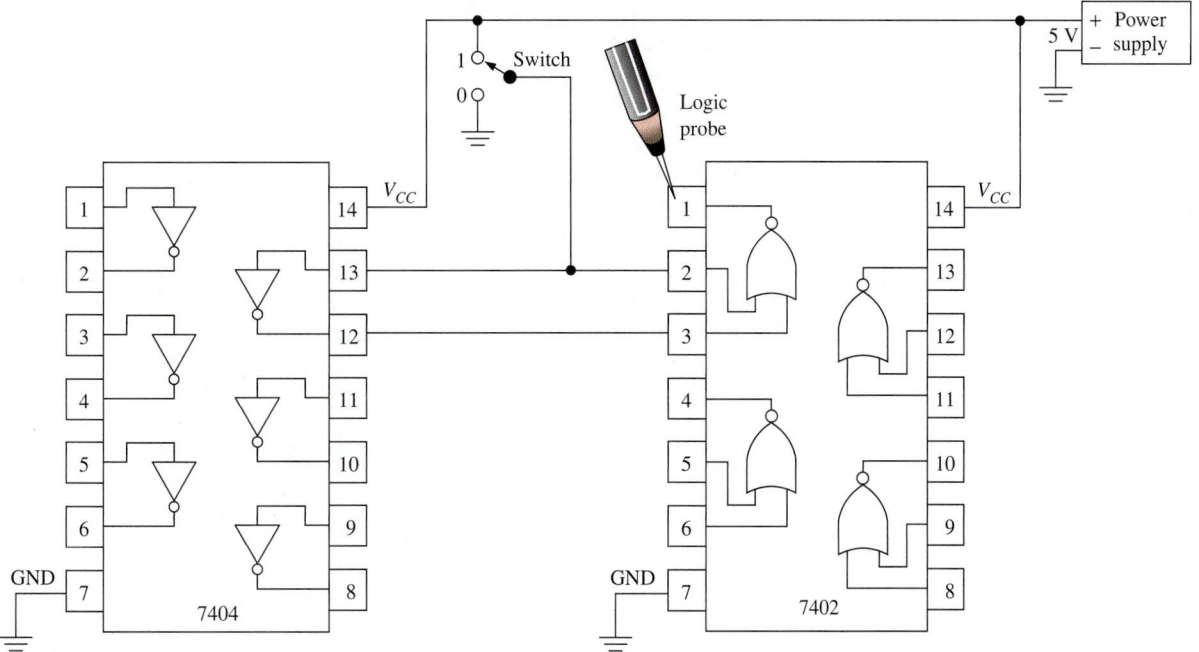

Figure P3–45

C T   **3–46.** Another circuit constructed the same way as Figure P3–45 causes the logic probe to come on when the switch is in the down (0) position. Further testing with the probe shows that pins 2 and 3 of the NOR IC are both LOW. Is anything wrong? If so, where is the fault?

T   **3–47.** Your company has purchased several of the 7430 eight-input NANDs shown in Figure P3–47. List the steps that you would follow to determine if they are all good ICs.

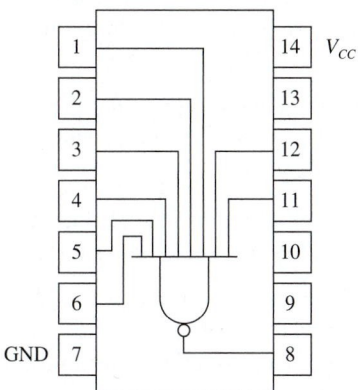

Figure P3–47

T

3–48. The previous data table was built by putting a logic probe on every pin of the 7427 NOR IC shown in Figure P3–48 while it was connected in a digital circuit. Which gates, if any, are bad, and why?

Pin	Logic Level
1	LOW
2	LOW
3	LOW
4	LOW
5	LOW
6	HIGH
7	LOW
8	Flashing
9	HIGH
10	LOW
11	Flashing
12	HIGH
13	HIGH
14	HIGH

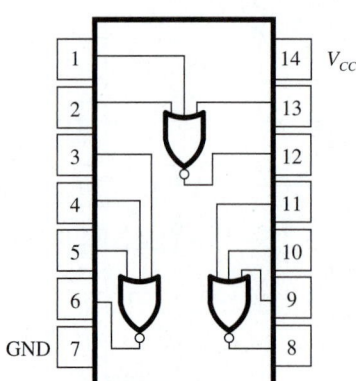

Figure P3–48

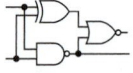

Schematic Interpretation Problems

See Appendix G for the schematic diagrams.

S

3–49. What are the component name and grid location of the two-input AND gate and the two-input OR gate in the Watchdog Timer schematic?

S

3–50. A logic probe is used to check the operation of the two-input AND and OR gates in the Watchdog Timer circuit. If the probe indicator is ON for pin 2 of both gates and flashing on pin 1, what will pin 3 be for (a) the AND gate and (b) the OR gate?

S

3–51. If you wanted to check the power supply connections for the 8031 IC (U8) on the 4096/4196 circuit, which pins would you check, and what level should they be?

S

3–52. On the 4096/4196 sheet 1 schematic, there are several gates labeled U1. Why are they all labeled the same?

S

3–53. Describe a method that you could use to check the operation of the inverter labeled U4:A of the Watchdog Timer. Assume that you have a dual-trace oscilloscope available for troubleshooting.

S

3–54. Locate the line labeled RAM_SL at location D8 of the HC11D0 schematic. To get a HIGH level on that line, what level must the inputs to U8 be?

C S

3–55. Locate the output pins labeled E and R/$\overline{\text{W}}$ on U1 of the HC11D0 schematic. During certain operations, line E goes HIGH and line R/$\overline{\text{W}}$ is then used to signify a READ operation if it is HIGH or a WRITE operation if it is LOW. For a READ operation, which line goes LOW: WE_B or OE_B?

Electronics Workbench®/MultiSIM® Exercises

E3–1. Load the circuit file for **Section 3–2.** Read the instructions in the *Description* window. The switches are used to input a 1 (up) or a 0 (down) to each gate input. Each switch can be moved by pressing the appropriate letter. The lamp connected to each gate output comes ON if the output is HIGH.

(a) What is the level at *X* and *Y* if all switches are up? Try it.

(b) What is the level at *X* and *Y* if all switches are down? Try it.

(c) Experimentally complete a truth table for each gate.

E3–2. Load the circuit file for **Section 3–3.** The Logic Analyzer shows the input waveforms *A* and *B,* and the output waveforms *X* and *Y*. Gate 1 and Gate 2 are hidden from your view; each is either an AND or an OR. Use the Logic Analyzer display to determine:

(a) What Gate 1 is, and

(b) What Gate 2 is.

E3–3. Load the circuit file for **Section 3–4.** This circuit is used to enable or disable the clock signal (*Cp*) from reaching the Logic Analyzer similar to Figures 3–17 and 3–18.

(a) Switch A must be in the _____ (up, down) position for the clock to be enabled.

(b) Switch B must be in the _____ (up, down) position for the clock to be enabled. Try both conditions.

C   **E3–4.** Load the circuit file for **Section 3–5.** All of the parts to build the clock enable circuit of Figure 3–20 are given. Make all of the necessary connections to make the circuit work and test its operation. What position must the Enable Switch be in to allow the receiving device to receive the clock pulses from *Cp*?

T   **E3–5.** Load the circuit file for **Section 3–6a.** This circuit is used to troubleshoot the number-4 gate of a 7408 Quad AND IC similar to Figure 3–24. Because this 7408 is working properly, the Logic Probe will flash when power is turned on. To troubleshoot the number-1 gate of the 7432 Quad NOR IC what should be connected to the following pins?

(a) Pin 1?

(b) Pin 2?

(c) Pin 3?

(d) Pin 7?

(e) Pin 14? Test your answers by moving the connections from the 7408 over to the 7402.

T   **E3–6.** Load the circuit file for **Section 3–6b.** There are one or more gates in each of the ICs shown that are bad. Use a Logic Pulser and Probe to find which gate or gates are bad (similar to Example 3–5).

(a) Which gate(s) are bad in the 7408?

(b) Which gate(s) are bad in the 7411?

(c) Which gate(s) are bad in the 7432?

E3–7. Load the circuit file for **Section 3–9a.** Read the instructions in the *Description* window. The switches are used to input a 1 (up) or a 0 (down) to each gate input. Each switch can be moved by pressing the appropriate letter. The lamp connected to each gate output comes ON if the output is HIGH.

(a) What is the level at X and Y if all switches are up? Try it.

(b) What is the level at X and Y if all switches are down? Try it.

(c) Experimentally complete a truth table for each gate.

E3–8. Load the circuit file for **Section 3–9b.** The Logic Analyzer shows the input waveforms A and B, and the output waveforms X and Y. Gate 1 and Gate 2 are hidden from your view, but each is either a NAND or a NOR. Use the Logic Analyzer display to determine:

(a) What Gate 1 is, and

(b) What Gate 2 is.

E3–9. Load the circuit file for **Section 3–10a.** This is the Johnson shift counter waveform generator from Figure 3–43. It is illustrated with A and B input to an AND gate.

(a) Is the output waveform correct?

(b) Write the Boolean equation at X.

(c) What is the time width of the X-waveform pulse?

E3–10. Load the circuit file for **Section 3–10b.** Change the inputs to the AND gate to A and C.

(a) What is the time at the rising edge, falling edge, and total pulse width of the X-output?

(b) Add Cp as a third input to the AND gate. How many positive pulses are output at X?

(c) What is the width of each positive pulse?

E3–11. Load the circuit file for **Section 3–10c.** The object here is to determine what gate is inside of the subcircuits labeled *gate 1* and *gate 2*. The output of gate 1 is displayed on the bottom trace. The next trace up is the output of gate 2.

(a) What is gate 1?

(b) What is gate 2?

E3–12. Load the circuit file for **Section 3–10d.** The object here is to determine what gate is inside of the subcircuits labeled *gate 3* and *gate 4*. The output of gate 4 is displayed on the bottom trace. The next trace up is the output of gate 3.

(a) What is gate 3?

(b) What is gate 4?

E3–13. Load the circuit file for **Section 3–10e.** Connect a logic gate to the Johnson outputs so that it will provide the following to the Logic Analyzer:

(a) The first three Cp pulses.

(b) A HIGH level from the 4 mS level to the 8 mS level.

Answers to Review Questions

3–1. True

3–2. To illustrate how the output level of a gate responds to all possible input-level combinations

3–3. To depict algebraically the operation of a logic gate

3–4. All inputs must be LOW.

3–5. To illustrate graphically how the output levels change in response to input-level changes

3–6. When the level of an input signal will have no effect on the output

3–7. HIGH

3–8. Positive power supply of 5 V to pin 14, ground at 0 V to pin 7

3–9. It uses an indicator lamp to tell you the digital level whenever it is placed in a circuit.

3–10. It provides digital pulses to the circuit being tested, which can be observed using a logic probe.

3–11. An inverter is used to complement or invert a digital signal.

3–12. A NAND gate is an AND gate with an inverter on its output.

3–13. HIGH

3–14. $X = \overline{A + B + C}$

3–15. HIGH

3–16. LOW

3–17. It is used as a repetitive waveform generator.

3–18. 7404; 4001

3–19. Triple, three-input NAND gates; triple, three-input NOR gates

4

Programmable Logic Devices: CPLDs and FPGAs with VHDL Design

OUTLINE

OBJECTIVES

Upon completion of this chapter, you should be able to:

- Explain the benefits of using PLDs.
- Describe the PLD design flow.
- Understand the differences between a PAL, PLA, SPLD, CPLD, and FPGA.
- Explain how a graphic editor and a VHDL text editor are used to define logic to a PLD.
- Interpret the output of a simulation file to describe logic operations.
- Interpret VHDL code for the basic logic gates.

INTRODUCTION

As you can imagine, stockpiling hundreds of different logic ICs to meet all the possible requirements of complex digital circuitry became very difficult. Besides having all of the possible logic on hand, another problem was the excessive amount of area on a printed-circuit board that was consumed by requiring a different IC for each different logic function. In many cases, only one or two gates on a quad or hex chip were used.

Then came "programmable logic"—the idea that implementing all logic designs using 7400- or 4000-series ICs is no longer needed. Instead, a company will purchase several user-configurable ICs that will be customized (i.e., programmed) to perform the specific logic operation that is required. These ICs are called **programmable logic devices (PLDs).**

4-1 PLD Design Flow

Samples of two PLDs are shown in Figure 4–1. They contain thousands of the basic logic gates plus advanced sequential logic functions inside a single package. This internal digital logic, however, is not yet configured to perform any particular function. One way to configure it is for the designer to first use PLD computer software to draw the logic that he or she needs implemented. This is called **CAD** (computer-aided design). The PLD software then performs a process called **schematic capture,** which reads the graphic drawing of the logic and converts (compiles) it to a binary file that accurately describes the logic to be implemented. This binary file is then used as an input to a programming process that electronically alters the internal PLD connections (synthesizes) to make it function specifically as required. Hundreds, or even thousands, of digital logic ICs will be replaced by a single PLD.

Another way to define the logic to be programmed into the PLD is to use a high-level language called Hardware Description Language (HDL). A specific form of HDL used by several manufacturers is called **VHDL,** which stands for VHSIC Hardware Description Language (where VHSIC stands for Very High-Speed Integrated Circuit). In this case, the inputs, outputs, and logic processes are defined using a programming language that looks a lot like C++. This method is somewhat more difficult to learn, but depending on the logic, it can be more a powerful—and simpler—tool with which to define complex or repetitive logic.

Figure 4–2 illustrates the design flow. First we need to define the digital logic problem that we want to solve. Once we have a good understanding of the problem, we can develop the equations to use in solving the logic operation that we want the circuit to perform.

(a)   (b)

Figure 4–1 Sample PLDs: (a) Altera EPM7128S; (b) Xilinx XC95108.

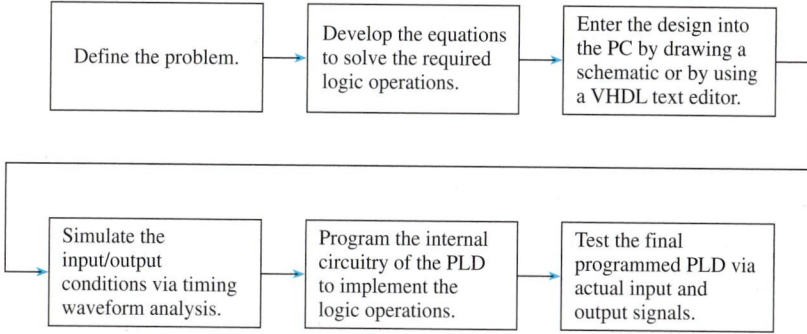

Figure 4–2 PLD product design flow.

After we have completed that work on paper, we will enter the design into a personal computer (PC) by drawing a schematic diagram using the CAD tools provided with the PLD software. In some cases, the design will instead be entered using the VHDL text editor provided. After the PC has analyzed the design, it will allow us to perform a simulation of the actual circuit to be implemented. To do this, we specify the input levels to our circuit, and we observe the resultant output waveforms on the PC screen using the waveform analysis tool provided.

If the computer simulation shows that our circuit works correctly, we can program the logic into a PLD chip that is connected by a cable to the back of our PC. The final step would be to connect actual inputs and outputs to the chip to check its performance in a real circuit.

To illustrate the power of a PLD, let's consider the logic circuit required to implement $X = \overline{A}B + \overline{B + C}$. Figure 4–3 shows the circuitry required to implement the logic using 7400-series ICs. As shown, we would need four different ICs to solve this equation. Wires are shown connecting one gate of each IC to one gate of the next IC until the logic requirements are met .

To solve this same logic using a PLD, we would draw the schematic or use VHDL to define the logic, then program that into a PLD. One possible PLD that could be used to implement this logic is the Altera EPM7128S (see Figure 4–4). After completing the steps listed in Figure 4–2, the internal circuitry of the PLD is configured (in this case) to input A, B, and C at pins 29, 30, and 31 and output to X at pin 73. The PLD software selected which pins to use, and as you can see, only a small portion of the PLD is actually used for this circuit .

This particular PLD is an 84-pin IC in a plastic leaded chip carrier (PLCC) package having 21 pins on a side. The notch signifies the upper left corner of the IC. Pin 1

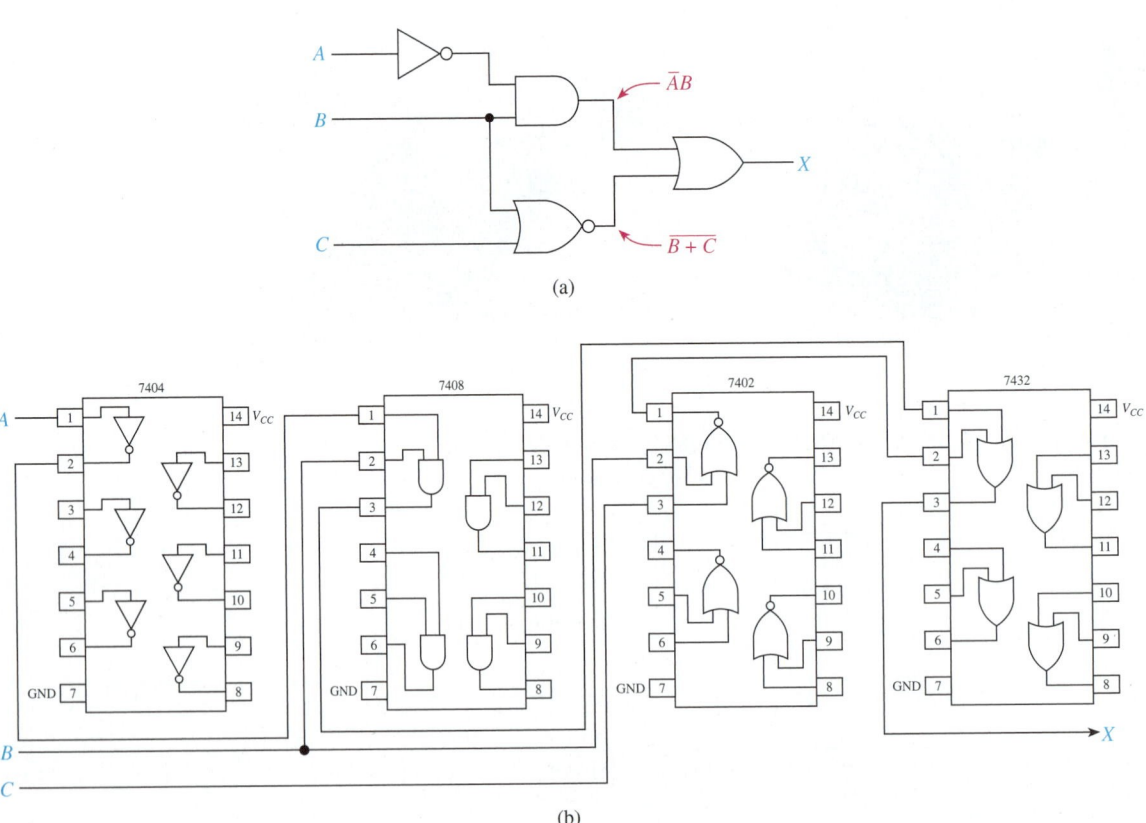

(a)

(b)

Figure 4–3   Implementing the equation $X = \overline{A}B + \overline{B + C}$ using 7400-series logic ICs: (a) Logic diagram; (b) Connections to IC chips.

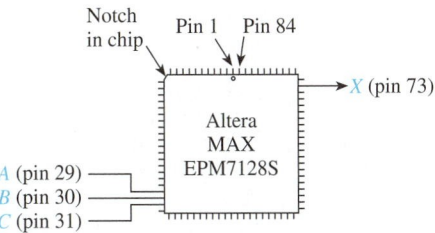

Figure 4–4 Implementing the equation $X = \overline{AB} + \overline{B} + C$ using a PLD.

is located in the middle of the upper row adjacent to a small indented circle; subsequent pin numbers are counted off counterclockwise from there. (A photograph of this particular chip is shown in Figure 4–1.)

As you may suspect, the price of a PLD is higher than a single 7400-series IC, but we've only used a small fraction of the PLD's capacity. We could enter and program hundreds of additional logic equations into the same PLD. The only practical limitation is the number of input and output pins that are available. Many PLDs are erasable and reprogrammable, allowing us to test many versions of our designs without ever changing ICs or the physical wiring of the gates.

We will learn design entry and waveform simulation in this chapter, and we will continue to explore PLD examples and problems throughout the remainder of this text.

One of the leading PLD manufacturers is Altera Corporation (www.altera.com), which has two software packages for programming its PLDs: MAX+PLUS® II and Quartus® II. Both packages are very similar, Quartus being the more advanced, covering their highest-level PLDs. MAX+PLUS® II software is more appropriate for entry- and intermediate-level PLD design and is included on the textbook CD along with solutions to all of the textbook examples.

PLD programmer boards that attach to the printer port on a PC can be purchased so that you can experience the programming and testing of actual PLD ICs. These programmer boards contain an Altera EPM7128S CPLD along with input switches, output LED indicators, and the interface circuitry to connect and communicate to your PC. Figure 4–5 shows photographs of two popular PLD programmer boards. The board in Figure 4–5 (a) is the RSR PLDT-2 from Electronics Express (www.elexp.com) and Figure 4–5 (b) is the Altera UP-1 (or UP-2) from the Altera University Program (www.altera.com). The Altera board also has an FPGA PLD, which can be used for more advanced projects. (The difference between CPLDs and FPGAs is explained in Section 4–2.)

4–2   PLD Architecture

Basically, there are three types of PLDs: simple programmable logic devices (SPLDs), complex programmable logic devices (CPLDs), and field-programmable gate arrays (FPGAs).

The SPLD

The **SPLD** is the most basic and least expensive form of programmable logic. It contains several configurable logic gates, programmable interconnection points, and may also have memory flip-flops. (Flip-flops are covered in Chapter 10.) To keep logic diagrams easy to read, a one-line convention has been adopted, as shown in Figure 4–6, which is just a small part of an SPLD, showing two inputs and four outputs. (A typical SPLD like the PAL in Figure 4–9 has 16 inputs plus their complements and 8 outputs.) As you can see in Figure 4–6, the A input is split into two different lines: A, and its complement $\overline{A}$. (The triangle symbol is a special type of inverter having two outputs: a true

(a)

(b)

Figure 4–5   PLD programmer boards: (a) RSR PLDT-2 (www.elexp.com); (b) UP-1 (University Program at www.altera.com).

and a complement.) The same goes for the B input and any others that are on the SPLD. The W, X, Y, and Z AND gates are programmable to have any of those four lines (A, $\overline{A}$, B, $\overline{B}$) as inputs.

The internal SPLD interconnect points are either made or not made by the PLD programming software. In Figure 4–6, the inputs to the W AND gate are connected to A and B. (The connections are shown by a dot.) The inputs to the X AND gate are connected to A and $\overline{B}$, and so on. The outputs of these AND gates are called the **product terms,** because X is the Boolean product of A and B and Y is the Boolean product of A and $\overline{B}$.

The product terms in Figure 4–6 are not very useful by themselves. The circuit is made more effective by adding an OR gate to the structure, as shown in Figure 4–7. This new configuration is the foundation for a **programmable array logic (PAL)**–type

SPLD. As Figure 4–7 shows, by OR-ing the four product terms together, we now have the Boolean sum of the four product terms, simply called the **Sum-of-Products (SOP).** The SOP is the most common form of Boolean equation used to represent digital logic. (For more on SOPs, see Section 5–6.)

The **programmable logic array (PLA)** goes one step further by providing *programmable* OR gates for combining the product terms. Figure 4–8 shows a small portion

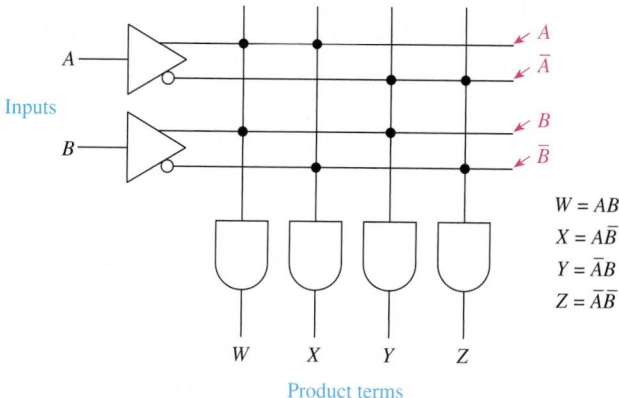

$$W = AB$$
$$X = A\bar{B}$$
$$Y = \bar{A}B$$
$$Z = \bar{A}\bar{B}$$

Figure 4–6   One-line convention for PLDs.

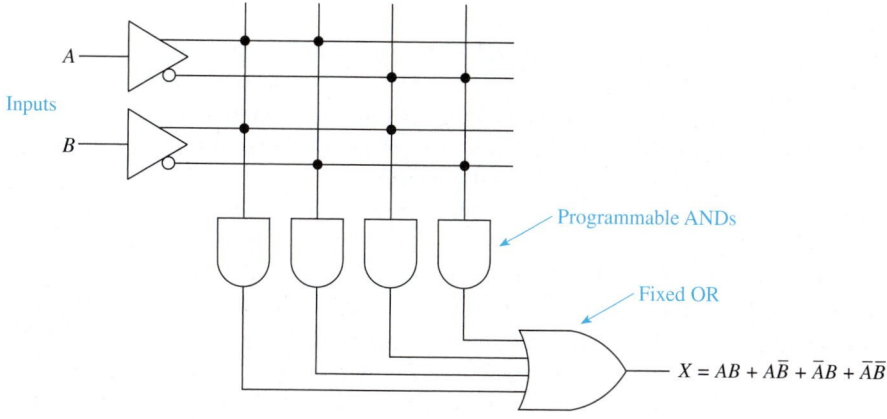

$$X = AB + A\bar{B} + \bar{A}B + \bar{A}\bar{B}$$

Figure 4–7   PAL architecture of an SPLD.

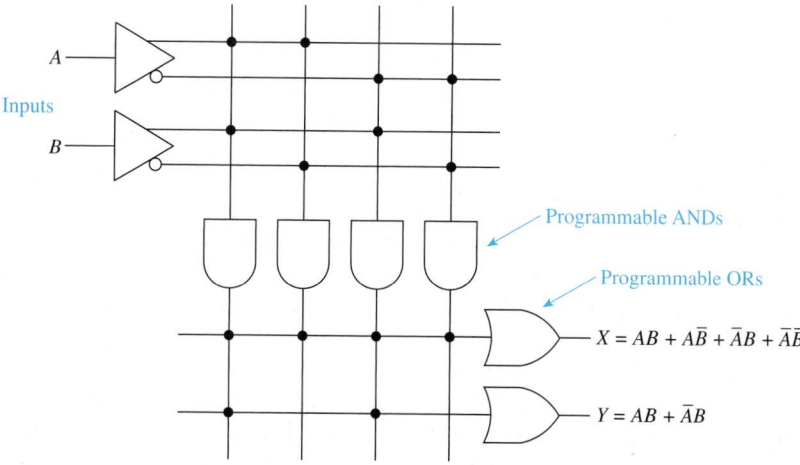

$$X = AB + A\bar{B} + \bar{A}B + \bar{A}\bar{B}$$
$$Y = AB + \bar{A}B$$

Figure 4–8   PLA architecture of an SPLD.

of a PLA. In this illustration, the PLA provides two SOP equations. The inputs to the first OR gate are programmed to connect to all four product terms ($X = AB + A\overline{B} + \overline{A}B + \overline{A}\,\overline{B}$). The inputs to the second OR gate are programmed to connect to only the first and third product terms ($Y = AB + \overline{A}B$).

Some SPLDs also contain a flip-flop memory section and data-steering circuitry. Flip-flop memory circuitry is used in a type of digital circuitry called *sequential logic*. This type of logic is a form of digital memory that changes states based on previous logic conditions and specific logic control inputs. (Sequential logic is covered in detail in Chapters 12 and 13.) The data-steering circuitry takes care of input and control signal interconnections and logic output destinations.

PAL16L8

A sample of a typical PAL device is the PAL16L8 shown in Figure 4–9. The number 16 in the part number signifies that it has 16 inputs. The 8 signifies 8 outputs and the letter L means that the outputs are active-LOW. An active-LOW output is one that goes LOW instead of HIGH when activated. Ten of the inputs in the figure are labeled with the letter I. Each of these can provide the true and the complement of the level placed on the pin. The other 6 inputs are labeled I/O. This means that they can be used as an input or an output. To come up with a total of 8 outputs, the other 2 dedicated outputs labeled O are provided on pins 12 and 19.

The CPLD

The **CPLD** is made by combining several PAL-type SPLDs into a single IC package, as shown in Figure 4–10. Each PAL-type structure is called a *macrocell*. Each macrocell has several I/O connection points, which go to the chips' external leads. The macrocells are all connected to control signals and to each other via the programmable interconnect matrix shown in the center of the structure.

Examples of CPLDs are the Altera MAX 7000S series and the Xilinx XC9500 series. (Photographs are shown in Figure 4–1, and data sheets are provided in Appendix B). These CPLDs perform the functions of thousands of individual logic gates. Both also feature a **nonvolatile** characteristic, meaning that when power is removed from the chip, they will remember their programmed logic and interconnections. (This type of memory is called EEPROM or Flash memory and is covered in Chapter 16.) These ICs can be repeatedly programmed to implement new designs or correct faulty ones, thus eliminating the need to rewire circuitry or buy new logic.

The FPGA

The **FPGA** differs from the CPLD in that, instead of solving the logic design by interconnecting logic gates, it uses a **look-up table** method to resolve the particular logic requirement. This technique allows PLD manufacturers to form a more streamlined design, with a much denser concentration of logic functions.

A look-up table is simply a truth table that lists all the possible input combinations with their desired output response. Figure 4–11 shows how an FPGA would use a look-up table to output the correct logic level for the equation $X = AB + A\overline{B}$. As you can see in the logic diagram, X will be HIGH if A and B are both HIGH, or if A is HIGH while B is LOW. Instead of forming the logic circuit, an FPGA builds the truth table shown in Figure 4–11(b). It then uses the logic levels received at its A and B input pins to pinpoint the location in the truth table to output a logic level at X. For example, Figure 4–11(c) shows the result if 1 is input at A and 0 is input at B. The FPGA will point to the result of the third look-up table entry and then output that level, which is a 1. Most look-up tables used in FPGAs utilize four inputs, resulting in 16 possible output results. Modern FPGAs have hundreds of look-up tables, configured with pro-

TIBPAL 16L8-15C
TIBPAL 16L8-20M
HIGH-PERFORMANCE *IMPACT*™ *PAL*® CIRCUITS
SRPS019A – FEBRUARY 1984 – REVISED APRIL 2000

logic diagram (positive logic)

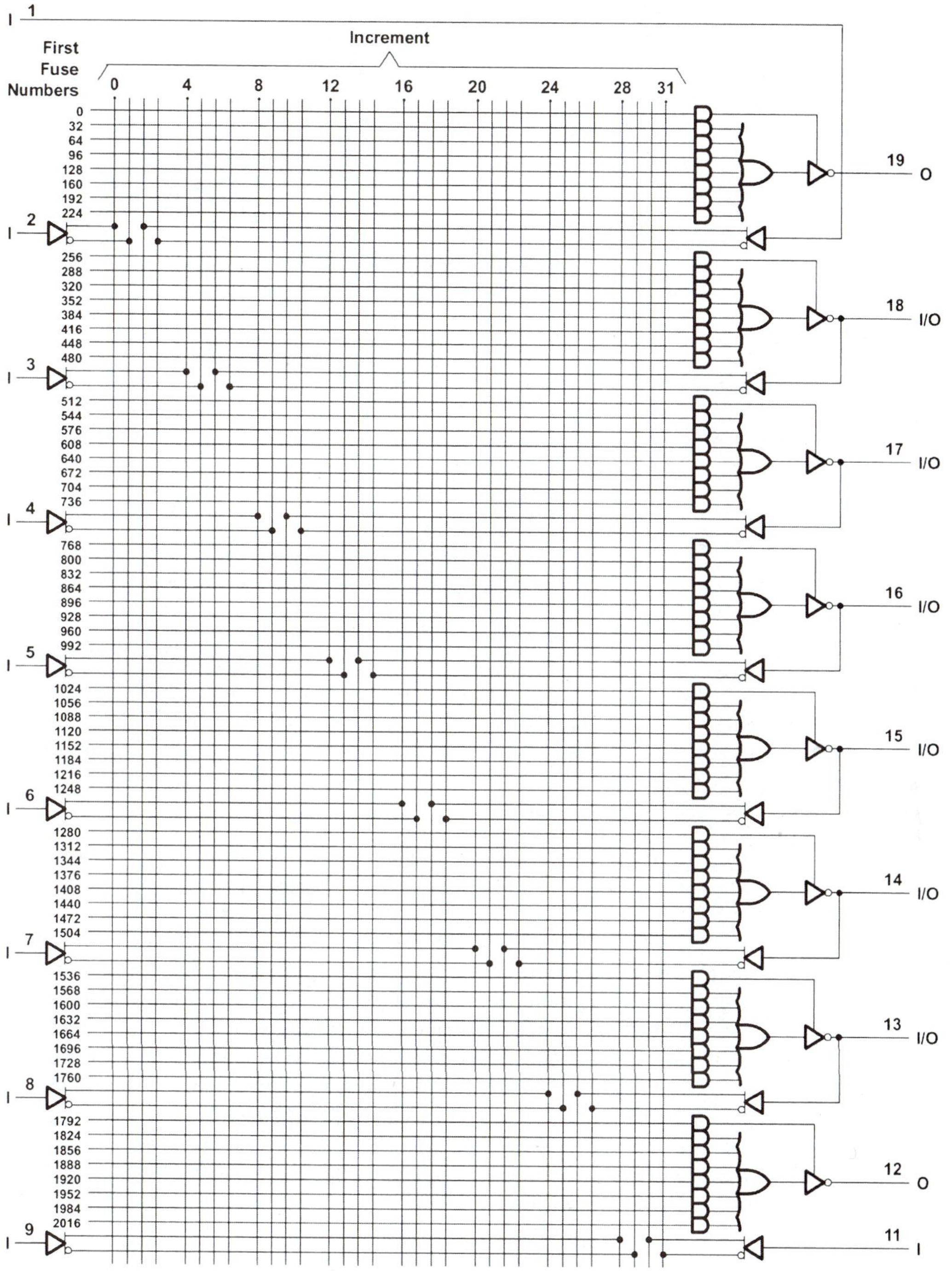

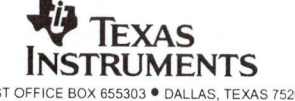

Fuse number = First fuse number + Increment

TEXAS INSTRUMENTS
POST OFFICE BOX 655303 • DALLAS, TEXAS 75265

Figure 4–9 The PAL16L8 SPLD logic diagram.

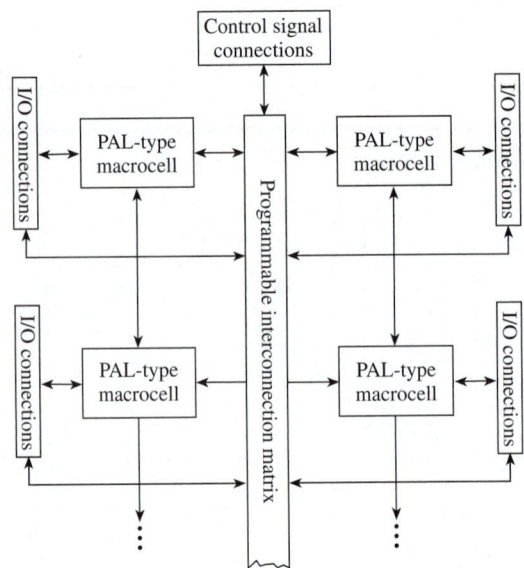

Figure 4–10 Internal structure of a CPLD.

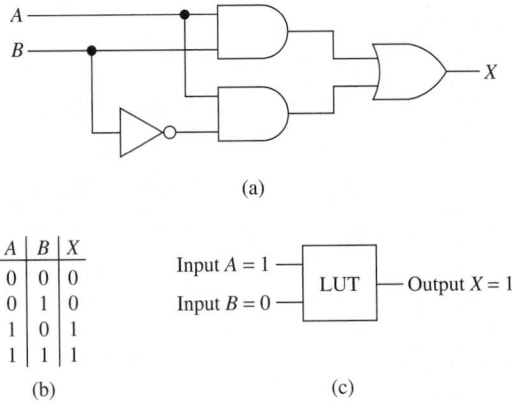

(a)

A	B	X
0	0	0
0	1	0
1	0	1
1	1	1

(b)

Input $A = 1$ — LUT — Output $X = 1$
Input $B = 0$ —

(c)

Figure 4–11 $X = AB + A\overline{B}$: (a) logic circuit representation; (b) FPGA look-up table method of resolving outputs; (c) look-up table results for $A = 1$; $B = 0$.

grammable interconnections feeding the I/O pins. This method of simply looking up 1's or 0's allows manufacturers to form a much denser PLD, with hundreds of times more equivalent logic gates than a CPLD contains.

Examples of FPGAs are the Xilinx XC4000 series and the Altera FLEX 10K series. Even though they generally hold a much higher amount of logic than CPLDs, their main drawback is that their memory uses SRAM technology, which is volatile. This means they lose their logic program once power is removed from the chip. (SRAMs are covered in Chapter 16.) Thus, their logic program must be loaded into the chip each time the system is powered up.

4–3 Using PLDs to Solve Basic Logic Designs

So, the next obvious question is "How do I design logic with a PLD?" We will use the MAX+PLUS® II software to design and simulate solutions modeled after Altera CPLDs. Then, if your laboratory has the PLD programmer boards like those shown in Figure 4–5, you can test the actual operation of the CPLD with switches and lights.

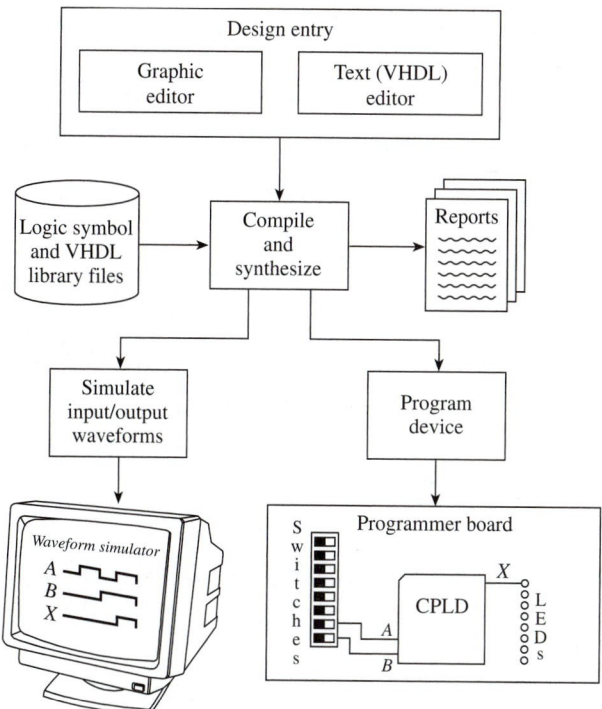

Figure 4–12   CPLD design flow.

Even without the boards, however, the design and simulation software is a great learning tool for digital logic.

Figure 4–12 shows the flow of operations required to design, simulate, and program a CPLD. Several methods are actually available to perform the design entry, but we will address the two most common: graphic, and VHDL. The **graphic editor** enables you to connect predefined logic symbols (AND, NAND, OR, etc.) together with inputs and outputs to define the logic operation that you need to implement. The **VHDL editor** is a text editor that helps you to define the logic in a programming language environment. In a text form, you specify the inputs, outputs, and logic operations that you need to implement.

The next step performed by the software is to compile and synthesize the design. A **compiler** is a language and symbol translation program that interprets VHDL statements and logic symbols, then translates them into a binary file that can be used to synthesize, then simulate and program the design into the CPLD IC. The compiler uses several symbol and VHDL library files to obtain the information needed to define the logic entered during the design entry stage. Report files are then generated that describe such things as I/O pin assignments, internal CPLD signal routing, and error messages. **Synthesizing** the design is the process the software completes to develop a model of the PLD's internal electrical connections, which will produce the actual logic functions that will later be simulated, then programmed into the PLD.

The **waveform simulator** provides a means to check the logic operation of your design. To use it, draw the input waveforms using the CAD tool provided, and the program will show the output response as if these inputs were applied to an actual CPLD. Finally, if you have a CPLD programmer board and the waveform simulation was accurate, you can program the CPLD and test it with actual inputs and outputs.

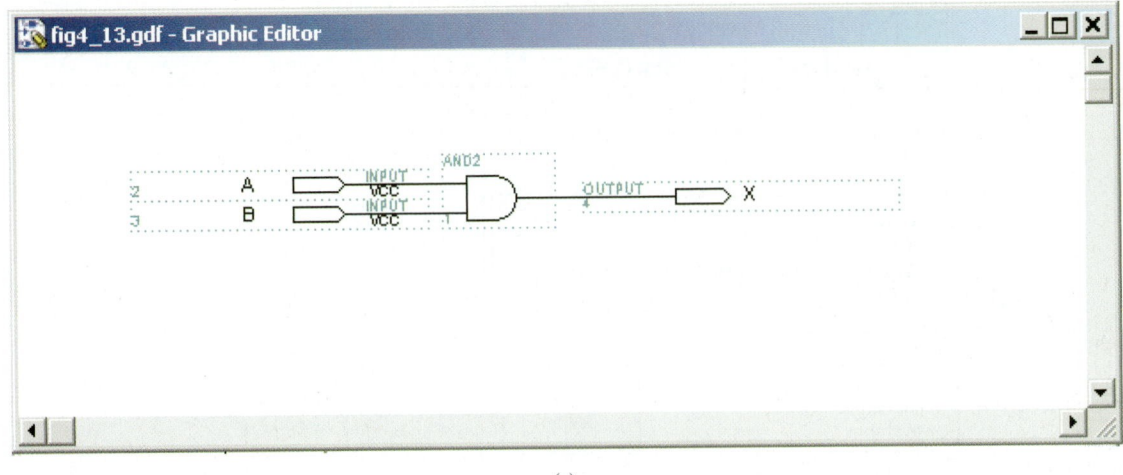

(a)

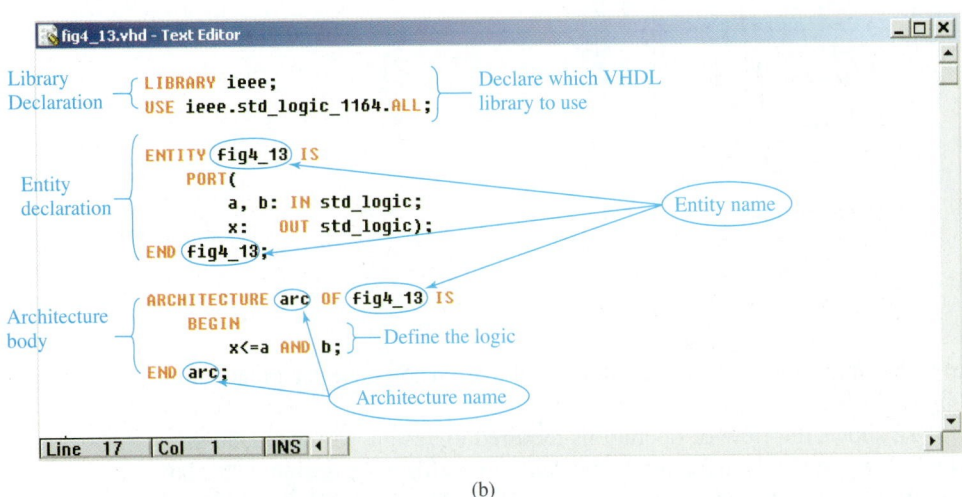

(b)

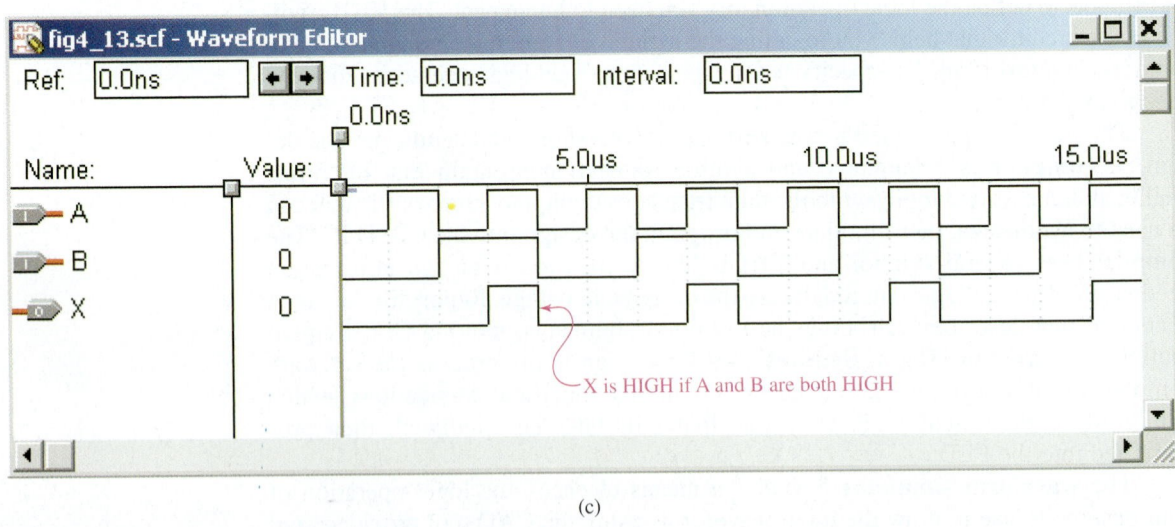

(c)

Figure 4–13   Computer screen displays generated by MAX+PLUS® II software for the design of a 2-input AND gate: (a) Graphic Editor file; (b) alternative method using the VHDL Text Editor file; (c) Waveform Editor file.

Max+Plus® II CPLD Software

Figures 4–13 (a), (b), and (c) are the actual computer screens that you will see when running the MAX+PLUS® II software to implement a simple 2-input AND gate following the design flow outlined in Figure 4–12. A tutorial on how to run the software appears in the Section 4–4.

Figure 4–13(a) is produced by the *Graphic Editor.* This method of design allows us to define the inputs, outputs, and circuit logic simply by drawing the logic diagram. This screen shows a 2-input AND gate with two input pins, *A* and *B,* and one output pin, *X.* This circuit was drawn by choosing each circuit component from a library of available symbols and then making each interconnection.

Figure 4–13(b) shows an alternate method of defining the same AND gate design using the *VHDL Text Editor.* The VHDL program is divided into three sections: **Library declaration, entity declaration,** and **architecture body.** As with most computer languages, the first statements of the program are used to declare the library source for resolving and translating the language within the body of the program. In VHDL this is called the *library declaration.* The IEEE standard library (*ieee.std_logic_1164.ALL*) is used most often by the VHDL compiler to translate references to the inputs, outputs, and logic statements used in the program.

The *entity declaration* defines the input (*a, b*) and output (*x*) ports to the CPLD. Note that the entity name (fig4_13) must match the file name (*fig4_13.vhd*) and it appears identical in three locations in the program listing. Also note the use of the underscore in the name because hyphens are not allowed.

The *architecture body* defines the internal logic operations ($x < = a$ AND b) that will be performed on those ports. (The symbol $< =$ means that output x receives the value of input a ANDed with input b.) The architecture name is arbitrary and it appears twice. The one used here is *arc.* As with the entity name, it cannot contain hyphens and it must start with a letter.

To make the reading of VHDL programs easier, a formatting convention has been established. Basically, all capitalized words are VHDL-reserved keywords, and all lower-case words and letters are variables. Even though VHDL is not case-sensitive, it is good practice for you to follow the convention presented in Figure 4–13(b). For example, writing the equation $x< = a$ AND b as $X< = A$ AND B would make no difference to VHDL, but it is harder to distinguish the keyword AND from the variables *A, B.*

You have probably guessed that for defining the action of a simple AND gate, VHDL design is more time-consuming than graphic entry, but we will see in later chapters that it is a much easier way to define logic when the circuits become more complex.

Figure 4–13 (c) shows the simulation of the circuit produced by the Waveform Editor. To produce that screen, the waveforms were first drawn for all possible combinations of *A* and *B* (like building a truth table). Then as the simulation is run, the software determines the logic state that would result at *X* for each combination of inputs and shows the result as the *X* waveform.

EXAMPLE 4–1

Figure 4–14 shows five computer screens generated by the MAX+ PLUS® II software. Each screen produces, or is the result of, a different logic circuit. Determine the Boolean equation that is being implemented in each case.

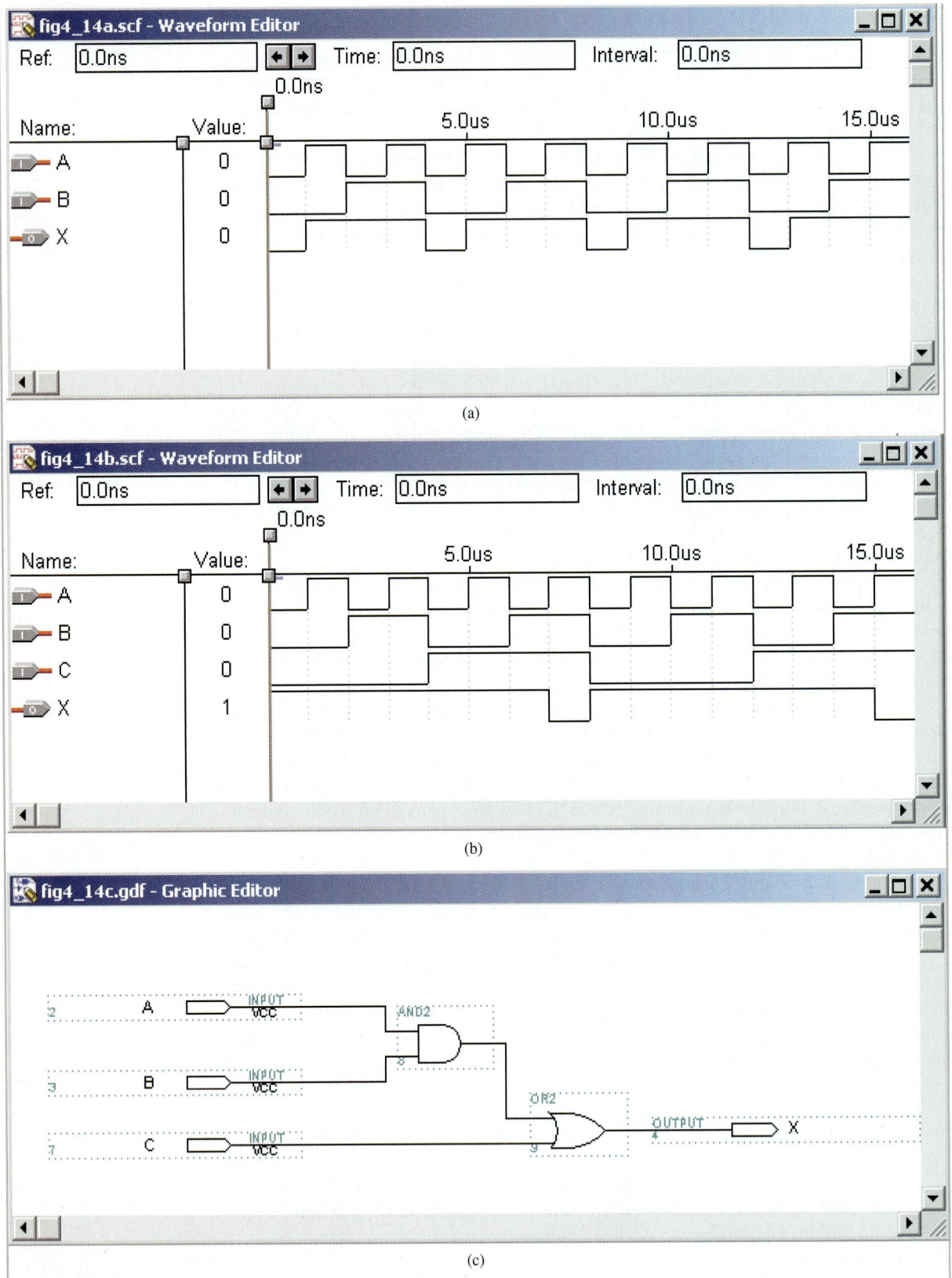

Figure 4–14 Computer screens generated by the MAX+PLUS® II software for Example 4–1.

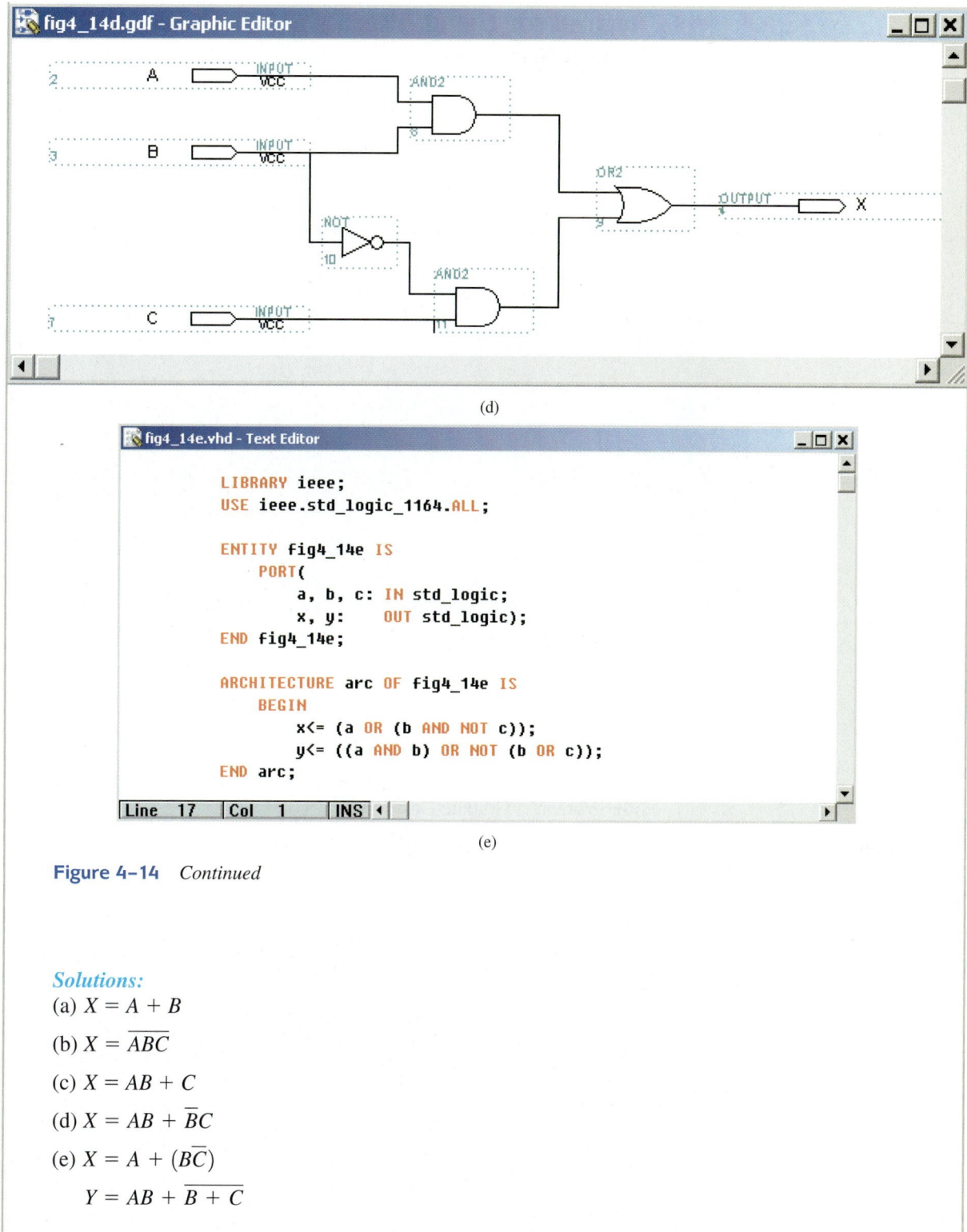

(d)

```
fig4_14e.vhd - Text Editor

        LIBRARY ieee;
        USE ieee.std_logic_1164.ALL;

        ENTITY fig4_14e IS
            PORT(
                a, b, c: IN std_logic;
                x, y:    OUT std_logic);
        END fig4_14e;

        ARCHITECTURE arc OF fig4_14e IS
            BEGIN
                x<= (a OR (b AND NOT c));
                y<= ((a AND b) OR NOT (b OR c));
        END arc;

Line  17   Col   1    INS
```

(e)

Figure 4–14 *Continued*

Solutions:

(a) $X = A + B$

(b) $X = \overline{ABC}$

(c) $X = AB + C$

(d) $X = AB + \overline{B}C$

(e) $X = A + (B\overline{C})$

$Y = AB + \overline{B + C}$

4–4 Tutorial for Using Altera's MAX+PLUS® II Design Software

Here, we will learn the basics of CPLD design and simulation by building a solution to the Boolean equation $X = AB + CD.$

Start the Altera MAX+PLUS® II program. The main screen is shown in Figure 4–15.

Figure 4–15   MAX+PLUS® II main screen.

Create a Graphic Design File (*.gdf)

1. To draw the logic circuit for our Boolean equation, we will use the graphic editor to create a Graphic Design File:

Choose **File > New > Graphic Editor file > .gdf**

Press **OK** (see Figure 4–16)

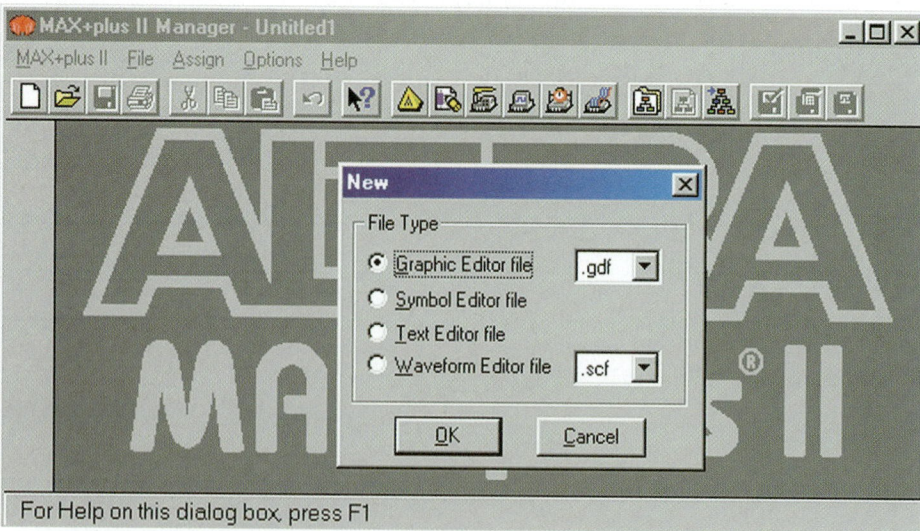

Figure 4–16   Creating a new Graphic Design File (.gdf).

2. You then want to assign a name to your file:

Choose **File > Save As**

File Name **C:\maxplus2\kleitz*boolean 1***

Press **OK** (see Figure 4–17)

When asked if OK to create new directory, press **YES**

> [The extension will be *.gdf*. Ask your class instructor, but you will probably want to use *your* name as the subdirectory in place of *kleitz*. The complete pathname **C:\maxplus2\kleitz\boolean1** creates the subdirectory *kleitz* and only needs to be given the first time that you save a file. Subsequent times will only require the file name (like *boolean1*) as long as you are careful to have the correct subdirectory name highlighted in the Directories column.]

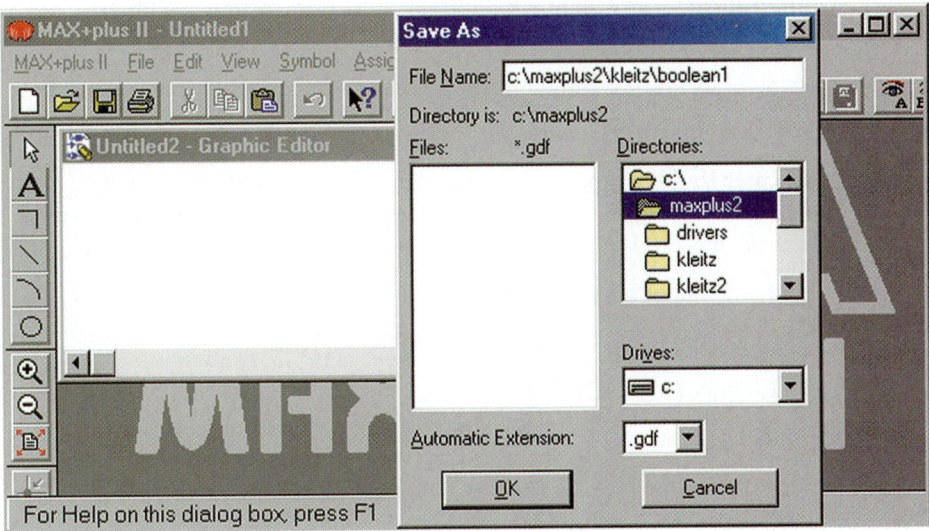

Figure 4–17 Saving the .gdf file in a new subdirectory.

3. Next, we need to specify a Project name:

Choose **File > Project > Set Project to Current File** (see Figure 4–18)

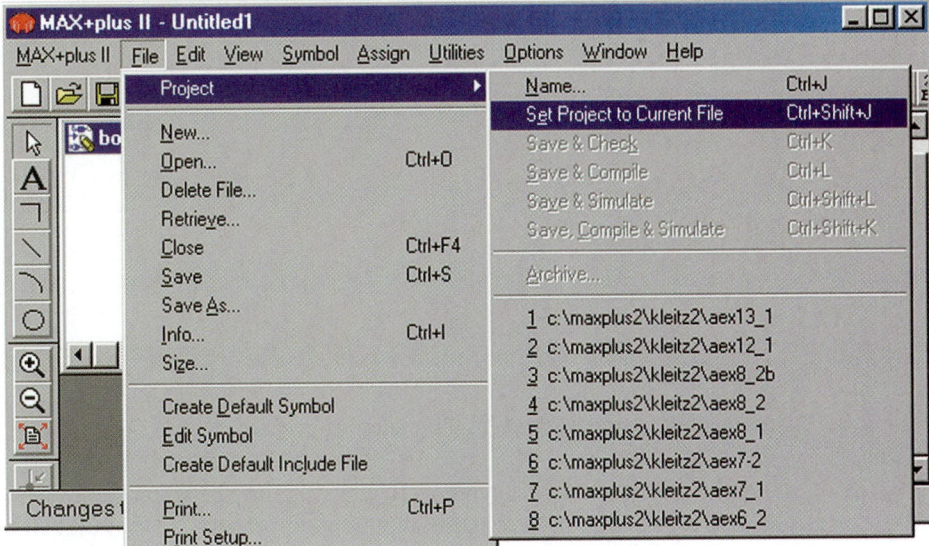

Figure 4–18 Setting the project to the current file.

4. Now you are ready to draw the logic circuit. To add a logic symbol to the graphic editor work space:

Double-click anywhere in the work space

Enter the symbol name for the first logic gate (a two-input AND gate) by typing **AND2,** then press **OK**

5. Repeat for the second AND gate and for the two-input OR gate **(OR2)**

[*Note:* To see all of the gates that are available for us to synthesize, try the following:

Double-click anywhere in the work space

In the **Enter Symbol** drop-down menu that appears, double-click on the **Symbol Library** entry ending with **\prim** (this is the primitives library). All of the available gates will appear under the Symbol Files heading.

To see all of the 7400-series chips available to synthesize, try the following:

Double-click anywhere in the work space

In the **Enter Symbol** drop-down menu that appears, double-click on the **Symbol Library** entry ending with **\mf** (this is the macrofunction library). All of the available 7400-series chips will appear under the Symbol Files heading.]

6. Input and output pins must now be added to the circuit:

Double-click somewhere near the input to the first gate

Type the symbol name **INPUT**

Press **OK**

Double-click on the **PIN_NAME,** and change it to **A**

7. Repeat for the other inputs: **B, C, D,** and the output **X.** (The results of steps 4–7 are shown in Figure 4–19.)

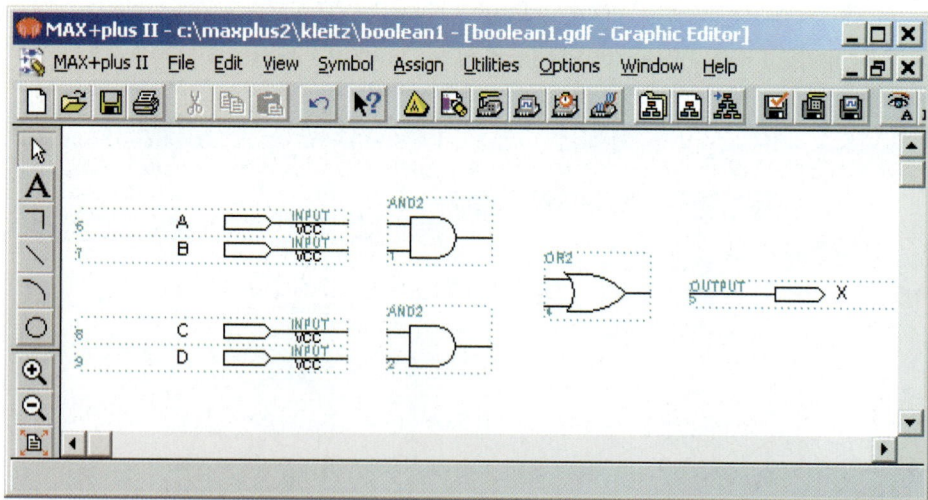

Figure 4–19   Adding logic symbols to the graphic editor work space.

8. Next, you want to connect the symbols by drawing lines between them. First, use the left mouse button to drag the input/output pinstubs so that they line them up with the appropriate logic gate. Then use the left button to draw a line from the end of each pinstub to the beginning of each corresponding logic gate input. If you make a mistake, highlight the line, and press the **Del** key (see Figure 4–20).

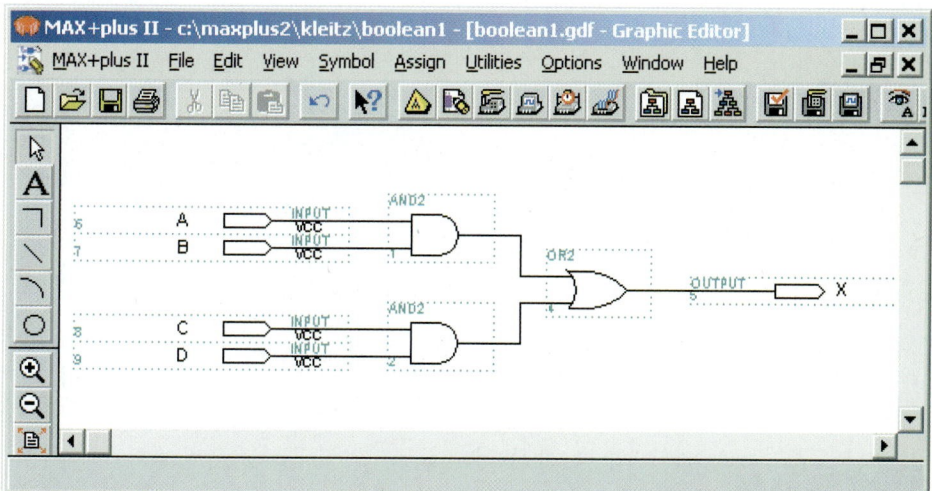

Figure 4–20   Adding connection lines.

9. We are done creating the graphic design file and can now assign an Altera device to be used to implement the design. To assign a device to the project:

Choose **Assign > Device**

The CPLD that we will use is in the Device Family **MAX7000S.** In the devices column highlight **EPM7128SLC84-7,** then press **OK.** Be sure that "Show Only Fastest Speed Grades" is *not* checked (see Figure 4–21).

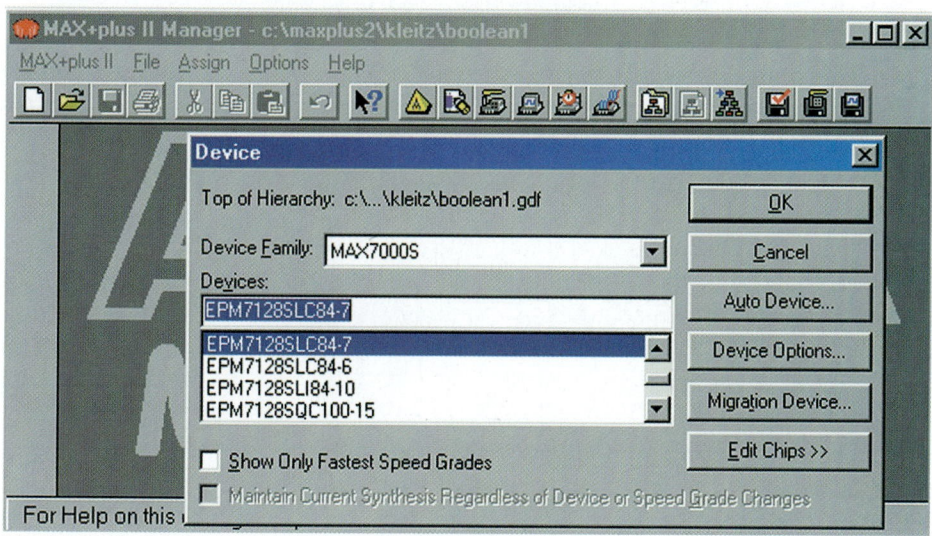

Figure 4–21   Assigning the part number to the device.

10. Finally, you want to save the project and compile it to check for errors:

Choose **File > Project > Save & Compile**

The compiler message should display 0 errors. Press **OK** (see Figure 4–22).

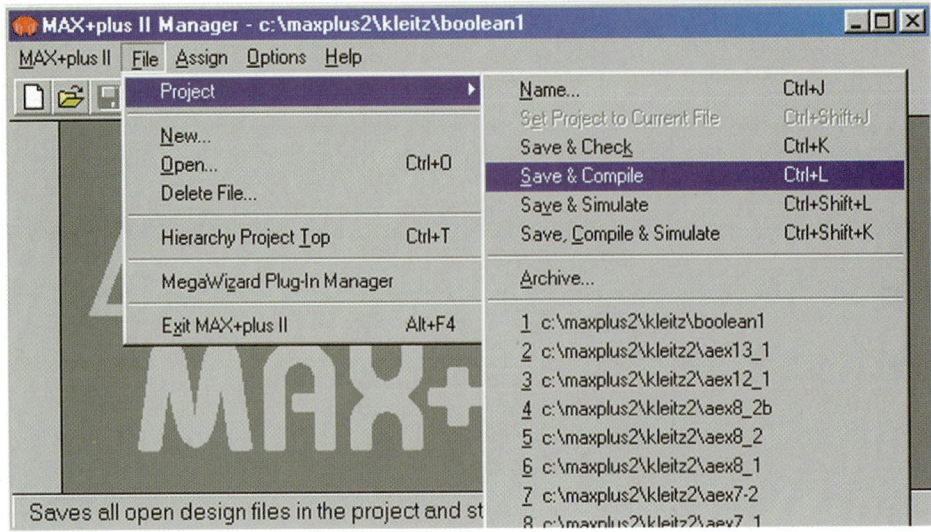

Figure 4–22 Saving and compiling the project.

Create a Simulator Channel File (.*scf*) for Waveform Simulation

Next, we will use the waveform editor to create a Simulator Channel File to test and simulate our logic design. To test our logic, we need to simulate all 16 possible combinations of inputs at **A, B, C,** and **D.**

1. To create a new Simulator Channel File:

Choose **File > New > Waveform Editor file > .scf**

Press **OK** (see Figure 4–23)

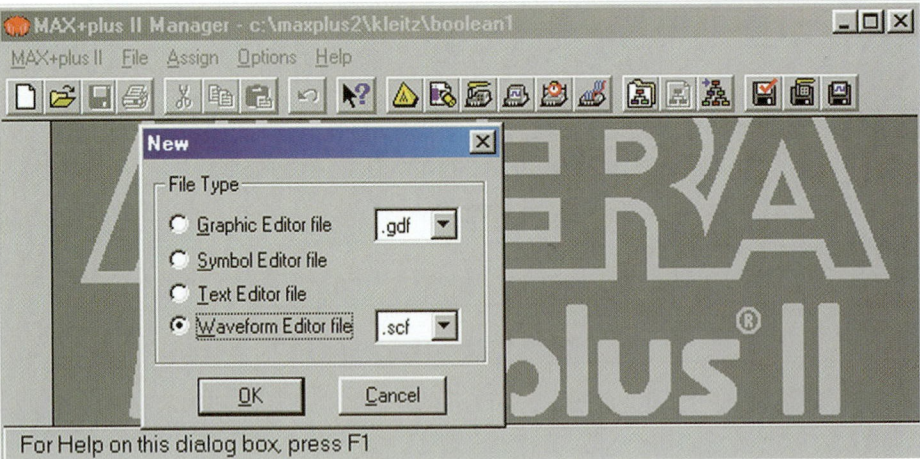

Figure 4–23 Creating a new Simulator Channel File (.*scf*).

2. You then want to assign the same name to this file:

Choose **File > Save as >** *Boolean1*

Press **OK**

[The file extension will be .*scf*. Again, be sure that the correct subdirectory name is highlighted first (see Figure 4–24).]

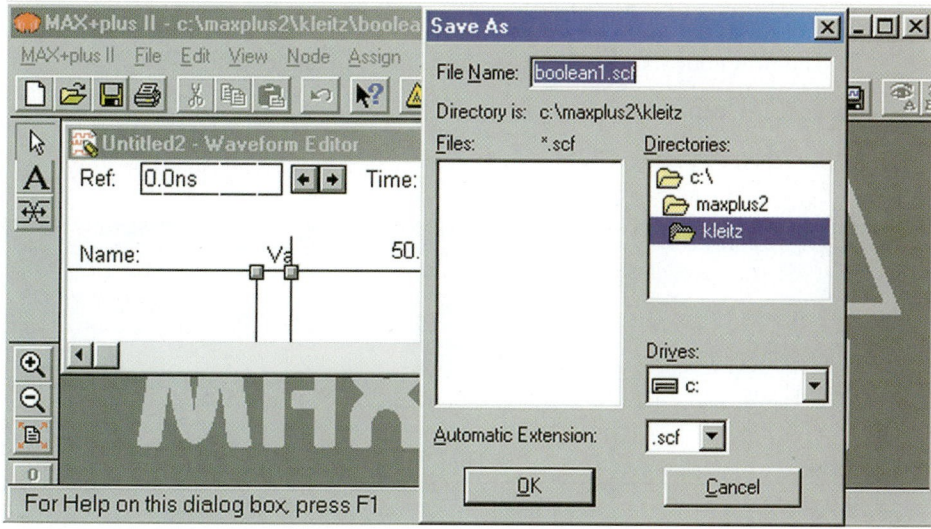

Figure 4–24   Saving the *.scf* file.

3. You should now be looking at a blank Waveform Editor screen. The first thing we need to do inside the waveform editor is create node names for the four inputs and one output. To do this, position the cursor in the area just below the word *Name:,* and then double-click. This brings up the Insert Node menu. Now, to enter the Node Name for the first input:

Enter **A**

Choose **Input Pin**

Press **OK** (see Figure 4–25)

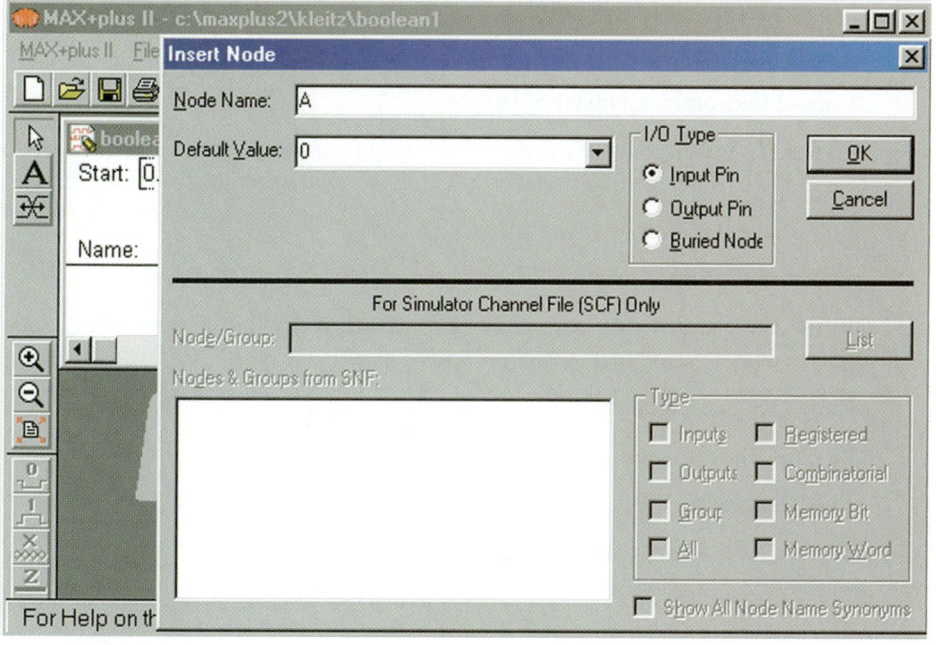

Helpful Hint

A shortcut for inserting node names is to choose Node>Enter Nodes from SNF. Press List and highlight the nodes that you want and press => to select them. Press OK to place them in your *.scf* file.

Figure 4–25   Creating node names in the *.scf* file.

4. Repeat step 3 for the other three inputs: **B, C,** and **D.**

5. The fifth node name will be for the output pin named **X:**

Enter **X > Output Pin**

Press **OK** (see Figure 4–26)

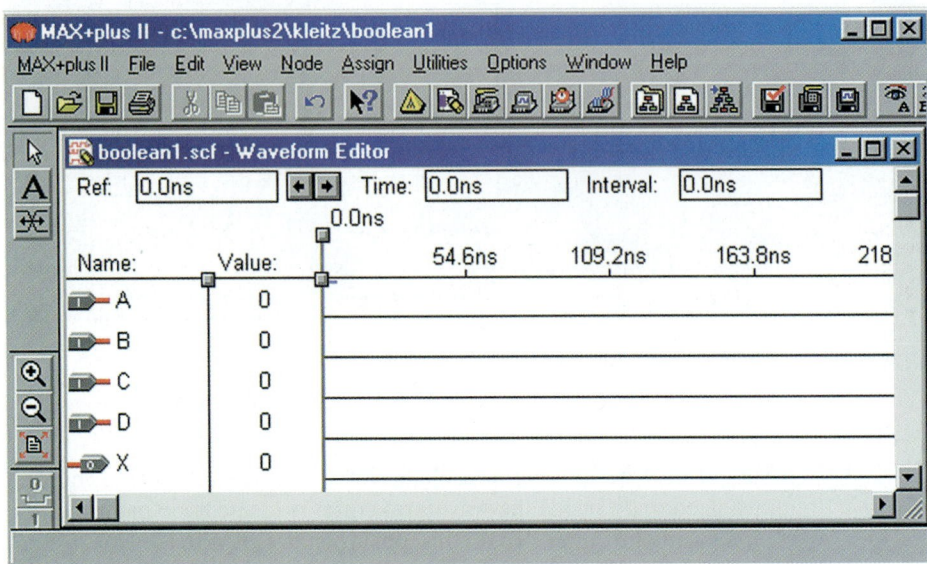

Figure 4–26 Naming all five input/output nodes.

6. Now we need to develop four input waveforms that will cover all 16 possible combinations of inputs. To set up these waveforms, perform the following steps (the parameters listed below are recommended to make your display match the figures in this tutorial):

Choose **File > Endtime > 16uS**

Choose **View > Fit In Window**

Choose **Options** Check **Show Grid**

Choose **Options > Gridsize > 1uS**

7. Next, we want to draw a sequence of waveforms that count from 0000 up to 1111. To draw the first waveform for the **A** input, make the following entries:

Left-click on node name **A**

Choose **Edit > Overwrite > Clock**

In the Overwrite Clock screen, choose **Multiply by 1** (this will draw clock pulses with a period of **1 X** 1µS)

Press **OK** (see Figure 4–27)

8. Draw the remaining waveforms:

Left-click on node **B**

Choose **Edit > Overwrite > Clock**

In the Overwrite Clock screen, enter **2** in the **multiply by** field (this will draw clock pulses with a period of **2 X** 1µS)

Left-click on node **C**

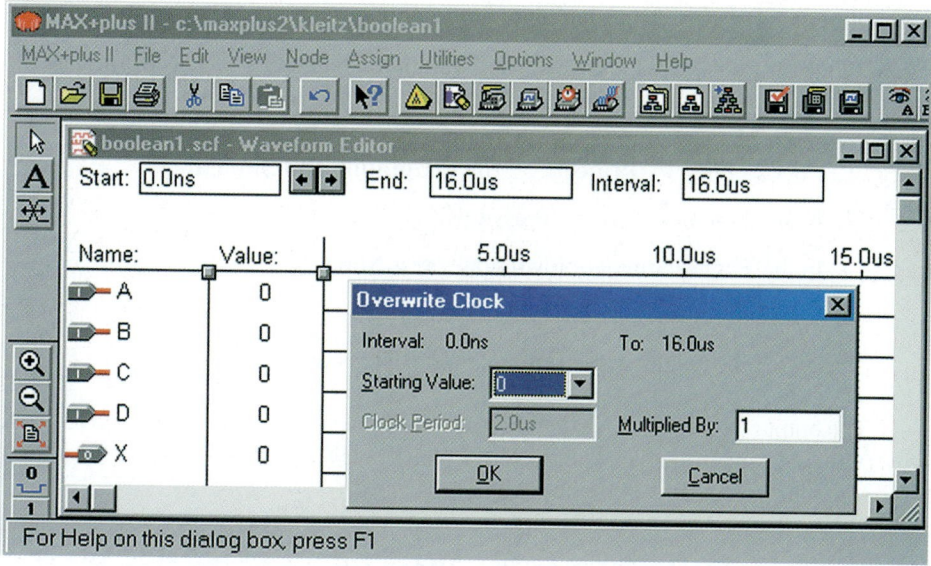

Figure 4–27 Drawing the input waveforms using the overwrite clock option.

Choose **Edit > Overwrite > Clock**

In the Overwrite Clock screen, enter **4** in the **multiply by** field (this will draw clock pulses with a period of **4 X** 1μS)

Left-click on node **D**

Choose **Edit > Overwrite > Clock**

In the Overwrite Clock screen, enter **8** in the **multiply by** field (this will draw clock pulses with a period of **8 X** 1μS)

9. The final Waveform Editor file will show waveforms counting from 0000 up to 1111. To save your file, press

File > Save (see Figure 4–28).

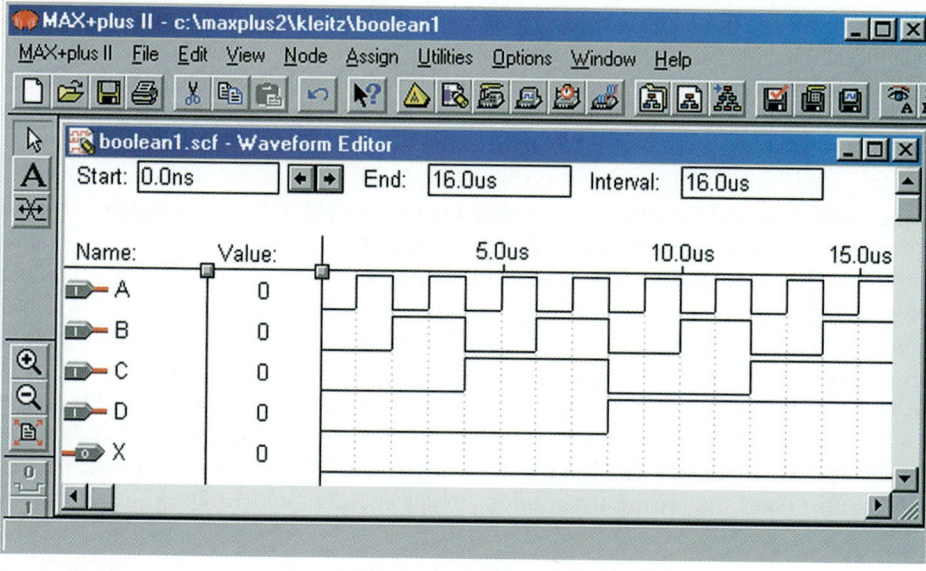

Figure 4–28 The four input wave forms counting from 0000 up to 1111.

Waveform Simulation

The waveform simulator will now use the Simulator Channel File (*.scf*) to simulate the inputs to the logic circuit in the Graphic Design File (*.gdf*) and produce the output at **X**:

1. Choose **File** > **Project** > **Save, Compile and Simulate.**

2. It should return *0 errors.* Press **OK.**

3. In the Timing Simulation window, press **Start.**

4. It should return *0 errors.* Press **OK.**

5. To see the results of the timing analysis press **Open SCF.**

The output waveform at **X** should satisfy the Boolean equation for **X = AB + CD.** Observing the waveforms you should verify that *X* is HIGH whenever *A* and *B* are both HIGH or whenever *C* and *D* are both HIGH.

Programming the PLD Using the Altera UP-1* or the RSR PLDT-2 Programmer Board

The UP-1 and the PLDT-2 programmer boards mentioned in Section 4–1 each have an Altera EPM7128S CPLD and all of the support circuitry required to download our logic designs so that they can run on the actual PLDs they were designed for. (The UP-1 also has an EPF10K20 FPGA.) You will need to refer to the UP-1 Users Guide or the RSR PLDT-2 Users Manual for all of the operating detail but its basic operations will be described in the paragraphs that follow.

In this section we will download the *boolean1.gdf* design to the EPM7128S CPLD. After the successful download we will test the operation of the logic programmed into the CPLD IC by connecting the on-board switches and LED to the IC and step through several combinations of input levels to be sure the output responds correctly.

The EPM7128S CPLD   This CMOS EEPROM-based PLD is a member of the Altera high-density MAX 7000 family. It is an 84-pin PLCC package containing 128 macrocells and has a total of 2500 usable gates. Each macrocell has a programmable-AND/fixed-OR array for implementing combinational logic. It also has a configurable register for implementing sequential circuits like counters and shift registers.

We will use this part for most of our designs. We will download our design to the IC and then test it in a stand-alone configuration. Because of its nonvolatile nature our design will remain on the IC even when power is removed and reapplied. Also, since it is an Electrically Erasable PROM (EEPROM), we can reprogram it over and over to test our other designs.

1(a) Configure and Connect Inputs and Outputs to the UP-1 Programmer Board (Skip this section if you are using the PLDT-2 board.)

The EPM7128S PLCC is socket-mounted on the programmer board. Female headers are on the board providing wiring access to each of the 84 IC pins. We will use solid 22- to 24-gauge hookup wire to connect to the 16 input switches and 16 output LEDs provided. The 16 switches are connected to provide logic 1 when the switch is in the OFF position (open) and logic 0 when in the ON position. The 16 LEDs on the UP-1 are all active-LOW, meaning that they illuminate when a LOW is placed on the female header associated with the LED. Refer to the UP-1 Users Guide for the header pin numbers and the wiring description of the switches and LEDs. (The theory of

---

*An upgraded version on the UP-1 is also available from the University Program at Altera. It is called the UP-2. Basically it is the same as the UP-1 except that it has a higher-density FPGA, the EPF10K70.

active-LOW and active-HIGH wiring of switches and LEDs in digital circuitry is covered in Chapter 11.)

To configure its operation, the UP-1 board has jumpers that need to be set according to the Users Guide instructions. Since the UP-1 has two PLDs, you will need to set the correct jumpers to access the EPM7128S CPLD for most experiments instead of the EPF10K20 FPGA. According to the Users Guide, this is done by placing the four configuration jumpers (TD1, TD0, Device, Board) in their top position.

A connection to the PC's parallel port is made with the cable included with the UP-1 board. DC power must be applied to the board via a user-supplied power supply. The UP-1 Users Guide provides all of the steps required to configure the board and make the hardware connections to the PC. A photo of the connections to a UP-1 board is shown in Figure 4–29.

1(b) Configure and Connect Inputs and Outputs to the PLDT-2 Programmer Board
(Skip this section if you are using the UP-1 board.)

The EPM7128S PLCC is socket-mounted on the programmer board. Female headers are on the board providing wiring access to each of the 84 IC pins. We could use solid 22- to 24-gauge hookup wire to connect to the 16 input switches and 16 output LEDs provided, but jumpers are provided on the board to make connections to 8 of the switches and 8 of the LEDs without using wire. The 16 switches are connected to provide logic 1 when the switch is in the OFF position (open) and logic 0 when in the ON position. The 16 LEDs on the PLDT-2 board are all active-HIGH, meaning that they illuminate when a HIGH is placed on the female header associated with the LED. Refer to the PLDT-2 Users Manual for the header pin numbers and the wiring description of the switches and LEDs. (The theory of active-LOW and active-HIGH wiring of switches and LEDs in digital circuitry is covered in Chapter 11.)

A connection to the PC's parallel port is made with the cable included with the PLDT-2 board. DC power must be applied to the board via a user-supplied power supply. The PLDT-2 Users Manual provides all of the steps required to configure the board and make the hardware connections to the PC. A photo of the connections to a PLDT-2 board is shown in Figure 4–30.

Figure 4–29 Connections to the UP-1 board.

Female headers for optional wiring to CPLD

Jumpers for connecting the switches to the CPLD

Jumpers for connecting the CPLD pins to the LED.

CPLD

Switch bank 1 showing sw1, 2, 3, 4 in their OFF position ('1')

Figure 4–30   Connections to the PLDT-2 board.

TABLE 4-1	EPM7128 CPLD pin assignments for connecting to the UP-1 and PLDT-2		
Input Switch Bank 1		**Output LED Bank 1**	
Switch number	**CPLD Pin Number**	**LED Number**	**CPLD Pin Number**
SW1 A	34	LED1 X	44
SW2 B	33	LED2	45
SW3 C	36	LED3	46
SW4 D	35	LED4	48
SW5	37	LED5	49
SW6	40	LED6	50
SW7	39	LED7	51
SW8	41	LED8	52

2. Connect the Switches and LED to Specific CPLD Pins   Up until now we didn't pay attention to which pins on the CPLD were going to be assigned for the inputs and outputs. Instead of allowing the software to make that determination, we will assign specific pin numbers and re-compile the *gdf* file before programming the IC. To keep our testing consistent, let's use the data in Table 4–1 for assigning the input and output pins.

The table shows that we need to wire SW1 of switch bank 1 to the female header corresponding to pin 34 of the EPM7128 CPLD IC. (If you are using the jumpers provided on the PLDT-2 board, you do not have to use any hookup wire. The jumpers are set up to align with the connections shown in Table 4–1.) SW2 of bank 1 connects to pin 33 and so on, as listed in the table. To implement the equation for the logic in the file *boolean1.gdf* ($X = AB + CD$), we will need to physically make connections for 4 inputs (*A, B, C, D*) and 1 output (*X*) to the CPLD. Then we need to make sure to assign the same pin numbers to our logic design using the MAX+PLUS® II software.

3. Assign pin numbers in the* boolean1.gdf *file   Assuming that the file *boolean1.gdf* has already been created and you have assigned a device to it (Figure 4–21), then you can now assign the pin numbers shown in blue in Table 4–1 to the logic circuit by performing the following steps:

Choose **File > Open >** *Boolean1.gdf* **> OK**

Choose **File > Project > Set Project to Current File**

Right-click on input *A* and select **Assign> Pin/Location/Chip** (See Figure 4–31)

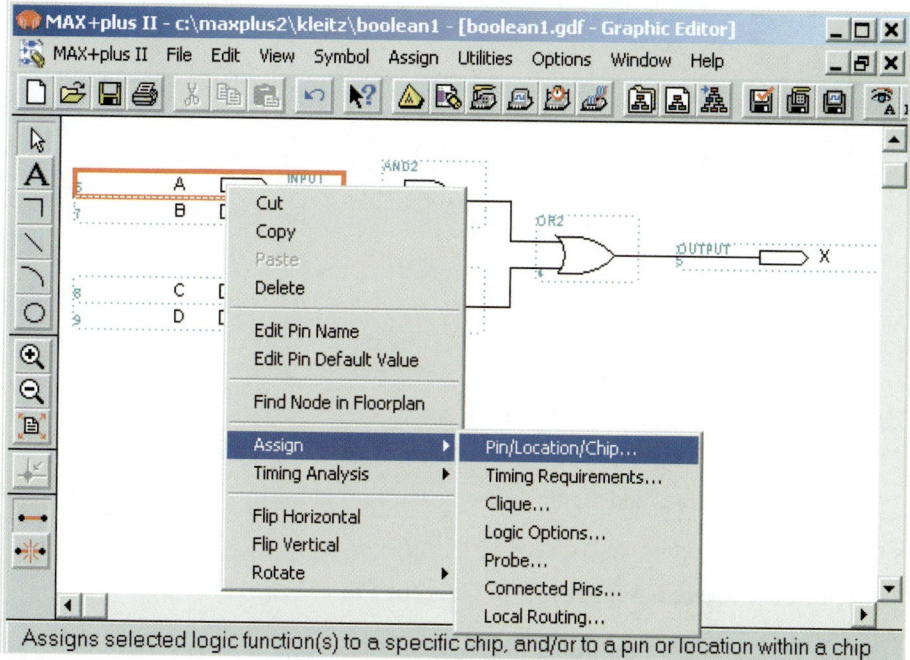

Figure 4–31 Assigning Pin numbers to the inputs and outputs.

Enter *A* for **Node Name,** enter *34* for **Pin,** enter *Input* for **Pin Type,** and Press **Add** (see Figure 4–32)

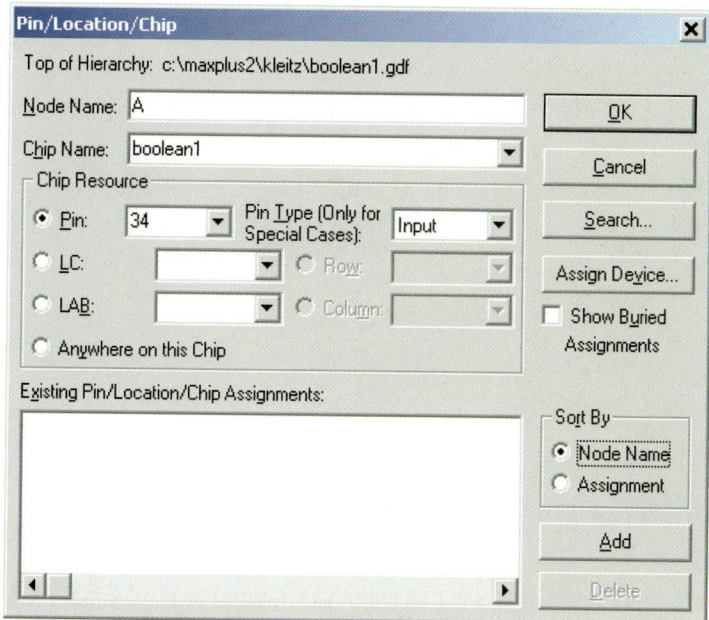

Figure 4–32 Adding pin definitions.

Repeat for **B, C, D** and **X** (Your file should now look like Figure 4–33 showing the assigned pin numbers.)

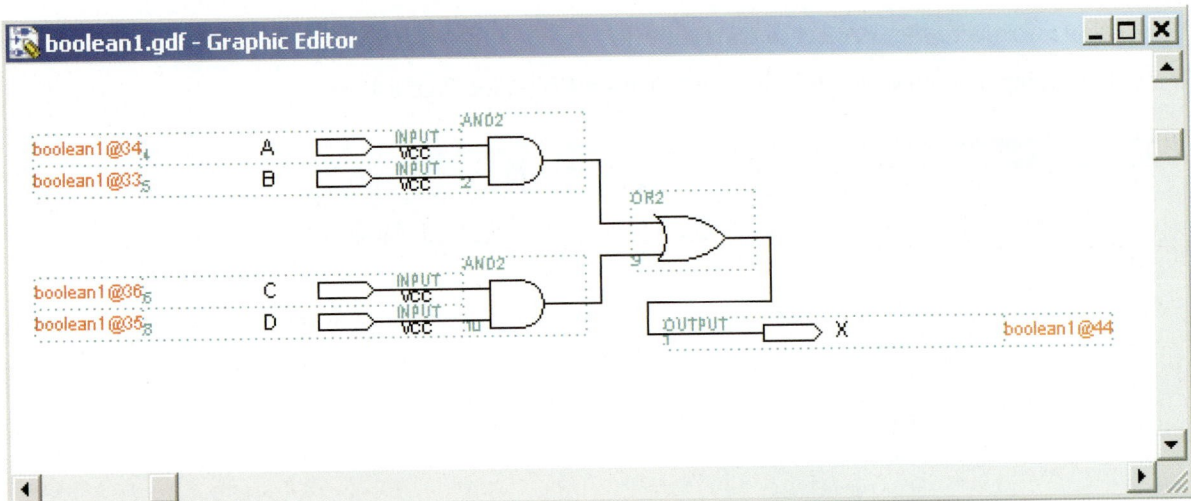

Figure 4–33   The *boolean1.gdf* file showing the assigned pin numbers.

Save these changes and set the project to the current file by performing these steps:

Choose **File > Save**

Choose **File > Project > Set Project to Current File**

4. Connect Wires to the Input Stimulus Switches and Output LED   The EPM7128S CPLD is an 84-pin IC. You will need to refer to the pin layout table provided for it in the UP-1 Users Guide to find the appropriate header to connect your inputs and output to. The layout for the headers is also shown in Figure 4–34.

```
        11  9  7  5  3  1  83  81  79  77  75
         x  10  8  6  4  2  84  82  80  78  76

    12  13                                        x   74
    14  15                                       73  72
    16  17                                       71  70
    18  19                                       69  68
    20  21             ALTERA                    67  66
    22  23             EPM7128S                   65  64
                       Headers
    24  25                                       63  62
    26  27                                       61  60
    28  29                                       59  58
    30  31                                       57  56
    32   x                                       55  54

        34  36  38  40  42  44  46  48  50  52   x
        33  35  37  39  41  43  45  47  49  51  53
```

Figure 4–34   Layout for the wiring headers on the CPLD.

According to Table 4–1, SW1 (*A*) should be wired to pin 34, SW2 (*B*) should be wired to pin 33, SW3 (*C*) should be wired to pin 36 and SW4 (*D*) should be wired to pin 35. The *X* output at pin 44 should be wired to LED1.

On the UP-1 board you must make the necessary connections from the header to the switches and LED using 22- or 24-gauge solid wire (see Figure 4–29). On the PLDT-2 board, use the jumpers provided to make those connections (see Figure 4–30).

5. Download (Program) Your Circuit Design to the CPLD   Before downloading:

(a) If you are using the UP-1 board, be sure that the four configuration jumpers are set for the EPM7128S CPLD according to the Users Guide (all four in the top position).
(b) Connect the parallel download cable to your PCs parallel port (Printer port, LPT1)
(c) Connect the DC power source.

To download the design:

Choose **MAX+PLUS® II > Programmer**

If this is the first time programming a device, a "Hardware Setup" window will appear.

Choose the correct **Hardware Type** (*ByteBlaster*) and **Port** (*LPT1*) for your programming board, then press **OK**

Back in the Programmer window press **Program**

After a successful program download (Programming complete) press **OK**

6. Test the Downloaded Design on the UP-1 or PLDT-2 Board   The logic circuit is now programmed into the EPM7128S CPLD. The connections from the switches and LED to the CPLD will allow you to test its logic operation. Remember, as the Users Guide states, putting a switch in the ON position places a LOW (0) at that input. Also remember that and the PLDT-2 board has active-HIGH LEDs and the UP-1 board has active-LOW LEDs. [*Note:* In the future, if you are using the UP-1 board, you could add an inverter (called a *NOT* gate) to the output signal to make it active-HIGH to make the truth table simpler to interpret.]

Exercise the CPLD by testing all possible combinations of inputs as you watch the output on the LED. If your design is working correctly, the results should match the truth table for $X = AB + CD$ shown in Table 4–2. This truth table shows that X is HIGH whenever A and B are both HIGH or whenever C and D are both HIGH.

TABLE 4–2	Truth Table for testing the CPLD program for $X = AB + CD$				
A (SW state)	B (SW state)	C (SW state)	D (SW state)	X (PLDT-2 LED state)	X (UP-1 LED state)
0 (ON)	0 (ON)	0 (ON)	0 (ON)	0 (INACTIVE)	0 (ACTIVE)
0 (ON)	0 (ON)	0 (ON)	1 (OFF)	0 (INACTIVE)	0 (ACTIVE)
0 (ON)	0 (ON)	1 (OFF)	0 (ON)	0 (INACTIVE)	0 (ACTIVE)
0 (ON)	0 (ON)	1 (OFF)	1 (OFF)	1 (ACTIVE)	1 (INACTIVE)
0 (ON)	1 (OFF)	0 (ON)	0 (ON)	0 (INACTIVE)	0 (ACTIVE)
0 (ON)	1 (OFF)	0 (ON)	1 (OFF)	0 (INACTIVE)	0 (ACTIVE)
0 (ON)	1 (OFF)	1 (OFF)	0 (ON)	0 (INACTIVE)	0 (ACTIVE)
0 (ON)	1 (OFF)	1 (OFF)	1 (OFF)	1 (ACTIVE)	1 (INACTIVE)
1 (OFF)	0 (ON)	0 (ON)	0 (ON)	0 (INACTIVE)	0 (ACTIVE)
1 (OFF)	0 (ON)	0 (ON)	1 (OFF)	0 (INACTIVE)	0 (ACTIVE)
1 (OFF)	0 (ON)	1 (OFF)	0 (ON)	0 (INACTIVE)	0 (ACTIVE)
1 (OFF)	0 (ON)	1 (OFF)	1 (OFF)	1 (ACTIVE)	1 (INACTIVE)
1 (OFF)	1 (OFF)	0 (ON)	0 (ON)	1 (ACTIVE)	1 (INACTIVE)
1 (OFF)	1 (OFF)	0 (ON)	1 (OFF)	1 (ACTIVE)	1 (INACTIVE)
1 (OFF)	1 (OFF)	1 (OFF)	0 (ON)	1 (ACTIVE)	1 (INACTIVE)
1 (OFF)	1 (OFF)	1 (OFF)	1 (OFF)	1 (ACTIVE)	1 (INACTIVE)

VHDL Design Entry

In this section, instead of designing the *boolean1* logic circuit using the graphic editor, we will use the VHDL Text Editor.

1. To define the boolean equation using VHDL:

Choose **File > New > Text Editor file**

Press **OK** (see Figure 4–35)

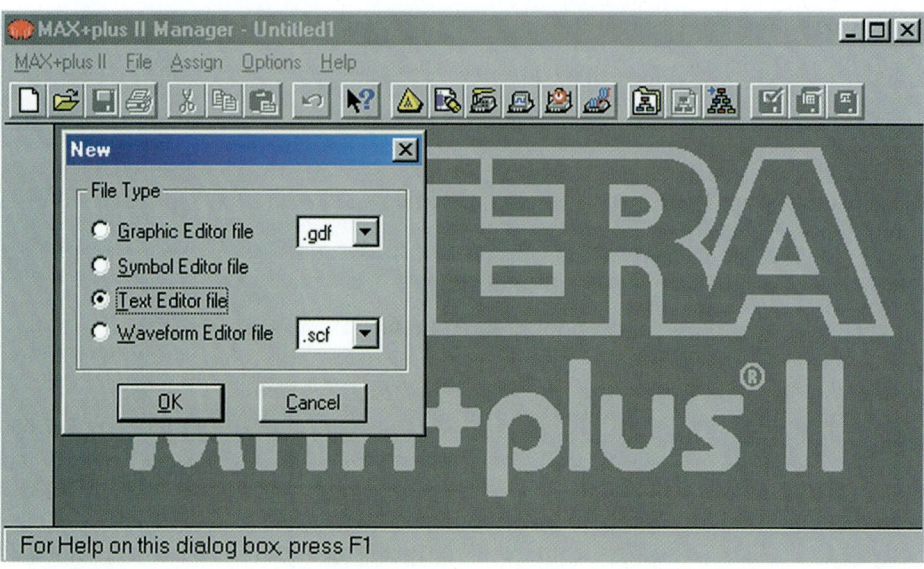

Figure 4–35 Creating a new VHDL file.

2. Type in the VHDL program for **X = AB + CD** shown in Figure 4–36.

```
LIBRARY ieee;
USE ieee.std_logic_1164.ALL;

ENTITY boolean1 IS
    PORT(
        a, b, c, d: IN std_logic;
        x:          OUT std_logic);
END boolean1;

ARCHITECTURE arc OF boolean1 IS
    BEGIN
        x<= (a AND b) OR (c AND d);
    END arc;
```

Figure 4–36 The VHDL program listing.

3. Save the VHDL file:

Choose **File > SaveAs >** *boolean1* **> Automatic Extension > .vhd**

(Be sure that the correct subdirectory is highlighted, as shown in Figure 4–37.)

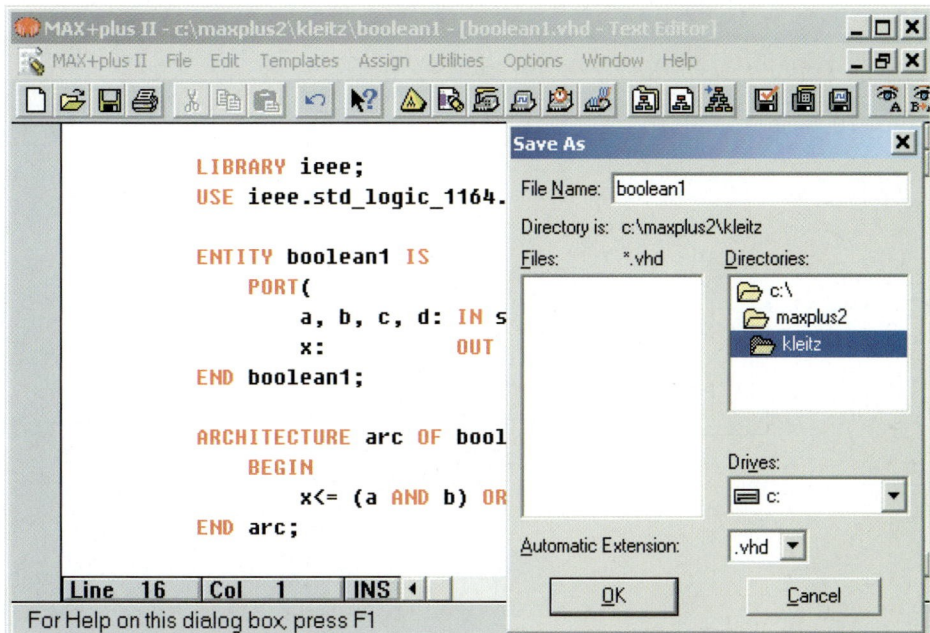

Figure 4–37   Saving the VHDL file.

4. To compile and test for errors in the VHDL program:

Choose **File > Project > Set Project to Current File**

Choose **File > Project > Save and Compile**

The compiler message should show *0 errors.* Now you can follow the steps previously outlined to build a Simulator Channel File (*.scf*) to test your design and then program the CPLD.

4–5   CPLD Applications

The logic design problems in this section will be solved using the tools provided in the MAX+PLUS® II software program. If you haven't already done so, you must work step by step through the tutorial instructions presented in Section 4–4. In each of the examples that follow, your goal is to design the logic circuit, perform a simulation of your circuit, and then, if you have a programmer board, you should download your results and test it on an actual CPLD with switches and LEDs.

EXAMPLE 4–2

Use Altera MAX+PLUS® II software to design the CPLD logic to implement the Boolean equation $X = \overline{A}B + \overline{A}B$.

a. Design the logic using the graphic editor to create a Graphic Design File (gdf) called *ex4_2.gdf.*

b. Test the operation of the CPLD logic by using the waveform editor to create a Simulator Channel File (scf) called *ex4_2.scf.* The simulation should show all possible combinations of inputs.

---

c. Build a truth table similar to Table 4–2 for use when exercising the inputs on a UP-1 or a PLDT-2 programmer board.

Solution: The results of the design are shown in Figures 4–38(a), (b), and (c). (The *.gdf* and *.scf* files can also be found on the CD included with this text.)

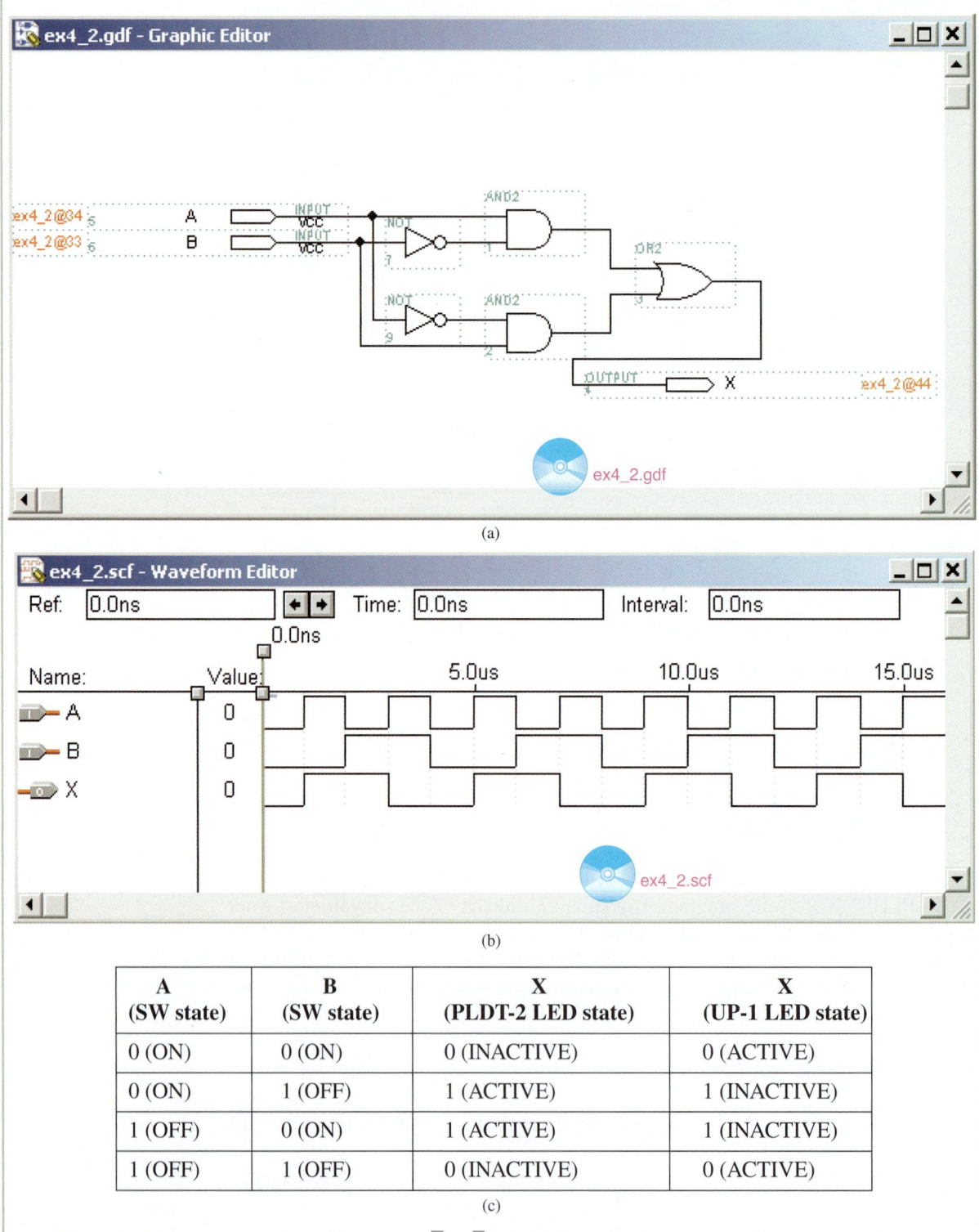

(a)

(b)

A (SW state)	B (SW state)	X (PLDT-2 LED state)	X (UP-1 LED state)
0 (ON)	0 (ON)	0 (INACTIVE)	0 (ACTIVE)
0 (ON)	1 (OFF)	1 (ACTIVE)	1 (INACTIVE)
1 (OFF)	0 (ON)	1 (ACTIVE)	1 (INACTIVE)
1 (OFF)	1 (OFF)	0 (INACTIVE)	0 (ACTIVE)

(c)

Figure 4–38 Solution to the equation $X = A\overline{B} + \overline{A}B$: (a) Graphic Design File; (b) Simulator Channel File; (c) Truth table.

EXAMPLE 4–3

Use Altera MAX+PLUS® II software to design the CPLD logic to implement the Boolean equation $X = AB\overline{C}$.

a. Design the logic using the graphic editor to create a Graphic Design File (gdf) called *ex4_3.gdf.*

b. Test the operation of the CPLD logic by using the waveform editor to create a Simulator Channel File (scf) called *ex4_3.scf.* The simulation should show all possible combinations of inputs.

c. Build a truth table similar to Table 4–2 for use when exercising the inputs on a UP-1 or a PLDT-2 programmer board.

Solution: The results of the design are shown in Figures 4–39(a), (b), and (c). (The *.gdf* and *.scf* files can also be found on the CD included with this text.)

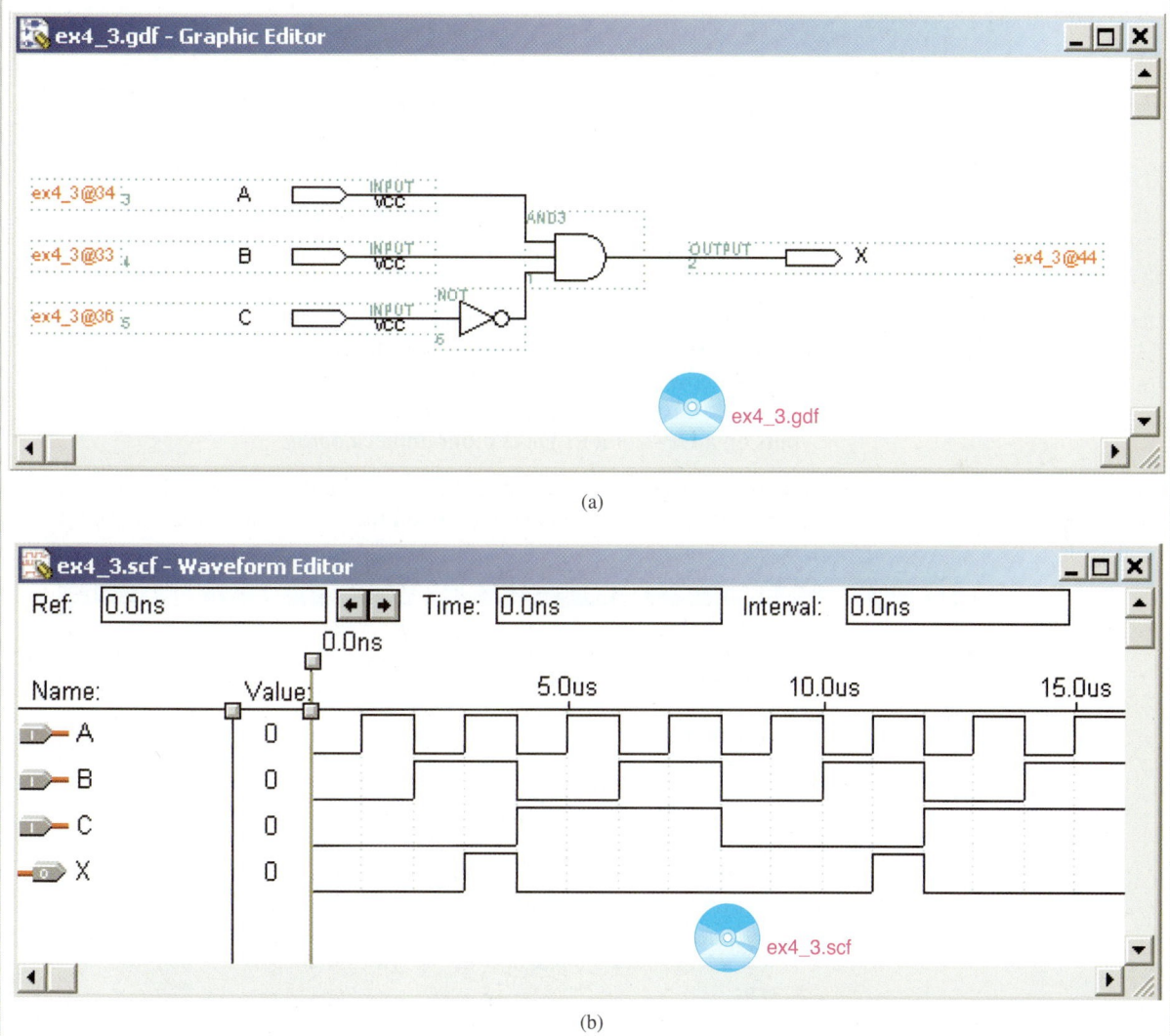

(a)

(b)

Figure 4–39 Solution to the equation $X = AB\overline{C}$: (a) Graphic Design File; (b) Simulator Channel File; (c) Truth table.

A (SW state)	B (SW state)	C (SW state)	X (PLDT-2 LED state)	X (UP-1 LED state)
0 (ON)	0 (ON)	0 (ON)	0 (INACTIVE)	0 (ACTIVE)
0 (ON)	0 (ON)	1 (OFF)	0 (INACTIVE)	0 (ACTIVE)
0 (ON)	1 (OFF)	0 (ON)	0 (INACTIVE)	0 (ACTIVE)
0 (ON)	1 (OFF)	1 (OFF)	0 (INACTIVE)	0 (ACTIVE)
1 (OFF)	0 (ON)	0 (ON)	0 (INACTIVE)	0 (ACTIVE)
1 (OFF)	0 (ON)	1 (OFF)	0 (INACTIVE)	0 (ACTIVE)
1 (OFF)	1 (OFF)	0 (ON)	1 (ACTIVE)	1 (INACTIVE)
1 (OFF)	1 (OFF)	1 (OFF)	0 (INACTIVE)	0 (ACTIVE)

(c)

Figure 4–39   *Continued*

EXAMPLE 4–4

Use Altera MAX+PLUS II® software to design the CPLD logic to implement the Boolean equation $X = \overline{A}BC + AB\overline{C}$.

   a. Design the logic using the text editor to create a VHDL File (vhd) called *ex4_4.vhd*.
   b. Test the operation of the CPLD logic by using the waveform editor to create a Simulator Channel File (scf) called *ex4_4.scf*. The simulation should show all possible combinations of inputs.
   c. Build a truth table similar to Table 4–2 for use when exercising the inputs on a UP-1 or a PLDT-2 programmer board.

Solution: The results of the design are shown in Figures 4–40(a), (b), and (c). (The vhd and scf files can also be found on the CD included with this text.)

```
ex4_4.vhd - Text Editor

LIBRARY ieee;
USE ieee.std_logic_1164.ALL;

ENTITY ex4_4 IS
    PORT(
        a, b, c: IN std_logic;
        x:       OUT std_logic);
END ex4_4;

ARCHITECTURE arc OF ex4_4 IS
    BEGIN
        x<= (NOT a AND b AND c) OR (a AND b AND NOT c);
END arc;
```
ex4_4.vhd

Line 19   Col 1   INS

(a)

Figure 4–40   Solution to the equation $X = \overline{A}BC + AB\overline{C}$: (a) VHDL program; (b) Simulator Channel File; (c) Truth table.

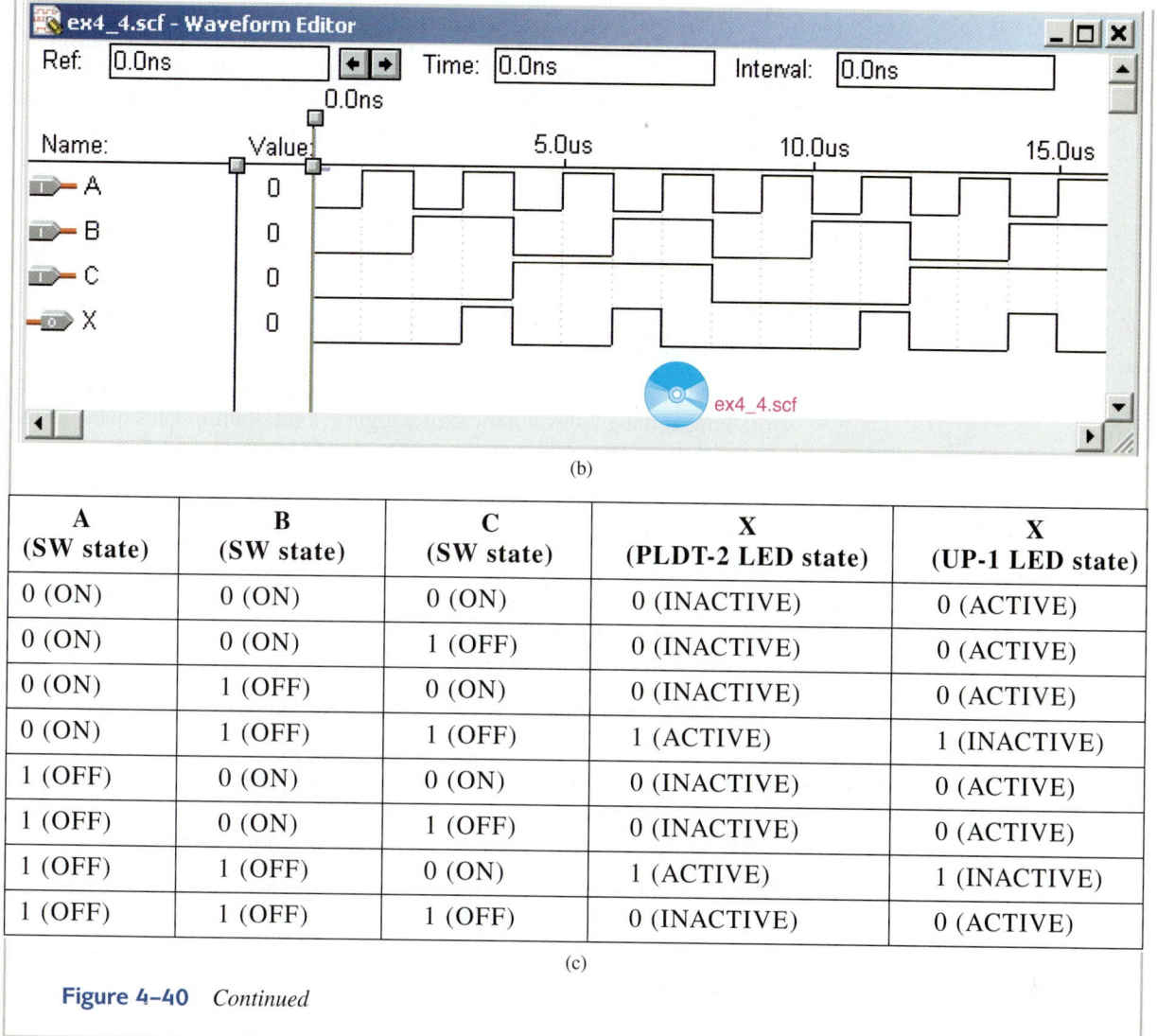

(b)

A (SW state)	B (SW state)	C (SW state)	X (PLDT-2 LED state)	X (UP-1 LED state)
0 (ON)	0 (ON)	0 (ON)	0 (INACTIVE)	0 (ACTIVE)
0 (ON)	0 (ON)	1 (OFF)	0 (INACTIVE)	0 (ACTIVE)
0 (ON)	1 (OFF)	0 (ON)	0 (INACTIVE)	0 (ACTIVE)
0 (ON)	1 (OFF)	1 (OFF)	1 (ACTIVE)	1 (INACTIVE)
1 (OFF)	0 (ON)	0 (ON)	0 (INACTIVE)	0 (ACTIVE)
1 (OFF)	0 (ON)	1 (OFF)	0 (INACTIVE)	0 (ACTIVE)
1 (OFF)	1 (OFF)	0 (ON)	1 (ACTIVE)	1 (INACTIVE)
1 (OFF)	1 (OFF)	1 (OFF)	0 (INACTIVE)	0 (ACTIVE)

(c)

Figure 4–40   *Continued*

Summary

In this chapter, we have learned that

1. PLDs can be used to replace 7400- and 4000-series ICs. They contain the equivalent of thousands of logic gates. CAD tools are used to configure them to implement the desired logic.

2. The two most common methods of PLD design entry are (graphic) entry and VHDL entry. To use graphic entry, the designer uses CAD tools to draw the logic to be implemented. To use VHDL entry, the designer uses a text editor to write program descriptions defining the logic to be implemented.

3. PLD design software usually includes a logic simulator. This feature allows the user to simulate levels to be input to the PLD, and it shows the output simulation to those input conditions.

4. Most PLDs are erasable and reprogrammable. This allows users to test many versions of their logic design without ever changing ICs.

5. Basically, there are three types of PLDs: SPLDs, CPLDs, and FPGAs. SPLDs use the PAL or PLA architecture. They consist of several multi-input AND gates whose outputs feed the inputs to OR gates and memory flip-flops. CPLDs consist of several interconnected SPLDs. FPGAs are the most dense form of PLD, solving logic using a look-up table to determine the desired output.

Glossary

Architecture Body: The section in a VHDL program defining the logic functions to be implemented.

CAD: Computer-Aided Design. This type of design uses a computer to aid in the drawing and logic development of a logic circuit. It eliminates many of the manual, time-consuming tasks once associated with logic design.

CPLD: Complex Programmable Logic Device. A PLD consisting of more than 100 interconnected SPLDs. A single chip can be programmed to implement hundreds of logic equations and operations.

Compiler: A language translation software module used by CPLD development systems to convert a schematic or VHDL code into a binary file to represent the digital logic to be implemented.

Entity Declaration: The section of a VHDL program defining the input and output ports.

FPGA: Field-Programmable Gate Array. The most dense form of PLD. It uses a look-up table to resolve its logic operations. Its main disadvantage is that most FPGAs are volatile, losing their memory when power is removed.

Graphic Editor: A software tool provided as part of the PLD development package. It provides a way to enter designs by drawing a schematic.

Library Declaration: The section of a VHDL program declaring the software libraries to be included in the program. These libraries are used by the compiler to resolve references to the various program commands.

Look-Up Table: Used by FPGA logic to determine the output level of a circuit based on the combinations of logic levels at its inputs. It is constructed as a truth table except that its outputs are only HIGH for specific combinations of inputs solving the given logic product terms.

Nonvolatile: Internal memory is maintained even when power is removed from the IC.

PAL: Programmable Array Logic. Its basic structure contains multiple inputs to several AND gates, the outputs of which are connected to a series of fixed ORs.

PLA: Programmable Logic Array. Its basic structure contains multiple inputs to several AND gates, the outputs of which are connected to a series of programmable ORs.

PLD: Programmable Logic Device. An IC containing thousands of undefined logic functions. A software development tool is used to specify (i.e., program) the specific logic to be implemented by the IC. PLD is the general term used to represent PLAs, PALs, SPLDs, CPLDs, and FPGAs.

Product Terms: Input variables that are ANDed together (e.g., ABC, $A\overline{BC}$:).

Schematic Capture: A method used by PLD software to input a design that is defined by a schematic.

SPLD: Simple Programmable Logic Devices. A programmable, digital logic IC containing several PAL or PLA structures with internal interconnections and memory registers.

Sum-of-Products SOP: Two or more product terms that are ORed together (e.g., $A\overline{B}C + A\overline{C}\,\overline{D} + BCD$).

Synthesize: The creation of a model of the PLD's internal electrical connections that will produce the actual logic functions defined by the user.

VHDL: VHSIC (Very High Speed Integrated Circuit) Hardware Description Language. A programming language used by PLD software to define a logic design by specifying a series of I/O definitions and logic equations.

VHDL Editor: A software program facilitating entry of text-based instructions comprising the VHDL program.

Waveform Simulator: The part of a PLD software development tool that allows users to simulate the input of several signals to a logic circuit and observe its response.

Problems

Section 4–1

4–1. How does programmable logic differ from discrete digital logic like the 7400 series?

4–2. What are two common ways to configure or define logic to PLD programming software?

4–3. What does HDL stand for in the acronym VHDL?

4–4. List the six steps in the PLD design flow.

4–5. How many different ICs would it take to implement the following equations?
(a) $X = AB + \overline{BC}$
(b) $Y = \overline{AB} + B\overline{C} + \overline{C + D}$

4–6. How is pin 1 identified in the PLCC package style used for the PLD in Figure 4–4?

4–7. What is the purpose of the PLD programmer boards shown in Figure 4–5?

Section 4–2

4–8. How many product terms are in the following equations?
(a) $X = A\overline{C} + BC + \overline{A}C$
(b) $Y = A\overline{B}C + B\overline{C}$
(c) $Z = AB\overline{C} + ACD + \overline{B}C\overline{D}$

4–9. How does a PLA differ from a PAL?

4–10. Redraw the PLA circuitry of Figure 4–8 to implement the following SOP equations:
(a) $X = \overline{A}B + \overline{A}\,\overline{B} + A\overline{B}$
(b) $Y = AB + \overline{A}B$

4–11. Why is it advantageous to use a CPLD that is nonvolatile?

4–12. Refer to the data sheets in Appendix B (or the manufacturers web site) to determine the number of usable gates and macrocells in each of the following CPLDs:

(a) Altera MAX EPM7128S

(b) Xilinx XC95108

4–13. Instead of interconnecting logic gates, the FPGA solves its logic requirements by using what method?

4–14. Draw a look-up table similar to Figure 4–10(b) for the equation $X = \overline{A}\,\overline{B} + A\overline{B}$.

4–15. Because most FPGAs are volatile, what must be done each time they are powered up?

Section 4–3

4–16. What are the two most common methods of design entry for CPLD development software?

4–17. What is the function of the compiler in CPLD development software?

4–18. What is the purpose of the three pin stubs in the *.gdf* file shown in Figure 4–13(a).

4–19. VHDL allows the user to enter the logic design via a _____ editor.

4–20. Define the purpose of the following three VHDL program segments:

(a) Library

(b) Entity

(c) Architecture

4–21. Write the VHDL entity declare for a three-input AND gate.

4–22. Write the VHDL architecture for a three-input AND gate.

4–23. Draw the logic circuit to be implemented by the following VHDL architecture body:

ARCHITECTURE arc OF p4_23 IS

BEGIN

x < = (a AND (b OR c));

y < = (a OR NOT b) AND NOT (b AND c);

z < = NOT (b AND c) OR NOT (a OR c);

END arc;

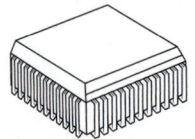

CPLD Problems

The following problems will be solved using the Altera MAX+PLUS® II software provided on the textbook CD. You will be asked to solve the design using the graphic design entry method or the VHDL design entry method. In either case you will demonstrate the circuit operation by producing a Simulator Channel File (*.scf*) that exercises all possible inputs to your circuit. The final step, if you have a programmer board like the UP-1 or PLDT-2, is to download your design to a CPLD and demonstrate its operation to your instructor.

Section 4–4

C4–1. Use a CPLD to implement the following Boolean equation: $X = \overline{AB}$.

(a) Create a Graphic Design File called *prob_c4_1.gdf* to define the logic circuit.

(b) Create a Simulator Channel File called *prob_c4_1.scf* to test the operation of your design by showing the output waveform for all possible input conditions.

(c) Build a truth table similar to Table 4–2 for your logic circuit and your programming board.

(d) Download the design to the CPLD on your programmer board and demonstrate its operation by monitoring the output LED as you step through all switch combinations shown in your truth table from part (c).

C4–2. Use a CPLD to implement the following Boolean equation: $X = AB + \overline{A}\,\overline{B}$.

(a) Create a Graphic Design File called *prob_c4_2.gdf* to define the logic circuit.

(b) Create a Simulator Channel File called *prob_c4_2.scf* to test the operation of your design by showing the output waveform for all possible input conditions.

(c) Build a truth table similar to Table 4–2 for your logic circuit and your programming board.

(d) Download the design to the CPLD on your programmer board and demonstrate its operation by monitoring the output LED as you step through all switch combinations shown in your truth table from part (c).

C4–3. Use a CPLD to implement the following Boolean equation: $X = A\overline{B}C$.

(a) Create a Graphic Design File called *prob_c4_3.gdf* to define the logic circuit.

(b) Create a Simulator Channel File called *prob_c4_3.scf* to test the operation of your design by showing the output waveform for all possible input conditions.

(c) Build a truth table similar to Table 4–2 for your logic circuit and your programming board.

(d) Download the design to the CPLD on your programmer board and demonstrate its operation by monitoring the output LED as you step through all switch combinations shown in your truth table from part (c).

C4–4. Use a CPLD to implement the following Boolean equation: $X = A\overline{B}C + \overline{A}\,\overline{B}C$.

(a) Create a VHDL File called *prob_c4_4.vhd* to define the logic circuit.

(b) Create a Simulator Channel File called *prob_c4_4.scf* to test the operation of your design by showing the output waveform for all possible input conditions.

(c) Build a truth table similar to Table 4–2 for your logic circuit and your programming board.

(d) Download the design to the CPLD on your programmer board and demonstrate its operation by monitoring the output LED as you step through all switch combinations shown in your truth table from part (c).

C4–5. Use a CPLD to implement the following Boolean equation: $X = A\overline{B} + \overline{C}\,\overline{D}$.

(a) Create a VHDL File called *prob_c4_5.vhd* to define the logic circuit.

(b) Create a Simulator Channel File called *prob_c4_5.scf* to test the operation of your design by showing the output waveform for all possible input conditions.

(c) Build a truth table similar to Table 4–2 for your logic circuit and your programming board.

(d) Download the design to the CPLD on your programmer board and demonstrate its operation by monitoring the output LED as you step through all switch combinations shown in your truth table from part (c).

5

Boolean Algebra and Reduction Techniques

OUTLINE

OBJECTIVES

Upon completion of this chapter, you should be able to:

- Write Boolean equations for combinational logic applications.
- Utilize Boolean algebra laws and rules for simplifying combinational logic circuits.
- Apply De Morgan's theorem to complex Boolean equations to arrive at simplified equivalent equations.
- Design single-gate logic circuits by utilizing the universal capability of NAND and NOR gates.
- Troubleshoot combinational logic circuits.
- Write VHDL programs to implement Boolean logic equations.
- Use the report file in MAX+PLUS II to determine simplified equations.
- Enter a truth table in VHDL using vectors and the selected signal assignment.
- Implement sum-of-products expressions utilizing AND–OR–INVERT gates.
- Utilize the Karnaugh mapping procedure to systematically reduce complex Boolean equations to their simplest form.
- Describe the steps involved in solving a complete system design application.

INTRODUCTION

Generally, you will find that the simple gate functions AND, OR, NAND, NOR, and INVERT are not enough by themselves to implement the complex requirements of digital systems. The basic gates will be used as the building blocks for the more complex logic that is implemented by using combinations of gates called *combinational logic*.

5–1   Combinational Logic

Combinational logic employs the use of two or more of the basic logic gates to form a more useful, complex function. For example, let's design the logic for an automobile warning buzzer using combinational logic. The criterion for the activation of the warning buzzer is as follows: The buzzer activates if the headlights are on *and* the driver's door is opened, *or* if the key is in the ignition *and* the door is opened.

The logic function for the automobile warning buzzer is illustrated symbolically in Figure 5–1. The figure illustrates a *combination* of logic functions that can be written as a Boolean equation in the form

$$B = K \text{ and } D \text{ or } H \text{ and } D$$

which is also written as

$$B = KD + HD$$

This equation can be stated as "B is HIGH if K and D are HIGH *or* if H and D are HIGH."

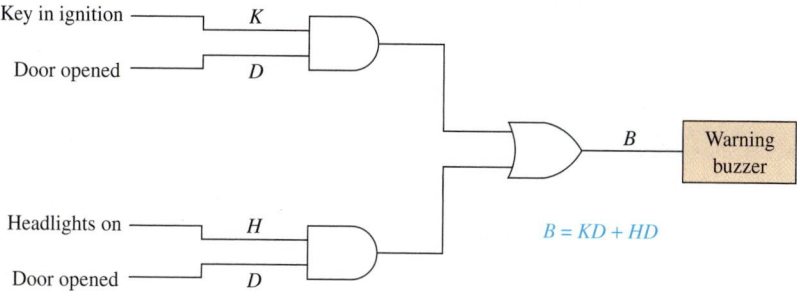

Figure 5–1   Combinational logic requirements for an automobile warning buzzer.

When you think about the operation of the warning buzzer, you may realize that it is activated whenever the door is opened *and* either the key is in the ignition *or* the headlights are on. If you can realize that, you have just performed your first **Boolean reduction** using Boolean algebra. (The systematic reduction of logic circuits is performed using Boolean algebra, named after the 19th-century mathematician George Boole.)

The new Boolean equation becomes $B = D$ and (K or H), also written as $B = D(K + H)$. (Notice the use of parentheses. Without them, the equation would imply that the buzzer activates if the door is opened with the key in the ignition or any time the headlights are on, which is invalid. $B \neq DK + H$. Parentheses are always required when an OR gate is input to an AND gate.) The new equation represents the same logic operation but is a simplified implementation, because it requires only two logic gates, as shown in Figure 5–2.

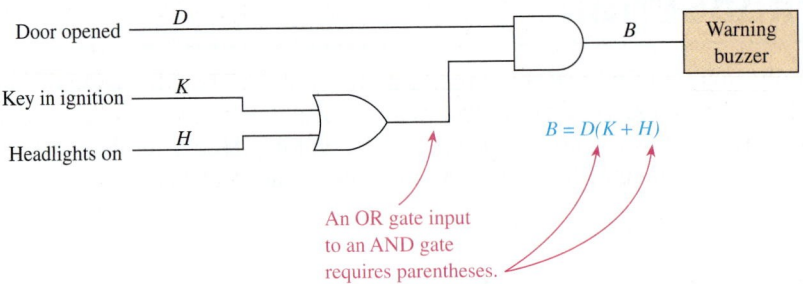

Figure 5–2 Reduced logic circuit for the automobile buzzer.

VHDL Proof of the Automobile Buzzer Circuit Reduction

An easy way to prove to yourself that the reduced circuit of Figures 5–2 is equivalent to the original circuit in Figure 5–1 is to describe each equation in a VHDL program and then run a simulation of all possible input conditions. The VHDL program is listed in Figure 5–3. The original circuit is described using the variable name "*b_original*" and the variable name for the reduced circuit is "*b_reduced*". As mentioned in Chapter 4, VHDL is not case-sensitive, but it is common practice to use a formatting scheme that capitalizes keywords like BEGIN, AND, OR and NOT and uses lower case for variables like *k, d* and *h*. Also, since VHDL equations have no order of precedence, it is mandatory to use parentheses to maintain proper grouping. The double hyphen (--) in the program is used to begin a comment. Comments are used for program documentation and are ignored by the VHDL compiler.

```
fig5_3.vhd - Text Editor                                      _ □ ×

        LIBRARY ieee;                      --Automobile Buzzer
        USE ieee.std_logic_1164.ALL;

        ENTITY fig5_3 IS
            PORT(
                k, d, h                    : IN std_logic;
                b_original, b_reduced    : OUT std_logic);
        END fig5_3;

        ARCHITECTURE arc OF fig5_3 IS
            BEGIN
                b_original<=(k AND d) OR (h AND d);
                b_reduced<= d AND (k OR h);
        END arc;
                                          fig5_3.vhd

Line  15   Col   9    INS
```

Figure 5–3 VHDL Program for comparing the two forms of the automobile buzzer circuit.

The simulation file is shown in Figure 5–4. Notice in the simulation that the waveform for *b_reduced* is identical to *b_original,* proving equality.

CHAPTER 5 ∣ BOOLEAN ALGEBRA AND REDUCTION TECHNIQUES

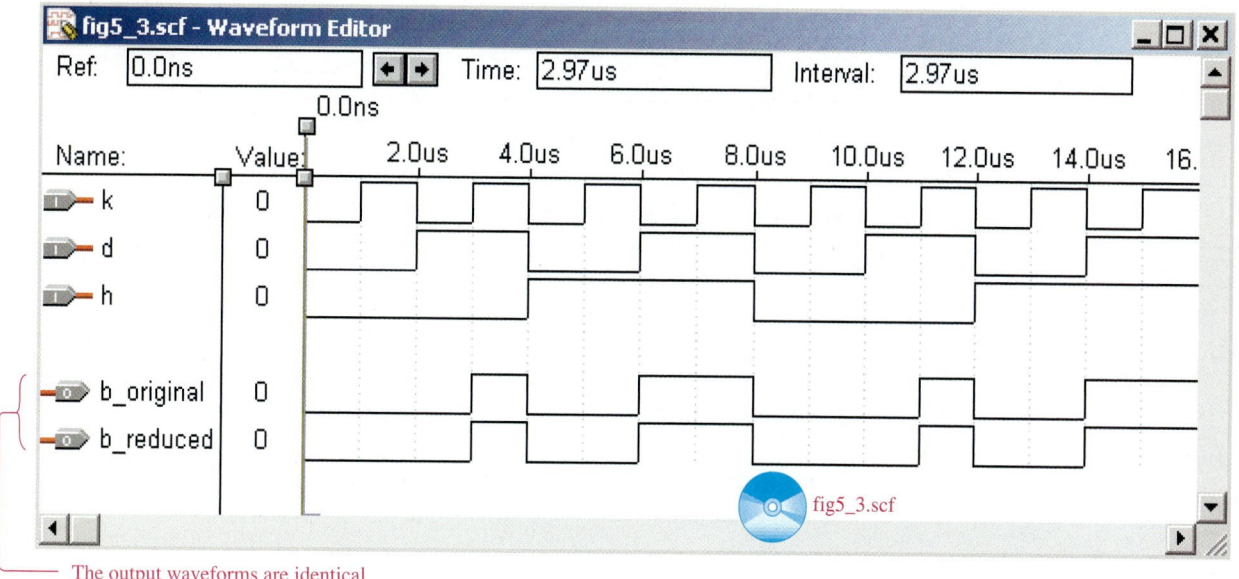

The output waveforms are identical

Figure 5–4 The simulator file for *fig5_3.vhd* showing that the two equations are identical.

EXAMPLE 5–1

Write the Boolean logic equation, and draw the logic circuit that represents the following function: A bank burglar alarm (*A*) is to activate if it is after banking hours (*H*) *and* the front door (*F*) is opened *or* if it is after banking hours (*H*) and the vault door is opened (*V*).

Solution: $A = HF + HV$. The logic circuit is shown in Figure 5–5.

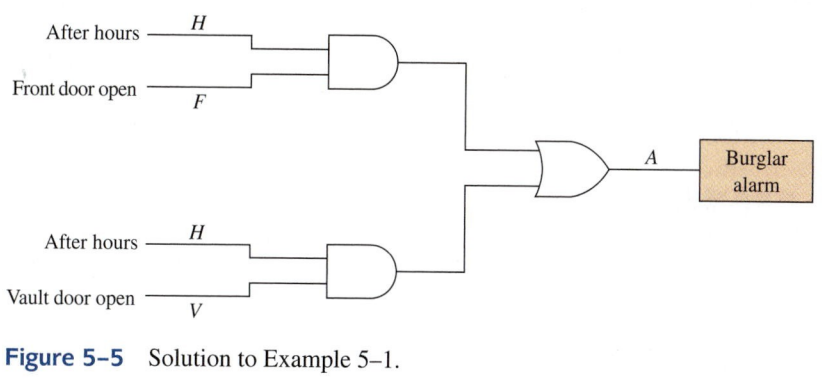

Figure 5–5 Solution to Example 5–1.

Team Discussion

What other applications of Boolean logic can you think of in the home, automobile, industry, and so on?

EXAMPLE 5–2

Using common reasoning, reduce the logic function described in Example 5–1 to a simpler form.

Solution: The alarm is activated if it is after banking hours *and* if either the front door is opened *or* the vault door is opened (see Figure 5–6). The simplified equation is written as

$$A = H(F + V) \qquad \text{(Notice the use of parentheses.)}$$

Team Discussion

How would this answer change if the parentheses were dropped?

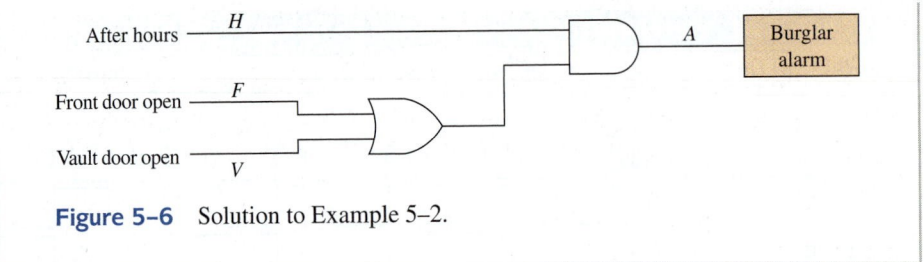

After hours ——— H

Front door open ——— F

Vault door open ——— V

A

Burglar alarm

Figure 5-6 Solution to Example 5-2.

EXAMPLE 5-3

Draw the logic circuit that could be used to implement the following Boolean equation:

$$X = AB + C(M + N)$$

Solution: The logic circuit is shown in Figure 5-7.

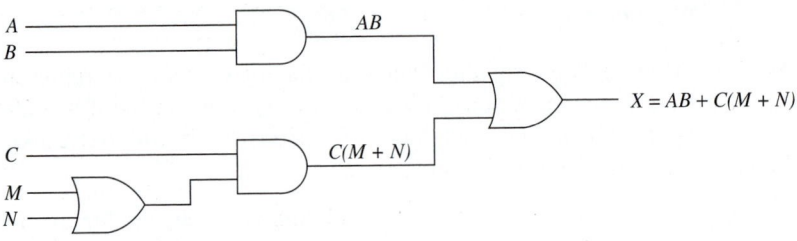

A
B
AB

C
M
N
C(M + N)

$X = AB + C(M + N)$

Figure 5-7 Solution to Example 5-3.

EXAMPLE 5-4

Write the Boolean equation for the logic circuit shown in Figure 5-8.

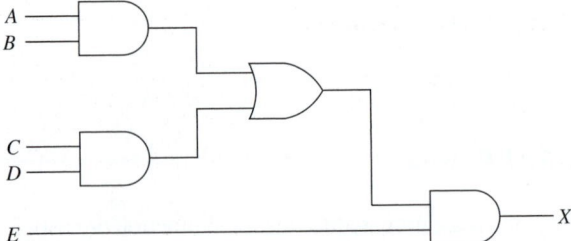

A
B

C
D

E

X

Figure 5-8 Combinational logic circuit for Example 5-4.

Solution: $X = (AB + CD)E$

5–2   Boolean Algebra Laws and Rules

Boolean algebra uses many of the same laws as those of ordinary algebra. The OR function ($X = A + B$) is *Boolean addition*, and the AND function ($X = AB$) is *Boolean multiplication*. The following three laws are the same for Boolean algebra as they are for ordinary algebra:

1. *Commutative law of addition: $A + B = B + A$, and multiplication: $AB = BA$.* These laws mean that the order of ORing or ANDing does not matter.

2. *Associative law of addition: $A + (B + C) = (A + B) + C$, and multiplication: $A(BC) = (AB)C$.* These laws mean that the grouping of several variables ORed or ANDed together does not matter.

3. *Distributive law: $A(B + C) = AB + AC$, and $(A + B)(C + D) = AC + AD + BC + BD$.* These laws show methods for expanding an equation containing ORs and ANDs.

These three laws hold true for any number of variables. For example, the commutative law can be applied to $X = A + BC + D$ to form the equivalent equation $X = BC + A + D$.

You may wonder when you will need to use one of the laws. Later in this chapter, you will see that by using these laws to rearrange Boolean equations, you will be able to change some combinational logic circuits to simpler **equivalent circuits** using fewer gates. You can gain a better understanding of the application of these laws by studying Figures 5–9 to 5–14.

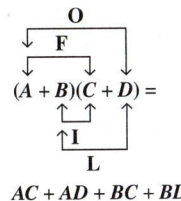
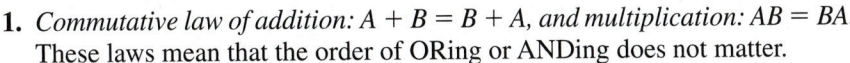

Figure 5–9   Using the commutative law of addition to rearrange an OR gate.

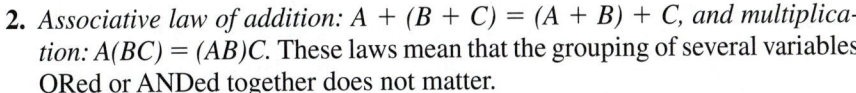

Figure 5–10   Using the commutative law of multiplication to rearrange an AND gate.

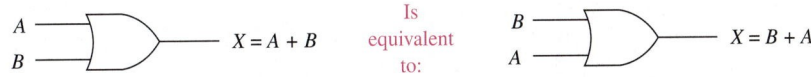

Figure 5–11   Using the associative law of addition to rearrange the grouping of OR gates.

Figure 5–12   Using the associative law of multiplication to rearrange the grouping of AND gates.

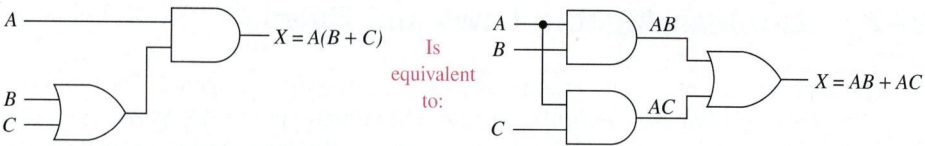

Figure 5–13   Using the distributive law to form an equivalent circuit.

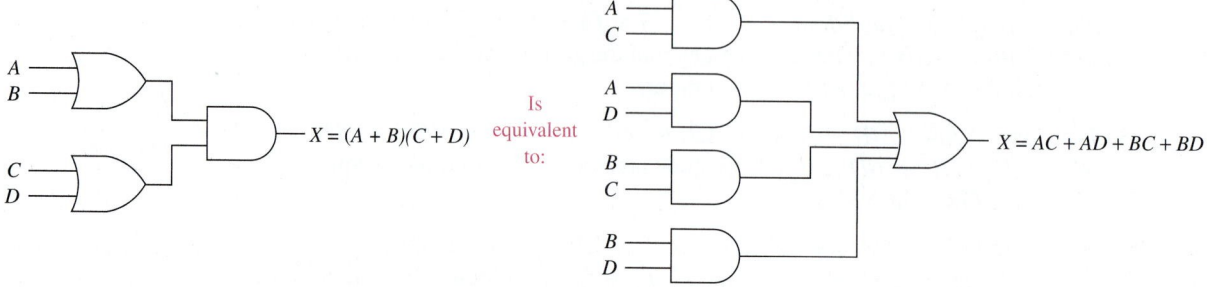

Figure 5–14   Using the distributive law to form an equivalent circuit (FOIL method).

Besides the three basic laws, several rules concern Boolean algebra. The rules of Boolean algebra allow us to combine or eliminate certain variables in the equation to form simpler equivalent circuits.

The following example illustrates the use of the first Boolean rule, which states that anything ANDed with a 0 will always output a 0.

═══════ **EXAMPLE 5–5** ═══════

A bank burglar alarm (B) will activate if it is after banking hours (A) and someone opens the front door (D). The logic level of the variable A is 1 after banking hours and 0 during banking hours. Also, the logic level of the variable D is 1 if the door sensing switch is opened and 0 if the door sensing switch is closed. The Boolean equation is, therefore, $B = AD$. The logic circuit to implement this function is shown in Figure 5–15(a).

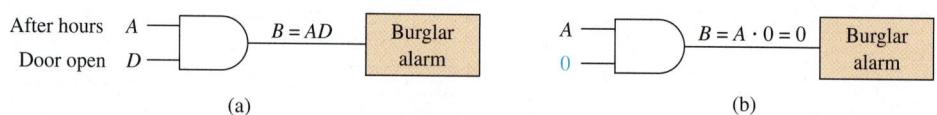

Figure 5–15   (a) Logic circuit for a simple burglar alarm: (b) disabling the burglar alarm by making $D = 0$.

Later, a burglar comes along and puts tape on the door-sensing switch, holding it closed so that it always puts out a 0 logic level. Now the Boolean equation ($B = AD$) becomes $B = A \cdot 0$ because the door-sensing switch is always 0. The alarm will never sound in this condition because one input to the AND gate is always 0. The burglar must have studied the Boolean rules and realized that anything ANDed with a 0 will output a 0, as shown in Figure 5–15(b).

Example 5–5 helped illustrate the reasoning for Boolean Rule 1. The other nine rules can be derived using common sense and knowing basic gate operation.

Helpful Hint

You should *make sense* out of these 10 rules—not simply memorize them.

Rule 1: Anything ANDed with a 0 is equal to 0 ($A \cdot 0 = 0$).

Rule 2: Anything ANDed with a 1 is equal to itself ($A \cdot 1 = A$). From Figure 5–16, we can see that, with one input tied to a 1, if the A input is 0, the X output is 0; if A is 1, X is 1. Therefore, X is equal to whatever the logic level of A is ($X = A$).

$$X = A \cdot 1 = A$$

A	1	X
0	1	0
1	1	1

X equals A

Figure 5–16 Logic circuit and truth table illustrating Rule 2.

Rule 3: Anything ORed with a 0 is equal to itself ($A + 0 = A$). In Figure 5–17, because one input is always 0, if $A = 1$, $X = 1$, and if $A = 0$, $X = 0$. Therefore, X is equal to whatever the logic level of A is ($X = A$).

$$X = A + 0 = A$$

A	0	X
0	0	0
1	0	1

X equals A

Figure 5–17 Logic circuit and truth table illustrating Rule 3.

Rule 4: Anything ORed with a 1 is equal to 1 ($A + 1 = 1$). In Figure 5–18, because one input to the OR gate is always 1, the output is always 1, no matter what A is ($X = 1$).

$$X = A + 1 = 1$$

A	1	X
0	1	1
1	1	1

X equals 1

Figure 5–18 Logic circuit and truth table illustrating Rule 4.

Rule 5: Anything ANDed with itself is equal to itself ($A \cdot A = A$). In Figure 5–19, because both inputs to the AND gate are A, if $A = 1$, 1 and 1 equals 1, and if $A = 0$, 0 and 0 equals 0. Therefore, X is equal to whatever the logic level of A is ($X = A$).

$$X = A \cdot A = A$$

A	A	X
0	0	0
1	1	1

X equals A

Figure 5–19 Logic circuit and truth table illustrating Rule 5.

Rule 6: Anything ORed with itself is equal to itself ($A + A = A$). In Figure 5–20, because both inputs to the OR gate are A, if $A = 1$, 1 or 1 equals 1, and if $A = 0$, 0 or 0 equals 0. Therefore, X is equal to whatever the logic level of A is ($X = A$).

$X = A + A = A$

A	A	X
0	0	0
1	1	1

$\}$ *X equals A*

Figure 5–20 Logic circuit and truth table illustrating Rule 6.

Rule 7: Anything ANDed with its own complement equals 0. In Figure 5–21, because the inputs are complements of each other, one of them is always 0. With a zero at the input, the output is always 0 ($X = 0$).

$X = A \cdot \overline{A} = 0$

A	$\overline{A}$	X
0	1	0
1	0	0

$\}$ *X equals 0*

Figure 5–21 Logic circuit and truth table illustrating Rule 7.

Rule 8: Anything ORed with its own complement equals 1. In Figure 5–22, because the inputs are complements of each other, one of them is always 1. With a 1 at the input, the output is always 1 ($X = 1$).

$X = A + \overline{A} = 1$

A	$\overline{A}$	X
0	1	1
1	0	1

$\}$ *X equals 1*

Figure 5–22 Logic circuit and truth table illustrating Rule 8.

Rule 9: A variable that is complemented twice will return to its original logic level. As shown in Figure 5–23, when a variable is complemented once, it changes to the opposite logic level. When it is complemented a second time, it changes back to its original logic level ($\overline{\overline{A}} = A$).

$X = \overline{\overline{A}} = A$

A	$\overline{A}$	$\overline{\overline{A}}$	X
0	1	0	0
1	0	1	1

$\}$ *X equals A*

Figure 5–23 Logic circuit and truth table illustrating Rule 9.

Rule 10: $A + \overline{A}B = A + B$ and $\overline{A} + AB = \overline{A} + B$. This rule differs from the others because it involves two variables. It is useful because, when an equation is in this form, one or more variables in the second term can be eliminated. The validity of these two equations is proven in Table 5–1. In each case, equivalence is demonstrated by showing that the truth table derived from the expression on the left side of the equation matches that on the right side.

TABLE 5–1		Using Truth Tables to Prove the Equations in Rule 10						
A	*B*	$A + \overline{A}B$	$A + B$	*A*	*B*	$\overline{A} + AB$	$\overline{A} + B$
0	0	0	0	0	0	1	1
0	1	1	1	0	1	1	1
1	0	1	1	1	0	0	0
1	1	1	1	1	1	1	1
		↑	↑			↑	↑
		Equivalent outputs				Equivalent outputs	

CHAPTER 5 I BOOLEAN ALGEBRA AND REDUCTION TECHNIQUES

Table 5–2 summarizes the laws and rules that relate to Boolean algebra. By using them, we can reduce complicated combinational logic circuits to their simplest form, as we will see in the next sections. The letters used in Table 5–2 are variables and were chosen arbitrarily. For example, $C + \overline{C}D = C + D$ is also a valid use of Rule 10(a).

TABLE 5–2	Boolean Laws and Rules for the Reduction of Combinational Logic Circuits
Laws	
1	$A + B = B + A$
	$AB = BA$
2	$A + (B + C) = (A + B) + C$
	$A(BC) = (AB)C$
3	$A(B + C) = AB + AC$
	$(A + B)(C + D) = AC + AD + BC + BD$
Rules	
1	$A \cdot 0 = 0$
2	$A \cdot 1 = A$
3	$A + 0 = A$
4	$A + 1 = 1$
5	$A \cdot A = A$
6	$A + A = A$
7	$A \cdot \overline{A} = 0$
8	$A + \overline{A} = 1$
9	$\overline{\overline{A}} = A$
10 (a)	$A + \overline{A}B = A + B$
(b)	$\overline{A} + AB = \overline{A} + B$

Review Questions

5–1. How many gates are required to implement the following Boolean equations?

(a) $X = (A + B)C$

(b) $Y = AC + BC$

(c) $Z = (ABC + CD)E$

5–2. Which Boolean law is used to transform each of the following equations?

(a) $B + (D + E) = (B + D) + E$

(b) $CAB = BCA$

(c) $(B + C)(A + D) = BA + BD + CA + CD$

5–3. The output of an AND gate with one of its inputs connected to 1 will always output a level equal to the level at the other input. True or false?

5–4. The output of an OR gate with one of its inputs connected to 1 will always output a level equal to the level at the other input. True or false?

5–5. If one input to an OR gate is connected to 0, the output will always be 0 regardless of the level on the other input. True or false?

5–6. Use one of the forms of Rule 10 to transform each of the following equations:

(a) $\overline{B} + AB = ?$

(b) $B + \overline{B}C = ?$

5–3 Simplification of Combinational Logic Circuits Using Boolean Algebra

Often in the design and development of digital systems, a designer will start with simple logic gate requirements but add more and more complex gating, making the final design a complex combination of several gates, with some having the same inputs. At that point, the designer must step back and review the combinational logic circuit that has been developed and see if there are ways of reducing the number of gates without changing the function of the circuit. If an equivalent circuit can be formed with fewer gates or fewer inputs, the cost of the circuit is reduced and its reliability is improved. This process is called the *reduction* or *simplification of combinational logic circuits* and is performed by using the laws and rules of Boolean algebra presented in the preceding section.

The following examples illustrate the use of Boolean algebra and present some techniques for the simplification of logic circuits.

EXAMPLE 5–6

The logic circuit shown in Figure 5–24 is used to turn on a warning buzzer at *X* based on the input conditions at *A, B,* and *C*. A simplified equivalent circuit that will perform the same function can be formed by using Boolean algebra. Write the equation of the circuit in Figure 5–24, simplify the equation, and draw the logic circuit of the simplified equation.

Helpful Hint

As a beginner, you should write the Boolean terms at each input to each gate, as shown here.

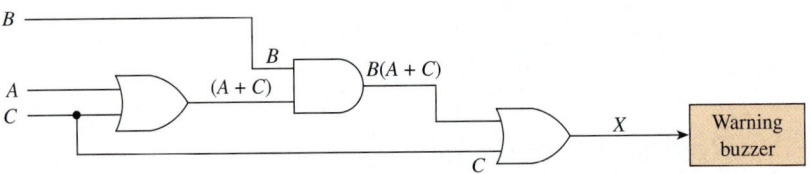

Figure 5–24 Logic circuit for Example 5–6.

Solution: The Boolean equation for *X* is

$$X = B(A + C) + C$$

To simplify, first apply *Law 3* [$B(A + C) = BA + BC$]:

$$X = BA + BC + C$$

Next, factor a *C* from terms 2 and 3:

$$X = BA + C(B + 1)$$

Apply *Rule 4* ($B + 1 = 1$):

$$X = BA + C \cdot 1$$

Apply *Rule 2* ($C \cdot 1 = C$):

$$X = BA + C$$

Apply *Law 1* ($BA = AB$):

$$X = AB + C \leftarrow simplified\ equation$$

The logic circuit of the simplified equation is shown in Figure 5–25.

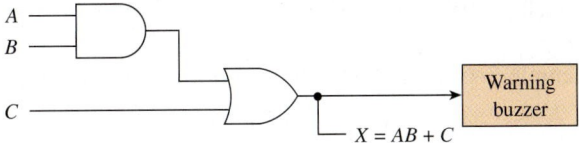

Figure 5–25 Simplified logic circuit for Example 5–6.

EXAMPLE 5–7

Repeat Example 5–6 for the logic circuit shown in Figure 5–26.

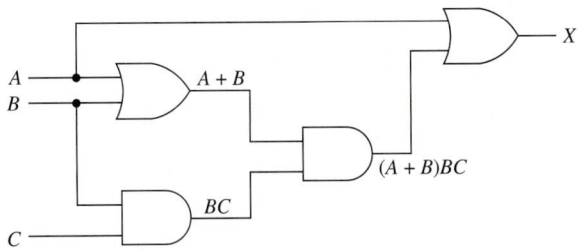

Figure 5–26 Logic circuit for Example 5–7.

Common Misconception

Without the parentheses in the first equation, the logic is invalid.

Solution: The Boolean equation for X is

$$X = (A + B)BC + A$$

To simplify, first apply *Law 3* [$(A + B)BC = ABC + BBC$]:

$$X = ABC + BBC + A$$

Apply *Rule 5* ($B \cdot B = B$):

$$X = ABC + BC + A$$

Factor a *BC* from terms 1 and 2:

$$X = BC(A + 1) + A$$

Apply *Rule 4* ($A + 1 = 1$):

$$X = BC \cdot 1 + A$$

Apply *Rule 2* ($BC \cdot 1 = BC$):

$$X = BC + A \leftarrow simplified\ equation$$

The logic circuit for the simplified equation is shown in Figure 5–27.

Figure 5–27 Simplified logic circuit for Example 5–7.

EXAMPLE 5–8

Repeat Example 5–6 for the logic circuit shown in Figure 5–28.

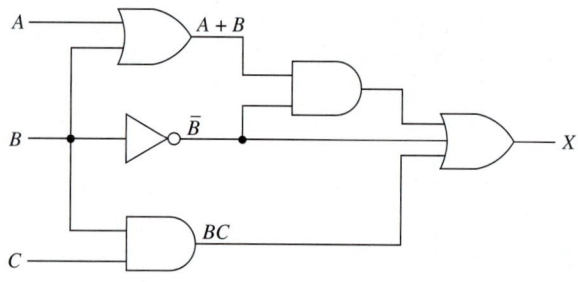

Figure 5–28   Logic circuit for Example 5–8.

Solution: The Boolean equation for X is

$$X = (A + B)\overline{B} + \overline{B} + BC$$

To simplify, first apply *Law 3* $[(A + B)\overline{B} = A\overline{B} + B\overline{B}]$:

$$X = A\overline{B} + B\overline{B} + \overline{B} + BC$$

Apply *Rule 7* $(B\overline{B} = 0)$:

$$X = A\overline{B} + 0 + \overline{B} + BC$$

Apply *Rule 3* $(A\overline{B} + 0 = A\overline{B})$:

$$X = A\overline{B} + \overline{B} + BC$$

Factor a $\overline{B}$ from terms 1 and 2:

$$X = \overline{B}(A + 1) + BC$$

Apply *Rule 4* $(A + 1 = 1)$:

$$X = \overline{B} \cdot 1 + BC$$

Apply *Rule 2* $(\overline{B} \cdot 1 = \overline{B})$:

$$X = \overline{B} + BC$$

Apply *Rule 10(b)* $(\overline{B} + BC = \overline{B} + C)$:

$$X = \overline{B} + C \leftarrow simplified\ equation$$

The logic circuit of the simplified equation is shown in Figure 5–29.

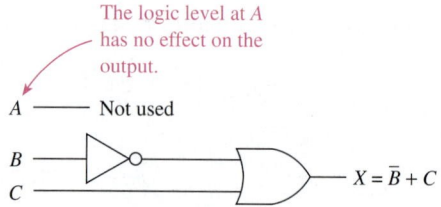

The logic level at A has no effect on the output.

Figure 5–29   Simplified logic circuit for Example 5–8.

EXAMPLE 5–9

Repeat Example 5–6 for the logic circuit shown in Figure 5–30.

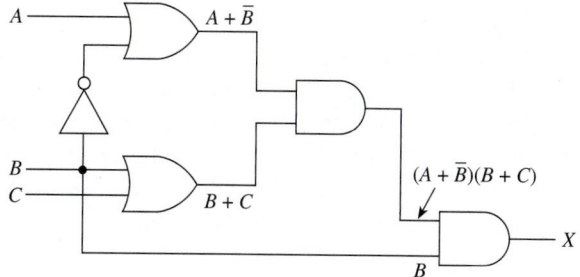

Figure 5–30 Logic circuit for Example 5–9.

Solution: The Boolean equation for *X* is

$$X = [(A + \overline{B})(B + C)]B$$

To simplify, first apply *Law 3:*

$$X = (AB + AC + \overline{B}B + \overline{B}C)B$$

The $\overline{B}B$ team can be eliminated using *Rule 7* and then *Rule 3:*

$$X = (AB + AC + \overline{B}C)B$$

Apply *Law 3* again:

$$X = ABB + ACB + \overline{B}CB$$

Apply *Law 1:*

$$X = ABB + ABC + \overline{B}BC$$

Apply *Rules 5 and 7:*

$$X = AB + ABC + 0 \cdot C$$

Apply *Rule 1:*

$$X = AB + ABC$$

Factor an *AB* from both terms:

$$X = AB(1 + C)$$

Apply *Rule 4* and then *Rule 2:*

$$X = AB \leftarrow simplified\ equation$$

The logic circuit of the simplified equation is shown in Figure 5–31.

A ——⊐
B ——⊐ — *X = AB*

C ——— Not used

Figure 5–31 Simplified logic for Example 5–9.

5–4 Using MAX+PLUS II To Determine Simplified Equations

Part of the compilation process performed by the MAX+PLUS II software is to determine the simplest form of the circuit before it synthesizes its logic. This eliminates unnecessary inputs and minimizes the number of gates used in the CPLD. If we redo Example 5–9 using MAX+PLUS II, the software will warn us of unused inputs and will also give us the final simplified equation. The VHDL program, *ex5_9.vhd,* is given in Figure 5–32(a). The original Boolean equation $X = [(A + \overline{B})(B + C)]B$ is entered in VHDL as x<= ((a OR NOT b) AND (b OR c))AND b;. The Waveform Editor was used to create the simulation file (*ex5_9.scf*) shown in Figure 5–32(b). If you study the

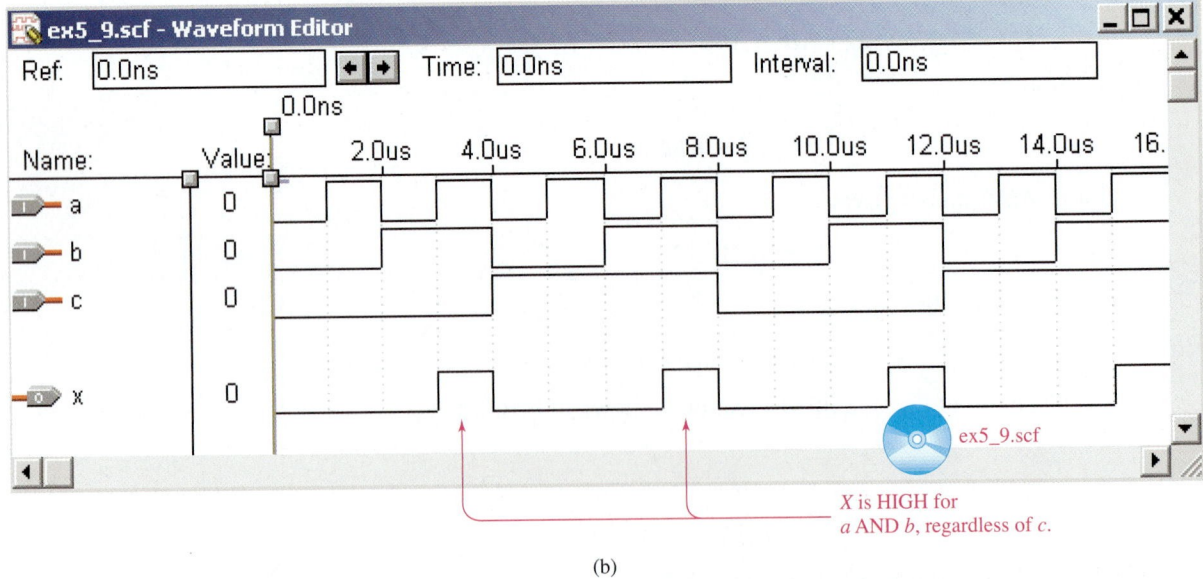

(a)

(b)

Figure 5–32 MAX+PLUS II solution to Example 5–9: (a) VHDL listing; (b) simulation file.

CHAPTER 5 | BOOLEAN ALGEBRA AND REDUCTION TECHNIQUES

results carefully you will see that *x* only goes HIGH when *a* and *b* are both HIGH, regardless of *c*.

When the *ex5_9.vhd* program was compiled, the compiler produced the warning message shown in Figure 5–33. As you can see, it tells us that after simplifying the equation, the *c* input was determined to be unnecessary and will therefore be ignored in the final design.

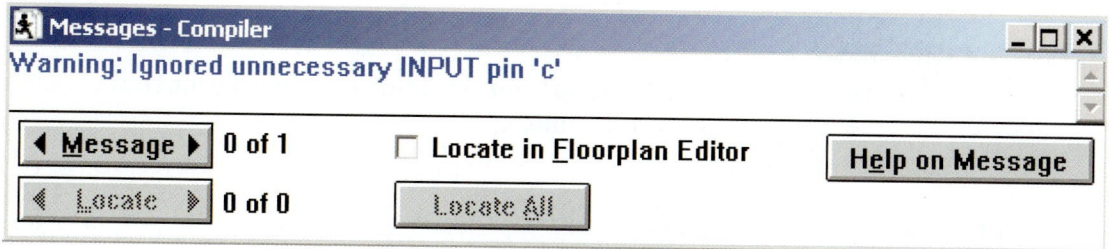

Figure 5–33 The compiler message noting the unnecessary input at *c*.

The report file, *ex5_9.rpt,* is something we haven't seen before but we will use it now to view the simplified equation. It can be loaded by using the following procedure:

Choose   **File Open** then to just list the report files,

Select   **Text Editor files *.rpt** Highlight the *rpt* file that you want to read and press **OK.** (See Figure 5–34)

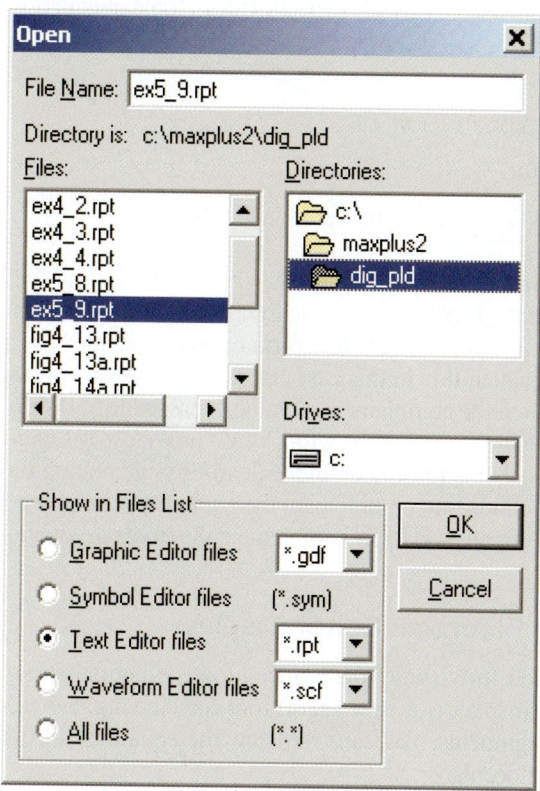

Figure 5–34 Opening the report file, *ex5_9.rpt.*

The report file gives all of the details on how the project is implemented. For now we will only concern ourselves with the Equations Section shown in Figure 5–35. This

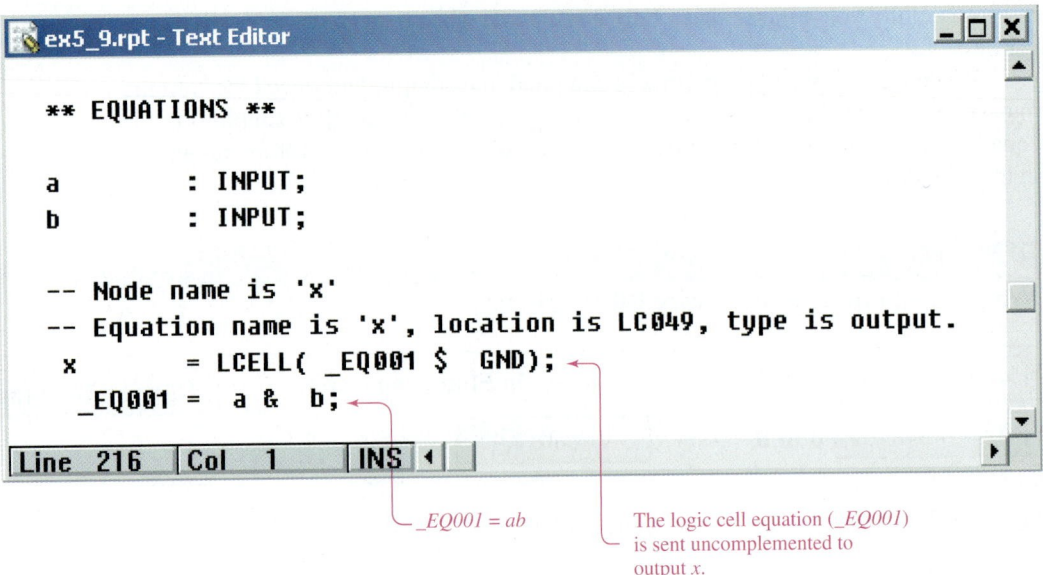

Figure 5–35 The Equations Section of the Report File.

part of the report shows us that a and b are the inputs, x is the output. The last line shows that the final equation is $a \& b$ (where "&" is defined as AND). Recall from Example 5–9 that we came up with the same equation ($X = AB$) by using Boolean algebra rules. Arithmetic operators used by MAX+PLUS II for Boolean equations are as follows:

& AND operator.

! NOT operator.

OR operator.

$ Exclusive-OR operator. (covered in Chapter 6)

The next-to-last line, $x = LCELL (\_EQ001 \$ GND)$; describes the disposition of the internal equation ($\_EQ001$) within the *LCELL* logic cell (or macrocell). This line is used to move the equation to x in its true form or complemented form. If $\_EQ001$ is followed by $ *GND* (which it is in this case) then the true form is passed to x. If it is followed by $ *VCC*, then the complemented value is passed to x. (We'll learn in Chapter 6 that Exclusive ORing a function with VCC provides its complemented value.) MAX+PLUS II does this to provide more flexibility in creating the most streamlined internal logic. If it chooses to complement the equation, we can use De Morgan's theorem (presented in Section 5–5) to un-complement the Boolean equation into its true form.

MAX+PLUS II Floorplan Editor Display

An alternate method for viewing the reduced logic equation is in the **Floorplan Editor Display.** This display is a very useful tool for viewing and modifying device pin and macrocell assignments. You can also view the equations section of the report file here. To view the floorplan:

Choose   **MAX+PLUS II**   –>   **Floorplan Editor**

Under the **Layout** tab be sure to check the same items as shown in Figure 5–36(a). The floorplan for *ex5_9* is shown in Figure 5–36(b). The LAB view that appears shows rout-

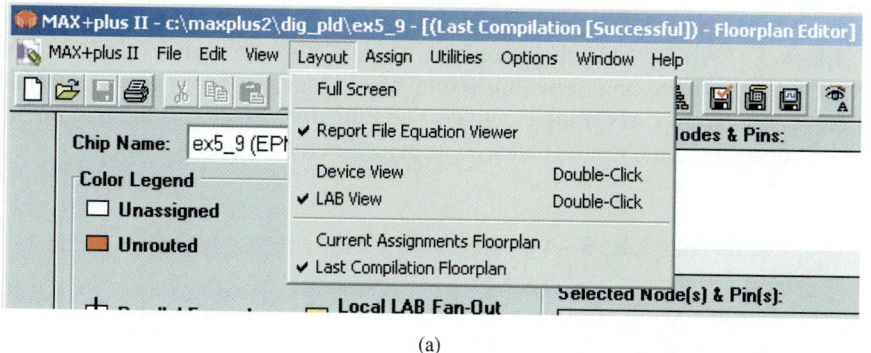

(a)

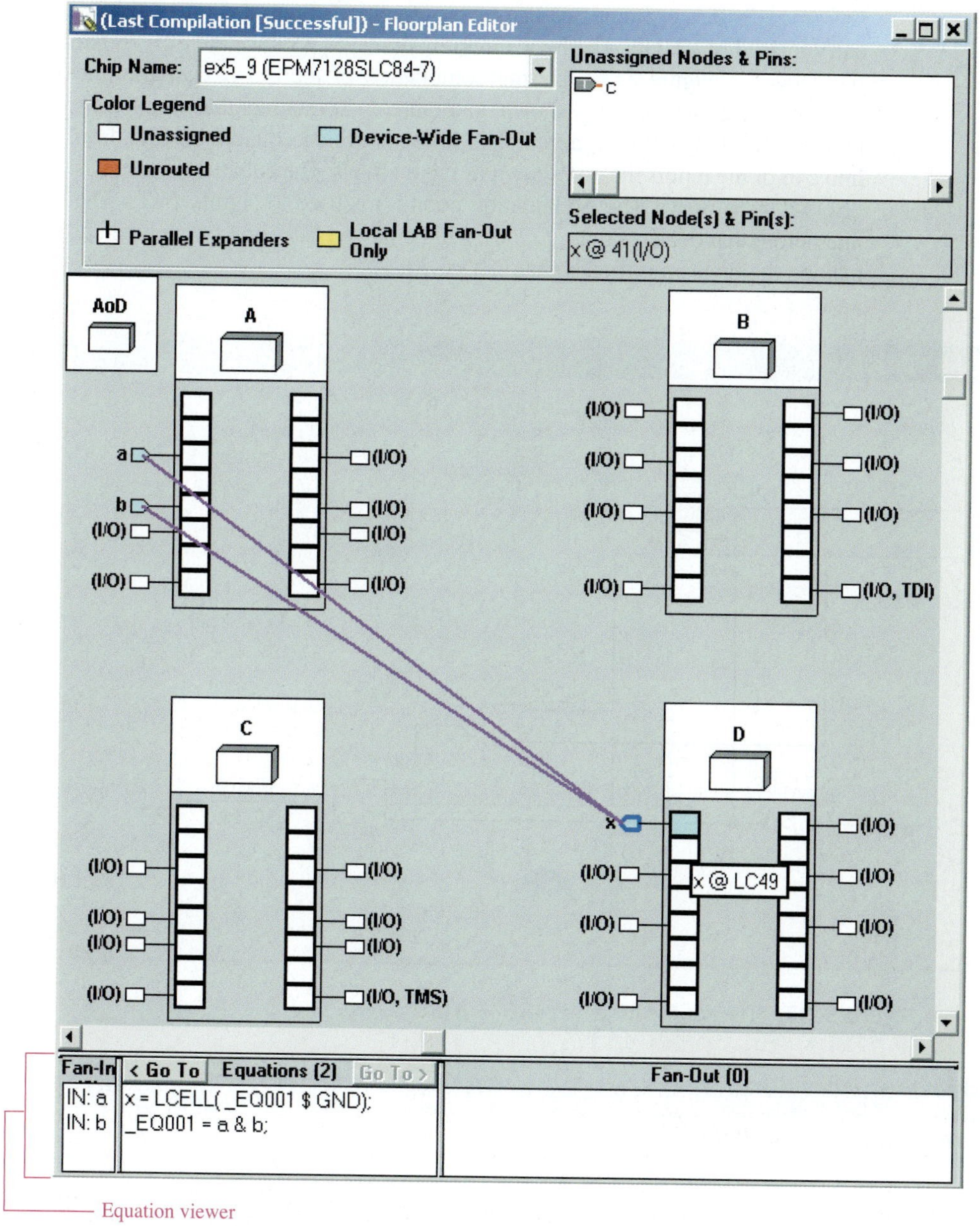

(Last Compilation [Successful]) - Floorplan Editor

Chip Name: ex5_9 (EPM7128SLC84-7)

Unassigned Nodes & Pins:

c

Color Legend

☐ Unassigned   ☐ Device-Wide Fan-Out

🟧 Unrouted

⊥ Parallel Expanders   ☐ Local LAB Fan-Out Only

Selected Node(s) & Pin(s):

x @ 41(I/O)

AoD

A

a

b
(I/O)
(I/O)
(I/O)
(I/O)
(I/O)

B

(I/O)
(I/O)
(I/O)
(I/O)

(I/O)
(I/O)
(I/O)
(I/O, TDI)

C

(I/O)
(I/O)
(I/O)
(I/O)
(I/O)
(I/O, TMS)

D

x
x @ LC49

(I/O)
(I/O)
(I/O)
(I/O)
(I/O)
(I/O)
(I/O)

Fan-In	< Go To	Equations (2)	Go To >	Fan-Out (0)
IN: a	x = LCELL(_EQ001 $ GND);			
IN: b	_EQ001 = a & b;			

Equation viewer

(b)

Figure 5–36 MAX+PLUS II Floorplan Editor: (a) layout selections; (b) LAB view showing interconnections and equations.

ing of the inputs and outputs. The *a* and *b* inputs are shown connected to **Logic Array Block** A (LAB A). This block contains **Logic Cells** 1 through 16 (LC1-LC16). There are 8 LABs on this chip providing a total of 128 logic cells (or macrocells). The output at *x* comes from LAB D logic cell 49. The Equations Section at the bottom of the figure shows the same results that were provided in the report file.

EXAMPLE 5–10

Use MAX+PLUS II to determine the simplified equation for
$X = (AB\overline{C} + B)B\overline{C}$.

Solution: In this example we'll use the Graphic Editor method of design entry instead of VHDL. The logic circuit is drawn to produce the *ex5_10.gdf* file shown in Figure 5–37. During compilation of the *ex5_10.gdf* file, the software gave the warning shown in Figure 5–38 stating that input *A* is unnecessary and will be ignored. The final simplified equation is found on line 246 of the report file and shown in Figure 5–39. The equation is given as _EQ001 = B & !C. The exclamation point (!) is used to signify NOT. The line before that declares that _EQ001 will be passed to *X*, un-complemented, making the final simplified equation $X = B\overline{C}$.

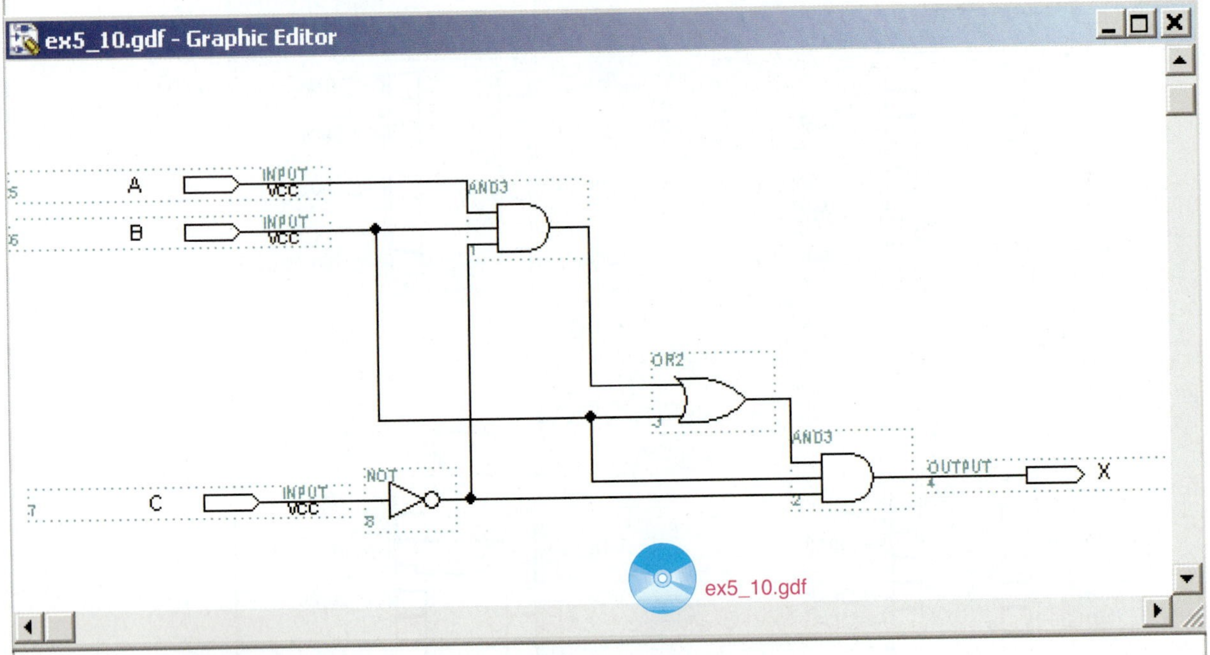

Figure 5–37 The gdf file for $X = (AB\overline{C} + B)B\overline{C}$.

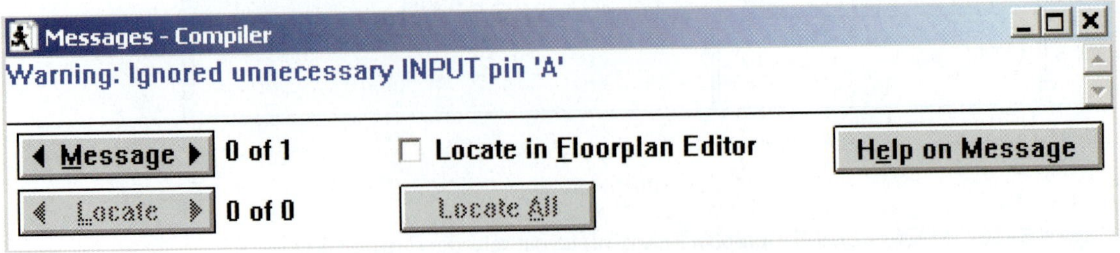

Figure 5–38 The compiler message noting the unused input.

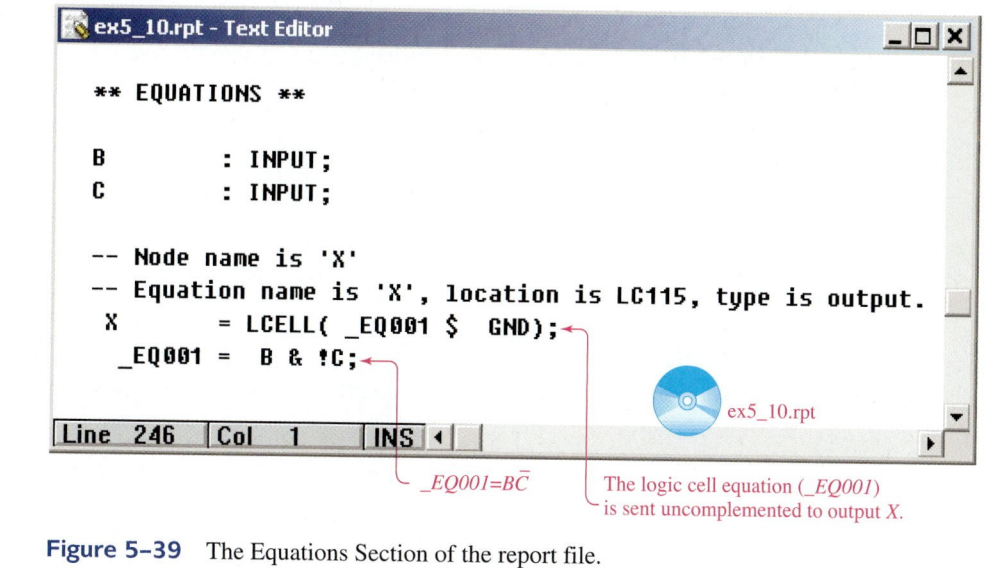

Figure 5–39 The Equations Section of the report file.

5–5 De Morgan's Theorem

You may have noticed that we did not use NANDs or NORs in any of the logic circuits in Section 5–3. To simplify circuits containing NANDs and NORs, we need to use a theorem developed by the mathematician Augustus De Morgan. This theorem allows us to convert an expression having an inversion bar over two or more variables into an expression having inversion bars over single variables only. This allows us to use the rules presented in the preceding section for the simplification of the equation.

In the form of an equation, **De Morgan's theorem** is stated as follows:

$$\overline{A \cdot B} = \overline{A} + \overline{B}$$

$$\overline{A + B} = \overline{A} \cdot \overline{B}$$

Also, for three or more variables,

$$\overline{A \cdot B \cdot C} = \overline{A} + \overline{B} + \overline{C}$$

$$\overline{A + B + C} = \overline{A} \cdot \overline{B} \cdot \overline{C}$$

Basically, to use the theorem, you break the bar over the variables and either change the AND to an OR or change the OR to an AND.

To prove to ourselves that this works, let's apply the theorem to a NAND gate and then compare the truth table of the equivalent circuit to that of the original NAND gate. As you can see in Figure 5–40, to use De Morgan's theorem on a NAND gate, first break the bar over the $A \cdot B$, then change the AND symbol to an OR. The new equation becomes $X = \overline{A} + \overline{B}$. Notice that **inversion bubbles** are used on the OR gate instead of inverters. By observing the truth tables of the two equations, we can see that the result in the X column is the same for both, which proves that they provide an equivalent output result.

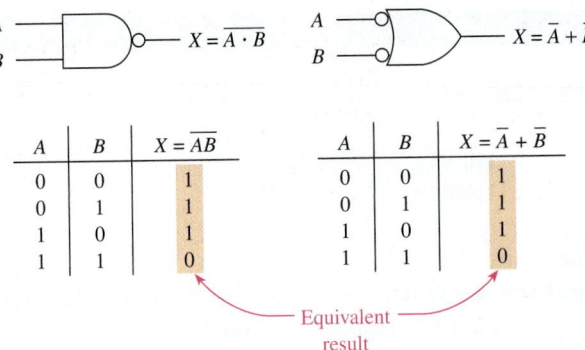

A
B
$X = \overline{A \cdot B}$
A
B
$X = \overline{A} + \overline{B}$

A	B	$X = \overline{AB}$
0	0	1
0	1	1
1	0	1
1	1	0

A	B	$X = \overline{A} + \overline{B}$
0	0	1
0	1	1
1	0	1
1	1	0

Equivalent result

Figure 5–40 De Morgan's theorem applied to NAND gate produces two identical truth tables.

A
B
$X = \overline{A + B}$
A
B
$X = \overline{A} \cdot \overline{B}$

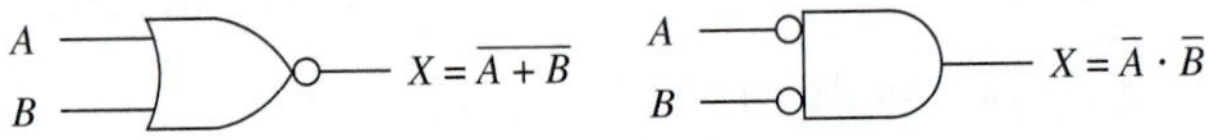

A	B	$X = \overline{A + B}$
0	0	1
0	1	0
1	0	0
1	1	0

A	B	$X = \overline{A} \cdot \overline{B}$
0	0	1
0	1	0
1	0	0
1	1	0

Equivalent result

(a)

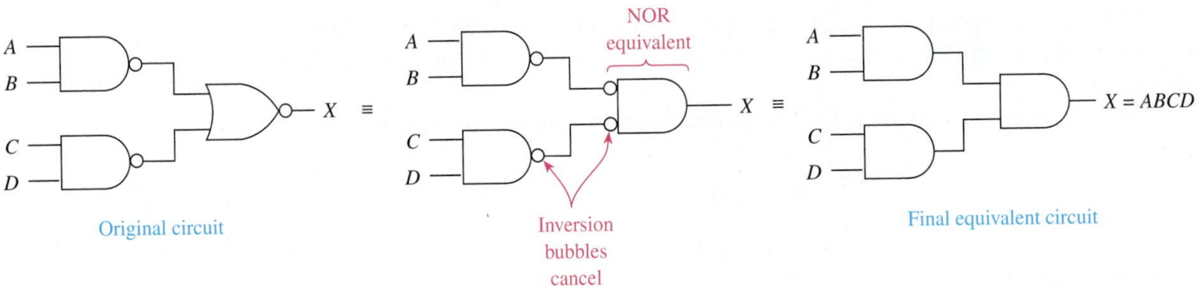

A
B
C
D
X
≡
NOR equivalent
A
B
C
D
X
≡
A
B
C
D
X = ABCD
Original circuit
Inversion bubbles cancel
Final equivalent circuit

(b)

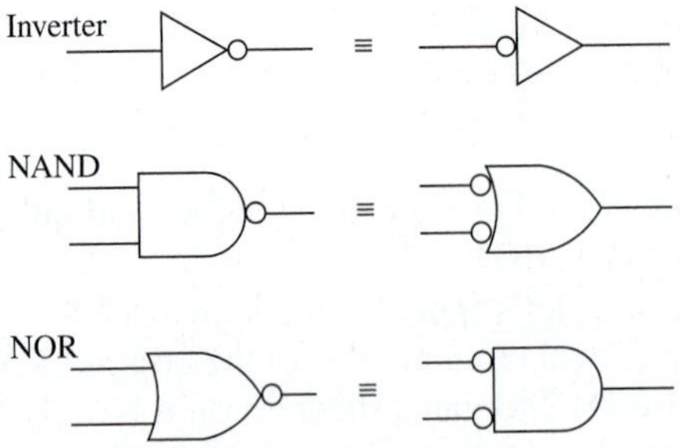

Inverter
≡
NAND
≡
NOR
≡

(c)

Figure 5–41 (a) De Morgan's theorem applied to NOR gate produces two identical truth tables; (b) using the alternative NOR symbol eases circuit simplification; (c) summary of alternative gate symbols.

Also, by looking at the two circuits, we can say that *an AND gate with its output inverted is equivalent to an OR gate with its inputs inverted.* Therefore, the OR gate with inverted inputs is sometimes used as an alternative symbol for a NAND gate.

By applying De Morgan's theorem to a NOR gate, we will also produce two identical truth tables, as shown in Figure 5–41(a). Therefore, we can also think of an OR gate with its output inverted as being equivalent to an AND gate with its inputs inverted. The inverted input AND gate symbol is also sometimes used as an alternative to the NOR gate symbol.

When you write the equation for an AND gate with its inputs inverted, be careful to keep the inversion bar over each individual variable (not both) because $\overline{A \cdot B}$ is not equal to $\overline{A} \cdot \overline{B}$. (Prove that to yourself by building a truth table for both.) Also, $\overline{A + B}$ is not equal to $\overline{A} + \overline{B}$.

The question always arises: Why would a designer ever use an *inverted-input OR gate* symbol instead of a NAND? Or why use an *inverted-input AND gate* symbol instead of a NOR? In complex logic diagrams, you will see both the inverted-input and the inverted-output symbols being used. The designer will use whichever symbol makes more sense for the particular application.

For example, referring to Figure 5–40, let's say you need a HIGH output level whenever either *A* or *B* is LOW. It makes sense to think of that function as an OR gate with inverted *A* and *B* inputs, but you could save two inverters by just using a NAND gate.

Also, referring to Figure 5–41(a), let's say you need a HIGH output whenever both *A* and *B* are LOW. You would probably use the *inverted-input AND gate* for your logic diagram because it makes sense logically, but you would use a NOR gate to actually implement the circuit because you could eliminate the inverters.

The alternative methods of drawing NANDs and NORs are also useful for the simplification of logic circuits. Take, for example, the circuit of Figure 5–41(b). By changing the NOR gate to an *inverted-input AND gate,* the inversion bubbles cancel, and the equation becomes simply $X = ABCD$. Figure 5–41(c) summarizes the alternative representations for the inverter, NAND, and NOR gates.

The following examples illustrate the application of De Morgan's theorem for the simplification of logic circuits.

EXAMPLE 5–11

Use MAX + PLUS II to prove the validity of the De Morgan's theorem circuits of Figures 5–40 and 5–41. Draw the circuits using the Graphic Editor and prove equivalence by performing a simulation with all possible input conditions.

Solution: The NAND and NOR circuits of Figures 5–40 and 5–41 are duplicated in the gdf file shown in Figure 5–42. *W* is the output of a NAND while *X* is the output of an inverted-input OR gate that is supposed to be equivalent. *Y* is the output of a NOR while *Z* is the output of an inverted-input AND gate that is supposed to be equivalent.

The waveform simulation file in Figure 5–43 shows every combination of input for *A, B* and *C, D*. By studying the resultant waveforms you can see that the output at *W* is identical to *X* and the output at *Y* is identical to *Z,* proving De Morgan's Theorem.

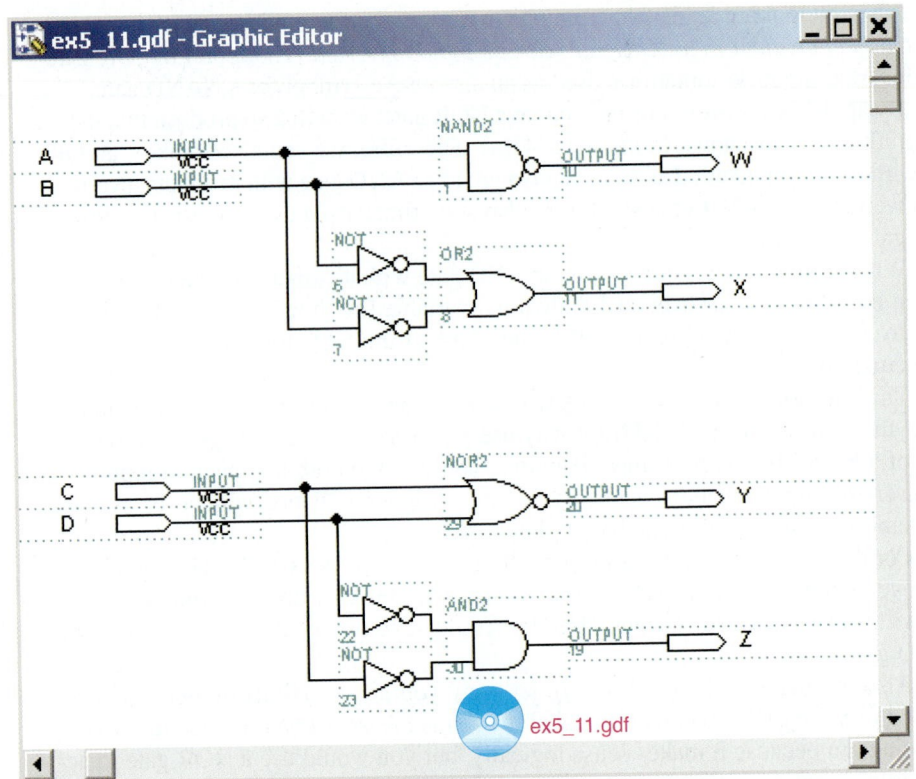

Figure 5–42 The gdf file of circuits used to prove De Morgan's theorem.

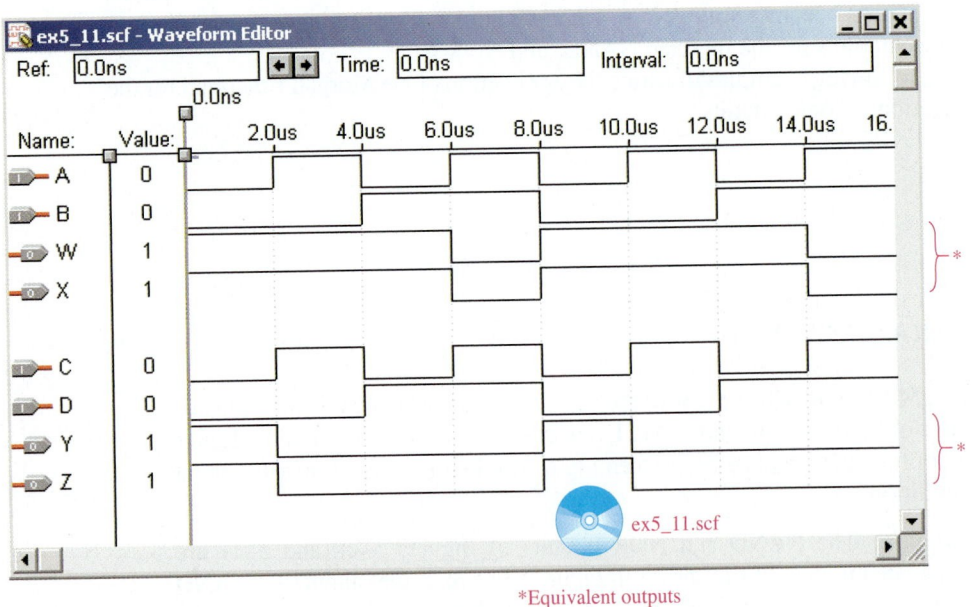

*Equivalent outputs

Figure 5–43 The waveform simulation demonstrating equivalent outputs.

CHAPTER 5 | BOOLEAN ALGEBRA AND REDUCTION TECHNIQUES

EXAMPLE 5–12

Write the Boolean equation for the circuit shown in Figure 5–44. Use De Morgan's theorem and then Boolean algebra rules to simplify the equation. Draw the simplified circuit.

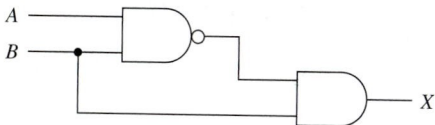

Figure 5–44

Helpful Hint

You must use parentheses to maintain proper grouping whenever you break the bar over a NAND or if an OR gate is input to an AND gate.

Solution: The Boolean equation at X is

$$X = \overline{AB} \cdot B$$

Applying De Morgan's theorem produces

$$X = (\overline{A} + \overline{B}) \cdot B$$

(Notice the use of parentheses to maintain proper grouping. *Rule:* Whenever you break the bar over a NAND you must use parentheses.) Using Boolean algebra rules produces

$$X = \overline{A}B + \overline{B}B$$
$$= \overline{A}B + 0$$
$$= \overline{A}B \quad \leftarrow simplified\ equation$$

The simplified circuit is shown in Figure 5–45.

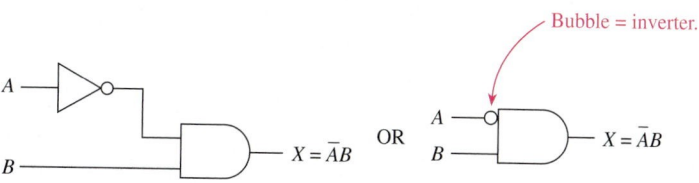

Figure 5–45 Simplified logic circuit for Example 5–12.

EXAMPLE 5–13

Repeat Example 5–12 for the circuit shown in Figure 5–46.

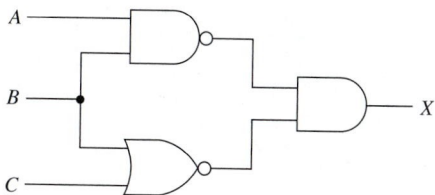

Figure 5–46

Solution: The Boolean equation at X is

$$X = \overline{AB} \cdot \overline{B + C}$$

Applying De Morgan's theorem produces

$$X = (\overline{A} + \overline{B}) \cdot \overline{B}\,\overline{C}$$

(Notice the use of parentheses to maintain proper grouping.) Using Boolean algebra rules produces

$$X = \overline{A}\,\overline{B}\,\overline{C} + \overline{B}\,\overline{B}\,\overline{C}$$
$$= \overline{A}\,\overline{B}\,\overline{C} + \overline{B}\,\overline{C}$$
$$= \overline{B}\,\overline{C}(\overline{A} + 1)$$
$$= \overline{B}\,\overline{C} \quad \leftarrow simplified\ equation$$

The simplified circuit is shown in Figure 5–47.

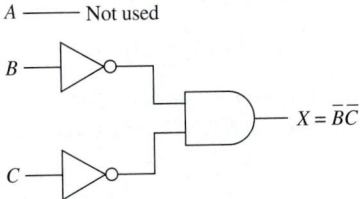

Figure 5–47 Simplified logic circuit for Example 5–13.

Also remember from Figure 5–41(a) that an AND gate with inverted inputs is equivalent to a NOR gate. Therefore, an equivalent solution to Example 5–13 would be a NOR gate with B and C as inputs, as shown in Figure 5–48.

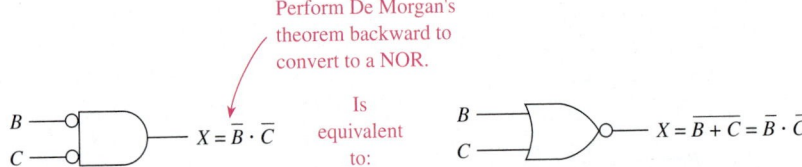

Figure 5–48 Equivalent solution to Example 5–13.

EXAMPLE 5-14

Repeat Example 5–12 for the circuit shown in Figure 5–49.

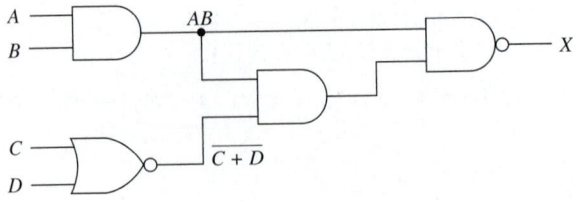

Figure 5–49

CHAPTER 5 | BOOLEAN ALGEBRA AND REDUCTION TECHNIQUES

Solution:

$$X = \overline{(AB \cdot \overline{C + D}) \cdot AB}$$
$$= \overline{AB \cdot \overline{C + D}} + \overline{AB}$$
$$= \overline{AB} + \overline{\overline{C + D}} + \overline{AB}$$
$$= \overline{A} + \overline{B} + C + D + \overline{A} + \overline{B}$$
$$= \overline{A} + \overline{B} + C + D \quad \leftarrow simplified\ equation$$

The simplified circuit is shown in Figure 5–50.

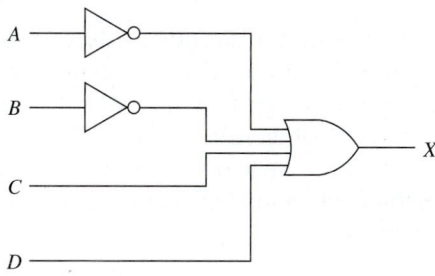

Figure 5–50 Simplified logic circuit for Example 5–14.

EXAMPLE 5–15

Use De Morgan's theorem and Boolean algebra on the circuit shown in Figure 5–51 to develop an equivalent circuit that has inversion bars covering only single variables.

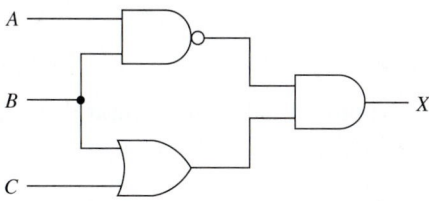

Figure 5–51

Solution: The Boolean equation at *X* is

$$X = \overline{AB} \cdot (B + C)$$

Applying De Morgan's theorem produces

$$X = (\overline{A} + \overline{B}) \cdot (B + C)$$

(Notice the use of parentheses to maintain proper grouping.) Using Boolean algebra rules produces

$$X = \overline{A}B + \overline{A}C + \overline{B}B + \overline{B}C$$
$$= \overline{A}B + \overline{A}C + \overline{B}C \quad \leftarrow final\ equation\ (sum\text{-}of\text{-}products\ form)$$

The equivalent circuit is shown in Figure 5–52.

Team Discussion

The final circuit in this example is actually *more* complicated than the original. As you will see later, it is in the form for implementation using AND-OR-INVERT gates and programmable logic devices. Besides, it is much easier to fill in a truth table from a sum of products (SOP). Build a truth table from the original equation and then from the final SOP to prove the point.

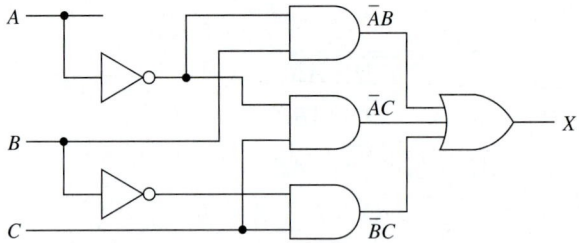

Figure 5–52 Logic circuit equivalent for Example 5–15.

Notice that the final equation actually produces a circuit that is more complicated than the original. In fact, if a technician were to build a circuit, he or she would choose the original because it is simpler and has fewer gates. However, the final equation is in a form called the **sum-of-products (SOP) form.** This form of the equation was achieved by using Boolean algebra and is very useful for building truth tables and Karnaugh maps, which are covered in Section 5–8.

EXAMPLE 5–16

Using De Morgan's theorem and Boolean algebra, prove that the two circuits shown in Figure 5–53 are equivalent.

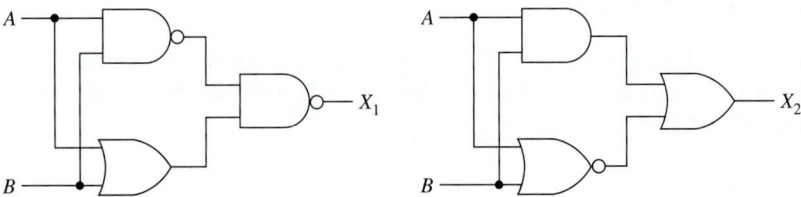

Figure 5–53

Solution: They can be proved to be equivalent if their simplified equations match.

$$X_1 = \overline{\overline{AB} \cdot (A + B)} \qquad X_2 = AB + \overline{A + B}$$

$$= \overline{\overline{AB}} + \overline{A + B} \qquad = AB + \overline{A}\,\overline{B}$$

$$= AB + \overline{A}\,\overline{B} \qquad \text{Equivalent}$$

EXAMPLE 5–17

Use MAX+PLUS II to simplify the equations:

$$X = \overline{AB + (\overline{B} + C)}$$

$$Y = \overline{AB} + \overline{B + C}$$

Solution: The logic for X and Y can be entered using the Graphic Editor or the VHDL Text Editor. VHDL entry was used in this example.

Figure 5–54 shows the VHDL program with the equations for *X* and *Y* appearing in the Architecture block.

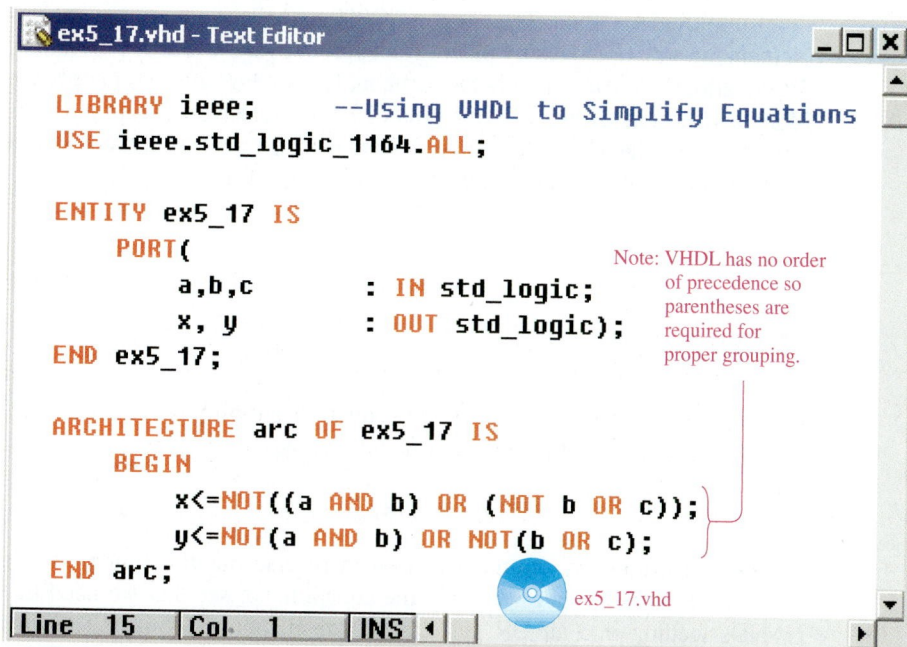

Figure 5–54   VHDL program for Example 5–17.

After performing a save and compile, the report file lists the simplified equations as shown in Figure 5–55. The simplified equation for *X* as determined by MAX + PLUS II is: $X = \overline{A}B\overline{C}$.

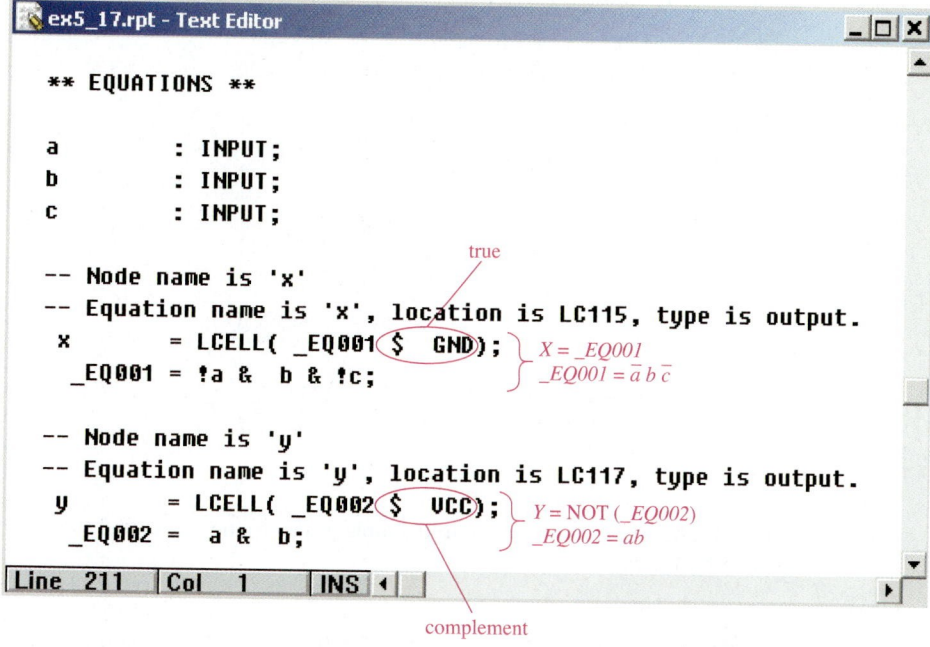

Figure 5–55   The Equations Section of the report file for Example 5–17.

The simplified equation for Y is the complement of that shown. As discussed in Section 5–4, the software sometimes lists the complement of the equation by using $ VCC in the LCELL expression. The LCELL expression for Y has the $ VCC so _EQ002 is actually the complement of the final simplified equation. To change _EQ002 to the true equation just complement the equation back by placing an overbar over the a & b. This makes the equation $Y = \overline{AB}$ which is the same as $Y = \overline{A} + \overline{B}$ after you apply De Morgan's theorem. (*Note:* "$" is the symbol for an Exclusive-OR gate and VCC is used to represent a logic "1." As we'll learn in Chapter 6, Exclusive-ORing a logic function with "1" produces the complement of that logic function.)

EXAMPLE 5–18

Draw the logic circuit for the following equation, simplify the equation, and construct a truth table for the simplified equation

$$X = \overline{A \cdot \overline{B}} + \overline{A \cdot (\overline{A} + C)}$$

Solution: To draw the circuit, we have to reverse our thinking from the previous examples. When we study the equation, we see that we need two NANDs feeding into an OR gate, as shown in Figure 5–56(a). Then we have to provide the inputs to the NAND gates, as shown in Figure 5–56(b).

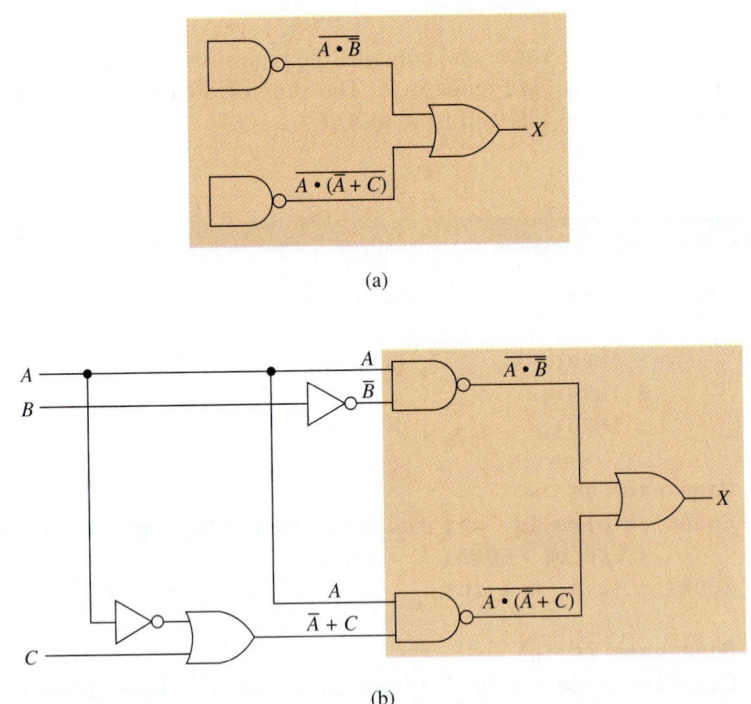

(a)

(b)

Figure 5–56 (a) Partial solution to Example 5–18; (b) logic circuit of the equation for Example 5–18.

Next, we use De Morgan's theorem and Boolean algebra to simplify the equation:

$$X = \overline{A \cdot \overline{B}} + \overline{A \cdot (\overline{A} + C)}$$

$$= (\overline{A} + \overline{\overline{B}}) + (\overline{A} + \overline{\overline{A} + C})$$

$$= \overline{A} + B + \overline{A} + \overline{\overline{A}} \cdot \overline{C}$$

$$= \overline{A} + \overline{A} + A\overline{C} + B$$

$$= \overline{A} + A\overline{C} + B$$

Apply *Rule 10:*

$$X = \overline{A} + \overline{C} + B \quad \leftarrow simplified\ equation$$

This equation can be interpreted as: X is HIGH if A is LOW or C is LOW or B is HIGH. Now, to construct a truth table (Table 5–3), we need three input columns (A, B, C) and eight entries ($2^3 = 8$), and we fill in a 1 for X when $A = 0$, $C = 0$, or $B = 1$.

TABLE 5–3	Truth Table for Example 5–18		
A	**B**	**C**	$\mathbf{X = \overline{A} + \overline{C} + B}$
0	0	0	1
0	0	1	1
0	1	0	1
0	1	1	1
1	0	0	1
1	0	1	0
1	1	0	1
1	1	1	1

EXAMPLE 5–19

Repeat Example 5–18 for the following equation:

$$X = \overline{A\overline{B} \cdot (A + C)} + \overline{\overline{A}B \cdot A} + \overline{B} + \overline{\overline{C}}$$

Solution: The required logic circuit is shown in Figure 5–57. The Boolean equation simplification is

$$X = \overline{A\overline{B} \cdot (A + C)} + \overline{\overline{A}B \cdot A} + \overline{B} + \overline{\overline{C}}$$

$$= \overline{A\overline{B}} + \overline{A + C} + \overline{\overline{A}B} \cdot (\overline{A} \cdot \overline{\overline{B}} \cdot \overline{\overline{C}})$$

$$= (\overline{A} + \overline{\overline{B}}) + \overline{A} \cdot \overline{C} + \overline{A}\,\overline{\overline{A}}BBC$$

$$= \overline{A} + B + \overline{A}\,\overline{C} + \overline{A}BC$$

$$= \overline{A}(1 + \overline{C}) + B + \overline{A}BC$$

$$= \overline{A} + B + \overline{A}BC$$

$$= \overline{A} + B(1 + \overline{A}C)$$

$$= \overline{A} + B \quad \leftarrow simplified\ equation$$

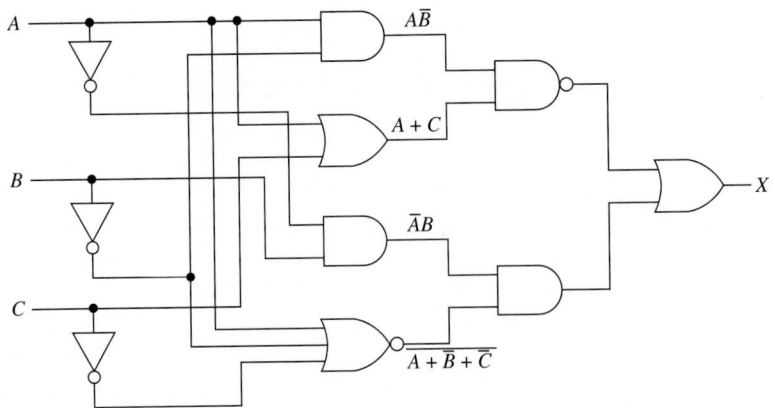

Figure 5–57 Logic circuit for the equation of Example 5–19.

TABLE 5-4			Truth Table for Example 5–19
A	**B**	**C**	$X = \overline{A} + B$
0	0	0	1
0	0	1	1
0	1	0	1
0	1	1	1
1	0	0	0
1	0	1	0
1	1	0	1
1	1	1	1

$X = 1$ if $A = 0$ or $B = 1$

Three columns are used in the truth table (Table 5–4) because the original equation contained three variables (*A*, *B*, *C*). *C* is considered a **don't care,** however, because it does not appear in the final equation and it does not matter whether it is 1 or 0.

From the simplified equation ($X = \overline{A} + B$), we can determine that $X = 1$ when *A* is 0 or when *B* is 1, and we fill in the truth table accordingly.

EXAMPLE 5–20

Complete the truth table and timing diagram for the following simplified Boolean equation:

$$X = AB + B\overline{C} + \overline{A}\,\overline{B}C$$

Solution: The required truth table and timing diagram are shown in Figure 5–58. To fill in the truth table for *X*, we first put a 1 for *X* when $A = 1$, $B = 1$. Then $X = 1$ for $B = 1$, $C = 0$. Then $X = 1$ for $A = 0$, $B = 0$, $C = 1$. All other entries for *X* are 0.

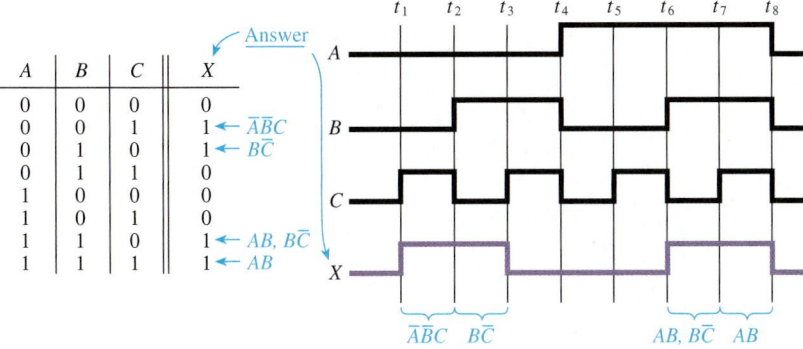

A	B	C	X
0	0	0	0
0	0	1	1 ← $\overline{A}\overline{B}C$
0	1	0	1 ← $B\overline{C}$
0	1	1	0
1	0	0	0
1	0	1	0
1	1	0	1 ← $AB, B\overline{C}$
1	1	1	1 ← AB

Figure 5–58   Truth table and timing diagram depicting the logic levels at X for all combinations of inputs.

The timing diagram performs the same function as the truth table, except it is a more graphic illustration of the HIGH and LOW logic levels of X as the A, B, and C inputs change over time. The logic levels at X are filled in the same way as they were for the truth table.

EXAMPLE 5–21

Repeat Example 5–20 for the following simplified equation:

$$X = A\overline{B}\,\overline{C} + \overline{A}\,\overline{B}C + ABC$$

Solution: The required truth table and timing diagram are shown in Figure 5–59.

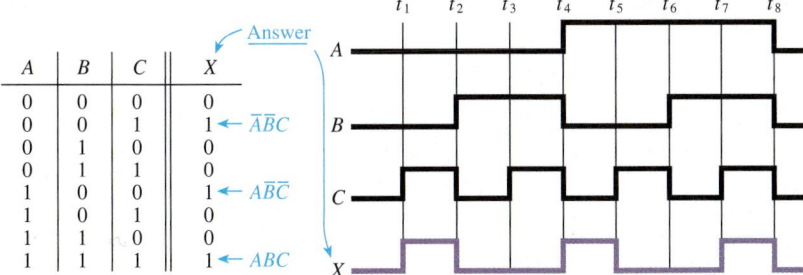

A	B	C	X
0	0	0	0
0	0	1	1 ← $\overline{A}\overline{B}C$
0	1	0	0
0	1	1	0
1	0	0	1 ← $A\overline{B}\overline{C}$
1	0	1	0
1	1	0	0
1	1	1	1 ← ABC

Figure 5–59   Truth table and timing diagram depicting the logic levels at X for all combinations of inputs.

Bubble Pushing

A shortcut method of forming equivalent logic circuits, based on De Morgan's theorem, is called **bubble pushing** and is illustrated in Figure 5–60. As you can see, to form the equivalent logic circuit, you must

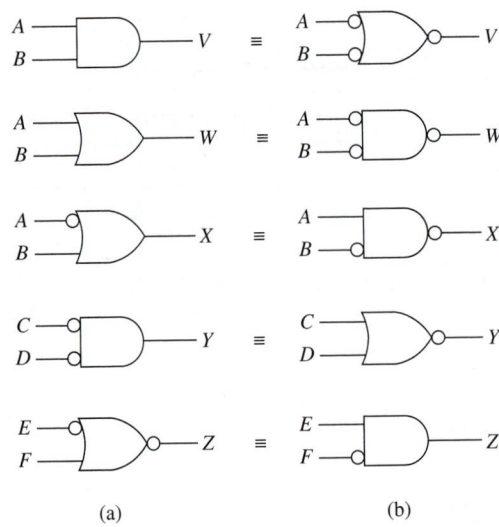

Figure 5–60 (a) Original logic circuits; (b) equivalent logic circuits.

1. Change the logic gate (AND to OR or OR to AND).

2. Add bubbles to the inputs and outputs where there were none, and remove the original bubbles.

Prove to yourself that this method works by comparing the truth table of each original circuit to its equivalent.

Notice in Figure 5–60 that we have equivalent logic circuits for the AND and OR gates (V and W). It is worth pointing out here that you will be seeing these two equivalents often when studying data memory ICs and microprocessor circuitry (Chapters 16 and 17). Figure 5–61 shows part of the gating circuitry that is often used to access microprocessor memory. Microprocessor control signals are usually **active-LOW,** meaning that they issue a LOW when they want to perform their specified task. *Also,* for the microprocessor to activate the block labeled Memory, the line labeled $\overline{MA}$ (memory access) must be made LOW. (The overbars on the variables signify that they are active-LOW.)

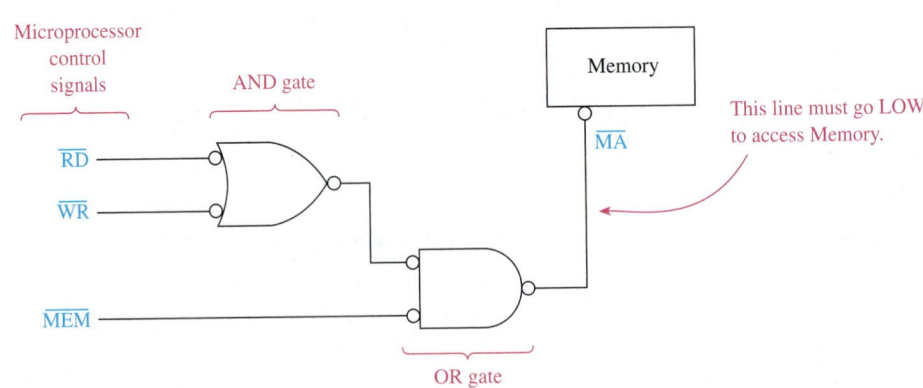

Figure 5–61 Typical gating circuitry used for microprocessor memory access.

The gating shown in Figure 5–61 will provide the LOW at $\overline{MA}$ if $\overline{MEM}$ is LOW *and* either $\overline{WR}$ is LOW *or* $\overline{RD}$ is LOW. The control signals from the microprocessor meet these conditions whenever the microprocessor is reading ($\overline{RD}$) or writing ($\overline{WR}$) from memory ($\overline{MEM}$). For example, if the microprocessor is to read from memory, it will make the $\overline{RD}$ line go LOW to signify that it wants to read, and it will make the $\overline{MEM}$ line go LOW to signify that it wants to read its information from memory. With these two lines LOW, MA is LOW, which activates the block labeled Memory. (When working with circuitry like this, it is better not to think of the bubbles as inverters; instead, think of that line as a part of the circuit that requires a LOW to "do its thing" or satisfy that input).

The OR gate with three bubbles outputs a LOW if either input is LOW. This symbol makes the logic easy to understand, but to actually implement the circuit, its equivalent (the 7408 AND gate) would be used. Also, the AND gate with three bubbles would actually be an OR gate (the 7432).

Review Questions

5–7. Why is De Morgan's theorem important in the simplification of Boolean equations?

5–8. Using De Morgan's theorem, you can prove that a NOR gate is equivalent to an _____ (OR, AND) gate with inverted inputs.

5–9. Using the bubble-pushing technique, an AND gate with one of its inputs inverted is equivalent to a _____ (NAND, NOR) gate with its other input inverted.

5–10. Using bubble pushing to convert an inverted-input OR gate will yield a(n) _____ (AND, NAND) gate.

5–6   Entering a Truth Table in VHDL Using a Vector Signal

Suppose we wanted to implement the logic for the truth table shown in Table 5–5. One method would be to write the Boolean equation for X by listing each combination of ABC that produces a HIGH at X, then simplify the equation and build the logic circuit. We could also write the equation for X as a VHLD architecture statement, and let the software synthesize it in a CPLD. However, in this section we will use techniques that employ several new concepts important to VHDL programmers.

TABLE 5–5	Truth Table to be Entered Using a Vector Data Type as an Internal Signal		
Inputs			**Output**
A	*B*	*C*	*X*
0	0	0	1
0	0	1	0
0	1	0	1
0	1	1	0
1	0	0	1
1	0	1	1
1	1	0	1
1	1	1	0
↑	↑	↑	
input(2)	input(1)	input(0)	

The first thing that we need to do is to define an internal **signal** to represent the three inputs. This internal signal will group the three inputs together as a 3-bit **vector.** Let's call this new internal vector signal "input." The following signal declare is placed within the architecture body, *just before* the BEGIN statement:

SIGNAL input: STD_LOGIC_VECTOR(2 downto 0);

This vector signal named *input* is similar to an array with three elements called input(2), input(1), and input(0). The specification **(2 downto 0)** defines three elements starting with element (2), then (1), then (0). To assign values to the three elements, the following assignment statements are placed *just after* the BEGIN statement:

input(2)<= a; --Move *a* to element 2 of the internal vector signal

input(1)<= b; --Move *b* to element 1 of the internal vector signal

input(0)<= c; --Move *c* to element 0 of the internal vector signal

[*Note:* The text following the double hyphen (--) is treated as a comment by VHDL. Comments are ignored by the VHDL compiler but are very useful for documenting our programs so that when you look at the program listing three months from now, you'll have a little help remembering why you did something the way you did.]

The final step is to assign the desired outputs for *X* for each input combination. We do this with the **Selected Signal Assignment** as follows:

WITH input SELECT

x<= '1' WHEN "000", -- x equals 1 when input equals "000"

'0' WHEN "001", -- x equals 0 when input equals "001"

'1' WHEN "010", -- x equals 1 when input equals "010"

'0' WHEN "011", -- x equals 0 when input equals "011"

'1' WHEN "100", -- x equals 1 when input equals "100"

'1' WHEN "101", -- x equals 1 when input equals "101"

'1' WHEN "110", -- x equals 1 when input equals "110"

'0' WHEN "111", -- x equals 0 when input equals "111"

'1' WHEN others;

The selected signal assignment is built to look just like the truth table entries. The last assignment uses the term *others*. This is required because the *std_logic* **type declaration** allows for many other bit states besides 1 and 0. [For example, a hyphen (-) can be used to specify "don't care" and a Z can be used to specify "High impedance (or Float)". A complete list of the *std_logic* data types is in Appendix E.] The "others" assignment will never be made because we will be inputting 1's and 0's to *a*, *b* and *c* but VHDL requires us to include it to cover all possibilities known to the language.

Also note that when making assignments, *single quotes* are used for making bit assignments and *double quotes* are used for making vector assignments. The complete VHDL program listing is shown in Figure 5–62.

An easy way to test the results of the program is to run a simulation and compare the waveforms with the original truth table. This is done in Figure 5–63.

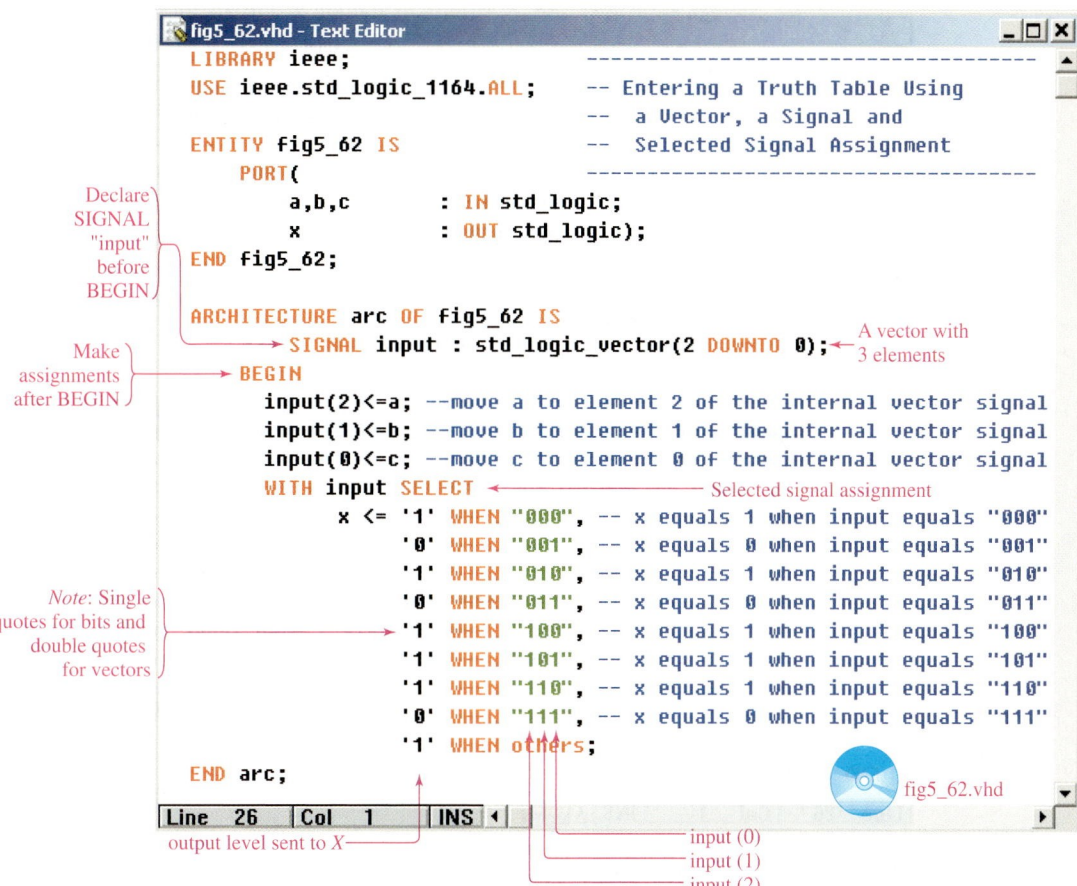

Declare SIGNAL "input" before BEGIN

Make assignments after BEGIN

Note: Single quotes for bits and double quotes for vectors

```
fig5_62.vhd - Text Editor                                                    _ □ ×
LIBRARY ieee;                        ------------------------------------
USE ieee.std_logic_1164.ALL;         -- Entering a Truth Table Using
                                     --   a Vector, a Signal and
ENTITY fig5_62 IS                    --   Selected Signal Assignment
    PORT(                            ------------------------------------
        a,b,c        : IN std_logic;
        x            : OUT std_logic);
END fig5_62;

ARCHITECTURE arc OF fig5_62 IS
        SIGNAL input : std_logic_vector(2 DOWNTO 0);    A vector with
BEGIN                                                   3 elements
    input(2)<=a; --move a to element 2 of the internal vector signal
    input(1)<=b; --move b to element 1 of the internal vector signal
    input(0)<=c; --move c to element 0 of the internal vector signal
    WITH input SELECT                    Selected signal assignment
            x <= '1' WHEN "000", -- x equals 1 when input equals "000"
                 '0' WHEN "001", -- x equals 0 when input equals "001"
                 '1' WHEN "010", -- x equals 1 when input equals "010"
                 '0' WHEN "011", -- x equals 0 when input equals "011"
                 '1' WHEN "100", -- x equals 1 when input equals "100"
                 '1' WHEN "101", -- x equals 1 when input equals "101"
                 '1' WHEN "110", -- x equals 1 when input equals "110"
                 '0' WHEN "111", -- x equals 0 when input equals "111"
                 '1' WHEN others;
END arc;                                                     fig5_62.vhd
Line  26   Col  1    INS
```

output level sent to *X*

input (0)
input (1)
input (2)

Figure 5–62 Program listing for entering a truth table in VHDL using a vector, a signal, and the selected signal assignment.

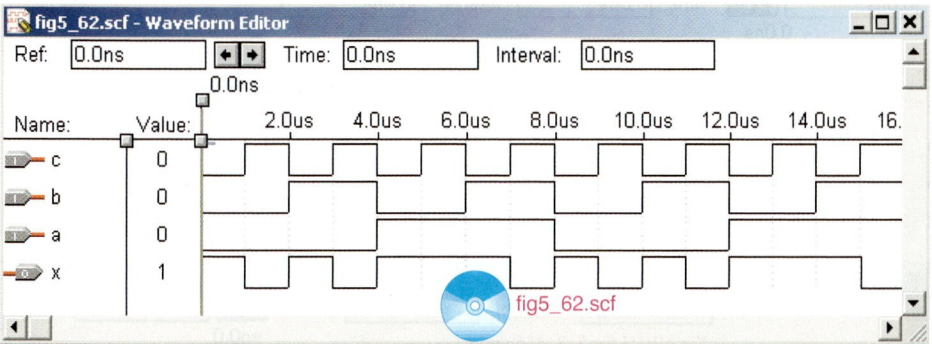

Figure 5–63 Waveform display used to check the simulation with the original truth table.

EXAMPLE 5–22

Design a logic circuit that can be used to tell when a 3-bit binary number is within the range of 2 (010_2) to 6 (110_2) inclusive. Use the VHDL selected signal assignment method discussed previously. Perform a simulation of your design by creating an scf file that steps through the entire range of input possibilities 000_2 to 111_2.

Solution: The VHDL program is shown in Figure 5–64 and the waveform simulation is shown in Figure 5–65.

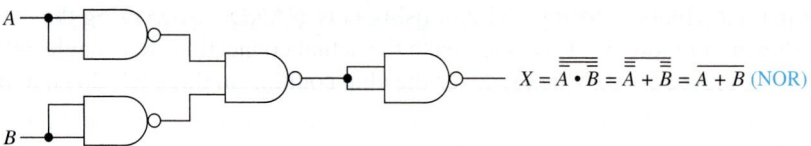

Figure 5–74   Forming a NOR from four NANDs.

The procedure for converting NOR gates into an inverter, OR, AND, or NAND is similar to the conversions just discussed for NAND gates. For example, to form an inverter from a NOR gate, just connect the inputs as shown in Figure 5–75.

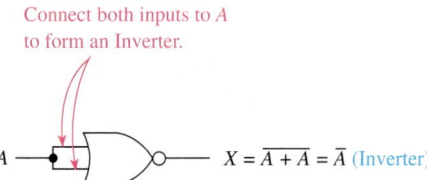

Figure 5–75   Forming an inverter from a NOR gate.

Take some time now to try to convert NORs to an OR, NORs to an AND, and NORs to a NAND. Prove to yourself that your solution is correct by using De Morgan's theorem and Boolean algebra.

EXAMPLE 5–24

Make the external connections to a 4001 CMOS NOR IC to implement the function $X = \overline{A} + B$.

Solution: We will need an inverter and an OR gate to provide the function for X. An inverter can be made from a NOR by connecting the inputs, and an OR can be made by inverting the output of a NOR, as shown in Figure 5–76.

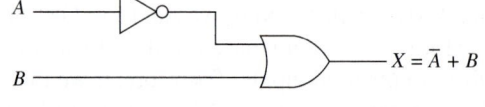

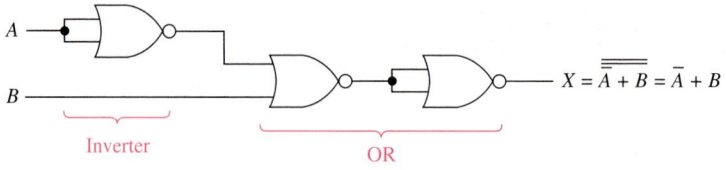

Figure 5–76   Implementing the function $X = \overline{A} + B$ using only NOR gates.

The pin configuration for the 4001 CMOS quad NOR can be found in a CMOS data book. Figure 5–77 shows the pin configuration and external connections to implement $X = \overline{A} + B$.

First, let's redraw the logic circuit using only NANDs. Now, using the configuration shown in Figure 5–71, we can make the actual connections to a single 7400 IC, as shown in Figure 5–72, which reduces the chip count from three ICs down to one.

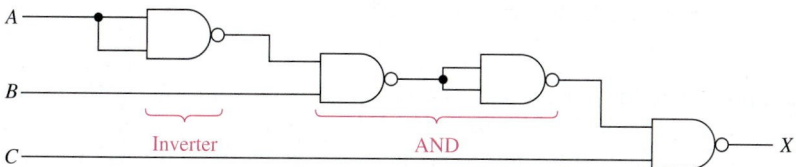

Figure 5–71 Equivalent logic circuit using only NANDs.

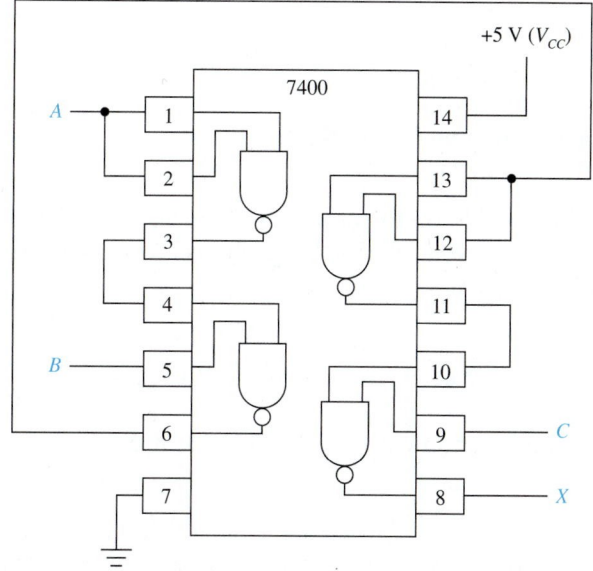

Figure 5–72 External connections to a 7400 TTL IC to form the circuit of Figure 5–71.

Besides forming inverters and ANDs from NANDs, we can form ORs and NORs from NANDs. Remember from De Morgan's theorem that an AND with an inverted output (NAND) is equivalent to an OR with inverted inputs. Therefore, if we invert the inputs to a NAND, we should find that it is equivalent to an OR, as shown in Figure 5–73.

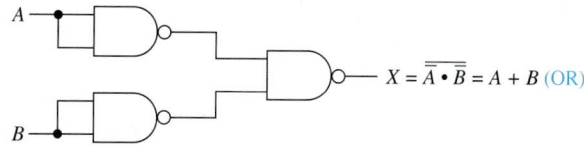

$$X = \overline{\overline{A} \cdot \overline{B}} = A + B \text{ (OR)}$$

Figure 5–73 Forming an OR from three NANDs.

Now, to form a NOR from NANDs, all we need to do is invert the output of Figure 5–73, as shown in Figure 5–74.

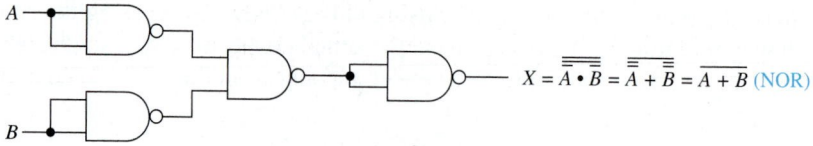

Figure 5–74   Forming a NOR from four NANDs.

The procedure for converting NOR gates into an inverter, OR, AND, or NAND is similar to the conversions just discussed for NAND gates. For example, to form an inverter from a NOR gate, just connect the inputs as shown in Figure 5–75.

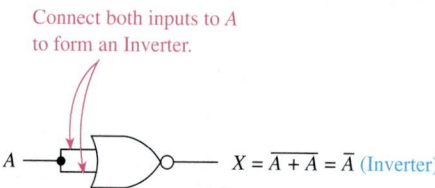

Figure 5–75   Forming an inverter from a NOR gate.

Take some time now to try to convert NORs to an OR, NORs to an AND, and NORs to a NAND. Prove to yourself that your solution is correct by using De Morgan's theorem and Boolean algebra.

EXAMPLE 5–24

Make the external connections to a 4001 CMOS NOR IC to implement the function $X = \overline{A} + B$.

Solution: We will need an inverter and an OR gate to provide the function for X. An inverter can be made from a NOR by connecting the inputs, and an OR can be made by inverting the output of a NOR, as shown in Figure 5–76.

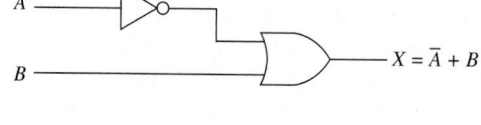

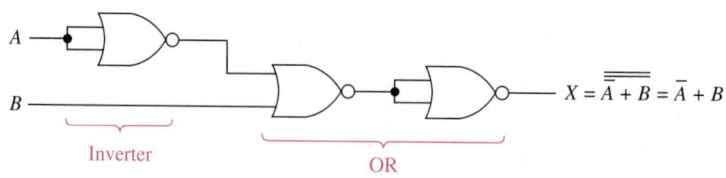

Figure 5–76   Implementing the function $X = \overline{A} + B$ using only NOR gates.

The pin configuration for the 4001 CMOS quad NOR can be found in a CMOS data book. Figure 5–77 shows the pin configuration and external connections to implement $X = \overline{A} + B$.

5–7   The Universal Capability of NAND and NOR Gates

NAND and NOR gates are sometimes referred to as **universal gates** because, by utilizing a combination of NANDs, all the other logic gates (inverter, AND, OR, NOR) can be formed. Also, by utilizing a combination of NORs, all the other logic gates (inverter, AND, OR, NAND) can be formed.

This principle is useful because you often may have extra NANDs available but actually need some other logic function. For example, let's say that you designed a circuit that required a NAND, an AND, and an inverter. You would probably purchase a 7400 quad NAND TTL IC. This chip has four NANDs in a single package. One of the NANDs will be used directly in your circuit. The AND requirement could actually be fulfilled by connecting the third and fourth NANDs on the chip to form an AND. The inverter can be formed from the second NAND on the chip. How do we convert a NAND into an inverter and two NANDs into an AND? Let's see.

An inverter can be formed from a NAND simply by connecting both NAND inputs, as shown in Figure 5–68. Both inputs to the NAND are, therefore, connected to A. The equation at X is $X = \overline{A \cdot A} = \overline{A}$, which is the inverter function.

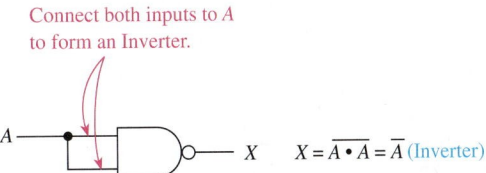

Figure 5–68   Forming an inverter from a NAND.

The next task is to form an AND from two NANDs. Do you have any ideas? What is the difference between a NAND and an AND? If we invert the output of a NAND, it will act like an AND, as shown in Figure 5–69.

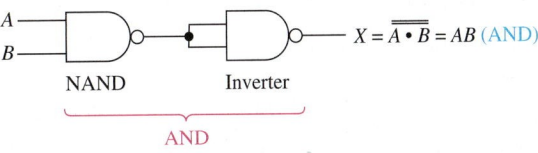

Figure 5–69   Forming an AND from two NANDs.

Now back to the original problem; we wanted to form a circuit requiring a NAND, an AND, and an inverter using a single 7400 quad NAND TTL IC. Let's make the external connections to the 7400 IC to form the circuit of Figure 5–70, which contains a NAND, an AND, and an inverter.

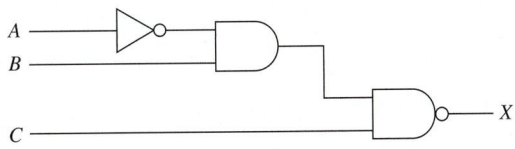

Figure 5–70   Logic circuit to be implemented using only NANDs.

Solution: The program listing is shown in Figure 5–66. The three tanks are grouped together as a vector instead of having three different variable names. This simplifies the program because now we don't have to define an internal vector signal and assign three variables to the signal like we did in Example 5–22.

The simulation in Figure 5–67 shows the alarm is HIGH whenever two or more tanks are HIGH. Notice, in the scf file, when listing the node names of a vector, you don't list the elements with parentheses like you do in VHDL. Instead, vector element tank(2) is named tank2, tank(1) is named tank1, and tank(0) is named tank0 as shown.

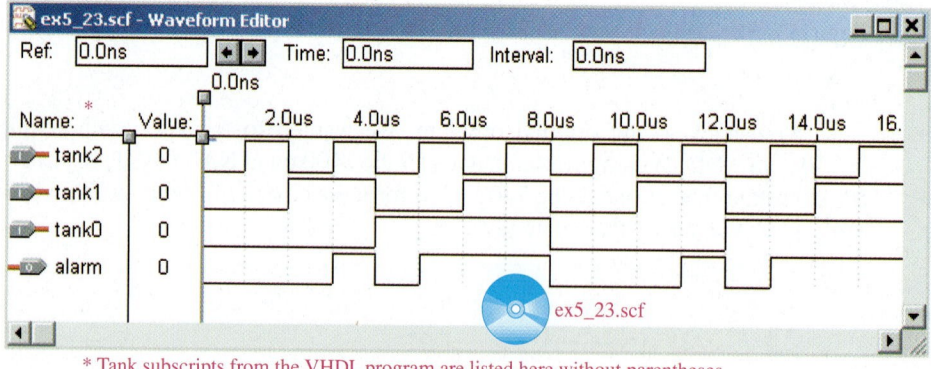

```
ex5_23.vhd - Text Editor                                    _ |□| x|
  LIBRARY ieee;                  ---------------------------
  USE ieee.std_logic_1164.ALL;   -- Chemical Tank Monitoring
                                 ---------------------------

  ENTITY ex5_23 IS
     PORT(
         tank               : IN std_logic_vector(2 downto 0);
         alarm              : OUT std_logic);
  END ex5_23;


  ARCHITECTURE arc OF ex5_23 IS          ┌──── tank (2)
     BEGIN                               ├──── tank (1)
        WITH tank SELECT                 └──── tank (0)
          alarm <= '0' WHEN "000", --------------------------------
                   '0' WHEN "001", -- alarm is HIGH for         --
                   '0' WHEN "010", --    any combination of      --
                   '1' WHEN "011", --    two or more tanks HIGH. --
                   '0' WHEN "100", --------------------------------
                   '1' WHEN "101",
                   '1' WHEN "110",
                   '1' WHEN "111",
                   '0' WHEN others;
  END arc;                              ◎  ex5_23.vhd
Line   22   Col   1    INS ◄ |                                   ► |
```

Figure 5–66 VHDL listing for Example 5–23.

```
ex5_23.scf - Waveform Editor                                _ |□| x|
 Ref: |0.0ns     | |◄|►|  Time: |0.0ns    |  Interval: |0.0ns    |
                  └ 0.0ns
 Name:  *    Value:    2.0us   4.0us   6.0us   8.0us  10.0us  12.0us  14.0us  16.
 ▥─ tank2      0
 ▥─ tank1      0
 ▥─ tank0      0
 ─◁▷ alarm     0
                               ◎  ex5_23.scf
◄| |                                                            |►|
```
* Tank subscripts from the VHDL program are listed here without parentheses.

Figure 5–67 Simulation file for Example 5–23.

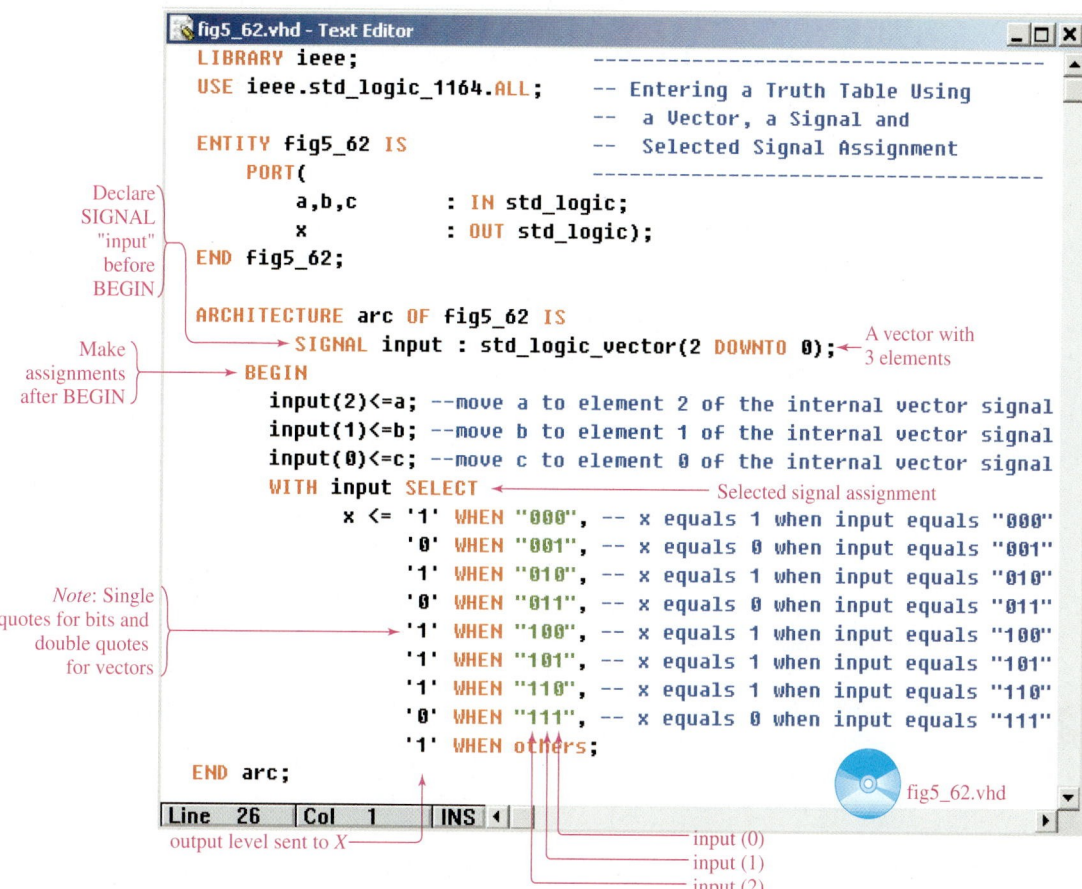

Figure 5–62 Program listing for entering a truth table in VHDL using a vector, a signal, and the selected signal assignment.

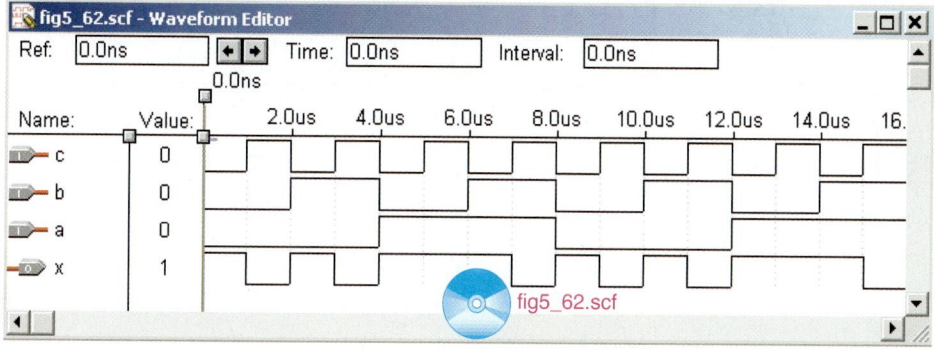

Figure 5–63 Waveform display used to check the simulation with the original truth table.

EXAMPLE 5–22

Design a logic circuit that can be used to tell when a 3-bit binary number is within the range of 2 (010_2) to 6 (110_2) inclusive. Use the VHDL selected signal assignment method discussed previously. Perform a simulation of your design by creating an scf file that steps through the entire range of input possibilities 000_2 to 111_2.

Solution: The VHDL program is shown in Figure 5–64 and the waveform simulation is shown in Figure 5–65.

```
ex5_22.vhd - Text Editor                                          _ □ ×

LIBRARY ieee;                      -------------------------------------
USE ieee.std_logic_1164.ALL; -- Using Vector, Signal and
                             --    Selected Signal Assignment

ENTITY ex5_22 IS                   -------------------------------------
    PORT(
        a,b,c        : IN std_logic;
        x            : OUT std_logic);
END ex5_22;


ARCHITECTURE arc OF ex5_22 IS
        SIGNAL input : std_logic_vector(2 DOWNTO 0);
    BEGIN
      input(2)<=a; --move a to element 2 of the internal vector signal
      input(1)<=b; --move b to element 1 of the internal vector signal
      input(0)<=c; --move c to element 0 of the internal vector signal
Make X  WITH input SELECT
HIGH for   x <= '0' WHEN "000", -- x equals 0 when input equals "000"
2, 3, 4, 5         '0' WHEN "001", -- x equals 0 when input equals "001"
and 6             '1' WHEN "010", -- x equals 1 when input equals "010"
                  '1' WHEN "011", -- x equals 1 when input equals "011"
                  '1' WHEN "100", -- x equals 1 when input equals "100"
                  '1' WHEN "101", -- x equals 1 when input equals "101"
                  '1' WHEN "110", -- x equals 1 when input equals "110"
                  '0' WHEN "111", -- x equals 0 when input equals "111"
                  '0' WHEN others;
    END arc;                                      ⊙ ex5_22.vhd

Line  26   Col   1     INS ◄
```

Figure 5–64   VHDL solution for Example 5–22.

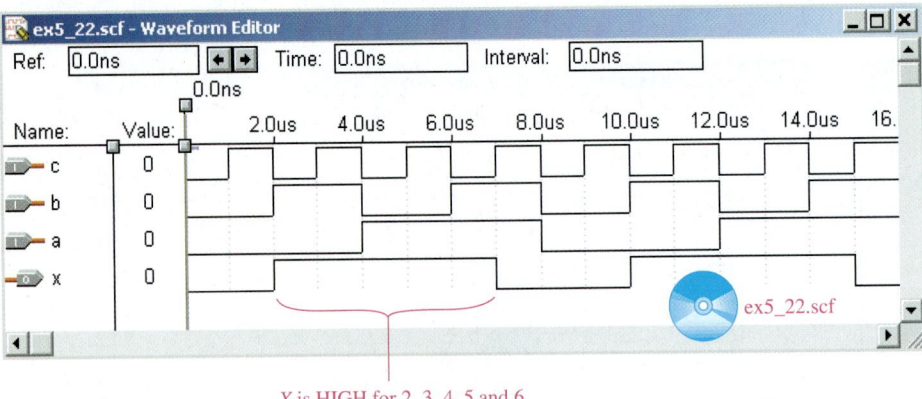

Figure 5–65   Waveform simulation of Example 5–22.

═══════════ **EXAMPLE 5–23** ═══════════

A chemical manufacturing facility needs to have a warning system to monitor its three chemical overflow holding tanks. Each tank has a HIGH/LOW level sensor. Design a system that activates a warning alarm whenever two or more tank levels are HIGH.

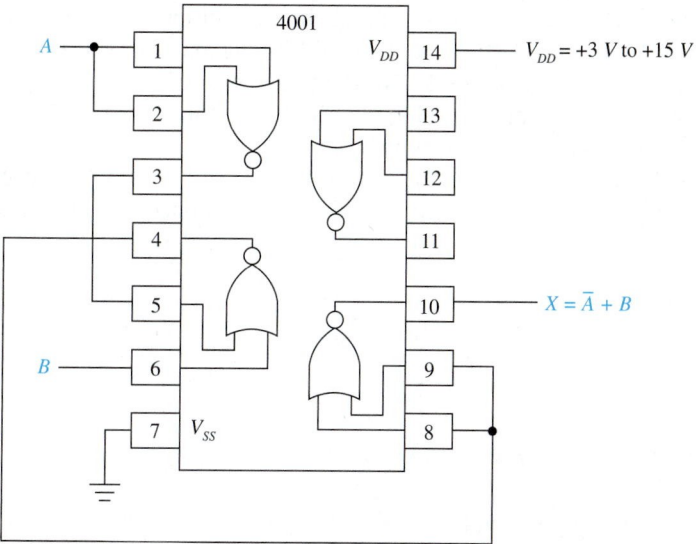

Figure 5–77 External connections to a 4001 CMOS IC to implement the circuit of Figure 5–76.

EXAMPLE 5–25

Troubleshooting

You have connected the circuit of Figure 5–77 and want to test it. Because the Boolean equation is $X = \overline{A} + B$, you first try $A = 0$, $B = 1$ and expect to get a 1 output at X, *but you don't.* V_{DD} is set to $+5$ V, and V_{SS} is connected to ground. Using a logic probe, you record the results shown in Table 5–6 at each pin. Determine the trouble with the circuit.

TABLE 5–6	Logic Probe Operation[a]
Probe on Pin	**Indicator Lamp**
1	Off
2	Off
3	On
4	Off
5	On
6	On
7	Off
8	Dim
9	Off
10	Off
11	On
12	Dim
13	Dim
14	On

[a]Lamp off, 0; lamp on, 1; lamp dim, float.

Solution: Because $A = 0$, pins 1 and 2 should both be 0, which they are. Pin 3 is a 1, because 0–0 into a NOR will produce a 1 output. Pin 6 is 1, because it is connected to the 1 at B. Pin 5 matches pin 3, as it is supposed to. Pin 4 sends a 0 to pins 8 and 9, but pin 8 is floating (not 0 or 1). That's it! The connection to pin 8 must be broken.

To be sure that the circuit operates properly, the problem at pin 8 should be corrected, and all four combinations of inputs at A and B should be tested.

EXAMPLE 5–26

(a) Write the simplified equation that will produce the output waveform at X, given the inputs at A, B, and C shown in Figure 5–78.

(b) Draw the logic circuit for this equation.

(c) Redraw the logic circuit using only NAND gates.

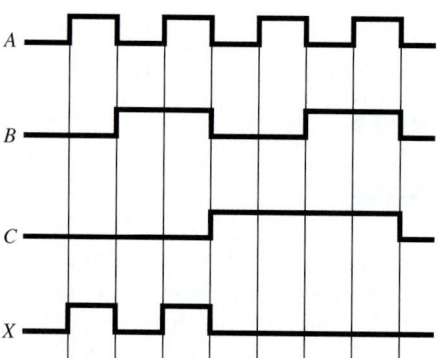

Figure 5–78

Solution: **(a)** The first HIGH pulse at X is produced for $A = 1$, $B = 0$, $C = 0$ $(A\overline{B}\,\overline{C})$. The second HIGH pulse at X happens when $A = 1$, $B = 1$, $C = 0$ $(AB\overline{C})$. Therefore, X is 1 for $A\overline{B}\,\overline{C}$ or $AB\overline{C}$.

$$X = A\overline{B}\,\overline{C} + AB\overline{C}$$

Simplifying yields

$$X = A\overline{C}(\overline{B} + B)$$
$$= A\overline{C}(1)$$
$$= A\overline{C} \;\leftarrow simplified\ equation$$

(b) The logic circuit is shown in Figure 5–79(a).

(c) Redrawing the same circuit using only NANDs produces the circuit shown in Figure 5–79(b).

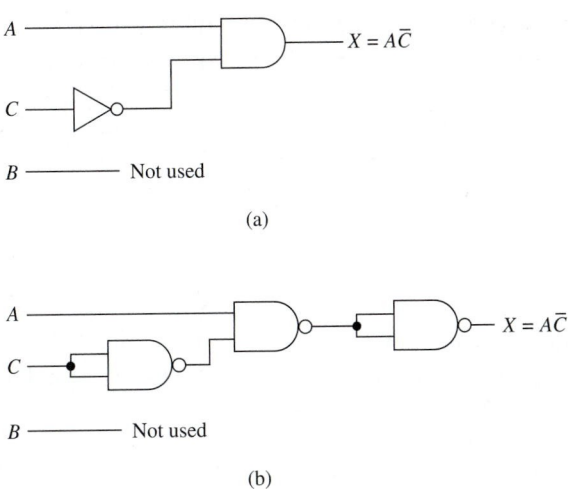

(a)

(b)

Figure 5–79 (a) Logic circuit that yields the waveform at X; (b) circuit of part (a) redrawn using only NANDs.

<hr/>

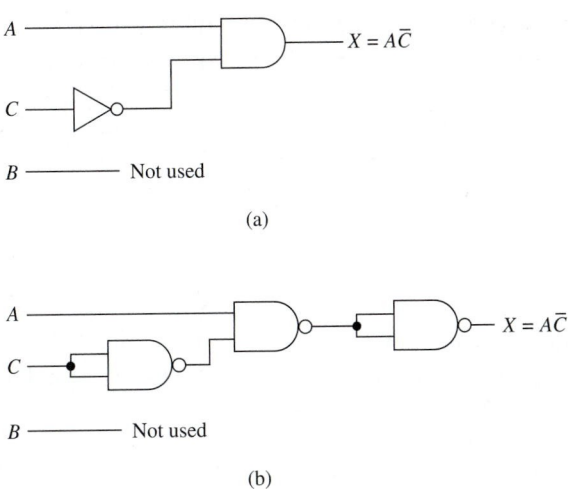

Review Questions

5–11. Why are NAND gates and NOR gates sometimes referred to as *universal* gates?

5–12. Why would a designer want to form an AND gate from two NAND gates?

5–13. How many inverters could be formed using a 7400 quad NAND IC?

5–8 AND–OR–INVERT Gates for Implementing Sum-of-Products Expressions

Most Boolean reductions result in an equation in one of two forms:

1. **Product-of-sums (POS) form**

2. **Sum-of-products (SOP) form**

The POS expression usually takes the form of two or more ORed variables within parentheses ANDed with two or more other variables within parentheses. Examples of POS expressions are

$$X = (A + \bar{B}) \cdot (B + C)$$

$$X = (B + \bar{C} + \bar{D}) \cdot (BC + \bar{E})$$

$$X = (A + \bar{C}) \cdot (\bar{B} + E) \cdot (C + B)$$

The SOP expression usually takes the form of two or more variables ANDed together ORed with two or more other variables ANDed together. Examples of SOP expressions are

$$X = A\overline{B} + AC + \overline{A}BC$$

$$X = AC\overline{D} + \overline{C}D + B$$

$$X = B\overline{C}\overline{D} + A\overline{B}DE + CD$$

The SOP expression is used most often because it lends itself nicely to the development of truth tables and timing diagrams. SOP circuits can also be constructed easily using a special combinational logic gate called the **AND–OR–INVERT gate.**

For example, let's work with the equation

$$X = \overline{A\overline{B} + \overline{C}D}$$

Using De Morgan's theorem yields

$$X = \overline{A\overline{B}} \cdot \overline{\overline{C}D}$$

Using De Morgan's theorem again puts it into a POS format:

$$X = (\overline{A} + B) \cdot (C + \overline{D}) \quad \leftarrow \text{POS}$$

Using the distributive law produces an equation in the SOP format:

$$X = \overline{A}C + \overline{A}\,\overline{D} + BC + B\overline{D} \quad \leftarrow \text{SOP}$$

Now, let's fill in a truth table for X (Table 5–7). Using the SOP expression, we put a 1 at X for $A = 0$, $C = 1$; for $A = 0$, $D = 0$; for $B = 1$, $C = 1$; and for $B = 1$, $D = 0$. That wasn't hard, was it?

However, if we were to use the POS expression, it would be more difficult to visualize. We would put a 1 at X for $A = 0$ or $B = 1$ whenever $C = 1$ or $D = 0$. Confusing? Yes, it is much more difficult to deal intuitively with POS expressions.

TABLE 5–7	Truth Table Completed Using the SOP Expression			
A	*B*	*C*	*D*	*X*
0	0	0	0	1
0	0	0	1	0
0	0	1	0	1
0	0	1	1	1
0	1	0	0	1
0	1	0	1	0
0	1	1	0	1
0	1	1	1	1
1	0	0	0	0
1	0	0	1	0
1	0	1	0	0
1	0	1	1	0
1	1	0	0	1
1	1	0	1	0
1	1	1	0	1
1	1	1	1	1

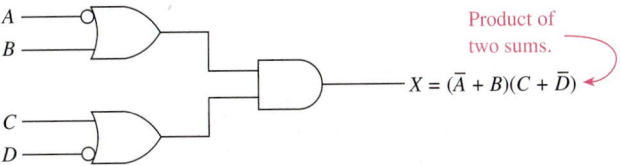

Figure 5–80 Logic circuit for the POS expression.

Drawing the logic circuit for the POS expression involves using OR gates feeding into an AND gate, as shown in Figure 5–80. Drawing the logic circuit for the SOP expression involves using AND gates feeding into an OR gate, as shown in Figure 5–81. The logic circuit for the SOP expression used more gates for this particular example, but the SOP form *is* easier to deal with and, besides, there is an IC gate specifically made to simplify the implementation of SOP circuits.

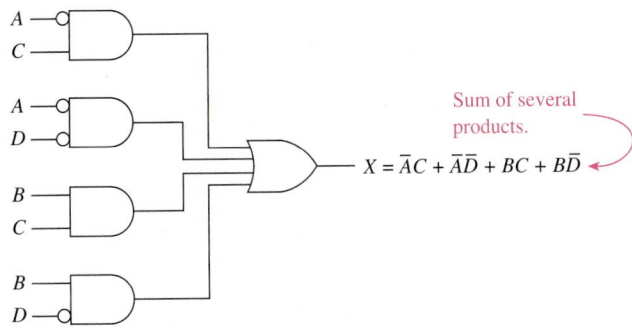

Figure 5–81 Logic circuit for the SOP expression.

That gate is the *AND–OR–INVERT* (AOI). AOIs are available in several different configurations within the TTL or CMOS families. Skim through your TTL and CMOS data books to identify some of the available AOIs. One AOI that is particularly well suited for implementing the logic of Figure 5–81 is the 74LS54 TTL IC. The pin configuration and logic symbol for the 74LS54 are shown in Figure 5–82.

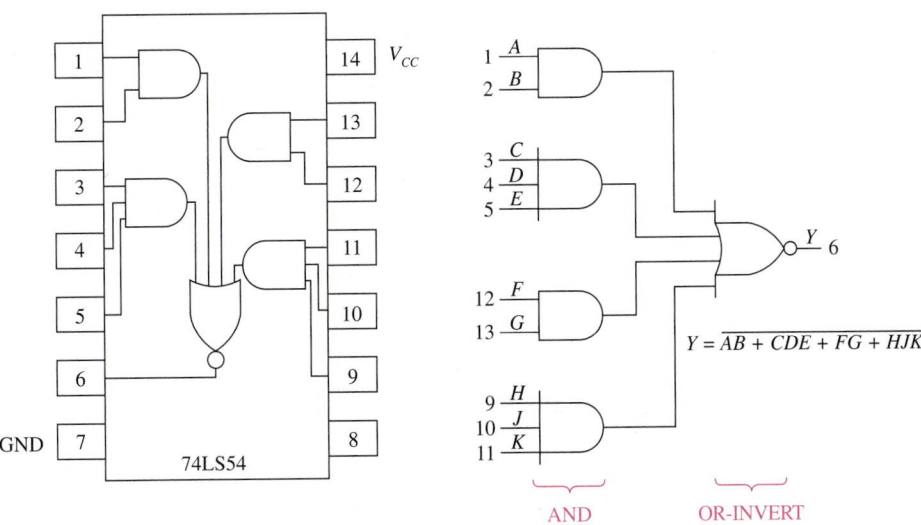

Figure 5–82 Pin configuration and logic symbol for the 74LS54 AOI gate.

Notice that the output at Y is inverted, so we have to place an inverter after Y. Also, two of the AND gates have *three* inputs instead of just the two-input gates that we need, so we just connect the unused third input to a 1. Figure 5–83 shows the required connections to the AOI to implement the SOP logic circuit of Figure 5–81. Omitting the inverter from Figure 5–83 would provide an active-LOW output function, which may be acceptable, depending on the operation required. (The new equation would be $\overline{X} = \overline{A}C + \overline{A}\,\overline{D} + BC + B\overline{D}$.)

Common Misconception

Students often forget the inverter, which makes the output active-LOW. The equations so far have been active-HIGH, but in later chapters, you will see why active-LOW is so common.

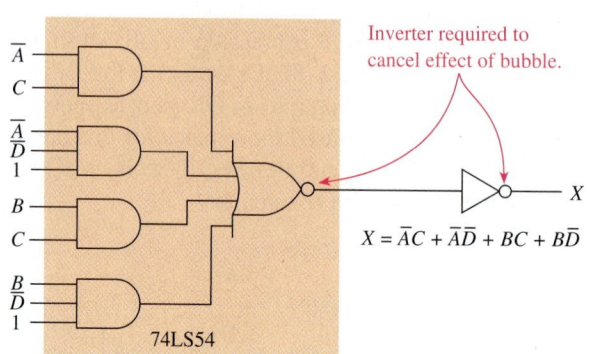

Figure 5–83 Using an AOI IC to implement an SOP equation.

EXAMPLE 5–27

Simplify the circuit shown in Figure 5–84 down to its SOP form, then draw the logic circuit of the simplified form using a 74LS54 AOI gate.

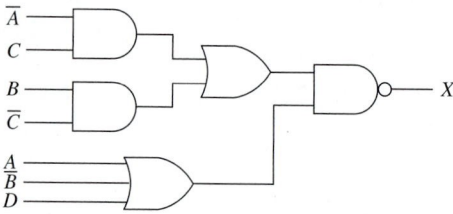

Figure 5–84 Original circuit for Example 5–27.

Solution:

$$X = \overline{(\overline{A}C + B\overline{C}) \cdot (A + \overline{B} + D)}$$

$$= \overline{\overline{A}C + B\overline{C}} + \overline{A + \overline{B} + D}$$

$$= \overline{\overline{A}C} \cdot \overline{B\overline{C}} + \overline{A}B\overline{D}$$

$$= (A + \overline{C})(\overline{B} + C) + \overline{A}B\overline{D}$$

$$= A\overline{B} + AC + \overline{B}\,\overline{C} + \overline{C}C + \overline{A}B\overline{D}$$

$$= A\overline{B} + AC + \overline{B}\,\overline{C} + \overline{A}B\overline{D} \quad \leftarrow \text{SOP}$$

The simplified circuit is shown in Figure 5–85.

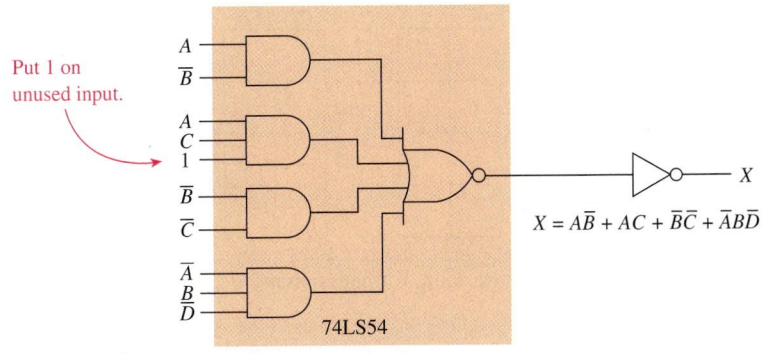

Team Discussion

What other options are available instead of inputting a 1 to the second AND gate?

Team Discussion

How could you create the AND–OR logic function using 5 NAND gates? (*Hint:* Use bubble pushing.)

Figure 5–85 Using an AOI IC to implement the simplified SOP equation for Example 5–27.

In the figure, inputs to the 74LS54 AOI IC are labeled:
- A, $\overline{B}$ (with "Put 1 on unused input." annotation)
- A, C, 1
- $\overline{B}$, $\overline{C}$
- $\overline{A}$, B, $\overline{D}$

$$X = A\overline{B} + AC + \overline{B}\,\overline{C} + \overline{A}B\overline{D}$$

▬▬▬ **Review Questions** ▬▬▬

5–14. Which form of Boolean equation is better suited for completing truth tables and timing diagrams, SOP or POS?

5–15. AOI ICs are used to implement _____ (SOP, POS) expressions.

5–16. The equation $X = AB + BCD + DE$ has only three product terms. If a 74LS54 AOI IC is used to implement the equation, what must be done with the three inputs to the unused fourth AND gate?

5–9 Karnaugh Mapping

We learned in previous sections that by using Boolean algebra and De Morgan's theorem, we can minimize the number of gates that are required to implement a particular logic function. This is very important for the reduction of circuit cost, physical size, and gate failures. You may have found that some of the steps in the Boolean reduction process require ingenuity on your part and a lot of practice.

Karnaugh mapping, named for its originator, Maurice Karnaugh who in 1953 developed another method of simplifying logic circuits. It still requires that you reduce the equation to an SOP form, but from there, you follow a *systematic approach,* which will always produce the simplest configuration possible for the logic circuit.

A **Karnaugh map** (K-map) is similar to a truth table in that it graphically shows the output level of a Boolean equation for each of the possible input variable combinations. Each output level is placed in a separate **cell** of the K-map. K-maps can be used to simplify equations having two, three, four, five, or six different input variables. Solving five- and six-variable K-maps is extremely cumbersome; they can be more practically solved using advanced computer techniques. In this book, we will solve two-, three-, and four-variable K-maps.

Determining the number of cells in a K-map is the same as finding the number of combinations or entries in a truth table. A two-variable map requires $2^2 = 4$ cells. A three-variable map requires $2^3 = 8$ cells. A four-variable map requires $2^4 = 16$ cells. The three different K-maps are shown in Figure 5–86.

Each cell within the K-map corresponds to a particular combination of the input variables. For example, in the two-variable K-map, the upper left cell corresponds to $\overline{A}\,\overline{B}$, the lower left cell is $A\overline{B}$, the upper right cell is $\overline{A}B$, and the lower right cell is AB.

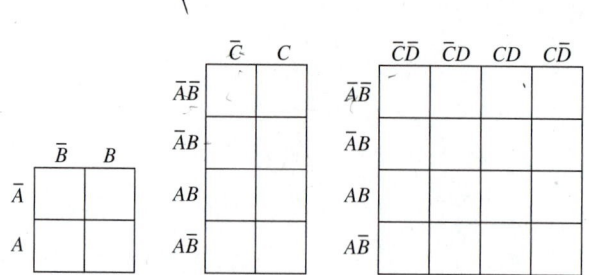

Figure 5–86   Two-, three-, and four-variable Karnaugh maps.

Also notice that when moving from one cell to an **adjacent cell,** only one variable changes. For example, look at the three-variable K-map. The upper left cell is $\overline{A}\,\overline{B}\,\overline{C}$; the adjacent cell just below it is $\overline{A}B\overline{C}$. In this case, the $\overline{A}\,\overline{C}$ remained the same and only the $\overline{B}$ changed, to B. The same holds true for each adjacent cell.

To use the K-map reduction procedure, you must perform the following steps:

1. Transform the Boolean equation to be reduced into an SOP expression.

2. Fill in the appropriate cells of the K-map.

3. Encircle adjacent cells in groups of two, four, or eight. (The more adjacent cells encircled, the simpler the final equation is; adjacent means a side is touching, not diagonal.)

4. Find each term of the final SOP equation by determining which variables remain constant within each circle.

Now, let's consider the equation

$$X = \overline{A}(\overline{B}C + \overline{B}\,\overline{C}) + \overline{A}B\overline{C}$$

First, transform the equation to an SOP expression:

$$X = \overline{A}\,\overline{B}C + \overline{A}\,\overline{B}\,\overline{C} + \overline{A}B\overline{C}$$

The terms of that SOP expression can be put into a truth table and then transferred to a K-map, as shown in Figure 5–87. Working with the K-map, we now encircle adjacent 1's in groups of two, four, or eight. We end up with two circles of two cells each, as shown in Figure 5–88. The first circle surrounds the two 1's at the top of the K-map, and the second circle surrounds the two 1's in the left column of the K-map.

Once the circles have been drawn encompassing all the 1's in the map, the final simplified equation is obtained by determining *which variables remain the same within each circle.* Well, the first circle (across the top) encompasses $\overline{A}\,\overline{B}\,\overline{C}$ and $\overline{A}\,\overline{B}C$. The variables that remain the same within the circle are $\overline{A}\,\overline{B}$. Therefore, $\overline{A}\,\overline{B}$ becomes one of the terms in the final SOP equation. The second circle (left column) encompasses $\overline{A}\,\overline{B}\,\overline{C}$ and $\overline{A}B\overline{C}$. The variables that remain the same within that circle are $\overline{A}\,\overline{C}$. Therefore, the second term in the final equation is $\overline{A}\,\overline{C}$.

A	B	C	X
0	0	0	1
0	0	1	1
0	1	0	1
0	1	1	0
1	0	0	0
1	0	1	0
1	1	0	0
1	1	1	0

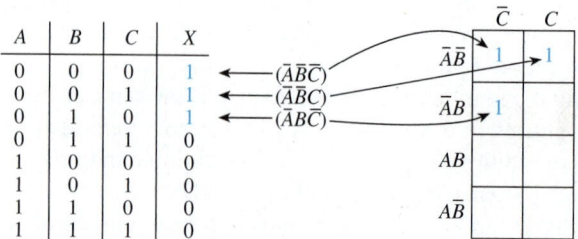

Figure 5–87   Truth table and Karnaugh map of $X = \overline{A}\,\overline{B}\,\overline{C} + \overline{A}\,\overline{B}C + \overline{A}B\overline{C}$.

CHAPTER 5 | BOOLEAN ALGEBRA AND REDUCTION TECHNIQUES

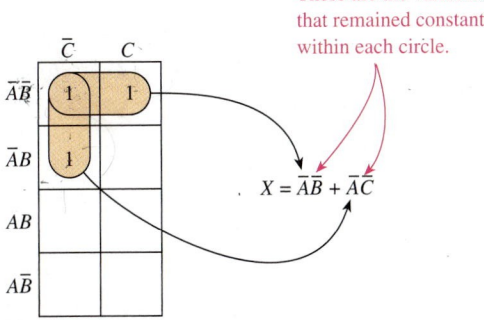

Figure 5–88 Encircling adjacent cells in a Karnaugh map.

Because the final equation is always written in the SOP format, the answer is $X = \overline{A}\,\overline{B} + \overline{A}\,\overline{C}$. Actually, the original equation was simple enough that we could have reduced it using standard Boolean algebra. Let's do it just to check our answer:

$$X = \overline{A}\,\overline{B}C + \overline{A}\,\overline{B}\,\overline{C} + \overline{A}B\overline{C}$$
$$= \overline{A}\,\overline{B}(C + \overline{C}) + \overline{A}B\overline{C}$$
$$= \overline{A}\,\overline{B} + \overline{A}B\overline{C}$$
$$= \overline{A}(\overline{B} + B\overline{C})$$
$$= \overline{A}(\overline{B} + \overline{C})$$
$$= \overline{A}\,\overline{B} + \overline{A}\,\overline{C} \;\surd$$

There are several other points to watch out for when applying the Karnaugh mapping technique. The following examples will be used to illustrate several important points in filling in the map, determining adjacencies, and obtaining the final equation. Work through these examples carefully so that you do not miss any special techniques.

EXAMPLE 5–28

Simplify the following SOP equation using the Karnaugh mapping technique:

$$X = \overline{A}B + \overline{A}\,\overline{B}\,\overline{C} + \overline{A}B\overline{C} + \overline{A}\overline{B}\,\overline{C}$$

Solution:
1. Construct an eight-cell K-map (see Figure 5–89), and fill in a 1 in each cell that corresponds to a term in the original equation. (Notice that $\overline{A}B$ has no C variable in it. Therefore, $\overline{A}B$ is satisfied whether C is HIGH or LOW, so $\overline{A}B$ will fill in two cells: $\overline{A}BC + \overline{A}B\overline{C}$.)

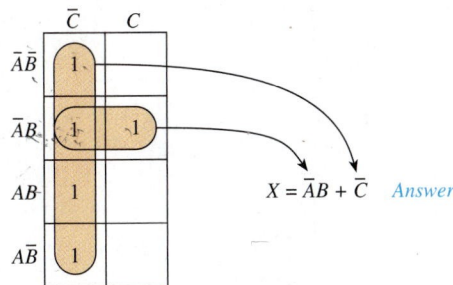

Figure 5–89 Karnaugh map and final equation for Example 5–28.

2. Encircle adjacent cells in the largest group of two or four or eight.

3. Identify the variables that remain the same within each circle, and write the final simplified SOP equation by ORing them together.

EXAMPLE 5–29

Simplify the following equation using the Karnaugh mapping procedure:

$$X = \overline{A}B\overline{C}D + A\overline{B}\,\overline{C}D + \overline{A}\,\overline{B}\,\overline{C}D + AB\overline{C}D + ABC\,\overline{D} + ABCD$$

Solution: Because there are four different variables in the equation, we need a 16-cell map ($2^4 = 16$), as shown in Figure 5–90.

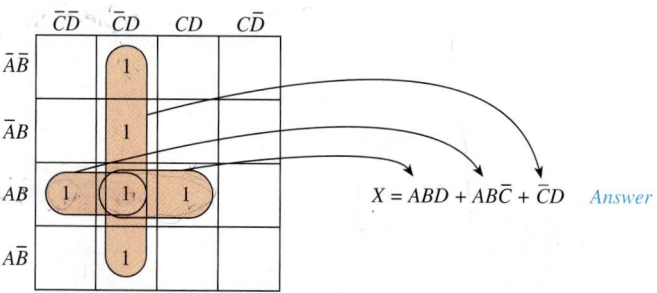

$$X = ABD + AB\overline{C} + \overline{C}D \quad \textit{Answer}$$

Figure 5–90 Solution to Example 5–29.

EXAMPLE 5–30

Simplify the following equation using the Karnaugh mapping procedure:

$$X = B\overline{C}\,\overline{D} + \overline{A}B\overline{C}D + AB\overline{C}D + \overline{A}BCD + ABCD$$

Solution: Notice in Figure 5–91 that the $B\overline{C}\,\overline{D}$ term in the original equation fills in *two* cells: $AB\overline{C}\,\overline{D} + \overline{A}BC\,\overline{D}$. Also notice in Figure 5–91 that we could have encircled four cells and then two cells, but that would not have given us the simplest final equation. By encircling four cells and then another four cells, we are sure to get the simplest final equation. (Always encircle the largest number of cells possible, even if some of the cells have already been encircled in another group.)

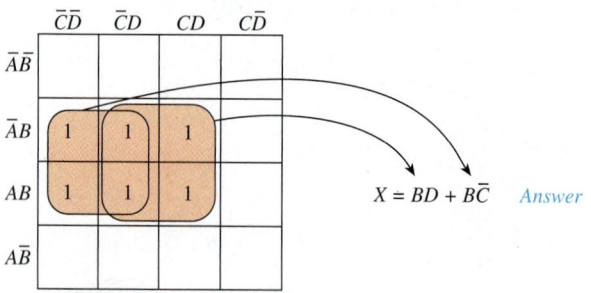

$$X = BD + B\overline{C} \quad \textit{Answer}$$

Figure 5–91 Solution to Example 5–30.

CHAPTER 5 | BOOLEAN ALGEBRA AND REDUCTION TECHNIQUES

EXAMPLE 5–31

Simplify the following equation using the Karnaugh mapping procedure:

$$X = \overline{A}\,\overline{B}\,\overline{C} + A\overline{C}\,\overline{D} + A\overline{B} + ABC\overline{D} + \overline{A}\,\overline{B}C$$

Solution: Notice in Figure 5–92 that a new technique called **wraparound** is introduced. You have to think of the K-map as a continuous cylinder in the horizontal direction, like the label on a soup can. This makes the left row of cells adjacent to the right row of cells. Also, in the vertical direction, a continuous cylinder like a soup can lying on its side makes the top row of cells adjacent to the bottom row of cells. In Figure 5–92, for example, the four top cells are adjacent to the four bottom cells, to combine as eight cells having the variable $\overline{B}$ in common.

Another circle of four is formed by the wraparound adjacencies of the lower left and lower right pairs combining to have $A\overline{D}$ in common. The final equation becomes $X = \overline{B} + A\overline{D}$. Compare that simple equation with the original equation that had five terms in it.

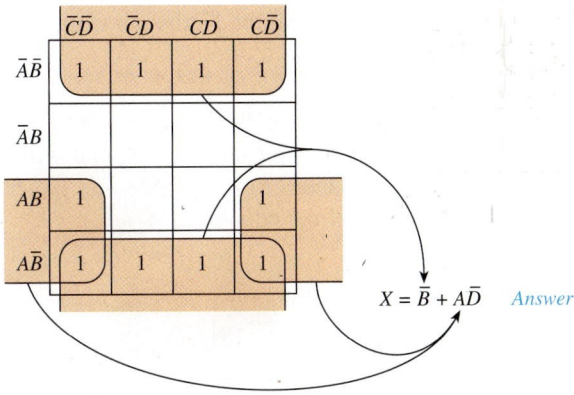

$X = \overline{B} + A\overline{D}$ *Answer*

Figure 5–92   Solution to Example 5–31 illustrating the wraparound feature.

EXAMPLE 5–32

Simplify the following equation using the Karnaugh mapping procedure:

$$X = \overline{B}(CD + \overline{C}) + C\overline{D}(\overline{\overline{A} + B} + AB)$$

Solution: Before filling in the K-map, an SOP expression must be formed:

$$X = \overline{B}CD + \overline{B}\,\overline{C} + C\overline{D}(\overline{A}\,\overline{B} + AB)$$
$$= \overline{B}CD + \overline{B}\,\overline{C} + \overline{A}\,\overline{B}C\overline{D} + ABC\overline{D}$$

The group of four 1's can be encircled to form $\overline{A}\,\overline{B}$, as shown in Figure 5–93. Another group of four can be encircled using wraparound to form $\overline{B}\,\overline{C}$. That leaves two 1's that are not combined with any others. The unattached 1 in the bottom row can be combined within a group of four, as shown, to form $\overline{B}D$.

The last 1 is not adjacent to any other, so it must be encircled by itself to form $ABC\overline{D}$. The final simplified equation is

$$X = \overline{A}\,\overline{B} + \overline{B}\,\overline{C} + \overline{B}D + ABC\overline{D}$$

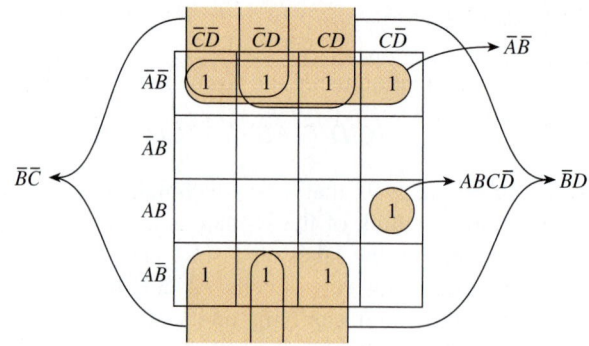

Figure 5–93 Solution to Example 5–32.

EXAMPLE 5–33

Simplify the following equation using the Karnaugh mapping procedure:

$$X = \overline{A}\,\overline{D} + A\overline{B}\,\overline{D} + \overline{A}\,\overline{C}D + \overline{A}CD$$

Solution: First, the group of eight cells can be encircled, as shown in Figure 5–94. $\overline{A}$ is the only variable present in each cell within the circle, so the circle of eight simply reduces to $\overline{A}$. (Notice that larger circles will reduce to fewer variables in the final equation.)

Team Discussion

What is the final equation of a map that has *all* cells filled in?

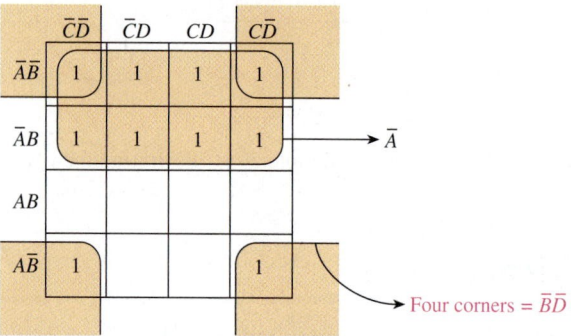

Figure 5–94 Solution to Example 5–33.

Also, all four corners are adjacent to each other because the K-map can be wrapped around in both the vertical *and* horizontal directions. Encircling the four corners results in $\overline{B}\,\overline{D}$. The final equation is

$$X = \overline{A} + \overline{B}\,\overline{D}$$

EXAMPLE 5–34

Simplify the following equation using the Karnaugh mapping procedure:

$$X = \overline{A}\,\overline{B}\,\overline{D} + A\overline{C}\,\overline{D} + \overline{A}B\overline{C} + AB\overline{C}D + A\overline{B}C\overline{D}$$

Solution: Encircling the four corners forms $\overline{B}\,\overline{D}$, as shown in Figure 5–95. The other group of four forms $B\overline{C}$. You may be tempted to encircle the $\overline{C}\,\overline{D}$ group of four as shown by the dotted line, but that would be a **redundancy** because each of those 1's is already contained within an existing circle. Therefore, the final equation is

$$X = \overline{B}\,\overline{D} + B\overline{C}$$

Team Discussion

So what's wrong with being redundant?

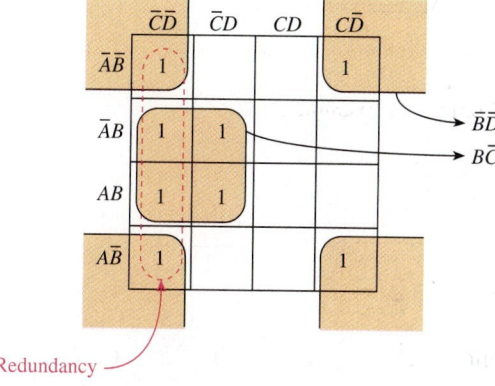

Figure 5–95   Solution to Example 5–34.

5–10   System Design Applications

Let's summarize the entire chapter now by working through two complete design problems. The following examples illustrate practical applications of a K-map to ensure that when we implement the circuit using an AOI, we will have the simplest possible solution.

SYSTEM DESIGN 5–1

Design a circuit that can be built using an AOI and inverters that will output a HIGH (1) whenever the 4-bit hexadecimal input is an odd number from 0 to 9.

Team Discussion

The LSB (variable A) is always HIGH for an odd number. Why can't we just say "odd number = A"?

TABLE 5–8	Hex Truth Table Used to Determine the Equation for Odd Numbers[a] from 0 to 9			
D	**C**	**B**	**A**	**DEC**
0	0	0	0	0
0	0	0	1	1 ← $\overline{A}\overline{B}\,\overline{C}\,\overline{D}$
0	0	1	0	2
0	0	1	1	3 ← $AB\overline{C}\,\overline{D}$
0	1	0	0	4
0	1	0	1	5 ← $\overline{A}BC\overline{D}$
0	1	1	0	6
0	1	1	1	7 ← $ABC\overline{D}$
1	0	0	0	8
1	0	0	1	9 ← $A\overline{B}\,\overline{C}D$

[a]Odd number $= A\overline{B}\,\overline{C}\,\overline{D} + AB\overline{C}\,\overline{D} + \overline{A}BC\overline{D} + ABC\overline{D} + A\overline{B}\,CD$.

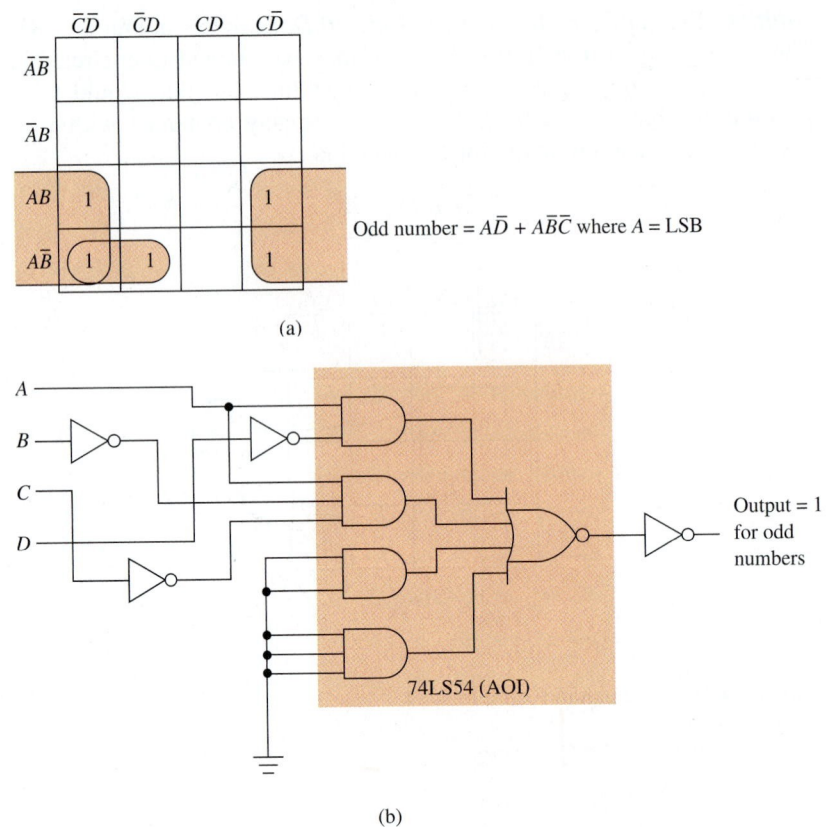

(a)

(b)

Figure 5–96   (a) Simplified equation derived from a Karnaugh map; (b) implementation of the odd-number decoder using an AOI.

Solution: First, build a truth table (Table 5–8) to identify which hex codes from 0 to 9 produce odd numbers. (Use the variable A to represent the 2^0 hex input, B for 2^1, C for 2^2, and D for 2^3.) Next, reduce this equation into its simplest form by using a Karnaugh map, as shown in Figure 5–96(a). Finally, using an AOI with inverters, the circuit can be constructed as shown in Figure 5–96(b).

SYSTEM DESIGN 5–2

A chemical plant needs a microprocessor-driven alarm system to warn of critical conditions in one of its chemical tanks. The tank has four HIGH/LOW (1/0) switches that monitor temperature (T), pressure (P), fluid level (L), and weight (W). Design a system that will notify the microprocessor to activate an alarm when any of the following conditions arise:

1. High fluid level with high temperature and high pressure
2. Low fluid level with high temperature and high weight
3. Low fluid level with low temperature and high pressure
4. Low fluid level with low weight and high temperature

Solution: First, write in Boolean equation form the conditions that will activate the alarm:

$$alarm = LTP + \overline{L}TW + \overline{L}\,\overline{T}P + \overline{L}WT$$

Next, factor the equation into its simplest form by using a Karnaugh map, as shown in Figure 5–97(a). Finally, using an AOI with inverters, the circuit can be constructed as shown in Figure 5–97(b).

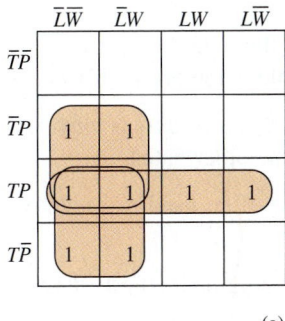

Alarm = $TP + P\overline{L} + T\overline{L}$

(a)

Team Discussion

By rereading conditions 2 and 4, can you logically explain why the weight is irrelevant and doesn't appear in the final equation?

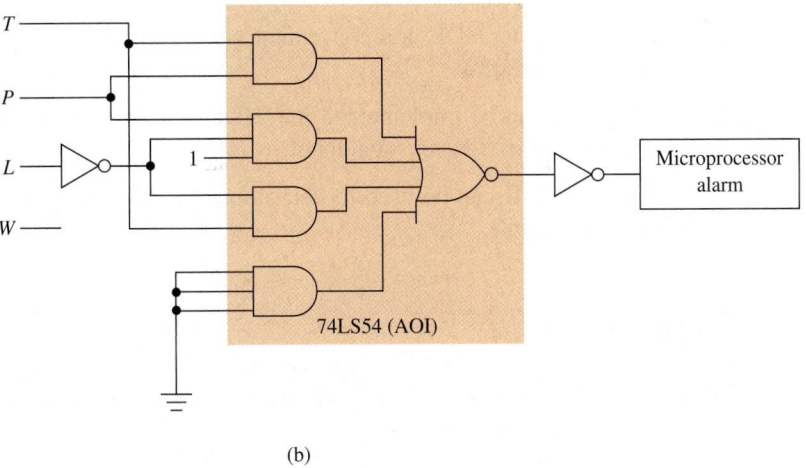

(b)

Figure 5–97 (a) Simplified equation derived from a Karnaugh map; (b) implementation of the chemical tank alarm using an AOI.

Review Questions

5–17. The number of cells in a Karnaugh map is equal to the number of entries in a corresponding truth table. True or false?

5–18. The order in which you label the rows and columns of a Karnaugh map does not matter as long as every combination of variables is used. True or false?

5–19. Adjacent cells in a Karnaugh map are encircled in groups of 2, 4, 6, or 8. True or false?

5–20. Which method of encircling eight adjacent cells in a Karnaugh map produces the simplest equation; two groups of four, or one group of eight?

Summary

In this chapter, we have learned that

1. Several logic gates can be connected together to form combinational logic.

2. There are several Boolean laws and rules that provide the means to form equivalent circuits.

3. Boolean algebra is used to reduce logic circuits to simpler equivalent circuits that function identically to the original circuit.

4. De Morgan's theorem is required in the reduction process whenever inversion bars cover more than one variable in the original Boolean equation.

5. NAND and NOR gates are sometimes referred to as *universal gates* because they can be used to form any of the other gates.

6. AND–OR–INVERT (AOI) gates are often used to implement sum-of-products (SOP) equations.

7. Karnaugh mapping provides a systematic method of reducing logic circuits.

8. Combinational logic designs can be entered into a computer using graphic design software or VHDL.

9. Using *vectors* in VHDL is a convenient way to group like signals together similar to an array.

10. Truth tables can be implemented in VHDL using vector signals with the *selected signal assignment* statement.

11. MAX+PLUS II can be used to determine the simplified equation of combinational circuits.

Glossary

Active-LOW: An output of a logic circuit that is LOW when activated, or an input that needs to be LOW to be activated.

Adjacent Cell: Cells within a Karnaugh map that border each other on one side or the top or bottom of the cell.

AND–OR–INVERT (AOI) Gate: An integrated circuit containing combinational logic consisting of several AND gates feeding into an OR gate and then an inverter. It is used to implement logic equations in the SOP format.

Boolean Reduction: An algebraic technique that follows specific rules to convert a Boolean equation into a simpler form.

Bubble Pushing: A shortcut method of forming equivalent circuits based on De Morgan's theorem.

Cell: Each box within a Karnaugh map. Each cell corresponds to a particular combination of input variable logic levels.

Combinational Logic: Logic circuits formed by combining several of the basic logic gates to form a more complex function.

De Morgan's Theorem: A Boolean law used for equation reduction that allows the user to convert an equation having an inversion bar over several variables into an equivalent equation having inversion bars over single variables only.

Don't Care: A variable appearing in a truth table or timing waveform that will have no effect on the final output regardless of the logic level of the variable. Therefore, don't-care variables can be ignored.

Equivalent Circuit: A simplified version of a logic circuit that can be used to perform the exact logic function of the original complex circuit.

Floorplan Editor Display: A MAX+PLUS II display that is used to view and modify the layout and configuration of a CPLD.

Inversion Bubbles: The bubble (or circle) can appear at the input or output of a logic gate. It indicates inversion (1 becomes 0; 0 becomes 1).

Karnaugh Map: A two-dimensional table of Boolean output levels used as a tool to perform a systematic reduction of complex logic circuits into simplified equivalent circuits.

Logic Array Block (LAB): Several logic cells put together as a group. The Altera EPM7128SLC CPLD has 8 LABs, each containing 16 logic cells.

Logic Cell: Also known as a macrocell, which is an array of AND-OR logic and I/O registers.

Product-of-Sums (POS) Form: A Boolean equation in the form of a group of ORed variables ANDed with another group of ORed variables [e.g., $X = (A + \overline{B} + C)(B + D)(\overline{A} + \overline{C})$].

Redundancy: Once all filled-in cells in a Karnaugh map are contained within a circle, the final simplified equation can be written. Drawing another circle around a different group of cells is needless (redundant).

Selected Signal Assignment: A VHDL statement that executes specific assignments based on the value of the specified signal used in the statement.

Sum-of-Products (SOP) Form: A Boolean equation in the form of a group of ANDed variables ORed with another group of ANDed variables (e.g., $X = ABC + \overline{B}DE + \overline{A}\,\overline{D}$).

Universal Gates: The NOR and NAND logic gates are sometimes called universal gates because any of the other logic gates can be formed from them.

Vector: A grouping of like signals similar to an array.

Wraparound: The left and right cells and the top and bottom cells of a Karnaugh map are actually adjacent to each other by means of the wraparound feature.

Section 5–1

5–1. Write the Boolean equation for each of the logic circuits shown in Figure P5–1.

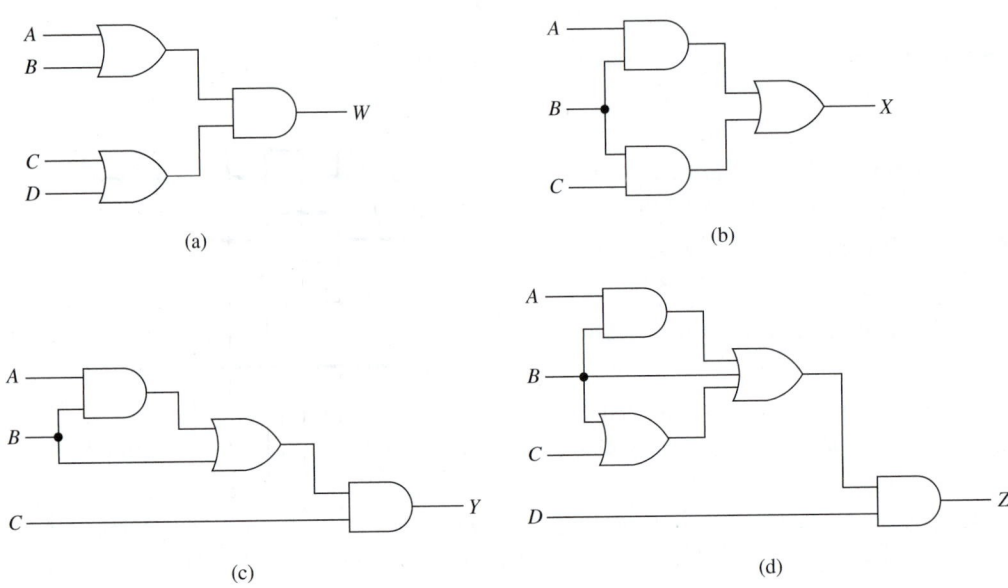

(a)

(b)

(c)

(d)

Figure P5–1

Section 5–2

5–2. Draw the logic circuit that would be used to implement the following Boolean equations.

(a) $M = (AB) + (C + D)$

(b) $N = (A + B + C)D$

(c) $P = (AC + BC)(A + C)$

(d) $Q = (A + B)BCD$

(e) $R = BC + D + AD$

(f) $S = B(A + C) + AC + D$

5–3. Construct a truth table for each of the equations given in Problem 5–2. (*Hint:* Where applicable, apply Law 3 to the equation first. *Do not* simplify the equation for this problem.)

5–4. Write the Boolean equation and then complete the timing diagram at W, X, Y, and Z for the logic circuits shown in Figure P5–4.

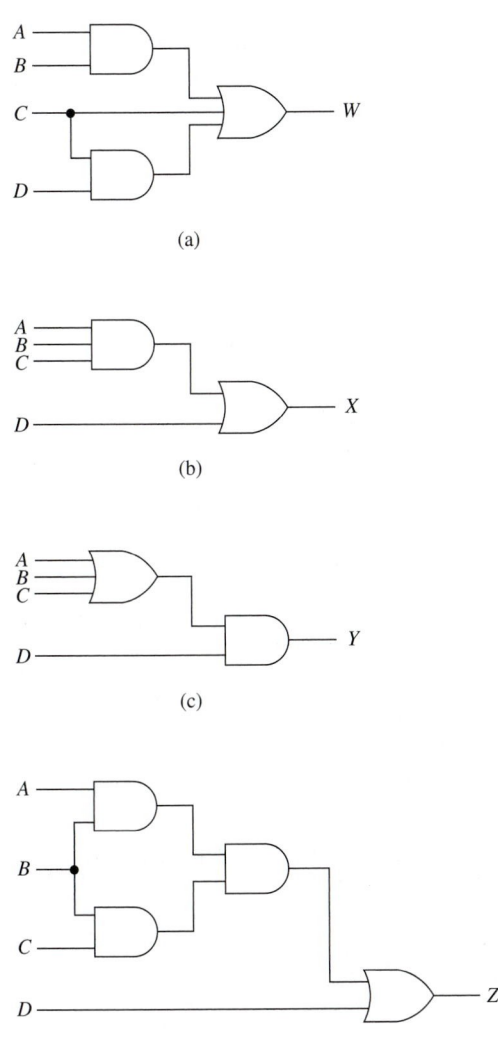

(a)

(b)

(c)

(d)

Figure P5–4

5–5. State the Boolean law that makes each of the equivalent circuits shown in Figure P5–5 valid.

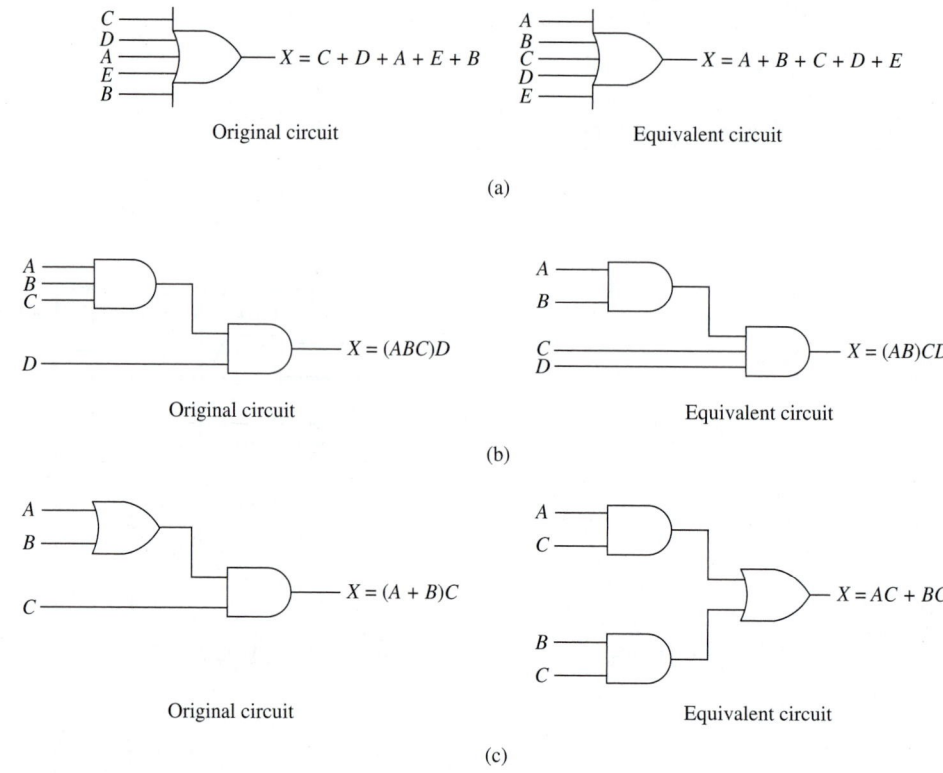

Figure P5–5

5–6. Using the 10 Boolean rules presented in Table 5–2, determine the outputs of the logic circuits shown in Figure P5–6.

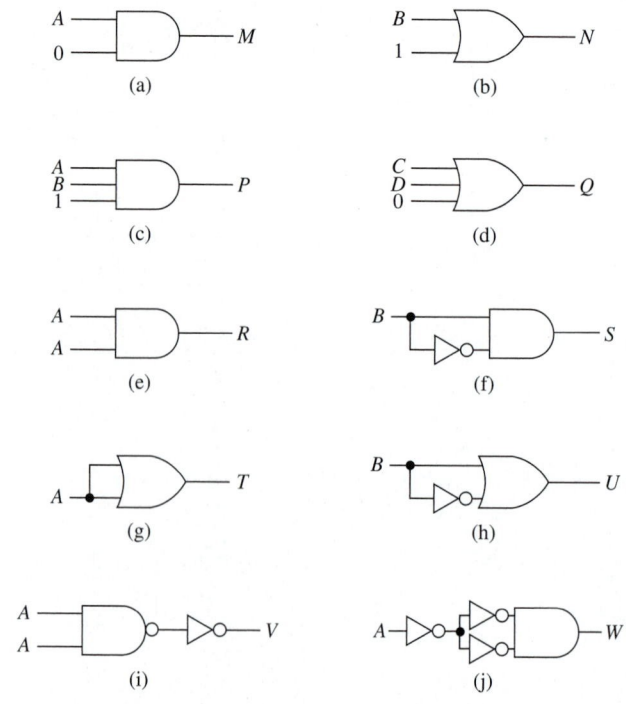

Figure P5–6

Section 5–3

5–7. Write the Boolean equation for the circuits of Figure P5–7. Simplify the equations, and draw the simplified logic circuit.

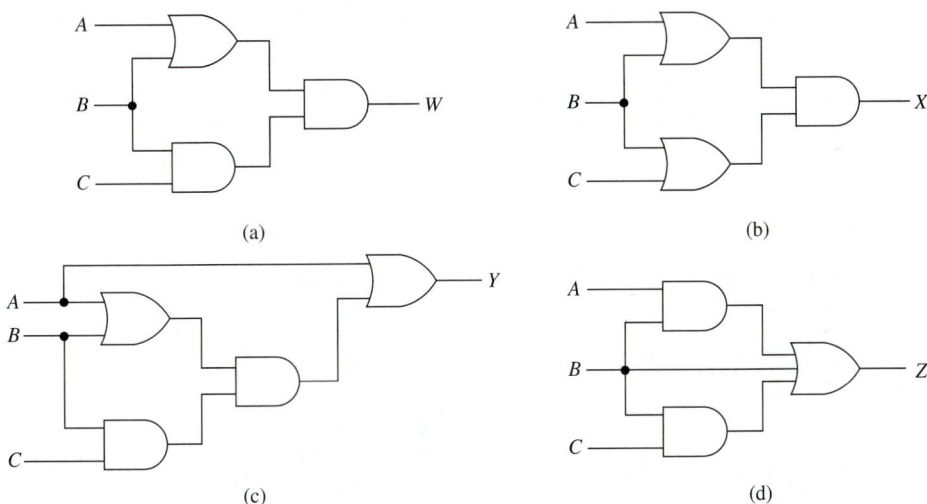

(a)

(b)

(c)

(d)

Figure P5–7

5–8. Repeat Problem 5–7 for the circuits shown in Figure P5–8.

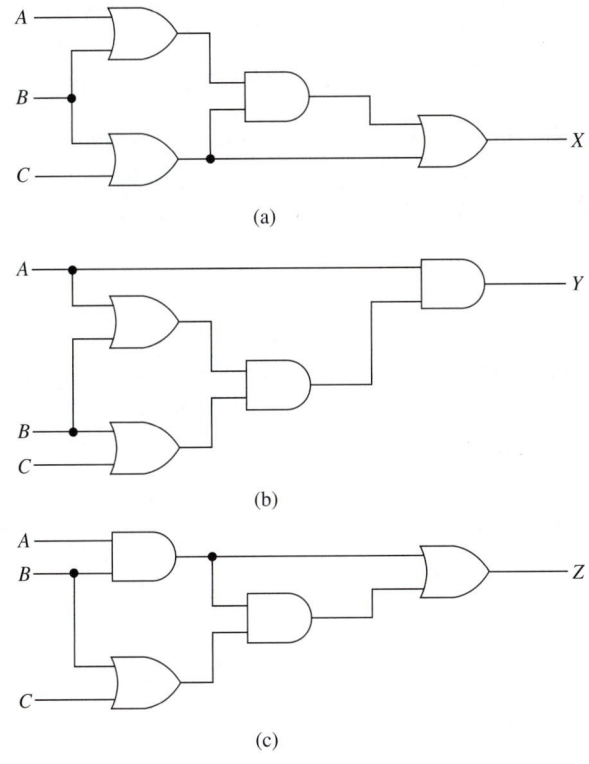

(a)

(b)

(c)

Figure P5–8

5–9. Draw the logic circuit for the following equations. Simplify the equations, and draw the simplified logic circuit.

(a) $V = AC + ACD + CD$

(b) $W = (BCD + C)CD$

(c) $X = (B + D)(A + C) + ABD$

(d) $Y = AB + BC + ABC$

(e) $Z = ABC + CD + CDE$

5–10. Construct a truth table for each of the simplified equations of Problem 5–9.

5–11. The pin layouts for a 74HCT08 CMOS AND gate and a 74HCT32 CMOS OR gate are given in Figure P5–11. Make the external connections to the chips to implement the following logic equation. (Simplify the logic equation first.)

$$X = (A + B)(D + C) + ABD$$

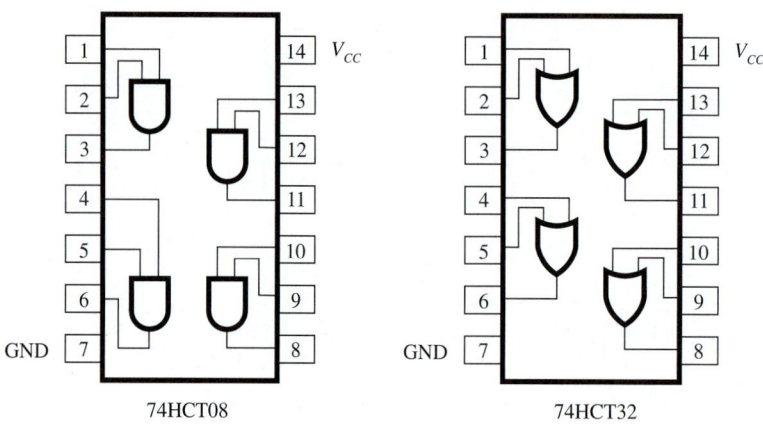

Figure P5–11

5–12. Repeat Problem 5–11 for the following equation.

$$Y = AB(C + BD) + BD$$

Section 5–4

5–13. Write a sentence describing how De Morgan's theorem is applied in the simplification of a logic equation.

5–14. (a) De Morgan's theorem can be used to prove that an OR gate with inverted inputs is equivalent to what type of gate?

(b) An AND gate with inverted inputs is equivalent to what type of gate?

5–15. Which two circuits in Figure P5–15 produce equivalent output equations?

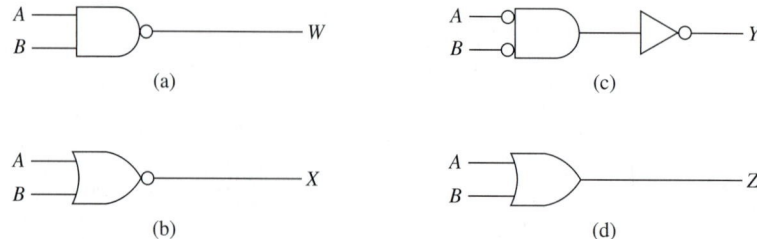

Figure P5–15

5–16. Use De Morgan's theorem to prove that a NOR gate with inverted inputs is equivalent to an AND gate.

5–17. Draw the logic circuit for the following equations. Apply De Morgan's theorem and Boolean algebra rules to reduce them to equations having inversion bars over single variables only. Draw the simplified circuit.

(a) $W = \overline{\overline{AB} + \overline{A}} + C$

(b) $X = \overline{A\overline{B} + C} + \overline{BC}$

(c) $Y = \overline{(AB) + C} + B\overline{C}$

(d) $Z = \overline{AB + (\overline{A} + C)}$

5–18. Write the Boolean equation for the circuits of Figure P5–18. Use De Morgan's theorem and Boolean algebra rules to simplify the equation. Draw the simplified circuit.

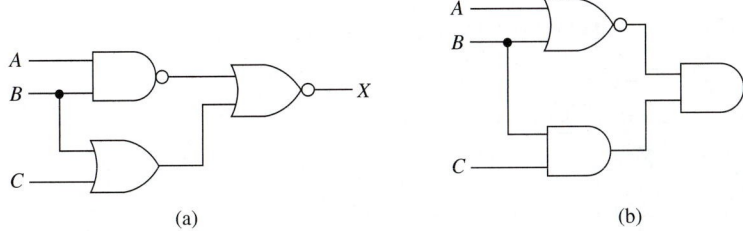

(a)   (b)

Figure P5–18

5–19. Repeat Problem 5–17 for the following equations.

(a) $W = \overline{\overline{\overline{AB} + CD}} + \overline{AC\overline{D}}$

(b) $X = \overline{\overline{\overline{A} + B} \cdot BC} + \overline{BC}$

(c) $Y = \overline{AB\overline{C} + D} + \overline{A\overline{B}} + \overline{B\overline{C}}$

(d) $Z = \overline{(C + D)\overline{A\overline{C}D}}(\overline{A}C + \overline{D})$

5–20. Repeat Problem 5–18 for the circuits of Figure P5–20.

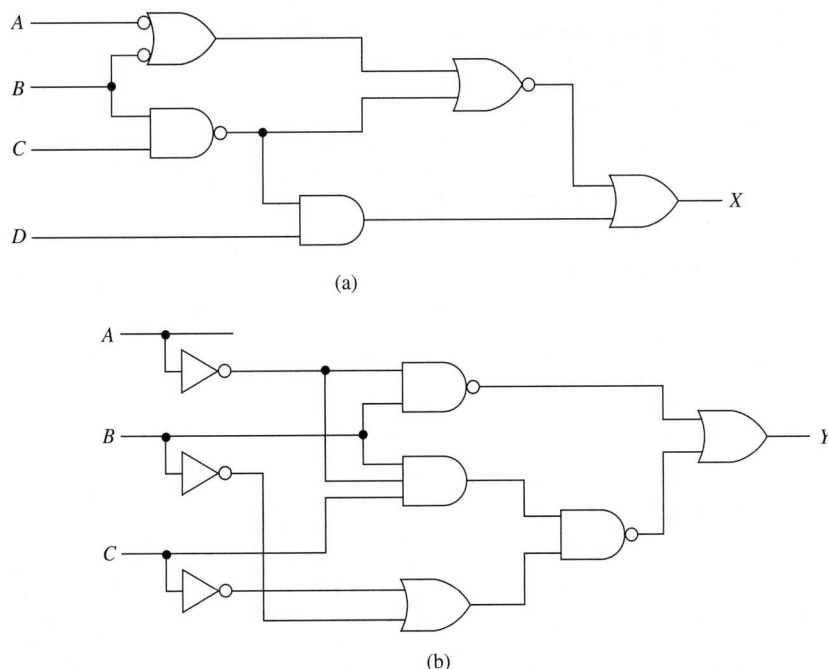

(a)

(b)

Figure 5–20

D *   **5–21.** Design a logic circuit that will output a 1 (HIGH) only if A and B are both 1 while either C or D is 1.

D   **5–22.** Design a logic circuit that will output a 0 only if A or B is 0.

D   **5–23.** Design a logic circuit that will output a LOW only if A is HIGH or B is HIGH while C is LOW or D is LOW.

C D   **5–24.** Design a logic circuit that will output a HIGH if only one of the inputs A, B, or C is LOW.

C D   **5–25.** Design a circuit that outputs a 1 when the binary value of $ABCD$ ($D = $ LSB) is > 11.

C D   **5–26.** Design a circuit that outputs a LOW when the binary value of $ABCD$ ($D = $ LSB) is > 7 and < 10.

5–27. Complete a truth table for the following simplified Boolean equations.

(a) $W = A\overline{B}\,\overline{C} + \overline{B}C + \overline{A}B$

(b) $X = \overline{A}\,\overline{B} + A\overline{B}C + B\overline{C}$

(c) $Y = \overline{C}D + \overline{A}\,\overline{B}\,\overline{C}\,\overline{D} + BCD + \overline{A}C\overline{D}$

(d) $Z = \overline{A}BC\overline{D} + \overline{A}C + C\overline{D} + \overline{B}\,\overline{C}$

5–28. Complete the timing diagram in Figure P5–28 for the following simplified Boolean equations.

(a) $X = \overline{A}\,\overline{B}\,\overline{C} + ABC + A\overline{C}$

(b) $Y = \overline{B} + \overline{A}B\overline{C} + AC$

(c) $Z = B\overline{C} + A\overline{B} + \overline{A}BC$

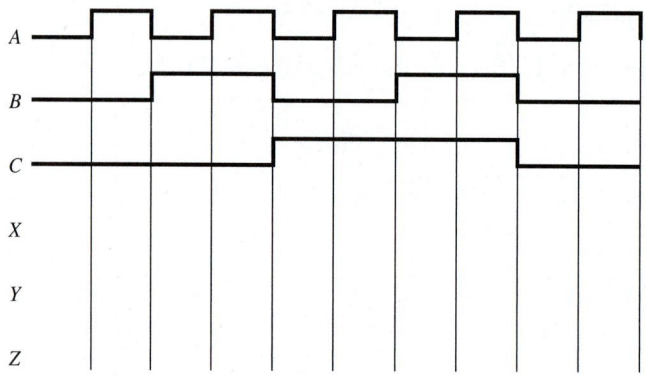

Figure P5–28

5–29. Use the bubble-pushing technique to convert the gates in Figure P5–29.

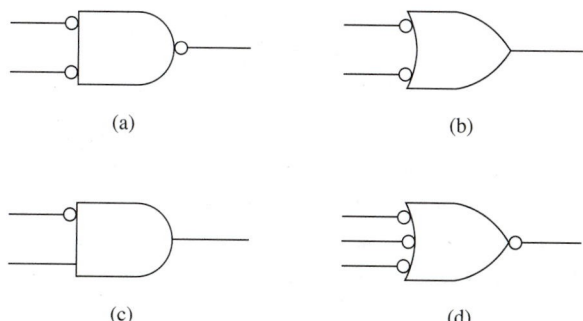

(a)   (b)

(c)   (d)

Figure P5–29

D C **5–30.** Some computer systems have two disk drives, commonly called drive A and drive B, for storing and retrieving data. Assume that your computer has four control signals provided by its internal microprocessor to enable data to be read and written to either drive. Design a gating scheme similar to that provided in Figure 5–61 to supply an active-LOW drive select signal to drive A $(\overline{DS_a})$ or to drive B $(\overline{DS_b})$ whenever they are read or written to. The four control signals are also active-LOW and are labeled $\overline{RD}$ (Read), $\overline{WR}$ (Write), $\overline{DA}$ (drive A), and $\overline{DB}$ (drive B).

Section 5–5

5–31. Draw the connections required to convert

(a) A NAND gate into an inverter

(b) A NOR gate into an inverter

5–32. Draw the connections required to construct

(a) An OR gate from two NOR gates

(b) An AND gate from two NAND gates

(c) An AND gate from several NOR gates

(d) A NOR gate from several NAND gates

5–33. Redraw the logic circuits of Figure P5–33 to their equivalents *using only* NOR gates.

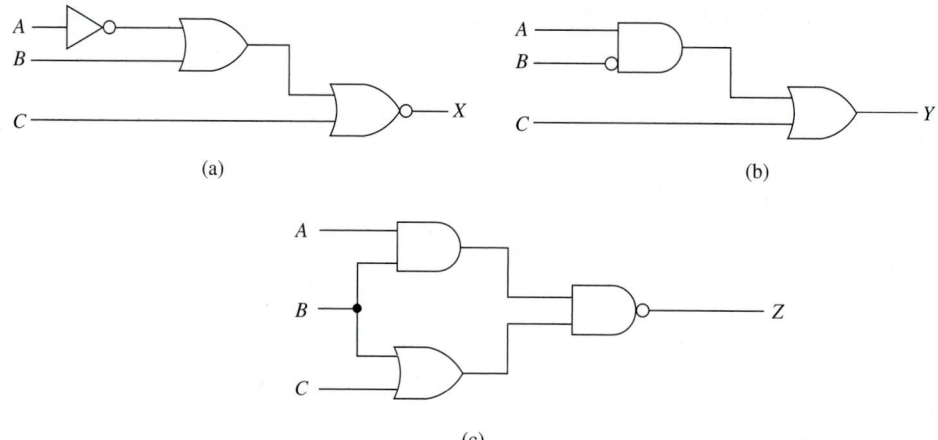

(a)

(b)

(c)

Figure P5–33

C   **5–34.** Convert the circuits of Figure P5–34 to their equivalents *using only* NAND gates. Next, make the external connections to a 7400 quad NAND to implement the new circuit. (Each new equivalent circuit is limited to *four* NAND gates.)

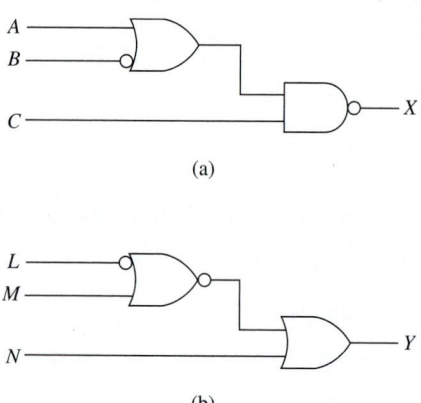

(a)

(b)

Figure P5–34

Section 5–6

5–35. Identify each of the following Boolean equations as a POS expression, a SOP expression, or both.

(a) $U = A\overline{B}C + BC + \overline{A}C$

(b) $V = (A + C)(\overline{B} + \overline{C})$

(c) $W = A\overline{C}(\overline{B} + C)$

(d) $X = AB + \overline{C} + BD$

(e) $Y = (A\overline{B} + D)(A + \overline{C}D)$

(f) $Z = (A + \overline{B})(BC + A) + \overline{A}B + CD$

5–36. Simplify the circuit of Figure P5–36 down to its SOP form, then draw the logic circuit of the simplified from implemented using a 74LS54 AOI gate.

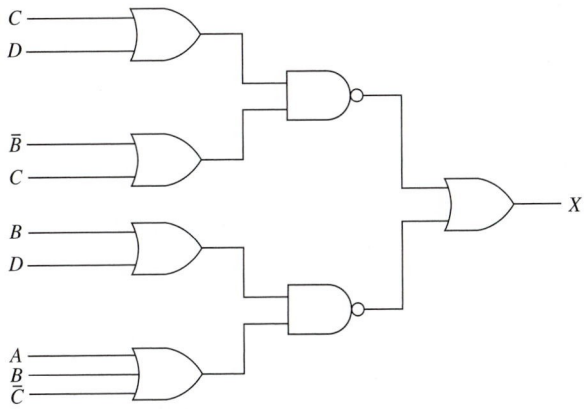

Figure P5–36

Section 5–7

5–37. Using a Karnaugh map, reduce the following equations to a minimum form.

(a) $X = AB\overline{C} + \overline{A}B + \overline{A}\,\overline{B}$

(b) $Y = BC + \overline{A}\,\overline{B}C + B\overline{C}$

(c) $Z = ABC + A\overline{B}\,\overline{C} + \overline{A}\,BC + AB\overline{C}$

5–38. Using a Karnaugh map, reduce the following equations to a minimum form.

(a) $W = \overline{B}(C\overline{D} + \overline{A}D) + \overline{B}\,\overline{C}(A + \overline{A}\,D)$

(b) $X = \overline{A}\,\overline{B}\,\overline{D} + B(\overline{C}\,\overline{D} + ACD) + A\overline{B}\,\overline{D}$

(c) $Y = A(C\overline{D} + \overline{C}\,\overline{D}) + A\overline{B}D + \overline{A}\,\overline{B}CD$

(d) $Z = \overline{B}\,\overline{C}D + B\overline{C}D + \overline{C}\,\overline{D} + C\overline{D}(B + \overline{A}\,\overline{B})$

C   **5–39.** Use a Karnaugh map to simplify the circuits in Figure P5–39.

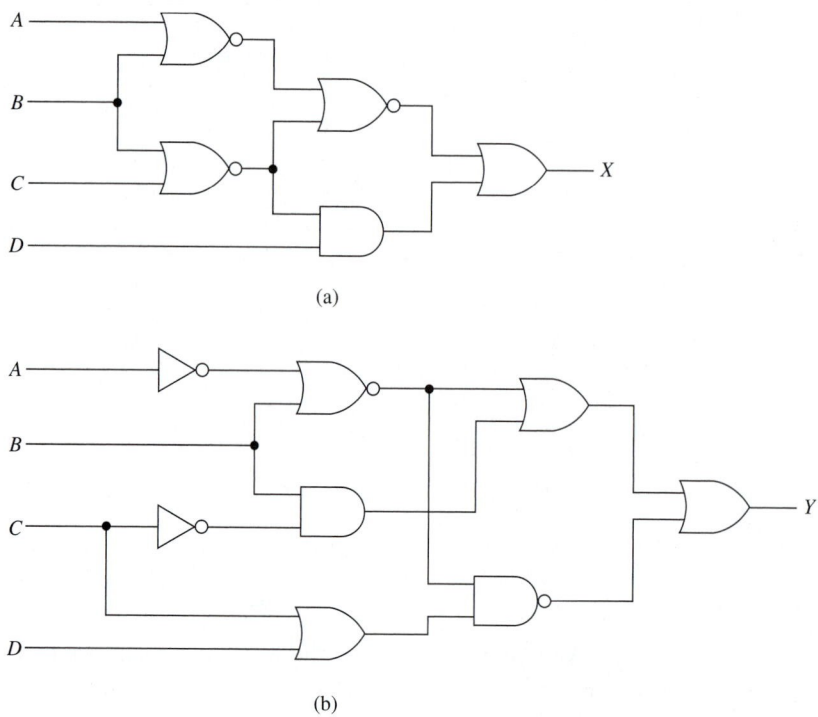

(a)

(b)

Figure P5–39

Section 5–8

C   **5–40.** Seven-segment displays are commonly used in calculators to display each decimal digit. Each segment of a digit is controlled separately, and when all seven of the segments are on, the number 8 is displayed. The upper right segment of the display comes on when displaying the numbers 0, 1, 2, 3, 4, 7, 8, and 9. (The numerical designation for each of the digits 0 to 9 is shown in Figure P5–40 and described in more detail in Section 12–6.) Design a circuit that outputs a HIGH (1) whenever a 4-bit BCD code translates to a number that uses the upper right segment. Use variable A to represent the 2^3 BCD input. Implement your design with an AOI and inverters.

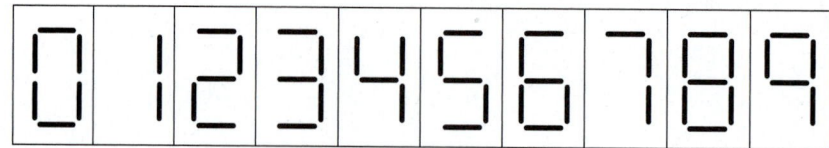

Figure P5–40

C D   **5–41.** Repeat Problem 5–40 for the lower left segment of a seven-segment display (0, 2, 6, 8).

T   **5–42.** The logic circuit of Figure P5–42(a) is implemented by making connections to the 7400 as shown in Figure P5–42(b). The circuit is not working properly. The problem is in the IC connections or in the IC itself. The data table in Figure P5–42(c) is completed by using a logic probe at each pin. Identify the problem.

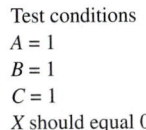

Test conditions
A = 1
B = 1
C = 1
X should equal 0

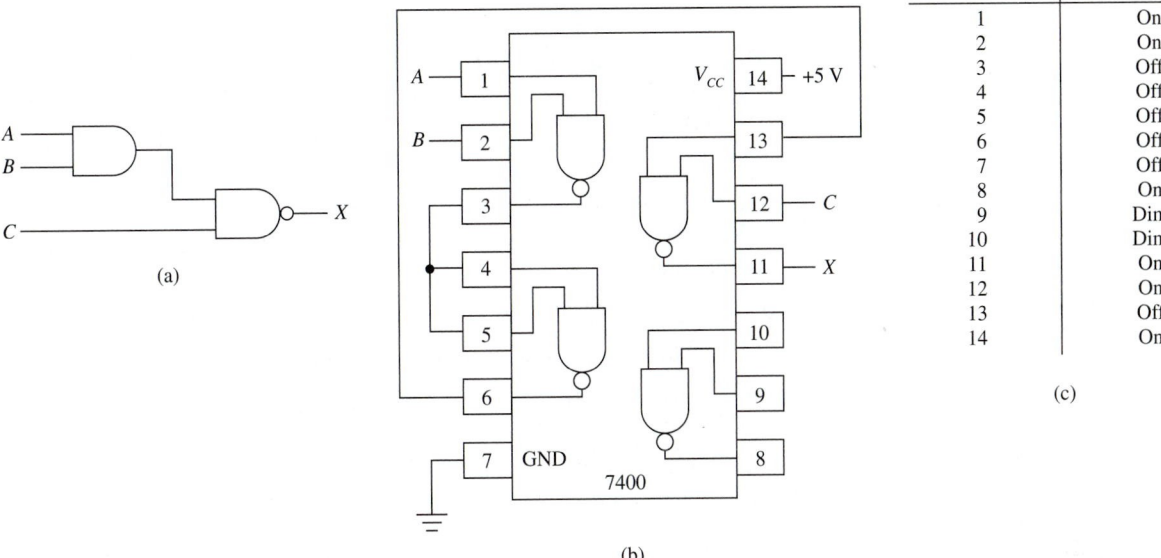

Probe on pin:	Indicator lamp
1	On
2	On
3	Off
4	Off
5	Off
6	Off
7	Off
8	On
9	Dim
10	Dim
11	On
12	On
13	Off
14	On

(c)

(a)

(b)

Figure P5–42

T   **5–43.** Repeat Problem 5–42 for the circuit shown in Figure P5–43.

Test conditions
A = 0
B = 1
C = 1
X should equal 0

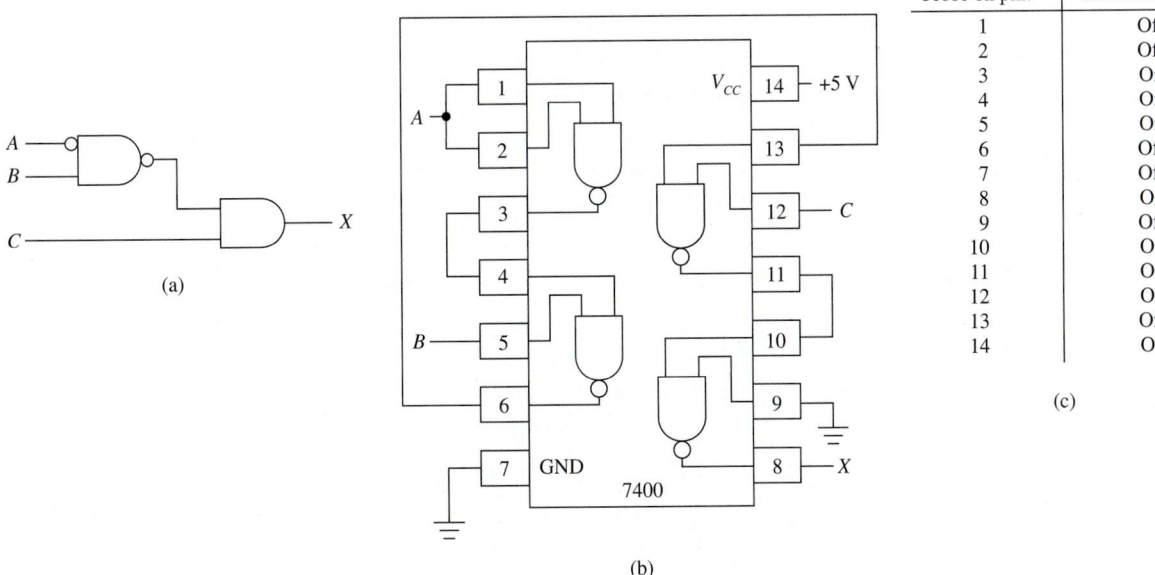

Probe on pin:	Indicator lamp
1	Off
2	Off
3	On
4	On
5	On
6	Off
7	Off
8	On
9	Off
10	On
11	On
12	On
13	Off
14	On

(c)

(a)

(b)

Figure 5–43

Schematic Interpretation Problems

See Appendix G for the schematic diagrams.

S   **5–44.** Find U8 in the HC11D0 schematic. Pins 11 and 12 are unused so they are connected to V_{CC}. What if they were connected to ground instead?

S   **5–45.** Find U1:A in the Watchdog Timer schematic. This device is called a flip-flop and is explained in Chapter 10. It has two inputs, D and CLK, and two outputs, Q_A and $\overline{Q_A}$ Write the Boolean equation at the output (pin 3) of U2:A.

S   **5–46.** Write the Boolean equation at the output (pin 3) of U12:A in the Watchdog Timer schematic. (*Hint:* Use the information given in Problem 5–45.)

C S   **5–47.** Locate the U14 gates in the 4096/4196 schematic.

(a) Write the Boolean equation of the output at pin 6 of U14.

(b) What kind of gate does it turn into if you use the bubble-pushing technique?

(c) This is a 74HC08. What kind of logic gate is that?

(d) Complete the following sentence: Pin 3 of U14:A goes LOW if _____ OR if _____.

C S   **5–48.** U10 of the 4096/4196 schematic is a RAM memory IC. Its operation is discussed in Chapter 16. To enable the chip to work, the Chip Enable input at pin 20 must be made LOW. Write a sentence describing the logic operation that makes that line go LOW. (*Hint:* Pin 20 of U10 goes LOW if _____.)

Electronics Workbench®/MultiSIM® Exercises

E5–1. Load the circuit file for **Section 5–1a.** This circuit is an automobile warning system used to warn you if leave your key in the ignition or leave your headlights on as you leave your car.

(a) Write the Boolean equation at B. Test your Boolean equation by moving the appropriate switches.

(b) The equation and the circuit can be reduced to a simpler form using just two gates and three switches to perform the same operation. What is the reduced equation? Test your reduced equation by building the new circuit.

E5–2. Load the circuit file for **Section 5–1b.**

(a) Create a truth table using the Logic Converter. How many different input combinations produce a 1 in the output?

(b) Use the Logic Converter to find the simplified equation at X. What is the simplified equation?

E5–3. Load the circuit file for **Section 5–3a.**

(a) Create a truth table using the Logic Converter. How many different input combinations produce a 1 in the output?

(b) Use the Logic Converter to find the simplified equation at X. What is the simplified equation?

E5–4. Load the circuit file for **Section 5–3b.**

(a) What is the Boolean equation at X?

(b) Create a truth table using the Logic Converter. How many different input combinations produce a 1 in the output?

(c) Use the Logic Converter to find the simplified equation at X. What is the simplified equation?

E5–5. Load the circuit file for **Section 5–3c.** Use the gates that are provided to draw the logic circuit for the following equation: $X = (ABC + B)BC$.

(a) Create a truth table using the Logic Converter. How many different input combinations produce a 1 in the output?

(b) Use the Logic Converter to find the simplified equation at X. What is the simplified equation?

(c) Draw the simplified circuit using the Logic Converter.

E5–6. Load the circuit file for **Section 5–3d.** Use the gates that are provided to draw the logic circuit for the following equation: $X = ABD + CD + CDE$.

(a) Create a truth table using the Logic Converter. How many different input combinations produce a 1 in the output?

(b) Use the Logic Converter to find the simplified equation at X. What is the simplified equation?

(c) Draw the simplified circuit using the Logic Converter.

E5–7. Load the circuit file for **Section 5–3e.** The Combinational logic circuit inside of the box labeled "COMBO1" produces an output at X. Use the waveforms shown on the Logic Analyzer to determine the Boolean logic that is inside circuit "COMBO1." Write the equation at X.

E5–8. Load the circuit file for the **Section 5–3f.** The combinational logic circuit inside of the box labeled "COMBO2" produces an output at X. Study the waveforms shown on the Logic Analyzer to determine the Boolean logic that is inside circuit "COMBO2." Write the equation at X.

E5–9. Load the circuit file for **Section 5–4a.** The circuit shown has a Boolean equation of $X = (AB)' (A + B)'$. The apostrophe (′) is used instead of an overbar.

(a) Create a truth table using the Logic Converter. How many different input combinations produce a 1 in the output?

(b) Use the Logic Converter to find the simplified equation at X. What is the simplified equation?

E5–10. Load the circuit file for **Section 5–4b.** The circuit shown is a Combinational logic circuit.

(a) What is the Boolean equation at X?

(b) Create a truth table using the Logic Converter. How many different input combinations produce a 1 in the output?

(c) Use the Logic Converter to find the simplified equation at X. What is the simplified equation?

C **E5–11.** Load the circuit file for **Section 5–4c.** Use the gates that are provided to draw the logic circuit for the following equation: $X = A(B + C)' + (BC)'$.

(a) Create a truth table using the Logic Converter. How many different input combinations produce a 1 in the output?

(b) Use the Logic Converter to find the simplified equation at X. What is the simplified equation?

(c) Draw the simplified circuit using the Logic Converter.

C

E5–12. Load the circuit file for **Section 5–4d.** Use the gates that are provided to draw the logic circuit for the following equation: $X = (ABC' + D)' + (AB' + BC')'$.

(a) Create a truth table using the Logic Converter. How many different input combinations produce a 1 in the output?

(b) Use the Logic Converter to find the simplified equation at X. What is the simplified equation?

(c) Draw the simplified circuit using the Logic Converter.

E5–13. Load the circuit file for **Section 5–4e.** On a separate piece of paper use the "Bubble-Pushing" technique to convert the gates connected to X and Y.

(a) What logic gate could be used to provide the logic at X?

(b) What logic gate could be used to provide the logic at Y? Check your answer by observing the output at X and Y on the Logic Analyzer.

C D

E5–14. Load the circuit file for **Section 5–4f.** The Word Generator is set up to output a binary up-counter waveform similar to the one commonly used in the textbook. Design a circuit that will output a HIGH if only one of the inputs A, B, or C is LOW. Connect the output of your design to the Logic Analyzer. Study the four waveforms to see if your design worked.

D

E5–15. Load the circuit file for **Section 5–4g.** The Word Generator is set up to output a binary up-counter waveform similar to the one commonly used in the textbook. Design a circuit that will output a HIGH when the binary value of ABCD (D = LSB) is greater than 11. Connect the output of your design to the Logic Analyzer. Study the five waveforms to see if your design worked.

D

E5–16. Load the circuit file for **Section 5–4h.** The Word Generator is set up to output a binary up-counter waveform similar to the one commonly used in the textbook. Design a circuit that will output a LOW when the binary value of ABCD (D = LSB) is greater than 7 and less than 10. Connect the output of your design to the Logic Analyzer. Study the five waveforms to see if your design worked.

C T

E5–17. Load the circuit file for **Section 5–4i.** The 7400 shown is a Quad NAND.

(a) If no other ICs are available, how many gates on the 7400 are required to implement the equation $X = A'B$?

(b) One of the gates on this 7400 is bad. Use the Logic Analyzer to determine which one.

(c) With the three remaining good gates, connect the circuit for $X = A'B$. Route its output to the Logic Analyzer to check its operation.($X = 1$ if A = 0 AND B = 1).

T

E5–18. Load the circuit file for **Section 5–4j.** The 7400 shown is a Quad NAND.

(a) On a separate piece of paper write the Boolean equation for the circuit shown.

(b) Simplify the equation.

(c) Use the Logic Analyzer to observe the waveforms. Are they what you expect? If not, troubleshoot the circuit using the Logic Analyzer.

(d) Is one of the gates bad? Substitute gate-4 for the bad gate and retest the circuit.

The following problems are solved using the Altera MAX+PLUS II software. In each case the design is completed by building a graphic design file (gdf) or a VHDL file (vhd) and then proving the results by producing a simulation (scf) file. [*Note:* if you build a vhd file having the same name as the gdf file be sure to choose **Set Project To Current File** (Under **File > Project**). This will ensure that the compiler uses the current file to synthesize and simulate your design. Also, you can use the same simulation (scf) file for either design method. The simulation will be performed on whichever project file is currently set.]

A final step that can be performed is to download the design to a CPLD on a programmer board like the UP-1 or PLDT-2 and demonstrate it to your instructor.

Section 5-1

C5–1. Prove that the reduced circuit for the bank alarm in Figure 5–6 is equivalent to its original in Figure 5–5. Call the output of the original circuit *A_original* and call the output of the reduced circuit *A_reduced.*

(a) Enter the logic circuit for the original circuit and for the reduced circuit in the same graphic design file called *prob_c5_1.gdf.* Prove that the equations produce identical results by building a waveform simulator channel file called *prob_c5_1.scf* that tests all possible input conditions at *H, F,* and *V.*

(b) Enter the logic equation for the original circuit and for the reduced circuit in the same VHDL file called *prob_c5_1.vhd.* Prove that the equations produce identical results by building a waveform simulator channel file called *prob_c5_1.scf* that tests all possible input conditions at *H, F,* and *V.*

(c) Download your design to a CPLD IC. Discuss your observations of the alarm LED (*A_reduced*) with your instructor as you try various combinations of the switches representing banking hours (H), vault door (V), and front door (F).

C5–2. Design the logic to implement the following Boolean equation (do not reduce):

$X = AB + BC + CD$

(a) Enter the logic circuit for the equation as a graphic design file called *prob_c5_2.gdf.* Simulate the results of your design by building a waveform simulator channel file called *prob_c5_2.scf* that tests all possible input conditions at *A, B, C,* and *D.*

(b) Enter the logic circuit for the equation as a VHDL file called *prob_c5_2.vhd.* Simulate the results of your design by building a waveform simulator channel file called *prob_c5_2.scf* that tests all possible input conditions at *A, B, C,* and *D.*

(c) Download your design to a CPLD IC. Discuss your observations of the output LED *(X)* with your instructor as you try various combinations of the switches representing *A, B, C,* and *D.*

C5–3. Repeat problem C5–2 (a), (b), and (c) for the following equation:

$$Y = ABC + AD + BD$$

Section 5–2

C5–4. Ten rules for Boolean reduction were given in Table 5–2. The 10th rule states that:

1) $A + \overline{A}B = A + B$ and

2) $\overline{A} + AB = \overline{A} + B$

(a) Create a graphic design file (*prob_c5_4.gdf*) and a simulator channel file (*prob_c5_4.scf*) to prove that both equations in (1) and both equations in (2) are equivalent.

(b) Create a VHDL file (*prob_c5_4.vhd*) and a simulator channel file (*prob_c5_4.scf*) to prove that both equations in (1) and both equations in (2) are equivalent.

Section 5–4

C5–5. Use the MAX+PLUS II software to determine the simplified form of the following Boolean equation:

$$X = \overline{A}B(B + \overline{A}BC)$$

Enter the circuit design using the Graphic Editor to create a file called *prob_c5_5.gdf*. Determine the simplified equation by reviewing the Equations Section of the report file (*prob_c5_5.rpt*) using the technique shown in Example 5–10.

C5–6. Repeat Problem C5–5 for the following equations:

(a) $Y = BC(\overline{A}BC + AB)$

(b) $Z = A(AB\overline{C} + \overline{B}\,\overline{C})$

Section 5-5

C5–7. Use the MAX+PLUS II software to determine the simplified form of the following Boolean equations:

(a) $Y = \overline{\overline{A + C} + \overline{B}C}$

(b) $Z = \overline{(A + \overline{C}) + BC}$

Enter the circuit design using the VHDL text editor to create a file called *prob_c5_7.vhd*. Determine the simplified equation by reviewing the Equations Section of the report file (*prob_c5_7.rpt*) using the technique shown in Example 5–17.

C5–8. Design the logic to implement the circuit in Example 5–13 (do not reduce):

(a) Enter the logic circuit given in the example as a graphic design file called *prob_c5_8.gdf*. Simulate the results of your design by building a waveform simulator channel file called *prob_c5_8.scf* that tests all possible input conditions at *A, B,* and *C*.

(b) Enter the logic circuit for the equation as a VHDL file called *prob_c5_8.vhd*. Simulate the results of your design by building a waveform simulator channel file called *prob_c5_8.scf* that tests all possible input conditions at *A, B,* and *C*.

(c) Download your design to a CPLD IC. Discuss your observations of the output LED *(X)* with your instructor as you try various combinations of the switches representing *A, B,* and *C*.

C5–9. Repeat problem C5-8 (a), (b), and (c) for Example 5–14.

C5–10. Design the logic using the VHDL text editor to implement the following Boolean equations:

(a) $X = \overline{\overline{AB}(A + B)}$

(b) $Y = \overline{\overline{A + B} + AB}$

(c) $Z = AB + \overline{A + B}$

Enter all three equations in the same architecture section of the program (*prob_c5_10.vhd*). Determine which two of those equations yield equivalent outputs by studying their waveforms in the simulator channel file (*prob_c5_10.scf*).

C5–11. A chemical processing plant has four HIGH/LOW sensors on each of its chemical tanks. [Temperature (T), Pressure (P), Fluid Level (L), and Weight (W)]. Several different combinations of sensor levels need to be constantly monitored. Design a CPLD solution using a VHDL program (*prob_c5_11.vhd*) that will tell the circuit to turn on any of the three indicator lights [Emergency (E), Warning (W), or Check (C)] if the listed conditions are met:

1) (E) *Emergency:* Shut down and drain system if any of the following exists:
   (a) High T, high P with low W
   (b) High T, high P with low L
   (c) High T, low P, with (low W or low L)
2) (W) *Warning:* Check controls and perform corrections if any of the following exists:
   (a) High P, high L with low W
   (b) High P, high W with low L
   (c) High P, low L, with low T
3) (C) *Check:* Read gauges and report if any of the following exists:
   (a) Any two levels are high
   (b) Any time W is high

Build a simulator channel file (*prob_c5_11.scf*) to simulate the operation of all three indicator lights and then download the program to a CPLD to demonstrate its complete operation to your instructor.

C5–12. MAX+PLUS II provides active-LOW input, active-LOW output gates called BNAND2 and BNOR2 in the primitive symbols library of the Graphic Editor. Use those gates in a graphic design file (*prob_c5_12.gdf*) to implement the microprocessor memory gating scheme presented in Figure 5–61. Exercise the design by creating a simulator channel file (*prob_c5_12.scf*) that illustrates the following sequence of events:

(a) Read from memory

(b) Wait (all control signals HIGH)

(c) Write to memory

(d) Wait

(e) Repeat (a)–(d) once again

[*Hint:* Specialized (nonrepetitive) control waveforms can be created by highlighting areas of a waveform and selecting a HIGH level or LOW level from the left side menu.] After a successful simulation, download the design to a CPLD and discuss your observations with your instructor as you physically simulate read/write operations with the on-board switches.

C5–13. Create a graphic design file (*prob_c5_13.gdf*) using BNAND and BNOR gates to implement the computer disk drive controller explained in problem 5–30. Exercise the design by creating a simulator channel file (*prob_c5_13.scf*) that illustrates the following sequence of events:

(a) Read from disk A

(b) Wait (all control signals HIGH)

(c) Write to disk drive B

(d) Wait

(e) Repeat (a)–(d) once again

[*Hint:* Specialized (nonrepetitive) control waveforms can be created by highlighting areas of a waveform and selecting a HIGH level or LOW level from the left side menu.] After a successful simulation, download the design to a CPLD and discuss your observations with your instructor as you physically simulate read/write operations with the on-board switches.

Section 5–6

C5–14. Design a logic circuit using VHDL (*prob_c5_14.vhd*) that can be used to tell when a 4-bit binary number is odd and within the range of 6 (0110_2) to 14 (1110_2) inclusive. Use the VHDL selected signal assignment method shown in Example 5–22. Perform a simulation of your design by creating a simulator channel file (*prob_c5_14.scf*) that steps through the entire range of input possibilities 0000_2 to 1111_2. After a successful simulation, download the design to a CPLD and discuss your observations with your instructor as you physically count through all possibilities on the on-board switches.

C5–15. A chemical processing facility needs to have a warning system to monitor an overflow condition in their four chemical holding tanks. Each tank has a HIGH/LOW level sensor. The tanks are labeled T3, T2, T1, and T0. Design a system that activates a warning alarm whenever the two odd-numbered tanks (T3 and T1) are both HIGH or whenever the two even-numbered tanks (T2 and T0) are both HIGH. Write a VHDL program (*prob_c5_15.vhd*) that groups the tanks together as a vector and uses the selected signal assignment similar to the one used in Example 5–23. Perform a simulation of your design by creating a simulator channel file (*prob_c5_15.scf*) that steps through the entire range of input possibilities 0000_2 to 1111_2. After a successful simulation, download the design to a CPLD and discuss your observations with your instructor as you physically test all possibilities on the on-board switches.

Answers to Review Questions

5–1. (a) 2

  (b) 3

  (c) 4

5–2. (a) Associative law of addition

  (b) commutative law of multiplication

  (c) distributive law

5–3. True

5–4. False

5–5. False

5–6. (a) $A + \overline{B}$

  (b) $B + C$

5–7. Because it enables you to convert an expression having an inversion bar over more than one variable into an expression with inversion bars over single variables only

5–8. AND

5–9. NOR

5–10. NAND

5–11. Because by utilizing a combination of these gates, all other gates can be formed

5–12. Because in designing a circuit you may have extra NAND gates available and can avoid using extra ICs

5–13. 4

5–14. SOP

5–15. SOP

5–16. They must be connected to ground.

5–17. True

5–18. False

5–19. False

5–20. One group of 8

6

Exclusive-OR and Exclusive-NOR Gates

OUTLINE

OBJECTIVES

Upon completion of this chapter, you should be able to:

- Describe the operation and use of exclusive-OR and exclusive-NOR gates.
- Construct truth tables and draw timing diagrams for exclusive-OR and exclusive-NOR gates.
- Simplify combinational logic circuits containing exclusive-OR and exclusive-NOR gates.
- Design odd- and even-parity generator and checker systems.
- Explain the operation of a binary comparator and a controlled inverter.
- Implement circuits in CPLD ICs using VHDL.

INTRODUCTION

We have seen in the previous chapters that by using various combinations of the basic gates, we can form almost any logic function that we need. Often, a particular combination of logic gates provides a function that is especially useful for a wide variety of tasks. The AOI discussed in Chapter 5 is one such circuit. In this chapter, we learn about and design systems using two new combinational logic gates: the exclusive-OR and the exclusive-NOR.

6–1 The Exclusive-OR Gate

Remember, a two-input OR gate provides a HIGH output if one input or the other input is HIGH *or if both inputs are HIGH.* The **exclusive-OR,** on the other hand, provides a HIGH output if one input or the other input is HIGH, *but not both.* This point is made more clear by comparing the truth tables for a two-input OR gate versus an exclusive-OR gate, as shown in Table 6–1.

TABLE 6–1	Truth Tables for an OR Gate Versus an Exclusive-OR Gate				
A	*B*	*X*	*A*	*B*	*X*
0	0	0	0	0	0
0	1	1	0	1	1
1	0	1	1	0	1
1	1	1	1	1	0
(OR)			**(Exclusive-OR)**		

The Boolean equation for the Ex-OR function is written $X = \overline{A}B + A\overline{B}$ and can be constructed using the combinational logic shown in Figure 6–1. By experimenting and using Boolean reduction, we can find several other combinations of the basic gates that provide the Ex-OR function. For example, the combination of AND, OR, and NAND gates shown in Figure 6–2 will reduce to the "one or the other but not both" (Ex-OR) function.

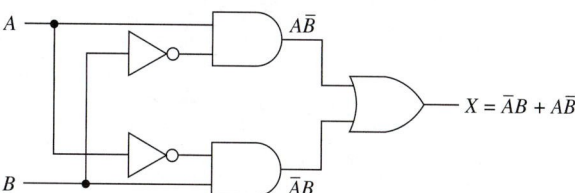

Figure 6–1 Logic circuit for providing the exclusive-OR function.

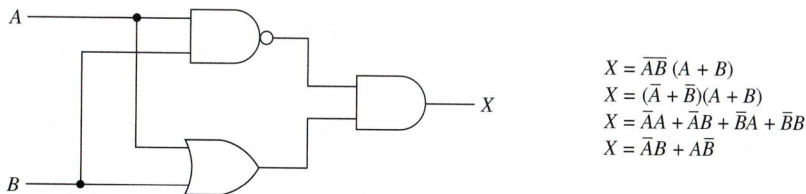

$$X = \overline{AB}\,(A + B)$$
$$X = (\overline{A} + \overline{B})(A + B)$$
$$X = \overline{A}A + \overline{A}B + \overline{B}A + \overline{B}B$$
$$X = \overline{A}B + A\overline{B}$$

Figure 6–2 Exclusive-OR built with an AND–OR–NAND combination.

The exclusive-OR gate is common enough to deserve its own logic symbol and equation, as shown in Figure 6–3. (Note the shorthand method of writing the Boolean equation is to use a plus sign with a circle around it.)

$$X = A \oplus B = \overline{A}B + A\overline{B}$$

Figure 6–3 Logic symbol and equation for the exclusive-OR.

6–2 The Exclusive-NOR Gate

The **exclusive-NOR** is the complement of the exclusive-OR. A comparison of the truth tables in Table 6–2 illustrates this point.

TABLE 6–2	Truth Tables of the Exclusive-NOR Versus the Exclusive-OR				
$X = AB + \overline{A}\,\overline{B}$			$X = \overline{A}B + A\overline{B}$		
A	B	X	A	B	X
0	0	0	0	0	0
0	1	1	0	1	1
1	0	1	1	0	1
1	1	1	1	1	0
Exclusive-NOR			Exclusive-OR		

The truth table for the Ex-NOR shows a HIGH output for both inputs LOW or both inputs HIGH. The Ex-NOR is sometimes called the *equality gate* because both inputs must be equal to get a HIGH output. The basic logic circuit and symbol for the Ex-NOR are shown in Figure 6–4.

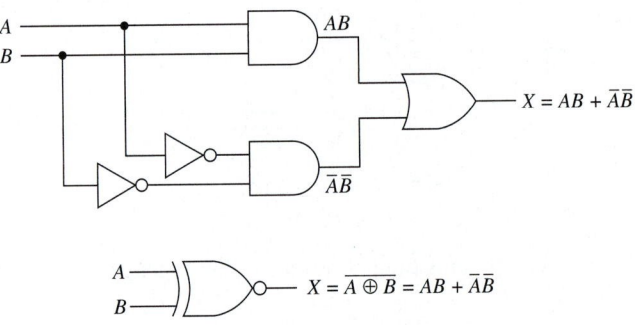

Figure 6–4   Exclusive-NOR logic circuit and logic symbol.

Summary

The exclusive-OR and exclusive-NOR gates are two-input logic gates that provide a very important, commonly used function that we will see in upcoming examples. Basically, the gates operate as follows:

The exclusive-OR gate provides a HIGH output for one or the other inputs HIGH, but not both ($X = \overline{A}B + A\overline{B}$).

The exclusive-NOR gate provides a HIGH output for both inputs HIGH or both inputs LOW ($X = AB + \overline{A}\,\overline{B}$).

Also, the Ex-OR and Ex-NOR gates are available in both TTL and CMOS IC packages. For example, the 7486 is a TTL quad Ex-OR and the 4077 is a CMOS quad Ex-NOR.

EXAMPLE 6–1

Determine for each circuit shown in Figure 6–5 if its output provides the Ex-OR function, the Ex-NOR function, or neither.

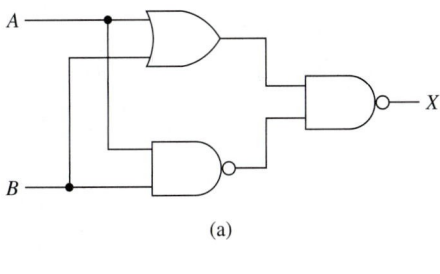

(a)

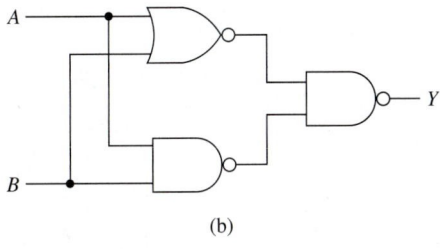

(b)

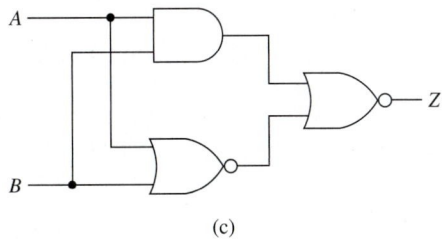

(c)

Figure 6–5

Solution:

(a) $X = \overline{(A + B)\overline{\overline{AB}}}$

$\qquad = \overline{A + B} + \overline{\overline{\overline{AB}}}$

$\qquad = \overline{A}\,\overline{B} + AB$ $\leftarrow$ Ex-NOR

(b) $Y = \overline{\overline{A + B}\;\overline{AB}}$

$\qquad = \overline{\overline{A + B}} + \overline{\overline{AB}}$

$\qquad = A + B + AB$

$\qquad = A + B(1 + A)$

$\qquad = A + B$ $\leftarrow$ neither (OR function)

(c) $Z = \overline{AB + \overline{A + B}}$

$\qquad = \overline{AB}\;\overline{\overline{A + B}}$

$\qquad = (\overline{A} + \overline{B})(A + B)$

$\qquad = \overline{A}B + \overline{A}A + \overline{B}A + \overline{B}B$

$\qquad = \overline{A}B + A\overline{B}$ $\leftarrow$ Ex–OR

EXAMPLE 6–2

Write the Boolean equation for the circuit shown in Figure 6–6 and simplify.

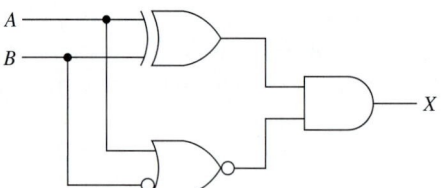

Figure 6–6

Solution:

$$X = (\overline{AB} + A\overline{B})\overline{A + \overline{B}}$$
$$= (\overline{AB} + A\overline{B})\overline{A}\,\overline{\overline{B}}$$
$$= \overline{AB}\,\overline{A}B + A\overline{B}\,\overline{A}B$$
$$= \overline{A}B$$

EXAMPLE 6–3

Write the Boolean equation for the circuit shown in Figure 6–7 and simplify.

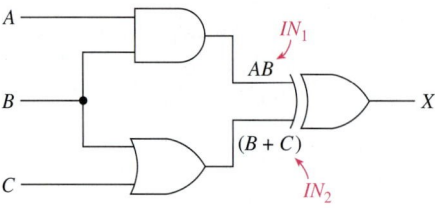

Figure 6–7

Solution:

$$X = \overline{AB}(B + C) + AB(\overline{B + C})$$
$$= (\overline{A} + \overline{B})(B + C) + AB\overline{B}\,\overline{C}$$
$$= \overline{A}B + \overline{A}C + \overline{B}B + \overline{B}C$$
$$= \overline{A}B + \overline{A}C + \overline{B}C$$

Hint:

$$X = \overline{IN_1}IN_2 + IN_1\overline{IN_2}$$

Review Questions

6–1. The exclusive-OR gate is the complement (or inverse) of the OR gate. True or false?

6–2. The exclusive-OR gate is the complement of the exclusive-NOR gate. True or false?

6–3. Write the Boolean equation for an exclusive-NOR gate.

6-3   Parity Generator/Checker

Now let's look at some digital systems that use the Ex-OR and Ex-NOR gates. We start by studying the parity generator.

In the **transmission** of binary information from one digital device to another, it is possible for external **electrical noise** or other disturbances to cause an error in the digital signal. For example, if a 4-bit digital system is transmitting a BCD 5 (0101), electrical noise present on the line during the transmission of the LSB may change a 1 to a 0. If so, the receiving device on the other end of the transmission line would receive a BCD 4 (0100), which is wrong. If a parity system is used, this error would be recognized, and the receiving device would signal an error condition or ask the transmitting device to retransmit.

Parity systems are defined as either *odd parity* or *even parity*. The parity system adds an extra bit to the digital information being transmitted. A 4-bit system will require a fifth bit, an 8-bit system will require a ninth bit, and so on.

In a 4-bit system such as BCD or hexadecimal, the fifth bit is the parity bit and will be a 1 or 0, depending on what the other 4 bits are. In an *odd-parity* system, the parity bit that is added must make the *sum of all 5 bits odd*. In an *even-parity* system, the parity bit makes the *sum of all 5 bits even*.

The parity generator is the circuit that creates the parity bit. On the receiving end, a parity checker determines if the 5-bit result is of the right parity. The type of system (odd or even) must be agreed on beforehand so that the parity checker knows what to look for (this is called *protocol*). Also, the parity bit can be placed next to the MSB *or* LSB as long as the device on the receiving end knows which bit is parity and which bits are data.

Let's look at the example of transmitting the BCD number 5 (0101) in an odd-parity system.

As shown in Figure 6–8, the transmitting device puts a 0101 on the BCD lines. The parity generator puts a 1 on the parity-bit line, making the sum of the bits odd $(0 + 1 + 0 + 1 + 1 = 3)$. The parity checker at the receiving end checks to see that the 5 bits are odd and, if so, assumes that the BCD information is valid.

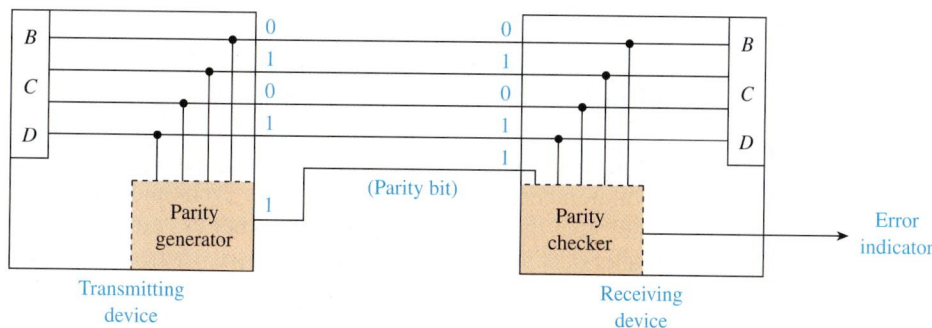

Figure 6–8 Odd-parity generator/checker system.

Helpful Hint

Typically, the *error indicator* is actually a signal that initiates a retransmission of the original signal or produces an error message on a computer display.

If, however, the data in the LSB were changed due to electrical noise somewhere in the transmission cable, the parity checker would detect that an even-parity number was received and would signal an error condition on the **error indicator** output.

This scheme detects only errors that occur to 1 bit. If 2 bits were changed, the parity checker would think everything is okay. However, the likelihood of 2 bits being affected is highly unusual. An error occurring to even 1 bit is unusual.

EXAMPLE 6-4

Add a parity bit next to the LSB of the following hexadecimal codes to form even parity: 0111, 1101, 1010, 1111, 1000, 0000.

Solution:

01111
11011
10100
11110
10001
00000

↑— parity bit

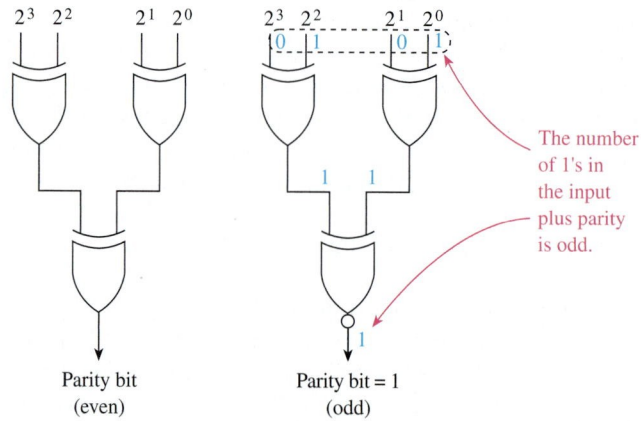

Figure 6–9   Even- and odd-parity generators.

The parity generator and checker can be constructed from exclusive-OR gates. Figure 6–9 shows the connections to form a 4-bit even- and a 4-bit odd-parity generator. The odd-parity generator has the BCD number 5 (0101) at its inputs. If you follow the logic through with these bits, you will see that the parity bit will be a 1, just as we want. Try some different 4-bit numbers at the inputs to both the even- and odd-parity generators to prove to yourself that they work properly. Computer systems generally transmit 8 or 16 bits of parallel data at a time. An 8-bit even-parity generator can be constructed by adding more gates, as shown in Figure 6–10.

A parity checker is constructed in the same way as the parity generator, except that in a 4-bit system, there must be five inputs (including the parity bit), and the output is used as the error indicator (1 = error condition). Figure 6–11 shows a 5-bit even-parity checker. The BCD 6 with even parity is input. Follow the logic through the diagram to prove to yourself that the output will be 0, meaning "no error."

IC Parity Generator/Checker

You may have guessed by now that parity generator and checker circuits are available in single IC packages. One popular 9-bit parity generator/checker is the 74280 TTL IC (or 74HC280 CMOS IC). The logic symbol and **function table** for the 74280 are given in Figure 6–12.

CHAPTER 6 ı EXCLUSIVE-OR AND EXCLUSIVE-NOR GATES

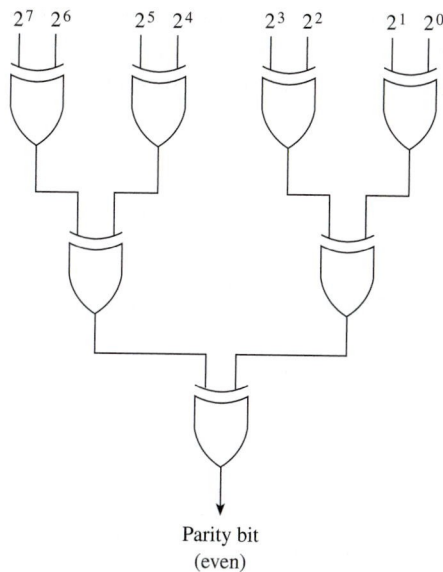

Figure 6–10 Eight-bit even-parity generator.

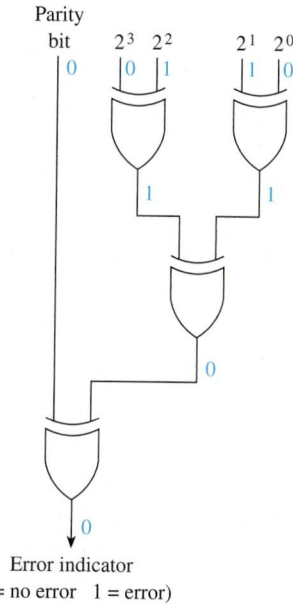

Figure 6–11 Five-bit even-parity checker.

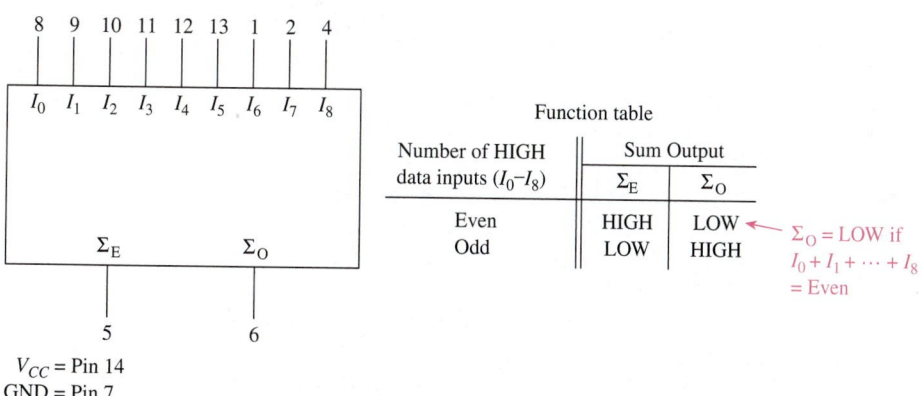

Function table		
Number of HIGH data inputs (I_0–I_8)	Sum Output	
	Σ_E	Σ_O
Even	HIGH	LOW
Odd	LOW	HIGH

Σ_O = LOW if $I_0 + I_1 + \cdots + I_8$ = Even

V_{CC} = Pin 14
GND = Pin 7

Figure 6–12 Logic symbol and function table for the 74280 9-bit parity generator/checker.

Inside Your PC

One of the most prevalent uses of parity is in the main RAM memory in a PC. Most systems use a 9-bit memory scheme (8 bits data, with 1 parity bit). The extra bits add one-ninth to the cost of the memory, and parity checking slightly increases the memory access time. However, it is well worth the expense to ensure data integrity.

The 74280 has nine inputs. If used as a parity checker, the first eight inputs would be the data input, and the ninth would be the parity-bit input. If your system is looking for even parity, the sum of the nine inputs should be even, which will produce a HIGH at the Σ_E output and a LOW at the Σ_O output.

6–4 System Design Applications

EXAMPLE 6–5

Parity Error-Detection System

Common Misconception

Students often have a hard time understanding why we use the sum-*odd* (Σ_O) output in an even system. The key to understanding that reasoning is found in the function table for the 74280 in Figure 6–12.

Using 74280s, design a complete parity generator/checking system. It is to be used in an 8-bit, even-parity computer configuration.

Solution: Parity generator: Because the 74280 has nine inputs, we have to connect the unused ninth input (I_8) to ground (0) so that it will not affect our result. The 8-bit input data are connected to I_0 through I_7.

Now, the generator sums bits I_0 through I_7 and puts out a LOW on Σ_O and a HIGH on Σ_E if the sum is even. Therefore, the parity bit generated should be taken from the Σ_O output because we want the sum of all 9 bits sent to the receiving device to be even.

Parity checker: The checker will receive all 9 bits and check if their sum is even. If their sum *is* even, the Σ_E line goes HIGH. We will use the Σ_O output because it will be LOW for "no error" and HIGH for "error." The complete circuit design is shown in Figure 6–13.

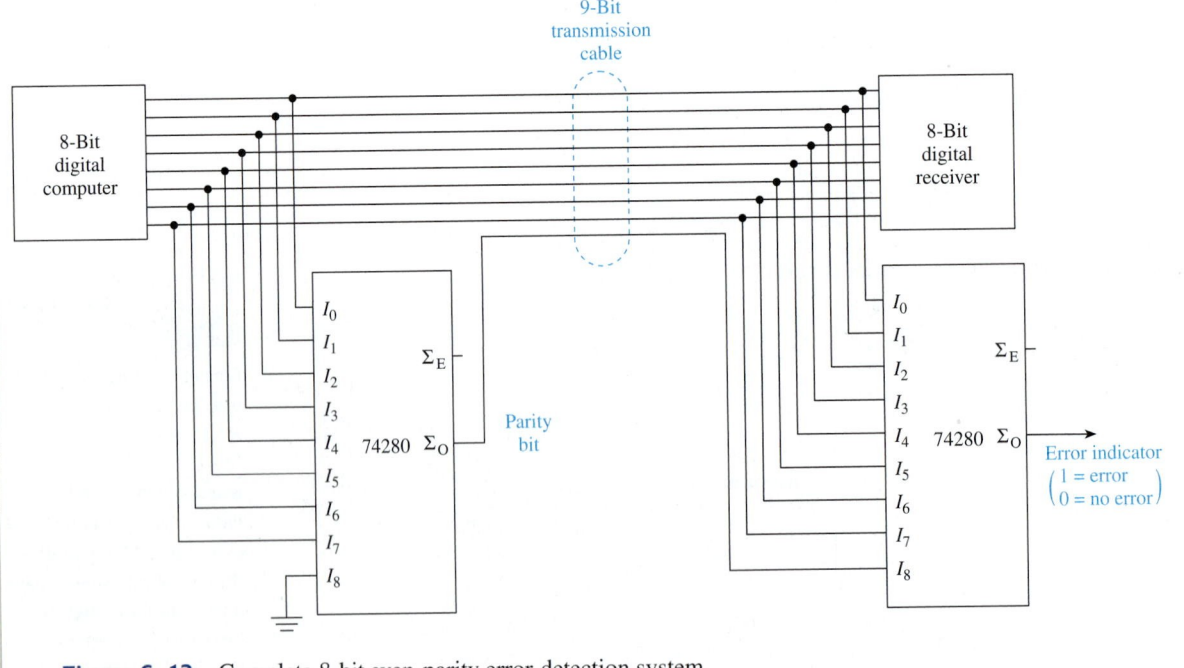

Figure 6–13 Complete 8-bit even-parity error-detection system.

EXAMPLE 6-6

Parallel Binary Comparator

Design a system—called a parallel binary **comparator**—that compares the 4-bit **binary string** A to the 4-bit binary string B. If the strings are exactly equal, provide a HIGH-level output to drive a warning buzzer.

Solution: Using four exclusive-NOR gates, we can compare string A to string B, bit by bit. Remember, if both inputs to an exclusive-NOR are the same (0–0 or 1–1), it outputs a 1. If all four Ex-NOR gates are outputting a 1, the 4 bits of string A must match the 4 bits of string B. The complete circuit design is shown in Figure 6–14.

Team Discussion

Test your *bubble-pushing* skills by determining what the AND gate must be converted to if ex-ORs were used instead of ex-NORs.

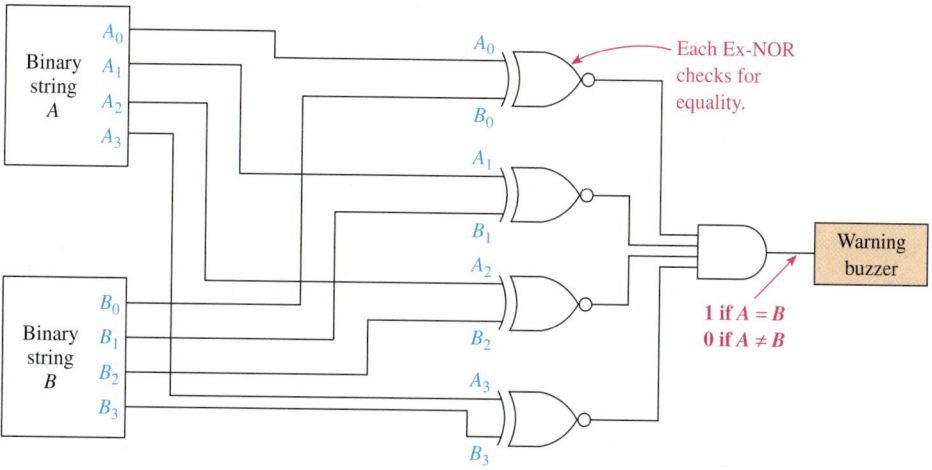

Figure 6–14   Binary comparator system.

EXAMPLE 6-7

Controlled Inverter

Often in binary arithmetic circuits, we need to have a device that complements an entire binary string when told to do so by some control signal. Design an 8-bit **controlled inverter** (complementing) circuit. The circuit will receive a control signal that, if HIGH, causes the circuit to complement the 8-bit string and, if LOW, does not.

Solution: The circuit shown in Figure 6–15 can be used to provide the complementing function. If the control signal (C) is HIGH, each of the input data bits is complemented at the output. If the control signal is LOW, the data bits pass through to the output uncomplemented. Two 7486 quad exclusive-OR ICs could be used to implement this design.

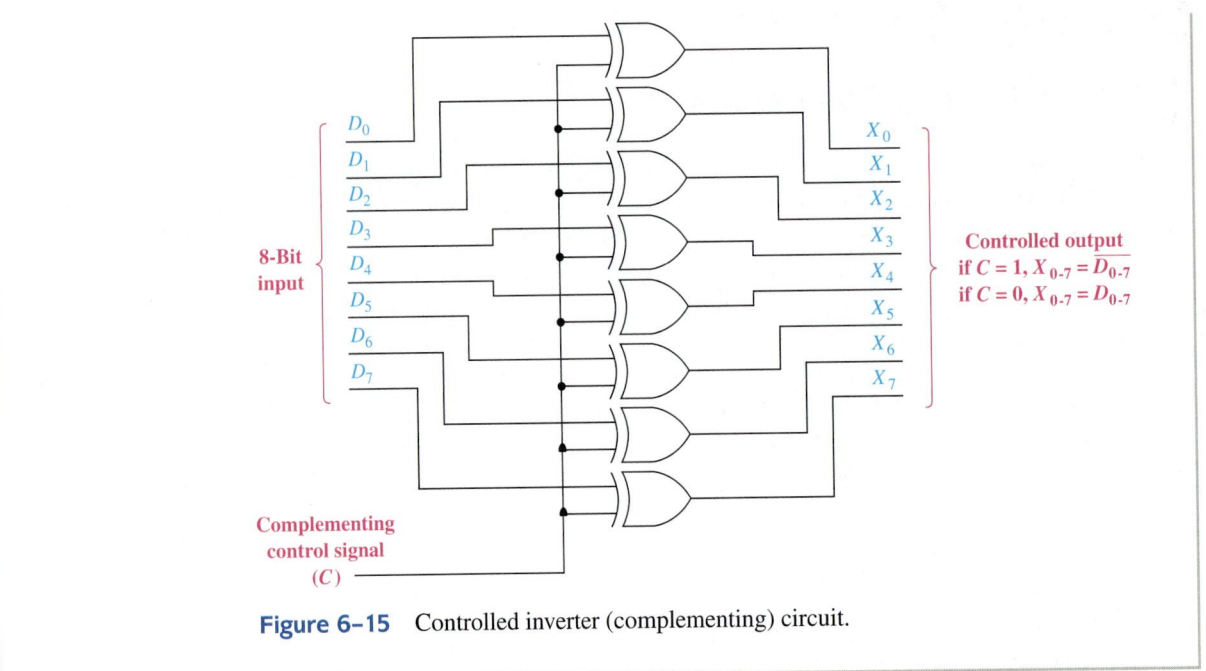

Figure 6–15 Controlled inverter (complementing) circuit.

In the figure:

8-Bit input: D_0, D_1, D_2, D_3, D_4, D_5, D_6, D_7

Outputs: X_0, X_1, X_2, X_3, X_4, X_5, X_6, X_7

Controlled output
if $C = 1$, $X_{0-7} = \overline{D_{0-7}}$
if $C = 0$, $X_{0-7} = D_{0-7}$

Complementing control signal (C)

Review Questions

6–4. An odd parity generator produces a 1 if the sum of its inputs is odd. True or false?

6–5. In an 8-bit parallel transmission system, if one or two of the bits are changed due to electrical noise, the parity checker will detect the error. True or false?

6–6. Which output of the 74280 parity generator is used as the parity bit in an *odd* system?

6–7. If all nine inputs to a 74280 are HIGH, the output at Σ_E will be _____ (HIGH, LOW)?

6–5 CPLD Design Applications with VHDL

In this section we will design circuits related to Ex-ORs and Ex-NORs by building graphic design files and VHDL programs. Several new concepts related to CPLDs will be introduced, including the use of 7400-series **macro-functions, grouping** nodes into a common **bus,** changing a group's **radix,** and creating a VHDL **Process** and **For Loop.**

Example 6–8 examines the characteristics of odd and even parity by using the pre-defined macro-function for the 74280 parity generator. Examples 6–8, 6–9, and 6–10 will group common inputs and outputs together as a bus. These groups can be displayed in the Waveform Editor in any of four different radixes: bin, hex, oct, or dec. Example 6–10 introduces the concept of using loops in VHDL to perform repetitive operations.

EXAMPLE 6-8

The 74280 Parity Generator Using the MAX+PLUS II Macro-function

Demonstrate the operation of the 74280 parity generator by building a graphic design file and a simulator channel file. While creating the gdf file, when in the **enter symbol** mode, type: 74280b. The 74280b supports bus operations. This symbol will be found in the macro-function symbol library subdirectory "\mf". Provide a binary count on the 9-bit input so that several combinations of odd and even parity are generated.

Helpful Hint

If you draw the connection line from the 74280b to D[8..0] (instead of the other direction), a bus line style (thick line) is chosen automatically.

Solution: The *ex6_8.gdf* file is shown in Figure 6–16. The 9 data input bits are grouped together as a 9-bit bus by specifying the input name as D[8..0]. This shortcut is used instead of specifying D8, D7, and so on, down to D0. The connecting line from the input pin to the parity symbol must be defined as a bus. This is done by right-clicking on the line and choosing **line style > bus** (the thick line).

The *ex6_8.scf* file in Figure 6–17 shows the result of a simulation. The D input waveform is set up as a bus that counts up from 0. The binary radix is selected for the D waveform by right-clicking on D[8..0] and choosing: **Enter Group > Bin > OK.** The two output waveforms prove the operation of the 74280 as specified back in Figure 6–12. The Sum_even waveform is HIGH whenever the sum of the HIGH input bits is even. The Sum_odd waveform is HIGH whenever the sum of the HIGH input bits is odd.

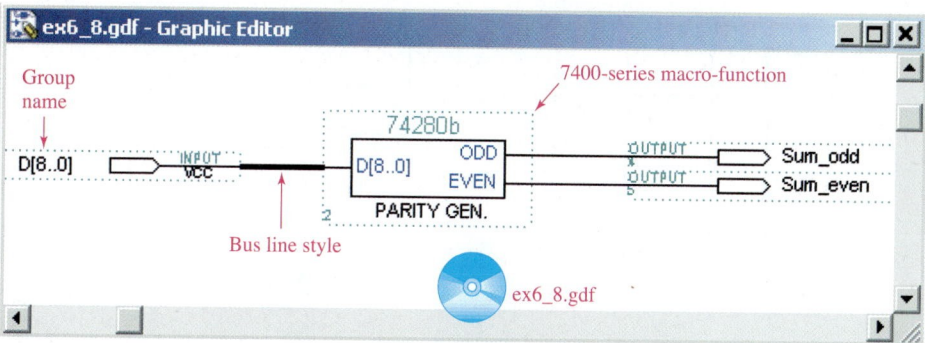

Figure 6-16 The graphic design file for Example 6–8.

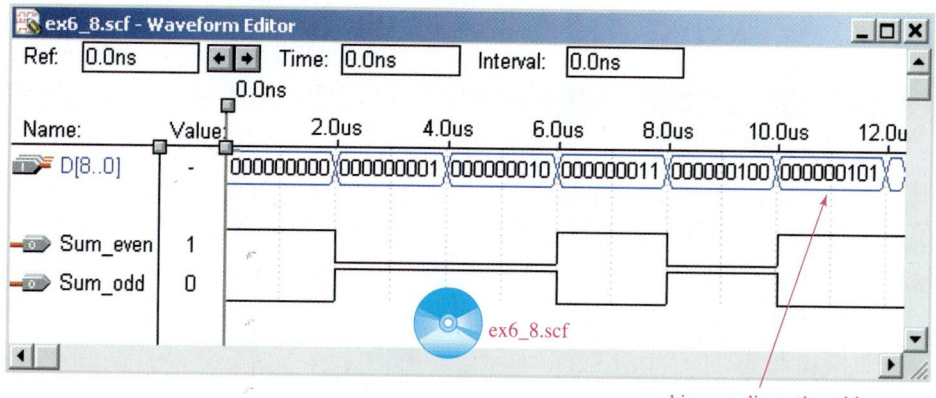

Figure 6-17 The simulator channel file for Example 6–8.

EXAMPLE 6-9

CPLD Parallel Binary Comparator

Reproduce the parallel binary comparator of Example 6–6 using MAX+PLUS II software tools. Complete the circuit using both design entry methods: gdf and VHDL. Test its operation by building an scf file that inputs several 4-bit input combinations at *A[3..0]* and *B[3..0]*. (Make some equal and some not.)

Solution: The graphic design method (*ex6_9.gdf*) is shown in Figure 6–18. All four bits of the A-string are grouped together as a common bus *A[3..0]*. The same with *B[3..0]*.

The simulation file (*ex6_9.scf*) is shown in Figure 6–19. The A and B inputs were initially set up as counters with a hexadecimal radix. Then several of the B values were changed to force inequality. To do this, highlight the hex number that you wish to change, then right-click on it and choose: **overwrite > group value**, then enter a new number and press **OK**. The proof that the circuit works can be seen by noting that the output at *W* goes HIGH whenever the A-bits equal the B-bits.

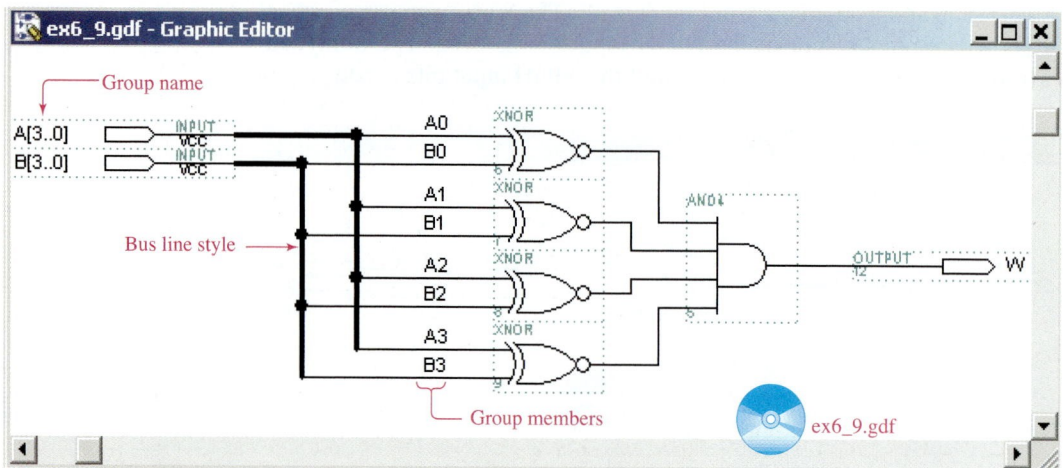

Figure 6–18 The graphic design file for Example 6–9.

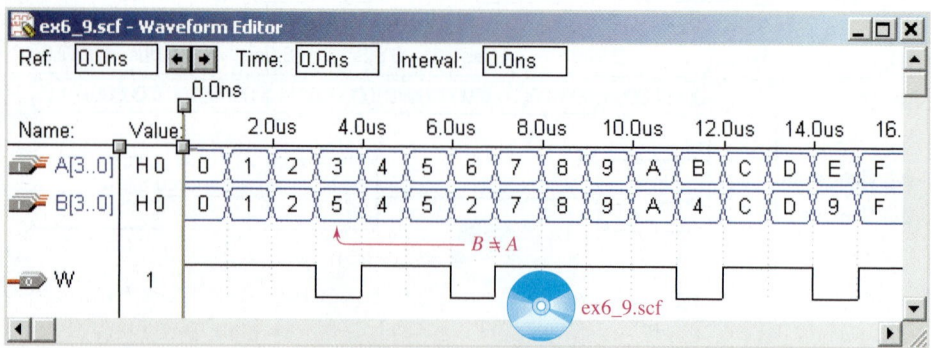

Figure 6–19 The simulator channel file for Example 6–9.

The VHDL design entry method (*ex6_9.vhd*) is shown in Figure 6–20. The results of this design must also be tested by performing a simulation. (*Note:* Be sure that the simulation is being performed on the results of the VHDL design by choosing **File > Project > Set Project To Current File** after you compile the VHDL program.)

```
ex6_9.vhd - Text Editor                              _ □ ×
   LIBRARY ieee;                    --------------------
   USE ieee.std_logic_1164.ALL;     -- Parallel Binary --
                                    --   Comparator    --
   ENTITY ex6_9 IS                  --------------------
      PORT(
         a              : IN std_logic_vector (3 DOWNTO 0);
         b              : IN std_logic_vector (3 DOWNTO 0);
         w              : OUT std_logic);
   END ex6_9 ;


   ARCHITECTURE arc OF ex6_9 IS
   BEGIN
   w<=(a(0) XNOR b(0)) AND (a(1) XNOR b(1)) AND
      (a(2) XNOR b(2)) AND (a(3) XNOR b(3));
   END arc;
                                                  ex6_9.vhd
 Line  15   Col  1    INS ◄
```

Describes the circuit of Figure 6-18

Figure 6–20 The VHDL design file for Example 6–9.

═══ **EXAMPLE 6–10** ═══

CPLD Controlled Inverter

Reproduce the controlled inverter of Example 6–7 using MAX+PLUS II software tools. Complete the circuit using both design entry methods: gdf and VHDL. Test its operation by building an scf file that inputs a count on the data input *d[3..0]* while the control input, *c* randomly goes LOW then HIGH to complement the bits.

Solution: The *ex6_10.gdf* file is shown in Figure 6–21. Note that the data inputs *d[3..0]* and the controlled output *x[3..0]* are grouped together as a bus for simplicity.

The simulation file (*ex6_10.scf*) is shown in Figure 6–22. Notice that when the complementing control signal *c* is LOW, the data bits are passed out to *x* uncomplemented, but when *c* is HIGH, the data bits at *x* are complemented.

The VHDL design entry method (*ex6_10.vhd*) is shown in Figure 6–23. This is our first introduction to sequential process loops. The loop control is useful whenever you need to perform repetitive operations or assignments. In this case we are XORing each data bit input with the complementing control

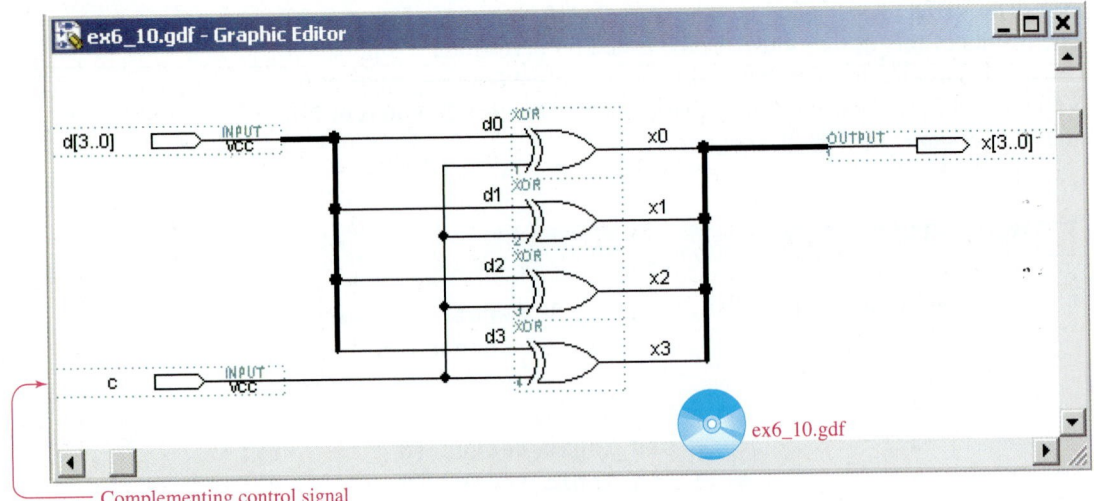

Figure 6–21 The graphic design file for Example 6–10.

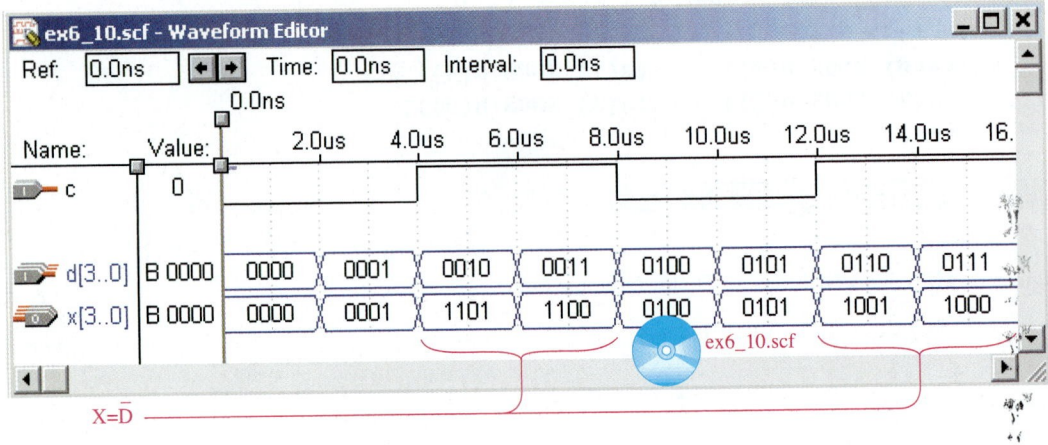

Figure 6–22 The simulator channel file for Example 6–10.

signal to assign each *x* output. This is considered to be a **sequential** operation. This means that when executing the program, *x(3)* is assigned before *x(2)*, and *x(2)* is assigned before *x(1)*, and so on. If, instead of using the process loop, we assigned each output with separate statements we would be making **concurrent** assignments. This way, *x(3)* will receive its logic level concurrently (at the same time) with *x(2)*, *x(1)*, and *x(0)*. The concurrent assignments would be made using the following program segment in place of the process loop:

$$x(3) <= d(3) \; XOR \; c;$$

$$x(2) <= d(2) \; XOR \; c;$$

$$x(1) <= d(1) \; XOR \; c;$$

$$x(0) <= d(0) \; XOR \; c;$$

In this program either method works just as well, but as we will learn, sequential statements will play a much more important role when we design sequential circuits like counters and shift registers in Chapters 12 and 13.

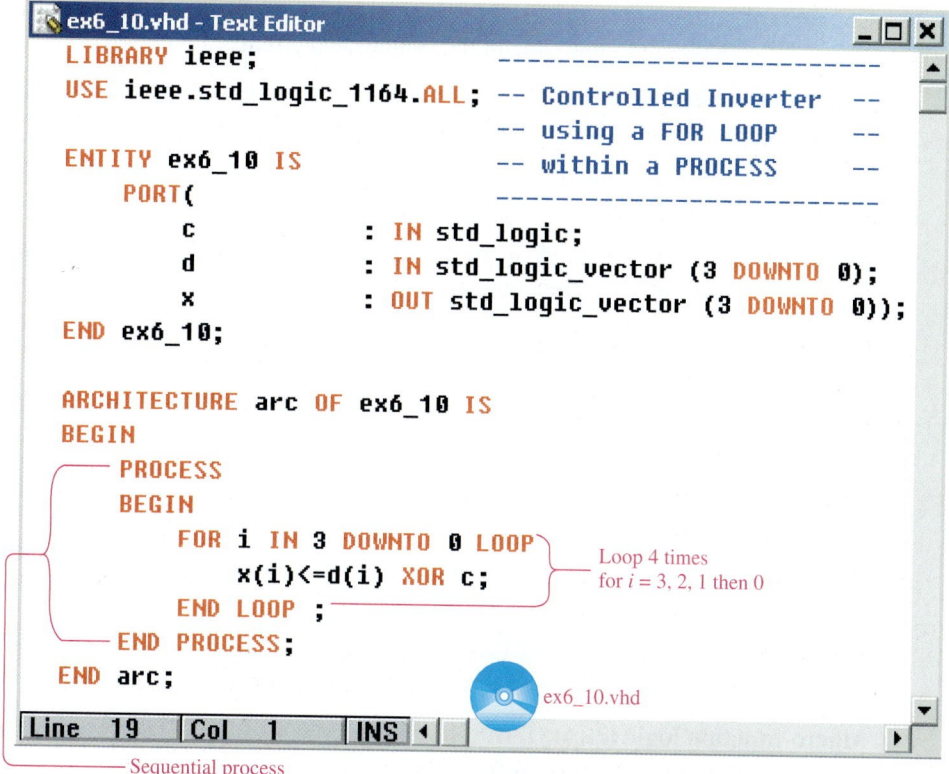

```
ex6_10.vhd - Text Editor                                    _ □ ✕
    LIBRARY ieee;                     --------------------------- ▲
    USE ieee.std_logic_1164.ALL; -- Controlled Inverter  --
                                     -- using a FOR LOOP  --
    ENTITY ex6_10 IS                 -- within a PROCESS  --
        PORT(                        ---------------------------
            c              : IN std_logic;
            d              : IN std_logic_vector (3 DOWNTO 0);
            x              : OUT std_logic_vector (3 DOWNTO 0));
    END ex6_10;


    ARCHITECTURE arc OF ex6_10 IS
    BEGIN
        PROCESS
        BEGIN
            FOR i IN 3 DOWNTO 0 LOOP          Loop 4 times
                x(i)<=d(i) XOR c;            for i = 3, 2, 1 then 0
            END LOOP ;
        END PROCESS;
    END arc;                                          ex6_10.vhd

 Line  19   Col  1     INS ◄                                     ▶
```

Sequential process

Figure 6–23 The VHDL design file for Example 6–10.

Summary

In this chapter, we have learned that

1. The exclusive-OR gate outputs a HIGH if one or the other inputs, but not both, is HIGH.

2. The exclusive-NOR gate outputs a HIGH if both inputs are HIGH or if both inputs are LOW.

3. A parity bit is commonly used for error detection during the transmission of digital signals.

4. Exclusive-OR and NOR gates are used in applications such as parity checking, binary comparison, and controlled complementing circuits.

5. CPLDs can be used to implement circuits containing the exclusive gates.

Glossary

Binary String: Two or more binary bits used collectively to form a meaningful binary representation.

Bus: A group of inputs or outputs having a common use such as bits in a binary string.

Comparator: A device or system that identifies an equality between two quantities.

Concurrent statements: In VHDL, concurrent statements are those that are all executed at the same time in the synthesized circuit.

Controlled Inverter: A digital circuit capable of complementing a binary string of bits based on an external control signal.

Electrical Noise: Unwanted electrical irregularities that can cause a change in a digital logic level.

Error Indicator: A visual display or digital signal that is used to signify that an error has occurred within a digital system.

Exclusive-NOR: A gate that produces a HIGH output for both inputs HIGH or both inputs LOW.

Exclusive-OR: A gate that produces a HIGH output for one or the other input HIGH, but not both.

For Loop: In VHDL, the For Loop allows the programmer to specify multiple iterations of program statements like assignments or circuit definitions.

Function Table: A chart that illustrates the input/output operating characteristics of an integrated circuit.

Group: Inputs or outputs having common characteristics such as bits in a binary string can be put together as a "Group" and referred to as a single name.

Macro-function logic library: A library in the MAX+PLUS II software containing most of the 7400-series fixed-function logic.

Parity: An error-detection scheme used to detect a change in the value of a bit.

Process statement: In VHDL, the Process statement is used to declare the beginning of a series of sequential operations.

Radix: A number system such as: binary, hexadecimal, octal, or decimal.

Sequential statements: In VHDL, sequential statements are those that are all executed one after another in the synthesized circuit.

Transmission: The transfer of digital signals from one location to another.

Problems

Sections 6–1 and 6–2

6–1. Describe in words the operation of an exclusive-OR gate and of an exclusive-NOR gate.

6–2. Describe in words the difference between

(a) An exclusive-OR and an OR gate

(b) An exclusive-NOR and an AND gate

6–3. Complete the timing diagram in Figure P6–3 for the exclusive-OR and the exclusive-NOR.

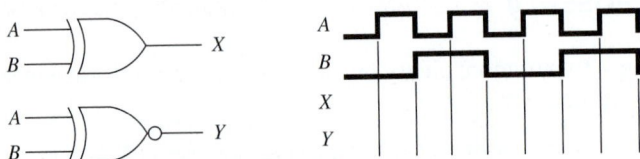

Figure P6-3

6–4. Write the Boolean equations for the circuits in Figure P6–4. Simplify the equations and determine if they function as an Ex-OR, Ex-NOR, or neither.

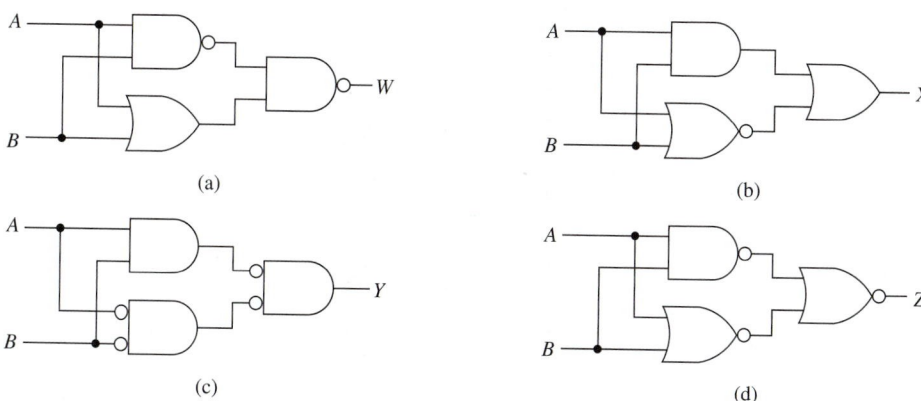

Figure P6-4

D   **6–5.** Design an exclusive-OR gate constructed from all NOR gates.

D   **6–6.** Design an exclusive-NOR gate constructed from all NAND gates.

6–7. Write the Boolean equations for the circuits of Figure P6–7. Reduce the equations to their simplest form.

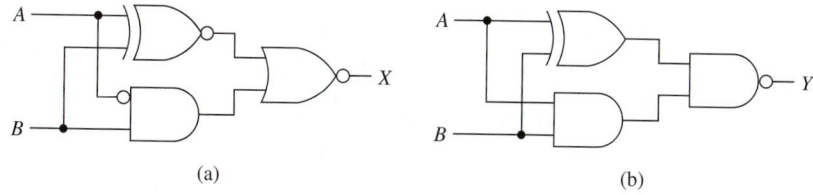

Figure P6-7

C   **6–8.** Repeat Problem 6–7 for the circuits of Figure P6–8.

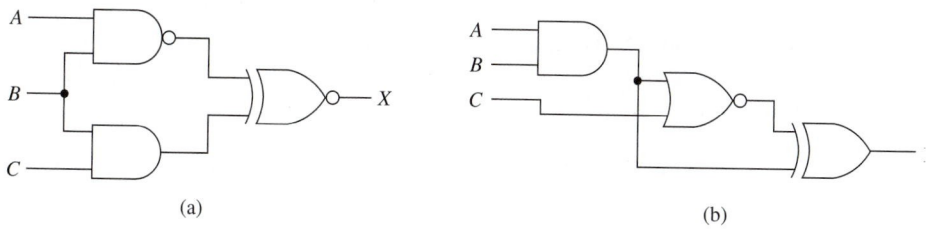

Figure P6-8

Section 6–3

6–9. Convert the following hexadecimal numbers to their 8-bit binary code. Add a parity bit next to the LSB to form odd parity.

A7   4C   79   F3   00   FF

6–10. The pin configuration of the 74HC86 CMOS quad exclusive-OR IC is given in Figure P6–10. Make the external connections to the IC to form a 4-bit even-parity generator.

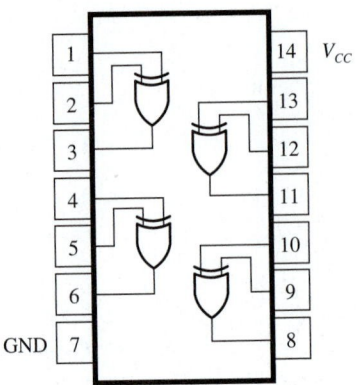

Figure P6-10

6–11. Repeat Problem 6–10 for a 5-bit even-parity checker. Use the pin configuration shown in Figure P6–11.

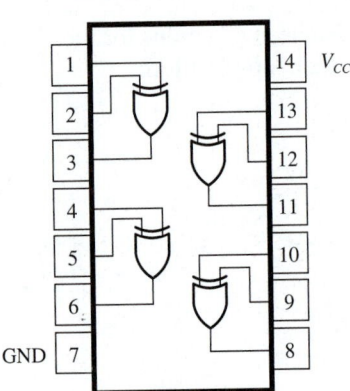

Figure P6-11

Section 6–4

6–12. Figure P6–12 shows another design used to form a 4-bit parity generator. Determine if the circuit will function as an odd- or even-parity generator.

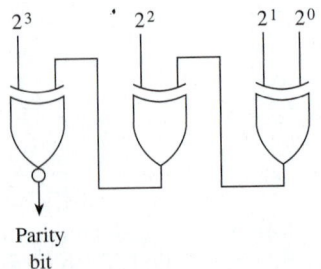

Figure P6-12

C D   **6–13.** Referring to Figure 6–13, design and sketch a 4-bit odd-parity error-detection system. Use two 74280 ICs and a five-line transmission cable between the sending and receiving devices.

C D   **6–14.** Design a binary comparator system similar to Figure 6–14 using exclusive-ORs instead of exclusive-NORs.

C   **6–15.** If the exclusive-ORs in Figure 6–15 are replaced by exclusive-NORs, will the circuit still function as a controlled inverter? If so, should *C* be HIGH or LOW to complement?

Schematic Interpretation Problems

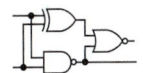

See Appendix G for the schematic diagrams.

C D S   **6–16.** Find Port 1 (P1.7–P1.0) of U8 in the 4096/4196 schematic. On a separate piece of paper, draw an 8-bit controlled inverter for that output port. The inverting function is to be controlled by the P3.5 output (pin 15).

C D S   **6–17.** Find Port 2 (P2.7–P2.0) of U8 in the 4096/4196 schematic. This port outputs the high-order address bits for the system (A8–A15). (Microcontroller addresses are discussed further in Chapter 16.) On a separate piece of paper, draw a binary comparator that compares the 4 bits A8–A11 to the 4 bits A12–A15. The HIGH output for an equal comparison is to be input to P3.4 (pin 14) of U8.

Electronics Workbench®/MultiSIM® Exercises

E6–1. Load the circuit file for **Section 6–2a.** The switches are used to input a 1(up) or a 0(down) to each gate input. The lamp connected to each gate output comes ON if the output is HIGH.

(a) What is the level at X and Y if all switches are up? Try it.

(b) What is the level at X and Y if all switches are down? Try it.

(c) Experimentally complete a truth table for each gate.

E6–2. Load the circuit file for **Section 6–2b.** The Logic Analyzer shows the input waveforms A and B and the output waveforms X and Y. Gate 1 and Gate 2 are hidden from your view, but each is either an Ex-OR or an Ex-NOR. Use the Logic Analyzer display to determine the following:

(a) What is Gate 1, and

(b) What is Gate 2?

T   **E6–3.** Load the circuit file for **Section 6–2c.** This circuit is used to troubleshoot the number-4 gate of a 7486 Quad Ex-OR IC. Because that gate is working OK, the Logic Probe will flash.

(a) What if the unused input (Pin13) was tied to ground instead of Vcc, would the Logic Probe still flash? Why? Try it.

(b) Test the remaining three Ex-OR gates on the chip. Are any bad?

E6–4. Load the circuit file for **Section 6–2d.** Write the Boolean equation at X. Connect the circuit to the Logic Converter and check your answer.

E6–5. Load the circuit file for **Section 6–2e.** Write the simplified Boolean equation at X. Connect the circuit to the Logic Converter and check your answer.

E6–6. Load the circuit file for **Section 6–2f.** Write the simplified Boolean equation at X. Connect the circuit to the Logic Converter and check your answer.

E6–7. Load the circuit file for **Section 6–3.** On a piece of paper, make up a chart for the even parity bit that would be generated for the binary count from 0000 to 1111 (0 to 15). Check all 16 of your answers by pressing "step" on the Word Generator repeatedly as you compare your parity bit with the Even Parity Light. *Note:* The number 1 is an odd number, and the number 2 is even. Why do they both generate an even parity bit?

D

E6–8. Load the circuit file for **Section 6–4.** This is a Parallel Binary Comparator similar to Figure 6–14. Two 4-bit binary strings are provided by the Word Generator.

(a) What type of Word Generator numbers turn the light ON?

(b) Let's say that when you go to build the circuit in lab, you can't find any Ex-NORs but have four Ex-ORs. To get the same circuit function, what must the AND gate be changed to? Try it.

CPLD Problems

The following problems are solved using the Altera MAX+PLUS II software. In each case the design is completed by building a graphic design file (gdf) or a VHDL file (vhd) and then proving the results by producing a simulation (scf) file. [*Note:* If you build a vhd file having the same name as the gdf file be sure to choose **Set Project To Current File** (Under **File > Project**). This will ensure that the compiler uses the current file to synthesize and simulate your design. Also, you can use the same simulation (scf) file for either design method. The simulation will be performed on whichever project file is currently set.]

A final step that can be performed is to download the design to a CPLD on a programmer board like the UP-1 or PLDT-2 and demonstrate it to your instructor.

C6–1. Use the macro-function library to test a parity circuit like in Example 6–8. Use the 74280 (not the 74280b) to determine the odd/even parity for several 1-digit hexadecimal numbers.

(a) Build a gdf file called *prob_c6_1.gdf* using the 74280 macro-function. Use a 4-bit group called *D[3..0]* to provide the hex digit input and include both the *sum_odd* and *sum_even* outputs. Since you will only use four inputs, just ground (gnd) the five unused bits.

(b) Simulate the operation by entering the following hex digits into the *D[3..0]* group of an scf file named *prob_c6_1.scf*: AF19714C. (See Example 6–9 for entering specific group numbers into the waveform.)

(c) Download your design to a CPLD IC. Discuss your observations of the odd and even LEDs with your instructor as you use the switches to step through the eight hex inputs.

C6–2. Redesign the binary comparator of Example 6–9 using Ex-ORs instead of Ex-NORs. Bubble-push the original circuit to determine which gate is required now instead of the AND.

(a) Build a gdf file (*prob_c6_2.gdf*) and run a simulation (*prob_c6_2.scf*) of the circuit with some equal, and some unequal inputs at *A[3..0]* and *B[3..0]*.

(b) Build a VHDL file (*prob_c6_2.vhd*) and run a simulation (*prob_c6_2.scf*) of the circuit with some equal and some unequal inputs at *A[3..0]* and *B[3..0]*.

(c) Download your design to a CPLD IC. Discuss your observations of the *W* output LED with your instructor as you use the switches to step through several combinations of equal and unequal inputs.

C6–3. Redo problem C6-2 (a), (b), and (c) for an 8-bit comparator.

C6–4. MAX+PLUS II provides an 8-bit bus oriented magnitude comparator named **8mcompb.** It compares an A-string with a B-string and provides three outputs indicating less-than, greater-than, and equal. Build a gdf file to exercise this macro-function. Simulate its operation by entering several 2-digit hex numbers as you monitor all three output waveforms.

C6–5. Redo Example 6–10 for an 8-bit controlled inverter.

(a) Build a gdf file and then perform a simulation to observe the invert/non-invert function.

(b) Build a VHDL file and redo the simulation with the VHDL file set as the current project.

(c) Download your design to a CPLD IC. Discuss your observations of the output LEDs with your instructor as you enter a binary string on the switches and use a push-button to control the complementing action.

Answers to Review Questions

6–1. False

6–2. True

6–3. $X = AB + \overline{A}\,\overline{B}$

6–4. False

6–5. False

6–6. Σ_E

6–7. LOW

7 Arithmetic Operations and Circuits

OUTLINE

OBJECTIVES

Upon completion of this chapter, you should be able to:

- Perform the four binary arithmetic functions: addition, subtraction, multiplication, and division.
- Convert positive and negative numbers to signed two's-complement notation.
- Perform two's-complement, hexadecimal, and BCD arithmetic.
- Explain the design and operation of a half-adder and a full-adder circuit.
- Utilize full-adder ICs to implement arithmetic circuits.
- Explain the operation of a two's-complement adder/subtractor circuit and a BCD adder circuit.
- Explain the function of an arithmetic/logic unit (ALU).
- Implement arithmetic functions in CPLDs using VHDL.

INTRODUCTION

An important function of digital systems and computers is the execution of arithmetic operations. In this chapter, we will see that there is no magic in taking the sum of two numbers electronically. Instead, there is a basic set of logic-circuit building blocks, and the arithmetic operations follow a step-by-step procedure to arrive at the correct answer. All the "electronic arithmetic" will be performed using digital input and output levels with basic combinational logic circuits or medium-scale-integration (MSI) chips.

7–1 Binary Arithmetic

Before studying the actual digital electronic requirements for arithmetic circuits, let's look at the procedures for performing the four basic arithmetic functions: addition, subtraction, multiplication, and division.

Addition

The procedure for adding numbers in binary is similar to adding in decimal, except that the binary sum is made up of only 1's and 0's. When the binary **sum** exceeds 1, you must carry a 1 to the next-more-significant column, as in regular decimal addition.

The four possible combinations of adding two binary numbers can be stated as follows:

$$0 + 0 = 0 \text{ carry } 0$$

$$0 + 1 = 1 \text{ carry } 0$$

$$1 + 0 = 1 \text{ carry } 0$$

$$1 + 1 = 0 \text{ carry } 1$$

The general form of binary addition in the least significant column can be written

$$A_0 + B_0 = \Sigma_0 + C_{out}$$

The sum output is given by the *summation* symbol (Σ), called sigma, and the **carry-out** is given by C_{out}. The truth table in Table 7–1 shows the four possible conditions when adding two binary digits.

TABLE 7–1	Truth Table for Addition of Two Binary Digits in the Least Significant Column		
A_0	B_0	Σ_0	C_{out}
0	0	0	0
0	1	1	0
1	0	1	0
1	1	0	1

If a carry-out *is* produced, it must be added to the next-more-significant column as a **carry-in** (C_{in}). Figure 7–1 shows this operation and truth table. In the truth table, the C_{in} term comes from the value of C_{out} from the previous addition. Now, with three possible inputs, there are eight combinations of outputs ($2^3 = 8$). Review the truth table to be sure that you understand how each sum and carry were determined.

A_1	B_1	C_{in}	Σ_1	C_{out}
0	0	0	0	0
0	0	1	1	0
0	1	0	1	0
0	1	1	0	1
1	0	0	1	0
1	0	1	0	1
1	1	0	0	1
1	1	1	1	1

Figure 7–1 Addition in the more significant columns requires including C_{in} with $A_1 + B_1$.

Now let's perform some binary additions. We represent all binary numbers in groups of 8 or 16 because that is the standard used for arithmetic in most digital computers today.

EXAMPLE 7–1

Perform the following decimal additions. Convert the original decimal numbers to binary and add them. Compare answers.(a) $5 + 2$; (b) $8 + 3$;(c) $18 + 2$;(d) $147 + 75$;(e) $31 + 7$.

Solution:

	Decimal	**Binary**
(a)	5 +2 — 7	0000 0101 +0000 0010 ———— 0000 0111 $= 7_{10}$ ✓
(b)	8 + 3 — 11	0000 1000 +0000 0011 ———— 0000 1011 $= 11_{10}$ ✓
(c)	18 + 2 — 20	0001 0010 +0000 0010 ———— 0001 0100 $= 20_{10}$ ✓
(d)	147 + 75 — 222	1001 0011 +0100 1011 ———— 1101 1110 $= 222_{10}$ ✓
(e)	31 + 7 — 38	0001 1111 +0000 0111 ———— 0010 0110 $= 38_{10}$ ✓

Subtraction

The four possible combinations of subtracting two binary numbers can be stated as follows:

$$0 - 0 = 0 \text{ borrow } 0$$

$$0 - 1 = 1 \text{ borrow } 1$$

$$1 - 0 = 1 \text{ borrow } 0$$

$$1 - 1 = 0 \text{ borrow } 0$$

The general form of binary subtraction in the least significant column can be written

$$A_0 - B_0 = R_0 + B_{out}$$

The difference, or **remainder,** from the subtraction is R_0, and if a **borrow** is required, B_{out} is 1. The truth table in Table 7–2 shows the four possible conditions when subtracting two binary digits.

If a borrow *is* required, the A_0 must borrow from A_1 in the next-more-significant column. When A_0 borrows from its left, A_0 increases by 2 (just as in decimal subtraction, where the number increases by 10). For example, let's subtract $2 - 1$ ($10_2 - 01_2$).

TABLE 7–2	Truth Table for Subtraction of Two Binary Digits in the Least Significant Column		
A_0	B_0	R_0	B_{out}
0	0	0	0
0	1	1	1
1	0	1	0
1	1	0	0

$\leftarrow$ Borrow required because $A_0 < B_0$

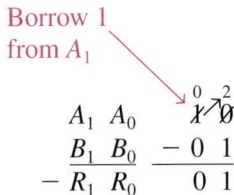

Because A_0 was 0, it borrowed 1 from A_1. A_1 becomes a 0, and A_0 becomes 2 (2_{10} or 10_2). Now the subtraction can take place: in the LS column, $2 - 1 = 1$, and in the MS column, $0 - 0 = 0$.

As you can see, the second column and all more significant columns first have to determine if A was borrowed from before subtracting $A - B$. Therefore, they have three input conditions, for a total of eight different possible combinations, as illustrated in Figure 7–2.

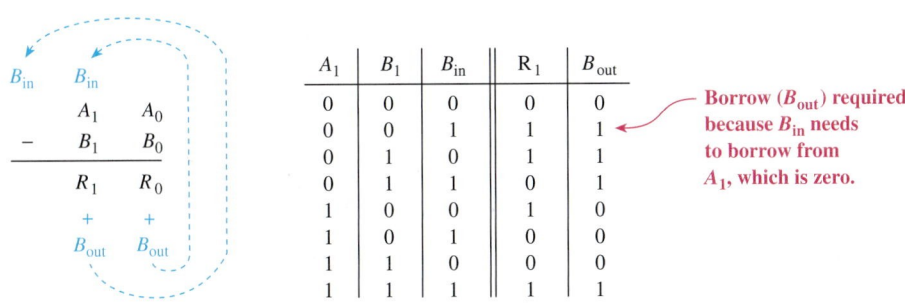

$\leftarrow$ Borrow (B_{out}) required because B_{in} needs to borrow from A_1, which is zero.

A_1	B_1	B_{in}	R_1	B_{out}
0	0	0	0	0
0	0	1	1	1
0	1	0	1	1
0	1	1	0	1
1	0	0	1	0
1	0	1	0	0
1	1	0	0	0
1	1	1	1	1

Figure 7–2 Subtraction in the more significant columns.

Helpful Hint

This table is difficult for most students. It helps to remind yourself where B_{in} comes from and what causes B_{out} to be 1.

The outputs in the truth table in Figure 7–2 are a little more complicated to figure out. To help you along, let's look at the subtraction $4 - 1$ ($0100_2 - 0001_2$):

$$
\begin{array}{ll}
4_{10} & A_3A_2A_1A_0 \\
- 1_{10} & A_3A_2A_1A_0 \\
\hline
3_{10} & R_3R_2R_1R_0
\end{array}
\qquad
\begin{array}{cccc}
0 & 1 & 0 & 0 \\
-0 & 0 & 0 & 1 \\
\hline
0 & 0 & 1 & 1 = 3_{10}
\end{array}
$$

To subtract $0100 - 0001$, A_0 must borrow from A_1, but A_1 is 0. Therefore, A_1 must first borrow from A_2, making A_2 a 0. Now A_1 is a 2. A_0 borrows from A_1, making A_1 a 1 and A_0 a 2. Now we can subtract to get 0011 (3_{10}). Actually, the process is very similar to the process you learned many years ago for regular decimal subtraction. Work through each entry in the truth table (Figure 7–2) to determine how it was derived.

Fortunately, as we will see in Section 7–2, digital computers use a much easier method for subtracting binary numbers, called *two's complement*. We do, however, need to know the standard method for subtracting binary numbers. Work through the following example to better familiarize yourself with the binary subtraction procedure.

EXAMPLE 7-2

Perform the following decimal subtractions. Convert the original decimal numbers to binary and subtract them. Compare answers. (a) $27 - 10$; (b) $9 - 4$; (c) $172 - 42$; (d) $154 - 54$; (e) $192 - 3$.

Solution:

	Decimal	**Binary**
(a)	27 $-$ 10 17	0001 1011 $-$ 0000 1010 0001 0001 = 17_{10} ✓
(b)	9 $-$ 4 5	0000 1001 $-$ 0000 0100 0000 0101 = 5_{10} ✓
(c)	172 $-$ 42 130	1010 1100 $-$ 0010 1010 1000 0010 = 130_{10} ✓
(d)	154 $-$ 54 100	1001 1010 $-$ 0011 0110 0110 0100 = 100_{10} ✓
(e)	192 $-$ 3 189	1100 0000 $-$ 0000 0011 1011 1101 = 189_{10} ✓

Multiplication

Binary multiplication is like decimal multiplication, except you deal only with 1's and 0's. Figure 7–3 illustrates the procedure for multiplying 13×11.

Decimal	*Binary*
13 $\times 11$ 13 13 143	0000 1101 (multiplicand) $\times$ 0000 1011 (multiplier) 0000 1101 00001 101 000000 00 0000110 1 0001000 1111 (product)

8-bit answer = 1000 1111 = 143_{10} ✓

Figure 7–3 Binary multiplication procedure.

The procedure for the multiplication in Figure 7–3 is as follows:

1. Multiply the 2^0 bit of the multiplier times the multiplicand.

2. Multiply the 2^1 bit of the multiplier times the multiplicand. Shift the result one position to the left before writing it down.

CHAPTER 7 | ARITHMETIC OPERATIONS AND CIRCUITS

3. Repeat step 2 for the 2^2 bit of the multiplier. Because the 2^2 bit is a 0, the result is 0.

4. Repeat step 2 for the 2^3 bit of the multiplier.

5. Repeating step 2 for the four leading 0's in the multiplier will have no effect on the answer, so don't bother.

6. Take the sum of the four partial products to get the final **product** of 143_{10}. (Written as an 8-bit number, the product is $1000\ 1111_2$.)

EXAMPLE 7–3

Perform the following decimal multiplications. Convert the original decimal numbers to binary and multiply them. Compare answers. **(a)** 5×3; **(b)** 45×3; **(c)** 15×15; **(d)** 23×9.

Solution:

Common Misconception

Most errors in binary multiplication occur when students are careless in the vertical alignment of the addition columns.

```
        Decimal              Binary
(a)        5               0000  0101
        ×  3             × 0000  0011
          ──               ─────────
          15                0000  0101
                         + 00000  101
                           ──────────
                           00000  1111 = 0000  1111 = 15₁₀ ✓
```

```
(b)       45               0010  1101
        ×  3             × 0000  0011
         ───               ─────────
         135                0010  1101
                         + 00101  101
                           ──────────
                           01000  0111 = 1000  0111 = 135₁₀ ✓
```

```
(c)       15                 0000  1111
        × 15              × 0000  1111
         ───               ──────────
          75                 0000  1111
        + 15                00001  111
         ───               000011  11
         225             + 0000111  1
                           ───────────
                           0001110 0001 = 1110  0001 = 225₁₀ ✓
```

```
(d)       23               0001  0111
        ×  9             × 0000  1001
         ───               ─────────
         207                0001  0111
                           00000  000
                          000000  00
                          0001011  1
                           ──────────
                          0001100  1111    = 1100  1111 = 207₁₀ ✓
```

Team Discussion

Develop a method to determine the value to carry when adding columns with several 1's in them, such as those encountered when multiplying 15×15.

Division

Binary division uses the same procedure as decimal division. Example 7–4 illustrates this procedure.

Helpful Hint

It is beneficial to review the procedure for base 10 long division that you learned in grade school.

EXAMPLE 7–4

Perform the following decimal divisions. Convert the original decimal numbers to binary and divide them. Compare answers. **(a)** $9 \div 3$; **(b)** $35 \div 5$; **(c)** $135 \div 15$; **(d)** $221 \div 17$.

Solution:

Decimal   **Binary**

(a)
$$3\overline{)9} \quad\quad 3$$
$$-9$$
$$\overline{0}$$

$$0000\ 0011\overline{)0000\ 1001} \quad 11 = 3_{10}\ \checkmark$$
$$\quad\quad\quad -\ \ 11$$
$$\quad\quad\quad \overline{\quad 11}$$
$$\quad\quad\quad -\ 11$$
$$\quad\quad\quad \overline{\quad 0}$$

(b)
$$5\overline{)35} \quad\quad 7$$
$$-35$$
$$\overline{0}$$

$$0000\ 0101\overline{)0010\ 0011} \quad 111 = 7_{10}\ \checkmark$$
$$\quad\quad\quad -\ 1\ 01$$
$$\quad\quad\quad \overline{\quad 111}$$
$$\quad\quad\quad -\ 101$$
$$\quad\quad\quad \overline{\quad 101}$$
$$\quad\quad\quad -\ 101$$
$$\quad\quad\quad \overline{\quad 0}$$

(c)
$$15\overline{)135} \quad\quad 9$$
$$-135$$
$$\overline{0}$$

$$0000\ 1111\overline{)1000\ 0111} \quad 1001 = 9_{10}\ \checkmark$$
$$\quad\quad\quad -\ 111\ 1$$
$$\quad\quad\quad \overline{\quad 1111}$$
$$\quad\quad\quad -\ 1111$$
$$\quad\quad\quad \overline{\quad 0}$$

(d)
$$17\overline{)221} \quad\quad 13$$
$$-17$$
$$\overline{51}$$
$$51$$
$$\overline{0}$$

$$0001\ 0001\overline{)1101\ 1101} \quad 1101 = 13_{10}\ \checkmark$$
$$\quad\quad\quad -\ 1000\ 1$$
$$\quad\quad\quad \overline{\quad 101\ 01}$$
$$\quad\quad\quad -\ 100\ 01$$
$$\quad\quad\quad \overline{\quad 1\ 0001}$$
$$\quad\quad\quad -\ 1\ 0001$$
$$\quad\quad\quad \overline{\quad 0}$$

Review Questions

7–1. Binary addition in the least significant column deals with how many inputs and how many outputs?

7–2. In binary subtraction, the borrow-out of the least significant column becomes the borrow-in of the next-more-significant column. True or false?

7–3. Binary multiplication and division are performed by a series of additions and subtractions. True or false?

7–2   Two's-Complement Representation

The most widely used method of representing binary numbers and performing arithmetic in computer systems is by using the **two's-complement method.** With this method, both positive and negative numbers can be represented using the same format, and binary subtraction is greatly simplified.

All along we have seen representing binary numbers in groups of eight for a reason. Most computer systems are based on 8- or 16-bit numbers. In an 8-bit system, the total number of different combinations of bits is 256 (2^8); in a 16-bit system, the number is 65,536 (2^{16}).

To be able to represent both positive *and* negative numbers, the two's-complement format uses the most significant bit (MSB) of the 8- or 16-bit number to signify whether the number is positive or negative. The MSB is, therefore, called the **sign bit** and is defined as 0 for positive numbers and 1 for negative numbers. *Signed two's-complement numbers are shown in Figure 7–4.*

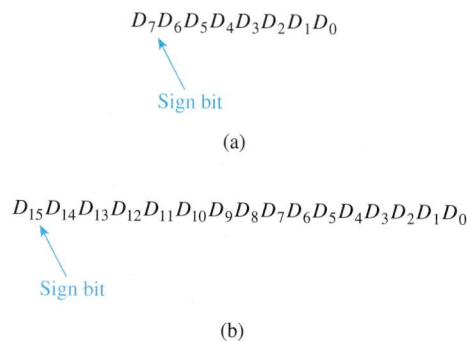

(a)

(b)

Figure 7–4   Two's-complement numbers: (a) 8-bit number; (b) 16-bit number.

The *range of positive numbers* in an 8-bit system is 0000 0000 to 0111 1111 (0 to 127). The *range of negative numbers* is 1111 1111 to 1000 0000 (-1 to -128). In general, the maximum positive number is equal to $2^{N-1} - 1$, and the maximum negative number is $-(2^{N-1})$, where N is the number of bits in the number, including the sign bit (e.g., for an 8-bit positive number, $2^{8-1} - 1 = 127$).

A table of two's-complement numbers can be developed by starting with some positive number and continuously subtracting 1. Table 7–3 shows the signed two's-complement numbers from $+7$ to -8.

Converting a decimal number to two's complement, and vice versa, is simple and can be done easily using logic gates, as we will see later in this chapter. For now, let's deal with 8-bit numbers; however, the procedure for 16-bit numbers is exactly the same.

Steps for Decimal-to-Two's-Complement Conversion

1. If the decimal number is positive, the two's-complement number is the true binary equivalent of the decimal number (e.g., $+18 = 0001\ 0010$).

2. If the decimal number is negative, the two's-complement number is found by
   (a) Complementing each bit of the true binary equivalent of the decimal number (this is called the **one's complement**).
   (b) Adding 1 to the one's-complement number to get the magnitude bits. (The sign bit will always end up being 1.)

Team Discussion

Try to represent the number 160_{10} in two's-complement for an 8-bit system. Why doesn't it work?

TABLE 7–3	Signed Two's-Complement Numbers + 7 Through − 8

Decimal	Two's Complement
+ 7	0000 0111
+ 6	0000 0110
+ 5	0000 0101
+ 4	0000 0100
+ 3	0000 0011
+ 2	0000 0010
+ 1	0000 0001
0	0000 0000
− 1	1111 1111
− 2	1111 1110
− 3	1111 1101
− 4	1111 1100
− 5	1111 1011
− 6	1111 1010
− 7	1111 1001
− 8	1111 1000

Steps for Two's-Complement-to-Decimal Conversion

1. If the two's-complement number is positive (sign bit = 0), do a regular binary-to-decimal conversion.

2. If the two's-complement number is negative (sign bit = 1), the decimal sign will be −, and the decimal number is found by
   (a) Complementing the entire two's-complement number, bit by bit.
   (b) Adding 1 to arrive at the true binary equivalent.
   (c) Doing a regular binary-to-decimal conversion to get the decimal numeric value.

The following examples illustrate the conversion process.

Common Misconception

As soon as some students see the phrase "convert to two's complement," they go ahead with the procedure for negative numbers whether the original number is positive or negative.

EXAMPLE 7–5

Convert $+ 35_{10}$ to two's complement.

Solution:

$$\text{True binary} = 0010\ 0011$$
$$\text{Two's complement} = 0010\ 0011 \quad \textit{Answer}$$

EXAMPLE 7–6

Convert $- 35_{10}$ to two's complement.

Solution:

$$\text{True binary} = 0010\ 0011$$
$$\text{One's complement} = 1101\ 1100$$
$$\text{Add 1} = \qquad +1$$
$$\text{Two's complement} = 1101\ 1101 \quad \textit{Answer}$$

EXAMPLE 7-7

Convert 1101 1101 two's complement back to decimal.

Solution: The sign bit is 1, so the decimal result will be negative.

$$\begin{array}{rl}
\text{Two's complement} = & 1101\ 1101 \\
\text{Complement} = & 0010\ 0010 \\
\text{Add 1} = & +1 \\
\hline
\text{True binary} = & 0010\ 0011 \\
\text{Decimal complement} = & -35 \quad \textit{Answer}
\end{array}$$

EXAMPLE 7-8

Convert -98_{10} to two's complement.

Solution:

$$\begin{array}{rl}
\text{True binary} = & 0110\ 0010 \\
\text{One's complement} = & 1001\ 1101 \\
\text{Add 1} = & +1 \\
\hline
\text{Two's complement} = & 1001\ 1110 \quad \textit{Answer}
\end{array}$$

EXAMPLE 7-9

Convert 1011 0010 two's complement to decimal.

Solution: The sign bit is 1, so the decimal result will be negative.

$$\begin{array}{rl}
\text{Two's complement} = & 1011\ 0010 \\
\text{Complement} = & 0100\ 1101 \\
\text{Add 1} = & +1 \\
\hline
\text{True binary} = & 0100\ 1110 \\
\text{Decimal complement} = & -78 \quad \textit{Answer}
\end{array}$$

Review Questions

7-4. Which bit in an 8-bit two's-complement number is used as the sign bit?

7-5. Are the following two's-complement numbers positive or negative?
   (a) 1010 0011
   (b) 0010 1101
   (c) 1000 0000

7-3 Two's-Complement Arithmetic

All four of the basic arithmetic functions involving positive *and* negative numbers can be dealt with very simply using two's-complement arithmetic. Subtraction is done by *adding* the two two's-complement numbers. Thus, the same digital circuitry can be used for additions *and* subtractions, and there is no need always to subtract the smaller number from the larger number. We must be careful, however, not to exceed the *maximum range* of the two's-complement number: $+127$ to -128 for 8-bit systems, and $+32,767$ to $-32,768$ for 16-bit systems ($+2^{N-1} - 1$ to -2^{N-1}).

When *adding* numbers in the two's-complement form, simply perform a regular binary addition to get the result. When *subtracting* numbers in the two's-complement form, convert the number being subtracted to a *negative* two's-complement number, and perform a regular binary addition [e.g., $5 - 3 = 5 + (-3)$]. The result will be a two's-complement number, and if the result is negative, the sign bit will be 1.

Work through the following examples to familiarize yourself with the addition and subtraction procedure.

EXAMPLE 7–10

Add $19 + 27$ using 8-bit two's-complement arithmetic.

Solution:

$$
\begin{aligned}
19 &= 0001\ 0011 \\
27 &= \underline{0001\ 1011} \\
\text{Sum} &= 0010\ 1110 = 46_{10}
\end{aligned}
$$

EXAMPLE 7–11

Perform the following subtractions using 8-bit two's-complement arithmetic.

(a) $18 - 7$;

(b) $21 - 13$;

(c) $118 - 54$;

(d) $59 - 96$.

Solution:

(a) $18 - 7$ is the same as $18 + (-7)$, so just add 18 to negative 7.

$$
\begin{aligned}
+18 &= 0001\ 0010 \\
-7 &= \underline{1111\ 1001} \\
\text{Sum} &= 0000\ 1011 = 11_{10}
\end{aligned}
$$

Note: The carry-out of the MSB is ignored. (It will always occur for positive sums.) The 8-bit answer is 0000 1011.

(b)
$$
\begin{aligned}
+21 &= 0001\ 0101 \\
-13 &= \underline{1111\ 0011} \\
\text{Sum} &= 0000\ 1000 = 8_{10}
\end{aligned}
$$

(c)
$$
\begin{aligned}
+118 &= 0111\ 0110 \\
-54 &= \underline{1100\ 1010} \\
\text{Sum} &= 0100\ 0000 = 64_{10}
\end{aligned}
$$

(d)
$$
\begin{aligned}
+59 &= 0011\ 1011 \\
-96 &= \underline{1010\ 0000} \\
\text{Sum} &= 1101\ 1011 = -37_{10}
\end{aligned}
$$

7–6. Which of the following decimal numbers cannot be converted to 8-bit two's-complement notation?

(a) 89

(b) 135

(c) − 107

(d) − 144

7–7. The procedure for subtracting numbers in two's-complement notation is exactly the same as for adding numbers. True or false?

7–8. When subtracting a smaller number from a larger number in two's complement, there will always be a carry-out of the MSB, which will be ignored. True or false?

7–4 Hexadecimal Arithmetic

Hexadecimal representation, as discussed in Chapter 1, is a method of representing groups of 4 bits as a single digit. Hexadecimal notation has been widely adopted by manufacturers of computers and microprocessors because it simplifies the documentation and use of their equipment. Eight- and 16-bit computer system data, program instructions, and addresses use hexadecimal to make them easier to interpret and work with than their binary equivalents.

Hexadecimal Addition

Remember, hexadecimal is a base 16 numbering system, meaning that it has 16 different digits (as shown in Table 7–4). Adding 3 + 6 in hex equals 9, and 5 + 7 equals C. But, adding 9 + 8 in hex equals a sum greater than F, which will create a carry. The sum of 9 + 8 is 17_{10}, which is 1 larger than 16, making the answer 11_{16}.

TABLE 7–4	Hexadecimal Digits with Their Equivalent Binary and Decimal Values	
Hexadecimal	**Binary**	**Decimal**
0	0000	0
1	0001	1
2	0010	2
3	0011	3
4	0100	4
5	0101	5
6	0110	6
7	0111	7
8	1000	8
9	1001	9
A	1010	10
B	1011	11
C	1100	12
D	1101	13
E	1110	14
F	1111	15

The procedure for adding hex digits is as follows:

1. Add the two hex digits by working with their decimal equivalents.

2. If the decimal sum is less than 16, write down the hex equivalent.

3. If the decimal sum is 16 or more, subtract 16, write down the hex result in that column, and carry 1 to the next-more-significant column.

Work through the following examples to familiarize yourself with this procedure.

EXAMPLE 7–12

Add 9 + C in hex.

Solution: C is equivalent to decimal 12.

$$12 + 9 = 21$$

Because 21 is greater than 16: **(a)** subtract $21 - 16 = 5$, and **(b)** carry 1 to the next-more-significant column. Therefore,

$$9 + C = 15_{16} \quad \textit{Answer}$$

EXAMPLE 7–13

Add 4F + 2D in hex.

Solution:

$$
\begin{array}{r}
4\,F \\
+\ 2\,D \\
\hline
7\,C
\end{array}
\quad \textit{Answer}
$$

Explanation: $F + D = 15 + 13 = 28$, which is 12 with a carry ($28 - 16 = 12$). The 12 is written down as C; $4 + 2 + \text{carry} = 7$.

EXAMPLE 7–14

Add A7C5 + 2DA8 in hex.

Solution:

$$
\begin{array}{r}
A\,7\,C\,5 \\
+\ 2\,D\,A\,8 \\
\hline
D\,5\,6\,D
\end{array}
\quad \textit{Answer}
$$

Explanation: $5 + 8 = 13$, which is D, $C + A = 22$, which is 6 with a carry. $7 + D + \text{carry} = 21$, which is 5 with a carry. $A + 2 + \text{carry} = 13$, which is D.

Alternative Method: An alternative method of hexadecimal addition, which you might find more straightforward, is to convert the hex numbers to binary and perform a regular binary addition. The binary sum is then converted back to hex. For example:

$$\begin{array}{r} 4\,F \\ + \;\underline{2\,D} \end{array} \Rightarrow \begin{array}{r} 0100\;\;1111_2 \\ + \;\underline{0010\;\;1101_2} \\ 0111\;\;1100_2 = 7\,C_{16}\;\checkmark \end{array}$$

Hexadecimal Subtraction

Subtraction of hexadecimal numbers is similar to decimal subtraction, except that when you borrow 1 from the left, the borrower increases in value by 16. Consider the hexadecimal subtraction $24 - 0C$.

$$\begin{array}{r} 24 \\ - \;\underline{0C} \\ 18 \end{array}$$

Explanation: We cannot subtract C from 4, so the 4 borrows 1 from the 2. This changes the 2 to a 1, and the 4 increases in value to 20 $(4 + 16 = 20)$. Now, $20 - C = 20 - 12 = 8$, and $1 - 0 = 1$. Therefore,

$$24 - 0C = 18$$

The next two examples illustrate hexadecimal subtraction.

EXAMPLE 7–15

Subtract $D7 - A8$ in hex.

Solution:

$$\begin{array}{r} D\,7 \\ - \;\underline{A\,8} \\ 2\,F \end{array} \quad \textit{Answer}$$

Explanation: 7 borrows from the D, which increases its value to 23 $(7 + 16 = 23)$, and $23 - 8 = 15$, which is an F. D becomes a C, and $C - A = 12 - 10 = 2$.

EXAMPLE 7–16

Subtract $A05C - 24CA$ in hex.

Solution:

$$\begin{array}{r} A\,0\,5\,C \\ - \;\underline{2\,4\,C\,A} \\ 7\,B\,9\,2 \end{array} \quad \textit{Answer}$$

Explanation: $C - A = 12 - 10 = 2$. The 5 borrows from the 0, which borrows from the A $(5 + 16 = 21)$; $21 - C = 21 - 12 = 9$. The 0 borrowed from the A, but it was also borrowed from, so it is now a 15; $15 - 4 = 11$, which is a B. The A was borrowed from, so it is now a 9; $9 - 2 = 7$.

7–9. Why is hexadecimal arithmetic commonly used when working with 8-, 16-, and 32-bit computer systems?

7–10. When adding two hex digits, if the sum is greater than _____ (9, 15, 16), the result will be a two-digit answer.

7–11. When subtracting hex digits, if the least significant digit borrows from its left, its value increases by _____ (10, 16).

7–5 BCD Arithmetic

If human beings had 16 fingers and toes, we probably would have adopted hexadecimal as our primary numbering system instead of decimal, and dealing with microprocessor-generated numbers would have been so much easier. (Just think how much better we could play a piano, too!) But, unfortunately, we normally deal in base 10 decimal numbers. Digital electronics naturally works in binary, and we have to group four binary digits together to get enough combinations to represent the 10 different decimal digits. This 4-bit code is called *binary-coded decimal (BCD)*.

So what we have is a 4-bit code that is used to represent the decimal digits that we need when reading a display on calculators or computer output. The problem arises when we try to add or subtract these BCD numbers. For example, digital circuitry would naturally like to add the BCD numbers 1000 + 0011 to get 1011, but 1011 is an invalid BCD result. (In Chapter 1, we described the range of valid BCD numbers as 0000 to 1001.) Therefore, when adding BCD numbers, we have to build extra circuitry to check the result to be certain that each group of 4 bits is a valid BCD number.

BCD Addition

Addition is the most important operation because subtraction, multiplication, and division can all be done by a series of additions or two's-complement additions.

The procedure for BCD addition is as follows:

1. Add the BCD numbers as regular true binary numbers.

2. If the sum is 9 (1001) or less, it is a valid BCD answer; leave it as is.

3. If the sum is greater than 9 or there is a carry-out of the MSB, it is an invalid BCD number; do step 4.

4. If it is invalid, add 6 (0110) to the result to make it valid. Any carry-out of the MSB is added to the next-more-significant BCD number.

5. Repeat steps 1 to 4 for each group of BCD bits.

Use this procedure for the following example.

EXAMPLE 7–17

Convert the following decimal numbers to BCD and add them. Convert the result back to decimal to check your answer.

(a) 8 + 7;

(b) 9 + 9;

(c) $52 + 63$;

(d) $78 + 69$.

Solution:

(a) $8 = 1000$
      $+\ 7 = \underline{0111}$
      Sum $= \overline{1111}$ (invalid BCD, so add six)
      Add $6 = \underline{0110}$
          $1\quad \overline{0101} = 0001\ 0101_{BCD} = 15_{10}$ √

(b) $9 = \quad 1001$
      $+\ 9 = \quad \underline{1001}$
      Sum $= 1\ \overline{0010}$ (invalid because of carry)
              ↖ cy
      Add $6 = \quad \underline{0110}$
          $1\ \ \overline{1000} = 0001\ 1000_{BCD} = 18_{10}$ √

(c) $52 = \ 0101\ 0010$
      $+\ 63 = \ \underline{0110\ 0011}$
      Sum $= \ 1011\ 0101$
      Add $6 = \ \underline{0110} \qquad$ invalid
          $1\ 0001\ 0101 = 0001\ 0001\ 0101 = 115_{10}$ √

(d) $78 = \ 0111\ 1000$
      $+\ 69 = \ \underline{0110\ 1001}$
      Sum $= \ 1110\ 0001 \qquad$ (Both groups of 4
              ↖ cy $\qquad\qquad$ BCD bits are invalid)
      Add $6 = \ \quad\quad \underline{0110}$
          $1110\ 0111$
      Add $6 = \ \underline{0110}$
          $1\ 0100\ 0111 = 0001\ 0100\ 0111 = 147_{10}$ √

When one of the numbers being added is negative (such as in subtraction), the procedure is much more difficult, but it basically follows a complement-then-add procedure, which is not covered in this book but is similar to that introduced in Section 7–3.

Now that we understand the more common arithmetic operations that take place within digital equipment, we are ready for the remainder of the chapter, which explains the actual circuitry used to perform these operations.

Review Questions

7–12. When adding two BCD digits, the sum is invalid and needs correction if it is _____ or if _____.

7–13. What procedure is used to correct the result of a BCD addition if the sum is greater than 9?

7–6 Arithmetic Circuits

All the arithmetic operations and procedures covered in the previous sections can be implemented using adders formed from the basic logic gates. For a large number of digits we can use **medium-scale-integration (MSI)** circuits, which actually have several adders within a single integrated package.

Basic Adder Circuit

By reviewing the truth table in Figure 7–5, we can determine the input conditions that produce each combination of sum and carry output bits. Figure 7–5 shows the addition of two 2-bit numbers. This could easily be expanded to cover 4-, 8-, or 16-bit addition. Notice that addition in the least-significant-bit (LSB) column requires analyzing only two inputs (A_0 plus B_0) to determine the output sum (Σ_0) and carry (C_{out}), but any more significant columns (2^1 column and up) require the inclusion of a third input, which is the carry-in (C_{in}) from the column to its right. For example, the carry-out (C_{out}) of the 2^0 column becomes the carry-in (C_{in}) to the 2^1 column. Figure 7–5(c) shows the inclusion of a third input for the truth table of the more significant column additions.

Half-Adder

Designing logic circuits to automatically implement the desired outputs for these truth tables is simple. Look at the LSB truth table; for what input conditions is the Σ_0 bit HIGH? The answer is *A or B* HIGH but *not both* (exclusive-OR function). For what input condition is the C_{out} bit HIGH? The answer is *A and B* HIGH (AND function). Therefore, the circuit design to perform addition in the LSB column can be implemented using an exclusive-OR and an AND gate. That circuit is called a **half-adder** and is shown in Figure 7–6. If the exclusive-OR function in Figure 7–6 is implemented using an AND–NOR–NOR configuration, we can tap off the AND gate for the carry, as shown in Figure7–7. [The AND–NOR–NOR configuration is an Ex-OR, as proved in Figure 6–5(c).]

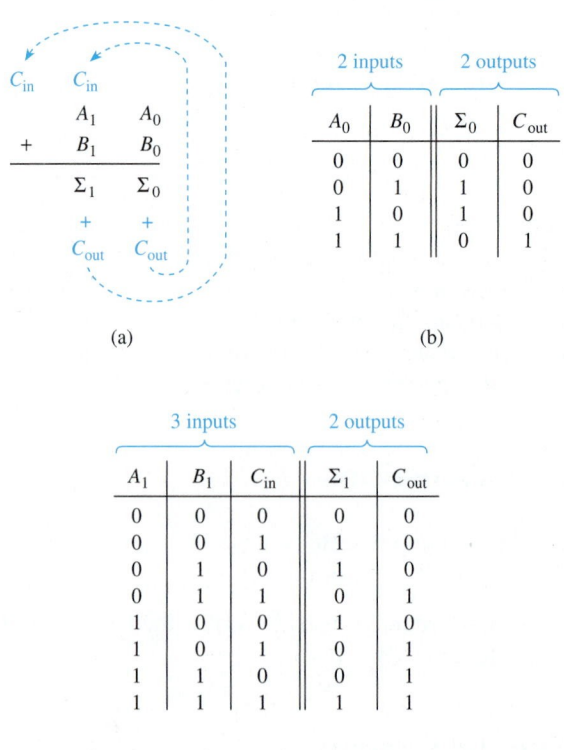

(a)

2 inputs		2 outputs	
A_0	B_0	Σ_0	C_{out}
0	0	0	0
0	1	1	0
1	0	1	0
1	1	0	1

(b)

3 inputs			2 outputs	
A_1	B_1	C_{in}	Σ_1	C_{out}
0	0	0	0	0
0	0	1	1	0
0	1	0	1	0
0	1	1	0	1
1	0	0	1	0
1	0	1	0	1
1	1	0	0	1
1	1	1	1	1

(c)

Figure 7–5 (a) Addition of two 2-bit binary numbers; (b) truth table for the LSB addition; (c) truth table for the more significant column.

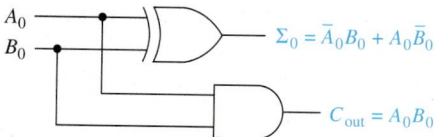

Figure 7–6 Half-adder circuit for addition in the LSB column.

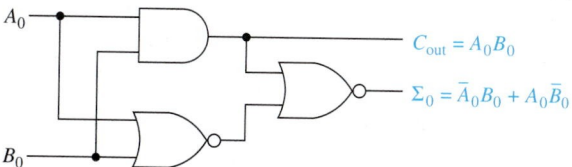

Figure 7–7 Alternative half-adder circuit built from an AND-NOR-NOR configuration.

Full-Adder

As you can see in Figure 7–5, addition in the 2^1 (or higher) column requires three inputs to produce the sum (Σ_1) and carry (C_{out}) outputs. Look at the truth table [Figure 7–5(c)]; for what input conditions is the sum output (Σ_1) HIGH? The answer is that the Σ_1 bit is HIGH whenever the three inputs (A_1, B_1, C_{in}) are *odd*. From Chapter 6, you may remember that an even-parity generator produces a HIGH output whenever the sum of the inputs is odd. Therefore, we can use an even-parity generator to generate our Σ_1 output bit, as shown in Figure 7–8.

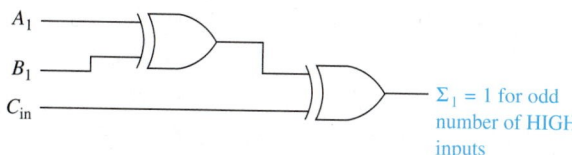

Figure 7–8 The sum (Σ_1) function of the full-adder is generated from an even-parity generator.

How about the carry-out (C_{out}) bit? What input conditions produce a HIGH at C_{out}? The answer is that C_{out} is HIGH whenever any two of the inputs are HIGH. Therefore, we can take care of C_{out} with three ANDs and an OR, as shown in Figure 7–9.

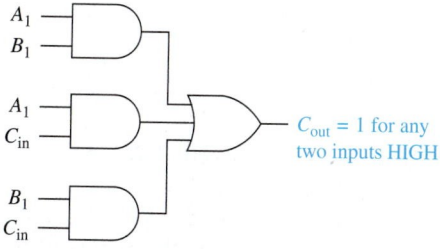

Figure 7–9 Carry-out (C_{out}) function of the full-adder.

The two parts of the full-adder circuit shown in Figures 7–8 and 7–9 can be combined to form the complete *full-adder* circuit shown in Figure 7–10. In the figure, the Σ_1 function is produced using the same logic as that in Figure 7–8 (an Ex-OR feeding an Ex-OR). The C_{out} function comes from A_1B_1 or C_{in} ($A_1\overline{B_1} + \overline{A_1}B_1$).

Prove to yourself that the Boolean equation at C_{out} will produce the necessary result. [*Hint*: Write the equation for C_{out} from the truth table in Figure 7–5(c).] Also, Example 7–18 will help you better understand the operation of the full-adder.

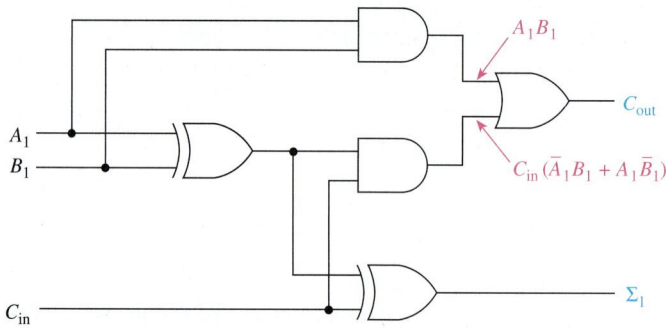

Figure 7–10 Logic diagram of a full-adder.

Helpful Hint

Wow, you should be getting excited about this! We have actually designed and demonstrated a circuit that adds two numbers. We are developing the fundamental building block for the modern computer.

EXAMPLE 7–18

Apply the following input bits to the full-adder of Figure 7–10 to demonstrate its operation ($A_1 = 0$, $B_1 = 1$, $C_{in} = 1$).

Solution: The full-adder operation is shown in Figure 7–11.

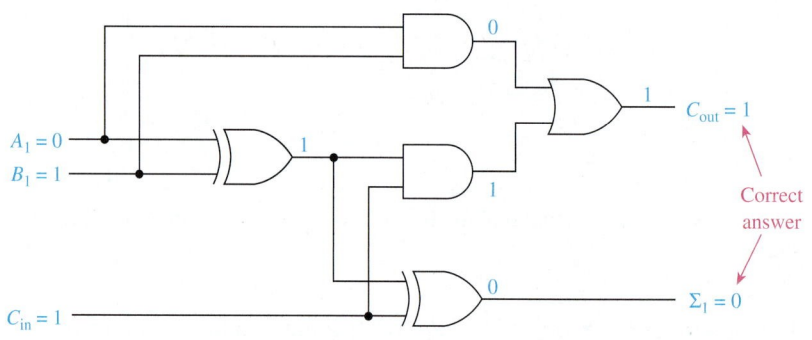

Figure 7–11 Full-adder operation for Example 7–18.

EXAMPLE 7–19

VHDL Description of a Full-Adder

Write the VHDL statements required to implement the full-adder of Figure 7–10. Run a simulation to check the results of the Σ_1 and C_{out} bits. Compare the simulator output to Figure 7–5(c).

Solution: The VHDL program is shown in Figure 7–12. Two equations are in the architecture of the program depicting the Boolean equation for the sum and carry. These are called "**concurrent**" statements because they synthesize two logic circuits that will be executed concurrently (at the same time) as soon at the inputs to the logic (a_1, b_1, and c_{in}) are provided.

The simulation of the circuitry is shown in Figure 7–13. As you can see, the sum bit *sum1* is HIGH for any odd input and the carry *cout* is HIGH whenever any two or more input bits are HIGH.

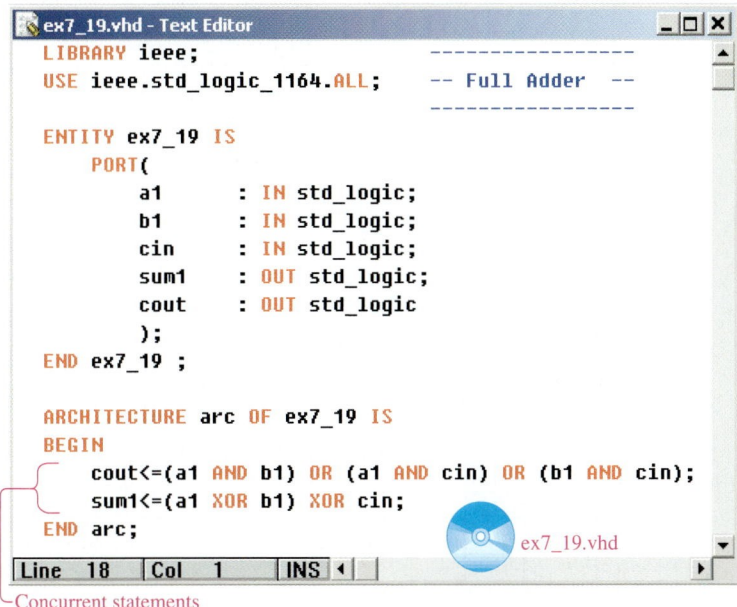

Figure 7–12 The graphic design file for the full-adder.

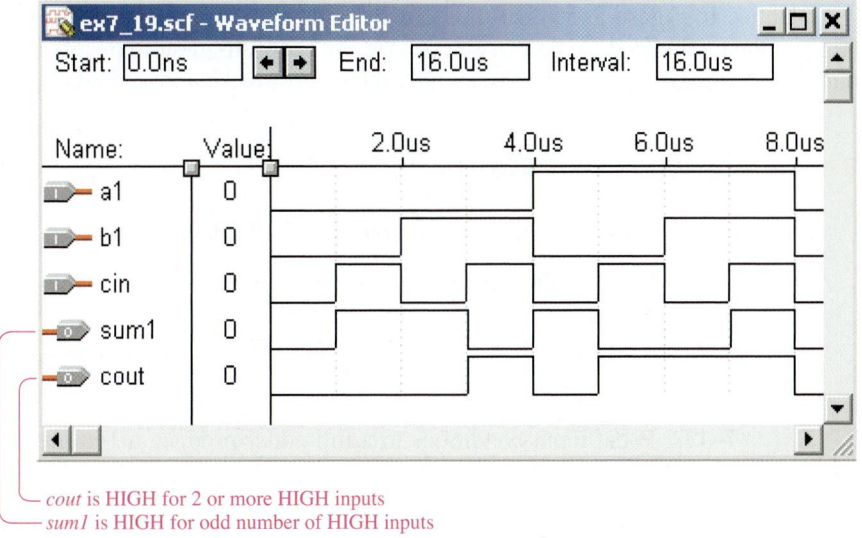

cout is HIGH for 2 or more HIGH inputs
sum1 is HIGH for odd number of HIGH inputs

Figure 7–13 The simulation proving the operation of the full-adder.

Block Diagrams

Now that we know the construction of half-adder and full-adder circuits, we can simplify their representation by just drawing a box with the input and output lines, as shown in Figure 7–14. When drawing multibit adders, a **block diagram** is used to represent the addition in each column. For example, in the case of a 4-bit adder, the 2^0 column needs only a half-adder because there will be no carry-in. Each of the more significant columns requires a full-adder, as shown in Figure 7–15.

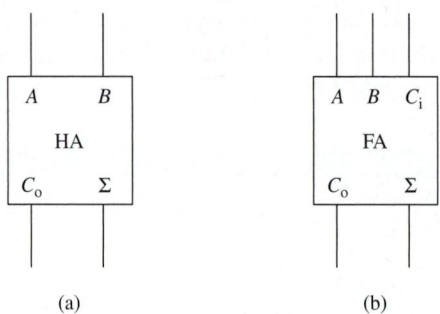

(a)   (b)

Figure 7–14 Block diagrams of (a) half-adder; (b) full-adder.

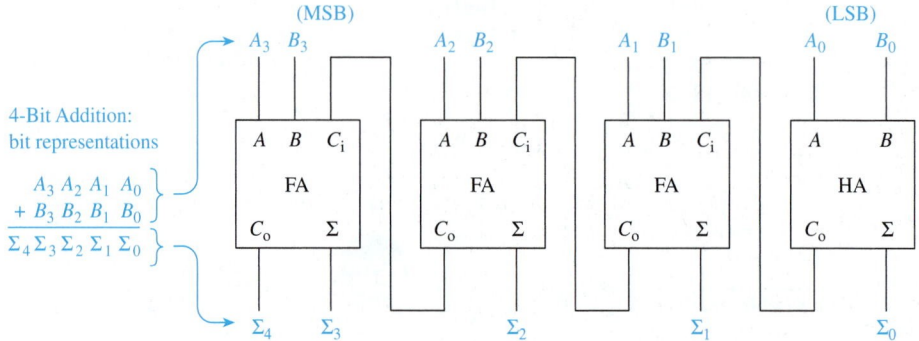

Figure 7–15 Block diagram of a 4-bit binary adder.

Notice in Figure 7–15 that the LSB half-adder has no carry-in. The carry-out (C_{out}) of the LSB becomes the carry-in (C_{in}) to the next full-adder to its left. The carry-out (C_{out}) of the MSB full-adder is actually the highest-order sum output (Σ_4).

Review Questions

7–14. Name the inputs and outputs of a half-adder.

7–15. Why are the input requirements of a full-adder different from those of a half-adder?

7–16. The sum output (Σ) of a full-adder is 1 if the sum of its three inputs is _____ (odd, even).

7–17. What input conditions to a full-adder produce a 1 at the carry-out (C_{out})?

7–7   Four-Bit Full-Adder ICs

Medium-scale-integration (MSI) ICs are available with four full-adders in a single package. Table 7–5 lists the most popular adder ICs. Each adder in the table contains four full-

TABLE 7–5	MSI Adder ICs	
Device	**Family**	**Description**
7483	TTL	4-Bit binary full-adder, fast carry
74HC283	CMOS	4-Bit binary full-adder, fast carry
4008	CMOS	4-Bit binary full-adder, fast carry

adders, and all are functionally equivalent. However, their pin layouts differ (refer to your data manual for the pin layouts). They each will add two 4-bit **binary words** plus one incoming carry. The binary sum appears on the sum outputs (Σ_1 to Σ_4) and the outgoing carry.

Figure 7–16 shows the functional diagram, the logic diagram, and the logic symbol for the 7483. In the figure, the least significant binary inputs (2^0) come into the A_1B_1 terminals, and the most significant (2^3) come into the A_4B_4 terminals. (Be careful; depending on which manufacturer's data manual you are using, the inputs may be labeled A_1B_1 to A_4B_4 or A_0B_0 to A_3B_3). The carry-out (C_{out}) from each full-adder is *internally connected* to the carry-in of the next full-adder. The carry-out of the last full-adder is brought out to a terminal to be used as the sum_5 (Σ_5) output or to be used as a carry-in (C_{in}) to the next full-adder IC if more than 4 bits are to be added (as in Example 7–20).

Something else that we have not seen before is the **fast-look-ahead carry** [see Figure 7–16(a)]. This is very important for speeding up the arithmetic process. For example, if we were adding two 8-bit numbers using two 7483s, the fast-look-ahead carry evaluates the four **low-order** inputs (A_1B_1 to A_4B_4) to determine if they are going to produce a carry-out of the fourth full-adder to be passed on to the next-higher-order adder IC (see Example 7–20). In this way, the addition of the **high-order** bits (2^4 to 2^7) can take place concurrently with the low-order (2^0 to 2^3) addition *without having to wait* for the carries to propagate, or **ripple,** through FA_1 to FA_2 to FA_3 to FA_4 to become available to the high-order addition. A discussion of the connections for the addition of two 8-bit numbers using two 7483s is presented in the following example.

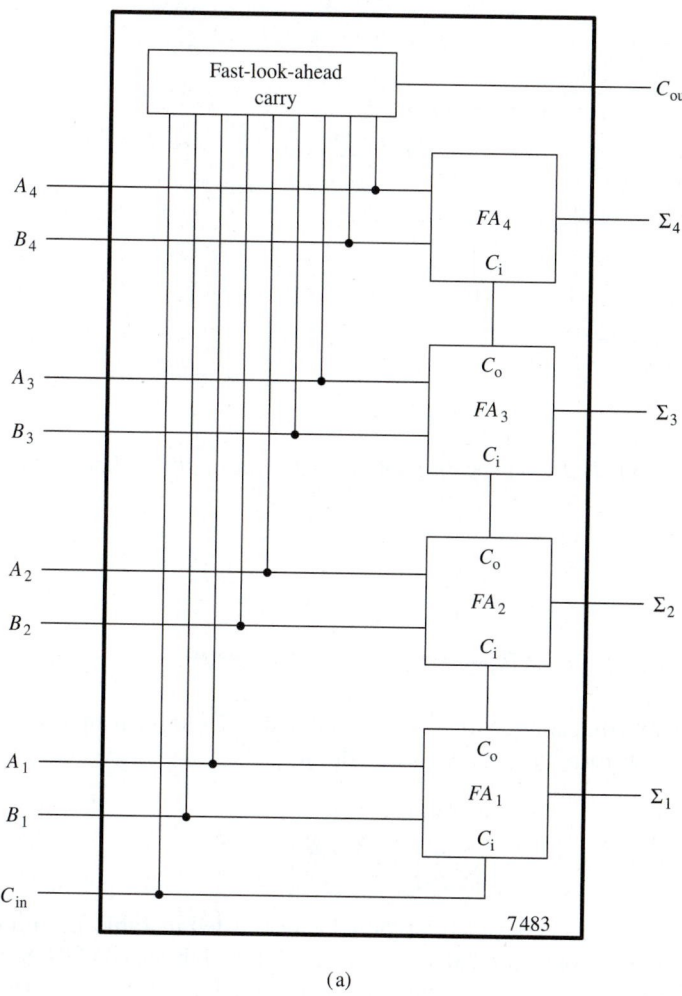

(a)

Figure 7–16 The 7483 4-bit full-adder: (a) functional diagram;

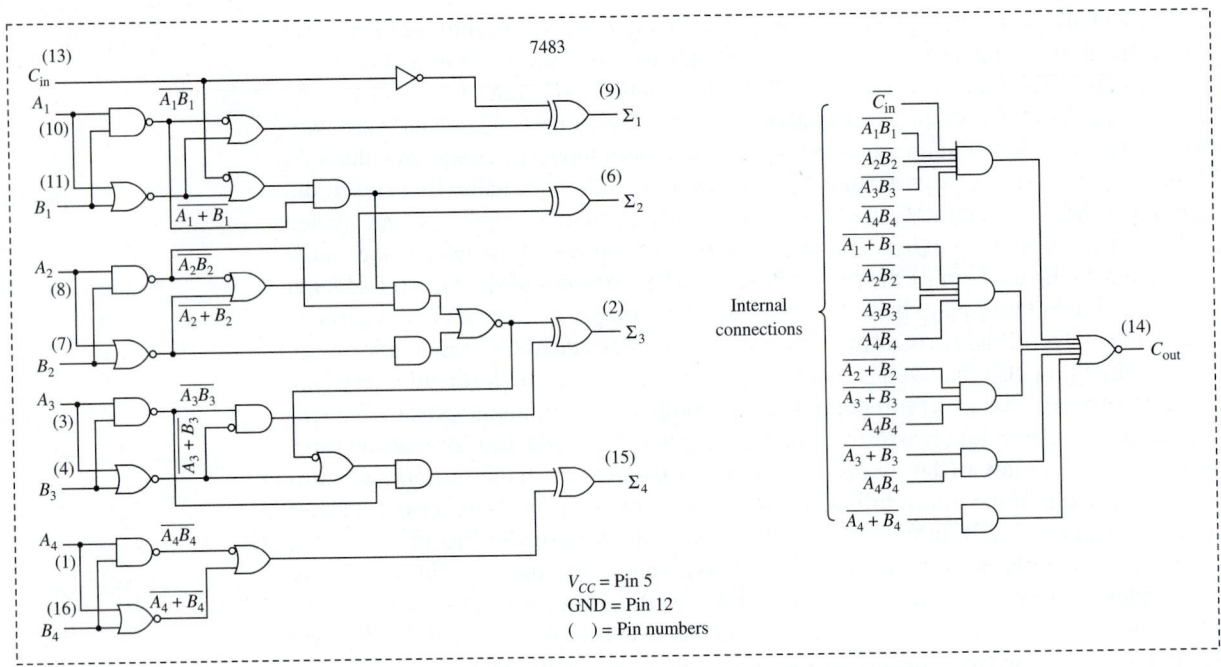

(b)

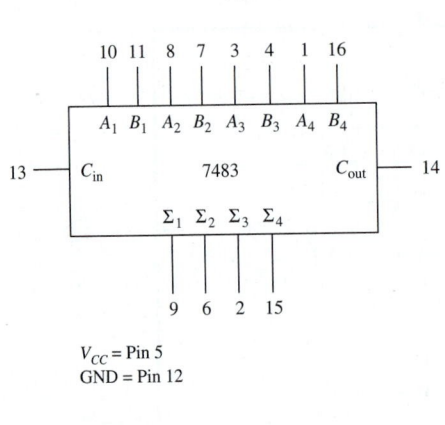

(c)

Figure 7–16 *(Continued)* (b) logic diagram; (c) logic symbol. [(b) Courtesy of Philips Semiconductors.]

EXAMPLE 7–20

Show the external connections to two 4-bit adder ICs to form an 8-bit adder capable of performing the following addition:

$$A_7A_6A_5A_4A_3A_2A_1A_0$$
$$+ \ B_7B_6B_5B_4B_3B_2B_1B_0$$
$$\overline{\Sigma_8\Sigma_7\Sigma_6\Sigma_5\Sigma_4\Sigma_3\Sigma_2\Sigma_1\Sigma_0}$$

Solution: We can choose any of the IC adders listed in Table 7–5 for our design. Let's choose the 74HC283, which is the high-speed CMOS version of the 4-bit adder (it has the same logic symbol as the 7483). As you can see in Figure 7–17, the two 8-bit numbers are brought into the A_1B_1-

---

CHAPTER 7 | ARITHMETIC OPERATIONS AND CIRCUITS

to-A_4B_4 inputs of each chip, and the sum output comes out of the Σ_4-to-Σ_1 outputs of each chip.

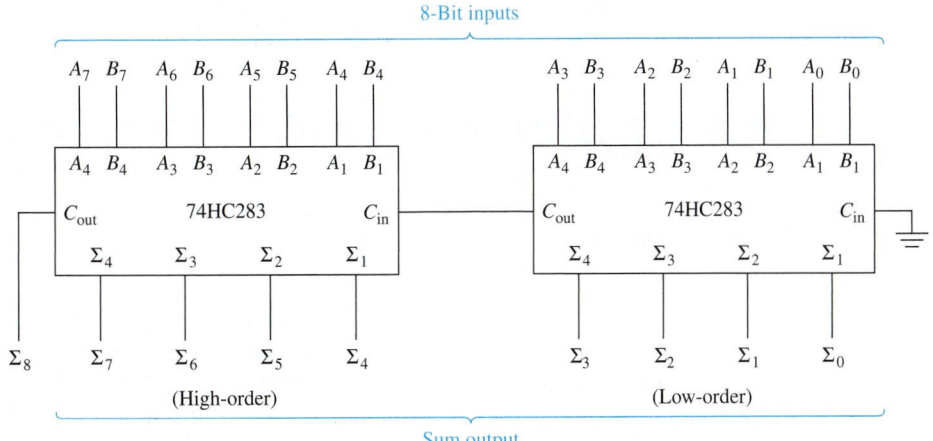

Figure 7-17 Eight-bit binary adder using two 74HC283 ICs.

The C_{in} of the least significant addition ($A_0 + B_0$) is grounded (0) because there is no carry-in (it acts like a half-adder), and if it were left floating, the IC would not know whether to assume a 1 state or 0 state.

The carry-out (C_{out}) from the addition of $A_3 + B_3$ must be connected to the carry-in (C_{in}) of the $A_4 + B_4$ addition, as shown. The fast-look-ahead carry circuit ensures that the carry-out (C_{out}) signal from the low-order addition is provided in the carry-in (C_{in}) of the high-order addition within a very short period of time so that the $A_4 + B_4$ addition can take place without having to wait for all the internal carries to propagate through all four of the low-order additions first. (The actual time requirements for the sum and carry outputs are discussed in Chapter 9, when we look at IC specifications.)

Team Discussion

What if you only wanted to add two 6-bit numbers? How could you get at the internal carry to output to E_6?

Review Questions

7-18. All the adders in the 7483 4-bit adder are full-adders. What is done with the carry-in (C_{in}) to make the first adder act like a half-adder?

7-19. What is the purpose of the fast-look-ahead carry in the 7483 IC?

7-8   VHDL Adders Using Integer Arithmetic

The VHDL language allows us to describe the addition process as an arithmetic expression using the **arithmetic operator** and a new data type called **integer**. Previously we declared inputs and outputs as *std_logic* or *std_logic_vector*. We used that data type to represent a 1 or a 0, or a vector of 1's and 0's (array). The integer data type allows us to specify inputs and outputs as numeric values other than 1 and 0 and perform arithmetic operations on them.

When declaring an input or output as an integer, you must also specify the range of the value. For example, if the inputs are for a 4-bit adder, the range of each number will be 0 to 15 (0000_2 to 1111_2). The result of a 4-bit addition will be 5-bit sum having a range of 0 to 31 (00000_2 to 11111_2). When synthesizing the circuit, the software determines how many input and output bits will be required and assigns the correct number of pins to satisfy the range requested in the integer declare. For example, if the range is 0 to 15, the software knows to allocate four individual input pins for that input name. Figure 7–18 shows a VHDL program that uses the integer type to form a 4-bit binary adder. The assignment statement in the architecture adds the *astring* plus the *bstring* with the *cin*.

```
adder_4b.vhd - Text Editor                              _ □ ×
  LIBRARY ieee;                      --------------------------
  USE ieee.std_logic_1164.all; -- 4-bit Binary Adder using --
                               --    Integer Arithmetic      --
  ENTITY adder_4b IS           --------------------------
      PORT
      (
          cin             : IN        integer RANGE 0 TO 1;
          astring         : IN        integer RANGE 0 TO 15;
          bstring         : IN        integer RANGE 0 TO 15;
          sum_string      : OUT       integer RANGE 0 TO 31
      );
  END adder_4b ;


  ARCHITECTURE arc OF adder_4b IS
  BEGIN
      sum_string<=astring+bstring+cin;
  END arc;                                      adder_4b.vhd
Line  17   Col  1    INS
```

Figure 7–18   Using the integer data type in a VHDL program to form a 4-bit adder.

To verify the circuit operation, the simulation file shown in Figure 7–19 was created. The values used for *astring, bstring,* and *cin* are arbitrary, and the radix used for the string values is hexadecimal. Notice the additional output called *sum_string4*. *Sum_string4* is the fifth bit of the sum, which would have to be used as a carry-out if this was to feed the carry-in of another 4-bit adder like we did in Figure 7–17. It is also used to indicate that the sum exceeded (overflowed) the maximum value of a 4-bit number.

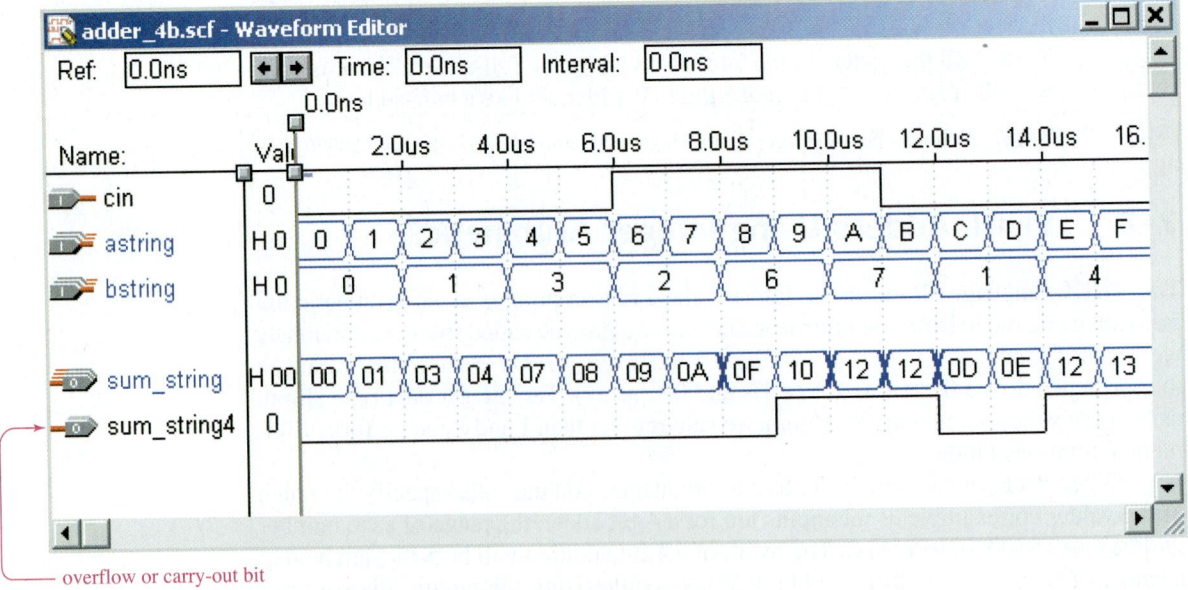

— overflow or carry-out bit

Figure 7–19   The simulation file for the 4-bit adder of Figure 7–18.

It is informative to look at the pin assignments chosen by the MAX+PLUS II software for the integer input and output strings. The easiest way to do this is to look at the device view layout in the Floorplan Editor as shown in Figure 7–20.

Choose **MAX+PLUS II > Floorplan Editor > Layout > Last Compilation** and **Device View** checked.

As you can see, the *astring* and *bstring* integers were each assigned four input pins and the *sum_string* output was assigned five.

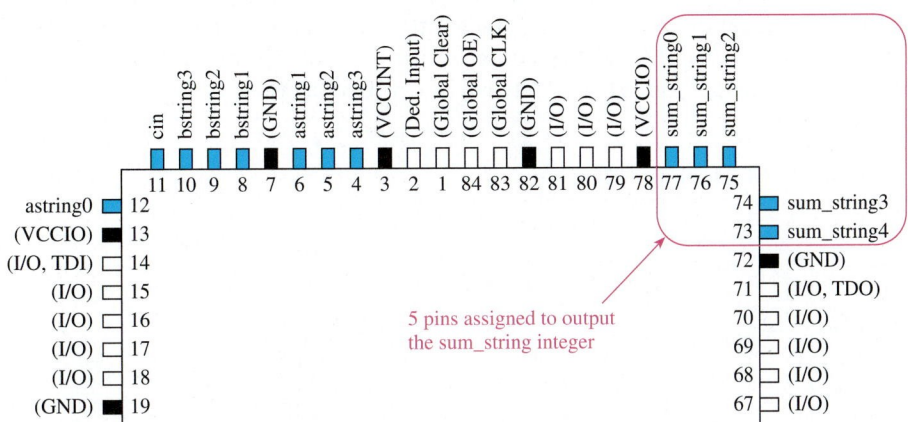

Figure 7–20 Partial floorplan view showing the pin assignments on the EPM7128SLC CPLD for the 4-bit adder of Figure 7–18.

7–9   System Design Applications

Each arithmetic operation discussed in Sections 7–1 through 7–5 can be performed by using circuits built from IC adders and logic gates. First, we will design a circuit to perform two's-complement arithmetic, and next, we will design a BCD adder.

Two's-Complement Adder/Subtractor Circuit

A quick review of Section 7–3 reminds us that positive two's-complement numbers are exactly the same as regular true binary numbers and can be added using regular binary addition. Also, subtraction in two's-complement arithmetic is performed by converting the number to be subtracted to a *negative* number in the two's-complement form and then using regular binary addition. Therefore, once our numbers are in two's-complement form, we can use a binary adder to get the answer whether we are adding *or* subtracting.

For example, to subtract $18 - 9$, we would first convert 9 to a negative two's-complement number by complementing each bit and then adding 1. We would then add $18 + (-9)$:

$$
\begin{array}{rl}
\text{Two's complement of } 18 = & 0001\ 0010 \\
+\ \text{Two's complement of } -9 = & \underline{1111\ 0111} \\
\text{Sum} = & \overline{0000\ 1001} = +9_{10}\ \textit{Answer}
\end{array}
$$

So it looks like all we need for a combination adder/subtractor circuit is an input switch or signal to signify addition or subtraction so that we will know whether to form a positive or a negative two's complement of the second number. Then we will just use a binary adder to get the final result.

To form negative two's complement, we can use the controlled inverter circuit presented in Figure 6–17 and add 1 to its output. Figure 7–21 shows the complete circuit used to implement a two's-complement adder/subtractor using two 4008 CMOS adders. The 4008s are CMOS 4-bit binary adders. The 8-bit number on the A inputs (A_7 to A_0) is brought directly into the adders. The other 8-bit binary number comes in on the B_7 to B_0 lines. If the B number is to be subtracted, the *complementing switch* will be in the up (1) position, causing each bit in the B number to be complemented (one's complement). At the same time, the low-order C_{in} receives a 1, which has the effect of adding a 1 to the already complemented B number, making it a negative two's-complement number.

Now the 4008s perform a regular binary addition. If the *complementing switch* is up, the number on the B inputs is subtracted from the number on the A inputs. If it is down, the sum is taken. As discussed in Section 7–3, the C_{out} of the MSB is ignored. The result can range from 0111 1111 ($+ 127$) to 1000 0000 ($- 128$).

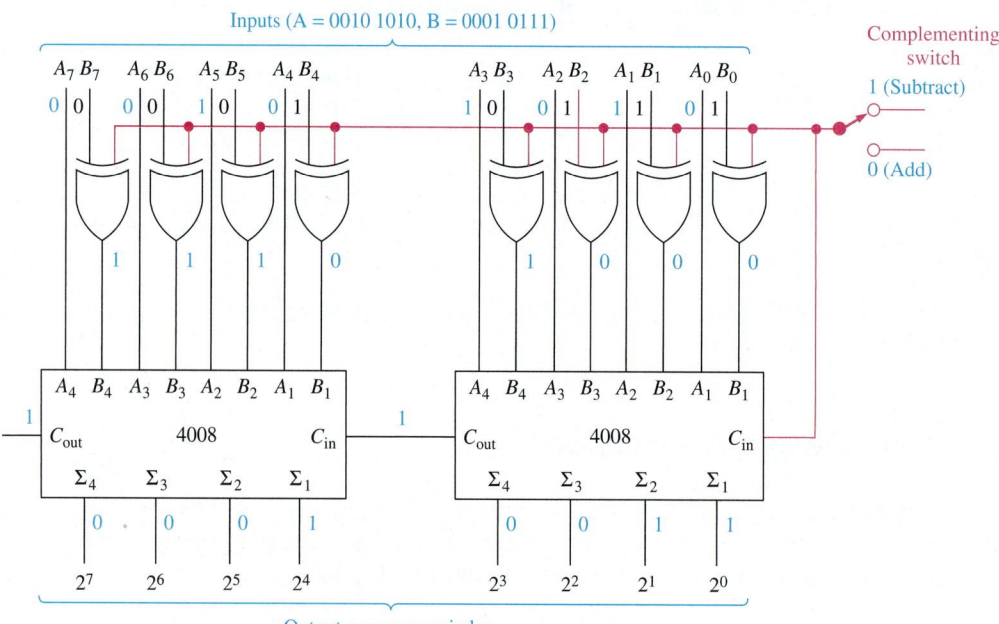

Figure 7–21 Eight-bit two's-complement adder/subtractor illustrating the subtraction $42 - 23 = 19$.

EXAMPLE 7-20

Prove that the subtraction $42 - 23$ produces the correct answer at the outputs by labeling the input and output lines on Figure 7–21.

Solution: $42 - 23$ should equal 19 (0001 0011). Convert the decimal input numbers to regular binary, and label Figure 7–21 (42 = 0010 1010, 23 = 0001 0111). The B input number is complemented, the LSB C_{in} is 1, and the final answer is 0001 0011, which proves that the circuit works for that number.

Try adding and subtracting some other numbers to better familiarize yourself with the operation of the circuit of Figure 7–21.

BCD Adder Circuit

BCD adders can also be formed using the IC 4-bit binary adders. The problem, as you may remember from Section 7–5, is that when any group-of-four BCD sum exceeds 9, or when there is a carry-out, the number is invalid and must be corrected by adding 6 to the invalid answer to get the correct BCD answer. (The valid range of BCD numbers is 0000 to 1001.)

For example, adding $0111_{BCD} + 0110_{BCD}$ (7 + 6) gives us an invalid result:

$$
\begin{array}{r}
0111 \\
+\ 0110 \\
\hline
1101 \quad \longleftarrow \text{ invalid BCD} \\
+\ 0110 \quad \longleftarrow \text{ add 6 to correct} \\
\hline
1 \quad 0011 \\
\end{array}
$$

carry to next BCD digit

The corrected answer is $0001\ 0011_{BCD}$, which equals 13.

Checking for a sum greater than 9, or a carry-out, can be done easily using logic gates. Then, when an invalid sum occurs, it can be corrected by adding 6 (0110) via the connections shown in Figure 7–22. The upper 7483 performs a basic 4-bit addition. If its sum is greater than 9, the Σ_4 (2^3) output *and* either the Σ_3 *or* Σ_2 (2^2 or 2^1) output will be HIGH. A sum greater than 9 *or* a carry-out will produce a HIGH out of the left OR gate, placing a HIGH–HIGH at the A_3 and A_2 inputs of the correction adder, which has the effect of adding a 6 to the original addition. If there is no carry and the original sum is not greater than 9, the correction adder adds 0000.

Common Misconception

Students typically have a hard time seeing where the error correction number 6 (0110) is input to the correction adder at A_4-A_1.

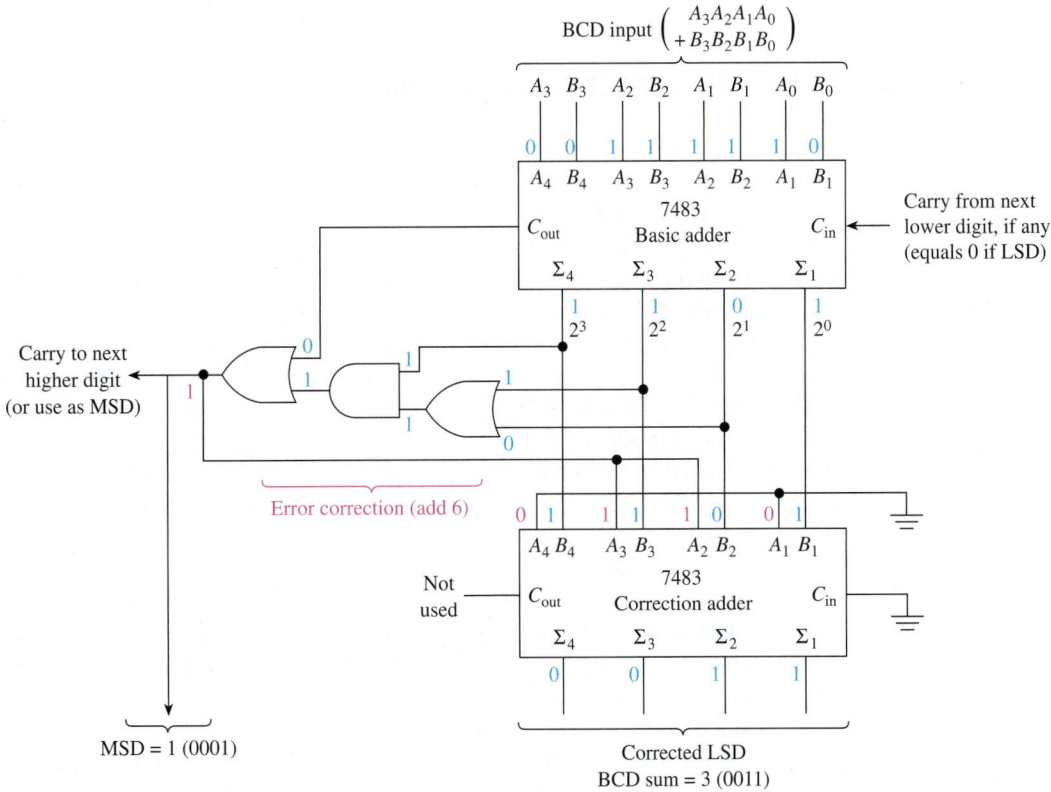

Figure 7–22 BCD adder illustrating the addition 7 + 6 = 13 (0111 + 0110 = 0001 0011 BCD).

EXAMPLE 7-22

Prove that the BCD addition $0111 + 0110$ $(7 + 6)$ produces the correct answer at the outputs by labeling the input and output lines on Figure 7-22.

Solution: The sum out of the basic adder is 13 (1101). Because the 2^3 bit and the 2^2 bit are both HIGH, the error correction OR gate puts out a HIGH, which is added to the next more significant BCD digit and also puts a HIGH–HIGH at A_3, A_2 of the correction adder, which adds 6. The correct answer has a 3 for the least significant digit (LSD) and a 1 in the next more significant digit, for the correct answer of 13.

Familiarize yourself with the operation of Figure 7-22 by testing the addition of several other BCD numbers.

BCD Adder IC: A 4-bit BCD adder is available in a single IC package. The 74HCT583 IC has internal correction circuitry to add two 4-bit numbers and produce a corrected 4-bit answer with carry-out. Refer to a high-speed CMOS data book for an in-depth description of the chip.

Review Questions

7-20. The complementing switch in Figure 7-21 is placed in the 1 position to subtract B from A. Explain how this position converts the binary number on the B inputs into a signed two's-complement number.

7-21. What is the purpose of the AND and OR gates in the BCD adder circuit of Figure 7-22?

7-10 Arithmetic/Logic Units

Arithmetic/logic units (ALUs) are available in large-scale IC packages (LSI). The LSI circuits are generally considered to be ICs containing from 100 to 10,000 gate equivalents. Typically, an **ALU** is a multipurpose device capable of providing several different arithmetic and logic operations. The specific operation to be performed is chosen by the user by placing a specific binary code on the mode select inputs. Microprocessors may also have ALUs built in as one of their many operational units. In such cases, the specific operation to be performed is chosen by software instructions.

The ALU that we learn to use in this section is the 74181 (TTL) or 74HC181 (CMOS). The 74181 is a 4-bit ALU that provides 16 arithmetic plus 16 logic operations. Its logic symbol and function table are given in Figure 7-23. The mode control input (M) is used to set the mode of operation as either *logic (M = H) or arithmetic (M = L)*. When M is HIGH, all internal carries are disabled, and the device performs *logic operations* on the individual bits (A_0 to A_3, B_0 to B_3), as indicated in the function table.

When M is LOW, the internal carries are enabled, and the device performs *arithmetic operations* on the two 4-bit binary inputs. Ripple carry output is provided at $\overline{C_{N+4}}$, and fast-look-ahead carry is provided at G and P for high-speed arithmetic operations. The carry-in and carry-out terminals are each active-LOW (as signified by the bubble), which means that a 0 signifies a carry.

Once the **mode control** (M) is set, you have 16 choices within either the logic or arithmetic categories. The specific function you want is selected by applying the appropriate binary code to the **function select** inputs (S_3 to S_0).

For example, with $M = H$ and $S_3 S_2 S_1 S_0 = LLLL$, the F outputs will be equal to the complement of A (see the function table). This means that $F_0 = \overline{A_0}, F_1 = \overline{A_1}, F_2 = \overline{A_2}$ and

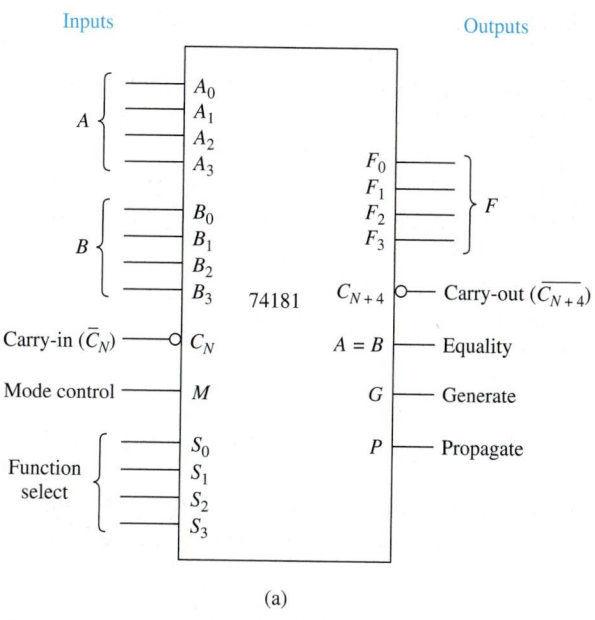

Inputs

Outputs

74181

Carry-in ($\overline{C}_N$)

Mode control

Function select

Carry-out ($\overline{C_{N+4}}$)

$A = B$ — Equality

G — Generate

P — Propagate

(a)

Mode select				Logic functions	Arithmetic operations
S_3	S_2	S_1	S_0	(M = H)	(M = L)($\overline{C}_n$ = H)
L	L	L	L	$F = \overline{A}$	$F = A$
L	L	L	H	$F = \overline{A + B}$	$F = A + B$
L	L	H	L	$F = \overline{A}B$	$F = A + \overline{B}$
L	L	H	H	$F = 0$	F = minus 1 (2's comp.)
L	H	L	L	$F = \overline{AB}$	$F = A$ plus $A\overline{B}$
L	H	L	H	$F = \overline{B}$	$F = (A + B)$ plus $A\overline{B}$
L	H	H	L	$F = A \oplus B$	$F = A$ minus B minus 1
L	H	H	H	$F = A\overline{B}$	$F = A\overline{B}$ minus 1
H	L	L	L	$F = \overline{A} + B$	$F = A$ plus AB
H	L	L	H	$F = \overline{A \oplus B}$	$F = A$ plus B
H	L	H	L	$F = B$	$F = (A + \overline{B})$ plus AB
H	L	H	H	$F = AB$	$F = AB$ minus 1
H	H	L	L	$F = 1$	$F = A$ plus A*
H	H	L	H	$F = A + \overline{B}$	$F = (A + B)$ plus A
H	H	H	L	$F = A + B$	$F = (A + \overline{B})$ plus A
H	H	H	H	$F = A$	$F = A$ minus 1

$F = A$ means:
$F_0 = A_0, F_1 = A_1, F_2 = A_2, F_3 = A_3$

*Each bit is shifted to the next-more-significant position.

(b)

Figure 7–23   The 74181 ALU: (a) logic symbol; (b) function table.

$F_3 = \overline{A}_3$. Another example is with $M = H$ and $S_3S_2S_1S_0 = HHHL$; the F outputs will be equal to $A + B$ *(A or B)*. This means that $F_0 = A_0 + B_0, F_1 = A_1 + B_1, F_2 = A_2 + B_2$, and $F_3 = A_3 + B_3$.

From the function table we can see that other logic operations (AND, NAND, NOR, Ex-OR, Ex-NOR, and several others) are available.

The function table in Figure 7–23(b) also shows the result of the 16 different *arithmetic operations* available when $M = L$. Note that the results listed are with carry-in ($\overline{C_N}$) equal to H (no carry). For $\overline{C_N} = L$, just add 1 to all results. All results produced by the device are in two's-complement notation. Also, in the function table, note that the $+$ *sign* means *logical-OR* and the word "PLUS" means *arithmetic-SUM*.

For example, to subtract B from A ($A_3A_2A_1A_0 - B_3B_2B_1B_0$), set $M = L$ and $S_3S_2S_1S_0 = LHHL$. The result at the F outputs will be the two's complement of A minus B minus *1*; therefore, to get just A minus B, we need to add for 1. (This can be done automatically by setting $\overline{C_N} = 0$.) Also, as discussed earlier for two's-complement subtraction, a carry-out (borrow) is generated ($\overline{C_{N+4}} = 0$) when the result is positive or zero. Just ignore it.

Read through the function table to see the other 15 arithmetic operations that are available.

EXAMPLE 7–23

Show the external connections to a 74181 to form a 4-bit subtractor. Label the input and output pins with the binary states that occur when subtracting $13 - 7$ ($A = 13, B = 7$).

Solution: The 4-bit subtractor is shown in Figure 7–24. The ALU is set in the subtract mode by setting $M = 0$ and $S_3S_2S_1S_0 = 0110$ (*LHHL*). 13 (1101) is input at A, and 7 (0111) is input at B.

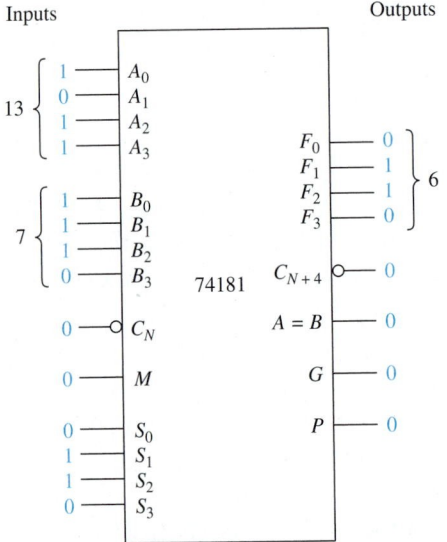

Figure 7–24   Four-bit binary subtractor using the 74181 ALU to subtract 13–7.

By setting $\overline{C_N} = 0$, the output at F_0, F_1, F_2, F_3 will be A minus B instead of A minus B minus 1, as shown in the function table [Figure 7–23(b)]. The result of the subtraction is a positive 6 (0110) with a carry-out ($\overline{C_{N+4}} = 0$). (As before, with two's-complement subtraction, there is a carry-out for any positive or zero answer, which is ignored.)

EXAMPLE 7–23

Place the following values at the inputs of a 74181: $A_3 A_2 A_1 A_0 = 1001$ $B_3 B_2 B_1 B_0 = 0011$ $S_3 S_2 S_1 S_0 = 1101$ and $C_N = 1$.

(a) With $M = 1$, determine the output at F ($F_3 F_2 F_1 F_0$).

(b) Change M to 0 and determine the output at F ($F_3 F_2 F_1 F_0$).

Solution: (a) From the chart in Figure 7–23(b), the logic function chosen is $F = A + \overline{B}$ (F equals A ORed with the complement of B).

$$A = 1001$$
$$B = 0011$$
$$\overline{B} = 1100$$
$$F = A + \overline{B} = 1101 \quad \textit{Answer}$$

(b) With $M = 0$, the arithmetic operation is (A ORed with B) with the result added to A.

$$A = 1001$$
$$B = 0011$$
$$A \text{ OR } B = 1011$$
$$A \text{ OR } B \text{ PLUS } A = 0100 \quad \textit{Answer}$$

Review Questions

7–22. What is the purpose of the *mode control* input to the 74181 arithmetic/logic unit?

7–23. If $M = H$ and $S_3, S_2, S_1, S_0 = L, L, H, H$ on the 74181, then F_3, F_2, F_1, F_0 will be set to L, L, L, L. True or false?

7–24. The arithmetic operations of the 74181 include both $F = A + B$ and $F = A$ PLUS B. How are the two designations different?

7–11 CPLD Applications with VHDL and LPMs

In this section we will duplicate several of the arithmetic circuits covered earlier in this chapter using CPLD implementation created with macrofunctions, VHDL and a new form of design entry that uses a built-in **Library of Parameterized Modules (LPM)**. LPMs are provided in the MAX+PLUS II software to ease the design process for commonly used systems like adders and ALUs.

EXAMPLE 7–25

Build a graphic design file for a 4-bit adder using the macro-function for the 74283 fixed-function IC. Group the A inputs, B inputs, and SUM outputs as busses. Simulate several different additions as you monitor the results on the SUM bits and C_{out}.

Solution: The gdf file utilizing the 74283 macrofunction is shown in Figure 7–25. The A and B inputs and the SUM outputs are grouped together as 4-bit busses at the terminal pins by giving them the names $A[3..0]$, $B[3..0]$ and $SUM[3..0]$. Each element of the group must then be broken out by specifying the element name ($A0$, $B0$, $A1$, and so on) on the line entering or leaving the symbol.

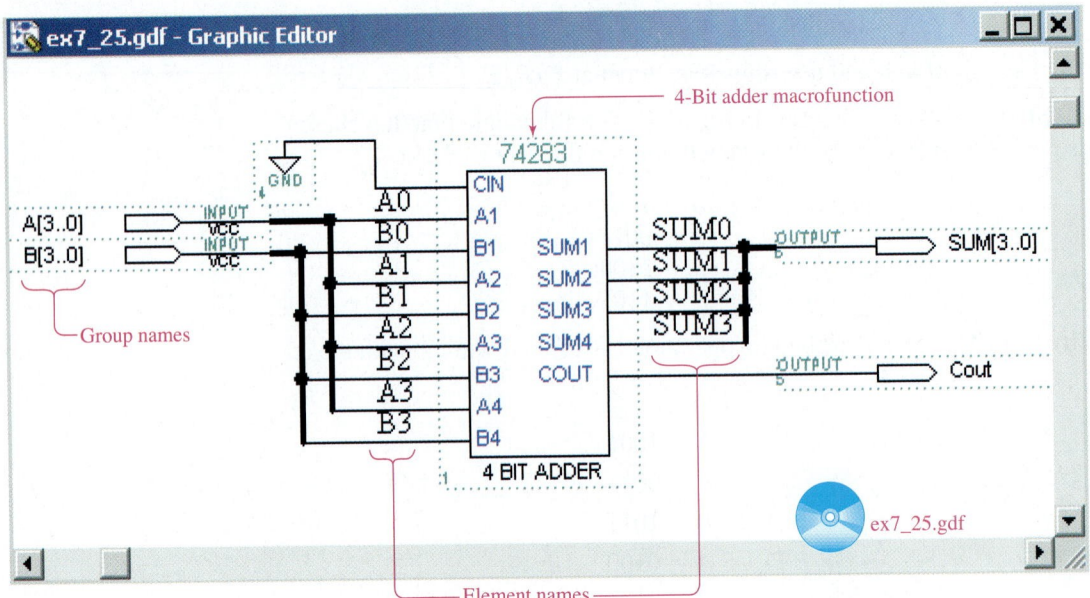

Figure 7–25 The 74283 4-bit adder macrofunction for Example 7–25.

The simulation in Figure 7–26 shows a variety of numbers that were chosen for A and B to exercise the adder and test the C_{out} line. As you can see, the numbers are all listed with a hexadecimal radix. For example, in the first addition, instead of adding $0011 + 0110 = 1001$, the simulation shows $3 + 6 = 9$. The third addition, $6 + 6$ would equal 12, but 12 in hex is C. The fifth, sixth, and seventh additions cause a carry-out because they are greater than 15.

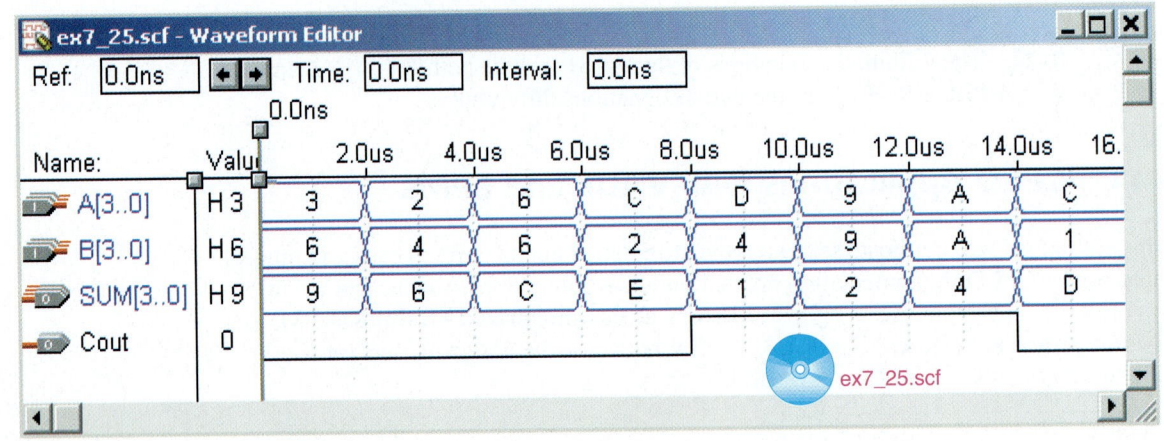

Figure 7–26 The simulation of the 4-bit adder for Example 7–25.

EXAMPLE 7–26

Design an 8-bit adder/subtractor in VHDL using the $+$ and $-$ arithmetic operators. Back in Figure 7–18 we used integer data types for performing arithmetic. Instead, in this example, use std_logic vectors for the input/output data types, and use a WHEN-ELSE **conditional signal assignment** to select whether to add or subtract the numbers. Create a simulation to test several 8-bit additions and subtractions.

Solution: Figure 7–27 shows the VHDL listing for the adder/subtractor. The first thing that you may notice is that there is a new library declare statement called *ieee.std_logic_signed*. This is required whenever you are using vectors in signed arithmetic. **Signed numbers** are those where the MSB is used to represent the sign as is the case with two's-complement notation. The *add_sub* input will be used to tell the logic whether to add the input numbers or to subtract them. Since this is an 8-bit system we use a vector size of (7 DOWNTO 0). The WHEN-ELSE statement is called a "conditional signal assignment" because it performs a specific assignment based on the condition listed after the WHEN command. When *add_sub* = '0' the inputs are added. Otherwise (ELSE) they are subtracted.

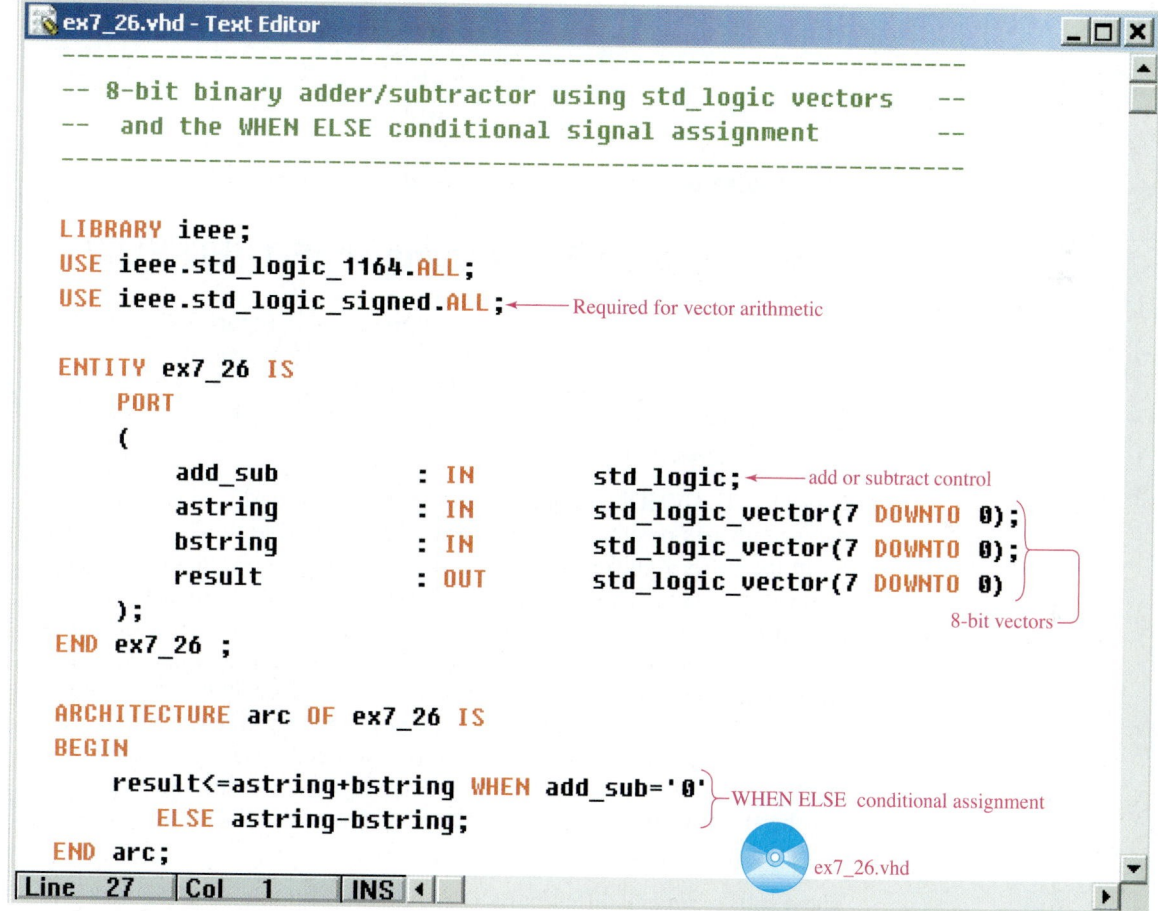

Figure 7–27   An 8-bit adder/subtractor employing the WHEN-ELSE assignment.

The simulation in Figure 7–28 shows the results of several different additions and subtractions. The first six operations are additions showing the hexadecimal sum of the *astring* plus the *bstring*. The next six operations are subtractions because the *add_sub* line is HIGH. The *result* output is in signed two's-complement notation. The results of the first two subtractions are positive but the next two are negative because *bstring* is larger than *astring*. Study the third subtraction and prove to yourself that 08_{16} minus $0C_{16}$ equals FC_{16}.

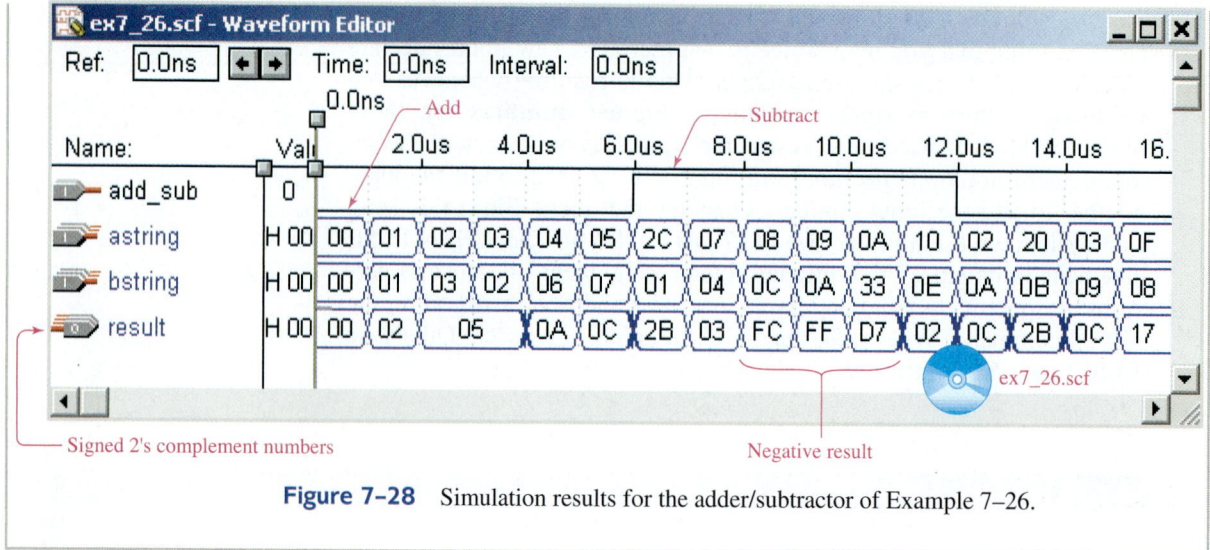

Figure 7–28   Simulation results for the adder/subtractor of Example 7–26.

EXAMPLE 7–27

BCD Correction Adder Using an IF-THEN-ELSE

Use VHDL to reproduce the BCD adder presented in Figure 7–22. Assume that the 4-bit inputs on *a* and *b* will always be valid positive BCD numbers. When the two BCD numbers are added together, use an **IF-THEN-ELSE** statement to determine if the sum is greater than 9. If so, then that sum must be corrected by adding 6.

Solution: The VHDL program is shown in Figure 7–29. Since the numbers are always positive, we don't want the program to think that a HIGH bit in the MSB signifies a negative result. Using *ieee.std_logic_unsigned* (instead of *signed*) will ensure that all of the numbers are treated as positives (**unsigned number**). As you saw in Figure 7–22, there is an intermediate sum produced after the initial addition before it goes to the correction adder. We will call this *bin_result* and declare it as a signal because it is an internal interconnection point, not an input or output. The IF-THEN-ELSE statement is **sequential** and needs to be put inside of a PROCESS as shown. Since the *bin_result* is a 5-bit string, the carry-out will be the fifth bit, so checking for a value greater than 9 will also be checking for a carry-out.

The results of several different BCD additions are simulated in Figure 7–30. A binary radix is chosen for all numbers. The *bcd result* was set up as eight bits to accommodate for two BCD digits that will occur for sums greater than 9. Study the simulation carefully to ensure that you believe that all of the *bcd_result* values are correct.

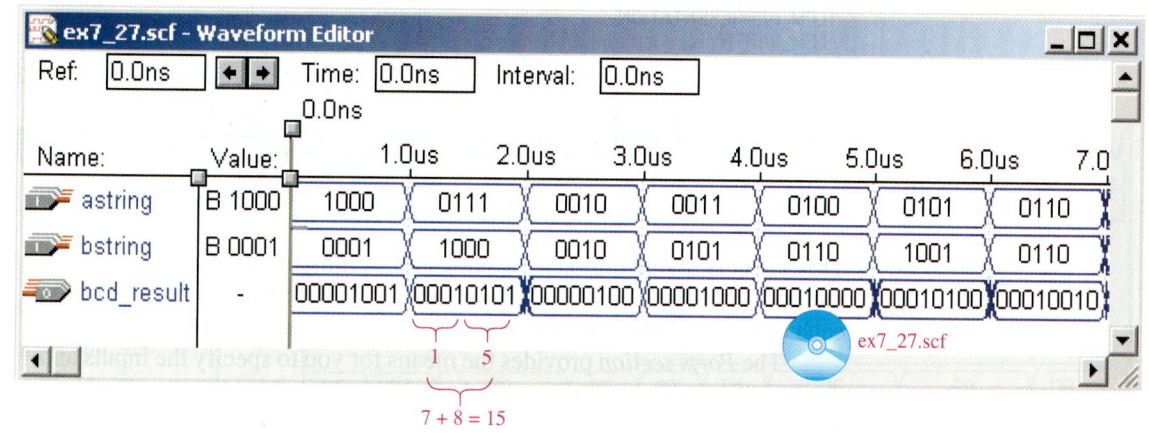

```
ex7_27.vhd - Text Editor                                          _ □ ✕
    LIBRARY ieee;                        ------------------------
    USE ieee.std_logic_1164.ALL;         -- BCD Correction Adder  --
    USE ieee.std_logic_unsigned.ALL;     --  using IF-THEN-ELSE   --
                                         ------------------------
    ENTITY ex7_27 IS
        PORT
        (
            astring         : IN        std_logic_vector(3 DOWNTO 0);
            bstring         : IN        std_logic_vector(3 DOWNTO 0);
            bcd_result      : OUT       std_logic_vector(7 DOWNTO 0)
        );
    END ex7_27 ;


    ARCHITECTURE arc OF ex7_27 IS
    SIGNAL bin_result         :          std_logic_vector(7 DOWNTO 0);
    BEGIN                                ── Internal interconnection signal
        bin_result<=astring+bstring;
        PROCESS
        BEGIN
            IF bin_result>"01001"
                THEN bcd_result<=bin_result+"0110";
                ELSE bcd_result<=bin_result;
            END IF;
        END PROCESS;
    END arc;                                              ex7_27.vhd
Line  25   Col  2    INS ◀                                      ▶
                 └── IF-THEN-ELSE sequential statements
```

Figure 7–29   VHDL implementation of a BCD adder for Example 7–27.

```
ex7_27.scf - Waveform Editor                                          _ □ ✕
Ref:  0.0ns   ◀ ▶   Time:  0.0ns    Interval:  0.0ns
               0.0ns
Name:        Value:        1.0us   2.0us   3.0us   4.0us   5.0us   6.0us   7.0
  astring    B 1000    1000  0111  0010  0011  0100  0101  0110
  bstring    B 0001    0001  1000  0010  0101  0110  1001  0110
  bcd_result   -     00001001 00010101 00000100 00001000 00010000 00010100 00010010
                                 1    5
                                                              ex7_27.scf
                              7 + 8 = 15
```

Figure 7–30   Simulation of BCD addition for Example 7–27.

Sign Bit: The leftmost, or MSB, in a two's-complement number, used to signify the sign of the number (1 = negative, 0 = positive).

Signed number: Positive and negative numbers represented by two's-complement notation where the MSB represents the sign bit.

Sum: The result of the addition of numbers.

Two's Complement: A binary numbering representation that simplifies arithmetic in digital systems.

Unsigned number: Positive numbers where the MSB is part of the number (not a sign bit).

WHEN-ELSE: A clause used to direct the operation of a conditional signal assignment.

Problems

Section 7–1

7–1. Perform the following decimal additions, convert the original decimal numbers to binary, and add them. Compare answers.

(a)   6
    + 3

(b)   8
    + 7

(c)   22
    + 6

(d)   29
    + 37

(e)   134
    + 66

(f)   254
    + 36

(g)   208
    + 127

(h)   196
    + 156

7–2. Repeat Problem 7–1 for the following subtractions.

(a)   15
    − 4

(b)   22
    − 11

(c)   84
    − 36

(d)   66
    − 31

(e)   126
    − 64

(f)   113
    − 88

(g)   109
    − 60

(h)   111
    − 104

7–3. Repeat Problem 7–1 for the following multiplications.

(a)   7
    × 3

(b)   6
    × 7

(c)   12
    × 5

(d)   39
    × 7

(e)   63
    × 125

(f)   127
    × 15

(g)   31
    × 13

(h)   255
    × 127

7–4. Repeat Problem 7–1 for the following divisions.

(a) 4)12   (b) 3)15   (c) 12)48   (d) 5)25
(e) 5)125   (f) 14)294   (g) 15)195   (h) 12)228

Section 7–2

7–5. Produce a table of 8-bit two's-complement numbers from + 15 to − 15.

7–6. Convert the following decimal numbers to 8-bit two's-complement notation.

(a) 7   (b) − 7   (c) 14   (d) 36   (e) − 36
(f) 66   (g) − 48   (h) 112   (i) − 112   (j) − 125

CHAPTER 7 | ARITHMETIC OPERATIONS AND CIRCUITS

```
ex7_27.vhd - Text Editor                                          _ □ ✕

   LIBRARY ieee;                        ----------------------------
   USE ieee.std_logic_1164.ALL;         -- BCD Correction Adder  --
   USE ieee.std_logic_unsigned.ALL;     --  using IF-THEN-ELSE   --
                                        ----------------------------
   ENTITY ex7_27 IS
      PORT
      (
          astring           : IN      std_logic_vector(3 DOWNTO 0);
          bstring           : IN      std_logic_vector(3 DOWNTO 0);
          bcd_result        : OUT     std_logic_vector(7 DOWNTO 0)
      );
   END ex7_27 ;

   ARCHITECTURE arc OF ex7_27 IS
   SIGNAL bin_result         :         std_logic_vector(7 DOWNTO 0);
   BEGIN                               ─── Internal interconnection signal
       bin_result<=astring+bstring;
       PROCESS
       BEGIN
         IF bin_result>"01001"
             THEN bcd_result<=bin_result+"0110";
             ELSE bcd_result<=bin_result;
         END IF;
       END PROCESS;
     END arc;                                          ◉ ex7_27.vhd

Line  25   Col  2    INS ◀                                        ▶
```

IF-THEN-ELSE sequential statements

Figure 7–29   VHDL implementation of a BCD adder for Example 7–27.

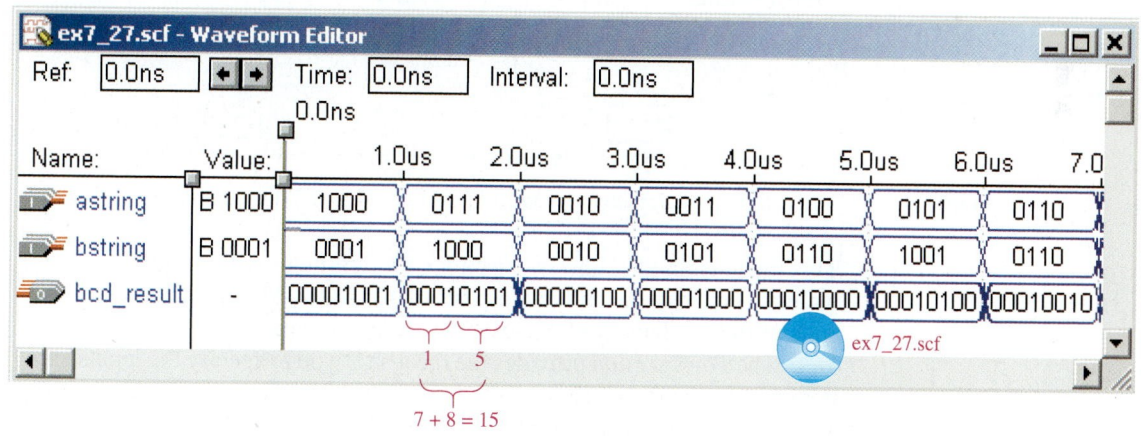

Figure 7–30   Simulation of BCD addition for Example 7–27.

EXAMPLE 7–28

LPM Adder/Subtractor

Rather than "reinvent the wheel" for each digital design, MAX+PLUS II software provides a library of pre-designed complex logic functions commonly used in digital systems. It is called the **Library of Parameterized Modules** and is found in the *mega_lpm* subdirectory. Enter the symbol for the *lpm_add_sub* in the Graphic Editor screen. Right click on the symbol and choose **Edit Ports/Parameters.** The pop-up menu shown in Figure 7–31 appears.

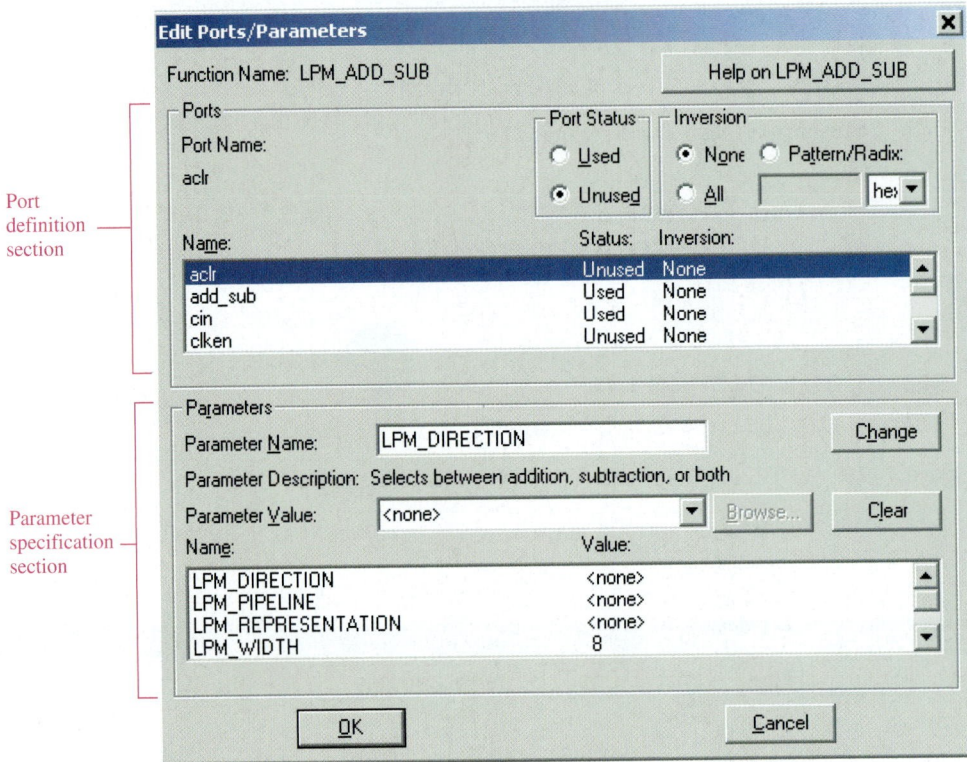

Figure 7–31 The Edit Ports/Parameters menu for the ADD_SUB LPM used in Example 7–28.

This menu is divided into two sections for defining the Ports and the Parameters. (Pressing the button **Help on LPM_ADD_SUB** provides detail on each of the available port names and parameter names.)

The *Ports section* provides the means for you to specify the inputs and outputs that are important to your design. For this example select the **Used** status for the following ports:

add_sub (This input is HIGH for addition, LOW for subtraction)

cin (This input is added to the result during addition, and its complement is subtracted from the result during subtraction.)

cout (This output is HIGH after adding numbers that overflow the bit width. It is LOW after subtracting numbers that have a negative result.)

dataa[ ] (This input receives the *a* number.)

datab[ ] (This input receives the *b* number.)

result[ ] (This output receives the result of the operation.)

The *Parameters section* provides the means for you to specify the operating characteristics of the module. For this example the only parameter that we need to specify is the LPM_WIDTH. Set it to 8 so that we can perform 8-bit arithmetic.

The LPM symbol with the input and output names added is shown in Figure 7–32. The 8-bit inputs and outputs are given *group names* and connected with the *bus* line style. The simulation in Figure 7–33 exercises the circuit by adding and then subtracting several combinations of input values. Carefully study the hex value in the *result* and *cout* waveforms to prove to yourself that the arithmetic answers are correct.

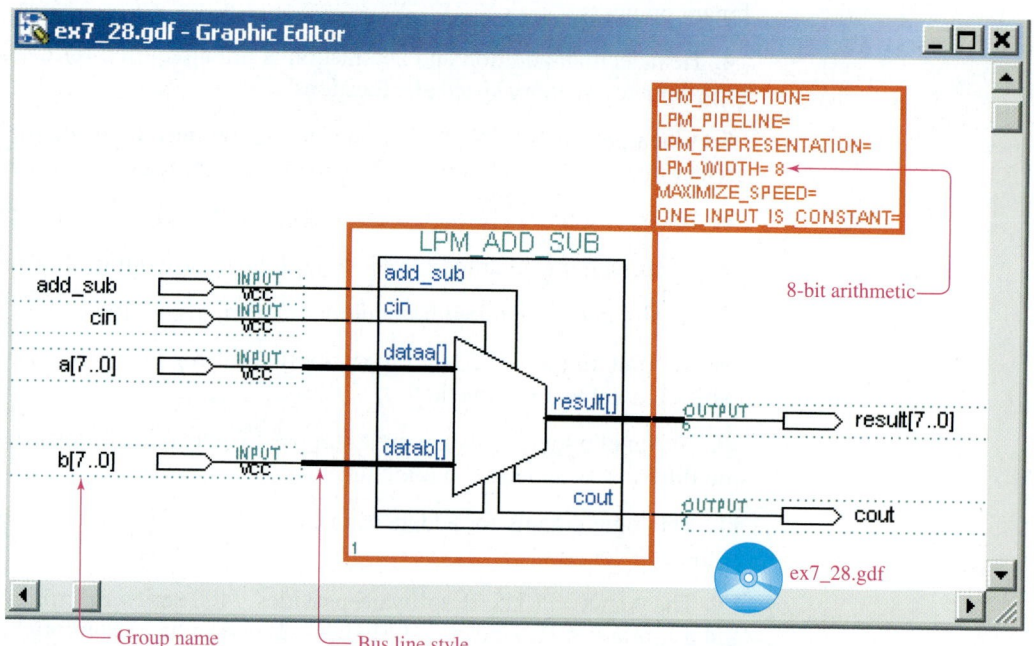

Figure 7–32 Inputs and outputs connected to the ADD_SUB LPM used in Example 7–28.

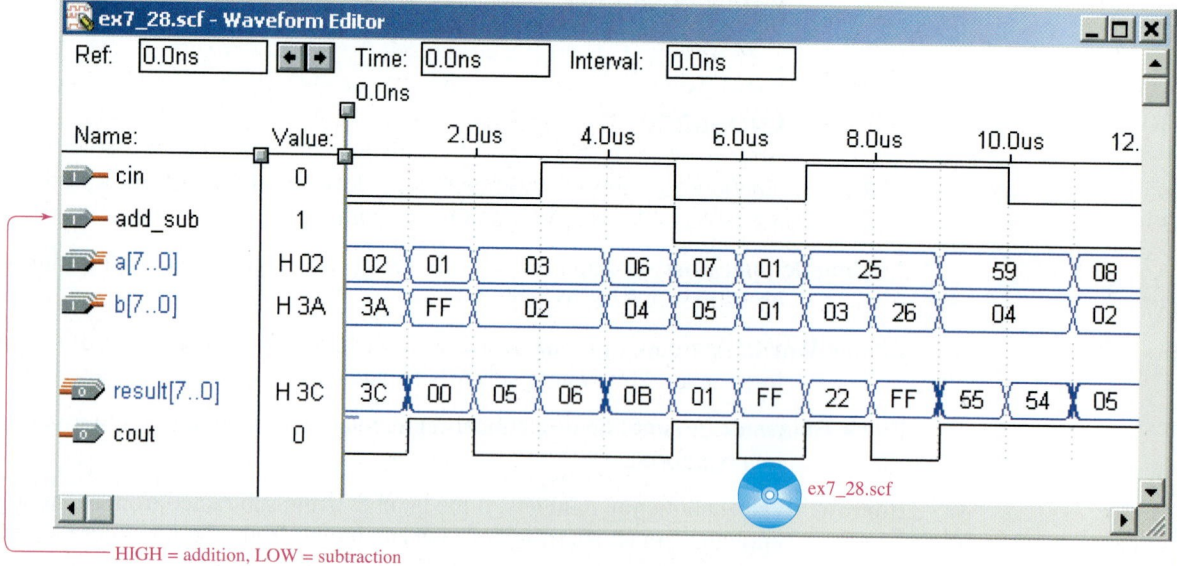

Figure 7–33 Addition and subtraction simulation for Example 7–28.

Summary

In this chapter, we have learned that

1. The binary arithmetic functions of addition, subtraction, multiplication, and division can be performed bit by bit using several of the same rules of regular base 10 arithmetic.

2. The two's-complement representation of binary numbers is commonly used by computer systems for representing positive and negative numbers.

3. Two's-complement arithmetic simplifies the process of subtraction of binary numbers.

4. Hexadecimal addition and subtraction is often required for determining computer memory space and locations.

5. When performing BCD addition a correction must be made for sums greater than 9 or when a carry to the next more significant digit occurs.

6. Binary adders can be built using simple combinational logic circuits.

7. A half-adder is required for addition of the least significant bits.

8. A full-adder is required for addition of the more significant bits.

9. Multibit full-adder ICs are commonly used for binary addition and two's-complement arithmetic.

10. Arithmetic/logic units are multipurpose ICs capable of providing several different arithmetic and logic functions.

11. The logic circuits for adders can be described in VHDL using integer arithmetic.

12. The MAX+PLUS II software provides 7400-series macrofunctions and a Library of Parameterized Modules to ease in the design of complex digital systems.

13. Conditional assignments can be made using the IF-THEN-ELSE or the WHEN-ELSE VHDL statements.

Glossary

ALU: Arithmetic/logic unit. A multifunction IC device used to perform a variety of user-selectable arithmetic and logic operations.

Arithmetic operator: Symbols used in the VHDL language to represent arithmetic operations that are to be performed.

Binary Word: A group, or string, of binary bits. In a 4-bit system, a word is 4 bits long. In an 8-bit system, a word is 8 bits long, and so on.

Block Diagram: A simplified functional representation of a circuit or system drawn in a box format.

Borrow: When subtracting numbers, if the number being subtracted from is not large enough, it must "borrow," or take an amount from, the next-more-significant digit.

Carry-In: An amount from a less-significant-digit addition that is applied to the current addition.

Carry-Out: When adding numbers, when the sum is greater than the amount allowed in that position, part of the sum must be applied to the next-more-significant position.

Concurrent statements: VHDL statements that are executed at the same time. In the synthesized circuit, the assignments associated with these statements will occur at the same time also.

Conditional signal assignment: In VHDL, an assignment that is made to a signal or an output based on the condition of another signal. The WHEN-ELSE clause is used for this purpose.

Fast-Look-Ahead Carry: When cascading several full-adders end to end, the carry-out of the last full-adder cannot be determined until each of the previous full-adder additions is completed. The internal carry must ripple or propagate through each of the lower-order adders before reaching the last full-adder. A fast-look-ahead carry system is used to speed up the process in a multibit system by reading all the input bits simultaneously to determine ahead of time if a carry-out of the last full-adder is going to occur.

Full-Adder: An adder circuit having three inputs, used to add two binary digits plus a carry. It produces their sum and carry as outputs.

Function Select: On an ALU, the pins used to select the actual arithmetic or logic operation to be performed.

Half-Adder: An adder circuit used in the LS position when adding two binary digits with no carry-in to consider. It produces their sum and carry as outputs.

High Order: In numbering systems, the high-order positions are those representing the larger magnitudes.

IF-THEN-ELSE: A clause preceding a series of sequential assignments that will be executed based on the results of the condition.

Integer: A VHDL data type that accommodates whole numbers.

Library of Parameterized Modules (LPM): A group of pre-defined modules that are the implementation of commonly used complex digital logic.

Low Order: In numbering systems, the low-order positions are those representing the smaller magnitudes.

Mode Control: On an ALU, the pin used to select either the arithmetic or the logic mode of operation.

MSI: Medium-scale integration. An IC chip containing combinational logic that is packed more densely than a basic logic gate IC (small-scale integration, SSI) but not as dense as a microprocessor IC (large-scale integration, LSI). MSI circuits are generally considered to be ICs containing, from 12 to 100 gate equivalents.

One's Complement: A binary number that is a direct (true) complement, bit by bit, of some other number.

Product: The result of the multiplication of numbers.

Remainder: The result of the subtraction of numbers.

Ripple Carry: *See* fast-look-ahead carry.

Sequential statements: VHDL statements that are executed one after the other in the order they appear. In the synthesized circuit, the assignments associated with these statements will occur one after the other also.

Sign Bit: The leftmost, or MSB, in a two's-complement number, used to signify the sign of the number (1 = negative, 0 = positive).

Signed number: Positive and negative numbers represented by two's-complement notation where the MSB represents the sign bit.

Sum: The result of the addition of numbers.

Two's Complement: A binary numbering representation that simplifies arithmetic in digital systems.

Unsigned number: Positive numbers where the MSB is part of the number (not a sign bit).

WHEN-ELSE: A clause used to direct the operation of a conditional signal assignment.

▬▬ Problems ▬▬

Section 7–1

7–1. Perform the following decimal additions, convert the original decimal numbers to binary, and add them. Compare answers.

(a) $\begin{array}{r} 6 \\ + 3 \\ \hline \end{array}$ (b) $\begin{array}{r} 8 \\ + 7 \\ \hline \end{array}$ (c) $\begin{array}{r} 22 \\ + 6 \\ \hline \end{array}$ (d) $\begin{array}{r} 29 \\ + 37 \\ \hline \end{array}$

(e) $\begin{array}{r} 134 \\ + 66 \\ \hline \end{array}$ (f) $\begin{array}{r} 254 \\ + 36 \\ \hline \end{array}$ (g) $\begin{array}{r} 208 \\ + 127 \\ \hline \end{array}$ (h) $\begin{array}{r} 196 \\ + 156 \\ \hline \end{array}$

7–2. Repeat Problem 7–1 for the following subtractions.

(a) $\begin{array}{r} 15 \\ - 4 \\ \hline \end{array}$ (b) $\begin{array}{r} 22 \\ - 11 \\ \hline \end{array}$ (c) $\begin{array}{r} 84 \\ - 36 \\ \hline \end{array}$ (d) $\begin{array}{r} 66 \\ - 31 \\ \hline \end{array}$

(e) $\begin{array}{r} 126 \\ - 64 \\ \hline \end{array}$ (f) $\begin{array}{r} 113 \\ - 88 \\ \hline \end{array}$ (g) $\begin{array}{r} 109 \\ - 60 \\ \hline \end{array}$ (h) $\begin{array}{r} 111 \\ - 104 \\ \hline \end{array}$

7–3. Repeat Problem 7–1 for the following multiplications.

(a) $\begin{array}{r} 7 \\ \times 3 \\ \hline \end{array}$ (b) $\begin{array}{r} 6 \\ \times 7 \\ \hline \end{array}$ (c) $\begin{array}{r} 12 \\ \times 5 \\ \hline \end{array}$ (d) $\begin{array}{r} 39 \\ \times 7 \\ \hline \end{array}$

(e) $\begin{array}{r} 63 \\ \times 125 \\ \hline \end{array}$ (f) $\begin{array}{r} 127 \\ \times 15 \\ \hline \end{array}$ (g) $\begin{array}{r} 31 \\ \times 13 \\ \hline \end{array}$ (h) $\begin{array}{r} 255 \\ \times 127 \\ \hline \end{array}$

7–4. Repeat Problem 7–1 for the following divisions.

(a) $4\overline{)12}$ (b) $3\overline{)15}$ (c) $12\overline{)48}$ (d) $5\overline{)25}$

(e) $5\overline{)125}$ (f) $14\overline{)294}$ (g) $15\overline{)195}$ (h) $12\overline{)228}$

Section 7–2

7–5. Produce a table of 8-bit two's-complement numbers from + 15 to − 15.

7–6. Convert the following decimal numbers to 8-bit two's-complement notation.

(a) 7 (b) − 7 (c) 14 (d) 36 (e) − 36
(f) 66 (g) − 48 (h) 112 (i) − 112 (j) − 125

7–7. Convert the following two's-complement numbers to decimal.

(a) 0001 0110 (b) 0000 1111

(c) 0101 1100 (d) 1000 0110

(e) 1110 1110 (f) 1000 0001

(g) 0111 1111 (h) 1111 1111

Section 7–3

7–8. What is the maximum positive-to-negative range of a two's-complement number in each?

(a) An 8-bit system (b) A 16-bit system

7–9. Convert the following decimal numbers to two's-complement form and perform the operation indicated.

(a) $\begin{array}{r} 5 \\ + 7 \end{array}$ (b) $\begin{array}{r} 12 \\ - 6 \end{array}$ (c) $\begin{array}{r} 32 \\ + 18 \end{array}$ (d) $\begin{array}{r} 32 \\ - 18 \end{array}$

(e) $\begin{array}{r} -28 \\ + 38 \end{array}$ (f) $\begin{array}{r} 125 \\ - 66 \end{array}$ (g) $\begin{array}{r} 36 \\ - 48 \end{array}$ (h) $\begin{array}{r} -36 \\ - 48 \end{array}$

Section 7–4

7–10. Build a table similar to Table 7–4 for hex digits 0C to 22.

7–11. Add the following hexadecimal numbers.

(a) $\begin{array}{r} A \\ + 4 \end{array}$ (b) $\begin{array}{r} 7 \\ + 6 \end{array}$ (c) $\begin{array}{r} 0B \\ + 16 \end{array}$ (d) $\begin{array}{r} 23 \\ + A7 \end{array}$

(e) $\begin{array}{r} 8A \\ + 82 \end{array}$ (f) $\begin{array}{r} A7 \\ + BB \end{array}$ (g) $\begin{array}{r} A049 \\ + 0AFC \end{array}$ (h) $\begin{array}{r} 0FFF \\ + 9001 \end{array}$

7–12. Subtract the following hexadecimal numbers.

(a) $\begin{array}{r} A \\ - 4 \end{array}$ (b) $\begin{array}{r} 8 \\ - 2 \end{array}$ (c) $\begin{array}{r} 1B \\ - 06 \end{array}$ (d) $\begin{array}{r} A7 \\ - 18 \end{array}$

(e) $\begin{array}{r} 2A \\ - 07 \end{array}$ (f) $\begin{array}{r} A7 \\ - 1D \end{array}$ (g) $\begin{array}{r} 4A2D \\ - 1A2F \end{array}$ (h) $\begin{array}{r} 8BB0 \\ - 4AC8 \end{array}$

7–13. Memory locations in personal computers are usually given in hexadecimal. If a computer programmer writes a program that requires 100 memory locations, determine the last memory location that is used if the program starts at location 2C8DH (H = base 16 hexadecimal).

7–14. A particular model of personal computer indicates in its owner's manual that the following memory locations are used for the storage of operating system subroutines: 07A4BH to 0BD78H inclusive and 02F80H to 03000H inclusive. Determine the total number of memory locations used for that purpose (in hex).

Section 7–5

7–15. Which of the following bit strings cannot be valid BCD numbers?

(a) 0111 1001 (b) 0101 1010

(c) 1110 0010 (d) 0100 1000

(e) 1011 0110 (f) 0100 1001

7–16. Convert the following decimal numbers to BCD and add them. Convert the result back to decimal to check your answer.

(a) $\begin{array}{r} 8 \\ + \ 3 \\ \hline \end{array}$ (b) $\begin{array}{r} 12 \\ + \ 16 \\ \hline \end{array}$ (c) $\begin{array}{r} 43 \\ + \ 72 \\ \hline \end{array}$ (d) $\begin{array}{r} 47 \\ + \ 38 \\ \hline \end{array}$

(e) $\begin{array}{r} 12 \\ + \ 89 \\ \hline \end{array}$ (f) $\begin{array}{r} 36 \\ + \ 22 \\ \hline \end{array}$ (g) $\begin{array}{r} 99 \\ + \ 11 \\ \hline \end{array}$ (h) $\begin{array}{r} 80 \\ + \ 23 \\ \hline \end{array}$

Section 7–6

7–17. Under what circumstances would you use a half-adder instead of a full-adder?

7–18. Reconstruct the half-adder circuit of Figure 7–7 using only NOR gates.

C   **7–19.** The circuit in Figure P7–19 is an attempt to build a half-adder. Will the C_{out} and Σ_0 function properly? (*Hint:* Write the Boolean equation at C_{out} and Σ_0.)

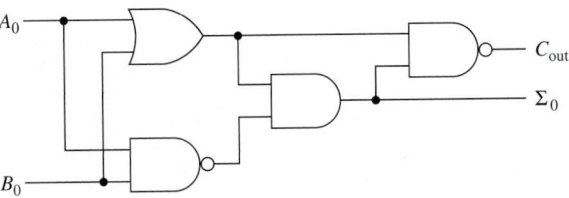

Figure P7–19

C   **7–20.** Use a Karnaugh map to prove that the C_{out} function of the full-adder whose truth table is given in Figure 7–5(c) can be implemented using the circuit given in Figure 7–9.

Section 7–7

7–21. Draw the block diagram of a 4-bit full-adder using *four full-adders.*

7–22. In Figure 7–17, the C_{in} to the first adder is grounded; explain why. Also, why isn't the C_{in} to the second adder grounded?

D   **7–23.** Design and draw a 6-bit binary adder similar to Figure 7–17 using two 7483 4-bit adders.

7–24. The 7483 has a fast-look-ahead carry. Explain why that is beneficial in some adder designs.

D   **7–25.** Design and draw a 16-bit binary adder using four 4008 CMOS 4-bit adders.

Section 7–8

7–26. What changes would have to be made to the adder/subtractor circuit of Figure 7–21 if exclusive-NORs are to be used instead of exclusive-ORs?

T   **7–27.** Figure P7–27 is a 4-bit two's-complement adder/subtractor. We are attempting to subtract $9 - 3$ ($1001 - 0011$) but keep getting the wrong answer of 10 (1010). Each test node in the circuit is labeled with the logic state observed using a logic probe. Find the two faults in the circuit.

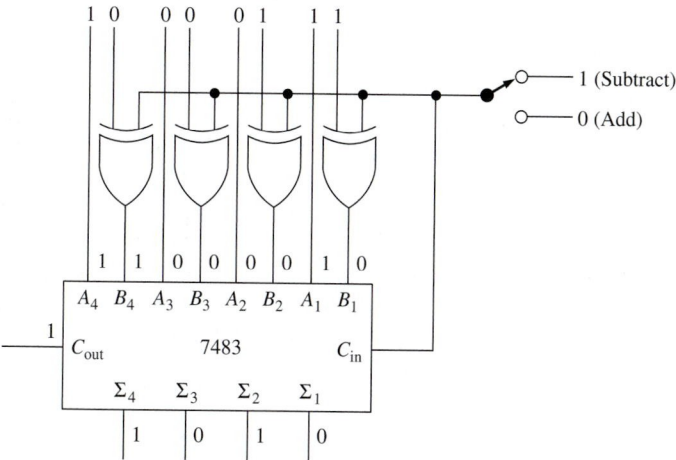

Figure P7–27

Section 7–9

T   **7–28.** Figure P7–28 is supposed to be set up as a one-digit hexadecimal adder. To test it, the values C and 2 (1100 + 0010) are input to the A and B inputs. The answer should be C + 2 = E (1110), but it is not! The figure is labeled with the states observed with a logic probe. Find the problem(s)!

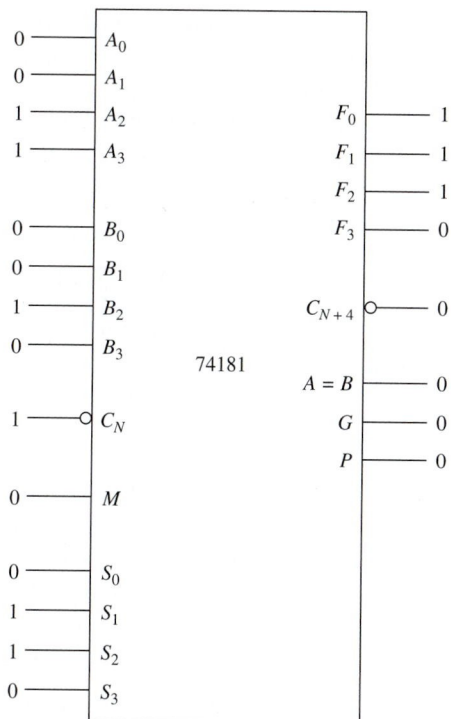

Figure P7–28

C   **7–29.** Re-solve Example 7–23 (a) and (b) for $S_3 - S_0 = 0100$.

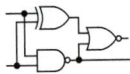

See Appendix G for the schematic diagrams.

S D **7–30.** Find U9 and U10 of the Watchdog Timer schematic. These are counter ICs that output their 4-bit binary count to $Q_0 - Q_3$. On a separate piece of paper, draw the connections that you would make to add the output of U9 to the output of U10 using a 74HC283 4-bit adder IC.

S C D **7–31.** Find Port E (PE0–PE7) of U1 in the HC11D0 schematic. Assume that the low-order bits (PE0–PE3) contain the 4-bit binary number A and the high-order bits (PE4–PE7) contain the 4-bit binary number B. On a separate piece of paper, make the connections necessary to subtract A minus B using a two's-complement adder/subtractor circuit design. The result of the subtraction is to be read by Port A, bits PA4–PA7.

S C D **7–32.** Repeat Problem 7–31 using a 74181 ALU.

Electronics Workbench®/MultiSIM® Exercises

E7–1. Load the circuit file for **Section 7–6a.** This circuit functions like the Full Adder described in Section 7–6.

(a) How many inputs and how many outputs does it have?

(b) The bits to be added are input via the A, B, and C switches. On a piece of paper, produce a truth table by stepping through all eight possible combinations of switch settings.

(c) Complete the following sentences: "The Sum output goes HIGH if. . . . " "The C_{out} output goes HIGH if. . . . "

C D **E7–2.** Load the circuit file for **Section 7–6b.** This circuit is a full-adder. To perform an addition of two 4-bit numbers as in Figure 7–15, we need to duplicate the circuit four times to create four full-adder subcircuits. Drag the four full-adder subcircuits onto the screen and connect them as a 4-bit adder.

(a) What must be done with the C_{in} of the first FA?

(b) Test your circuit operation by performing the addition: $7 + 6 = 13$ $(0111 + 0110 = 1101)$. Demonstrate your results to your teacher.

D **E7–3.** Load the circuit file for **Section 7–7.** The 7483s shown are subcircuits, each containing four full-adders. Make the necessary connections to form an 8-bit adder similar to Figure 7–17.

(a) Test your design by adding $37 + 22$. What is the result of the addition?

(b) Repeat part (a) for the addition $67 + 45$.

C D **E7–4.** Load the circuit file for **Section 7–8a.** All of the components necessary to construct a two's-complement adder/subtractor are provided in the circuit window. Make the connections similar to those shown in Figure 7–21. Demonstrate the operation of your design to your teacher by performing the following operations:

(a) $75 + 50 = 125$ (Which output summation LEDs are ON?)

(b) $75 - 50 = 25$ (Which output summation LEDs are ON?)

D **E7–5.** Load the circuit file for Section 7–8b. All of the components necessary to construct a BCD adder are provided in the circuit window. Make the connections similar to those shown in Figure 7–22. Demonstrate the operation of your design to your teacher by performing the following additions:

(a) 7 + 6 (Which output summation LEDs are ON?)

(b) 9 + 8 (Which output summation LEDs are ON?)

CPLD Problems

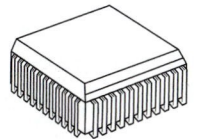

C7–1. The VHDL full-adder equations given in Figure 7–12 were based on the *sum* equation derived in Figure 7–8 and the C_{out} equation derived from Figure 7–9. Figure 7–10 combines the two into one circuit.

(a) Use the same equation for *sum1* that was given in Figure 7–12, but rewrite the VHDL equation for C_{out} to match the circuit in Figure 7–10. Save this program as *prob_c7_1.vhd*.

(b) Test the new VHDL design by simulating all possible inputs at *a1*, *b1*, and *cin*.

(c) Download your design to a CPLD IC. Discuss your observations of the *sum1* and C_{out} LEDs with your instructor as you use the switches to step through all eight input combinations.

C7–2. The VHDL program in Figure 7–18 is the implementation of a 4-bit adder.

(a) Make the necessary changes to make it an 8-bit adder. Save this program as *prob_c7_2.vhd*.

(b) Test its operation by performing the following hex additions with *cin*= 0, then *cin*= 1: 44+ 22; 20+ 4D; AA+ 22; FF+ 01.

(c) Use the Floorplan Editor to determine the pin assignments for all nine *sum_string* outputs.

C7–3. Figure 7–25 uses a 74283 macrofunction to form a 4-bit adder.

(a) Use a second 74283 to form an 8-bit adder similar to Figure 7–17. Save this program as *prob_c7_3.gdf*.

(b) Test its operation by performing the following hex additions: 33+ 14; 17+ 03; 7A+ 22; BB+ 16; FF+ 01; DD+ 66.

C7–4. Figure 7–21 shows an 8-bit adder/subtractor built with adder ICs and XOR gates.

(a) Use the Graphic Editor to design a 4-bit adder/subtractor using a 74283 macrofunction and four XOR gates. Save this program as *prob_c7_4.gdf*.

(b) Test its operation by simulating its 4-bit result as you perform the following hex arithmetic: 2+ 6; 7+ 6; A+ 2; 1+ 7; 9− 6; C− 1; F− 7; 6− 8.

(c) Download your design to a CPLD IC. Discuss your observations of its 4-bit *result* LEDs with your instructor as you use the switches to step through the four additions and four subtractions.

C7–5. Figure 7–27 shows a VHDL implementation of an 8-bit adder/subtractor using std_logic vectors and the WHEN-ELSE conditional signal assignment.

(a) Rewrite the program to implement a 4-bit adder/subtractor. Save this program as *prob_c7_5.vhd*.

(b) Test its operation by simulating its 4-bit result as you perform the following hex arithmetic: 2+ 6; 7+ 6, A+ 2; 1+ 7; 9+ 6; C− 1; F− 7; 6− 8.

(c) Download your design to a CPLD IC. Discuss your observations of its 4-bit *result* LEDs with your instructor as you use the switches to step through the four additions and four subtractions.

C7–6. Figure 7–29 (Example 7–27) shows the VHDL implementation of a BCD correction adder using the IF-THEN-ELSE statement.

(a) Use Windows Explorer to copy the file *ex7_27.vhd* located on the textbook CD over to the *maxplus2* subdirectory on your hard drive assigned for your projects. Then in the MAX+PLUS II program, open that file and choose: **File > Project > Set Project to Current File** and then **File > Project > Save and Compile.**

(b) Create a simulation file to test the following BCD additions: 4 + 4; 4 + 7; 3 + 7; 9 + 9; 7 + 7; 9 + 4. Check that the 8-bit BCD results are correct.

(c) Download your design to a CPLD IC. Discuss your observations of its 8-bit BCD *result* displayed on the LEDs with your instructor as you use the switches to step through the six additions.

C7–7. Figure 7–32 (Example 7–28) uses an LPM module to perform additions and subtractions.

(a) Create a file called *prob_c7_7.gdf* using the LPM_ADD_SUB to perform 16-bit additions.

(b) Test its operation by simulating the following hex additions: 20A7 + 1111; 00BB + 2012; 4AFC + 1322; 7FFF + 0001.

C7–8. MAX+PLUS II software provides an LPM module called LPM_MULT that can be used to multiply two numbers.

(a) Enter the LPM_MULT module into a new Graphic Editor window. Set it up to multiply two 4-bit numbers called *a[3..0]* and *b[3..0]* to form a product called *p[7..0]*.

(b) Create a simulation file that demonstrates the multiplication of the following decimal numbers (Set all radixes to decimal): 2×2; 2×8; 4×4, 9×9; 12×8; 15×15.

(c) Download your design to a CPLD IC. Discuss your observations of its 8-bit product displayed on the LEDs with your instructor as you use the switches to step through the six multiplications.

Answers to Review Questions

7–1. 2 inputs, 2 outputs

7–2. True

7–3. True

7–4. D_7

7–5. (a) Negative
  (b) positive
  (c) negative

7–6. b, d

7–7. True

7–8. True

7–9. Because it simplifies the documentation and use of the equipment

7–10. 15

7–11. 16

7–12. Greater than 9; there is a carry out of the MSB.

7–13. Add 6 (0110).

7–14. Inputs: A_0, B_0; outputs: Σ_0, C_{out}

7–15. Because it needs a carry-in from the previous adder

7–16. Odd

7–17. When any two of the inputs are HIGH

7–18. Connect it to zero.

7–19. To speed up the arithmetic process

7–20. It provides a C_{in}, and it puts a 1 on the inputs of the X-OR gates, which inverts B.

7–21. They check for a sum greater than 9 and provide a C_{out}.

7–22. It sets the mode of operation for either logic or arithmetic.

7–23. True

7–24. + means logical OR; *plus* means arithmetic sum.

8

Code Converters, Multiplexers, and Demultiplexers

OUTLINE

OBJECTIVES

Upon completion of this chapter, you should be able to:

- Utilize an IC magnitude comparator to perform binary comparisons.
- Describe the function of a decoder and an encoder.
- Design the internal circuitry for encoding and decoding.
- Utilize manufacturers' data sheets to determine the operation of IC decoder and encoder chips.
- Explain the procedure involved in binary, BCD, and Gray code converting.
- Explain the operation of code-converter circuits built from SSI and MSI ICs.
- Describe the function and uses of multiplexers and demultiplexers.
- Design circuits that employ multiplexer and demultiplexer ICs.

INTRODUCTION

Information, or data, that is used by digital devices comes in many formats. The mechanisms for the conversion, transfer, and selection of data are handled by combinational logic ICs.

In this chapter, we first take a general approach to the understanding of data-handling circuits and then deal with the specific operation and application of practical data-handling MSI chips. The MSI chips covered include comparators, decoders, encoders, code converters, multiplexers, and demultiplexers.

8-1 Comparators

Often in the evaluation of digital information, it is important to compare two binary strings (or binary words) to determine if they are exactly equal. This comparison process is performed by a digital **comparator.**

The basic comparator evaluates two binary strings bit by bit and outputs a 1 if they are exactly equal. An exclusive-NOR gate is the easiest way to compare the equality of bits. If both bits are equal (0–0 or 1–1), the Ex-NOR puts out a 1.

To compare more than just 2 bits, we need additional Ex-NORs, and the output of all of them must be 1. For example, to design a comparator to evaluate two 4-bit numbers, we need four Ex-NORs. To determine total equality, connect all four outputs into an AND gate. That way, if all four outputs are 1's, the AND gate puts out a 1. Figure 8–1 shows a comparator circuit built from exclusive-NORs and an AND gate.

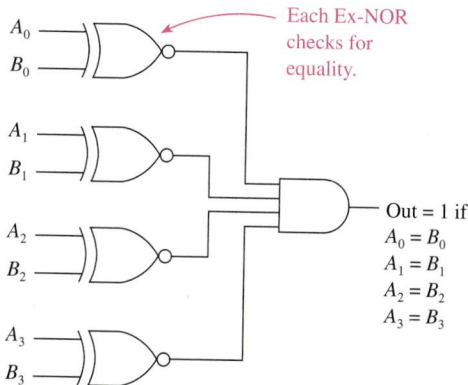

Figure 8–1 Binary comparator for comparing two 4-bit binary strings.

Studying Figure 8–1, you should realize that if A_0–B_0 equals 1–1 or 0–0, the top Ex-NOR will output a 1. The same holds true for the second, third, and fourth Ex-NOR gates. If all of them output a 1, the AND gate outputs a 1, indicating equality.

EXAMPLE 8–1

Referring to Figure 8–1, determine if the following pairs of input binary numbers will output a 1.

(a) $A_3A_2A_1A_0 = 1\ 0\ 1\ 1$

   $B_3B_2B_1B_0 = 1\ 0\ 1\ 1$

(b) $A_3A_2A_1A_0 = 0\ 1\ 1\ 0$

   $B_3B_2B_1B_0 = 0\ 1\ 1\ 1$

Solution: **(a)** When the A and B numbers are applied to the inputs, each of the four Ex-NORs will output 1's, so the output of the AND gate will be 1 (equality).

(b) For this case, the first three Ex-NORs will output 1's, but the last Ex-NOR will output a 0 because its inputs are not equal. The AND gate will output a 0 (inequality).

Integrated-circuit *magnitude comparators* are available in both the TTL and CMOS families. A magnitude comparator not only determines if A equals B but also if A is *greater than B* or A is *less than B*.

The 7485 is a TTL 4-bit magnitude comparator. The pin configuration and logic symbol for the 7485 are given in Figure 8–2. The 7485 can be used just like the basic comparator of Figure 8–1 by using the A inputs, B inputs, and the equality output ($A = B$). The 7485 has the additional feature of telling you which number is larger if the equality is not met. The $A > B$ output is 1 if A is larger than B, and the $A < B$ output is 1 if B is larger than A.

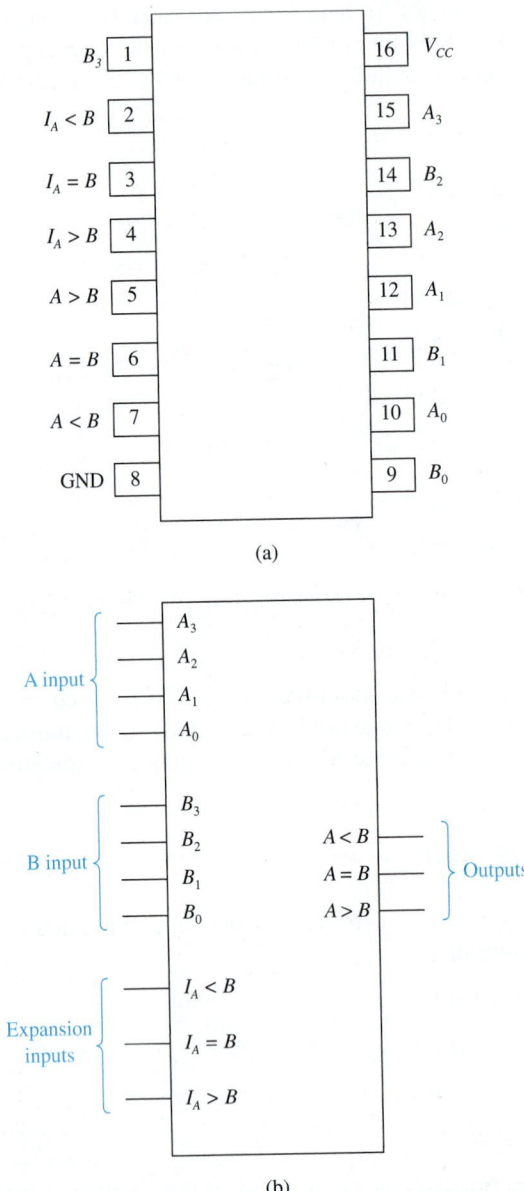

Figure 8–2   The 7485 4-bit magnitude comparator: (a) pin configuration; (b) logic symbol.

The expansion inputs $I_A < B$, $I_A = B$, and $I_A > B$ are used for expansion to a system capable of comparisons greater than 4 bits. For example, to set up a circuit capable of comparing two *8-bit words*, two 7485s are required. The $A > B$, $A = B$, and $A < B$ outputs of the low-order (least significant) comparator are connected to the expansion

inputs of the high-order comparator. That way, the comparators act together, comparing two entire 8-bit words and outputting the result from the high-order comparator outputs. For proper operation, the expansion inputs to the low-order comparator should be tied as follows: $I_A > B = $ LOW, $I_A = B = $ HIGH, and $I_A < B = $ LOW. Expansion to greater than 8 bits using multiple 7485s is also possible. Figure 8–3 shows the connections for magnitude comparison of two 8-bit binary strings. If the high-order A inputs are equal to the high-order B inputs, then the expansion inputs are used as a tiebreaker.

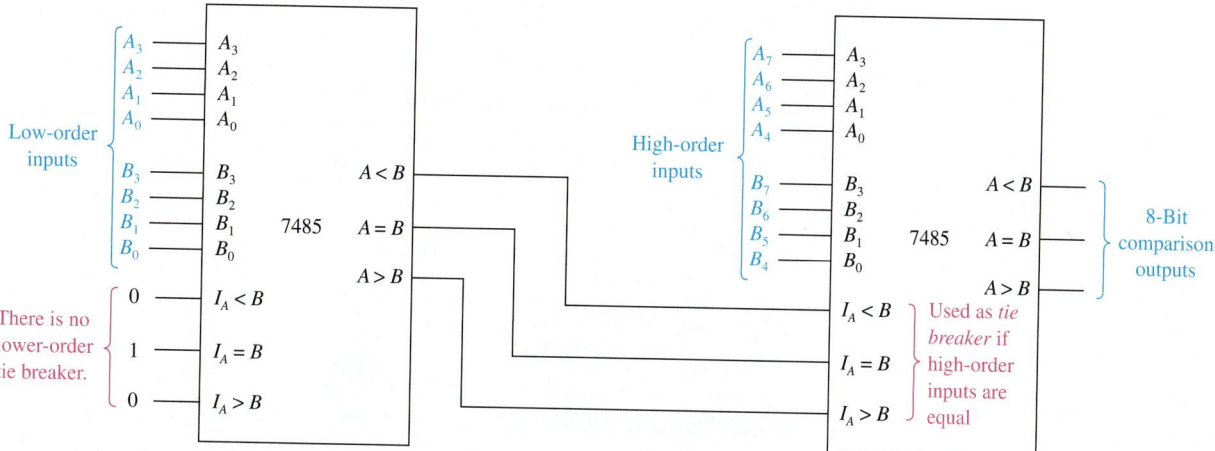

Figure 8–3 Magnitude comparison of two 8-bit binary strings (or binary words).

Review Questions

8–1. More than one output of the 7485 comparator can be simultaneously HIGH. True or false?

8–2. If all inputs to a 7485 comparator are LOW except for the $I_A < B$ input, what will the output be?

8–2   VHDL Comparator Using IF-THEN-ELSE

In this section we'll see how easy it is to design the 8-bit comparator system of Figure 8–3 using VHDL. The program listing is shown in Figure 8–4.

The entity section declares the a and b inputs as 8-bit vectors. During the simulation stage, we will enter several 2-digit hex numbers into those a and b inputs to compare their magnitude. Three separate outputs are declared to signify magnitude: a greater than b (*agb*); a equal b (*aeb*); and a less than b (*alb*).

In the beginning of the architecture section a vector called *result* is declared as an internal *signal*. This 3-bit vector will receive the result of the IF-THEN-ELSE comparisons. After the comparisons load this *result* vector, three individual assignments must be made to output the vector elements to their associated outputs at *agb*, *aeb*, and *alb*. The IF comparison statements need to be incorporated into a PROCESS. The PROCESS needs the sensitivity list (a,b). This sensitivity list tells VHDL to execute the statements inside of this process *only when there is a change in a or b*.

As soon as there is a change in the sensitivity list items, all of the statements in the process will be executed sequentially as shown in the flowchart in Figure 8–5.

```
compare_8b.vhd - Text Editor                                              _ □ X

LIBRARY ieee;                        ----------------------------
USE ieee.std_logic_1164.ALL;         -- 8-bit Comparator using  --
                                     -- IF-THEN-ELSE and ELSIF  --
                                     ----------------------------
ENTITY compare_8b IS
    PORT(a, b              : IN    std_logic_vector(7 DOWNTO 0);
         agb, aeb, alb     : OUT   std_logic);
END compare_8b;

ARCHITECTURE arc OF compare_8b IS
    SIGNAL result              : std_logic_vector(2 DOWNTO 0);  ◄──┐  3-bit internal
BEGIN                                                              │  SIGNAL result
    PROCESS (a,b)  ──── Sensitivity list
      BEGIN                                              ─── alb
        IF    a<b THEN
                result     <=   "001";       ──── aeb
        ELSIF a=b THEN
                result     <=   "010";   ─── agb
        ELSIF a>b THEN
                result     <=   "100";
        ELSE
                result     <=   "000";
        END IF;
        agb <=  result(2); ─┐
        aeb <=  result(1);  │  Assign vector elements
        alb <=  result(0); ─┘  to outputs
      END PROCESS;                              🔵  compare_8b.vhd
END arc;
                                                                         ▼
Line  27  |Col  1  | INS ◄|                                            ►
```

Figure 8–4   VHDL program for an 8-bit comparator similar to Figure 8–3.

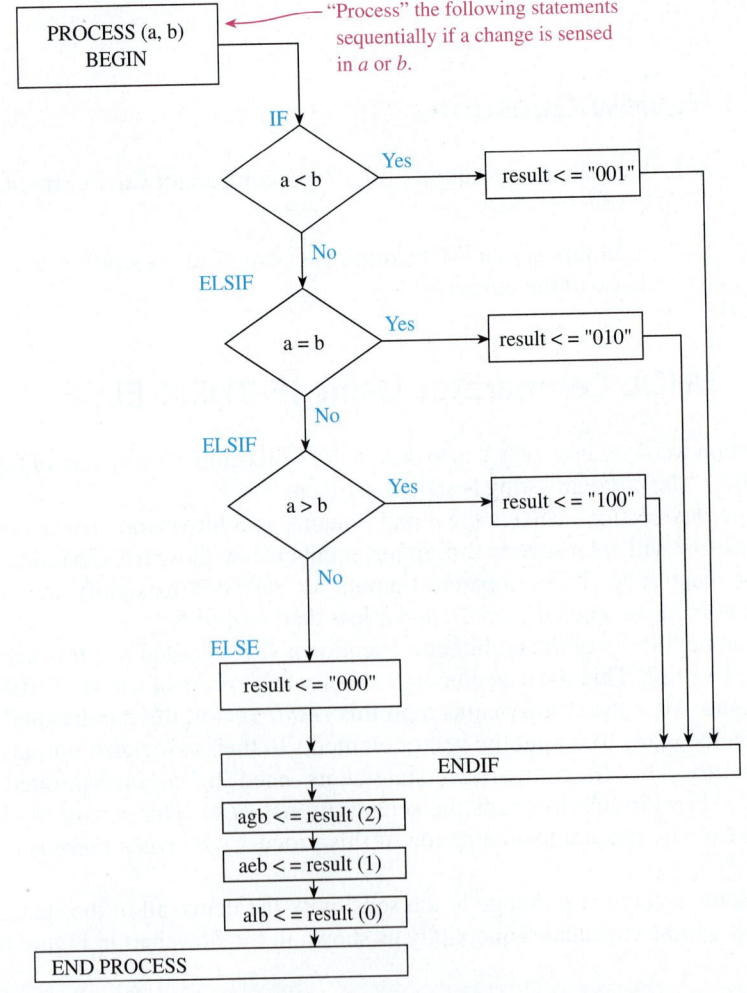

Figure 8–5   Flowchart showing the sequential execution within the PROCESS.

296

The first IF statement determines if *a* is less than *b*. If it is, then the three bits "001" are moved to the *result* vector and program execution passes to the END IF clause. The 1 in the rightmost position corresponds to *result* element number 0 *result(0)*, which will later be moved to output pin *alb* (*a* less than *b*). If the answer to the first IF is NO, then control passes to the ELSIF a = b statement, where a YES/NO determination is made, and so on for the ELSIF a > b. As you can see in the flowchart, if the answer is NO for all three IF statements, then the *result* is set to 000. This may happen because as mentioned before, std_logic has many other states besides just 1 and 0. (HIGH impedance and FLOAT are two other common states.)

After the END IF statement, each individual *result* element is assigned to its corresponding output pin.

The simulation in Figure 8–6 shows the results of several comparisons between *a* and *b*. For example, the first comparison (*a* = 05, *b* = 05) results in making the *aeb* (*a* equals *b*) output HIGH and all others LOW. The next comparison (*a* = AA, *b* = A7) produces a HIGH at *agb*, and so on. Check over the other six comparisons to be sure that they make sense to you.

One peculiar occurrence that you may notice is the short duration **glitch** on the *alb* line. This false switching is an undesirable trait that sometimes occurs because of internal CPLD circuit switching delays. Figure 8–7 shows an expanded view of the first glitch.

Notice that not only is there a false HIGH on the *alb* line, but the *aeb* line doesn't immediately go LOW when the *a*, *b* inputs switch. Also, the *agb* line takes some additional time before it goes HIGH to signify *a* greater than *b*. All of these can cause timing problems in high-speed switching applications. Notice that the time denoted by the vertical cursor is 1.0315 μs. This is 0.0315 μs after the 1.0 μs switching point. 1.0315 μs is the point where all three lines are valid. This is an important piece of information that the simulator gives us. It means that when using this particular CPLD IC for this purpose, you must wait at least 0.0315 μs (31.5 ns) after the *a* and *b* numbers switch before the outputs are valid to read.

(*Note:* The expanded view shown in Figure 8–7 is displayed by moving the vertical reference cursor to the 1.0 μs mark by pressing the left/right arrows located between the **Ref:** and **Time:** boxes in the Waveform Editor window. Then press **View > Zoom In** several times to expand the view as shown.)

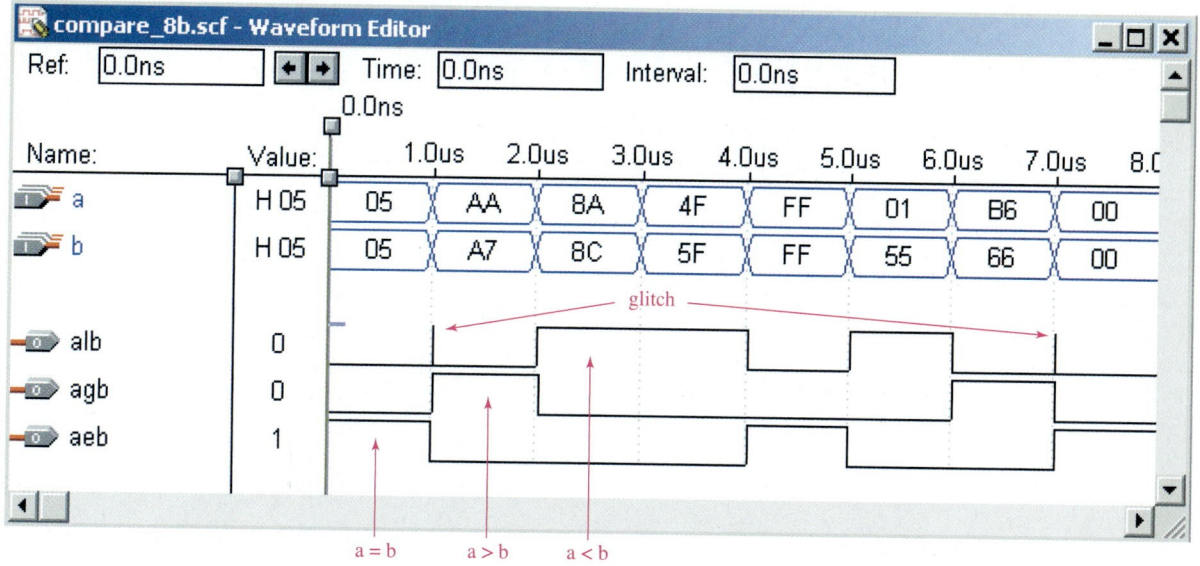

Figure 8–6   Simulation of the 8-bit comparator of Figure 8–4.

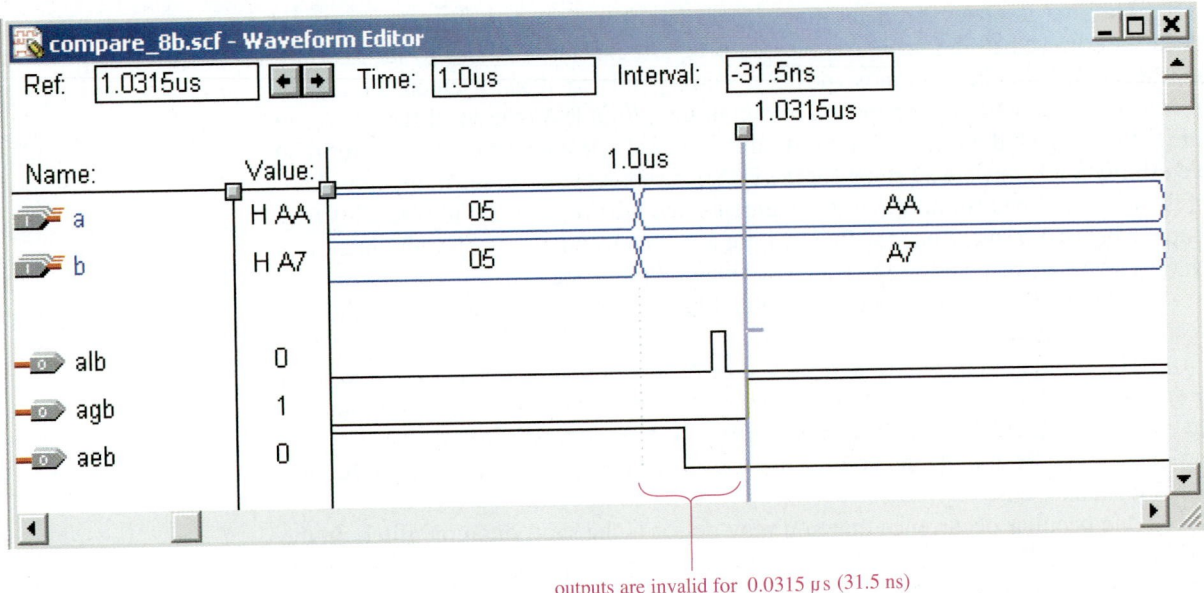

outputs are invalid for 0.0315 μs (31.5 ns)

Figure 8–7 Expanded view of the first glitch in the simulation display.

8–3 Decoding

Decoding is the process of converting some code (such as binary, BCD, or hex) into a singular active output representing its numeric value. Take, for example, a system that reads a 4-bit BCD code and converts it to its appropriate decimal number by turning on a decimal indicating lamp. Figure 8–8 illustrates such a system. This **decoder** is made up of a combination of logic gates that produces a HIGH at one of the 10 outputs, based on the levels at the four inputs.

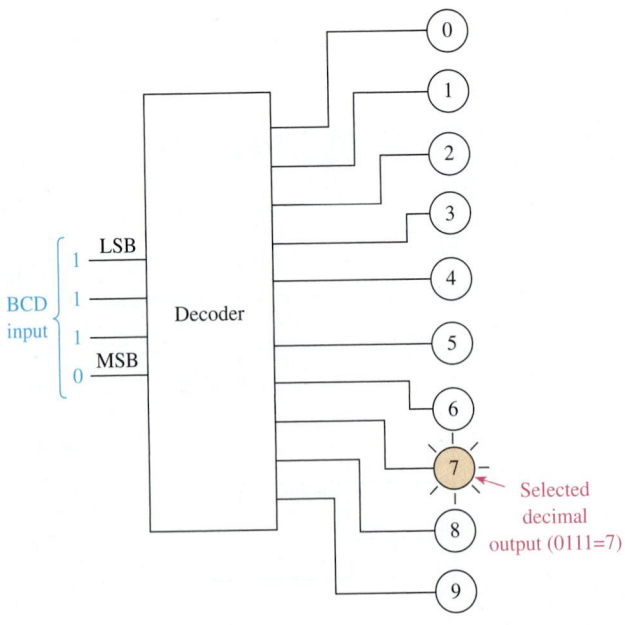

Figure 8–8 A BCD decoder selects the correct decimal-indicating lamp based on the BCD input.

In this section, we learn how to use decoder ICs by first looking at the combinational logic that makes them work and then by selecting the actual decoder IC and making the appropriate pin connections.

3-Bit Binary-to-Octal Decoding

To design a decoder, it is useful first to make a truth table of all possible input/output combinations. An octal decoder must provide eight outputs, one for each of the eight different combinations of inputs, as shown in Table 8–1.

Before the design is made, we must decide if we want an *active-HIGH-level* output or an *active-LOW-level* output to indicate the value selected. For example, the *active-HIGH* truth table in Table 8–1(a) shows us that, for an input of 011 (3), output 3 is HIGH, and all other outputs are LOW. The *active-LOW* truth table is just the opposite (output 3 is LOW, and all other outputs are HIGH).

Therefore, we have to know whether the indicating lamp (or other receiving device) requires a HIGH level to activate or a LOW level. We will learn in Chapter 9 that most devices used in digital electronics are designed to activate from a LOW-level signal, so most decoder designs use *active-LOW* outputs, as shown in Table 8–1(b). The combinational logic requirements to produce a LOW at output 3 for an input of 011 are shown in Figure 8–9.

To design the complete octal decoder, we need a separate NAND gate for each of the eight outputs. The input connections for each of the NAND gates can be determined by referring to Table 8–1(b). For example, the NAND gate 5 inputs are connected to the $2^2 - \overline{2^1} - 2^0$ input lines, NAND gate 6 is connected to the $2^2 - 2^1 - \overline{2^0}$ input lines, and so on. The complete circuit is shown in Figure 8–10. Each NAND gate in Figure 8–10

TABLE 8–1 | Truth Tables for an Octal Decoder

(a) Active-HIGH Outputs

Input			Output							
2^2	2^1	2^0	0	1	2	3	4	5	6	7
0	0	0	1	0	0	0	0	0	0	0
0	0	1	0	1	0	0	0	0	0	0
0	1	0	0	0	1	0	0	0	0	0
0	1	1	0	0	0	1	0	0	0	0
1	0	0	0	0	0	0	1	0	0	0
1	0	1	0	0	0	0	0	1	0	0
1	1	0	0	0	0	0	0	0	1	0
1	1	1	0	0	0	0	0	0	0	1

← *Note:* the selected output goes HIGH.

(b) Active-LOW Outputs

Input			Output							
2^2	2^1	2^0	0	1	2	3	4	5	6	7
0	0	0	0	1	1	1	1	1	1	1
0	0	1	1	0	1	1	1	1	1	1
0	1	0	1	1	0	1	1	1	1	1
0	1	1	1	1	1	0	1	1	1	1
1	0	0	1	1	1	1	0	1	1	1
1	0	1	1	1	1	1	1	0	1	1
1	1	0	1	1	1	1	1	1	0	1
1	1	1	1	1	1	1	1	1	1	0

← *Note:* the selected output goes LOW.

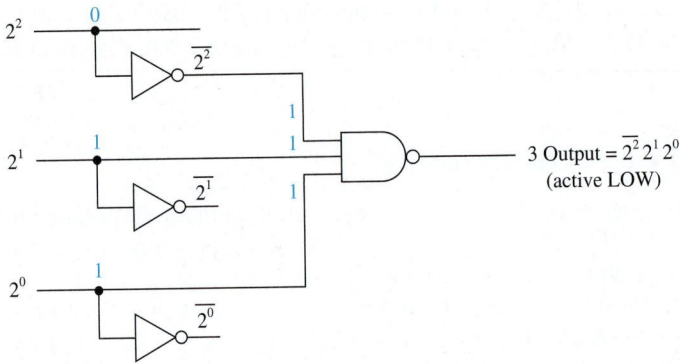

Figure 8–9 Logic requirements to produce a LOW at output 3 for a 011 input.

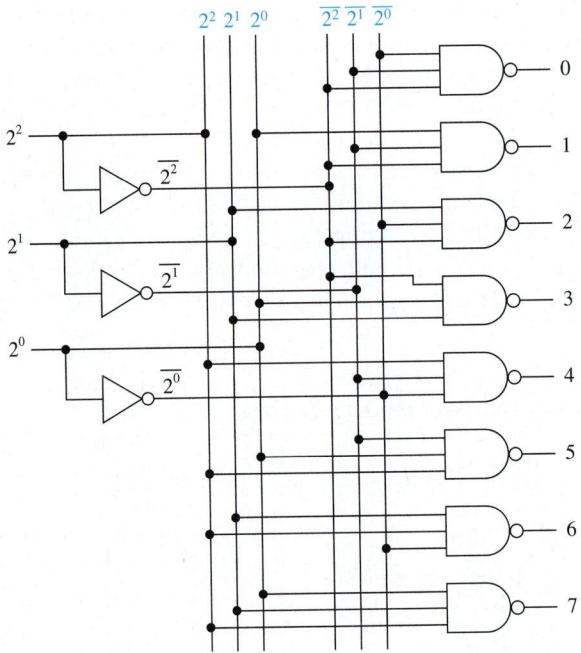

Figure 8–10 Complete circuit for an active-LOW output octal (1-of-8) decoder.

is wired so that its output goes LOW when the correct combination of input levels is present at its input. BCD and hexadecimal decoders can be designed in a similar manner.

The octal decoder is sometimes referred to as a *1-of-8 decoder* because, based on the input code, one of the eight outputs will be active. It is also known as a *3-line-to-8-line decoder* because it has three input lines and eight output lines.

Integrated-circuit decoder chips provide basic decoding as well as several other useful functions. Manufacturers' data books list several decoders and give function tables illustrating the input/output operation and special functions. Rather than designing decoders using combinational logic, it is much more important to be able to use a data book to find the decoder that you need and to determine the proper pin connections and operating procedure to perform a specific decoding task. Table 8–2 lists some of the more popular TTL decoder ICs. (Equivalent CMOS ICs are also available.)

TABLE 8–2	Decoder ICs

Device Number	Function
74138	1-of-8 octal decoder (3-line-to-8-line)
7442	1-of-10 BCD decoder (4-line-to-10-line)
74154	1-of-16 hex decoder (4-line-to-16-line)
7447	BCD-to-seven-segment decoder (covered in Chapter 12)

Octal Decoder IC

The 74138 is an octal decoder capable of decoding the eight possible octal codes into eight separate active-LOW outputs, just like our combinational logic design. It also has three enable inputs for additional flexibility. Figure 8–11 shows information presented in a data book for the 74138.

Helpful Hint

In most cases, the 74LS and 74HC series are interchangeable with the standard 74 series of ICs.

Helpful Hint

Decoding the *A* inputs in Figure 8–11(c) could have been done using just three inverters, similar to Figure 8–10. This is a good time to start thinking about gate loading, which is covered in Chapter 9. (Using six inverters ensures that each *A* input drives only one gate load.)

Helpful Hint

This is a good time to begin realizing the meaning of overbars in schematics. Don't develop the bad habit of thinking that $\overline{E_1}$ and $\overline{E_2}$ are inverted as they enter the IC. Instead, realize that $\overline{E_1}$ and $\overline{E_2}$ require a LOW to be satisfied. Also, the eight outputs each become active by going LOW.

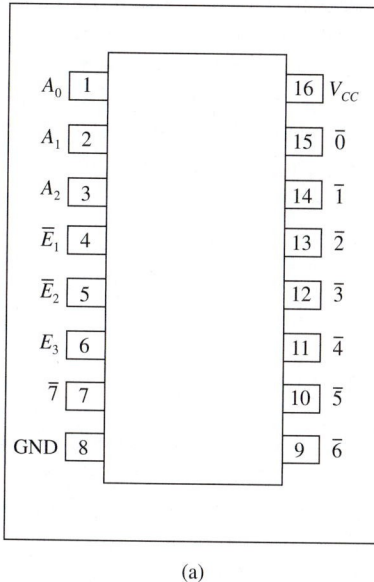

(a)

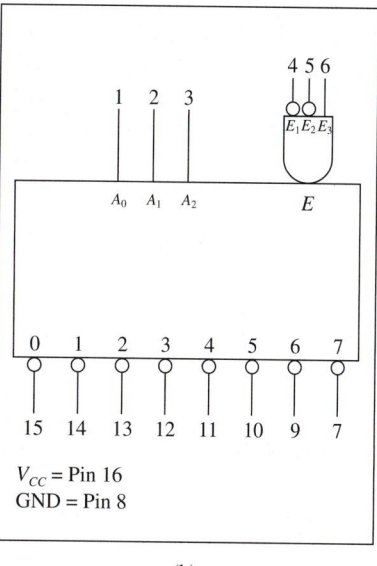

(b)

	Inputs						Outputs							
$\overline{E_1}$	$\overline{E_2}$	E_3	A_0	A_1	A_2	$\overline{0}$	$\overline{1}$	$\overline{2}$	$\overline{3}$	$\overline{4}$	$\overline{5}$	$\overline{6}$	$\overline{7}$
H	X	X	X	X	X	H	H	H	H	H	H	H	H
X	H	X	X	X	X	H	H	H	H	H	H	H	H
X	X	L	X	X	X	H	H	H	H	H	H	H	H
L	L	H	L	L	L	L	H	H	H	H	H	H	H
L	L	H	H	L	L	H	L	H	H	H	H	H	H
L	L	H	L	H	L	H	H	L	H	H	H	H	H
L	L	H	H	H	L	H	H	H	L	H	H	H	H
L	L	H	L	L	H	H	H	H	H	L	H	H	H
L	L	H	H	L	H	H	H	H	H	H	L	H	H
L	L	H	L	H	H	H	H	H	H	H	H	L	H
L	L	H	H	H	H	H	H	H	H	H	H	H	L

Notes
H = HIGH voltage level
L = LOW voltage level
X = Don't care
A_2 = MSB

(c)

(d)

Figure 8–11 The 74138 octal decoder: (a) pin configuration; (b) logic symbol; (c) logic diagram; (d) function table. (Courtesy of Philips Semiconductors.)

Just by looking at the logic symbol [Figure 8–11(b)] and function table [Figure 8–11(d)], we can figure out the complete operation of the chip. First, the inversion bubbles on the decoded outputs indicate active-LOW operation. The three inputs $\overline{E_1}$, $\overline{E_2}$, and E_3 are used to *enable* the chip. The function table shows that the chip is disabled (all outputs HIGH) *unless* $\overline{E_1}$ = LOW *and* $\overline{E_2}$ = LOW *and* E_3 = HIGH. The enables are useful for go/no-go operation of the chip based on some external control signal.

When the chip is *disabled*, the ×'s in the binary input columns A_0, A_1, and A_2 indicate **don't-care** levels, meaning the outputs will all be HIGH no matter at what level A_0, A_1, and A_2 are. When the chip is *enabled*, the binary inputs A_0, A_1, and A_2 are used to select which output goes LOW. In this case, A_0 is the least significant bit (LSB) input. Be aware that some manufacturers label the inputs *A*, *B*, *C* instead of A_0, A_1, A_2 and assume that *A* is the LSB.

The logic diagram in Figure 8–11(c) shows the actual internal combinational logic required to perform the decoding. The extra inverters on the inputs are required to prevent excessive loading of the driving source(s). These internal inverters supply the driving current to the eight NAND gates instead of the driving source(s) having to do it. (Gate loading is discussed in Chapter 9.) The three enable inputs ($\overline{E_1}$, $\overline{E_2}$, and E_3) are connected to an AND gate, which can disable all the output NANDs by sending them a LOW input level if $\overline{E_1}$, $\overline{E_2}$, and E_3 are not 001. Example 8–2 shows a waveform analysis of the 74138, and Section 8–7 discusses its use in a microcomputer application.

EXAMPLE 8–2

Sketch the output waveforms of the 74138 in Figure 8–12. Figure 8–13 shows the input waveforms to the 74138.

Solution:

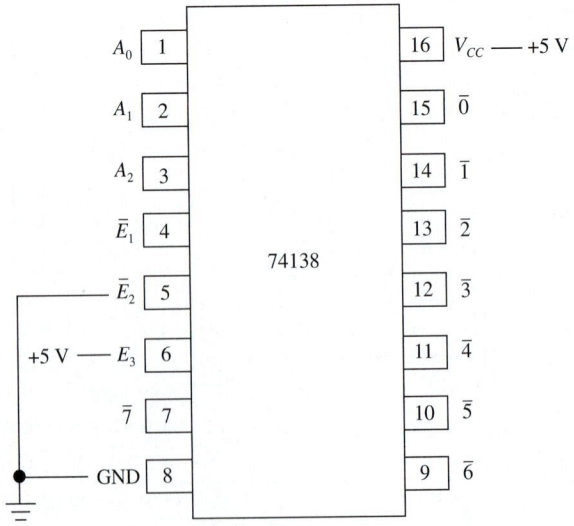

Figure 8–12   The 74138 connections for Example 8–2.

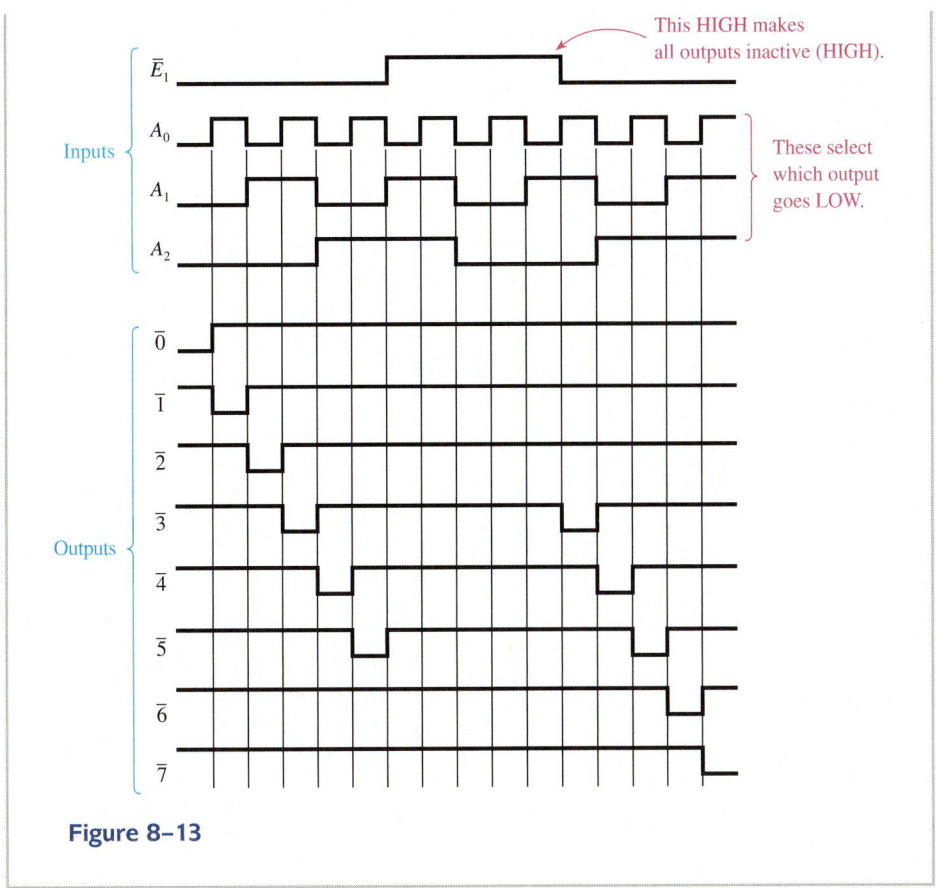

Figure 8–13

BCD Decoder IC

The 7442 is a BCD-to-decimal (1-of-10) decoder. It has four pins for the BCD input bits (0000 to 1001) and 10 active-LOW outputs for the decoded decimal numbers. Figure 8–14 gives the operational information for the 7442 from a manufacturer's data book.

Hexadecimal 1-of-16 Decoder IC

The 74154 is a 1-of-16 decoder. It accepts a 4-bit binary input (0000 to 1111), decodes it, and provides an active-LOW output to one of the 16 output pins. It also has a two-input active-LOW enable gate for disabling the outputs. If either enable input ($\overline{E_0}$ or $\overline{E_1}$) is made HIGH, the outputs are forced HIGH regardless of the A_0 to A_3 inputs. The operational information for the 74154 is given in Figure 8–15.

The logic diagram in Figure 8–15(c) shows the actual combinational logic circuit that is used to provide the decoding. The inverted-input AND gate is used in the circuit to disable all output NAND gates if either $\overline{E_0}$ or $\overline{E_1}$ is made HIGH. Follow the logic levels through the circuit for several combinations of inputs to A_0 through A_3 to prove its operation.

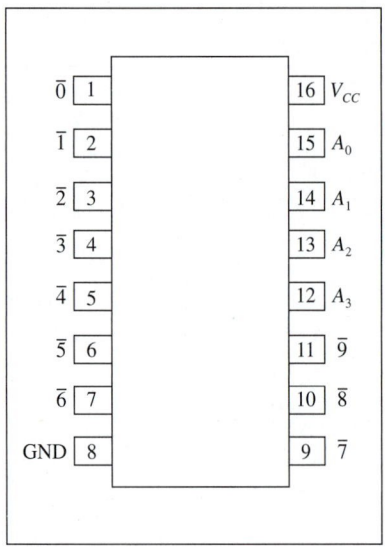

(a)

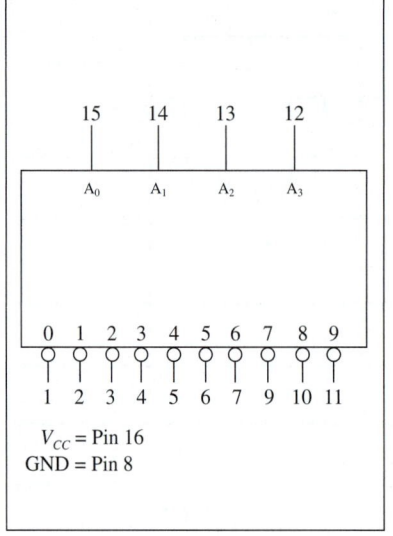

(b)

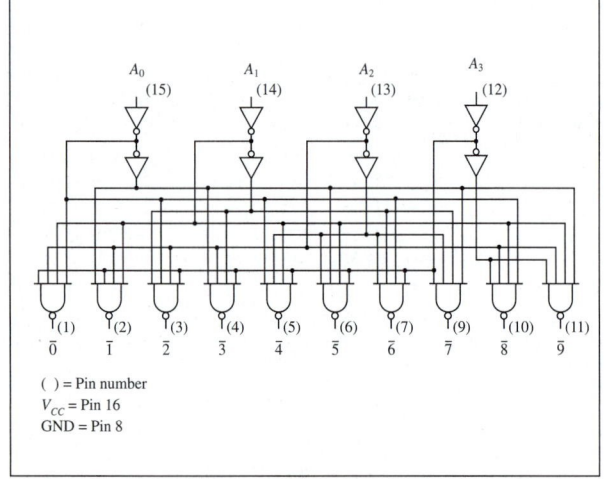

(c)

A_3	A_2	A_1	A_0	$\overline{0}$	$\overline{1}$	$\overline{2}$	$\overline{3}$	$\overline{4}$	$\overline{5}$	$\overline{6}$	$\overline{7}$	$\overline{8}$	$\overline{9}$
L	L	L	L	L	H	H	H	H	H	H	H	H	H
L	L	L	H	H	L	H	H	H	H	H	H	H	H
L	L	H	L	H	H	L	H	H	H	H	H	H	H
L	L	H	H	H	H	H	L	H	H	H	H	H	H
L	H	L	L	H	H	H	H	L	H	H	H	H	H
L	H	L	H	H	H	H	H	H	L	H	H	H	H
L	H	H	L	H	H	H	H	H	H	L	H	H	H
L	H	H	H	H	H	H	H	H	H	H	L	H	H
H	L	L	L	H	H	H	H	H	H	H	H	L	H
H	L	L	H	H	H	H	H	H	H	H	H	H	L
H	L	H	L	H	H	H	H	H	H	H	H	H	H
H	L	H	H	H	H	H	H	H	H	H	H	H	H
H	H	L	L	H	H	H	H	H	H	H	H	H	H
H	H	L	H	H	H	H	H	H	H	H	H	H	H
H	H	H	L	H	H	H	H	H	H	H	H	H	H
H	H	H	H	H	H	H	H	H	H	H	H	H	H

H = HIGH voltage level
L = LOW voltage level

(d)

Team Discussion

What happens when you enter an invalid BCD string [see Figure 8–10(d)]?

Figure 8–14 The 7442 BCD-to-DEC decoder: (a) pin configuration; (b) logic symbol; (c) logic diagram; (d) function table. (Courtesy of Philips Semiconductors.)

Review Questions

8–3. A BCD-to-decimal decoder has how many inputs and how many outputs?

8–4. An octal decoder with active-LOW outputs will output seven LOWs and one HIGH for each combination of inputs. True or false?

8–5. A hexadecimal decoder is sometimes called a 4-line-to-10-line decoder. True or false?

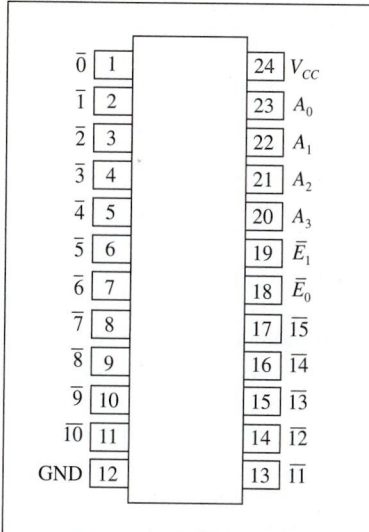

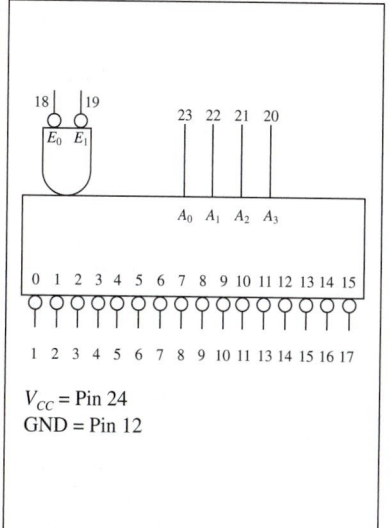

| (a) | | | (b) | | |

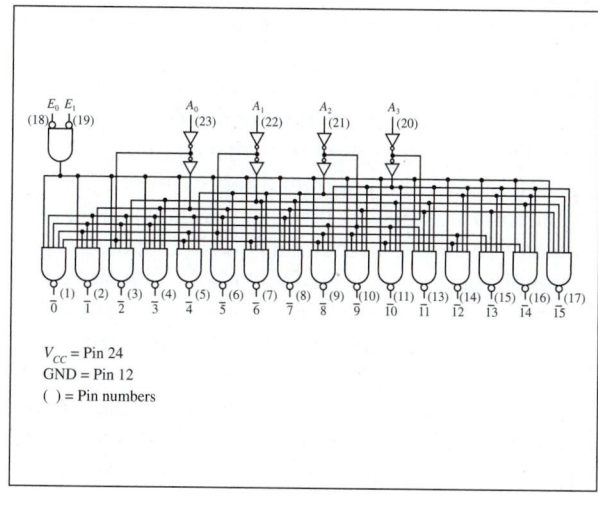

V_{CC} = Pin 24
GND = Pin 12
( ) = Pin numbers

(c)

Inputs						Outputs																
$\overline{E}_1$	$\overline{E}_2$	A_3	A_2	A_1	A_0	$\overline{0}$	$\overline{1}$	$\overline{2}$	$\overline{3}$	$\overline{4}$	$\overline{5}$	$\overline{6}$	$\overline{7}$	$\overline{8}$	$\overline{9}$	$\overline{10}$	$\overline{11}$	$\overline{12}$	$\overline{13}$	$\overline{14}$	$\overline{15}$
L	H	X	X	X	X	H	H	H	H	H	H	H	H	H	H	H	H	H	H	H	H
H	L	X	X	X	X	H	H	H	H	H	H	H	H	H	H	H	H	H	H	H	H
H	H	X	X	X	X	H	H	H	H	H	H	H	H	H	H	H	H	H	H	H	H
L	L	L	L	L	L	L	H	H	H	H	H	H	H	H	H	H	H	H	H	H	H
L	L	L	L	L	H	H	L	H	H	H	H	H	H	H	H	H	H	H	H	H	H
L	L	L	L	H	L	H	H	L	H	H	H	H	H	H	H	H	H	H	H	H	H
L	L	L	L	H	H	H	H	H	L	H	H	H	H	H	H	H	H	H	H	H	H
L	L	L	H	L	L	H	H	H	H	L	H	H	H	H	H	H	H	H	H	H	H
L	L	L	H	L	H	H	H	H	H	H	L	H	H	H	H	H	H	H	H	H	H
L	L	L	H	H	L	H	H	H	H	H	H	L	H	H	H	H	H	H	H	H	H
L	L	L	H	H	H	H	H	H	H	H	H	H	L	H	H	H	H	H	H	H	H
L	L	H	L	L	L	H	H	H	H	H	H	H	H	L	H	H	H	H	H	H	H
L	L	H	L	L	H	H	H	H	H	H	H	H	H	H	L	H	H	H	H	H	H
L	L	H	L	H	L	H	H	H	H	H	H	H	H	H	H	L	H	H	H	H	H
L	L	H	L	H	H	H	H	H	H	H	H	H	H	H	H	H	L	H	H	H	H
L	L	H	H	L	L	H	H	H	H	H	H	H	H	H	H	H	H	L	H	H	H
L	L	H	H	L	H	H	H	H	H	H	H	H	H	H	H	H	H	H	L	H	H
L	L	H	H	H	L	H	H	H	H	H	H	H	H	H	H	H	H	H	H	L	H
L	L	H	H	H	H	H	H	H	H	H	H	H	H	H	H	H	H	H	H	H	L

H = HIGH voltage level
L = LOW voltage level
X = Don't care   (d)

Figure 8–15   The 74154 1-of-16 decoder: (a) pin configuration; (b) logic symbol; (c) logic diagram; (d) function table. (Courtesy of Philips Semiconductors.)

 **Team Discussion**

Why are the *A* inputs listed as don't cares in the first three entries of the function table?

8–6. Only one of the three *enable* inputs must be satisfied to enable the 74138 decoder IC. True or false?

8–7. The 7442 BCD decoder has active-_____ (LOW, HIGH) inputs and active-_____ (LOW, HIGH) outputs.

8–4   Decoders Implemented in the VHDL Language

The decoders described in the previous section can all be described using the VHDL language. More importantly, they can be customized to meet specific needs instead of having to work around the fixed-function characteristics of the 7400-series of ICs. In this section we will use an octal decoder to illustrate the flexibility VHDL provides. Figure 8–16(a) shows the block diagram of a basic octal decoder, also known as a 3-line-to-8-line decoder. Figure 8–16(b) shows the addition of an enable signal to enable/disable the outputs. As discussed previously, the inputs, outputs, and enable could be active-HIGH or active-LOW. Figure 8–16 shows all active-HIGH signals.

The first VHDL method, shown in Figure 8–17(a) implements the function table for a decoder as a *series of Boolean equations*. The first equation states that *y0* will be HIGH if *a2* = 0 AND *a1* = 0 AND *a0* = 0. Notice that the Boolean equations form a binary counter that specifies all combinations on the *a*-inputs from 000 up to 111. To make the *y*-outputs active-LOW use parentheses to NOT the whole quantity to the right of the equal sign. The simulation waveforms in Figure 8–17(b) show the HIGH outputs at *y* that correspond to each combination of *a*-inputs.

Figure 8–18 shows an alternate method that produces the same results. In this method, the inputs and outputs are each grouped as vectors and a *selected signal assignment* statement is used to set the appropriate *y*-output HIGH based on the 3-bit code at the *a*-inputs.

The next two VHDL listings include an *enable* feature like the one shown in Figure 8–16(b). In Figure 8–19 you can see that the enable input (*en*) is included as an additional entity port declaration. The program is written so that *en* must be HIGH for any output to be selected to go HIGH. For the program to check the *en* bit, it is first combined with the 3-bit *a* input forming a new 4-bit string. This is called **concatenation**. This takes two new statements. First, an internal SIGNAL named *inputs* is declared as a 4-bit vector. The signal *inputs* is then loaded with the concatenation of *en* with the three *a*-input bits forming the 4-bit vector (& is the symbol used to concatenate). This 4-bit vector is now used in the selected signal assignment statements to select the appropriate *y* output to go HIGH. Notice that whenever *en* (the leftmost bit) is LOW, control drops to the "others" clause, setting all outputs LOW.

The octal decoder shown in Figure 8–20(a) uses an IF-THEN-ELSE clause and a CASE statement to select the appropriate *y* output to go HIGH. In some ways this makes the most sense logically because it asks "IF *en* equals '1' THEN perform the CASE assignments, ELSE set output vector *y* to all zeros". IF statements are sequential in nature and therefore must be placed within a PROCESS. Also, the CASE method of assignment is chosen because the selected signal assignment method is not allowed with IF statements. The PROCESS sensitivity list consists of *a* and *en*. Whenever either of these changes, the PROCESS is executed.

The waveform simulation in Figure 8–20(b) shows how the *a*-input vector dictates which *y* output goes HIGH as long as *en* is HIGH. When *en* is LOW all outputs are made LOW.

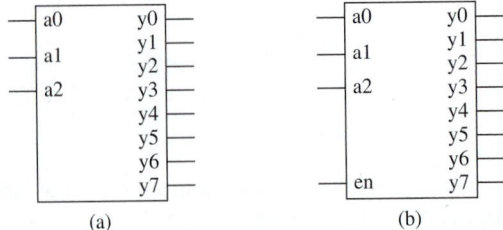

        (a)   (b)

Figure 8–16 Decoder block diagrams: (a) octal decoder; (b) octal decoder with an active-HIGH enable control input.

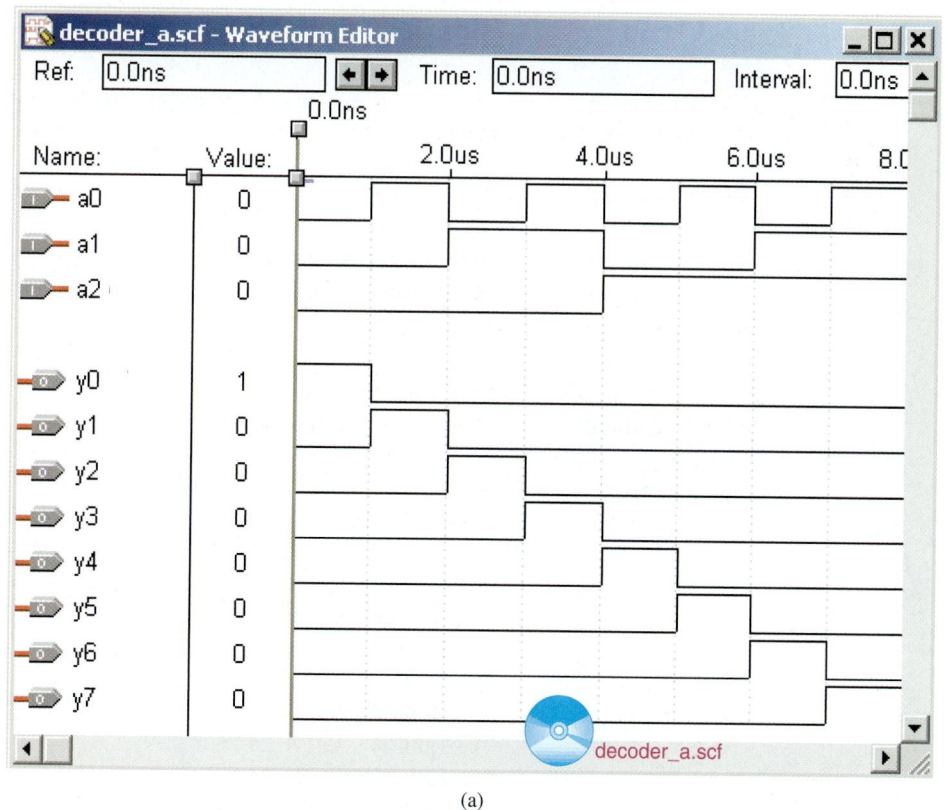

```
decoder_a.vhd - Text Editor                                    _ □ X
   LIBRARY ieee;    -- Octal decoder using Boolean equations--
   USE ieee.std_logic_1164.ALL;

   ENTITY decoder_a IS
       PORT(a0,a1,a2                  : IN  std_logic;
            y0,y1,y2,y3,y4,y5,y6,y7: OUT std_logic);
   END decoder_a ;

   ARCHITECTURE arc OF decoder_a IS
   BEGIN
       y0  <=  (NOT a2) AND (NOT a1) AND (NOT a0);
       y1  <=  (NOT a2) AND (NOT a1) AND (    a0);
       y2  <=  (NOT a2) AND (    a1) AND (NOT a0);
       y3  <=  (NOT a2) AND (    a1) AND (    a0);
       y4  <=  (    a2) AND (NOT a1) AND (NOT a0);
       y5  <=  (    a2) AND (NOT a1) AND (    a0);
       y6  <=  (    a2) AND (    a1) AND (NOT a0);
       y7  <=  (    a2) AND (    a1) AND (    a0);
   END arc;
Line   19   Col   1      INS
```

$yo = \bar{a}_2 \bar{a}_1 \bar{a}_0$

decoder_a.vhd

(a)

(a)

Figure 8–17 Octal decoder: (a) VHDL program using Boolean equations; (b) simulation of the decoded waveforms.

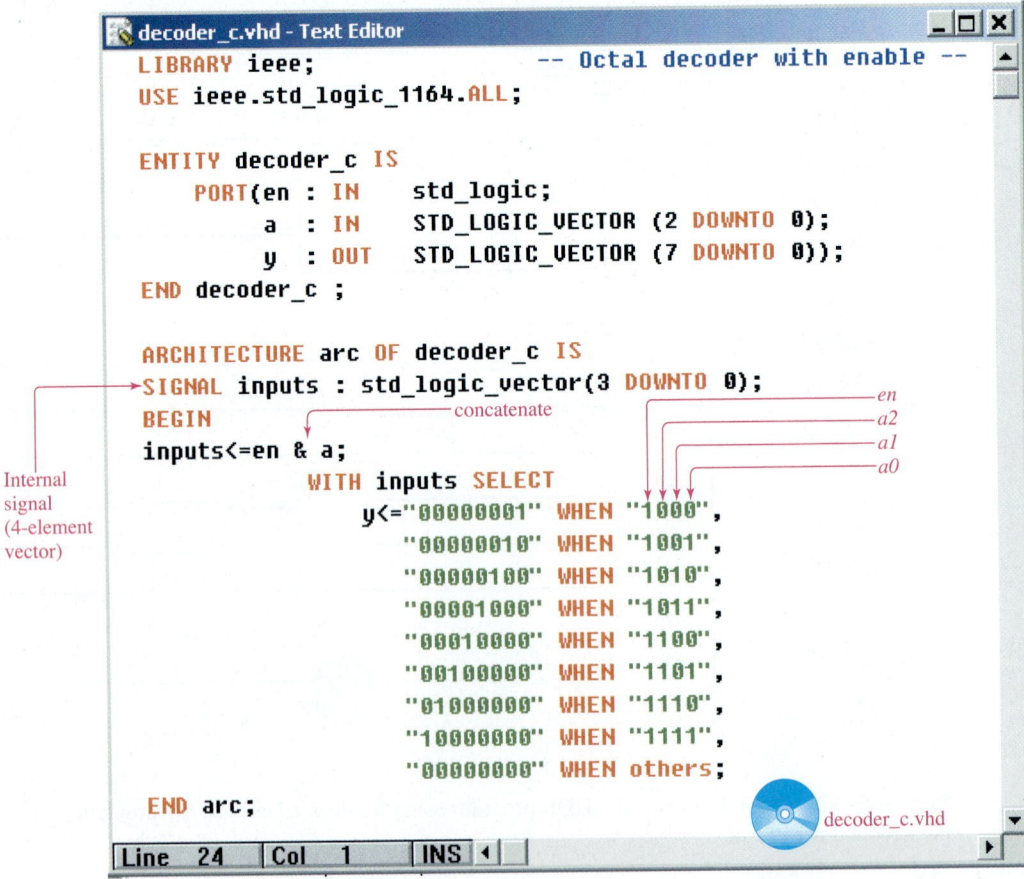

```
decoder_b.vhd - Text Editor                                      _ □ X
    LIBRARY ieee;                     -- Octal decoder using vectors
    USE ieee.std_logic_1164.ALL;      --  and
                                      -- selected signal assignment

    ENTITY decoder_b IS
        PORT(a  : IN     STD_LOGIC_VECTOR (2 downto 0);
             y  : OUT    STD_LOGIC_VECTOR (7 downto 0));
    END decoder_b ;


    ARCHITECTURE arc OF decoder_b IS                    y7
    BEGIN                                               y0
        WITH a SELECT                                          a2  a0
                    y<="00000001" WHEN "000",
                       "00000010" WHEN "001",
                       "00000100" WHEN "010",
                       "00001000" WHEN "011",
                       "00010000" WHEN "100",
                       "00100000" WHEN "101",
                       "01000000" WHEN "110",
                       "10000000" WHEN "111",
                       "00000000" WHEN others;
                                                          decoder_b.vhd
    END arc;
Line   21    Col   1    INS
```

Figure 8–18 A VHDL octal decoder implemented with vectors and the selected signal assignment.

```
decoder_c.vhd - Text Editor                                      _ □ X
    LIBRARY ieee;                        -- Octal decoder with enable --
    USE ieee.std_logic_1164.ALL;


    ENTITY decoder_c IS
        PORT(en : IN     std_logic;
             a  : IN     STD_LOGIC_VECTOR (2 DOWNTO 0);
             y  : OUT    STD_LOGIC_VECTOR (7 DOWNTO 0));
    END decoder_c ;


    ARCHITECTURE arc OF decoder_c IS
    SIGNAL inputs : std_logic_vector(3 DOWNTO 0);           en
    BEGIN                       concatenate                 a2
    inputs<=en & a;                                         a1
                                                            a0
                WITH inputs SELECT
                    y<="00000001" WHEN "1000",
                       "00000010" WHEN "1001",
                       "00000100" WHEN "1010",
                       "00001000" WHEN "1011",
                       "00010000" WHEN "1100",
                       "00100000" WHEN "1101",
                       "01000000" WHEN "1110",
                       "10000000" WHEN "1111",
                       "00000000" WHEN others;
    END arc;                                              decoder_c.vhd
Line   24    Col   1    INS
```

Internal signal (4-element vector)

Figure 8–19 VHDL listing of an octal decoder with an enable input.

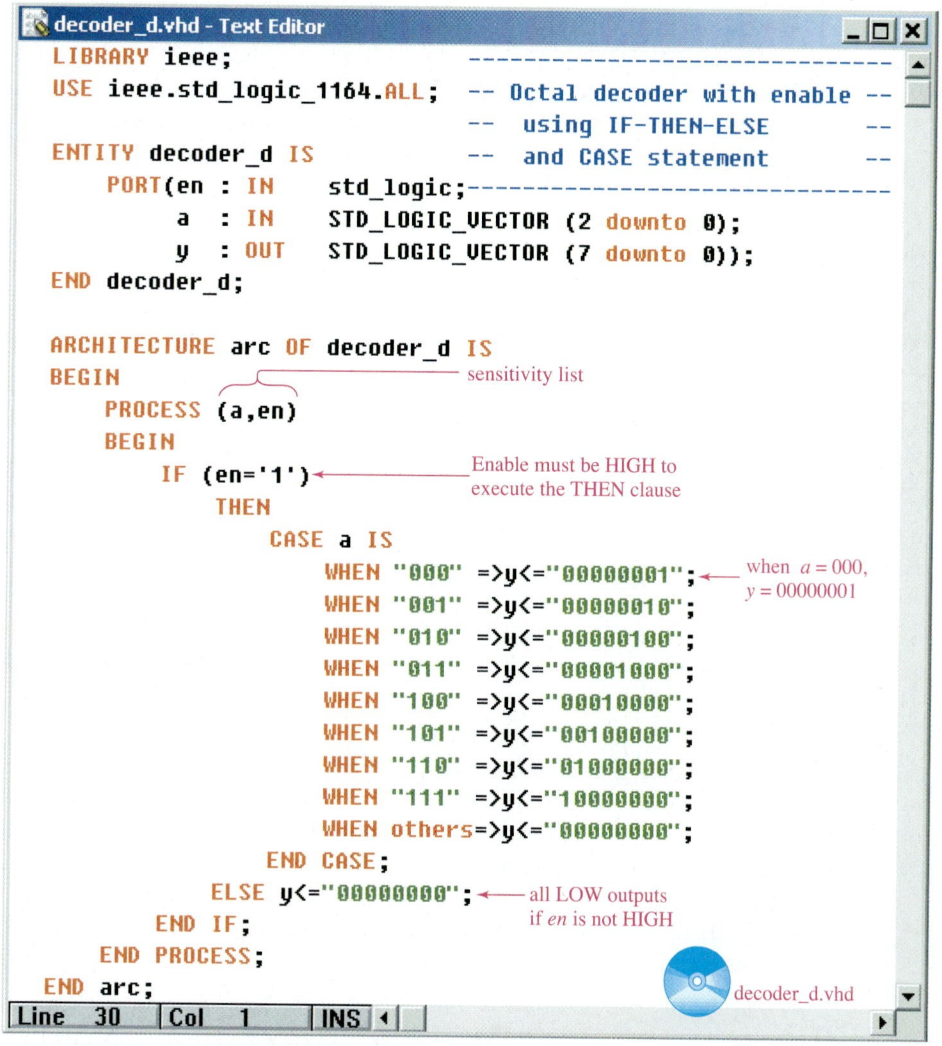

```
decoder_d.vhd - Text Editor                                      _ □ ×

    LIBRARY ieee;                     -------------------------------
    USE ieee.std_logic_1164.ALL;    -- Octal decoder with enable --
                                     --   using IF-THEN-ELSE     --
    ENTITY decoder_d IS              --   and CASE statement     --
        PORT(en : IN     std_logic;-------------------------------
             a  : IN     STD_LOGIC_VECTOR (2 downto 0);
             y  : OUT    STD_LOGIC_VECTOR (7 downto 0));
    END decoder_d;


    ARCHITECTURE arc OF decoder_d IS
    BEGIN                                    ⟵ sensitivity list
        PROCESS (a,en)
        BEGIN
            IF (en='1')              ⟵ Enable must be HIGH to
                THEN                    execute the THEN clause
                    CASE a IS
                        WHEN "000" =>y<="00000001";  ⟵ when a = 000,
                        WHEN "001" =>y<="00000010";      y = 00000001
                        WHEN "010" =>y<="00000100";
                        WHEN "011" =>y<="00001000";
                        WHEN "100" =>y<="00010000";
                        WHEN "101" =>y<="00100000";
                        WHEN "110" =>y<="01000000";
                        WHEN "111" =>y<="10000000";
                        WHEN others=>y<="00000000";
                    END CASE;
                ELSE y<="00000000";   ⟵ all LOW outputs
            END IF;                       if en is not HIGH
        END PROCESS;
    END arc;                                          decoder_d.vhd

Line   30   Col   1    INS ◄                                      ►
```

(a)

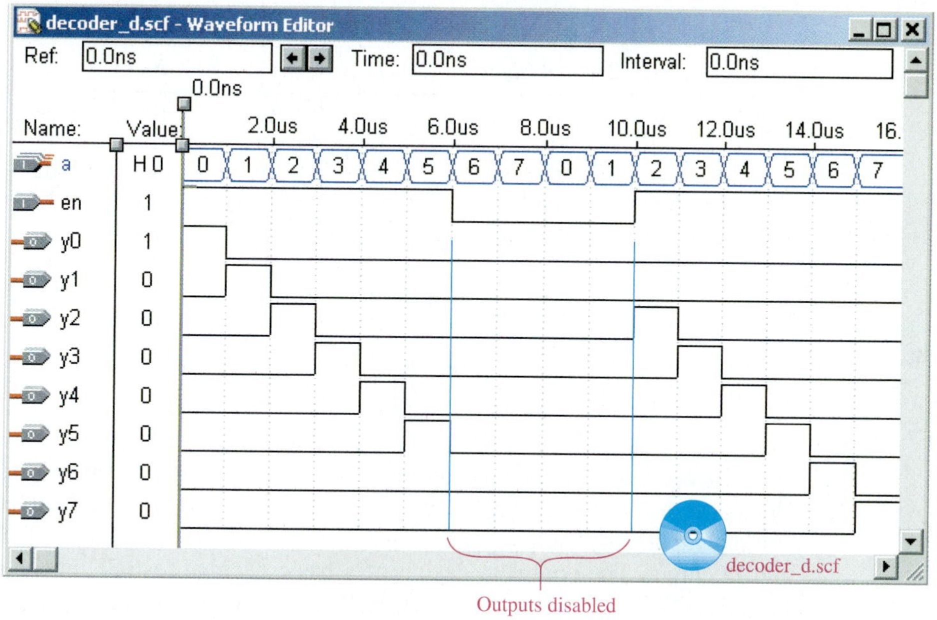

(b)

Figure 8–20   Octal decoder: (a) VHDL program using the IF-THEN-ELSE and CASE statements; (b) simulation waveforms of the octal decoder with enable control.

309

8–5 Encoding

Encoding is the opposite process from decoding. Encoding is used to generate a coded output (such as BCD or binary) from a singular active numeric input line. For example, Figure 8–21 shows a typical block diagram for a decimal-to-BCD **encoder** and an octal-to-binary encoder.

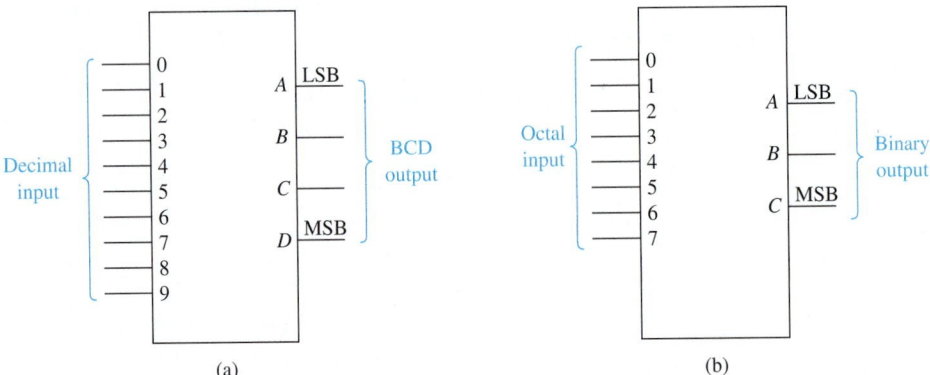

Figure 8–21 Typical block diagrams for encoders: (a) decimal-to-BCD encoder; (b) octal-to-binary encoder.

The design of encoders using combinational logic can be done by reviewing the truth table (see Table 8–3) for the operation to determine the relationship each output has with the inputs. For example, by studying Table 8–3 for a decimal-to-BCD encoder, we can see that the A output (2^0) is HIGH for all odd decimal input numbers (1, 3, 5, 7, and 9). The B output (2^1) is HIGH for decimal inputs 2, 3, 6, and 7. The C output (2^2) is HIGH for decimal inputs 4, 5, 6, and 7, and the D output (2^3) is HIGH for decimal inputs 8 and 9.

TABLE 8–3	Decimal-to-BCD Encoder Truth Table			
	BCD Output			
Decimal Input	**D**	**C**	**B**	**A**
0	0	0	0	0
1	0	0	0	1
2	0	0	1	0
3	0	0	1	1
4	0	1	0	0
5	0	1	0	1
6	0	1	1	0
7	0	1	1	1
8	1	0	0	0
9	1	0	0	1

Now, from what we have just observed, it seems that we can design a decimal-to-BCD encoder with just four OR gates; the A output OR gate goes HIGH for any odd decimal input, the B output goes HIGH for 2 *or* 3 *or* 6 *or* 7, and so on, for the C output and D output. The complete design of a basic decimal-to-BCD encoder is given in Figure 8–22. The design for an octal-to-binary encoder uses the same procedure, but, of course, these encoders are available in IC form: the 74147 decimal to BCD and the 74148 octal to binary.

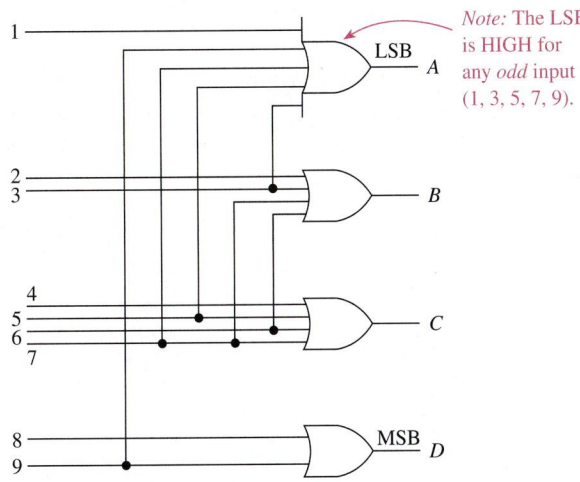

Note: The LSB is HIGH for any *odd* input (1, 3, 5, 7, 9).

LSB — A

B

C

MSB — D

Figure 8–22 Basic decimal-to-BCD encoder.

The 74147 Decimal-to-BCD Encoder

The 74147 operates similarly to our basic design from Figure 8–22 except for two major differences:

1. The inputs *and* outputs are all active-LOW [see the bubbles on the logic symbol, Figure 8–23(a)].

2. The 74147 is a *priority* encoder, which means that if more than one decimal number is input, the highest numeric input has *priority* and will be encoded to the output [see the function table, Figure 8–23(b)]. For example, looking at the second line in the function table, if $\overline{I_9}$ is LOW (decimal 9), all other inputs are don't care (could be HIGH *or* LOW), and the BCD output is 0110 (active-LOW BCD-9).

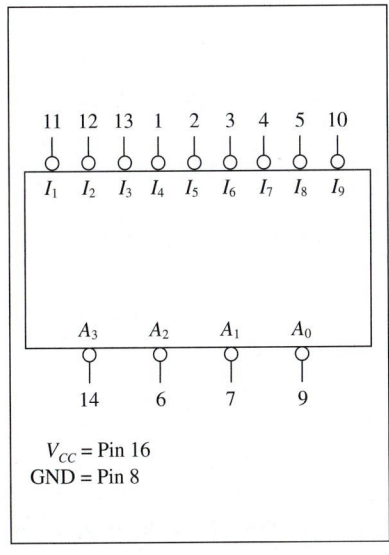

11 12 13 1 2 3 4 5 10

I_1 I_2 I_3 I_4 I_5 I_6 I_7 I_8 I_9

A_3 A_2 A_1 A_0

14   6   7   9

V_{CC} = Pin 16
GND = Pin 8

(a)

Input									Output			
$\overline{I_1}$	$\overline{I_2}$	$\overline{I_3}$	$\overline{I_4}$	$\overline{I_5}$	$\overline{I_6}$	$\overline{I_7}$	$\overline{I_8}$	$\overline{I_9}$	$\overline{A_3}$	$\overline{A_2}$	$\overline{A_1}$	$\overline{A_0}$
H	H	H	H	H	H	H	H	H	H	H	H	H
X	X	X	X	X	X	X	X	L	L	H	H	L
X	X	X	X	X	X	X	L	H	L	H	H	H
X	X	X	X	X	X	L	H	H	H	L	L	L
X	X	X	X	X	L	H	H	H	H	L	L	H
X	X	X	X	L	H	H	H	H	H	L	H	L
X	X	X	L	H	H	H	H	H	H	L	H	H
X	X	L	H	H	H	H	H	H	H	H	L	L
X	L	H	H	H	H	H	H	H	H	H	L	H
L	H	H	H	H	H	H	H	H	H	H	H	L

H = HIGH voltage level
L = LOW voltage level
X = Don't care

(b)

Figure 8–23 The 74147 decimal-to-BCD (10-line-to-4-line) encoder: (a) logic symbol; (b) function table.

EXAMPLE 8-3

For simplicity, the 74147 IC shown in Figure 8–24 is set up for encoding just three of its inputs (7, 8, and 9). Using the function table from Figure 8–23(b), sketch the outputs at $\overline{A}_0$, $\overline{A}_1$, $\overline{A}_2$, and $\overline{A}_3$ as the $\overline{I}_7$, $\overline{I}_8$, and $\overline{I}_9$ inputs are switching as shown in Figure 8–25.

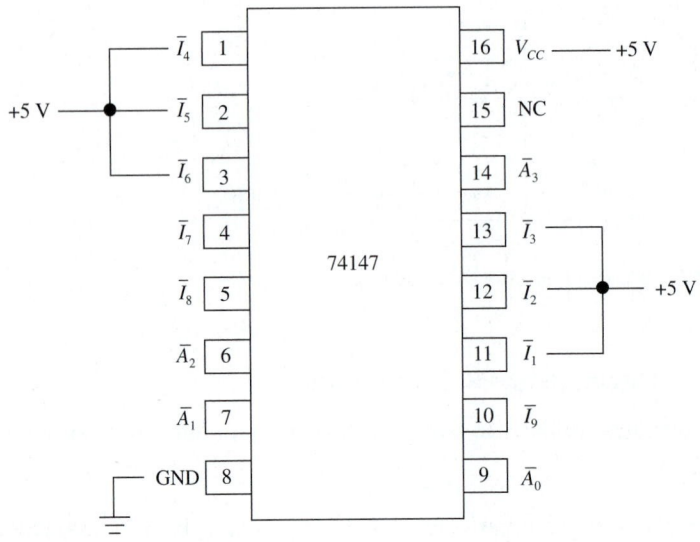

Figure 8–24

Solution:

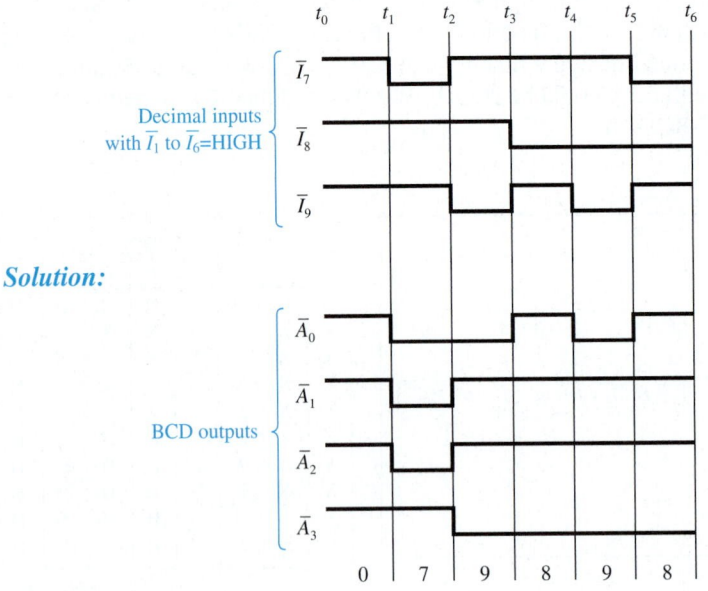

Figure 8–25

Explanation: The $\overline{I}_1$ to $\overline{I}_6$ inputs are all tied HIGH and have no effect on the output.

t_0–t_1: Decimal inputs are all HIGH; BCD outputs represent a 0.

t_1–t_2: $\overline{I}_7$ is LOW; BCD outputs represent a 7. (Active-LOW 7 = 1000.)

t_2–t_3: $\overline{I_9}$ is LOW; BCD outputs represent a 9.

t_3–t_4: $\overline{I_8}$ is LOW; BCD outputs represent an 8.

t_4–t_5: $\overline{I_8}$ *and* $\overline{I_9}$ are LOW; $\overline{I_9}$ has priority; BCD outputs represent a 9.

t_5–t_6: $\overline{I_7}$ *and* $\overline{I_8}$ are LOW; $\overline{I_8}$ has priority; BCD outputs represent an 8.

The 74148 Octal-to-Binary Encoder

The 74148 encoder accepts data from eight active-LOW inputs and provides a binary representation on three active-LOW outputs. It is also a **priority** encoder, so when two or more inputs are active simultaneously, the input with the highest priority is represented on the output, with input line $\overline{I_7}$ having the highest priority. The logic symbol and function table in Figure 8–26 give us some other information as well.

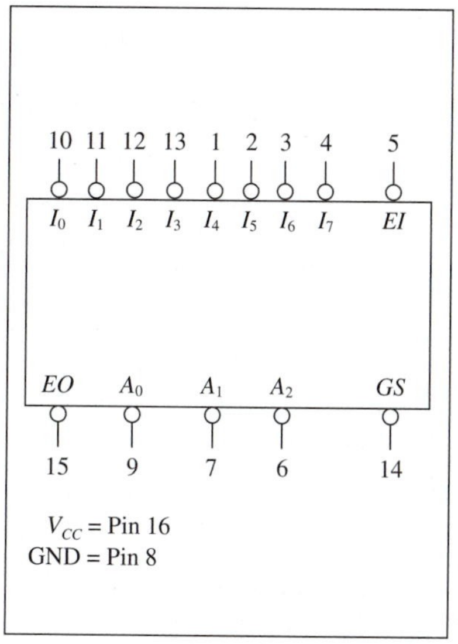

	Inputs									Outputs				
$\overline{EI}$	$\overline{I_0}$	$\overline{I_1}$	$\overline{I_2}$	$\overline{I_3}$	$\overline{I_4}$	$\overline{I_5}$	$\overline{I_6}$	$\overline{I_7}$	$\overline{GS}$	$\overline{A_0}$	$\overline{A_1}$	$\overline{A_2}$	$\overline{EO}$
H	X	X	X	X	X	X	X	X	H	H	H	H	H
L	H	H	H	H	H	H	H	H	H	H	H	H	L
L	X	X	X	X	X	X	X	L	L	L	L	L	H
L	X	X	X	X	X	X	L	H	L	H	L	L	H
L	X	X	X	X	X	L	H	H	L	L	H	L	H
L	X	X	X	X	L	H	H	H	L	H	H	L	H
L	X	X	X	L	H	H	H	H	L	L	L	H	H
L	X	X	L	H	H	H	H	H	L	H	L	H	H
L	X	L	H	H	H	H	H	H	L	L	H	H	H
L	L	H	H	H	H	H	H	H	L	H	H	H	H

H = HIGH voltage level
L = LOW voltage level
X = Don't care

(b)

(a)

Figure 8–26   The 74148 octal-to-binary (8-line-to-3-line) encoder: (a) logic symbol; (b) functional table. (Courtesy of Philips Semiconductors.)

The 74148 can be expanded to any number of inputs by using several 74148s and their $\overline{EI}$, $\overline{EO}$, and $\overline{GS}$ pins. These special pins are defined as follows:

$\overline{EI}$ Active-LOW enable input: a HIGH on this input forces all outputs ($\overline{A_0}$ to $\overline{A_2}$, $\overline{EO}$, $\overline{GS}$) to their inactive (HIGH) state.

$\overline{EO}$ Active-LOW enable output: this output pin goes LOW when all inputs ($\overline{I_0}$ to $\overline{I_7}$) are inactive (HIGH) and $\overline{EI}$ is LOW.

$\overline{GS}$ Active-LOW group signal output: this output pin goes LOW whenever any of the inputs ($\overline{I_0}$ to $\overline{I_7}$) are active (LOW) and $\overline{EI}$ is LOW.

The following example illustrates the use of these pins.

EXAMPLE 8-4

Sketch the output waveforms for the 74148 connected as shown in Figure 8–27. The input waveforms to $\overline{I_6}$, $\overline{I_7}$, and $\overline{EI}$ are given in Figure 8–28. (Inputs $\overline{I_0}$ to $\overline{I_5}$ are tied HIGH for simplicity.)

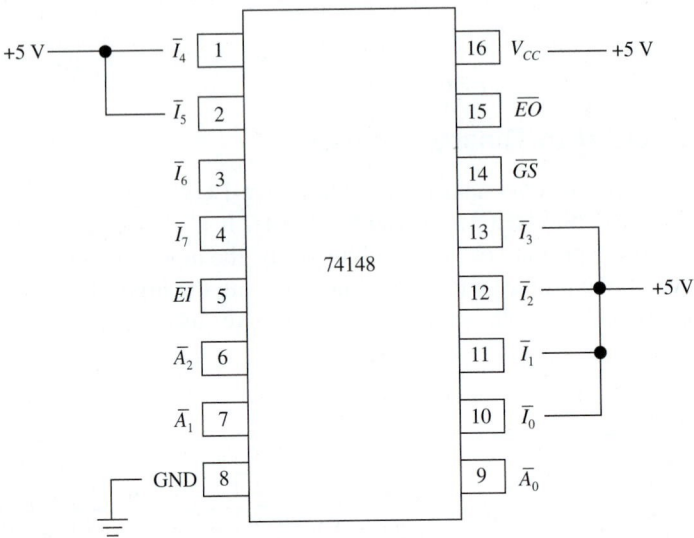

Figure 8–27 The 74148 connections for Example 8–4.

Solution:

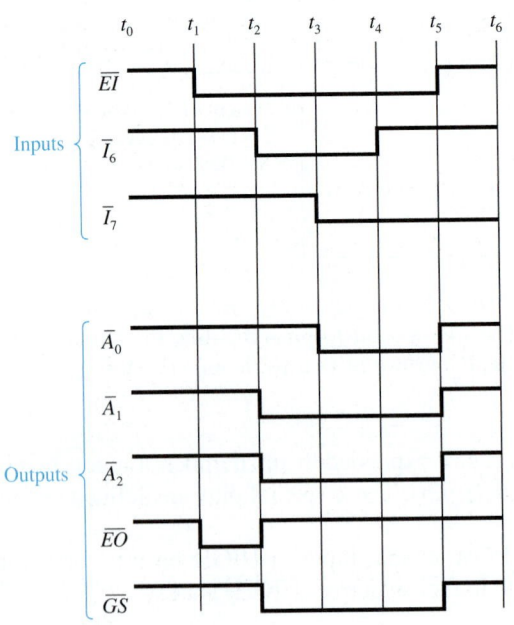

Figure 8–28

t_0–t_1: All outputs are forced HIGH by the HIGH on $\overline{EI}$.
t_1–t_2: $\overline{EI}$ is LOW to enable the inputs, but $\overline{I_0}$ to $\overline{I_7}$ are all HIGH (inactive), so $\overline{EO}$ goes LOW.

t_2–t_3: $\overline{GS}$ goes LOW because one of the inputs ($\overline{I_6}$) is active; the active-LOW binary output is equal to 6.

t_3–t_4: $\overline{I_7}$ and $\overline{I_6}$ are LOW; $\overline{I_7}$ has priority; output = 7.

t_4–t_5: $\overline{I_7}$ is LOW; output = 7.

t_5–t_6: All outputs are forced HIGH by the HIGH on $\overline{EI}$.

EXAMPLE 8–5

VHDL Octal Priority Encoder

Use VHDL to design an active-HIGH input, active-HIGH output, octal priority encoder. Create a waveform simulation file to test that both individual and multiple active inputs are encoded correctly.

Solution: The program listing is given in Figure 8–29. The **Conditional Signal Assignment** statement was used instead of a Selected Signal Assignment because it operates on a *priority* basis. For example, the first line checks for i(7) = '1'. [*i(7)* is the leftmost bit in the *i* input vector.] If it is TRUE then "111" is moved to output vector *a* and control passes to the end because all ELSE clauses will be skipped. If i(7) = '1' is FALSE, then the next line is executed for i(6) = '1', and so on. This imposes a priority, guaranteeing that the highest input will be encoded if multiple inputs are HIGH.

```
ex8_5.vhd - Text Editor                                      _ □ ✕
  LIBRARY ieee;                    -- Octal priority encoder
  USE ieee.std_logic_1164.ALL;     -- using
                                   -- conditional signal assignment
  ENTITY ex8_5 IS
      PORT(i  : IN    std_logic_vector (7 DOWNTO 0);
           a  : OUT   std_logic_vector (2 DOWNTO 0));
  END ex8_5 ;

  ARCHITECTURE arc OF ex8_5 IS
  BEGIN
                a<="111" WHEN i(7)='1' ELSE
                   "110" WHEN i(6)='1' ELSE
                   "101" WHEN i(5)='1' ELSE      The first clause
                   "100" WHEN i(4)='1' ELSE      that is TRUE has
                   "011" WHEN i(3)='1' ELSE      priority, all
                   "010" WHEN i(2)='1' ELSE      others are
                   "001" WHEN i(1)='1' ELSE      skipped.
                   "000" WHEN i(0)='1' ELSE
                   "000";
  END arc;

Line  20   Col  1    INS ◄                                         ►
```

Figure 8–29 VHDL program for the octal encoder of Example 8–5.

Proof that the encoder is working properly is shown in Figure 8–30. Notice that at the 2.0μs mark, *i5* AND *i4* are *both* HIGH. Since 5 is larger than 4, it gets encoded at the *a* output as shown. Carefully review the remainder of the waveform to convince yourself that all results are valid.

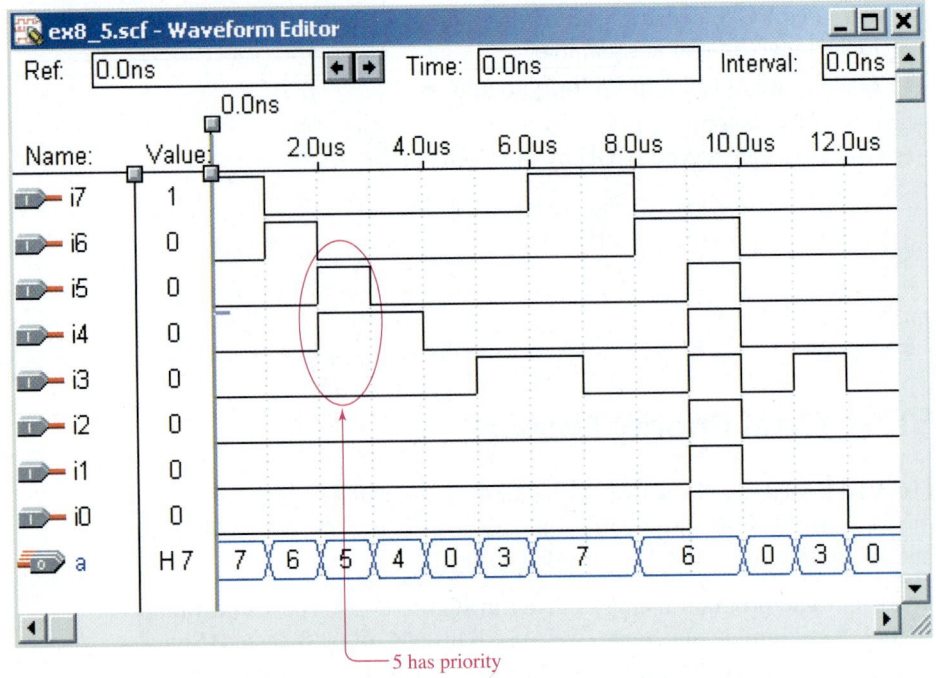

Figure 8–30 Simulated waveforms for the octal encoder of Example 8–5.

8–8. How does an encoder differ from a decoder?

8–9. If more than one input to a *priority* encoder is active, which input will be encoded?

8–10. (a) If all inputs to a 74147 encoder are HIGH, what will the A_3–A_0 outputs be?

    (b) Repeat part (a) for all inputs being LOW.

8–11. What are the five outputs of the 74148? Are they active-LOW or active-HIGH?

8–6   Code Converters

Often it is important to convert a coded number into another form that is more usable by a computer or digital system. The prime example of this is with binary-coded decimal (BCD). We have seen that BCD is very important for visual display communication between a computer and human beings. But BCD is very difficult to deal with arithmetically. Algorithms, or procedures, have been developed for the conversion of BCD to binary by computer programs (**software**) so that the computer will be able to perform all arithmetic operations in binary.

Another way to convert BCD to binary, the **hardware** approach, is with MSI ICs. Additional circuitry is involved, but it is much faster to convert using hardware rather than software. We look at both methods for the conversion of BCD to binary.

BCD-to-Binary Conversion

If you were going to convert BCD to binary using software program statements, you would first have to develop a procedure, or algorithm, for the conversion. Take, for example, the number 26_{10} in BCD.

$$\overbrace{2}^{} \qquad \overbrace{6}^{}$$
$$0010 \qquad 0110$$

If you simply apply regular binary weighting to each bit, you would come up with 38 ($2^1 + 2^2 + 2^5 = 38$). You must realize that the second group of BCD positions has a new progression of powers of 2 but with a **weighting factor** of 10, as shown in Figure 8–31. Now, if we go back and apply the proper weighting factors to 26_{10} in BCD, we should get the correct binary equivalent.

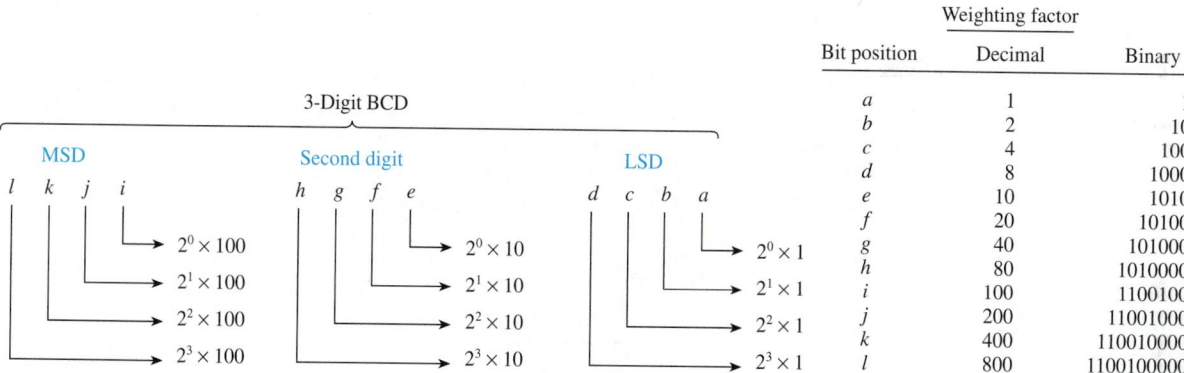

Figure 8–31   Weighting factors for BCD bit positions.

Bit position	Weighting factor	
	Decimal	Binary
a	1	1
b	2	10
c	4	100
d	8	1000
e	10	1010
f	20	10100
g	40	101000
h	80	1010000
i	100	1100100
j	200	11001000
k	400	110010000
l	800	1100100000

EXAMPLE 8–6

Using the weighting factors given in Figure 8–31, convert the BCD equivalent of 26_{10} to binary.

Solution:

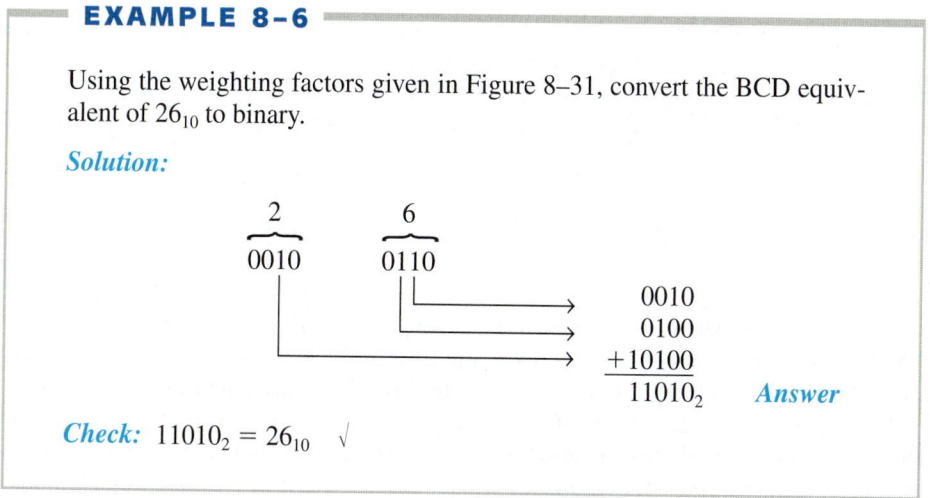

Check: $11010_2 = 26_{10}$ √

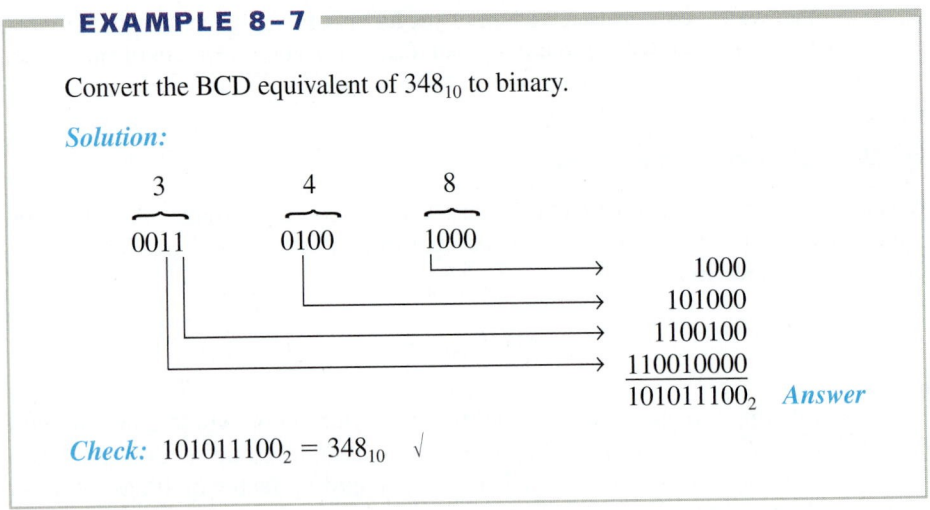

EXAMPLE 8-7

Convert the BCD equivalent of 348_{10} to binary.

Solution:

$$
\begin{array}{ccc}
3 & 4 & 8 \\
\overbrace{0011} & \overbrace{0100} & \overbrace{1000}
\end{array}
$$

$$
\begin{array}{r}
1000 \\
101000 \\
1100100 \\
110010000 \\
\hline
101011100_2 \quad \textit{Answer}
\end{array}
$$

Check: $101011100_2 = 348_{10}$ ✓

Conversion of BCD to Binary Using the 74184

Examples 8–6 and 8–7 illustrate one procedure of conversion that can be used as an algorithm for a computer program (software). The hardware approach using the 74184 IC is another way to accomplish BCD-to-binary conversion.

The logic symbol in Figure 8–32 shows eight active-HIGH binary outputs. Y_1 to Y_5 are outputs for regular BCD-to-binary conversion. Y_6 to Y_8 are used for a special BCD code called nine's complement and ten's complement.

The active-HIGH BCD bits are input on A through E. The $\overline{G}$ is an active-LOW enable input. When $\overline{G}$ is HIGH, all outputs are forced HIGH.

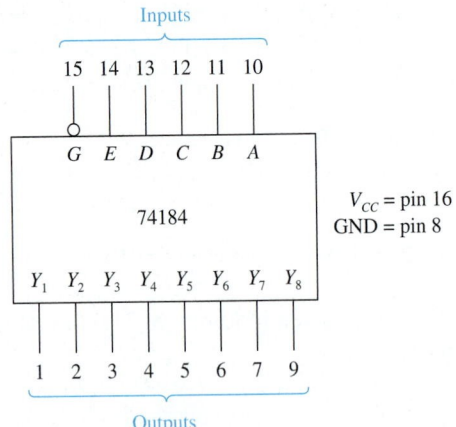

Figure 8-32   Logic symbol for the 74184 BCD-to-binary converter.

Figure 8–33 shows the connections to form a 6-bit BCD converter. Because the LSB of the BCD input is always equal to the LSB of the binary output, the connection is made straight from input to output. The other BCD bits are connected to the A to E inputs. They have the weighting of $A = 2, B = 4, C = 8, D = 10$, and $E = 20$. Because only 2 bits are available for the MSD BCD input, the largest BCD digit in that position will be 3 (11). More useful setups, providing for the input of two or three complete BCD digits, are shown in Figure 8–34(a) and (b).

A companion chip, the 74185, is used to work the opposite way, binary to BCD. Figure 8–34(c) and (d) shows the 74185 used to perform binary-to-BCD conversions.

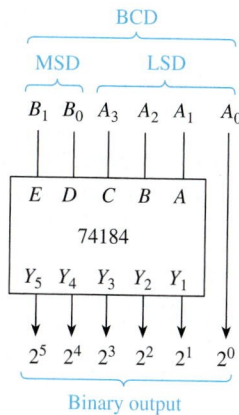

Figure 8–33 Six-bit BCD-to-binary converter.

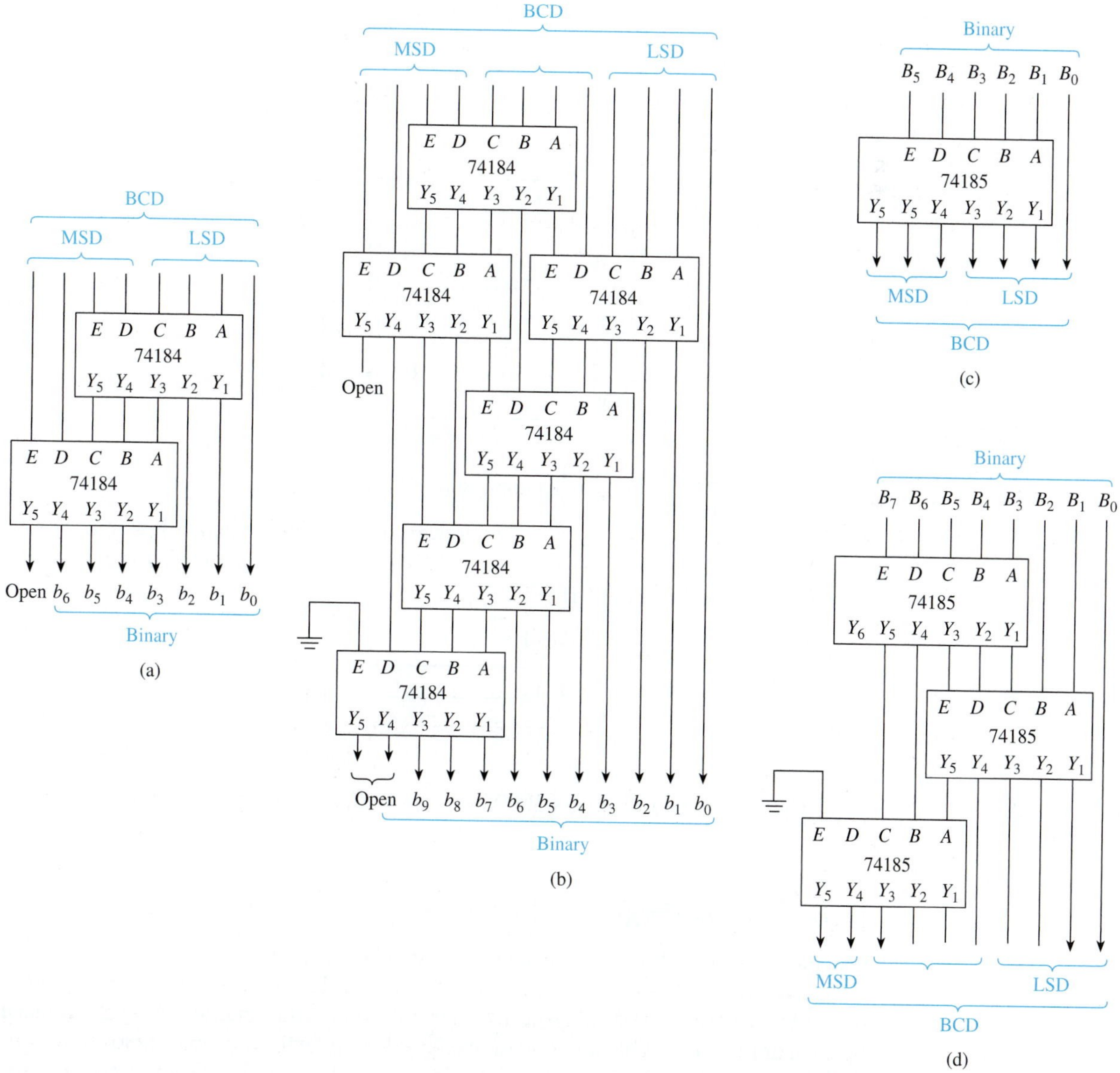

Figure 8–34 BCD-to-binary conversions using the 74184 and binary-to-BCD conversions using the 74185: (a) BCD-to-binary converter for two BCD decades; (b) BCD-to-binary converter for three BCD decades; (c) 6-bit binary-to-BCD converter; (d) 8-bit binary-to-BCD converter. (Courtesy of Texas Instruments, Inc.)

EXAMPLE 8–8

Show how the BCD code of the number 65 is converted by the circuit of Figure 8–34(a) by placing 1's and 0's at the inputs and outputs.

Solution: The BCD-to-binary conversion is shown in Figure 8–35. The upper 74184 is used to convert the least significant 6 bits, which are 100101. Using the proper binary weighting, 100101 becomes 011001.

```
100101
   └────────────→    1
   └────────────→   100
   └────────────→ 10100
                  ───────
                  011001
```

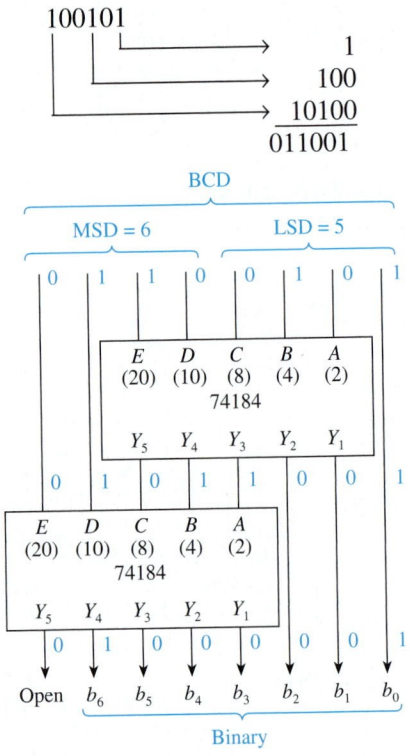

Figure 8–35 Solution to Example 8–8.

The lower 74184 is used to convert the most significant 2 bits plus 4 bits from the upper 74184, which are 010110. Using proper binary weighting, 010110 becomes 010000.

```
010110
   └────────────→    10
   └────────────→   100
   └────────────→  1010
                  ───────
                  010000
```

The final binary result is 01000001, which checks out to be equal to 65_{10}.

BCD-to-Seven-Segment Converters

Calculators and other devices with numeric displays use another form of code conversion involving BCD-to-seven-segment conversion. The term *seven segment* comes from the fact that these displays utilize seven different illuminating segments to make up each of the 10 possible numeric digits. A **code converter** must be employed to convert the 4-bit BCD into a 7-bit code to drive each digit. A commonly used BCD-to-seven-segment converter is the 7447 IC. An in-depth discussion of displays and display converter/drivers will be given in Section 12–6 after you have a better understanding of the circuitry used to create the numeric data to be displayed.

Gray Code

The **Gray code** is another useful code used in digital systems. It is used primarily for indicating the angular position of a shaft on rotating machinery, such as automated lathes and drill presses. This code is like binary in that it can have as many bits as necessary, and the more bits, the more possible combinations of output codes (number of combinations = 2^N). A 4-bit Gray code, for example, has $2^4 = 16$ different representations, giving a resolution of 1 out of 16 possible angular positions at 22.5° each ($360°/16 = 22.5°$).

The difference between the Gray code and the regular binary code is illustrated in Table 8–4. Notice in the table that the Gray code varies by only 1 bit from one entry to the next and from the last entry (15) back to the beginning (0). Now, if each Gray code represents a different position on a rotating wheel, as the wheel turns, the code read from one position to the next will vary by only 1 bit (see Figure 8–36).

If the same wheel were labeled in *binary,* as the wheel turned from 7 to 8, the code would change from 0111 to 1000. If the digital machine happened to be reading

TABLE 8–4	Four-Bit Gray Code	
Decimal	**Binary**	**Gray**
0	0000	0000
1	0001	0001
2	0010	0011
3	0011	0010
4	0100	0110
5	0101	0111
6	0110	0101
7	0111	0100
8	1000	1100
9	1001	1101
10	1010	1111
11	1011	1110
12	1100	1010
13	1101	1011
14	1110	1001
15	1111	1000

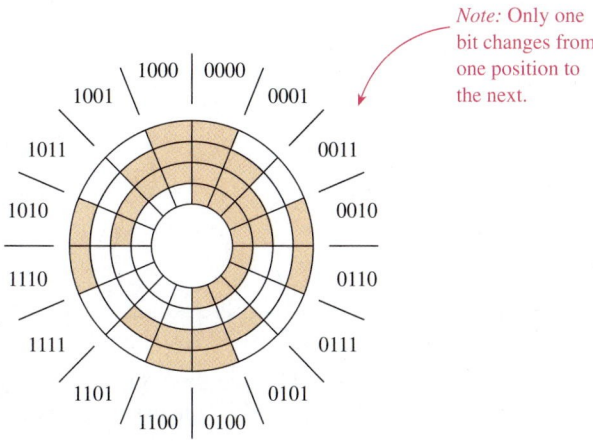

Note: Only one bit changes from one position to the next.

Figure 8–36   Gray code wheel.

the shaft position just as the code was changing, it might see 0111 or 1000, but because all 4 bits are changing (0 to 1 or 1 to 0), the code that it reads may be anything from 0000 to 1111. Therefore, the potential for an error using the regular binary system is great.

With the Gray code wheel, on the other hand, when the position changes from 7 to 8, the code changes from 0100 to 1100. The MSB is the only bit that changes, so if a read is taken right on the border between the two numbers, either a 0100 is read or a 1100 is read (no problem).

Gray Code Conversions

The determination of the Gray code equivalents and the conversions between Gray code and binary code are done very simply with exclusive-OR gates, as shown in Figures 8–37 and 8–38.

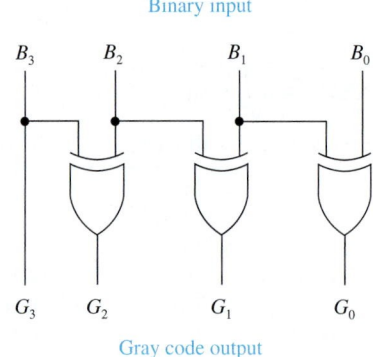

Figure 8–37   Binary–to–Gray code converter.

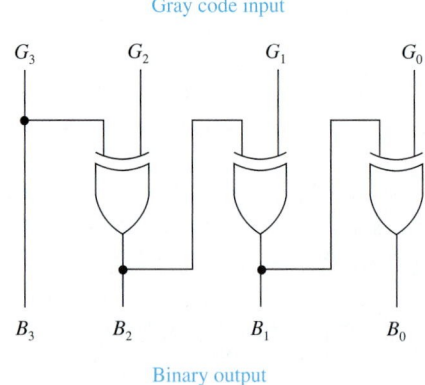

Figure 8–38   Gray code–to–binary converter.

EXAMPLE 8–9

Test the operation of the binary–to–Gray code converter of Figure 8–37 by labeling the inputs and outputs with the conversion of binary 0110 to Gray code.

Solution: The operation of the converter is shown in Figure 8–39.

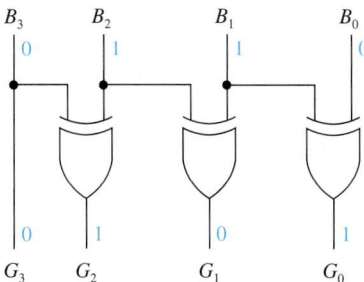

Figure 8–39

EXAMPLE 8–10

Repeat Example 8–9 for the Gray code–to–binary converter of Figure 8–38 by converting a Gray code 0011 to binary.

Solution: The operation of the converter is shown in Figure 8–40.

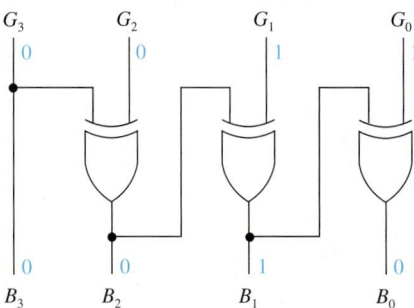

Figure 8–40

Review Questions

8–12. What is the binary weighting factor of the MSB of a two-digit (8-bit) BCD number?

8–13. How many 74184 ICs are required to convert a three-digit BCD number to binary?

8–14. Why is Gray code used for indicating the shaft position of rotating machinery rather than regular binary code?

8–7   Multiplexers

A **multiplexer** is a device capable of funneling several data lines into a single line for transmission to another point. The multiplexer has two or more digital input signals connected to its input. Control signals are also input to tell which data-input line to select for transmission (data selection). Figure 8–41 illustrates the function of a multiplexer.

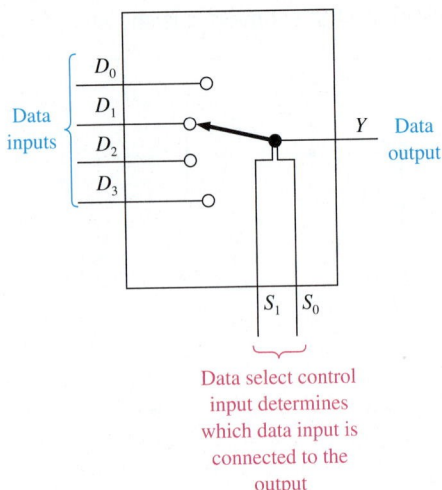

Data select control input determines which data input is connected to the output

Figure 8–41 Functional diagram of a four-line multiplexer.

The multiplexer is also known as a *data selector*. Figure 8–41 shows that the *data select control inputs* (S_1, S_0) are responsible for determining which data input (D_0 to D_3) is selected to be transmitted to the data-output line (Y). The S_1, S_0 inputs will be a binary code that corresponds to the data-input line that you want to select. If $S_1 = 0$, $S_0 = 0$, then D_0 is selected; if $S_1 = 0$, $S_0 = 1$, then D_1 is selected; and so on. Table 8–5 lists the codes for input data selection.

TABLE 8–5	Data Select Input Codes for Figure 8–41

Data Select Control Inputs		
S_1	S_0	**Data Input Selected**
0	0	D_0
0	1	D_1
1	0	D_2
1	1	D_3

A sample four-line multiplexer built from SSI logic gates is shown in Figure 8–42. The control inputs (S_1, S_0) take care of enabling the correct AND gate to pass just one of the data inputs through to the output. In Figure 8–42, 1's and 0's were placed on the diagram to show the levels that occur when selecting data input D_1. Notice that AND gate 1 is enabled, passing D_1 to the output, whereas all other AND gates are disabled.

Two-, 4-, 8-, and 16-input multiplexers are readily available in MSI packages. Table 8–6 lists some popular TTL and CMOS multiplexers. (H-CMOS, or high-speed CMOS, is compared with the other logic families in detail in Chapter 9.)

The 74151 Eight-Line Multiplexer

The logic symbol and logic diagram for the 74151 are given in Figure 8–43. Because the 74151 has eight lines to select from (I_0 to I_7), it requires three data select inputs (S_2, S_1, S_0) to determine which input to choose ($2^3 = 8$). True (Y) and complemented ($\overline{Y}$) outputs are provided. The active-LOW enable input ($\overline{E}$) disables all inputs when it is HIGH and forces Y LOW regardless of all other inputs.

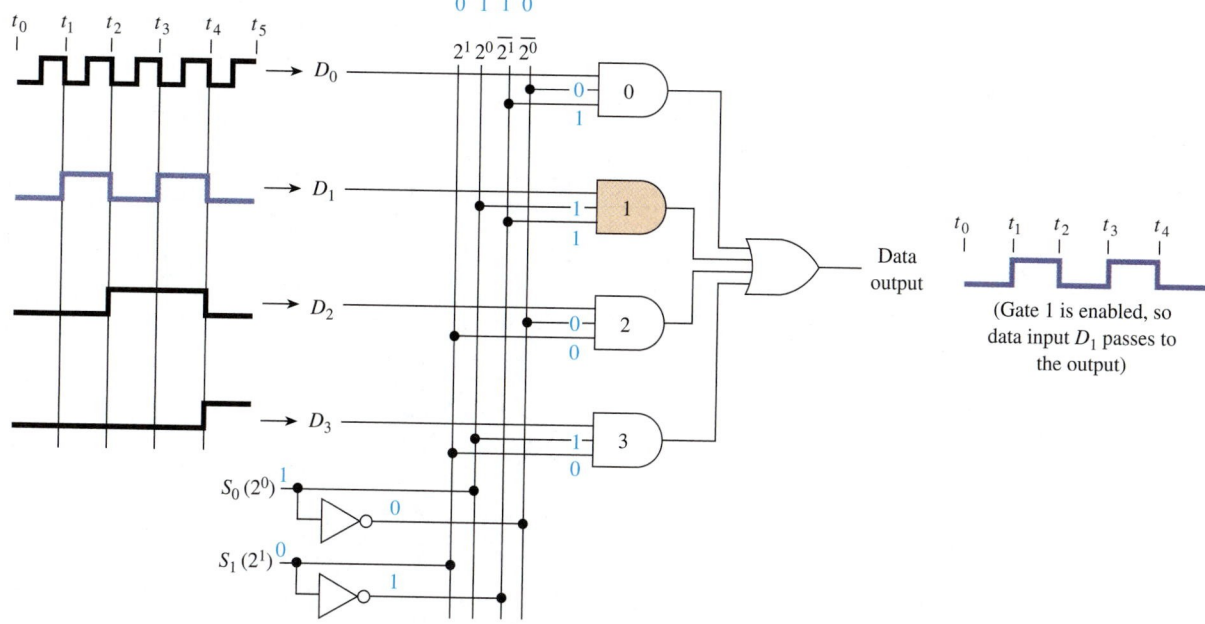

Figure 8–42   Logic diagram for a four-line multiplexer.

TABLE 8–6	TTL and CMOS Multiplexers	
Function	**Device**	**Logic Family**
Quad two-input	74157	TTL
	74HC157	H-CMOS
	4019	CMOS
Dual eight-input	74153	TTL
	74HC153	H-CMOS
	4539	CMOS
Eight-input	74151	TTL
	74HC151	H-CMOS
	4512	CMOS
Sixteen-input	74150	TTL

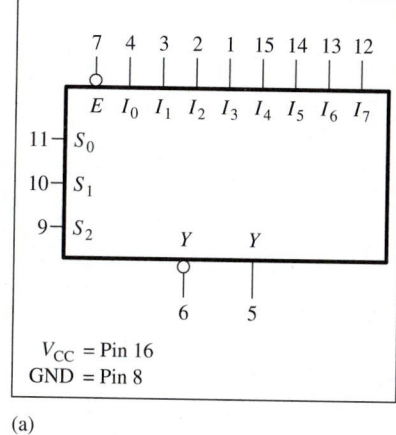

(a)

Figure 8–43   The 74151 eight-line multiplexer: (a) logic symbol.

Common Misconception

Students often question the operation of the *Y* outputs. The trial of several different inputs on the logic diagram [Figure 8–43(b)] will reinforce your understanding of the IC. Use the logic diagram to determine the output at *Y* when $\overline{E}$ is HIGH.

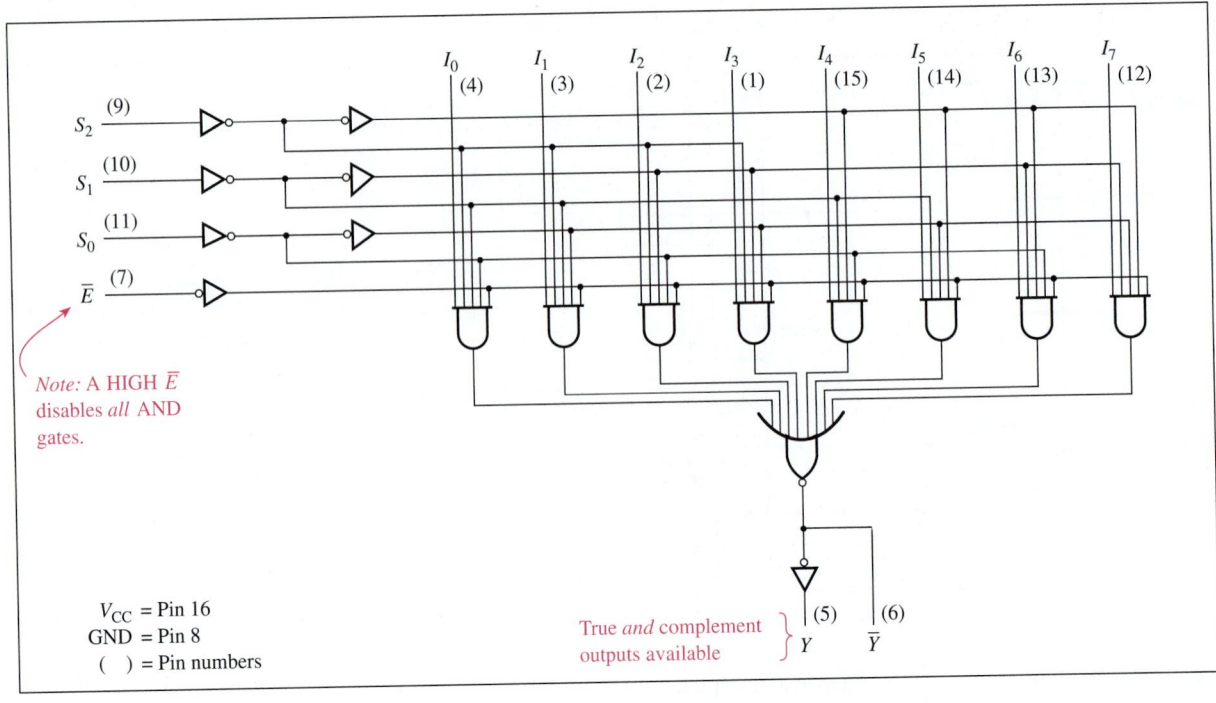

V_{CC} = Pin 16
GND = Pin 8
( ) = Pin numbers

Note: A HIGH $\overline{E}$ disables *all* AND gates.

True *and* complement outputs available $\}$ (5) (6)
Y $\overline{Y}$

(b)

Figure 8–43 *(Continued)* (b) logic diagram. (Courtesy of Philips Semiconductors.)

EXAMPLE 8–11

Sketch the output waveforms at Y for the 74151 shown in Figure 8–44. For this example, the eight input lines (I_0 to I_7) are each connected to a constant level, and the data select lines (S_0 to S_2) and input enable ($\overline{E}$) are given as input waveforms.

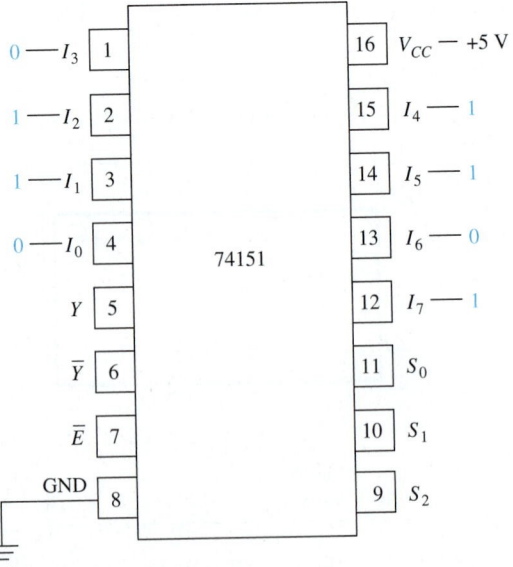

Figure 8–44 The 74151 multiplexer pin connections for Example 8–11.

See Figure 8–45. From t_0 to t_8, the waveforms at S_0, S_1, and S_2 form a binary counter from 000 to 111. Therefore, the output at Y will be selected from I_0, then I_1, then I_2, and so on, up to I_7. From t_8 to t_9, the S_0, S_1, and S_2 inputs are back to 000, so I_0 will be selected for output. From t_9 to t_{11}, the $\overline{E}$ enable line goes HIGH, disabling all inputs and forcing Y LOW.

Solution:

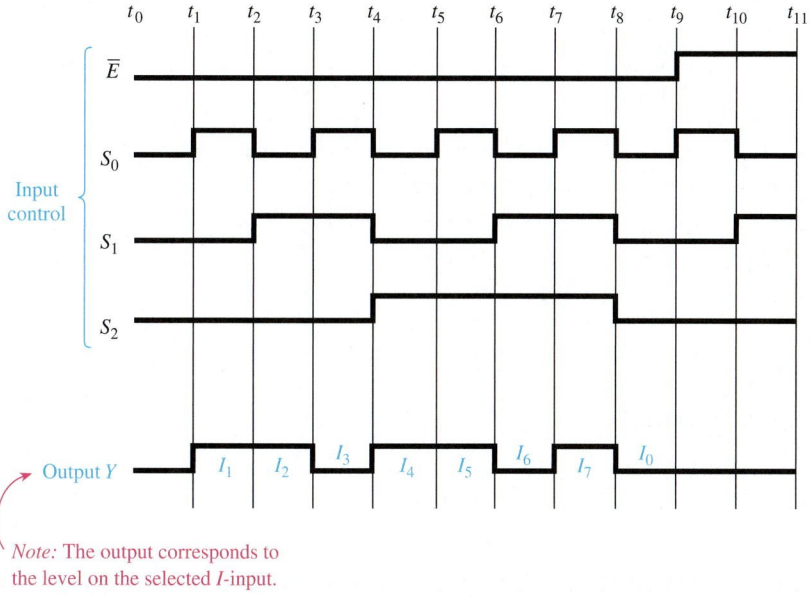

Note: The output corresponds to the level on the selected *I*-input.

Figure 8–45

EXAMPLE 8–12

Using two 74151s, design a 16-line multiplexer controlled by four data select control inputs.

Solution: The multiplexer is shown in Figure 8–46. Because there are 16 data input lines, we must use four data select inputs ($2^4 = 16$). (A is the LSB data select line, and D is the MSB.)

When the data select is in the range from 0000 to 0111, the D line is 0, which enables the low-order (left) multiplexer selecting the D_0 to D_7 inputs and disables the high-order (right) multiplexer.

When the data select inputs are in the range from 1000 to 1111, the D line is 1, which disables the low-order multiplexer and enables the high-order multiplexer, allowing D_8 to D_{15} to be selected. Because the Y output of a disabled multiplexer is 0, an OR gate is used to combine the two outputs, allowing the output from the enabled multiplexer to pass through.

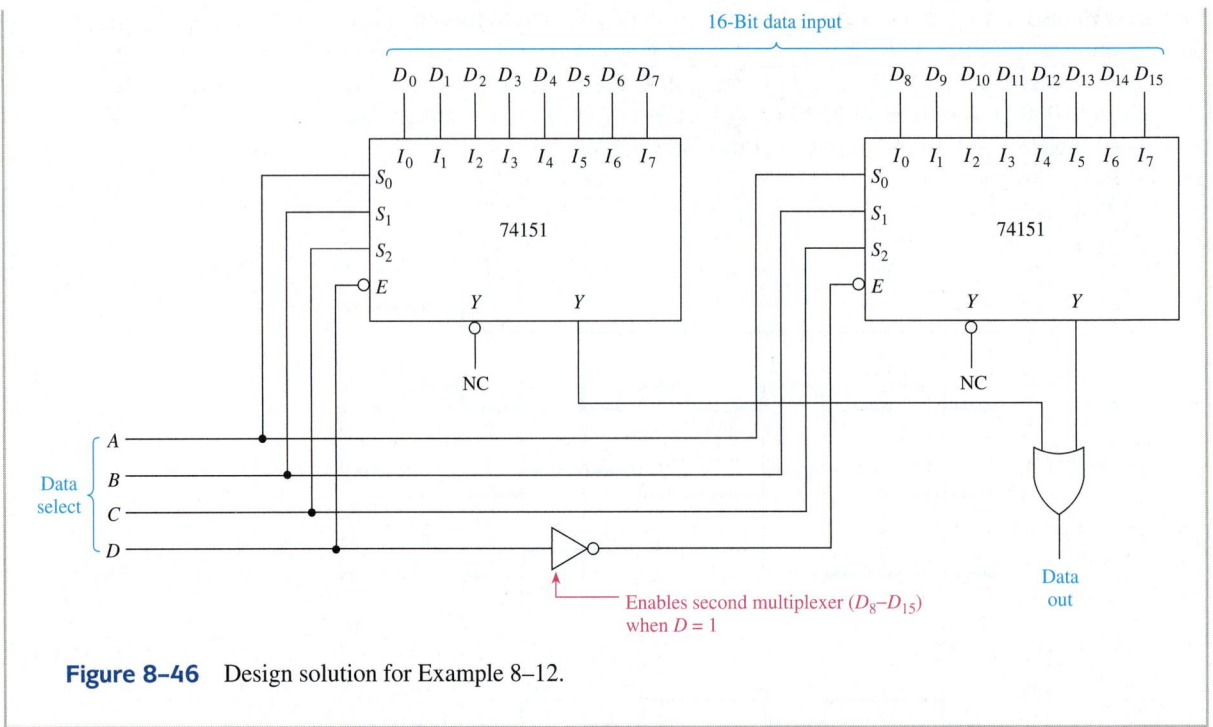

16-Bit data input

D_0 D_1 D_2 D_3 D_4 D_5 D_6 D_7 D_8 D_9 D_{10} D_{11} D_{12} D_{13} D_{14} D_{15}

74151

74151

NC

NC

Data select

A

B

C

D

Enables second multiplexer (D_8–D_{15}) when $D = 1$

Data out

Figure 8–46 Design solution for Example 8–12.

Providing Combination Logic Functions with a Multiplexer

Multiplexers have many other uses besides functioning as data selectors. Another important role of a multiplexer is for implementing combinational logic circuits. One multiplexer can take the place of several Small-Scale Integration (SSI) logic gates, as shown in the following example.

EXAMPLE 8–13

Use a multiplexer to implement the function

$$X = \overline{A}\,\overline{B}\,\overline{C}D + A\overline{B}\,\overline{C}D + AB\overline{C}\,\overline{D} + \overline{A}BC + \overline{A}\,\overline{B}C$$

Solution: The equation is in the sum-of-products form. Each term in the equation, when fulfilled, will make $X = 1$. For example, when $A = 0$, $B = 0$, $C = 0$, and $D = 1$ ($\overline{A}\,\overline{B}\,\overline{C}D$), X will receive a 1. Also, if $A = 1$, $B = 0$, $C = 0$, and $D = 1$ ($A\overline{B}\,\overline{C}D$), X will receive a 1, and so on.

If the A, B, C, and D variables are used as the data input selectors of a 16-line multiplexer (four input variables can have 16 possible combinations) and the appropriate digital levels are placed at the multiplexer data inputs, we can implement the function for X. We will use the 74150 16-line multiplexer. A 1 must be placed at each data input that satisfies any term in the Boolean equation. The truth table and the 74150 connections to implement the function are given in Figure 8–47.

The logic symbol for the 74150 is similar to the 74151 except that it has 16 input data lines, four data select lines, and only the complemented output. To test the operation of the circuit, let's try some entries from the truth table to see that X is valid. For example, if $A = 0$, $B = 0$, $C = 0$, and $D = 1$ ($\overline{A}\,\overline{B}\,\overline{C}D$), the multiplexer will select D_1, which is 1. It

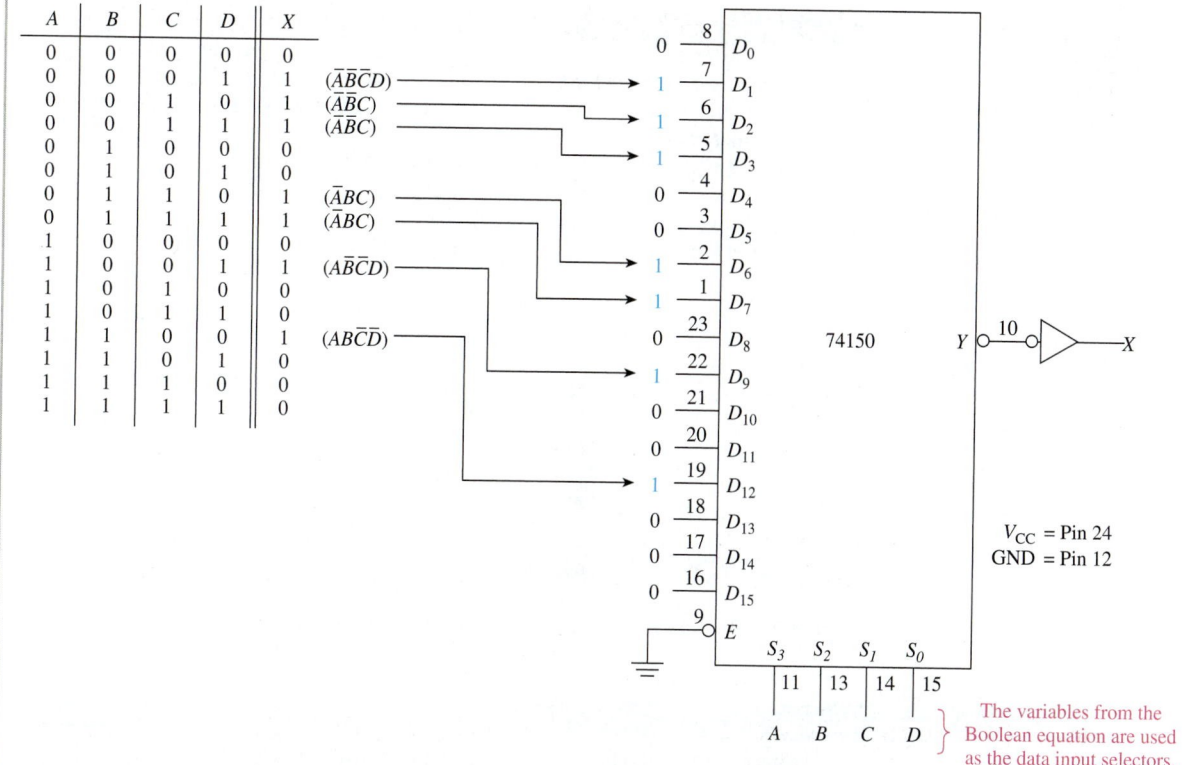

A	B	C	D	X
0	0	0	0	0
0	0	0	1	1
0	0	1	0	1
0	0	1	1	1
0	1	0	0	0
0	1	0	1	0
0	1	1	0	1
0	1	1	1	1
1	0	0	0	0
1	0	0	1	1
1	0	1	0	0
1	0	1	1	0
1	1	0	0	1
1	1	0	1	0
1	1	1	0	0
1	1	1	1	0

Figure 8–47 Truth table and solution for the implementation of the Boolean equation. $X = \overline{A}\,\overline{B}\,CD + A\overline{B}\,\overline{C}D + ABC\overline{D} + \overline{A}BC + \overline{A}\,\overline{B}C$ using a 16-line multiplexer.

gets inverted twice before reaching X, so X receives a 1, which is correct. Work through the rest of them yourself, and you will see that the Boolean function is fulfilled.

VHDL 4-line Multiplexer

Use VHDL to design a 4-line multiplexer similar to Figure 8–41. Test its operation by creating a waveform simulation file that alternately routes each of the inputs to the y output.

Solution: The program listing is given in Figure 8–48. The individual d inputs are grouped together as a 4-bit vector. The data selector bits (s) are grouped as a 2-bit vector, and y is a **scalar** (individual) quantity. A selected signal assignment statement is used to route the appropriate d input to the y output.

Proof that the multiplexer is working properly is shown in Figure 8–49. During the first 16 μs the data select inputs ($s1$ and $s0$) are set to 0-0 causing waveform $d0$ to be routed to y. From 16 μs to 32 μs, the data select inputs are set to 0-1 causing waveform $d1$ to be routed to y, and so on.

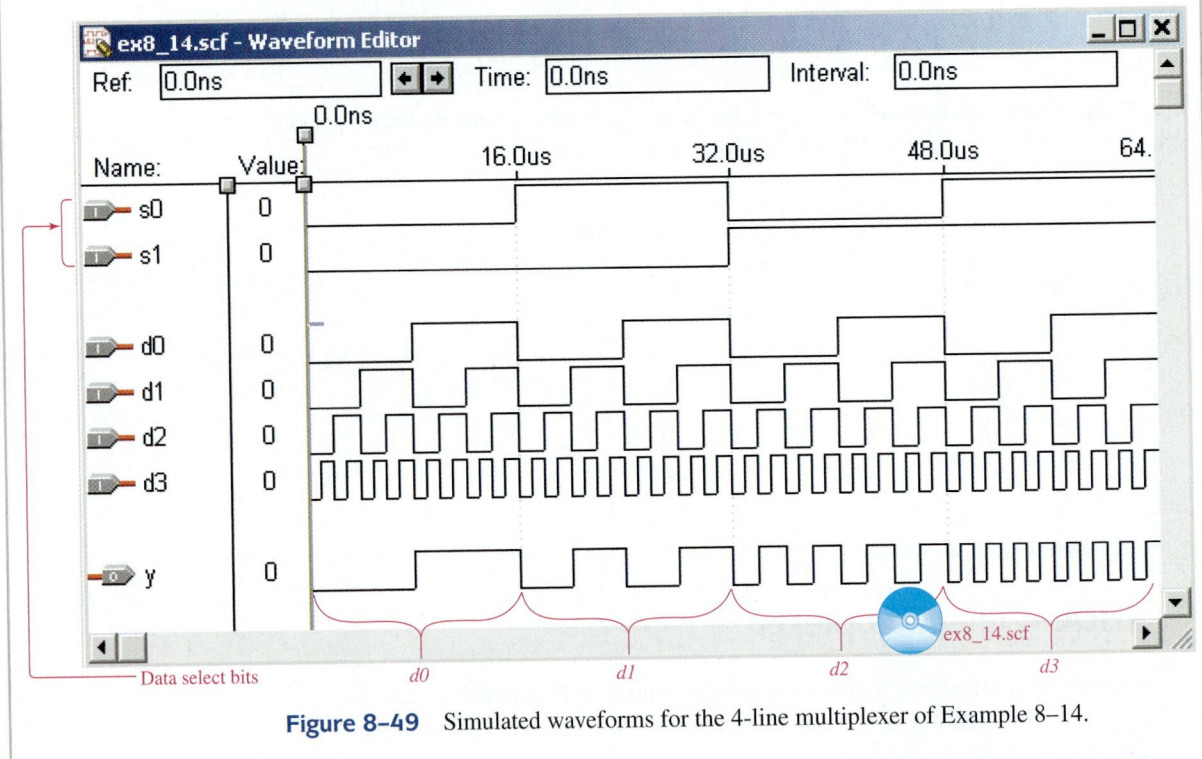

Figure 8–48 VHDL program for the 4-line multiplexer of Example 8–14.

Figure 8–49 Simulated waveforms for the 4-line multiplexer of Example 8–14.

8–8   Demultiplexers

Demultiplexing is the opposite procedure from multiplexing. We can think of a **demultiplexer** as a *data distributor*. It takes a single input data value and routes it to one of several outputs, as illustrated in Figure 8–50(a).

The results of a simulation are shown in Figure 8–50(b). The object of this demultiplexer simulation is to route the *D* waveform to one of the *Y* outputs selected by

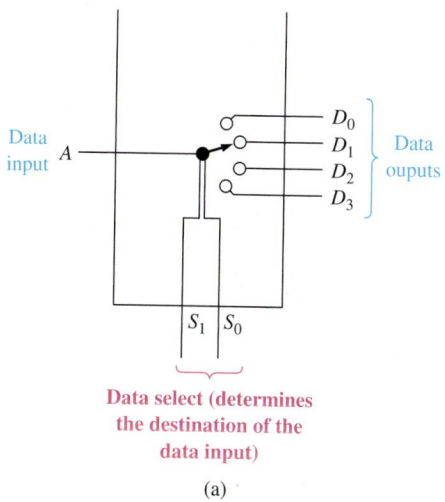

Data input A

D_0
D_1
D_2
D_3

Data ouputs

S_1 S_0

Data select (determines the destination of the data input)

(a)

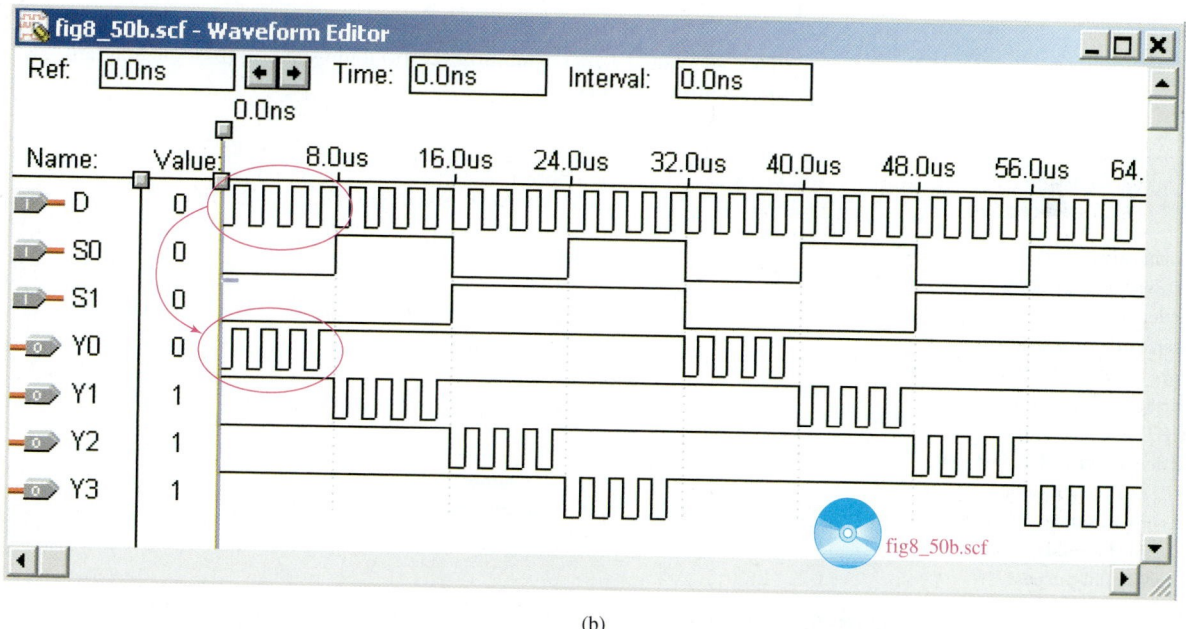

(b)

Figure 8–50 4-line demultiplexer: (a) functional diagram; (b) waveform simulation.

the two data select inputs, $S0$ and $S1$. Notice that during the first 8.0 μs, D is routed to $Y0$ because the data select inputs are set to 0-0. From 8.0 to 16 μs D is routed to $Y1$ because the data select inputs are 0-1, and so on.

Integrated-circuit demultiplexers come in several configurations of inputs/outputs. The two that we discuss in this section are the 74139 dual 4-line demultiplexer and the 74154 16-line demultiplexer.

The logic diagram and logic symbol for the 74139 are given in Figure 8–51. Notice that the 74139 is divided into two equal sections. By looking at the logic diagram, you will see that the schematic is the same as that of a 2-line-to-4-line decoder. Decoders and demultiplexers are the same, except with a decoder you hold the $\overline{E}$ enable line LOW and enter a code at the A_0A_1 inputs. As a demultiplexer, the A_0A_1 inputs are used to select the destination of input data. The input data are brought in via the $\overline{E}$ line. The 74138 3-line-to-8-line decoder that we covered earlier in this chapter can also function as an 8-line demultiplexer.

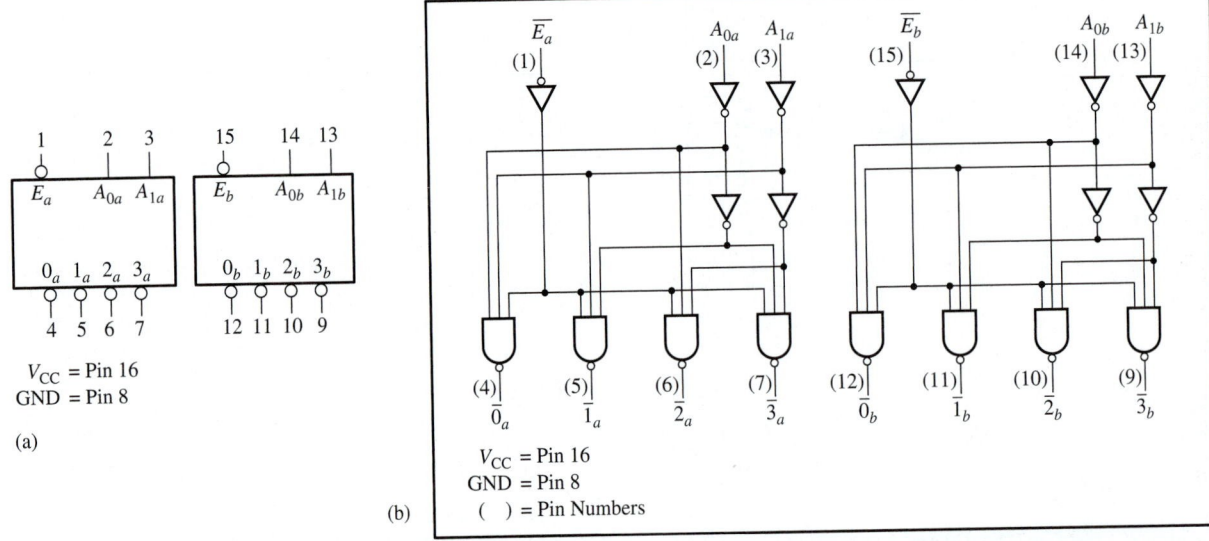

Figure 8–51 The 74139 dual 4-line demultiplexer: (a) logic symbol; (b) logic diagram. (Courtesy of Philips Semiconductors.)

Common Misconception

Students are sometimes confused when we use a decoder IC as a demultiplexer. For example, in Figure 8–52, the 74139 is described as a demultiplexer, but it could also function as a decoder if we use $\overline{E}_a$ as an enable and A_0, A_1 as binary inputs. Discuss how a 74138 could also be used as a dual-purpose chip.

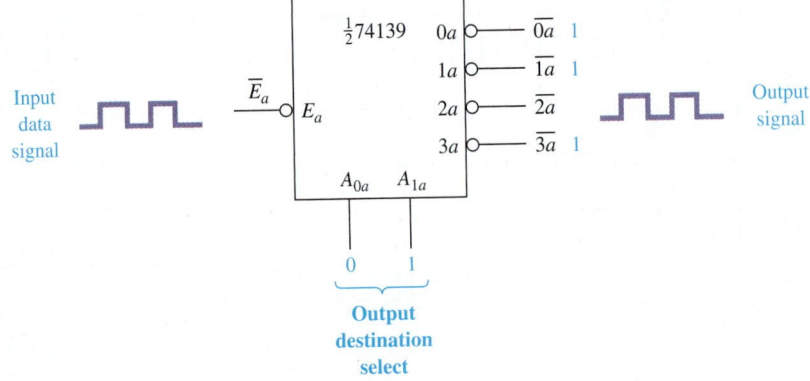

Figure 8–52 Connections to route an input data signal to the $\overline{2}_a$ output of a 74139 demultiplexer.

To use the 74139 as a demultiplexer to route some input data signal to, let's say, the $\overline{2}_a$ output, the connections shown in Figure 8–52 would be made. In the figure, the destination $\overline{2}_a$ is selected by making $A_{1a} = 1$ and $A_{0a} = 0$. The input signal is brought into the enable line $(\overline{E}_a)$. When $\overline{E}_a$ goes LOW, the selected output line goes LOW; when $\overline{E}_a$ goes HIGH, the selected output line goes HIGH. (Because the outputs are active-LOW, all nonselected lines remain HIGH continuously.)

The 74154 was used earlier in the chapter as a 4-line-to-16-line hexadecimal decoder. It can also be used as a 16-line demultiplexer. Figure 8–53 shows how it can be connected to route an input data signal to the $\overline{5}$ output.

Analog Multiplexer/Demultiplexer

Several analog multiplexers/demultiplexers are available in the CMOS family. The 4051, 4052, and 4053 are combination multiplexer *and* demultiplexer CMOS ICs. (High-speed CMOS versions, such as the 74HCT4051, are also available.) They can function in either configuration because their inputs and outputs are **bidirectional,**

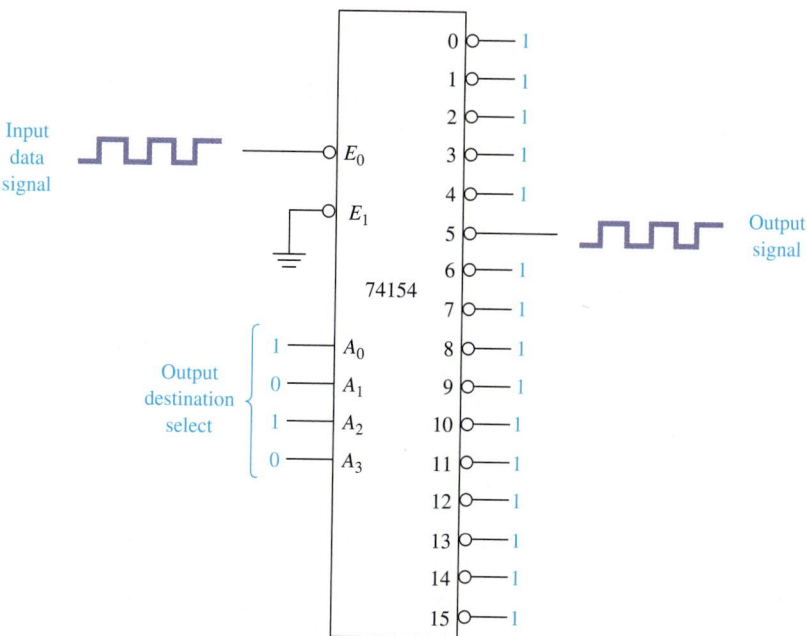

Figure 8–53 The 74154 demultiplexer connections to route an input signal to the $\overline{5}$ output.

meaning that the flow can go in either direction. Also, they are *analog,* meaning that they can input and output levels other than just 1 and 0. The input/output levels can be any analog voltage between the positive and negative supply levels.

The functional diagram for the 4051 eight-channel multiplexer/demultiplexer is given in Figure 8–54. The eight square boxes in the functional diagram represent the bidirectional I/O lines. Used as a multiplexer, the analog levels come in on the Y_0 to Y_7 lines, and the decoder selects which of these inputs are output to the Z line. As a demultiplexer, the connections are reversed, with the input coming into the Z line and the output going out on one of the Y_0 to Y_7 lines.

Review Questions

8–15. Why is a *multiplexer* sometimes called a data selector?

8–16. Why is a *demultiplexer* sometimes called a data distributor?

8–17. What is the function of the S_0, S_1, and S_2 pins on the 74151 multiplexer?

8–18. What is the function of the A_0, A_1, A_2, and A_3 pins on the 74154 demultiplexer?

8–9 System Design Applications

Microprocessor Address Decoding

The 74138 and its CMOS version, the 74HCT138, are popular choices for decoding the address lines in **microprocessor** circuits. A typical 8-bit microprocessor such as the Intel 8085A or the Motorola 6809 has 16 address lines (A_0–A_{15}) for designating

Helpful Hint

The applications that follow demonstrate how the MSI chips covered in this chapter interface with practical microprocessor-based systems. They are meant to provide a positive experience for you by showing that you can comprehend circuits of higher complexity now that you have mastered some of the building blocks.

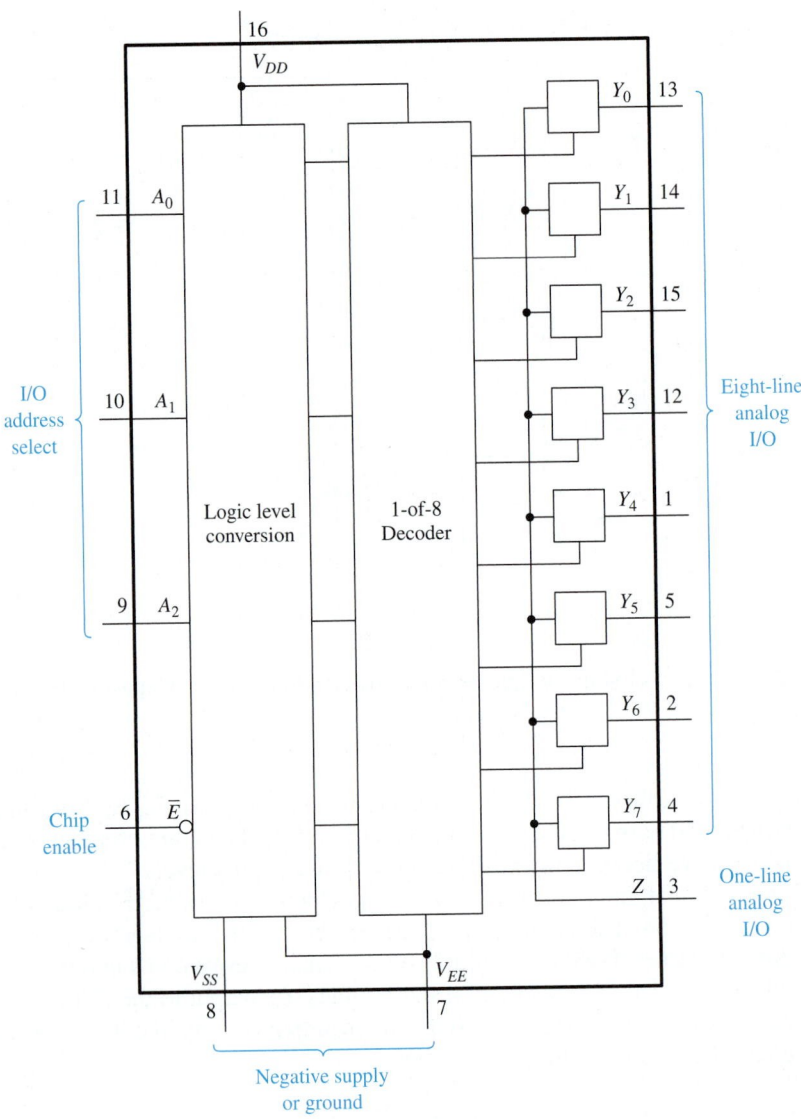

I/O address select

Eight-line analog I/O

One-line analog I/O

Chip enable

Negative supply or ground

Figure 8–54 The 4051 CMOS analog multiplexer/demultiplexer. (Courtesy of Philips Semiconductors.)

unique **addresses** (locations) for all the peripheral devices and memory connected to it. When a microprocessor-based system has a large amount of memory connected to it, a designer often chooses to set the memory up in groups, called *memory banks*. For example, Figure 8–55 shows a decoding scheme that can be used to select one of eight separate memory banks within a microprocessor-based system. The high-order bits of the address (A_{12}–A_{15}) are output by the microprocessor to designate which memory bank is to be accessed. In this design, A_{15} must be LOW for the decoder IC to be enabled. The three other high-order bits, A_{12}–A_{14}, are then used to select the designated memory bank.

The 8085A also outputs control signals that are used to enable/disable memory operations. First, if we are performing a memory operation, we must be doing a read ($\overline{\text{RD}}$) or a write ($\overline{\text{WR}}$). The inverted-input OR gate (NAND) provides the HIGH to the E_3 enable if it receives a LOW $\overline{\text{RD}}$ *or* a LOW $\overline{\text{WR}}$. The other control signal, IO/$\overline{\text{M}}$, is used by the 8085A to distinguish between input/output (IO) to peripheral devices versus memory operations. If IO/$\overline{\text{M}}$ is HIGH, an IO operation is to take place, and if it is LOW, a

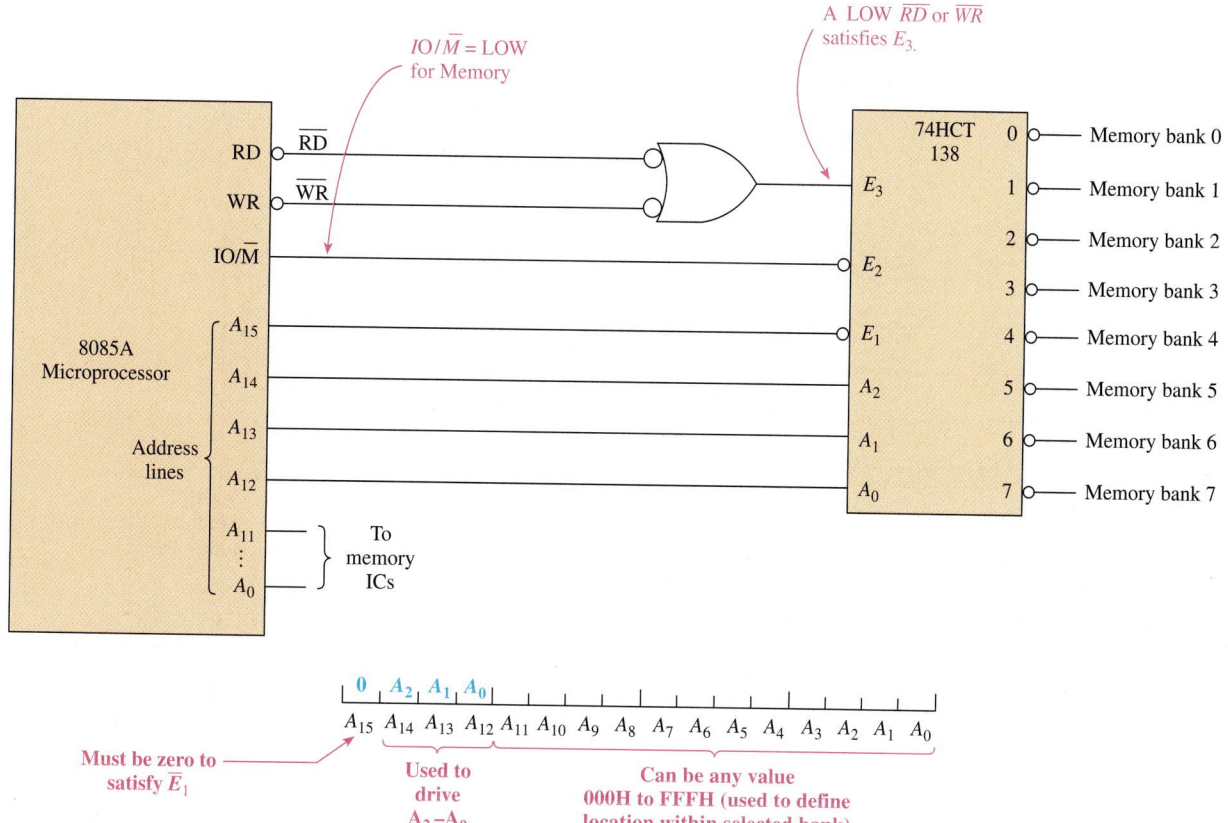

Figure 8–55 Using the 74HCT138 for a memory address decoder in an 8085A microprocessor system.

memory operation is to occur. Therefore, one of the memory banks will be selected if IO/M̄ is LOW and address line A_{15} is LOW while either R̄D̄ is LOW or W̄R̄ is LOW.

EXAMPLE 8–14

What is the range of addresses that can be specified by the 8085A in Figure 8–55 to access memory within bank 2?

Solution: Referring to the chart in Figure 8–55, address bit A_{13} must be HIGH to make A_1 in the decoder HIGH to select bank 2. The other address bits, A_0–A_{11}, are not used by the decoder IC and can be any value. Therefore, any address within the range of 2000H through 2FFFH will select memory bank 2.

Alarm Encoder for a Microcontroller

This design application uses a 74148 encoder to monitor the fluid level of eight chemical tanks. If any level exceeds a predetermined height, a sensor outputs a LOW level to the input of the encoder. The encoder encodes the active input into a 3-bit binary code to be read by a **microcontroller.** This way, the microcontroller needs to use only three input lines to monitor eight separate points. Figure 8–56 shows the circuit connections.

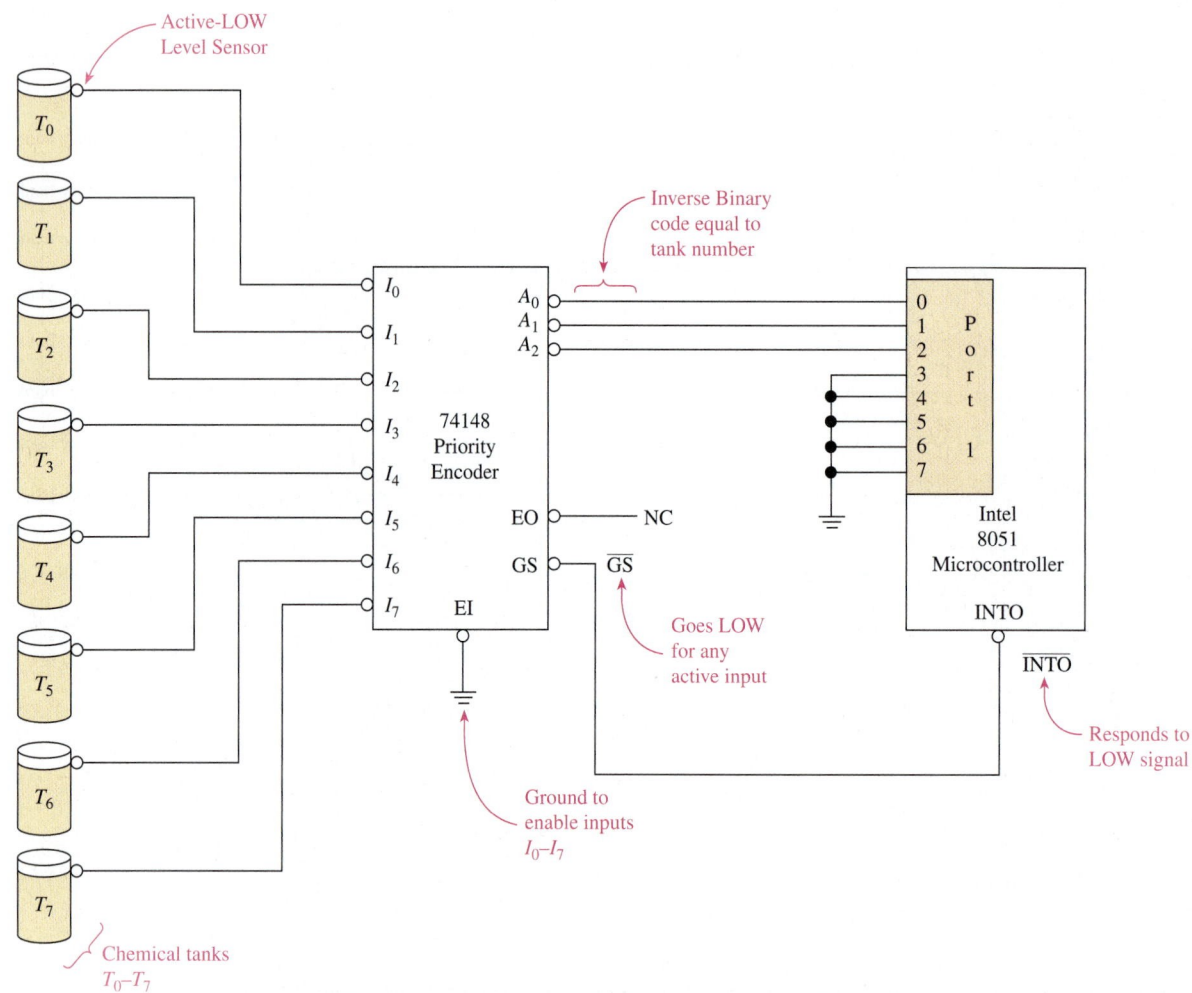

Figure 8–56 Using a 74148 to encode an active alarm to be monitored by a microcontroller.

A microcontroller differs from a microprocessor in that it has several input/output ports and memory built into its architecture, making it better suited for monitoring and control applications. The microcontroller used here is the Intel 8051. We will use one of its 8-bit ports to read the encoded alarm code, and we will use its interrupt input, $\overline{\text{INT0}}$, to receive notification that an alarm has occurred. The 8051 will be programmed to be in a HALT mode (or, in some versions, a low-power SLEEP mode) until it receives an interrupt signal to take a specific action. In this case, it will perform the desired response to the alarm when it receives a LOW at $\overline{\text{INT0}}$. This LOW interrupt signal is provided by $\overline{\text{GS}}$, which goes LOW whenever any of the 74148 inputs becomes active.*

Serial Data Multiplexing for a Microcontroller

Multiplexing and demultiplexing are very useful for data communication between a computer system and serial data terminals. The advantage is that only one serial receive line and one serial transmit line are required by the computer to communicate with several data terminals. A typical configuration is shown in Figure 8–57.

---

*For an in-depth study of the 8085A microprocessor and the 8051 microcontroller, refer to William Kleitz, *Digital and Microprocessor Fundamentals*. Fourth Edition, Prentice Hall, Upper Saddle River, N.J., 2003.

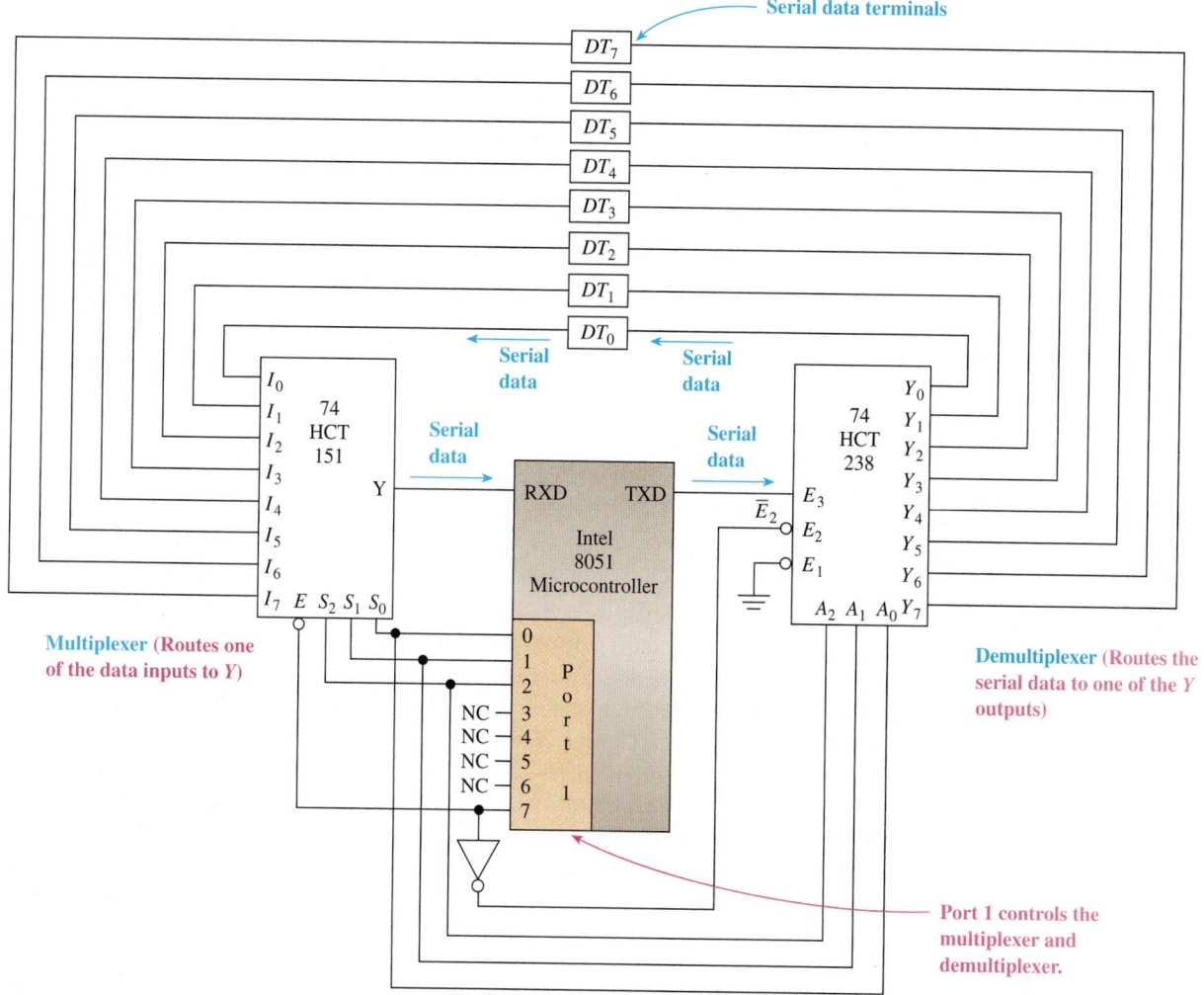

Figure 8–57 Using a multiplexer, demultiplexer, and microcontroller to provide communication capability to several serial data terminals.

Again, the 8051 is used because of its built-in control and communication capability. Its RXD and TXD pins are designed to receive (RXD) and transmit (TXD) serial data at a speed and in a format dictated by a computer program written by the user.

The selected data terminal to be read from is routed through the 74HCT151 multiplexer to the microcontroller's serial input terminal, RXD. First, the computer program writes the appropriate hex code to Port 1 to enable the 74HCT151 to route the serial data stream from the selected data terminal through to its Y output. The 8051 then reads the serial data at its RXD input and performs the desired action.

To output to one of the data terminals, the 8051 must first output the appropriate hex code to Port 1 to select the correct data terminal; then, it outputs serial data on its TXD pin. The 74HCT238 is used as a demultiplexer (data distributor) in this application. The 74HCT238 is identical to the 74138 decoder/demultiplexer except the 74HCT238 has noninverting outputs. The HCT versions are used to match the high speed of the 8051 and keep the power requirements to a minimum (more on speed and power requirements in Chapter 9). The hex code output at Port 1 must provide a LOW to $\overline{E}_2$ and the proper data-routing select code to $A_2 - A_0$. The selected Y output then duplicates the HIGH/LOW levels presented at the E_3 pin.

EXAMPLE 8–15

Determine the correct hex codes that must be output to Port 1 in Figure 8–57 to accomplish the following action: (a) read from data terminal 3 (DT3); (b) write to data terminal 6 (DT6).

Solution The 0, 1, and 2 outputs of Port 1 are used to control the S_0, S_1, and S_2 input-select pins of the 74HCT151 and the A_0, A_1, and A_2 output-select pins of the 74HCT238. Output 7 of Port 1 is used to determine which IC is enabled; 0 enables the 74HCT151 multiplexer, and 1 enables the 74HCT238 demultiplexer. (a) Assuming that the NC (no connection) lines are zeros, the hex code to read from data terminal 3 is 03H (0000 0011). (b) The hex code to write to data terminal 6 is 86H (1000 0110).

Analog Multiplexer Application

The 4051 is very versatile for controlling analog voltages with digital signals. One use is in the design of multitrace oscilloscope displays for displaying as many as eight traces on the same display screen. To do that, each input signal to be displayed must be **superimposed** on (added to) a different voltage level so that each trace will be at a different Y-axis level on the display screen.

The 4051 (and its high-speed version, the 74HCT4051) can be set up to sequentially output eight different voltage levels repeatedly if connected as shown in Figure 8–58. The resistor voltage-divider network in Figure 8–58 is set up to drop 0.5 V across each 100-Ω resistor. This will put 0.5 V at Y_0, 1.0 V at Y_1, and so on. The binary counter is a device that outputs a binary progression from 000 up to 111 at its 2^0, 2^1, and 2^2 outputs, which causes each of the Y_0 to Y_7 inputs to be successively selected for Z out, one at a time, in order. The result is the staircase waveform shown in Figure 8–58, which can superimpose a different voltage level on each of eight separate digital input signals that are brought in via the 74151 8-line *digital* multiplexer (not shown) driven by the same binary counter.

Multiplexed Display Application

Figure 8–59 shows a common method of using multiplexing to reduce the cost of producing a multidigit display in a computer or digital device like a calculator or wristwatch. Multiplexing multidigit displays reduces circuit cost and failure rate by *sharing* common ICs, components, and conductors. The seven-segment digit displays, decoders, and drivers are covered in detail in Chapter 12. For now, we need to know that a decoding process must take place to convert the BCD digit information to a recognizable digit display. We will also assume that the "arithmetic circuitry" takes care of loading the four digit registers with the proper data.

The digit **bus** and display bus are each just a common set of conductors *shared by* the digit storage registers and display segments. The four digit-registers are, therefore, multiplexed into a single-digit bus, and the display bus is demultiplexed into the four-digit displays.

The 74139 four-line demultiplexer takes care of sequentially accessing each of the four digits. It first outputs a LOW on the $\overline{0}$ line. This enables the LS digit register *and* the LS digit display. The LS BCD information travels down the digit bus to the decoder/driver, which *decodes* the BCD into the special seven-segment code used by the LS digit display and *drives* the LS digit display.

Next, the second digit register and display are enabled, then the third, and then the fourth. This process continues repeatedly, each digit being on one-fourth of the

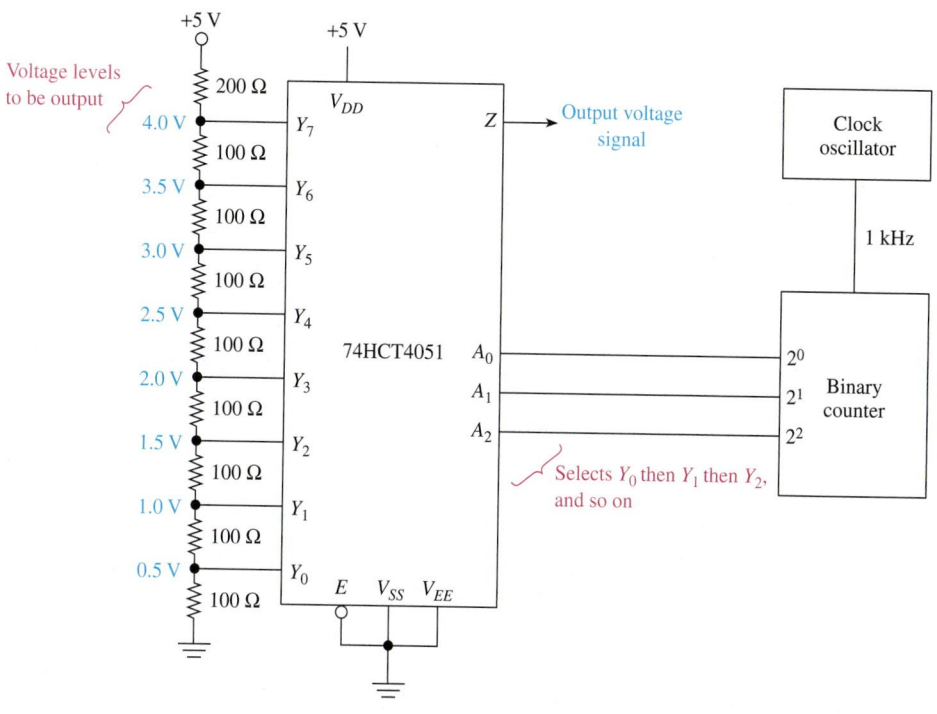

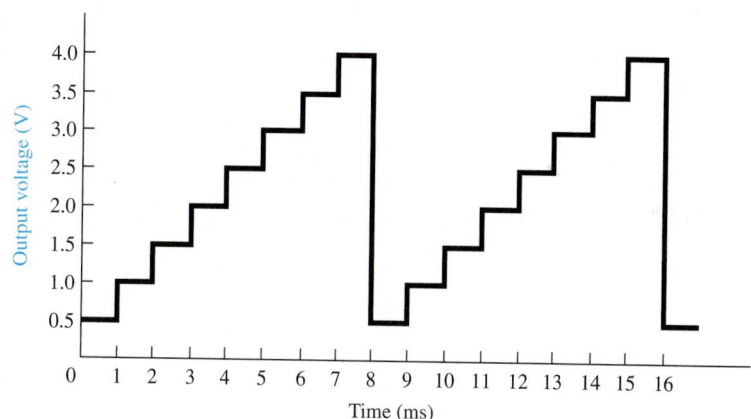

Figure 8–58 The 74HCT4051 analog multiplexer used as a staircase generator.

time. The circulation is set up fast enough (1 kHz or more) that it appears that all four digits are on at the same time. The external arithmetic circuitry is free to change the display at any time simply by reloading the temporary digit registers.

Review Questions

8–19. In the address decoding circuit of Figure 8–55, the A_{15} address bit must be HIGH to access memory bank 7. True or false?

8–20. What IC would be used to implement the inverted-input OR gate in Figure 8–55: a 7400, a 7402, a 7408, or a 7432?

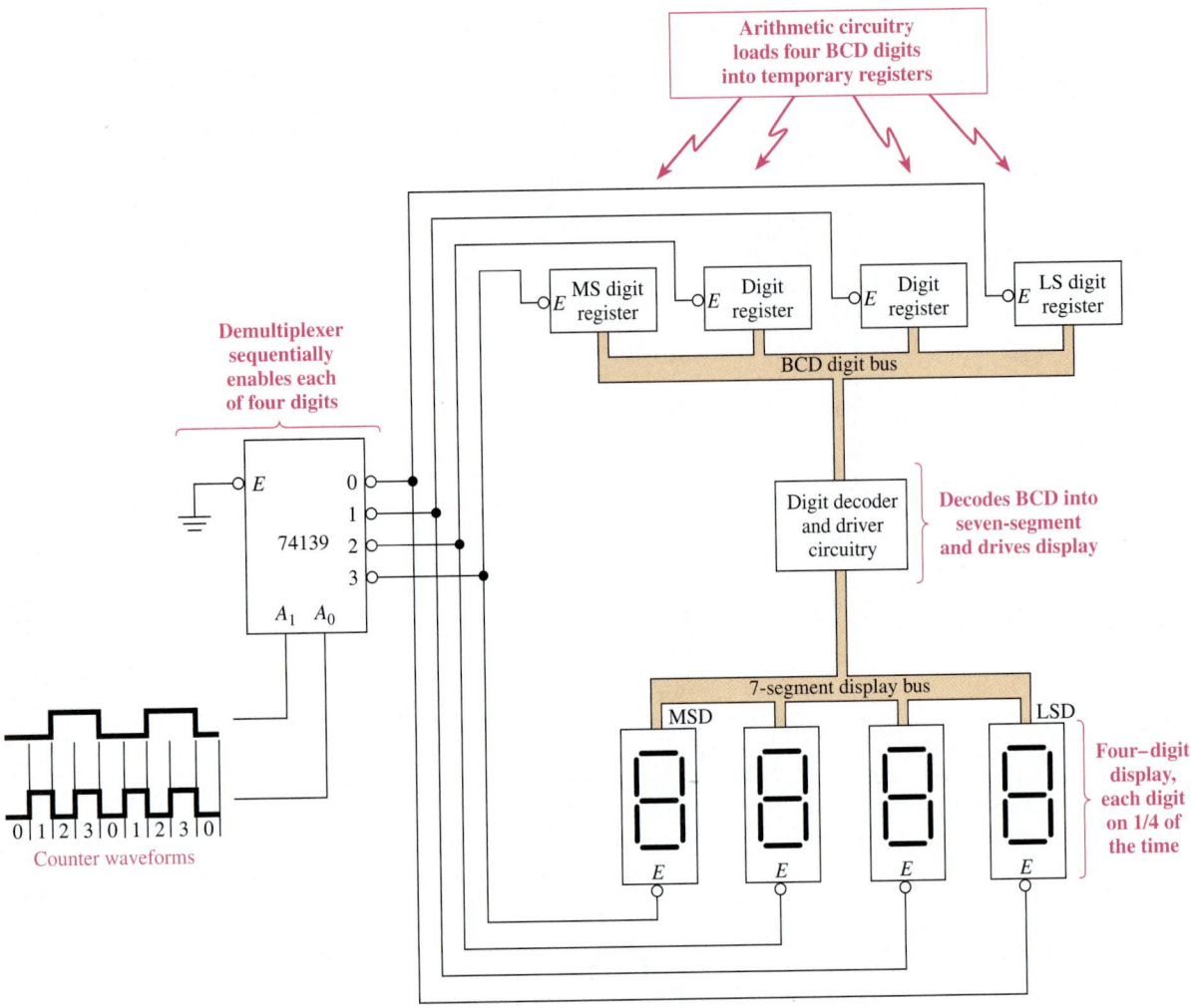

Figure 8–59 Multiplexed four-digit display block diagram.

8–21. To read from memory bank 0 in Figure 8–55, the microprocessor will output _____, _____, and _____ on the $\overline{RD}$, $\overline{WR}$, and $IO/\overline{M}$ lines.

8–22. In Figure 8–56, how is the microcontroller notified of a high fluid level at one of the chemical tanks?

8–23. The circuit of Figure 8–56 would work properly but have inverted outputs at $A_0 - A_2$ if the level sensor outputs were active-HIGH instead of active-LOW. True or false?

8–24. The circuit of Figure 8–57 does not allow for both transmitting and receiving serial data simultaneously. True or false?

8–25. In Figure 8–57, the 74HCT151 could be switched with the 74HCT238 and still work properly. True or false?

8–26. What is the purpose of the binary counter in Figure 8–58?

8–27. Describe the circuit operation required to display the number 5 in the MSD position of the display in Figure 8–59.

8–10   CPLD Design Applications Using LPMs

The Library of Parameterized Modules provides design solutions to many of the functions covered in this chapter. The next three examples will illustrate the power of these LPMs and firm up your understanding of comparators, decoders, and multiplexers.

EXAMPLE 8–16

LPM Comparator

Create a graphic design file that uses an LPM module to compare two 8-bit strings.

Solution: The module named LPM_COMPARE is found in the mega_lpm subdirectory. To get there, choose: **File > New > Graphic Editor File.** Then double-click in the work area and select the **mega_lpm** subdirectory. Find the symbol file called LPM_COMPARE. The symbol that appears is similar to that shown in Figure 8–60.

Three of the outputs are not needed for this design. To hide them, right-click on the symbol and select "edit Ports/Parameters" and check the "Unused" box for *ageb, aleb,* and *aneb.* Then in the "Parameters" section, set the LPM_WIDTH to 8. Since inputs *a* and *b* are vectors, connect them to the symbol using a bus line style as shown.

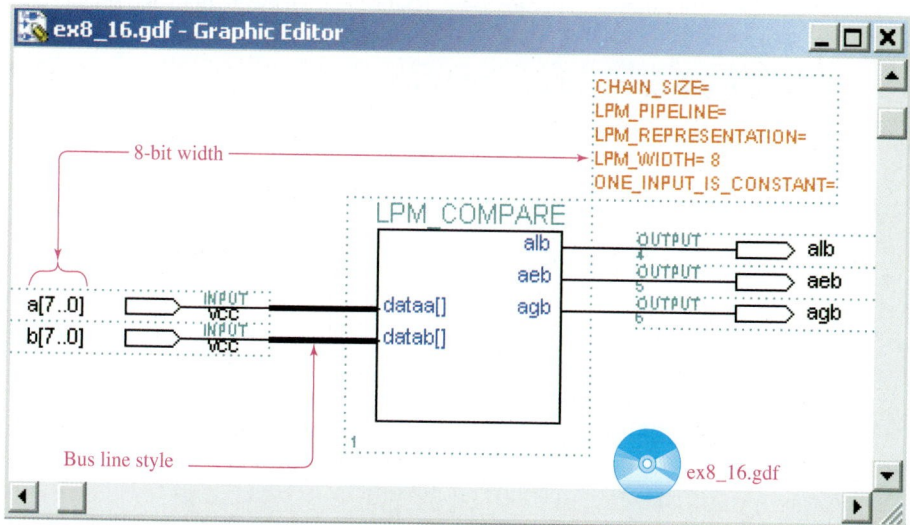

Figure 8–60   LPM module for an 8-bit comparator.

The waveform simulation file is shown in Figure 8–61. Carefully check all three outputs for each combination of inputs to verify proper operation.

SECTION 8–10 | CPLD DESIGN APPLICATIONS USING LPMS

341

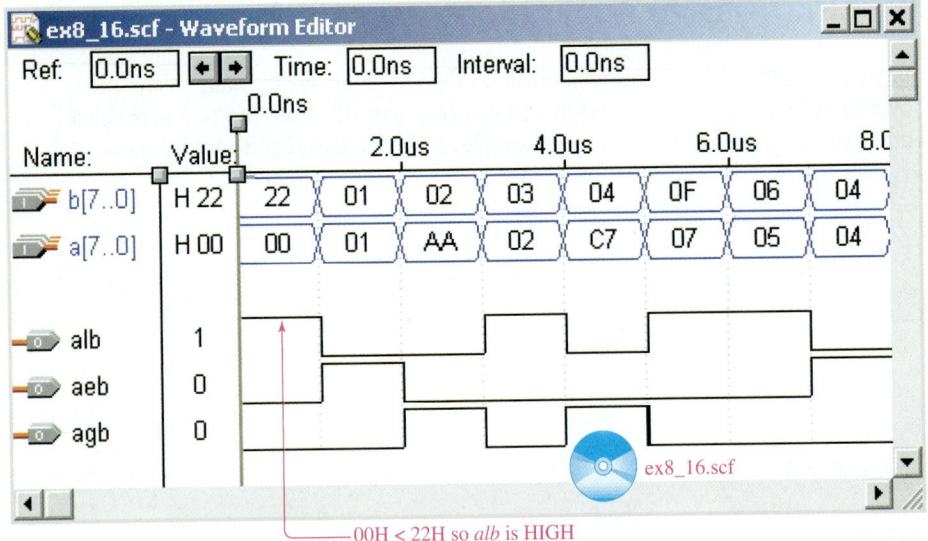

Figure 8–61 Simulation waveforms for the LPM comparator.

EXAMPLE 8–17

LPM Decoder

Create a graphic design file that uses an LPM module to form a 3-line-to-8-line (octal) decoder.

Solution: The module named LPM_DECODE is found in the mega_lpm subdirectory. To get there, choose: **File > New > Graphic Editor File.** Then double-click in the work area and select the **mega_lpm** subdirectory. Find the symbol file called LPM_DECODE. The symbol that appears is similar to that shown in Figure 8–62. Right-click on the symbol and select "edit

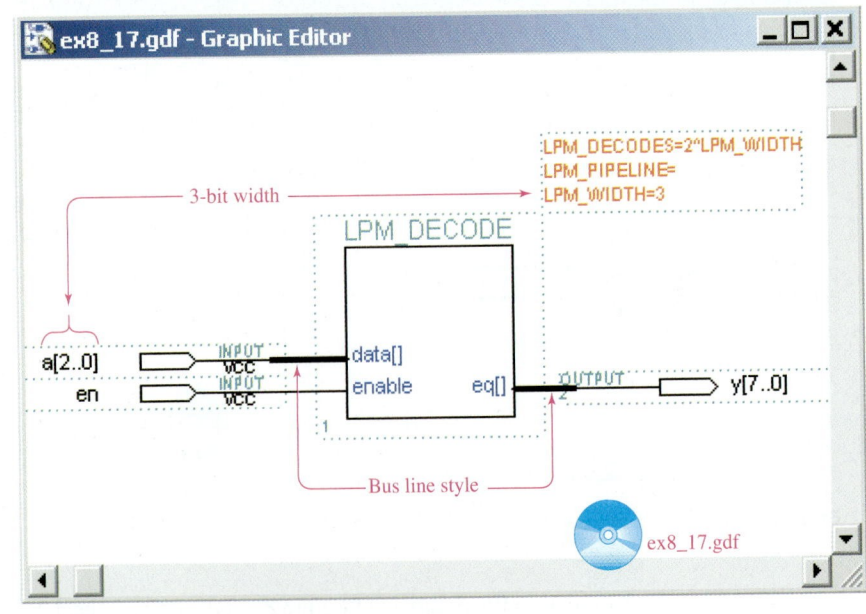

Figure 8–62 LPM module for an octal decoder.

Ports/Parameters". Then in the "Parameters" section, set the LPM_WIDTH to 3. Since inputs *a* and outputs *y* are vectors, connect them to the symbol using a bus line style as shown.

The waveform simulation file is shown in Figure 8–63. Notice that as long as *en* is HIGH, the 3-bit number on the *a* inputs will determine which *y* output goes HIGH. Also, when *en* is LOW, all outputs are kept LOW. Carefully check all seven outputs for each combination of inputs to verify proper operation.

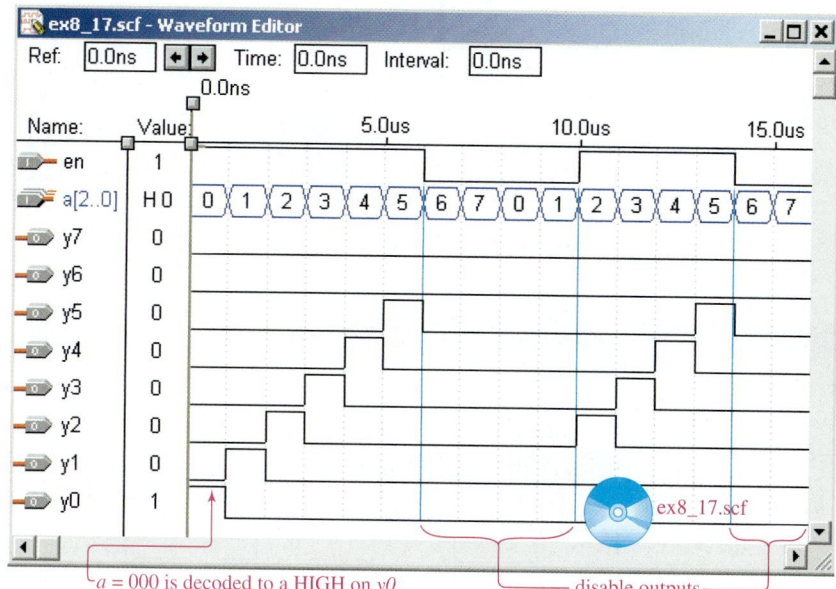

Figure 8–63 Simulation waveforms for the LPM decoder.

EXAMPLE 8–18

LPM Multiplexer

Create a graphic design file that uses an LPM module to form an 8-line multiplexer.

Solution: The module named MUX is found in the mega_lpm subdirectory. To get there choose: **File > New > Graphic Editor File.** Then double-click in the work area and select the **mega_lpm** subdirectory. Find the symbol file called MUX. (Don't confuse this with the module called LPM_MUX, which is a *bus* multiplexer.) The symbol that appears is similar to that shown in Figure 8–64. Right-click on the symbol and select "edit Ports/Parameters". Then in the "Parameters" section, set the WIDTH to 8. Since inputs *d* and *s* are vectors, connect them to the symbol using a bus line style as shown.

The waveform simulation file is shown in Figure 8–65. Several irregular waveforms were arbitrarily drawn to test the operation. For each new value of the selection bits (*s*), the corresponding *d* is routed to the *y* output. Carefully check *y* to verify that it duplicates the corresponding section of the *d* waveform for all eight values of *s*.

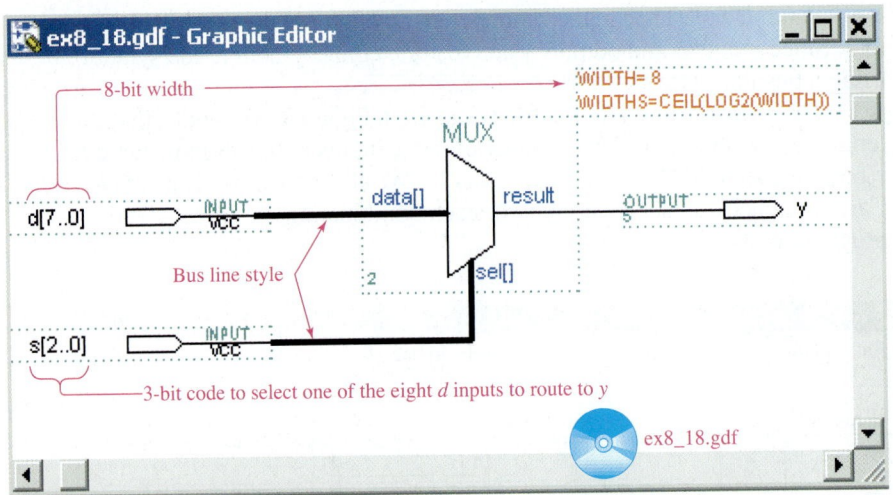

Figure 8–64   LPM module for an 8-line multiplexer.

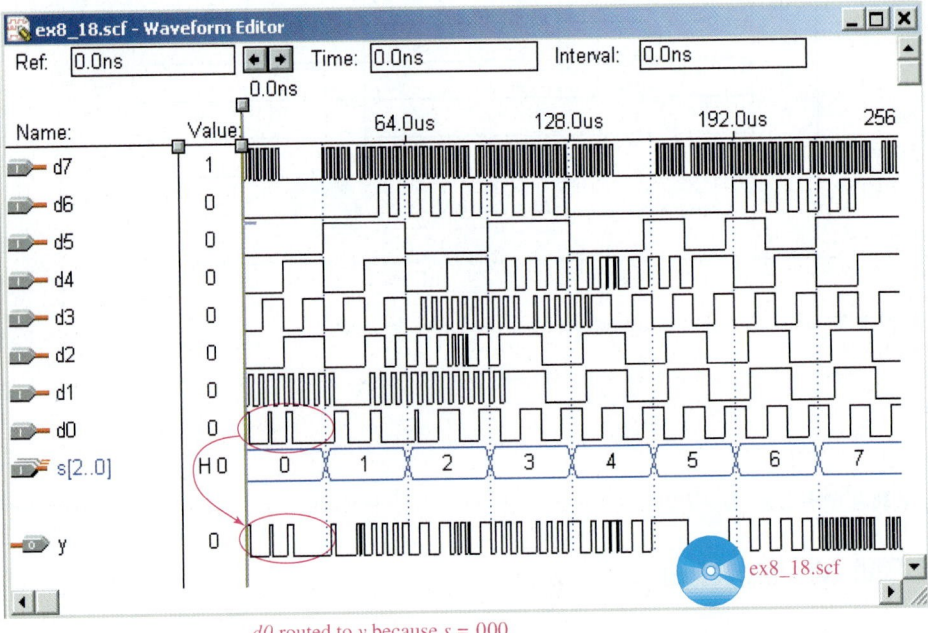

d0 routed to *y* because *s* = 000

Figure 8–65   Simulation waveforms for the LPM multiplexer.

Summary

In this chapter, we have learned that

1. Comparators can be used to determine equality or which of two binary strings is larger.

2. Decoders can be used to convert a binary code into a singular active output representing its numeric value.

3. Encoders can be used to generate a coded output from a singular active numeric input line.

4. ICs are available to convert BCD to binary and binary to BCD.

5. The Gray code is useful for indicating the angular position of a shaft on a rotating device, such as a motor.

6. Multiplexers are capable of funneling several data lines into a single line for transmission to another point.

7. Demultiplexers are used to take a single data value or waveform and route it to one of several outputs.

8. The logic functions for comparators, decoders, and multiplexers can be described in VHDL.

9. The MAX+PLUS II software provides a Library of Parameterized Modules to ease in the design of comparators, decoders, and multiplexers.

Glossary

Address: A unique binary value that is used to distinguish the location of individual memory bytes or peripheral devices.

Bidirectional: A device capable of functioning in either of two directions, thus being able to reverse its input/output functions.

Bus: A group of conductors that have a common purpose and are shared by several devices or ICs.

Code Converter: A device that converts one type of binary representation to another, such as BCD to binary or binary to Gray code.

Comparator: A device used to compare the magnitude or size of two binary bit strings or words.

Decoder: A device that converts a digital code such as hex or octal into a single output representing its numeric value.

Demultiplexer: A device or circuit capable of routing a single data-input line to one of several data-output lines; sometimes referred to as a *data distributor*.

Don't Care (×): A variable that is signified in a function table as a don't care, or ×, can take on either value, HIGH *or* LOW, without having any effect on the output.

Encoder: A device that converts a weighted numeric input line to an equivalent digital code, such as hex or octal.

Glitch: False switching seen on a waveform due to internal CPLD circuit switching delays.

Gray Code: A binary coding system used primarily in rotating machinery to indicate a shaft position. Each successive binary string within the code changes by only 1 bit.

Hardware/Software: Sometimes solutions to digital applications can be done using hardware *or* software. The *software* approach uses computer program statements to solve the application, whereas the *hardware* approach uses digital electronic devices and ICs.

Microcontroller: Sometimes referred to as "a computer on a chip," it is especially well suited for data acquisition and control applications. In a single IC package, it will contain a microprocessor, memory, I/O ports, and communication capability, among other features.

Microprocessor: A large-scale integration (LSI) IC that is the fundamental building block of a digital computer. It is controlled by software programs that allow it to do all digital arithmetic, logic, and I/O operations. It does not have the I/O control capability of a microcontroller but is better suited for executing complex software programs, such as word processing or spreadsheets.

Multiplexer: A device or circuit capable of selecting one of several data input lines for output to a single line; sometimes referred to as a *data selector.*

Priority: When more than one input to a device is active and only one can be acted on, the one with the highest priority will be acted on.

Scalar: VHDL variable types that are used to represent singular quantities usually declared as std_logic.

Superimpose: Combining two waveforms together such that the result is the sum of their levels at each point in time.

Weighting Factor: The digit within a numeric string of data is worth more or less depending on which position it is in. A weighting factor is applied to determine its worth.

Problems

Section 8–1

D
8–1. Design a binary comparator circuit using exclusive-ORs and a NOR gate that will compare two 8-bit binary strings.

8–2. Label all the lines in your design for Problem 8–1 with the digital levels that will occur when comparing $A = 1101\ 1001$ and $B = 1101\ 1001$.

8–3. Label the digital levels on all the lines in Figure 8–3 that would occur when comparing the two 8-bit strings $A = 1011\ 0101$ and $B = 1100\ 0011$.

Section 8–2

8–4. Write a two-sentence description of the function of a decoder.

8–5. Construct a truth table similar to Table 8–1 for an active-LOW output BCD (1-of-10) decoder.

8–6. What state must the inputs, $\overline{E_1}$, $\overline{E_2}$, and E_3 be in to *enable* the 74138 decoder?

8–7. What does the $\times$ signify in the function table for the 74138?

8–8. Describe the difference between active-LOW outputs and active-HIGH outputs.

8–9. Sketch the output waveforms ($\overline{0}$ to $\overline{7}$) given the inputs shown in Figure P8–9(b) to the 74138 of Figure P8–9(a).

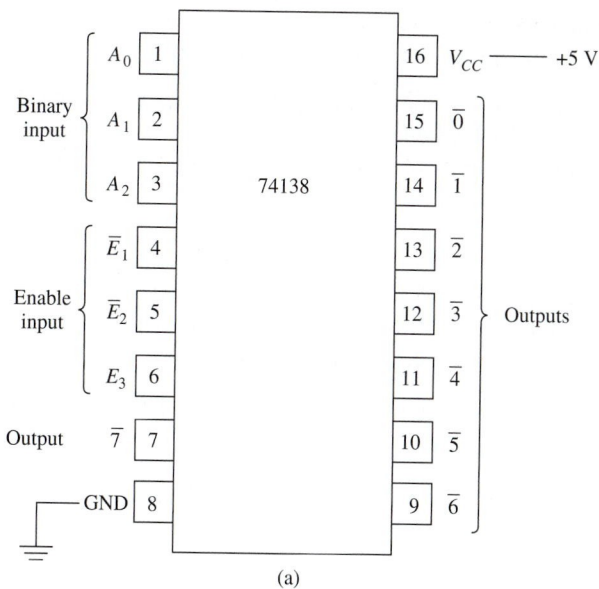

(a)

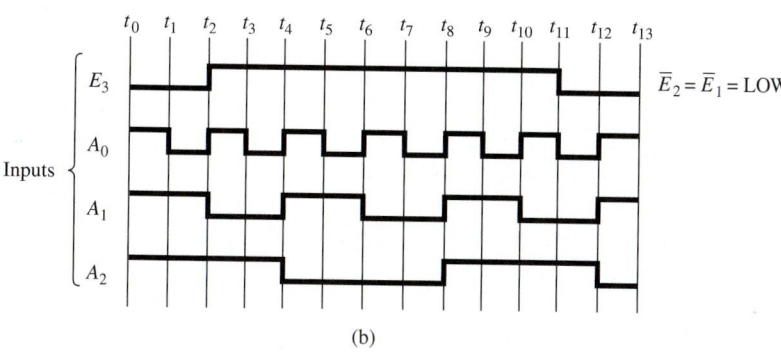

$\overline{E}_2 = \overline{E}_1 = LOW$

Inputs

(b)

Figure P8–9

8–10. Repeat Problem 8–9 for the input waveforms shown in Figure P8–10.

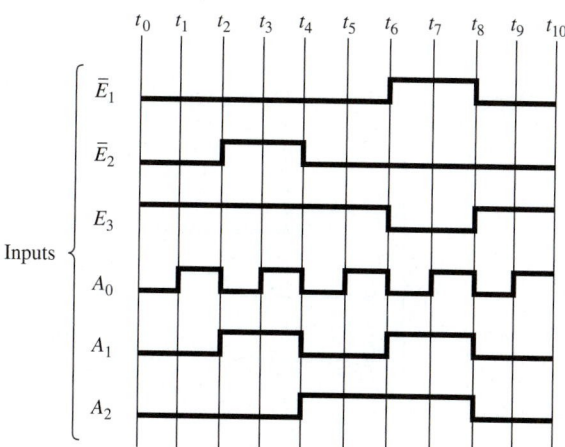

Inputs

Figure P8–10

8–11. What state do the outputs of a 7442 BCD decoder go to when an invalid BCD number (10 to 15) is input to A_0 to A_3?

D   **8–12.** Design a circuit, based on a 74154 4-line-to-16-line decoder, that will output a HIGH whenever the 4-bit binary input is greater than 12. (When the binary input is less than or equal to 12, it will output a LOW.)

Section 8–3

8–13. With the 74147 priority encoder, if two different decimal numbers are input at the same time, which will be encoded?

8–14. A 74147 is connected with $\overline{I_1} = \overline{I_2} = \overline{I_3} = $ LOW and $\overline{I_4} = \overline{I_5} = \overline{I_6} = \overline{I_7} = \overline{I_8} = \overline{I_9} = $ HIGH. Determine $\overline{A_0}, \overline{A_1}, \overline{A_2},$ and $\overline{A_3}$.

8–15. Sketch the output waveforms $(\overline{A_0}, \overline{A_1}, \overline{A_2}, \overline{EO}, \overline{GS})$ given the inputs shown in Figure P8–15(b) to the 74148 of Figure P8–15(a).

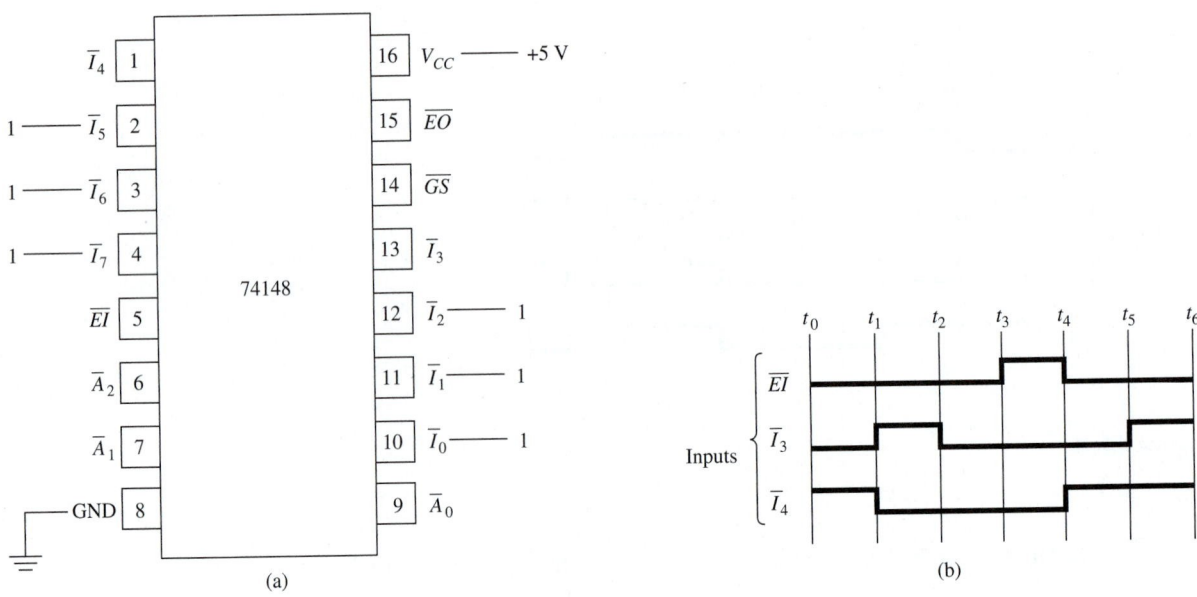

Figure P8–15

8–16. Repeat Problem 8–15 for the waveforms shown in Figure P8–16.

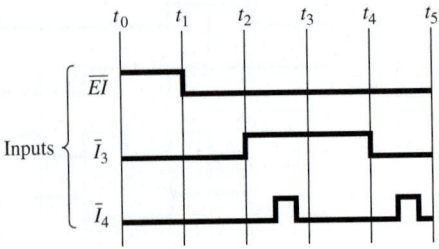

Figure P8–16

     CHAPTER 8 | CODE CONVERTERS, MULTIPLEXERS, AND DEMULTIPLEXERS

C   **8–17.** Two 74148s are connected in Figure P8–17 to form an active-LOW input, active-LOW output hexadecimal (16-line-to-4-line) priority encoder. Show the logic levels on each line in Figure P8–17 for encoding an input hexadecimal C (12) to an output binary 1100 (active-LOW 0011).

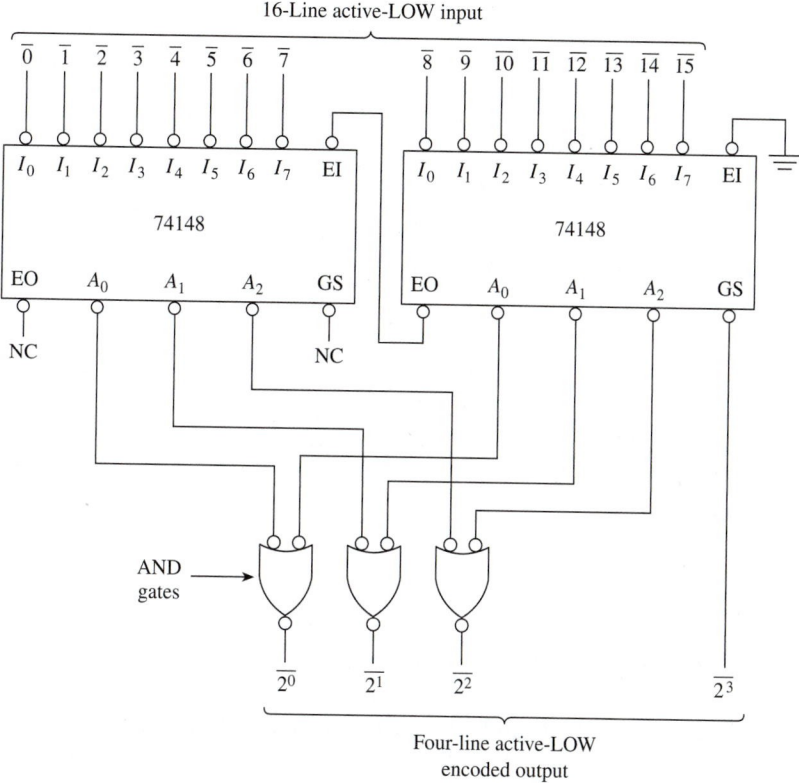

Figure P8–17

C   **8–18.** Repeat Problem 8–17 for encoding an input hexadecimal 6 to an output binary six (active-LOW 1001).

Section 8–4

8–19. Using the weighting factors given in Figure 8–31, convert the following decimal numbers to BCD and then to binary.

(a) 32   (c) 55

(b) 46   (d) 68

C   **8–20.** Figure P8–20 is a two-digit BCD-to-binary converter. Show how the number 49 ($0100\ 1001_{BCD}$) is converted to binary by placing 1's and 0's at the inputs and outputs.

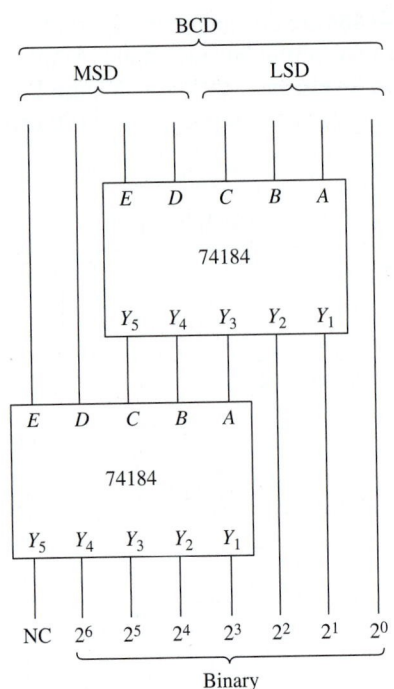

Figure P8-20

C

8–21. Repeat Problem 8–20 for the number 73.

8–22. Convert the following Gray codes to binary using the circuit of Figure 8–38.

(a) 1100   (c) 1110
(b) 0101   (d) 0111

8–23. Convert the following binary numbers to Gray code using the circuit of Figure 8–37.

(a) 1010   (c) 0011
(b) 1111   (d) 0001

Section 8–5

8–24. The connections shown in Figure P8–24 are made to the 74151 8-line multiplexer. Determine Y and $\overline{Y}$.

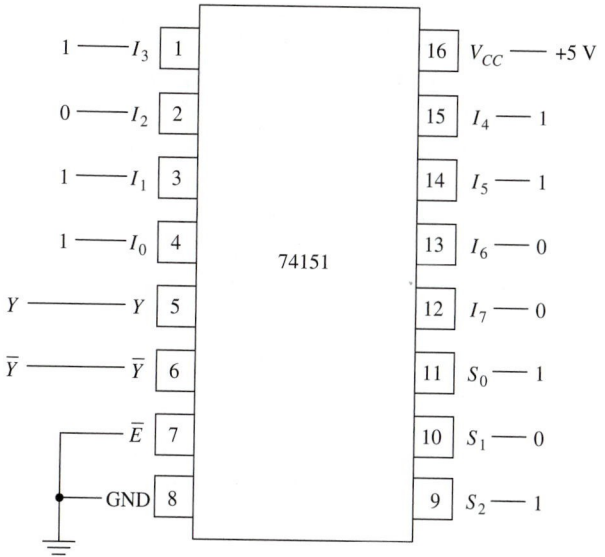

Figure P8–24

C D   **8–25.** Using a technique similar to that presented in Figure 8–46, design a 32-bit multiplexer using four 74151's.

C D   **8–26.** Design a circuit that will output a LOW whenever a month has 31 days. The month number (1 to 12) is input as a 4-bit binary number (January = 0001, and so on). (*Hint:* Use a 74150.)

Section 8–6

C D   **8–27.** Design an 8-bit demultiplexer using one 74139.

C D   **8–28.** Design a 16-bit demultiplexer using two 74138's.

T   **8–29.** There is a malfunction in a digital system that contains several multiplexer and demultiplexer ICs. A reading was taken at each pin with a logic probe, and the results were recorded in Table 8–7. Which IC or ICs are not working correctly?

Section 8–7

8–30. Which memory bank is accessed in Figure 8–55 for each of the following hex addresses output by the microprocessor?

(a) 3000H   (c)   507CH

(b) 6000H   (d)   8001H

TABLE 8-7	IC Logic States for Troubleshooting Problem 8–29						
74150		**74151**		**74139**		**74154**	
Pin	Level	Pin	Level	Pin	Level	Pin	Level
1	0	1	1	1	0	1	1
2	1	2	0	2	1	2	1
3	1	3	0	3	0	3	1
4	0	4	1	4	0	4	1
5	1	5	1	5	1	5	1
6	0	6	0	6	1	6	1
7	1	7	0	7	1	7	1
8	1	8	0	8	0	8	1
9	0	9	0	9	0	9	1
10	0	10	0	10	0	10	1
11	0	11	0	11	1	11	1
12	0	12	0	12	0	12	0
13	1	13	1	13	1	13	1
14	1	14	0	14	0	14	1
15	1	15	0	15	1	15	1
16	0	16	1	16	1	16	1
17	1					17	1
18	0					18	1
19	1					19	0
20	1					20	0
21	1					21	0
22	0					22	1
23	1					23	1
24	1					24	1

8–31. A logic analyzer was used to monitor the 19 microprocessor lines shown in Figure 8–55. Describe the operation that was taking place if the following levels were observed from top to bottom.

(a) 100 0101 0000 0000 0011

(b) 010 0110 0000 1100 0111

8–32. What hex number will be read by port 1 of the 8051 microcontroller in Figure 8–56 if the fluid level is high in the following chemical tanks?

(a) Tank 1

(d) All tanks at a high fluid level

(b) Tank 6

(e) No tanks at a high fluid level

(c) Tanks 2 and 7

8–33. What hex code must be output by port 1 of the 8051 in Figure 8–57 to perform the following operations?

(a) Read from data terminal 5

(b) Read from data terminal 7

(c) Write to data terminal 2

(d) Write to data terminals 1 and 2 at the same time

8–34. Determine the voltage level of the output voltage signal in Figure 8–58 when the binary count reaches 110 (6_{10}).

8–35. What are the output levels at $\overline{0}, \overline{1}, \overline{2}$, and $\overline{3}$ of the demultiplexer in Figure 8–59 when the circuit is displaying the number 3 in the LSD position?

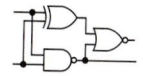

Schematic Interpretation Problems

See Appendix G for the schematic diagrams.

S **8–36.** Find the two 4-bit-magnitude comparators, U7 and U8, in the Watchdog Timer schematic. Which IC receives the high-order binary data, U7 or U8? (*Hint:* The bold lines in that schematic represent a *bus,* which is a group of conductors that are shared by several ICs. It simplifies the diagram by showing a single bold line instead of several separate lines. When the individual lines are taken off the bus they are labeled, appropriately, 0–1–2–3 and 4–5–6–7 in this application.)

S **8–37.** Where is the final output of the comparison made by U7, U8 used in the Watchdog Timer schematic?

S **8–38.** Find the octal decoder U5 in the HC11D0 schematic. Determine the levels on AS, AD13, AD14, and AD15 required to provide an active-LOW signal on the line labeled MON_SL.

S C **8–39.** Locate the address decoder section (U3:C, U5, and U9) in the HC11D0 schematic. The octal decoder (U9) is used to determine if the LCD (LCD_SL) or the keyboard (KEY_SL) is to be active.

Determine the levels on AD3-5, AD11-15, and AS required to select the LCD.

Repeat for selecting the keyboard.

S C **8–40.** Find the decoders U28 and U29 on sheet 2 of the 4096/4196 schematic. They are cascaded together to form a 1-of-18 decoder for the lines labeled ICS1–ICS18.

Determine the levels on pins 2, 5, 6, 9, and 12 of U31 to provide an active-LOW output at ICS5.

Repeat for ICS18.

S C D **8–41.** Design a circuit using a 74151 multiplexer that will allow you to input the status of the four temperatures and four pressures in the chemical monitoring system in Figure 1–6 (Chapter 1). Assume that the 68HC11 microcontroller in the HC11D0 schematic will be used to read each of the status bits, one at a time (P_D, then T_D, then P_C, and so on). Also assume that the only input and output pins available for use are at port D, bits PD2, PD3, PD4, and PD5.

Electronics Workbench®/MultiSIM® Exercises

D **E8–1.** Load the circuit file for **Section 8–1a.** This is a binary comparator similar to Figure 8–1. Test its operation by observing the Equality Indicator status for various 4-bit inputs at the A-inputs and B-inputs.

(a) What input conditions are required for the indicator to come ON?

(b) Redesign the circuit using Ex-ORs instead of Ex-NORs. What must the AND gate be changed to to get the same output result? Test your design with several different input combinations.

C D **E8–2.** Load the circuit file for **Section 8–1b.** Design a comparator that determines equality and, if not equal, also tells us which input is larger. This new comparator will have eight inputs (A0-A3, B0-B3) and three outputs (A = B, A > B, A < B). For example, if the A-inputs equal the B-inputs, then the A = B output goes HIGH and the other outputs go LOW, and so on. The key component in your design should be the 7483 connected as a binary subtractor similar to the circuit designed in the file "sec 7–8a" (or Figure 7–21 in the textbook). Before attempting to complete the design determine the levels at the outputs of a binary subtractor for inputs where A = B, then A > B, and then A < B. What is the level of the five outputs of the subtractor for the following three conditions:

(a) A = 7, B = 7

(b) A = 7, B = 5

(c) A = 5, B = 7?

Now you should realize what external gating is required to implement your design. Build the circuit on the screen, and test your design. Demonstrate your working design to your teacher.

D

E8–3. Load the circuit file for **Section 8–1c.** This circuit simulates a control circuit that is used to fill a chemical tank. Design a circuit using a binary comparator to turn on a buzzer when the tank reaches level 5. Demonstrate your design to your teacher.

E8–4. Load the circuit file for **Section 8–2a.** The 74138 shown is a 3-line-to-8-line decoder. The Word Generator is used to exercise the chip by outputting a binary count from 000 to 111 repeatedly to the A, B, C inputs (A = LSB). The two active-LOW enables are grounded, but the active-HIGH enable (G1) is connected to a switch.

(a) What position (up, down) must the switch be in for the chip to be enabled?

(b) If the binary input to A, B, C is 101(5) what is the level of the eight output LEDs if the enable switch is up?

(c) Repeat **(b)** for the enable switch down. Turn the power switch ON, and check your answers.

E8–5. Load the circuit file for **Section 8–2b.** In this exercise the Logic Analyzer will be used to observe the output waveforms of the 74138 similar to those seen in Figure 8–13.

(a) The HIGH-order bit of the Word Generator is used to create the waveform that is used for the G2A′ enable input on the 74138. That active-LOW enable input is displayed as the top trace. Describe the state of the outputs when it is HIGH.

(b) Describe what you would expect to observe if that signal was used to drive the active-HIGH enable input (G1) instead. Try it.

(c) Expand the Word Generator, and study the four-digit hex words that are to be output at the 16 binary output pins. Five of the entries begin with an 8 instead of 0. Why?

(d) How would you modify the four-digit hex words in the Word Generator to disable the chip whenever the #7 is to be output. Test your answer.

C D

E8–6. Load the circuit file for **Section 8–2c.** This circuit simulates a control circuit that is used to fill a chemical tank. Design a circuit using the 7442 BCD-to-decimal decoder and other logic gates to turn on a buzzer when:

(a) The tank is filled to the middle levels 4 or 5 or 6.

(b) The tank exceeds level 6 or drops below level 3.

(c) When the tank overflows (beyond level 9). Demonstrate your design to your teacher.

D **E8–7.** Load the circuit file for **Section 8–3a.** The circuit shown is a decimal counter. Connect a 74147 decimal-to-BCD encoder so that it displays on LEDs the corresponding BCD code of the selected decimal LED. Connect a seven-segment display to the BCD LEDs to observe the ten decimal digits. Demonstrate the working design to your teacher.

C D **E8–8.** Load the circuit file for **Section 8–3b.** This circuit simulates a chemical processing plant. To add chemicals to a tank, press the corresponding key. For example, to increase the level in Tank 0, press 0 repeatedly. When it reaches a dangerous level near the top, a warning LED comes on.

(a) Design a circuit that sounds a warning buzzer if any of the tanks are too high.

(b) Add a seven-segment display to your design of part (a) that displays the decimal number of the tank that went too high. (*Hint:* You will need an encoder to achieve this). Demonstrate your design to your teacher.

E8–9. Load the circuit file for **Section 8–5a.** This is a four-line multiplexer similar to Figure 8–42. By moving switches S1 and S0 you can pick which square wave frequency is to be routed to the output.

(a) The Output Data LED will flash fastest for S1 = _____, S0 = _____?

(b) The Output Data LED will flash slowest for S1 = _____, S0 = _____? Try it.

C T **E8–10.** Load the circuit file for **Section 8–5b.** This is a four-line multiplexer similar to Figure 8–42. By moving switches S1 and S0, you can pick which square wave frequency is to be routed to the output. As you test the four different combinations of switch settings, there should be four different LED flashing speeds, but there are not.

(a) Describe the problem that you observe when you test the four combinations of switch settings.

(b) Troubleshoot the circuit with a logic probe, and list the fault that you find. Fix it, and test your corrected circuit.

C T **E8–11.** Load the circuit file for **Section 8-5c.** This is a four-line multiplexer similar to Figure 8–42. By moving switches S1 and S0, you can pick which square wave frequency is to be routed to the output. As you test the four different combinations of switch settings, there should be four different LED flashing speeds, but there are not.

(a) Describe the problem that you observe when you test the four combinations of switch settings.

(b) Troubleshoot the circuit with a logic probe, and list the fault that you find. Fix it, and test your corrected circuit.

C **E8–12.** Load the circuit file for **Section 8–6a.** The Demultiplexer is capable of distributing the Input Data Signal to one of the eight possible Output Destination LEDs shown by selecting the proper settings on the Data Select Switches.

(a) The switches 2,1,0 should be set to _____, _____, _____ to route the Input Data Signal to the 6' output. Make the necessary connections, and test your answer.

(b) The Input Data Signal has a 25% duty cycle, which means that it is HIGH for 25% of the period. Do you think that the signal coming out of the 6' terminal will still be a 25% duty cycle (Yes or No)? Check your answer by connecting Channel-1 of the oscilloscope to the Input Data Signal and Channel-2 to the 6′ output of the DeMux to observe the two signals.

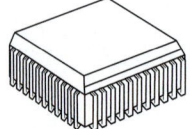

CPLD Problems

C8–1. The VHDL program in Figure 8–4 is the implementation of an 8-bit comparator.

(a) Make the necessary changes to make it a 4-bit comparator. Save this program as *prob_c8_1.vhd.*

(b) Test its operation by creating waveform simulations comparing the magnitude of the following hex numbers: a = F, A, 8, 2, 0, D, 5, 1 compared to: b = D, F, A, 2, 0, 9, 6, 2.

(c) Download your design to a CPLD IC. Discuss your observations of the *alb, aeb,* and *agb* output LEDs with your instructor as you use the switches to step through all eight input combinations.

C8–2. The VHDL program in Figure 8–17(a) is an octal decoder implemented with Boolean equations. It has active-HIGH inputs and outputs.

(a) Make the necessary changes to make it an active-LOW output decoder. Save this program as *prob_c8_2.vhd.*

(b) Test its operation by creating waveform simulations that monitor all eight outputs as the *a* inputs count from 000 up to 111.

(c) Download your design to a CPLD IC. Discuss your observations of the eight output LEDs with your instructor as you use the switches to step through all eight input combinations.

C8–3. The VHDL program in Figure 8–18 is an octal decoder implemented with selected signal assignments. It has active-HIGH inputs and outputs.

(a) Make the necessary changes to make it an active-LOW output decoder. Save this program as *prob_c8_3.vhd.*

(b) Test its operation by creating waveform simulations that monitor all eight outputs as the *a* inputs count from 000 up to 111.

(c) Download your design to a CPLD IC. Discuss your observations of the eight output LEDs with your instructor as you use the switches to step through all eight input combinations.

C8–4. The VHDL program in Figure 8–19 is an octal decoder with an active-HIGH enable (*en*) input. It has active-HIGH *a*-inputs and *y*-outputs.

(a) Make the necessary changes to make the enable input active-LOW. Save this program as *prob_c8_4.vhd.*

(b) Test its operation by creating waveform simulations that monitor all eight outputs as the *a* inputs count from 000 up to 111 with *en* LOW. Repeat for *en* HIGH.

(c) Download your design to a CPLD IC. Discuss your observations of the eight output LEDs with your instructor as you use the switches to step through all 16 input combinations.

C8–5. The VHDL program in Figure 8–20(a) is an octal decoder with an active-HIGH enable (*en*) input. It has active-HIGH *a*-inputs and *y*-outputs and uses an IF-THEN-ELSE statement to determine if the enable input is satisfied.

(a) Make the necessary changes to make it function exactly like a 74138 fixed-function IC (three enables and eight active-LOW outputs). Save this program as *prob_c8_5.vhd*.

(b) Test its operation by creating waveform simulations that monitor all eight outputs as the *a* inputs count from 000 up to 111 with various combinations of *en1, en2,* and *en3*.

(c) Download your design to a CPLD IC. Discuss your observations of the eight output LEDs with your instructor as you use the switches to step through all eight input combinations with the enables satisfied (001). Then with the *y7* output active, demonstrate various ways to disable/enable it with the three *en* lines.

C8–6. The VHDL program in Figure 8–29 is an octal priority encoder implemented with conditional signal assignments. It has active-HIGH *i*-inputs and *a*-outputs.

(a) Make the necessary changes to make the inputs active-LOW. Save this program as *prob_c8_6.vhd*.

(b) Test its operation by creating waveform simulations that monitor the three *a* outputs as individual, then multiple, *i* inputs go LOW.

(c) Download your design to a CPLD IC. Discuss your observations of the three output LEDs with your instructor as you use the switches to step through several combinations of individual and multiple active inputs.

C8–7. The VHDL program in Figure 8–48 is a 4-line multiplexer implemented with selected signal assignments.

(a) Make the necessary changes to make it an 8-line multiplexer. Save this program as *prob_c8_7.vhd*.

(b) A waveform simulator for an 8-bit multiplexer was developed in Figure 8–65 for Example 8–18. This simulator file is on the textbook CD named *ex8_18.scf*. Test your 8-bit multiplexer design using those waveforms. (*Hint:* After loading *ex8_18.scf* from the CD, if must be saved as *prob_c8_7.scf* to match up with the name of your vhd file.)

C8–8. The graphic design file in Figure 8–60 is an 8-bit comparator implemented with an LPM.

(a) Add additional gating to the circuit to provide a HIGH output if the two input strings are equal and odd. Call this new output *oddeq*. Do the same for equal and even (called *eveneq*). Save this program as *prob_c8_8.gdf*. (*Hint:* the LSB member of the *a*-input group can be identified as *a0* in the Graphic Editor and is HIGH for odd numbers. Figure 6–18 shows how a line can be labeled as an individual member of a group.)

(b) Test its operation by creating waveform simulations that monitor all five outputs as you compare the following eight pairs of numbers. a = 11, 99, AA, 25, A7, 2A, 24, 88; b = 11, 88, AA, 26, A7, 2F, 22, 88.

C8–9. The graphic design file in Figure 8–62 is an octal decoder implemented with an LPM.

(a) Add additional gating to the circuit to make it function exactly like a 74138 fixed-function IC with active-LOW outputs and a triple-enable input. The outputs can be made active-LOW in the *edit ports/parameters* section of the LPM. Save this program as *prob_c8_9.gdf*.

(b) Test its operation by creating waveform simulations that monitor all eight outputs as the *a* inputs count from 000 up to 111 with various combinations of *en1, en2,* and *en3.*

(c) Download your design to a CPLD IC. Discuss your observations of the eight output LEDs with your instructor as you use the switches to step through all eight input combinations with the enables satisfied (001). Then with the *y7* output active, demonstrate various ways to disable/enable it with the three *en* lines.

C8–10. The graphic design file in Figure 8–64 is an 8-line multiplexer implemented with an LPM.

(a) Make the necessary changes to make it a 4-line multiplexer. Save this program as *prob_c8_10.gdf.*

(b) Test its operation by creating waveform simulations that alternately route each of the four *d* inputs to the *y* output.

Answers to Review Questions

8–1. False

8–2. $A < B = 1$

8–3. 4 inputs, 10 outputs

8–4. False

8–5. False

8–6. False

8–7. HIGH, LOW

8–8. An encoder creates a coded output from a singular active numeric input line. A decoder is the opposite, creating a single output from a numeric input code.

8–9. The highest numeric input

8–10. (a) All HIGH (b) $A_3 = L$, $A_2 = H, A_1 = H, A_0 = L$

8–11. A_0, A_1, A_2: binary outputs; EO: enable output; GS: group signal output. All outputs are active LOW.

8–12. 80

8–13. 6

8–14. Because it varies by only 1 bit when the shaft is rotated from one position to the next

8–15. Because it selects which data input is to be sent to the data output

8–16. Because it takes a single input data line and routes it to one of several outputs

8–17. They select which one of the input lines is sent to the output.

8–18. They select which one of the output lines the input data are sent to.

8–19. False

8–20. 7400

8–21. 0, 1, 0

8–22. $\overline{GS}$ goes LOW.

8–23. False

8–24. True

8–25. False

8–26. It outputs a binary progression, allowing each of the Y_0 to Y_7 inputs to appear at the Z output in a staircase fashion.

8–27. The number 5 is loaded into the MSD register. When the A_0 and A_1 lines of the 74139 reach 1–1, the number 3 line outputs a LOW, enabling the MSD register and the MSD of the display. The number 5 is then transferred to the display by the decoder/driver circuitry.

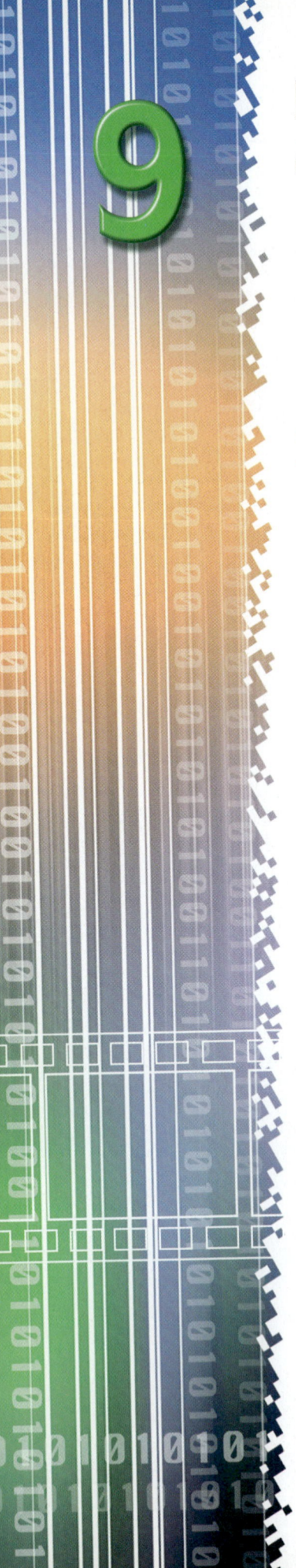

Logic Families and Their Characteristics

OBJECTIVES

Upon completion of this chapter, you should be able to:

- Analyze the internal circuitry of a TTL NAND gate for both the HIGH and LOW output states.
- Determine IC input and output voltage and current ratings from the manufacturer's data manual.
- Explain gate loading, fan-out, noise margin, and time parameters.
- Design wired-output circuits using open-collector TTL gates.
- Discuss the differences and proper use of the various subfamilies within both the TTL and CMOS lines of ICs.
- Describe the reasoning and various techniques for interfacing between the TTL, CMOS, and ECL families of ICs.

INTRODUCTION

Integrated-circuit logic gates (small-scale integration, SSI), combinational logic circuits (medium-scale integration, MSI), and microprocessor systems (large-scale integration and very large-scale integration, LSI and VLSI) are readily available from several manufacturers through distributors and electronic parts suppliers. Basically, there are three commonly used families of digital IC logic: **TTL** (transistor–transistor logic), **CMOS** (complementary metal oxide semiconductor), and **ECL** (emitter-coupled logic). Within each family, several, subfamilies (or series) of logic types are available, with different ratings for speed, power consumption, temperature range, voltage levels, and current levels.

Fortunately, the different manufacturers of digital logic ICs have standardized a numbering scheme so that basic part numbers are the same regardless of the manufacturer. The prefix of the part number, however, will differ because it is the manufacturer's abbreviation. For example, a typical TTL part number might be S74F08N. The 7408 is the basic number used by all manufacturers for a quad *AND* gate. The F stands for the *FAST* TTL subfamily, and the S prefix is the manufacturer's code for Signetics. National Semiconductor uses the prefix DM, and Texas Instruments uses the prefix SN. The N suffix at the end of the part number is used to specify the package type. N is used for the plastic dual-in-line (DIP), W is used for the ceramic flatpack, and D is used for the surface-mounted SO plastic package. The best sources of information on available package styles and their dimensions are the manufacturers' data manuals. Most data manuals list the 7408 as 5408/7408. The 54XX series is the military version, which has less stringent power supply requirements and an extended temperature range of $-55°$ to $+125°C$, whereas the 74XX is the commercial version, with a temperature range of $0°$ to $+70°C$ and strict power supply requirements.

For the purposes of this text, reference is usually made to the 74XX commercial version, and the manufacturer's prefix code and package-style suffix code are ignored. The XX is used in this book to fill the space normally occupied by the actual part number. For example, the part number for an inverter in the 74XX series is 7404.

9–1   The TTL Family

The standard 74XX TTL IC family has evolved through several stages since the late 1960s. Along the way, improvements have been made to reduce the internal time delays and power consumption. At the same time, each manufacturer has introduced chips with new functions and applications.

The fundamental operation of a TTL chip can be explained by studying the internal circuitry of the basic two-input 7400 NAND gate shown in Figure 9–1. (The

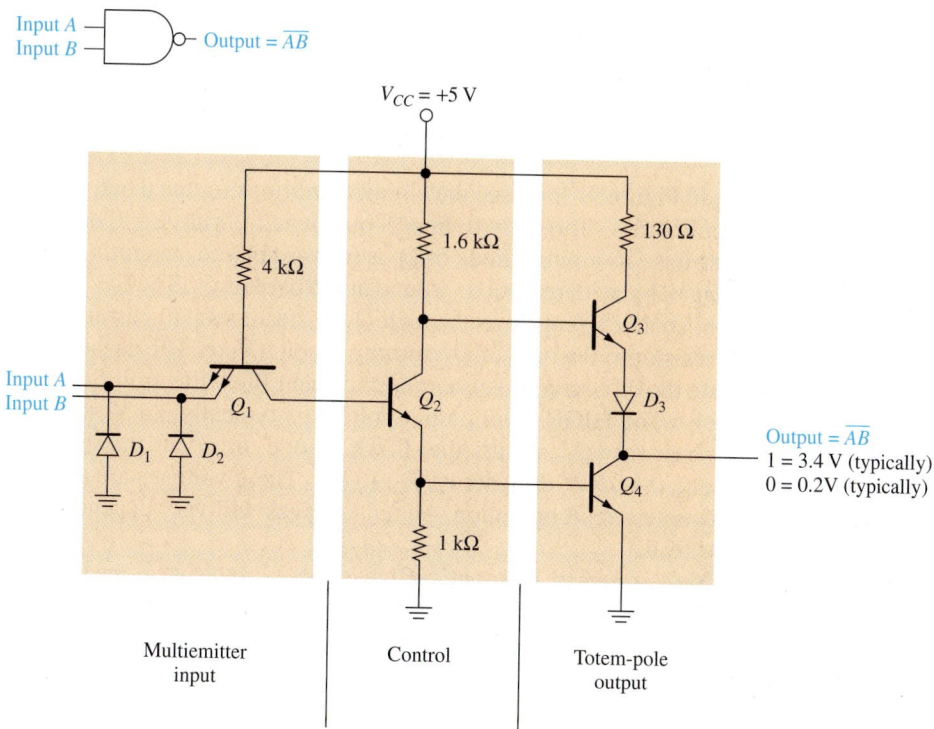

Figure 9–1   Internal circuitry of a 7400 two-input NAND gate.

NAND is the simplest of the gates, requiring the least amount of circuitry to implement.) The diodes D_1 and D_2 are negative clamping diodes used to protect the inputs from any short-term negative input voltages. The input transistor, Q_1, acts like an AND gate and is usually fabricated with a *multiemitter* transistor, which characterizes TTL technology. (To produce two-, three-, four-, and eight-input NAND gates, the manufacturer uses two-, three-, four-, and eight-emitter transistors.) Q_2 provides control and current boosting to the totem-pole output stage.

The reasoning for the totem-pole setup was discussed in Chapter 2. Basically, when the output is HIGH (1), Q_4 is OFF (open) and Q_3 is ON (short). When the output is LOW (0), Q_4 is ON and Q_3 is OFF. Because one or the other transistor is always OFF, the current flow from V_{CC} to ground in that section of the circuit is minimized.

To study the operation of the circuit in more detail, let's first review some basic electronics. An *NPN* transistor is basically two diodes: a *P* to *N* from base to emitter, and another *P* to *N* from base to collector, as shown in Figure 9–2. The base-to-emitter diode is forward biased by applying a positive voltage on the base with respect to the emitter. A forward-biased base-to-emitter diode will have 0.7 V across it and cause the collector-to-emitter junction to become almost a short circuit, with approximately 0.3 V across it.

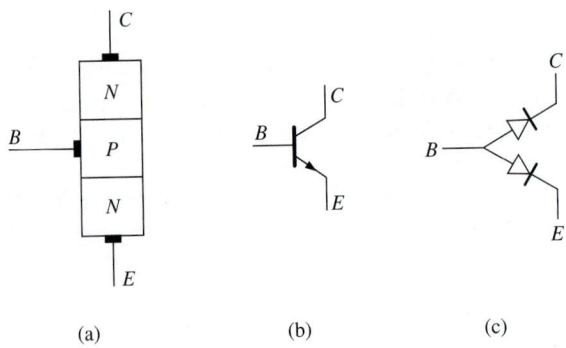

(a) (b) (c)

Figure 9–2 *NPN* transistor: (a) physical layout; (b) symbol; (c) diode equivalent.

Now, referring to Figure 9–3, we see the circuit conditions for the 0 output state and 1 output state. In Figure 9–3(a) ($A = 0$, $B = 0$, output $= 1$), with $A = 0$ or $B = 0$ or both equal to 0, the base-to-emitter diode of Q_1 is forward biased, saturating (turning on) Q_1 and placing 0.3 V with respect to ground at the base of Q_2. The 0.3 V is not enough to turn Q_2 on, so no current flows through Q_2; instead, a small current flows through the 1.6-kΩ resistor to the base of Q_3, turning Q_3 on. (*Note:* The dashed lines in Figure 9–3 indicate the direction of *conventional* current flow. Electron flow is in the opposite direction.) The HIGH-level output voltage is typically 3.4 V, which is the 4.8 V at the base of Q_3 minus the 0.7-V diode drop at the base-to-emitter diode of Q_3 and the 0.7-V drop across D_3. (*Note:* These voltages are approximations used to illustrate circuit operation. Actual voltages will vary depending on the connected output load.)

In Figure 9–3(b) ($A = 1$, $B = 1$, output $= 0$), with $A = 1$ *and* $B = 1$, the base-to-emitter diode of Q_1 is reverse biased, *but* the base-to-collector diode [see Figure 9–2(c)] of Q_1 is forward biased. Current will flow down through the base to collector of Q_1, turning Q_2 on with a positive base voltage and turning Q_4 on with a positive base voltage. The output voltage will be approximately 0.3 V. Q_3 is kept off because there is not enough voltage between the base of Q_3 (1.0 V) to the cathode of D_3 (0.3 V) to overcome the two 0.7-V diode drops required to allow current flow.

CHAPTER 9 | LOGIC FAMILIES AND THEIR CHARACTERISTICS

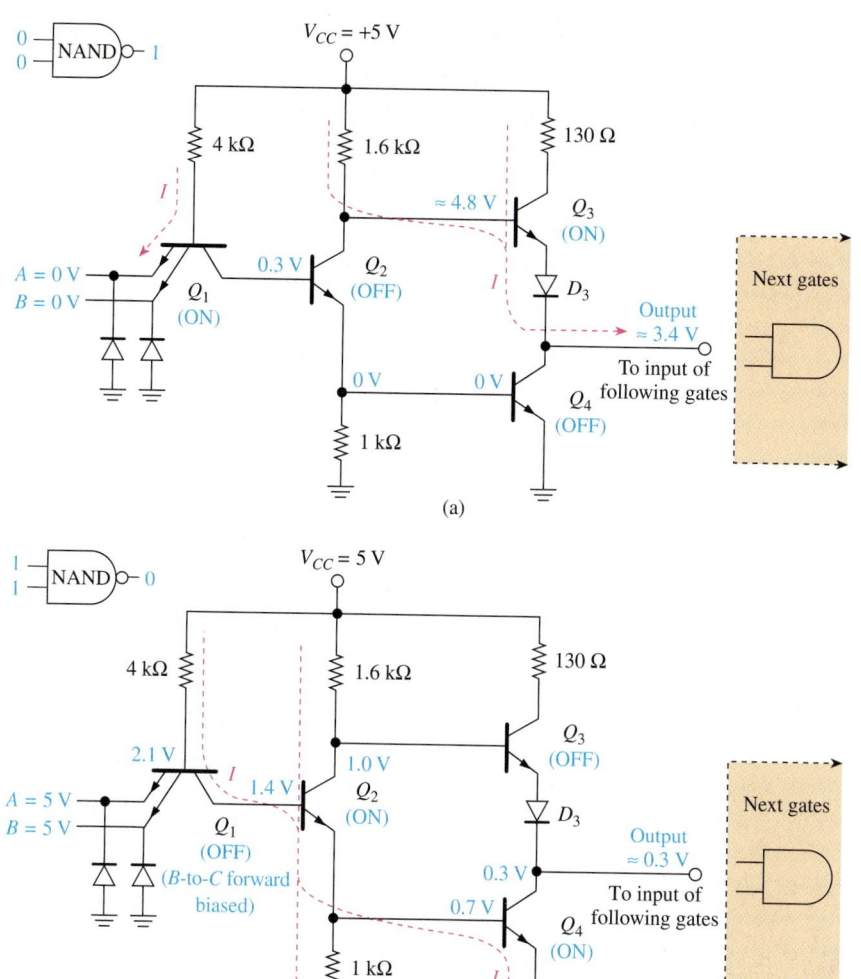

Team Discussion

Discuss why V_{out} drops as gate loads are added to the output of Figure 9–3(a).

Common Misconception

You have probably never seen a transistor on its side, like Q_1. Just realize that it is still turned ON the same way, by applying a positive current flow from base to emitter.

Team Discussion

Think about what happens to the voltage level and current demand on the V_{CC} supply at the switching point if both Q_3 and Q_4 are momentarily ON at the same time.

Figure 9–3 Equivalent circuits for a TTL NAND in the (a) HIGH output state; (b) LOW output state (I = conventional current flow).

Review Questions

9–1. The part number for a basic logic gate varies from manufacturer to manufacturer. True or false?

9–2. The input signal to a TTL NAND gate travels through three stages of internal circuitry: *input, control,* and _____.

9–3. A forward-biased NPN transistor will have approximately _____ volts across its base–emitter junction and _____ volts across its collector–emitter junction.

9–2   TTL Voltage and Current Ratings

Basically, we like to think of TTL circuits as operating at 0- and 5-V levels, but, as you can see in Figure 9–3, that just is not true. As we draw more and more current out of the HIGH-level output, the output voltage drops lower and lower, until finally, it will not be recognized as a HIGH level anymore by the other TTL gates that it is feeding.

Input/Output Current and Fan-Out

The **fan-out** of a subfamily is defined as the number of gate inputs of the same subfamily that can be connected to a single output without exceeding the current ratings of the gate. (A typical fan-out for most TTL subfamilies is 10.) Figure 9–4 shows an example of fan-out with 10 gates driven from a single gate.

Common Misconception

Students often think that I_{OL} (or I_{OH}) is the *actual* output current, when really it is the *maximum limit not to be exceeded.*

Team Discussion

Describe the difference between the I_{out} of a 7400 that is feeding five inverter inputs versus one that is feeding five inverters connected end to end.

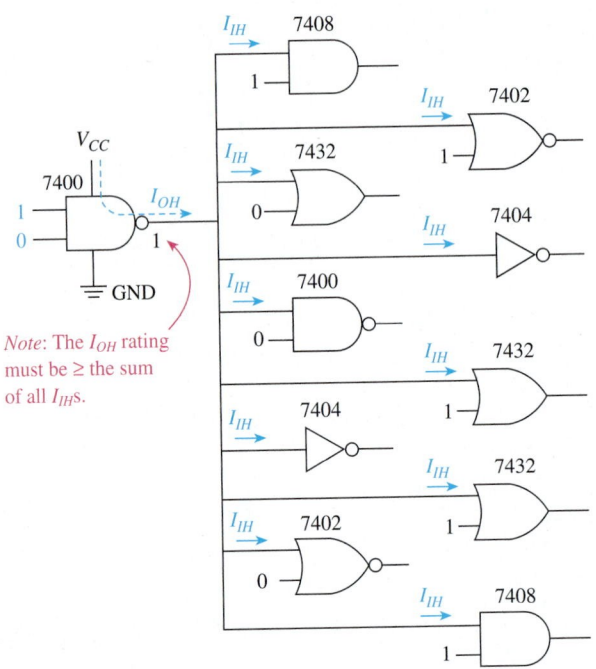

Figure 9–4 Ten gates driven from a single source.

To determine fan-out, you must know how much input current a gate load draws (I_I) and how much output current the driving gate can supply (I_O). In Figure 9–4, the single 7400 is the driving gate, supplying current to 10 other gate loads. The output current capability for the HIGH condition is abbreviated I_{OH} and is called a **source current.** I_{OH} for the 7400 is $-400\ \mu A$ maximum. (The minus sign signifies conventional current *leaving* the gate.)

The input current requirement for the HIGH condition is abbreviated I_{IH} and for the 74XX subfamily is equal to $40\ \mu A$ maximum. To find the fan-out, divide the source current ($-400\ \mu A$) by the input requirements for a gate ($40\ \mu A$). The fan-out is $400\ \mu A / 40\ \mu A = 10$.

For the LOW condition, the maximum output current for the 74XX subfamily is 16 mA, and the input requirement for each 74XX gate is -1.6 mA maximum, also for a fan-out of 10. The fan-out is usually the same for both the HIGH and LOW conditions for the 74XX gates; if not, we use the lower of the two.

Because a LOW output level is close to 0 V, the current actually flows into the output terminal and sinks down to ground. This is called a **sink current** and is illustrated in Figure 9–5. In the figure, two gates are connected to the output of gate 1. The total current that gate 1 must sink in this case is 2×1.6 mA = 3.2 mA. Because the maximum current a gate can sink in the LOW condition (I_{OL}) is 16 mA, gate 1 is well within its maximum rating of I_{OL}. (Gate 1 could sink the current from as many as *10* gate inputs.) As gates are added to the output of gate 1, its output voltage will rise above its 0 V level. It will still register as a LOW, however, as long as the number of gates is less than 10.

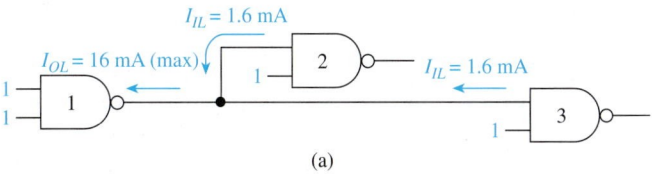

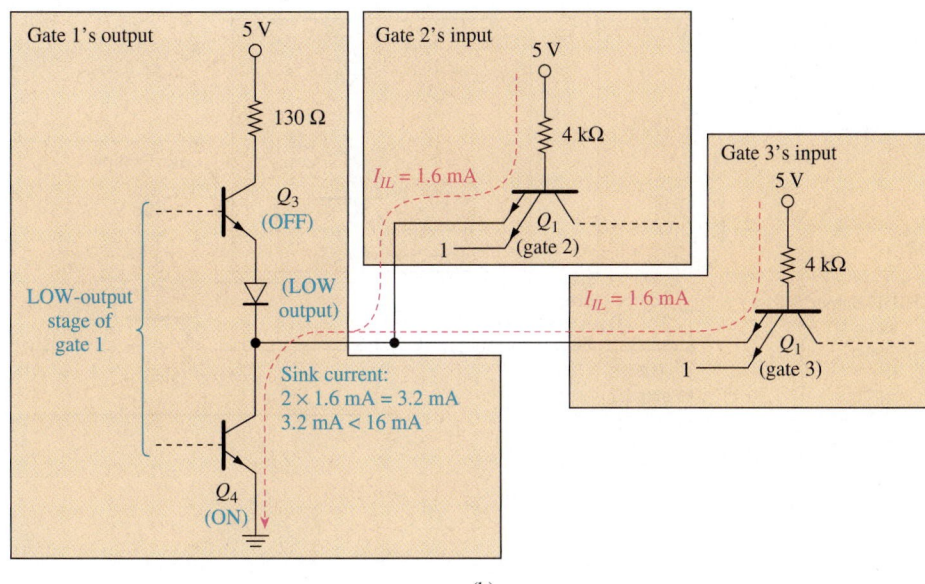

(b)

Figure 9–5 Totem-pole LOW output of a TTL gate sinking the input currents from two gate inputs: (a) logic gate symbols; (b) logic gate internal circuitry.

For the HIGH-output condition, the circuitry is the same, but the current flow is reversed, as shown in Figure 9–6. In the figure, you can see that the 40 μA going into each input is actually a small reverse leakage current flowing against the emitter arrow. In this case, the output of gate 1 is sourcing $-80\ \mu$A to the inputs of gates 2 and 3. The $-80\ \mu$A is well below the maximum allowed HIGH-output current rating of $-400\ \mu$A.

Summary of Input/Output Current and Fan-Out

1. The maximum current that an input to a *standard* (that is, a 74XX) TTL gate can sink or source is

    I_{IL}—low-level input current $= -1.6$ mA ($-1600\ \mu$A)
    I_{IH}—high-level input current $= 40\ \mu$A

    (The minus sign signifies current *leaving* the gate.)

2. The maximum current that the output of a *standard* TTL gate can sink or source is

    I_{OL}—low-level output current $= 16$ mA (16,000 μA)
    I_{OH}—high-level output current $= -400\ \mu$A ($-800\ \mu$A for some)

    (*Note:* This is *not* the actual amount of current leaving or entering a gate's output; rather, it is the *maximum capability* of the gate to sink or source current. The *actual* output current that flows depends on the number and type of loads connected.)

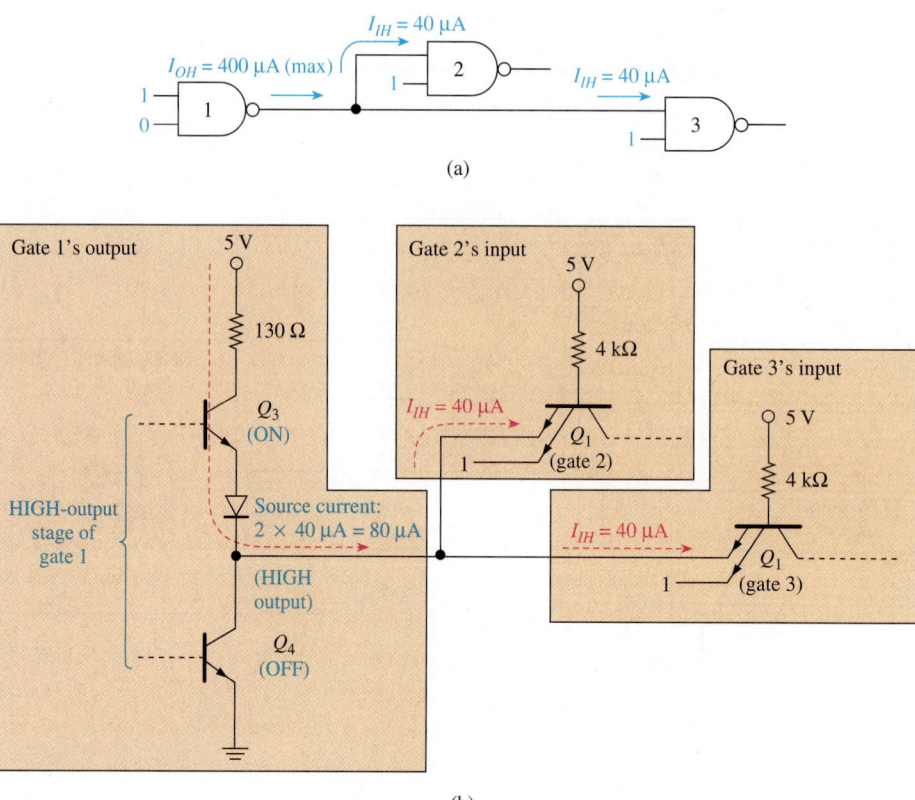

Figure 9–6 Totem-pole HIGH output of a TTL gate sourcing current to two gate inputs: (a) logic gate symbols; (b) logic gate internal circuitry.

3. The maximum number of gate inputs that can be connected to a *standard* TTL gate output is 10 (fan-out = 10). Fan-out is determined by taking the smaller result of I_{OL}/I_{IL} or I_{OH}/I_{IH}.

Input/Output Voltages and Noise Margin

We must also concern ourselves with the specifications for the acceptable input and output voltage levels. For the *LOW output condition,* the lower transistor (Q_4) in the totem-pole output stage is saturated (ON), and the upper one (Q_3) is cut off (OFF). V_{out} for the LOW condition (V_{OL}) is the voltage across the saturated Q_4, which has a typical value of 0.2 V and a maximum value of 0.4 V, as specified in the manufacturer's data manual.

For the *HIGH output condition,* the upper transistor (Q_3) is saturated, and the lower transistor (Q_4) is cut off. The voltage that reaches the output (V_{OH}) is V_{CC} minus the drop across the 130-Ω resistor, minus the C-E drop, minus the diode drop. Manufacturers' data sheets specify that the HIGH-level output is typically 3.4 V, and they guarantee that the worst-case minimum value will be 2.4 V. This means that the next gate input must interpret any voltage from 2.4 V up to 5.0 V as a HIGH level. Therefore, we must also consider the *input* voltage-level specifications (V_{IH}, V_{IL}).

Manufacturers guarantee that any voltage between a minimum of 2.0 V up to 5.0 V will be interpreted as a HIGH (V_{IH}). Also, any voltage from a maximum of 0.8 V down to 0 V will be interpreted as a LOW (V_{IL}).

These values leave us a little margin for error, what is called the **noise margin.** For example, V_{OL} is guaranteed not to exceed 0.4 V, and V_{IL} can be as high as 0.8 V and still be interpreted as a LOW. Therefore, we have 0.4 V (0.8 V − 0.4 V) of leeway (noise margin), as illustrated in Figure 9–7(a) and (b). Input voltages that fall within the "uncertain region" in Figure 9–7(b) will produce unpredictable results.

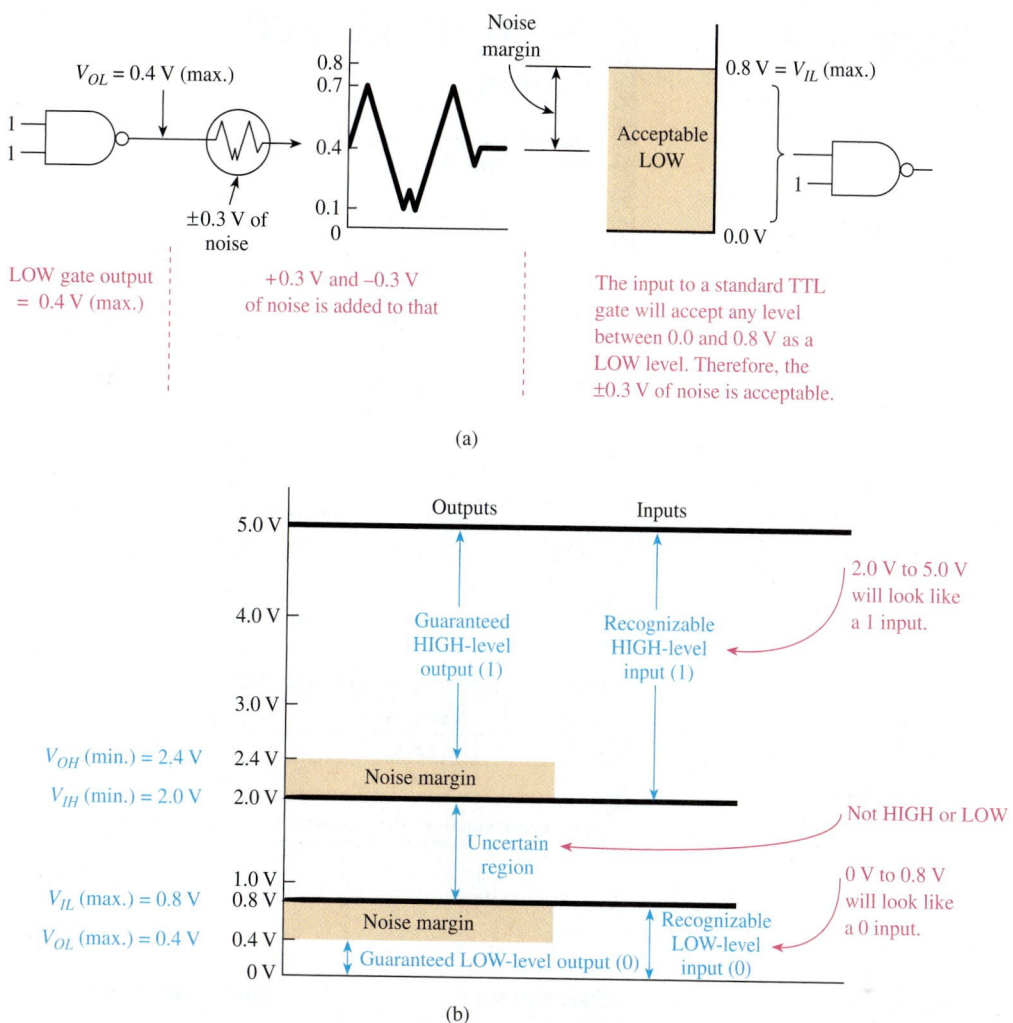

Figure 9–7 (a) Adding noise to a LOW-level output; (b) graphical illustration of the input/output voltage levels for the standard 74XX TTL series.

Parameter	Minimum	Typical	Maximum	
TABLE 9–1	**Standard 74XX Series Voltage Levels**			
V_{OL}		0.2 V	0.4 V	} Noise margin
V_{IL}			0.8 V	= 0.4 V
V_{OH}	2.4 V	3.4 V		} Noise margin
V_{IH}	2.0 V			= 0.4 V

Noise margin (HIGH) = V_{OH} (min) − V_{IH} (min)
Noise margin (LOW) = V_{IL} (max) − V_{OL} (max)

Table 9–1 is a summary of input/output voltage levels and noise margins for the standard family of TTL ICs. These numbers are the most common, but be sure that you can locate these values on a data sheet. The data sheet for a 7400 NAND gate is given in Figure 9–8. Study the data sheet to be sure that you can locate all of the voltage and current ratings covered so far. The prudent designer will always assume worst-case values to ensure that his or her design will always work for any conditions that may arise.

Signetics

7400, LS00, S00
Gates

Quad Two-Input NAND Gate
Product Specification

Logic Products

Typical switching speed

Typical power supply requirements

TYPE	TYPICAL PROPAGATION DELAY	TYPICAL SUPPLY CURRENT (TOTAL)
7400	9ns	8mA
74LS00	9.5ns	1.6mA
74S00	3ns	15mA

Gives part numbers for various package styles.

ORDERING CODE

PACKAGES	COMMERCIAL RANGE $V_{CC} = 5V \pm 5\%$; $T_A = 0°C$ to $+70°C$	
Plastic DIP	N7400N, N74LS00N, N74S00N	
Plastic SO	N74LS00D, N74S00D	

NOTE:
For information regarding devices processed to Military Specifications, see the Signetics Military Products Data Manual.

Shows every Input/Output combination

FUNCTION TABLE

INPUTS		OUTPUT
A	B	Y
L	L	H
L	H	H
H	L	H
H	H	L

H = HIGH voltage level
L = LOW voltage level

INPUT AND OUTPUT LOADING AND FAN-OUT TABLE

PINS	DESCRIPTION	74	74S	74LS
A, B	Inputs	1ul	1Sul	1LSul
Y	Output	10ul	10Sul	10LSul

NOTE:
Where a 74 unit load (ul) is understood to be 40µA I_{IH} and −1.6mA I_{IL}, a 74S unit load (Sul) is 50µA I_{IH} and −2.0mA I_{IL}, and 74LS unit load (LSul) is 20µA I_{IH} and −0.4mA I_{IL}.

Means that the output can drive 10 unit load inputs of the same family (fan-out = 10)

Dependency Notation symbol

Traditional Logic symbol

Gives IC wiring information

PIN CONFIGURATION

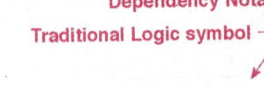

LOGIC SYMBOL

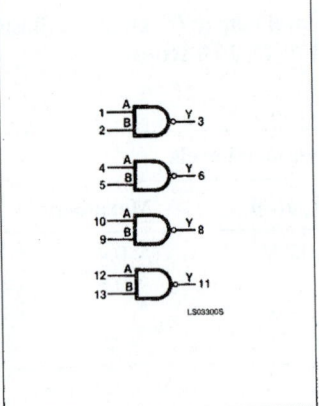

LOGIC SYMBOL (IEEE/IEC)

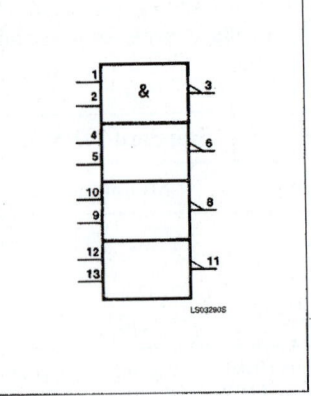

Figure 9–8 The 7400 data sheet.

CHAPTER 9 | LOGIC FAMILIES AND THEIR CHARACTERISTICS

Gates

7400, LS00, S00

Range not to be exceeded

ABSOLUTE MAXIMUM RATINGS (Over operating free-air temperature range unless otherwise noted.)

	PARAMETER	74	74LS	74S	UNIT
V_{CC}	Supply voltage	7.0	7.0	7.0	V
V_{IN}	Input voltage	−0.5 to +5.5	−0.5 to +7.0	−0.5 to +5.5	V
I_{IN}	Input current	−30 to +5	−30 to +1	−30 to +5	mA
V_{OUT}	Voltage applied to output in HIGH output state	−0.5 to +V_{CC}	−0.5 to +V_{CC}	−0.5 to +V_{CC}	V
T_A	Operating free-air temperature range		0 to 70		°C

Normal range to be used

Specs for each 7400 series

RECOMMENDED OPERATING CONDITIONS

	PARAMETER	74			74LS			74S			UNIT
		Min	Nom	Max	Min	Nom	Max	Min	Nom	Max	
V_{CC}	Supply voltage	4.75	5.0	5.25	4.75	5.0	5.25	4.75	5.0	5.25	V
V_{IH}	HIGH-level input voltage	2.0			2.0			2.0			V
V_{IL}	LOW-level input voltage			+0.8			+0.8			+0.8	V
I_{IK}	Input clamp current			−12			−18			−18	mA
I_{OH}	HIGH-level output current			−400			−400			−1000	μA
I_{OL}	LOW-level output current			16			8			20	mA
T_A	Operating free-air temperature	0		70	0		70	0		70	°C

Input voltage specs

Output current specs

Stay within this range but use 5.0 V nominal

TEST CIRCUITS AND WAVEFORMS

Shows the results of an input pulse applied to a 7400 Device Under Test (DUT)

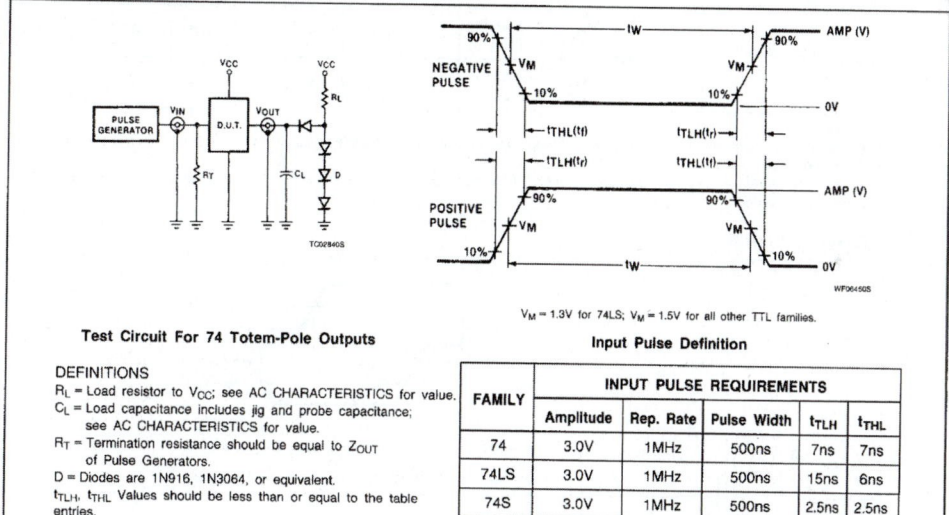

$V_M = 1.3V$ for 74LS; $V_M = 1.5V$ for all other TTL families.

Test Circuit For 74 Totem-Pole Outputs

Input Pulse Definition

DEFINITIONS
R_L = Load resistor to V_{CC}; see AC CHARACTERISTICS for value.
C_L = Load capacitance includes jig and probe capacitance; see AC CHARACTERISTICS for value.
R_T = Termination resistance should be equal to Z_{OUT} of Pulse Generators.
D = Diodes are 1N916, 1N3064, or equivalent.
t_{TLH}, t_{THL} Values should be less than or equal to the table entries.

FAMILY	INPUT PULSE REQUIREMENTS				
	Amplitude	Rep. Rate	Pulse Width	t_{TLH}	t_{THL}
74	3.0V	1MHz	500ns	7ns	7ns
74LS	3.0V	1MHz	500ns	15ns	6ns
74S	3.0V	1MHz	500ns	2.5ns	2.5ns

Figure 9–8 *(Continued)*

Gates

The Min and Max are guaranteed limits but you can expect the typ (typical)

DC ELECTRICAL CHARACTERISTICS (Over recommended operating free-air temperature range unless otherwise noted.)

	PARAMETER	TEST CONDITIONS[1]	7400			74LS00			74S00			UNIT
			Min	Typ[2]	Max	Min	Typ[2]	Max	Min	Typ[2]	Max	
V_{OH}	HIGH-level output voltage	V_{CC} = MIN, V_{IH} = MIN, V_{IL} = MAX, I_{OH} = MAX	2.4	3.4		2.7	3.4		2.7	3.4		V
V_{OL}	LOW-level output voltage	V_{CC} = MIN, V_{IH} = MIN	I_{OL} = MAX		0.2	0.4		0.35	0.5			0.5	V
			I_{OL} = 4mA (74LS)					0.25	0.4				V
V_{IK}	Input clamp voltage	V_{CC} = MIN, I_I = I_{IK}			−1.5			−1.5			−1.2	V
I_I	Input current at maximum input voltage	V_{CC} = MAX	V_I = 5.5V			1.0						1.0	mA
			V_I = 7.0V						0.1				mA
I_{IH}	HIGH-level input current	V_{CC} = MAX	V_I = 2.4V			40							μA
			V_I = 2.7V						20			50	μA
I_{IL}	LOW-level input current	V_{CC} = MAX	V_I = 0.4V			−1.6			−0.4				mA
			V_I = 0.5V									−2.0	mA
I_{OS}	Short-circuit output current[3]	V_{CC} = MAX		−18		−55	−20		−100	−40		−100	mA
I_{CC}	Supply current (total)	V_{CC} = MAX	I_{CCH} Outputs HIGH		4	8		0.8	1.6		10	16	mA
			I_{CCL} Outputs LOW		12	22		2.4	4.4		20	36	mA

Output voltage specs

Input current specs

NOTES:
1. For conditions shown as MIN or MAX, use the appropriate value specified under recommended operating conditions for the applicable type.
2. All typical values are at V_{CC} = 5V, T_A = 25°C.
3. I_{OS} is tested with V_{OUT} = +0.5V and V_{CC} = V_{CC} MAX + 0.5V. Not more than one output should be shorted at a time and duration of the short circuit should not exceed one second.

AC WAVEFORM

Waveform shows definitions for propagation time specs

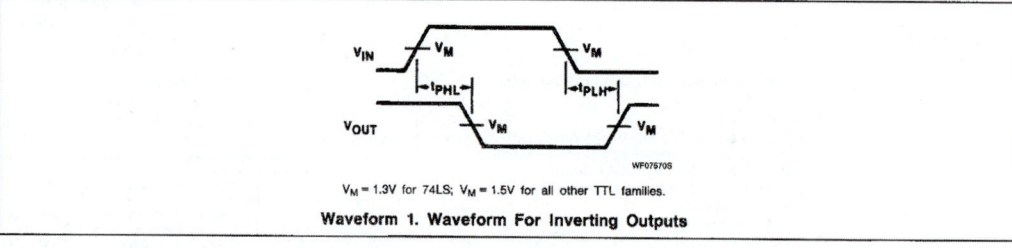

V_M = 1.3V for 74LS; V_M = 1.5V for all other TTL families.

Waveform 1. Waveform For Inverting Outputs

AC ELECTRICAL CHARACTERISTICS T_A = 25°C, V_{CC} = 5.0V

	PARAMETER	TEST CONDITIONS	74		74LS		74S		UNIT
			C_L = 15pF, R_L = 400Ω		C_L = 15pF, R_L = 2kΩ		C_L = 15pF, R_L = 280Ω		
			Min	Max	Min	Max	Min	Max	
t_{PLH}	Propagation delay	Waveform 1		22		15		4.5	ns
t_{PHL}				15		15		5.0	

Propagation delay specs

Figure 9–8 *(Continued)*

The following examples illustrate the use of the current and voltage ratings for establishing acceptable operating conditions for TTL logic gates.

EXAMPLE 9–1

Find the voltages and currents that are asked for in Figure 9–9 if the gates are all standard (74XX) TTLs.

(a) Find V_a and I_a for Figure 9–9(a).

(b) Find V_a, V_b, and I_b for Figure 9–9(b).

(c) Find V_a, V_b, and I_b for Figure 9–9(c).

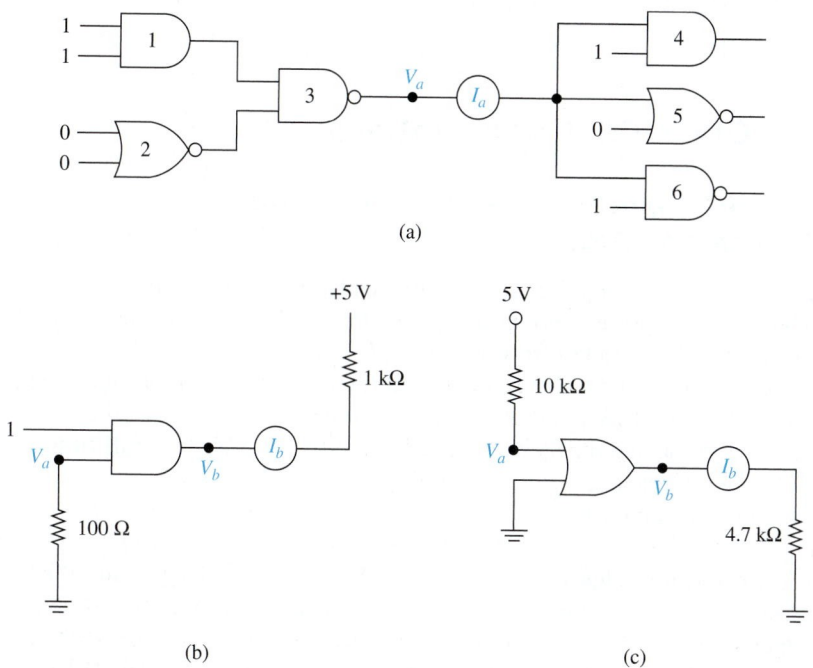

(a)

(b)   (c)

Figure 9–9   Voltage and current ratings.

Solution: **(a)** The input to gate 3 is a 1–1, so the output will be LOW. Using the *typical* value, $V_a = 0.2$ V. Because gate 3 is LOW, it will be sinking current from the three other gates: 4, 5, and 6. The typical value for each I_{IL} is -1.6 mA; therefore, $I_a = -4.8$ mA (-1.6 mA $-$ 1.6 mA $-$ 1.6 mA).

(b) The 100-Ω resistor to ground will place a LOW level at that input. I_{IL} typically is -1.6 mA, which flows down through the 100-Ω resistor, making $V_a = 0.16$ V (1.6 mA $\times$ 100 Ω). The 0.16 V at V_a will be recognized as a LOW level ($V_{IL} = 0.8$ V max.), so the AND gate will output a LOW level; $V_b = 0.2$ V (typ.). The AND gate will sink current from the 1-kΩ resistor; $I_b = 4.8$ mA [(5 V $-$ 0.2 V)/1 kΩ]. The 4.8 mA is well below the maximum allowed current of 16 mA (I_{OL}), so the AND gate will not burn out.

(c) I_{IH} into the OR gate is 40 μA; therefore, the voltage at $V_a = 4.6$ V [5 V $-$ (10 kΩ $\times$ 40 μA)]. The output level of the OR gate will be HIGH (V_{OH}), making $V_b = 3.4$ V and $I_b = 3.4$ V/4.7 k$\Omega = 723$ μA. The 723 μA is below the maximum rating of the OR gate ($I_{OH} = -800$ μA max.). Therefore, the OR gate will not burn out.

9–4. Why aren't the HIGH/LOW output levels of a TTL gate exactly 5.0 and 0 V?

9–5. Describe what is meant by fan-out.

9–6. List the names and abbreviations of the four input and output currents of a digital IC.

9–7. Describe the difference between *sink* and *source* output current.

9–8. Determine if the following input voltages will be interpreted as HIGH, LOW, or undetermined logic levels in a standard TTL IC.

(a) 3.0 V (c) 1.0 V

(b) 2.2 V (d) 0.6 V

9–3   Other TTL Considerations

Pulse-Time Parameters: Rise Time, Fall Time, and Propagation Delay

We have been using ideal pulses for the input and output waveforms up until now. Actually, however, the pulse is not perfectly square; it takes time for the digital level to rise from 0 up to 1 and to fall from 1 down to 0.

As shown in Figure 9–10(a), the **rise time** (t_r) is the length of time it takes for a pulse to rise from its 10% point up to its 90% point. For a 5-V pulse, the 10% point is 0.5 V (10% $\times$ 5 V), and the 90% point is 4.5 V (90% $\times$ 5 V). The **fall time** (t_f) is the length of time it takes to fall from the 90% point to the 10% point.

Not only are input and output waveforms sloped on their rising and falling edges, but there is also a delay time for an input wave to propagate through an IC to the output, called the **propagation delay** (t_{PLH} and t_{PHL}). The propagation delay is due to limitations in transistor switching speeds caused by undesirable internal capacitive stored charges.

Figure 9–10(b) shows that it takes a certain length of time for an input pulse to reach the output of an IC gate. A specific measurement point (1.5 V for the standard TTL series)

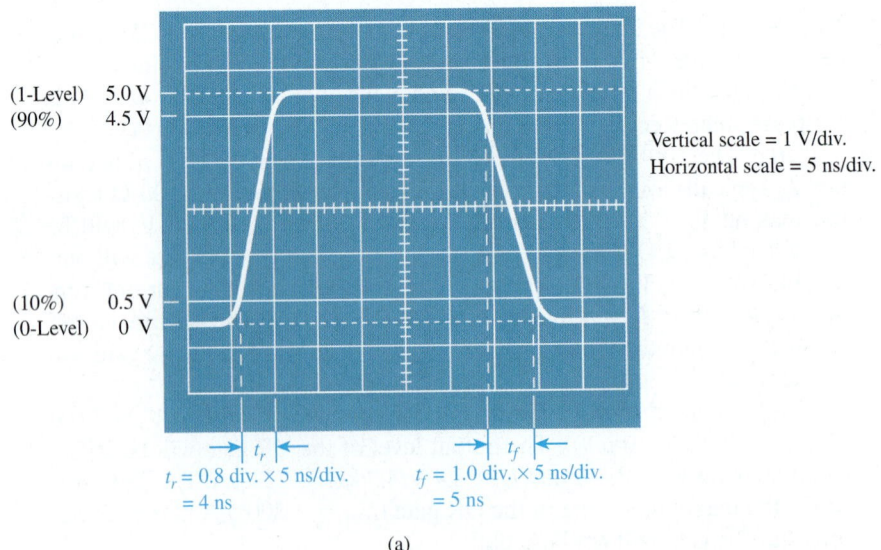

Vertical scale = 1 V/div.
Horizontal scale = 5 ns/div.

(1-Level) 5.0 V
(90%)   4.5 V

(10%)   0.5 V
(0-Level) 0 V

t_r

t_r = 0.8 div. × 5 ns/div.
   = 4 ns

t_f

t_f = 1.0 div. × 5 ns/div.
   = 5 ns

(a)

Figure 9–10   Oscilloscope displays: (a) pulse rise and fall times;

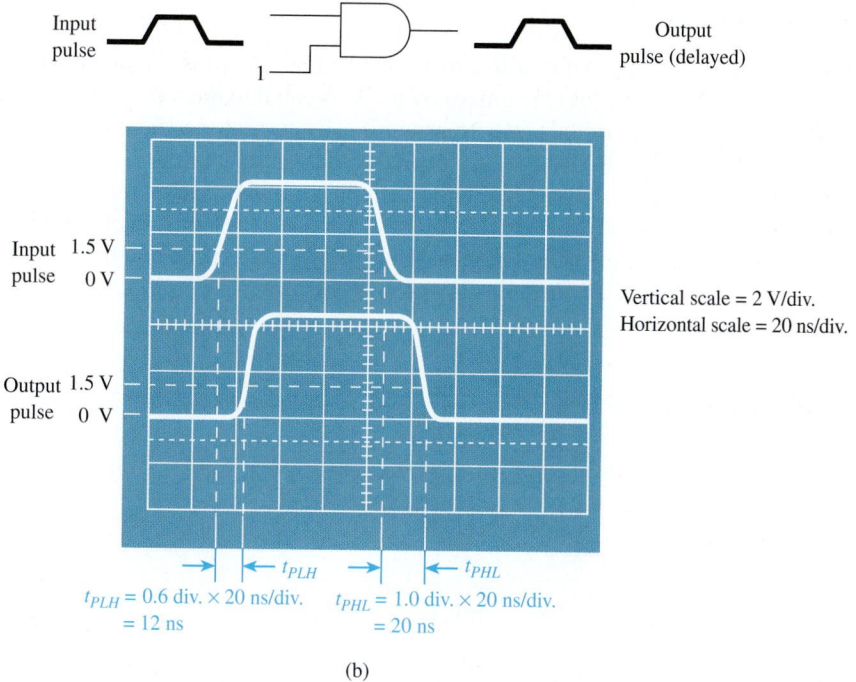

Vertical scale = 2 V/div.
Horizontal scale = 20 ns/div.

t_{PLH} = 0.6 div. × 20 ns/div.
= 12 ns

t_{PHL} = 1.0 div. × 20 ns/div.
= 20 ns

(b)

Figure 9–10 *(Continued)* (b) propagation delay times.

is used as a reference. The propagation delay time for the *output* to respond in the LOW-to-HIGH direction is labeled t_{PLH}, and in the HIGH-to-LOW direction, it is labeled t_{PHL}.

EXAMPLE 9–2

The propagation delay times for the 7402 NOR gate shown in Figure 9–11 are listed in a TTL data manual as $t_{PLH} = 22$ ns and $t_{PHL} = 15$ ns. Sketch and label the input and output pulses to a 7402.

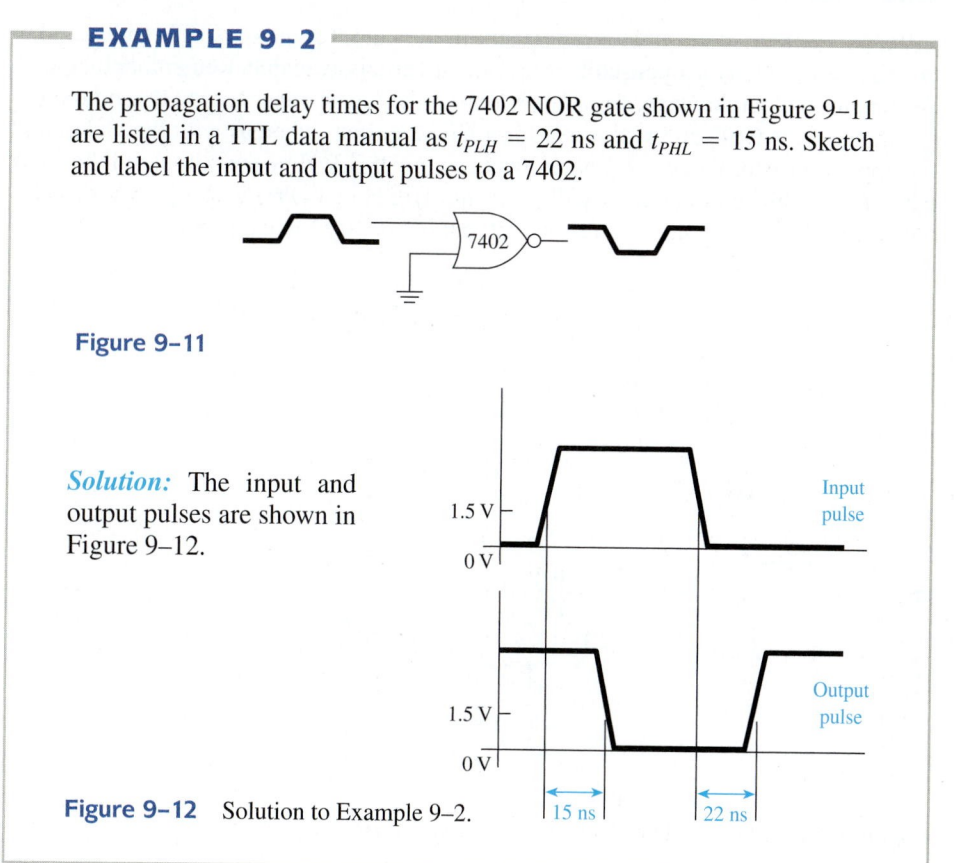

Figure 9–11

Solution: The input and output pulses are shown in Figure 9–12.

Figure 9–12 Solution to Example 9–2.

Common Misconception

The subscripts *LH* and *HL* pertain to the output, not the input.

Helpful Hint

It is very important that you can find these specs, as well as others, in a data book. (Data sheets for some common ICs are provided in Appendix B.)

Power Dissipation

Another operating characteristic of ICs that must be considered is the **power dissipation.** The power dissipated (or consumed) by an IC is equal to the total power supplied to the IC power supply terminals (V_{CC} to ground). The current that enters the V_{CC} supply terminal is called I_{CC}. Two values are given for the supply current: I_{CCH} and I_{CCL} for use when the outputs are HIGH or when the outputs are LOW. Because the outputs are usually switching between HIGH and LOW, if we assume a 50% duty cycle (HIGH half of the time, LOW half of the time), then an average I_{CC} can be used and the power dissipation determined from the formula $P_D = V_{CC} \times I_{CC}$ (av.).

EXAMPLE 9–3

The total supply current for a 7402 NOR IC is given as $I_{CCL} = 14$ mA, $I_{CCH} = 8$ mA. Determine the power dissipation of the IC.

Solution:

$$P_D = V_{CC} \times I_{CC} \text{ (av.)}$$

$$= 5.0 \text{ V} \times \frac{14 \text{ mA} + 8 \text{ mA}}{2} = 55 \text{ mW}$$

Open-Collector Outputs

Instead of using a totem-pole arrangement in the output stage of a TTL gate, another arrangement, called the **open-collector** (OC) **output,** is available. Remember that, with the totem-pole output stage, for a LOW output the lower transistor is ON and the upper transistor is OFF, and vice versa for a HIGH output, whereas with the OC output the upper transistor is *removed,* as shown in Figure 9–13. Now the output will be *LOW* when Q_4 is ON, and the output will *float* (not HIGH or LOW) when Q_4 is OFF. This means that an OC output can sink current, but it *cannot* source current.

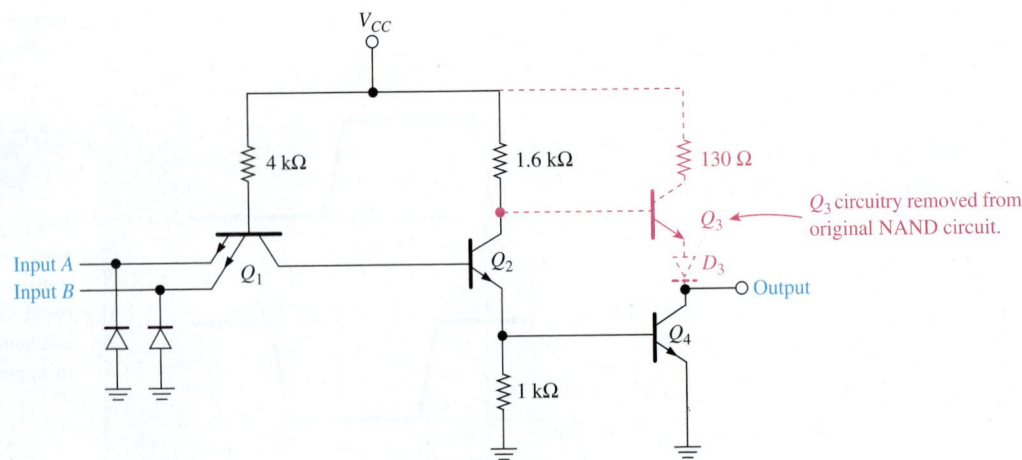

Figure 9–13   TTL NAND with an open-collector output.

CHAPTER 9 | LOGIC FAMILIES AND THEIR CHARACTERISTICS

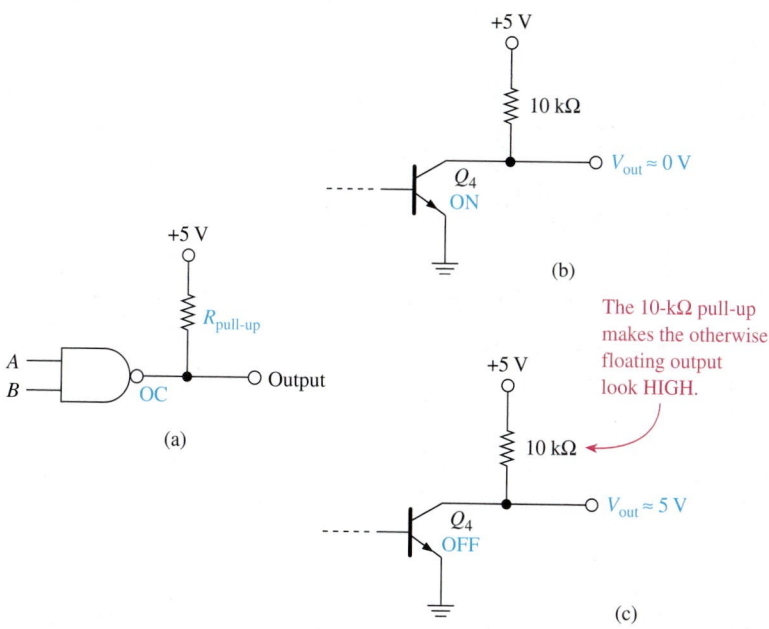

Figure 9–14 Using a pull-up resistor with an open-collector output: (a) adding a pull-up resistor to a NAND gate; (b) when Q_4 inside the NAND is ON, $V_{out} \approx 0$ V; (c) when Q_4 is OFF, the pull-up resistor provides ≈ 5 V to V_{out}.

To get an OC output to produce a HIGH, an external resistor (called a **pull-up resistor**) must be used, as shown in Figure 9–14. Now when Q_4 is OFF (open), the output is approximately 5 V (HIGH), and when Q_4 is ON (short), the output is approximately 0 V (LOW). The optimum size for a pull-up resistor depends on the size of the output's load and the leakage current through Q_4 (I_{OH}) when it is OFF. Usually, a good size for a pull-up resistor is 10 kΩ: 10 kΩ is not too small to allow excessive current flow when Q_4 is ON, and it is not too large to cause an excessive voltage drop across itself when Q_4 is OFF.

Open-collector **buffer**/driver ICs are available for output loads requiring large sink currents, such as displays, relays, or motors. The term *buffer/driver* signifies the ability to provide high output currents to drive heavy loads. Typical ICs of this type are the 7406 OC inverter buffer/driver and the 7407 OC buffer/driver. They are each capable of sinking up to 40 mA, which is 2.5 times greater than the 16-mA capability of the standard 7404 inverter.

Wired-Output Operation

The main use of the OC gates is when the outputs from two or more gates or other devices have to be tied together. Using the regular totem-pole output gates, if a gate having a HIGH output (5 V) is connected to another gate having a LOW output (0 V), you would have a direct short circuit, causing either or both gates to burn out.

Using OC gates, outputs can be connected without worrying about the 5 V–0 V conflict. When connected, they form **wired-AND** logic, as shown in Figure 9–15. The 7405 IC has six OC inverters in a single package. By tying their outputs together, as shown in Figure 9–15(a), we have in effect *ANDed* all the inverters. The outputs of all six inverters must be floating (all inputs must be LOW) to get a HIGH output ($X = 1$ if $A = 0$ AND $B = 0$ AND $C = 0$, and so on). If any of the inverter output transistors (Q_4) turn on, the output will go LOW. The result of this wired-AND connection is the six-input NOR function shown in Figure 9–15(c).

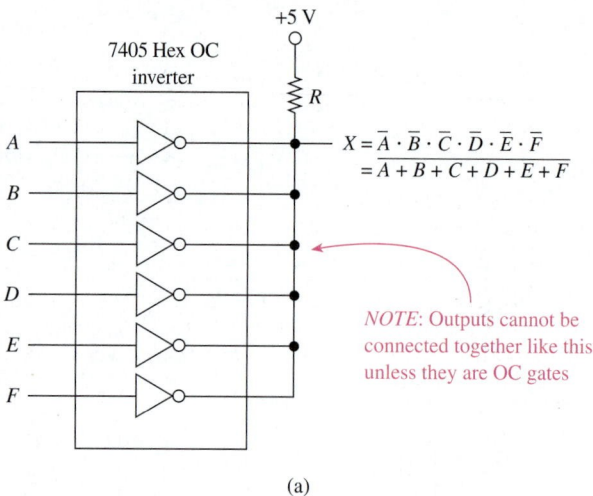

$$X = \overline{A} \cdot \overline{B} \cdot \overline{C} \cdot \overline{D} \cdot \overline{E} \cdot \overline{F}$$
$$= \overline{A + B + C + D + E + F}$$

NOTE: Outputs cannot be connected together like this unless they are OC gates

(a)

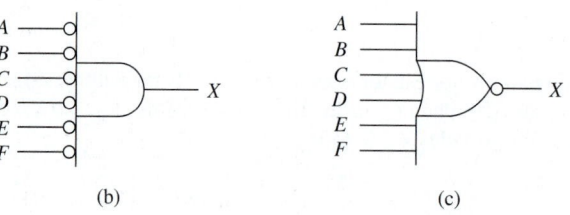

(b)   (c)

Figure 9–15 (a) Wired-AND connections to a hex OC inverter to form a six-input NOR gate; (b) AND gate representation; (c) alternative NOR gate representation.

EXAMPLE 9–4

Write the Boolean equation at the output of Figure 9–16(a).

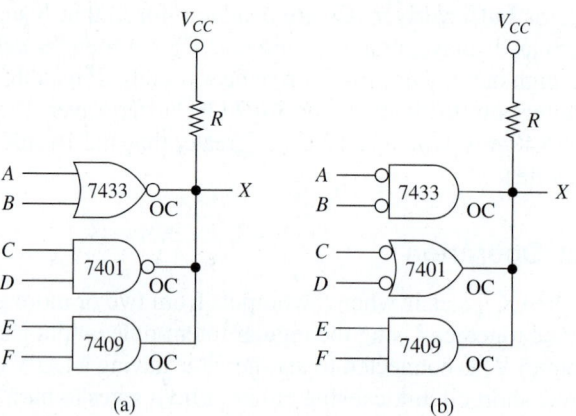

(a)   (b)

Figure 9–16 Wired-ANDing of open-collector gates for Example 9–4: (a) original circuit; (b) alternative gate representations used for clarity.

Solution: The output of all three gates in either circuit must be floating to get a HIGH output at *X*. Using Figure 9–16(b),

$$X = \overline{A}\,\overline{B} \cdot (\overline{C} + \overline{D}) \cdot EF$$

Disposition of Unused Inputs and Unused Gates

Electrically, *open inputs* degrade ac noise immunity as well as the switching speed of a circuit. For example, if two inputs to a three-input NAND gate are being used and the third is allowed to float, unpredictable results will occur if the third input picks up electrical noise from surrounding circuitry.

Unused inputs on AND and NAND gates should be tied HIGH, and on OR and NOR gates, they should be tied to ground. An example of this is a three-input AND gate that is using only two of its inputs.

Also, the outputs of *unused gates* on an IC should be forced HIGH to reduce the I_{CC} supply current and, thus, reduce power dissipation. To do this, tie AND and OR inputs HIGH, and tie NAND and NOR inputs LOW. An example of this is a quad NOR IC, where only three of the NOR gates are being used.

Power Supply Decoupling

In digital systems, there are heavy current demands on the main power supply. TTL logic tends to create spikes on the main V_{CC} line, especially at the logic-level transition point (LOW to HIGH or HIGH to LOW). At the logic-level transition, there is a period of time that the conduction in the upper and lower **totem-pole output** transistors overlaps. This drastically changes the demand for I_{CC} current, which causes sharp high-frequency spikes to occur on the V_{CC} (power supply) line. These spikes cause false switching of other devices connected to the same power supply line and can also induce magnetic fields that radiate electromagnetic interference (**EMI**).

Decoupling of IC power supply spikes from the main V_{CC} line can be accomplished by placing a 0.01- to 0.1-μF capacitor directly across the V_{CC}-to-ground pins on each IC in the system. The capacitors tend to hold the V_{CC} level at each IC constant, thus reducing the amount of EMI radiation that is emitted from the system and the likelihood of false switching. Locating these small capacitors close to the IC ensures that the current spike will be kept local to the chip instead of radiating through the entire system back to the power supply.

Review Questions

9–9. The *rise time* is the length of time required for a digital signal to travel from 0 V to its HIGH level. True or false?

9–10. The letters L and H in the abbreviation t_{PLH} refer to the transition in the _____ (input, output) signal.

9–11. Describe the function of a pull-up resistor when it is used with an *open-collector* TTL output.

9–4   Improved TTL Series

Integrated-circuit design engineers have constantly worked to improve the standard TTL series. In fact, very few new designs will incorporate the use of the original 74 series technology. A simple improvement that was made early on was simply reducing all the internal resistor values of the standard TTL series. This increased the power consumption (or dissipation), which was bad, but it reduced the internal $R \times C$ time constants that cause propagation delays. The result was the 74HXX series, which has almost half the propagation delay time but almost double the power consumption of the standard TTL series. The product of delay time $\times$ power (the speed–power product), which is a figure of merit for IC families, remained approximately the same, however.

Another series, the 74LXX, was developed using just the opposite approach. The internal resistors were increased, thus reducing power consumption, but the propagation delay increased, keeping the speed–power product about the same. The 74HXX and 74LXX series have, for the most part, been replaced now by the Schottky TTL and CMOS series of ICs.

Schottky TTL

The major speed limitation of the standard TTL series is due to the capacitive charge in the base region of the transistors. The transistors basically operate at either cutoff or saturation. When the transistor is saturated, charges build up at the base region, and when it is time to switch to cutoff, the stored charges must be dissipated, which takes time, thus causing propagation delay.

Schottky logic overcomes the saturation and stored charges problem by placing a Schottky diode across the base-to-collector junction, as shown in Figure 9–17. With the Schottky diode in place, any excess charge on the base is passed on to the collector, and the transistor is held just below saturation. The Schottky diode has a special metal junction that minimizes its own capacitive charge and increases its switching speed. Using Schottky-clamped transistors and decreased resistor values, the propagation delay is reduced by a factor of 4, and power consumption is only doubled. Therefore, the speed–power product of the 74SXX TTL series is improved to approximately half that of the 74XX TTL series (the lower, the better).

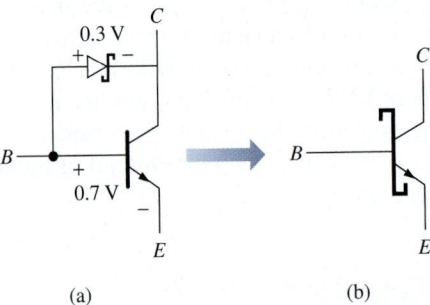

Figure 9–17   Schottky-clamped transistor: (a) Schottky diode reduces stored charges; (b) symbol.

Low-Power Schottky (LS):   (See Figure 9–8 for the 7400/74LS00 data sheet.) By using different integration techniques and increasing the values of the internal resistors, the power dissipation of the Schottky TTL is reduced significantly. The speed–power product of the 74LSXX TTL series is approximately one-third that of the 74SXX series and one-fifth that of the 74XX series.

Advanced Low-Power Schottky (ALS):   Further improvement of the 74LSXX series reduced the propagation delay time from 9 to 4 ns and the power dissipation from 2 to 1 mW per gate. The 74ALSXX and 74LS series rapidly replaced the standard 74XX and 74SXX series because of the speed and power improvements. However, as with any new technology, they were slightly more expensive.

Fast (F)

It was long clear to TTL IC design engineers that new processing technology was needed to improve the speed of the LS series. A new process of integration, called *oxide isolation* (also used by the ALS series), has reduced the propagation delay in the

74FXX series to below 3 ns. In this process, transistors are isolated from each other, not by a reverse-biased junction but by an actual channel of oxide. This dramatically reduces the size of the devices, which in turn reduces their associated capacitances and, thus, reduces propagation delay.

9–5   The CMOS Family

The CMOS family of integrated circuits differs from TTL by using an entirely different type of transistor as its basic building block. The TTL family uses **bipolar transistors** (*NPN* and *PNP*). CMOS (complementary metal oxide semiconductor) uses complementary pairs of transistors (*N* type and *P* type) called **MOSFETs** (metal oxide semiconductor field-effect transistors). MOSFETs are also used in other families of MOS ICs, including **PMOS** and **NMOS,** which are most commonly used for large-scale memories and microprocessors in the LSI and VLSI (large-scale and very large-scale integration) category. One advantage that MOSFETs have over bipolar transistors is that the input to a MOSFET is electrically isolated from the rest of the MOSFET [see Figure 9–18 (a) and (b)], giving it a high input impedance, which reduces the input current and power dissipation.

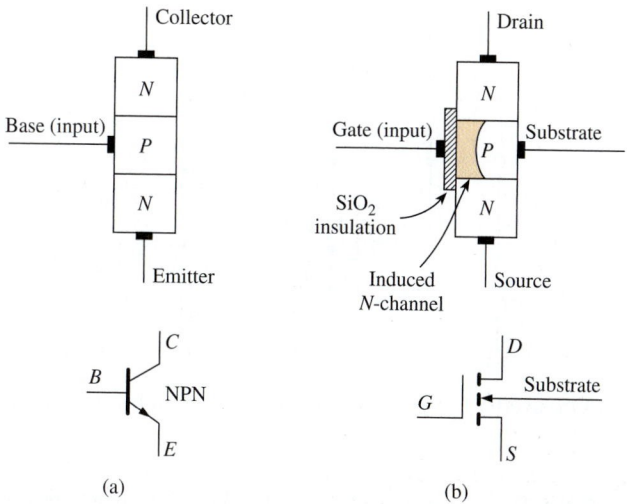

Figure 9–18   Simplified diagrams of bipolar and field-effect transistors: (a) *NPN* bipolar transistor used in TTL ICs; (b) *N*-channel MOSFET used in CMOS ICs.

The *N*-channel MOSFET is similar to the *NPN* bipolar transistor in that it is two back-to-back *N*–*P* junctions, and current will not flow down through it until a positive voltage is applied to the gate. The silicon dioxide (SiO_2) layer between the gate material and the *P* **substrate** (base) of the MOSFET prevents any gate current from flowing, which provides a high input impedance and low power consumption.

The MOSFET shown in Figure 9–18(b) is a normally OFF device because there are no negative carriers in the *P* material for current flow to occur. However, conventional current will flow down from drain to source if a positive voltage is applied to the gate with respect to the substrate. This voltage induces an electric field across the SiO_2 layer, which repels enough of the positive charges in the *P* material to form a channel of negative charges on the left side of the *P* material. This allows electrons to flow from source to drain (conventional current flows from drain to source). The channel that is formed is called an *N* channel because it contains negative carriers.

P-channel MOSFETs are just the opposite, constructed from *P–N–P* materials. The channel is formed by placing a *negative* voltage at the gate with respect to the substrate.

There are three major MOS technology families: PMOS (made up of *P*-channel MOSFETs), NMOS (made up of *N*-channel MOSFETs), and CMOS (made up of complementary *P*-channel and *N*-channel MOSFETs). MOS technology provides a higher packing density than bipolar TTL and, therefore, allows IC manufacturers to provide thousands of logic functions on a single IC chip (VLSI circuitry). For example, it is common to find computer systems with MOS memory ICs containing millions of memory cells per chip and MOS microcontroller ICs containing the combined logic of more than 10 LSI ICs. Between NMOS and PMOS, NMOS was historically more widely used because of its higher speed and packing density.

Fabricating both *P*- and *N*-channel transistors in the same package previously made CMOS slightly more expensive and limited its use in VLSI circuits. However, recent advances in fabrication techniques and mass production have reduced its price. Today, CMOS competes very favorably with the best bipolar TTL circuitry in the SSI and MSI arena and is by far the most popular technology in LSI and VLSI memory and microprocessor ICs.

Using an *N*-channel MOSFET with its complement, the *P*-channel MOSFET, a simple complementary MOS (CMOS) inverter can be formed, as shown in Figure 9–19.

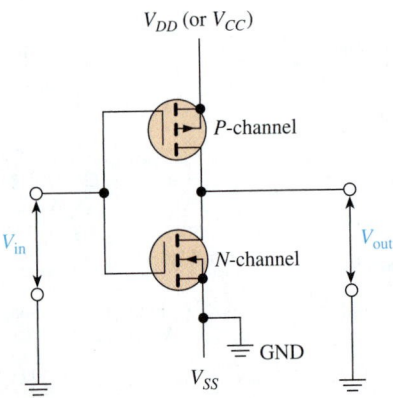

Figure 9–19   CMOS inverter formed from complementary *N*-channel/*P*-channel transistors.

TABLE 9–2	Basic MOSFET Switching Characteristics	
Gate Level[a]	***N*-Channel**	***P*-Channel**
1	ON	OFF
0	OFF	ON

[a] $1 \equiv V_{DD}$ (or V_{CC}); $0 \equiv V_{SS}$ (Gnd).

We can think of MOSFETs as ON/OFF switches, just as we did for bipolar transistors. Table 9–2 summarizes the ON/OFF operation of *N*- and *P*-channel MOSFETs. We can use Table 9–2 to prove that the circuit of Figure 9–19 operates as an inverter. With $V_{in} = 1$, the *N*-channel transistor is ON and the *P*-channel transistor is OFF, so $V_{out} = 0$. With $V_{in} = 0$, the *N*-channel transistor is OFF and the *P*-channel transistor is ON, so $V_{out} = 1$. Therefore, $V_{out} = \overline{V_{in}}$. Notice that this complementary action is very similar to the TTL totem-pole output stage, but much simpler to understand.

The other basic logic gates can also be formed using complementary MOSFET transistors. The operation of CMOS NAND and NOR gates can easily be understood by studying the schematics and data tables presented in Figure 9–20(a) and (b).

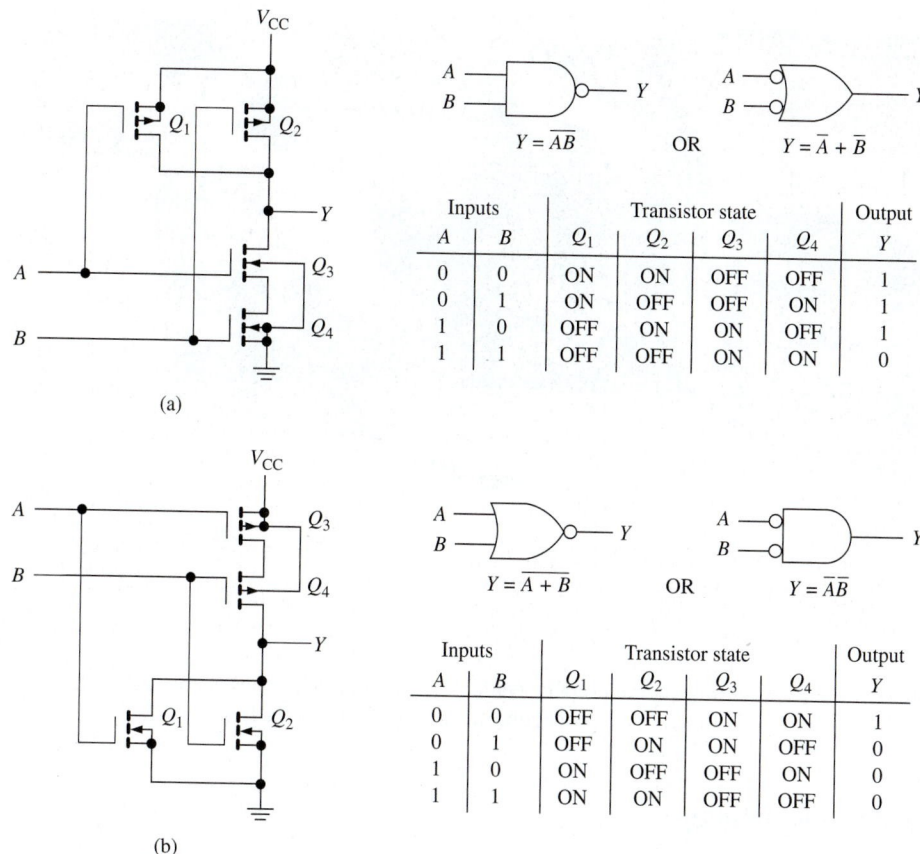

Inputs		Transistor state				Output
A	B	Q_1	Q_2	Q_3	Q_4	Y
0	0	ON	ON	OFF	OFF	1
0	1	ON	OFF	OFF	ON	1
1	0	OFF	ON	ON	OFF	1
1	1	OFF	OFF	ON	ON	0

$$A \;B \rightarrow Y = \overline{AB} \quad OR \quad Y = \overline{A} + \overline{B}$$

(a)

$$A \;B \rightarrow Y = \overline{A + B} \quad OR \quad Y = \overline{A}\,\overline{B}$$

Inputs		Transistor state				Output
A	B	Q_1	Q_2	Q_3	Q_4	Y
0	0	OFF	OFF	ON	ON	1
0	1	OFF	ON	ON	OFF	0
1	0	ON	OFF	OFF	ON	0
1	1	ON	ON	OFF	OFF	0

(b)

Figure 9–20   CMOS gate schematics: (a) NAND; (b) NOR.

Handling MOS Devices to Avoid Electrostatic Discharge

The silicon dioxide layer that isolates the gate from the substrate is so thin that it is very susceptible to burn-through from electrostatic discharge (ESD). ESD occurs when a static charge moves from one surface to another, such as from a human finger to an IC. You must be very careful and use the following guidelines when handling MOS devices:

1. Store the ICs in a conductive foam, or leave them in their original container.

2. Work on a conductive surface (e.g., a metal tabletop) that is properly grounded.

3. Ground all test equipment and soldering irons.

4. Wear a wrist strap to connect your wrist to ground with a length of wire and a 1-MΩ series resistor (see Figure 9–21).

5. Do not connect signals to the inputs while the device power supply is off.

6. Connect all unused inputs to V_{DD} or ground.

7. Don't wear electrostatic-prone clothing, such as wool, silk, or synthetic fibers.

8. Don't remove or insert an IC with the power on.

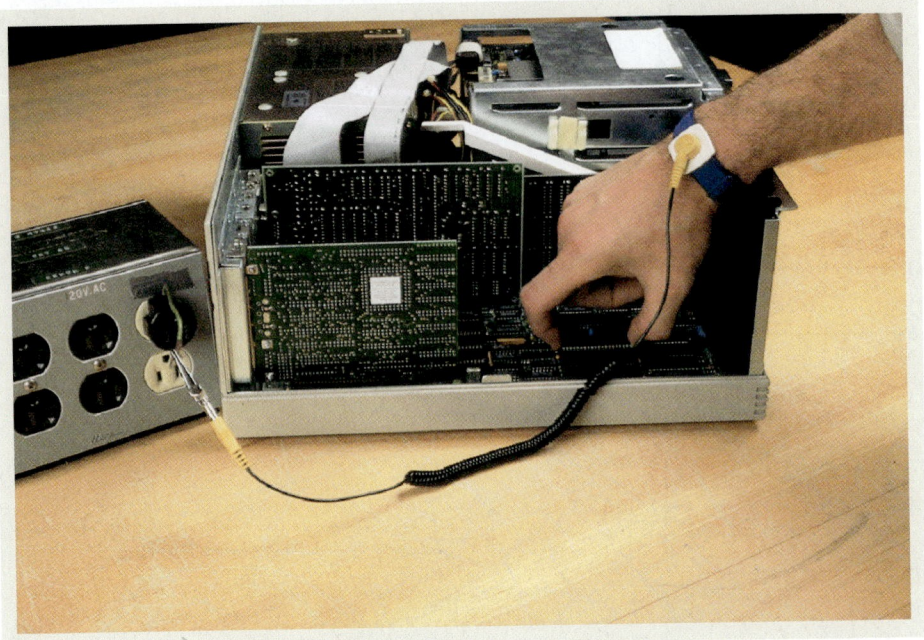

Figure 9–21 Wearing a commercially available wrist strap dissipates static charges from the technician's body to a ground connection while handling CMOS ICs.

CMOS Availability*

The CMOS family of ICs provides almost all the same functions that are available in the TTL family, plus CMOS provides several special-purpose functions not provided by TTL. Like TTL, the CMOS family has evolved into several different subfamilies, or series, each having better performance specifications than the previous one.

4000 Series: The 4000 series (or the improved 4000B) is the original CMOS line (see Figure 3–61). It became popular because it offered very low power consumption and could be used in battery-powered devices. It is much slower than any of the TTL series and has a low level of electrostatic discharge protection. The power supply voltage to the IC can range anywhere from $+3$ to $+15$ V, with the minimum 1-level input equal to $\frac{2}{3}V_{CC}$ and the maximum 0-level input equal to $\frac{1}{3}V_{CC}$.

40H00 Series: This series was designed to be faster than the 4000 series. It did overcome some of the speed limitations, but it is still much slower than LSTTL.

74C00 Series: This series was developed to be pin compatible with the TTL family, making interchangeability easier. It uses the same numbering scheme as TTL, except that it begins with 74C. It has a low-power advantage over the TTL family, but it is still much slower.

74HC00 and 74HCT00 Series: (See Appendix B for the data sheet.) The 74HC00 (high-speed CMOS) and 74HCT00 (high-speed CMOS, TTL compatible) offer a vast improvement over the original 74C00 series. The HC/HCT series are as speedy as the LSTTL series and still consume less power, depending on the operating frequency. They are pin compatible (the HCT is also input/output voltage-level compatible) with

---

*Upon publication of this textbook, data sheets for most of the TTL and CMOS logic families were available on the Internet. (See Appendix A for their World Wide Web [www] addresses.)

the TTL family, yet they offer greater noise immunity and greater voltage and temperature operating ranges. Further improvements to the HC/HCT series have led to the Advanced CMOS Logic (ACL) and Advanced CMOS Technology (ACT) series, which have even better operating characteristics.

74-BiCMOS Series: (See Appendix B for the 74ABT data sheet). Several IC manufacturers have developed technology that combines the best features of bipolar transistors and CMOS transistors, forming **BiCMOS** logic. The high-speed characteristics of bipolar *P–N* junctions are integrated with the low-power characteristics of CMOS to form an extremely low-power, high-speed family of digital logic. Each manufacturer uses different suffixes to identify its BiCMOS line. For example, Texas Instruments uses 74BCTXXX, Harris uses 74FCTXXX, and Signetics (Philips) uses 74ABTXXX.

The product line is especially well suited for and is mostly limited to microprocessor bus interface logic. This logic is mainly available in octal (8-bit) configurations used to interface 8-, 16-, and 32-bit microprocessors with high-speed peripheral devices such as memories and displays. An example is the 74ABT244 octal buffer from Philips. Its logic is equivalent to the 74244 of other families, but it has several advanced characteristics. It has TTL-compatible input and output voltages, gate input currents less than 0.01 μA, and output sink and source current capability of 64 and -32 mA, respectively. It is extremely fast, having a typical propagation delay of 2.9 ns.

One of the most desirable features of these bus-interface ICs is the fact that, when their outputs are inactive ($\overline{OE} = 1$) or HIGH, the current draw from the power supply (I_{CCZ} or I_{CCH}) is only 0.5 μA. Because interface logic spends a great deal of its time in an inactive (idle) state, this can translate into a power dissipation as low as 2.5 μW! The actual power dissipation depends on how often the IC is inactive and on the HIGH/LOW duty cycle of its outputs when it is active.

74-Low Voltage Series: (See Appendix B for the 74LV data sheet.) A new series of logic using a nominal supply voltage of 3.3 V (and lower) has been developed to meet the extremely low power design requirements of battery-powered and hand-held devices. These ICs are being designed into the circuits of notebook computers, mobile radios, hand-held video games, telecom equipment, and high-performance workstation computers. Some of the more common Low-Voltage families are identified by the following suffixes:

LV—Low-Voltage HCMOS

LVC—Low-Voltage CMOS

LVT—Low-Voltage Technology

ALVC—Advanced Low-Voltage CMOS

HLL—High-speed Low-power Low-voltage

The power consumption of CMOS logic ICs decreases approximately with the square of power supply voltage. The propagation delay increases slightly at this reduced voltage, but the speed is restored, and even increased, by using finer geometry and submicron CMOS technology that is tailored for low-power and low-voltage applications.

The supply voltage of LV logic can range from 1.2 to 3.6 V, which makes it well suited for battery-powered applications. When operated between 3.0 and 3.6 V, it can be interfaced directly with TTL levels. The switching speed of LV logic is extremely fast, ranging from approximately 9 ns for the LV series down to 2.1 ns for the ALVC. Like BiCMOS logic, the power dissipation of LV logic is negligible in the idle state or at low frequencies. At higher frequencies, the power dissipation is down to half as much as BiCMOS, depending on the power supply voltage used on the LV logic. Another key benefit of LV logic is its high output drive capability. The highest capability is provided by the LVT series, which can sink up to 64 mA and source up to 32 mA.

74AHC and 74AHCT Series: The advanced, high-speed CMOS is an enhanced version of the 74HC and 74HCT series. Designers who previously upgraded to the 74HC/HCT series can take the next step and migrate to this advanced version. It provides superior speed and low power consumption, and it has a broad product selection. 74AHC has half the static power consumption, one-third the propagation delay, high-output drive current, and can operate at a V_{CC} of 3.3 or 5 V.

Two new forms of packaging have emerged with this series: *Single-gate logic* and *Widebus*™. Single-gate logic has a lower pin count and takes up less area on a printed-circuit board by having only a single gate on the IC instead of the two-, four-, or six-gate versions. For example, the 74AHC1G00 is the single-gate version of the 74AHC00 quad NAND. It is a five-pin IC containing a single NAND gate.

The *Widebus*™ version* of logic is an extension of the octal ICs commonly found in microprocessor applications. They provide 16-bit I/O capability in a single IC package. For example, the 74AHC16244 is the *Widebus*™ version of the 74AHC244 octal buffer. It provides 16 buffers in a 48-pin IC package.

74AVC (Advanced Very Low-Voltage) CMOS Logic: Most modern internal PC bus interface circuitry must run at over 100 Mhz. That means a clock period of less than 10 ns! The AVC family was developed specifically to meet these faster speed requirements. It is designed to operate at the very low voltages (3.3, 2.5, 1.8, 1.5, and 1.2V) used in modern electronic circuitry. The maximum propagation delay is less than 2 ns, which allows a high-speed microprocessor to communicate with peripheral devices without having to enter wait states for the interface logic to catch up. Another key feature of the AVC family is called "Dynamic Output Control." This internal circuitry automatically adjusts the output impedance during logic-level transitions to minimize both the overshoot and undershoot that show up on the output as noise in high-speed switching circuits.

9–6   Emitter-Coupled Logic

Another family designed for extremely high-speed applications is emitter-coupled logic (ECL). ECL comes in two series, ECL 10K and ECL 100K. ECL is extremely fast, with propagation delay times as low as 0.8 ns. This speed makes it well suited for large mainframe computer systems that require a high number of operations per second, but that are not as concerned about an increase in power dissipation.

The high speed of ECL is achieved by never letting the transistors saturate; in fact, the whole basis for HIGH and LOW levels is determined by which transistor in a **differential amplifier** is conducting more.

Figure 9–22 shows a simplified diagram of the differential amplifier used in ECL circuits. The HIGH and LOW logic-level voltages (-0.8 and -1.7 V, respectively) are somewhat unusual and cause problems when interfacing to TTL and CMOS logic.

An ECL IC uses a supply voltage of -5.2 V at V_{EE} and 0 V at V_{CC}. The reference voltage on the base of Q_3 is set up by internal circuitry and determines the threshold between HIGH and LOW logic levels. In Figure 9–22(a), the base of Q_3 is at a more positive potential with respect to the emitter than Q_1 and Q_2 are. This causes Q_3 to conduct, placing a LOW at V_{out}.

If *either* input A or B is raised to -0.8 V (HIGH), the base of Q_1 or Q_2 will be at a higher potential than the base of Q_3, and Q_3 will stop conducting, making V_{out} HIGH. Figure 9–22(b) shows what happens when -0.8 V is placed on the A input.

Widebus is a registered trademark of Texas Instruments, Inc.

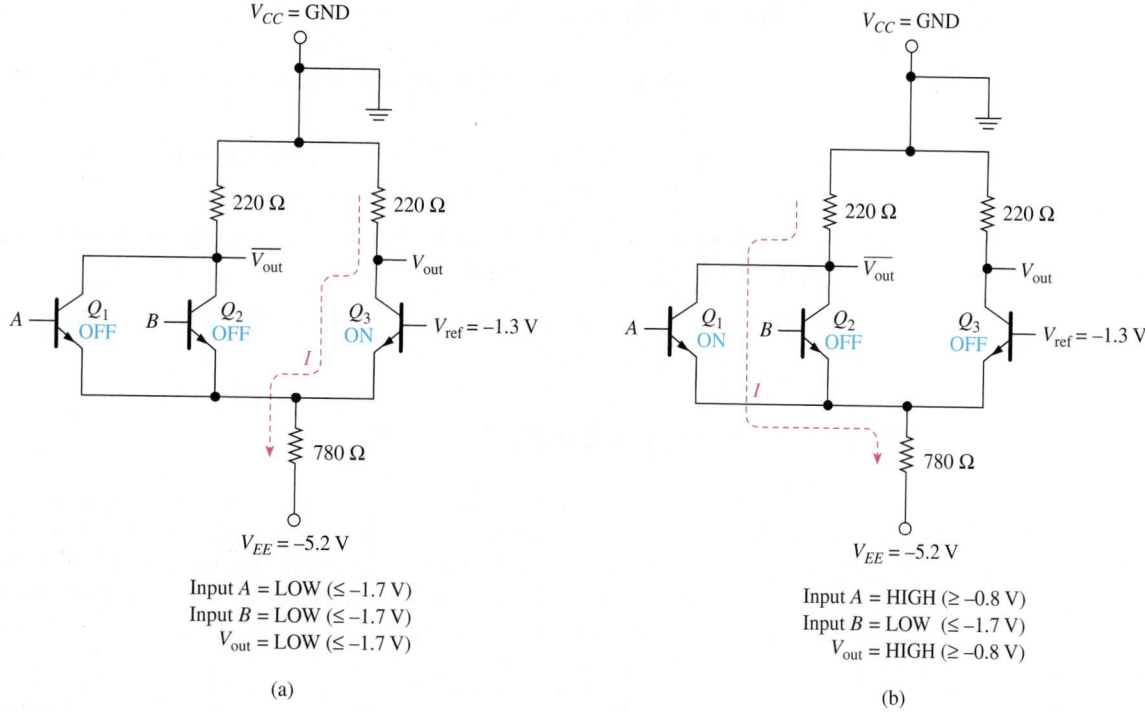

$V_{CC} = \text{GND}$

$220\ \Omega$ $220\ \Omega$

$\overline{V_{\text{out}}}$

V_{out}

A Q_1 OFF B Q_2 OFF Q_3 ON $V_{\text{ref}} = -1.3\ \text{V}$

I

$780\ \Omega$

$V_{EE} = -5.2\ \text{V}$

Input A = LOW (≤ -1.7 V)
Input B = LOW (≤ -1.7 V)
V_{out} = LOW (≤ -1.7 V)

(a)

$V_{CC} = \text{GND}$

$220\ \Omega$ $220\ \Omega$

$\overline{V_{\text{out}}}$

V_{out}

A Q_1 ON B Q_2 OFF Q_3 OFF $V_{\text{ref}} = -1.3\ \text{V}$

I

$780\ \Omega$

$V_{EE} = -5.2\ \text{V}$

Input A = HIGH (≥ -0.8 V)
Input B = LOW (≤ -1.7 V)
V_{out} = HIGH (≥ -0.8 V)

(b)

Figure 9–22 Differential amplifier input stage to an ECL OR/NOR gate: (a) LOW output; (b) HIGH output.

A
B
$V_{\text{out}} = A + B$
$\overline{V_{\text{out}}} = \overline{A + B}$

Inputs		Outputs	
A	B	V_{out}	$\overline{V_{\text{out}}}$
0	0	0	1
0	1	1	0
1	0	1	0
1	1	1	0

Figure 9–23 ECL OR/NOR symbol and truth table.

In any case, the transistors never become saturated, so capacitive charges are not built up on the base of the transistors to limit their switching speed. Figure 9–23 shows the logic symbol and truth table for the OR/NOR ECL gate.

Developing New Digital Logic Technologies[*]

The quest for logic devices that can operate at even higher frequencies and can be packed more densely in an IC package is a continuing process. Designers have high hopes for other new technologies, such as integrated injection logic (I^2L), silicon-on-sapphire (SOS), gallium arsenide (GaAs), and Josephen junction circuits. Eventually, propagation delays will be measured in picoseconds, and circuit densities will enable the supercomputer of today to become the desktop computer of tomorrow.

---

[*]New IC families are introduced every year. To keep up with the latest technology, please visit some of the Internet sites listed in Appendix A for the manufacturers of digital logic.

9–12. What effect did the Schottky-clamped transistor have on the operation of the standard TTL IC?

9–13. The earlier 4000 series of CMOS ICs provided what advantage over earlier TTL ICs? What was their disadvantage?

9–14. The BiCMOS family of ICs is fabricated using both CMOS transistors and _____ transistors.

9–15. The high speed of ECL ICs is achieved by fully saturating the ON transistor. True or false?

9–7   Comparing Logic Families

Throughout the years, system designers have been given a wide variety of digital logic to choose from. The main parameters to consider include speed, power dissipation, availability, types of functions, noise immunity, operating frequency, output-drive capability, and interfacing. First and foremost, however, are the basic speed and power concerns. Table 9–3 shows the propagation delay, power dissipation, and speed–power product for the most popular families.

TABLE 9–3	Typical Single-Gate Performance Specifications		
Family	**Propagation Delay (ns)**	**Power Dissipation (mW)**	**Speed–Power Product pW-s (picowatt-seconds)**
74	10	10	100
74S	3	20	60
74LS	9	2	18
74ALS	4	1	4
74F	2.7	4	11
4000B (CMOS)	105	1 at 1 MHz	105
74HC (CMOS)	10	1.5 at 1 MHz	15
74BCT (BiCMOS)	2.9	0.0003 to 7.5	0.00087 to 22
100K (ECL)	0.8	40	32

The speed–power product is a type of figure of merit, but it does not necessarily tell the ranking within a specific application. For example, to say that the speed–power product of 15 pW-s for the 74HC family is better than 32 pW-s for the 100K ECL family totally ignores the fact that ECL is a better choice for ultrahigh-speed applications.

Another way to view the speed–power relationships is with the graph shown in Figure 9–24. From the graph, you can see the wide spectrum of choices available. 4000B CMOS and 100K ECL are at opposite ends of the spectrum of speed versus power, whereas 74ALS and 74F seem to offer the best of both worlds.

The operating frequency for CMOS devices is critical for determining power dissipation. At very low frequencies, CMOS devices dissipate very little power, but at higher switching frequencies, charging and discharging the gate capacitances draws a heavy current from the power supply (I_{CC}) and, thus, increases the power dissipation ($P_D = V_{CC} \times I_{CC}$), as shown in Figure 9–25. We see that at high frequencies the power dissipations of 74HC CMOS and 74LS TTL are comparable. At today's microprocessor clock rates, 74HC CMOS ICs actually dissipate more power than 74LS or 74ALS. However, in typical systems, only a fraction of the gates are connected to switch as fast as the clock rate, so significant power savings can be realized by using the 74HC CMOS series.

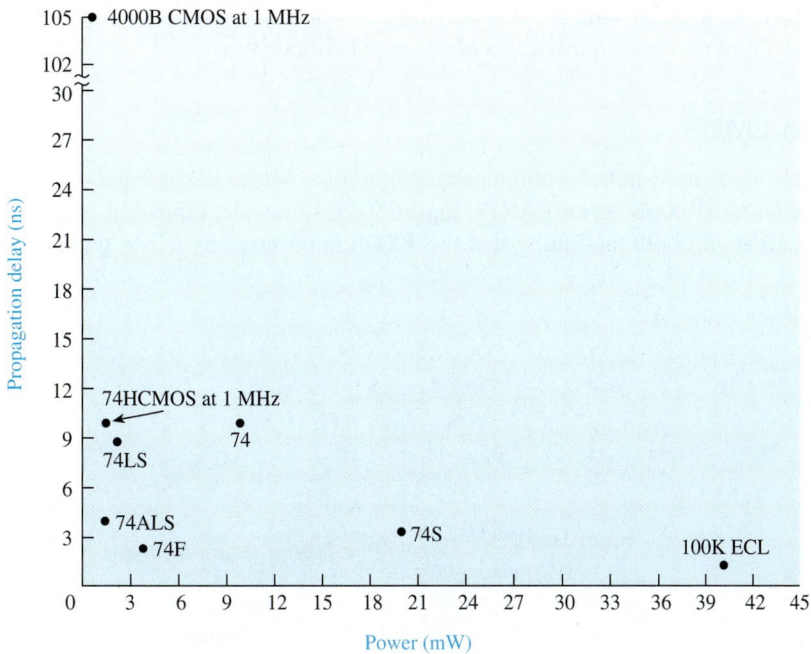

Figure 9–24   Graph of propagation delay versus power. (Courtesy of Philips Semiconductors.)

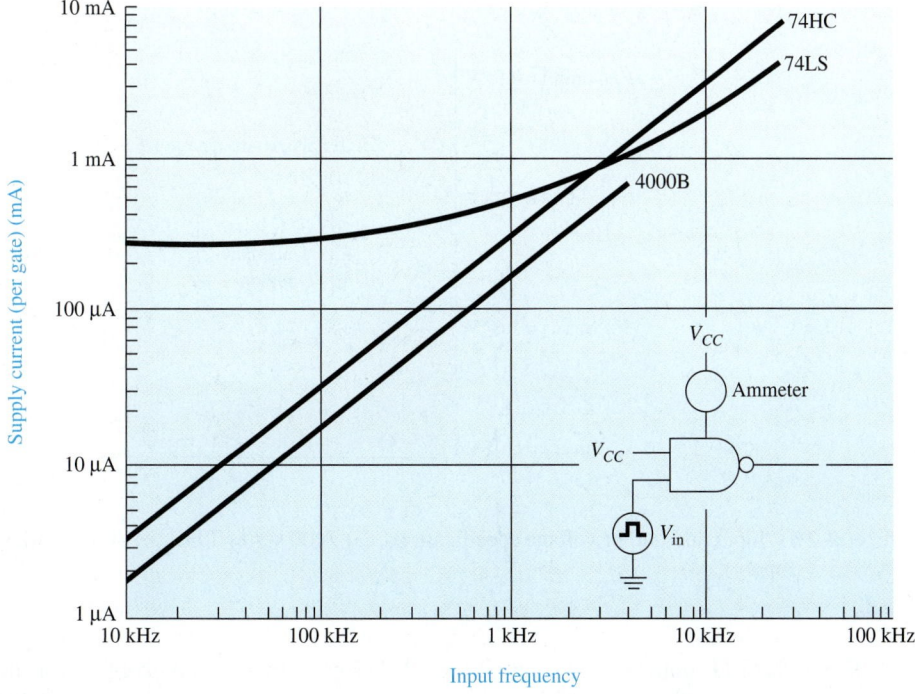

Figure 9–25   Power supply current versus frequency. (Courtesy of Philips Semiconductors.)

9–8   Interfacing Logic Families

Often, the need arises to interface (connect) between the various TTL and CMOS families. You have to make sure that a HIGH out of a TTL gate looks like a HIGH to the input of a CMOS gate, and vice versa. The same holds true for the LOW logic levels.

You also have to make sure that the driving gate can sink or source enough current to meet the input current requirements of the gate being driven.

TTL to CMOS

Let's start by looking at the problems that might arise when interfacing a standard 7400 series TTL to a 4000B series CMOS. Figure 9–26 shows the input and output voltage specifications for both, assuming that the 4000B is powered by a 5-V supply.

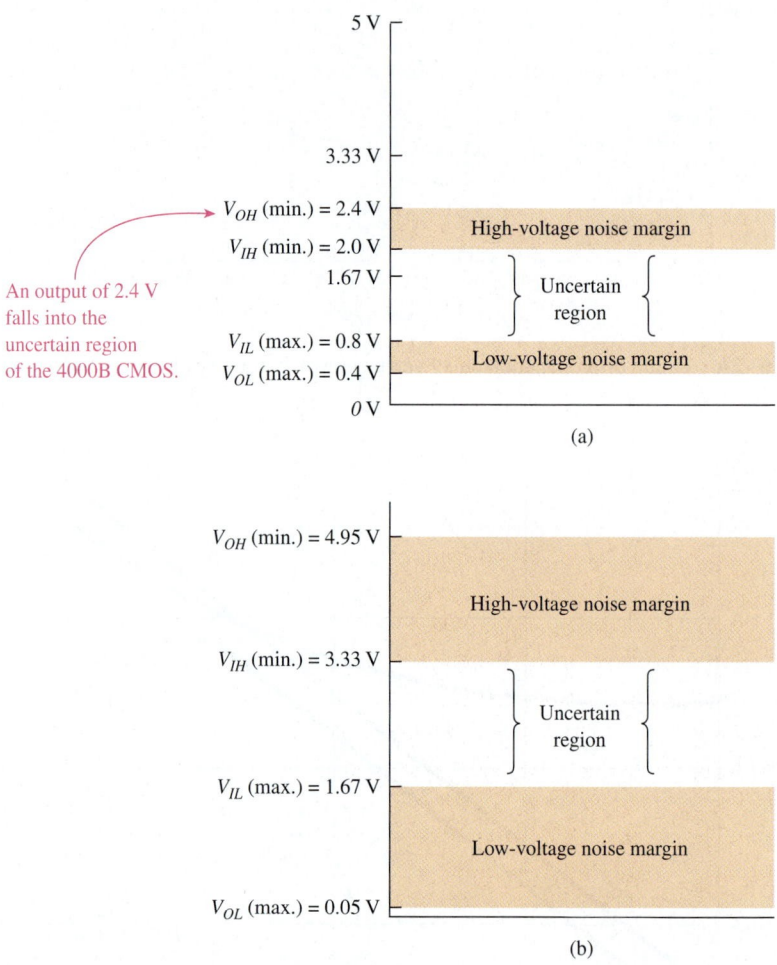

Figure 9–26 Input and output voltage specifications: (a) 7400 series TTL; (b) 4000B series CMOS (5-V supply).

When the TTL gate is used to drive the CMOS gate, there is no problem for the LOW-level output because the TTL guarantees a maximum LOW-level output of 0.4 V and the CMOS will accept any voltage up to 1.67 V ($\frac{1}{3}V_{CC}$) as a LOW-level input.

But, for the HIGH level, the TTL may output as little as 2.4 V as a HIGH. The CMOS expects at least 3.33 V as a HIGH-level input. Therefore, 2.4 V is unacceptable because it falls within the uncertain region. However, a resistor can be connected between the CMOS input to V_{CC}, as shown in Figure 9–27, to solve the HIGH-level input problem.

In Figure 9–27, with V_{out1} LOW, the 7404 will sink current from the 10-kΩ resistor and the I_{IL} from the 4069B, making V_{out2} HIGH. With V_{out1} HIGH, the 10-kΩ re-

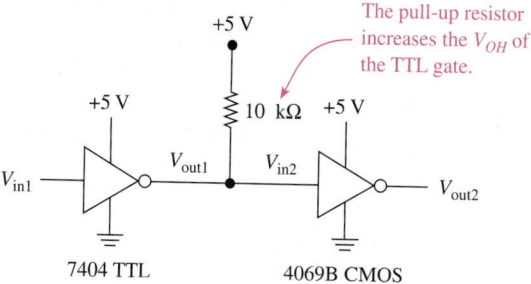

Figure 9–27 Using a pull-up resistor to interface TTL to CMOS.

sistor will pull the voltage at V_{in2} up to 5.0 V, causing V_{out2} to go LOW. The 10-kΩ resistor is called a *pull-up resistor* and is used to raise the output of the TTL gate closer to 5 V when it is in a HIGH output state. With V_{out1} HIGH, the voltage at V_{in2} will be almost 5 V because current into the 4069B is so LOW (≈ 1 μA) that the voltage drop across the 10 kΩ is insignificant, leaving almost 5.0 V at V_{in2} (V_{in2} = 5 V − 1 μA × 10 kΩ = 4.99 V).

The other thing to look at when interfacing is the current levels of all gates that are involved. In this case, the 7404 can sink (I_{OL}) 16 mA, which is easy enough for the I_{IL} of the 4069B (1 μA) plus the current from the 10-kΩ resistor (5 V/10 kΩ = 0.5 mA). I_{OH} of the 7404 (−400 μA) is no problem either because, with the pull-up resistor, the 7404 will not have to source current.

CMOS to TTL

When driving TTL from CMOS, the voltage levels are no problem because the CMOS will output approximately 4.95 V for a HIGH and 0.05 V for a LOW, which is easily interpreted by the TTL gate.

But, the current levels can be a real concern because 4000B CMOS has severe output-current limitations. (The 74C and 74HC series have much better output-current capabilities, however.) Figure 9–28 shows the input/output currents that flow when interfacing CMOS to TTL.

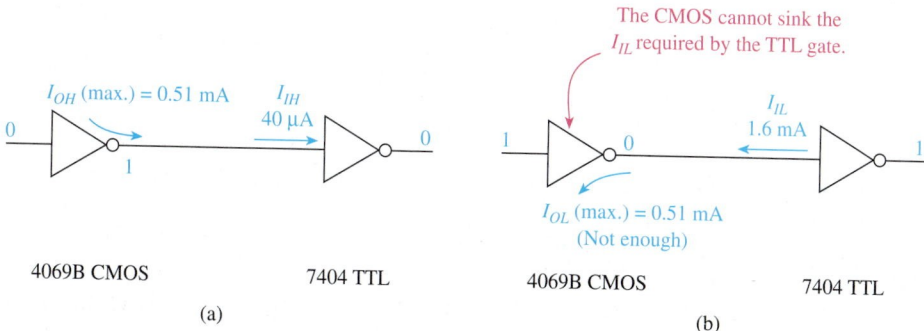

Figure 9–28 Current levels when interfacing CMOS to TTL: (a) CMOS I_{OH}; (b) CMOS I_{OL}.

For the HIGH output condition [Figure 9–28(a)], the 4069B CMOS can source a maximum current of 0.51 mA, which is enough to supply the HIGH-level input current (I_{IH}) to one 7404 inverter. But, for the LOW output condition, the 4069B can also sink only 0.51 mA, which is not enough for the 7404 LOW-level input current (I_{IL}).

Most of the 4000B series has the same problem of low-output drive current capability. To alleviate the problem, two special gates, the 4050 buffer and the 4049 inverting buffer, are specifically designed to provide high output current to solve many interfacing problems. They have drive capabilities of $I_{OL} = 4.0$ mA and $I_{OH} = -0.9$ mA, which is enough to drive two 74TTL loads, as shown in Figure 9–29.

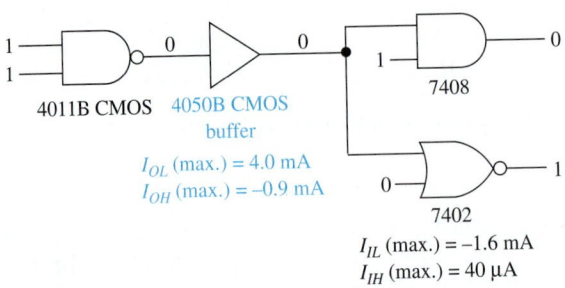

Figure 9–29 Using the 4050B CMOS buffer to supply sink and source current to two standard TTL loads.

Team Discussion

Discuss several attributes that make the HCT family an excellent choice for interfacing situations.

If the CMOS buffer were used to drive another TTL series, let's say, the 74LS series, we would have to refer to a TTL data book to determine how many loads could be connected without exceeding the output current limits. (The 4050B can actually drive *10* 74LS loads.) Table 9–4 summarizes the input/output voltage and current specifications of some popular TTL and CMOS series, which easily enables us to determine interface parameters and family characteristics.

TABLE 9–4	Worst-Case Values for Interfacing Considerations[a]					
Parameter	4000B CMOS	74HCMOS	74HCTMOS	74TTL	74LSTTL	74ALSTTL
V_{IH} (min.) (V)	3.33	3.5	2.0	2.0	2.0	2.0
V_{IL} (max.) (V)	1.67	1.0	0.8	0.8	0.8	0.8
V_{OH} (min.) (V)	4.95	4.9	4.9	2.4	2.7	2.7
V_{OL} (max.) (V)	0.05	0.1	0.1	0.4	0.4	0.4
I_{IH} (max.) (μA)	1	1	1	40	20	20
I_{IL} (max.) (μA)	-1	-1	-1	-1600	-400	-100
I_{OH} (max.) (mA)	-0.51	-4	-4	-0.4	-0.4	-0.4
I_{OL} (max.) (mA)	0.51	4	4	16	8	4

[a]All values are for $V_{supply} = 5.0$ V.

By reviewing Table 9–4, we can see that the 74HCMOS has relatively low input-current requirements compared to the bipolar TTL series. Its HIGH output can source 4 mA, which is 10 times the capability of the TTL series. Also, the noise margin for the 74HCMOS is much wider than any of the TTL series (1.4 V HIGH, 0.9 V LOW).

Because of the low input-current requirements, any of the TTL series can drive several of the 74HCMOS loads. An interfacing problem occurs in the voltage level, however. The 74HCMOS logic expects 3.5 V at a minimum for a HIGH-level input. The worst case (which we must always assume could happen) for the HIGH output level of a 74LSTTL is 2.7 V, so we will need to use a pull-up resistor at the 74LSTTL output to ensure an adequate HIGH level for the 74HCMOS input as shown in Figure 9–30.

The combinations of interfacing situations are extensive (74HCMOS to 74ALSTTL, 74TTL to 74LSTTL, and so on). In each case, reference to a data book must be made to

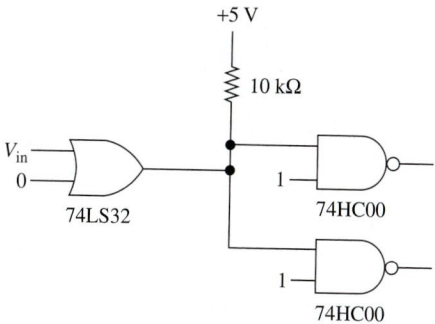

Figure 9–30 Interfacing 74LSTTL to 74HCMOS.

check the worst-case voltage and current parameters, as we will see in upcoming examples. In general, a pull-up resistor is required when interfacing TTL to CMOS to bring the HIGH-level TTL output up to a suitable level for the CMOS input. (The exception to the rule is when using *74HCTMOS,* which is designed for TTL voltage levels.) The disadvantages of using a pull-up resistor are that it takes up valuable room on a printed-circuit board and that it dissipates power in the form of heat.

Different series within the TTL family and the TTL-compatible 74HCTMOS series can be interfaced directly. The main concern then is determining how many gate loads can be connected to a single output.

Level Shifting

Another problem arises when you interface families that have different supply voltages. For example, the 4000B series can use anywhere from $+3$ to $+15$ V for a supply, and the ECL series uses -5.2 V for a supply.

This problem is solved by using **level-shifter** ICs. The 4049B and 4050B buffer ICs that were introduced earlier are also used for voltage-level shifting. Figure 9–31 shows the connections for interfacing 15-V CMOS to 5-V TTL.

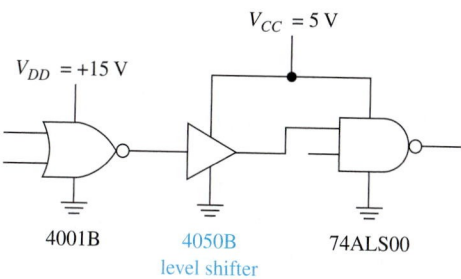

Figure 9–31 Using a level shifter to convert 0-V/15-V logic to 0-V/5-V logic.

The 4050B level-shifting buffer is powered from a 5-V supply and can actually accept 0-V/15-V logic levels at the input and then output the corresponding 0-V/5-V logic levels at the output. For an inverter function, use the 4049B instead of the 4050B.

The reverse conversion, 5-V TTL to 15-V CMOS, is accomplished with the 4504B CMOS level shifter, as shown in Figure 9–32. The 4504B level-shifting buffer requires two power supply inputs: the 5-V V_{CC} supply to enable it to recognize the 0-V/5-V input levels, and the 15-V supply to enable it to provide 0-V/15-V output levels.

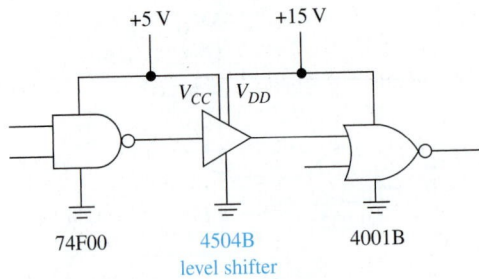

Figure 9–32   Level shifting 0-V/5-V TTL logic to 0-V/15-V CMOS logic.

ECL Interfacing

Interfacing 0-V/5-V logic levels to -5.2-V/0-V ECL circuitry requires another set of level shifters (or translators): the ECL 10125 and the ECL 10124, whose connections are shown in Figure 9–33.

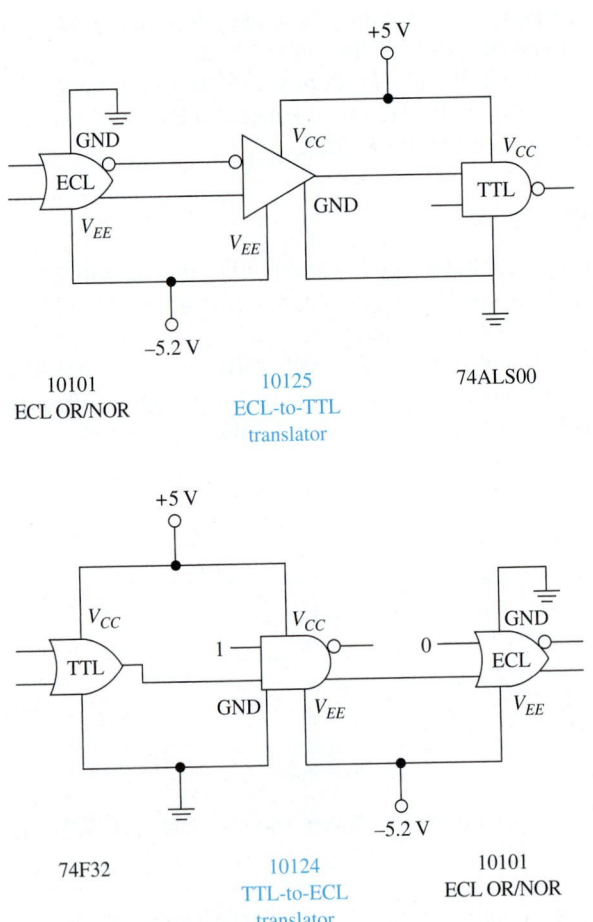

Figure 9–33   Circuit connections for translating between TTL and ECL levels.

EXAMPLE 9–5

Determine from Table 9–4 how many 74LSTTL logic gates can be driven by a single 74TTL logic gate.

CHAPTER 9 | LOGIC FAMILIES AND THEIR CHARACTERISTICS

Solution: The output voltage levels (V_{OL}, V_{OH}) of the 74TTL series are compatible with the input voltage levels (V_{IL}, V_{IH}) of the 74LSTTL series. The voltage noise margin for the HIGH level is 0.4 V (2.4 − 2.0), and for the LOW level it is 0.4 V (0.8 − 0.4).

The HIGH-level output current (I_{OH}) for the 74TTL series is −400 μA. Each 74LSTTL gate draws 20 μA of input current for the HIGH level (I_{IH}), so one 74TTL gate can drive 20 74LSTTL loads in the HIGH state (400 μA/20 μA = 20).

For the LOW state, the 74TTL I_{OL} is 16 mA and the 74LSTTL I_{IL} is −400 μA, meaning that, for the LOW condition, one 74TTL can drive 40 74LSTTL loads (16 mA/400 μA = 40). Therefore, considering both the LOW and HIGH conditions, a single 74TTL can drive 20 74LSTTL gates.

EXAMPLE 9–6

One 74HCT04 inverter is to be used to drive one input to each of the following gates: 7400 (NAND), 7402 (NOR), 74LS08 (AND), and 74ALS32 (OR). Draw the circuit, and label input and output worst-case voltages and currents. Will there be total voltage and current compatibility?

Solution: The circuit is shown in Figure 9–34. Figure 9–34(a) shows the worst-case HIGH-level values. If you sum all the input currents, the total that the 74HCT04 must supply is 120 μA (40 μA + 40 μA + 20 μA + 20 μA), which is well below the −4-mA maximum source capability of the 74HCT04. Also, the 4.9-V output voltage of the 74HCT04 *is* compatible with the 2.0-V minimum requirement of the TTL inputs, leaving a noise margin of 2.9 V (4.9 V − 2.0 V).

Figure 9–34(b) shows the worst-case LOW-level values. The sum of all the TTL input currents is 3.7 mA (1.6 mA + 1.6 mA + 400 μA + 100 μA), which is less than the 4-mA maximum sink capability of the 74HCT04. Also, the 0.1-V output of the 74HCT04 *is* compatible with the 0.8-V maximum requirement of the TTL inputs, leaving a noise margin of 0.7 V (0.8 V − 0.1 V).

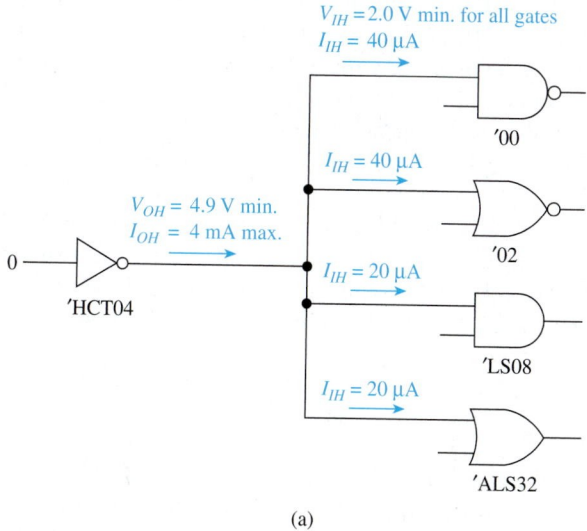

(a)

Figure 9–34 Interfacing a 74HCTMOS to several different TTL series: (a) HIGH-level values;

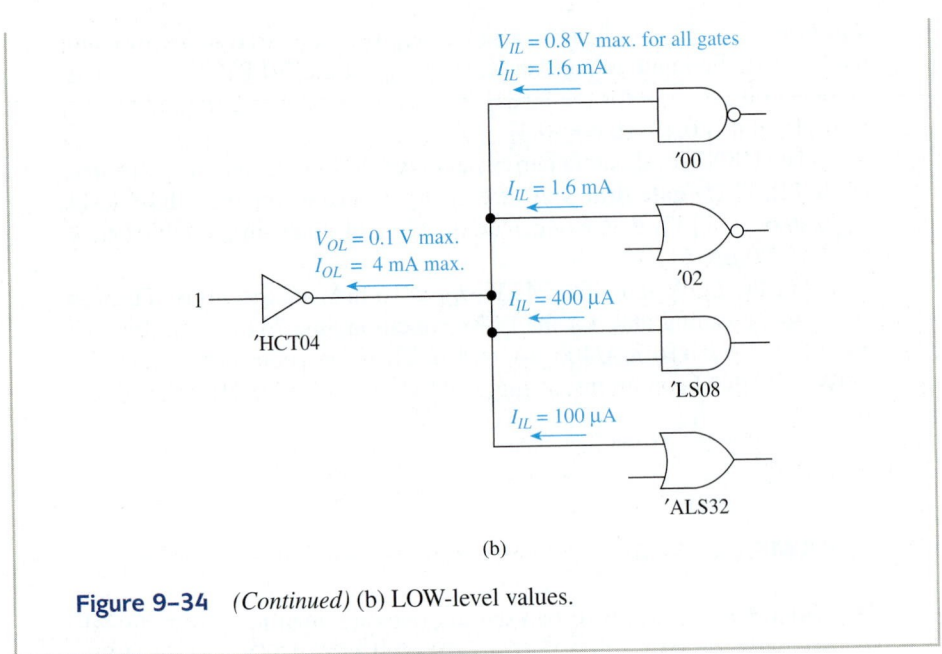

$V_{IL} = 0.8$ V max. for all gates
$I_{IL} = 1.6$ mA
'00

$I_{IL} = 1.6$ mA
'02

$V_{OL} = 0.1$ V max.
$I_{OL} = 4$ mA max.
'HCT04

$I_{IL} = 400$ µA
'LS08

$I_{IL} = 100$ µA
'ALS32

(b)

Figure 9–34 *(Continued)* (b) LOW-level values.

9–9 CPLD Electrical Characteristics

Each manufacturer of programmable logic will develop ICs with their own specific electrical characteristics. Now that we know how to interpret specs like input/output voltage and current, and speed power values, it's simply a matter of viewing the data sheet from the manufacturer's Web site and looking at the electrical characteristics section. The ICs that we are most concerned with in this text are the Altera MAX 7000 family of CPLDs. These data sheets are found in the literature section of the Altera Web site at www.altera.com.

MAX 7000 CPLD Specifications

MAX 7000 devices are fabricated with advanced CMOS technology and can interface with 5-, 3.3-, 2.5- or 1.8-V devices. They have one set of V_{CC} pins for internal operations and input buffers (V_{CCINT}) and another set for I/O drivers (V_{CCIO}).

The EPM7128S CPLD used on the UP-1 programmer board is part of the MAX7000S series. It can interface with 5-V and 3.3-V devices as shown in Figure 9–35. V_{CCINT} is always connected to 5 V, and V_{CCIO} must be connected to 5 V or 3.3 V, depending on the circuitry it is interfacing with.

MAX 7000S devices also provide an optional open-drain (functionally equivalent to open-collector) output for each I/O pin. Using an external pull-up resistor to 5 V insures HIGH-level voltage compatibility with both TTL and CMOS devices.

The DC operating conditions listed in the data sheet show that the MAX 7000 family has complete TTL and CMOS compatibility. For example, devices with V_{CCIO} set at 5 V have the following specs: $V_{OH} = 2.4$ V (min) at $I_{OH} = -4$mA and $V_{OL} = 0.45$ V (max) at

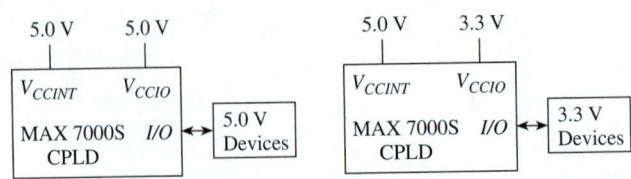

Figure 9–35 Interfacing to multiple voltages.

$I_{OL} = 12$mA. These voltage and current ratings will provide a substantial margin to interface to both TTL and CMOS devices.

MAX 7000-family CPLDs are extremely fast, having propagation delay times as low as 3.5 ns. A speed power optimization feature is also provided. Speed-critical macrocell portions of a design can be programmed to run at high speed/full power (turbo bit set HIGH), while the remaining portions run at reduced speed/low power (turbo bit set LOW). This can result in as much as a 50% reduction in power consumption. The power dissipation for the device can be calculated using the equation $P_D = I_{CCINT} \times V_{CCINT} + PIO$. PIO depends on the output load connected to the device, and I_{CCINT} for the EPM7128S CPLD can be approximated from the chart in Figure 9–36. The chart shows that the typical I_{CC} increases with operating frequency, but it is halved when programmed to the Low Power mode.

EPM7128S

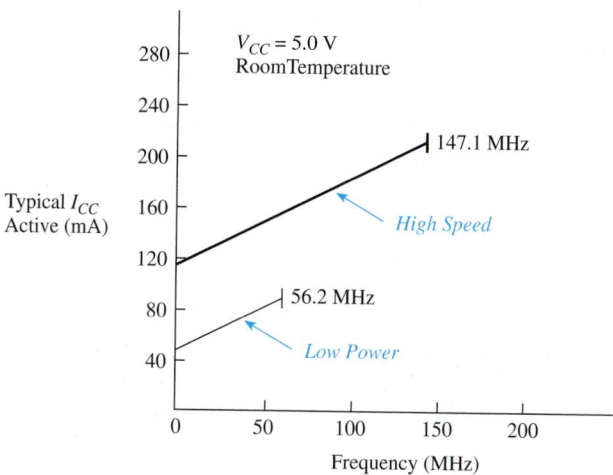

Figure 9–36 I_{CC} versus frequency for high speed and low power configurations.

Review Questions

9–16. Which logic family is faster, the 74ALS or 74HC?

9–17. Which logic family has a lower power dissipation, the 100K ECL or 74LS?

9–18. In the graph of Figure 9–24, the IC families plotted closest to the origin have the best speed–power products. True or false?

9–19. Use the graph in Figure 9–25 to determine which logic family has a lower power dissipation at 100 kHz, the 74HC CMOS or 74LS TTL?

9–20. What is the function of a pull-up resistor when interfacing a TTL IC to a CMOS IC?

9–21. What problem arises when interfacing 4000 series CMOS ICs to standard TTL ICs?

9–22. Which family has more desirable output voltage specifications, the 74HCTMOS or 74ALSTTL? Why?

9–23. What IC specifications are used to determine how many gates of one family can be driven from the output of another family?

Summary

In this chapter, we have learned that

1. There are basically three stages of internal circuitry in a TTL (transistor–transistor logic) IC: input, control, and output.

2. The input current (I_{IL} or I_{IH}) to an IC gate is a constant value specified by the IC manufacturer.

3. The output current of an IC gate depends on the size of the load connected to it. Its value cannot exceed the maximum rating of the chip, I_{OL} or I_{OH}.

4. The HIGH- and LOW-level output voltages of the standard TTL family are *not* 5 and 0 V but typically are 3.4 and 0.2 V.

5. The propagation delay is the length of time that it takes for the output of a gate to respond to a stimulus at its input.

6. The rise and fall times of a pulse describe how long it takes for the voltage to travel between its 10% and 90% levels.

7. Open-collector outputs are required whenever logic outputs are connected to a common point.

8. Several improved TTL families are available and continue to be introduced each year, providing decreased power consumption and decreased propagation delay.

9. The CMOS family uses complementary metal oxide semiconductor transistors instead of the bipolar transistors used in TTL ICs. Traditionally, the CMOS family consumed less power but was slower than TTL. However, recent advances in both technologies have narrowed the differences.

10. The BiCMOS family combines the best characteristics of bipolar technology and CMOS technology to provide logic functions that are optimized for the high-speed, low-power characteristics required in microprocessor systems.

11. Emitter-coupled logic provides the highest-speed ICs. Its drawback is its very high power consumption.

12. A figure of merit of IC families is the product of their propagation delay and power consumption, called the speed–power product (the lower, the better).

13. When interfacing logic families, several considerations must be made. The output voltage level of one family must be high and low enough to meet the input requirements of the receiving family. Also, the output current capability of the driving gate must be high enough for the input draw of the receiving gate or gates.

Glossary

BiCMOS: A logic family that is fabricated from a combination of bipolar transistors and complementary MOSFETs. It is an extremely fast low-power family.

Bipolar Transistor: Three-layer *N–P–N* or *P–N–P* junction transistor.

Buffer: A device placed between two other devices that provides isolation and current amplification. The input logic level is equal to the output logic level.

CMOS: Complementary metal oxide semiconductor.

Decoupling: A method of isolating voltage irregularities on the V_{CC} power supply line from an IC V_{CC} input.

Differential Amplifier: An amplifier that basically compares two inputs and provides an output signal based on the *difference* between the two input signals.

ECL: Emitter-coupled logic.

EMI: Electromagnetic interference. Undesirable radiated energy from a digital system caused by magnetic fields induced by high-speed switching.

Fall Time: The time required for a digital pulse to fall from 90% down to 10% of its maximum voltage level.

Fan-Out: The number of logic gate inputs that can be driven from a single gate output of the same subfamily.

Level Shifter: A device that provides an interface between two logic families having different power supply voltages.

MOSFET: Metal oxide semiconductor field-effect transistor.

NMOS: A family of ICs fabricated with N-channel MOSFETs.

Noise Margin: The voltage difference between the guaranteed output voltage level and the required input voltage level of a logic gate.

Open-Collector Output: A special output stage of the TTL family that has the upper transistor of a totem-pole configuration removed.

PMOS: A family of ICs fabricated with P-channel MOSFETs.

Power Dissipation: The electrical power (watts) that is consumed by a device and given off (dissipated) in the form of heat.

Propagation Delay: The time required for a change in logic level to travel from the input to the output of a logic gate.

Pull-Up Resistor: A resistor with one end connected to V_{CC} and the other end connected to a point in a logic circuit that needs to be raised to a voltage level closer to V_{CC}.

Rise Time: The time required for a digital pulse to rise from 10% up to 90% of its maximum voltage level.

Sink Current: Current entering the output or input of a logic gate.

Source Current: Current leaving the output or input of a logic gate.

Substrate: The silicon supporting structure or framework of an integrated circuit.

Totem-Pole Output: The output stage of the TTL family having two opposite-acting transistors, one above the other.

TTL: Transistor–transistor logic.

Wired-AND: The AND function that results from connecting several open-collector outputs together.

Problems

Section 9–1

9–1. What is the purpose of diodes D_1 and D_2 in Figure 9–1?

9–2. In Figure 9–1, when input A is connected to ground (0 V), calculate the approximate value of emitter current in Q_1.

9–3. In Figure 9–1, when the output is HIGH, how do you account for the output voltage being only about 3.4 V instead of 5.0 V?

9–4. In Figure 9–1, describe the state (ON or OFF) of Q_3 and Q_4 for

(a) Both inputs A and B LOW

(b) Both inputs A and B HIGH

Section 9–2

9–5. What does the negative sign in the rating of source current (e.g., $I_{OH} = -400\ \mu A$) signify?

9–6. For TTL outputs, which is higher, the source current or the sink current?

C   **9–7. (a)** Find V_a and I_a in the circuits of Figure P9–7 using the following specifications:

$$I_{IL} = -1.6\ \text{mA} \qquad I_{IH} = 40\ \mu A$$
$$V_{IL} = 0.8\ \text{V max} \qquad V_{IH} = 2.0\ \text{V min}$$
$$I_{OL} = 16\ \text{mA} \qquad I_{OH} = -400\ \mu A$$
$$V_{OL} = 0.2\ \text{V typ.} \qquad V_{OH} = 3.4\ \text{V typ.}$$

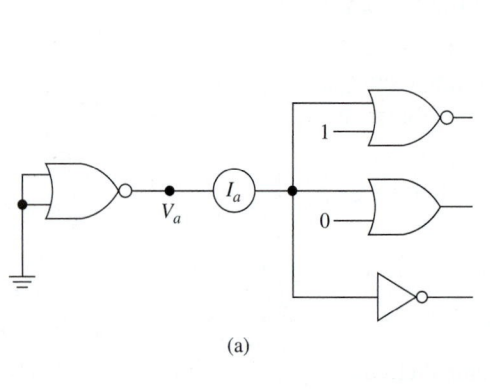

(a)

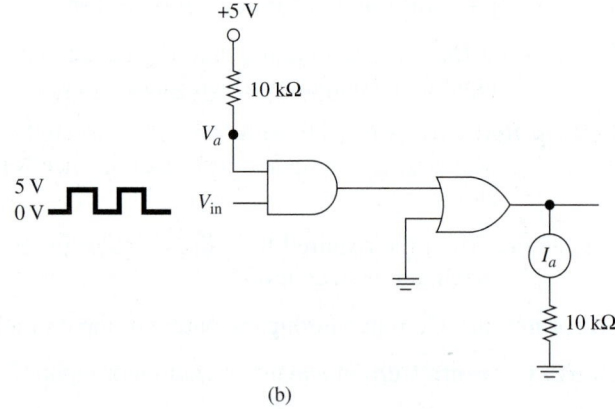

(b)

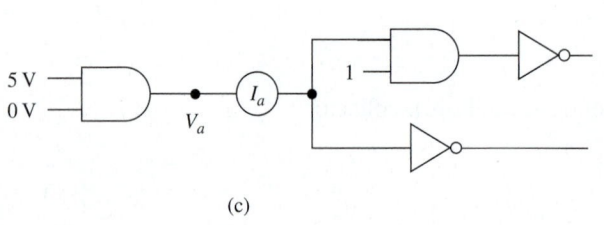

(c)

(d)

Figure P9–7

(b) Repeat part (a) using input/output specifications that you gather from a TTL data book, assuming that all gates are 74LSXX series.

Section 9–3

9–8. The input and output waveforms to an OR gate are given in Figure P9–8. Determine:

(a) The period and frequency of V_{in}

(b) The rise and fall times (t_r, t_f) of V_{in}

(c) The propagation delay times of (t_{PLH}, t_{PHL}) of the OR gate

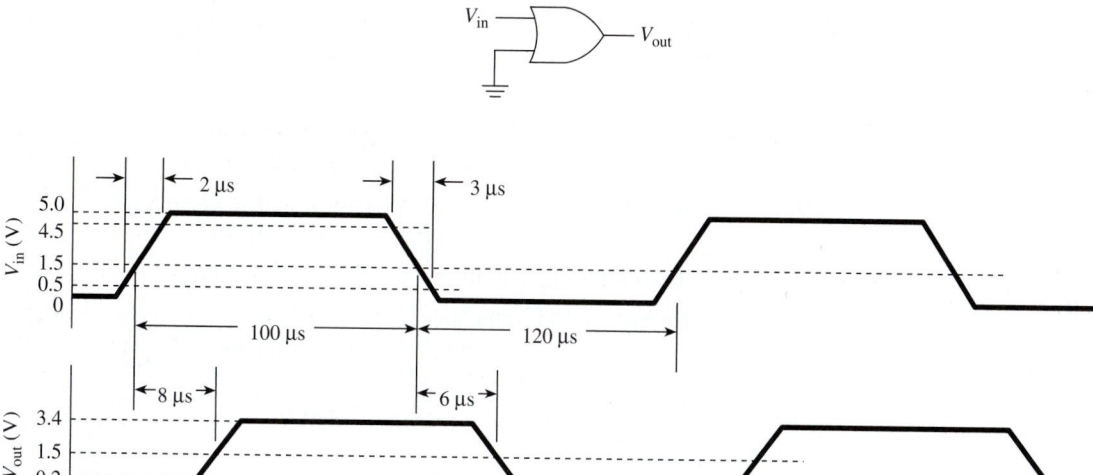

Figure P9–8

9–9. The propagation delay times for a 74LS08 AND gate (Figure P9–9) are $t_{PLH} = 15$ ns, $t_{PHL} = 20$ ns, and for a 7402 NOR gate, they are $t_{PLH} = 22$ ns, $t_{PHL} = 15$ ns. Sketch V_{out1} and V_{out2} showing the effects of propagation delay. (Assume 0 ns for the rise and fall times.)

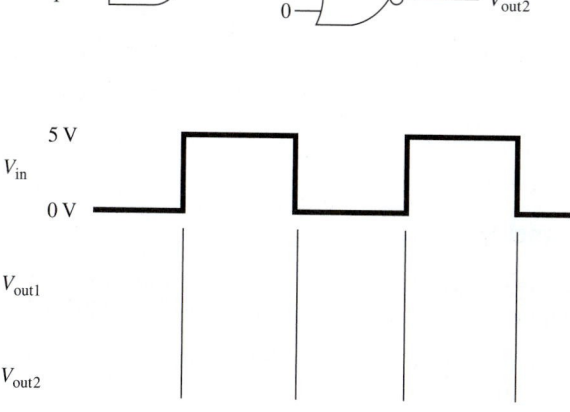

Figure P9–9

9–10. (a) Repeat Problem 9–9 for the circuit of Figure P9–10(a).

(b) Repeat Problem 9–9 for V_a, V_b, V_c, and V_d in Figure P9–10(b). (Use a TTL data book to determine the propagation delay times.)

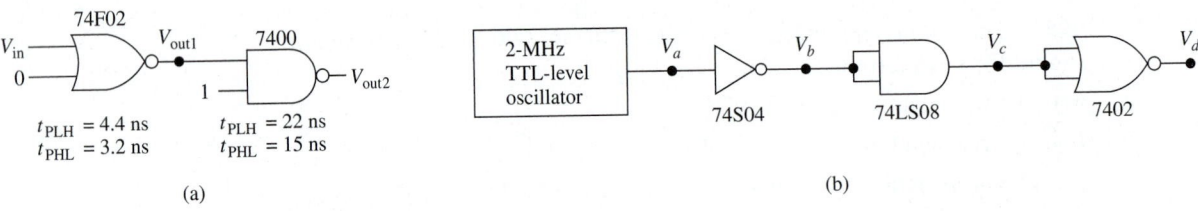

(a)   (b)

Figure P9–10

9–11. Refer to a TTL data book or the data sheets in Appendix B. Use the total supply current (I_{CCL}, I_{CCH}) to compare the power dissipation of a 7400 versus a 74LS00.

9–12. Refer to a TTL data sheet to compare the typical LOW-level output voltage (V_{OL}) at maximum output current for a 7400 versus a 74LS00.

9–13. (a) Refer to a TTL data sheet to determine the noise margins for the HIGH and LOW states of both the 7400 and 74LS00.

(b) Which has better noise margins, the 7400 or 74LS00?

9–14. (a) Refer to a TTL data sheet (or Appendix B) to determine which can sink more current at its output, the commercial 74LS00 or the military 54LS00.

(b) Which has a wider range of recommended V_{CC} supply voltage, the 7400 or the 5400?

9–15. Why is a pull-up resistor required at the output of an open-collector gate to achieve a HIGH-level output?

9–16. The wired-AND circuits in Figure P9–16 use all open-collector gates. Write the simplified Boolean equations at X and Y.

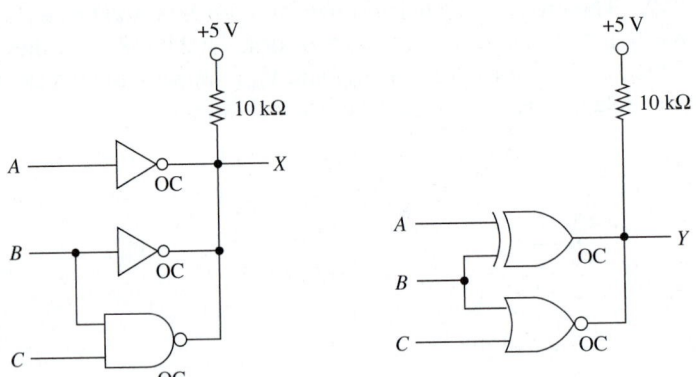

Figure P9–16

Sections 9–4 and 9–5

9–17. Make a general comparison of both the switching speed and power dissipation of the 7400 TTL series versus the 4000B CMOS series.

9–18. Which type of transistor, bipolar or field effect, is used in TTL ICs? In CMOS ICs?

9–19. Why is it important to store MOS ICs in antistatic conductive foam?

9–20. What is the principal reason that ECL ICs reach such high switching speeds?

9–21. Table 9–3 shows that the speed–power product of the 74ALS family is much better than the 100K ECL family. Why, then, are some large mainframe computers based on ECL technology?

9–22. The graph in Figure 9–24 shows the 4000B CMOS family in the opposite corner from the 100K ECL family. What is the significance of this?

9–23. Referring to Figure 9–25, which logic family dissipates less power at low frequencies, the 74LS or 74HC?

Section 9–8

C

9–24. (a) Using the data in Table 9–4, draw a graph of input and output specifications similar to Figure 9–26 for the 74HCMOS and the 74ALSTTL IC series.

(b) From your graphs of the two IC series, compare the HIGH- and LOW-level noise margins.

(c) From your graphs, can you see a problem in *directly* interfacing:

    (1) The 74HCMOS to the 74ALSTTL?

    (2) The 74ALSTTL to the 74HCMOS?

9–25. Refer to Table 9–4 to determine which of the following interfacing situations (driving gate-to-gate load) will require a pull-up resistor, and why?

(a) 74TTL to 74ALSTTL

(b) 74HCMOS to 74TTL

(c) 74TTL to 74HCMOS

(d) 74LSTTL to 74HCTMOS

(e) 74LSTTL to 4000B CMOS

9–26. Of the interfacing situations given in Problem 9–25, will any of the driving gates have trouble sinking or sourcing current to a single connected gate load?

9–27. From Table 9–4, determine:

(a) How many 74LSTTL loads can be driven by a single 74HCTMOS gate?

(b) How many 74HCTMOS loads can be driven by a single 74LSTTL gate?

Schematic Interpretation Problems

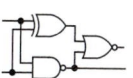

See Appendix G for the schematic diagrams.

S

9–28. Assume that the inverter U4:A in the Watchdog Timer schematic has the following propagation delay times: t_{PHL} = 7.0 nS, t_{PLH} = 9.0 nS. Also assume that WATCHDOG_CLK is a 10-MHz square wave. Sketch the waveforms at WATCHDOG_CLK and the input labeled CLK on U1:B on the same time axis.

S   **9–29.** Repeat Problem 9–28 with a 7404 used in place of the 74HC04. Assume that the 7404 has the following propagation delay times: $t_{PHL} = 15.0$ nS, $t_{PLH} = 22.0$ nS.

S C D **9–30.** Find U9 in the HC11D0 schematic. LCD_SL and KEY_SL are active-LOW outputs that signify that either the LCD is selected or the keyboard is selected. Add a logic gate to this schematic that outputs a LOW level called I/O_SEL whenever either the LCD *or* the keyboard is selected.

Electronics Workbench®/MultiSIM® Exercises

E9–1. Load the circuit file for **Section 9–1a.** This is a two-input NAND similar to Figure 9–1.

(a) The output should always be HIGH unless A AND B are both _____ (HIGH or LOW?). Check your answer by changing the A and B inputs. (Q3 and Q4 form the "totem-pole" arrangement explained in Section 9–1.)

(b) When the output is HIGH, Q3 should be _____, and Q4 should be _____ (ON or OFF?).

(c) When the output is LOW, Q3 should be _____, and Q4 should be _____ (ON or OFF?). Check your answers by measuring the voltage C-to-E on Q3 and Q4 for both conditions.

E9–2. Load the circuit file for **Section 9–3a.** Connect Channel-1 of the oscilloscope to the input of the LS-family AND gate. Connect Channel-2 to its output.

(a) Measure tplh and tphl (*Hint:* See Figure 9–10.)

(b) Repeat for the 4000-family AND gate.

E9–3. Load the circuit file for **Section 9–3b.**

(a) Use the oscilloscope to determine the rise and fall time of the pulse applied to the AND gate.

(b) Use the dual-trace feature of the oscilloscope to measure tphl and tplh of the AND gate (Use 1.5 V as the measurement reference level.)

E9–4. Load the circuit file for **Section 9–3c.** This is a two-input NAND with a totem-pole output.

(a) Measure Voh and Vol.

(b) Convert it to an open-collector NAND by deleting the appropriate comments. Measure Voh and Vol.

(c) What external component must be added to this new circuit to make Voh look like a HIGH? Try it.

CPLD Problems

(Data sheets for Altera devices can be downloaded from the literature section at www.altera.com).

C9–1. Find the chart that provides the device DC operating conditions for the MAX 7000 family.

(a) What is the minimum to maximum range of recognizable voltages for V_{IH} and V_{IL}?

(b) What is the worst-case minimum and maximum output voltages (V_{OH}, V_{OL}) for the device operating in a 5.0 V TTL system?

C9–2. Study the features section in the data sheets for the following three families: MAX 7000, MAX 7000A, and MAX 7000B.

(a) List the I/O voltage interface compatibility for each.

(b) List the nanosecond pin-to-pin logic delays for each.

(c) List the maximum counter frequencies for each.

Answers to Review Questions

9–1. False

9–2. Totem-pole output

9–3. 0.7, 0.3

9–4. Because of the voltage drops across the internal transistors of the gate

9–5. It is the number of gates of the same subfamily that can be connected to a single ouput without exceeding the current rating of the gate.

9–6. Input current HIGH condition (I_{IH}), input current LOW condition (I_{IL}), output current HIGH condition (I_{OH}), output current LOW condition (I_{OL})

9–7. Sink current flows into the gate and goes to ground. Source current flows out of the gate and supplies the other gates.

9–8. **(a)** HIGH

(b) HIGH

(c) undetermined

(d) LOW

9–9. False

9–10. Output

9–11. It pulls the output of the gate up to 5 V when the output transistor is off (float).

9–12. It reduces the propagation delay to achieve faster speeds.

9–13. Lower power dissipation, slower speeds.

9–14. Bipolar

9–15. False

9–16. 74ALS

9–17. 74LS

9–18. True

9–19. 74HC CMOS

9–20. To raise the output level of the TTL gate so it is recognized as a HIGH by the CMOS gate

9–21. The CMOS gate has severe output current limitations.

9–22. 74HCTMOS, because its voltage is almost perfect at 4.9 V HIGH and 0.1 V LOW.

9–23. Input current requirements (I_{IL}, I_{IH}) and output current capability (I_{OL}, I_{OH})

Flip-Flops and Registers

OUTLINE

OBJECTIVES

Upon completion of this chapter, you should be able to:

- Explain the internal circuit operation of *S-R* and gated *S-R* flip-flops.
- Compare the operation of *D* latches and *D* flip-flops by using timing diagrams.
- Describe the difference between pulse-triggered and edge-triggered flip-flops.
- Explain the theory of operation of master–slave devices.
- Connect IC *J-K* flip-flops as toggle and *D* flip-flops.
- Use timing diagrams to illustrate the synchronous and asynchronous operation of *J-K* flip-flops.
- Use VHDL to design flip-flops for CPLD implementation.

INTRODUCTION

The logic circuits that we have studied in the previous chapters have consisted mainly of logic gates (AND, OR, NAND, NOR, INVERT) and **combinational logic.** Starting in this chapter, we will deal with data storage circuitry that will **latch** on to (remember) a digital state (1 or 0).

This new type of digital circuitry is called **sequential logic,** because it is controlled by and used for controlling other circuitry in a specific sequence dictated by a control clock or enable/disable control signals.

The simplest form of data storage is the Set–Reset (*S-R*) flip-flop. These circuits are called **transparent latches** because the outputs respond immediately to changes at the input, and the input state will be remembered, or latched onto. The latch will sometimes have an enable input, which is used to control the latch to accept or ignore the *S-R* input states.

More sophisticated flip-flops use a clock as the control input and are used wherever the input and output signals must occur within a particular sequence.

10–1   *S-R* Flip-Flop

The *S-R* **flip-flop** is a data storage circuit that can be constructed using the basic gates covered in previous chapters. Using a cross-coupling scheme with two NOR gates, we can form the flip-flop shown in Figure 10–1.

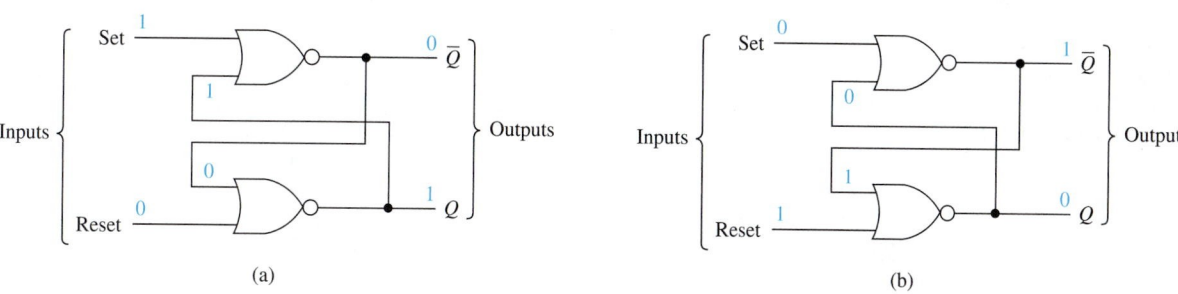

Figure 10–1   Cross-NOR *S-R* flip-flop: (a) Set condition; (b) Reset condition.

Let's start our analysis by placing a 1 (HIGH) on the **Set** and a 0 (LOW) on **Reset** [Figure 10–1(a)]. This is defined as the Set condition and should make the Q output 1 and $\overline{Q}$ output 0. A HIGH on the Set will make the output of the upper NOR equal 0 ($\overline{Q} = 0$), and that 0 is fed down to the lower NOR, which together with a LOW on the Reset input will cause the lower NOR's output to equal a 1 ($Q = 1$). [Remember, a NOR gate is always 0 output, except when *both* inputs are 0 (Chapter 3).]

Now, when the 1 is removed from the Set input, the flip-flop should remember that it is Set (i.e., $Q = 1$, $\overline{Q} = 0$). So with Set = 0, Reset = 0, and $Q = 1$ from previously being Set, let's continue our analysis. The upper NOR has a 0–1 at its inputs, making $\overline{Q} = 0$, whereas the lower NOR has a 0–0 at its inputs, keeping $Q = 1$. Great— the flip-flop remained Set even after the Set input was returned to 0 (called the *Hold* condition).

Now we should be able to Reset the flip-flop by making $S = 0$, $R = 1$ [Figure 10–1(b)]. With $R = 1$, the lower NOR will output a 0 ($Q = 0$), placing a 0–0 on the upper NOR, making its output 1 ($\overline{Q} = 1$); thus, the flip-flop "flipped" to its Reset state.

The only other input condition is when both S and R inputs are HIGH. In this case, both NORs will put out a LOW, making Q *and* $\overline{Q}$ equal 0, which is a condition that is not used. (Why would anyone want to Set *and* Reset at the same time, anyway!) Also, when you return to the Hold condition from $S = 1$, $R = 1$, you will get unpredictable results unless you know which input returned LOW last.

From the previous analysis, we can construct the *S-R* flip-flop **function table** shown in Table 10–1, which lists all input and output conditions.

Common Misconception

You may feel that you can't solve this circuit because one input to each NOR is unknown. Remember that you don't need to know both inputs. If one input is HIGH, the output is LOW.

TABLE 10–1	Function Table for Figure 10–1			
S	*R*	*Q*	$\overline{Q}$	**Comments**
0	0	*Q*	$\overline{Q}$	*Hold* condition (no change)
1	0	1	0	Flip-flop Set
0	1	0	1	Flip-flop Reset
1	1	0	0	Not used

An *S-R* flip-flop can also be made from cross-NAND gates, as shown in Figure 10–2. Prove to yourself that Figure 10–2 will produce the function table shown in Table 10–2. (Start with $S = 1$, $R = 0$, and remember that a NAND is LOW out only when *both* inputs are HIGH.) The symbols used for an *S-R* flip-flop are shown in Figure 10–3. The symbols show that both *true*, and **complemented** Q outputs are available. The second symbol is technically more accurate, but the first symbol is found most often in manufacturers' data manuals and throughout this book.

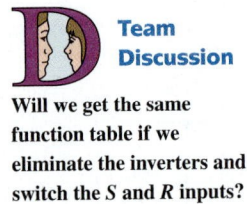

Team Discussion

Will we get the same function table if we eliminate the inverters and switch the *S* and *R* inputs?

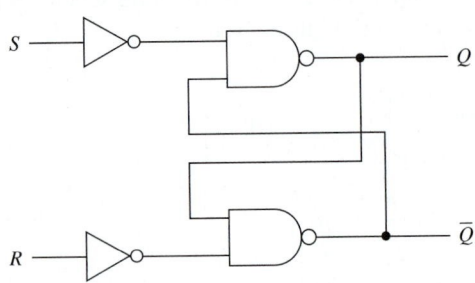

Figure 10–2 Cross-NAND *S-R* flip-flop.

TABLE 10–2		Function Table for Figure 10–2		
S	*R*	*Q*	$\overline{Q}$	**Comments**
0	0	*Q*	$\overline{Q}$	Hold condition
1	0	1	0	Flip-flop Set
0	1	0	1	Flip-flop Reset
1	1	1	1	Not used

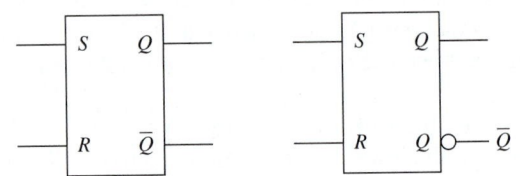

Figure 10–3 Symbols for an *S-R* flip-flop.

Now let's get practical and find an IC TTL NOR gate and draw the actual wiring connections to form a cross-NOR like Figure 10–1 so that we may check it in the lab.

The TTL data manual shows a quad NOR gate 7402. Looking at its pin layout in conjunction with Figure 10–1, we can draw the circuit of Figure 10–4. To check out the operation of Figure 10–4 in the lab, apply 5 V to pin 14 and ground pin 7. Set the flip-flop by placing a HIGH (5 V) to the Set input and a LOW (0 V, ground) to the Reset input. A logic probe attached to the Q output should register a HIGH. When the *S-R* inputs are returned to the 0–0 state, the Q output should remain latched in the 1 state. The Reset function can be checked using the same procedure.

S-R Timing Analysis

By performing a timing analysis on the *S-R* flip-flop, we can see why it is called transparent and also observe the latching phenomenon.

CHAPTER 10 I FLIP-FLOPS AND REGISTERS

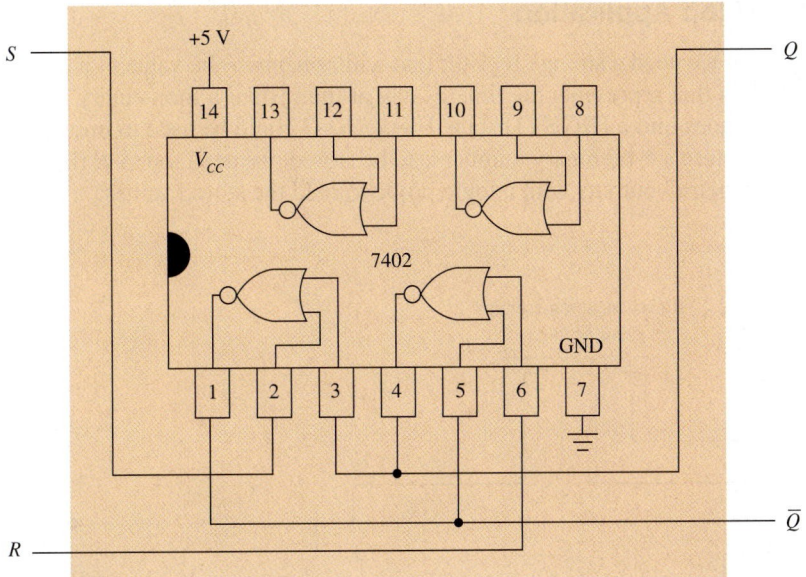

Figure 10–4 *S-R* flip-flop connections using a 7402 TTL IC.

EXAMPLE 10–1

To the *S-R* flip-flop shown in Figure 10–5, we connect the *S* and *R* wave-forms given in Figure 10–6. Sketch the *Q* output waveform that will result.

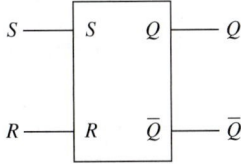

Figure 10–5

Solution:

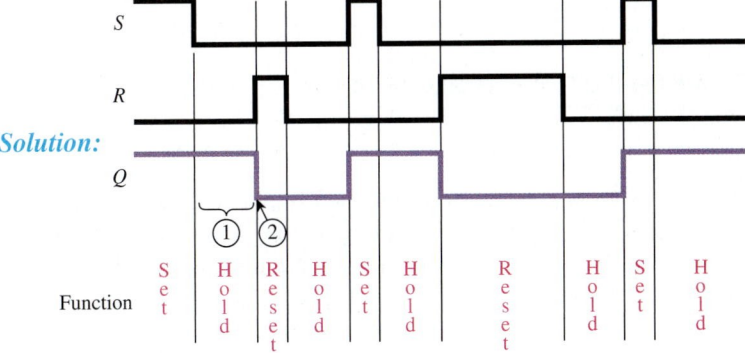

Figure 10–6

Notes:

1. The flip-flop is latched (held) in the Set condition even after the HIGH is removed from the *S* input.

2. The flip-flop is considered transparent because the *Q* output responds immediately to input changes.

S-R Flip-Flop Application

Let's say that we need a storage register that will *remember* the value of a binary number ($2^3 2^2 2^1 2^0$) that represents the time of day at the instant a momentary temperature limit switch goes into a HIGH (1) state. Figure 10–7 could be used to implement such a circuit. Because a 4-bit binary number is to be stored, we need four *S-R* flip-flops. We will look at their Q outputs with a logic probe to read the stored values.

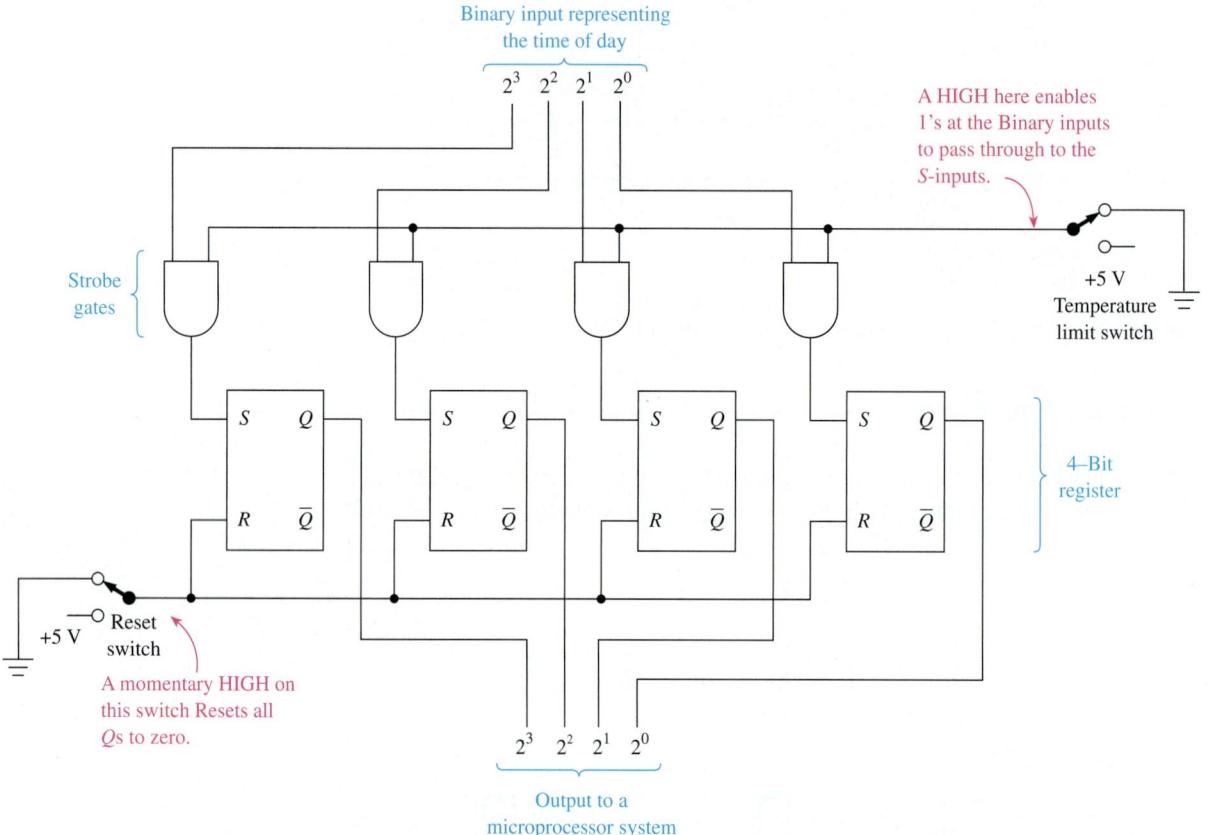

Figure 10–7 *S-R* flip-flop used as a storage register.

Team Discussion

Discuss some of the problems that may occur if the circuit is operated improperly by forgetting to reset or by allowing multiple temperature switch closures.

With the Reset switch in the up position, the R inputs will be zero. With the temperature limit switch in the up position, one input to each AND gate is grounded, keeping the S inputs at 0 also. To start the operation, first the Reset switch is momentarily pressed down, placing 5 V (1) on all four R inputs and, thus, resetting all flip-flops to 0.

Meanwhile, the binary input number is not allowed to reach the S inputs because a 0 is at the other input of each AND gate. (Gates used in this method are referred to as **strobe gates** because they let information pass only when they are **enabled**.)

When the temperature limit switch momentarily goes down, 5 V (1) will be placed at each strobe gate, allowing the binary number (1's and 0's) to pass through to the S inputs, thus setting the appropriate flip-flops. (Assume that the switch will go down only once.)

The binary input number representing the time of day that the temperature switch went down will be stored in the 4-bit register and can later be read by a logic probe or automatically by a microprocessor system.

Review Questions

10–1. A flip-flop is different from a basic logic gate because it *remembers* the state of the inputs after they are removed. True or false?

10–2. What levels must be placed on *S* and *R* to Set an *S-R* flip-flop?

10–3. What effect do $S = 0$ and $R = 0$ have on the output level at *Q*?

10–2 Gated *S-R* Flip-Flop

Simple gate circuits, combinational logic, and transparent *S-R* flip-flops are called **asynchronous** (not synchronous) because the output responds immediately to input changes. **Synchronous** circuits operate sequentially, in step, with a control input. To make an *S-R* flip-flop synchronous, we add a gated input to enable and disable the *S* and *R* inputs. Figure 10–8 shows the connections that make the cross-NOR *S-R* flip-flop into a gated *S-R* flip-flop.

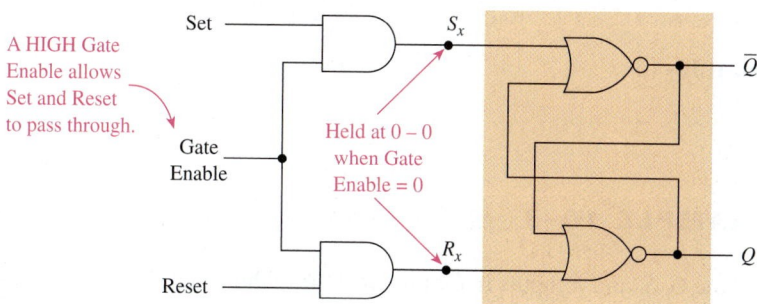

Figure 10–8 Gated *S-R* flip-flop.

The S_x and R_x lines in Figure 10–8 are the original Set and Reset inputs. With the addition of the AND gates, however, the S_x and R_x lines will be kept LOW–LOW (Hold condition) as long as the Gate Enable is LOW. The flip-flop will operate normally while the Gate Enable is HIGH. The function chart [Figure 10–9(b)] and Example 10–2 illustrate the operation of the gated *S-R* flip-flop.

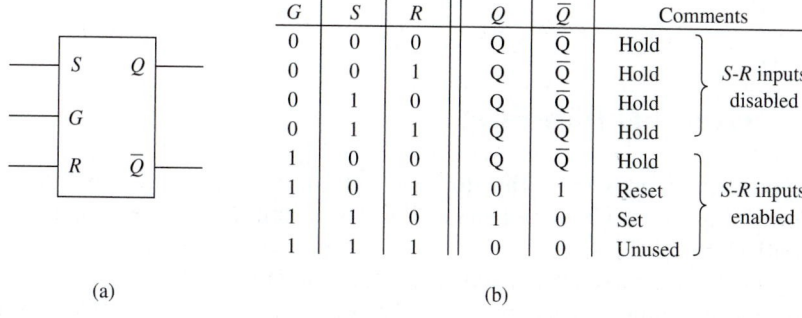

G	S	R	Q	$\overline{Q}$	Comments	
0	0	0	Q	$\overline{Q}$	Hold	
0	0	1	Q	$\overline{Q}$	Hold	*S-R* inputs
0	1	0	Q	$\overline{Q}$	Hold	disabled
0	1	1	Q	$\overline{Q}$	Hold	
1	0	0	Q	$\overline{Q}$	Hold	
1	0	1	0	1	Reset	*S-R* inputs
1	1	0	1	0	Set	enabled
1	1	1	0	0	Unused	

(a)   (b)

Figure 10–9 Function table and symbol for the gated *S-R* flip-flop of Figure 10–8.

EXAMPLE 10–2

Feed the G, S, and R inputs in Figure 10–10 into the gated S-R flip-flop, sketch the output wave at Q, and list the flip-flop functions.

Solution:

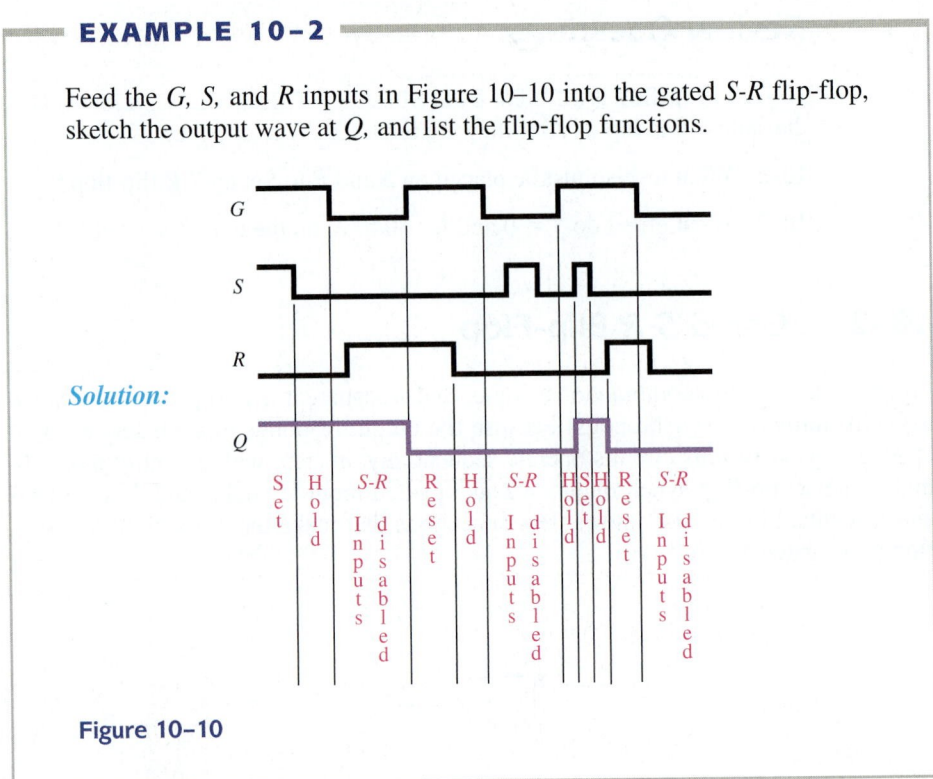

Figure 10–10

EXAMPLE 10–3

Feed the G, S, and R inputs in Figure 10–11 into the gated S-R flip-flop, and sketch the output wave at Q.

Solution:

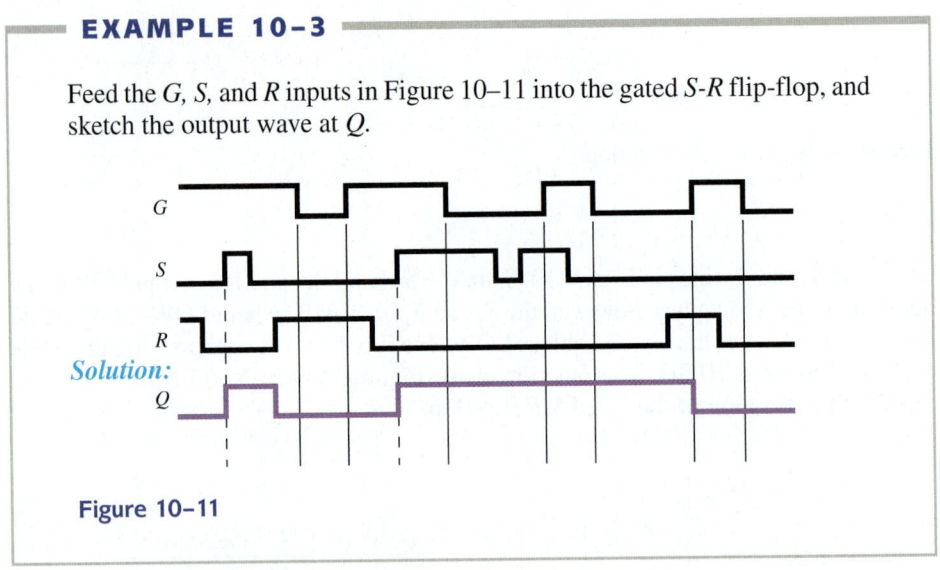

Figure 10–11

10–3 Gated *D* Flip-Flop

Another type of flip-flop is the D flip-flop (*Data* flip-flop). It can be formed from the gated S-R flip-flop by the addition of an inverter. This enables just a single input (D) to both Set *and* Reset the flip-flop.

In Figure 10–12, we see that S and R will be complements of each other, and S is connected to a single line labeled D (Data). The operation is such that Q will be the same as D while G is HIGH, and Q will remain latched when G goes LOW. (*Latched* means that Q remains constant regardless of changes in D.)

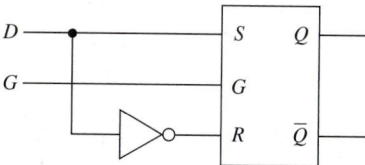

Figure 10–12 Gated *D* flip-flop.

Sketch the output waveform at *Q* for the inputs at *D* and *G* of the gated *D* flip-flop in Figure 10–13.

Solution:

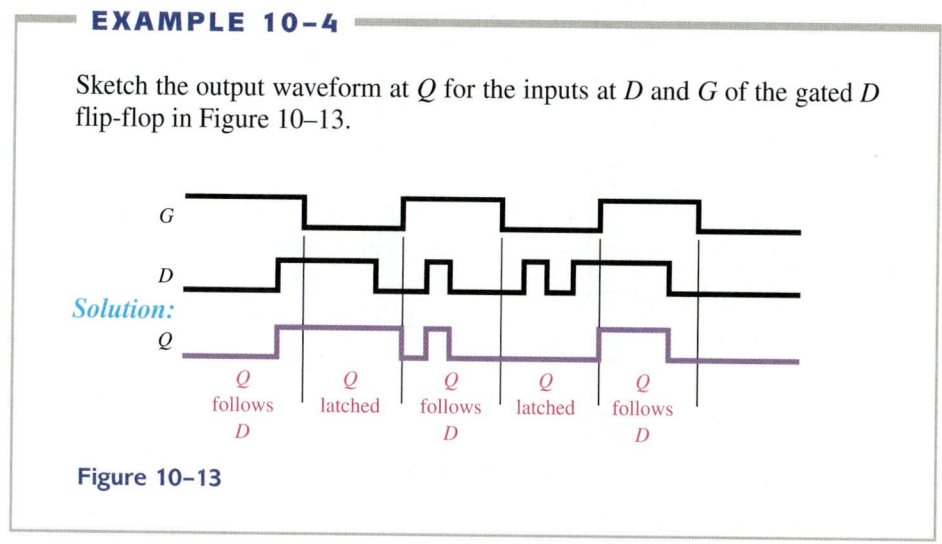

Figure 10–13

━━━ **Review Questions** ━━━

10–4. Explain why the *S-R* flip-flop is called asynchronous and the *gated S-R* flip-flop is called synchronous.

10–5. Changes in *S* and *R* while a gate is enabled have no effect on the *Q* output of a gated *S-R* flip-flop. True or false?

10–6. What procedure would you use to Reset the *Q* output of a gated *D* flip-flop?

10–4   *D* Latch: 7475 IC; VHDL Description

7475 IC

The 7475 is an example of an IC *D* latch (also called a *bistable latch*). It contains *four* transparent *D* latches. (See Section 13–10 for the 8-bit version, the 74LS373.) Its logic symbol and pin configuration are given in Figure 10–14. Latches 0 and 1 share a common enable (E_{0-1}), and latches 2 and 3 share a common enable (E_{2-3}). The enables act just like the *G*-input in the previous section.

From the function table (Table 10–3) we can see that the *Q* output will follow *D* (transparent) as long as the enable line *(E)* is HIGH (called active-HIGH enable). When *E* goes LOW, the *Q* output will become *latched* to the value that *D* was just before the HIGH-to-LOW transition of E.

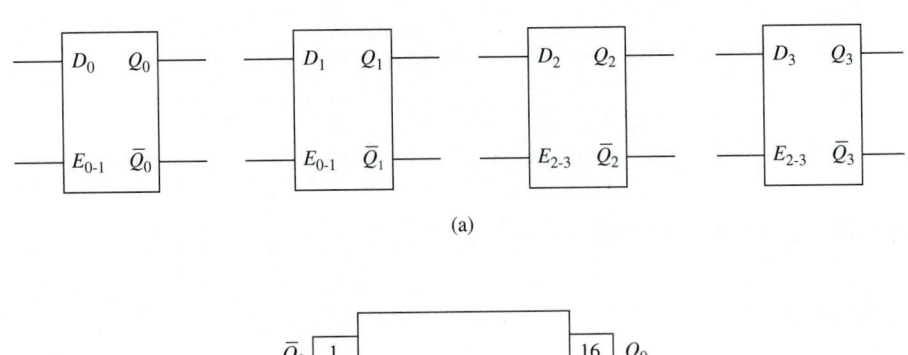

(a)

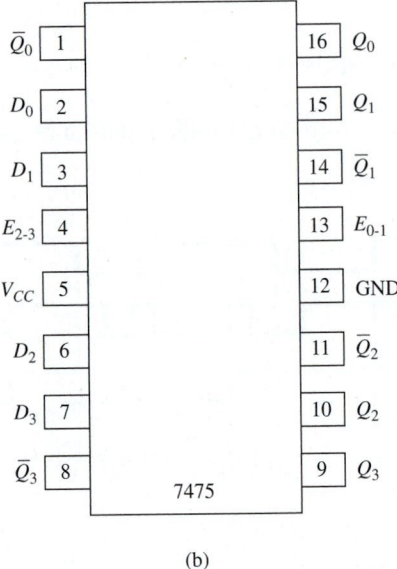

(b)

Figure 10–14 The 7475 quad bistable D latch: (a) logic symbol; (b) pin configuration.

Helpful Hint

Some students use the function table as a crutch. This can become a dangerous habit as the functions get more complex. You must get into the habit of understanding the *description* provided by the manufacturer. (In this case, what is meant by *transparent* and *latched*?)

TABLE 10–3 Function Table for a 7475[a]

	Inputs		Outputs	
Operating Mode	**E**	**D**	**Q**	**$\bar{Q}$**
Data enabled	H	L	L	H
Data enabled	H	H	H	L
Data latched	L	x	q	$\bar{q}$

[a]q = state of Q before the HIGH-to-LOW edge of E; x = don't care.

EXAMPLE 10–5

For the inputs at D_0 and E_{0-1} for the 7475 D latch shown in Figure 10–15, sketch the output waveform at Q_0 in Figure 10–16.

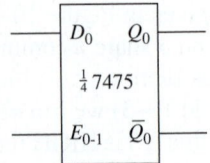

Figure 10–15

CHAPTER 10 | FLIP-FLOPS AND REGISTERS

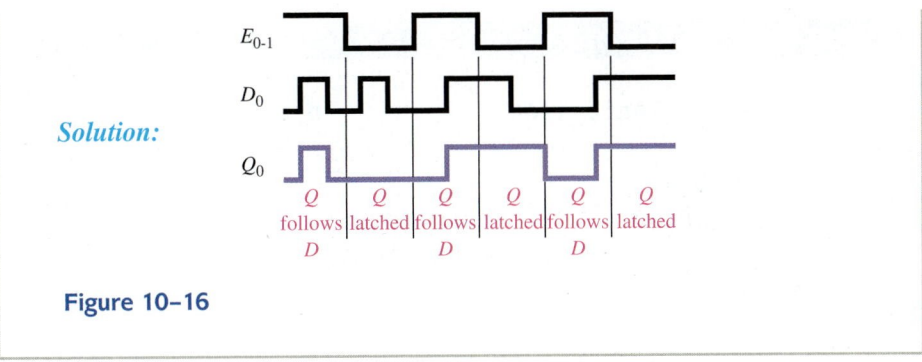

Figure 10–16

VHDL Description of a *D* Latch

The function of a *D* latch is to pass the binary state on its *D* input out to *Q* (transparent operation) while the enable is HIGH, and then store that value at *Q* (latch operation) when the enable goes LOW. This function can be implemented in a CPLD as a graphic design file (gdf) or as a VHDL program (vhd). Example 10–6 shows both methods and provides a simulation of the latch.

EXAMPLE 10–6

D Latch

Create a *D* latch using the graphic design method. A symbol called *LATCH* is found in the *prim* (primitives) subdirectory. Test the design by creating a simulation file. Repeat using the VHDL design entry method.

Solution: Figure 10–17 shows the solution as a gdf file using the primitive symbol *LATCH*. *d0* and *e0* are applied as inputs and *q0* is the output.

The VHDL solution is shown in Figure 10–18. Notice that *d0* and *e0* are listed in the sensitivity list of the PROCESS statement. This way, whenever either of them changes levels, the process is executed. The IF clause sets *q0* equal to *d0* when *e0* is HIGH. (If *e0* is not HIGH, *q0* remains unchanged.)

The simulation file is shown in Figure 10–19. This simulation must

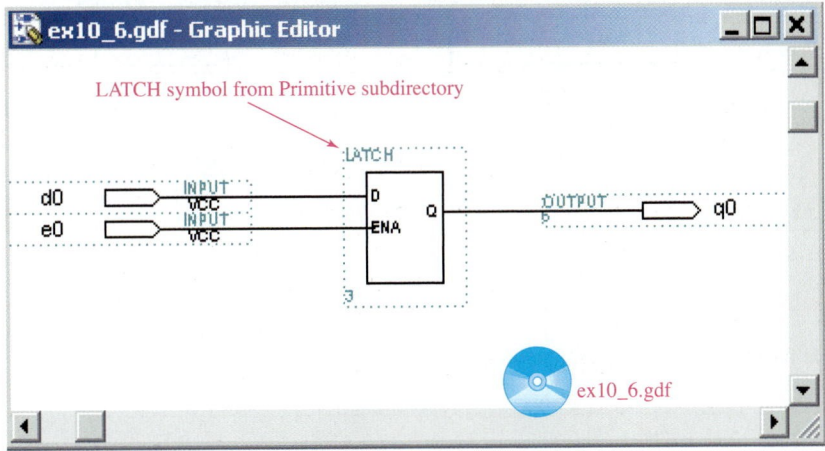

Figure 10–17 Graphic design file for the *D* latch of Example 10–6.

```
┌─────────────────────────────────────────────────────────────────────┐
│ 🗎 ex10_6.vhd – Text Editor                              _ ❑ ✕        │
├─────────────────────────────────────────────────────────────────────┤
│   LIBRARY ieee;                        ---------------        ▲       │
│   USE ieee.std_logic_1164.ALL;         -- D Latch  --                 │
│                                        ---------------                │
│                                                                       │
│   ENTITY ex10_6 IS                                                    │
│       PORT (e0,d0      : IN std_logic;                                │
│               q0       : OUT std_logic);                              │
│   END ex10_6 ;                                                        │
│                                                                       │
│                                                                       │
│   ARCHITECTURE arc OF ex10_6 IS                                       │
│   BEGIN                                  ── Sensitivity list          │
│       PROCESS (d0,e0)                                                 │
│       BEGIN                              ── Condition to load q0 with d0│
│           IF e0 = '1' THEN                                            │
│               q0<=d0;                                                 │
│           END IF;                                                     │
│       END PROCESS;                                                    │
│     END arc;                                                          │
│                                                       ex10_6.vhd      │
│                                                                    ▼  │
├─────────────────────────────────────────────────────────────────────┤
│ │Line   1  ││Col   1  ││INS│◄│ │                                   ►  │
└─────────────────────────────────────────────────────────────────────┘
```

Figure 10–18 VHDL design file for the *D* latch of Example 10–6.

be run for both design methods (graphic and VHDL) to check the validity of both implementations. (Be sure to select **Set Project To Current File** before running the simulation for each design method.) Notice in the simulation that *q0* follows the state of *d0* while *e0* is HIGH, then it becomes latched when *e0* goes LOW.

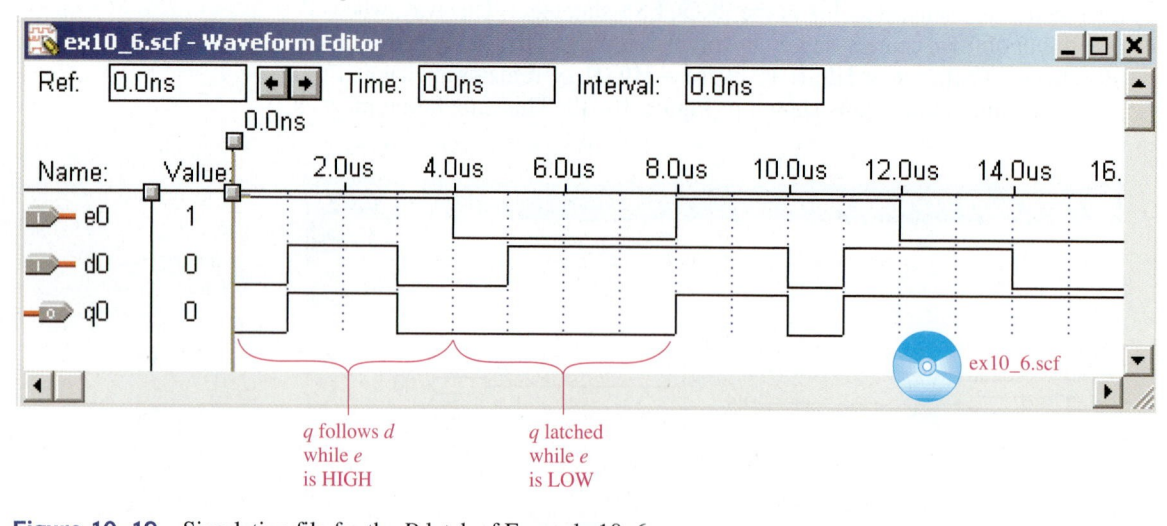

Figure 10–19 Simulation file for the *D* latch of Example 10–6.

10–7. The 7475 IC contains how many D latches?

10–8. The Q output of the 7475 D latch follows the level on the D input as long as E is _____ (HIGH or LOW).

10–9. Changes to D are ignored by the 7475 while E is LOW. True or false?

10–5   *D* Flip-Flop: 7474 IC; VHDL Description

7474 IC

The 7474 D flip-flop differs from the 7475 D latch in several ways. Most important, the 7474 is an **edge-triggered** device. This means that **transitions** in Q occur only at the edge of the input **trigger** pulse. The trigger pulse is usually a **clock** or timing signal instead of an enable line. In the case of the 7474, the trigger point is at the **positive edge** of C_p (LOW-to-HIGH transition). The small triangle on the D flip-flop symbol [Figure 10–20(a)] is used to indicate that it is edge triggered.

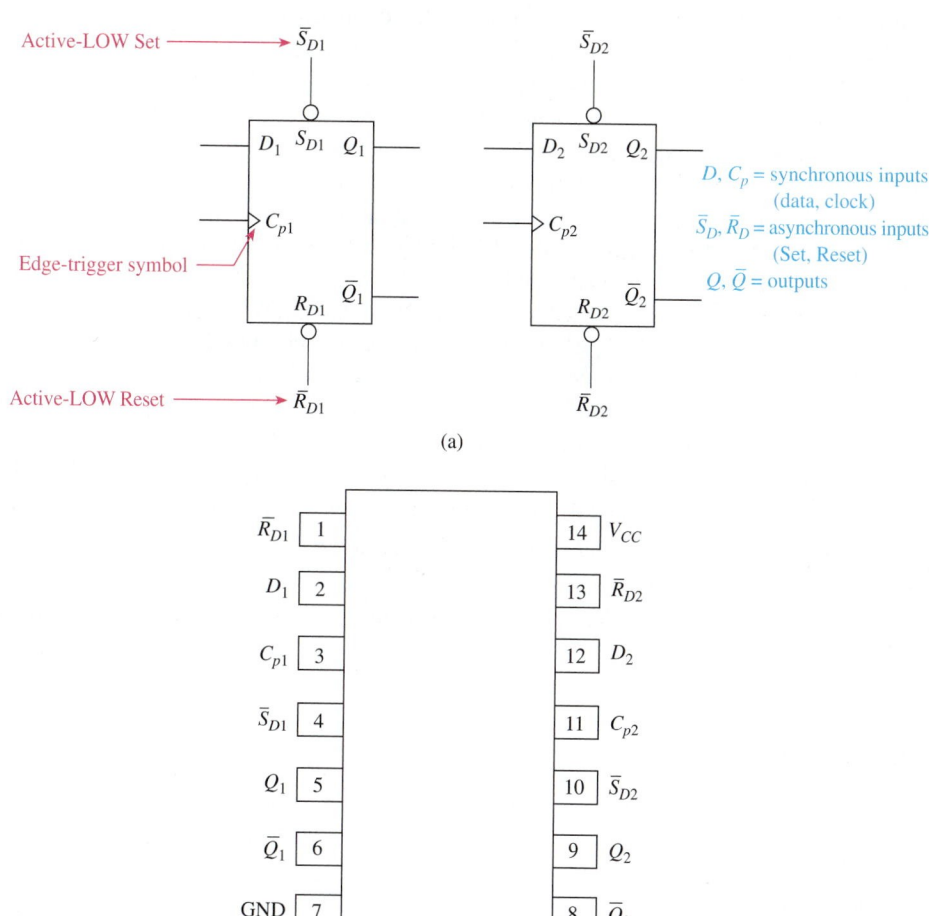

(a)

(b)

Figure 10–20   The 7474 dual D flip-flop: (a) logic symbol; (b) pin configuration.

Common Misconception

Don't mistakenly think that you can leave the asynchronous inputs disconnected if they are not to be used (they must be tied HIGH).

Common Misconception

Students often mistakenly think that applying a LOW to $\overline{S}_D$ will make Q = LOW.

Inside Your PC

The D latch and D flip-flop are also available as octal devices (8 flip-flops in a single IC). The 74LS373 and 74LS374 octal ICs are often used to interface to microprocessors. (See Sections 10–9 and 13–10.)

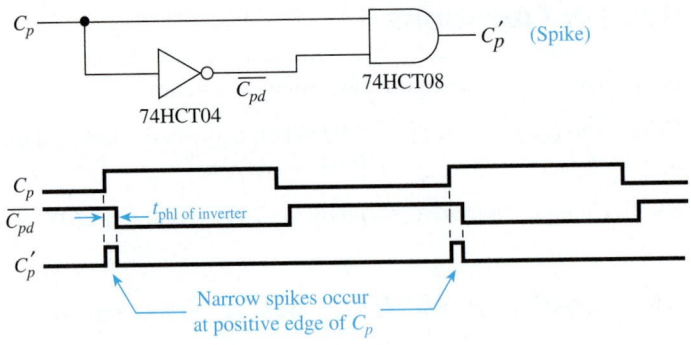

Figure 10-21 Positive edge-detection circuit and waveforms.

Edge-triggered devices are made to respond to only the *edge* of the clock signal by converting the positive clock input pulse into a single, narrow spike. Figure 10–21 shows a circuit similar to that inside the 7474 to convert the rising edge of C_p into a positive spike. This is called a *positive edge-detection circuit.*

In Figure 10–21, the original clock, C_p, is input to an inverter whose purpose is to invert the signal and delay it by the propagation delay time of the inverter (t_{phl}). This inverted, delayed signal, $\overline{C_{pd}}$, is then fed into the AND gate along with the original clock, C_p. By studying the waveforms, you can see that the output waveform, C_p', is a very narrow pulse (called a *spike*) that lines up with the positive edge of C_p. This is now used as the trigger signal inside the D flip-flop. Therefore, even though a very wide pulse is entered at C_p of the 7474, the edge-detection circuitry converts it to a spike so that the D flip-flop reacts only to data entered at D at the positive edge of C_p.

The 7474 has two distinct types of inputs: synchronous and asynchronous. The *synchronous inputs* are the D (Data) and C_p (Clock) inputs. The state at the D input will be transferred to Q at the positive edge of the input trigger (LOW-to-HIGH edge of C_p). The *asynchronous inputs* are $\overline{S_D}$ (Set) and $\overline{R_D}$ (Reset), which operate independently of D and C_p. Being asynchronous means that they are *not* in sync with the clock pulse, and the Q outputs will respond *immediately* to input changes at $\overline{S_D}$ and $\overline{R_D}$. The little circle at S_D and R_D means that they are **active-LOW** inputs, and because the circles act like inverters, the external pin on the IC is labeled as the complement of the internal label.

This all sounds complicated, but it really is not. Just realize that *a LOW on $\overline{S_D}$ will immediately Set the flip-flop, and a LOW on $\overline{R_D}$ will immediately Reset the flip-flop, regardless of the states at the synchronous (D, C_p) inputs.*

The function table (Table 10–4) and following examples illustrate the operation of the 7474 D flip-flop.

Team Discussion

Discuss how the 7474 might be used to remember that a pedestrian had pressed a crosswalk push button. (*Hint:* See Figure 12–43.)

TABLE 10-4	**Function Table for a 7474 D Flip-Flop[a]**					
	Inputs				**Outputs**	
Operating Mode	$\overline{S_D}$	$\overline{R_D}$	C_p	D	Q	$\overline{Q}$
Asynchronous Set	L	H	x	x	H	L
Asynchronous Reset	H	L	x	x	L	H
Not used	L	L	x	x	H	H
Synchronous Set	H	H	↑	h	H	L
Synchronous Reset	H	H	↑	1	L	H

X means that we "don't care" about these inputs because S_D or R_D are active.

[a]↑ = positive edge of clock; H = HIGH; h = HIGH level one setup time prior to positive clock edge; L = LOW; 1 = LOW level one setup time before positive clock edge; x = don't care.

The lowercase h in the D column indicates that, to do a synchronous Set, the D must be in a HIGH state at least one setup time before the positive edge of the clock. The same rules apply for the lowercase 1 (Reset).

The **setup time** for this flip-flop is 20 ns, which means that if D is changing while C_p is LOW, that's okay, but D must be held stable (HIGH or LOW) at least 20 ns *before* the LOW-to-HIGH transition of C_p. (We discuss setup time in greater detail in Chapter 11.) Also realize that the only digital level on the D input that is used is the level that is present at the positive edge of C_p.

We have learned a lot of new terms in regard to the 7474 (active-LOW, edge-triggered, asynchronous, and others). These terms are important because they apply to almost all the ICs that are used in the building of sequential circuits.

EXAMPLE 10–7

Sketch the output waveform at Q for the 7474 D flip-flop shown in Figure 10–22(a) whose input waveforms are as given in Figure 10–22(b).

(a)

Solution:

(b) AS = asynchronous Set
AR = asynchronous Reset
SS = synchronous Set
SR = synchronous Reset

Figure 10–22

Helpful Hint

Two ways that you can remember that it takes a LOW to asynchronously Set the flip flop are (1) the overbar on $\overline{S_D}$ and (2) the bubble on $\overline{S_D}$.

EXAMPLE 10-8

Sketch the output waveforms at Q for the 7474 D flip-flops shown in Figure 10–23 whose input waveforms are given in Figure 10–24.

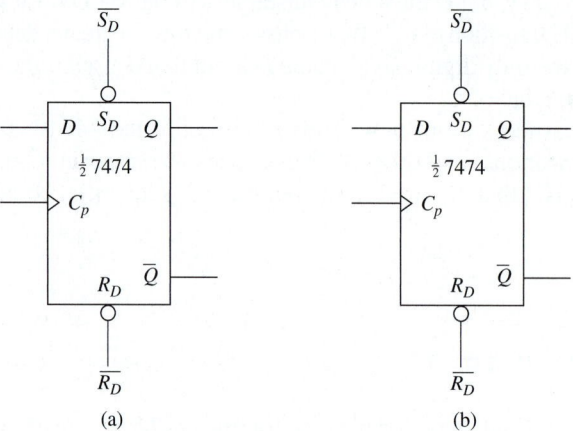

Figure 10–23

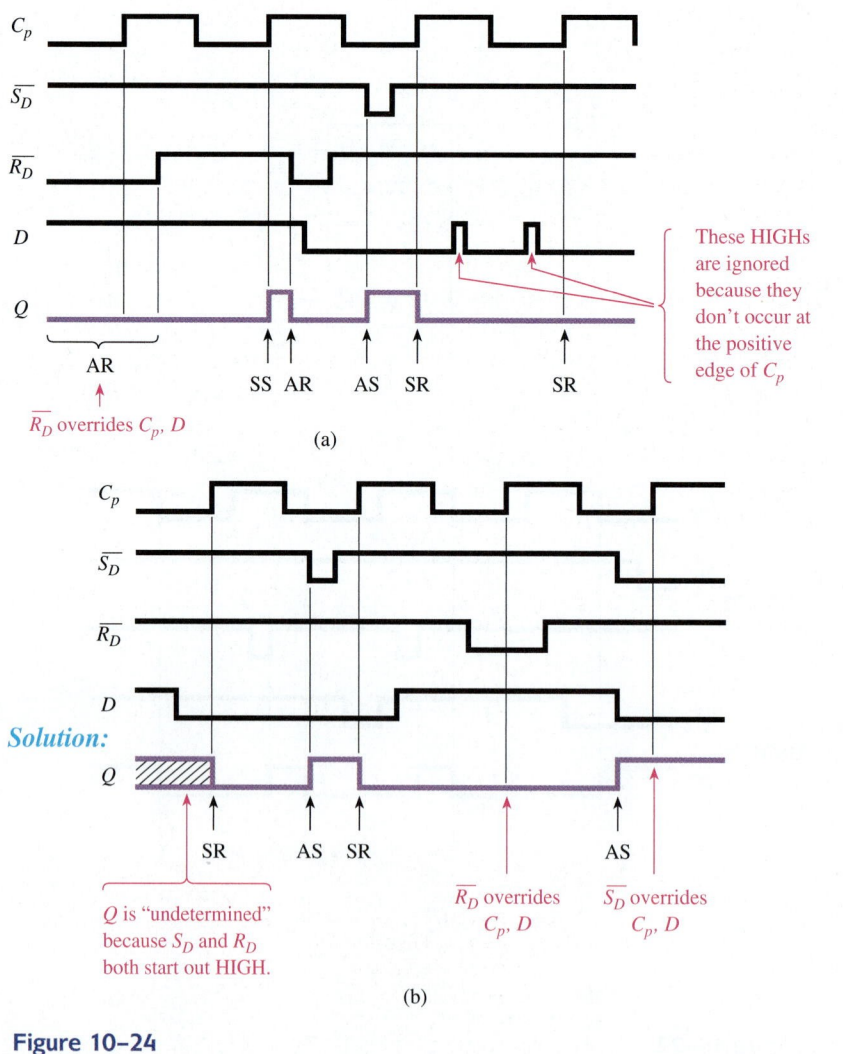

Figure 10–24

10–10. The 7474 is an edge-triggered device. What does this mean?

10–11. Which are the synchronous and which are the asynchronous inputs to the 7474 D flip-flop?

10–12. To perform an asynchronous Set, the $\overline{S_D}$ line must be made HIGH. True or false?

10–13. The purpose of the inverter in the edge-detection circuit of Figure 10–18 is to _____ and _____ the signal from C_p.

VHDL Description of a D Flip-Flop

The function of a D flip-flop is to remember the logic level on D at the instant that the input clock trigger makes a state transition (LOW to HIGH or HIGH to LOW). Flip-flops can also have asynchronous Set and Reset inputs. The flip-flop function can be implemented in a CPLD as a graphic design file (gdf) or as a VHDL program (vhd). Examples 10–9 and 10–10 show both methods and provide a simulation of the flip flops.

EXAMPLE 10–9

D Flip-Flop

Create a D flip-flop using the graphic design method. A symbol called *DFF* is found in the *prim* (primitives) subdirectory. Test the design by creating a simulation file. Repeat using the VHDL design entry method.

Solution: Figure 10–25 shows the solution as a gdf file using the primitive symbol *DFF. d, cp,* and *n_reset* are applied as inputs and *q* is the output. The name *n_reset* is chosen to represent the active-LOW asynchronous Reset because *reset* cannot be typed with the overbar. You will also notice that inside of the symbol, Altera uses the terms *Preset (PRN)* instead of *Set (Sd)* and *Clear (CLRN)* instead of *Reset (Rd)*.

The VHDL solution is shown in Figure 10–26(a). Within the PROCESS block, the first check is to determine if *n_reset* is LOW. If it is LOW, it overrides any synchronous operations on *cp* and *d* and resets *q* to '0'. If it

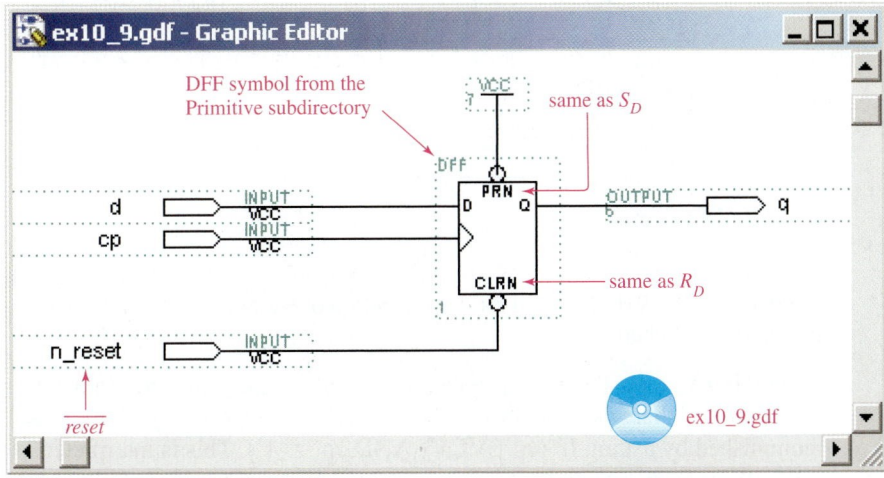

Figure 10–25 Graphic design file for the D flip-flop of Example 10–9.

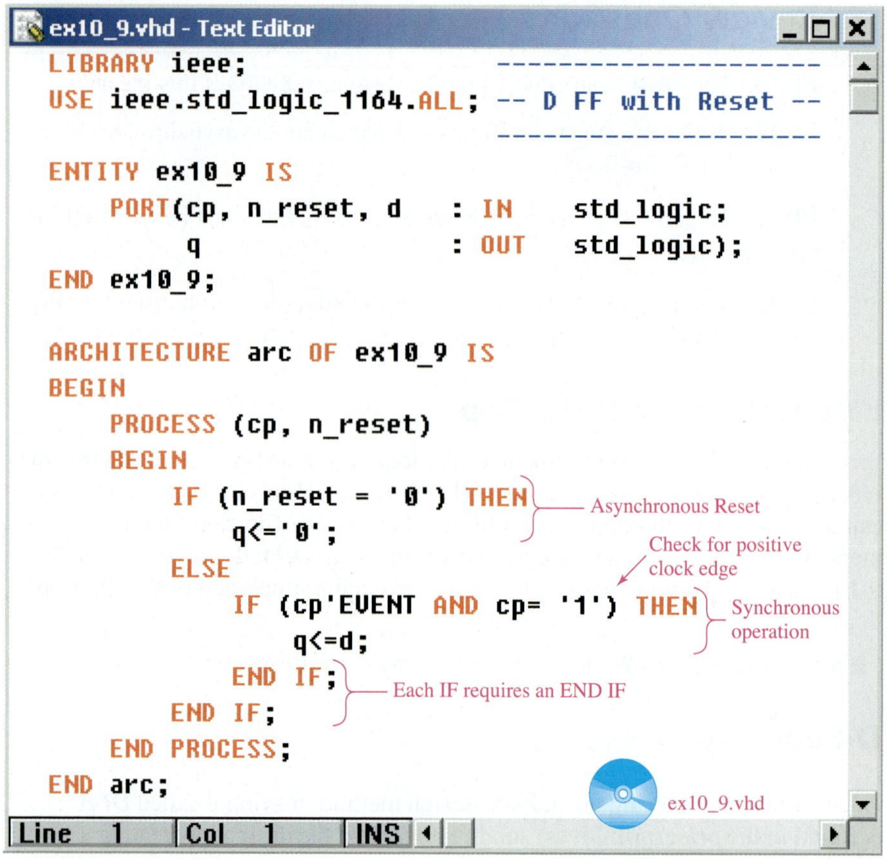

```
LIBRARY ieee;                          ----------------------
USE ieee.std_logic_1164.ALL;          -- D FF with Reset --
                                       ----------------------

ENTITY ex10_9 IS
    PORT(cp, n_reset, d    : IN    std_logic;
              q            : OUT   std_logic);
END ex10_9;

ARCHITECTURE arc OF ex10_9 IS
BEGIN
    PROCESS (cp, n_reset)
    BEGIN
        IF (n_reset = '0') THEN
            q<='0';
        ELSE
            IF (cp'EVENT AND cp= '1') THEN
                q<=d;
            END IF;
        END IF;
    END PROCESS;
END arc;
```


ex10_9.vhd - Text Editor
Asynchronous Reset
Check for positive clock edge
Synchronous operation
Each IF requires an END IF
ex10_9.vhd
Line 1
Col 1
INS

(a)

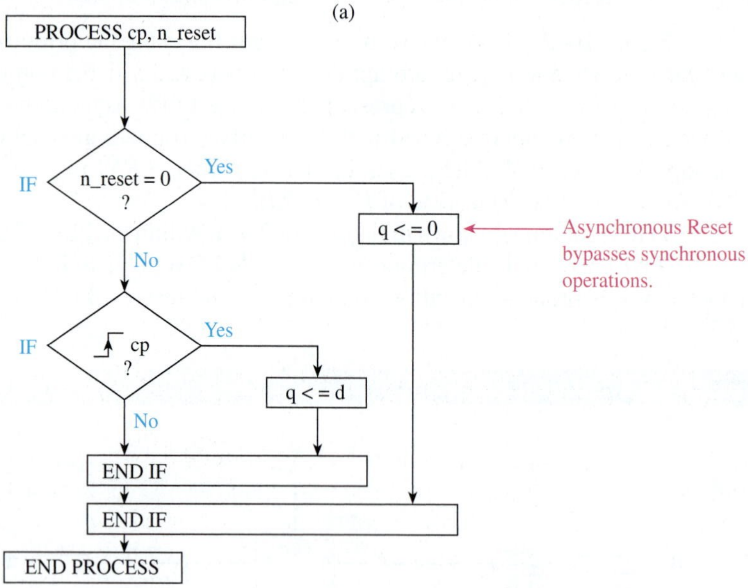

PROCESS cp, n_reset
IF
n_reset = 0 ?
Yes
No
q < = 0
Asynchronous Reset bypasses synchronous operations.
IF
cp ?
Yes
No
q < = d
END IF
END IF
END PROCESS

(b)

Figure 10–26 VHDL design for the *D* flip-flop of Example 10–9: (a) VHDL listing; (b) flow chart.

is not LOW, then the ELSE clause is executed. Here the program must determine if there is a positive (LOW to HIGH) edge on *cp*. This is accomplished by asking IF (cp'EVENT AND cp = '1'). This is interpreted as "Was there an event (change of state) on *cp* and is *cp* now HIGH?" To check for a negative (HIGH-to-LOW) clock edge, *cp* = '1' would be changed to

$cp = `0'$. So, if it is determined that there is a positive edge on cp, then q is loaded with the logic level of d. The VHDL program ends with an END IF for each IF statement and an END PROCESS for the PROCESS statement. Notice the use of indentation to delineate the beginning and end of the PROCESS and each IF block.

Whenever there are IF-THEN-ELSE statements in a program, a flow-chart like Figure 10–26(b) is helpful to visualize program branching. This makes it obvious that the asynchronous input n_reset has priority over the synchronous cp. If n_reset is LOW, q is reset to 0 and the check for an active clock edge is skipped.

The simulation file is shown in Figure 10–27. This simulation must be run for both design methods (graphic and VHDL) to check the validity of both implementations. (Be sure to select **Set Project to Current File** before running the simulation for each design method.) Notice in the simulation that starting n_reset LOW makes q start out LOW. Then after the 10 μs mark it resets q again. Also notice that at each positive edge of cp, q is set to the level of d. If cp is not on a positive edge, then d is ignored.

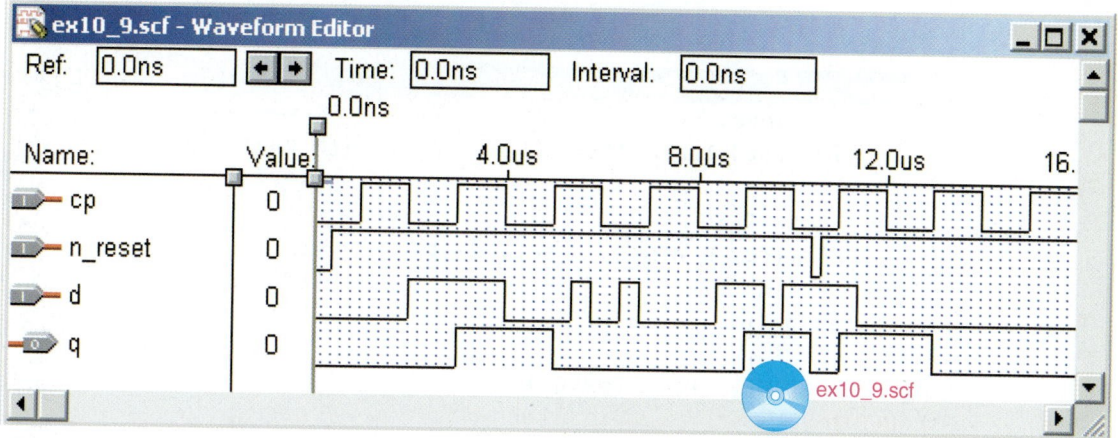

Figure 10–27   Simulation file for the D flip-flop of Example 10–9.

EXAMPLE 10–10

D Flip-Flop with Asynchronous Set and Reset

Create a D flip-flop using the graphic design method having both asynchronous Set and Reset inputs. A symbol called *DFF* is found in the *prim* (primitives) subdirectory. Test the design by creating a simulation file. Repeat using the VHDL design entry method.

Solution: Figure 10–28 shows the solution as a gdf file using the primitive symbol *DFF*. d, cp, n_set, and n_reset are applied as inputs, and q is the output. Notice that inside of the symbol, Altera uses *Present (PRN)* instead of *Set (Sd)* and *Clear (CLRN)* instead of *Reset (Rd)*.

The VHDL solution is shown in Figure 10–29(a). Within the PROCESS block the operation of the IF clauses are similar to that explained in Example 10–9 except, instead of using two statements for ELSE and IF, they are combined as ELSIF. Also notice that ELSIF does not require an END IF statement.

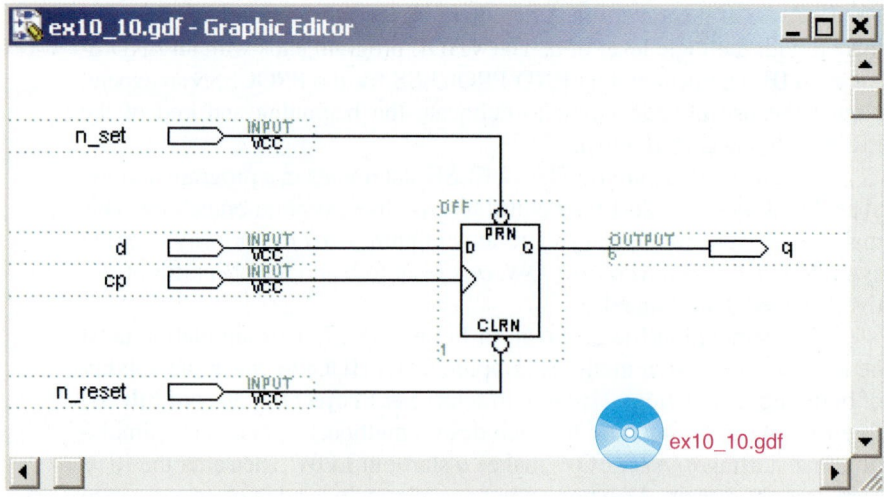

Figure 10–28   Graphic design file for the *D* flip-flop of Example 10–10.

```
ex10_10.vhd - Text Editor                                    _ □ X
LIBRARY ieee;                    ----------------------------
USE ieee.std_logic_1164.ALL; -- D FF with Reset and Set--
                                 ----------------------------
ENTITY ex10_10 IS
    PORT(cp, n_reset, n_set, d  : IN     std_logic;
         q                      : OUT    std_logic);
END ex10_10;

ARCHITECTURE arc OF ex10_10 IS
BEGIN
    PROCESS (cp, n_reset, n_set)
    BEGIN
        IF (n_reset = '0'AND n_set='1') THEN      Asynchronous
            q<='0';                               Reset
        ELSIF (n_reset = '1'AND n_set='0') THEN   Asynchronous
            q<='1';                               Set
        ELSIF (cp'EVENT AND cp= '1') THEN         Synchronous
            q<=d;                                 operation
        END IF;
                                          Positive clock edge
    END PROCESS;
END arc;
                                              ex10_10.vhd
Line    1    Col    1    INS ◄
```

(a)

Figure 10–29   VHDL design for the *D* flip-flop of Example 10–10: (a) VHDL listing;

The flowchart in Figure 10–29(b) helps visualize the program branching due to the IF THEN ELSE statements. As you can see, if either asynchronous input is active, *q* is Set or Reset and the check for the positive clock edge is skipped.

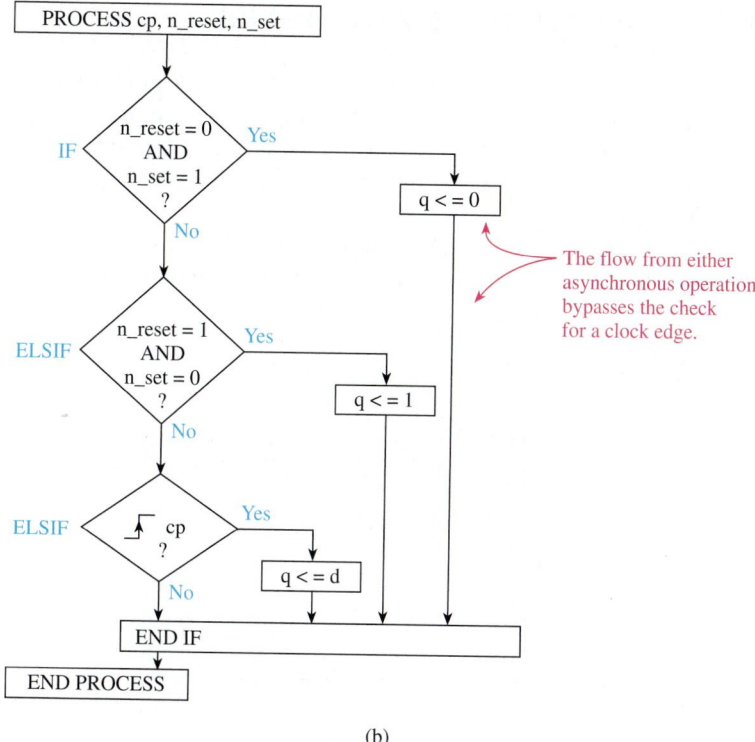

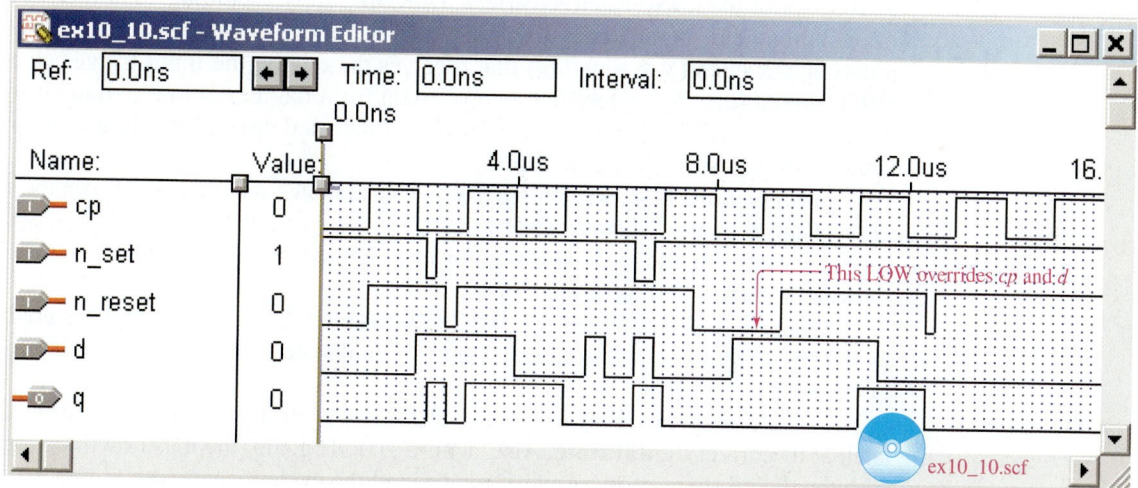

Figure 10–29 VHDL design for the *D* flip-flop of Example 10–10: (b) flowchart.

Figure 10–30 Simulation file for the *D* flip-flop of Example 10–10.

The simulation file is shown in Figure 10–30. This simulation must be run for both design methods (graphic and VHDL) to check the validity of both implementations. (Be sure to select **Set Project To Current File** before running the simulation for each design method.) This simulation illustrates asynchronous Set and Reset in several locations. Be sure that you can identify each one. Also study the synchronous inputs (*cp* and *d*) carefully to see that *q* responds to the level of *d* only at the positive clock edge. Notice at the 9 μs positive *cp* edge, *q* does not go HIGH because *n_reset* is active and overrides the synchronous inputs.

10–6 Master–Slave *J-K* Flip-Flop

Another type of flip-flop is the *J-K* flip-flop. It differs from the *S-R* flip-flop in that it has one new mode of operation, called **toggle.** Toggle means that Q and $\overline{Q}$ will switch to their *opposite* states at the active clock edge. (Q will switch from a 1 to 0 or from a 0 to a 1.) The synchronous inputs to the *J-K* flip-flop are labeled J, K, and C_p. J acts like the S input to an *S-R* flip-flop, and K acts like the R input to an *S-R* flip-flop. The toggle mode is achieved by making *both* J and K HIGH before the active clock edge. Table 10–5 shows the four synchronous operating modes of *J-K* flip-flops.

Toggle means Q flips to opposite state.

J acts like *Set.*
K acts like *Reset.*

TABLE 10–5	Synchronous Operating Modes of a *J-K* Flip-Flop	
Operating Mode	**J**	**K**
Hold	0	0
Set	1	0
Reset	0	1
Toggle	1	1

A number of the older flip-flops (74H71, 7472, 7473, 7476, 7478, 74104, 74105) are of the **master–slave** variety. They are rarely used today, but their theory of operation helps in understanding the newer varieties. They consist of two latches: a master *S-R* latch (*S-R* flip-flop) that receives data while the input trigger clock is HIGH, and a slave *S-R* latch that receives data from the master and outputs it when the clock goes LOW. Figure 10–31 shows a simplified equivalent circuit and logic symbol for a master–slave *J-K* flip-flop.

From Figure 10–31, we can see that the master latch will be loaded with the state of the J and K inputs, whereas AND gates 1 and 2 are enabled by a HIGH C_p (i.e., the *master* is loaded while C_p is HIGH). For now, let's ignore the feedback connections shown in Figure 10–31 as dashed lines.

When C_p goes LOW, gates 1 and 2 are **disabled,** but gates 3 and 4 are **enabled** by the HIGH from the inverter, allowing the digital state at the master to pass through to the slave latch inputs.

When C_p goes HIGH again, gates 3 and 4 will be disabled, thus keeping the slave latch at its current **digital state.** Also, with C_p HIGH again, the master will be loaded with the digital states of the J and K inputs, and the cycle repeats (see Figure 10–32). Master–slave flip-flops are called **pulse-triggered** or **level-triggered** devices because *input data are read during the entire time that the clock pulse is at a HIGH level.*

If you analyze the logic in Figure 10–31, *including* the dashed lines, you will see how the *toggle* operation occurs. With $J = 1$ and $K = 1$, let's assume that $Q = 1$ ($\overline{Q} = 0$). The dashed feedback connection from $Q = 1$ will enable gate 2 (gate 1 is disabled by the 0 on $\overline{Q}$), allowing the master to get reset when C_p goes HIGH. Therefore, Q (of the slave) will toggle to a 0 when C_p returns LOW.

With J and K still 1 and $Q = 0$, the next time C_p is HIGH, gate 1 will be enabled because $\overline{Q} = 1$. This will set the master. Then when C_p returns LOW, the Q output of the slave will toggle to a 1. In other words, the feedback connections allow only the *opposite* state to enter the master when $J = 1$ and $K = 1$. Notice that even

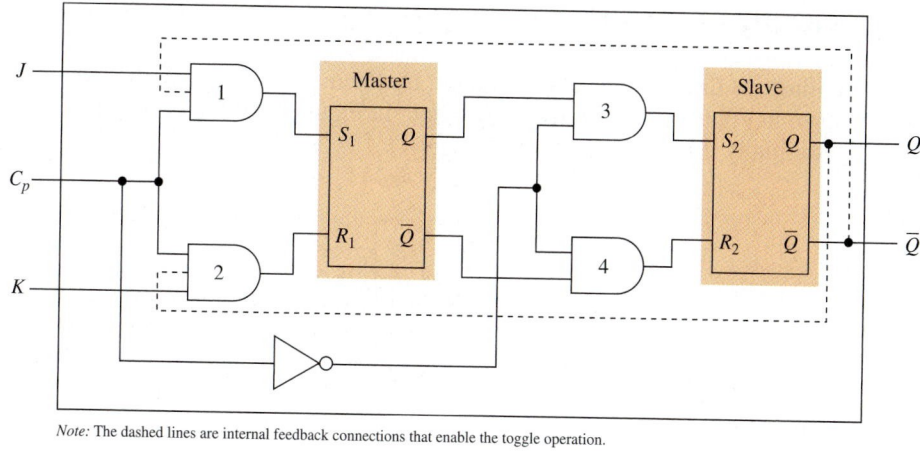

Note: The dashed lines are internal feedback connections that enable the toggle operation.

(a)

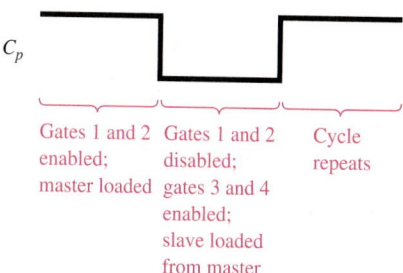

(b)

Figure 10–31   Positive pulse-triggered master–slave J-K flip-flop: (a) equivalent circuit; (b) logic symbol.

C_p

Gates 1 and 2   Gates 1 and 2   Cycle
enabled;   disabled;   repeats
master loaded   gates 3 and 4
                enabled;
                slave loaded
                from master

Figure 10–32   Enable/disable operation of the C_p line of a master–slave flip-flop.

if J and K are only *momentarily* made HIGH or if they are pulsed HIGH at different times while C_p is HIGH, the master will toggle and pass the toggle on to the slave when C_p goes LOW.

Occasionally, unwanted pulses or short glitches caused by electrostatic **noise** appear on J and K while C_p is HIGH. This phenomenon of interpreting unwanted signals on J and K while C_p is HIGH is called **ones catching** and is eliminated by the newer J-K flip-flops, which use an edge-triggering technique instead of pulse triggering.

EXAMPLE 10–11

To illustrate the master–slave operation, for the master–slave *J-K* flip-flop shown in Figure 10–33, draw the *Q* output in Figure 10–34. (Assume that *Q* is initially 0.)

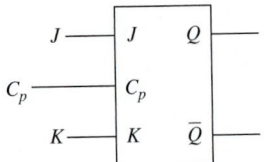

Figure 10–33

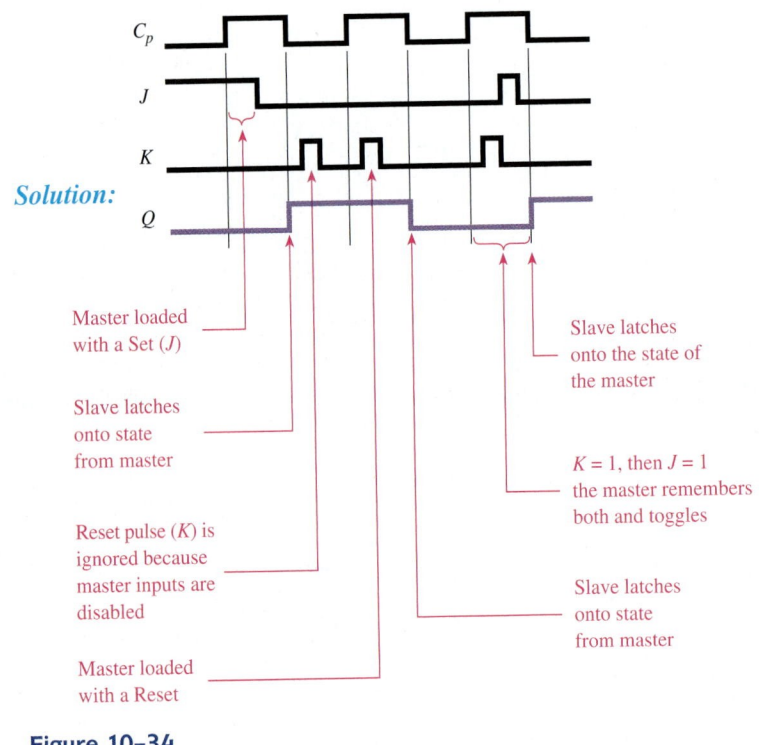

Solution:

Master loaded with a Set (*J*)

Slave latches onto state from master

Reset pulse (*K*) is ignored because master inputs are disabled

Master loaded with a Reset

Slave latches onto the state of the master

K = 1, then *J* = 1 the master remembers both and toggles

Slave latches onto state from master

Figure 10–34

EXAMPLE 10–12

For the master–slave *J-K* flip-flop shown in Figure 10–35, sketch the waveform at *Q* in Figure 10–36. (Assume that *Q* is initially 0.)

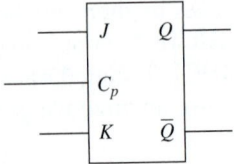

Figure 10–35

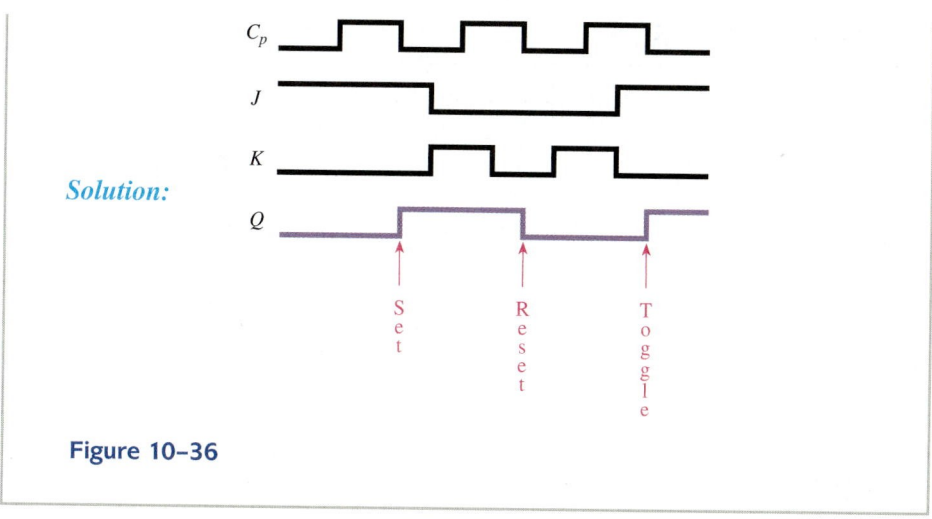

Solution:

Figure 10–36

10–7 Edge-Triggered *J-K* Flip-Flop with VHDL Model

With edge triggering, the flip-flop accepts data only on the *J* and *K* inputs that are present at the active clock edge (either the HIGH-to-LOW edge of C_p or the LOW-to-HIGH edge of C_p). This gives the design engineer the ability to accept input data on *J* and *K* at a precise instant in time. Transitions of the level *J* and *K* before or after the active clock trigger edge are ignored. The logic symbols for edge-triggered flip-flops use a small triangle at the clock input to signify that it is an edge-triggered device (see Figure 10–37).

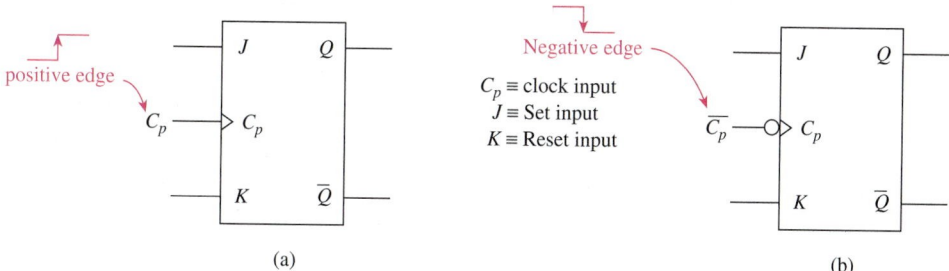

Figure 10–37 Symbols for edge-triggered *J-K* flip-flops: (a) positive edge triggered; (b) negative edge triggered.

Transitions of the *Q* output for the positive edge-triggered flip-flop shown in Figure 10–37(a) will occur when the C_p input goes from LOW to HIGH (positive edge). Figure 10–37(b) shows a negative edge-triggered flip-flop. The input clock signal will connect to the IC pin labeled $\overline{C_p}$. The small circle indicates that transitions in the output will occur at the HIGH-to-LOW edge (**negative edge**) of the $\overline{C_p}$ input.

The function table for a negative edge-triggered *J-K* flip flop is shown in Figure 10–38.

The downward arrow in the $\overline{C_p}$ column indicates that the flip-flop is triggered by the HIGH-to-LOW transition (negative edge) of the clock.

Operating mode	Inputs			Outputs	
	$\overline{C}_p$	J	K	Q	$\overline{Q}$
Hold	↓	0	0	No change	
Set	↓	1	0	1	0
Reset	↓	0	1	0	1
Toggle	↓	1	1	Opposite state	

↓ ≡ HIGH- to-LOW

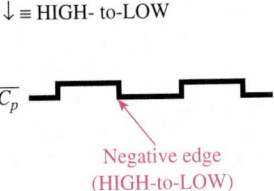

Negative edge
(HIGH-to-LOW)

Figure 10–38   Function table for a negative edge-triggered *J-K* flip-flop.

EXAMPLE 10–13

To illustrate edge triggering, for the negative edge-triggered *J-K* flip-flop shown in Figure 10–39, let's draw the *Q* output in Figure 10–40. (Assume that *Q* is initially 0.)

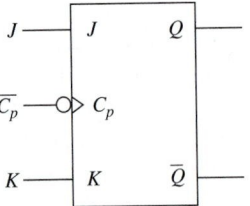

Figure 10–39

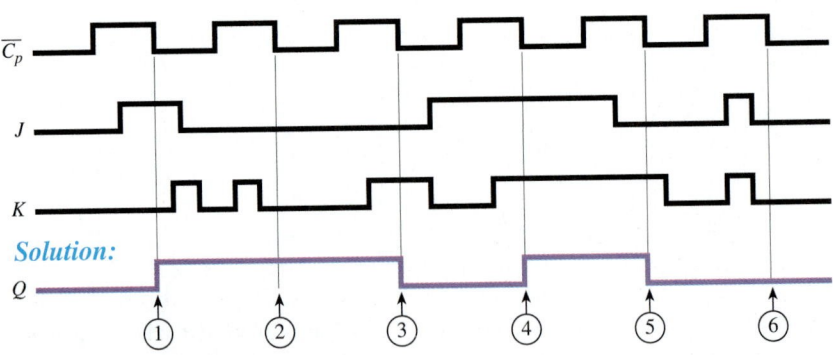

Solution:

① $J = 1, K = 0$ at the negative clock edge; *Q* is Set

② $J = 0, K = 0$ at the negative clock edge; *Q* is held
(transitions in *K* before the edge are ignored)

③ $J = 0, K = 1$ at the negative clock edge; *Q* is Reset

④ $J = 1, K = 1$ at the negative clock edge; *Q* toggles

⑤ $J = 0, K = 1$ at the negative clock edge; *Q* is Reset

⑥ $J = 0, K = 0$ at the negative clock edge; *Q* is held

Figure 10–40

VHDL Description of an Edge-triggered *J-K* Flip-Flop

A *J-K* flip-flop has four synchronous operations: Hold, Set, Reset, and Toggle. When an output toggles, it changes to the opposite state of what it was before the input trigger was applied. Because it has four operations, the VHDL solution can be made more streamlined by using the **CASE** statement instead of multiple IF-THEN-ELSE statements. *J-K* flip-flops can be designed for CPLDs using a graphic design method or as a VHDL program as shown in the following example.

EXAMPLE 10–14

J-K Flip-Flop

Create a *J-K* flip-flop using the graphic design method. A symbol called *JKFF* is found in the *prim* (primitives) subdirectory. Test the design by creating a simulation file. Repeat using the VHDL design entry method.

Solution: Figure 10–41 shows the solution as a gdf file using the primitive symbol *JKFF*. This symbol is a positive-edge flip-flop so an inverter was added to the clock line to make it a negative edge trigger and *n_cp* is used to represent $\overline{C_p}$.

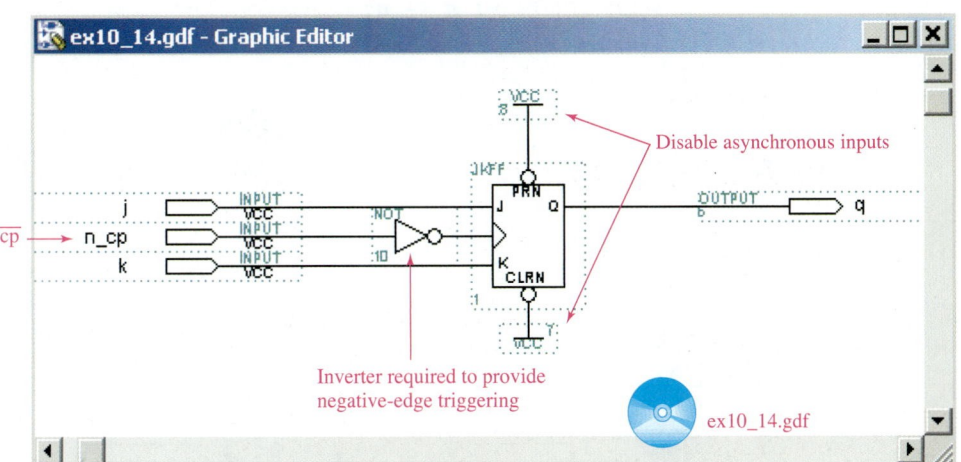

Figure 10–41   Graphic design file for the *J-K* flip-flop of Example 10–14.

The VHDL solution is shown in Figure 10–42(a). Notice in the entity that *q* is declared as a BUFFER to enable it to be used as both an input and an output. This is because the output assigned to *q* is sometimes determined by the state that is input to it from the previous value of *q*. [i.e. for the Hold condition, *q* becomes the state that *q* was before the clock transition (no change), and for Toggle, it becomes the opposite state]. The CASE statement is going to be checking a two-bit vector called *jk* so a vector SIGNAL must be declared, and the individual *j* and *k* inputs must be **concatenated** together using the & symbol to form a two-bit vector. The IF statement checks for a negative edge on *cp*. If it is a negative edge, then the CASE block is executed. All four flip-flop conditions are listed as the CASE conditions. The OTHERS clause is required to cover the other states allowed by *std_logic* besides 1 and 0.

The flowchart for the process is shown in Figure 10–42(b). This makes it obvious that the CASE statement is only executed if there is a negative edge on *cp*. If not, *q* is unaffected and the process ends.

```
ex10_14.vhd - Text Editor                                              _ □ X

LIBRARY ieee;                        ----------------------
USE ieee.std_logic_1164.ALL;  --      J-K FF       --
                                     ----------------------

ENTITY ex10_14 IS
    PORT(n_cp, j, k    : IN    std_logic;
         q             : BUFFER std_logic);
END ex10_14;
                          └─ Buffer declares q as output and input

ARCHITECTURE arc OF ex10_14 IS
SIGNAL jk : std_logic_vector (1 DOWNTO 0);
BEGIN          ┌─ Concatenate j and k into a 2-bit vector
    jk<=j&k;
    PROCESS (n_cp, j, k)
    BEGIN
        IF (n_cp'EVENT AND n_cp= '0') THEN  --Neg edge trigger
            CASE jk IS          ┌─ Double quote required for vector quantities
                WHEN "00" => q <= q;        --Hold
                WHEN "01" => q <= '0';      --Reset
                WHEN "10" => q <= '1';      --Set
                WHEN "11" => q <= NOT q;    --Toggle
                WHEN OTHERS => q <= q;
            END CASE;
        END IF;
    END PROCESS;
END arc;                                         ⊙ ex10_14.vhd

Line   1    Col   1    INS ◄
```

(a)

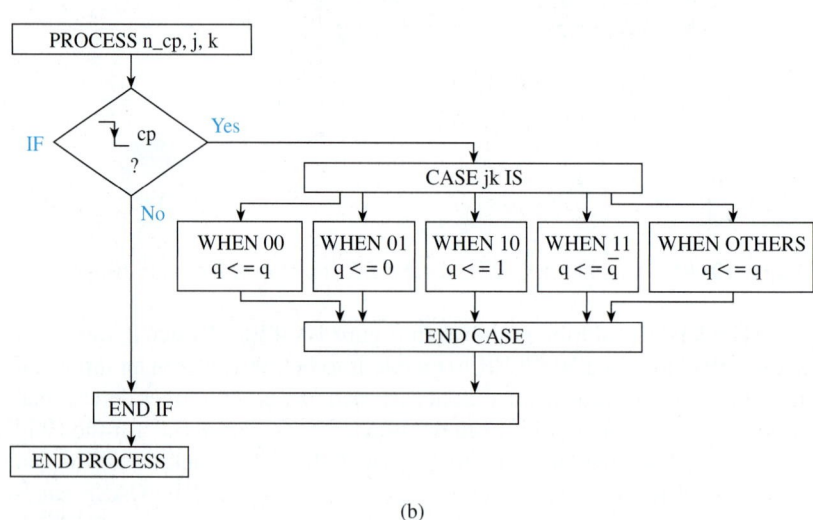

(b)

Figure 10–42 VHDL design file for the *J-K* flip-flop of Example 10–14:
(a) VHDL listing; (b) flowchart.

The simulation file is shown in Figure 10–43. This simulation must be run for both design methods (graphic and VHDL) to check the validity of both implementations. (Be sure to select **Set Project To Current File** before running the simulation for each design method.) This simulation illustrates the four operations of a *J-K* flip-flop. At the first negative clock edge, *j* and *k* are both LOW (Hold) so *q* remains unchanged. At the next edge, *j* is HIGH, *k* is LOW, so *q* is Set. At the next edge, *j* is LOW, *k* is HIGH, so *q* is Reset. For the last two negative clock edges, *j* and *k* are both HIGH so *q* toggles to the opposite state each time.

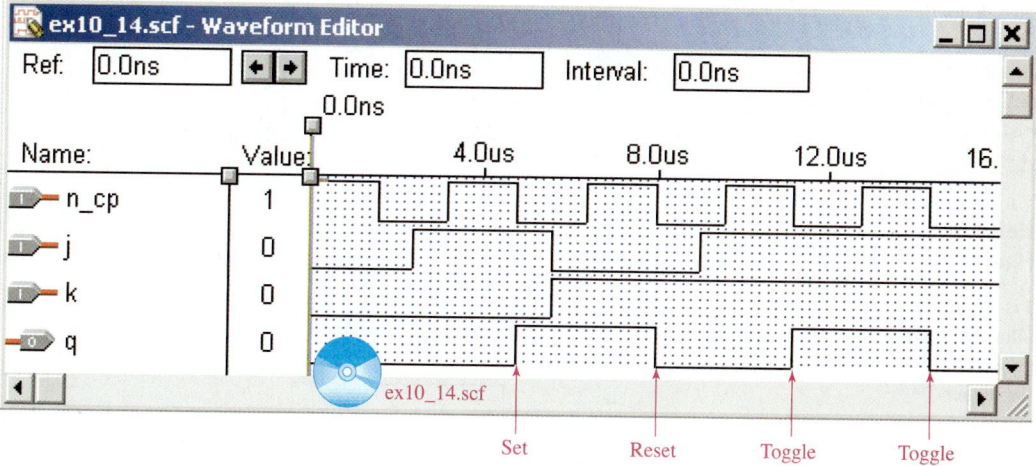

Figure 10–43 Simulation file for the *J-K* flip-flop of Example 10–14.

Review Questions

10–14. Describe why master–slave flip-flops are called *ones catching*.

10–15. The *Set* input to a *J-K* flip-flop is _____ *(J, K)* and the *Reset* input is _____ *(J, K)*.

10–16. The *edge-triggered J-K* flip-flop looks only at the *J-K* inputs that are present during the active clock edge on C_p. True or false?

10–17. What effect does the *toggle* operation of a *J-K* flip-flop have on the *Q* output?

10–8   Integrated-Circuit *J-K* Flip-Flop (7476, 74LS76)

Now let's take a look at actual *J-K* flip-flop ICs. The 7476 and 74LS76 are popular *J-K* flip-flops because they are both dual flip-flops (two flip-flops in each IC package) and they have asynchronous inputs ($\overline{R_D}$ and $\overline{S_D}$) as well as synchronous inputs (C_p, *J*, *K*). The 7476 is a positive pulse-triggered (master–slave) flip-flop, and the 74LS76 is a negative edge-triggered flip-flop, a situation that can trap the unwary technician who attempts to replace the 7476 with the 74LS76!

From Figure 10–44(a) and Table 10–6, we see that the asynchronous inputs $\overline{S_D}$ and $\overline{R_D}$ are *active-LOW*. That is, a LOW on $\overline{S_D}$ (Set) will Set the flip-flop (*Q* = 1), and a LOW on $\overline{R_D}$ will Reset the flip-flop (*Q* = 0). Remember, the asynchronous inputs will cause the flip-flop to respond immediately *without* regard to the clock trigger input.

For synchronous operations using *J*, *K*, and $\overline{C_p}$, the asynchronous inputs must be disabled by putting a HIGH level on both $\overline{S_D}$ and $\overline{R_D}$. The *J* and *K* inputs are read one setup time before the HIGH-to-LOW edge of the clock ($\overline{C_p}$). One setup time for the 74LS76 is 20 ns. This means that the state of *J* and *K* 20 ns *before* the negative edge of the clock is used to determine the synchronous operation to be performed. (Of course, the 7476 master–slave will read the state of *J* and *K* during the entire positive clock pulse.)

Helpful Hint

The 7476 master–slave can be demonstrated in lab to be a ones catcher. The 74LS76, on the other hand, will ignore *J* and *K* except at the trigger edge. Try it.

Helpful Hint

The 74LS112 is another dual *J-K* flip-flop. It is very popular because it is available in the higher-speed families, such as 74F112, 74ALS112, and 74HC112.

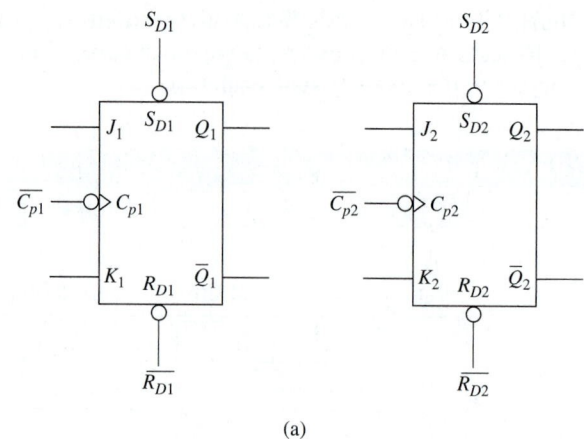

(a)

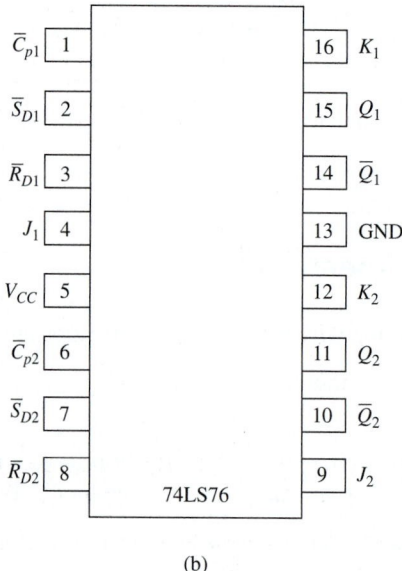

(b)

Figure 10–44 The 74LS76 negative edge-triggered flip-flop: (a) logic symbol; (b) pin configuration.

TABLE 10–6 | Function Table for the 74LS76[a]

	Inputs					Outputs	
Operating Mode	$\overline{S_D}$	$\overline{R_D}$	$\overline{C_p}$	*J*	*K*	*Q*	$\overline{Q}$
Asynchronous Set	L	H	x	x	x	H	L
Asynchronous Reset	H	L	x	x	x	L	H
Synchronous Hold	H	H	↓	l	l	q	$\overline{q}$
Synchronous Set	H	H	↓	h	l	H	L
Synchronous Reset	H	H	↓	l	h	L	H
Synchronous Toggle	H	H	↓	h	h	$\overline{q}$	q

[a]H = HIGH-voltage steady state; L = LOW-voltage steady state; h = HIGH voltage one setup time before negative clock edge; l = LOW voltage one setup time before negative clock edge; x = don't care; q = state of *Q* before negative clock edge; ↓ = HIGH-to-LOW (negative) clock edge.

Also notice that in the toggle mode ($J = K = 1$), after a negative clock edge, Q becomes whatever $\overline{Q}$ was before the clock edge, and vice versa (i.e., if $Q = 1$ before the negative clock edge, then $Q = 0$ after the negative clock edge).

Now let's work through several timing analysis examples to be sure that we fully understand the operation of *J-K* flip-flops.

EXAMPLE 10–15

Sketch the Q waveform for the 74LS76 negative edge-triggered *J-K* flip-flop shown in Figure 10–45, with the input waveforms given in Figure 10–46.

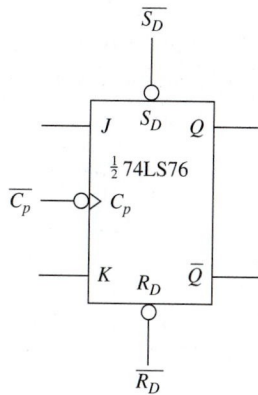

Figure 10–45

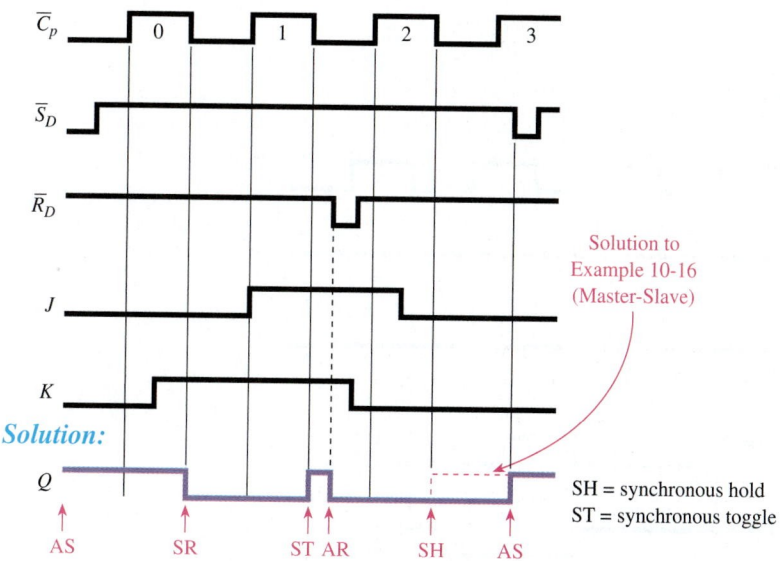

Solution to
Example 10-16
(Master-Slave)

SH = synchronous hold
ST = synchronous toggle

Figure 10–46

Note: Q changes only on the negative edge of $\overline{C_p}$, except when asynchronous operations $(\overline{S_D}, \overline{R_D})$ are taking place.

EXAMPLE 10–16

How would the Q waveform of Example 10–11 be different if we used a 7476 pulse-triggered master–slave flip-flop instead of the 74LS76?

Solution: During positive pulse 2, J is HIGH for a short time. The master latch within the 7476 will remember that and cause the flip-flop to do a synchronous Set ($Q = 1$) when $\overline{C_p}$ returns LOW. (See Figure 10–46.)

EXAMPLE 10–17

Sketch the Q waveform for the 7476 positive pulse-triggered flip-flop shown in Figure 10–47 with the input waveforms given in Figure 10–48.

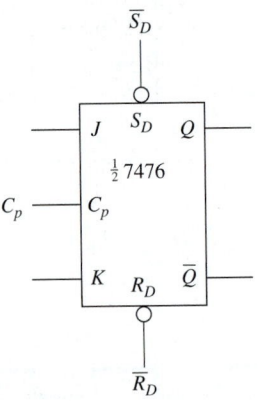

Figure 10–47

Solution:

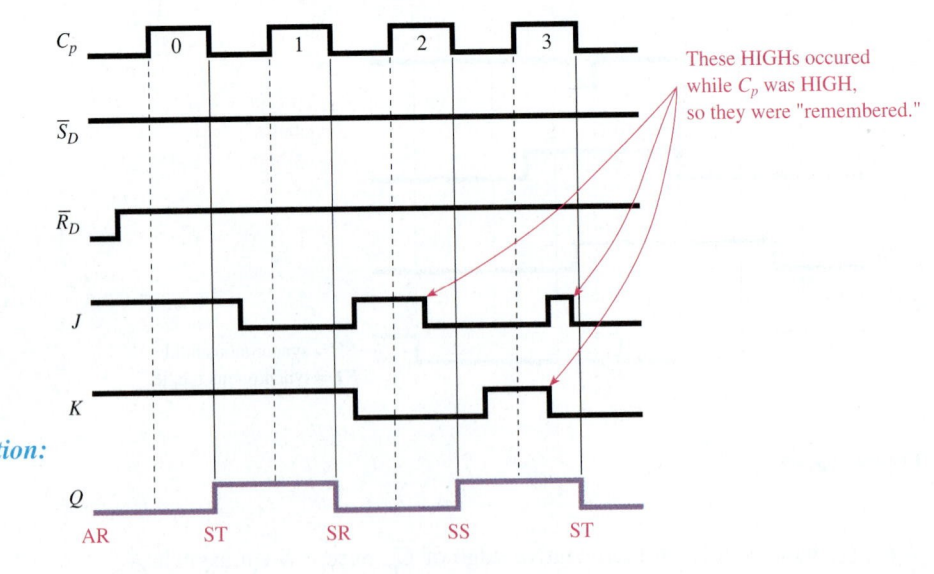

These HIGHs occured while C_p was HIGH, so they were "remembered."

Figure 10–48

EXAMPLE 10-18

The 74109 is a positive edge-triggered J-$\overline{K}$ flip-flop. The logic symbol (Figure 10–49) and input waveforms (Figure 10–50) are given; sketch Q.

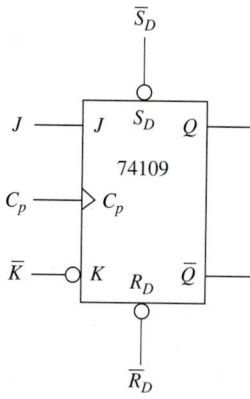

Figure 10–49

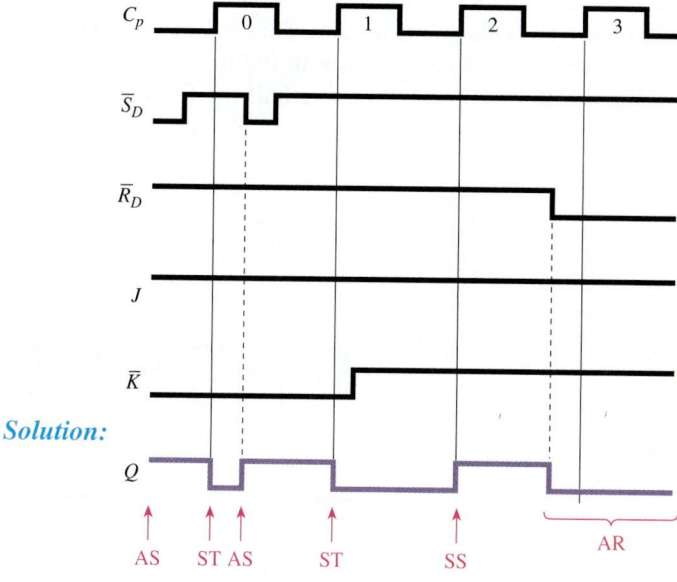

Solution:

Figure 10–50

Notes:

1. *Positive* edge triggering.

2. $\overline{K}$ instead of K; therefore, for a toggle, $J = 1, \overline{K} = 0$.

The J-K flip-flop can be used to form other flip-flops by making the appropriate external connections. For example, to form a D flip-flop, add an inverter between the J and K inputs, and bring the data into the J input, as shown in Figure 10–51.

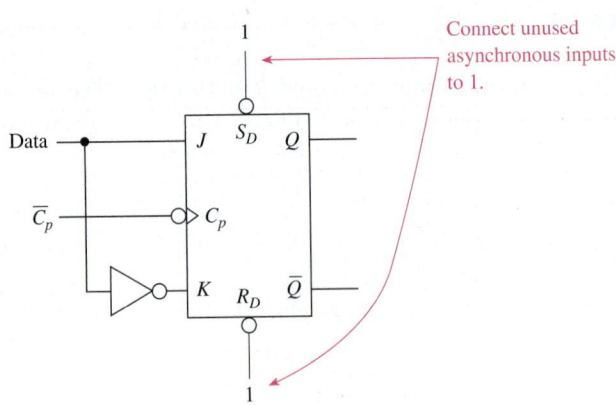

Figure 10–51   *D* flip-flop made from a *J-K* flip-flop.

The flip-flop in Figure 10–51 will operate as a *D* flip-flop because the data are brought in on the *J* terminal and its complement is at the *K*. So, if Data = 1, the flip-flop will be Set after the clock edge; if Data = 0, the flip-flop will be Reset after the clock edge. (*Note:* You lose the toggle mode and hold mode using this configuration.)

Also, it is often important for a flip-flop to operate in the toggle mode. This can be done simply by connecting both *J* and *K* to 1. This will cause the flip-flop to change states at each active clock edge, as shown in Figure 10–52. Notice that the frequency of the output waveform at *Q* will be one-half the frequency of the input waveform at $\overline{C_p}$.

Helpful Hint

It is interesting for you to see an application of the *D* flip-flop (check out the shift register in Figure 13–1) and the toggle flip-flop (check out the counter in Figure 12–9).

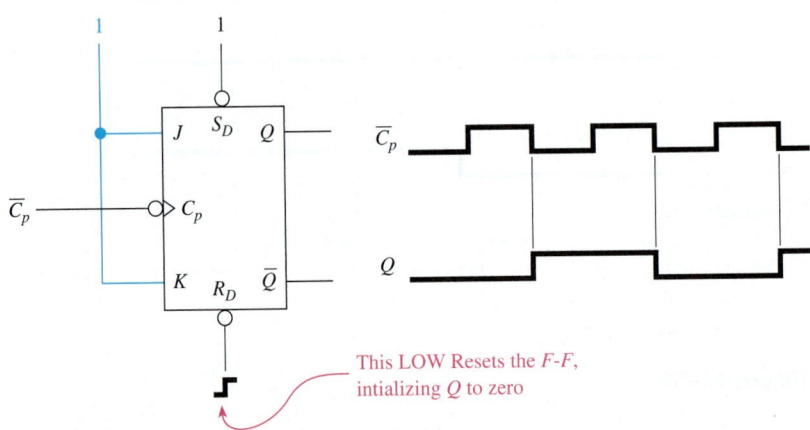

Figure 10–52   *J-K* connected as a toggle flip-flop.

Figure 10–53 shows the test apparatus used to display the input and output of a toggle flip-flop. The oscilloscope display is used to accurately show the timing and frequency relationship between the two waveforms.

As we have seen, there is a variety of flip-flops, each with its own operating characteristics. In Chapters 11 through 13, we learn how to use these ICs to perform sequential operations such as counting, data shifting, and sequencing.

First, let's summarize what we have learned about flip-flops by utilizing four common flip-flops in the same circuit and then supplying input signals and sketching the *Q* output of each (see Example 10 –19).

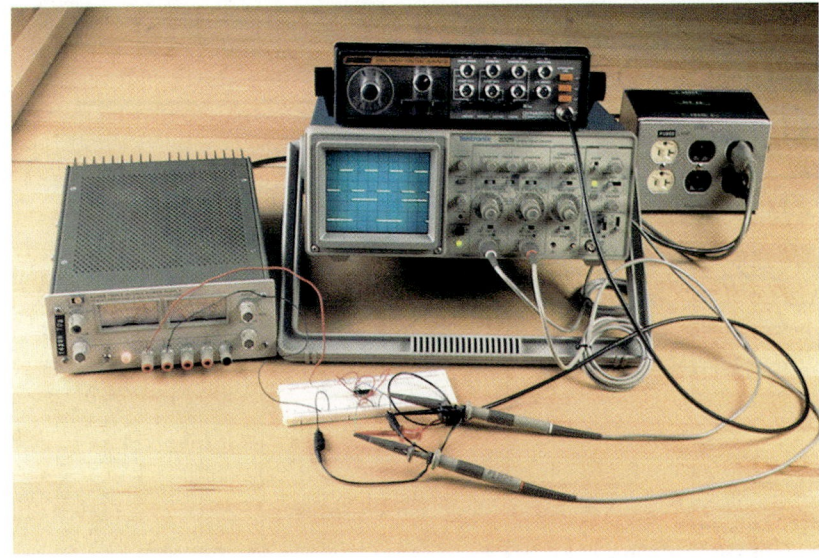

Figure 10–53 Test apparatus used to analyze the input and output of a toggle flip-flop.

EXAMPLE 10–19

For each of the flip-flops shown in Figure 10–54, sketch the Q outputs in Figure 10–55.

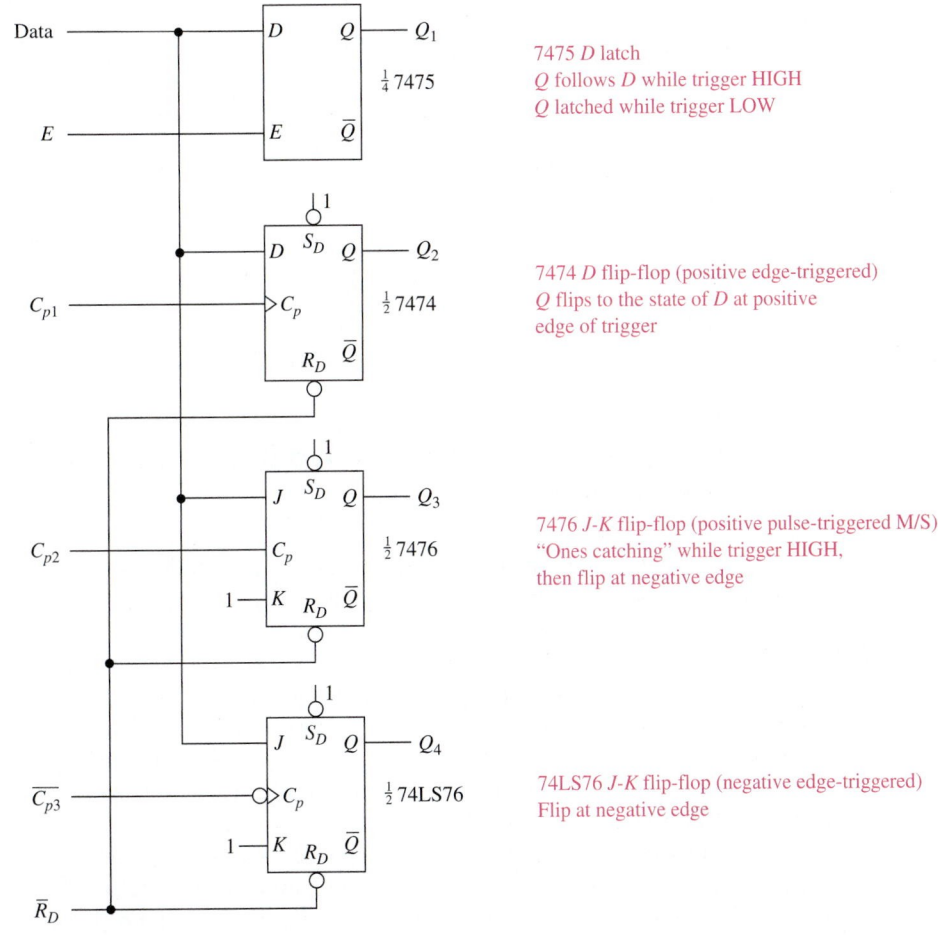

7475 D latch
Q follows D while trigger HIGH
Q latched while trigger LOW

7474 D flip-flop (positive edge-triggered)
Q flips to the state of D at positive edge of trigger

7476 J-K flip-flop (positive pulse-triggered M/S)
"Ones catching" while trigger HIGH, then flip at negative edge

74LS76 J-K flip-flop (negative edge-triggered)
Flip at negative edge

Figure 10–54

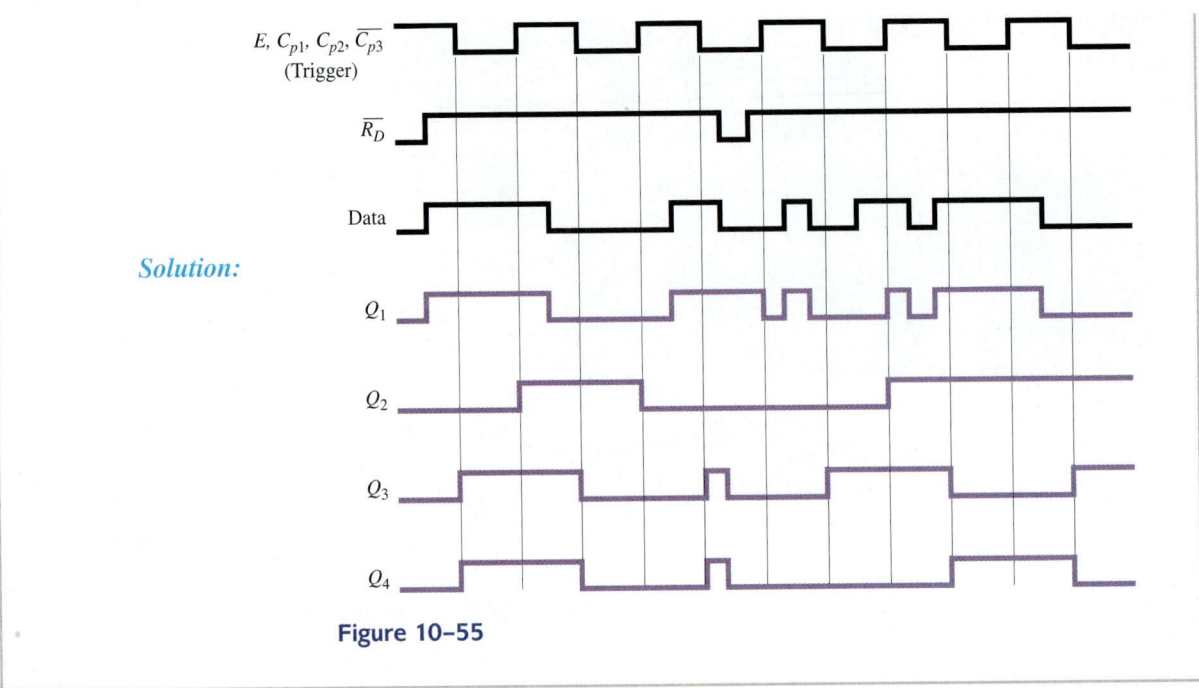

Figure 10–55

Solution:

Review Questions

10–18. How do you *asynchronously* Reset the 74LS76 flip-flop?

10–19. The synchronous inputs to the 74LS76 override the asynchronous inputs. True or false?

10–20. To operate a 74LS76 flip-flop synchronously, the $\overline{S_D}$ and $\overline{R_D}$ inputs must be held _____ (HIGH, LOW).

10–21. What is the distinction between uppercase and lowercase letters when used in the function table for the 74LS76 flip-flop?

10–9 Using an Octal *D* Flip-Flop in a Microcontroller Application

Most of the basic latches and flip-flops are also available as **octal** ICs. In this configuration, there are eight latches or flip-flops in a single IC package. If all eight latches or flip-flops are controlled by a common clock, it is called an 8-bit **register.** An example of an 8-bit *D* flip-flop register is the high-speed CMOS 74HCT273 (also available in the TTL LS and S families). The '273 contains eight *D* flip-flops, all controlled by a common edge-triggered clock, C_p (see Figure 10–56). At the positive edge of C_p, the 8 bits of data at D_0 through D_7 are stored in the eight *D* flip-flops and output at Q_0 through Q_7. The '273 also has an active-LOW master reset ($\overline{MR}$), which provides asynchronous Reset capability to all eight flip-flops.

An application of the '273 octal *D* flip-flop is shown in Figure 10–57. Here it is used as an *update and hold* register. Every 10 s, it receives a clock pulse from the Motorola 68HC11 microcontroller. The data that are on D_0–D_7 at each positive clock edge are stored in the register and output at Q_0–Q_7.

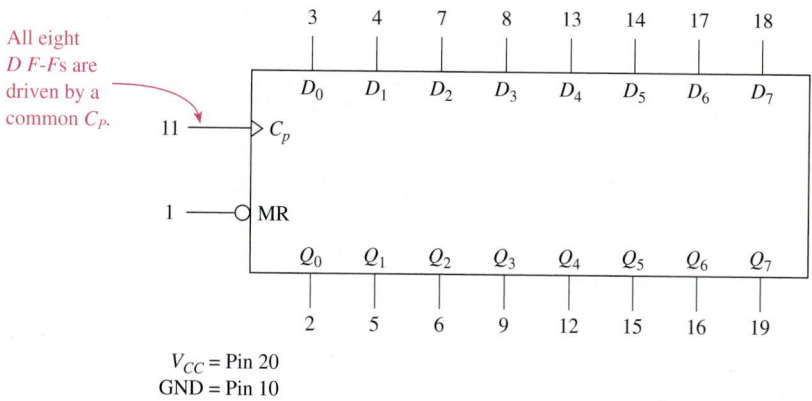

Figure 10–56   Logic diagram for a 74HCT273 octal *D* flip-flop.

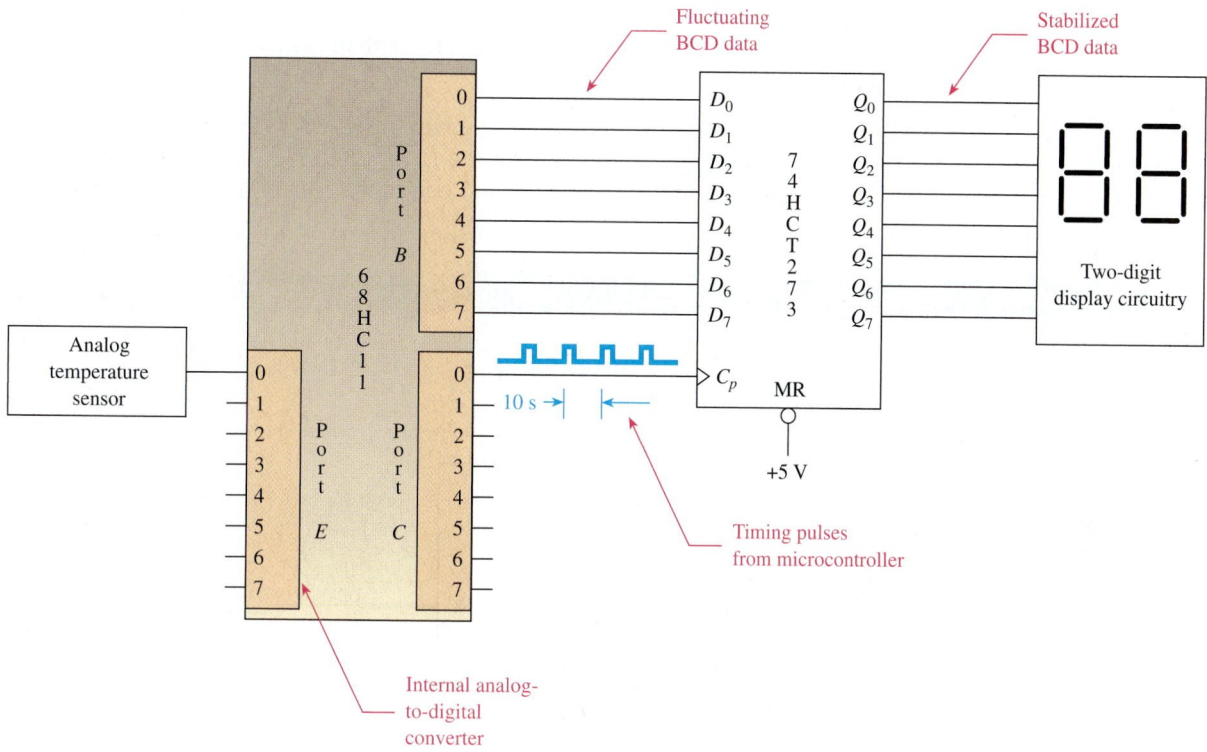

Figure 10–57   Using an octal *D* flip-flop to interface a display to a microcontroller.

The analog temperature sensor is designed to output a voltage that is proportional to degrees centigrade. (See Section 15–12 for the design of temperature sensors.) The 68HC11 microcontroller has the capability to read analog voltages and convert them into their equivalent digital value. A software program is written for the microcontroller to translate this digital string into a meaningful two-digit BCD output for the display.

The BCD output of the 68HC11 is constantly changing as the temperature fluctuates. One way to stabilize this fluctuating data is to use a **storage register** like the 74HCT273. Because the '273 only accepts the BCD data every 10 s, it will hold the two-digit display constant for that length of time, making it easier to read.

10–10 Using Altera's LPM Flip-Flop

MAX+PLUS II software provides a general-purpose flip-flop called *LPM_FF* in their Library of Parameterized Modules subdirectory called *\mega_lpm*. It has several different parameters available for the user to work with in the design of flip-flop circuits. A help screen describing all of its features can be viewed by right-clicking on the LPM symbol and selecting **Edit Ports/Parameters.** Pressing the **Help** button, which appears in the top right of the drop-down window, will provide a description of all of the inputs, outputs, and parameters that are available. The next two examples will illustrate some of its basic operating features with its resulting simulation waveforms.

═══ **EXAMPLE 10–20** ═══

LPM Flip-Flop

Use the LPM_FF to implement an octal *D* flip-flop with synchronous *clock* and *data* inputs and an asynchronous *clear.* Test its operation by producing a set of simulation waveforms that exercise both the synchronous and asynchronous inputs.

Solution: Create a graphic design file with the *LPM_FF* symbol. Make it an octal device by specifying *8* for the *LPM_WIDTH.* Change the *enable* port to *unused.* Attach the inputs and outputs as shown in Figure 10–58.

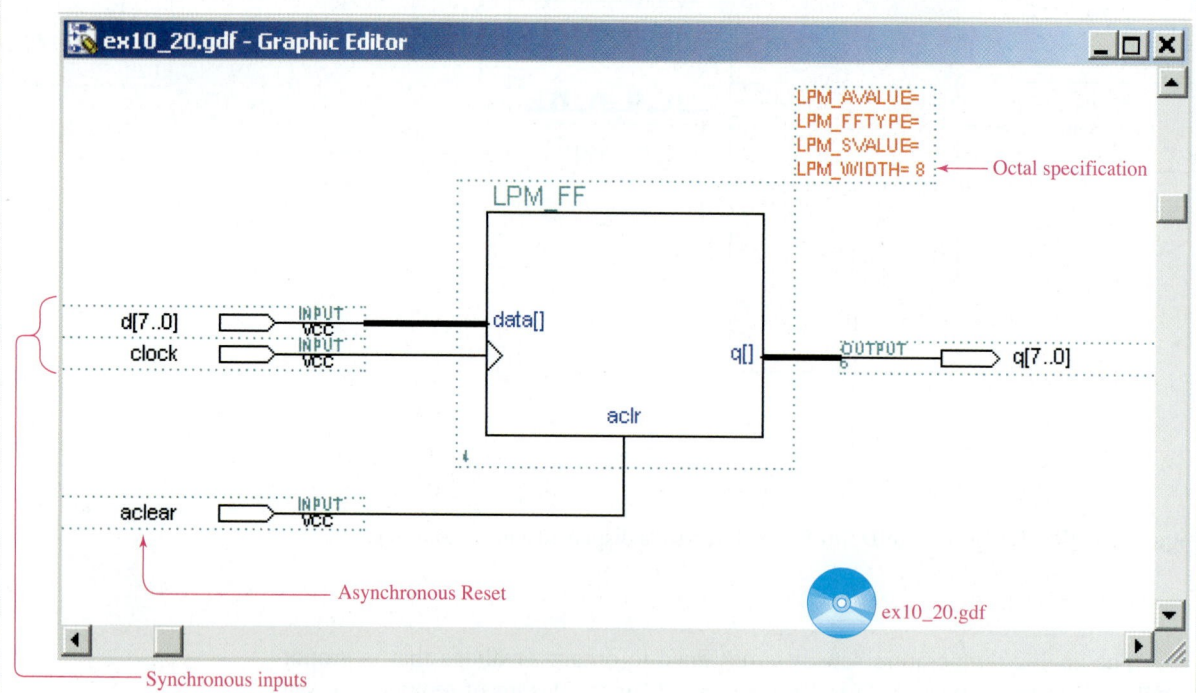

Figure 10–58 The LPM_FF used for Example 10–20.

Several waveforms are made up for the simulation to exercise the features of the LPM_FF (See Figure 10–59.) The initial HIGH on *aclear* resets all eight *q* outputs to 0's (00H). The first positive edge clock occurs at 1.0 μs. This stores the 8 bits from the *d*-inputs (FFH) into the register, which then appear on the *q* outputs just the way that an individual 7474 *D* flip-flop

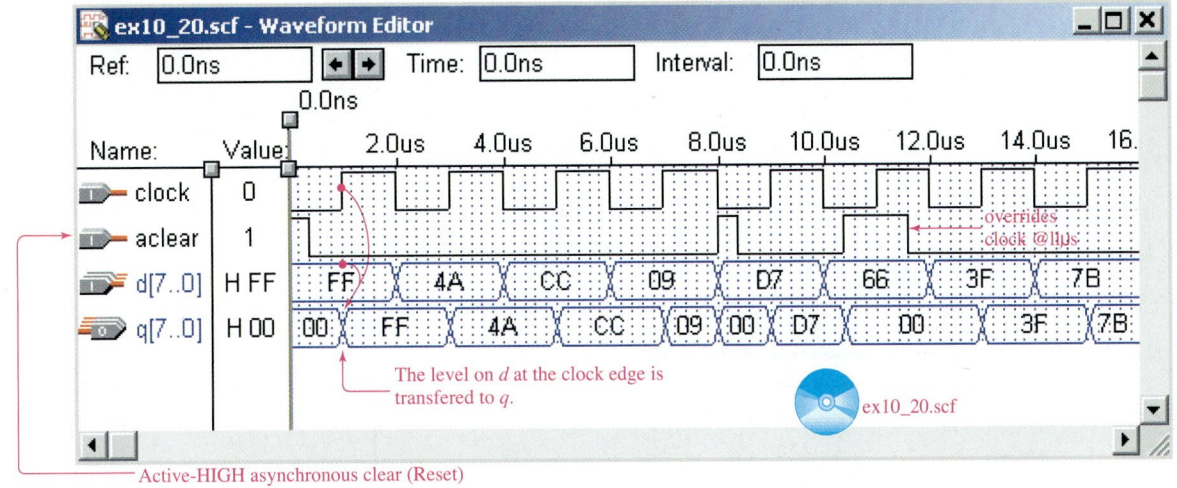

Figure 10–59 The simulation file for Example 10–20.

would. The same occurs at the 3-, 5-, and 7-μs marks. Then an asynchronous *aclear* is asserted at 8.0 μs. (It is an active-HIGH signal, but it could have been specified as active-LOW in the edit ports/parameters window.) This forces the *q*-outputs to 00H. The next *aclear* is asserted at 10.4 μs, resetting the *q* outputs. It is still HIGH beyond the 11 μs positive clock edge. Since *aclear* has priority, the synchronous operation of loading the *d* inputs (66H) will not occur.

EXAMPLE 10–21

LPM Flip-flop with asynchronous control

Use the LPM_FF to implement an octal *D* flip-flop with asynchronous clear, set, and load. Include an enable to control the clock. Test its operation by producing a set of simulation waveforms that exercise both the asynchronous and synchronous inputs.

Solution: Add the LPM_FF symbol to a graphic design file as shown in Figure 10–60. In the **Edit Ports/Parameters** drop-down window, specify *used* for all of the ports shown in Figure 10–59. Make it an octal device by specifying *8* for the *LPM_WIDTH*. Connect all of the inputs and outputs as shown.

The simulation file in Figure 10–61 is designed to exercise all of the synchronous and asynchronous inputs. As specified in the **Help** screen, the *enable* line must be HIGH to enable the clock to be read. (Notice that the positive clock transition at 5.0 μs is ignored because *enable* is LOW.) The *q* outputs initially start out at FFH because the asynchronous set (*aset*) is asserted HIGH right at the beginning of the waveforms. After that, the *q* outputs take on the value of the *d* inputs at each positive clock edge. At 8.4 μs the *aload* signal is asserted. This asynchronously loads the *d* inputs (33H) into the device regardless of the state of the clock. The last asynchronous operation occurs at 13.8 μs where *aclear* resets the *q* outputs to 00H.

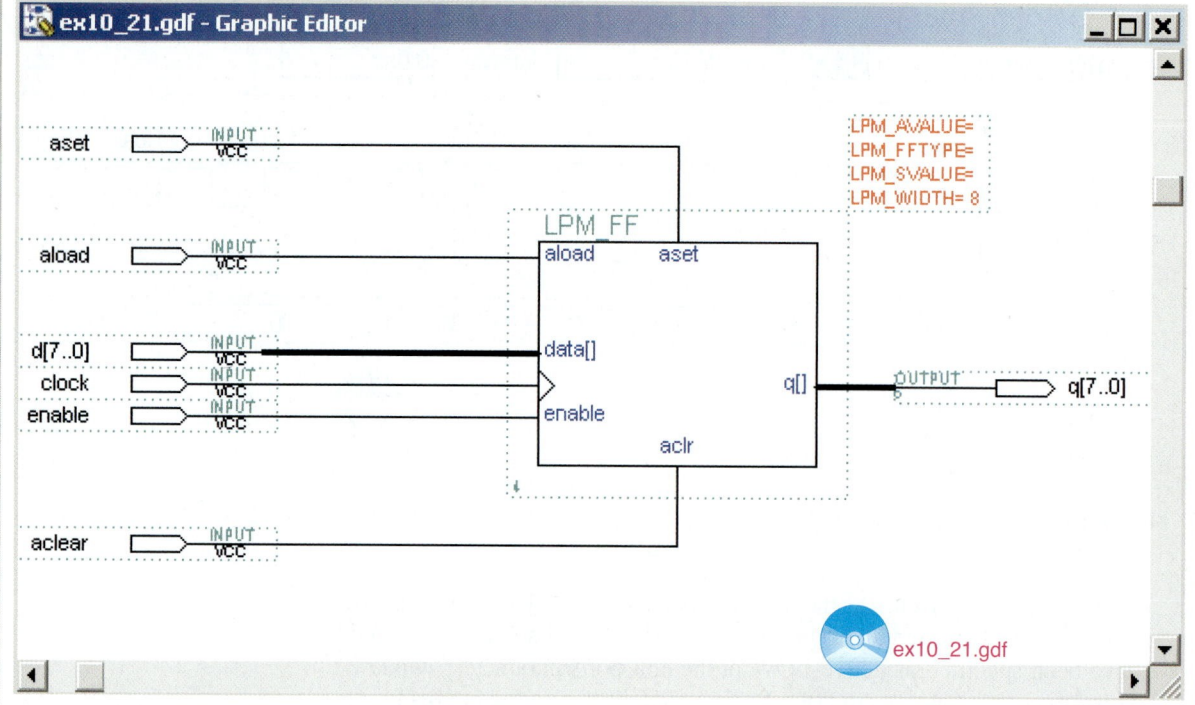

Figure 10–60 The LPM_FF used for Example 10–21.

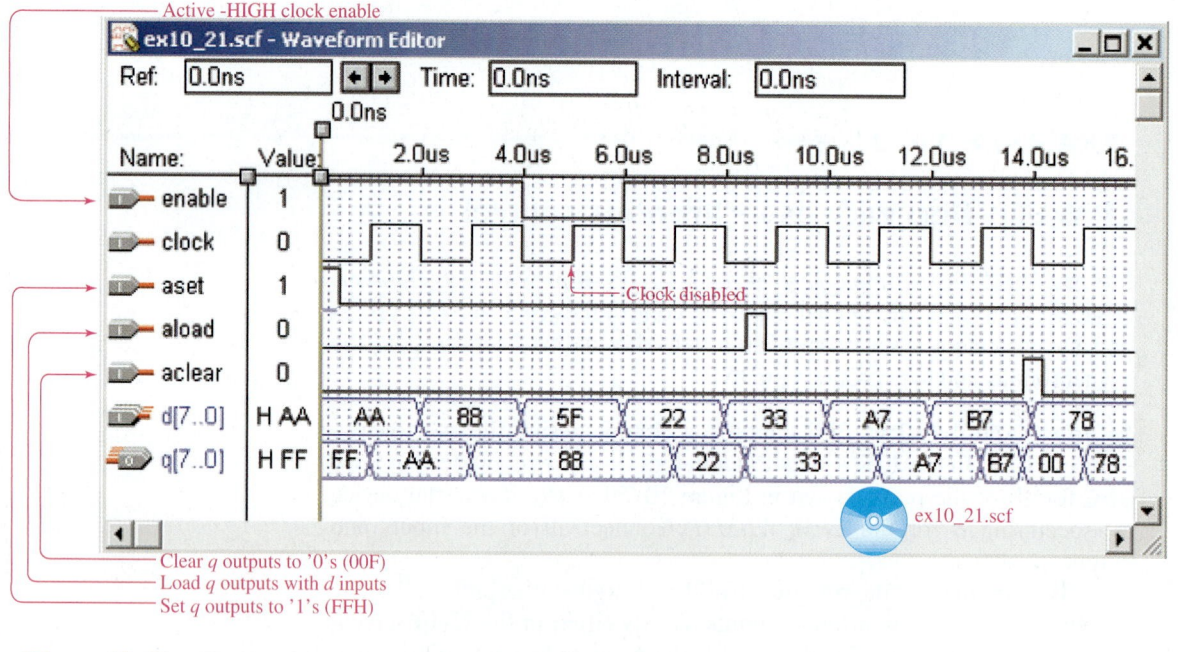

Figure 10–61 The simulation file for Example 10–21.

Summary

In this chapter, we have learned that

1. The *S-R* flip-flop is a single-bit data storage circuit that can be constructed using basic gates.

2. Adding gate enable circuitry to the *S-R* flip-flop makes it *synchronous*. This means that it will operate only under the control of a clock or enable signal.

3. The *D* flip-flop operates similar to the *S-R*, except it has only a single data input, *D*.

4. The 7475 is an IC *D* latch. The output *(Q)* follows *D* while the enable *(E)* is HIGH. When *E* goes LOW, *Q* remains latched.

5. The 7474 is an IC *D* flip-flop. It has two *synchronous* inputs, D and C_P, and two *asynchronous* inputs, $\overline{S_D}$ and $\overline{R_D}$. Q changes to the level of D at the *positive edge* of C_p. Q responds *immediately* to the asynchronous inputs regardless of the synchronous operations.

6. The *J-K* flip-flop differs from the *S-R* flip-flop because it can also perform a *toggle* operation. Toggling means that Q flips to its opposite state.

7. The master–slave *J-K* flip-flop consists of two latches: a *master* that receives data while the clock trigger is HIGH, and a *slave* that receives data from the master and outputs it to Q when the clock goes LOW.

8. The 74LS76 is an edge-triggered *J-K* flip-flop IC. It has synchronous and asynchronous inputs. The 7476 is similar, except it is a pulse-triggered master–slave type.

9. The 74HCT273 is an example of an *octal D* flip-flop. It has *eight D* flip-flops in a single IC package, making it ideal for microprocessor applications.

10. *D* latches, *D* flip-flops and *J-K* flip-flops can be described in VHDL and implemented in CPLDs.

11. The MAX+PLUS II software provides a general-purpose flip-flop in their LPM subdirectory that can be used to implement multi-bit *D* and toggle flip-flops.

Glossary

Active-LOW: Means that the input to or the output from a terminal must be LOW to be enabled, or "active."

Asynchronous: (Not synchronous.) A condition in which the output of a device will switch states instantaneously as the inputs change without regard to an input clock signal.

Clock: A device used to produce a periodic digital signal that repeatedly switches from LOW to HIGH and back at a predetermined rate.

Combinational Logic: The use of several of the basic gates (AND, OR, NOR, NAND) together to form more complex logic functions. The output of a combinational logic circuit is determined by the *present* logic levels at its inputs.

Complement: Opposite digital state (i.e., the complement of 0 is 1, and vice versa).

Concatenate: Combine two or more values end-to-end and treat them as one.

Digital State: The logic levels within a digital circuit (HIGH level = 1 state and LOW level = 0 state).

Disabled: The condition in which a digital circuit's inputs or outputs are not allowed to accept or transmit digital states.

Edge Triggered: The term given to a digital device that can accept inputs and change outputs only on the positive or negative *edge* of some input control signal or clock.

Enabled: The condition in which a digital circuit's inputs or outputs are allowed to accept or transmit digital states normally.

Flip-Flop: A circuit capable of storing a digital 1 or 0 level based on sequential digital levels input to it.

Function Table: A table that illustrates all the possible combinations of input and output states for a given digital IC or device.

Latch: The ability to *hold* onto a particular digital state. A latch circuit will hold the level of a digital pulse even after the input is removed.

Level Triggered: *See* Pulse triggered.

Master–Slave: A storage device consisting of two sections: the master section, which accepts input data while the clock is HIGH; and the slave section, which receives the data from the master when the clock goes LOW.

Negative Edge: The edge on a clock or trigger pulse that is making the transition from HIGH to LOW.

Noise: Any fluctuations in power supply voltages, switching surges, or electrostatic charges will cause irregularities in the HIGH- and LOW-level voltages of a digital signal. These irregularities or fluctuations in voltage levels are called electrical noise and can cause false readings of digital levels.

Octal: A group of eight. An octal flip-flop IC has eight flip-flops in a single package.

Ones Catching: A feature of the master–slave flip-flop that allows the master section to latch on to any 1 level that is felt at the inputs at any time while the input clock pulse is HIGH and then transfer those levels to the slave when the clock goes LOW.

Positive Edge: The edge on a clock or trigger pulse that is making the transition from LOW to HIGH.

Pulse Triggered: The term given to a digital device that can accept inputs during an entire positive or negative pulse of some input control signal or clock. (Also called level triggered.)

Register: A group of several flip-flops or latches that is used to store a binary string and is controlled by a common clock or enable signal.

Reset: A condition that produces a digital LOW (0) state.

Sequential Logic: Digital circuits that involve the use of a sequence of timing pulses to synchronize the reading of their input data. Storage devices such as flip-flops and latches and functional ICs such as counters and shift registers are sequential logic.

Set: A condition that produces a digital HIGH (1) state.

Setup Time: The length of time before the active edge of a trigger pulse (control signal) that the inputs of a digital device must be in a stable digital state. [That is, if the setup time of a device is 20 ns, the inputs must be held stable (and will be read) 20 ns before the trigger edge.]

Storage Register: Two or more data storage circuits (such as flip-flops or latches) used in conjunction with each other to hold several bits of information.

Strobe Gates: A control gate used to enable or disable inputs from reaching a particular digital device.

Synchronous: A condition in which the output of a device will operate only in synchronization with (in step with) a specific HIGH or LOW timing pulse or trigger signal.

Toggle: In a flip-flop, a toggle is when Q changes to the level of $\overline{Q}$ and $\overline{Q}$ changes to the level of Q.

Transition: The instant of change in digital state from HIGH to LOW or LOW to HIGH.

Transparent Latch: An asynchronous device whose outputs will hold onto the most recent digital state of the inputs. The outputs immediately follow the state of the inputs without regard to trigger input and remain in that state even after the inputs are removed or disabled.

Trigger: The input control signal to a digital device that is used to specify the instant that the device is to accept inputs or change outputs.

▬ Problems ▬

Section 10–1

10-1. Make the necessary connections to the 7400 quad NAND gate IC in Figure P10–1 to form the cross-NAND *S-R* flip-flop of Figure 10–2. [Remember that an inverter can be formed from a NAND (Chapter 3).]

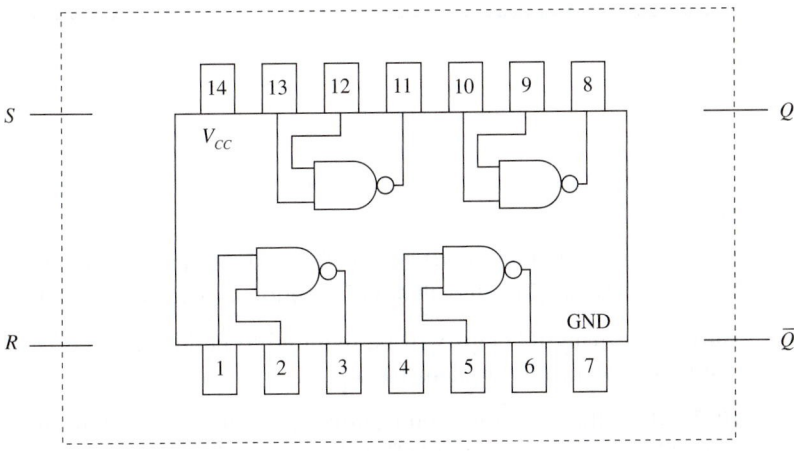

Figure P10–1

(c) Download your design to a CPLD IC. Discuss your observations of the *q* output LED with your instructor as you use the switches to step through the combinations of *e0* and *d0* simulated in (b).

C10-2. The VHDL program in Figure 10–26(a) is the implementation of a *D* flip-flop with an asynchronous Reset.

(a) Make the necessary changes to provide an asynchronous *Set* instead of *Reset*. Save this program as *prob_c10_2.vhd*.

(b) Test its operation by creating waveform simulations that demonstrate its synchronous (*cp* and *d*) and asynchronous (*n_set*) operations.

(c) Download your design to a CPLD IC. Discuss your observations of the *q* output LED with your instructor as you use the switches or pushbuttons to step through the combinations of *n_set*, *cp*, and *d* simulated in (b).

C10-3. The Primitives subdirectory contains a *D* flip-flop with a clock enable signal. It is called *DFFE*.

(a) Build a graphic design file containing this flip flop with I/O ports labeled similar the gdf file shown in Figure 10–28. Save this program as *prob_c10_3.gdf*.

(b) Test its operation by creating waveform simulations that demonstrate its synchronous and asynchronous operations as well as its clock enable/disable feature. (Use the *help* menu to find information on the operation of the *DFFE* primitive.)

(c) Download your design to a CPLD IC. Discuss your observations of the *q* output LED with your instructor as you use the switches or pushbuttons to step through the combinations of *n_set*, *n_reset*, *clock_enable*, *cp*, and *d* simulated in (b).

C10-4. The VHDL program in Figure 10–29(a) is the implementation of a *D* flip-flop with asynchronous Set and Reset.

(a) Make the necessary changes to provide a clock enable similar to that provided by the *DFFE* primitive used in Problem C10–3. Save this program as *prob_c10_4.vhd*.

(b) Test its operation by creating waveform simulations that demonstrate its synchronous and asynchronous operations as well as its clock enable/disable feature.

(c) Download your design to a CPLD IC. Discuss your observations of the *q* output LED with your instructor as you use the switches or pushbuttons to step through the combinations of *n_set*, *n_reset*, *clock_enable*, *cp*, and *d* simulated in (b).

C10-5. The graphic design file in Figure 10–41 is the implementation of a *J-K* flip-flop using the primitive symbol called *JKFF*.

(a) Make the necessary changes to provide input pins for the active-LOW Set and Reset which are currently connected to V_{CC}. Save this program as *prob_c10_5.gdf*.

(b) Test its operation by creating waveform simulations that demonstrate its asynchronous operations using *n_set* and *n_reset* as well as its synchronous operations using *j*, *k*, and *n_cp*.

(c) Download your design to a CPLD IC. Discuss your observations of the *q* output LED with your instructor as you use the switches or pushbuttons to step through the combinations of *n_set*, *n_reset*, *j*, *k*, and *n_cp* simulated in (b). [*Note:* The switches and pushbuttons on the UP-1 and UP-2 boards exhibit a phenomenon called *switch bounce*. This causes

E10-4. Load the circuit file for **Section 10–5a.** The 7474 D flip-flop has synchronous inputs (D, Cp) and asynchronous inputs (Sd', Rd').

(a) List the steps that you need to perform with the switches D and C to synchronously Reset Q.

(b) List the steps that you need to perform with the switches D and C to synchronously Set Q.

(c) If the asynchronous inputs (Sd', Rd') are allowed to float (not connected to Vcc), what happens to the operation of the D flip-flop?

(d) Connect the switches S and R to the asynchronous inputs. List the steps that you need to follow to Set and then to Reset the flip-flop.

C   **E10-5.** Load the circuit file for **Section 10–5b.** The Word Generator is used to inject the signal Cp, D, Sd', and Rd' into the D flip-flop shown. On a piece of paper, carefully sketch those four waveforms displayed on the Logic Analyzer, and then sketch what you think Q will look like. Check your answer by connecting Q to one of the Logic Analyzer inputs.

C   **E10-6.** Load the circuit file for **Section 10–5c.** The Word Generator is used to inject the signals Cp, D, Sd', and Rd' into the D flip-flop shown. On a piece of paper, carefully sketch those four waveforms displayed on the Logic Analyzer, and then sketch what you think Q will look like. Check your answer by connecting Q to one of the Logic Analyzer inputs.

E10-7. Load the circuit file for **Section 10–8a.** The 7476 J-K flip flop has synchronous inputs (J, K, Cp) and asynchronous inputs (Sd', Rd').

(a) List the steps that you need to perform to synchronously Reset Q.

(b) List the steps that you need to perform to synchronously Set Q.

(c) Make J = 1, K = 1. What happens if you now continuously press C?

(d) If the asynchronous inputs (Sd', Rd') are allowed to float (not connected to Vcc), what happens to the operation of the flip-flop?

(e) Connect the switches S and R to the asynchronous inputs. List the steps that you need to follow to Set and then to Reset the flip-flop.

C   **E10-8.** Load the circuit file for **Section 10–8b.** The Word generator is used to inject the signals Cp', J, K, Sd', and Rd' into the J-K flip-flop shown. On a piece of paper, carefully sketch those five waveforms displayed on the Logic Analyzer, and then sketch what you think Q will look like. Check your answer by connecting Q to one of the Logic Analyzer inputs.

C   **E10-9.** Load the circuit file for **Section 10–8c.** The Word Generator is used to inject the signals Cp', J, K, Sd', and Rd' into the J-K flip-flop shown. On a piece of paper, carefully sketch those five waveforms displayed on the Logic Analyzer, and then sketch what you think Q will look like. Check your answer by connecting Q to one of the Logic Analyzer inputs.

CPLD Problems

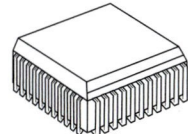

C10–1. The VHDL program in Figure 10–18 is the implementation of a *D* Latch.

(a) Make the necessary changes to make the enable active-LOW so that *q0* follows *d0* while *e0* is LOW. Save this program as *prob_c10_1.vhd.*

(b) Test its operation by creating waveform simulations that demonstrate its *transparent* operation and its *latching* feature.

(c) Download your design to a CPLD IC. Discuss your observations of the q output LED with your instructor as you use the switches to step through the combinations of *e0* and *d0* simulated in (b).

C10-2. The VHDL program in Figure 10–26(a) is the implementation of a *D* flip-flop with an asynchronous Reset.

(a) Make the necessary changes to provide an asynchronous *Set* instead of *Reset*. Save this program as *prob_c10_2.vhd.*

(b) Test its operation by creating waveform simulations that demonstrate its synchronous (*cp* and *d*) and asynchronous (*n_set*) operations.

(c) Download your design to a CPLD IC. Discuss your observations of the q output LED with your instructor as you use the switches or pushbuttons to step through the combinations of *n_set, cp,* and *d* simulated in (b).

C10-3. The Primitives subdirectory contains a *D* flip-flop with a clock enable signal. It is called *DFFE.*

(a) Build a graphic design file containing this flip flop with I/O ports labeled similar the gdf file shown in Figure 10–28. Save this program as *prob_c10_3.gdf.*

(b) Test its operation by creating waveform simulations that demonstrate its synchronous and asynchronous operations as well as its clock enable/disable feature. (Use the *help* menu to find information on the operation of the *DFFE* primitive.)

(c) Download your design to a CPLD IC. Discuss your observations of the q output LED with your instructor as you use the switches or pushbuttons to step through the combinations of *n_set, n_reset, clock_enable, cp,* and *d* simulated in (b).

C10-4. The VHDL program in Figure 10–29(a) is the implementation of a *D* flip-flop with asynchronous Set and Reset.

(a) Make the necessary changes to provide a clock enable similar to that provided by the *DFFE* primitive used in Problem C10–3. Save this program as *prob_c10_4.vhd.*

(b) Test its operation by creating waveform simulations that demonstrate its synchronous and asynchronous operations as well as its clock enable/disable feature.

(c) Download your design to a CPLD IC. Discuss your observations of the q output LED with your instructor as you use the switches or pushbuttons to step through the combinations of *n_set, n_reset, clock_enable, cp,* and *d* simulated in (b).

C10-5. The graphic design file in Figure 10–41 is the implementation of a *J-K* flip-flop using the primitive symbol called *JKFF.*

(a) Make the necessary changes to provide input pins for the active-LOW Set and Reset which are currently connected to V_{CC}. Save this program as *prob_c10_5.gdf.*

(b) Test its operation by creating waveform simulations that demonstrate its asynchronous operations using *n_set* and *n_reset* as well as its synchronous operations using *j, k,* and *n_cp.*

(c) Download your design to a CPLD IC. Discuss your observations of the q output LED with your instructor as you use the switches or pushbuttons to step through the combinations of *n_set, n_reset, j, k,* and *n_cp* simulated in (b). [*Note:* The switches and pushbuttons on the UP-1 and UP-2 boards exhibit a phenomenon called *switch bounce.* This causes

Schematic Interpretation Problems

See Appendix G for the schematic diagrams.

S C   **10-35.** Find U1:A of the Watchdog Timer schematic. Assume that initially, WATCHDOG_EN-LOW and CPU_RESET is pulsed LOW.

(a) What is the output level of U2:A?

(b) When WATCHDOG_EN goes HIGH, does the output of U2:A go LOW?

(c) What must happen to U1:A to make the output of U2:A go LOW?

S   **10-36.** In the Watchdog Timer schematic, both U14 flip-flops are Reset when there is a LOW /CPU_RESET _____ (and, or) a LOW $\overline{Q}$ from U14:B.

S   **10-37.** After being Reset, U14:A will be Set as soon as _____.

S   **10-38.** U5 and U6 are octal *D* flip-flops in the Watchdog Timer schematic. They provide two stages of latching for the 8-bit data bus labeled D(7:0).

(a) How are they initially Reset? (*Hint:* CLR is the abbreviation for CLEAR, which is the same as Master Reset.)

(b) What has to happen for the *Q*-outputs of U5 to receive the value of the data bus?

(c) What has to happen for the *Q*-outputs of U6 to receive the value of the U5 outputs?

Electronics Workbench®/MultiSIM® Exercises

E10–1. Load the circuit file for **Section 10–1a.** This is a cross-NAND S-R flip-flop similar to Figure 10–2.

(a) To "Set" Q to ON, you must make S = _____ and R = _____? Try it.

(b) To "Reset" Q to OFF, you must make S = _____ and R = _____? Try it.

(c) To "Hold" the last value of Q, you must make S = _____ and R = _____? Try it.

E10-2. Load the circuit file for **Section 10–1b.** This circuit is a cross-NAND S-R flip-flop similar to Figure 10–2. It can be constructed in lab using a single 7400 Quad NAND IC. Make the necessary connections to the 7400 to implement a cross-NAND S-R flip-flop. How did you make the inverters out of NANDs? Demonstrate Setting and Resetting Q to your instructor.

T   **E10-3.** Load the circuit file for **Section 10–1c.** Only one of the flip-flop circuits shown is working properly.

(a) What is the faulty component (if any) in ckt-1?

(b) What is the faulty component (if any) in ckt-2?

(c) What is the faulty component (if any) in ckt-3?

(d) What is the faulty component (if any) in ckt-4? Fix all of the faults and retest.

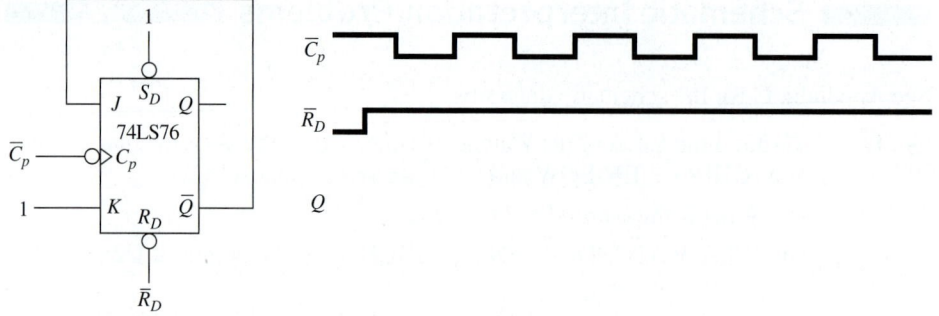

Figure P10-30

C   **10-31.** Sketch the output waveform at Q for Figure P10–31.

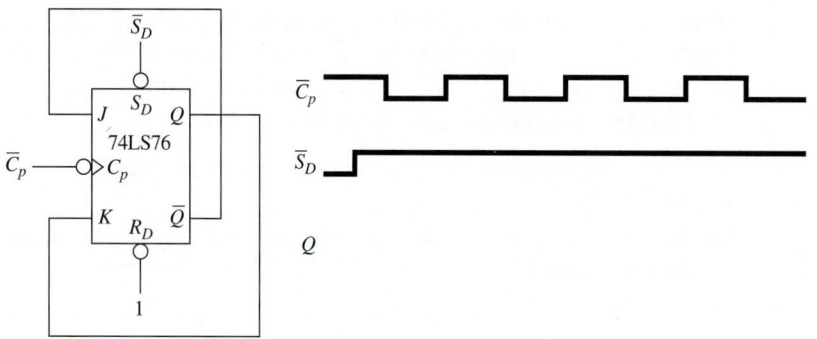

Figure P10-31

C   **10-32.** Sketch the output waveform at Q for Figure P10–32.

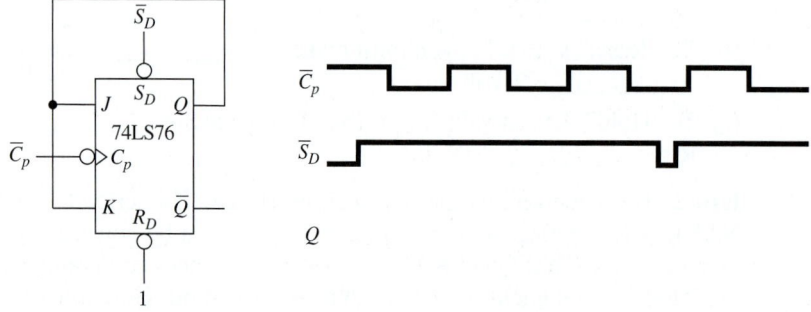

Figure P10-32

Section 10–9

C D   **10-33.** The 74HCT373 (or 74LS373) is an octal transparent latch. Refer to a data book (CMOS or TTL) to review its operation. Discuss why it can or cannot be used to replace the 273 in Figure 10–57.

C T   **10-34.** A designer decides to change the timing pulse increment in Figure 10–57 from 10 s to 10 ms. When she does, the least significant digit always displays the number 8. Explain why.

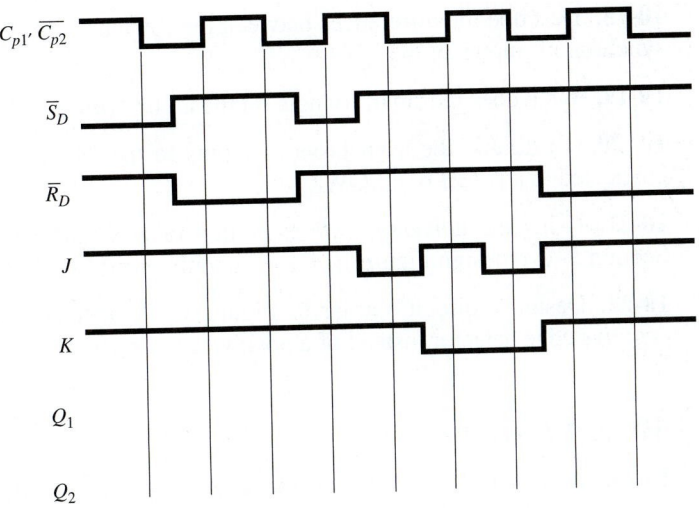

Figure P10–27

10-28. Sketch the output waveform at Q for Figure P10–28.

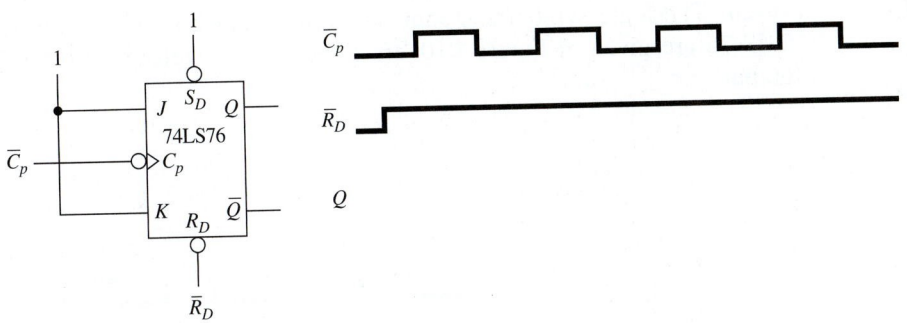

Figure P10–28

10-29. Sketch the output waveform at Q for Figure P10–29.

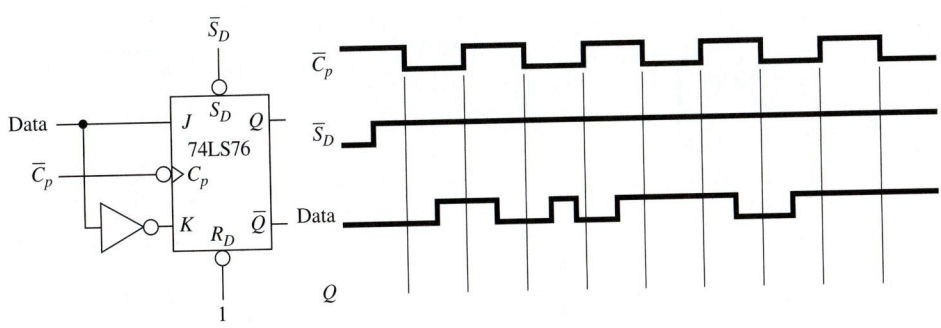

Figure P10–29

C   **10-30.** Sketch the output waveform at Q for Figure P10–30.

10-18. Describe the differences between the asynchronous inputs and the synchronous inputs of the 7474.

10-19. What does the small triangle on the C_p line of the 7474 indicate?

10–20. To disable the asynchronous inputs to the 7474, should they be connected to a HIGH or a LOW?

D **10-21.** Using the *universal gate* capability of a NAND gate covered in Section 5–7, redesign Figure 10–21 using only one 74HCT00 NAND IC.

D **10-22.** Design a circuit similar to Figure 10–21 that can be used as a *negative* edge detector instead of a positive edge detector.

Sections 10–6, 10–7, and 10–8

10-23. What is the one additional synchronous operating mode that the *J-K* flip-flop has that the *S-R* flip-flop does not have?

10-24. What are the asynchronous inputs to the 7476 *J-K* flip-flop? Are they active LOW or active HIGH?

10-25. The 7476 is called a *pulse-triggered master–slave* flip-flop, whereas the 74LS76 is called an *edge-triggered* flip-flop. Describe the differences between them.

C **10-26.** The logic symbol and input waveforms for both the 7476 and 74LS76 are given in Figure P10–26. Sketch the waveform at each Q output.

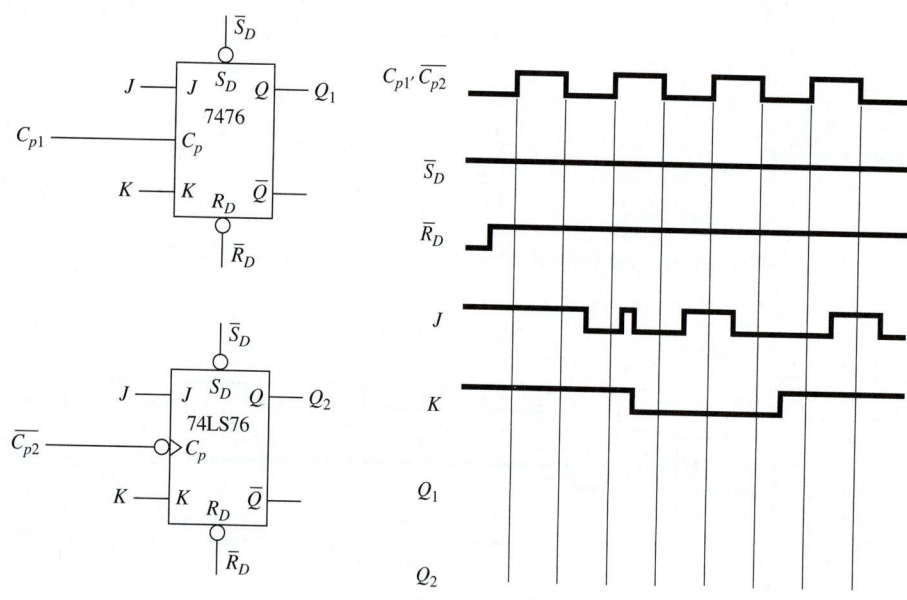

Figure P10–26

C **10-27.** Repeat Problem 10–26 for the input waveforms shown in Figure P10–27.

10-12. Explain why the 7475 is called *transparent* and why it is called a *latch*.

10-13. The 7475 is transparent while the E input is _____ (LOW or HIGH), and it is latched while E is _____ (LOW or HIGH).

Section 10–5

10-14. (a) What are the asynchronous inputs to the 7474 D flip-flop?

(b) What are the synchronous inputs to the 7474 D flip-flop?

10-15. The logic symbol for one-half of a 7474 dual D flip-flop is given in Figure P10–15(a). Sketch the Q output wave given the inputs at C_p, D, $\overline{S}_D$, and $\overline{R}_D$ shown in Figure P10–15(b).

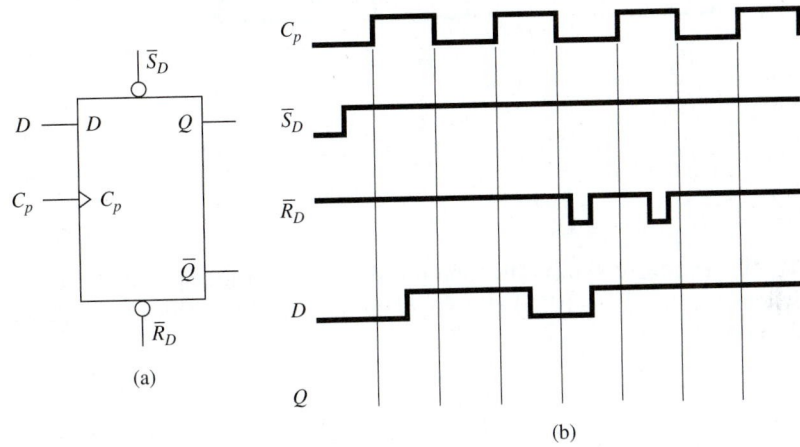

Figure P10–15

10-16. Repeat Problem 10–15 for the input waves shown in Figure P10–16.

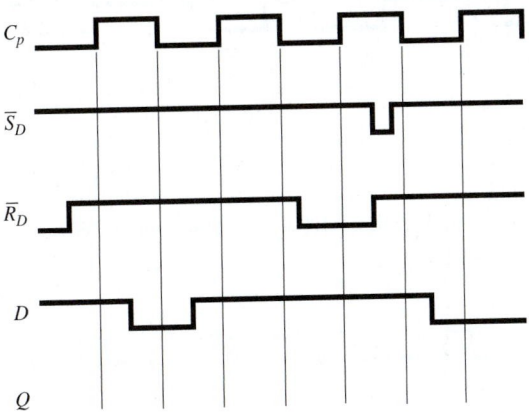

Figure P10–16

10-17. Describe several differences between the 7474 D flip-flop and the 7475 D latch.

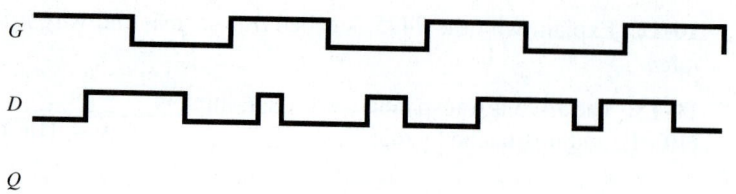

G

D

Q

Figure P10–8

10-9. Repeat Problem 10–8 for the G and D inputs shown in Figure P10–9.

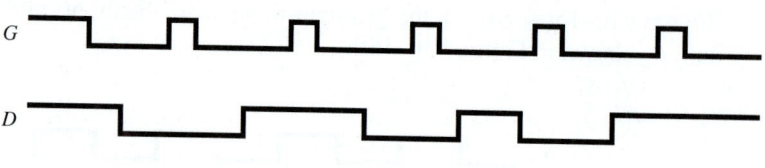

G

D

Q

Figure P10–9

Section 10–4

10-10. The logic symbol for one-fourth of a 7475 transparent D latch is given in Figure P10–10. Sketch the Q output waveform given the inputs at E and D.

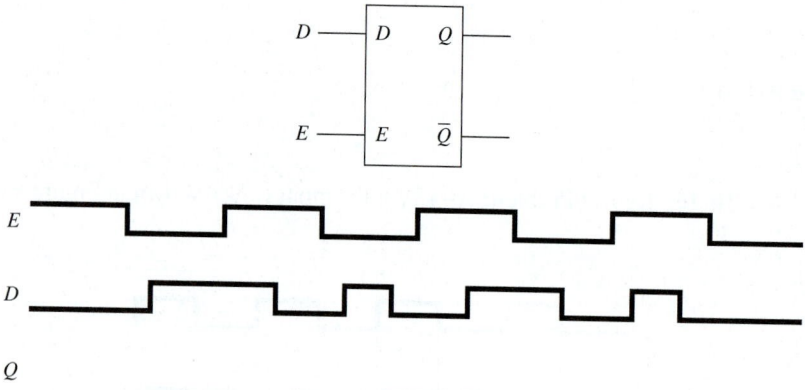

E

D

Q

Figure P10–10

10-11. Repeat Problem 10–10 for the waveforms at E and D shown in Figure P10–11.

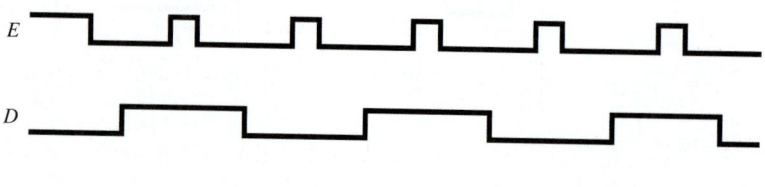

E

D

Q

Figure P10–11

PROBLEMS

Set: A condition that produces a digital HIGH (1) state.

Setup Time: The length of time before the active edge of a trigger pulse (control signal) that the inputs of a digital device must be in a stable digital state. [That is, if the setup time of a device is 20 ns, the inputs must be held stable (and will be read) 20 ns before the trigger edge.]

Storage Register: Two or more data storage circuits (such as flip-flops or latches) used in conjunction with each other to hold several bits of information.

Strobe Gates: A control gate used to enable or disable inputs from reaching a particular digital device.

Synchronous: A condition in which the output of a device will operate only in synchronization with (in step with) a specific HIGH or LOW timing pulse or trigger signal.

Toggle: In a flip-flop, a toggle is when Q changes to the level of $\overline{Q}$ and $\overline{Q}$ changes to the level of Q.

Transition: The instant of change in digital state from HIGH to LOW or LOW to HIGH.

Transparent Latch: An asynchronous device whose outputs will hold onto the most recent digital state of the inputs. The outputs immediately follow the state of the inputs without regard to trigger input and remain in that state even after the inputs are removed or disabled.

Trigger: The input control signal to a digital device that is used to specify the instant that the device is to accept inputs or change outputs.

Problems

Section 10–1

10-1. Make the necessary connections to the 7400 quad NAND gate IC in Figure P10–1 to form the cross-NAND S-R flip-flop of Figure 10–2. [Remember that an inverter can be formed from a NAND (Chapter 3).]

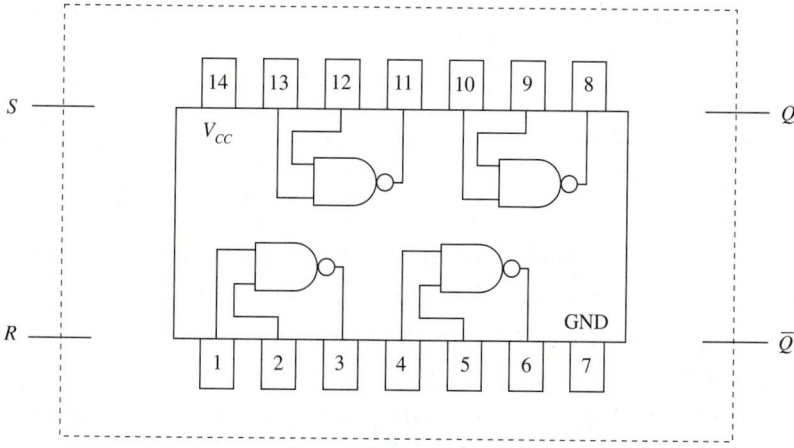

Figure P10–1

10-2. Sketch the Q output waveform for a gated S-R flip-flop (Figure 10–8), given the inputs at S, R, and G shown in Figure P10–2.

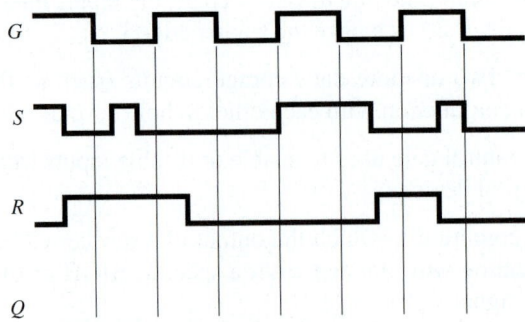

Figure P10–2

10-3. Repeat Problem 10–2 for the input waves shown in Figure P10–3.

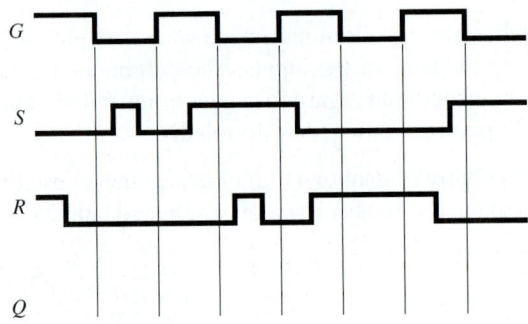

Figure P10–3

10-4. Problem 10–2 for the input waves shown in Figure P10–4.

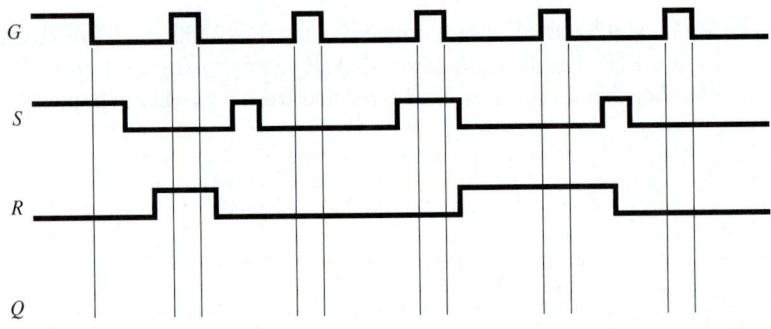

Figure P10–4

Section 10–3

D **10-5.** Referring to Figures 10–8 and 10–12, sketch the logic diagram using NORs, ANDs, and inverters that will function as a gated D flip-flop.

10-6. How many IC chips will be required to build the gated D flip-flop that you sketched in Problem 10–5?

10-7. Make the necessary connections to a 7402 quad NOR and a 7408 quad AND to form the gated D flip-flop of Problem 10–5.

10-8. Sketch the Q output waveform for the gated D flip-flop of Figure 10–12 given the D and G inputs shown in Figure P10–8.

each change in switch position to be interpreted as several HIGH/LOW pulses instead of one. This will produce unpredictable results in q when bouncing occurs on the clock input when j and k are in the *toggle* mode. The PLDT-2 board has four electronically de-bounced switches that eliminate this problem when used on the clock input *(n_cp).*]

C10-6. The VHDL program in Figure 10–42(a) is the implementation of a *J-K* flip-flop.

(a) Make the necessary program additions to provide active-LOW asynchronous *Set* and *Reset.* Save this program as *prob_c10_6.vhd.*

(b) Test its operation by creating waveform simulations that demonstrate its synchronous and asynchronous operations.

(c) Download your design to a CPLD IC. Discuss your observations of the q output LED with your instructor as you use the switches or pushbuttons to step through the combinations of *n_set, n_reset, j, k,* and *cp* simulated in (b).

C10-7. The LPM model in Figure 10–58 is the implementation of an octal *D* flip-flop with asynchronous clear.

(a) Use the **Edit Ports/Parameters** feature to specify the *aclear* as active-LOW and the clock as negative-edge triggered. Save this program as *prob_c10_7.gdf.*

(b) Test its operation by creating waveform simulations that demonstrate its synchronous and asynchronous operations. (Make the simulation similar to Figure 10–59, but with different values.)

(c) Download your design to a CPLD IC. Discuss your observations of the q output LED with your instructor as you use the switches or pushbuttons to step through the combinations of *aclear, clock,* and *d[7..0]* simulated in (b).

C10-8. The LPM model in Figure 10–60 is the implementation of an octal *D* flip-flop with asynchronous set, clear, and load.

(a) Use the **Edit Ports/Parameters** feature to specify the *aset, aclear, aload,* and *enable* as active-LOW and the *clock* as negative-edge triggered. Save this program as *prob_c10_8.gdf.*

(b) Test its operation by creating waveform simulations that demonstrate its synchronous and asynchronous operations. (Make the simulation similar to Figure 10–61, but with different values.)

(c) Download your design to a CPLD IC. Discuss your observations of the q output LED with your instructor as you use the switches or pushbuttons to step through the combinations of *aset, aclear, aload, enable, clock,* and *d[7..0]* simulated in (b).

Answers to Review Questions

10-1. True

10-2. $S = 1, R = 0$

10-3. None

10-4. An *S-R* flip-flop is asynchronous because the output responds immediately to input changes. The gated *S-R* is synchronous because it operates sequentially with the control input at the gate.

10-5. False

10-6. $D = 0, G = 1$

10-7. 4

10-8. HIGH

10-9. True

10-10. Transitions at Q will occur only at the edge of the input trigger pulse.

10-11. C_p and D are synchronous. $\overline{S}_D$ and $\overline{R}_D$ are asynchronous.

10-12. False

10-13. Invert, delay

10-14. Because the master will latch onto any HIGH inputs while the clock pulse is HIGH and transfer them to the slave when the clock goes LOW.

10-15. J, K

10-16. True

10-17. It switches the Q and $\overline{Q}$ outputs to their opposite states.

10-18. Apply a LOW on the $\overline{R}_D$ input.

10-19. False

10-20. HIGH

10-21. The uppercase letters mean steady state. The lowercase letters are used for levels one setup time before the negative clock edge.

11 Practical Considerations for Digital Design

OUTLINE

OBJECTIVES

Upon completion of this chapter, you should be able to:

- Describe the causes and effects of a race condition on synchronous flip-flop operation.
- Use manufacturers' data sheets to determine IC operating specifications such as setup time, hold time, propagation delay, and input/output voltage and current specifications.
- Perform worst-case analysis on the time-dependent operations of flip-flops and sequential circuitry.
- Design a series RC circuit to provide an automatic power-up reset function.
- Describe the wave-shaping capability and operating characteristics of Schmitt trigger ICs.
- Describe the problems caused by switch bounce and how to eliminate its effects.
- Calculate the optimum size for a pull-up resistor.

INTRODUCTION

We now have the major building blocks required to form sequential circuits. There are a few practical time and voltage considerations that we have to deal with first before we connect ICs to form sequential logic.

For instance, ideally, a 74LS76 flip-flop switches on the negative edge of the input clock, but actually, it could take the output as long as 30 ns to switch. Thirty nanoseconds (30×10^{-9} s) does not sound like much, but when you cascade several flip-flops end to end or any time you have combinational logic with flip-flops that rely on a high degree of accurate timing, the IC delay times could cause serious design problems.

Digital ICs have to keep up with the high speed of the microprocessors used in modern computer systems. For example, a microprocessor operating at a clock frequency of 20 MHz will have a clock period of 50 ns. The slower IC families that have switching speeds of 20 to 30 ns would create timing problems that would lead to misinterpretation of digital levels in a system operating that fast. To ensure reliability in high-speed systems, designers must consider the worst-case timing scenario. They cannot simply play it safe and build in a large margin of error because that would mean that a slower operating system would be developed, which would not compete favorably in the modern marketplace.

In this chapter, we look at the *actual* operating characteristics of digital ICs as they relate to output delay times, input setup requirements, and input/output voltage and current levels. With a good knowledge of the practical aspects of digital ICs, we then develop the external circuitry needed to interface to digital logic.

11–1 Flip-Flop Time Parameters

There are several time parameters listed in IC manufacturers' data manuals that require careful analysis. For example, let's look at Figure 11–1, which uses a 74LS76 flip-flop with the J and $\overline{C_p}$ inputs brought in from some external circuit.

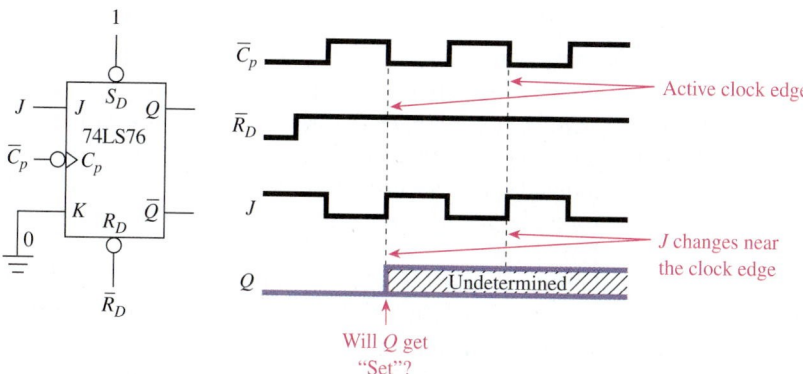

Figure 11–1 A possible race condition on a *J-K* flip-flop creates an undetermined result at *Q*.

The waveform shown for J and $\overline{C_p}$ will create a **race condition.** *Race* is the term used when the inputs to a triggerable device (like a flip-flop) are changing at the same time that the active trigger edge of the input clock is making its transition. In the case of Figure 11–1, the J waveform is changing from LOW to HIGH exactly at the negative edge of the clock, so what is J at the negative edge of the clock, LOW or HIGH?

Now when you look at Figure 11–1, you should ask the question, Will Q ever get Set? Remember from Chapter 10 that J must be HIGH at the negative edge of $\overline{C_p}$ to set the flip-flop. Actually, J must be HIGH one *setup time* before the negative edge of the clock.

The **setup time** is the length of time before the active clock edge that the flip-flop looks back to determine the levels to use at the inputs. In other words, for Figure 11–1, the flip-flop will look back one setup time before the negative clock edge to determine the levels at J and K.

The setup time for the 74LS76 is 20 ns, so we must ask, were J and K HIGH or LOW 20 ns before the negative clock edge? Well, K is tied to ground, so it was LOW, and depending on when J changed from LOW to HIGH, the flip-flop may have Set ($J = 1$, $K = 0$) or Held ($J = 0$, $K = 0$).

In a data manual, the manufacturer will provide **ac waveforms** that illustrate the measuring points for all the various time parameters. The illustration for setup time will look something like Figure 11–2.

The active transition (trigger point) of the $\overline{C_p}$ input (clock) occurs when $\overline{C_p}$ goes from above to below the 1.3-V level.

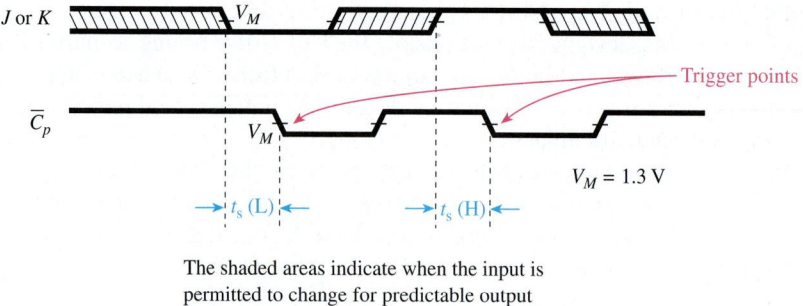

Figure 11–2 Setup time waveform specifications for a 74LS76.

Setup time (LOW), t_s (L), is given as 20 ns. This means that *J* and *K* can be changing states 21 ns or more before the active transition of $\overline{C_p}$; but to be interpreted as a LOW, they must be 1.3 V *or less* at 20 ns *before* the active transition of $\overline{C_p}$.

Setup time (HIGH), t_s (H), is also given as 20 ns. This means that *J* and *K* can be changing states 21 ns or more before the active edge of $\overline{C_p}$; but to be interpreted as a HIGH, they must be 1.3 V *or more* at 20 ns *before* the active transition of $\overline{C_p}$.

Did you follow all that? If not, go back and read it again! Sometimes, material like this has to be read over and over again, carefully, to be fully understood.

Not only does the input have to be set up some definite time *before* the clock edge, but it also has to be *held* for a definite time after the clock edge. This time is called the **hold time** [t_h (L) and t_h (H)].

Fortunately, the hold time for the 74LS76 (and most other flip-flops) is given as 0 ns. This means that the desired levels at *J* and *K* must be held 0 ns *after* the **active clock edge.** In other words, the levels do not have to be held beyond the active clock edge for most flip-flops. In the case of the 74LS76, the desired level for *J* and *K* must be present from 20 ns before the negative clock edge to 0 ns after the clock edge.

For example, for a 74LS76 to have a LOW level on *J* and *K,* the waveforms in Figure 11–3 illustrate the *minimum* setup and hold times allowed to still have the LOW reliably interpreted as a LOW. Figure 11–3 shows us that *J* and *K* are allowed to change states any time greater than 20 ns before the negative clock edge, and because the hold time is zero, they are permitted to change immediately after the negative clock edge.

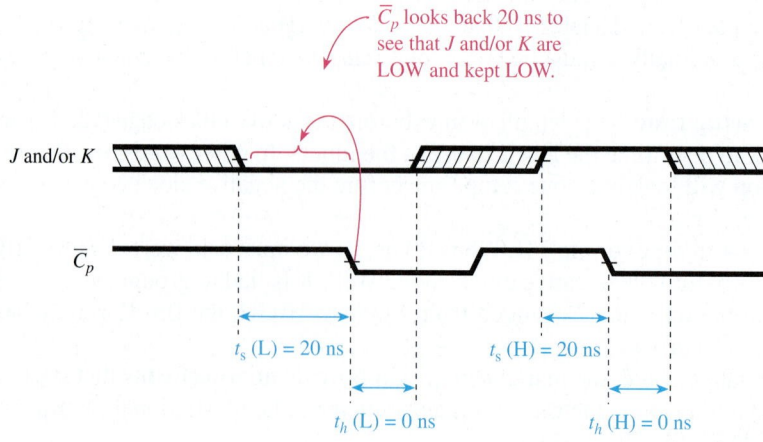

Figure 11–3 Setup and hold parameters for a 74LS76 flip-flop.

Do you notice in Examples 11–1 and 11–2 that the Q output changes *exactly* on the negative clock edge? Do you really think that it will? It won't! Electrical charges that build up inside any digital logic circuit won't allow it to change states instantaneously as the inputs change. This delay from input to output is called **propagation delay.** There are propagation delays from the synchronous inputs to the output and also from the asynchronous inputs to the output.

EXAMPLE 11–1

Following the rules for setup and hold times, for the 74H106 shown in Figure 11–4, sketch the waveform at Q in Figure 11–5 [t_s (L) = 13 ns, t_s (H) = 10 ns, t_h (L) = t_h (H) = 0 ns].

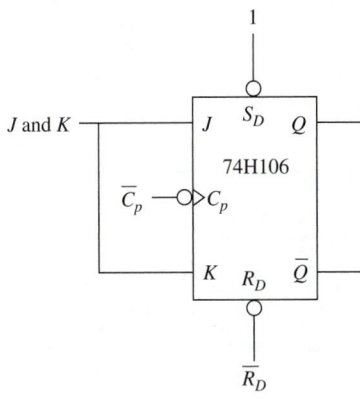

Figure 11–4

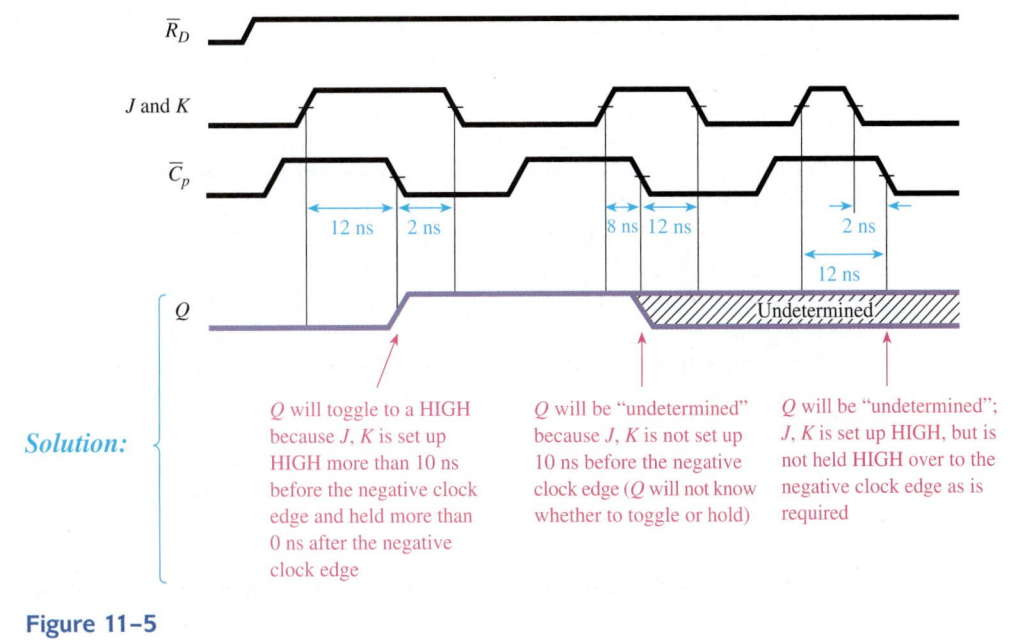

Solution:

Q will toggle to a HIGH because J, K is set up HIGH more than 10 ns before the negative clock edge and held more than 0 ns after the negative clock edge

Q will be "undetermined" because J, K is not set up 10 ns before the negative clock edge (Q will not know whether to toggle or hold)

Q will be "undetermined"; J, K is set up HIGH, but is not held HIGH over to the negative clock edge as is required

Figure 11–5

EXAMPLE 11–2

Sketch the Q output for the 74H106 shown in Figure 11–6, with the input waveforms given in Figure 11–7 [t_s (L) = 13 ns, t_s (H) = 10 ns, t_h (L) = t_h (H) = 0 ns].

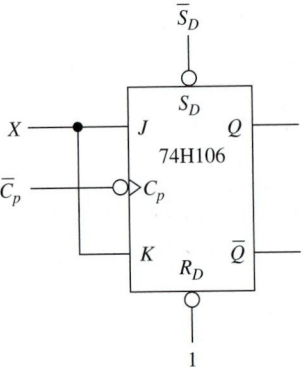

Figure 11–6

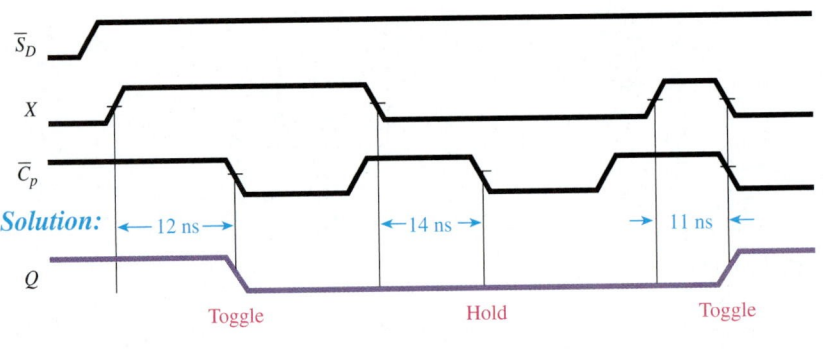

Figure 11–7

For example, there is a propagation delay period from the instant that $\overline{R_D}$ or $\overline{S_D}$ goes LOW until the Q output responds accordingly. The data manual shows a *maximum* propagation delay for $\overline{S_D}$ to Q of 20 ns and for $\overline{R_D}$ to Q of 30 ns. Because a LOW on $\overline{S_D}$ causes Q to go *LOW to HIGH,* the propagation delay is abbreviated t_{PLH}. A LOW on $\overline{R_D}$ causes Q to go *HIGH to LOW;* therefore, use t_{PHL} for that propagation delay, as illustrated in Figure 11–8.

The propagation delay from the clock trigger point to the Q output is also called t_{PLH} or t_{PHL}, depending on whether the Q output is going LOW to HIGH or HIGH to LOW. For the 74LS76 clock to output, t_{PLH} = 20 ns, and t_{PHL} = 30 ns. Figure 11–9 illustrates the synchronous propagation delays.

Besides setup, hold, and propagation delay times, the manufacturer's data manual will also give

1. Maximum frequency (f_{max}), or the maximum frequency allowed at the clock input. Any frequency above this limit will yield unpredictable results.

2. Clock pulse width (LOW) [t_w(L)], or the minimum width (in nanoseconds) that is allowed at the clock input during the LOW level for reliable operation.

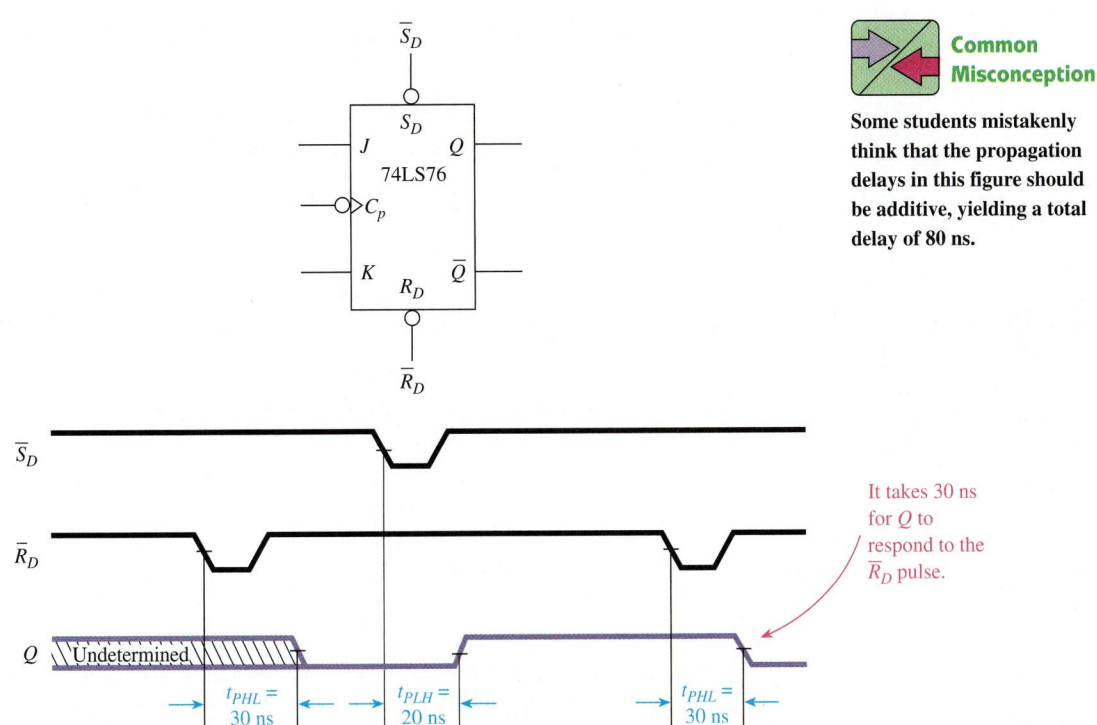

It takes 30 ns
for Q to
respond to the
$\overline{R}_D$ pulse.

Figure 11–8 Propagation delay for the asynchronous input to Q output for the 74LS76.

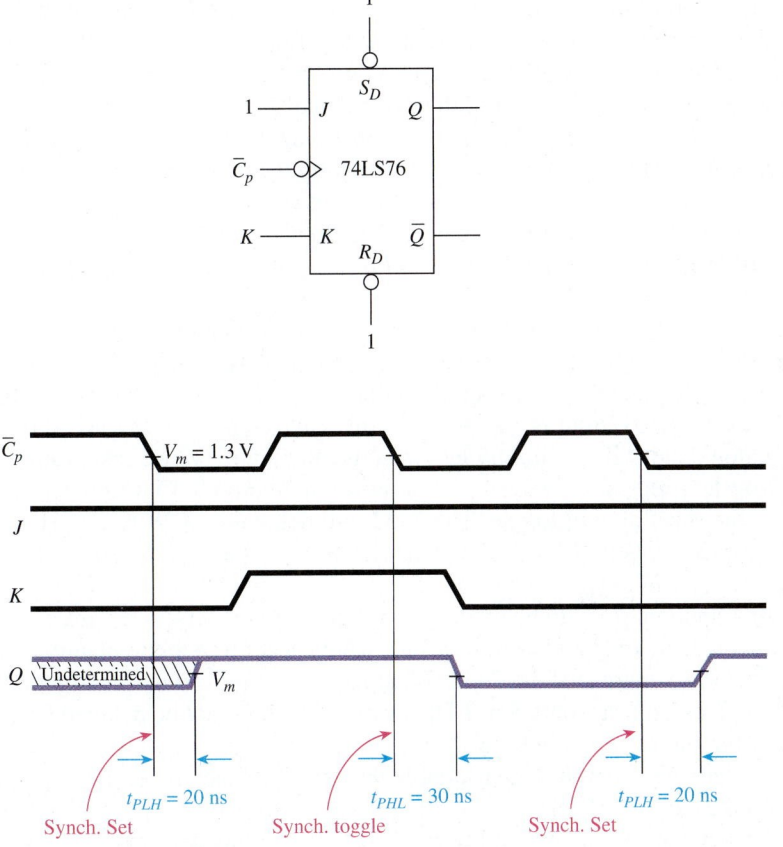

Figure 11–9 Propagation delay for the clock to output of the 74LS76.

3. Clock pulse width (HIGH) [t_w(H)], or the minimum width (in nanoseconds) that is allowed at the clock input during the HIGH level for reliable operation.

4. Set or Reset pulse width (LOW) [t_w(L)], or the minimum width (in nanoseconds) of the LOW pulse at the Set $\overline{S_D}$ or Reset $\overline{R_D}$ inputs.

Figure 11–10 shows the measurement points for these specifications.

Team Discussion

Why do manufacturers give *maximum* propagation delay times but *minimum* pulse widths?

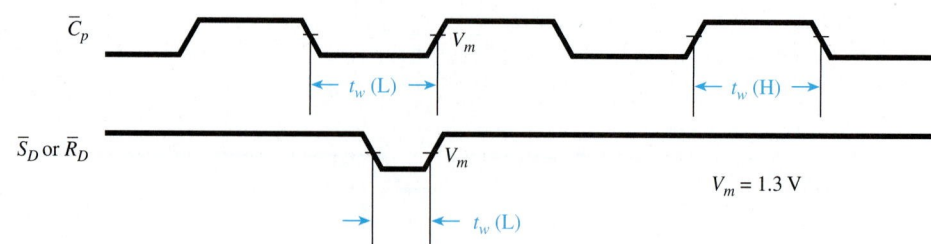

Figure 11–10   Minimum pulse-width specifications.

Helpful Hint

It is very important that you learn how to interpret a data manual. New devices and ICs are constantly being introduced. Often, the only way to learn their operation and features is to study a data sheet.

If these specifications are not met, the flip-flop may enter a **metastable** state. In this state, the Q-output voltage will become an invalid level (neither HIGH nor LOW) for a short time (metastable), then stabilize HIGH or LOW. For example, if Q is HIGH and you apply too short of an $\overline{R_D}$ pulse to an LS TTL IC, Q may become 1.5 V (invalid) for a few milliseconds, then either return HIGH (> 2.4 V) or go LOW (< 0.4 V). This condition also occurs if you do not meet the setup or hold time requirements.

Complete specifications for the 7476/74LS76 flip-flop are given in Figure 11–11. Can you locate all the specifications that we have discussed so far? If you have your own data manual, look at some of the other flip-flops, and see how they compare. In the front of the manual, you will find a section that describes all the IC specifications and abbreviations used throughout the data manual.

Now that we understand most of the operating characteristics of digital ICs, let's examine why they are so important and what implications they have on our design of digital circuits.

To get the flip-flop in Example 11–3 to toggle, we have to move C_p to the right by at least 10 ns so that J is HIGH 10 ns *before* the positive edge of $\overline{C_p}$. By saying "move it to the right," we mean "delay it by at least 10 ns."

Helpful Hint

Data sheets for ICs are provided at manufacturers' www sites. A suggested list is given in Appendix A. Compare the data sheet in Figure 11–11 to ones that you find for a similar JK, the 74F112, 74ALS112, or 74HC112.

One common way to introduce delay is to insert one or more IC gates in the C_p line, as shown in Figure 11–14, so that their combined propagation delay is greater than 10 ns. From the manufacturer's specifications, we can see that the propagation delay for a 7432 input to output is t_{PHL} = 22 ns max. and t_{PLH} = 15 ns max. [typically, the propagation delay will be slightly less than the maximum (worst case) rating].

Now let's redraw the waveforms as shown in Figure 11–15 with the delayed clock to see if the flip-flop will toggle. The 7432 will delay the LOW-to-HIGH edge of the clock by approximately 15 ns, so J will be HIGH one setup time before the trigger point on C_p; thus, Q will toggle.

An important point to be made here is that in Figure 11–14 we are relying on the propagation delay of the 7432 to be 15 ns, which according to the manufacturer is the worst case (maximum) for the 7432. What happens if the *actual* propagation delay is less than 15 ns, let's say only 8 ns? The clock (C_p) would not be delayed far enough to the right for J to be set up in time.

Special *delay-gate* ICs are available that provide exact, predefined delays specifically for the purpose of delaying a particular signal to enable proper time relationships. One such delay gate is shown in Figure 11–16. To use this delay gate, you would connect the signal that you want delayed to the C_p input terminal. You then select the

FLIP-FLOPS

54/7476, LS76

Dual J-K Flip-Flop

DESCRIPTION

The '76 is a dual J-K flip-flop with individual J, K, Clock, Set, and Reset inputs. The 7476 is positive pulse-triggered. JK information is loaded into the master while the Clock is HIGH and transferred to the slave on the HIGH-to-LOW Clock transition. The J and K inputs must be stable while the Clock is HIGH for conventional operation.

The 74LS76 is a negative edge-triggered flip-flop. The J and K inputs must be stable only one setup time prior to the HIGH-to-LOW Clock transition.

The Set ($\overline{S}_D$) and Reset ($\overline{R}_D$) are asynchronous active LOW inputs. When LOW, they override the Clock and Data inputs, forcing the outputs to the steady state levels as shown in the Function Table.

TYPE	TYPICAL f_{MAX}	TYPICAL SUPPLY CURRENT (Total)
7476	20MHz	10mA
74LS76	45MHz	4mA

Gives part numbers for various package types, V_{CC}, and temperature ranges.

ORDERING CODE

PACKAGES	COMMERCIAL RANGES $V_{CC} = 5V \pm 5\%$; $T_A = 0°C$ to $+70°C$		MILITARY RANGES $V_{CC} = 5V \pm 10\%$; $T_A = -55°C$ to $+125°C$	
Plastic DIP	N7476N	• N74LS76N		
Ceramic DIP			S5476F	• S54LS76F
Flatpack			S5476W	• S54LS76W

2 unit loads means that these inputs draw 2 times the I_I ratings in this family.

INPUT AND OUTPUT LOADING AND FAN-OUT TABLE

PINS	DESCRIPTION	54/74	54/74LS
$\overline{CP}$	Clock input	2ul	2LSul
$\overline{R}_D, \overline{S}_D$	Reset and Set inputs	2ul	2LSul
J, K	Data inputs	1ul	1LSul
Q, $\overline{Q}$	Outputs	10ul	10LSul

NOTE
Where a 54/74 unit load (ul) is understood to be 40μA I_{IH} and −1.6 mA I_{IL} and a 54/74LS unit load (LSul) is 20μA I_{IH} and −0.4mA I_{IL}.

10 unit loads means that these outputs can supply 10 times the amount of current required for a single unit input load (or fan-out = 10).

PIN CONFIGURATION

Notice the overbars indicating active-LOW terminals.

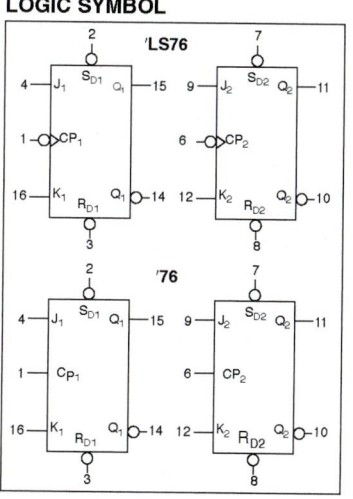

LOGIC SYMBOL

LOGIC SYMBOL (IEEE/IEC)

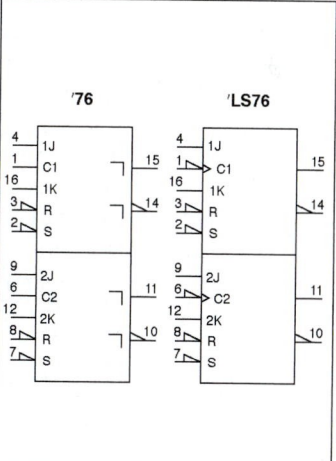

Note: In this figure, the abbreviation for Clock is $\overline{CP}$ instead of $\overline{C_p}$.

Figure 11–11 Typical data sheet for a 7476/74LS76. (Courtesy of Philips Semiconductors.)

FLIP-FLOPS

54/7476, LS76

— Gives the input requirements and output result for each operating mode.

LOGIC DIAGRAM

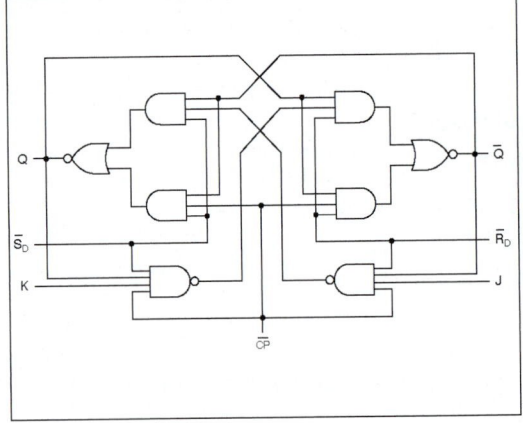

FUNCTION TABLE

OPERATING MODE	INPUTS					OUTPUTS	
	$\overline{S}_D$	$\overline{R}_D$	$\overline{CP}^{(b)}$	J	K	Q	$\overline{Q}$
Asynchronous Set	L	H	X	X	X	H	L
Asynchronous Reset (Clear)	H	L	X	X	X	L	H
Undetermined[a]	L	L	X	X	X	H	H
Toggle	H	H	⊓	h	h	$\overline{q}$	q
Load "0" (Reset)	H	H	⊓	l	h	L	H
Load "1" (Set)	H	H	⊓	h	l	H	L
Hold "no change"	H	H	⊓	l	l	q	$\overline{q}$

H = HIGH voltage level steady state.
h = HIGH voltage level one setup time prior to the HIGH-to-LOW Clock transition.[C]
L = LOW voltage level steady state.
l = LOW voltage level one setup time prior to the HIGH-to-LOW Clock transition.[C]
q = Lowercase letters indicate the state of the referenced output prior to the HIGH-to-LOW Clock transition.
X = Don't care.
⊓ = Positive Clock pulse.

NOTES
a. Both outputs will be HIGH while both $\overline{S}_D$ and $\overline{R}_D$ are LOW, but the output states are unpredictable if $\overline{S}_D$ and $\overline{R}_D$ go HIGH simultaneously.
b. The 74LS76 is edge triggered. Data must be stable one setup time prior to the negative edge of the Clock for predictable operation.
c. The J and K inputs of the 7476 must be stable while the Clock is HIGH for conventional operation.

ABSOLUTE MAXIMUM RATINGS (Over operating free-air temperature range unless otherwise noted.)

	PARAMETER	54	54LS	74	74LS	UNIT
V_{CC}	Supply voltage	7.0	7.0	7.0	7.0	V
V_{IN}	Input voltage	−0.5 to +5.5	−0.5 to +7.0	−0.5 to +5.5	−0.5 to +7.0	V
I_{IN}	Input current	−30 to +5	−30 to +1	−30 to +5	−30 to +1	mA
V_{OUT}	Voltage applied to output in HIGH ouput state	−0.5 to +V_{CC}	−0.5 to +V_{CC}	−0.5 to +V_{CC}	−0.5 to +V_{CC}	V
T_A	Operating free-air temperature range	−55 to +125		0 to 70		°C

RECOMMENDED OPERATING CONDITIONS

HIGH/LOW input voltage requirements.

Maximum output currents.

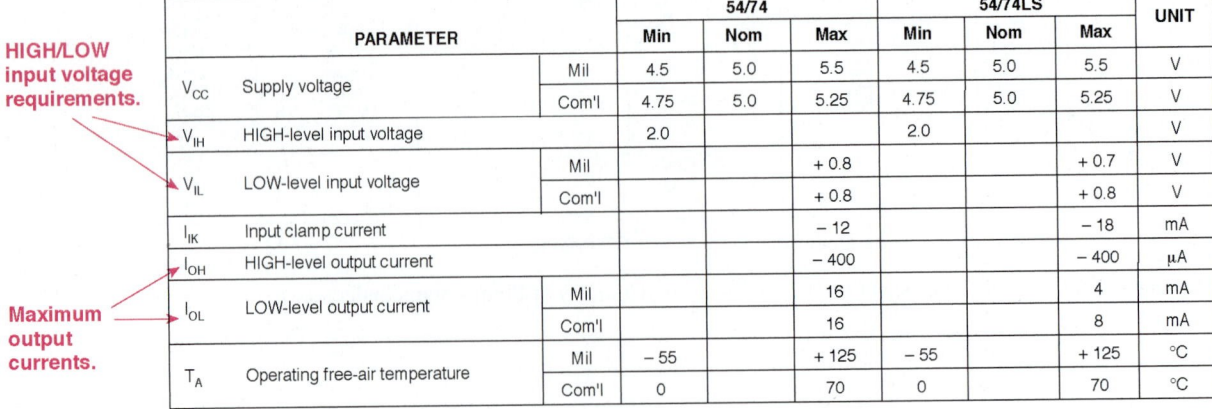

	PARAMETER		54/74			54/74LS			UNIT
			Min	Nom	Max	Min	Nom	Max	
V_{CC}	Supply voltage	Mil	4.5	5.0	5.5	4.5	5.0	5.5	V
		Com'l	4.75	5.0	5.25	4.75	5.0	5.25	V
V_{IH}	HIGH-level input voltage		2.0			2.0			V
V_{IL}	LOW-level input voltage	Mil			+0.8			+0.7	V
		Com'l			+0.8			+0.8	V
I_{IK}	Input clamp current				−12			−18	mA
I_{OH}	HIGH-level output current				−400			−400	μA
I_{OL}	LOW-level output current	Mil			16			4	mA
		Com'l			16			8	mA
T_A	Operating free-air temperature	Mil	−55		+125	−55		+125	°C
		Com'l	0		70	0		70	°C

Figure 11–11 *(Continued)*

CHAPTER 11 | PRACTICAL CONSIDERATIONS FOR DIGITAL DESIGN

FLIP-FLOPS

DC ELECTRICAL CHARACTERISTICS (Over recommended operating free-air temperature range unless otherwise noted.)

Typical and worst-case output voltages.

HIGH/LOW input current requirements.

PARAMETER		TEST CONDITIONS[1]			54/7476			54/74LS76			UNIT
					Min	Typ[2]	Max	Min	Typ[2]	Max	
V_{OH}	HIGH-level output voltage	V_{CC} = MIN, V_{IH} = MIN, V_{IL} = MAX, I_{OH} = MAX		Mil	2.4	3.4		2.5	3.4		V
				Com'l	2.4	3.4		2.7	3.4		V
V_{OL}	LOW-level output voltage	V_{CC} = MIN, V_{IL} = MAX, V_{IH} = MIN	I_{OL} = MAX	Mil		0.2	0.4		0.25	0.4	V
				Com'l		0.2	0.4		0.35	0.5	V
			I_{OL} = 4mA	74LS					0.25	0.4	V
V_{IK}	Input clamp voltage	V_{CC} = MIN, $I_I = I_{IK}$					−1.5			−1.5	V
I_I	Input current at maximum input voltage	V_{CC} = MAX	V_I = 5.5 V				1.0				mA
			V_I = 7.0V	J, K Inputs						0.1	mA
				$\overline{S}_D, \overline{R}_D$ Inputs						0.3	mA
				$\overline{CP}$ Inputs						0.4	mA
I_{IH}	HIGH-level input current	V_{CC} = MAX	V_I = 2.4V	J, K Inputs			40				μA
				$\overline{S}_D, \overline{R}_D$ Inputs			80				μA
				$\overline{CP}$ Inputs			80				μA
			V_I = 2.7V	J, K Inputs						20	μA
				$\overline{S}_D, \overline{R}_D$ Inputs						60	μA
				$\overline{CP}$ Inputs						80	μA
I_{IL}	LOW-level input current[5]	V_{CC} = MAX, V_I = 0.4V		J, K Inputs			−1.6			−0.4	mA
				$\overline{S}_D, \overline{R}_D$ Inputs			−3.2			−0.8	mA
				$\overline{CP}$ Inputs			−3.2			−0.8	mA
I_{OS}	Short-circuit output current[3]	V_{CC} = MAX		Mil	−20		−57	−20		−100	mA
				Com'l	−18		−57	−20		−100	mA
I_{CC}	Supply current[4] (total)	V_{CC} = MAX				10	40		4	8	mA

NOTES

1. For conditions shown as MIN or MAX, use the appropriate value specified under recommended operating conditions for the applicable type.
2. All typical values are at V_{CC} = 5V, T_A = 25° C.
3. I_{OS} is tested with V_{OUT} = + 0.5V and V_{CC} = V_{CC} MAX + 0.5V. Not more than one output should be shorted at a time and duration of the short circuit should not exceed one second.
4. With the Clock input grounded and all outputs open, I_{CC} is measured with the Q and $\overline{Q}$ outputs HIGH in turn.
5. $\overline{S}_D$ is tested with $\overline{R}_D$ HIGH, and $\overline{R}_D$ is tested with $\overline{S}_D$ HIGH.

Waveforms 1, 2, and 3 on the next page show the measurement points for frequency and propagation delay values.

AC CHARACTERISTICS T_A = 25° C, V_{CC} = 5.0V

PARAMETER		TEST CONDITIONS	54/74 C_L = 15pF, R_L = 400Ω		54/74LS C_L = 15pF, R_L = 2kΩ		UNIT
			Min	Max	Min	Max	
f_{MAX}	Maximum Clock frequency	Waveform 3	15		30		MHz
t_{PLH} t_{PHL}	Propagation delay Clock to output	Waveform 1, 'LS76 Waveform 3, '76		25 40		20 30	ns
t_{PLH} t_{PHL}	Propagation delay $\overline{S}_D$ or $\overline{R}_D$ to output	Waveform 2		25 40		20 30	ns

NOTE

Per industry convention, f_{MAX} is the worst case value of the maixmum device operating frequency with no constraints on t_r, t_f, pulse width, or duty cycle.

Figure 11-11 *(Continued)*

FLIP-FLOPS

AC SETUP REQUIREMENTS $T_A = 25°$ C, $V_{CC} = 5.0$V

Minimum pulse widths

PARAMETER		TEST CONDITIONS	54/74		54/74LS		UNIT
			Min	Max	Min	Max	
t_w(H)	Clock pulse width (HIGH)	Waveform 1	20		20		ns
t_w(L)	Clock pulse width (LOW)	Waveform 1	47				ns
t_w(L)	Reset pulse width (LOW)	Waveform 2	25		25		ns
t_s	Setup time J or K to Clock(C)	Waveform 1	0		20		ns
t_h	Hold time J or K to Clock	Waveform 1	0		0		ns

Setup and hold times

These waveforms define time measurement points.

AC WAVEFORMS

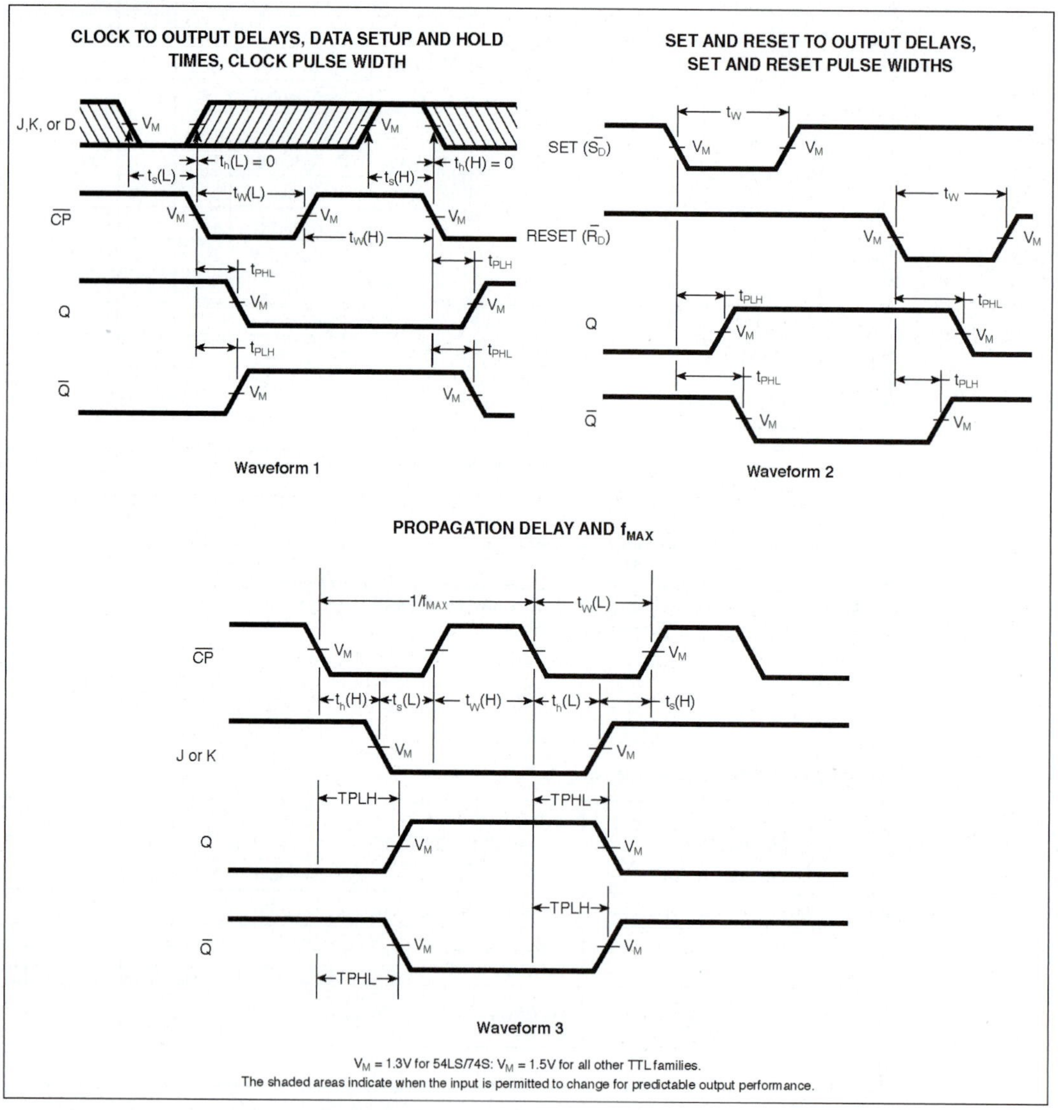

Figure 11–11   *(Continued)*

delayed output waveform that suits your needs. The output waveforms are identical to the input except delayed by 5, 10, 15, or 20 ns. Complemented, delayed waveforms are also available at the $\overline{5}$, $\overline{10}$, $\overline{15}$, and $\overline{20}$ outputs. Delay gates with various other delay intervals are also available.

EXAMPLE 11-3

The 74109 is a positive edge-triggered J-$\overline{K}$ flip-flop. If we attach the J input to the clock as shown in Figure 11–12, will the flip-flop's Q output toggle?

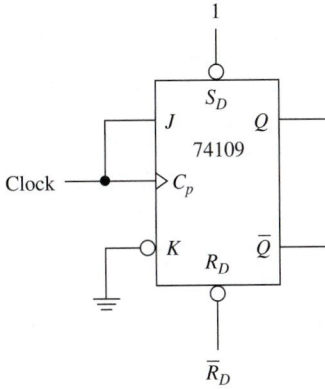

Figure 11-12

Solution: t_s for a 74109 is 10 ns, which means that J must be HIGH and $\overline{K}$ must be LOW 10 ns *before* the positive clock edge. When we draw the waveforms as shown in Figure 11–13, we see that J and C_p are exactly the same. Therefore, because J is not HIGH one setup time before the positive clock edge, Q will not toggle.

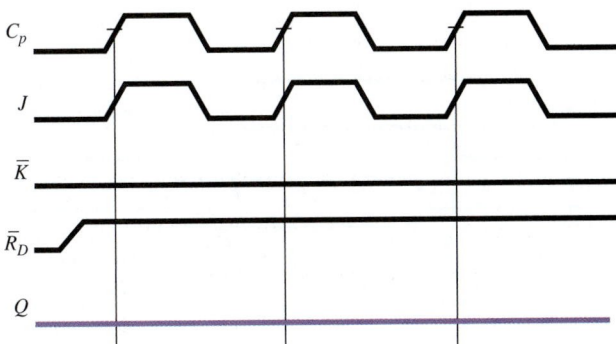

Figure 11-13

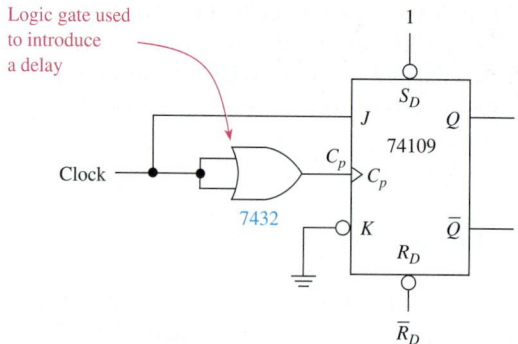

Logic gate used to introduce a delay

Figure 11–14 Modification of flip-flop in Example 11–3 to allow it to toggle.

Team Discussion

If manufacturers provided a minimum *and* a maximum propagation delay, which would you use?

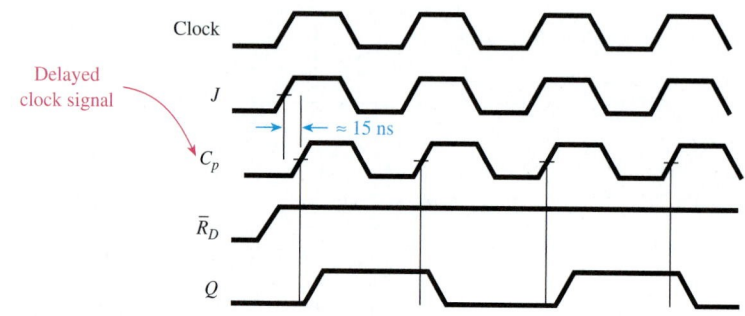

Delayed clock signal

Figure 11–15 Timing waveforms for Figure 11–14.

Helpful Hint

The circuitry inside a typical delay gate is given in Figure 16–13(a).

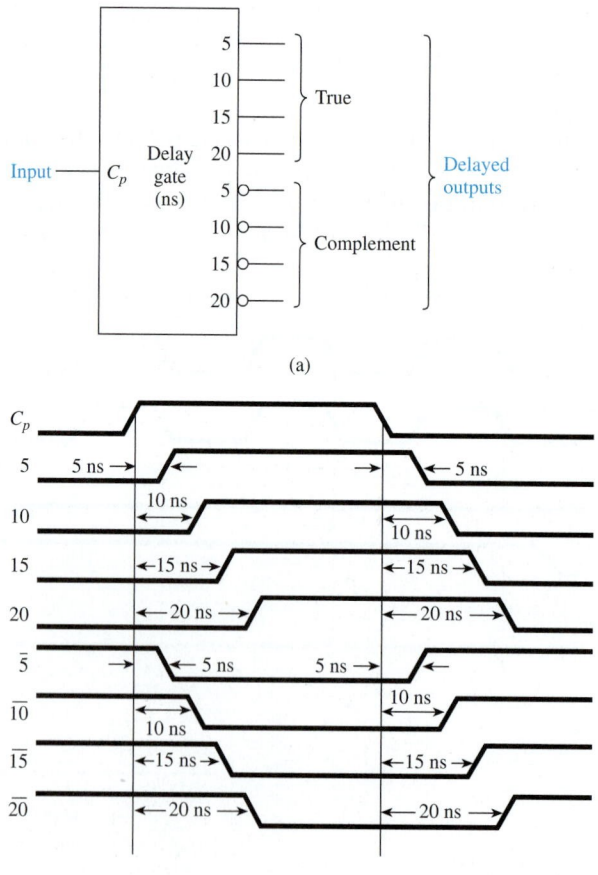

Figure 11–16 A 5-ns multitap delay gate: (a) logic symbol; (b) output waveforms.

EXAMPLE 11-4

Use the setup, hold, and propagation delay times from a data manual to determine if the 74109 J-$\overline{K}$ flip-flop in the circuit shown in Figure 11–17 will toggle. (Assume that the flip-flop is initially Reset, and remember that for a toggle, $J = 1$, $\overline{K} = 0$.)

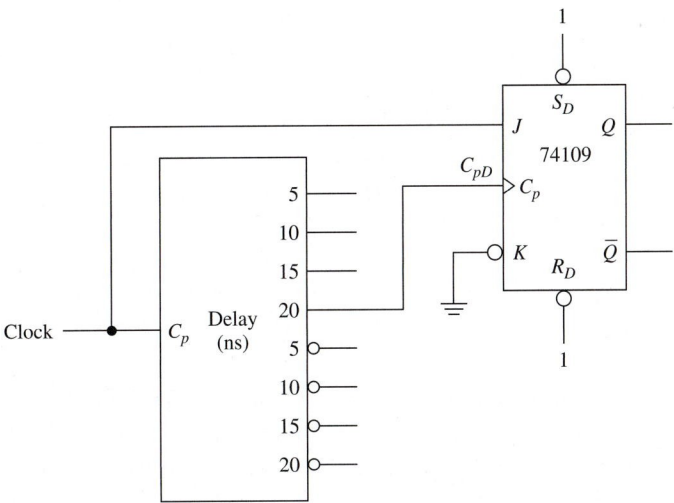

Figure 11–17

Solution: First, draw the waveforms as shown in Figure 11–18. C_{pD}, the delayed clock, makes its LOW-to-HIGH transition and triggers the flip-flop 20 ns *after J* makes its transition. Looking at the waveforms, *J is* HIGH 10 ns *before* the positive edge of C_{pD} and *is* held HIGH ≥ 6 ns *after* the positive edge of C_{pD}. Therefore, *the flip-flop will toggle* at each positive edge of C_{pD}.

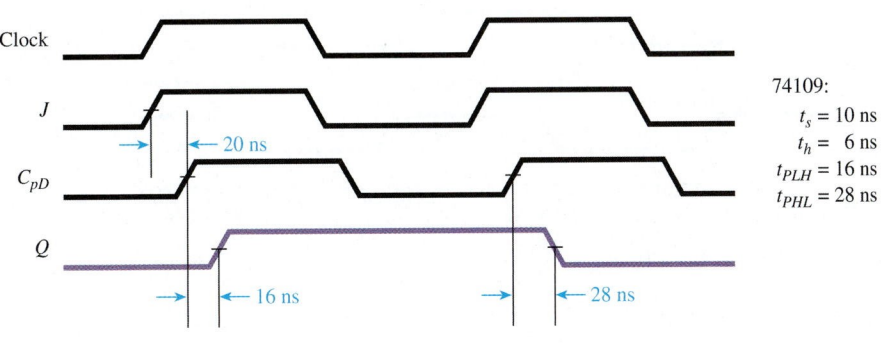

74109:
$t_s = 10$ ns
$t_h = 6$ ns
$t_{PLH} = 16$ ns
$t_{PHL} = 28$ ns

Figure 11–18

EXAMPLE 11-5

Use the specifications from a data manual to determine if the 74LS112 J-K flip-flop in the circuit shown in Figure 11–19 will toggle. (Assume that the flip-flop is initially Reset.)

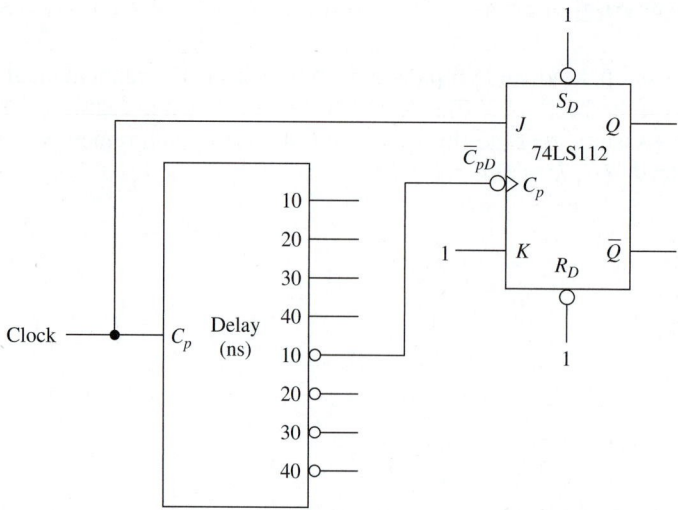

Figure 11–19

Solution: First, draw the waveforms as shown in Figure 11–20. $\overline{C_{pD}}$ is inverted and delayed by 10 ns from the Clock and *J* waveforms. Each negative edge of $\overline{C_{pD}}$ triggers the flip-flop. Looking at the waveforms, the *J* input *is not* set up HIGH ≥ 20 ns before the negative $\overline{C_{pD}}$ edge and, therefore, *is not* interpreted as a HIGH. The flip-flop output will be *undetermined* from then on because it cannot distinguish if *J* is a HIGH or a LOW at each negative $\overline{C_{pD}}$ edge.

To correct the problem, $\overline{C_{pD}}$ should be connected to the $\overline{30}$- ns tap on the delay gate instead of the $\overline{10}$- ns tap. This way, when the flip-flop "looks back" 20 ns from the negative edge of $\overline{C_{pD}}$, it will see a HIGH on *J*, allowing the toggle operation to occur.

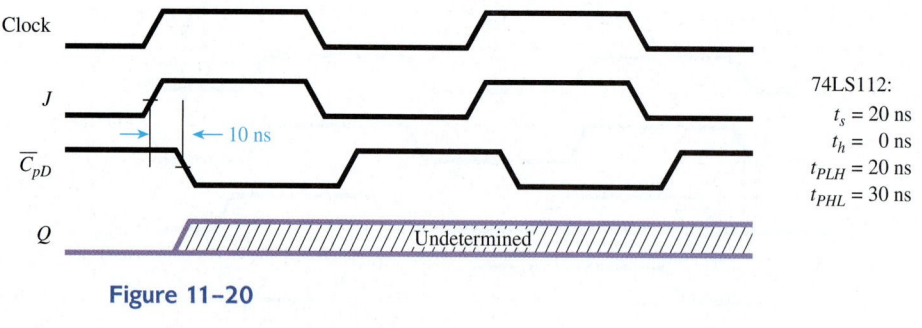

74LS112:
$t_s = 20$ ns
$t_h = 0$ ns
$t_{PLH} = 20$ ns
$t_{PHL} = 30$ ns

Figure 11–20

EXAMPLE 11–6

The repetitive waveforms shown in Figure 11–21(b) are input to the 7474 *D* flip-flop shown in Figure 11–21(a). Because of poor timing, *Q* never goes HIGH. Add a delay gate to correct the timing problem. (Assume that rise, fall, and propagation delay times are 0 ns.)

CHAPTER 11 I PRACTICAL CONSIDERATIONS FOR DIGITAL DESIGN

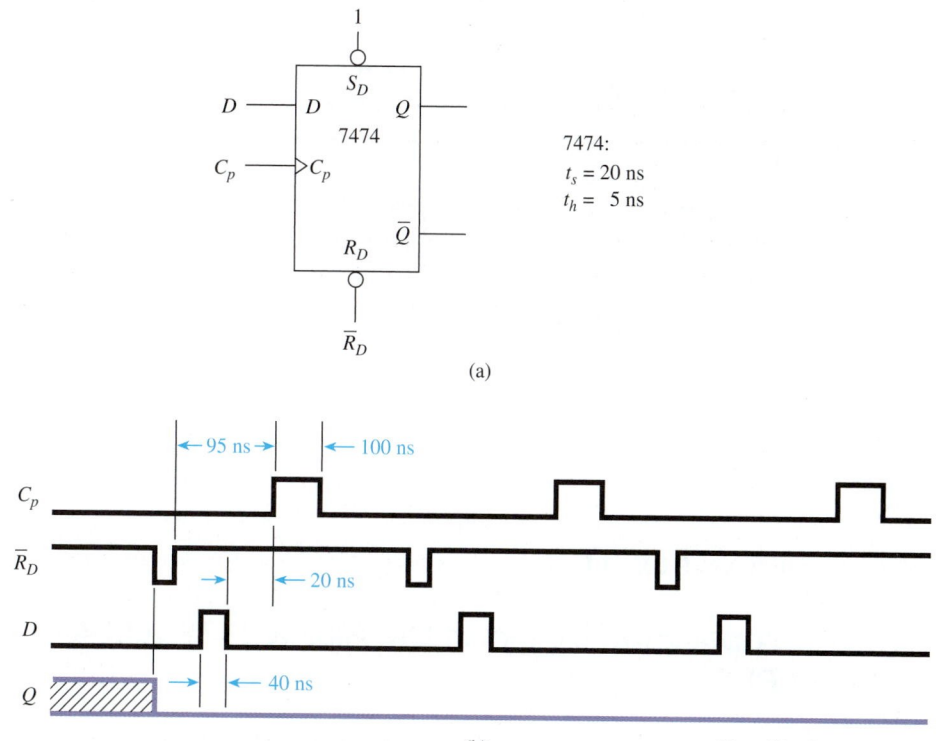

7474:
$t_s = 20$ ns
$t_h = 5$ ns

(a)

(b)   *Note:* Not drawn to scale.

Figure 11–21

Team Discussion

To obtain the fastest possible operating speed in a digital system, designers sometimes push the times to the absolute limits specified by the manufacturer. Discuss some of the considerations the designer must watch for and some of the problems that may occur.

Solution: Q never goes HIGH because the 40-ns HIGH pulse on D does not occur at the positive edge of C_p. Delaying the D waveform by 30 ns will move D to the right far enough to fulfill the necessary setup and hold times to allow the flip-flop to get Set at every positive edge of C_p, as shown in Figure 11–22. (D_D is the delayed D waveform, which has been shifted to the right by 30 ns to correct the timing problem.)

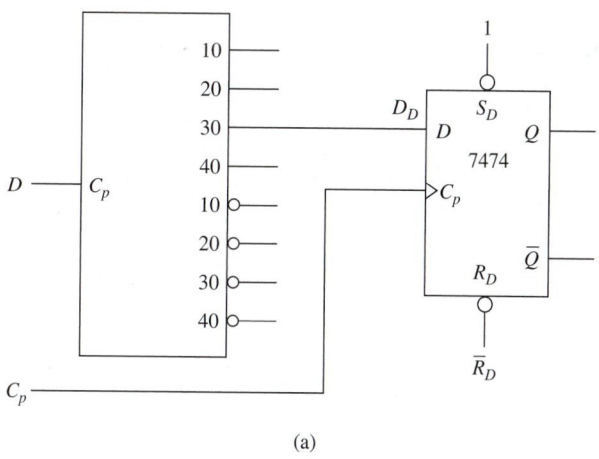

(a)

Figure 11–22

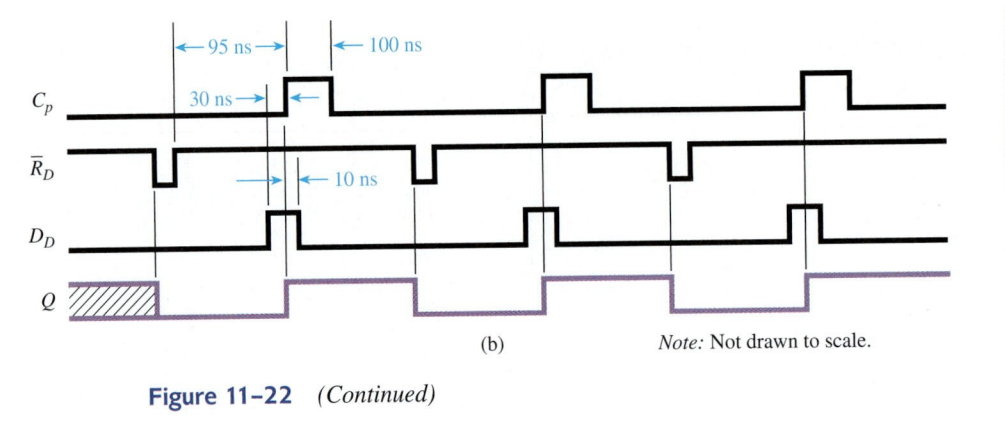

(b) *Note:* Not drawn to scale.

Figure 11–22 *(Continued)*

EXAMPLE 11–7

Propagation Delays in the Altera 7128SLC84-7 CPLD

Create a simulator channel file (scf) that exercises a D flip-flop in such a way to allow us to determine the propagation delays for its synchronous and asynchronous inputs to the q output. Include an AND gate at its output to show the delay for the output $x = dq$.

Solution: Figure 11–23 shows the DFF primitive with the required inputs and outputs. The scf file for the circuit is shown in Figure 11–24. Since the propagation times will be in the nanosecond range, the **End Time** (File menu) was set to 1 μs (1000 ns). The **Grid Size** (Option menu) was set to 20 ns to draw the *cp* and *d* waveforms and set to 2 ns to draw the narrow *n_sd* and *n_rd* pulses. If you look carefully, you can see a delay between the *n_sd* to *q* waveforms and from the *cp* to *q* waveforms.

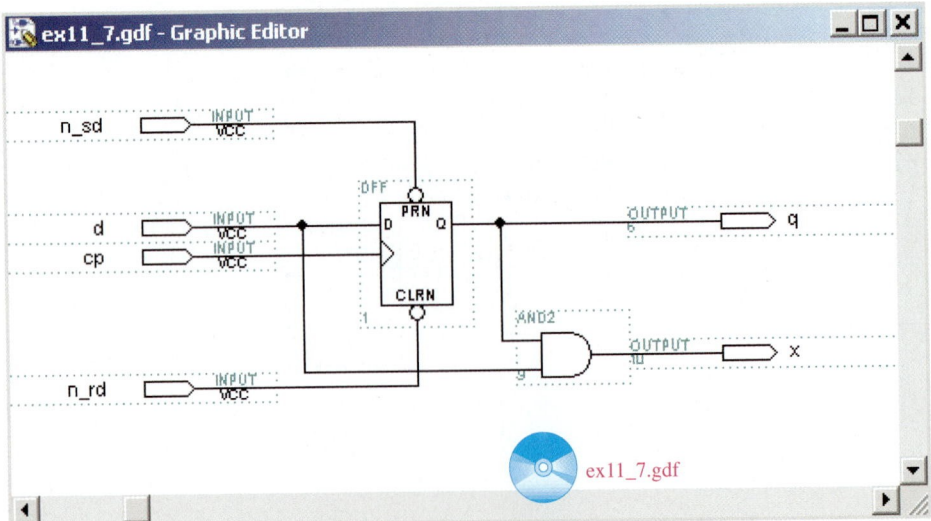

Figure 11–23 DFF Primitive used to illustrate CPLD propagation delays in Example 11–7.

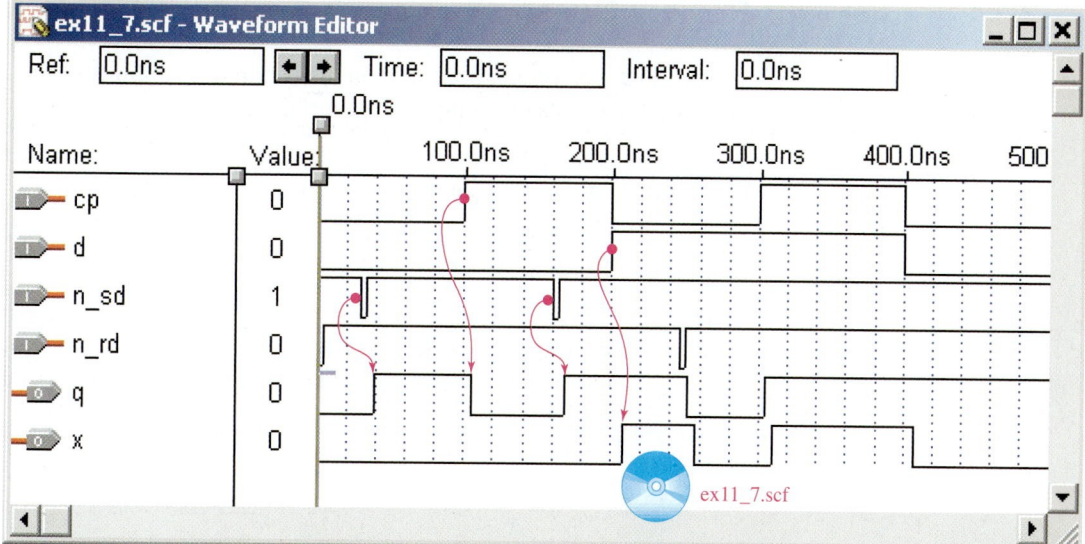

Figure 11–24 Waveform simulation used to exercise the CPLD of Example 11–7.

To accurately measure the time in nanoseconds, we need to **Zoom In** (View menu) to the waveform as shown in Figure 11–25. Then by clicking near the negative edge on *q,* the time reference cursor will appear as shown. Press the left and right cursor movement arrows to measure the time between each active edge in the scf file shown. By doing this, the following propagation delays were determined:

t_{PHL} (*cp*-to-*q*) = 4.5 ns

t_{PLH} (*n_sd*-to-*q*) = 8.5 ns

t_{PLH} (*d*AND*q*-to-*x*) = 7.5 ns

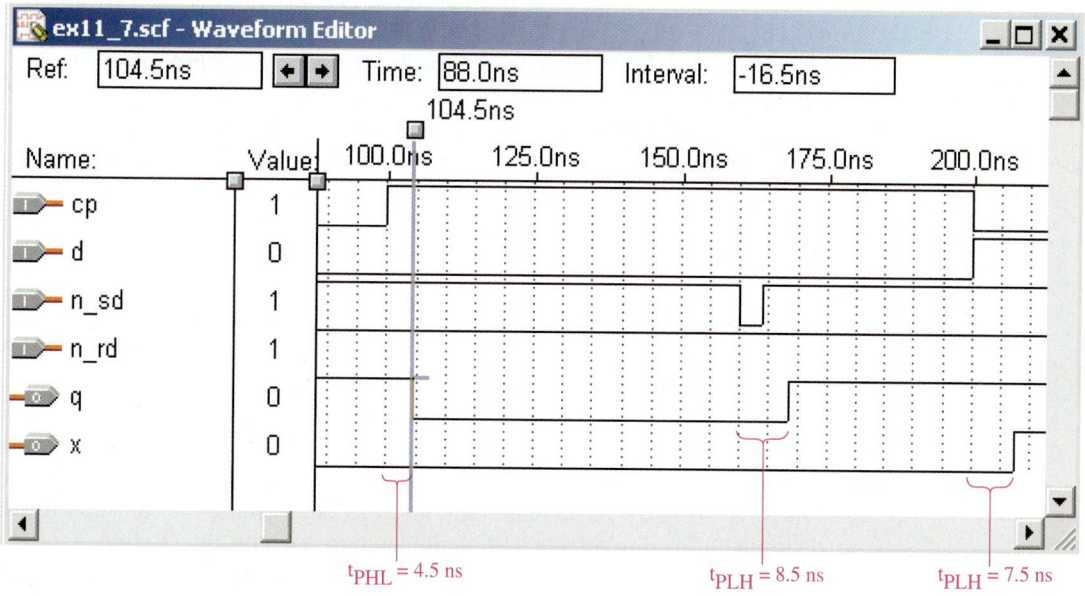

Figure 11–25 Zooming in on the 100 ns—to—200 ns section of the simulation for Example 11–7.

11–1. A *race condition* occurs when the Q output of a flip-flop changes at the same time as its clock input. True or false?

11–2. *Setup time* is the length of time that the clock input must be held stable before its active transition. True or false?

11–3. Describe what manufacturers mean when they specify a *hold time* of 0 ns for a flip-flop.

11–4. The abbreviation t_{PHL} is used to specify the _____ of an IC from input to output. The letters *HL* in the abbreviation refer to the _____ (input, output) waveform changing from HIGH to LOW.

11–5. Under what circumstances would a digital circuit design require a delay gate?

11–2 Automatic Reset

Often, it is advantageous to automatically Reset (or Set) all resettable (or settable) devices as soon as power is first applied to a digital circuit. In the case of resetting flip-flops, we want a LOW voltage level (0) present at the $\overline{R_D}$ inputs for a short duration immediately following **power-up,** but then after a short time (usually a few microseconds), we want the $\overline{R_D}$ line to return to a HIGH (1) level so that the flip-flops can start their synchronous operations.

To implement such an operation, we might use a series **RC circuit** to charge a capacitor that is initially discharged (0). A short time after the power-up voltage is applied to the *RC* circuit and the flip-flop's V_{CC}, the capacitor will charge up to a value high enough to be considered a HIGH (1) by the $\overline{R_D}$ input.

Basic electronic theory states that in a series *RC* circuit, the capacitor becomes almost fully charged after a time equal to the product *5RC*. This means that in Figure 11–26(a) the capacitor (which is initially discharged via the internal resistance of the $\overline{R_D}$ terminal) will begin to charge toward the 5-V level through R as soon as the switch is closed.

Before the capacitor reaches the HIGH-level threshold of the 74LS76 ($\approx$2.0 V), the temporary LOW on the $\overline{R_D}$ terminal will cause the flip-flop to Reset. As soon as the capacitor charges to above 2.0 V, the $\overline{R_D}$ terminal will see a HIGH, allowing the flip-flop to perform its normal synchronous operations. The waveforms that appear on the V_{CC} and $\overline{R_D}$ lines as the power switch is closed and opened are shown in Figure 11–26(b).

This automatic resetting scheme can be used in circuits employing single or multiple resettable ICs. Depending on the device being Reset, the length of time that the Reset line is at a LOW level will be approximately 1 μs.

As you add more and more devices to the Reset line, the time duration of the LOW will decrease because of the additional charging paths supplied by the internal circuitry of the ICs. Remember, there is a minimum allowable width for the LOW Reset pulse ($\approx$25 ns for a 74LS76). To increase the time, you can increase the capacitor to 0.01 μ F, or to eliminate loading effects and create a sharp edge on the $\overline{R_D}$ line, a 7407 buffer could be inserted in series with the $\overline{R_D}$ input. (A silicon diode can also be added from the top of the capacitor up to the V_{CC} line to discharge the capacitor more rapidly when power is removed.) We will utilize the **automatic Reset** feature several times in Chapters 12 and 13.

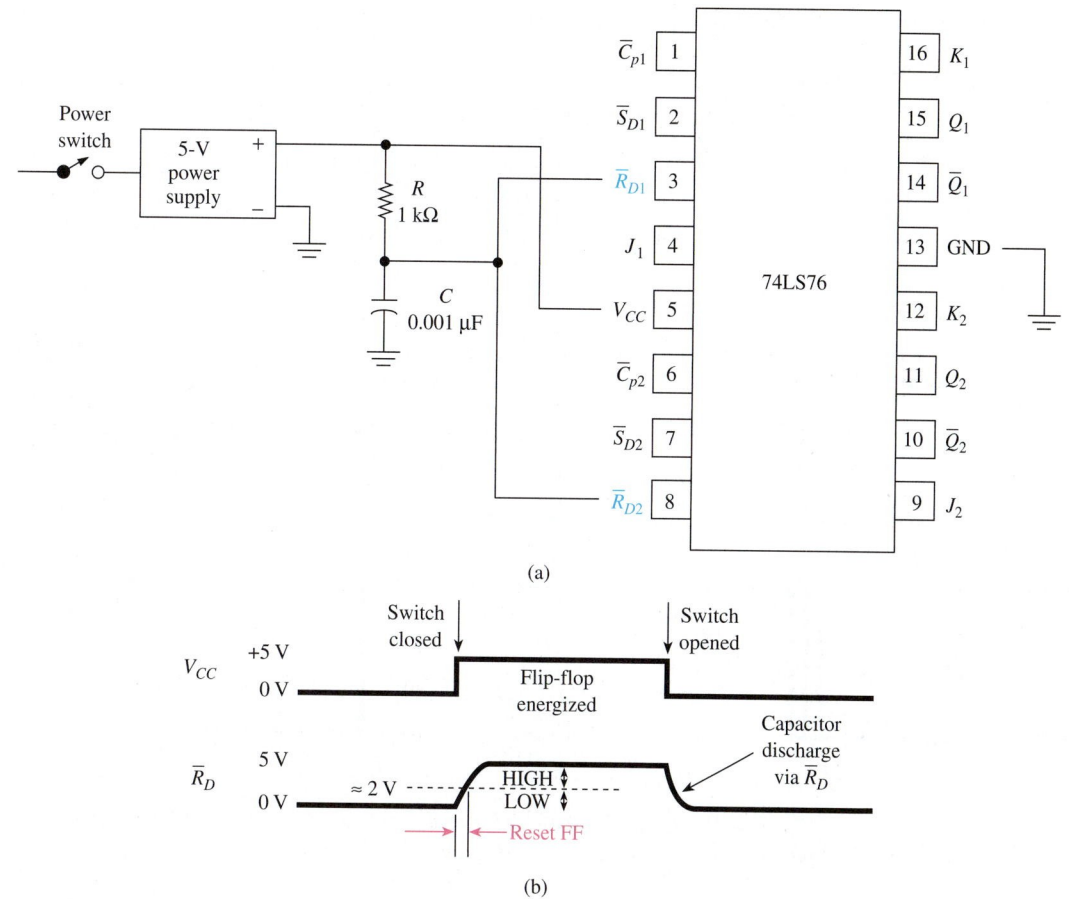

Figure 11–26 Automatic power-up Reset for a *J-K* flip-flop: (a) circuit connections; (b) waveforms.

11–3 Schmitt Trigger ICs

A **Schmitt trigger** is a special type of IC that is used to transform slowly changing waveforms into sharply defined, jitter-free output signals. They are useful for changing clock edges that may have slow rise and fall times into straight vertical edges.

The Schmitt trigger employs a technique called **positive feedback** internally to speed up the level transitions and also to introduce an effect called **hysteresis.** Hysteresis means that the switching threshold on a positive-going input signal is at a higher level than the switching threshold on a negative-going input signal [see Figure 11–27(b)]. This is useful for devices that have to ignore small amounts of **jitter,** or electrical noise on input signals. Notice in Figure 11–27(a) that when the positive- and negative-going thresholds are exactly the same, as with standard gates, and a small amount of noise causes the input to jitter slightly, the output will switch back and forth several times until the input level is far above the threshold voltage.

Figure 11–27 illustrates the difference in the output waveforms for a standard 7404 inverter and a 7414 Schmitt trigger inverter. As you can see in Figure 11–27(b), the output (V_{out2}) is an inverted, jitter-free pulse. On the other hand, just think if V_{out1} were fed into the $\overline{C}_p$ input of a 74LS76 hooked up as a toggle flip-flop; the flip-flop would toggle three times (three negative edges) instead of once as was intended.

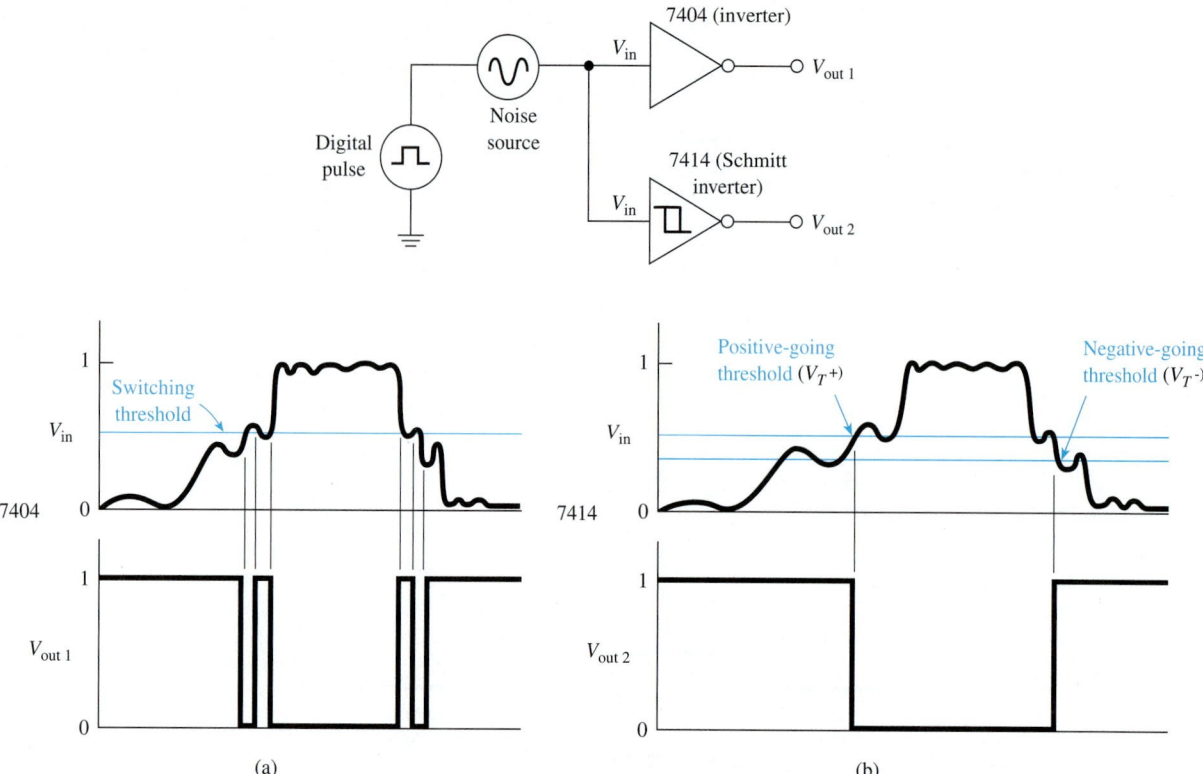

Figure 11-27   Switching characteristics of inverters: (a) false switching of the regular inverter; (b) jitter-free operation of a Schmitt trigger.

The difference between the positive- and negative-going thresholds is defined as the hysteresis voltage. For the 7414, the positive-going threshold (V_{T+}) is typically 1.7 V, and the negative-going threshold (V_{T-}) is typically 0.9 V, yielding a hysteresis voltage (ΔV_T) of 0.8 V. The small box symbol ($\sqcap$) inside the 7414 symbol is used to indicate that it is a Schmitt trigger inverter instead of a regular inverter.

The most important specification for Schmitt trigger devices is illustrated by use of a **transfer function** graph, which is a plot of V_{out} versus V_{in}. From the transfer function, we can determine the HIGH- and LOW-level output voltages (typically 3.4 and 0.2 V, the same as for most TTL gates), as well as V_{T+}, V_{T-}, and ΔV_T.

Common Misconception

Using an oscilloscope in the X-Y mode, you can view the transfer function as you apply a 0- to 5-V triangle wave to V_{in}. You may think that it is not working because the vertical lines are not apparent. They do not show because the output switches so fast. However, the threshold points are still obvious to see.

Team Discussion

How would the transfer function of a noninverting Schmitt differ from Figure 11–28?

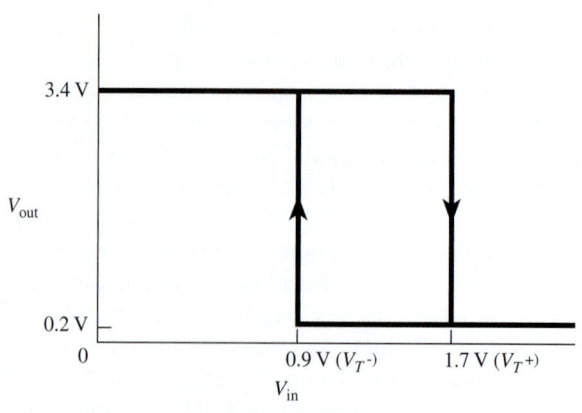

Figure 11-28   Transfer function for a 7414 Schmitt trigger inverter.

CHAPTER 11 | PRACTICAL CONSIDERATIONS FOR DIGITAL DESIGN

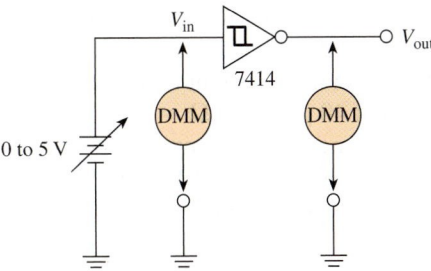

Figure 11–29 Circuit used to experimentally produce a Schmitt trigger transfer function.

Figure 11–28 shows the transfer function for the 7414. The transfer function graph is produced experimentally by using a variable voltage source at the input to the Schmitt and a DMM (digital multimeter) at V_{in} and V_{out}, as shown in Figure 11–29.

As the V_{in} of Figure 11–29 is increased from 0 V up toward 5 V, V_{out} will start out at approximately 3.4 V (1) and switch to 0.2 V (0) when V_{in} exceeds the positive-going **threshold** (≈ 1.7 V). The output transition from HIGH to LOW is indicated in Figure 11–28 by the downward arrow. As V_{in} is increased up to 5 V, V_{out} remains at 0.2 V (0).

As the input voltage is then decreased down toward 0 V, V_{out} will remain LOW until the negative-going threshold is passed (≈ 0.9 V). At that point, the output will switch up to 3.4 V (1), as indicated by the upward arrow in Figure 11–28. As V_{in} continues to 0 V, V_{out} remains HIGH at 3.4 V. The hysteresis in this case is 1.7 V − 0.9 V = 0.8 V.

EXAMPLE 11-8

Let's use the Schmitt trigger to convert a small-signal sine wave (E_s) into a square wave (V_{out}).

Solution: The diode is used to short the negative 4 V from E_s to ground to protect the Schmitt input, as shown in Figure 11–30(a). The 1-kΩ resistor will limit

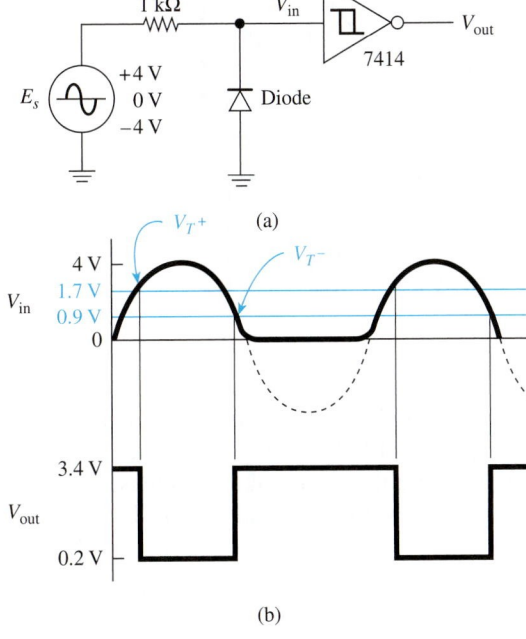

Figure 11–30

Helpful Hint

Actually, there will be a slight negative voltage of −0.7 V at V_{in} during the negative cycle.

Team Discussion

Why is the duty cycle of V_{out} always going to be greater than 50%?

Solution: The transfer function is shown in Figure 11–36.

Common Misconception

Students often draw the transfer function in Figure 11–36 backward, thinking that it is an *inverting* Schmitt.

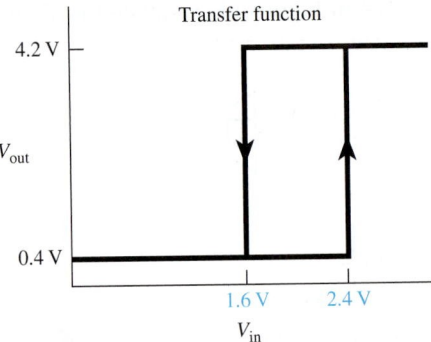

Figure 11–36

Review Questions

11–6. In an automatic power-up Reset *RC* circuit, the voltage across the _____ (resistor, capacitor) provides the initial LOW to the $\overline{R_D}$ inputs.

11–7. The input voltage to a Schmitt trigger IC has to cross two different switching points called the _____ and the _____. The voltage differential between these two switching points is called the _____.

11–8. A Schmitt trigger IC is capable of "cleaning up" a square wave that may have a small amount of noise on it. Briefly explain how it works.

11–9. The transfer function of a Schmitt trigger device graphically shows the relationship between the _____ and _____ voltages.

11–4 Switch Debouncing

Often, mechanical switches are used in the design of digital circuits. Unfortunately, however, most switches exhibit a phenomenon called **switch bounce.** Switch bounce is the action that occurs when a mechanical switch is opened or closed. For example, when the contacts of a switch are closed, the electrical and mechanical connection is first made, but due to a slight springing action of the contacts, they will bounce back open, then close, then open, then close, continuing repeatedly until they finally settle down in the closed position. This bouncing action will typically take place for as long as 50 ms.

A typical connection for a single-pole, single-throw **(SPST) switch** is shown in Figure 11–37. This is a poor design because, if we expect the toggle to operate only once when we close the switch, we will be out of luck because of switch bounce. Why do we say that? Let's look at the waveform at $\overline{C_p}$ to see what actually happens when a switch is closed.

Figure 11–38 shows that $\overline{C_p}$ will receive several LOW pulses each time the switch is closed instead of the single pulse that we expect. This is because as the switch is first closed, the contacts come together and then bounce apart several times before settling down together. [The 10-kΩ pull-up resistor in Figure 11–37 is necessary to hold the voltage level at $\overline{C_p}$ up close to +5 V while the switch is open. If the

EXAMPLE 11–10

Sketch V_{out} of the 7414 in Figure 11–33 given the V_{in} waveform shown in Figure 11–34.

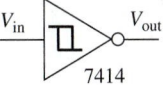

Figure 11–33

Solution:

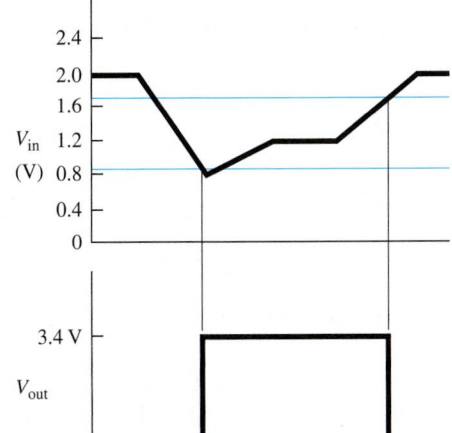

Figure 11–34

EXAMPLE 11–11

Draw and completely label the V_{out} versus V_{in} transfer function for the Schmitt trigger device whose V_{in} and V_{out} waveforms are given in Figure 11–35.

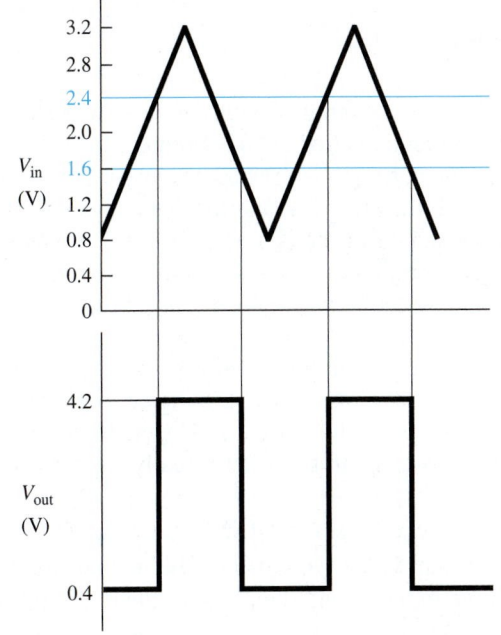

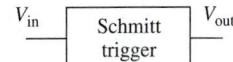

Figure 11–35

**Common
Misconception**

Students often draw the
transfer function in
Figure 11–36 backward,
thinking that it is an
inverting Schmitt.

Solution: The transfer function is shown in Figure 11–36.

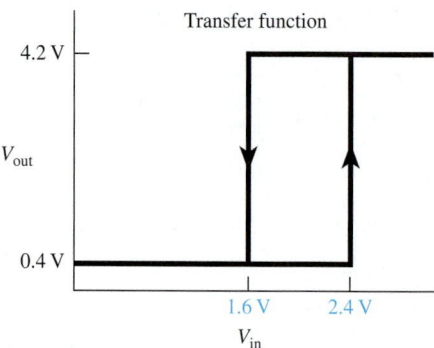

Figure 11–36

Review Questions

11–6. In an automatic power-up Reset *RC* circuit, the voltage across the _____ (resistor, capacitor) provides the initial LOW to the $\overline{R_D}$ inputs.

11–7. The input voltage to a Schmitt trigger IC has to cross two different switching points called the _____ and the _____. The voltage differential between these two switching points is called the _____.

11–8. A Schmitt trigger IC is capable of "cleaning up" a square wave that may have a small amount of noise on it. Briefly explain how it works.

11–9. The transfer function of a Schmitt trigger device graphically shows the relationship between the _____ and _____ voltages.

11–4   Switch Debouncing

Often, mechanical switches are used in the design of digital circuits. Unfortunately, however, most switches exhibit a phenomenon called **switch bounce.** Switch bounce is the action that occurs when a mechanical switch is opened or closed. For example, when the contacts of a switch are closed, the electrical and mechanical connection is first made, but due to a slight springing action of the contacts, they will bounce back open, then close, then open, then close, continuing repeatedly until they finally settle down in the closed position. This bouncing action will typically take place for as long as 50 ms.

A typical connection for a single-pole, single-throw **(SPST) switch** is shown in Figure 11–37. This is a poor design because, if we expect the toggle to operate only once when we close the switch, we will be out of luck because of switch bounce. Why do we say that? Let's look at the waveform at $\overline{C_p}$ to see what actually happens when a switch is closed.

Figure 11–38 shows that $\overline{C_p}$ will receive several LOW pulses each time the switch is closed instead of the single pulse that we expect. This is because as the switch is first closed, the contacts come together and then bounce apart several times before settling down together. [The 10-kΩ pull-up resistor in Figure 11–37 is necessary to hold the voltage level at $\overline{C_p}$ up close to +5 V while the switch is open. If the

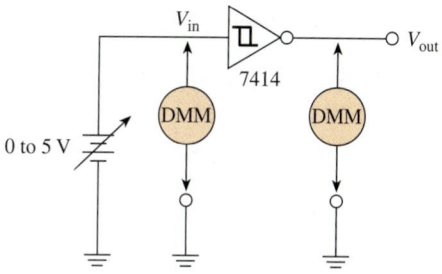

Figure 11–29   Circuit used to experimentally produce a Schmitt trigger transfer function.

Figure 11–28 shows the transfer function for the 7414. The transfer function graph is produced experimentally by using a variable voltage source at the input to the Schmitt and a DMM (digital multimeter) at V_{in} and V_{out}, as shown in Figure 11–29.

As the V_{in} of Figure 11–29 is increased from 0 V up toward 5 V, V_{out} will start out at approximately 3.4 V (1) and switch to 0.2 V (0) when V_{in} exceeds the positive-going **threshold** (≈ 1.7 V). The output transition from HIGH to LOW is indicated in Figure 11–28 by the downward arrow. As V_{in} is increased up to 5 V, V_{out} remains at 0.2 V (0).

As the input voltage is then decreased down toward 0 V, V_{out} will remain LOW until the negative-going threshold is passed (≈ 0.9 V). At that point, the output will switch up to 3.4 V (1), as indicated by the upward arrow in Figure 11–28. As V_{in} continues to 0 V, V_{out} remains HIGH at 3.4 V. The hysteresis in this case is 1.7 V $-$ 0.9 V $= 0.8$ V.

EXAMPLE 11–8

Let's use the Schmitt trigger to convert a small-signal sine wave (E_s) into a square wave (V_{out}).

Solution: The diode is used to short the negative 4 V from E_s to ground to protect the Schmitt input, as shown in Figure 11–30(a). The 1-kΩ resistor will limit

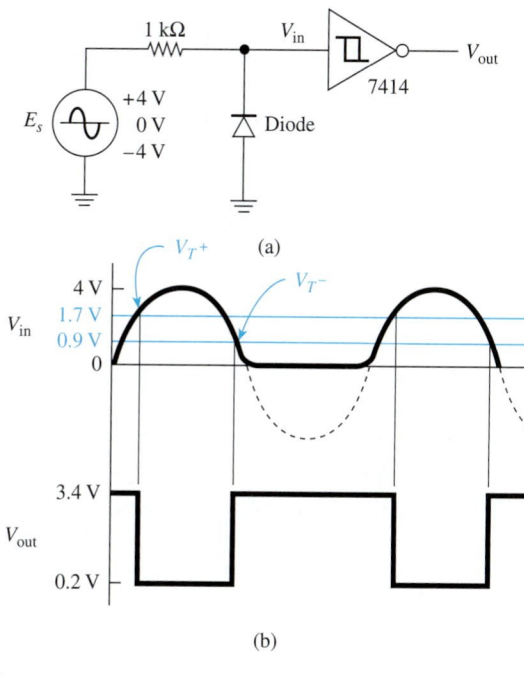

Figure 11–30

Helpful Hint

Actually, there will be a slight negative voltage of -0.7 V at V_{in} during the negative cycle.

Team Discussion

Why is the duty cycle of V_{out} always going to be greater than 50%?

the current through the diode when it is conducting. $[I_{\text{diode}} = (4 - 0.7 \text{ V})/1 \text{ k}\Omega = 3.3 \text{ mA}$, which is well within the rating of most silicon diodes.]

Also, the HIGH-level input current to the Schmitt (I_{IH}) is only 40 μA, causing a voltage drop of 40 μA $\times$ 1 kΩ = 0.04 V when V_{in} is HIGH. (We can assume that 0.04 V is negligible compared to + 4.0 V.)

The input to the Schmitt will, therefore, be a half-wave signal with a 4.0-V peak. The output will be a square wave, as shown in Figure 11–30(b).

EXAMPLE 11–9

The V_{in} waveform to the 74132 Schmitt trigger NAND gate in Figure 11–31 is given in Figure 11–32.

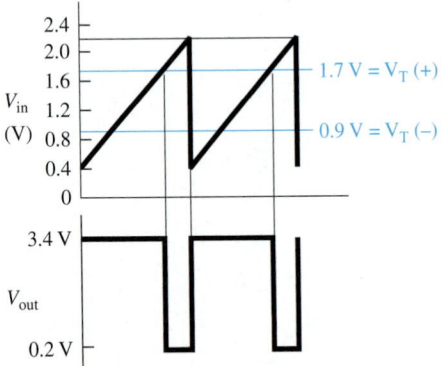

74132

V_{in} —[ ]— V_{out}

Figure 11–31

(a) Sketch the V_{out} waveform. (The 74132 has the same voltage specifications as the 7414.)
(b) Determine the duty cycle of the output waveform; the **duty cycle** is defined as

$$\frac{\text{time HIGH}}{\text{time HIGH} + \text{time LOW}} \times 100\%$$

Solution:

(a)   The V_{out} waveform is shown in Figure 11–32.

Figure 11–32

(b) V_{out} stays HIGH while V_{in} increases from 0.4 to 1.7 V, for a change of 1.3 V. V_{out} stays LOW while V_{in} increases from 1.7 to 2.2 V, for a change of 0.5 V. Because the input voltage increases linearly with respect to time, the change in V_{in} is proportional to time duration, so

$$\text{duty cycle} = \frac{1.3 \text{ V}}{1.3 \text{ V} + 0.5 \text{ V}} \times 100\% = 72.2\%$$

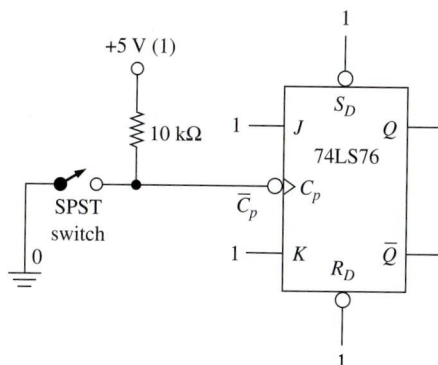

Team Discussion

When is switch bounce a problem? When doesn't it matter?

Figure 11–37   Switch used as a clock input to a toggle flip-flop.

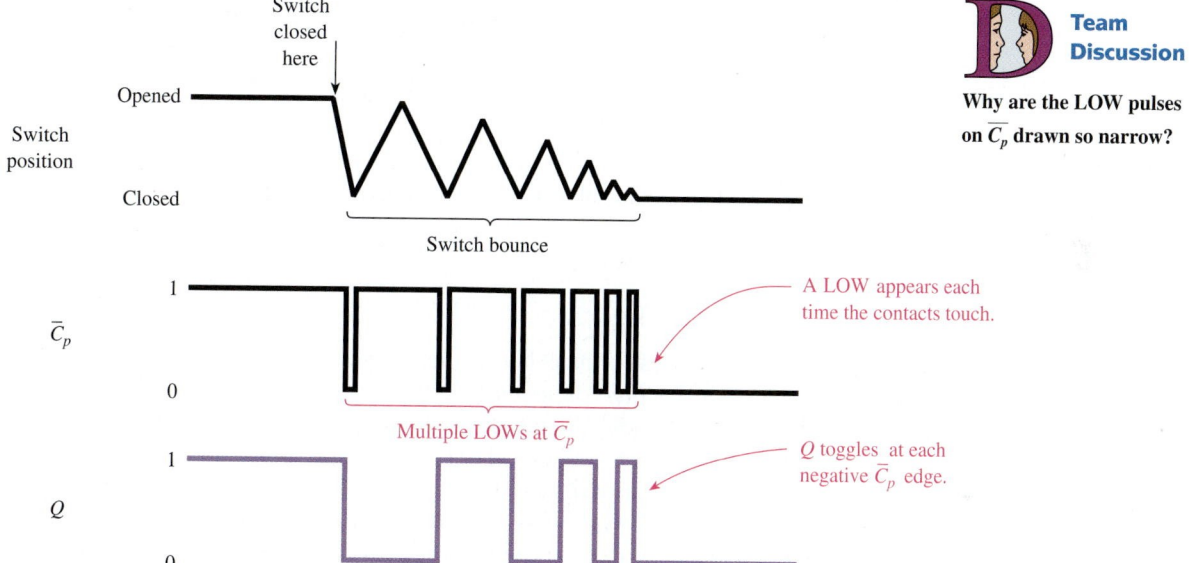

Team Discussion

Why are the LOW pulses on $\overline{C_p}$ drawn so narrow?

Figure 11–38   Waveform at point $\overline{C_p}$ for Figure 11–37.

pull-up resistor were not used, the voltage at $\overline{C_p}$ with the switch open would be undetermined, but *with* the 10 kΩ (and realizing that the current into the $\overline{C_p}$ terminal is negligible), the level at the $\overline{C_p}$ terminal will be held at approximately +5 V while the switch is open.]

There are several ways to eliminate the effects of switch bounce. If you need to debounce a single-pole, single-throw switch or push button, the Schmitt trigger scheme shown in Figure 11–39 can be used. With the switch open, the capacitor will be charged to +5 V (1), keeping $V_{out} = 0$. When the switch is closed, the capacitor will discharge rapidly to zero via the 100-Ω current-limiting resistor, making V_{out} equal to 1. Then, as the switch bounces, the capacitor will repeatedly try to charge slowly back up to a HIGH and then discharge rapidly to zero. The RC charging time constant (10 k × 0.47 μF) is long enough that the capacitor will not get the chance to charge up high enough (above V_{T^+}) before the switch bounces back to the closed position. This keeps V_{out} equal to 1.

When the switch is reopened, the capacitor is allowed to charge all the way up to +5 V. When it crosses V_{T^+}, V_{out} will switch to 0, as shown in Figure 11–39. The result

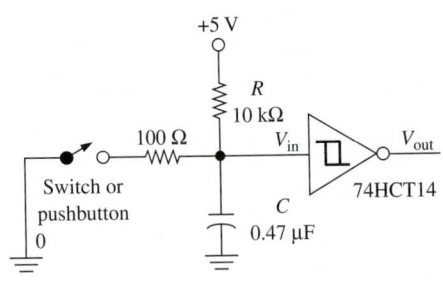

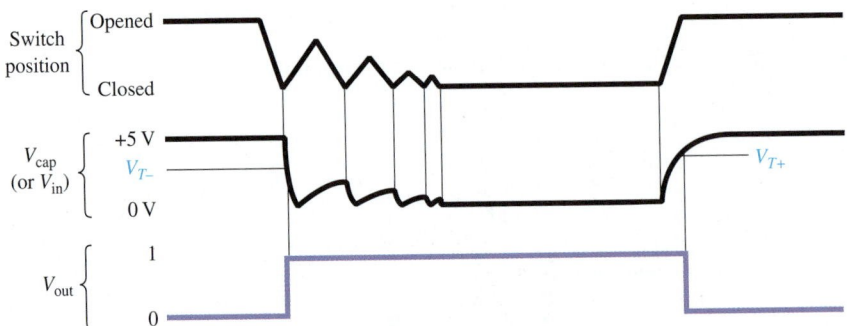

Figure 11–39 Schmitt method of debouncing a single-pole, single-throw switch.

is that by closing the switch or push button once *you will get only a single pulse at the output even though the switch is bouncing.*

To debounce single-pole, double-throw switches, a different method is required, as illustrated in Figures 11–40 and 11–41. The single-pole, double-throw switch shown in Figure 11–40(a) actually has three positions: (1) position A, (2) in between position A and position B while it is making its transition, and (3) position B. The cross-NAND debouncer works very similarly to the cross-NAND *S-R* flip-flop presented in Figure 10–2.

When the switch is in position A, OUT will be Set (1). When the switch is moved to position B, it bounces, causing OUT to Reset, Hold, Reset, Hold, Reset, Hold repeatedly until the switch stops bouncing and settles into position B. From the time the switch first touched position B until it is returned to position A, OUT will be Reset even though the switch is bouncing.

When the switch is returned to position A, it will bounce, causing OUT to be Set, Hold, Set, Hold, Set, Hold repeatedly until the switch stops bouncing. In this case, OUT will be Set and remain Set from the moment the switch first touched position A, even though the switch is bouncing.

Figure 11–41 shows another way to debounce a single-pole, double-throw switch using a 7474 *D* flip-flop. (Actually, any flip-flop with asynchronous $\overline{S_D}$ and $\overline{R_D}$ inputs can be used.)

The waveforms created from Figure 11–41 will look the same as Figure 11–40(b). As the switch is moved to position A but is still bouncing, the flip-flop will Set, Hold, Set, Hold, Set, Hold repeatedly until the switch settles into position A, keeping the flip-flop Set. When it is moved to position B, the flip-flop will Reset, Hold, Reset, Hold, and so on until it settles down, keeping the flip-flop Reset.

Figures 11–42(a) and (b) show the results of when using a *D* flip-flop to debounce multiple LOWs on its asynchronous inputs. Assume that *n_sd* and *n_rd* are to be connected to a SPDT switch that makes multiple closures to a LOW before it settles down.

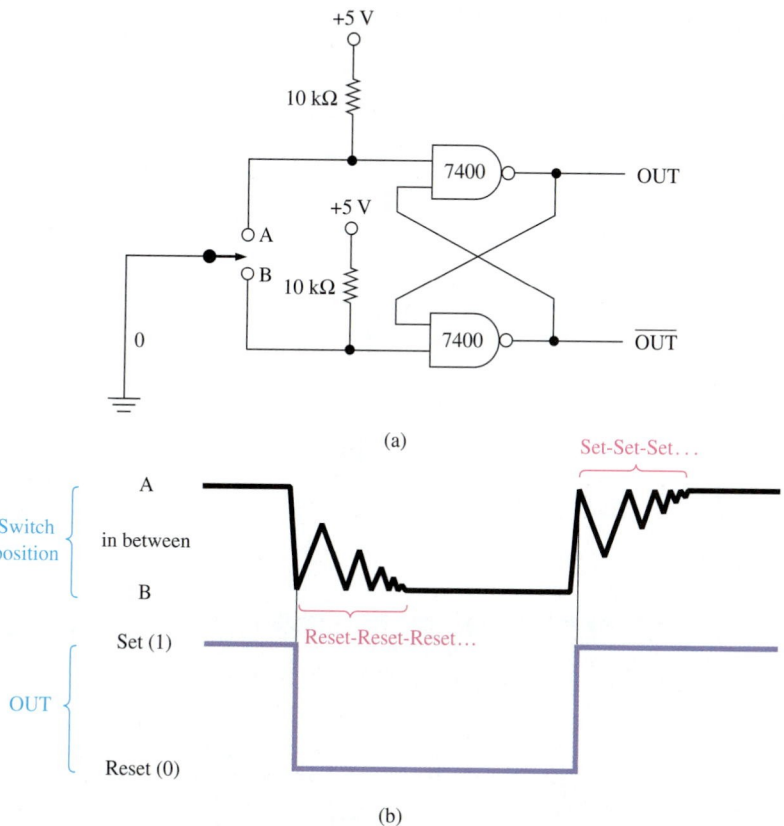

(a)

(b)

Figure 11–40 (a) Cross-NAND method of debouncing a single-pole, double-throw switch; (b) waveforms for part (a).

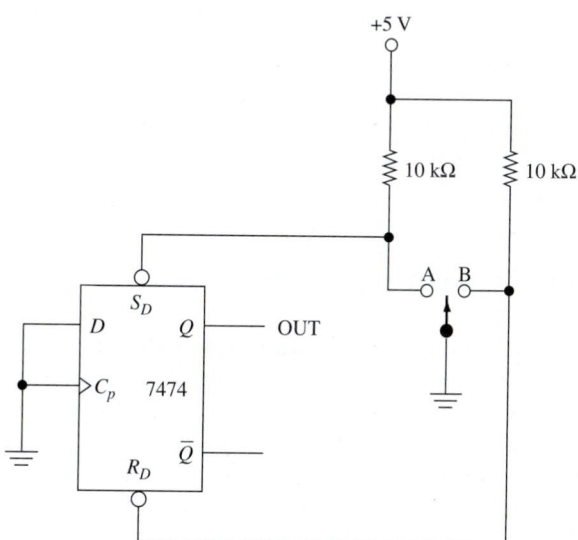

Figure 11–41 *D* flip-flop method of debouncing a single-pole, double-throw switch.

The simulation waveforms show that the first LOW closure on *n_sd* sets the *q* output and as the signal continues to bounce, it has no effect on *q* because it is already Set. At the 6 μs mark, *n_sd* is made HIGH and *n_rd* is allowed to make multiple LOW closures. As you can see, only the first LOW is used to reset *q* and the subsequent LOWs are ignored because *q* is already Reset.

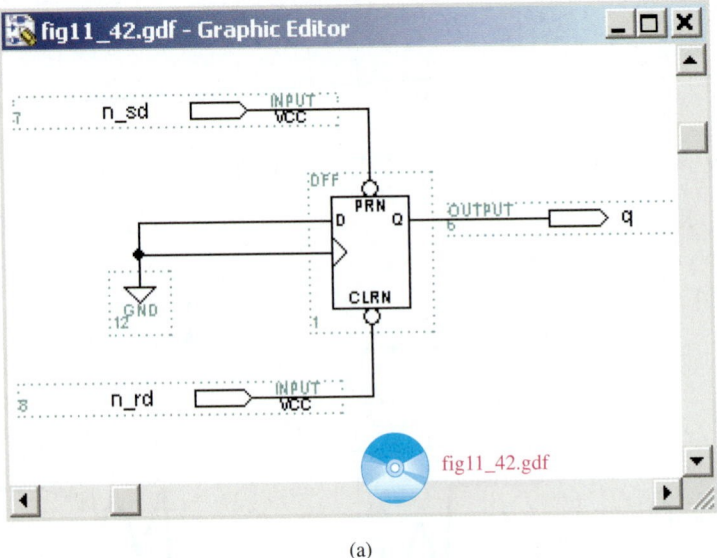

(a)

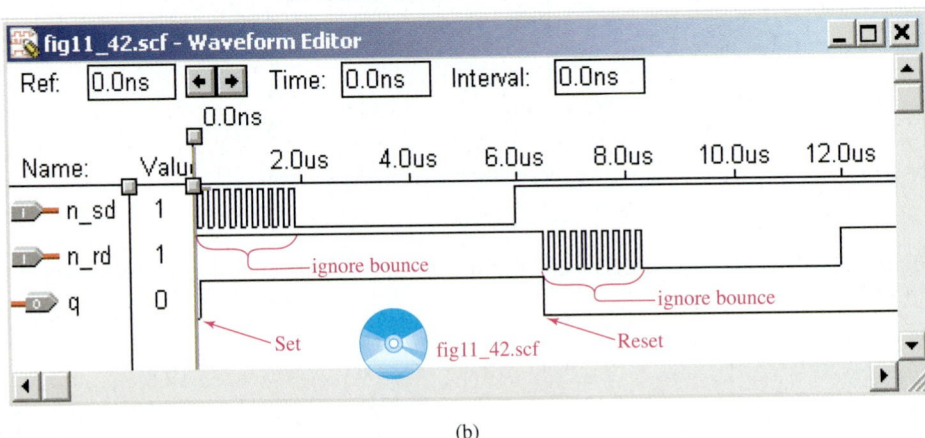

(b)

Figure 11–42 Using a D flip-flop to debounce multiple LOWs on its asynchronous inputs: (a) circuit diagram; (b) waveform simulation.

11–5 Sizing Pull-Up Resistors

By the way, how do we know what size **pull-up resistors** to use in circuits like the one in Figure 11–41? Remember, the object of the pull-up resistor is to keep a terminal at a HIGH level when it would normally be at a **float** (not 1 or 0) level. In Figure 11–41, when the switch is *between* points A and B, current will flow down through the 10-kΩ resistor to $\overline{S_D}$. I_{IH} for $\overline{S_D}$ is 80 µA. This causes a voltage drop of 80 µA $\times$ 10 kΩ = 0.8 V, leaving 4.2 V at $\overline{S_D}$, which is well within the HIGH specifications of the 7474.

You may ask, Why not just make the pull-up resistor very small to minimize the voltage drop across it? Well, when the switch is in position A or B, we have a direct connection to ground. If the resistor is too small, we will have excessive current and high power consumption. However, a 10-kΩ or larger resistor would work just fine. So check the I_{IH} and V_{IH} values in a data book and keep within their ratings. Usually, a 10-kΩ pull-up resistor is a good size for most digital circuits.

When you have to provide a **pull-down resistor** (to keep a floating terminal LOW), a much smaller resistor is required because I_{IL} is typically much higher than I_{IH}.

For example, if $I_{IL} = -1.6$ mA and a 100-Ω pull-down resistor is used, the voltage across the resistor is 0.160 V, which will be interpreted as a LOW. One concern of using a pull-down resistor is the high power dissipation of the resistor.

EXAMPLE 11–12

Determine the power dissipation of the 10-kΩ pull-up resistors used in Figure 11–40(a). Also, determine the HIGH-level voltage at the input to the NAND gates.

Solution: The specs for a 7400 NAND from a TTL data manual are

$$\left. \begin{array}{l} I_{IL} = 1.6 \text{ mA max.} \\ I_{IH} = 40 \text{ }\mu\text{A max.} \\ V_{IL} = 0.8 \text{ V max.} \\ V_{IH} = 2.0 \text{ V min.} \end{array} \right\} \begin{array}{l} \text{You can review these} \\ \text{terms in Chapter 9.} \end{array}$$

When the switch is *between* positions A and B, I_{IH} will flow from the +5 V, through the 10-kΩ resistor, into the NAND. The power dissipation is

$$P = I^2 \times R$$
$$= (40 \text{ }\mu\text{A})^2 \times 10 \text{ k}\Omega$$
$$= 16 \text{ }\mu\text{W}$$

The high-level input voltage

$$V = V_{CC} - I_{IH} \times R$$
$$= 5 \text{ V} - 40 \text{ }\mu\text{A} \times 10 \text{ k}\Omega$$
$$= 4.6 \text{ V}$$

The 4.6-V HIGH-level input voltage is above the 2.0-V V_{IH} limit given in the specs, and 16 μW is negligible for most applications.

When the switch is moved to either A or B, the power dissipation in the resistor is

$$P = \frac{V^2}{R}$$
$$= \frac{5 \text{ V}^2}{10 \text{ k}\Omega}$$
$$= 2.5 \text{ mW}$$

The value of 2.5 mW is still negligible for most applications. If not, increase the size of the 10-kΩ pull-up resistor and recalculate for the HIGH-level input voltage and power dissipation. As long as you keep the HIGH-level input voltage above the specified limit of 2.0 V, the circuit will operate properly.

Team Discussion

Sizing pull-up resistors for open-collector outputs requires that you also consider the size of the load resistor. For example, if the load is 10 kΩ, what happens if you use a 10-kΩ pull up?

11–6 Practical Input and Output Considerations

Before designing and building the practical digital circuits in the next few chapters, let's study some circuit designs for (1) a simple 5-V power supply, (2) a clock to drive synchronous trigger inputs, and (3) circuit connections for LED interfacing to the outputs of integrated-circuit chips.

A 5-V Power Supply

For now, we limit our discussion to the TTL family of ICs. From the data manual, we can see that TTL requires a constant supply voltage of 5.0 V $\pm$ 5%. Also, the total supply current requirement into the V_{CC} terminal ranges from 20 to 100 mA for most TTL ICs.

For the power supply, the 78XX series of IC **voltage regulators** is inexpensive and easy to use. To construct a regulated 5.0-V supply, we use the 7805 (the 05 designates 5 V; a 7808 would designate an 8.0-V supply). The 7805 is a three-terminal device (input, ground, output) capable of supplying 5.0 V $\pm$ 0.2% at 1000 mA. Figure 11–43 shows how a 7805 voltage regulator is used in conjunction with an ac-to-dc **rectifier** circuit.

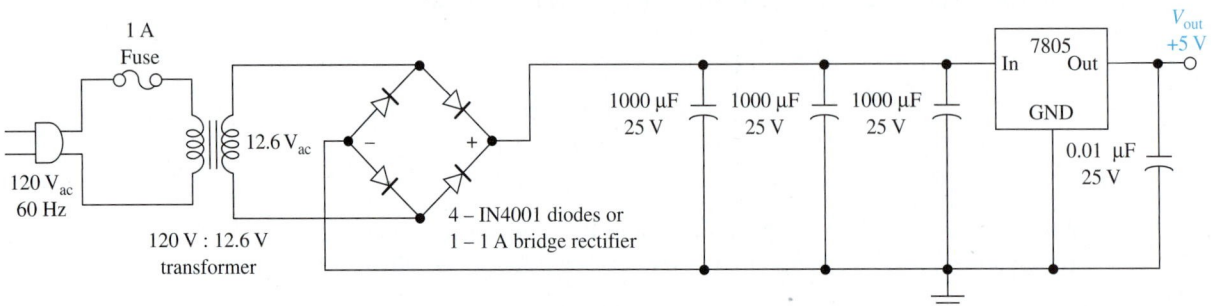

Figure 11–43   Complete 5-V, 1-A TTL power supply.

In Figure 11–43 the 12.6-V ac rms is rectified by the diodes (or a four-terminal bridge rectifier) into a full-wave dc of approximately 20 V. The 3000 µF of capacitance is required to hold the dc level into the 7805 at a high, steady level. The 7805 will automatically decrease the 20-V dc input to a solid, **ripple**-free 5.0-V dc output.

The 0.01-µF capacitor is recommended by TTL manufacturers for *decoupling* the power supply. Tantalum capacitors work best and should be mounted as close as possible to the V_{CC}-to-ground pins on every TTL IC used in your circuit. Their size should be between 0.01 and 0.1 µF, with a voltage rating $\geq$ 5 V. The purpose of the capacitor is to eliminate the effects of voltage spikes created from the internal TTL switching and electrostatic noise generated on the power and ground lines.

The 7805 will get very hot when your circuit draws more than 0.5 A. In that case, it should be mounted on a heat sink to help dissipate the heat.

A 60-Hz Clock

Figure 11–44 shows a circuit design for a simple 60-Hz TTL-level (0 to 5 V) clock that can be powered from the same transformer used in Figure 11–43 and used to drive the clock inputs to our synchronous ICs. Our electric power industry supplies us with accurate 60-Hz ac voltages. It is a simple task to reduce the voltage to usable levels and still maintain a 60-Hz [60-pulse-per-second (pps)] signal.

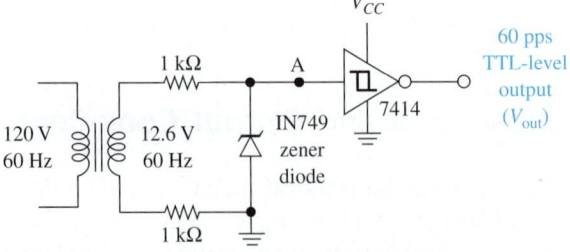

Figure 11–44   Accurate 60-Hz, TTL-level clock pulse generator.

From analog electronics courses, you may remember that a zener diode will conduct normally in the forward-biased direction, and in the reverse-biased direction, it will start conducting when a voltage level equal to its *reverse* **zener breakdown** rating is reached (4.3 V for the IN749).

The 1-kΩ resistors are required to limit the zener current to reasonable levels. Figure 11–44 shows the waveform that will appear at point A and V_{out} of Figure 11–45. The zener breaks down at 4.3 V, which is high enough for a one-level input to the Schmitt trigger but not too high to burn out the chip. The V_{out} waveform will be an accurate 60-pulse-per-second, approximately 50% duty cycle square wave. As we will see in Chapter 12, this frequency can easily be divided down to 1 pulse per second by using toggle flip-flops. One pulse per second is handy because it is slow enough to see on visual displays (like LEDs) and accurate enough to use as a trigger pulse on a digital clock.

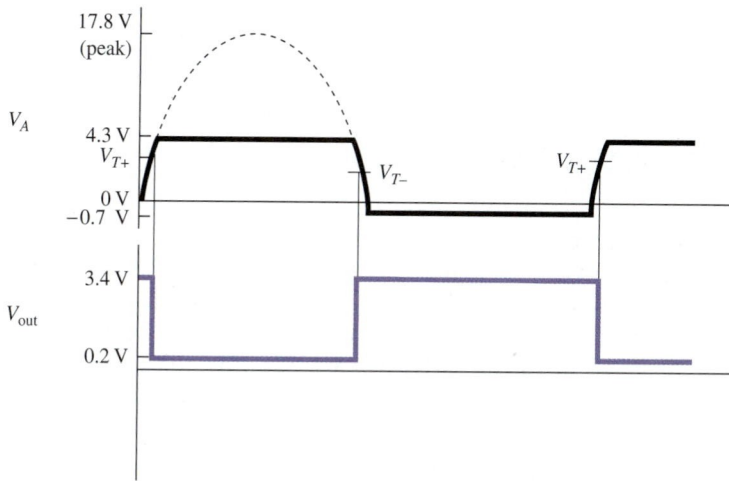

Figure 11–45   Voltage waveform at point *A* and V_{out} of Figure 11–44.

Driving Light-Emitting Diodes

Light-emitting diodes (LEDs) are good devices to visually display a HIGH (1) or LOW (0) digital state. A typical red LED will drop 1.7 V cathode to anode when forward biased (positive anode-to-cathode voltage) and will illuminate with 10 to 20 mA flowing through it. In the reverse-biased direction (zero or negative anode-to-cathode voltage), the LED will block current flow and not illuminate.

Because it takes 10 to 20 mA to illuminate an LED, we may have trouble driving it with a TTL output. From the TTL data manual, we can determine that most ICs can sink (0-level output) a lot more current than they can source (1-level output). Typically, the maximum sink current, I_{OL}, is 16 mA, and the maximum source current, I_{OH}, is only 0.4 mA. Therefore, we had better use a LOW level (0) to turn on our LED instead of a HIGH level.

Figure 11–46 shows how we can drive an LED from the output of a TTL circuit (a *J-K* flip-flop in this case). The *J-K* flip-flop is set up in the toggle mode so that *Q* will flip states once each second.

When *Q* is LOW (0 V), the LED is forward biased, and current will flow through the LED and resistor and sink into the *Q* output. The 330-Ω resistor is required to limit the series current to 10 mA [$I = (5 V - 1.7 V)/330 Ω = 10$ mA], and 10 mA into the LOW-level *Q* output will not burn out the flip-flop. If, however, we were trying to turn the LED on with a HIGH-level output, we would turn the LED around and connect the cathode to ground. But, 10 mA would exceed the limit of I_{OH} on the 7476 and either burn it out or just not illuminate the LED.

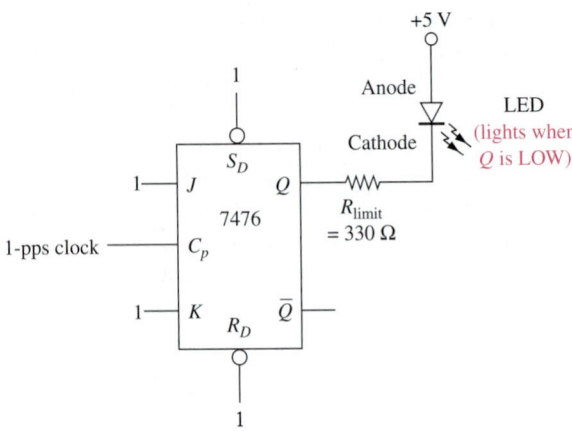

Figure 11–46 Driving an LED.

Connecting Multiple I/O to a CPLD or FPGA

Figure 11–47 shows how multiple inputs and outputs can be connected to a CPLD or FPGA. A single-pole, single-throw (SPST) dual-in-line package (DIP) switch is used to input HIGH/LOW levels into the device. With all eight switches open, the pull-up resistors provide all HIGH levels to the inputs (In_0 through In_7). When any switch is closed, a LOW will appear on that input. Debounced clock inputs are provided at Clk_0 and Clk_1 by the cross-NAND S-R flip flop connections shown. When either SPDT switch (or SPDT pushbutton) goes from its down position to its up position, the level on Clk changes from LOW to HIGH. This level change on Clk makes only one transition even as the switch bounces. This single transition is very important if the clock input is connected to logic that counts each time the switch is moved from LOW to HIGH because including the bounces would give an inaccurate count result.

The LEDs shown in Figure 11–47 are connected as active-HIGH logic indicators. Unlike most 7400-series ICs, CPLDs and FPGAs are capable of sinking and sourcing far more than the 10 mA required to illuminate an LED. Outputting a HIGH at any output (Out_0 through Out_7) will source current through the corresponding LED and 330-ohm limiting resistor, turning that LED ON. (The display can be made active-LOW by reversing the LEDs and changing the ground connection to V_{CC}.)

Phototransistor Input to a Latching Alarm System

A **phototransistor** is made to turn off and on by shining light on its base region. It is encased in clear plastic and is turned on when light strikes its base region or if the base connection is forward biased with an external voltage. The resistance from collector to emitter for an OFF transistor is typically 1 to 10 MΩ. An ON transistor will range from 1000 Ω to as low as 10 Ω depending on the light intensity. If even more sensitivity is required, a CDS photocell could be used in place of the phototransistor.

The circuit of Figure 11–48 uses a phototransistor in an alarm system. The phototransistor could be placed in a doorway and positioned so that light is normally striking it. This will keep its resistance low and the voltage at point A low. When a person walks through the doorway, the light is interrupted, making the voltage at point A momentarily high. The 74HCT14 Schmitt inverters will react by outputting a LOW-to-HIGH pulse. This creates the clock trigger to the D flip-flop, which will latch HIGH, turning on the alarm. The alarm will remain on until the Reset pushbutton is pressed.

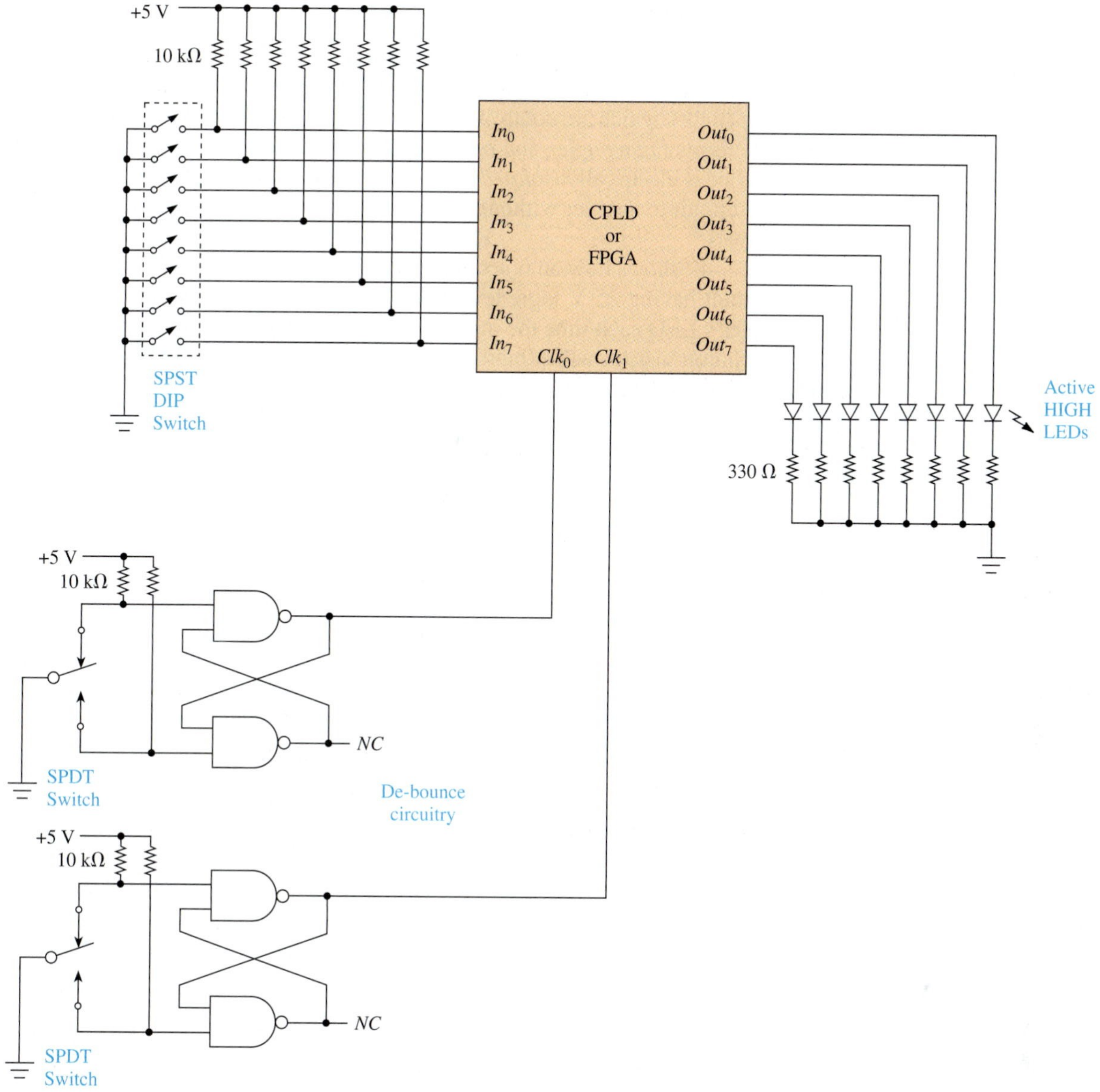

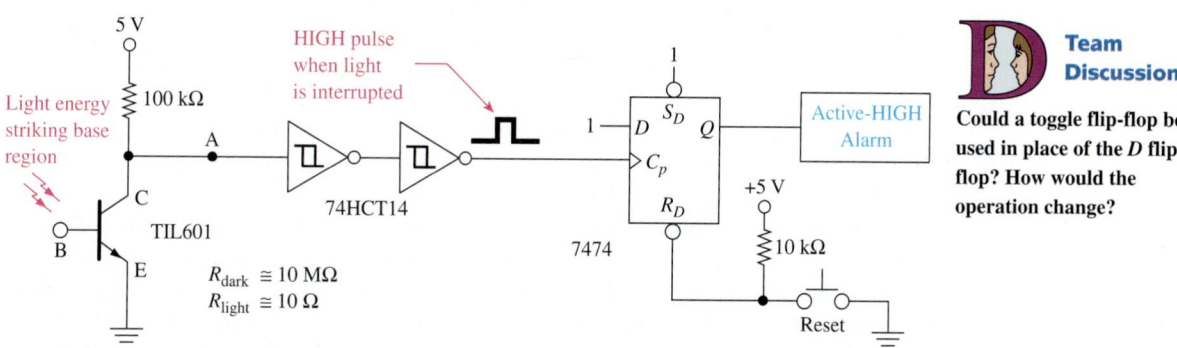

Figure 11–47 Multiple inputs and outputs connected to a CPLD or FPGA.

Figure 11–48 Phototransistor used as an input to a latching alarm system.

Team Discussion

Could a toggle flip-flop be used in place of the D flip-flop? How would the operation change?

Using an Optocoupler for Level Shifting

An **optocoupler** (or optoisolator) is an IC with an LED and phototransistor encased in the same package. The phototransistor has a very high OFF resistance (dark) and a low ON resistance (light), which are controlled by the amount of light striking its base from the LED. The terms *optocoupler* and *optoisolator* come from the fact that the output side of the device is electrically *isolated* from the input side and can, therefore, be used to *couple* one circuit to another without being concerned about incompatible or harmful voltage levels.

Figure 11–49 shows how an optocoupler can be used to transmit TTL-level data to another circuit having 25-V logic levels. The 7408 is used to sink current through the optocoupler's LED each time the input signal goes LOW. Each time the LED illuminates, the phototransistor will exhibit a low resistance, making V_{out} LOW. Therefore, the output signal will be in phase with the input signal, but its HIGH/LOW levels will be approximately 25 V/0 V. Notice that the 25-V circuit is totally separate from the 5-V TTL circuit, providing complete isolation from the potentially damaging higher voltage.

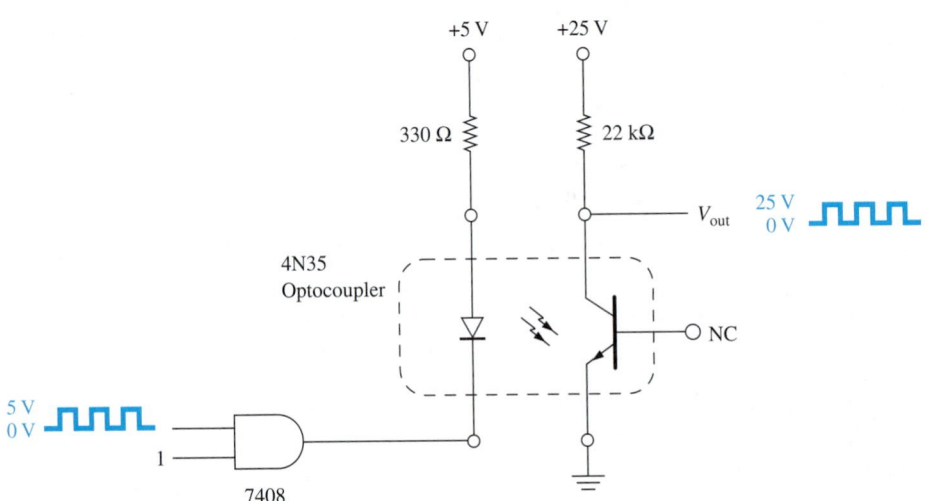

Figure 11–49 An optocoupler provides isolation in a level-shifting application.

A Power MOSFET Used to Drive a Relay and AC Motor

The output drive capability of digital logic severely limits the size of the load that can be connected. An LS-TTL buffer such as the 74LS244 can sink up to 24 mA, and the BiCMOS 74ABT244 can sink up to 64 mA. But, this is still far below the current requirements of some loads. A common way to boost the current capability is to use a *power MOSFET*, which is particularly well suited for these applications. This is a transistor specifically designed to have a very high input impedance to limit current draw into its gate and also be capable of passing a high current through its drain to source. (See Section 9–5 to review MOSFET operation.)

Figure 11–50 shows a circuit that could be used to drive a $\frac{1}{3}$-hp ac motor from a digital logic circuit. Because the starting current of a motor can be extremely high, we will use a relay with a 24-V dc coil and a 50-A contact rating. A relay of this size may require as much as 200 mA to energize the coil to pull in the contacts. A MOSFET such as the IRF130 can pass up to 12 A through its drain to source, so it can easily handle this relay coil requirement.

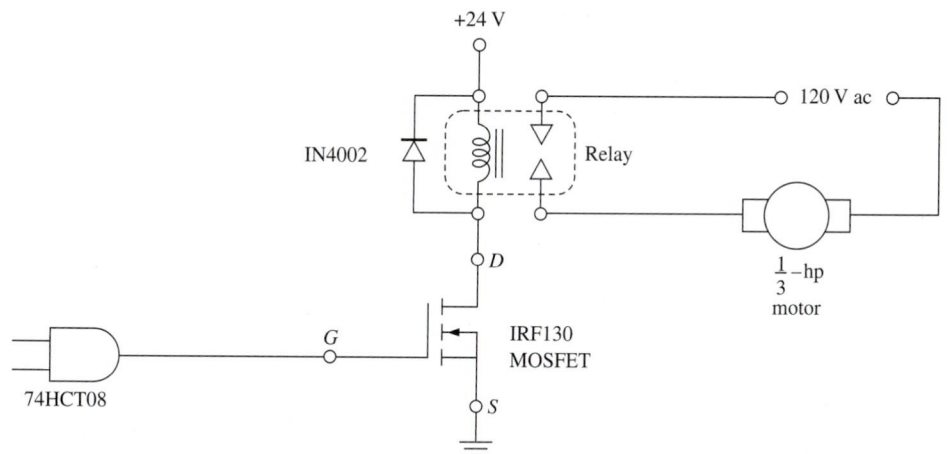

Team Discussion

The relay and the optocoupler are common means used to interface digital logic to the outside world. List several devices that might be driven from a digital or microprocessor circuit.

Figure 11–50   Using a power MOSFET to interface digital logic to high-power ac circuitry.

When the 74HCT08 outputs a HIGH (5 V) to the gate of the MOSFET, the drain to source becomes a short ($\approx 0.2\Omega$). This allows current to flow through the relay coil, creating the magnetic flux required to pull in the contacts. The motor will start. The 1N4002 diode provides arc protection across the coil when it is deenergized. (See Section 2–6 for a review of relays.)

Level Detecting with an Analog Comparator

Analog comparators such as the LM339 are commonly used to interface to digital circuitry. A comparison of the two analog voltages at the comparator's input are used to determine the device's output logic level (1 or 0). If the analog voltage at the $(+)$ input is greater than the voltage at the $(-)$ input, then the output is a logic 1. Otherwise, the output is a logic 0. The output of the LM339 acts like an open-collector gate, so a pull-up resistor is required to make the output HIGH.

The circuit in Figure 11–51 is used to detect when the temperature of a furnace exceeds 100°C. The LM35 is a temperature sensor used to monitor the furnace temperature. It outputs 10 mV for each degree Celcius. (For example, if it is 20°C, it will output 200 mV.)

This circuit is set up to sound an alarm when the temperature exceeds 100°C. The 10-k potentiometer can be set at any value as a reference for comparison. In this case, we want to set it at 1.00 V. When the temperature exceeds 100°C, the $(+)$ input becomes greater than the $(-)$, and the LM339 outputs a HIGH. This HIGH will find its way to the buzzer as long as the Enable switch is in the UP position. Piezo alarm

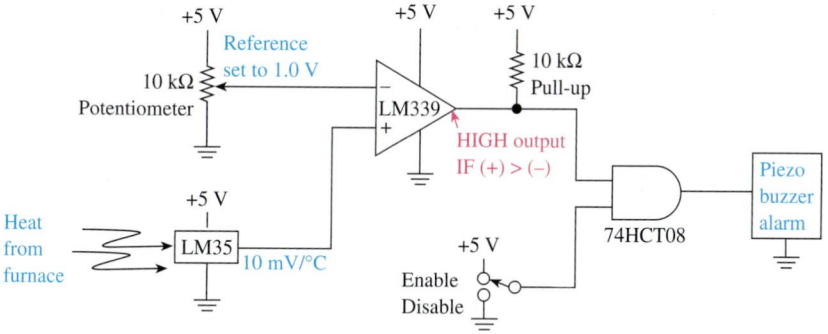

Figure 11–51   Using an LM339 analog comparator to interface to digital logic.

buzzers are noted for their very small current draw and can easily be driven by the digital logic gate.

Many other types of sensors could be monitored by the comparator instead of the temperature sensor. Sensors that output levels in the 0- to 5-V range are available for monitoring such quantities as pressure, velocity, light intensity, and displacement. The reference level set by the potentiometer allows you to select exactly what level triggers the alarm.

Review Questions

11–10. What is the cause of switch bounce, and why is it harmful in digital circuits?

11–11. What size resistor is better suited for a pull-up resistor, 10 kΩ or 100 Ω?

11–12. What 78XX series IC voltage regulator could be used to build an inexpensive 12-V power supply?

11–13. The zener diode serves two purposes in the pulse generator design in Figure 11–44. What are those purposes?

11–14. Why are LEDs usually connected as active-LOW indicator lights instead of active HIGH?

11–15. Is the phototransistor alarm in Figure 11–48 better suited for a home burglar alarm or as an alarm to announce when a customer has entered a store? Why?

11–16. How would the operation of the alarm in Figure 11–48 change if only one Schmitt inverter was used instead of two?

11–17. Why is an optocoupler sometimes referred to as an optoisolator?

11–18. In Figure 11–50, why can't you drive the relay directly with 74HCT08 instead of using the MOSFET?

Summary

In this chapter, we have learned that

1. Unpredictable results on IC logic can occur if strict timing requirements are not met.

2. A setup time is required to ensure that the input data to a logic circuit is present some definite time before the active clock edge.

3. A hold time is required to ensure that the input data to a logic circuit is held for some definite time after the active clock edge.

4. The propagation delay is the length of time it takes for the output of a logic circuit to respond to an input stimulus.

5. Delay gates are available to purposely introduce time delays when required.

6. The charging voltage on a capacitor in a series RC circuit can be used to create a short delay for a power-up reset.

7. The two key features of Schmitt trigger ICs are that they output extremely sharp edges and that they have two distinct input threshold voltages. The difference between the threshold voltages is called the hysteresis voltage.

8. Mechanical switches exhibit a phenomenon called switch bounce, which can cause problems in most kinds of logic circuits.

9. Pull-up resistors are required to make a normally floating input act like a HIGH. Pull-down resistors are required to make a normally floating input act like a LOW.

10. A practical, inexpensive 5-V power supply can be made with just a transformer, four diodes, some capacitors, and a voltage regulator.

11. A 60-pulse-per-second clock oscillator can be made using the power supply's transformer and a few additional components.

12. The resistance from collector to emitter of a phototransistor changes from approximately 10 MΩ down to approximately 1000 Ω when light shines on its base region.

13. An optocoupler provides electrical isolation from one part of a circuit to another.

14. Power MOSFETs are commonly used to increase the output drive capability of IC logic from less than 100 mA to more than 1A.

Glossary

Active Clock Edge: A clock edge is the point in time where the waveform is changing from HIGH to LOW (negative edge) or LOW to HIGH (positive edge). The *active* clock edge is the edge (either positive or negative) used to trigger a synchronous device to accept input digital states.

AC Waveforms: Test waveforms that are supplied by IC manufacturers for design engineers to determine timing sequence and measurement points for such quantities as setup, hold, and propagation delay times.

Automatic Reset: A scheme used to automatically Set or Reset all storage ICs (usually flip-flops) to a Set or Reset condition when power is first applied to them so that their starting condition can always be determined.

Duty Cycle: The ratio of the length of time a periodic wave is HIGH versus the total period of the wave.

Float: A condition in which an input or output line in a circuit is neither HIGH nor LOW because it is not directly connected to a high or low voltage level.

Hold Time: The length of time *after* the active clock edge that the input data to be recognized (usually *J* and *K*) must be held stable to ensure recognition.

Hysteresis: In digital Schmitt trigger ICs, hysteresis is the difference in voltage between the positive-going switching threshold and the negative-going switching threshold at the input.

Jitter: A term used in digital electronics to describe a waveform that has some degree of electrical noise on it, causing it to rise and fall slightly between and during level transitions.

Metastable: A logic-level transition that becomes neither HIGH nor LOW followed by a valid but undetermined state. It occurs briefly on the output of sequential logic circuits like flip-flops when certain input timing specifications are not met.

Optocoupler: A device having an LED and a phototransistor encased in the same package. Illuminating the LED turns the transistor on, providing optical coupling and isolation between two circuits.

Phototransistor: An optically sensitive transistor that is turned on when light strikes its base region.

Positive Feedback: A technique employed by Schmitt triggers that involves taking a small sample of the output of a circuit and feeding it back into the input of the same circuit to increase its switching speed and introduce hysteresis.

Power-Up: The term used to describe the initial events or states that occur when power is first applied to an IC or digital system.

Propagation Delay: The length of time that it takes for an input level change to pass through an IC and appear as a level change at the output.

Pull-Down Resistor: A resistor with one end connected to a LOW voltage level and the other end connected to an input or output line so that, when that line is in the float condition (not HIGH or LOW), the voltage level on that line will, instead, be pulled down to a LOW state.

Pull-Up Resistor: A resistor with one end connected to a HIGH voltage level and the other end connected to an input or output line so that, when that line is in a float condition (not HIGH or LOW), the voltage level on that line will, instead, be pulled up to a HIGH state.

Race Condition: The condition that occurs when a digital input level (1 or 0) is changing states at the same time as the active clock edge of a synchronous device, making the input level at that time undetermined.

_RC_ Circuit: A simple series circuit consisting of a resistor and a capacitor used to provide time delay.

Rectifier: An electronic device used to convert an ac voltage into a dc voltage.

Ripple: A small fluctuation in the output voltage of a power supply that is the result of poor filtering and regulation.

Schmitt Trigger: A circuit used in digital electronics to provide ultrafast level transitions and introduce hysteresis for improving jittery or slowly rising waveforms.

Setup Time: The length of time _before_ the active clock edge that the input data to be recognized (usually J and K) must be held stable to ensure recognition.

SPDT: The abbreviation for single pole, double throw. A SPDT switch switches a single line to one of two possible output lines.

SPST Switch: The abbreviation for single pole, single throw. A SPST switch is used to make or break contact in a single electrical line.

Switch Bounce: An undesirable characteristic of most switches when they physically make and break contact several times (bounce) each time they are opened or closed.

Threshold: The exact voltage level at the input to a digital IC that causes it to switch states. In Schmitt trigger ICs, there are two different threshold levels: the positive-going threshold (LOW to HIGH), and the negative-going threshold (HIGH to LOW).

Transfer Function: A plot of V_{out} versus V_{in} that is used to graphically determine the operating specifications of a Schmitt trigger.

Voltage Regulator: An electronic device or circuit that is used to adjust and control a voltage to remain at a constant level.

Zener Breakdown: The voltage across the terminals of a zener diode when it is conducting current in the reverse-biased direction.

■ Problems ■

Section 11–1

11–1. Sketch the Q output waveform for a 74LS76 given the input waveforms shown in Figure P11–1 [use t_s (L) = 20 ns, t_s (H) = 20 ns, t_h (L) = 0 ns, t_h (H) = 0 ns, t_{PLH} = 0 ns, t_{PHL} = 0 ns].

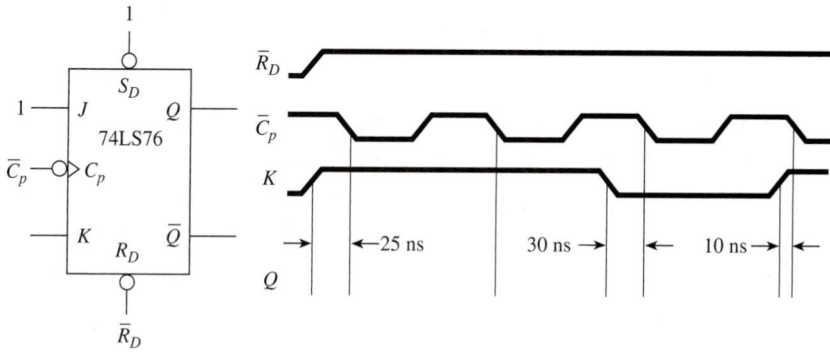

Figure P11–1

11–2. Repeat Problem 11–1 for the waveforms shown in Figure P11–2.

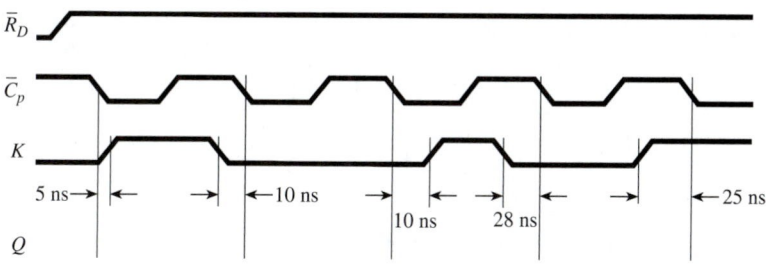

Figure P11–2

11–3. Using actual specifications for a 74LS76, label the propagation delay times on the waveforms shown in Figure P11–3.

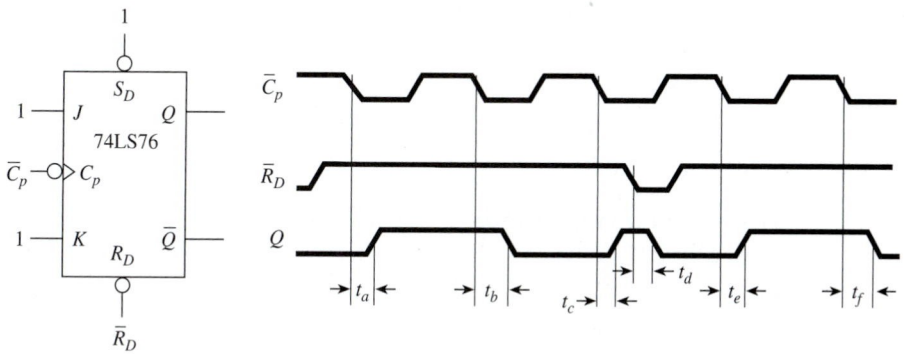

Figure P11–3

11–4. Repeat Problem 11–3 for the waveforms shown in Figure P11–4. Use specifications for a 74109 in the toggle mode ($J = 1$, $\overline{K} = 0$).

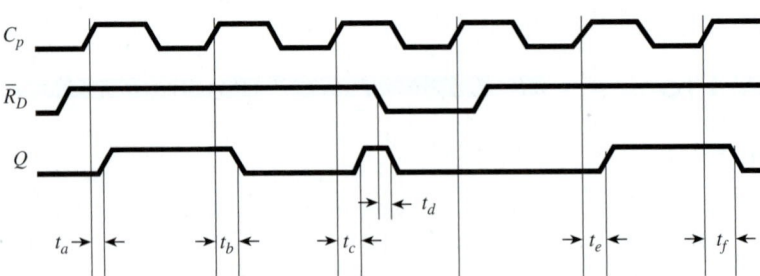

Figure P11–4

11–5. Describe the problem that may arise when using the 7432 OR gate to delay the clock signal into the flip-flop circuit of Figure 11–14.

C **11–6.** Sketch the output at $\overline{C_{pD}}$ and Q for the flip-flop circuit shown in Figure P11–6. (Ignore propagation delays in the 74LS76.)

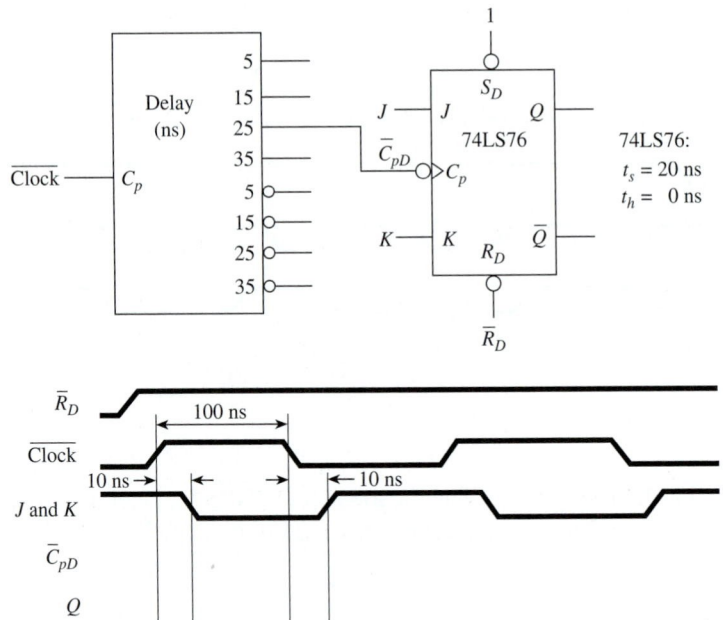

Figure P11–6

C **11–7.** Redraw the waveforms given in Problem 11–6 if the 35-ns delay tap is used instead of the 25-ns tap.

C **11–8.** (a) Sketch the output at D_D and Q for the flip-flop circuit shown in Figure P11–8. (Ignore propagation delays in the 7474.) (b) Connect D_D to the 30-ns tap, and repeat part (a). See Figure P11–8.

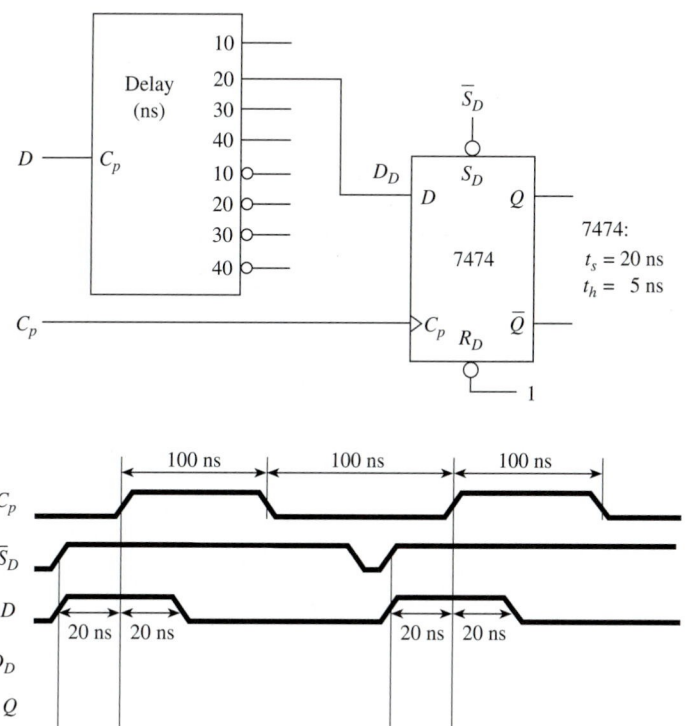

Figure P11–8

Sections 11–2 and 11–3

11–9. One particular Schmitt trigger inverter has a positive-going threshold of 1.9 V and a negative-going threshold of 0.7 V. Its V_{OH} (typical) is 3.6 V and V_{OL} (typical) is 0.2 V. Sketch the transfer function (V_{out} versus V_{in}) for this Schmitt trigger.

11–10. If the input waveform (V_{in}) shown in Figure P11–10 is fed into the Schmitt trigger described in Problem 11–9, sketch its output waveform (V_{out}).

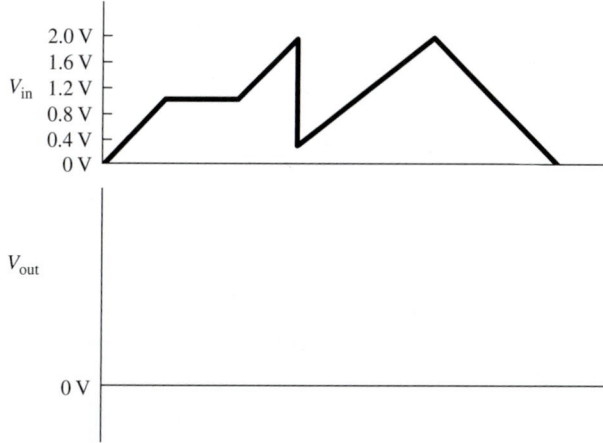

Figure P11–10

11–11. If the waveform shown in Figure P11–11 is fed into a 7414 Schmitt trigger inverter, sketch V_{out} and determine the duty cycle of V_{out}.

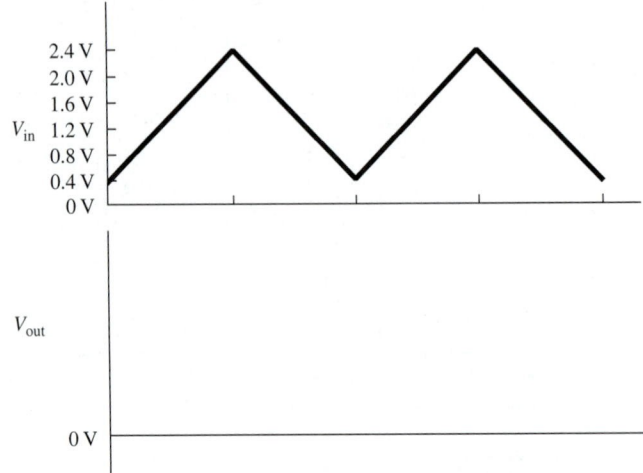

Figure P11–11

11–12. If the V_{in} and V_{out} waveforms shown in Figure P11–12 are observed on a Schmitt trigger device, determine its characteristics and sketch the transfer function (V_{out} versus V_{in}).

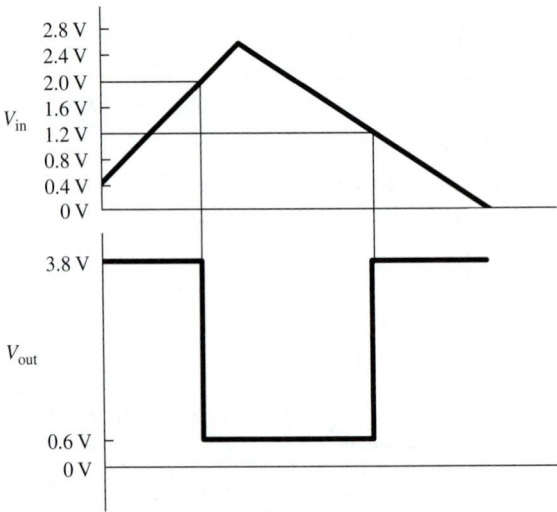

Figure P11–12

Section 11–4

T   **11–13.** The Q output of the 74LS76 in Figure 11–37 is used to drive an LED. Sometimes when the switch is closed, the LED toggles to its opposite state, but sometimes it does not. Discuss the probable cause and a solution to the problem.

Section 11–5

D   **11–14.** Occasionally, instead of using a pull-up resistor, a pull-down resistor is required because a floating connection must be held LOW, as shown in Figure P11–14. Why can't a 10-kΩ resistor be used for this purpose? Could a 100- resistor be used? How would the use of a 74HCT74 improve the situation?

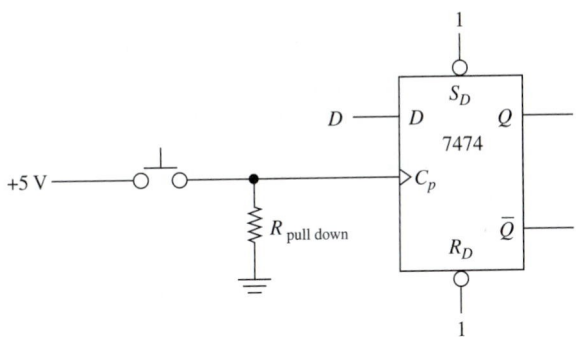

Figure P11–14

Section 11–6

T   **11–15.** A problem arises in a digital system that you have designed. Using a multimeter, you find that none of your ICs is receiving 5-V V_{CC} power. Your 5-V power supply is the one given in Figure 11–43. Outline a procedure that you would follow to troubleshoot your power supply.

D   **11–16.** Design a 60-pulse-per-second TTL-level pulse generator similar to that in Figure 11–44 using an optocoupler instead of the zener diode.

11–17. In Figure 11–48, assume that the phototransistor has an ON resistance of 1 kΩ and an OFF resistance of 1 MΩ. Determine the voltage at point A when the light is striking, and then not striking, the phototransistor.

C T   **11–18.** You are asked to troubleshoot the alarm circuit in Figure 11–48, which is not working. You find that the voltage at C_p is stuck at 0.2 V for both light and dark conditions. The voltage at point A is also stuck at approximately 1.0 V. You then take the inverters out of the circuit, test them, and determine that they are working. While the inverters are out, you notice that the voltage at point A jumped up to 4.8 V. When you shine a light on the phototransistor, it drops to 0.2 V! Looking further, you notice that a 7414 was substituted for the 74HCT14. What is the problem?

D   **11–19.** A good choice for an alarm in Figure 11–48 is a 5-V piezo buzzer. The problem is that it takes approximately 10 mA to operate the buzzer and the 7474 can only source 0.4 mA. Any ideas?

11–20. Assume that the ON and OFF resistances of the phototransistor in a 4N35 optocoupler are 1-kΩ ON, 1-MΩ OFF. Determine the actual values of V_{out} in Figure 11–49.

11–21. If the relay used in Figure 11–50 has a coil resistance of 100 Ω, determine the coil current when the MOSFET is ON. (Assume that $R_{DS(ON)} = 0.2 \ \Omega$.)

Schematic Interpretation Problems

See Appendix G for the schematic diagrams.

S **11–22.** Find the section of the watchdog timer schematic that shows U14:A, U15:A, and U14:B. Assume that pins 2 and 13 of U15:A are both HIGH and that U14:A is initially reset. Apply a positive pulse on the line labeled WATCHDOG_SEL.

(a) Discuss the possible setup time problems that may occur with U14:B.

(b) Discuss how the situation changes if pin 1 of U15:A is already HIGH and the positive pulse comes in on pin 2 instead.

S D **11–23.** On a separate piece of paper, draw the circuit connections to add a bank of eight LEDs with current-limiting resistors to the octal *D* flip-flop, U5, in the 4096/4196 schematic.

S C D **11–24.** On a separate piece of paper draw the connections to input the following values to port PA of the 68HC11 microcontroller in the HC11D0 master board schematic.

(a) Monitor light/no light conditions by using a light-sensitive phototransistor connected to PA1.

(b) Interface the 0-V/15-V (LOW/HIGH) levels from a 4050B CMOS buffer to PA3 via an optocoupler.

S C **11–25.** S2 in grid location B-1 in the HC11D0 schematic is a set of seven 10-kΩ pull-up resistors contained in a single DIP. They all have a common connection to V_{CC}, as shown. Explain their purpose as they relate to the U12 DIP-switch package and the MODA, MODB inputs to the 68HC11 microcontroller.

S D **11–26.** On a separate piece of paper, add the circuitry to provide +5 V on sheet 2 of the 4096/4196 schematic. (Tap off of the +UNREG signal provided.)

Electronics Workbench®/MultiSIM® Exercises

E11–1. Load the circuit file for **Section 11–3a.** The subcircuit labeled "Schmitt1" is a Schmitt trigger device similar to Figure 11–29 with unknown characteristics. Varying the potentiometer R will increase and decrease Vin so that you can determine the switching points Vt+ and Vt−. Determine the switching thresholds so that you can sketch and label the voltage levels on a transfer function similar to Figure 11–28.

E11–2. Load the circuit file for **Section 11–3b.** The subcircuit labeled "Schmitt1" is a Schmitt trigger device similar to Example 11–11 with unknown characteristics. The Function Generator is used to provide a varying Vin as the Oscilloscope monitors Vin and Vout.

(a) From the expanded oscilloscope display, determine Vt+, Vt−, Voh, and Vol, and sketch a transfer function.

(b) Press the B/A button in the Time Base Section to display ChB versus ChA. This is the transfer function. Does it match the results from part (a)?

C **E11–3.** Load the circuit file for **Section 11–3c.** The subcircuit labeled "Schmitt1" is a Schmitt trigger device similar to Example 11–11 with unknown characteristics.

(a) Connect a varying Voltage into Vin and measure Vin, Vout so that you can sketch a transfer function for the device. What are Vt+, Vt−, Voh, and Vol?

(b) Replace the varying voltage source with a triangle wave from a function generator. Use an oscilloscope to display the Vin, Vout waveforms and transfer function. Do the values match? Show your instructor.

Answers to Review Questions

11–1. False

11–2. False

11–3. It means that the input levels don't have to be held beyond the active clock edge.

11–4. Propagation delay, output

11–5. To enable proper setup and hold times

11–6. Capacitor

11–7. Positive-going threshold, negative-going threshold, hysteresis

11–8. The switching threshold on a positive-going input signal is at a higher level than the switching threshold on a negative-going input signal. This is called hysteresis. The output is steady as long as the input noise does not exceed the hysteresis voltage.

11–9. Input, output

11–10. It is caused by the springing action of the contacts, and it can cause false triggering of a digital circuit.

11–11. 10 kΩ

11–12. 7812

11–13. It cuts off the negative cycle of the sine wave and limits the positive cycle to 4.3 V.

11–14. Because the ICs to which they are connected can sink more current than they can source

11–15. Home alarm, because the alarm is latched in the ON state until the flip-flop is manually reset

11–16. The clock signal would be normally HIGH and drop LOW when interrupted, and the flip-flop would latch when the signal returned HIGH.

11–17. The output side is electrically isolated from the input side.

11–18. The current capability of the output is too low to trigger the relay.

12

Counter Circuits and VHDL State Machines

OBJECTIVES

Upon completion of this chapter, you should be able to:

- Use timing diagrams for the analysis of sequential logic circuits.
- Design any modulus ripple counter and frequency divider using *J-K* flip-flops.
- Describe the difference between ripple counters and synchronous counters.
- Solve various counter design applications using 4-bit counter ICs and external gating.
- Connect seven-segment LEDs and BCD decoders to form multidigit numeric displays.
- Cascade counter ICs to provide for higher counting and frequency division.

INTRODUCTION

Now that we understand the operation of flip-flops and latches, we can apply our knowledge to the design and application of sequential logic circuits. One common application of sequential logic arrives from the need to count events and time the duration of various processes. These applications are called **sequential** because they follow a predetermined sequence of digital states and are triggered by a timing pulse or clock.

     To be useful in digital circuitry and microprocessor systems, counters normally count in binary and can be made to stop or recycle to the beginning at any time. In a recycling counter, the number of different binary states defines the **modulus** (MOD) of the counter. For example, a counter that counts from 0 to 7 is called a MOD-8 counter. For a counter to count from 0 to 7, it must have three binary outputs and one clock trigger input, as shown in Figure 12–1.

Normally, each binary output will come from the Q output of a flip-flop. Flip-flops are used because they can hold, or remember, a binary state until the next clock or trigger pulse comes along. The count sequence of a 0 to 7 binary counter is shown in Table 12–1 and Figure 12–2.

Before studying counter circuits, let's analyze some circuits containing logic gates with flip-flops to get a feeling for the analytical process involved in determining the output waveforms of sequential circuits.

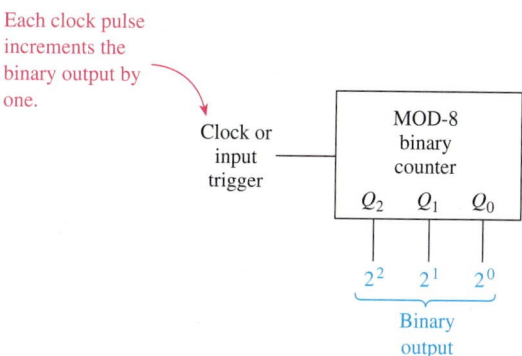

Figure 12–1   Simplified block diagram of a MOD-8 binary counter.

TABLE 12–1	Binary Count Sequence of a MOD-8 Binary Counter		
Q_2	Q_1	Q_0	Count
0	0	0	0
0	0	1	1
0	1	0	2
0	1	1	3
1	0	0	4
1	0	1	5
1	1	0	6
1	1	1	7
0	0	0	0
0	0	1	1
0	1	0	2
0	1	1	3
	and so on		

Eight different binary states

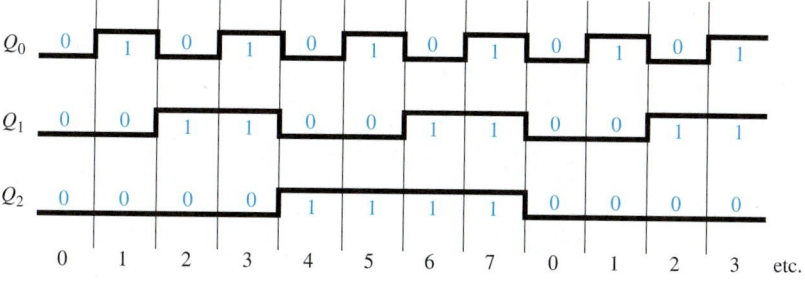

Figure 12–2   Waveforms for a MOD-8 binary counter.

12–1 Analysis of Sequential Circuits

To get our minds thinking in terms of sequential analysis, let's look at an example that mixes combinational logic gates with flip-flops and whose operation is dictated by a specific sequence of input waveforms, as shown in Example 12–1.

EXAMPLE 12–1

The 7474 shown in Figure 12–3 is a positive edge-triggered D flip-flop. The waveforms shown in Figure 12–4 are applied to the inputs at A and C_p. Sketch the resultant waveform at D, Q, $\overline{Q}$, and X.

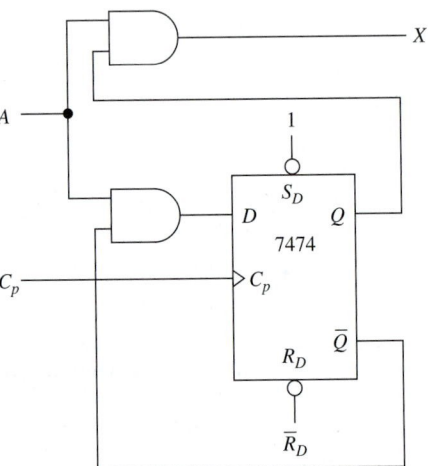

Figure 12–3

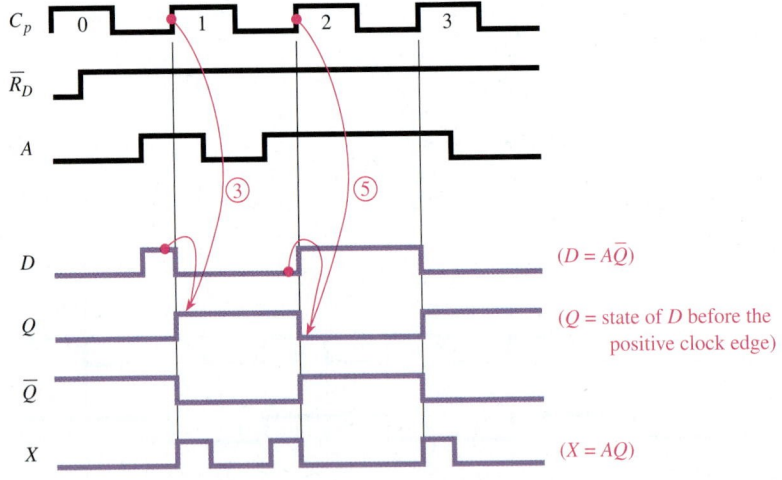

$(D = A\overline{Q})$

$(Q$ = state of D before the positive clock edge$)$

$(X = AQ)$

Figure 12–4

Solution:

1. $Q = 0, \overline{Q} = 1$ during the 0 period because of $\overline{R_D}$.
2. D is equal to $A\overline{Q}$; X is equal to AQ (therefore, the level at D and X will change whenever the inputs to the AND gates change, regardless of the state of the input clock).
3. At the positive edge of pulse 1, D is HIGH, so Q will go HIGH and $\overline{Q}$ will go LOW and remain there until the positive edge of pulse 2.
4. During period 1, D will equal $A\overline{Q}$ and X will equal AQ, as shown.
5. At the positive edge of pulse 2, D is LOW, so the flip-flop will Reset $(Q = 0; \overline{Q} = 1)$ and remain there until the positive edge of pulse 3.
6. At the positive edge of pulse 3, D is HIGH, so the flip-flop will Set $(Q = 1; \overline{Q} = 0)$.

The timing analysis in Examples 12–1 and 12–2 was done by observing the level on D before the positive clock edge and realizing that D follows the level of $A\overline{Q}$.

EXAMPLE 12–2

Using the same circuit of Example 12–1, sketch the waveforms at D, Q, $\overline{Q}$, and X, given the input waves shown in Figure 12–5.

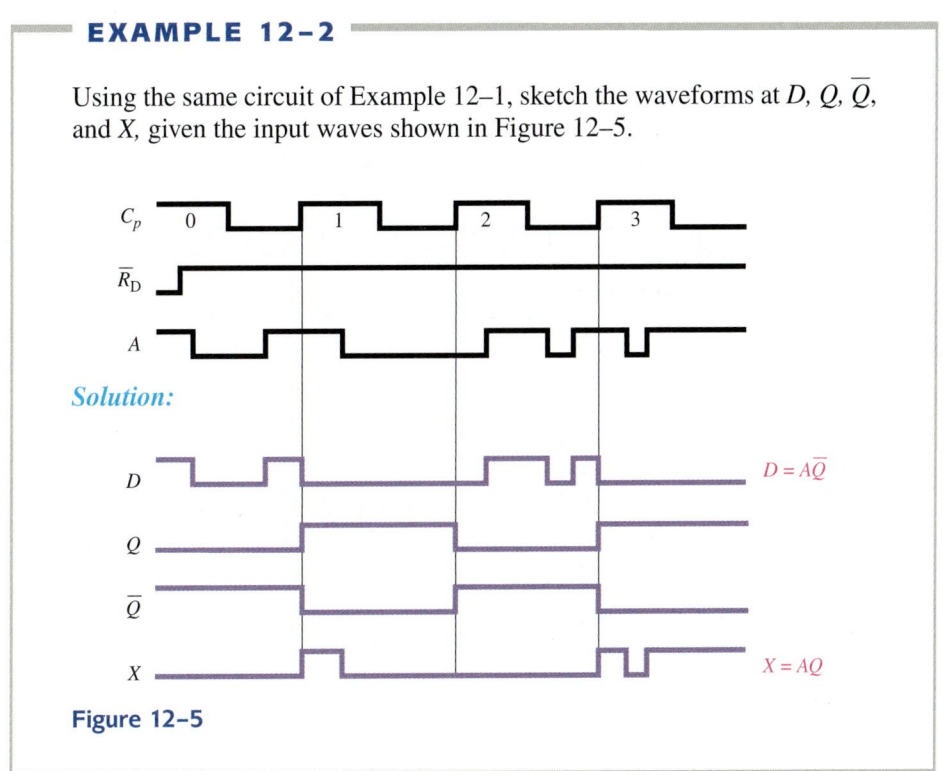

Solution:

Figure 12–5

When a *J-K* flip-flop is used, we have to consider the level at J and K at the active clock edge as well as any asynchronous operations that may be taking place. Examples 12–3 and 12–4 illustrate the timing analysis of sequential circuits utilizing *J-K* flip-flops.

EXAMPLE 12-3

The 74ALS112 shown in Figure 12–6 is a negative edge-triggered flip-flop. The waveforms in Figure 12–7 are applied to the inputs at A and $\overline{C}_{p0}$. Sketch the resultant waveforms at J_1, K_1, Q_0, and Q_1. Notice that the clock input to the second flip-flop comes from Q_0. Also, $J_1 = AQ_1$, $K_1 = AQ_0$.

Common Misconception

Students often think that *both* flip-flops are triggered from $\overline{C}_{p0}$.

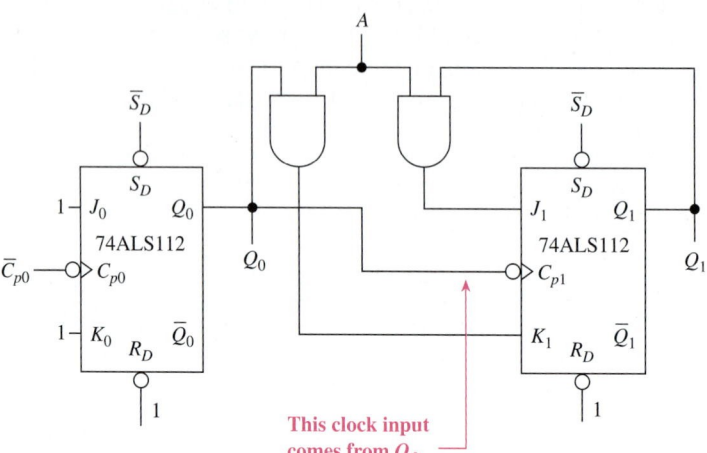

Figure 12-6

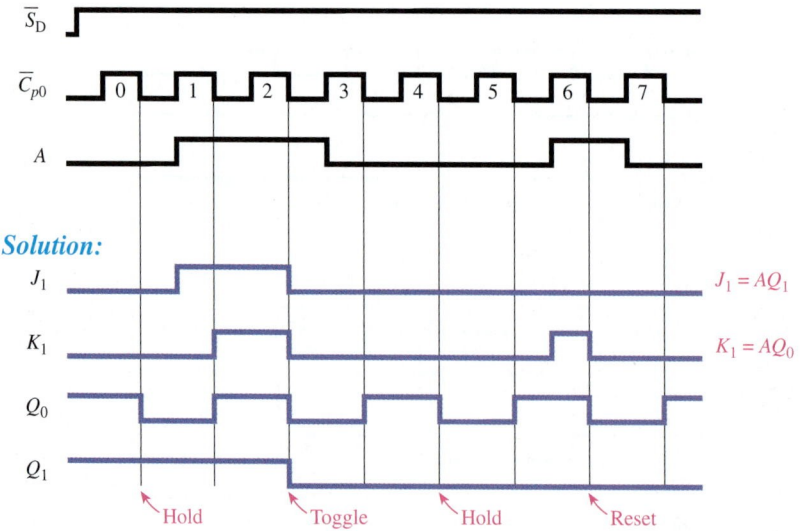

Figure 12-7

1. Because $J_0 = 1$ and $K_0 = 1$, then Q_0 will *toggle* at each negative edge of $\overline{C}_{p0}$.

2. The second flip-flop will be triggered at each negative edge of the Q_0 line.

3. The levels at J_1 and K_1 just before the negative edge of the Q_0 line will determine the synchronous operation of the second flip-flop.

4. After Q_0 and Q_1 are determined for each period, the new levels for J_1 and K_1 can be determined from $J_1 = AQ_1$ and $K_1 = AQ_0$.

EXAMPLE 12–4

Repeat Example 12–3 for the waveforms shown in Figure 12–8.

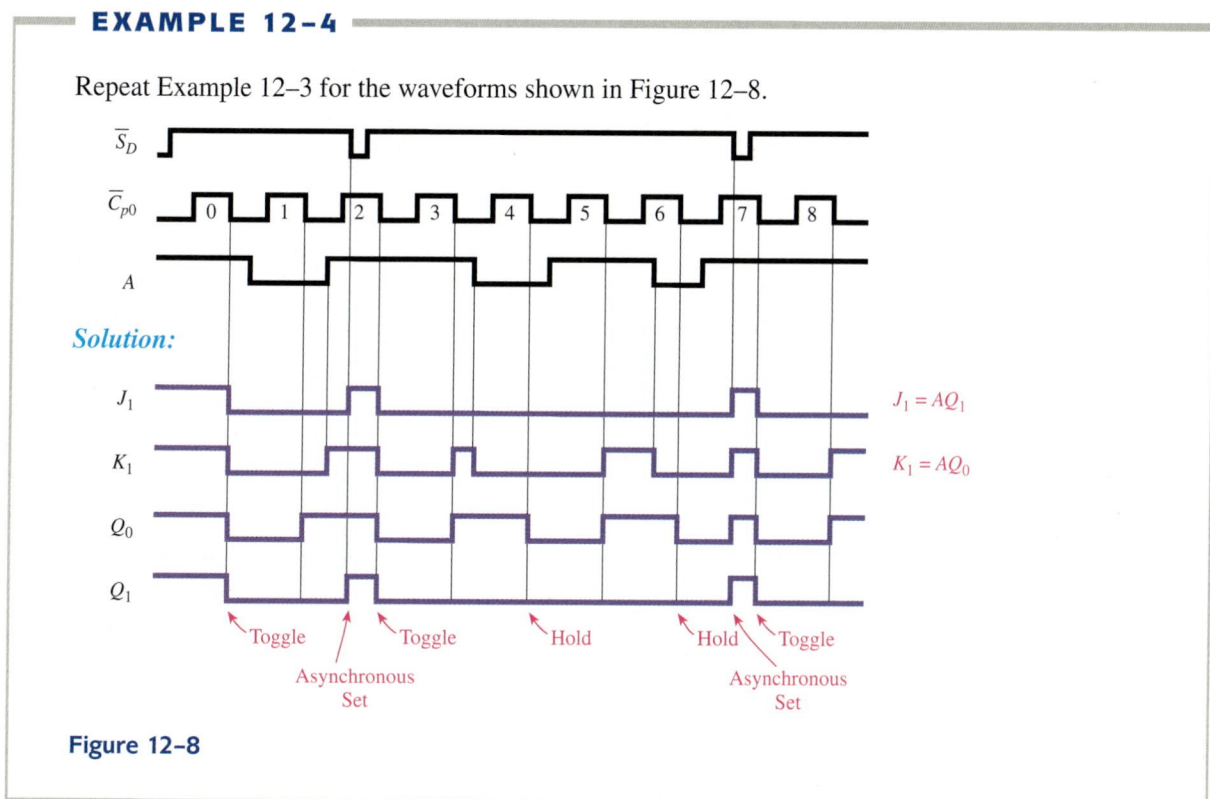

Figure 12–8

12–2 Ripple Counters: JK FFs and VHDL Description

Flip-flops can be used to form binary counters. The counter output waveforms discussed in the beginning of this chapter (Figure 12–2) could be generated by using three flip-flops cascaded together (**cascade** means to connect the Q output of one flip-flop to the clock input of the next). Three flip-flops are needed to form a 3-bit counter (each flip-flop will represent a different power of 2: 2^2, 2^1, 2^0). With three flip-flops, we can produce 2^3 different combinations of binary outputs ($2^3 = 8$). The eight different binary outputs from a 3-bit binary counter will be 000, 001, 010, 011, 100, 101, 110, and 111.

If we have a 4-bit binary counter, we would count from 0000 up to 1111, which is 16 different binary outputs. As it turns out, we can determine the number of different binary output states (modulus) by using the following formula:

$$\text{modulus} = 2^N, \qquad \text{where } N = \text{number of flip-flops}$$

To form a 3-bit binary counter, we cascade three *J-K* flip-flops, each operating in the *toggle mode,* as shown in Figure 12–9. The clock input used to increment the binary count comes into the $\overline{C_p}$ input of the first flip-flop. Each flip-flop will toggle every time its clock input receives a HIGH-to-LOW edge.

Now, with the knowledge that we have gained by analyzing the sequential circuits in Section 12–1, it should be easy to determine the output waveforms of the 3-bit binary **ripple counter** of Figure 12–9.

When we analyze the circuit and waveforms, we see that Q_0 toggles at each negative edge of $\overline{C_{p0}}$, Q_1 toggles at each negative edge of Q_0, and Q_2 toggles at each negative edge of Q_1. The result is that the outputs will "count" repeatedly from 000 up to 111 and then 000 to 111, as shown in Figure 12–10(a). The term *ripple* is derived from the fact that the input clock trigger is not connected to each flip-flop directly but, instead, has to propagate down through each flip-flop to reach the next.

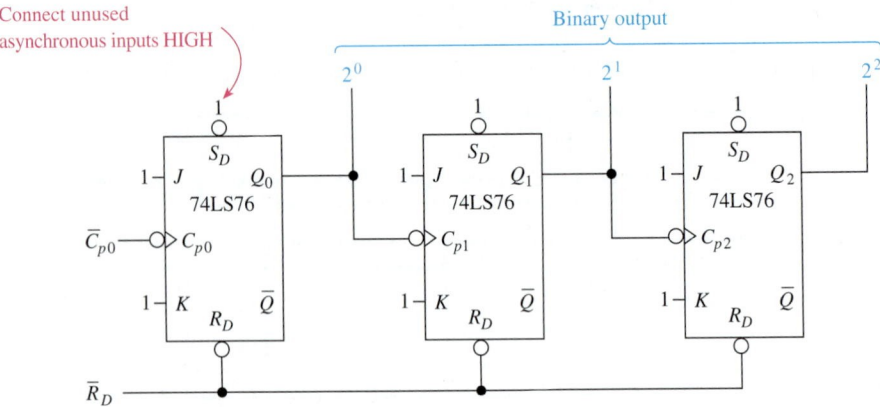

Figure 12–9 Three-bit binary ripple counter.

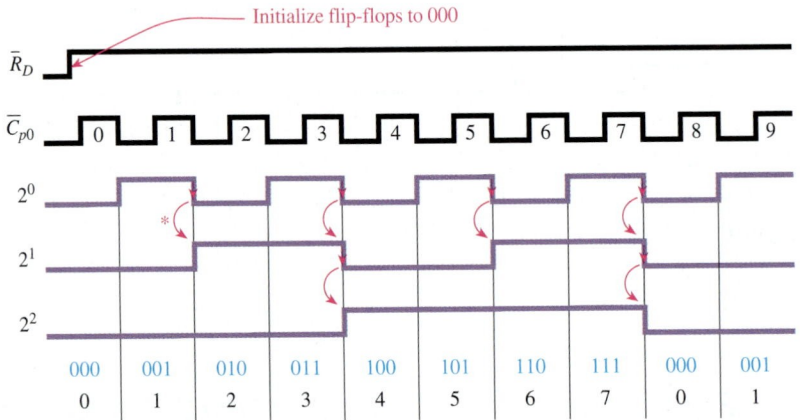

Figure 12–10 The 3-bit binary ripple counter: (a) waveforms; (b) state diagram.

For example, look at clock pulse 7. The negative edge of $\overline{C_{p0}}$ causes Q_0 to toggle LOW . . . which causes Q_1 to toggle LOW . . . which causes Q_2 to toggle LOW. There will definitely be a propagation delay between the time that $\overline{C_{p0}}$ goes LOW until Q_2 finally goes LOW. Because of this delay, ripple counters are called **asynchronous counters,** meaning that each flip-flop is not triggered at exactly the same time.

Another way to observe the count sequence is by using a **state diagram,** as shown in Figure 12–10(b). This drawing shows the Q output levels after each negative transition of $\overline{C_p}$.

Synchronous counters can be formed by driving each flip-flop's clock by the same clock input. Synchronous counters are more complicated, however, and will be covered after we have a thorough understanding of asynchronous ripple counters.

The propagation delay inherent in ripple counters places limitations on the maximum frequency allowed by the input trigger clock. The reason is that if the input clock has an active trigger edge before the previous trigger edge has propagated through all the flip-flops, you will get an erroneous binary output.

Let's look at the 3-bit counter waveforms in more detail, now taking into account the propagation delays of the 74LS76 flip-flops. In reality, the 2^0 waveform will be delayed to the right **(skewed)** by the propagation of the first flip-flop. The 2^1 waveform will be skewed to the right from the 2^0 waveform, and the 2^2 waveform will be skewed to the right from the 2^1 waveform. This is a cumulative effect that causes the 2^2 waveform to be skewed to the right of the original $\overline{C_{p0}}$ waveform by three propagation delays. [Remember, however, that the propagation delay, even for slow flip-flops like the 74LS76, is in the 20-ns range, which will not hurt us until the input clock period is very short, 100 to 200 ns (5 to 10 MHz).] Figure 12–11 illustrates the effect of propagation delay on the output waveform.

From Figure 12–11, we can see that the length of time that it takes to change from binary 011 to 100 (3 to 4) will be

$$t_{PHL1} + t_{PHL2} + t_{PLH3} = 30 \text{ ns} + 30 \text{ ns} + 20 \text{ ns} = 80 \text{ ns}$$

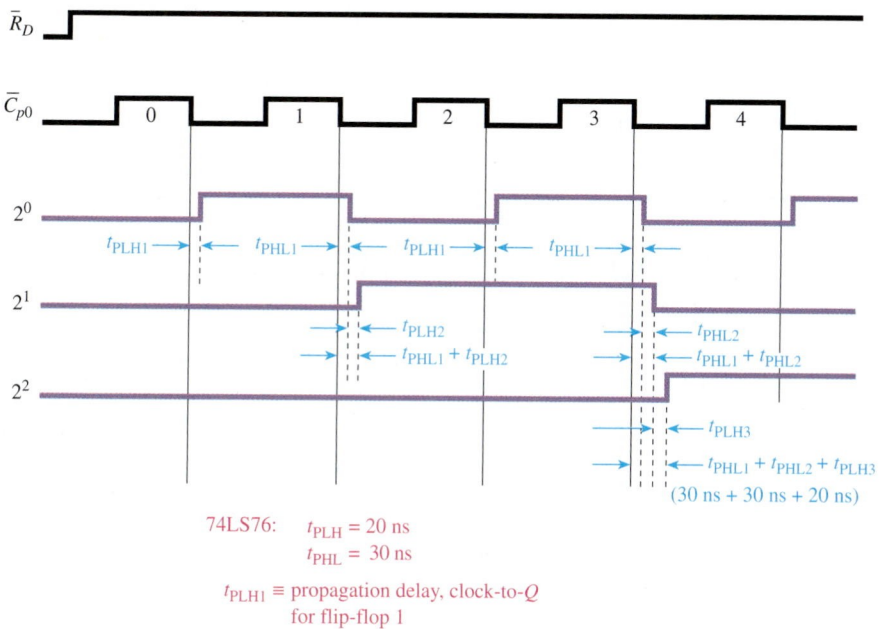

Team Discussion

For 80 ns during period 4, the output is invalid. How does this affect the minimum clock period that can be used?

Figure 12–11 Effect of propagation delay on ripple counter outputs.

As we cascade more and more flip-flops to form higher-modulus counters, the cumulative effect of the propagation delay becomes more of a problem. The approximate maximum frequency (f_{max}) of a ripple counter due to the accumulation of propagation delays can be determined using the following formula:

$$f_{max} = \frac{1}{N \times t_p}$$

where N = number of flip-flops, and

t_p = average propagation delay of each flip-flop (C_p-to-Q)

A MOD-16 ripple counter can be built using four ($2^4 = 16$) flip-flops. Figures 12–12 and 12–13 show the circuit design and waveforms for a MOD-16 ripple counter. From the waveforms, we can see that the 2^1 line toggles at every negative edge of the 2^0 line, the 2^2 line toggles at every negative edge of the 2^1 line, and so on, down through each successive flip-flop. When the count reaches 15 (1111), the next negative edge of $\overline{C_{p0}}$ causes all four flip-flops to toggle and changes the count to 0 (0000). Figure 12–13(b) is a photograph of the MOD-16 waveforms displayed on an eight-trace logic analyzer.

Down-Counters

On occasion, there is a need to count down in binary instead of counting up. To form a down-counter, simply take the binary outputs from the $\overline{Q}$ outputs instead of the Q outputs, as shown in Figure 12–14. The down-counter waveforms are shown in Figure 12–15.

Team Discussion

On paper, connect a four-input NOR gate to the 2^0–2^3 outputs in Figure 12–12. Sketch the output of the NOR gate, including *glitches* (short-duration error pulses) that occur due to propagation delay.

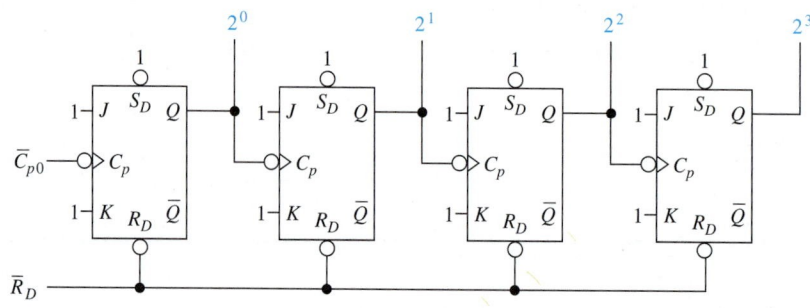

Figure 12–12 MOD-16 ripple counter.

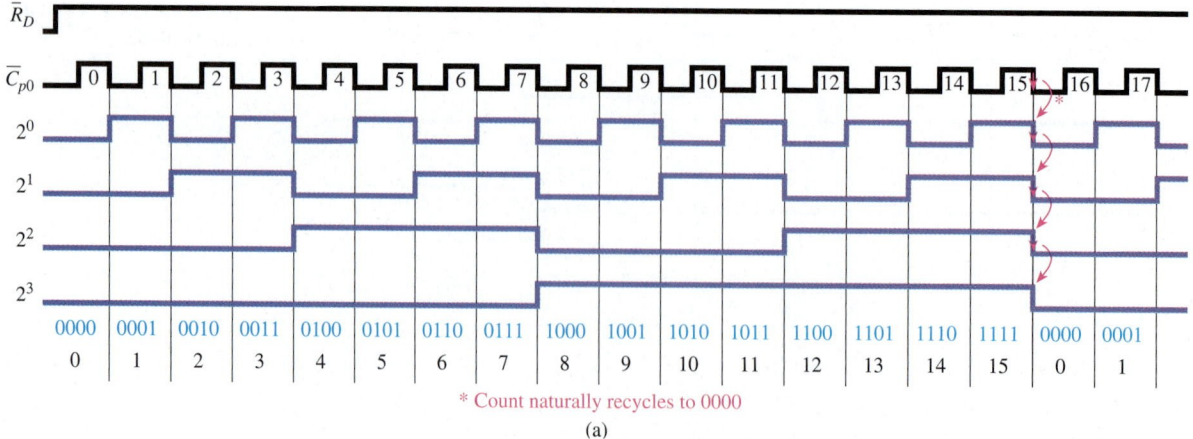

* Count naturally recycles to 0000

(a)

Figure 12–13 MOD-16 ripple counter waveforms: (a) theoretical counter waveforms;

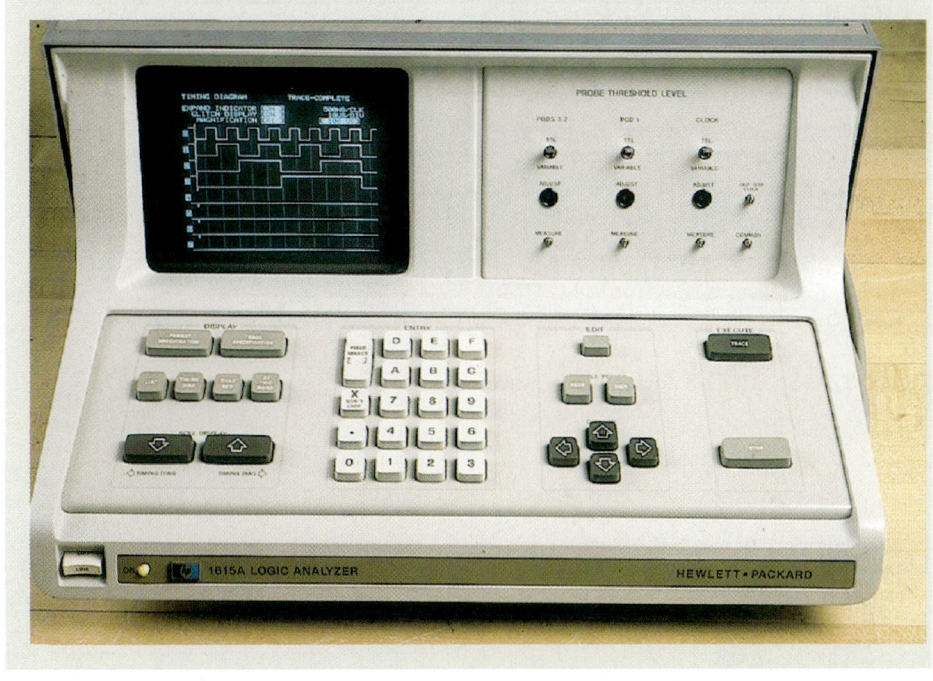

(b)

Figure 12–13 *(Continued)* (b) actual counter output display on a logic analyzer.

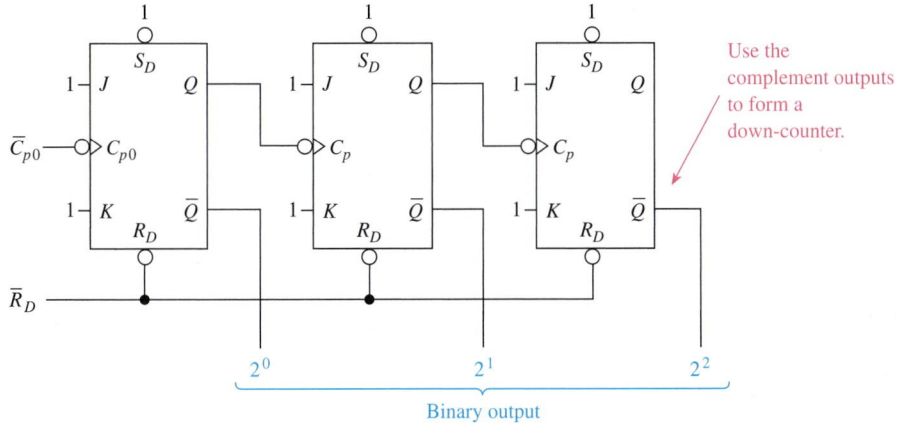

Use the complement outputs to form a down-counter.

Binary output

Figure 12–14 MOD-8 ripple down-counter.

When you compare the waveforms of the up-counter of Figure 12–10 to the down-counter of Figure 12–15, you can see that they are exact complements of each other. That is easy to understand because the binary output is taken from $\overline{Q}$ instead of Q.

VHDL Description of a Mod-16 Up Counter

Four-bit MOD-16 counters can be implemented in CPLDs using a VHDL description as shown in Figure 12–16. The inputs used here (n_cp and n_rd) are the same as Figure 12–12 ($\overline{C_p}$ and $\overline{R_D}$). The q output is declared as a BUFFER instead of an OUTPUT because q is used in an assignment statement ($q <= q + 1;$) where is it treated as an input on the right

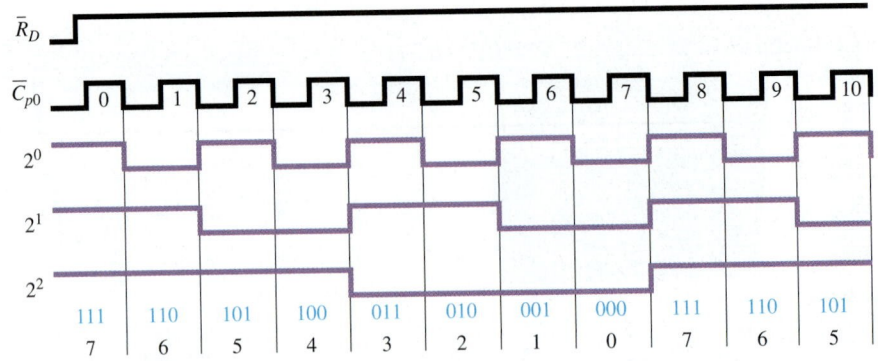

Figure 12–15 MOD-8 down-counter waveforms.

```
mod16up.vhd - Text Editor                                          _ □ X

    LIBRARY ieee;                        ---------------------------------
    USE ieee.std_logic_1164.ALL;         -- Mod-16 Up-Counter          --
                                          ---------------------------------
    ENTITY mod16up IS                                        4-bit output
        PORT(n_cp, n_rd        : IN     std_logic;
                q              : BUFFER integer RANGE 0 TO 15);
    END mod16up;
                                                    Input and output

    ARCHITECTURE arc OF mod16up IS
    BEGIN
        PROCESS (n_cp, n_rd)
        BEGIN
            IF  (n_rd='0') THEN              Asyncronous Reset has priority
                q <= 0;
                ELSIF (n_cp'EVENT AND n_cp='0') THEN
                    q <=q+1;                         Negative clock edge
            END IF;                   Input to equation
        END PROCESS;                  Output to CPLD pins
    END arc;
                                                        mod16up.vhd

Line  1    Col  1    INS ◄
```

Figure 12–16 VHDL listing for the MOD-16 up counter.

side and an output on the left side. Since the integer RANGE is 0 to 15, *q* will be assigned 4 output ports: *q0, q1, q2,* and *q3.*

The IF statement first checks to see if *n_rd* is LOW. If it is, *q* is set to 0 and control passes to the END IF statement. Else, if there is a negative clock edge, *q* is incremented by 1.

Figure 12–17 shows the results of a simulation of the synthesized program. After the initial reset, it counts 0 to 15 just as the MOD-16 constructed from four *J-K* flip-flops did.

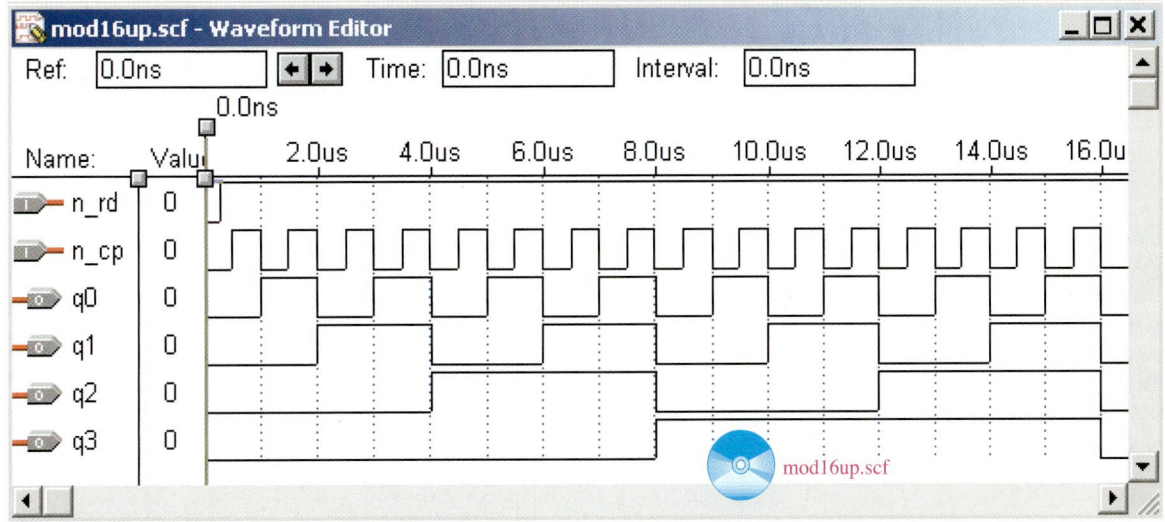

Figure 12–17 Simulation of the Mod-16 up counter.

Review Questions

12–1. When analyzing digital circuits containing basic gates combined with sequential logic like flip-flops, you must remember that gate outputs can change at any time, whereas sequential logic only changes at the active clock edges. True or false?

12–2. For a binary ripple counter to function properly, all J and K inputs must be tied _____ (HIGH, LOW), and all $\overline{S_D}$ and $\overline{R_D}$ inputs must be tied _____ (HIGH, LOW) to count.

12–3. What effect does propagation delay have on ripple counter outputs?

12–4. How can a ripple up-counter be converted to a down-counter?

12–3 Design of Divide-by-*N* Counters

Counter circuits are also used as frequency dividers to reduce the frequency of periodic waveforms. For example, if we study the waveforms generated by the MOD-8 counter of Figure 12–10, we can see that the frequency of the 2^2 output line is one-eighth of the frequency of the $\overline{C_{p0}}$ input clock line. This concept is illustrated in the block diagram of Figure 12–18, assuming that the input frequency is 24 kHz. So, as it turns out, a MOD-8 counter can be used as a divide-by-8 frequency divider, and a MOD-16 can be used as a divide-by-16 frequency divider. Notice that the duty cycle of each of the outputs in Figures 12–10 and 12–13 is 50%.

What if we need a divide-by-5 (MOD-5) counter? We can modify the MOD-8 counter so that when it reaches the number 5 (101) all flip-flops will be Reset. The new count sequence will be 0–1–2–3–4–0–1–2–3–4–0–, and so on. To get the counter to Reset at number 5 (binary 101), you will have to monitor the 2^0 and 2^2 lines and, when they are both HIGH, put out a LOW Reset pulse to all flip-flops. Figure 12–19 shows a circuit that can do this for us.

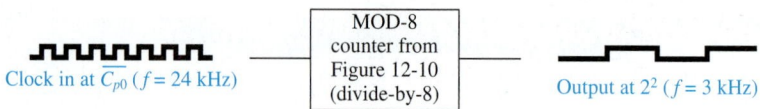

Figure 12–18   Block diagram of a divide-by-8 counter.

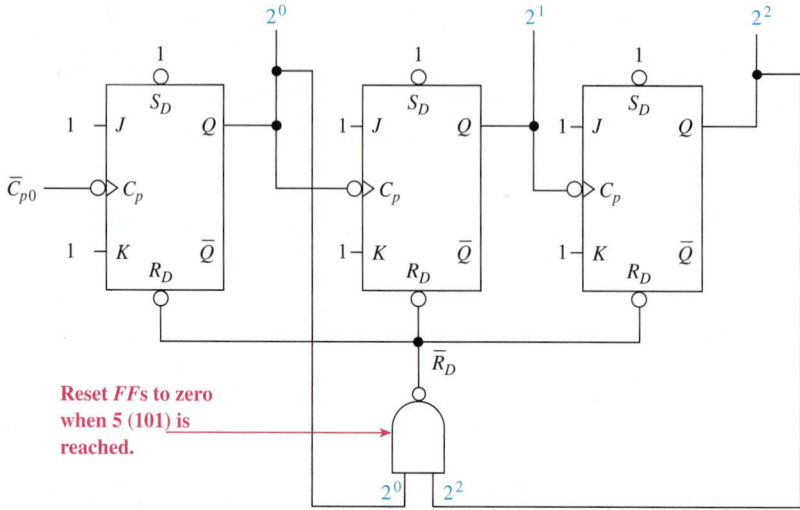

Reset *FF*s to zero when 5 (101) is reached.

Figure 12–19   Connections to form a divide-by-5 (MOD-5) binary counter.

As you can see, the inputs to the NAND gate are connected to the 2^0 and 2^2 lines, so when the number 5 (101) comes up, the NAND puts out a LOW level to Reset all flip-flops. The waveforms in Figure 12–20(a) and the state diagram of Figure 12–20(b) illustrate the operation of the MOD-5 counter of Figure 12–19.

As we can see in Figure 12–20(a), the number 5 will appear at the outputs for a short duration, just long enough to Reset the flip-flops. The resulting short pulse on the 2^0 line is called a **glitch.** Do you think you could determine how long the glitch is? (Assume that the flip-flop is a 74LS76 and the NAND gate is a 7400.)

Because t_{PHL} of the NAND gate is 15 ns, it takes that long just to drive the $\overline{R_D}$ inputs LOW. But, then it also takes 30 ns (t_{PHL}) for the LOW on $\overline{R_D}$ to Reset the Q output to LOW. Therefore, the total length of the glitch is 45 ns. If the input clock period is in the microsecond range, then 45 ns is insignificant, but at extremely high

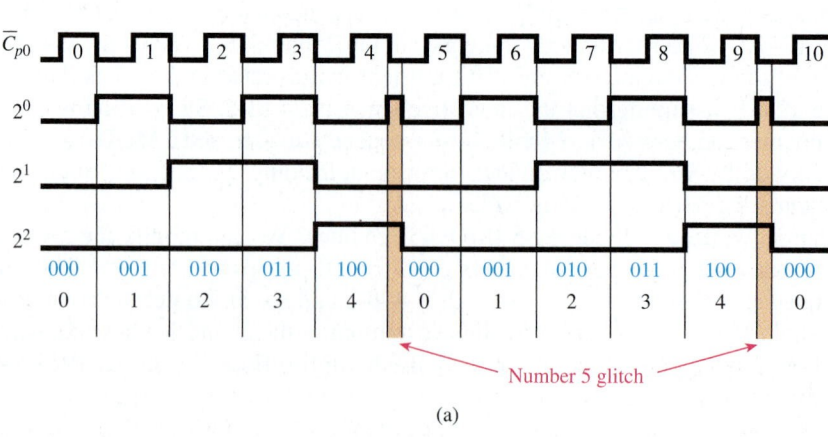

Number 5 glitch

(a)

Figure 12–20   The MOD-5 counter of Figure 12–19: (a) waveforms;

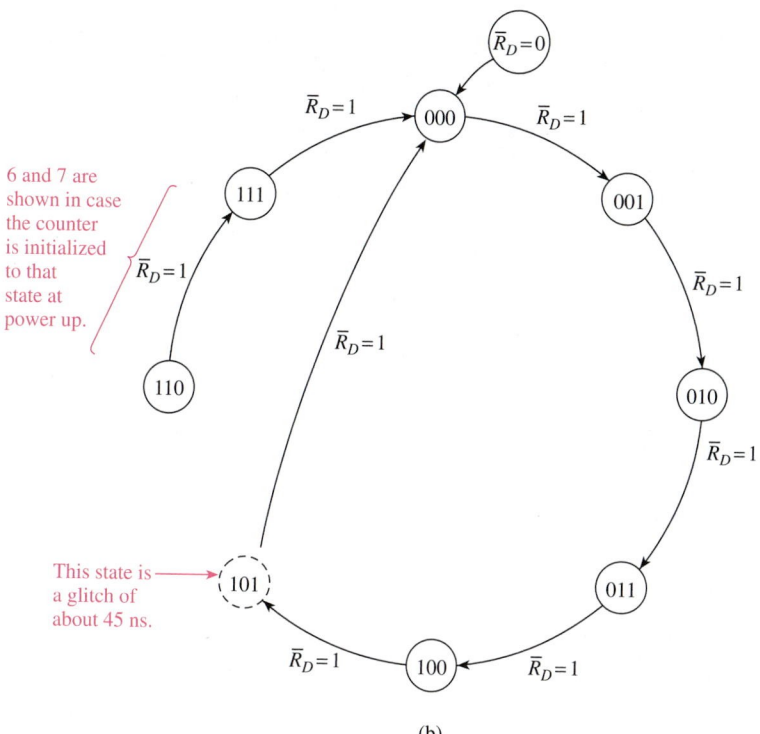

The state diagram shows states with transitions:

$\bar{R}_D=0$

$\bar{R}_D=1$ 000 $\bar{R}_D=1$

111   001

6 and 7 are shown in case the counter is initialized to that state at power up.

$\bar{R}_D=1$

110

$\bar{R}_D=1$

$\bar{R}_D=1$

010

$\bar{R}_D=1$

This state is a glitch of about 45 ns.   101

011

$\bar{R}_D=1$ 100 $\bar{R}_D=1$

(b)

Figure 12–20 *(Continued)* (b) state diagram.

clock frequencies, that glitch could give us erroneous results. Also notice that the duty cycle of each of the outputs is not 50% anymore.

Any modulus counter (**divide-by-N** counter) can be formed by using external gating to Reset at a predetermined number. The following examples illustrate the design of some other divide-by-N counters.

EXAMPLE 12–5

Design a MOD-6 ripple up-counter that can be manually Reset by an external push button.

Solution: The ripple up-counter is shown in Figure 12–21. The count sequence will be 0–1–2–3–4–5. When 6 (binary 110) is reached, the output of the AND gate will go HIGH, causing the NOR gate to put a LOW on the $\bar{R}_D$ line, resetting all flip-flops to zero.

As soon as all outputs return to zero, the AND gate will go back to a LOW output, causing the NOR and $\bar{R}_D$ to return to a HIGH, allowing the counter to count again.

This cycle continues to repeat until the manual Reset push button is pressed. The HIGH from the push button will also cause the counter to Reset. The 100-Ω pull-down resistor will keep the input to the NOR gate LOW when the push button is in the open position. [I_{IL} (NOR) $= -1.6$ mA, $V_{100\Omega} = 1.6$ mA $\times$ 100 $\Omega = 0.160$ V $\equiv$ LOW.]

SECTION 12–3 I DESIGN OF DIVIDE-BY-N COUNTERS

Team Discussion

Why do we need such a small value for the pull-down resistor? What if we use a 10-kΩ resistor?

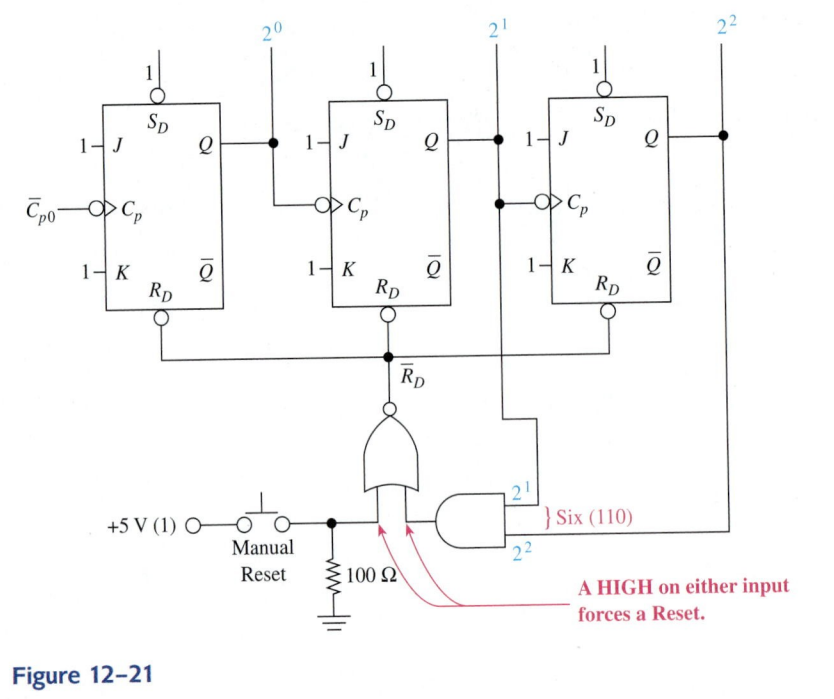

} Six (110)

A HIGH on either input forces a Reset.

Figure 12–21

EXAMPLE 12–6

Design a MOD-10 ripple up-counter with a manual push button Reset.

Solution: The ripple up-counter is shown in Figure 12–22. Four flip-flops are required to give us a possibility of $2^4 = 16$ binary states ($2^3 = 8$ would not be enough). We want to stop the count and automatically Reset when 10 (binary 1010) is reached. This is taken care of by the AND gate feeding into the NOR, making the $\overline{R_D}$ line go LOW when 10 is reached. The count sequence will be 0–1–2–3–4–5–6–7–8–9–0–1–, and so on, which is a MOD-10 up-counter. (The number 10 is only a glitch and is not considered to be part of the output count.)

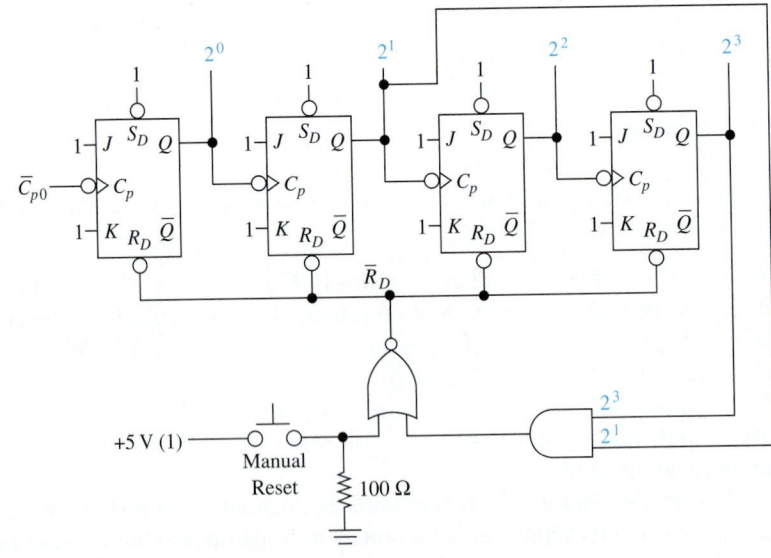

Figure 12–22

EXAMPLE 12-7

Design a MOD-6 down-counter with a manual push button Reset (the count sequence should be 7–6–5–4–3–2–7–6–5–, and so on).

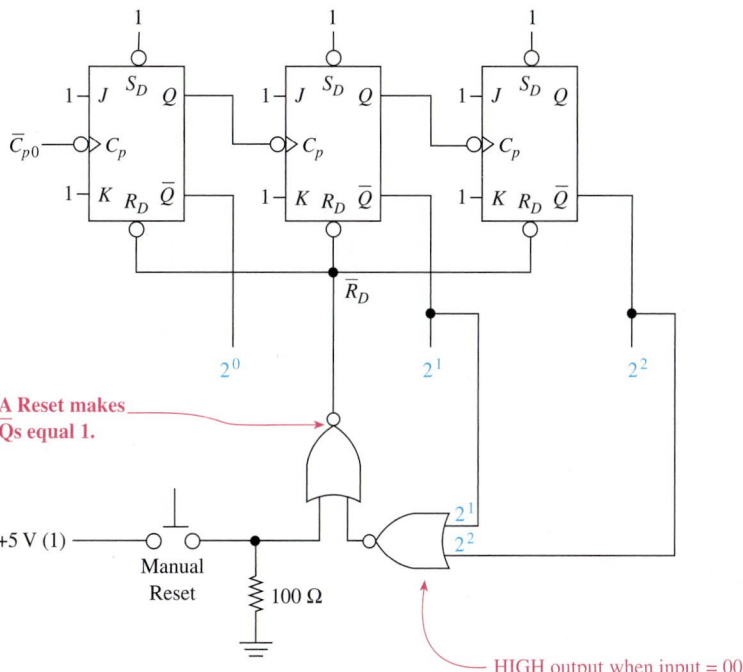

Common Misconception

A LOW on $\overline{R_D}$ normally makes the outputs 000, but in this case, it makes them 111.

Figure 12–23

Solution: The down-counter is shown in Figure 12–23. First, by pressing the manual Reset push button, all flip-flops will Reset, making the counter outputs, taken from the $\overline{Q}$'s, to be 1 1 1. The count sequence that we want is 7–6–5–4–3–2, then Reset to 7 again when 1 is reached (binary 001). When 1 is reached, that is the first time that 2^1 and 2^2 are both LOW. The NOR gate connected to 2^1 and 2^2 will give a HIGH output when both of its inputs are LOW, causing the $\overline{R_D}$ line to go LOW.

EXAMPLE 12-8

Design a MOD-5 up-counter that counts in the sequence 6–7–8–9–10–6–7–8–9–10–6–, and so on.

Solution: The up-counter is shown in Figure 12–24. By pressing the manual Preset push button, the 2^1 and 2^2 flip-flops get Set while the 2^0 and 2^3 flip-flops get Reset. This will give the number 6 (binary 0110) at the output. In the count mode, when the count reaches 11 (binary 1011), the output of the AND gates goes HIGH, causing the $\overline{Preset}$ line to go LOW and recycling the count to 6 again.

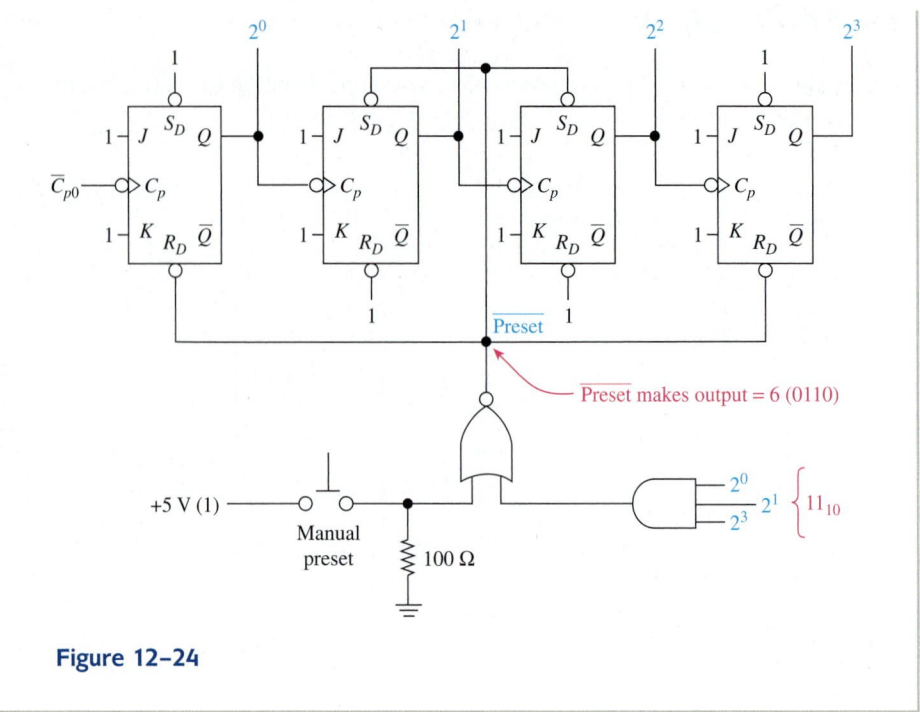

Figure 12–24

EXAMPLE 12–9

Design a down-counter that counts in the sequence 6–5–4–3–2–6–5–4–3–2–6–5–, and so on.

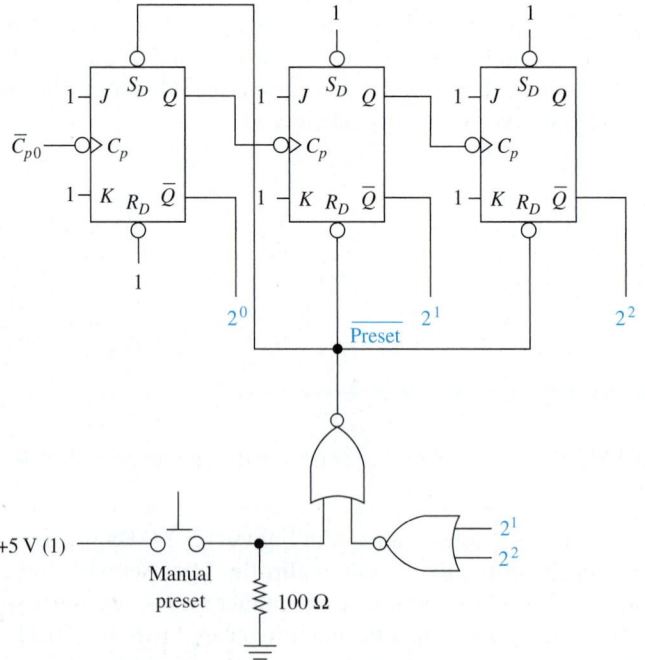

Figure 12–25

CHAPTER 12 | COUNTER CIRCUITS AND VHDL STATE MACHINES

Solution: The down-counter is shown in Figure 12–25. When the $\overline{\text{Preset}}$ line goes LOW, the 2^0 flip-flop is Set, and the other two flip-flops are Reset (this gives a 6 at the Q outputs). As the counter counts down toward zero, the 2^1 and 2^2 will both go LOW at the count of 1 (binary 001), and the $\overline{\text{Preset}}$ line will then go LOW again, starting the cycle over again.

EXAMPLE 12–10

Design a counter that counts 0–1–2–3–4–5 and then stops and turns on an LED. The process is initiated by pressing a start push button.

Solution: The required counter is shown in Figure 12–26. When power is first applied to the circuit (power-up), the capacitor will charge up toward 5 V. It starts out at a zero level, however, which causes the 7474 to Reset ($Q_D = 0$). The LOW at Q_D will remain there until the start button is pressed. With a LOW at Q_D, the three counter flip-flops are all held in the Reset state (binary 000). The output of the NAND gate is HIGH, so the LED is OFF.

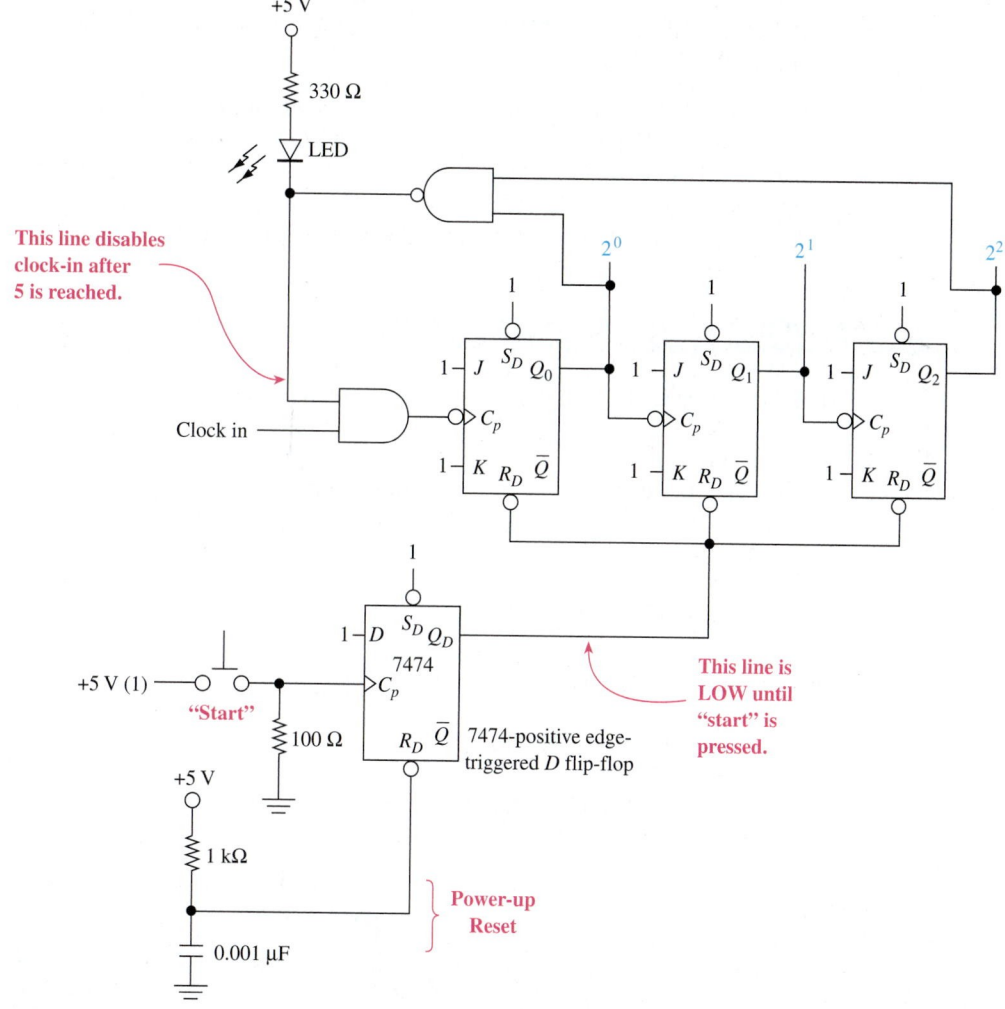

Figure 12–26

When the start button is pressed, Q_D goes HIGH and stays HIGH after the button starts bouncing and is released. With Q_D HIGH, the counter begins counting: 0–1–2–3–4–5. When 5 is reached, the output of the NAND gate goes LOW, turning on the LED. The current through the LED will be (5 V − 1.7 V)/330 Ω = 10 mA. The NAND gate can sink a maximum of 16 mA (I_{OL} = 16 mA), so 10 mA will not burn it out.

The LOW output of the NAND gate is also fed to the input of the AND gate, which will disable the clock input. Because the clock cannot get through the AND gate to the first flip-flop, the counter stays at 5, and the LED stays lit.

If you want to Reset the counter to zero again, you could put a push button in parallel across the capacitor so that, when it is pressed, Q_D will go LOW and stay LOW until the start button is pressed again.

EXAMPLE 12–11

VHDL Description of a MOD-10 Up-Counter

Make one change to the MOD-16 counter of Figure 12–16 to change it to a MOD-10.

Solution: The modified VHDL program is shown in Figure 12–27. Notice that the IF statement resets q to 0 if there is a LOW *n_rd* or if the count has reached 10. A simulation of the design is shown in Figure 12–28. Notice the glitch on the *q1* line at the 10 μs mark. This is because for one iteration of the PROCESS loop q equals 10, but then the next time around it is reset to 0.

```
ex12_11.vhd - Text Editor                              _ □ ×

LIBRARY ieee;                         --------------------------
USE ieee.std_logic_1164.ALL;          -- Mod-10 Up-Counter    --
                                      --------------------------

ENTITY ex12_11 IS
    PORT(n_cp, n_rd        : IN    std_logic;
            q              : BUFFER integer RANGE 0 TO 15);
END ex12_11;


ARCHITECTURE arc OF ex12_11 IS
BEGIN
    PROCESS (n_cp, n_rd)                    ——— Modification for Mod-10
    BEGIN
        IF   (n_rd='0' OR q=10) THEN
             q <= 0;
             ELSIF (n_cp'EVENT AND n_cp='0') THEN
                   q <=q+1;
        END IF;
    END PROCESS;
END arc;                                        ex12_11.vhd

Line   1    Col  1     INS ◄│
```

Figure 12–27   VHDL listing for the MOD-10 up-counter of Example 12–11.

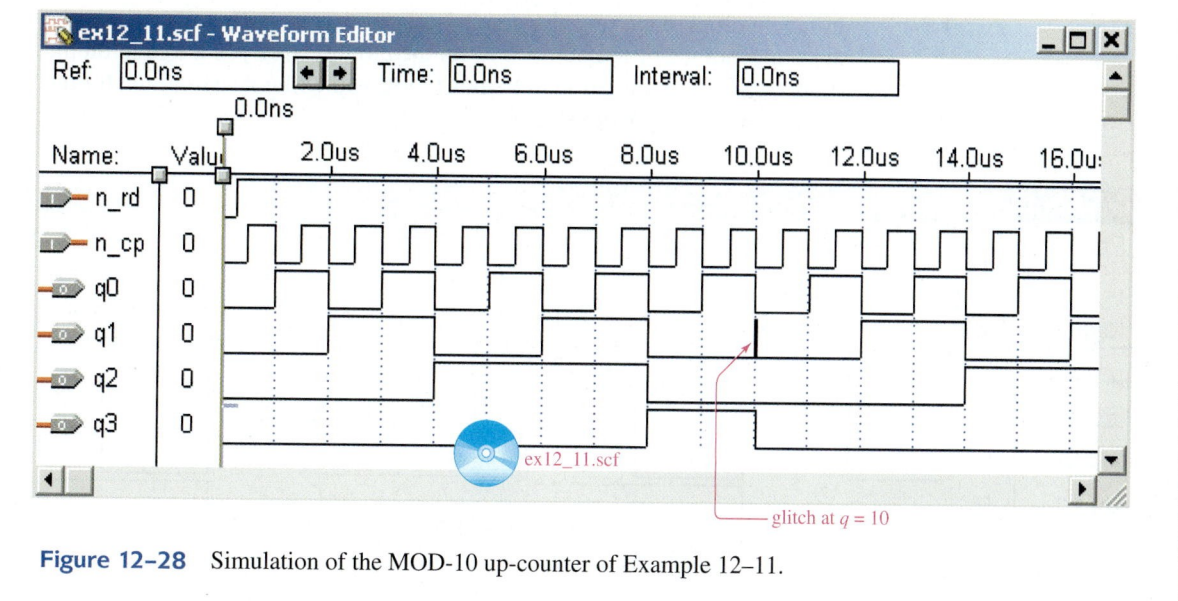

Figure 12–28 Simulation of the MOD-10 up-counter of Example 12–11.

EXAMPLE 12–12

VHDL Description of a Glitch-Free Counter

Change the MOD-10 counter so that the number 10 never appears at the output. This will eliminate the glitch on the *q1* output.

Solution: The modified VHDL program is shown in Figure 12–29. Instead of checking for 10 and then resetting, this program resets *q* when there is

```
ex12_12.vhd - Text Editor                                    _ □ ×
    LIBRARY ieee;                      ----------------------------------
    USE ieee.std_logic_1164.ALL;       -- Mod-10 Glitch-Free Up-Counter --
                                       ----------------------------------
    ENTITY ex12_12 IS
        PORT(n_cp, n_rd        : IN    std_logic;
                q              : BUFFER integer RANGE 0 TO 15);
    END ex12_12;

    ARCHITECTURE arc OF ex12_12 IS
    BEGIN
        PROCESS (n_cp, n_rd)
        BEGIN
            IF  (n_rd='0') THEN
                q<= 0;
                ELSIF (n_cp'EVENT AND n_cp='0' ) THEN      Reset q at the
                    IF (q=9) THEN                          end of #9 period
                        q<=0;
                        ELSE q<=q+1;
                    END IF;
            END IF;
        END PROCESS;
    END arc;                                    ex12_12.vhd
Line  1    Col  1      INS
```

Figure 12–29 VHDL listing for the glitch-free MOD-10 up-counter of Example 12–12.

an active clock edge and the count is currently on 9. This eliminates the short appearance of the number 10 on the q outputs. If you zoom in on the 10 μs area of the simulation in Figure 12–30 you will not see a glitch for the count goes directly from 9 to 0.

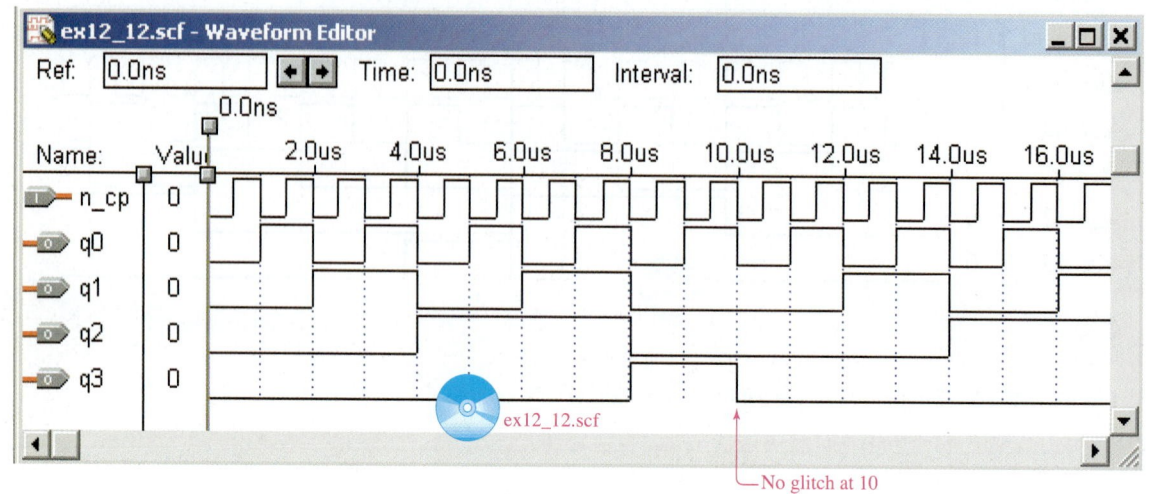

Figure 12–30 Simulation of the glitch-free MOD-10 up-counter of Example 12–12.

Review Questions

12–5. A MOD-16 counter can function as a divide-by-16 frequency divider by taking the output from the _____ output.

12–6. To convert a 4-bit MOD-16 counter to a MOD-12 counter, the flip-flops must be Reset when the counter reaches the number _____ (11, 12, 13).

12–7. Briefly describe the operation of the Manual Reset push-button circuitry used in the MOD-N counters in this section.

12–4 Ripple Counter ICs

Helpful Hint

The 74293 and 74290 are electrically identical to the 7493 and 7490, except V_{cc} and GND are moved to the outside corners of the chip. Also, the 74393 and 74390 are *dual* 4-bit counters. Check the Internet to determine what subfamilies are available.

Four-bit binary ripple counters are available in a single IC package. The most popular are the 7490, 7492, and 7493 TTL ICs.

Figure 12–31 shows the internal logic diagram for the 7493 4-bit binary ripple counter. The 7493 has four *J-K* flip-flops in a single package. It is divided into two sections: a divide-by-2, and a divide-by-8. The first flip-flop provides the divide-by-2 with its $\overline{C_{p0}}$ input and Q_0 output. The second group has three flip-flops cascaded to each other and provides the divide-by-8 via the $\overline{C_{p1}}$ input and $Q_1 Q_2 Q_3$ outputs. To get a divide-by-16, you can *externally* connect Q_0 to $\overline{C_{p1}}$ so that all four flip-flops are cascaded end to end, as shown in Figure 12–32. Notice that two Master Reset inputs (MR_1, MR_2) are provided to asynchronously Reset all four flip-flops. When MR_1 and MR_2 are both HIGH, all Q's will be Reset to 0. (MR_1 or MR_2 must be held LOW to enable the count mode.)

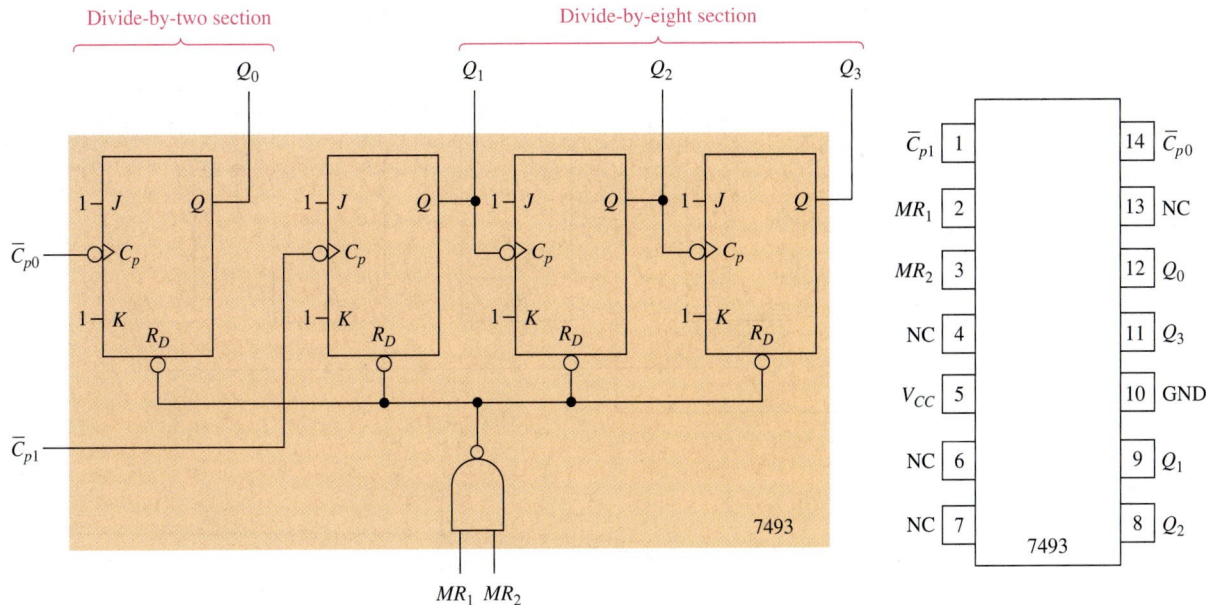

Figure 12–31 Logic diagram and pin configuration for a 7493 4-bit ripple counter IC.

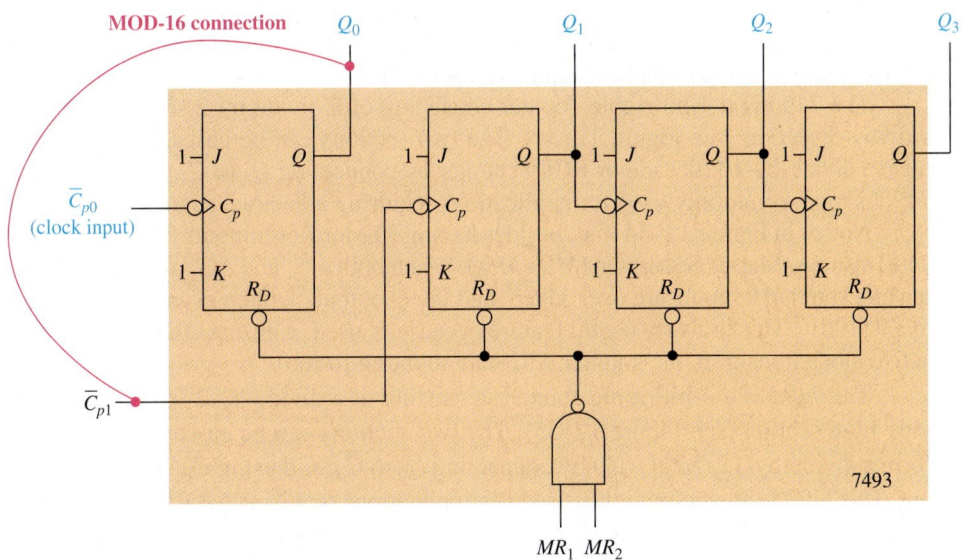

Figure 12–32 A 7493 connected as a MOD-16 ripple counter.

With the MOD-16 connection, the frequency output at Q_0 is equal to one-half the frequency input at $\overline{C_{p0}}$. Also, $f_{Q1} = \frac{1}{4} f_{\overline{Cp0}}$, $f_{Q2} = \frac{1}{8} f_{\overline{Cp0}}$, and $f_{Q3} = \frac{1}{16} f_{\overline{Cp0}}$.

The 7493 can be used to form any modulus counter less than or equal to MOD-16 by utilizing the MR_1 and MR_2 inputs. For example, to form a MOD-12 counter, simply make the external connections shown in Figure 12–33.

The count sequence of the MOD-12 counter will be 0–1–2–3–4–5–6–7–8–9–10–11–0–1, and so on. Each time 12 (1100) tries to appear at the outputs, a HIGH–HIGH is placed on MR_1–MR_2, and the counter resets to zero.

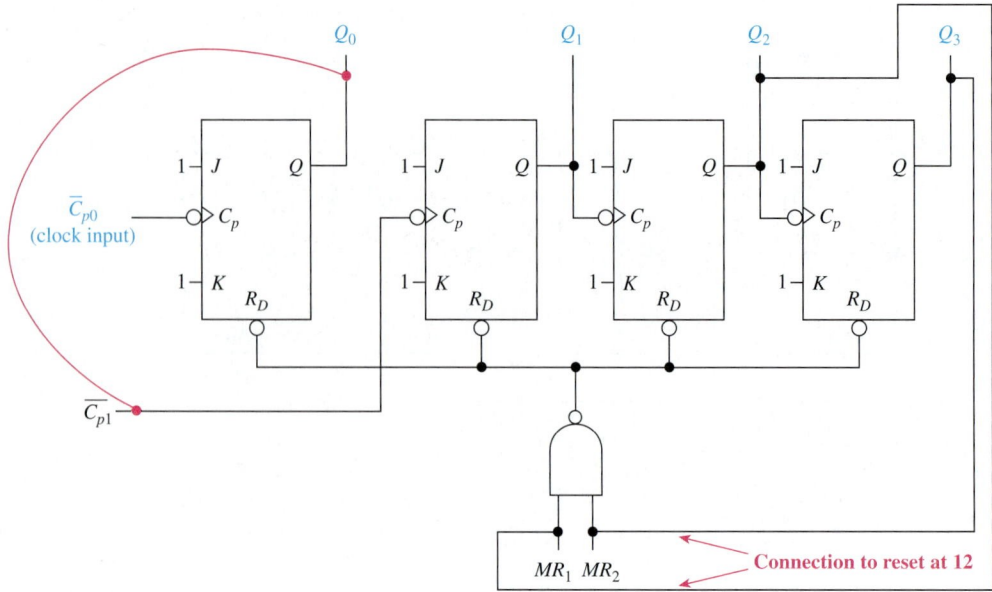

Figure 12–33 External connections to a 7493 to form a MOD-12 counter.

Two other common ripple counter ICs are the 7490 and 7492. They both have four internal flip-flops like the 7493, but through the application of internal gating, they automatically recycle to 0 after 9 and 11, respectively.

The 7490 is a 4-bit ripple counter consisting of a divide-by-2 section and a divide-by-5 section (see Figure 12–34). The two sections can be cascaded together to form a divide-by-10 (decade or BCD) counter by connecting Q_0 to $\overline{C_{p1}}$ externally. The 7490 is most commonly used for applications requiring a decimal (0 to 9) display.

Notice in Figure 12–34 that, besides having Master Reset inputs (MR_1–MR_2), the 7490 also has Master Set inputs (MS_1–MS_2). When both MS_1 and MS_2 are made HIGH, the clock and MR inputs are overridden, and the Q outputs will be asynchronously Set to a 9 (1001). This is a very useful feature because, if used, it ensures that *after* the first active clock transition, the counter will start counting from 0.

The 7492 is a 4-bit ripple counter consisting of a divide-by-2 section and a divide-by-6 section (see Figure 12–35). The two sections can be cascaded together to form a divide-by-12 (MOD-12) by connecting Q_0 to $\overline{C_{p1}}$ and using $\overline{C_{p0}}$ as the clock input. The 7492 is most commonly used for applications requiring MOD-12 and MOD-6 frequency dividing, such as in digital clocks. You can get a divide-by-6 frequency divider simply by ignoring the $\overline{C_{p0}}$ input of the first flip-flop and, instead, bringing the clock input into $\overline{C_{p1}}$, which is the input to the divide-by-6 section. (One peculiarity of the 7492 is that, when connected as a MOD-12, it does *not* count sequentially from 0 to 11. Instead, it counts from 0 to 13, skipping 6 and 7, but still functions as a divide-by-12.)

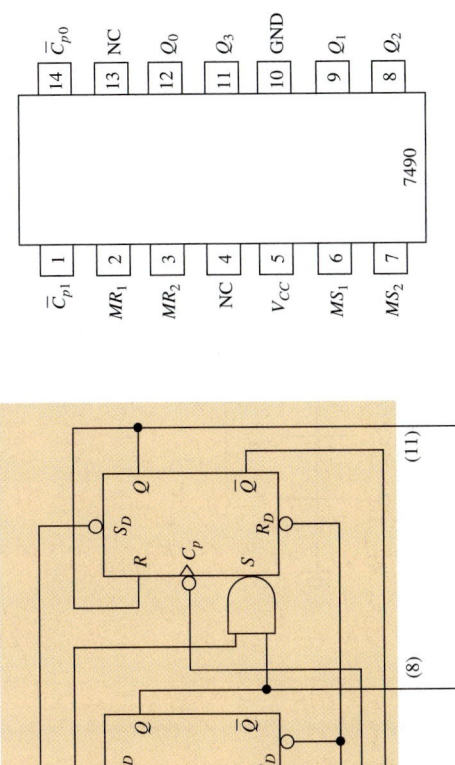

Figure 12–34 Logic diagram and pin configuration for a 7490 decade counter.

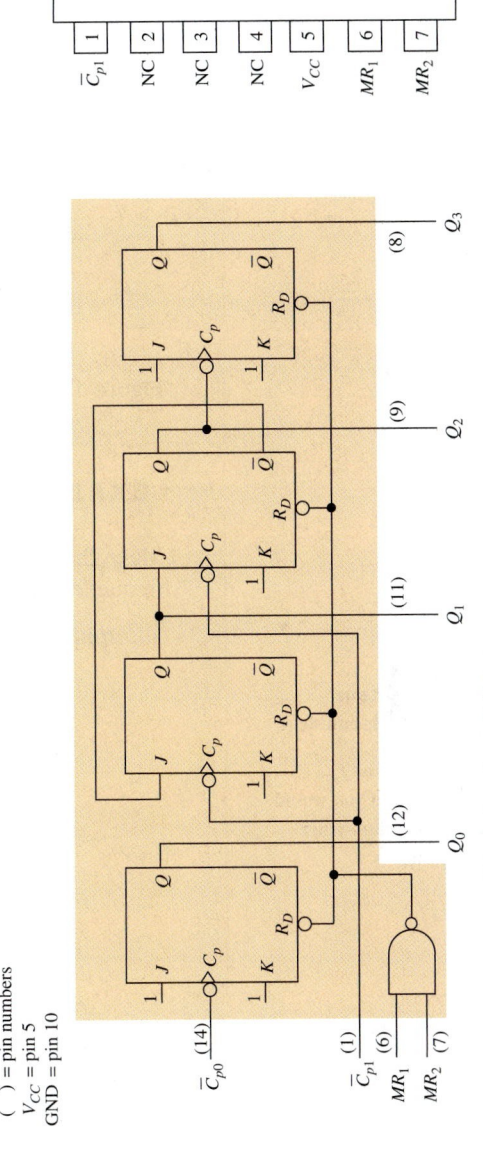

Figure 12–35 Logic diagram and pin configuration for a 7492 counter.

EXAMPLE 12-13

Make the necessary external connections to a 7490 to form a MOD-10 counter.

Solution: The MOD-10 counter is shown in Figure 12–36.

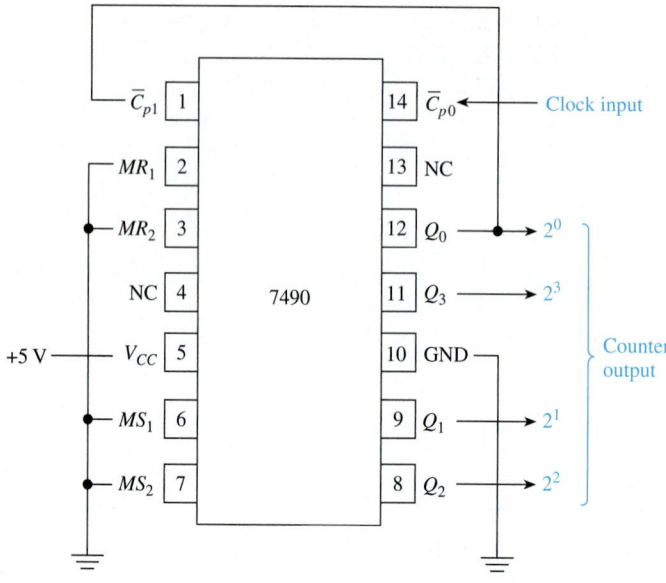

Figure 12–36

EXAMPLE 12-14

Make the necessary external connections to a 7492 to form a divide-by-6 frequency divider $(f_{out} = \frac{1}{6} f_{in})$.

Solution: The frequency divider is shown in Figure 12–37.

Team Discussion

Sketch the f_{in} and f_{out} waveforms that you would observe on a dual-trace oscilloscope.

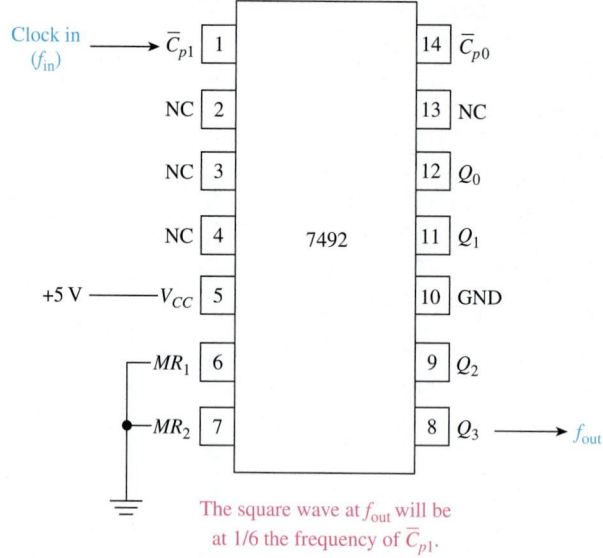

The square wave at f_{out} will be at 1/6 the frequency of $\bar{C}_{p1}$.

Figure 12–37

EXAMPLE 12-15

Make the necessary external connections to a 7490 to form a MOD-8 counter (0 to 7). Also, upon initial power-up, set the counter at 9 so that, after the first active input clock edge, the output will be 0 and the count sequence will proceed from there.

Solution: The MOD-8 counter is shown in Figure 12–38. The output of the 7414 Schmitt inverter will initially be HIGH when power is first turned on because the capacitor feeding its input is initially discharged to zero. This HIGH on MS_1 and MS_2 will Set the counter to 9. Then, as the capacitor charges up above 1.7 V, the Schmitt will switch to a LOW output, allowing the counter to start its synchronous counting sequence. The inverter on MR_1 is necessary to keep the counter from Resetting when the outputs are at 9 ($Q_0, Q_3 = 1, 1$).

Q_0 is connected to $\overline{C}_{p1}$ so that all four flip-flops are cascaded. When the counter reaches 8 (1000), the MR_1 and MR_2 lines will equal 1–1, causing the counter to Reset to 0. The counter will continue to count in the sequence 0–1–2–3–4–5–6–7–0–1–2, and so on, continuously.

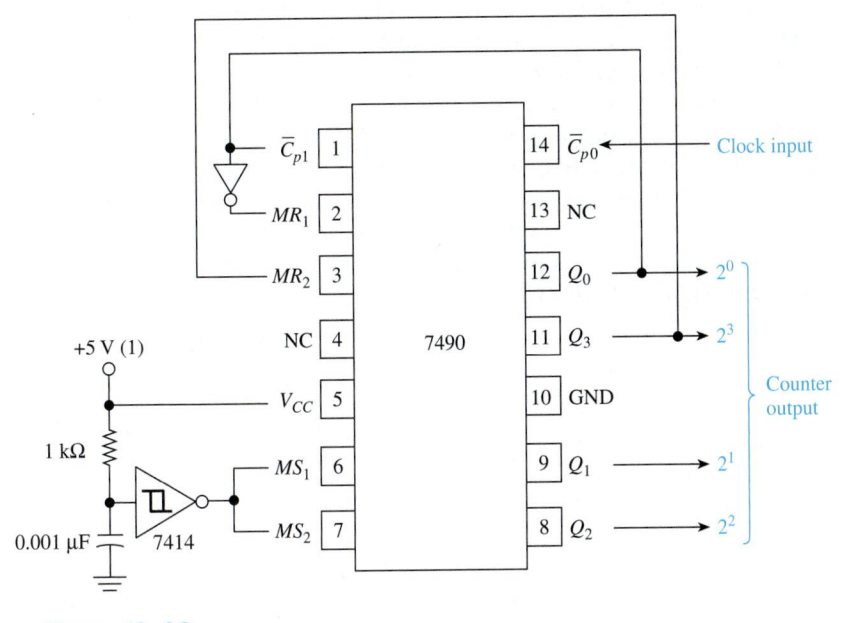

Figure 12–38

Review Questions

12–8. What is the highest modulus of each of the following counter ICs: 7490, 7492, 7493?

12–9. Why does the 7493 counter IC have *two* clock inputs?

12–10. What happens to the Q-outputs of the 7490 counter when you put 1's on the MS inputs?

12–5 System Design Applications

Integrated-circuit counter chips are used in a multitude of applications dealing with timing operations, counting, sequencing, and frequency division. To implement a complete system application, output devices such as LED indicators, seven-segment LED displays, relay drivers, and alarm buzzers must be configured to operate from the counter outputs. The synchronous and asynchronous inputs can be driven by such devices as a clock oscillator, a push-button switch, the output from another digital IC, or control signals provided by a microprocessor.

EXAMPLE 12–16

For example, let's consider an application that requires an LED indicator to illuminate for 1 s once every 13 s to signal an assembly line worker to perform some manual operation.

Solution: To solve this design problem, we first have to come up with a clock oscillator that produces 1 pulse per second (pps).

The first part of Figure 12–39(a), which is used to produce the 60-pps clock, was described in detail in Section 11–6. To divide the 60 pps down to 1 pps, we can cascade a MOD-10 counter with a MOD-6 counter to create a divide-by-60 circuit.

The 7490 connected as a MOD-10 is chosen for the divide-by-10 section. If you study the output waveforms of a MOD-10 counter, you can see that Q_3 will oscillate at a frequency one-tenth of the frequency at $\overline{C_{p0}}$. Then, if we use Q_3 to trigger the input clock of the divide-by-6 section, the overall effect will be a divide-by-60. [The 7492 is used for the divide-by-6 section simply by using $\overline{C_{p1}}$ as the input and taking the 1-pps output from Q_3, as shown in Figure 12–39(a).]

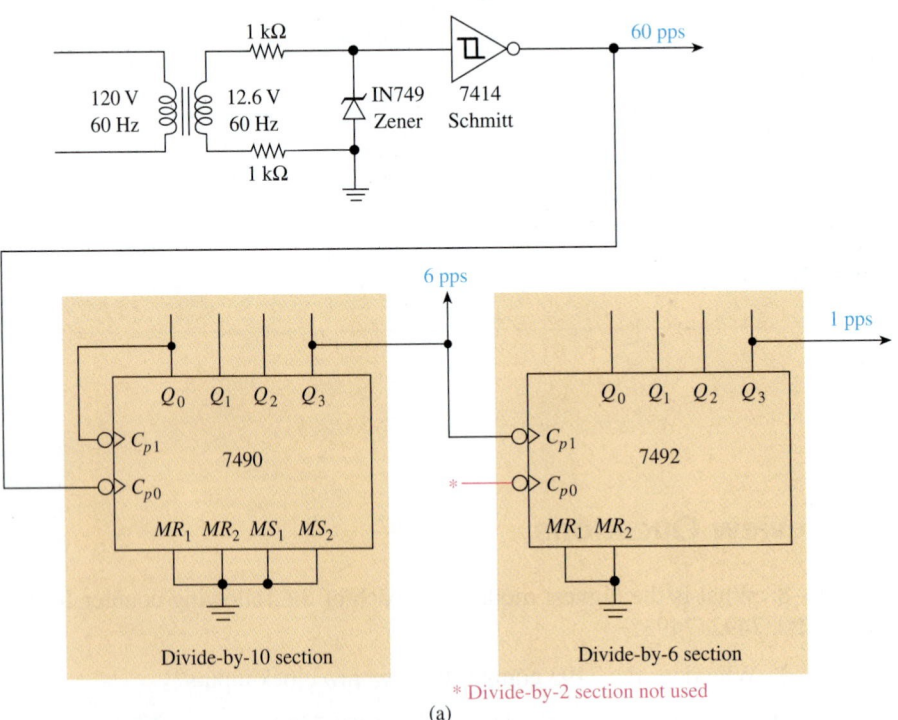

(a)

Figure 12–39 (a) Circuit used to produce 1 pps;

CHAPTER 12 | COUNTER CIRCUITS AND VHDL STATE MACHINES

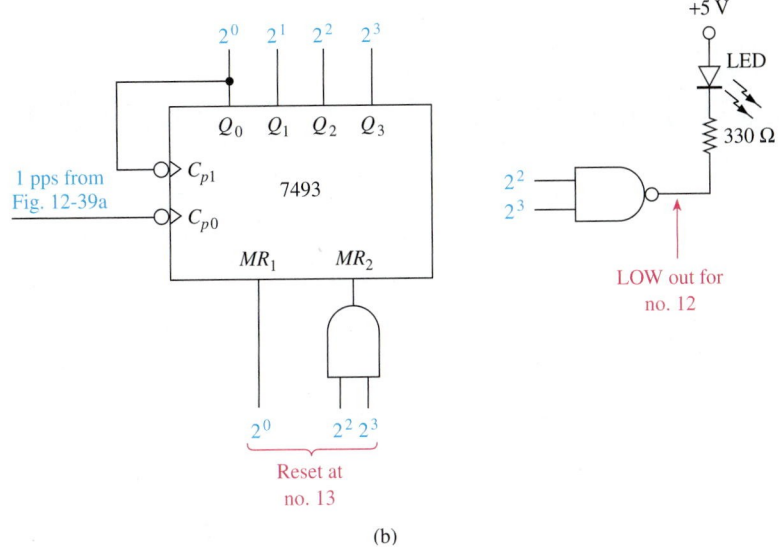

Figure 12–39 *(Continued)* (b) circuit used to illuminate an LED once every 13 s.

The next step in the system design is to use the 1-pps clock to enable a circuit to turn on an LED for 1 s once every 13 s. It sounds like we need a MOD-13 counter (0 to 12) and a gating scheme that turns on an LED when the count is on the number 12. A 7493 can be used for a MOD-13 counter, and a NAND gate can be used to sink the current from an LED when the number 12 ($Q_2 = 1$, $Q_3 = 1$) occurs. Figure 12–39(b) shows the necessary circuit connections.

Notice in Figure 12–39(b) that a MOD-13 is formed by connecting Q_0 to $\overline{C_{p1}}$ and resetting the counter when the number 13 is reached, resulting in a count of 0 to 12. Also, when the number 12 is reached, the NAND gate's output goes LOW, turning on the LED. [$I_{LED} = (5 \text{ V} - 1.7 \text{ V})/330 \text{ }\Omega = 10 \text{ mA}$].

EXAMPLE 12–17

Design a circuit to turn on an LED for 20 ms once every 100 ms. Assume that you have a 50-Hz (50-pps) clock available.

Solution: Because 20 ms is one-fifth of 100 ms, we should use a MOD-5 counter such as the one available in the 7490 IC. To determine which outputs to use to drive the LED, let's look at the waveforms generated by a 7490 connected as a MOD-5 counter.

Remember that the second section of a 7490 is a MOD-5 counter (0 to 4). If the input frequency is 50 Hz, each count will last for 20 ms (1/50 Hz = 20 ms), as shown in Figure 12–40(a).

Notice that the Q_3 line goes HIGH for 20 ms once every 100 ms. So if we just invert the Q_3 line and use it to drive the LED, we have the solution to our problem! Figure 12–40(b) shows the final solution.

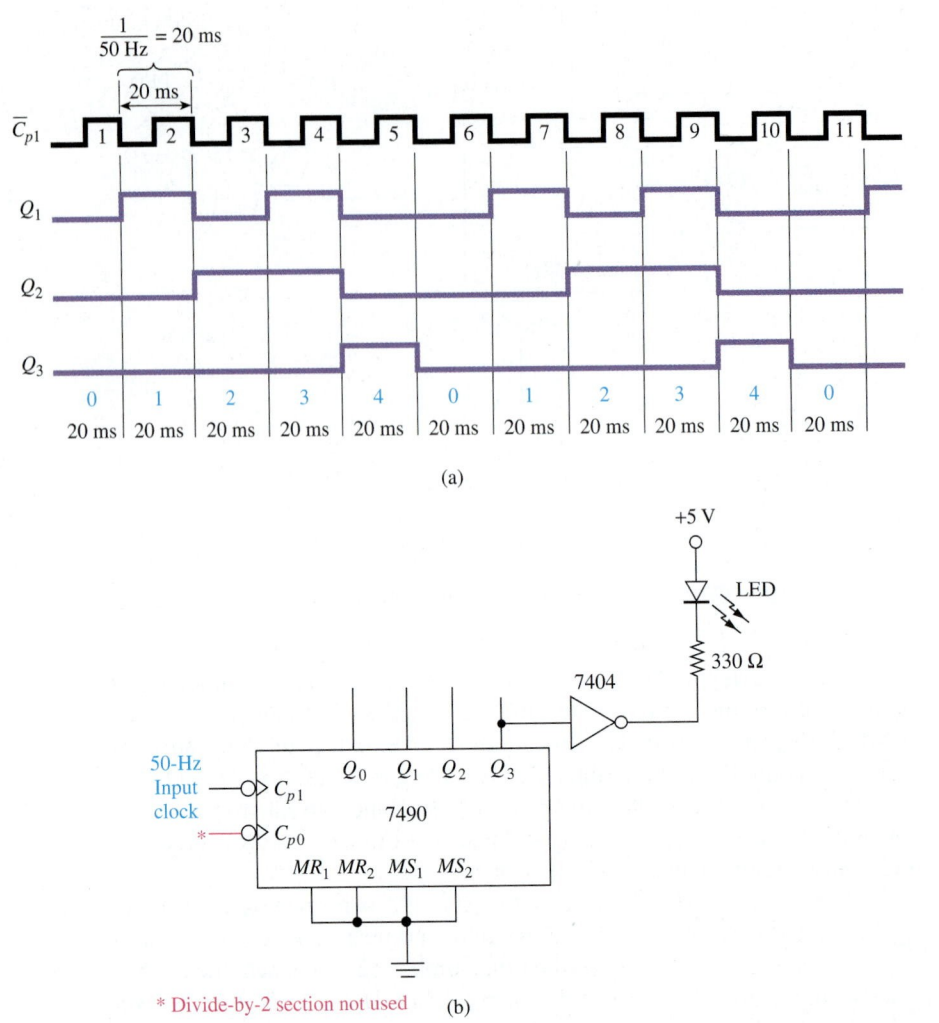

Figure 12–40 (a) Output waveforms from a MOD-5 counter driven by a 50-Hz input clock; (b) solution to Example 12–17.

EXAMPLE 12-18

Design a three-digit decimal counter that can count from 000 to 999.

Solution: We have already seen that a 7490 is a single-digit decimal (0 to 9) counter. If we cascade three 7490s together and use the low-order counter to trigger the second digit counter and the second digit counter to trigger the high-order-digit counter, they will count from 000 up to 999. (Keep in mind that the outputs will be binary-coded decimal in groups of 4. In Section 12–6, we will see how we can convert the BCD outputs into actual decimal digits.)

If you review the output waveforms of a 7490 connected as a MOD-10 counter, you can see that at the end of the cycle, when the count changes from 9 (1001) to 0 (0000), the 2^3 output line goes from HIGH to LOW. When cascading counters, you can use that HIGH-to-LOW transition to trigger the input to the next-highest-order counter. That will work out great

Helpful Hint

Review MOD-10 waveforms to see why the 2^3 signal is used to clock each successive BCD digit.

because we want the next-highest-order decimal digit to increment by 1 each time the lower-order digit has completed its 0-through-9 cycle (i.e., the transition from 009 to 010). The complete circuit diagram for a 000-to-999 BCD counter is shown in Figure 12–41.

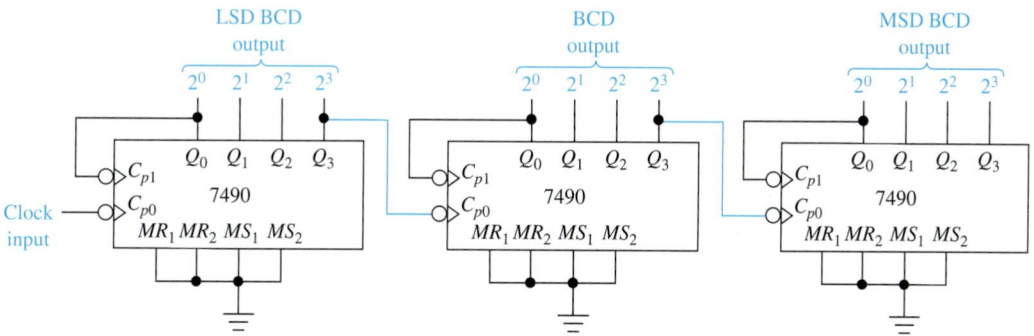

Figure 12–41 Cascading 7490s to form a 000-to-999 BCD output counter.

EXAMPLE 12–19

Design and sketch a block diagram of a digital clock capable of displaying hours, minutes, and seconds.

Solution: First, we have to design a 1-pps clock to feed into the least significant digit of the seconds counter. The seconds will be made up of two cascaded counters that count 00 to 59. When the seconds change from 59 to 00, that transition will be used to trigger the minutes digits to increment by 1. The minutes will also be made up of two cascaded counters that count from 00 to 59. When the minutes change from 59 to 00, that transition will be used to trigger the hours digits to increment by 1. Finally, when the hours reach 12, all counters should be Reset to 0. The digital clock will display the time from 00:00:00 to 11:59:59.

Figure 12–42 is the final circuit that could be used to implement a digital clock. A 1-pps clock (similar to the one shown in Figure 12–39) is used as the initial clock trigger into the least significant digit (LSD) counter of the seconds display. This counter is a MOD-10 constructed from a 7490 IC. Each second this counter will increment. When it changes from 9 to 0, the HIGH-to-LOW edge on the 2^3 line will serve as a clock pulse into the most significant digit (MSD) counter of the seconds display. This counter is a MOD-6 constructed from a 7492 IC.

After 59 s, the 2^2 output of the MOD-6 counter will go HIGH to LOW [once each minute (1 ppm)], triggering the MOD-10 of the minutes section. When the minutes exceed 59, the 2^2 output of that MOD-6 counter will trigger the MOD-10 of the hours section.

The MOD-2 of the hours section is just a single toggle flip-flop having a 1 or 0 output. The hours section is set up to count from 0 to 11. When 12 is reached, the AND gate resets both hours counters. The clock display will, therefore, be 00:00:00 to 11:59:59.

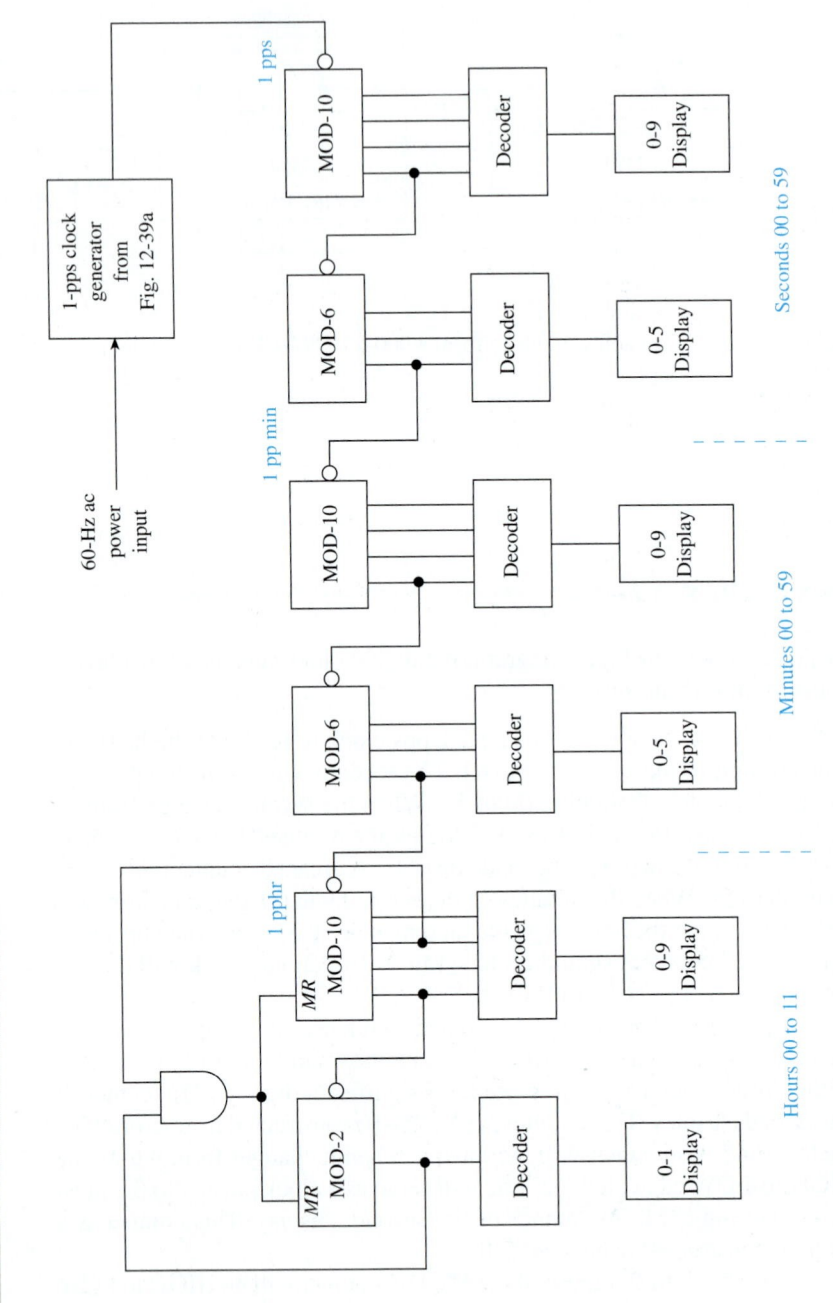

Figure 12–42 Block diagram for a digital clock.

If you want the clock to display 1:00:00 to 12:59:59 instead, you will have to check for a 13 in the hours section instead of 12. When 13 is reached, you will want to Reset the MOD-2 counter and *Preset* the MOD-10 counter to a 1. Presettable counters such as the 74192 are used in a case like this. (Presettable counters are covered later in this chapter.)

The decoders are required to convert the BCD from the counters into a special code that can be used by the actual display device. Digit displays and decoders are discussed in Section 12–6.

Team Discussion

Discuss ways that you might modify the clock generator circuit to provide a "fast-forward" feature for setting the time.

EXAMPLE 12–20

Design an egg-timer circuit. The timer will be started when you press a push button. After 3 minutes, a 5-V, 10-mA dc piezoelectric buzzer will begin buzzing.

Solution: The first thing to take care of is to divide the 1-pps clock previously designed in Figure 12–39(a) down to a 1-ppm clock. At 1 ppm, when the count reaches 3, the buzzer should be enabled and the input clock disabled. An automatic power-up Reset is required on all the counters so that the minute counter will start at zero. A D latch can be utilized for the push-button starter so that, after the push button is released, the latch remembers and will keep the counting process going.

The circuit of Figure 12–43 can be used to implement this design. When power is first turned on, the automatic Reset circuit will Reset all counter outputs and Reset the 7474, making $\overline{Q} = 1$. With $\overline{Q}$ HIGH, the OR gate will stay HIGH, disabling the clock from getting through to the first 7492.

When the start push button is momentarily depressed, $\overline{Q}$ will go LOW, allowing the 1-pps clock to reach $\overline{C_{p1}}$. The first two counters are connected as a MOD-6 and a MOD-10 to yield a divide-by-60, so we have 1 ppm available for the last counter, which serves as a minute counter. When the count reaches 3 in the last 7490, the AND gate goes HIGH, disabling the clock input. This causes the 7404 to go LOW, providing sink current for the buzzer to operate. The buzzer is turned off by turning off the main power supply.

Review Questions

12–11. How could you form a divide-by-60 using two IC counters?

12–12. When cascading several counter ICs end to end, which Q-output drives the clock input to each successive stage?

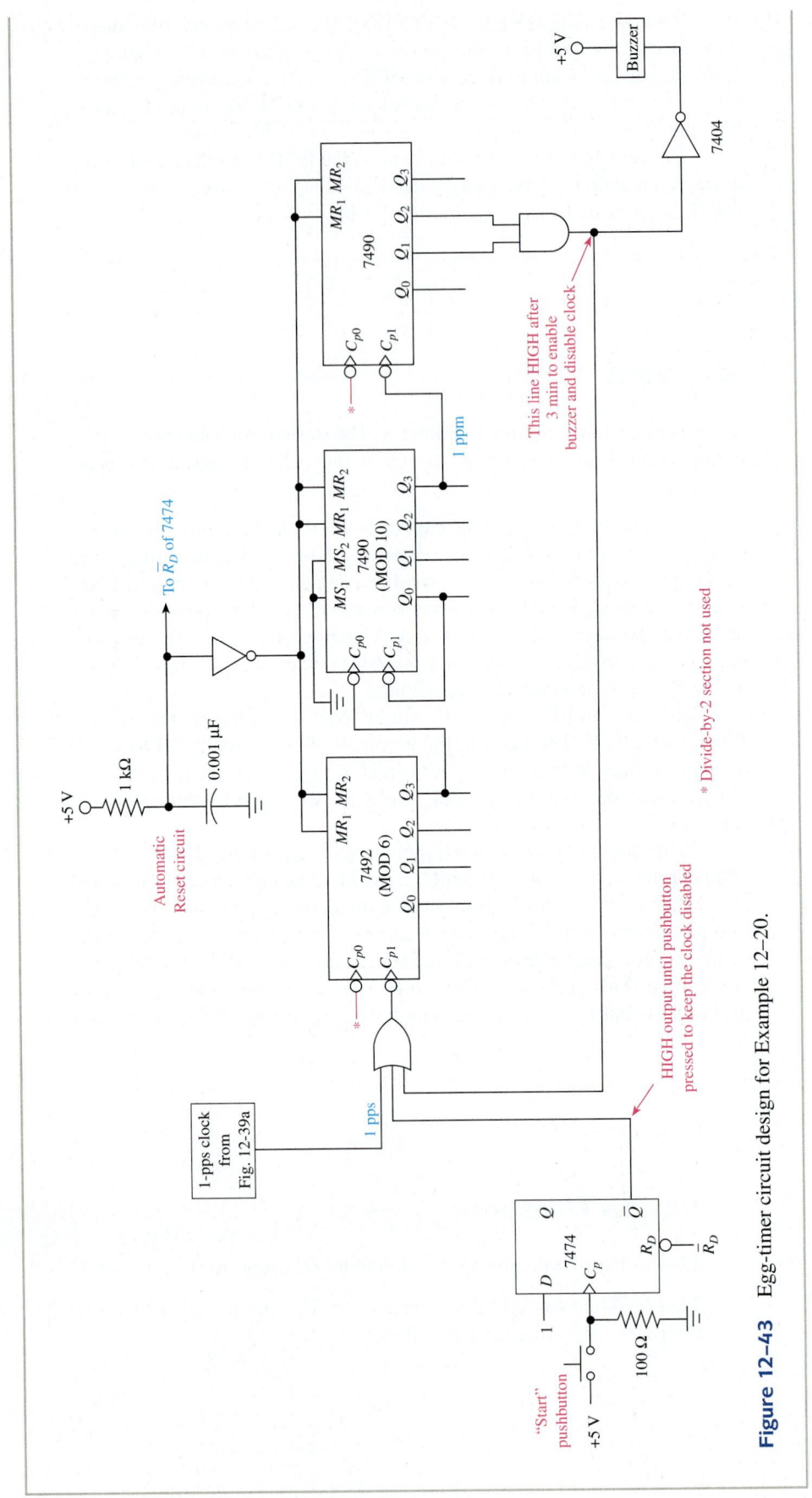

Figure 12–43 Egg-timer circuit design for Example 12–20.

12–6 Seven-Segment LED Display Decoders: The 7447 IC and VHDL Description

In Section 12–5, we discussed counter circuits that are used to display decimal (0 to 9) numbers. If a counter is to display a decimal number, the count on each 4-bit counter cannot exceed 9 (1001). In other words, the counters must be outputting binary-coded decimal (BCD). As described in Chapter 2, BCD is a 4-bit binary string used to represent the 10 decimal digits. To be useful, however, the BCD must be decoded by a decoder into a format that can be used to drive a decimal numeric display. The most popular display technique is the **seven-segment LED** display.

A seven-segment LED display is actually made up of seven separate light-emitting diodes in a single package. The LEDs are oriented so as to form an 8. Most seven-segment LEDs have an eighth LED used for a decimal point.

The job of the decoder is to convert the 4-bit BCD code into a seven-segment code that will turn on the appropriate LED segments to display the correct decimal digit. For instance, if the BCD is 0111 (7), the decoder must develop a code to turn on the top segment and the two right segments (7).

Common-Anode LED Display

The physical layout of a seven-segment LED display is shown in Figure 12–44. This figure shows that the anode of each LED (segment) is connected to the +5-V supply. Now, to illuminate an LED, its cathode must be grounded through a series-limiting resistor, as shown in Figure 12–45. The value of the limiting resistor can be found by knowing that the voltage drop across an LED is 1.7 V and that it takes approximately 10 mA to illuminate it. Therefore,

$$R_{\text{limit}} = \frac{5.0 \text{ V} - 1.7 \text{ V}}{10 \text{ mA}} = 330 \text{ }\Omega$$

Each segment in the display unit is illuminated in the same way. Figure 12–46 shows the numerical designations for the 10 allowable decimal digits.

Common-anode displays are *active-LOW* (LOW-enable) devices because it takes a LOW to turn on (illuminate) a segment. Therefore, the decoder IC used to drive a **common-anode LED** must have active-LOW outputs.

Common-cathode LEDs and decoders are also available but are not as popular because they are *active-HIGH*, and ICs typically cannot *source* (1 output) as much current as they can *sink* (0 output).

BCD-to-Seven-Segment Decoder/Driver ICs*

The 7447 is the most popular common-anode decoder/LED driver. Basically, the 7447 has a 4-bit BCD input and seven individual active-LOW outputs (one for each LED segment). As shown in Figure 12–47, it also has a *lamp test* ($\overline{LT}$) input for testing *all* segments, and it also has **ripple blanking** input and output.

To complete the connection between the 7447 and the seven-segment LED, we need seven 330-Ω resistors (eight if the decimal point is included) for current limiting. Dual-in-line package (DIP) *resistor networks* are available and simplify the wiring process because all seven (or eight) resistors are in a single DIP.

---

*Very versatile CMOS seven-segment decoders are the 4543 and its high-speed version, the 74HCT4543. The 4543 provides active-HIGH *or* active-LOW outputs and can drive LED displays as well as **liquid-crystal displays (LCDs).** LCDs are used in low-power battery applications such as calculators and watches. (See Application 16–2). Their segments don't actually emit light, but instead, the individual liquid-crystal segments will polarize to become either opaque (black) or transparent (white) to external light.

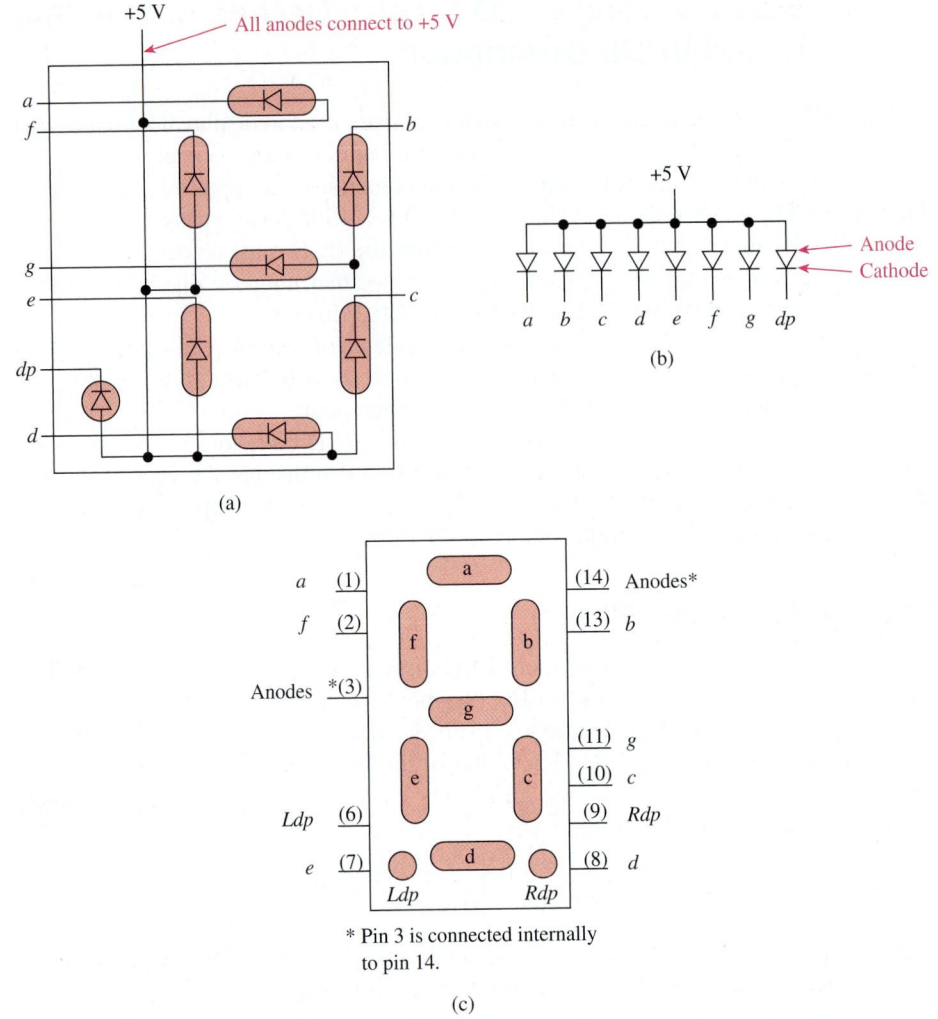

(a)

(b)

(c)

* Pin 3 is connected internally
to pin 14.

Figure 12–44 Seven-segment common-anode LED display: (a) physical layout;
(b) schematic; (c) pin configuration.

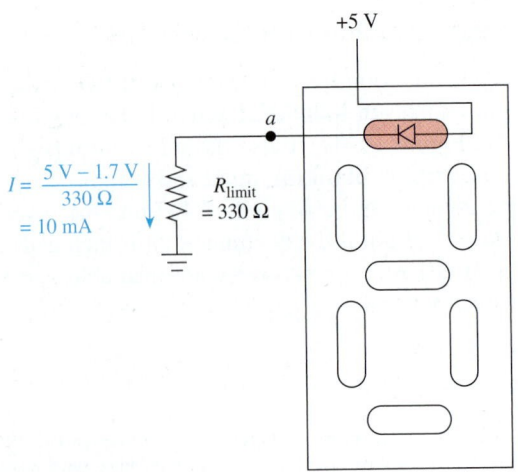

$$I = \frac{5\text{ V} - 1.7\text{ V}}{330\ \Omega}$$
$$= 10\text{ mA}$$

$R_{limit} = 330\ \Omega$

Figure 12–45 Illuminating the *a* segment.

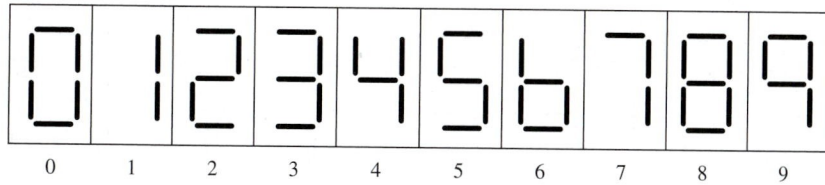

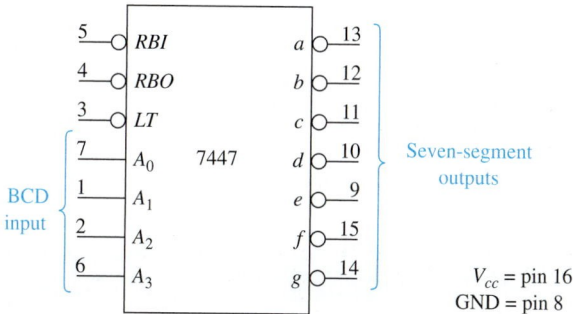

Figure 12–46 Numeric designations for a seven-segment LED.

Figure 12–47 Logic symbol for a 7447 decoder.

Figure 12–48 shows typical decoder–resistor–DIP–LED connections. As an example of how Figure 12–48 works, if a MOD-10 counter's outputs are connected to the BCD input and the count is at six (0110_{BCD}), the following will happen:

1. The decoder will determine that a 0110_{BCD} must send the $\overline{c}, \overline{d}, \overline{e}, \overline{f}, \overline{g}$ outputs LOW ($\overline{a}, \overline{b}$ will be HIGH for 6).

2. The LOW on those outputs will provide a path for the sink current in the appropriate LED segments via the 330-Ω resistors (the 7447 can sink up to 40 mA at each output).

3. The decimal number 6 will be illuminated, together with the decimal point if the *dp* switch is closed.

A complete three-digit decimal display system is shown in Figure 12–49. The three counters in the figure are connected as MOD-10 counters with the input clock oscillator connected to the least significant counter. The three counters are cascaded by connecting the Q_3 output of the first to the $\overline{C_{p0}}$ of the next, and so on.

Notice that the decimal point of the LSD is always on, so the counters will, therefore, count from .0 up to 99.9. If the clock oscillator is set at 10 pps, the LSD will indicate tenths of seconds. Also notice that the ripple blanking inputs and outputs ($\overline{RBI}$ and $\overline{RBO}$) are used in this design. They are active LOW and are used for *leading-zero suppression*. For example, if the display output is at 1.4, would you like it to read 01.4 or 1.4? To suppress the leading zero and make it a blank, ground the $\overline{RBI}$ terminal of the MSD decoder. How about if the output is at .6? Would you like it to read 00.6, 0.6, or .6? To suppress the second zero when the MSD is blank, simply connect the $\overline{RBO}$ of the MSD decoder to the $\overline{RBI}$ of the second digit decoder. The way this works is that, if the MSD is blank (zero suppressed), the MSD decoder puts a LOW out at $\overline{RBO}$. This LOW is connected to the $\overline{RBI}$ of the second decoder, which forces a blank output (zero suppression) if its BCD input is zero.

The $\overline{RBI}$ and $\overline{RBO}$ can also be used for zero suppression of trailing zeros. For example, if you have an eight-digit display, the $\overline{RBI}$'s and $\overline{RBO}$'s could be used to automatically suppress the number 0046.0910 to be displayed as 46.091.

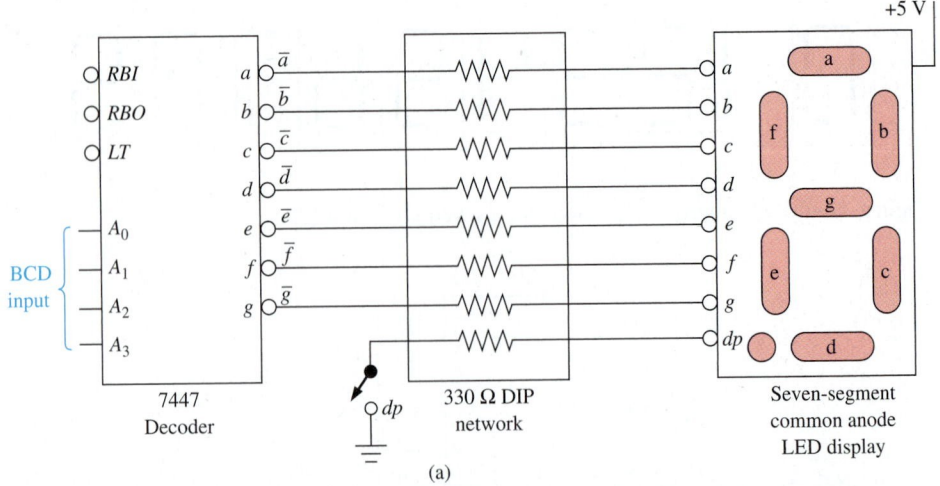

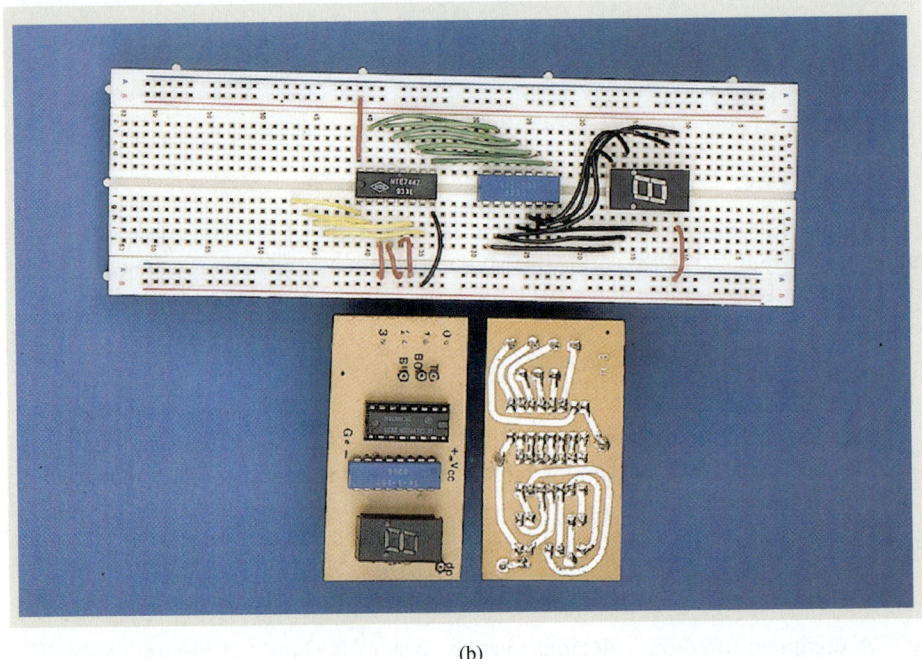

(b)

Figure 12–48 Driving a seven-segment LED display: (a) logic circuit connections; (b) photo of the actual circuit on a breadboard and a printed circuit.

Intelligent LED displays are also available. These displays contain an integrated logic circuit in the same package with the LEDs. For example, the TIL306 has a built-in BCD counter, a 4-bit data latch, a BCD-to-seven-segment decoder, and the drive circuitry along with the display LEDs. For displaying hexadecimal digits, the TIL311 can be used. It doesn't contain a counter like the TIL306, but it does have a latch, decoder, and driver along with the LED display. It accepts a 4-bit hex input and displays the 16 digits 0 through F.

Driving a Multiplexed Display with a Microcontroller

Multidigit LED or LCD displays are commonly used in microprocessor systems. To drive each digit of a six-digit display using separate, dedicated drivers would require six 8-bit I/O ports. Instead, a *multiplexing* scheme is usually used. Using the

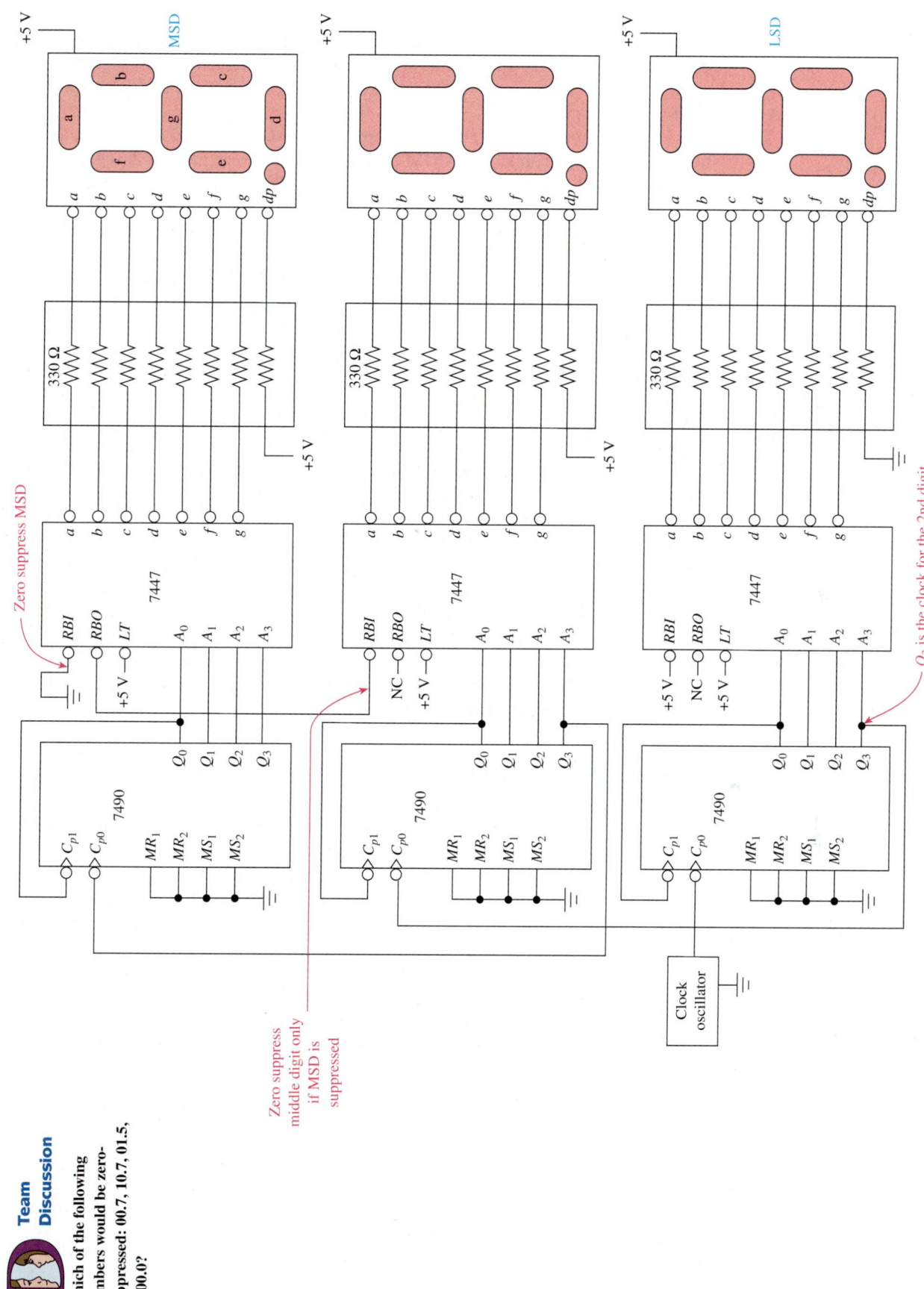

Figure 12–49 Complete three-digit decimal display system.

541

multiplexing technique, up to eight digits can be driven by using only two output ports. One output port is used to select which *digit* is to be active, whereas the other port is used to drive the appropriate *segments* within the selected digit. Figure 12–50 shows how two I/O ports of an 8051 microcontroller can be used to drive a six-digit multiplexed display.

The displays used in Figure 12–50 are common-cathode LEDs. To enable a digit to work, the connection labeled COM must be grounded. The individual segments are then illuminated by supplying +5 V via a 150-Ω limiting resistor to the appropriate segment.

It takes approximately 10 mA to illuminate a single segment. If all segments in one digit are on, as with the number 8, the current in the COM line will be 70 mA. The output ports of the 8051 can only sink 1.6 mA. This is why we need the *PNP* transistors set up as current buffers. When port 1 outputs a 0 on bit 0, the first *PNP* turns on, shorting the emitter to collector. This allows current to flow from the +5-V supply through the 150-Ω limiting resistor, to the *a* segments. None of the *a* segments will illuminate unless one of the digits' COM lines is brought LOW. To enable the LSD, port 2 will output a 0 on bit 0, which shorts the emitter to collector of that transistor. This provides a path for current to flow from the COM on the LSD to ground.

Notice that enabling both the segment and the digit requires an active-LOW signal. To drive all six digits, we have to *scan* the entire display repeatedly with the appropriate numbers to be displayed. For example, to display the number 123456, we need to turn on the segments for the number 1 (*b* and *c*) and then turn on the MSD. We then turn off all digits, turn on the segments for the number 2 (*a, b, g, e,* and *d*), and turn on the next digit. We then turn off all digits, turn on the segments for the number 3, and turn on the next digit. This process repeats until all six digits have been flashed on once. At that point, the MSD is cycled back on, followed by each of the next digits. By repeating this cycle over and over again, the number 123456 appears to be on all the time. This process of decoding the segments and scanning the digits is performed by software instructions written for the 8051 output ports.

VHDL Description of the Seven-Segment Decoder

The decoding feature of the 7447 IC can easily be described in VHDL using the Selected Signal Assignment or Case statement (see Chapter 8 for VHDL decoder techniques). The 7447 decodes a 4-bit BCD input into seven individual active-LOW outputs. The active-LOW outputs are then connected to a common-anode seven-segment LED to display the 10 decimal digits (0 to 9). Before designing a BCD decoder, it is helpful to draw a truth table to identify which outputs go LOW for each BCD input combination (see Table 12–2).

For example, comparing the table entry for the decimal number 0 to the numeric designations given back in Figure 12–46 shows us that all segments should be ON except the *g* segment. (Since we are connecting to a common-anode display, turning a segment ON requires a LOW.) Compare the remainder of Table 12–2 to the designations given in Figure 12–46.

The VHDL program in Figure 12–51 uses a Selected Signal Assignment statement to assign the appropriate output levels for each BCD input. Internal SIGNAL vectors are declared for *bcd_in* and *out_segs*. Before using *bcd_in* in the SELECT statement, it is assigned the concatenation of all four BCD bits (A3 & A2 & A1 & A0) input from the CPLD input ports. After the *out_segs* selected signal assignments are complete, they are separated out to the individual output pins (na, nb, nc, nd, ne, nf, and ng) using seven assignment statements.

Figure 12–52 shows a simulation of the BCD decoding operation.

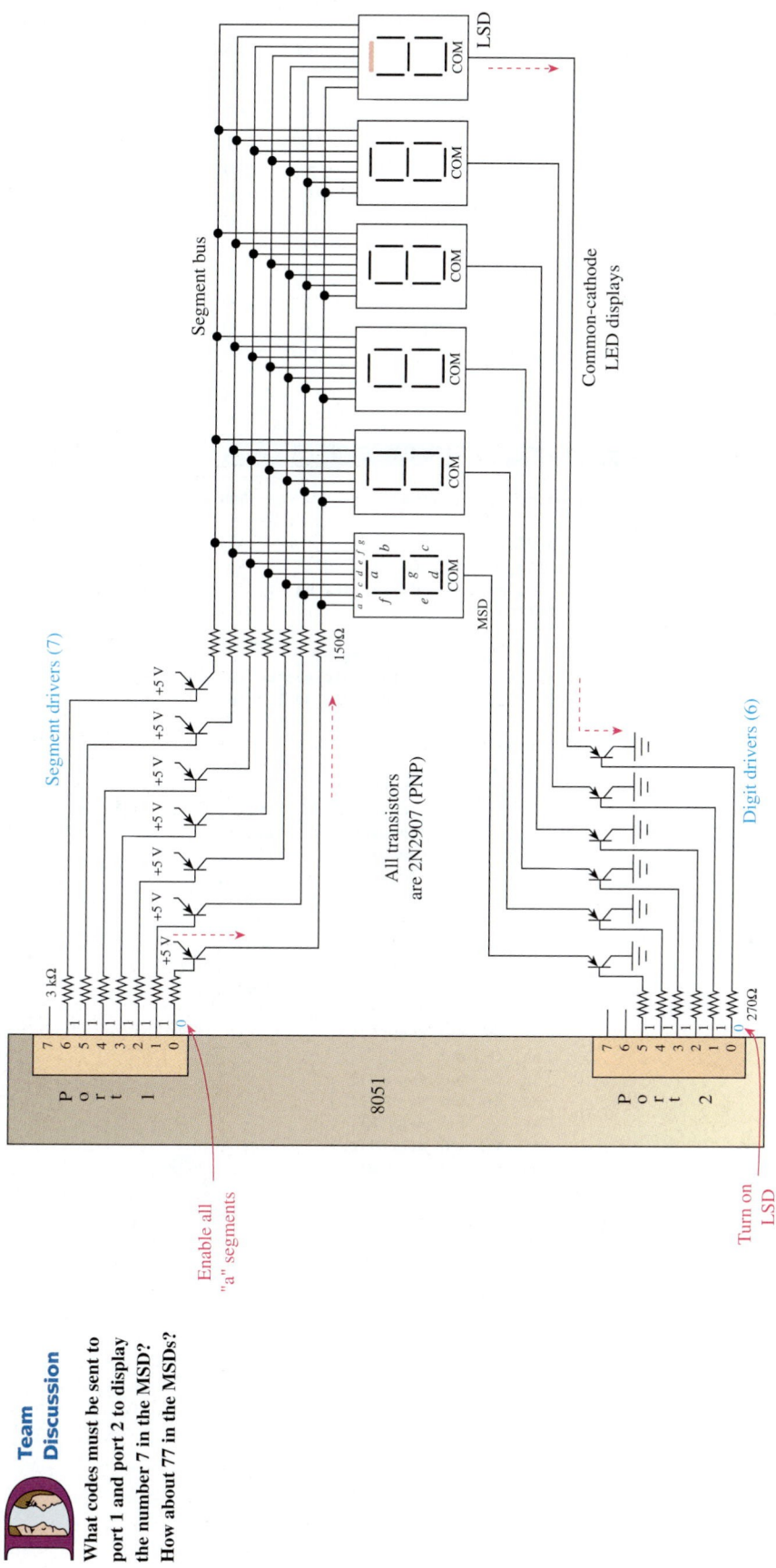

Figure 12–50 Multiplexed six-digit display with the *a* segment of the LSD illuminated. (From *Digital and Microprocessor Fundamentals*, Fourth Edition, by William Kleitz, Prentice Hall, Upper Saddle River, NJ., 2003.)

Team Discussion

What codes must be sent to port 1 and port 2 to display the number 7 in the MSD? How about 77 in the MSDs?

TABLE 12–2 | Truth Table for a BCD to Seven-Segment Decoder

A3	A2	A1	A0	na	nb	nc	nd	ne	nf	ng
0	0	0	0	0	0	0	0	0	0	1
0	0	0	1	1	0	0	1	1	1	1
0	0	1	0	0	0	1	0	0	1	0
0	0	1	1	0	0	0	0	1	1	0
0	1	0	0	1	0	0	1	1	0	0
0	1	0	1	0	1	0	0	1	0	0
0	1	1	0	1	1	0	0	0	0	0
0	1	1	1	0	0	0	1	1	1	1
1	0	0	0	0	0	0	0	0	0	0
1	0	0	1	0	0	0	1	1	0	0

```
bcdto7seg.vhd - Text Editor                                    _□×
LIBRARY ieee;                          ----------------------------
USE ieee.std_logic_1164.ALL;           -- BCD to 7 Segment Decoder --
                                       ----------------------------
ENTITY bcdto7seg IS
    PORT(A3, A2, A1, A0                : IN    std_logic;
         na, nb, nc, nd, ne, nf, ng    : OUT   std_logic);
END bcdto7seg;


ARCHITECTURE arc OF bcdto7seg IS
    SIGNAL bcd_in   : std_logic_vector (3 DOWNTO 0);
    SIGNAL out_segs : std_logic_vector (6 DOWNTO 0);
BEGIN
    bcd_in <= A3 & A2 & A1 & A0;        ← Concatenate 4 inputs
    WITH bcd_in SELECT                     into internal SIGNAL
       out_segs <= "0000001" WHEN "0000",  └ a segment (active-LOW)
                   "1001111" WHEN "0001",
                   "0010010" WHEN "0010",  ← g segment (active-LOW)
                   "0000110" WHEN "0011",
                   "1001100" WHEN "0100",
                   "0100100" WHEN "0101",
                   "1100000" WHEN "0110",
                   "0001111" WHEN "0111",
                   "0000000" WHEN "1000",
                   "0001100" WHEN "1001",
                   "1111111" WHEN others;
    na  <= out_segs(6);
    nb  <= out_segs(5);
    nc  <= out_segs(4);
    nd  <= out_segs(3);        Separate internal SIGNAL vector
    ne  <= out_segs(2);        into individual outputs
    nf  <= out_segs(1);
    ng  <= out_segs(0);
END arc;                                            bcdto7seg.vhd
Line   1   Col   1    INS ◄
```

Figure 12–51 VHDL program for a BCD to seven-segment decoder.

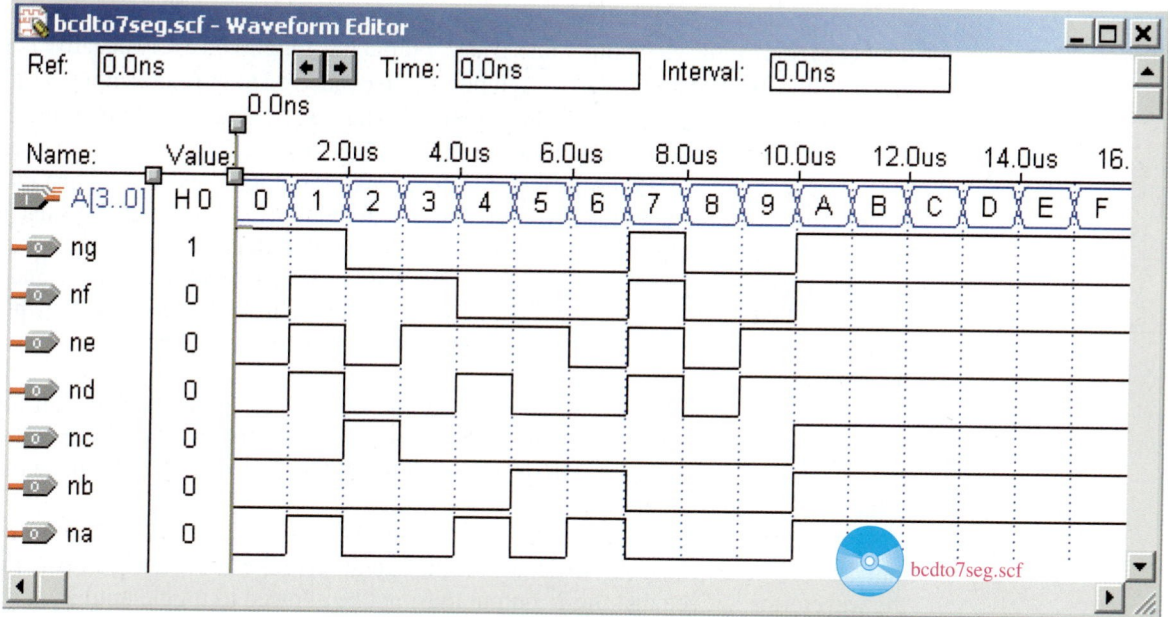

Figure 12–52 Simulation of the BCD to seven-segment decoder.

Review Questions

12–13. Seven-segment displays are either common anode or common cathode. What does this mean?

12–14. List the active segments, by letter, that form the following digits on a seven-segment display: 5, 0.

12–15. Why are series resistors required when driving a seven-segment LED display?

12–16. The 7447 IC is used to convert _____ data into _____ data for common- _____ LEDs.

12–17. Liquid crystal displays (LCDs) use more power but are capable of emitting a brighter light than LEDs. True or false?

12–18. What is the advantage of using a multiplexing scheme for multi-digit displays like the one shown in Figure 12–50?

12–7 Synchronous Counters

Remember the problems we discussed with ripple counters due to the accumulated propagation delay of the clock from flip-flop to flip-flop? (See Figure 12–11.) Well, synchronous counters eliminate that problem because all the clock inputs ($\overline{C_p}$'s) are tied to a common clock input line, so each flip-flop will be triggered at the same time (thus, any Q output transitions will occur at the same time).

If we want to design a 4-bit synchronous counter, we need four flip-flops, giving us a MOD-16 (2^4) binary counter. Keep in mind that, because all the $\overline{C_p}$ inputs receive a trigger at the same time, we must hold certain flip-flops from making output transitions until it is their turn. To design the connection scheme for the synchronous counter, let's first study the output waveforms of a 4-bit binary counter to determine which flip-flops are to be held from toggling, and when.

Common Misconception

Students sometimes mistakenly think that the propagation delay of the AND gates would keep the outputs from switching at the same time.

From the waveforms in Figure 12–53(a), we can see that the 2^0 output is a continuous toggle off the clock input line. The 2^1 output line toggles on every negative edge of the 2^0 line, but because the 2^1's $\overline{C_{p0}}$ input is also connected to the clock input, it must be held from toggling until the 2^0 line is HIGH. This can be done simply by tying the J and K inputs to the 2^0 line, as shown in Figure 12–53(b).

The same logic follows through for the 2^2 and 2^3 output lines. The 2^2 line must be held from toggling until the 2^0 *and* 2^1 lines are both HIGH. Also, the 2^3 line must be held from toggling until the 2^0 *and* 2^1 *and* 2^2 lines are all HIGH.

To keep the appropriate flip-flops in the *hold* or *toggle* condition, their J and K inputs are tied together and, through the use of additional AND gates, as shown in Figure 12–53(b), the J-K inputs will be both 0 or 1, depending on whether they are to be in the hold or toggle mode.

From Figure 12–53(b), we can see that the same clock input is driving all four flip-flops. The 2^1 flip-flop will be in the hold mode ($J_1 = K_1 = 0$) until the 2^0 output goes HIGH, which will force J_1-K_1 HIGH, allowing the 2^1 flip-flop to toggle when the next negative clock edge comes in.

Now, observe the output waveforms [Figure 12–53(a)] while you look at the circuit design [Figure 12–53(b)] to determine the operation of the last two flip-flops. From the waveforms, we see that the 2^2 output must not be allowed to toggle until 2^0 *and* 2^1 are both HIGH. Well, the first AND gate in Figure 12–53(b) takes care of that by holding J_2-K_2 LOW. The same method is used to keep the 2^3 output from toggling until the 2^0 *and* 2^1 *and* 2^2 outputs are *all* HIGH.

As you can see, the circuit is more complicated, but the cumulative effect of propagation delays through the flip-flops is not a problem as it was in ripple counters. This

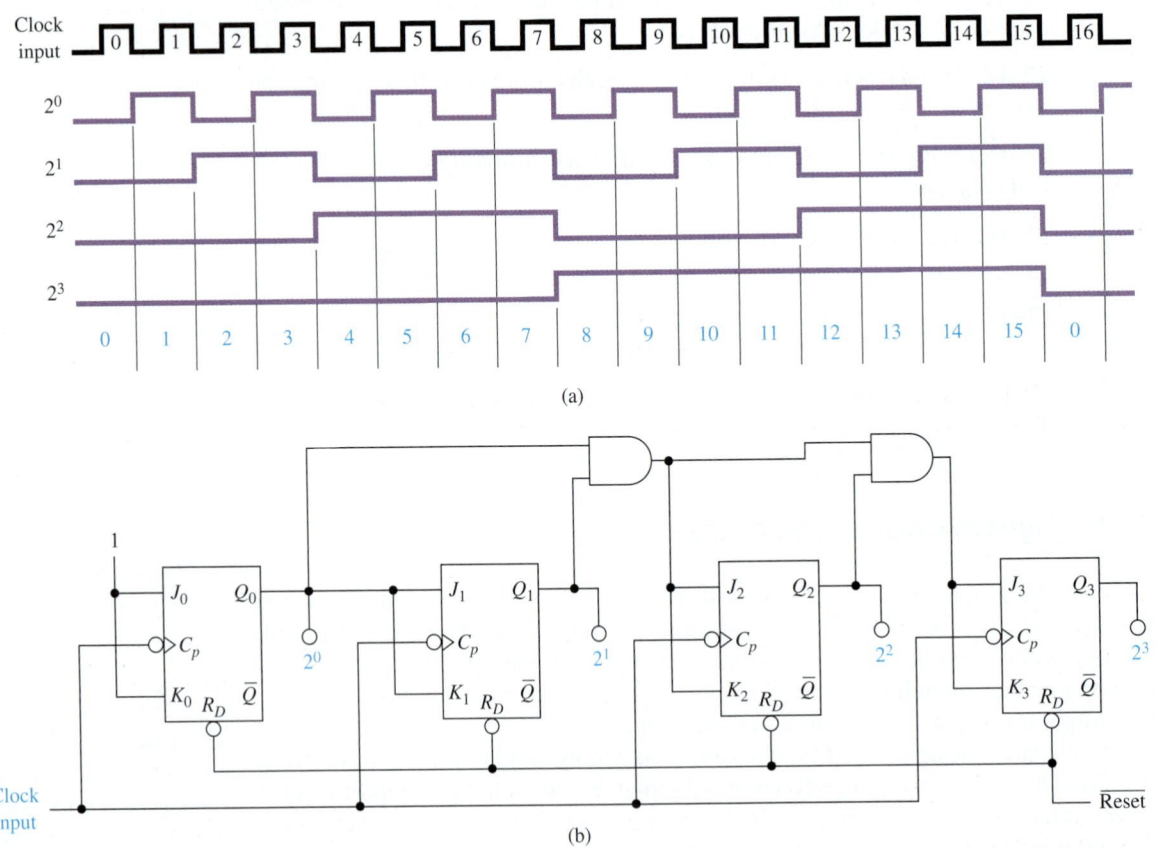

Figure 12–53 Four-bit MOD-16 synchronous counter: (a) output waveforms; (b) circuit connections.

is because all output transitions will occur at the same time, because all flip-flops are triggered from the same input line. (There *is* a propagation delay through the AND gates, but it will not affect the Q outputs of the flip-flops.)

As with ripple counters, synchronous counters can be used as down-counters by taking the output from the $\overline{Q}$ outputs and can form any modulus count by resetting the count to zero after some predetermined binary number has been reached.

EXAMPLE 12–21

Design a MOD-6 synchronous binary up-counter.

Solution: A MOD-6 counter will count 0–1–2–3–4–5–0–1–, and so on. To count to 5, we will need three flip-flops and will have to Reset the count to zero when the number 6 (110_2) is reached, as shown in the circuit in Figure 12–54.

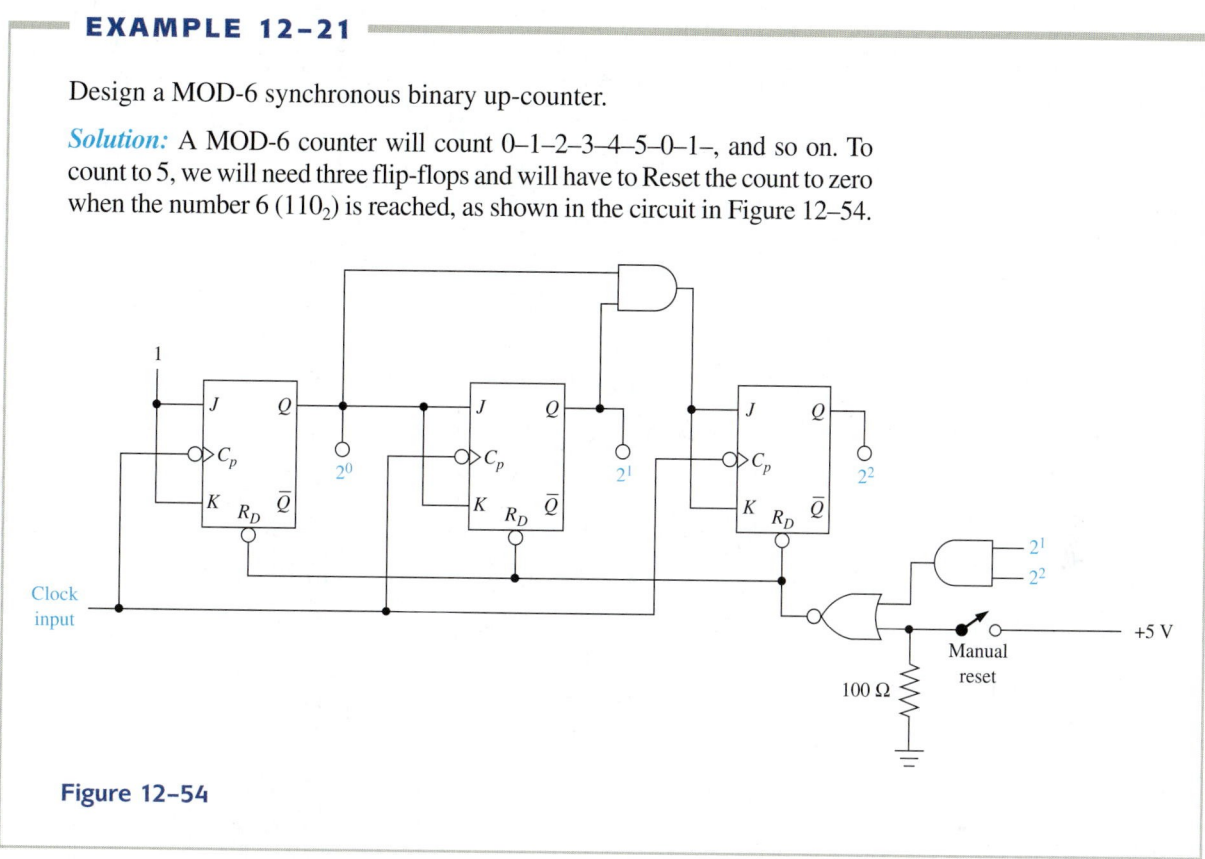

Figure 12–54

System Design Application

Synchronous binary counters have many applications in the timing and sequencing of digital systems. The following design will illustrate one of these applications.

EXAMPLE 12–22

Let's say that your company needs a system that will count the number of hours of darkness each day. The senior design engineer for your company will be connecting his microcontroller-based system to your counter outputs after you are sure that your system is working correctly. After the counter outputs are read, the microcontroller will issue a LOW Reset pulse to your counter to Reset it to all zeros sometime before sunset.

Solution: You decide to use a synchronous counter but realize that it may be dark outside for as many as 18 h per day. A 4-bit counter will not count high enough, so first you have to come up with the 5-bit synchronous counter design that is shown in Figure 12–55. That was not hard; you just had to add one more AND gate and a flip-flop to a 4-bit counter.

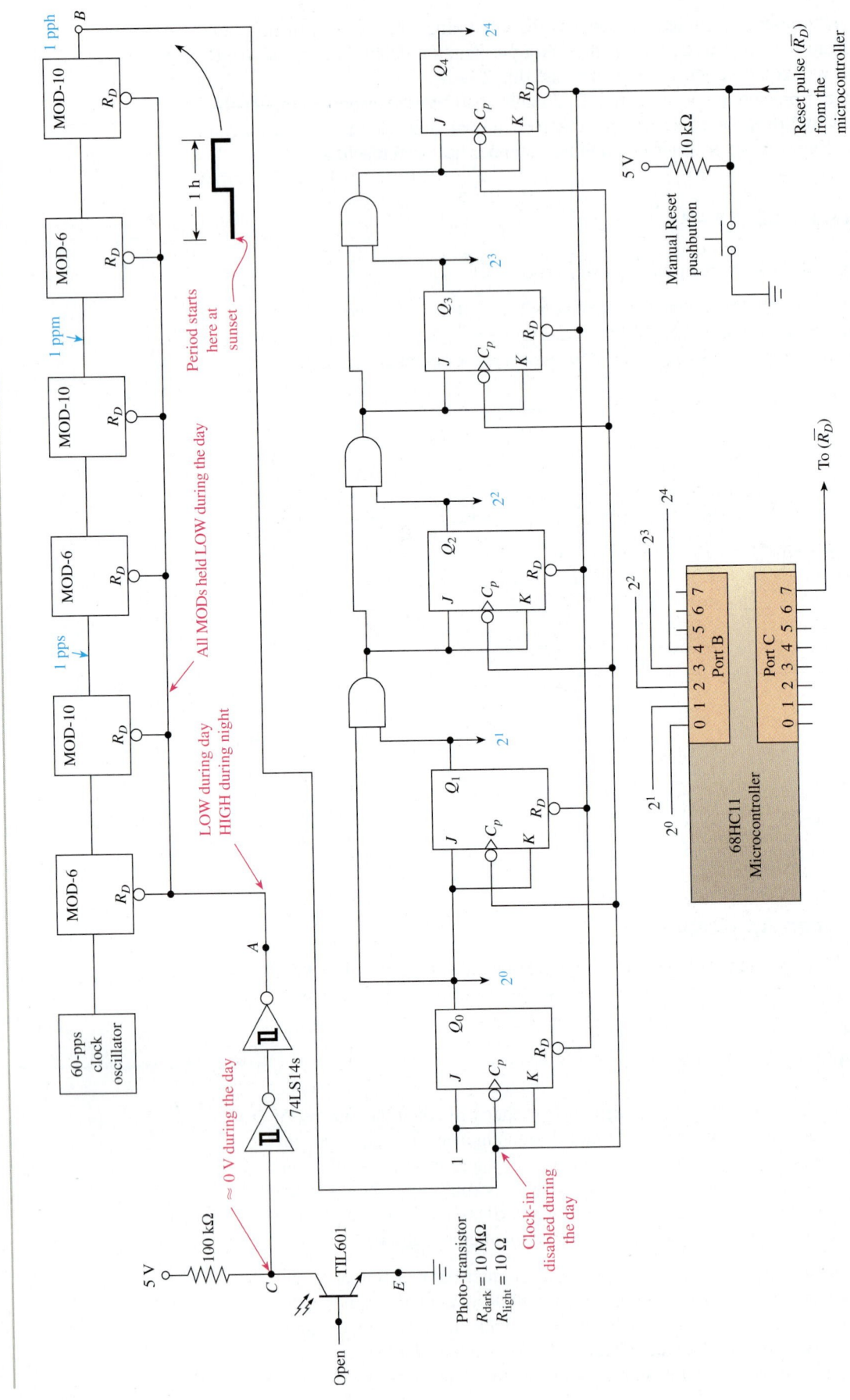

Figure 12–55 System design solution for the "hours of darkness counter."

From analog electronics, you remembered that a phototransistor has varying resistance from collector to emitter, depending on how much light strikes it. The **phototransistor** that you decide to use has a resistance of 10 MΩ when it is in the dark and 10 Ω when it is in the daylight. Your final circuit design is shown in Figure 12–55.

Explanation of Figure 12–55: Let's start with the 5-bit synchronous counter. With the addition of the last AND gate and flip-flop, it will be capable of counting from 0 up to 31 (MOD-32). The manual Reset push button, when depressed, will Reset the counter to zero.

The phototransistor collector-to-ground voltage will be almost zero during daylight because the collector-to-emitter resistance acts almost like a short. (Depending on the transistor used, the ON resistance may be as low as 10 Ω.) The Schmitt inverters are used to give a sharp HIGH-to-LOW and LOW-to-HIGH at sunset and sunrise to eliminate any false clock switching. Schmitt triggers are most commonly available as inverting functions, so two of them are necessary so that a LOW at the collector will come through as a LOW at point A.

The LOW at point A during the daylight will hold all the MOD counters (divide-by-N's) at zero so that at the beginning of sunset the waveform at point B will start out LOW and take one full hour before it goes HIGH to LOW, triggering the first transition at Q. During the nighttime, point B will **oscillate** at one pulse per hour, incrementing the counter once each hour. At sunrise, point A goes LOW, forcing all MODs LOW and disabling the clock. The counter outputs at Q_0 to Q_4 will be read by the microcontroller during the day and then Reset.

Review Questions

12–19. What advantage do synchronous counters have over ripple counters?

12–20. Because each flip-flop in a synchronous counter is driven by the same clock input, what keeps *all* flip-flops from toggling at each active clock edge?

12–21. The 5-bit synchronous counter in Figure 12–55 counts continuously, day and night, but is ignored by the microcontroller during the day. True or false?

12–8 Synchronous Up/Down-Counter ICs

Four-bit synchronous binary counters are available in a single IC package. Two popular synchronous IC counters are the 74192 and 74193. They both have some features that were not available on the ripple counter ICs. They can count *up or down* and can be *preset* to any count that you desire. The 74192 is a BCD decade **up/down-counter,** and the 74193 is a 4-bit binary up/down-counter. The logic symbol used for both counters is shown in Figure 12–56.

There are two separate clock inputs: C_{pU} for counting up, and C_{pD} for counting down. One clock must be held HIGH while counting with the other. The binary output count is taken from Q_0 to Q_3, which are the outputs from four internal J-K flip-flops. The Master Reset *(MR)* is an active-HIGH Reset for resetting the Q outputs to zero.

The counter can be preset by placing any binary value on the parallel data inputs $(D_0$ to $D_3)$ and then driving the Parallel Load $(\overline{PL})$ line LOW. The parallel load operation will change the counter outputs regardless of the conditions of the clock inputs.

Helpful Hint

The 74HC192 and 74HC193 are high-speed CMOS counters. They function the same but are faster and use less power.

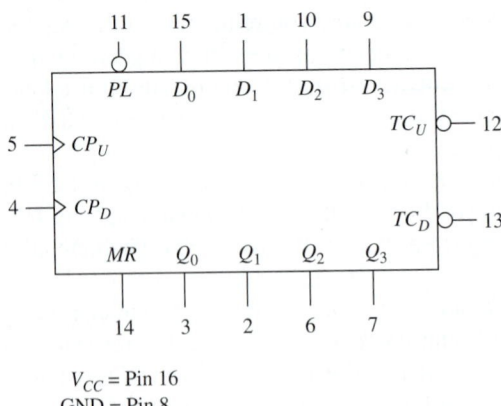

V_{CC} = Pin 16
GND = Pin 8

Figure 12–56   Logic symbol for the 74192 and 74193 synchronous counter ICs.

The **Terminal Count** Up $(\overline{TC_U})$ and Terminal Count Down $(\overline{TC_D})$ are normally HIGH. The $\overline{TC_U}$ is used to indicate that the maximum count is reached and the count is about to recycle to zero (carry condition). The $\overline{TC_U}$ line goes LOW for the 74193 when the count reaches 15 *and* the input clock (C_{pU}) goes HIGH to LOW. $\overline{TC_U}$ remains LOW until C_{pU} returns HIGH. This LOW pulse at $\overline{TC_U}$ can be used as a clock input to the next-higher-order stage of a multistage counter.

The $\overline{TC_U}$ output for the 74192 is similar, except that it goes LOW at 9 *and* a LOW C_{pU} (see Figure 12–57). The Boolean equations for $\overline{TC_U}$, therefore, are as follows:

$$\text{LOW at } \overline{TC_U} = Q_0 Q_1 Q_2 Q_3 \overline{C_{pU}} \quad (74193)$$

$$\text{LOW at } \overline{TC_U} = Q_0 Q_3 \overline{C_{pU}} \quad (74192)$$

The Terminal Count Down $(\overline{TC_D})$ is used to indicate that the minimum count is reached and the count is about to recycle to the maximum (15 or 9) count (borrow condition). Therefore, $\overline{TC_D}$ goes LOW when the down-count reaches zero and the input clock (C_{pD}) goes LOW (see Figure 12–59). The Boolean equation at $\overline{TC_D}$ is

$$\text{LOW at } \overline{TC_D} = \overline{Q_0}\,\overline{Q_1}\,\overline{Q_2}\,\overline{Q_3}\,\overline{C_{pD}} \quad (74192 \text{ and } 74193)$$

The function table shown in Table 12–3 can be used to show the four operating modes (Reset, Load, Count up, and Count down) of the 74192/74193.

The best way to illustrate how these chips operate is to exercise all their functions and observe the resultant waveforms, as shown in the following examples.

TABLE 12–3 | Function Table for the 74192/74193 Synchronous Counter IC[a]

Operating Mode	Inputs								Outputs					
	MR	$\overline{PL}$	C_{pU}	C_{pD}	D_0	D_1	D_2	D_3	Q_0	Q_1	Q_2	Q_3	$\overline{TC_U}$	$\overline{TC_D}$
Reset	H	×	×	L	×	×	×	×	L	L	L	L	H	L
	H	×	×	H	×	×	×	×	L	L	L	L	H	H
Parallel load	L	L	×	L	L	L	L	L	L	L	L	L	H	L
	L	L	×	H	L	L	L	L	L	L	L	L	H	H
	L	L	L	×	H	H	H	H	H	H	H	H	L	H
	L	L	H	×	H	H	H	H	H	H	H	H	H	H
Count up	L	H	↑	H	×	×	×	×	Count up				H	H
Count down	L	H	H	↑	×	×	×	×	Count down				H	H

[a]H = HIGH voltage level; L = LOW voltage level; × = don't care; ↑ = LOW-to-HIGH clock transition.

550   CHAPTER 12 | COUNTER CIRCUITS AND VHDL STATE MACHINES

Draw the input and output timing waveforms for a 74192 that goes through the following sequence of operation:

1. Reset all outputs to zero.
2. Parallel load a 7 (0111).
3. Count up five counts.
4. Count down five counts.

Solution: The timing waveforms are shown in Figure 12–57.

Common Misconception

Students often mistakenly draw the LOW at $\overline{TC}$ for the entire terminal count clock period instead of only while the clock is LOW.

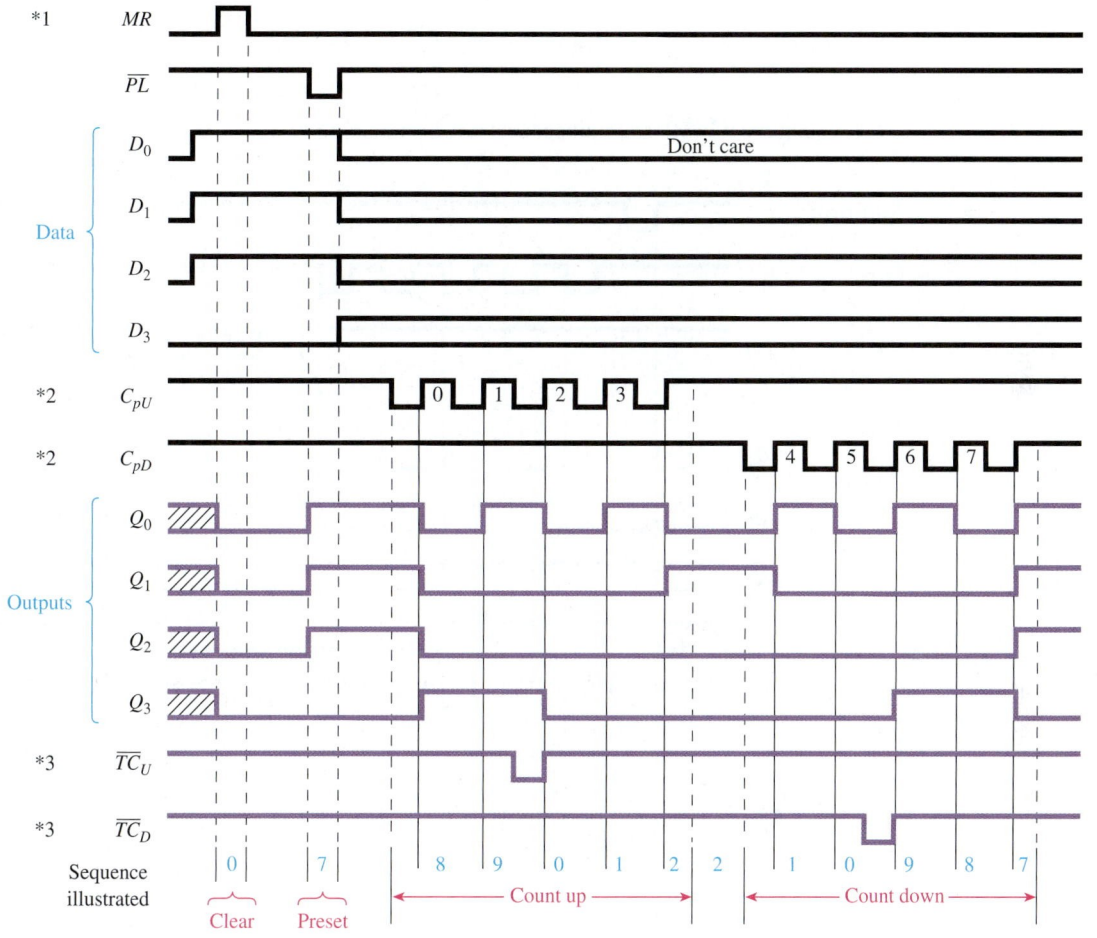

Notes
1. Clear overrides load data and count inputs.
2. When counting up, count-down input must be HIGH; when counting down, count-up input must be HIGH.
3. $\overline{TC}$ follows the clock during the terminal count.

Figure 12–57 Timing waveforms for the 74192 used in Example 12–23.

EXAMPLE 12-24

Draw the output waveforms for the 74193 shown in Figure 12–58, given the waveforms shown in Figure 12–59. (Initially, set $D_0 = 1$, $D_1 = 0$, $D_2 = 1$, $D_3 = 1$, and $MR = 0$.)

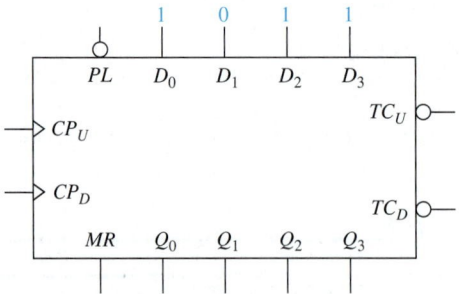

Figure 12–58 Circuit connections for Example 12–24.

Solution:

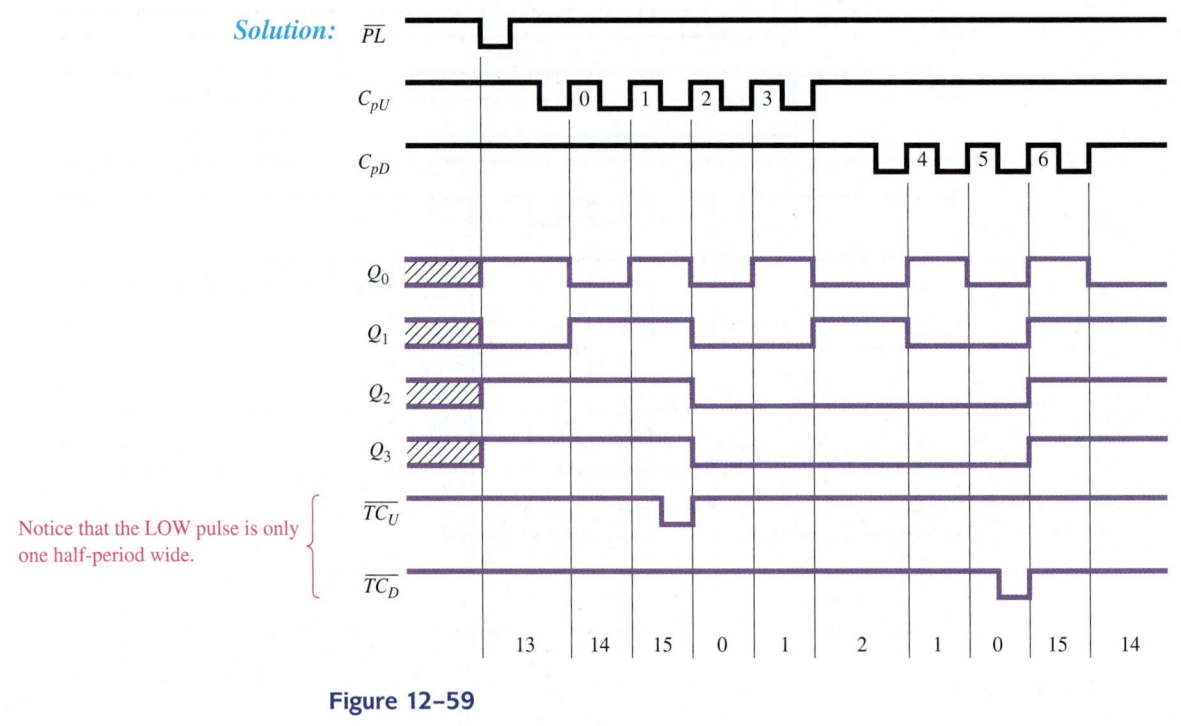

Notice that the LOW pulse is only one half-period wide.

Figure 12–59

EXAMPLE 12-25

Design a decimal counter that will count from 00 to 99 using two 74192 counters and the necessary drive circuitry for the two-digit display. (Display circuitry was explained in Section 12–5.)

Solution: The 74192s can be used to form a multistage counter by connecting the $\overline{TC_U}$ of the first counter to the C_{pU} of the second counter. $\overline{TC_U}$ will go LOW, then HIGH, when the first counter goes from 9 to 0 (carry). This LOW-to-HIGH edge can be used as the clock input to the second stage, as shown in Figure 12–60.

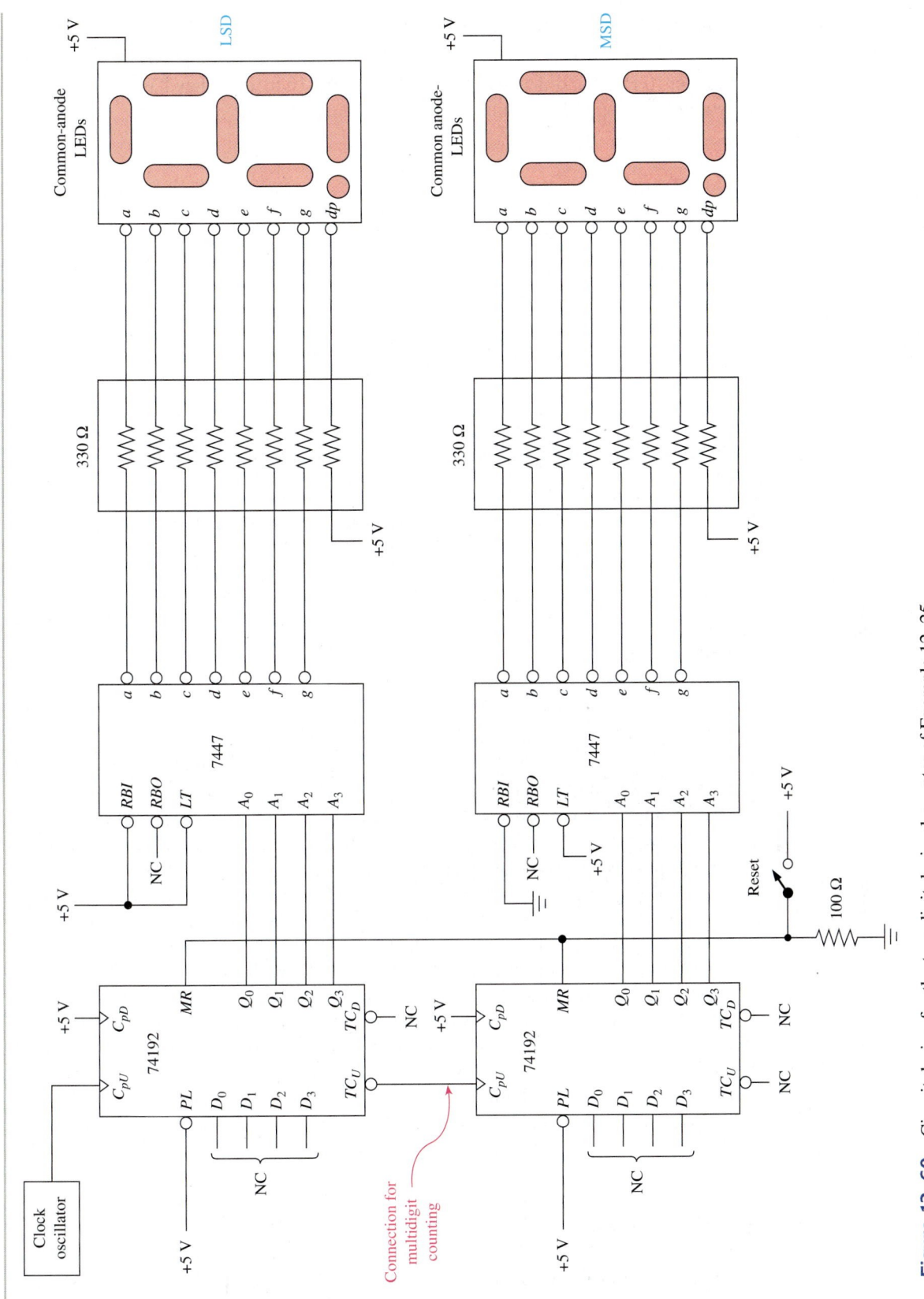

Figure 12-60 Circuit design for the two-digit decimal counter of Example 12–25.

74190/74191 Synchronous Counter ICs

Other forms of synchronous counters are the 74190 and 74191. The 74190 is a BCD counter (0 to 9), and the 74191 is a 4-bit counter (0 to 15). They have some different features and input/output pins, as shown in Figure 12–61.

Helpful Hint

The 74190- and 74160-series counters have several additional features and are, therefore, more complicated to use. If you have a data book or access to the Internet, this is a good opportunity for you to try to figure out the operation of the counter by reading the manufacturer's description instead of the textbook.

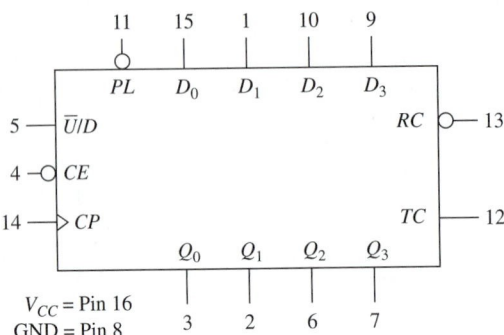

V_{CC} = Pin 16
GND = Pin 8

Figure 12–61 Logic symbol for the 74190/74191 synchronous counters.

The 74190/74191 can be preset to any count by using the **Parallel Load** $(\overline{PL})$ operation. It can count up or down by using the $\overline{U}/D$ input. With $\overline{U}/D = 0$, it will count up, and with $\overline{U}/D = 1$, it will count down. The Count Enable input $(\overline{CE})$ is an active-LOW input used to enable/inhibit the counter. With $\overline{CE} = 0$, the counter is enabled. With $\overline{CE} = 1$, the counter stops and holds the current states of the Q_0 to Q_3 outputs.

The Terminal Count output *(TC)* is normally LOW, but it goes HIGH when the counter reaches zero in the count-down mode and 15 (or 9) in the count-up mode. The ripple clock output $(\overline{RC})$ follows the input clock (C_p) whenever *TC* is HIGH. In other words, in the count-down mode, when zero is reached, $\overline{RC}$ will go LOW when C_p goes LOW. The $\overline{RC}$ output can be used as a clock input to the next higher stage of a multistage counter, just the way that the $\overline{TC}$ outputs of the 74192/74193 were used. In either case, however, the multistage counter will not be truly synchronous because of the small propagation delay from C_p to $\overline{RC}$ of each counter.

For a multistage counter to be *truly* synchronous, the C_p of each stage must be connected to the *same* clock input line. The 74190/74191 counters enable you to do this by using the *TC* output to inhibit each successive stage from counting until the previous stage is at its Terminal Count. Figure 12–62 shows how three 74191s can be connected to form a true 12-bit binary synchronous counter.

In Figure 12–62, we can see that each counter stage is driven by the same clock, making it truly synchronous. The second stage is inhibited from counting until the first stage reaches 15. The second stage will then increment by 1 at the next positive clock edge. Stage 1 will then inhibit stage 2 via the *TC*-to-$\overline{CE}$ connection while stage 1 is counting up to 15 again. The same operation between stages 2 and 3 also keeps stage 3 from incrementing until stages 1 and 2 both reach 15.

74160/61/62/63 Synchronous Counter ICs

Finally, another type of counter allows you to perform true synchronous counting without using external gates, as we had to in Figure 12–62. The 74160/74161/74162/74163 synchronous counter ICs have *two* Count Enable inputs *(CEP* and *CET)* and a Terminal Count output to facilitate high-speed synchronous counting. The logic symbol is given in Figure 12–63. From the logic symbol, we can see that this counter is similar to the previous synchronous counters, except that it has two active-HIGH Count

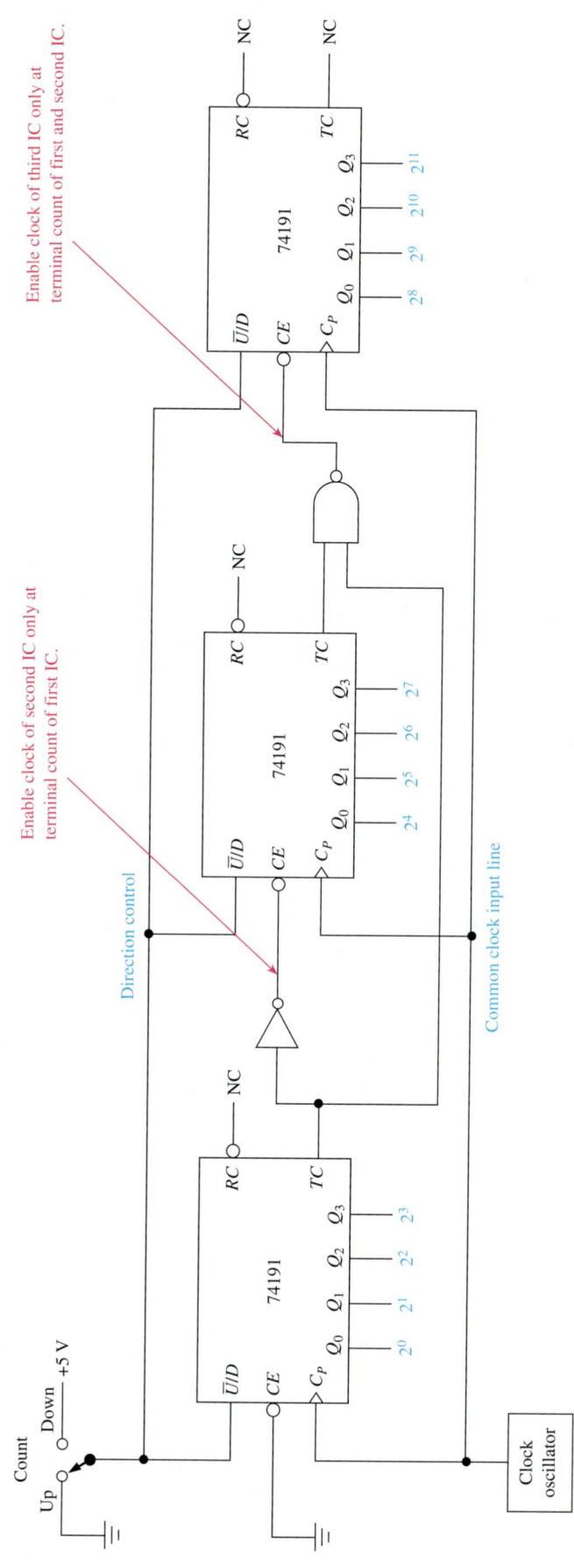

Figure 12–62 Twelve-bit synchronous binary counter using the Terminal Count and Count Enable features.

555

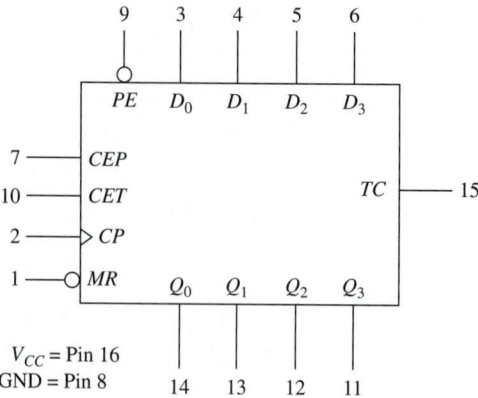

Figure 12–63 Logic symbol for the 74160/74161/74162/74163 synchronous counter.

Enable inputs *(CEP* and *CET)* and an active-HIGH Terminal Count *(TC)* output. (There are other differences between this and other synchronous counters, but we will leave it up to you to determine those from reading your TTL data manual.)

Both count enables *(CEP* and *CET)* must be HIGH to count. The Terminal Count output *(TC)* will go HIGH when the highest count is reached. *TC* will be forced LOW, however, when *CET* goes LOW, even though the highest count may be reached. This is an important feature that enables the multistage counter of Figure 12–64 to operate properly.

Review Questions

12–22. What is the function of the $\overline{TC_U}$ and $\overline{TC_D}$ output pins on the 74193 synchronous counter IC?

12–23. How do you change the 74190 from an up-counter to a down-counter?

12–24. The $\overline{CE}$ input to the 74190 synchronous counter is the *Chip Enable* used to enable/disable the *Q*-outputs. True or false?

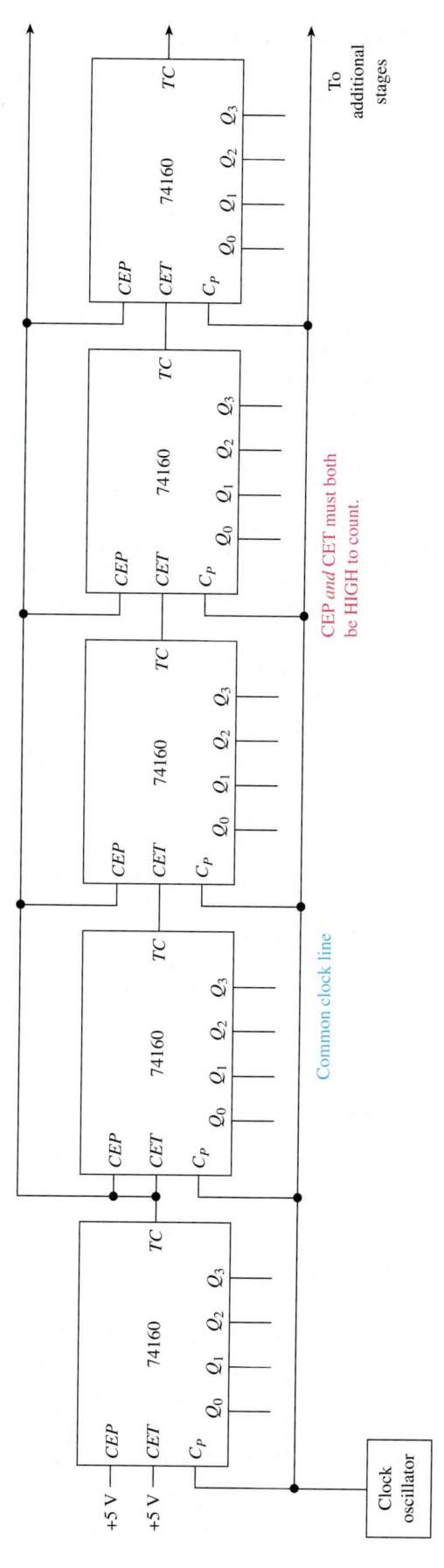

Figure 12–64 High-speed multistage synchronous counter.

12–9 Applications of Synchronous Counter ICs

The following applications will explain some useful design strategy and circuit operations using synchronous counter ICs.

Design a counter that will count up 0 to 9, then down 9 to 0, then up 0 to 9 repeatedly using a synchronous counter and various gates.

Solution: Because the count is 0 to 9, a BCD counter will work. Also, we want to go up, then down, then up, and so on, so it would be easy if we had a reversible counter like the 74190 and just toggled the $\overline{U}/D$ terminal each time the Terminal Count is reached. Figure 12–65 could be used to implement this circuit. When power is first applied, the 74190 will be Parallel Loaded with a 5 (0101), and the direction line will be 1. (The 5 is chosen arbitrarily because it is somewhere between the terminal counts 0 and 9.) The counter will count down to 0, at which time *TC* will go HIGH, causing the flip-flop to toggle and changing the direction to 0. With the clock oscillator still running, the counter will reverse and start counting up. When 9 is reached, *TC* goes HIGH, again changing the direction, and the cycle repeats.

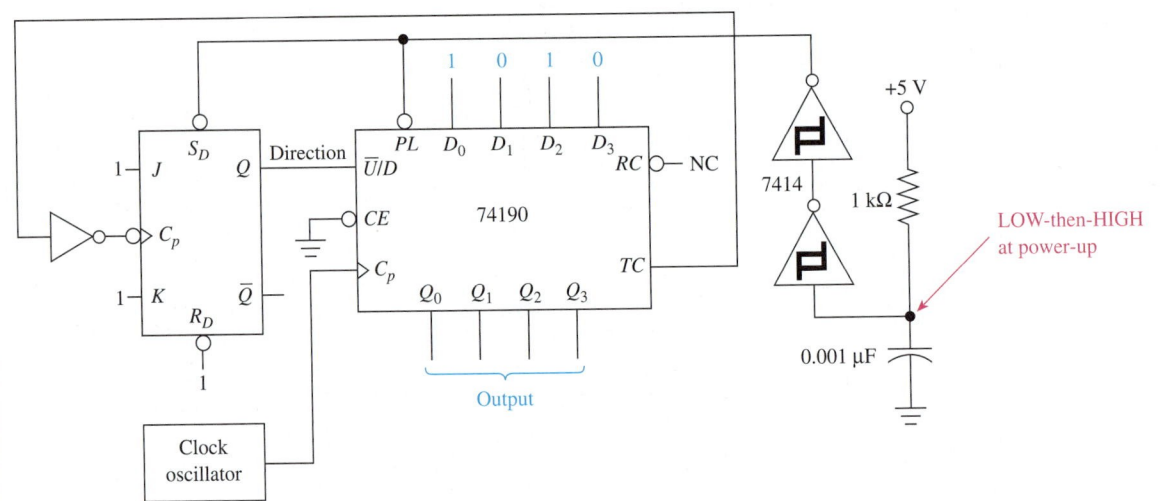

Figure 12–65 Self-reversing BCD counter (solution to Example 12–26).

Design and sketch the timing waveforms for a divide-by-9 frequency divider using a 74193 counter.

Solution: We can use the Parallel Load feature of the 74193 to set the counter at some initial value and then count down to zero. When we reach zero, we will have to Parallel Load the counter to its initial value and count down again, making sure the repetitive cycle repeats once every <u>nine</u> clock periods. Figure 12–66 could be used to implement such a circuit. TC_D is fed back into $\overline{PL}$. This means that, when the Terminal Count is reached, the

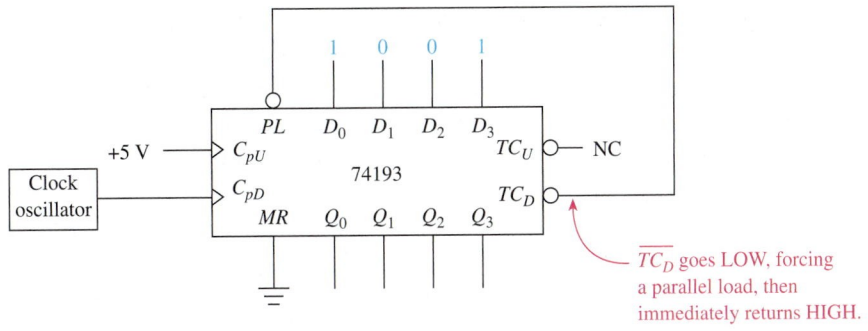

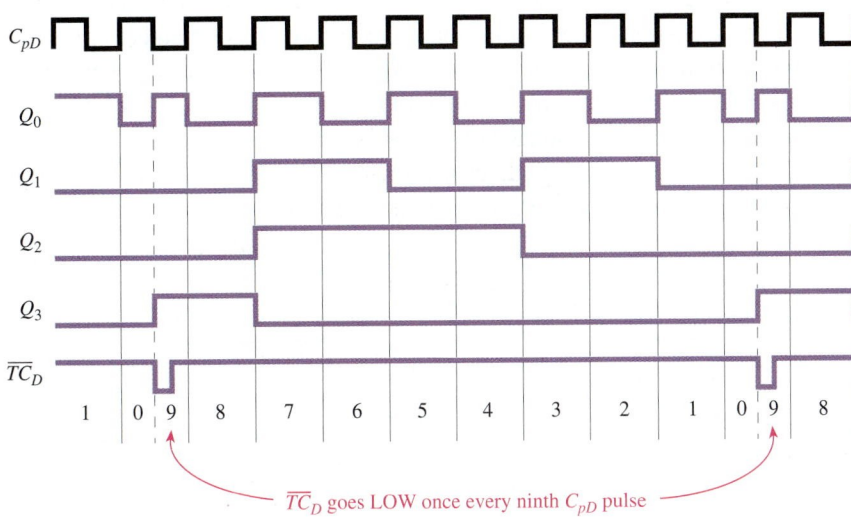

$\overline{TC_D}$ goes LOW, forcing a parallel load, then immediately returns HIGH.

$\overline{TC_D}$ goes LOW once every ninth C_{pD} pulse

Figure 12–66 Circuit design and timing waveforms for a divide-by-9 frequency divider.

LOW out of $\overline{TC_D}$ will enable the Parallel Load, making the outputs equal to the D_0 to D_3 inputs (1001).

The time waveforms arbitrarily start at 1 and count down. Notice at zero (Terminal Count) that $\overline{TC_D}$ goes LOW when C_{pD} goes LOW (remember that a LOW at $\overline{TC_D} = \overline{Q_0} \, \overline{Q_1} \, \overline{Q_2} \, \overline{Q_3} \, \overline{C_{pD}}$). As soon as $\overline{TC_D}$ goes LOW, the outputs return to 9, thus causing $\overline{TC_D}$ to go back HIGH again. Therefore, $\overline{TC_D}$ is a narrow pulse just long enough to perform the Parallel Load operation.

The down-counting resumes until zero is reached again, which causes the Parallel Load of 9 to occur again. The $\overline{TC_D}$ pulse occurs once every ninth C_{pD} pulse; thus, we have a divide-by-9. (A different duty-cycle divide-by-9 can be gotten from the Q_3 or Q_2 outputs.)

> **Helpful Hint**
>
> This circuit configuration works fine as a frequency divider, but not as a counter, because the 0 and 9 appear for only one-half of a period.

EXAMPLE 12–28

Design a divide-by-200 using synchronous counters.

Solution: The number 200 exceeds the maximum count of a single 4-bit counter. Two 4-bit counters can be cascaded together to form an 8-bit counter capable of counting 256 states ($2^8 = 256$).

The 74193 is a logical choice for a 4-bit counter. We can cascade two of them together to form an 8-bit down-counter. If we preload with the binary equivalent of the number 200 and count down to zero, we can use the borrow output ($\overline{TC_{D2}}$) to drive the Parallel Load ($\overline{PL}$) line LOW to recycle back to 200. Figure 12–67 shows the circuit connections to form this 8-bit divide-by-200 counter. The two 74193 counters will start out at some unknown value and start counting down toward zero. The borrow-out ($\overline{TC_{D2}}$) line will go LOW when the count reaches zero and C_{pD2} is LOW. As soon as $\overline{TC_{D2}}$ goes LOW, a Parallel Load of number 200 takes place, making $\overline{TC_{D2}}$ go back HIGH again. Therefore, $\overline{TC_{D2}}$ is just a short glitch, and the number zero will appear at the outputs for just one-half of a clock period, and the number 200 will appear the other one-half of the same clock period. The remainder of the numbers will follow a regular counting sequence (199 down to 1), giving us 200 complete clock pulses between the LOW pulses on $\overline{TC_{D2}}$. If the short glitch on $\overline{TC_{D2}}$ is not wide enough as a divide-by-200 output, it could be widened to produce any duty cycle without affecting the output frequency by using a one-shot multivibrator pulse stretcher. (One shots are discussed in Chapter 14.)

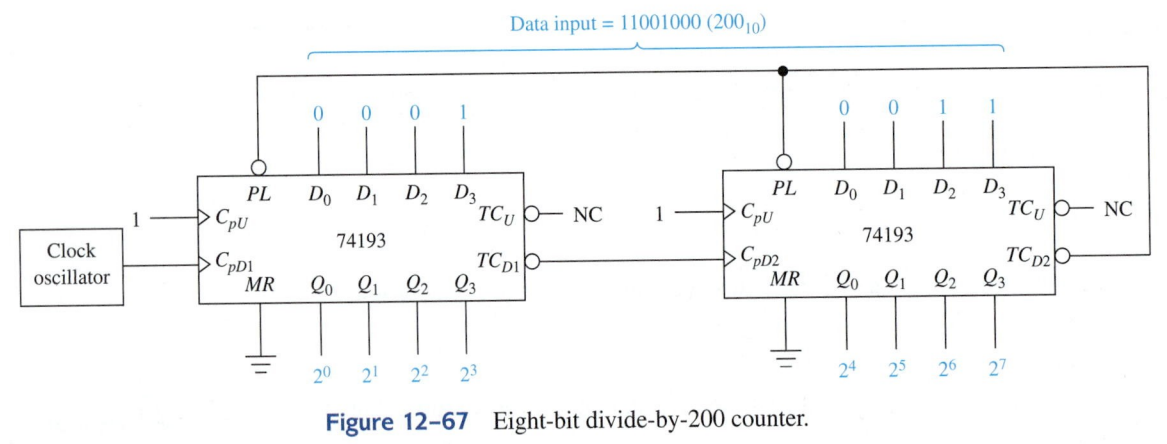

Figure 12–67 Eight-bit divide-by-200 counter.

EXAMPLE 12–29

Use a 74163 to form a MOD-7 synchronous up-counter. Sketch the timing waveforms.

Solution: The 74163 has a *synchronous* Reset feature—that is, a LOW level at the Master Reset ($\overline{MR}$) input will Reset all flip-flops (Q_0 to Q_3) at the next positive clock ($\overline{C_p}$) edge. Therefore, what we can do is bring the Q_1 and Q_2 (binary 6) lines into a NAND gate to drive the $\overline{MR}$ line LOW when the count is at 6. The next positive C_p edge would normally increase the count to 7 but, instead, will Reset the count to 0. The result is a count from 0 to 6, which is a MOD-7.

Remember that with previous MOD-N counters we would look for the number that was one greater than the last number to be counted, and when we reached it, we would Reset the count to zero because we went beyond the modulus required. That method of resetting after the fact works, but it lets a short-duration glitch (unwanted state) through to the outputs before resetting (see Figure 12–20). With the 74163, using a synchronous Reset, as shown in Figure 12–68, we can avoid that problem.

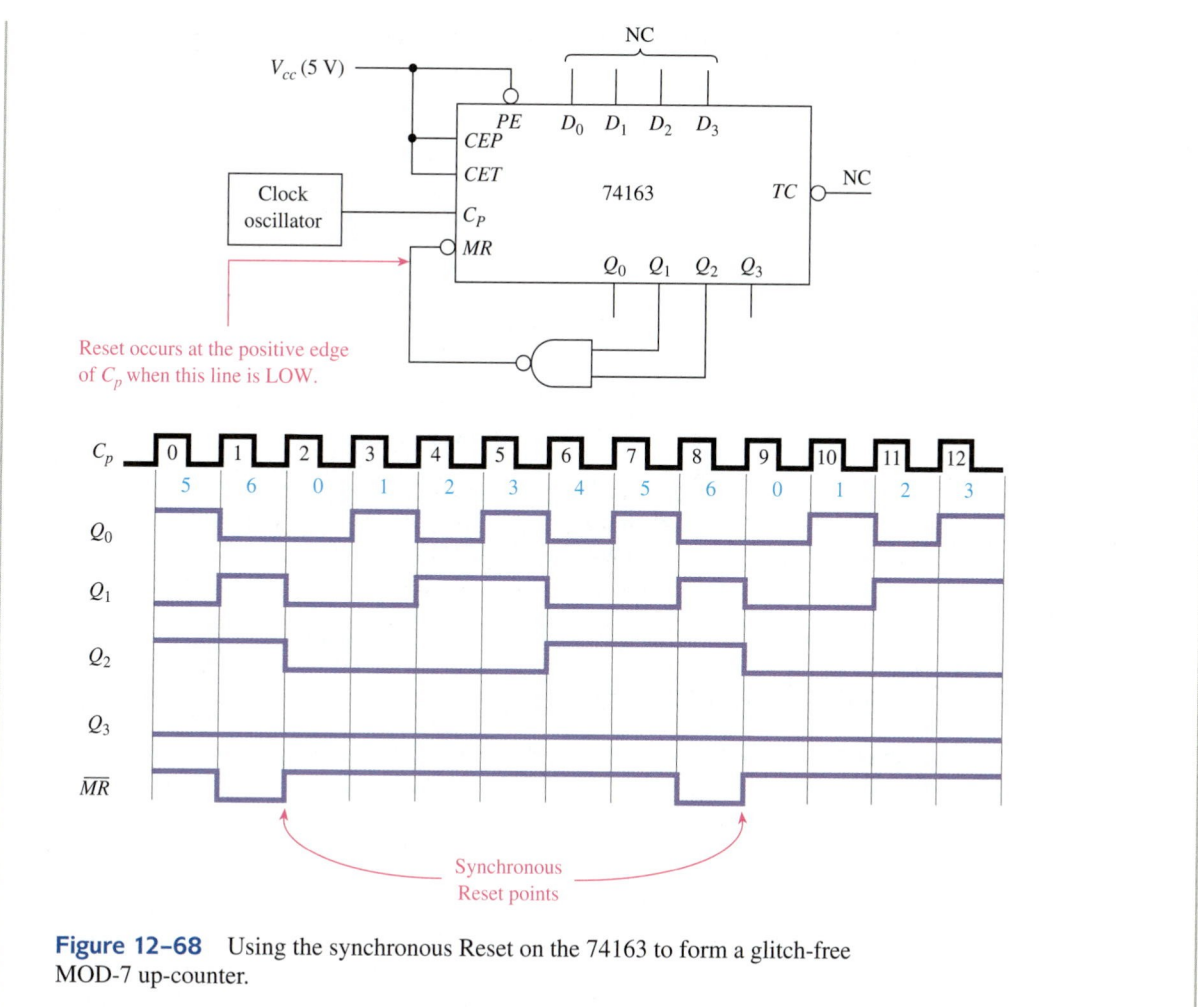

Figure 12–68   Using the synchronous Reset on the 74163 to form a glitch-free MOD-7 up-counter.

12–10 VHDL and LPM Counters

As seen is some of the previous sections, VHDL is an efficient method for designing counters. In this section we will use VHDL to duplicate many of the counter features that we've used with the fixed-function 7400-series ICs such as Parallel Load, up/down counting, and Count Enable. We will also learn to use the LPM counters that are available in the MAX+PLUS II software.

VHDL Up-Counter

Figure 12–69 gives the VHDL listing for a 4-bit up-counter that also provides asynchronous Reset and Parallel Load. It is designed to function similar to the 74193, featuring positive-edge clock triggering and asynchronous inputs that override the synchronous up-counting. The n_pl is an active-LOW Parallel Load that loads the 4-bit counter with the value input on the pl_data inputs. In the PORT assignments, notice that the pl_data input and the q output are declared as integers with a RANGE of 0 to 15. VHDL will automatically allocate a 4-bit data path for both to accommodate values up to $15(1111_2)$. The IF statements are set up so that if either asynchronous input is active it will override the synchronous counting of cp pulses. As you can see, if n_rd is active then q is reset to 0. If n_pl is active, then q is parallel loaded with pl_data. If neither are active, then q is incremented at each positive cp pulse.

```
counter_a.vhd - Text Editor                          _ □ X

LIBRARY ieee;                        --------------------------
USE ieee.std_logic_1164.ALL;         -- 4-bit Up-Counter with --
                                     -- Async Reset and Load  --
ENTITY counter_a IS                  --------------------------
    PORT(cp, n_rd, n_pl      : IN std_logic;
            pl_data          : IN integer RANGE 0 TO 15;
            q                : BUFFER integer RANGE 0 TO 15);
END counter_a;                                              4bit 0000 to 1111


ARCHITECTURE arc OF counter_a IS
BEGIN
    PROCESS (cp, n_rd, n_pl)
    BEGIN
        IF (n_rd='0' AND n_pl='1') THEN
            q<= 0;                                  ── Asynchronous has priority
        ELSIF (n_rd='1' AND n_pl='0') THEN
            q<=pl_data;
        ELSIF (cp'EVENT AND cp='1' ) THEN
            q<=q+1;                                 ── Positive edge trigger
        END IF;
    END PROCESS;
END arc;                                         counter_a.vhd

Line   1    Col   1    INS
```

Figure 12–69 VHDL program for a 4-bit up-counter with asynchronous Reset and Parallel Load.

The object of the simulation in Figure 12–70 is to exercise, or test, all of the operating features of the counter. It starts out with a reset pulse on *n_rd* to reset *q* to zero. (The *q* outputs are shown in the waveform as a group having a hexadecimal radix.) Notice that at each positive *cp* pulse, *q* is incremented by 1. This counting continues until *n_pl* is asserted LOW near the 5.25 μs mark. At that point, the positive clock edge is overridden and *q* is parallel loaded with *pl_data,* which at that point is a hexadecimal 1. The count-up continues from there until the next *n_rd* pulse at 8.25 μs resets *q* and then the *n_pl* pulse at 14.25 μs sets *q* to 4.

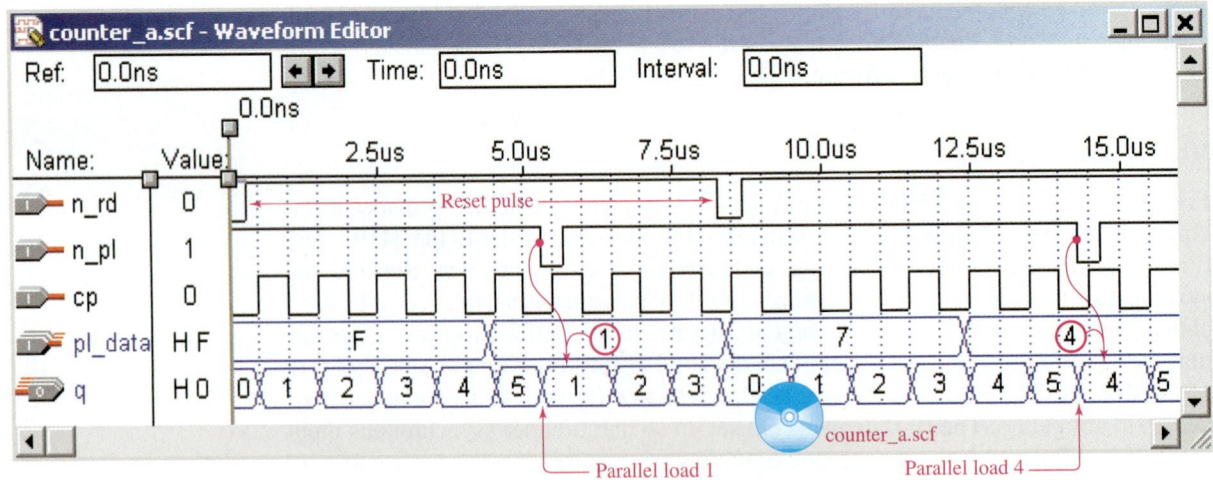

Figure 12–70 Simulation of the counter of Figure 12–69.

VHDL Up/Down-Counter

With the addition of a few statements, we can include a Count Enable feature and provide up/down direction control so that our counter functions like a 74191 IC counter. Figure 12–71 shows the VHDL program to accomplish this. We need to include two new control bits called n_ce for "count enable" and u_d for up/down control. Notice in the architecture section, if u_d is 0 then q is incremented (counts up), ELSE q is decremented. However, neither operation is performed unless the active-LOW n_ce count enable is 0.

Since this program has so many IF-ELSE and ELSIF statements, it is important to draw a flow chart to map out all of the branching that takes place (see Figure 12–72). The first and second diamond-shaped decision boxes show that if either asynchronous operation is performed, then control bypasses the synchronous check for a clock pulse and instead passes to the END IF and then the END PROCESS. If both asynchronous operations are NO, then the process checks for a positive cp edge, then a LOW n_ce, then the level of u_d.

The simulation in Figure 12–73 was carefully developed to ensure that all of the counter's features were tested. The Reset and Parallel Load operations are similar to the last counter, but the Count Enable and up/down control show the additional flexibility of the logic implementation. Notice that for the first 7 µs, q is counting up because u_d is LOW (except for the parallel-load at 4.25 µs). From 7 µs to 16 µs it counts down (except for the Reset and Parallel-Load operations). The counting is disabled with a HIGH on n_ce from 11.25 µs to 12.75 µs. During that span of time, q remains at 0 even though two positive clock edges occur.

```
counter_b.vhd - Text Editor                              _ □ X

  LIBRARY ieee;                       ---4-bit Up/Down-Counter    --
  USE ieee.std_logic_1164.ALL;        -- with Async Reset and Load --
                                      -- and clock enable          --
  ENTITY counter_b IS                 -----------------------------
      PORT(cp, n_rd, n_pl, n_ce, u_d     : IN std_logic;
           pl_data                       : IN integer RANGE 0 TO 15;
           q                             : BUFFER integer RANGE 0 TO 15);
  END counter_b;


  ARCHITECTURE arc OF counter_b IS
  BEGIN
      PROCESS (cp, n_rd, n_pl)
      BEGIN
          IF  (n_rd='0' AND n_pl='1') THEN
              q<= 0;                                    ⎤— Asyncronous
          ELSIF (n_rd='1' AND n_pl='0') THEN           ⎦
              q<=pl_data;
          ELSIF (cp'EVENT AND cp='1' ) THEN
              IF (n_ce='0') THEN   ←————— Clock enable
                  IF (u_d='0') THEN                  ⎤
                      q<=q+1;                        ⎬— Up/down control
                  ELSE                               ⎦
                      q<=q-1;
                  END IF;
              END IF;
          END IF;
      END PROCESS;
  END arc;                            (●) counter_b.vhd

Line  1    Col  1    INS ◄                                   ►
```

Figure 12–71 VHDL program for a 4-bit up/down-counter with asynchronous Reset, Parallel Load, and Clock Enable.

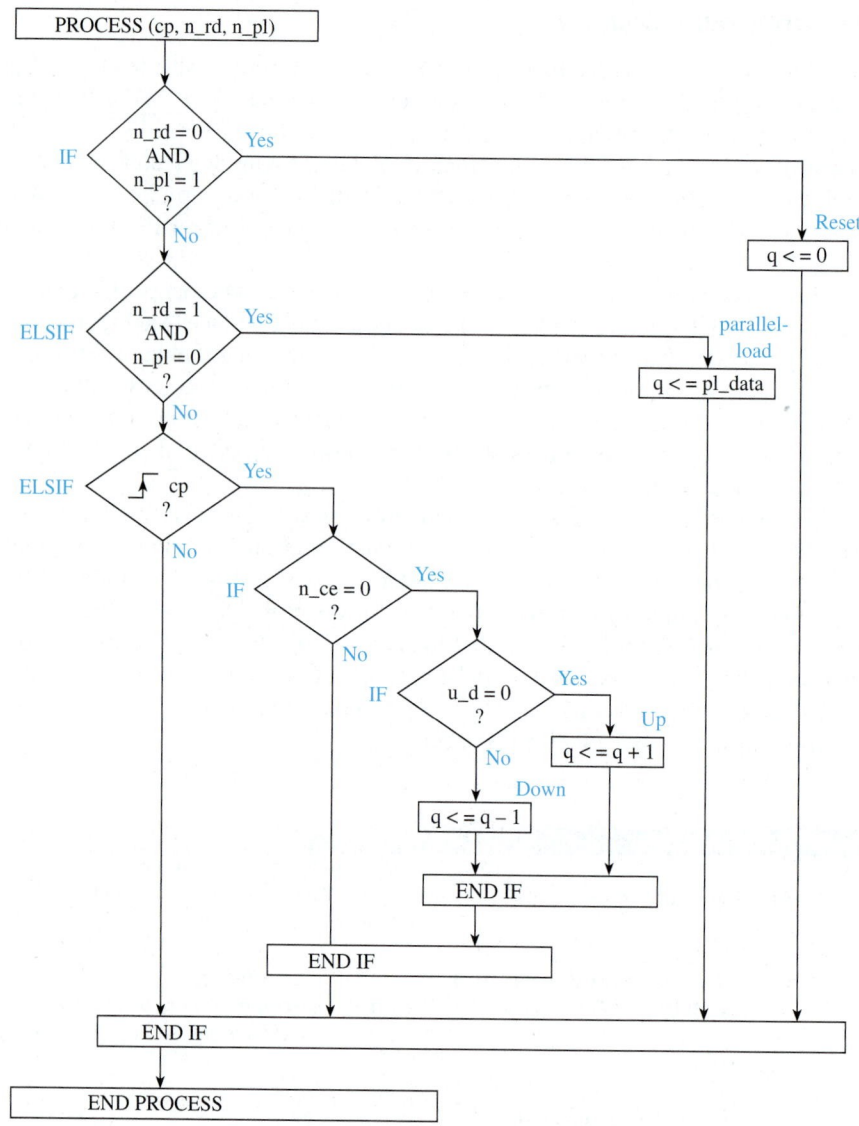

Figure 12–72 Flowchart for the counter of Figure 12–71.

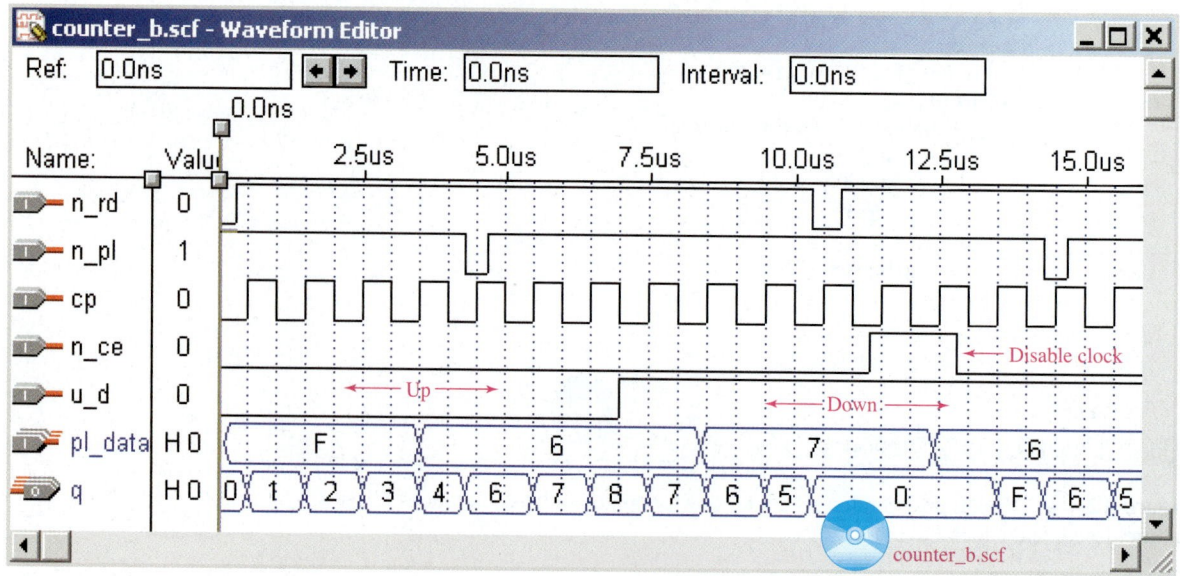

564   **Figure 12–73** Simulation of the counter of Figure 12–71.

LPM Counter

The Library of Parameterized Modules has a predefined counter called LPM_COUNTER. This gives us the maximum in flexibility by providing several asynchronous and synchronous inputs as well as allowing us to specify the LPM_WIDTH and LPM_MODULUS. To see the entire list of options, add the LPM_COUNTER symbol to a blank graphic design file (gdf), right-click on the symbol, and choose **Edit Ports/Parameters.** A Help screen is provided that lists comments for each of the counter ports and parameters.

Figure 12–74 shows an LPM up/down-counter with an asynchronous Set and Clear and a Count Enable. By reading the help menu, it is determined that port *updown* must be HIGH to count up and *cnt_en* must be HIGH to enable the count. Asynchronous set and clear (*aset* and *aclr*) are both active-HIGH. The parameter *LPM_AVALUE* contains the data value that will be loaded into the counter when *aset* is asserted HIGH. This example is set up as a MOD-10 so the *LPM_WIDTH* was assigned a value of 4 to accommodate the highest count, which will be $9(1001_2)$.

A simulation was developed in Figure 12–75 to exercise all of the features that are used in Figure 12–74. A HIGH pulse on *aclr* initially resets q to 0. The outputs then count up at each positive pulse until *aset* is asserted HIGH at 4 μs, loading the number 7 into the counter. As the count continues past 9, it rolls over to 0 because *LPM_MODULUS* is set to 10 (0 through 9). It continues to count up until *updown* goes LOW at 8 μs, reversing the count. Down counting continues until *cnt_en* goes LOW at 14 μs, disabling the count.

12–11 Implementing State Machines in VHDL

Many processes in digital electronics follow a predefined sequence of steps initiated by a series of clock pulses. These processes can be driven by a single clock input and have one or more outputs that respond in a particular order at each clock input pulse. Besides the clock trigger, these processes often have other external stimuli that have an effect on the state of the outputs. This sequence of events can be implemented in a logic system called a **state machine.** The outputs of a state machine follow a predictable sequence, triggered by a clock and other input stimulus.

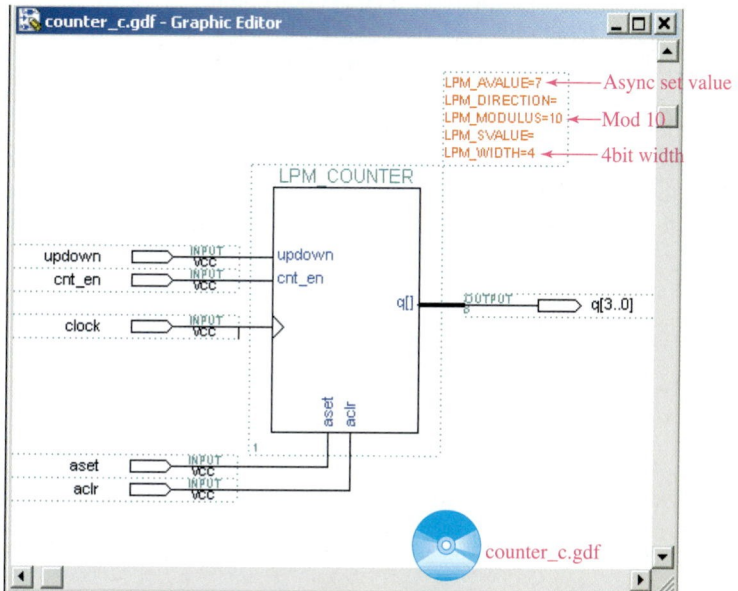

Figure 12–74 LPM up/down-counter with an asynchronous Set and Clear and a Count Enable.

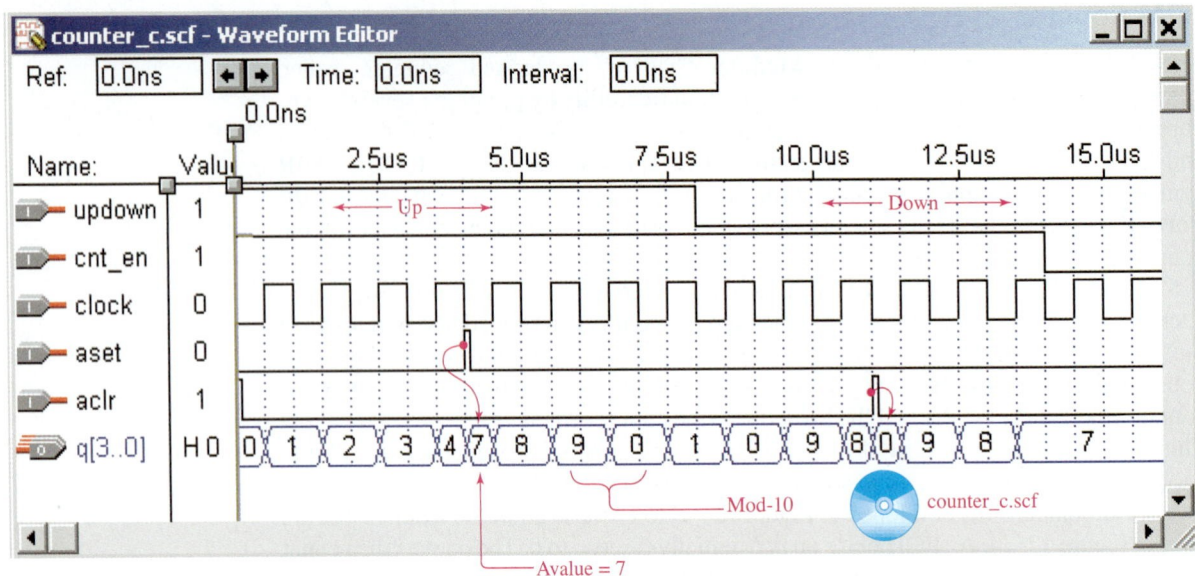

Figure 12–75   Waveform simulation of the LPM counter of Figure 12–74.

State Machine for a Gray Code Sequencer

State machines can be implemented in VHDL by defining the correct sequence of output states and then stepping through the states in a numerical order or in an order dictated by the level of one or more control inputs. A simple illustration of a state machine is for the implementation of a Gray code sequencer. The Gray code was introduced in Section 8–6 as a means to encode positions on a rotating wheel in such a way that as the wheel moves from one coded position to the next, only one bit in the code changes. Table 12–4 shows the Gray code in a 3-bit system having eight unique positions. Notice the order does not follow the same pattern as a binary counter, so conventional coding schemes will not work. Think of the chart as having eight states. State 0 (*s0*) is 000, state 1 (*s1*) is 001, state 2 (*s2*) is 011, and so on as shown in Table 12–4.

The VHDL program to implement a Gray code counter is shown in Figure 12–76. The only input is *clk,* which is used to trigger each state change in the *q* outputs. The outputs at *q* are a 3-bit vector, which will represent the Gray code from Table 12–4. The main function of the sate machine is to change from one state to the next as an input trigger (*clk* in this case) is applied.

In this program, the states are given the names *s0* through *s7* in the TYPE statement. This statement defines *state_type* as a new data type having the values *s0, s1, s2, s3, s4, s5, s6,* and *s7.* This is called an **enumeration type** because it is a type whose

TABLE 12–4	Three-bit Gray Code Chart
State	**Gray Code**
s0	000
s1	001
s2	011
s3	010
s4	110
s5	111
s6	101
s7	100

```vhdl
LIBRARY ieee;                    -------------------------------------
USE ieee.std_logic_1164.ALL;  -- State Machine Gray Code Counter --
                                 -------------------------------------
ENTITY state_gray IS
    PORT(clk    : IN    STD_LOGIC;
         q      : OUT   STD_LOGIC_VECTOR(2 downto 0));
END state_gray ;

ARCHITECTURE arc OF state_gray IS
    TYPE state_type IS (s0, s1, s2, s3, s4, s5, s6, s7);
    SIGNAL state: state_type;
BEGIN
    PROCESS (clk)
    BEGIN
        IF clk'EVENT AND clk = '1' THEN
            CASE state IS
                WHEN s0 => state <= s1;
                WHEN s1 => state <= s2;
                WHEN s2 => state <= s3;
                WHEN s3 => state <= s4;
                WHEN s4 => state <= s5;
                WHEN s5 => state <= s6;
                WHEN s6 => state <= s7;
                WHEN s7 => state <= s0;
            END CASE;
        END IF;
    END PROCESS;

    WITH state SELECT
        q   <=  "000"   WHEN    s0,
                "001"   WHEN    s1,
                "011"   WHEN    s2,
                "010"   WHEN    s3,
                "110"   WHEN    s4,
                "111"   WHEN    s5,
                "101"   WHEN    s6,
                "100"   WHEN    s7;
END arc;
```

Defines a new data type having the allowable values *s0-s7*.

Increment *state* to next state.

Assign binary to *q* based on value of *state*.

state_gray.vhd

Figure 12–76 VHDL program to create a Gray code counter.

allowable values are all user-defined. The next program statement defines an internal SIGNAL called *state,* which is declared as having the data type *state_type* that we just defined. (Notice the use of the colon preceding *state_type.* This is the same syntax that we have been using to declare other types like *std_logic* or *integer.*) Later in the program we will be assigning the data values *s0* through *s7* to the signal names *state.*

In the PROCESS loop, the CASE statement is executed after each positive clock edge. The case assignments are made in such a way that if the **present state** is *s0* then the **next state** is *s1*, and so on. This has the effect of making the signal *state* step to the next state at each positive clock edge. Notice that when the present state is *s7,* then the next state is *s0,* which will cause *state* to cycle back to the beginning of the sequence.

After each *state* has been assigned by the CASE assignment group, then a selected signal assignment is used to output the correct Gray code to *q* based on the value of *state.* For example, if *state* is *s2,* then *q* receives "011", if *state* is *s3,* then *q* receives "010"; and so on.

Figure 12–77 shows a simulation of the program. Notice that at each positive clock edge, the *q*-outputs sequence to the next Gray code.

State Machine for a Stepper Motor

Another common use for state machine design is for creating the logic circuitry to control stepper motors. Stepper motors are commonly found in computer disk drives and robotic mechanisms and are described in detail in Section 13–9. Basically a stepper motor takes small angular "steps" based on the binary code output to it. Table 12–5 shows the binary codes used to drive a stepper motor clockwise and counterclockwise.

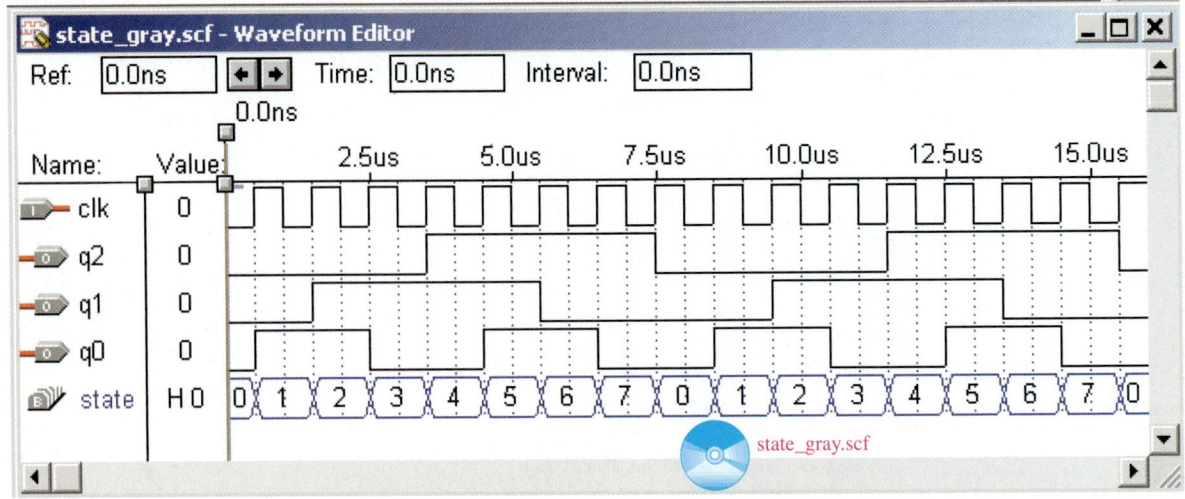

Figure 12–77 Simulation of the Gray code counter of Figure 12–76.

TABLE 12–5	Binary Codes for Clockwise and Counterclockwise Stepper Motor Rotation

CW	CCW
0001 (s0)	1000 (s3)
0010 (s1)	0100 (s2)
0100 (s2)	0010 (s1)
1000 (s3)	0001 (s0)
0001 (s0)	1000 (s3)
Etc.	Etc.

TABLE 12–6	State Changes for Clockwise and Counterclockwise Stepper Motor Rotation		

Clockwise rotation		Counterclockwise rotation	
Present state	**Next state**	**Present state**	**Next state**
s0	*s1*	*s3*	*s2*
s1	*s2*	*s2*	*s1*
s2	*s3*	*s1*	*s0*
s3	*s0*	*s0*	*s3*

For example, to step clockwise, the binary codes sent to the stepper will be 0001-0010-0100-1000-0001-, and so on.

In state machine design we need to refer to the states as present-state and next-state. For example, if counterclockwise rotation is desired and if the present state is *s0*, then the next state is *s1*. If the present state is *s1*, then the next state is *s2*, and so on. Table 12–6 lists the order of the states for the stepper motor.

Figure 12–78 provides a state diagram to show the state changes for the stepper motor sequencer. The direction control input is named *dir*. As you can see, when *dir* is HIGH, the states sequence from *s0* to *s1* to *s2* to *s3* to *s0*, and so on. This will be a clockwise rotation as defined in Tables 12–5 and 12–6. With *dir=0* the direction will be counterclockwise.

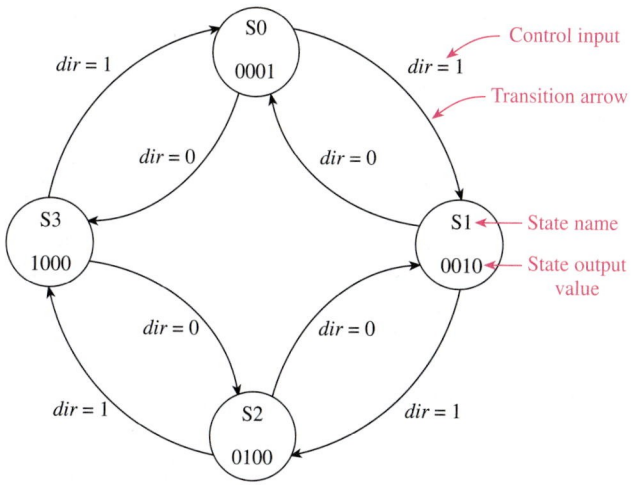

Figure 12–78 State diagram for a stepper motor sequencer with direction control.

```
state_stepper_b.vhd - Text Editor                              _ □ ×

LIBRARY ieee;                    ------------------------------------
USE ieee.std_logic_1164.ALL; -- State Machine Stepper Motor Sequencer --
                             --      with Direction Control          --
ENTITY state_stepper_b IS        ------------------------------------
    PORT(clk, dir  : IN    STD_LOGIC;
         q         : OUT   STD_LOGIC_VECTOR(3 downto 0));
END state_stepper_b ;

ARCHITECTURE arc OF state_stepper_b IS
    TYPE state_type IS (s0, s1, s2, s3);
    SIGNAL state: state_type;
BEGIN
    PROCESS (clk)
    BEGIN
        IF clk'EVENT AND clk = '1' THEN
            IF dir='1' THEN      -- Clockwise Steps
                CASE state IS
                    WHEN s0 => state <= s1;
                    WHEN s1 => state <= s2;     ⎤ states increment
                    WHEN s2 => state <= s3;     ⎦
                    WHEN s3 => state <= s0;
                END CASE;
            ELSE                 -- Counter-clockwise Steps
                CASE state IS
                    WHEN s0 => state <= s3;
                    WHEN s1 => state <= s0;     ⎤ states decrement
                    WHEN s2 => state <= s1;     ⎦
                    WHEN s3 => state <= s2;
                END CASE;
            END IF;
        END IF;
    END PROCESS;

    WITH state SELECT
        q   <=  "0001"  WHEN   s0,
                "0010"  WHEN   s1,
                "0100"  WHEN   s2,
                "1000"  WHEN   s3;
    END arc;                                    state_stepper_b.vhd
Line  1    Col  1    INS ◄
```

Figure 12–79 VHDL state machine for a stepper motor sequencer with direction control.

Figure 12–79 shows the VHDL program to implement a 4-bit stepper motor sequencer with direction control. The inputs are *clk* and *dir* and the output is a 4-bit vector named *q*. The *state_type* is defined with four values: *s0, s1, s2,* and *s3*. The internal signal, *state* is declared as TYPE *state_type* and will be assigned values in the CASE assignment group. The input *dir* is used to determine which group of CASE assignments are made (i.e., the ones for clockwise rotation or counterclockwise rotation.) Keep in mind that the PROCESS loop is only executed if there is a logic level change in *clk* because that is the only value provided in the sensitivity list. Once inside the PROCESS loop, the first IF statement checks for a positive *clk* edge. If there is a positive edge, one of the eight possible CASE assignments will be executed and then control passes out of the PROCESS loop. The selected signal assignment then uses the value of *state* to assign the next-state binary value to the *q* output vector.

Figure 12–80 shows a simulation of the stepper motor sequence sent to the *q* outputs. Notice that the individual HIGH travels from *q0* to *q1* to *q2* to *q3* repeatedly until *dir* is LOW, at which time the direction reverses.

State machines with multiple control inputs

Most computer peripherals require several control inputs and outputs to communicate with the computer's microprocessor. These handshaking signals ensure that data transfer between the device and the microprocessor goes smoothly without interfering with other devices. Typical control signal names are *read, write, ready-to-receive, ready-to-transmit, buffer-full, end-of-transmit,* and *parity-error.* A common solution to these complex timing and handshaking control signals is to implement a state machine in a CPLD.

An example of a device that has to handshake with a microprocessor is an analog-to-digital converter (ADC). ADCs are covered in detail in Section 15–10. To develop a state machine solution to the ADC interface controller we need to know the basic operation of the ADC. We'll use the 8-bit ADC in Figure 12–81 for this example. The object is to input an analog voltage to the ADC and have it determine what its 8-bit binary equivalent is. To complete this conversion, the following steps must be performed:

1. The microprocessor must issue a "trigger" (*t*) to the ADC controller (a CPLD) to start the process.

2. Upon receiving the trigger, the controller will issue a "start-convert" (*sc*) pulse to the ADC.

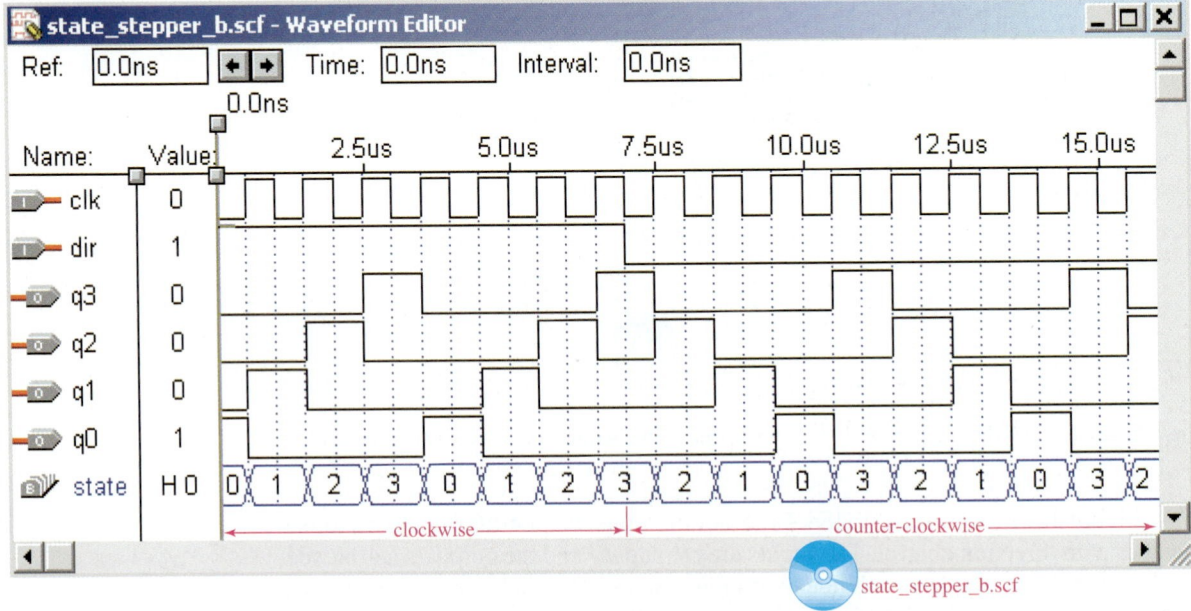

state_stepper_b.scf

Figure 12–80 Simulation of the stepper motor sequencer program of Figure 12–79.

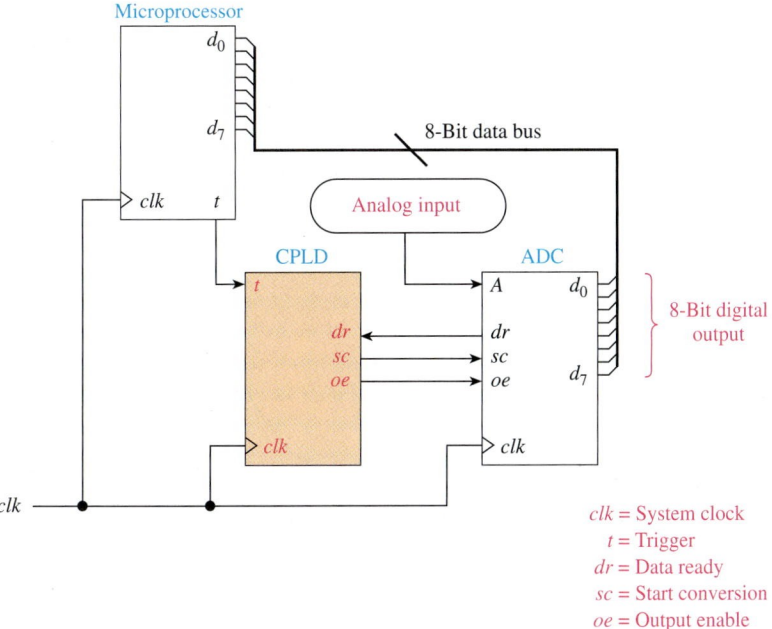

Figure 12–81 Block diagram of the state machine implementation of an ADC controller.

3. The ADC will read the analog input voltage and begin a process called "successive approximation" to determine the binary equivalent of that analog value.

4. When the ADC has completed its successive approximation, it will issue a "data-ready" (*dr*) pulse to the controller.

5. Upon receiving the data-ready pulse, the controller will issue an "output-enable" (*oe*) pulse to the ADC enabling the ADC to transmit its 8-bit result to the data bus. This completes the process.

The sequential process of the controller is illustrated in the state diagram of Figure 12–82. This diagram shows four states: *idle, start, waiting,* and *read.* Before the process begins and after the process is complete, the controller is in the *idle* state with *sc* = 0 and *oe* = 0. The loop outside of the *idle* state indicates that the process flow stays in that state as long as *t,dr* = *0,x.* This means that it stays in the *idle* state as long as the trigger (*t*) is 0 (*dr* is a don't-care). The **transition arrow** leaving the *idle* state indicates that the process moves to the *start* state when the trigger is applied as specified by *t,dr* = *1,x* (*t* = *1, dr* is still a don't-care). The *start* state issues the following outputs: *sc* = 1 and *oe* = 0. (The HIGH on *sc* performs a "start-conversion" in the ADC.)

The transition out of the *start* state specifies (*t,dr* = *x,x*). This is called an **unconditional transition** because both inputs are don't-cares. The process moves to the next state called *waiting.* The *waiting* state resets *sc* to 0 and keeps *oe* at 0. Since the external loop on this state says *t,dr* = *x,0* then the process remains here as long as *dr* = 0. (It is waiting for the ADC to issue a HIGH data-ready signal.)

As soon as the ADC completes its analog conversion, it outputs a HIGH on *dr.* According to the state diagram, as soon as *dr* = 1 the process transitions to the *read* state. The *read* state issues a HIGH on *oe* (allowing the ADC to output its binary data to the data bus) and then does an unconditional transition back to the *idle* state, where it remains until another trigger is applied.

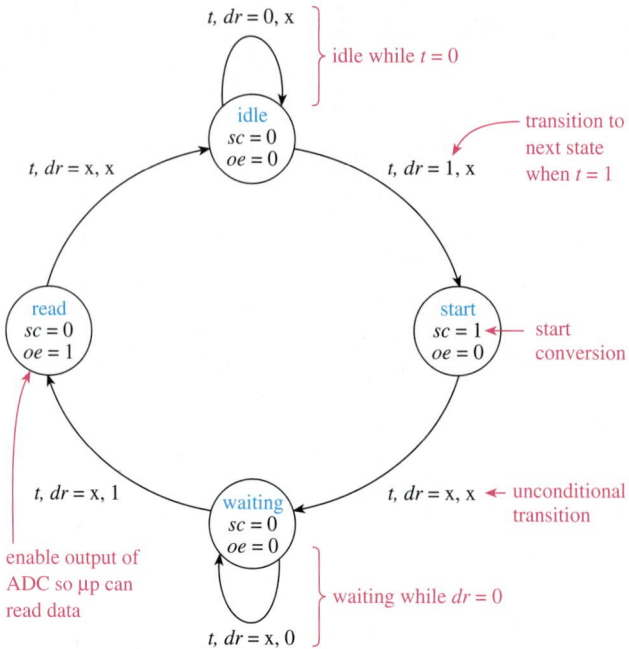

Figure 12–82 State diagram of the ADC controller process.

The VHDL program to implement this state machine is given in Figure 12–83. The CASE statement passes control to one of the four different states: *idle, start, waiting,* or *read.* The program logic is coded to match the state diagram exactly. For example, WHEN in the *idle* case, IF *t* is 0 then the state remains *idle,* and *sc* and *oe* are set to 0. ELSIF *t* is 1 (trigger applied) THEN the state is changed to *start* and *sc* is set to 1, starting the ADC conversion.

After any CASE assignments are made, program control passes to the END CASE, END IF, END PROCESS statements and awaits the next active edge of *clk.* As you can see, the CASE assignments for the *start* state and the *read* state are unconditional (no IF clause) and each changes the state to the next state name. The *waiting* state checks the status of *dr* and sets *oe* to 1 if *dr* is 1. (This means that the outputs are enabled if the data is ready.)

The simulation file for the ADC controller is shown in Figure 12–84. It shows that the ADC controller is triggered at the 1 μs mark and again at the 9 μs mark. When *t* is HIGH and the first positive clock edge arrives at 1.5 μs, *sc* goes HIGH and the state changes to state 1 (*start*). At the next positive clock edge there is an unconditional change to state 2 (*waiting*). The process now waits for the data ready (*dr*) to go HIGH, which it does at the 5 μs mark. When *dr* is HIGH and the first positive clock edge arrives at 5.5 μs, *oe* goes HIGH and the state changes to state 3 (*read*). Now the system has completed one complete analog-to-digital conversion and the microprocessor has received the 8-bit binary result via the data bus. At the next clock edge, the state changes to state 0 (*idle*), where it remains until the next trigger is applied at 9 μs.

CHAPTER 12 I COUNTER CIRCUITS AND VHDL STATE MACHINES

```
state_adc.vhd - Text Editor                                                    _ □ x
                                        ----------------------------
    LIBRARY ieee;                       -- State Machine Design of --
    USE ieee.std_logic_1164.ALL;        --   an ADC Controller      --
    ENTITY state_adc IS                 ----------------------------
        PORT(clk, t,dr  : IN    std_logic;
             sc,oe       : OUT   std_logic);
    END state_adc ;

    ARCHITECTURE arc OF state_adc IS
        TYPE state_type IS (idle,start,waiting,read);
        SIGNAL state: state_type;
    BEGIN
        PROCESS(clk)
        BEGIN
            IF (clk'EVENT and clk = '1') THEN
                CASE state IS
                    WHEN idle=>
                        IF t = '0' THEN
                            state         <= idle;            ⎫
                            sc            <='0';              ⎬  Stay in idle until t = 1
                            oe            <='0';              ⎭
                        ELSIF t = '1' THEN
                            state         <= start;           ⎫  Move to start
                            sc            <='1';              ⎬  and set sc = 1
                            oe            <='0';              ⎭
                        END IF;
                    WHEN start=>
                        state         <= waiting;     ◄── Uncondition state
                        sc            <='0';              change to waiting
                        oe            <='0';
                    WHEN waiting=>
                        IF dr = '0' THEN
                            state         <= waiting;         ⎫  Stay in waiting
                            sc            <='0';              ⎬  until dr = 1
                            oe            <='0';              ⎭
                        ELSIF dr = '1' THEN
                            state         <= read;            ⎫  Move to read
                            sc            <='0';              ⎬  and set oe = 1
                            oe            <='1';              ⎭
                        END IF;
                    WHEN read=>
                        state         <= idle;        ◄── Unconditional state
                        sc            <='0';              change to idle
                        oe            <='0';
                    WHEN OTHERS =>
                        state         <= idle;
                        sc            <='0';
                        oe            <='0';
                END CASE;
            END IF;
        END PROCESS;
    END arc;                                      ● state_adc.vhd

Line   1    Col   1    INS ◄
```

Figure 12–83   VHDL state machine implementation of an ADC controller.

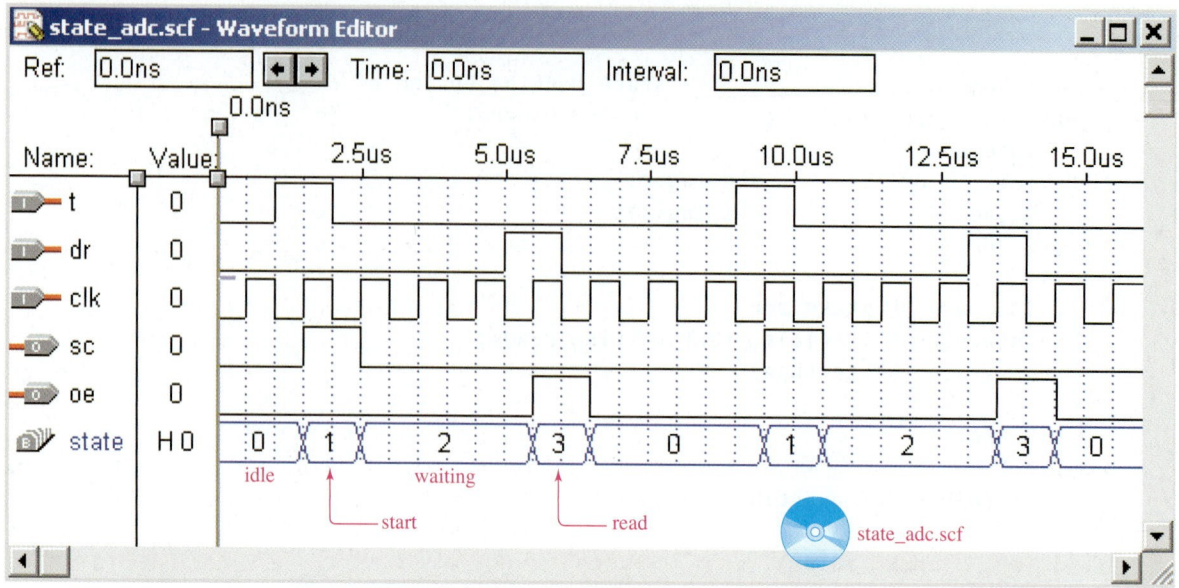

Figure 12–84 Simulation file for the ADC controller.

EXAMPLE 12–30

The handshaking and interface requirements of an automated teller machine (ATM) are often fulfilled by implementing a state machine in a CPLD. Figure 12–85 shows the block diagram of the interconnections between an ATM, a CPLD controller, and a thermal printer.

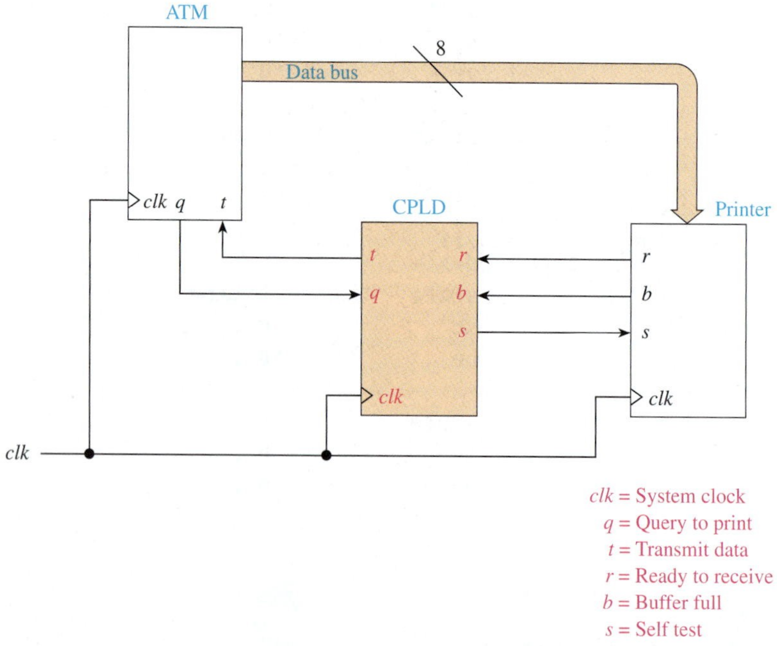

clk = System clock
q = Query to print
t = Transmit data
r = Ready to receive
b = Buffer full
s = Self test

Figure 12–85 Block diagram of an ATM-thermal printer interface.

The order of operation is as follows:

1. When the ATM has information to print, it makes the Query (q) line HIGH.

2. When the CPLD controller sees a HIGH q, it issues a HIGH Self-Test (s) to have the printer perform a self-test of its print mechanism.

3. When the self-test completes successfully, the printer returns a HIGH on Ready-to-Receive (r).

4. With r HIGH the controller issues a HIGH on Transmit-Data (t), which tells the ATM system to send the 8-bit parallel data to the printer. The ATM continues to hold q HIGH as long as there are data to print. When the stream of data on the data bus ends, the printer issues a LOW on r.

5. The printer has a Buffer-Full (b) signal that signifies that it has received too much data in its internal holding buffer and has to catch up on its printing duties before it can accept any more data. In this case, the ATM must wait until b goes back LOW before continuing to print.

6. While waiting for the buffer to clear, or while transmitting data, if r goes LOW, signifying that the printer is NOT Ready-to-Receive, the controller must return to the self-test state and postpone printing until r goes back HIGH.

Assignment:

(a) Draw a state diagram of the flow between the four states: *idle, selftest, transmit,* and *waiting.*

(b) Write the VHDL to implement the design.

(c) Create a simulation file that tests the normal sequence of printing and then a printing sequence that is interrupted by a Buffer-Full and then a LOW Ready-to-Receive.

Solution: The state diagram is in Figure 12–86, the VHDL program is in Figure 12–87, and its simulation is shown in Figure 12–88.

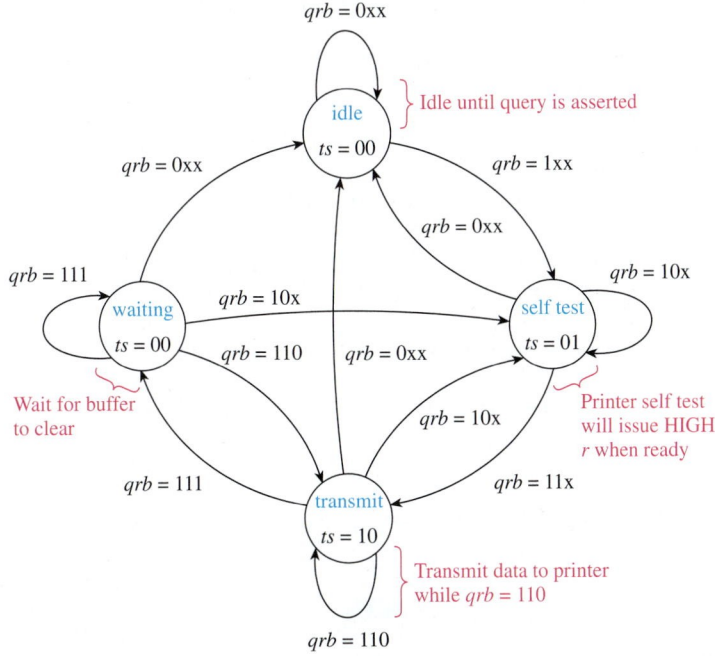

Figure 12–86 State diagram for the ATM-thermal printer interface.

```
state_atm.vhd - Text Editor                                    _ □ X
    LIBRARY ieee;                       -- State Machine Design of  --
    USE ieee.std_logic_1164.ALL;        -- an ATM Printer Controller --
    ENTITY state_atm IS                 ------------------------------
        PORT(clk,q,r,b : IN    std_logic;
             t,s       : OUT   std_logic);
    END state_atm;
    ARCHITECTURE arc OF state_atm IS
        TYPE state_type IS (idle,selftest,transmit,waiting);
        SIGNAL state: state_type;
    BEGIN
        PROCESS(clk)
        BEGIN
            IF (clk'EVENT and clk = '1') THEN
                CASE state IS
                    WHEN idle=>
                        IF q = '0' THEN
                            state<= idle;        t<='0'; s<='0';    Idle until
                        ELSIF q = '1' THEN                          q = 1
                            state<= selftest;    t<='0'; s<='1';
                        END IF;
                    WHEN selftest=>
                        IF q = '0' THEN                             Return to idle
                            state<= idle;        t<='0'; s<='0';    if q goes LOW
                        ELSIF q = '1' AND r = '0' THEN
                            state<= selftest;    t<='0'; s<='1';    Continue selftest
                        ELSIF q = '1' AND r = '1' THEN              until r = 1
                            state<= transmit;    t<='1'; s<='0';
                        END IF;
                    WHEN transmit=>
                        IF q = '0' THEN                             Return to idle
                            state<= idle;        t<='0'; s<='0';    if q goes LOW
                        ELSIF q = '1' AND r = '0' THEN              Return to selftest
                            state<= selftest;    t<='0'; s<='1';    if r goes LOW
                        ELSIF q = '1' AND r = '1' THEN
                            IF b = '0' THEN                         Transmit data
                                state<= transmit;    t<='1'; s<='0';  while b = 0
                            ELSE
                                state<= waiting;     t<='0'; s<='0';  Go to waiting
                            END IF;                                 if b goes HIGH
                        END IF;
                    WHEN waiting=>
                        IF q = '0' THEN                             Return to idle
                            state <= idle;       t<='0'; s<='0';    if q goes LOW
                        ELSIF q = '1' AND r = '0' THEN              Return to selftest
                            state<= selftest;    t<='0'; s<='1';    if r goes LOW
                        ELSIF q = '1' AND r = '1' THEN
                            IF b = '0' THEN                         Return to transmit
                                state<= transmit;    t<='1'; s<='0';  if buffer clears
                            ELSE
                                state<= waiting;     t<='0'; s<='0';  Keep waiting
                            END IF;                                 until buffer clears
                        END IF;
                    WHEN OTHERS =>
                        state <= idle;       t<='0'; s<='0';
                END CASE;
            END IF;
        END PROCESS;
    END arc;                                        state_atm.vhd
Line   1    Col  1    INS
```

Figure 12–87   VHDL program for the ATM-thermal printer interface.

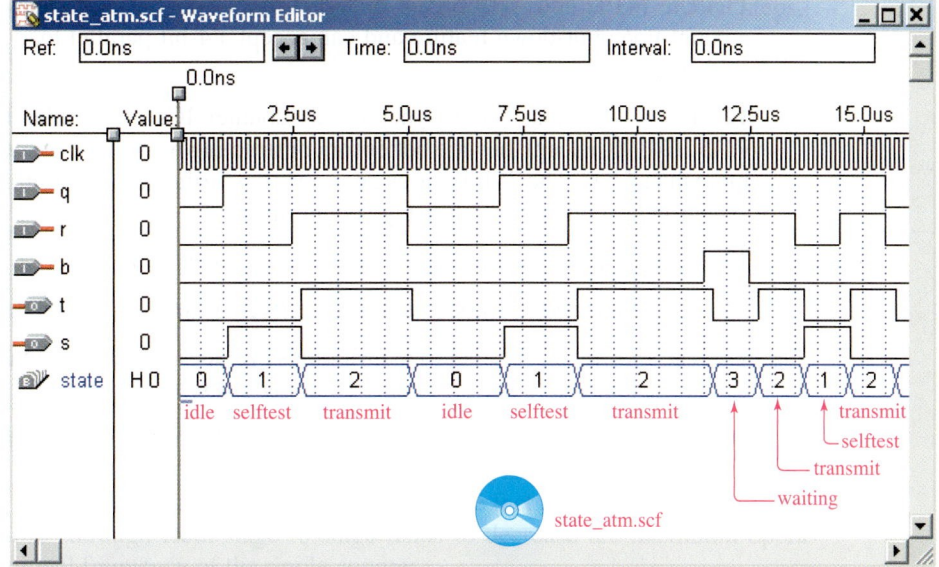

Figure 12–88 Simulation of the ATM-thermal printer interface.

Summary

In this chapter, we have learned that

1. Toggle flip-flops can be cascaded end to end to form ripple counters.

2. Ripple counters cannot be used in high-speed circuits because of the problem they have with the accumulation of propagation delay through all the flip-flops.

3. A down-counter can be built by taking the outputs from the $\overline{Q}$'s of a ripple counter.

4. Any modulus (or divide-by) counter can be formed by resetting the basic ripple counter when a specific count is reached.

5. A glitch is a short-duration pulse that may appear on some of the output bits of a counter.

6. Ripple counter ICs such as the 7490, 7492, and 7493 have four flip-flops integrated into a single package providing four-bit counter operations.

7. Four-bit counter ICs can be cascaded end to end to form counters with higher than MOD-16 capability.

8. Seven-segment LED displays choose between seven separate LEDs (plus a decimal point LED) to form the 10 decimal digits. They are constructed with either the anodes or the cathodes connected to a common pin.

9. LED displays require a decoder/driver IC such as the 7447 to decode BCD data into a seven-bit code to activate the appropriate segments to illuminate the correct digit.

10. Synchronous counters eliminate the problem of accumulated propagation delay associated with ripple counters by driving all four flip-flops with a common clock.

Section 12-1

12-1. How are sequential logic circuits different from combinational logic gate circuits?

12-2. The waveforms shown in Figure P12–2 are applied to the inputs at A, $\overline{R_D}$, and C_p. Sketch the resultant waveforms at D, Q, $\overline{Q}$, and X.

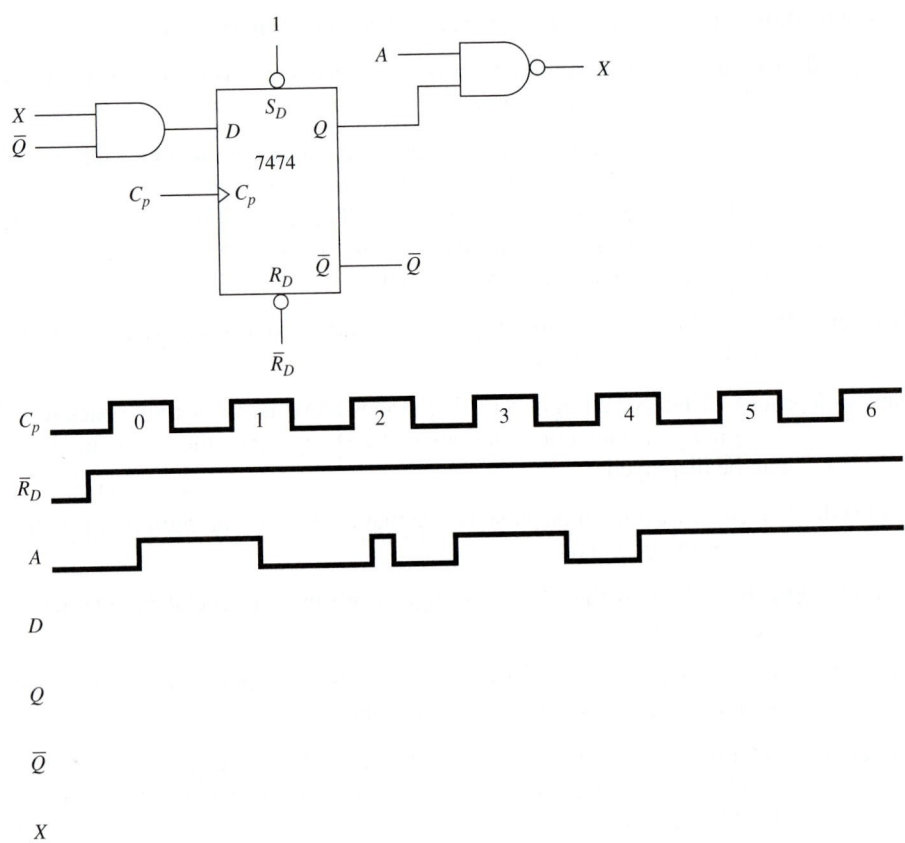

Figure P12-2

12-3. Repeat Problem 12–2 for the input waveforms shown in Figure P12–3.

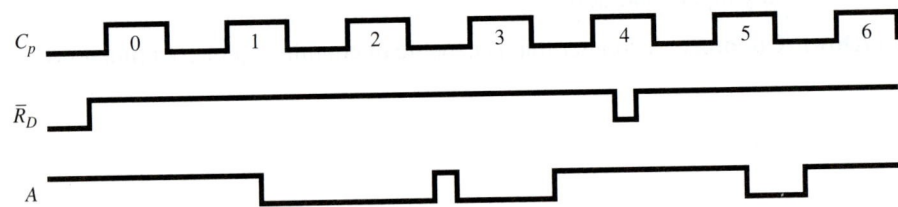

Figure P12-3

Modulus: In a digital counter, the modulus is the number of different counter steps.

Next State: In a state machine, this is the *next* state to be performed in the process.

Oscillate: Change digital states repeatedly (HIGH–LOW–HIGH–LOW–, and so on).

Parallel Load: A feature on some counters that allows you to load all 4 bits of a counter at the same time, asynchronously.

Phototransistor: A transistor whose collector-to-emitter current and resistance vary depending on the amount of light shining on its base junction.

Present State: In a state machine, this is the *current* state in the process.

Ripple Blanking: A feature supplied with display decoders to enable the suppression of leading and trailing zeros.

Ripple Counter: Asynchronous counter. A multibit counter whose clock input trigger is not connected to each flip-flop but, instead, has to propagate through each flip-flop to reach the input of the next. The fact that the clock has to "ripple" through from stage to stage tends to decrease the maximum operational frequency of the ripple counter.

Sequential: Operations that follow a predetermined sequence of digital states triggered by a timing pulse or clock.

Seven-Segment LED: Seven light-emitting diodes fabricated in a single package. By energizing various combinations of LED segments, the 10 decimal digits can be displayed.

Skewed: A skewed waveform or pulse is one that is offset to the right or left with respect to the time axis.

State Diagram: A drawing that shows the logic levels in a sequential circuit after each trigger event.

State Machine: A logic system whose outputs follow a predictable sequence, triggered by a clock and other input stimulus.

Synchronous Counter: A multibit counter whose clock input trigger is connected to each flip-flop so that each flip-flop will operate in step with the same input clock transition.

Terminal Count: The highest (or lowest) count in a multibit counting sequence.

Transition Arrow: In a state machine, this arrow points to the next state to be performed.

Unconditional Transition: In a state machine, this transition is made regardless of any input conditions.

Up/Down-Counter: A counter that is capable of counting up or counting down.

Section 12–1

12–1. How are sequential logic circuits different from combinational logic gate circuits?

12–2. The waveforms shown in Figure P12–2 are applied to the inputs at A, $\overline{R_D}$, and C_p. Sketch the resultant waveforms at D, Q, $\overline{Q}$, and X.

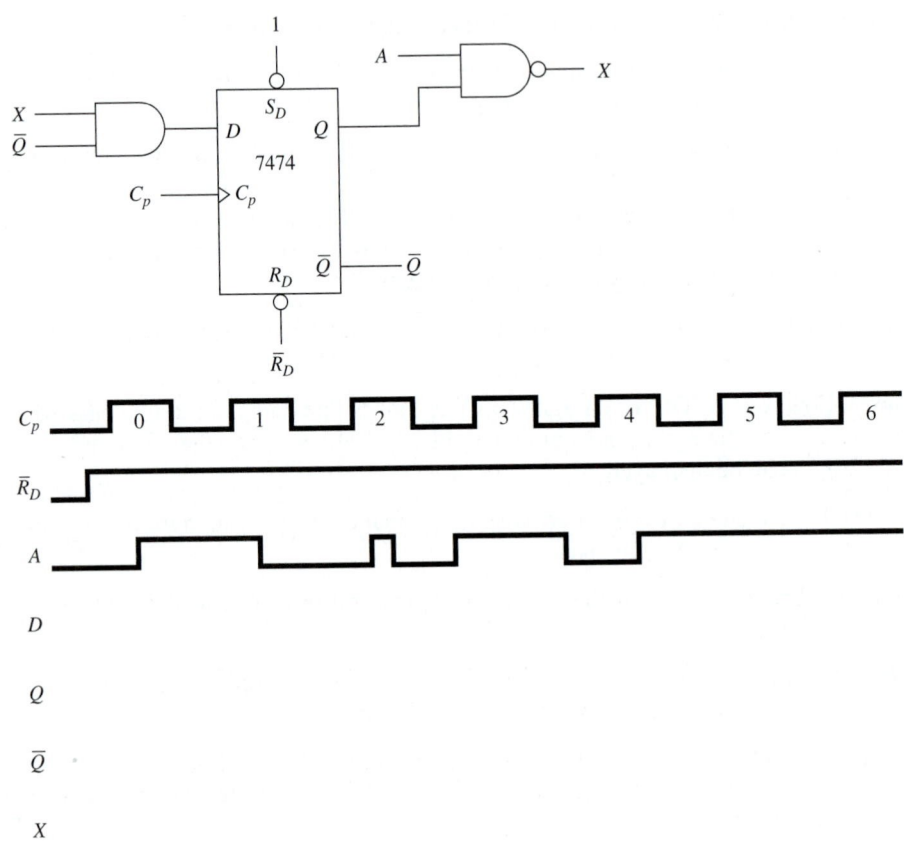

Figure P12–2

12–3. Repeat Problem 12–2 for the input waveforms shown in Figure P12–3.

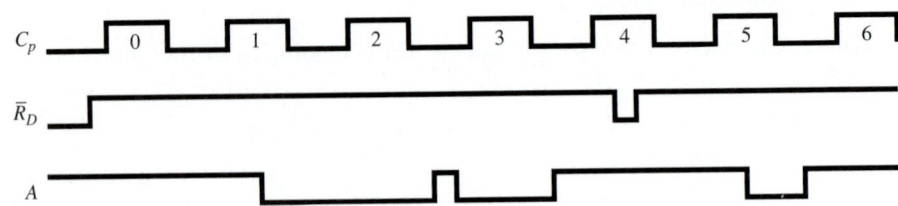

Figure P12–3

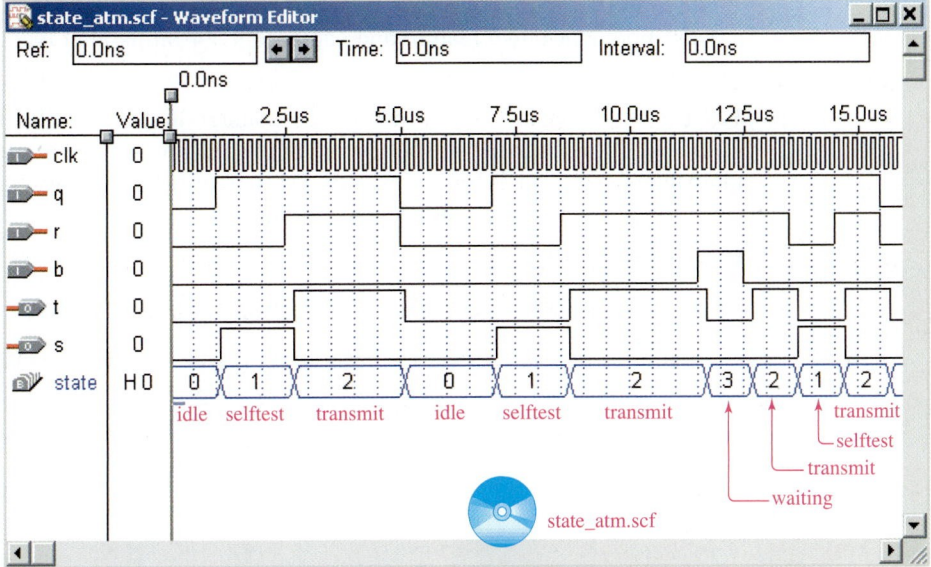

Figure 12–88 Simulation of the ATM-thermal printer interface.

Summary

In this chapter, we have learned that

1. Toggle flip-flops can be cascaded end to end to form ripple counters.

2. Ripple counters cannot be used in high-speed circuits because of the problem they have with the accumulation of propagation delay through all the flip-flops.

3. A down-counter can be built by taking the outputs from the $\overline{Q}$'s of a ripple counter.

4. Any modulus (or divide-by) counter can be formed by resetting the basic ripple counter when a specific count is reached.

5. A glitch is a short-duration pulse that may appear on some of the output bits of a counter.

6. Ripple counter ICs such as the 7490, 7492, and 7493 have four flip-flops integrated into a single package providing four-bit counter operations.

7. Four-bit counter ICs can be cascaded end to end to form counters with higher than MOD-16 capability.

8. Seven-segment LED displays choose between seven separate LEDs (plus a decimal point LED) to form the 10 decimal digits. They are constructed with either the anodes or the cathodes connected to a common pin.

9. LED displays require a decoder/driver IC such as the 7447 to decode BCD data into a seven-bit code to activate the appropriate segments to illuminate the correct digit.

10. Synchronous counters eliminate the problem of accumulated propagation delay associated with ripple counters by driving all four flip-flops with a common clock.

11. The 74192 and 74193 are 4-bit synchronous counter ICs. They have a count-up/count-down feature and can accept a 4-bit parallel load of binary data.

12. The 74190 and 74191 synchronous counter ICs are similar to the 74192/74193, except they are better for constructing multistage counters of more than 4 bits. The 74160 series goes one step further and allows for truly synchronous high-speed multistage counting.

13. VHDL can be used to implement MOD-N counters.

14. A seven-segment decoder can be effectively described in VHDL.

15. The Library of Parameterized Modules provides an LPM counter that can be customized to perform many counting tasks.

16. State machines can be implemented in VHDL.

Glossary

Asynchronous Counter: See *ripple counter*.

Cascade: In multistage systems, when the output of one stage is fed directly into the input of the next.

Common-Anode LED: A seven-segment LED display whose LED anodes are all connected to a common point and supplied with +5 V. Each LED segment is then turned on by supplying a LOW level (via a limiting resistor) to the appropriate LED cathode.

Divide-by-N: The Q outputs in counter operations will oscillate at a frequency that is at some multiple (N) of the input clock frequency. For example, in a divide-by-8 (MOD-8) counter the output frequency of the highest-order Q (Q_2) is one-eighth the frequency of the input clock.

Enumeration Type: In VLDL, this is a data type whose allowable values are all user-defined.

Glitch: A short-duration-level change in a digital circuit.

Liquid Crystal Display (LCD): A low-power display technology that creates an image by selectively making sections of its crystalline structure either opaque or transparent to incident light. This forms dark and light segments that, together, create alphanumeric images.

12–4. The waveforms shown in Figure P12–4 are applied to the inputs at A, $\overline{C_p}$, and $\overline{R_D}$. Sketch the resultant waveforms at J, K, Q, and $\overline{Q}$.

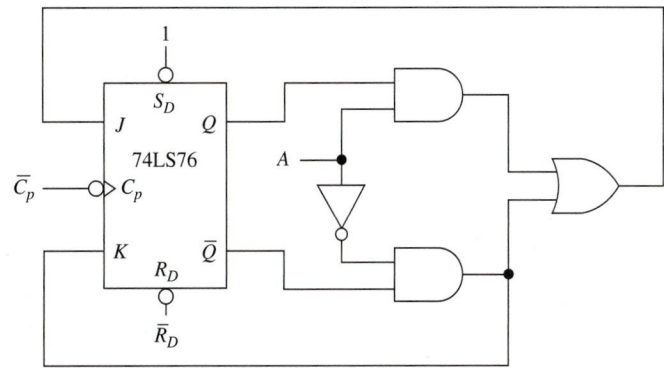

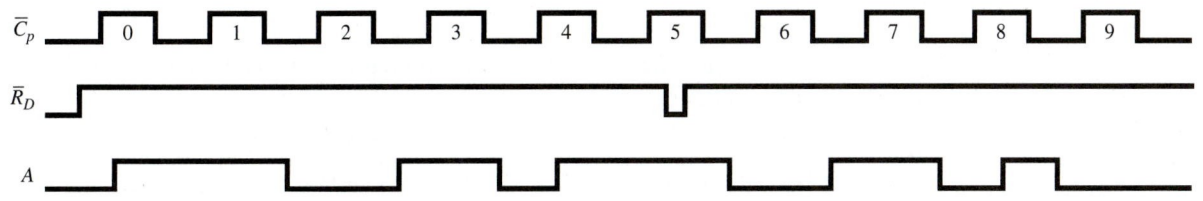

Figure P12–4

12–5. Repeat Problem 12–4 for the input waveforms shown in Figure P12–5.

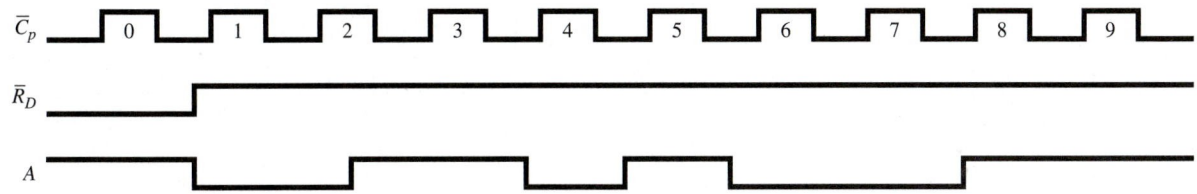

Figure P12–5

Section 12–2

12–6. What is the modulus of a counter whose output counts from

(a) 0 to 7?   **(d)** 10 to 0?

(b) 0 to 18?   **(e)** 2 to 15?

(c) 5 to 0?   **(f)** 7 to 3?

12–7. How may J-K flip-flops are required to construct the following counters?

(a) MOD-7   **(d)** MOD-20

(b) MOD-8   **(e)** MOD-33

(c) MOD-2   **(f)** MOD-15

12–8. If the input frequency to a 6-bit counter is 10 MHz, what is the frequency at the following output terminals?

(a) 2^0 **(d)** 2^3

(b) 2^1 **(e)** 2^4

(c) 2^2 **(f)** 2^5

12–9. Draw the timing waveforms at $\overline{C}_p$, 2^0, 2^1, and 2^2 for a 3-bit binary up-counter for 10 clock pulses.

12–10. Repeat Problem 12–9 for a binary down-counter.

12–11. What is the highest binary number that can be counted using the following number of flip-flops?

(a) 2   (c) 7

(b) 4   (d) 1

Section 12–3

12–12. In a 5-bit counter, the frequency at the following output terminals is what fraction of the input clock frequency?

(a) 2^0 (d) 2^3

(b) 2^1 (e) 2^4

(c) 2^2

12–13. How many flip-flops are required to form the following divide-by-N frequency dividers?

(a) Divide-by-4   (c) Divide-by-12

(b) Divide-by-15   (d) Divide-by-18

12–14. Explain why the propagation delay of a flip-flop affects the maximum frequency at which a ripple counter can operate.

C   **12–15.** Sketch the $\overline{C}_p$, 2^0, 2^1, and 2^2 output waveforms for the counter shown in Figure P12–15. (Assume that flip-flops are initially Reset.)

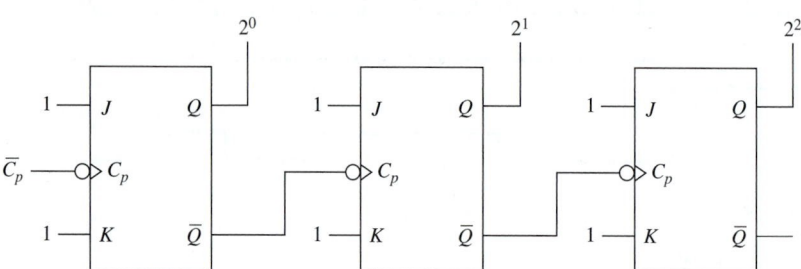

Figure P12–15

12–16. Is the counter of Problem 12–15 an up- or down-counter, and is it a MOD-8 or MOD-16?

12–17. Sketch the connections to a 3-bit ripple up-counter that can be used as a divide-by-6 frequency divider.

D   **12–18.** Design a circuit that will convert a 2-MHz input frequency into a 0.4-MHz output frequency.

D   **12–19.** Design and sketch a MOD-11 ripple up-counter that can be manually Reset by an external push button.

C D   **12–20.** Design and sketch a MOD-5 ripple down-counter with a manual Reset push button. (The count sequence should be 7–6–5–4–3–7–6–5–, and so on.)

C D   **12–21.** Repeat Problem 12–20 for a count sequence of 10–9–8–7–6–10–9–8–, and so on.

CHAPTER 12 | COUNTER CIRCUITS AND VHDL STATE MACHINES

C D **12–22.** Design a MOD-4 ripple up-counter that counts in the sequence 10–11–12–13–10–11–12–, and so on.

C D **12–23.** Redesign the Reset circuitry of Figure 12–21 using a 7401 open-collector NAND and an active-LOW push button with a 10-kΩ pull-up resistor. (Use no other gates.)

C T **12–24.** When you test your design for Problem 12–23, it works fine until you depress the push button. After that, it becomes a MOD-8 counter. When you check the ICs, you find that a 7400 was used instead of the 7401. What caused the MOD-6 counter to turn into a MOD-8?

C D **12–25.** The circuit in Figure P12–25 is being considered as a replacement for the Reset circuitry of Figure 12–21. Do you think that it will work? Why?

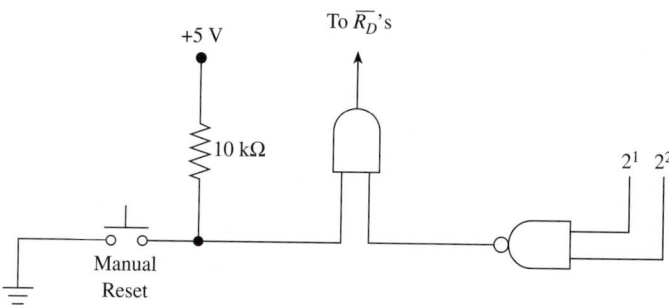

Figure P12–25

Section 12–4

12–26. Describe the major differences among the 7490, 7492, and the 7493 TTL ICs.

12–27. Assume that you have one 7490 and one 7492. Show the external connections that are required to form a divide-by-24.

12–28. Repeat Problem 12–27 using two 7492s to form a divide-by-36.

12–29. Using as many 7492s and 7490s as you need, sketch the external connections required to divide a 60-pps clock down to one pulse per day.

12–30. Make the necessary external connections to a 7493 to form a MOD-10 counter.

Section 12–5

D **12–31.** Design a ripple counter circuit that will flash an LED ON for 40 ms and OFF for 20 ms (assume that a 100-Hz clock oscillator is available). (*Hint*: Study the output waveforms from a MOD-6 counter.)

C D **12–32.** Design a circuit that will turn on an LED 6 s after you press a momentary push button. (Assume that a 60-pps clock is available.)

12–33. What modification to the egg timer circuit of Figure 12–43 could be made to allow you to turn off the buzzer without shutting off the power?

Section 12–6

12–34. Calculate the size of the series current-limiting resistor that could be used in Figure 12–45 to limit the LED current to 15 mA instead of 10 mA.

T **12–35.** In Figure 12–48, instead of using a resistor dip network, some designers use a single limiting resistor in series with the 5-V supply and connect the 7447 outputs directly to the LED inputs to save money. It works, but the display does not look as good. Can you explain why?

Section 12–7

12–36. What advantage does a synchronous counter have over a ripple counter?

12–37. Sketch the waveforms at $\overline{C_p}$, 2^0, 2^1, and 2^2 for 10 clock pulses for the 3-bit synchronous counter shown in Figure P12–37.

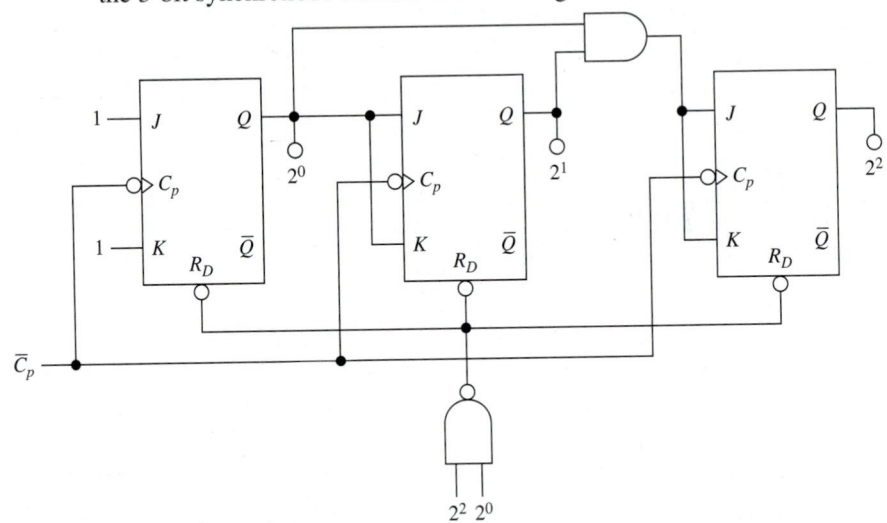

Figure P12–37

12–38. The duty cycle of a square wave is defined as the time the wave is HIGH, divided by the total time for one period. From the waveforms that you sketched for Problem 12–37, find the duty cycle for the 2^2 output wave.

Sections 12–8 and 12–9

C **12–39.** Sketch the timing waveforms at $\overline{TC_D}$, $\overline{TC_U}$, Q_0, Q_1, Q_2, and Q_3 for the 74192 counter shown in Figure P12–39.

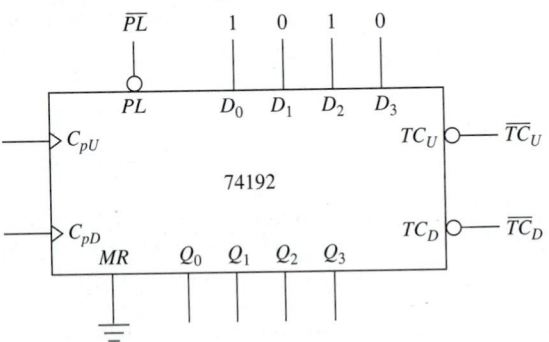

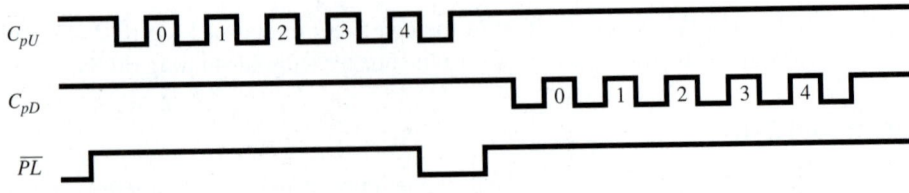

Figure P12–39

CHAPTER 12 I COUNTER CIRCUITS AND VHDL STATE MACHINES

C

12–40. Sketch the timing waveforms at $\overline{RC}$, TC, Q_0, Q_1, Q_2, and Q_3 for the 74191 counter shown in Figure P12–40.

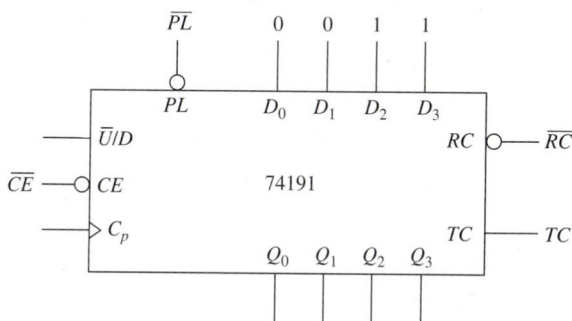

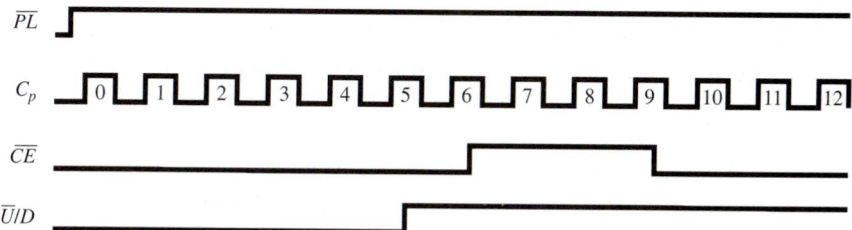

Figure P12–40

C D **12–41.** Make all the necessary pin connections to a 74193 without using external gating to form a divide-by-4 frequency divider. Make it an up-counter, and show the waveforms at C_{pU}, $\overline{TC_U}$, Q_0, Q_1, Q_2, and Q_3.

C D **12–42.** Using the synchronous Reset feature of the 74163 counter, make the necessary connections to form a glitch-free MOD-12 up-counter.

Schematic Interpretation Problems

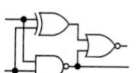

See Appendix G for the schematic diagrams.

S **12–43.** The 74161s in the Watchdog Timer schematic are used to form an 8-bit counter.

(a) Which is the HIGH-order and which is the LOW-order counter?

(b) Is the parallel-load feature being used on these counters?

(c) How are the counters reset in this circuit?

S D **12–44.** On a separate piece of paper, redesign the counter section of the Watchdog Timer schematic by replacing the 74161s with 74193s.

S C D **12–45.** The 68HC11 microcontroller in the HC11D0 master board schematic provides a clock output signal at the pin labeled E. This clock signal is used as the input to the LCD controller, M1 (grid location E-7). The frequency of this signal is 9.8304 MHz, as dictated by the crystal on the 68HC11. To experiment with different clock speeds on the LCD controller, you want to divide that frequency by 2, 4, 8, and 16 before inputting it to pins 6 and 10. Design a circuit using a 4-bit counter IC connected as a frequency divider and a multiplexer IC to select which counter output is sent to the LCD controller for its clock signal.

Electronics Workbench®/MultiSIM® Exercises

E12–1. Load the circuit file for **Section 12–2a.** This is a 4-bit binary counter. Press P several times to observe the counting action.

(a) Which is the least-significant output?

(b) Connect the seven-segment display to the binary output lines. List the digits that appear on the display as you press P repeatedly.

T

E12–2. Load the circuit file for **Section 12–2b.** This is a 4-bit binary counter. Press P several times to observe the counting action. There seems to be a problem. Any ideas? Fix the fault, and test your fix.

T

E12–3. Load the circuit file for **Section 12–2c.** This is a 4-bit binary counter. Press P several times to observe the counting action. There seems to be a problem. Any ideas? Fix the fault, and test your fix.

E12–4. Load the circuit file for **Section 12–2d.** The output of this 4-bit binary counter can be observed on the Logic Analyzer (LA). The Word Generator is used to perform an initial reset then apply clock pulses to the first flip-flop.

(a) Is the count incremented on each negative or positive edge of Cp'?

(b) Besides Cp', which waveform has the highest frequency?

(c) What is the modulus of this counter?

C D

E12–5. Load the circuit file for **Section 12–2e.** RO1, RO2 are the same as MR1, MR2)

(a) What is the modulus of this counter?

(b) What modification must be made to make it a MOD-12? Try it.

(c) You can't see it on the Logic Analyzer, but one of the lines of your Mod-12 will have a short-duration glitch on it. Which one will it be? (It can be observed with the Oscilloscope if you increase the Word Generator frequency to 1 MHZ.)

D

E12–6. Load the circuit file for **Section 12–4a.** This counter is driven by a two pulse-per-second clock.

(a) What is the modulus of this counter circuit?

(b) What modification needs to be done to form a Mod-12?

(c) What modification needs to be done to form a Mod-13? Try them.

C D

E12–7. Load the circuit file for **Section 12–4b.**

(a) Design and make the connections necessary to form a 00 to 99 counter using two 7490s and two seven-segment displays.

(b) Make a modification to your circuit that will provide a "fast-forward" feature that will change the count speed from 1 PPS to 10 PPS by throwing a switch. (Limit yourself to a single clock input source for your circuit.)

(c) Make another modification to your circuit to "stop the count" when it reaches a specific number, let's say 23.

D

E12–8. Load the circuit file for **Section 12–4c.** Design a circuit to demonstrate on a dual-trace oscilloscope the divide-by-six capability of a 7492.

C

E12–9. Load the circuit file for **Section 12–8a.** The Word Generator is used to inject signals into the 74192 up/down-counter. On a piece of paper, carefully sketch the four waveforms MR, PL, Cpu, and Cpd, and then sketch what you think Qa, Qb, Qc, Qd, Tcd′, and Tcu′ will look like. Check your answer by connecting those six outputs to the Logic Analyzer.

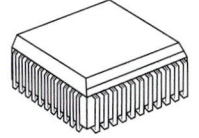

CPLD Problems

C12–1. The VHDL program in Figure 12–16 is the implementation of a MOD-16 up-counter. Load the files *mod16up.vhd* and *mod16up.scf* from the textbook CD. Save these files with the new names *prob_c12_1.vhd* and *prob_c12_1.scf.* Compile and simulate this program. (Remember, the Entity name must be changed to the new name in all three locations before attempting to compile.)

(a) Zoom the simulation display to determine the propagation delay from *n_cp to q0*.

(b) Add a reset pulse (*n_rd*) to the simulation at 5.0 μs and rerun the simulation.

(c) Change the counter to a down-counter and rerun the simulation.

(d) Download your design of part (c) to a CPLD IC. Discuss your observations of the *q* output LEDs with your instructor as you demonstrate the down-counting operation and the asynchronous reset. Use a de-bounced switch for *n_cp*. (The UP-1 board does not have de-bounced switches so you must build an off-board switch de-bouncer similar to the one explained in Figure 11–47.)

C12–2. The VHDL program in Figure 12–27 is the implementation of a MOD-10 up-counter.

(a) Make the necessary changes to make it a positive edge-triggered MOD-12 up-counter. Save this program as *prob_c12_2.vhd.*

(b) Test its operation by creating waveform simulations that demonstrate its up-counting and asynchronous reset operations.

(c) Zoom in on the *q2* waveform to determine the width of its glitch.

(d) Download your design to a CPLD IC. Discuss your observations of the *q* output LEDs with your instructor as you demonstrate the up-counting operation and the asynchronous reset. Use a de-bounced switch for *n_cp*. (The UP-1 board does not have de-bounced switches so you must build an off-board switch de-bouncer similar to the one explained in Figure 11–47.)

C12–3. Modify Problem C12–2 to eliminate the glitch on the *q2* waveform. (*Hint:* see Example 12–12.) Save this program as *prob_c12_3.vhd.* Demonstrate its glitch-free operation by creating simulation waveforms.

C12–4. The VHDL program in Figure 12–51 is a seven-segment decoder for common-anode displays.

(a) What changes need to be made to make it able to drive a common-cathode display?

(b) Make the changes specified in part (a) and run a simulation to demonstrate its operation.

(c) Add a new input to the decoder called *ca_cc*. When this line is HIGH, perform decoding for a common-anode display, and when it is LOW, perform decoding for a common-cathode display. Create a simulation file to demonstrate its operation.

C12–5. Figure 12–69 is a 4-bit up-counter with asynchronous Reset and Parallel Load features. Modify its simulation file to perform the following operations:

(a) Reset, then count up four pulses.

(b) Parallel Load an 8, and then count up four more pulses.

(c) Reset, then count up five more pulses.

(a) Parallel Load a 7, and then continue counting up for the remainder of the clock pulses.

[*Hint:* To draw narrow waveform pulses, change the grid size to 100 ns by choosing **Options -> Grid Size** and enter 100 ns. Next, highlight the area that you want to be LOW and press the **0** (LOW pulse) shortcut key.]

C12–6. Figure 12–71 is a 4-bit up/down-counter with a Clock Enable and asynchronous Reset and Parallel Load features. Modify its simulation file to perform the following operations:

(a) Reset, then count up four pulses.

(b) Parallel Load D_{16}, and then count up four more pulses.

(c) Count down eight pulses then disable the clock ($n\_ce = 1$) for the remainder of the clock pulses.

[*Hint:* To draw narrow waveform pulses, change the grid size to 100 ns by choosing **Option -> Grid Size** and enter 100 ns. Next highlight the area that you want to be LOW and press the **0** (LOW pulse) shortcut key.]

C12–7. Add a terminal-count (*tc*) feature to the VHDL counter in Figure 12–71. This output is similar to that provided by the 74191 of Section 12–8. [In the count-up mode *tc* goes HIGH when 15 (F_{16}) is reached, and in the count-down mode it goes HIGH when 0 is reached.] Create a simulation file that demonstrates its operation.

C12–8. Figure 12–74 shows the LPM_COUNTER set up as a MOD-10 up/down-counter with asynchronous Set (*aset*) and Clear (*aclr*). Test its operation by performing the following steps:

(a) *Aset* a 7, then count down 4 pulses.

(b) Then clear and count down 4 more pulses.

(c) Then count up 6 pulses, *aset* a 7, then continue counting up to the end.

C12–9. Use an LPM_COUNTER to design a MOD-12 down-counter. Develop a simulation file to demonstrate its operation.

C12–10. Use an LPM_COUNTER to design a divide-by-60 frequency divider. Develop a simulation file to demonstrate its operation (Show the clock input waveform and the single divide-by-60 output waveform.)

C12–11. Figure 12–76 shows a VHDL program that uses state machine design to create a Gray code counter.

(a) Use a technique similar to that to develop a sequencer that counts up for just the odd number (1-3-5-7-9-1-, etc.). Develop a simulation file to demonstrate its operation.

(b) Add a direction control input so that you can change it from an odd up-counter to an odd down-counter. Develop a simulation file to demonstrate its operation.

C12–12. Figure 12–79 shows a VHDL program that uses state machine design to create a stepper-motor driver. Instead of activating output 1, then 2, then 3, then 4, then 1, repeatedly as shown, motor rotation half-steps can be made by activating output 1, then (1 and 2), then 2, then (2 and 3), etc, repeatedly (i.e. 0001-0011-0010-0110-0100-1100-1000-1001-0001-, and so on).

(a) Rewrite the program to output half-steps for the stepper motor.

(b) Develop a simulation file to demonstrate its operation for both forward and reverse directions.

C12–13. Figures 12–81, 12–82, 12–83, and 12–84 show how an ADC interface to a microprocessor is implemented in VHDL using a state machine. For simplicity, the circuit used all active-HIGH signals but in reality the control signals used for ADCs are active-LOW.

(a) Modify Figures 12–81 and 12–82 assuming active-LOW signals for *dr, sc,* and *oe.*

(b) Rewrite the VHDL program in Figure 12–83 using those active-LOW signals.

(c) Redraw the simulation waveforms using those active-LOW signals.

C12–14. The state diagram in Figure 12–82 shows that as soon as data-ready (*dr*) goes HIGH, control passes from the *waiting* state to the *read* state. The purpose of the read state is to make *oe* LOW, which makes the ADC outputs active so that the microprocessor can read the 8-bit digital data. A possible problem arises because control then passes unconditionally to the *idle* state at the next active clock edge. Depending on the speed of the microprocessor, this may be too short of a time duration for the data to be read. One way to correct this problem is to monitor the trigger (*t*) signal from the microprocessor. Once this signal is HIGH, it does not go back LOW until the entire process is complete and the microprocessor has completed reading the data bus. Modify the state diagram and VHDL program so that the transition to idle does not occur until trigger (*t*) goes LOW. Perform a simulation of the modified program.

C12–15. Figures 12–85, 12–86, 12–87, and 12–88 show how an ATM interface to a printer is implemented in VHDL using a state machine. For simplicity, the circuit used all active-HIGH signals but in reality the control signals used for most printers are active-LOW.

(a) Modify all Figures 12–85 and 12–86 assuming active-LOW signals for *r, b,* and *s.*

(b) Rewrite the VHDL program in Figure 12–87 using those active-LOW signals.

(c) Redraw the simulation waveforms using those active-LOW signals.

C12–16. The state diagram for the ATM controller in Figure 12–86 shows that when the transmission of data is complete, *q* goes LOW and the state transitions to *idle.* Make the necessary modifications to the state diagram and the VHDL program so that when *q* goes LOW, a transition is made to a new state called *eject* that sends a HIGH signal to the printer telling it to cut and eject the paper. [The CPLD now has three outputs: *t, s,* and *e* (eject)]. Build a simulation file to demonstrate its operation.

C12–17. A chemical tank flushing system consists of a CPLD controlling the draining and refilling of a water tank that is used to dilute a chemical production by-product that needs to be flushed periodically.

Process definition:

(a) In the *standby* state the holding tank is full of water.

(b) When a flush signal (*f*) is asserted, transition is made to the *drain* state. The tank's drain valve (*dv*) is activated, emptying all of the water and chemical waste into a drain.

(c) When the tank is empty, a fluid-at-bottom signal (*b*) is issued from the tank. A transition is made to the *fill* state.

(d) The tank's fill-valve (*fv*) is activated to refill the tank with clean water until the fluid-at-top signal (*t*) is issued from the tank. A transition is made back to the *standby* state.

The CPLD has the following I/O:

(a) Active-HIGH "flush" signal input (*f*) from a push-button to initiate the process.

(b) Active-LOW "fluid-at-top" signal input (*t*) from the tank sensor to signify that the fluid is at the top of the tank.

(c) Active-LOW "fluid-at-bottom" signal input (*b*) from the tank sensor to signify that the fluid is totally drained from the tank.

(d) Active-HIGH "fill-valve" signal output (*fv*) to the tank to open the fill valve to allow the water to fill the tank.

(e) Active-HIGH "drain valve" signal output (*dv*) to the tank to open the drain valve to allow the fluid to drain from the tank.

Assignment: Complete a block diagram, state diagram, VHDL program, and simulation of the system.

Answers to Review Questions

12-1. True

12-2. HIGH, HIGH

12-3. It places limitations on the maximum frequency allowed by the input trigger clock because each output is delayed from the previous one.

12-4. By taking the binary output from the $\overline{Q}$ outputs

12-5. 2^3

12-6. 12

12-7. When the push button is pressed, 5 V is applied to the NOR gate, driving its output LOW. The 100-Ω pull-down resistor will keep the input LOW when the push button is in the open position.

12-8. MOD-10, MOD-12, MOD-16

12-9. One is for the divide-by-2 section, and the other is for the divide-by-8 section.

12-10. $Q_0 = 1, Q_1 = 0, Q_2 = 0, Q_3 = 1$

12-11. By cascading a divide-by-10 with a divide-by-6

12-12. The Q associated with the most significant bit

12-13. This means that all of the anodes or cathodes of the LED segments are connected together.

12-14. a c d f g, a b c d e f

12-15. To limit the current flowing through the LED segments

12-16. Active-HIGH, active-LOW, anode

12-17. False

12-18. It reduces both the number of I/O ports and data paths needed.

12-19. They don't have the problem of accumulated propagation delay.

12-20. J and K inputs are tied together and are controlled through the use of an AND gate.

12-21. False

12-22. They are the terminal count pins, which are used to indicate when the terminal count is reached and the count is about to recycle.

12-23. $\overline{U}/D = 1$

12-24. False

13

Shift Registers

OUTLINE

OBJECTIVES

Upon completion of this chapter, you should be able to:

- Connect *J-K* flip-flops as serial or parallel-in to serial or parallel-out multibit shift registers.
- Draw timing waveforms to illustrate shift register operation.
- Explain the operation and application of ring and Johnson shift counters.
- Make external connections to MSI shift register ICs to perform conversions between serial and parallel data formats.
- Explain the operation and application of three-state output buffers.
- Discuss the operation of circuit design applications that employ shift registers.

INTRODUCTION

Registers are required in digital systems for the temporary storage of a group of bits. **Data bits** (1's and 0's) traveling through a digital system sometimes have to be temporarily stopped, copied, moved, or even shifted to the right or left one or more positions.

A shift register facilitates this movement and storage of data bits. Most shift registers can handle parallel movement of data bits, as well as serial movement, and can also be used to convert from parallel to serial and from serial to parallel.

13–1   Shift Register Basics

Let's take a look at the contents of a 4-bit shift register as it receives 4 bits of parallel data and shifts them to the right four positions into some other digital device. The timing for the shift operations is provided by the input clock. The data bits will shift to the right by one position for each input clock pulse, as shown in Figure 13–1.

In the figure, the group of four boxes is four D flip-flops comprising the 4-bit shift register. The first step is to parallel load the register with a 1–0–0–0. *Parallel load*

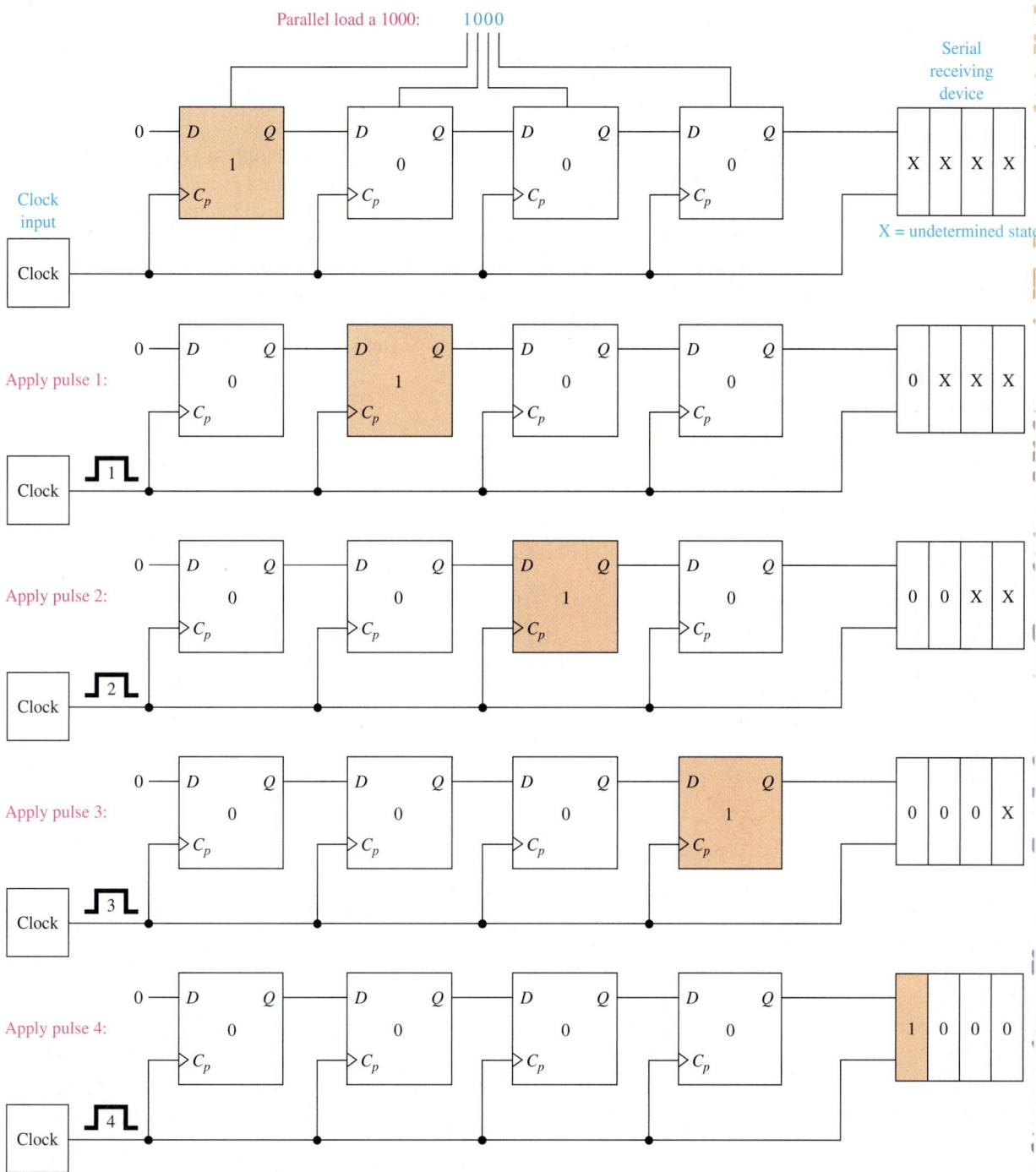

Figure 13–1   Block diagram of a 4-bit shift register for parallel-to-serial conversion.

means to load all four flip-flops at the same time. This is done by momentarily enabling the appropriate asynchronous Set $(\overline{S_D})$ and Reset $(\overline{R_D})$ inputs.

Next, the first clock pulse causes all bits to shift to the right by 1 because the input to each flip-flop comes from the Q output of the flip-flop to its left. Each successive pulse causes all data bits to shift one more position to the right.

At the end of the fourth clock pulse, all data bits have been shifted all the way across, and now all four original data bits appear, in the correct order, in the serial receiving device. The connections between the fourth flip-flop and the serial receiving device could be a three-conductor serial transmission cable (serial data, clock, and ground).

Figure 13–1 illustrated a parallel-to-serial conversion. Shift registers can also be used for serial-to-parallel, parallel-to-parallel, and serial-to-serial, as well as shift-right, shift-left operations, as indicated in Figure 13–2(a). Each of these configurations and an explanation of the need for a *recirculating line* are explained in upcoming sections.

Figure 13–2(b) shows how shift registers are commonly used in data communications systems. Computers operate on data internally in a *parallel format*. To communicate over a serial cable like the one used by the RS232 standard or a telephone line, the data must first be converted (**data conversion**) to the *serial format*. For example, for computer A to send data to computer B, computer A will parallel load 8 bits of data into shift register A and then apply eight clock pulses. The 8 data bits output from shift register A will travel across the serial communication line to shift register B, which is concurrently loading the 8 bits. After shift-register B has received all 8 data

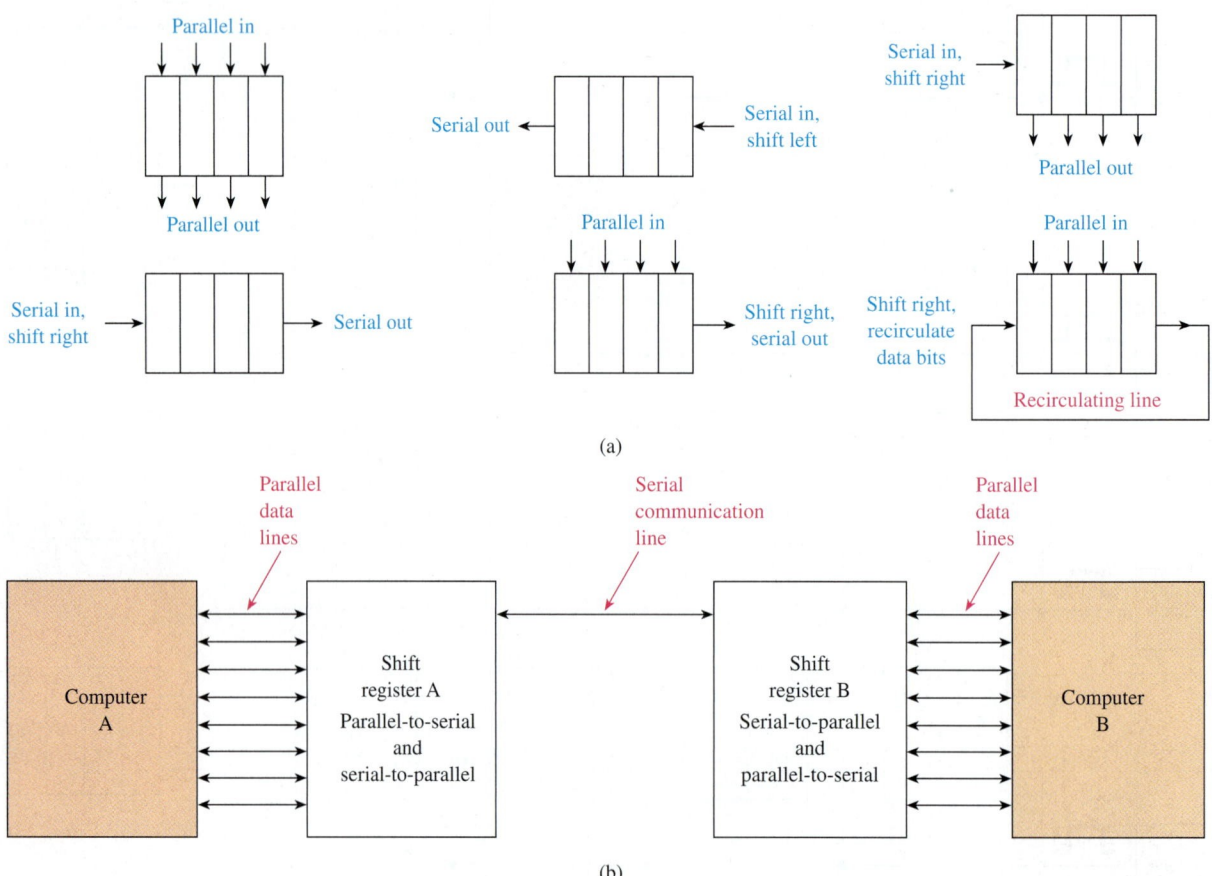

Figure 13–2 Shift register operations: (a) types of data movement and conversion; (b) using shift registers to provide serial communication between computers having parallel data.

bits, it will output them on its parallel output lines to computer B. This is a simplification of the digital communication* that takes place between computers, but it illustrates the heart of the system, the **shift register.**

13–2 Parallel-to-Serial Conversion

Now let's look at the actual circuit connections for a shift register. The data storage elements can be D flip-flops, S-R flip-flops, or J-K flip-flops. We are familiar with J-K flip-flops, so let's stick with them. Most J-K's are negative edge triggered (like the 74LS76) and will have an active-LOW asynchronous Set ($\overline{S_D}$) and Reset ($\overline{R_D}$).

Figure 13–3 shows the circuit connections for a 4-bit parallel-in, serial-out shift register that is first Reset, then parallel loaded with an active-LOW 7 (1000), and then shifted right four positions.

Notice in Figure 13–3(a) that all $\overline{C_p}$ inputs are fed from a common clock input. Each flip-flop will respond to its J-K inputs at every negative clock input edge. Because every J-K input is connected to the preceding stage output, then at each negative clock edge each flip-flop will change to the state of the flip-flop to its left. In other words, all data bits will be shifted one position to the right.

Now, looking at the timing diagram, in the beginning of period 1, $\overline{R_D}$ goes LOW, resetting Q_0 to Q_3 to zero. Next, the parallel data are input (parallel loaded) via the D_0 to D_3 input lines. (Because the $\overline{S_D}$ inputs are active LOW, the complement of the number to be loaded must be used. The $\overline{S_D}$ inputs must be returned HIGH before shifting can be initiated.)

At the first negative clock edge,

Q_0 takes on the value of Q_1

Q_1 takes on the value of Q_2

Q_2 takes on the value of Q_3

Q_3 is Reset by $J = 0, K = 1$

In effect, the bits have all shifted one position to the right. Next, the negative edges of periods 2, 3, and 4 will each shift the bits one more position to the right.

The serial output data comes out of the right-end flip-flop (Q_0). Because the LSB was parallel loaded into the rightmost flip-flop, the LSB will be shifted out first. The order of the parallel input data bits could have been reversed, and the MSB would have come out first. Either case is acceptable. It is up to the designer to know which is first, MSB or LSB, and when to sample (or read) the serial output data line.

13–3 Recirculating Register

Recirculating the rightmost data bits back into the beginning of the register can be accomplished by connecting Q_0 back to J_3 and $\overline{Q_0}$ back to K_3. This way, the original parallel-loaded data bits will never be lost. After every fourth clock pulse, the Q_3 to Q_0 outputs will contain the original 4 data bits. Therefore, with the addition of the recirculating lines to Figure 13–3(a), the register becomes a parallel-in, serial, *and* parallel-out.

---

*An IC called the UART (universal asynchronous receiver transmitter) is used to perform the shift register operation and to create the other control signals necessary for computer communication. That circuitry is used in a computer modem for communication over telephone lines and cable.

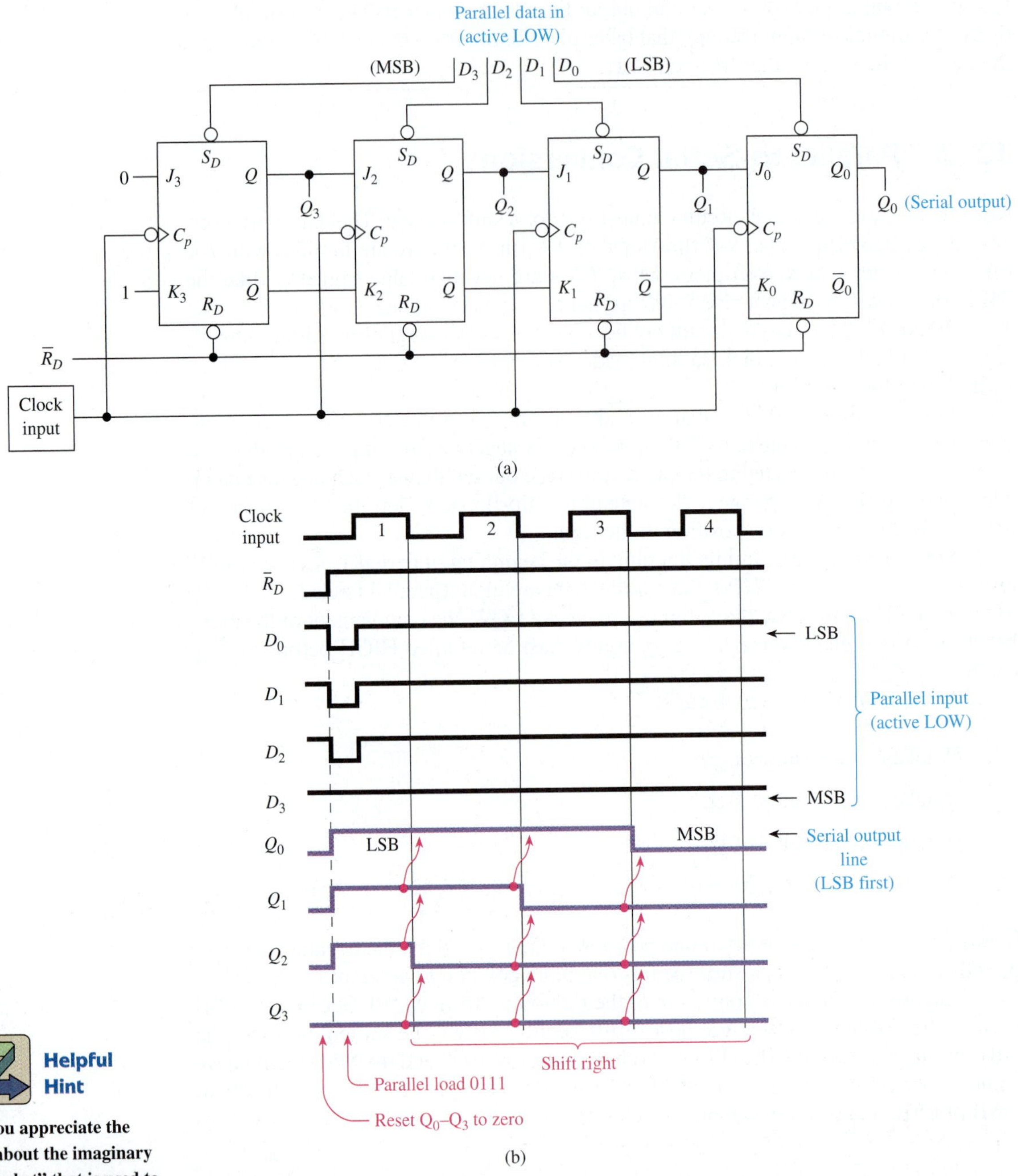

Parallel data in
(active LOW)

(MSB) D_3 D_2 D_1 D_0 (LSB)

(a)

Clock
input

$\overline{R}_D$

D_0 ← LSB

D_1

Parallel input
(active LOW)

D_2

D_3 ← MSB

Q_0 LSB MSB ← Serial output
line
(LSB first)

Q_1

Q_2

Q_3

Shift right

Parallel load 0111

Reset Q_0–Q_3 to zero

(b)

**Helpful
Hint**

Can you appreciate the
story about the imaginary
"bit bucket" that is used to
catch the bits that drop out
of the Q_0 output if the clock
is allowed to continue?

Figure 13–3   (a) Four-bit parallel-in, serial-out shift register using 74LS76 *J-K* flip-flops;
(b) waveforms produced by parallel loading a 7 (0 1 1 1) and shifting right by four clock
pulses. (An active-LOW 7 is 1000.)

Review Questions

13–1. All flip-flops within a shift register are driven by the same clock input. True or false?

13–2. What connections allow data to pass from one flip-flop to the next
in a shift register?

13–3. How are data parallel loaded into a shift register constructed from *J-K* flip-flops?

13–4. If a hexadecimal C is parallel loaded into the shift register of Figure 13–3 and four clock pulses are applied, what is the state of the *Q* outputs? If recirculating lines are connected and the same operation occurs, what is the state of the *Q* outputs?

13–4 Serial-to-Parallel Conversion

Serial-in, parallel-out shift registers can also be made up of *J-K* flip-flop storage and a shift-right operation. The idea is to put the serial data in on the serial input line, LSB first (or MSB first, depending on the direction of the shift) and clock the shift register four times (for a 4-bit register), stop, and then read the parallel data from the Q_0 to Q_3 outputs. Figure 13–4(a) shows a 4-bit serial-to-parallel shift register converter. The serial data are coming in on the left at D_S. The flip-flops are connected in a shift-right fashion. The inverter at D_S is required to ensure that if $D_S = 1$ then $J = 1$, $K = 0$, and the first flip-flop will Set. Each of the other flip-flops takes on the value of the flip-flop to its left at each negative clock edge.

Each bit of the serial input must be present on the D_S line before the corresponding negative clock edge. After four clock pulses, all 4 serial data bits have been shifted into their appropriate flip-flop. At that time, the parallel output data can be read by some other digital device.

If the clock were allowed to continue beyond four pulses, the data bits would continue shifting out of the right end of the register and be lost if you tried to read them again later. This problem is corrected by the **Strobe** line. It is used to Enable-then-Disable the clock at the appropriate time so that the shift-right process will stop. Using a Strobe signal is a popular technique in digital electronics to Enable or Disable some function during a specific time period.

13–5 Ring Shift Counter and Johnson Shift Counter

Two common circuits that are used to create sequential control waveforms for digital systems are the ring and Johnson **shift counters.** They are similar to a synchronous counter because the clock input to each flip-flop is driven by the same clock input. Their outputs do not count in true binary but, instead, provide a repetitive sequence of digital output levels. These shift counters are used to control a sequence of events in a digital system (**digital sequencer**).

In the case of a 4-bit *ring shift counter,* the output at each flip-flop will be HIGH for one clock period, then LOW for the next three, and then repeat, as shown in Figure 13–5(b). To form the ring shift counter of Figure 13–5(a), the Q-$\overline{Q}$ output of each stage is fed to the *J-K* input of the next stage, and the Q-$\overline{Q}$ output of the last stage is fed back to the *J-K* input of the first stage. Before applying clock pulses, the shift counter is preset with a 1–0–0–0.

Ring Shift Counter Operation

The *RC* circuit connected to the power supply will provide a LOW-then-HIGH as soon as the power is turned on, forcing a HIGH–LOW–LOW–LOW at Q_0–Q_1–Q_2–Q_3, which is the necessary preset condition for a ring shift counter. At the first negative clock input edge, Q_0 will go LOW because just before the clock edge J_0 was LOW (from Q_3) and K_0 was HIGH (from $\overline{Q_3}$). At that same clock edge, Q_1 will go HIGH because its *J-K* inputs are connected to Q_0-$\overline{Q_0}$, which were 1–0. The Q_2 and Q_3 flip-flops will remain Reset (LOW) because their *J-K* inputs see a 0–1 from the previous flip-flops.

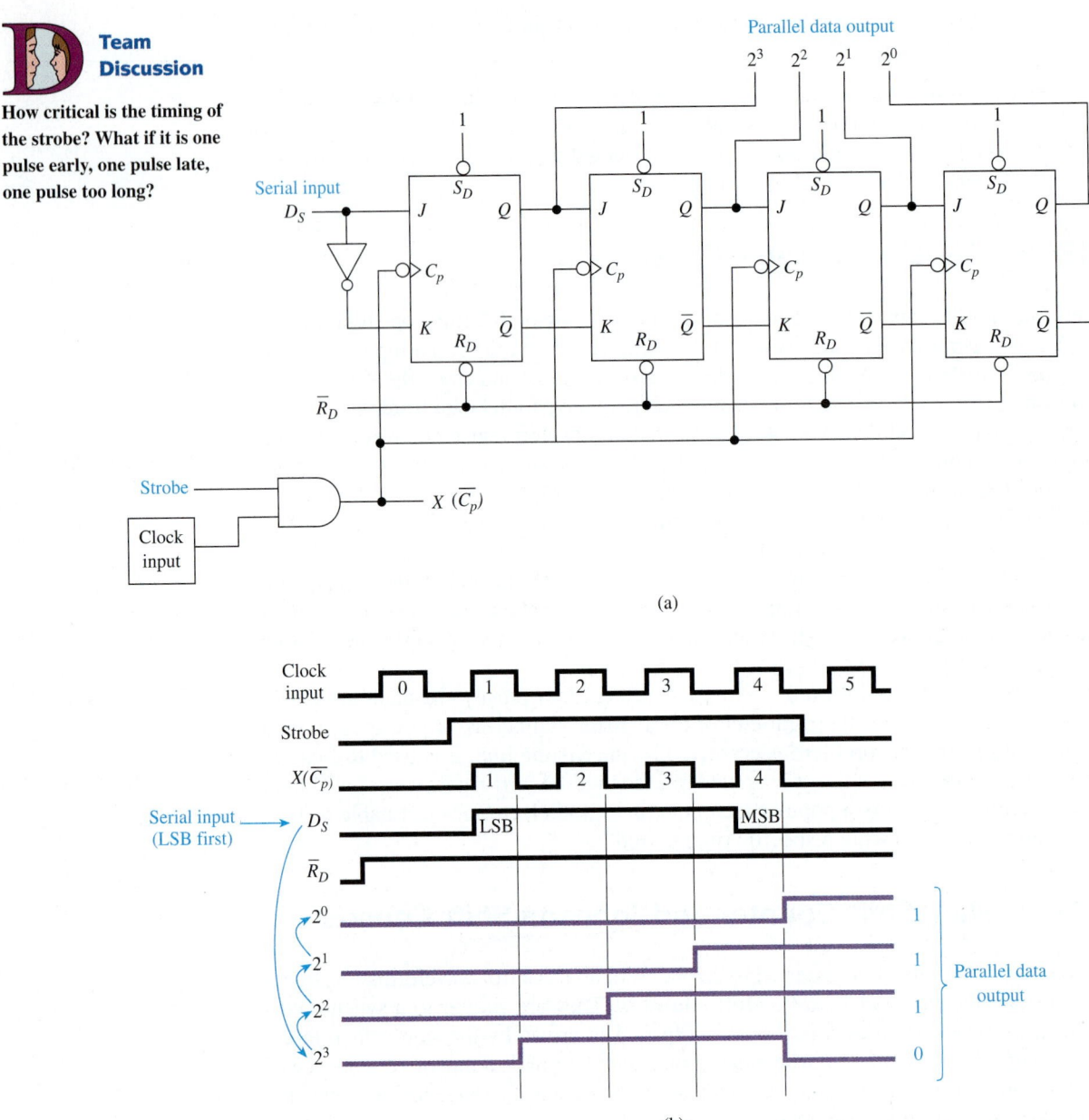

Team Discussion

How critical is the timing of the strobe? What if it is one pulse early, one pulse late, one pulse too long?

(a)

(b)

Figure 13–4 (a) Four-bit serial-to-parallel shift register; (b) waveforms produced by a serial-to-parallel conversion of the binary number 0 1 1 1.

Now the ring shift counter is outputting a 0–1–0–0 (period 2). At the negative edge of period 2, the flip-flop outputs will respond to whatever levels are present at their J-K inputs, the same as explained in the preceding paragraph. That is, because J_2-K_2 are looking back at (connected to) Q_1-$\overline{Q}_1$ (1–0), then Q_2 will go HIGH. All other flip-flops are looking back at a 0–1, so they will Reset (LOW). This cycle repeats continuously. The system acts like it is continuously "pushing" the initial HIGH level at Q_0 through the four flip-flops.

The *Johnson shift counter* circuit is similar to the ring shift counter except that the output lines of the last flip-flop are crossed (thus, an alternative name is *twisted ring counter*) before feeding back to the input of the first flip-flop, and *all* flip-flops are initially Reset, as shown in Figure 13–6.

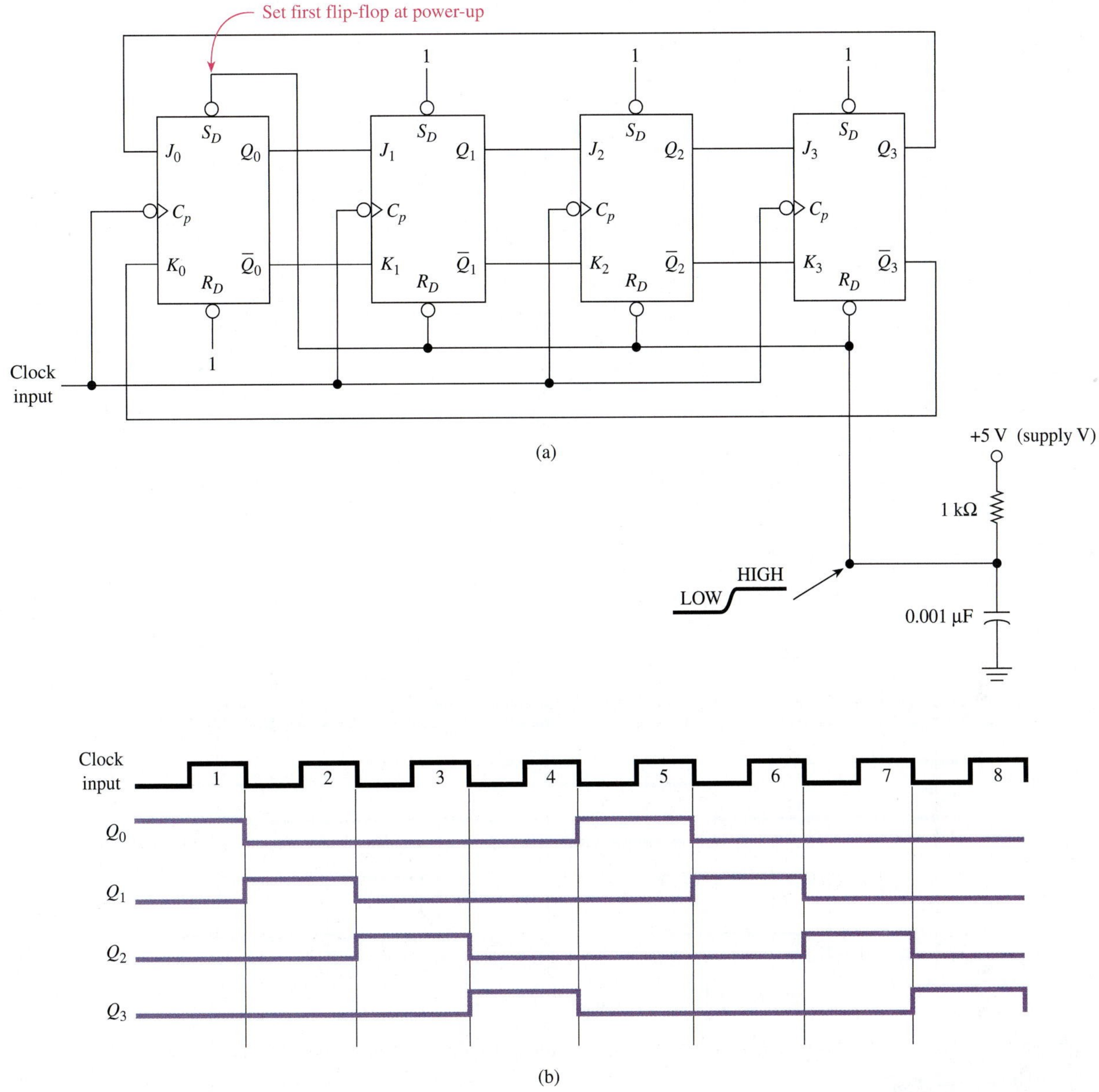

Set first flip-flop at power-up

(a)

+5 V (supply V)

1 kΩ

HIGH

LOW

0.001 μF

Clock input

Clock input

Q_0

Q_1

Q_2

Q_3

(b)

Figure 13–5   Ring shift counter: (a) circuit connections; (b) output waveforms.

Team Discussion

How would you implement this circuit using 7474 D flip-flops?

Johnson Shift Counter Operation

The RC circuit provides an automatic Reset to all four flip-flops, so the initial outputs will all be Reset (LOW). At the first negative clock edge, the first flip-flop will Set (HIGH) because J_0 is connected to $\overline{Q}_3$ (HIGH) and K_0 is connected to Q_3 (LOW). The Q_1, Q_2, and Q_3 outputs will follow the state of their preceding flip-flop because of their direct connection J-to-Q. Therefore, during period 2, the output is 1–0–0–0.

At the next negative clock edge, Q_0 remains HIGH because it takes on the *opposite* state of Q_3, Q_1 goes HIGH because it takes on the *same* state as Q_0, Q_2 stays LOW, and Q_3 stays LOW. Now the output is 1–1–0–0.

The sequence continues as shown in Figure 13–6. Notice that, during period 5, Q_3 gets Set HIGH. At the end of period 5, Q_0 gets Reset LOW because the outputs of Q_3 are crossed, so Q_0 takes on the opposite state of Q_3.

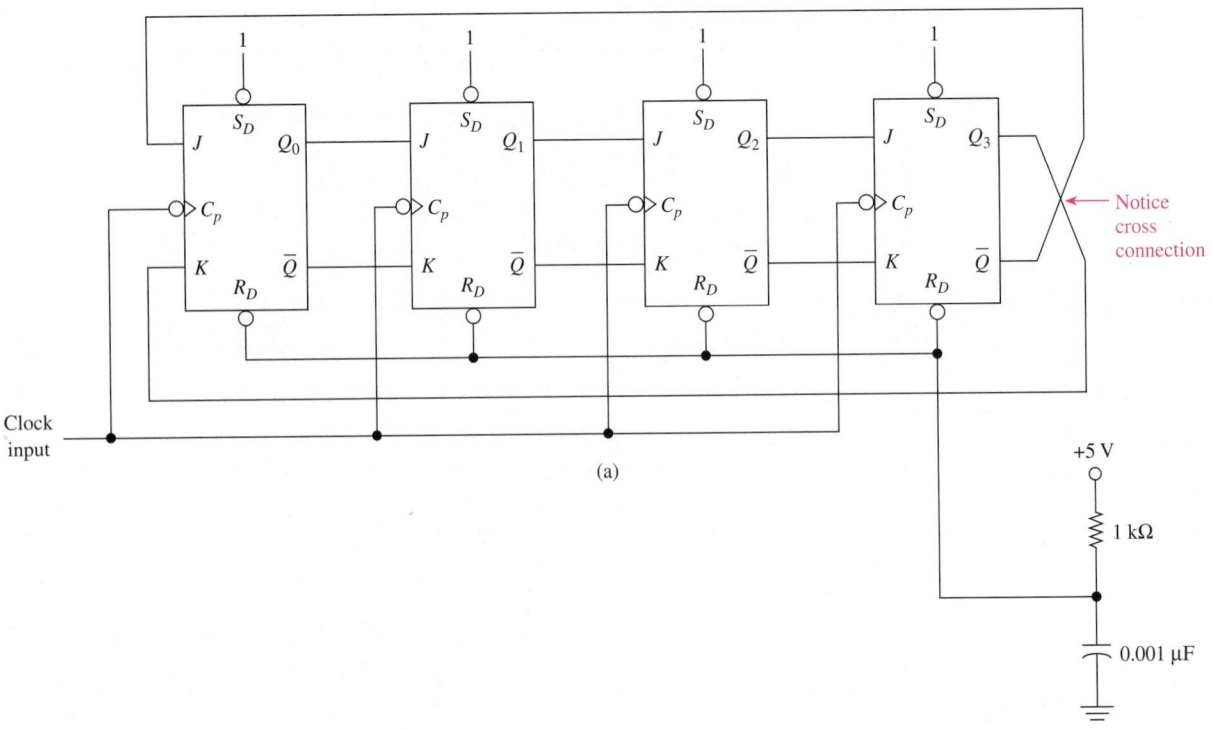

(a)

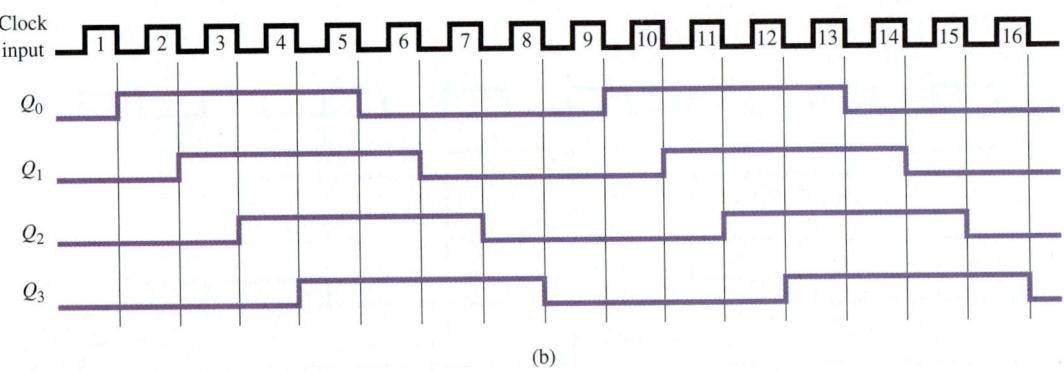

(b)

Figure 13–6 Johnson shift counter: (a) circuit connections; (b) output waveforms.

Team Discussion

To see if you really understand the circuit operation, try to sketch the waveforms if the cross-connection is made between the third and the fourth flip-flop instead of at the end.

Review Questions

13–5. What happens to the initial parallel-loaded data in the shift register of Figure 13–4 if the *Strobe* line is never disabled?

13–6. To operate properly, a ring shift counter must be parallel loaded with _____, and a Johnson shift counter must be parallel loaded with _____.

13–6 VHDL Description of Shift Registers

The VHDL language provides the ultimate in flexibility in designing shift registers. They can be defined as shift-left or shift-right simply by changing the order of the assignment statements in the architecture section. You can define parallel loading as synchronous or asynchronous by the placement of the IF statement in the program. Other features such as strobing the clock and recirculating the data bits can also be added to the basic design.

Figure 13–7 shows a 4-bit shift-right shift register. An internal SIGNAL register (*reg*) is declared and used to receive the shifted data at each clock pulse. Notice that the leftmost register element [*reg(3)*] is loaded with *ser_data,* and each subsequent element is loaded with the element from its left. This, in effect, shifts the data to the right. After the shift to the right is made for all four register elements, the *q*-output vector is loaded with the contents of *reg*. The simulation waveform shown in Figure 13–8 illustrates the data bits shifting to the right on each negative clock pulse. At the 2 μs clock edge, the *ser_data* line is HIGH, which sets *q3* HIGH. From then on, the data travels to each lower element number (shifting right).

Figure 13–9 shows a 4-bit shift-right shift register with a parallel-load feature. It is similar to the previous shift register but has an IF statement to check for a parallel-load operation. If *pl = '1'* then the internal register is loaded with the data present on the *par_data* inputs. Since this check is made before the check for the clock edge, then it is asynchronous, having priority over the synchronous shifting operation. The simulation waveforms shown in Figure 13–10 show the initial parallel loading of the number 9, then the data bits shifting to the right on each negative clock pulse.

13–7 Shift Register ICs

Four-bit and 8-bit shift registers are commonly available in IC packages. Depending on your needs, practically every possible load, shift, and conversion operation is available in a shift register IC.

```
shift_a.vhd - Text Editor                                    _ □ X
  LIBRARY ieee;                  --------------------------------
  USE ieee.std_logic_1164.ALL; -- 4-bit Serial-in Shift Right --
                               --   Shift Register            --
  ENTITY shift_a IS            --------------------------------
      PORT(n_cp, ser_data : IN std_logic;
              q               : OUT std_logic_vector(3 DOWNTO 0));
  END shift_a ;


  ARCHITECTURE arc OF shift_a IS
      SIGNAL reg : std_logic_vector (3 DOWNTO 0);
  BEGIN                          Declare an internal register that acts like four FFs
      PROCESS (n_cp)
      BEGIN
          IF (n_cp'EVENT AND n_cp='0' ) THEN
              reg(3)<= ser_data;
              reg(2)<= reg(3);          Shift each bit to the right
              reg(1)<= reg(2);          at each clock edge
              reg(0)<= reg(1);
          END IF;
          q<=reg;            Send results to output
      END PROCESS;
  END arc;                                         shift_a.vhd
  Line   1    Col   1      INS
```

Figure 13–7   VHDL description of a serial-in shift-right shift register.

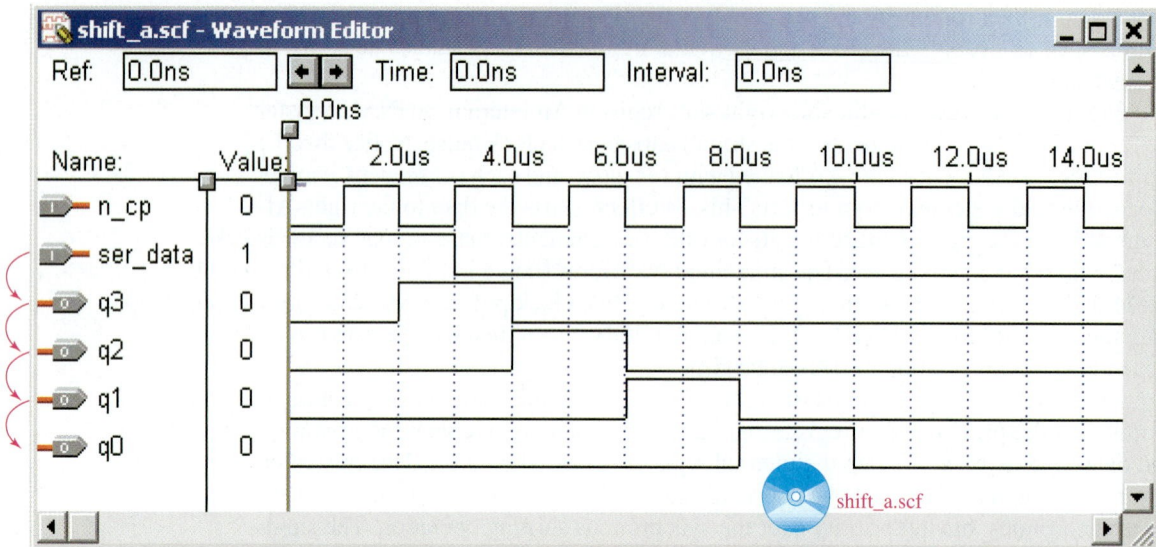

Figure 13–8 Simulation of the shift register defined in Figure 13–7.

```
shift_b.vhd - Text Editor                                    _ □ ×
LIBRARY ieee;                    ------------------------------
USE ieee.std_logic_1164.ALL; -- 4-bit Parallel-Load Shift  --
                             --    Right Shift Register     --
ENTITY shift_b IS            ------------------------------
    PORT(n_cp, pl      : IN std_logic;
         par_data      : IN std_logic_vector (3 DOWNTO 0);
         q             : OUT std_logic_vector(3 DOWNTO 0));
END shift_b ;


ARCHITECTURE arc OF shift_b IS
    SIGNAL reg : std_logic_vector (3 DOWNTO 0);
BEGIN
    PROCESS (n_cp,pl)
    BEGIN
        IF (pl='1') THEN                Parallel load overrides
            reg<=par_data;              synchronous n_cp operations
        ELSIF (n_cp'EVENT AND n_cp='0' ) THEN
            reg(3)<= '0';
            reg(2)<= reg(3);
            reg(1)<= reg(2);
            reg(0)<= reg(1);
        END IF;
        q<=reg;
    END PROCESS;
END arc;                                          shift_b.vhd
Line   1    Col   1    INS ◄
```

Figure 13–9 VHDL description of a parallel-load shift-right shift register.

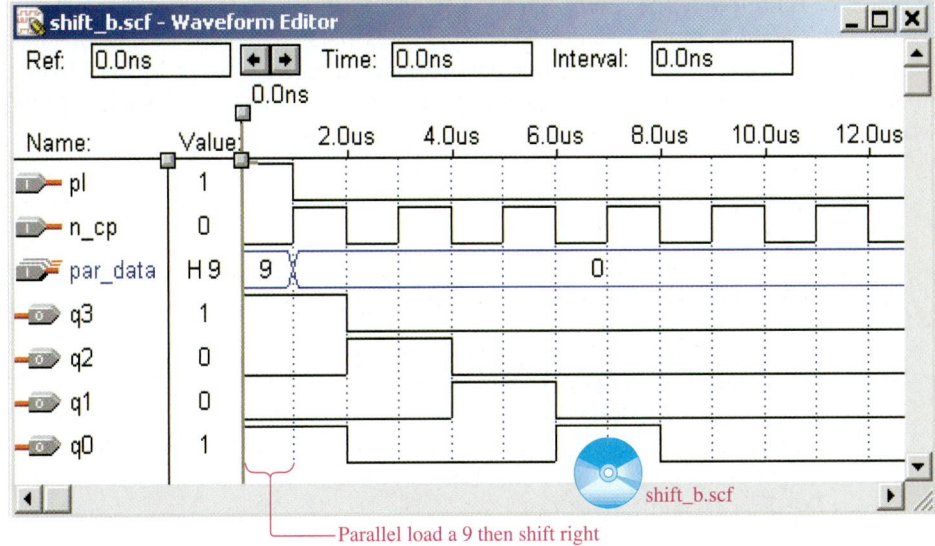

Parallel load a 9 then shift right

Figure 13–10 Simulation of the shift register defined in Figure 13–9.

Let's look at four popular shift register ICs to get familiar with using our data manuals and understanding the terminology and procedure for performing the various operations.

The 74164 8-Bit Serial-In, Parallel-Out Shift Register

By looking at the logic symbol and logic diagram for the 74164 (see Figure 13–11), we can see that it saves us the task of wiring together eight D flip-flops. The 74164 has two serial input lines (D_{Sa} and D_{Sb}), synchronously read in by a positive edge-triggered clock (C_p). The logic diagram [Figure 13–11(b)] shows both D_S inputs feeding into an

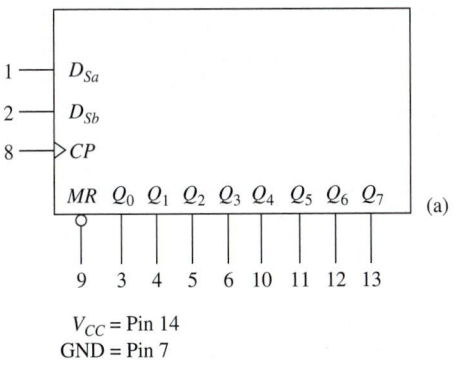

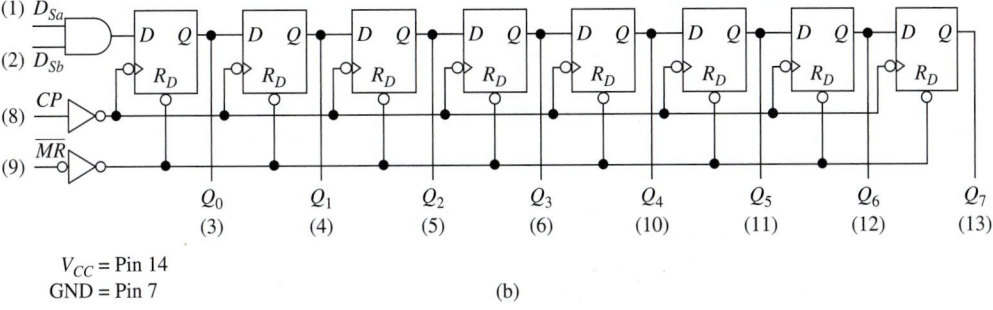

Figure 13–11 The 74164 shift register IC: (a) logic symbol; (b) logic diagram.

AND gate. Therefore, either input can be used as an active-HIGH enable for data entry through the other input. Each positive edge clock pulse will shift the data bits one position to the right. Therefore, the first data bit entered (either LSB or MSB) will end up in the far right D flip-flop (Q_7) after eight clock pulses. The $\overline{MR}$ is an active-LOW Master Reset that resets all eight flip-flops when pulsed LOW.

EXAMPLE 13–1

Draw the circuit connections and timing waveforms for the serial-to-parallel conversion of the binary number 11010010 using a 74164 shift register.

Solution: The serial-to-parallel conversion circuit and waveforms are shown in Figure 13–12. First, the register is cleared by a LOW on $\overline{MR}$, making Q_0–$Q_7 = 0$. The Strobe line is required to make sure that we get only eight clock pulses. The serial data are entered on the D_{Sb} line, MSB first. After eight clock pulses, the 8 data bits can be read at the parallel output pins (MSB at Q_7 and LSB at Q_0).

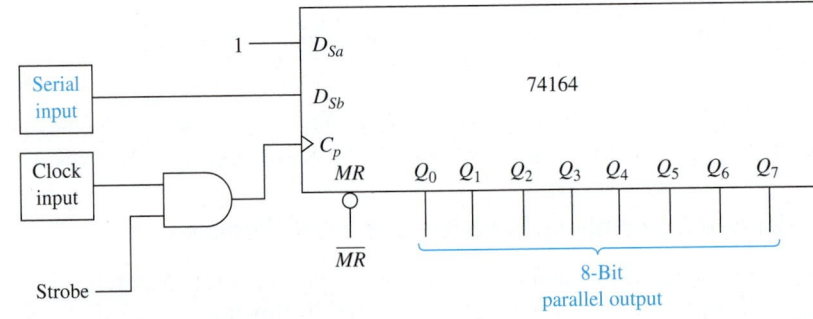

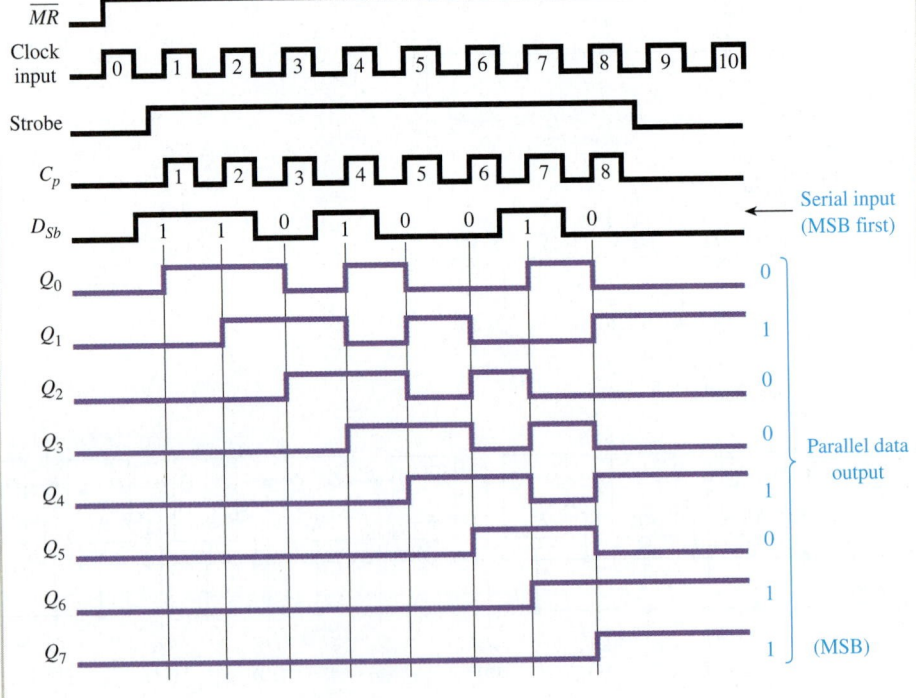

Figure 13–12 Serial-to-parallel conversion using the 74164 shift register IC.

The 74165

The next IC to consider is the 74165 8-bit serial *or* parallel-in, serial-out shift register. The logic symbol for the 74165 is given in Figure 13–13.

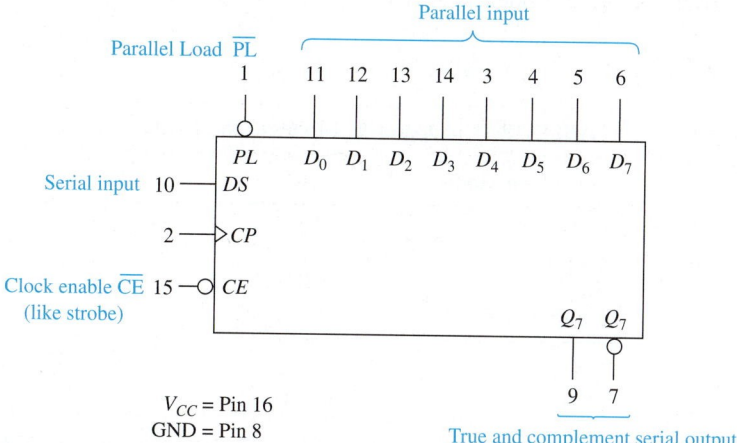

Figure 13–13 Logic symbol for the 74165 8-bit serial or parallel-in, serial-out register.

Just by looking at the logic symbol, you should be able to determine the operation of the 74165. The $\overline{PL}$ is an active-LOW terminal for performing a parallel load of the 8 parallel input data bits. The $\overline{CE}$ is an active-LOW clock enable for starting/stopping (shifting/holding) the shift operation by enabling/disabling the clock (same function as the Strobe in Example 13–1).

The clock input (C_p) is positive edge triggered, so after each positive edge, the data bits are shifted one position to the right. The serial output (Q_7) and its complement ($\overline{Q_7}$) are available from the rightmost flip-flop's outputs.

The 74194

Another shift register IC is the 74194 4-bit bidirectional universal shift register. It is called universal because it has a wide range of applications, including serial or parallel input, serial or parallel output, shift left or right, hold, and asynchronous Reset. The logic symbol for the 74194 is shown in Figure 13–14.

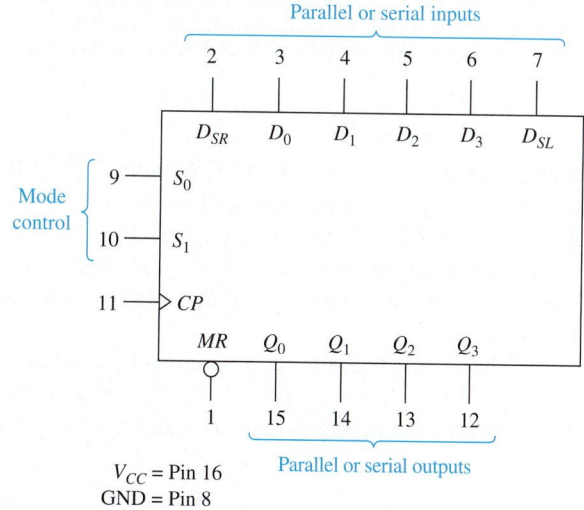

Figure 13–14 Logic symbol for the 74194 universal shift register.

The major differences with the 74194 are that there are separate serial inputs for shifting left or shifting right and that the operating mode is determined by the digital states of the **mode control** inputs, S_0 and S_1. S_0 and S_1 can be thought of as receiving a 2-bit binary code representing one of four possible operating modes ($2^2 = 4$ combinations). The four operating modes are shown in Table 13–1.

TABLE 13–1	Operating Modes of the 74194	
Operating Mode	S_1	S_0
Hold	0	0
Shift left	1	0
Shift right	0	1
Parallel load	1	1

A complete mode select-function table for the 74194 is shown in Table 13–2. Table 13–2 can be used to determine the procedure and expected outcome of the various shift register operations.

TABLE 13–2	Mode Select-Function Table for the 74194[a]										
	Inputs							**Outputs**			
Operating Mode	C_p	$\overline{MR}$	S_1	S_0	D_{SR}	D_{SL}	D_n	Q_0	Q_1	Q_2	Q_3
Reset (clear)	$\times$	L	$\times$	$\times$	$\times$	$\times$	$\times$	L	L	L	L
Hold (do nothing)	$\times$	H	1[b]	1[b]	$\times$	$\times$	$\times$	q_0	q_1	q_2	q_3
Shift left	$\uparrow$	H	h	1[b]	$\times$	1	$\times$	q_1	q_2	q_3	L
$(Q_N \leftarrow Q_{N+1}, Q_3 \leftarrow D_{SL})$	$\uparrow$	H	h	1[b]	$\times$	h	$\times$	q_1	q_2	q_3	H
Shift right	$\uparrow$	H	1[b]	h	1	$\times$	$\times$	L	q_0	q_1	q_2
$(D_{SR} \rightarrow Q_0, Q_N \rightarrow Q_{N+1})$	$\uparrow$	H	1[b]	h	h	$\times$	$\times$	H	q_0	q_1	q_2
Parallel load	$\uparrow$	H	h	h	$\times$	$\times$	d_n	d_0	d_1	d_2	d_3

[a]H = HIGH voltage level; h = HIGH voltage level one setup time before the LOW-to-HIGH clock transition; L = LOW voltage level; 1 = LOW voltage level one setup time before the LOW-to-HIGH clock transition; $d_n(q_n)$ = lowercase letters indicate the state of the referenced input (or output) one setup time before the LOW-to-HIGH clock transition; $\times$ = don't care; $\uparrow$ = LOW-to-HIGH clock transition.
[b]The HIGH-to-LOW transition of the S_0 and S_1 inputs on the 74194 should take place only while C_P is HIGH for conventional operation.

From the function table, we can see that a LOW input to the Master Reset ($\overline{MR}$) asynchronously resets Q_0 to Q_3 to 0. A parallel load is accomplished by making S_0, S_1 both HIGH and placing the parallel input data on D_0 to D_3. The register will then be parallel loaded synchronously by the first positive clock (C_p) edge. The 4 data bits can then be shifted to the right or left by making $S_0 - S_1$ 1–0 or 0–1 and applying an input clock to C_p.

A **recirculating** shift-right register can be set up by connecting Q_3 back into D_{SR} and applying a clock input (C_p) with $S_1 = 0$, $S_0 = 1$. Also, a recirculating shift-left register can be set up by connecting Q_0 into D_{SL} and applying a clock input (C_p) with $S_1 = 1$, $S_0 = 0$.

The best way to get a feel for the operation of the 74194 is to study the timing waveforms for a typical sequence of operations. Figure 13–15 shows the input data, control waveforms, and the output (Q_0 to Q_3) waveforms generated by a clear–load–shift right–shift left–inhibit–clear sequence. Study these waveforms carefully until you thoroughly understand the setup of the mode controls and the reason for each state change in the Q_0 to Q_3 outputs.

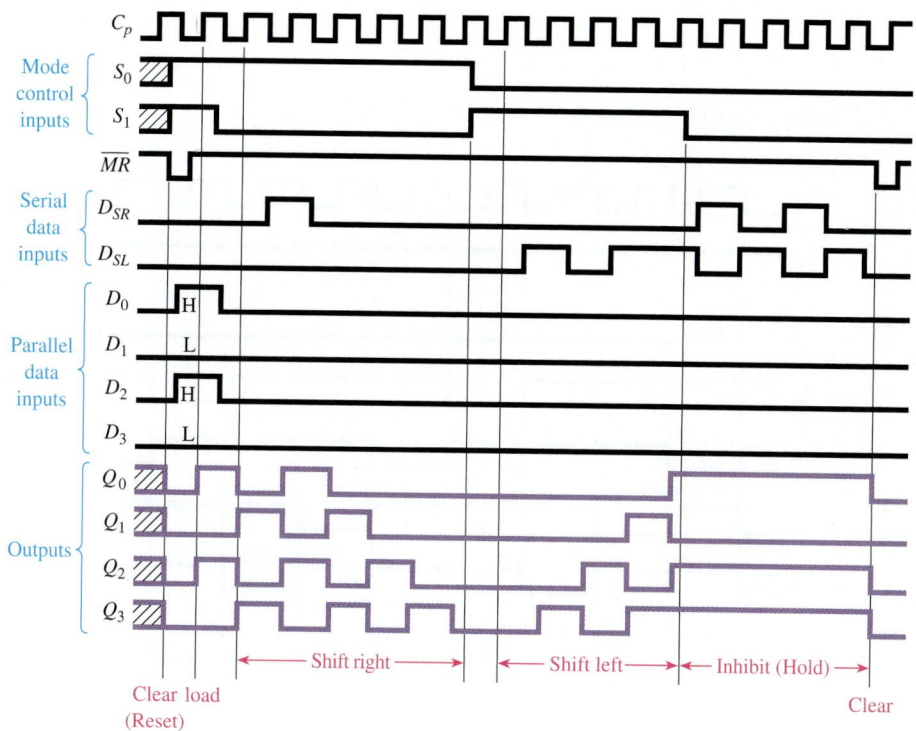

Figure 13–15 Typical clear–load–shift right–shift left–inhibit–clear sequence for a 74194.

EXAMPLE 13–2

Draw the circuit connection and timing waveforms for a recirculating shift-right register. The register should be loaded initially with a hexadecimal D (1101).

Solution: The shift-right register is shown in Figure 13–16. First, the S_0 and S_1 mode controls are set to 1–1 for parallel loading D_0 to D_3. When the first positive clock edge (pulse 0) comes in, the data present on D_0 to D_3 are loaded into Q_0 to Q_3. Next, S_0 and S_1 are made 1–0 to perform shift-right operations. At the positive edge of each successive clock edge, the data are shifted one position to the right ($Q_0 \rightarrow Q_1$, $Q_1 \rightarrow Q_2$, $Q_2 \rightarrow Q_3$, $Q_3 \rightarrow D_{SR}$, $D_{SR} \rightarrow Q_0$). The recirculating connection from Q_3 back to D_{SR} keeps the data from being lost. After each fourth clock pulse, the circulating data are back in their original position.

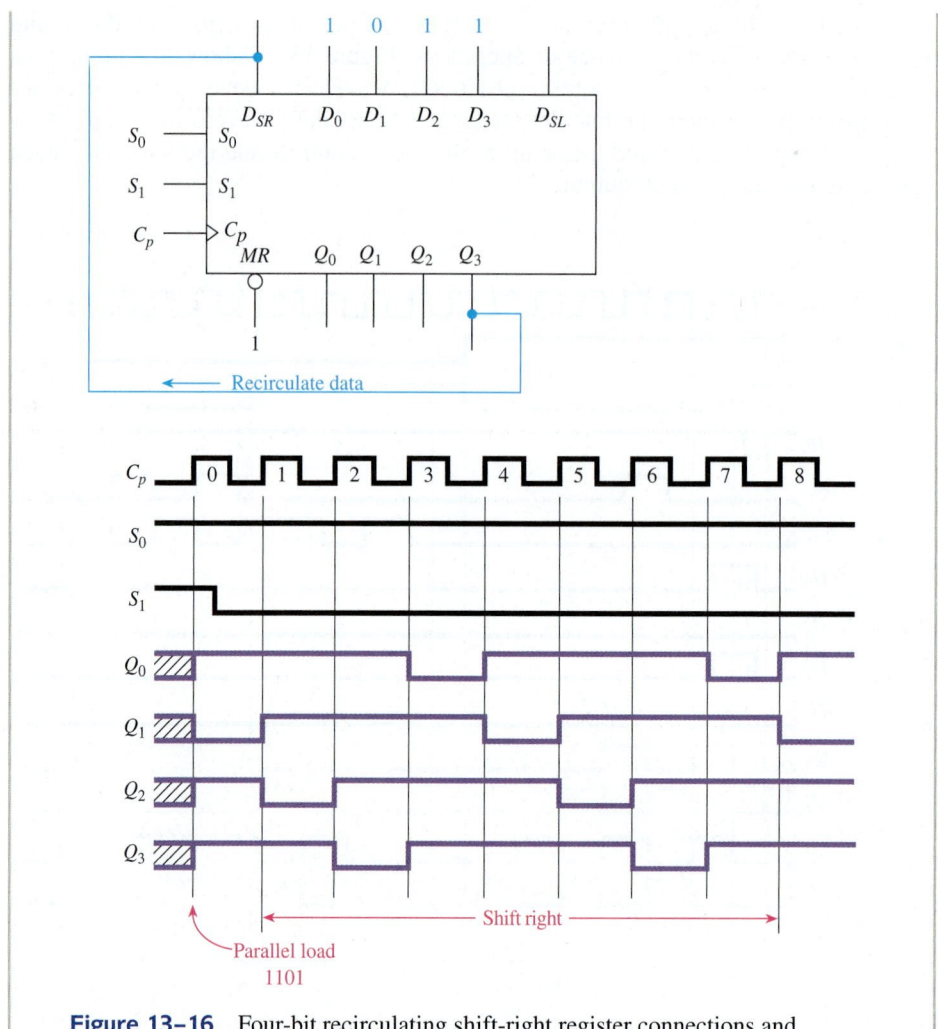

Figure 13–16 Four-bit recirculating shift-right register connections and waveforms using the 74194.

Three-State Outputs

Another valuable feature available on some shift registers is a **three-state output.** The term *three-state* (or *tristate*) is derived from the fact that the output can have one of three levels: HIGH, LOW, or float. The symbol and function table for a three-state output buffer are shown in Figure 13–17.

From Figure 13–17, we can see that the circuit acts like a straight buffer (output = input) when $\overline{OE}$ is LOW (active-LOW Output Enable). When the output is disabled $\overline{OE}$ = HIGH), the output level is placed in the **float,** or **high-impedance state.** In the high-impedance state, the output looks like an open circuit to anything else connected to it. In other words, in the float state the output is neither HIGH nor LOW and cannot sink nor source current.

Three-state outputs are necessary when you need to connect the output of more than one register to the same points. For example, if you have two 4-bit registers, one containing data from device 1 and the other containing different data from device 2, and you want to connect both sets of outputs to the same receiving device, one device must be in the float condition while the other's output is enabled, and vice versa. This way, only one set of outputs is connected to the receiving device at a time to avoid a conflict.

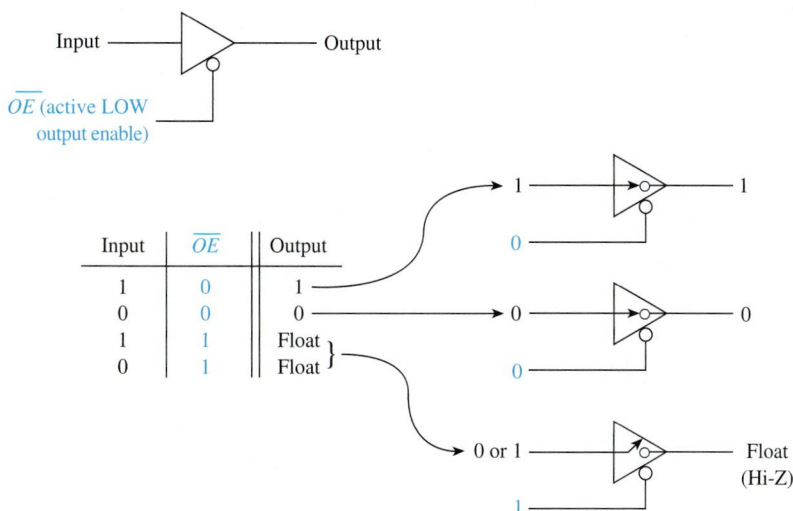

Input	$\overline{OE}$	Output
1	0	1
0	0	0
1	1	Float
0	1	Float

Figure 13–17 Three-state output buffer symbol and function table.

The 74395A

To further illustrate the operation of three-state buffers, let's look at a register that has three-state outputs and discuss a system design example that uses two 4-bit registers to feed a single receiving device. The 74395A is a 4-bit shift (right) register with three-state outputs, as shown in Figure 13–18. From the logic circuit diagram [Figure 13–18(b)], we can see that the Q_0 and Q_3 outputs are "three-stated" and will not be allowed to pass data unless a LOW is present at the Output Enable pin $(\overline{OE})$. Also, a non-three-stated output, Q'_3, is made available to enable the user to cascade with another register and shift data bits to the cascaded register whether the regular outputs (Q_0 to Q_3) are enabled or not (i.e., to cascade two 4-bit registers, Q'_3 would be connected to D_S of the second stage).

Otherwise, the chip's operation is similar to previously discussed shift registers. The **Parallel Enable** (*PE*) input is active-HIGH for enabling the parallel data input (D_0 to D_3) to be synchronously loaded on the negative clock edge. D_S is the serial data input line for synchronously loading serial data, and $\overline{MR}$ is an active-LOW Master Reset.

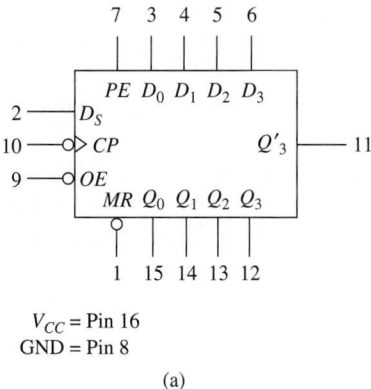

V_{CC} = Pin 16
GND = Pin 8

(a)

Figure 13–18 The 74395A 4-bit shift register with three-state outputs: (a) logic symbol

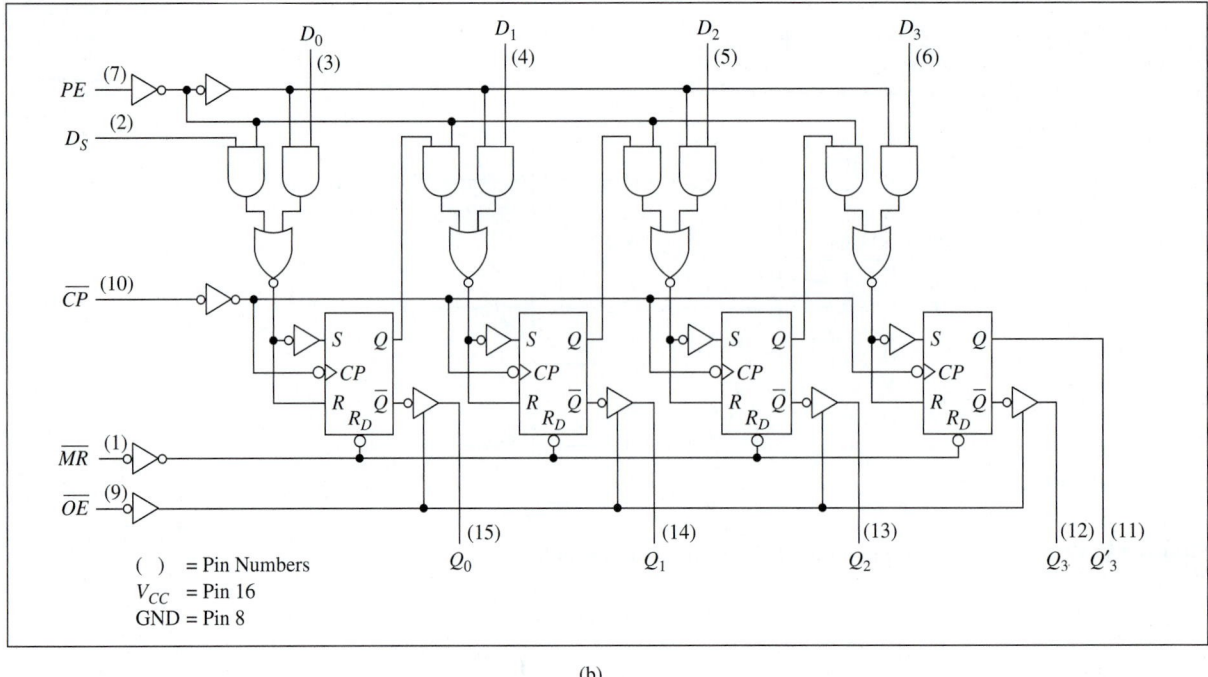

$(\ )$ = Pin Numbers
V_{CC} = Pin 16
GND = Pin 8

(b)

Figure 13-18 *(Continued)* The 74395A 4-bit shift register with three-state outputs: (b) logic circuit.

EXAMPLE 13-3

Sketch the circuit connections for a two-register system that alternately feeds one register and then the other into a 4-bit receiving device. Upon power-up, load register 1 with 0111 and register 2 with 1101.

Solution: The two-register system is shown in Figure 13–19. Notice that both sets of outputs go to a common point. The three-state outputs allow us to do this by only enabling one set of outputs at a time, so that there is no conflict between HIGH's and LOW's. One way to alternately enable one register and then the other is to use a toggle flip-flop and feed the Q output to the upper Output Enable $(\overline{OE})$ and the $\overline{Q}$ to the lower Output Enable, as shown in Figure 13–19.

Also, upon initial power-up, we want to parallel load both 4-bit registers. To do this, the D_0 to D_3 inputs contain the proper bit string to be loaded, and the Parallel Enable (*PE*) is held HIGH. Because the parallel-load function is synchronous (needs a clock trigger), we will supply a HIGH-then-LOW pulse to $\overline{C}_p$ via the *RC* Schmitt circuit.

CHAPTER 13 | SHIFT REGISTERS

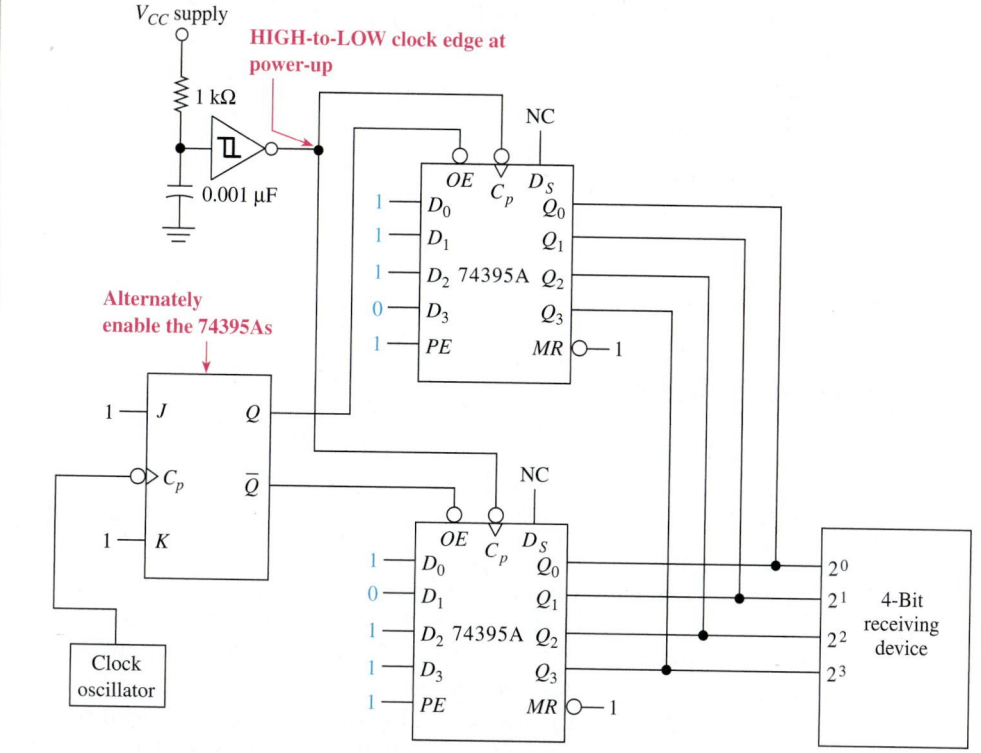

Figure 13–19 Two 4-bit, three-state output shift registers feeding a common receiving device.

Review Questions

13–7. To input serial data into the 74164 shift register, one D_S input must be held _____ (HIGH, LOW) while the other receives the serial data.

13–8. What is the function of the $\overline{CE}$ input to the 74165 shift register?

13–9. Why is the 74194 IC sometimes called a "universal" shift register?

13–10. List the steps that you would follow to parallel load a hexadecimal B into a 74194 shift register.

13–11. To make the 74194 act as a shift-left *recirculating* register, a connection must be made from _____ to _____ and S_0 and S_1 must be _____ _____.

13–12. How does the operation of the Parallel Enable (*PE*) on the 74395A shift register differ from the Parallel Load ($\overline{PL}$) of the 74165?

13–13. The outputs of the 74395A shift register are disabled by making ($\overline{OE}$) _____ (HIGH, LOW), which makes $Q_0 - Q_3$ _____ (HIGH, LOW, float).

13–8 System Design Applications for Shift Registers

Shift registers have many applications in digital sequencing, storage, and transmission of serial and parallel data. The following examples will illustrate some of these applications.

EXAMPLE 13–4

Using a ring shift counter as a sequencing device, design a traffic light controller that goes through the following sequence: green, 20 s; yellow, 10 s; red, 20 s. Also, at night, flash the yellow light on and off continuously.

Solution: By studying the waveforms from Figure 13–5, you will notice that, if we add one more flip-flop to that 4-bit ring shift counter and use a clock input of 1 pulse per 10 s, we could tap off the Q outputs using OR gates to get a 20–10–20 sequence. Also, we could use a phototransistor to determine night from day. During the night we want to stop the ring shift counter and flash the yellow light. Figure 13–20 shows a 5-bit ring counter that could be used as this traffic light controller. First, let's make sure that the green–yellow–red sequence will work properly during the daytime. During the daytime, outdoor light shines on the phototransistor, making its collector-to-emitter resistance LOW, placing a low voltage at the input to the first Schmitt inverter, and causing a LOW input to OR gate 4. The 1-pps clock oscillator will pass through OR gate 4 into the MOD-10, which divides the frequency down to one pulse per 10 s. The output from the MOD-10 is used to drive the clock input to the 5-bit ring shift counter, which will circulate a single HIGH level down through each successive flip-flop for 10 s at each Q output, as shown in the timing waveforms.

OR gates 1, 2, and 3 are connected to the ring counter outputs in such a way that the green light will be on if Q_0 or Q_1 is HIGH, which occurs for 20 s. The yellow light will come on next for 10 s due to Q_2 being on, and then the red light will come on for 20 s because either Q_3 or Q_4 is HIGH.

At nighttime, the phototransistor changes to a high resistance, placing a HIGH at the input to the first Schmitt inverter, which places a HIGH at OR gate 4. This makes its output HIGH, stopping the clock input oscillations to the ring counter.

Also at nighttime, the LOW output from the first Schmitt inverter is connected to the ring counter Resets, holding the Q outputs at 0. The HIGH output from the second Schmitt inverter enables the AND gate to pass the 1-pps clock oscillator on to OR gate 2, causing the yellow light to flash.

At sunrise, the output from the first Schmitt inverter changes from a LOW to a HIGH, allowing the ring counter to start again. This LOW-to-HIGH transition causes an instantaneous surge of current to flow through the RC circuit. That current will cause a HIGH at the input of the third Schmitt inverter, which places a LOW at $\overline{S_{D0}}$, setting Q_0 HIGH. When the surge current has passed (a few microseconds), $\overline{S_{D0}}$ returns to a HIGH, and the ring counter will proceed to rotate the HIGH level from Q_0 to Q_1 to Q_2 to Q_3 to Q_4 continuously, all day, as shown in the timing waveforms.

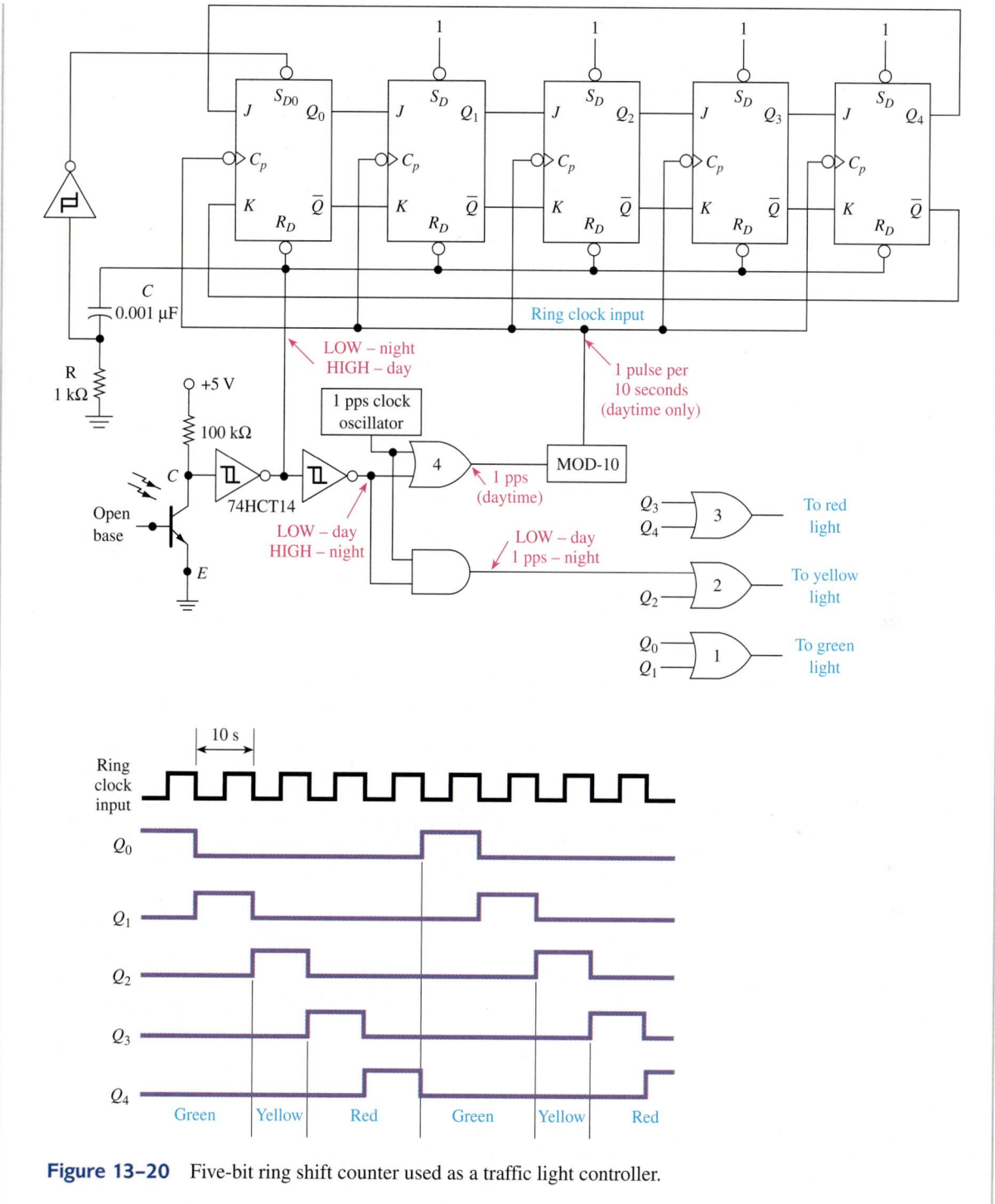

Figure 13–20   Five-bit ring shift counter used as a traffic light controller.

EXAMPLE 13–5

Design a 16-bit serial-to-parallel converter.

Solution: First, we have to look through a TTL data manual to see what is available. The 74164 is an 8-bit serial-in, parallel-out shift register. Let's cascade two of them together to form a 16-bit register. Figure 13–21 shows that the Q_7 output is fed into the serial input of the second 8-bit register. This way, as the data bits are shifted through the register, when they reach Q_7 the next shift will pass the data into Q_0 of the second register (via D_{Sa}), making it a 16-bit register. The second serial input (D_{Sb}) of each stage is internally ANDed with the D_{Sa} input so that it serves as an active-HIGH enable input.

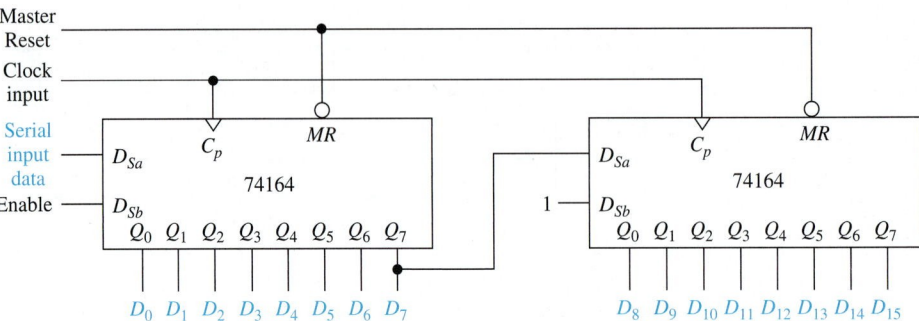

Figure 13–21   Sixteen-bit serial-to-parallel converter.

EXAMPLE 13-6

Design a circuit and provide the input waveforms required to perform a parallel-to-serial conversion. Specifically, a hexadecimal B (1011) is to be parallel loaded and then transmitted repeatedly to a serial device LSB first.

Solution: By controlling the mode control inputs ($S_0 - S_1$) of a 74194, we can perform a parallel load and then shift right repeatedly. The serial output data are taken from Q_3, as shown in Figure 13–22. The 74194 universal shift register is connected as a recirculating parallel-to-serial converter. Each time the Q_3 serial output level is sent to the serial device, it is also recirculated back into the left end of the shift register.

First, at the positive edge of clock pulse 0, the register is parallel loaded with a 1011 (B_{16}) because the mode controls (S_0-S_1) are HIGH–HIGH. (D_3 is loaded with the LSB because it will be the first bit out when we shift right.)

Next, the mode controls ($S_0 - S_1$) are changed to HIGH–LOW for a shift-right operation. Now, each successive positive clock edge will shift the

Common Misconception

This is a good circuit to simulate in the lab using push buttons, switches, and LEDs. Students often fail to realize the importance of the order of operation among C_p, $D_0 - D_3$, and $S_0 - S_1$.

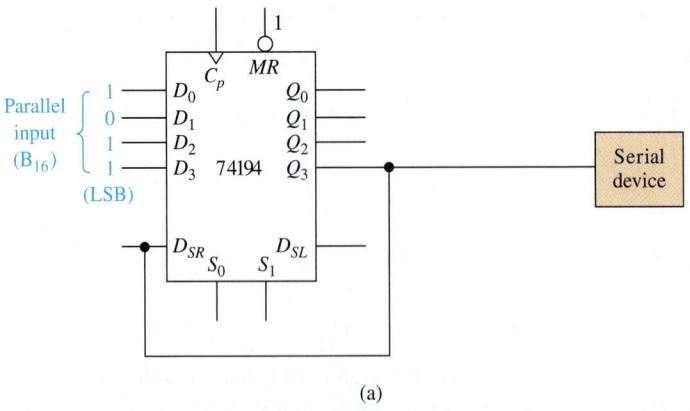

(a)

Figure 13–22   Four-bit parallel-to-serial converter: (a) circuit connections;

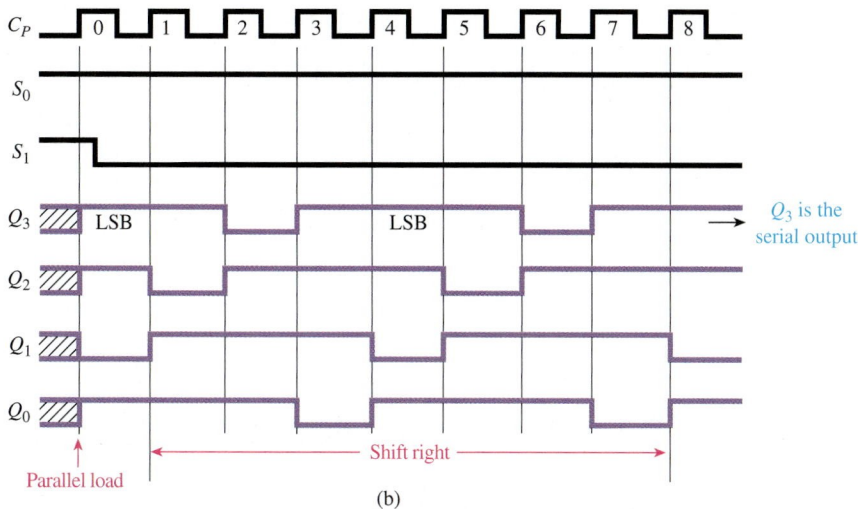

Figure 13-22 *(Continued)* Four-bit parallel-to-serial converter: (b) waveforms.

data bit one position to the right. The Q_3 output will continuously have the levels 1101–1101–1101–, and so on, which is a backward hexadecimal B (LSB first).

EXAMPLE 13-7

Design an interface to an 8-bit serial printer. Sketch the waveforms required to transmit the single ASCII code for an asterisk (*). *Note:* ASCII is the 7-bit code that was given in Chapter 1. The ASCII code for an asterisk is 010 1010. Let's make the unused eighth bit (MSB) a zero.

Solution: The circuit design and waveforms are shown in Figure 13–23. The 74165 is chosen for the job because it is an 8-bit register that can be parallel loaded and then shifted synchronously by the clock input to provide the serial output to the printer.

During pulse 0, the register is loaded with the ASCII code for an asterisk (the LSB is put into D_7 because we want it to come out first). The clock input is then enabled by a LOW on $\overline{CE}$ (**Clock Enable**). Each positive pulse on C_p from then on will shift the data bits one position to the right. After the eighth clock pulse (0 to 7), the printer will have received all 8 serial data bits. Then, the $\overline{CE}$ line is brought HIGH to disable the synchronous clock input. To avoid any racing problems, the printer will read the Q_7 line at each negative edge of C_p so that the level will definitely be a stable HIGH or LOW, as shown in Figure 13–23.

At this point, you may be wondering how, practically, we are going to electronically provide the necessary signals on the $\overline{CE}$ and $\overline{PL}$ lines. An exact degree of timing must be provided on these lines to ensure that the register–printer interface communicates properly. These signals will be provided by a microprocessor and are called the **handshaking** signals.

Microprocessor theory and programming are advanced digital topics and are not discussed in this book. For now, it is important for us to realize that these signals are required and to be able to sketch their timing diagrams.

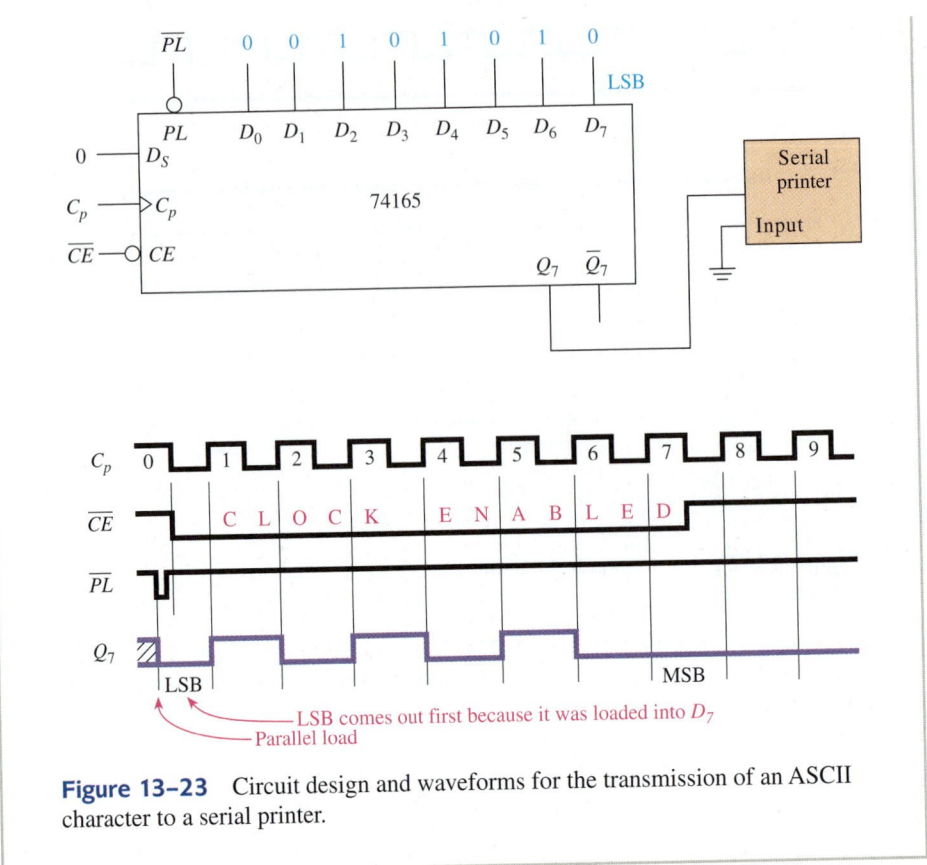

Figure 13–23 Circuit design and waveforms for the transmission of an ASCII character to a serial printer.

13–9 Driving a Stepper Motor with a Shift Register

A **stepper motor** makes its rotation in steps instead of a smooth, continuous motion as with conventional motors. Typical **stepping angles** are 15° or 7.5° per step, requiring 24 or 48 steps, respectively, to complete one revolution. The stepping action is controlled by digital levels that energize magnetic coils within the motor.

Because they are driven by sequential digital signals, it is a good application for shift registers. For example, a shift register circuit could be developed to cause the stepper motor to rotate at 100 rpm for 32 revolutions and then stop. This is useful for applications requiring exact positioning control without the use of **closed-loop feedback** circuitry to monitor the position. Typical applications are floppy disk Read/Write head positioning, printer type head and line feed control, and robotics.

There are several ways to construct a motor to achieve this digitally controlled stepping action. One such way is illustrated in Figure 13–24. This particular stepper motor construction uses four **stator** (stationary) **coils** set up as four **pole pairs.** Each stator pole is offset from the previous one by 45°. The directions of the windings are such that energizing any one coil will develop a north field at one pole and a south field at the opposite pole. The north and south poles created by energizing coil 1 are shown in Figure 13–24. The rotating part of the motor (the **rotor**) is designed with three ferromagnetic pairs spaced 60° apart from each other. (A **ferromagnetic** material is one that is attracted to magnetic fields). Because the stator poles are spaced 45° apart, this makes the next stator-to-rotor 15° out of alignment.

In Figure 13–24, the rotor has aligned itself with the **flux lines** created by the north–south stator poles of coil 1. To step the rotor 15° clockwise, coil 1 is deenergized,

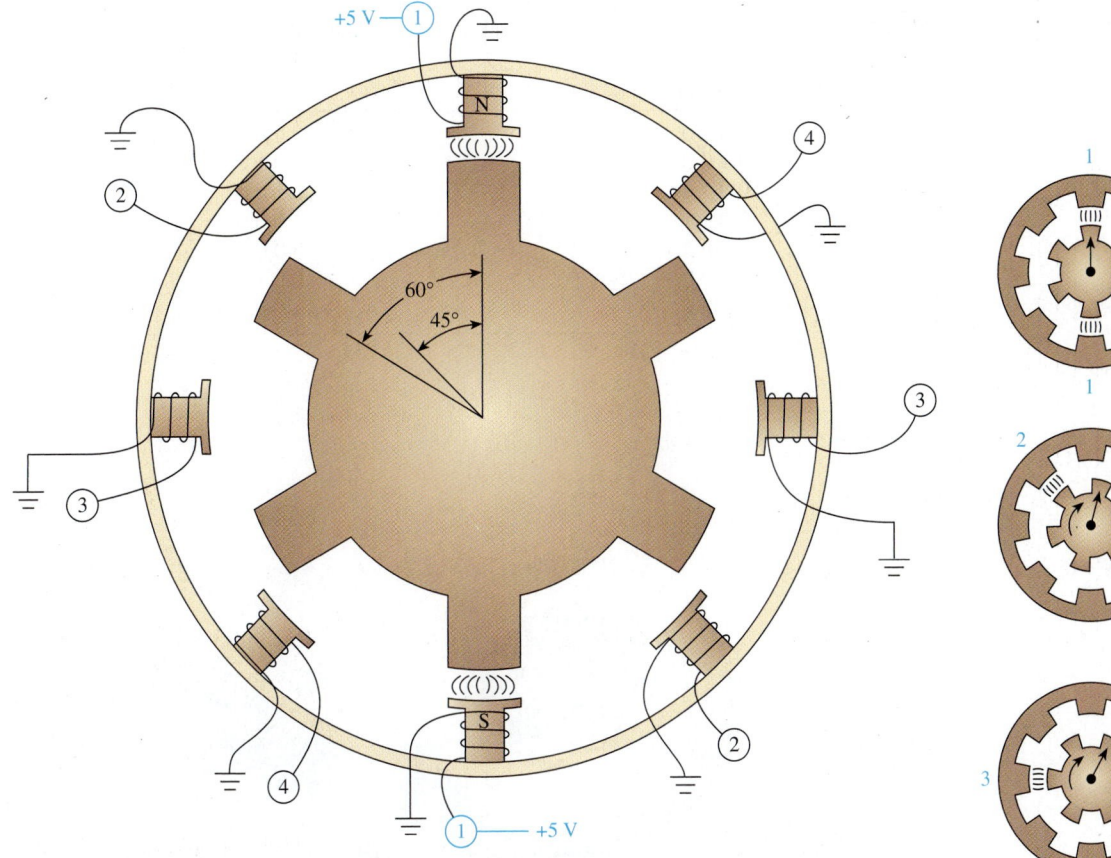

Figure 13–24 A four-coil stepper motor with stator coil 1 energized. (From *Digital and Microprocessor Fundamentals, Fourth Edition* by William Kleitz, Prentice Hall, Upper Saddle River, N.J. 2003.)

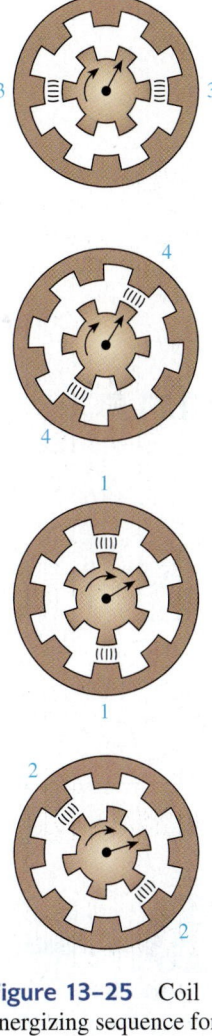

Figure 13–25 Coil energizing sequence for 15° clockwise steps.

and coil 2 is energized. The closest rotor pair to coil 2 will now line up with stator pole pair 2's flux lines. The next 15° steps are made by energizing coil 3, then 4, then 1, then 2, and so on, for as many steps as you require. Figure 13–25 shows the stepping action achieved by energizing each successive coil six times. Table 13–3 shows the digital codes that are applied to the stator coils for 15° clockwise and 15° counterclockwise rotation. Figure 13–26 shows a stepper motor with the motor removed to expose two of the stator coils.

The amount of current required to energize a coil pair is much higher than the capability of a 74194, so we will need some current-buffering circuitry similar to that shown in Figure 13–27.

The output of the upper 7406 inverting buffer in Figure 13–28 is LOW, forward biasing the base-emitter of the MJ2955 *PNP* power transistor. This causes the collector–emitter to short, allowing the large current to flow through the number 1 coils to ground. The IN4001 diodes protect the coils from arcing over when the current is stopped.

The 74194 is first parallel loaded with 0001 and then changed to a shift-right operation by making S_0, $S_1 = 1, 0$. Each positive clock edge shifts the ON bit one position to the right. The Q outputs will follow the clockwise pattern shown in Table 13–3, causing the motor to rotate. The speed of revolution is dictated by the period of C_p.

TABLE 13–3 | **Digital Codes for 15° Clockwise and Counterclockwise Rotation**

Clockwise Coil 1 2 3 4	Counterclockwise Coil 1 2 3 4
1 0 0 0	0 0 0 1
0 1 0 0	0 0 1 0
0 0 1 0	0 1 0 0
0 0 0 1	1 0 0 0
1 0 0 0	0 0 0 1
0 1 0 0	0 0 1 0
and so on	and so on

Figure 13–26 Stepper motor with the rotor removed to expose two of the stator coils.

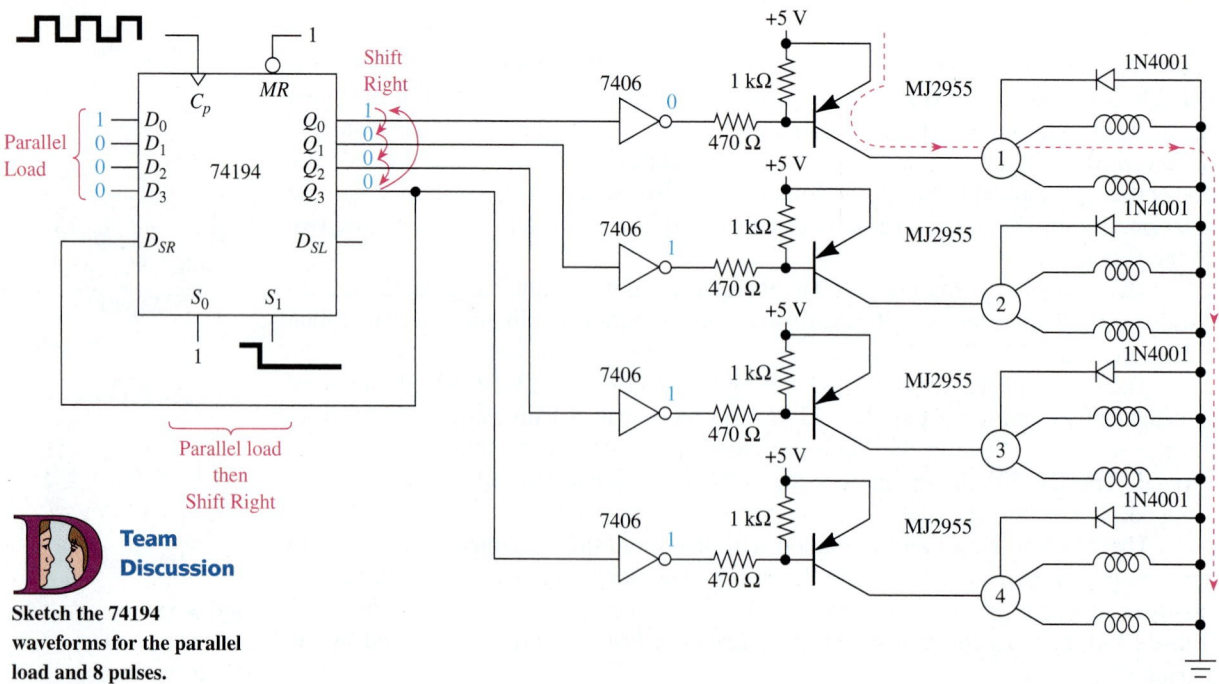

Team Discussion

Sketch the 74194 waveforms for the parallel load and 8 pulses.

Figure 13–27 Drive circuitry for a four-coil stepper motor showing the number 1 coils energized.

618

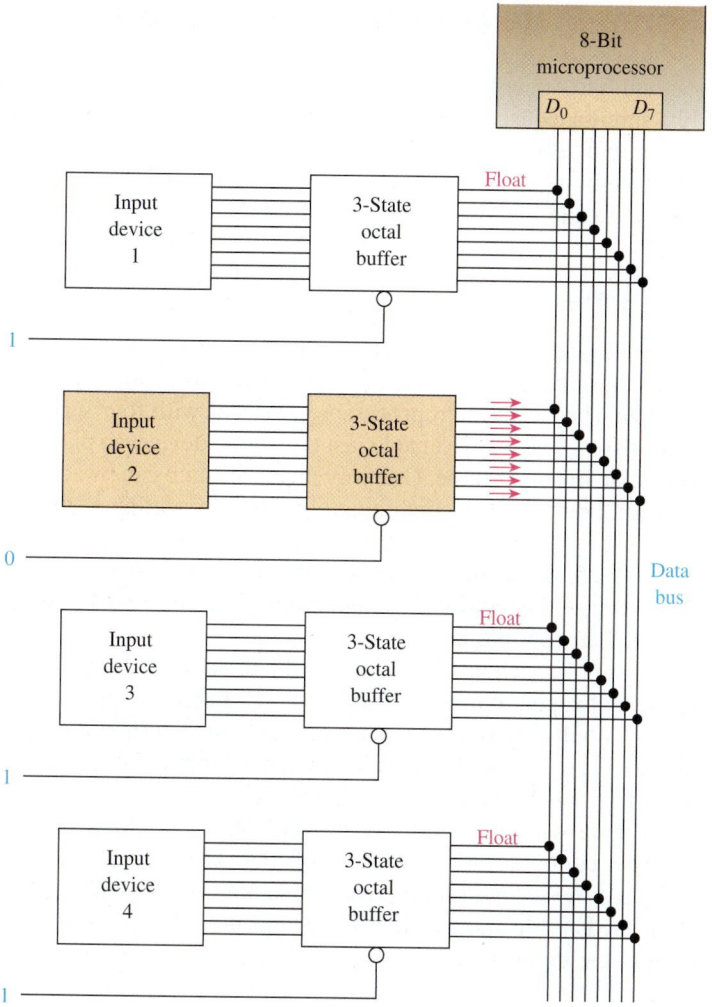

Inside Your PC

In a modern PC, the input devices needing access to the microprocessor might be: a DVD, a hard drive, a floppy, and a keyboard.

Figure 13–28 Using a three-state octal buffer to pass 8 data bits from input device 2 to the data bus.

Review Questions

13–14. The traffic light controller of Figure 13–20 flashes the yellow light at night because the level at the collector of the phototransistor is _____ (LOW, HIGH), which _____ (enables, disables) the AND gate.

13–15. What circuitry is responsible for parallel loading a 1 into the first flip-flop of Figure 13–20 at the beginning of each day?

13–16. The serial-to-parallel converter in Figure 13–21 could also be used for serial in to serial out. True or false?

13–17. How would the waveforms change in Figure 13–22 if Q_3 were connected to D_{SL} instead of D_{SR}?

13–18. How is the $\overline{CE}$ input used on the 74165 in Figure 13–23?

13–19. Is the stepper motor in Figure 13–27 turning clockwise or counterclockwise? How would you change its direction?

13–10 Three-State Buffers, Latches, and Transceivers

When we start studying microprocessor hardware, we'll see a need for transmitting a number of bits simultaneously as a group (**data transmission**). A single flip-flop will not suffice. What we need is a group of flip-flops, called a **register,** to facilitate the movement and temporary storage of binary information. The most commonly used registers are 8 bits wide and function as either a buffer, latch, or transceiver.

Three-State Buffers

In microprocessor systems, several input and output devices must share the same data lines going into the microprocessor IC. (These shared data lines are called the **data bus**). For example, if an 8-bit microprocessor interfaces with four separate 8-bit input devices, we must provide a way to enable just one of the devices to place its data on the data bus and disable the other three. One way this procedure can be accomplished is to use three-state octal buffers (ICs containing eight buffers similar to the one discussed in Figure 13–17).

In Figure 13–28, the second buffer is enabled, which allows the 8 data bits from Input Device 2 to reach the data bus. The other three buffers are disabled, thus keeping their outputs in a *float* condition.

A buffer is simply a device that, when enabled, passes a digital level from its input to its output unchanged. It provides isolation, or a **buffer,** between the input device and the data bus. A buffer also provides the sink or source current required by any devices connected to its output without loading down the input device. An octal buffer IC has eight individual buffers within a single package.

A popular three-state octal buffer is the 74LS244 shown in Figure 13–29. Notice that the buffers are configured in two groups of four. The first group (group a) is controlled by $\overline{OE}_a$, and the second group (group b) is controlled by $\overline{OE}_b$. OE is an abbreviation for **Output Enable** and is active-LOW, meaning that it takes a LOW to allow data to pass from the inputs (I) to the outputs (Y). Other features of the 74LS244 are

Helpful Hint

It is instructive for you to see the '244 in a practical microprocessor application, such as the one shown in Figure 17–6.

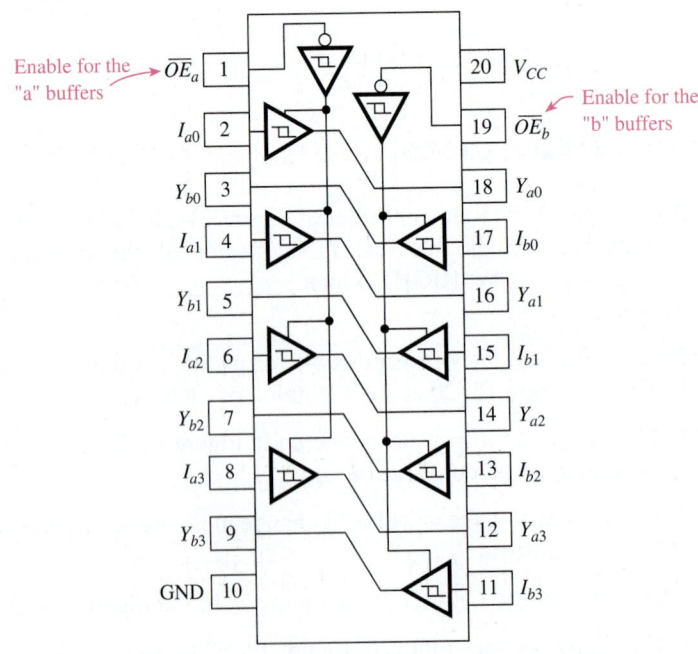

Figure 13–29 Pin configuration for the 74LS244 three-state octal buffer.

that it has Schmitt trigger hysteresis and very high sink and source current capabilities (24 and 15 mA, respectively).

Octal Latches/Flip-Flops

In microprocessor systems, we need latches and flip-flops to remember digital states that a microprocessor issues before it goes on to other tasks. Take, for example, a microprocessor system that drives two separate 8-bit output devices, as shown in Figure 13–30.

To send information to output device 1, the microprocessor first sets up the data bus ($D_0 - D_7$) with the appropriate data and then issues a LOW-to-HIGH pulse on line C_1. The positive edge of the pulse causes the data at $D_0 - D_7$ of the flip-flop to be stored at $Q_0 - Q_7$. Because $\overline{OE}$ is tied LOW, its data are sent on to output device 1. (The diagonal line with the number 8 above it is a shorthand method used to indicate eight separate lines or conductors.)

Next, the microprocessor sets up the data bus with data for output device 2 and issues a LOW-to-HIGH pulse on C_2. Now the second **octal** D flip-flop is loaded with valid data. The outputs of the D flip-flops will remain at those digital levels, thus allowing the microprocessor to go on to perform other tasks.

Earlier in this text, we studied the 7475 **transparent latch** and the 7474 D flip-flop. The 74LS373 and 74LS374 shown in Figure 13–31 operate similarly, except that they were developed to handle 8-bit data operations.

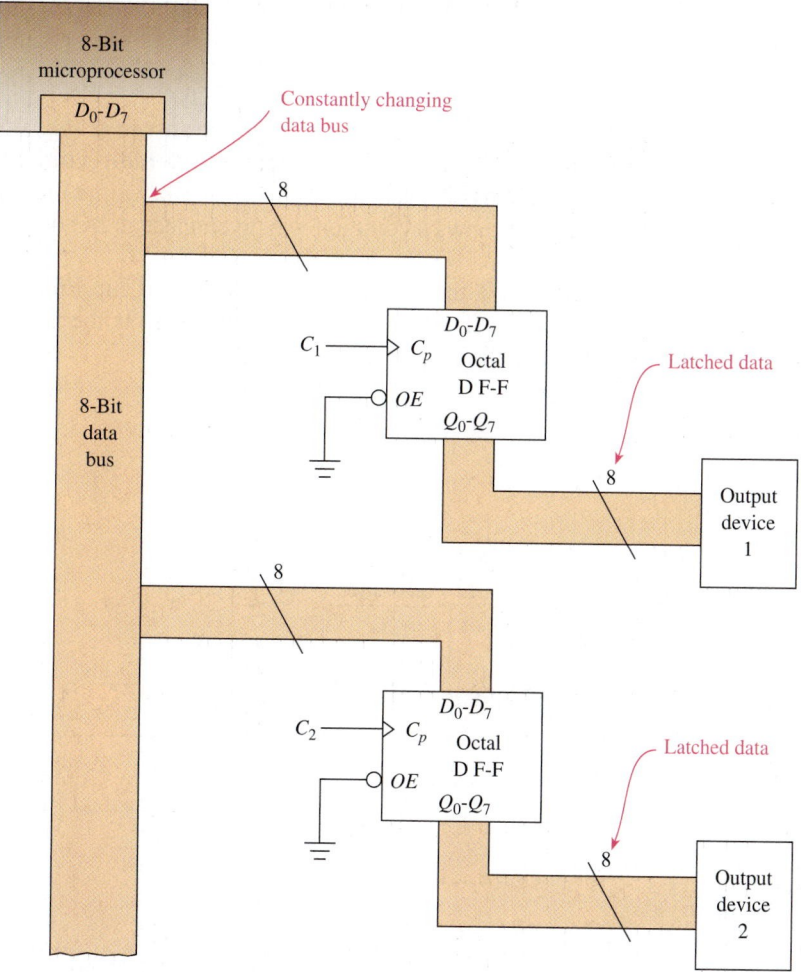

Figure 13–30 Using octal D flip-flops to capture data that appear momentarily on a microprocessor data bus.

Helpful Hint

The '374 is used in the microprocessor application in Figure 17–6.

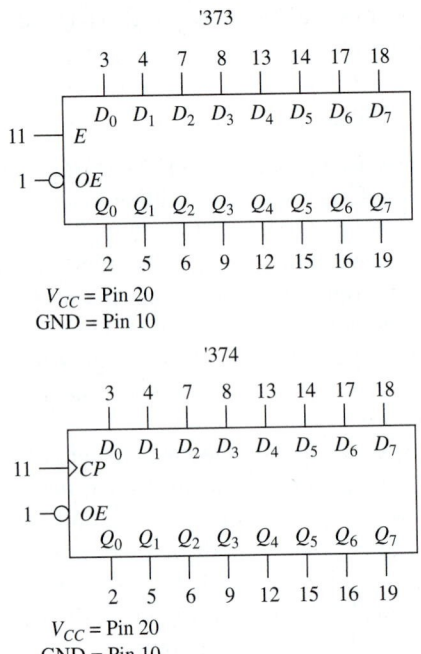

Figure 13–31   Logic symbol for the 74LS373 octal latch and the 74LS374 octal *D* flip-flop.

Transceivers

Another way to connect devices to a shared data bus is to use a **transceiver** (transmitter/receiver). The transceiver differs from a buffer or latch because it is **bidirectional.** This capability is necessary for interfacing devices that are used for *both input and output* to a microprocessor. Figure 13–32 shows a common way to connect an I/O device to a data bus via a transceiver.

To make input/output device 1 the active interface, the $\overline{CE}$ (Chip Enable) line must first be made LOW. If $\overline{CE}$ is HIGH, the transceiver disconnects the I/O device from the bus by making the connection float.

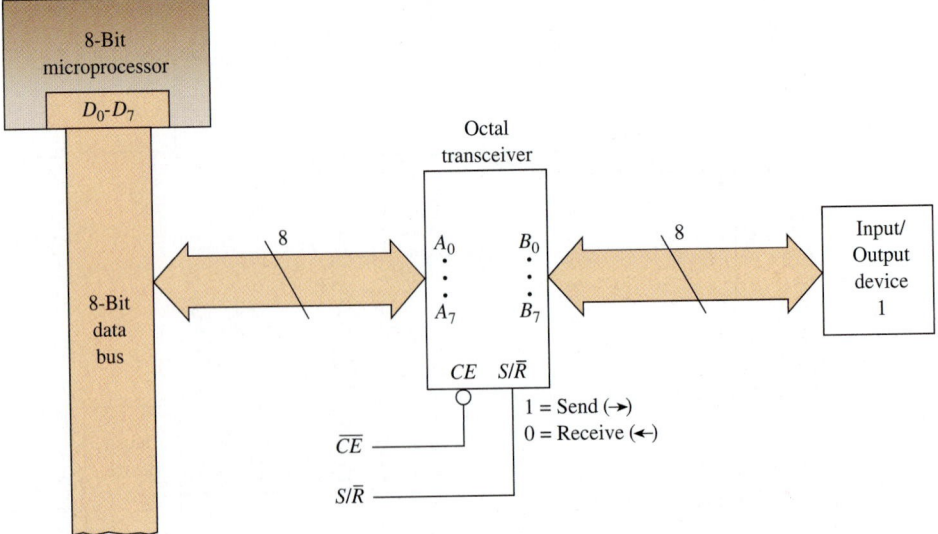

Figure 13–32   Using an octal transceiver to interface an input/output device to an 8-bit data bus.

After making $\overline{CE}$ LOW, the microprocessor then issues the appropriate level on the $S/\overline{R}$ line depending on whether it wants to *send data to* the I/O device or *receive data from* the I/O device. If $S/\overline{R}$ is made HIGH, the transceiver allows data to pass to the I/O device (from A to B). If $S/\overline{R}$ is made LOW, the transceiver allows data to pass to the microprocessor data bus (from B to A).

To see how a transceiver is able to both send and receive data, study the internal logic of the 74LS245 shown in Figure 13–33.

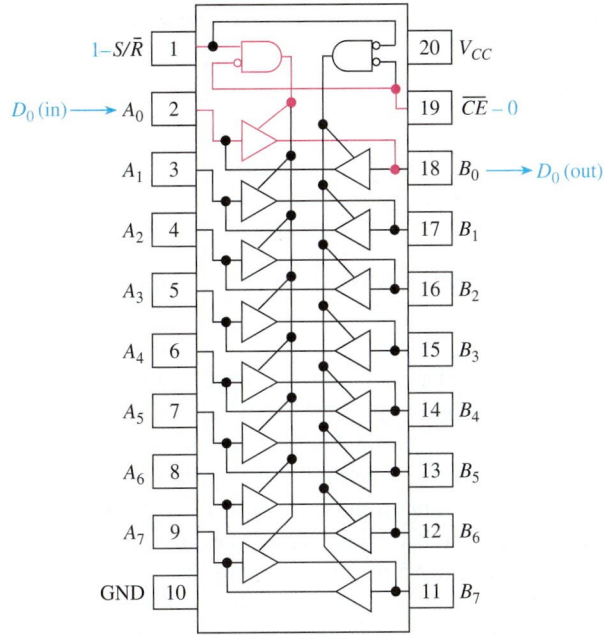

Figure 13–33 Pin configuration and internal logic of the 74LS245 octal three-state transceiver. A single data bit (D_0) is shown being sent from A_0 to B_0.

Review Questions

13–20. The 74LS244 provides buffering for a total of _____ signals. The outputs are all forced to their *float* state by making _____ and _____ HIGH.

13–21. The 74LS374 octal D flip-flop is a _____ device, whereas the 74LS244 octal buffer is a _____ device (synchronous, asynchronous).

13–22. A transceiver like the 74LS245 is *bidirectional*, allowing data to flow in either direction through it. True or false?

13–11 Using the LPM Shift Register and 74194 Macrofunction

The MAX+PLUS II software provides a general-purpose shift register in its Library of Parameterized Modules called *LPM_SHIFTREG*. When you study the **Edit Ports/ Parameters** pop-up screen, you'll see many useful features such as LPM_WIDTH, serial and parallel input and output, asynchronous and synchronous loading, and Clock Enable. Figure 13–34 shows a basic configuration set up for serial in, shift-right with

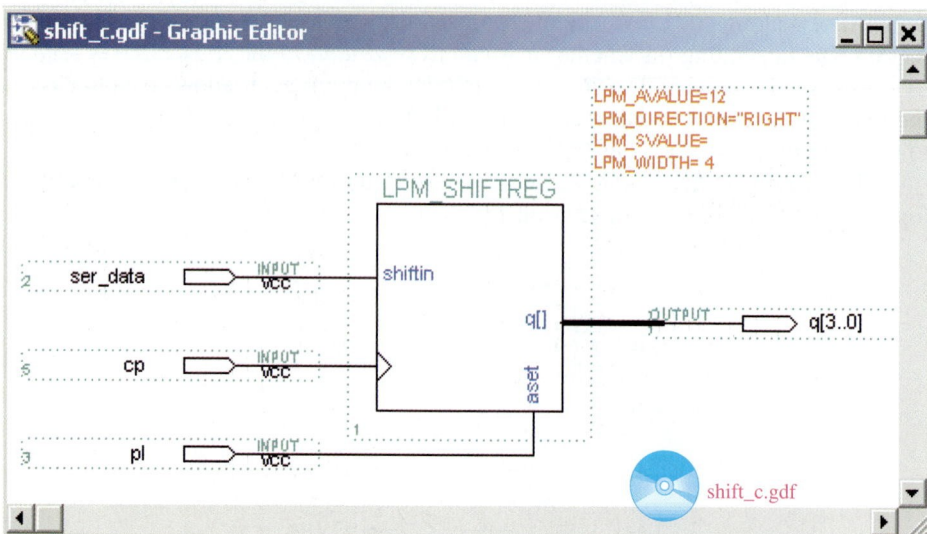

Figure 13–34 LPM shift register connections.

synchronous parallel loading. The *Parameters* section defines the width as *4,* the direction as *right* and the asynchronous value to be parallel loaded as *12.* By reading the help screen it is determined that if the terminal labeled *aset* is asserted during the active clock edge, the LPM_AVALUE will be loaded into the register.

Figure 13–35 shows a simulation of the shift register. At the first positive clock edge, the *ser_data* is HIGH and is loaded into *q3.* The data are then shifted right at each of the next four clock edges. At 9.5 µs, *pl* is asserted, parallel-loading the value 12 into the register. The data are then shifted right at each of the next clock edges.

Models for the 7400-series shift registers are provided by the MAX+PLUS II software as "macrofunctions" in the *mf* subdirectory. The 74194 is used in Figure 13–36 as a "universal" shift register. The left and right serial inputs (*SLSI* and *SRSI*) are

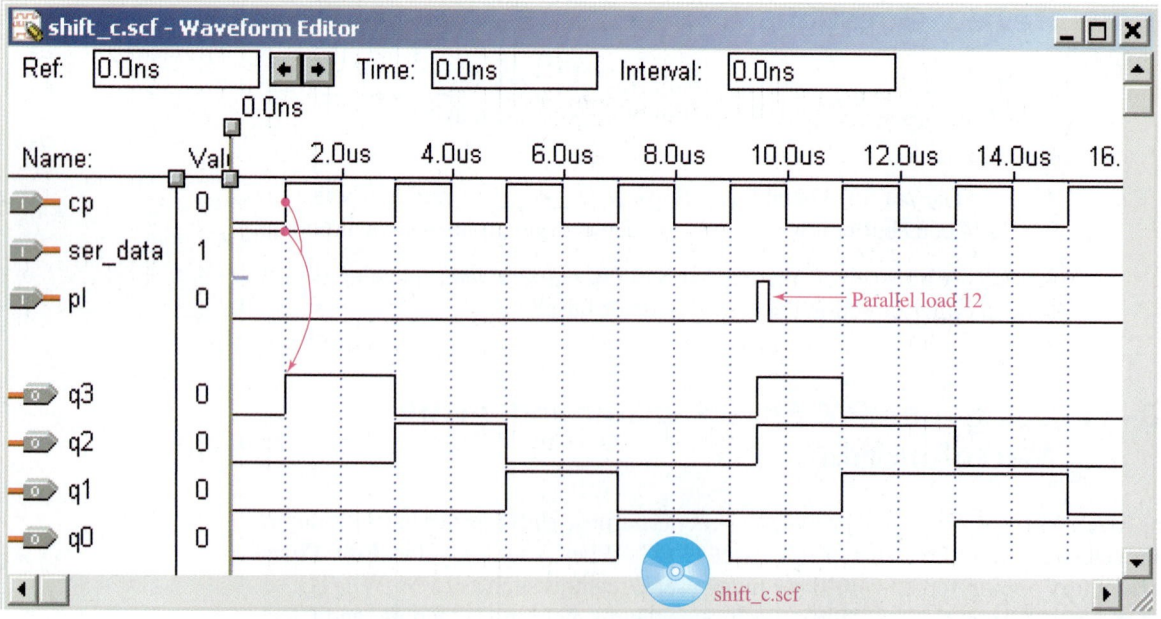

Figure 13–35 Simulation of the LPM shift register of Figure 13–34.

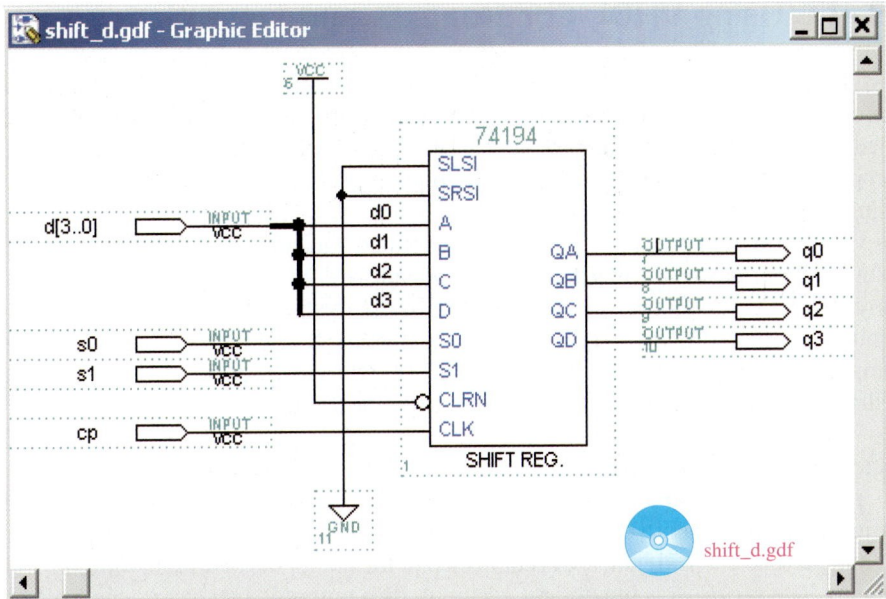

Figure 13–36   The 74194 macrofunction connected as a universal shift register.

both grounded so we will see '0' serially input in each case. The parallel data is brought in as a bus and then broken into individual elements *d0, d1, d2,* and *d3*. The outputs at *q* are left as individual elements so that we can observe the shifting of the waveforms.

The waveform simulations are shown in Figure 13–37. Since *s1-s0* are *1-1* the parallel data I_{16} (0001) is loaded into the register. For the next four pulses, *s1-s0* equals *0-1* causing the data to be shifted left at each positive clock edge. At 10.5 μs, C_{16} is parallel loaded into the register. The remainder of the waveforms show the data being shifted right.

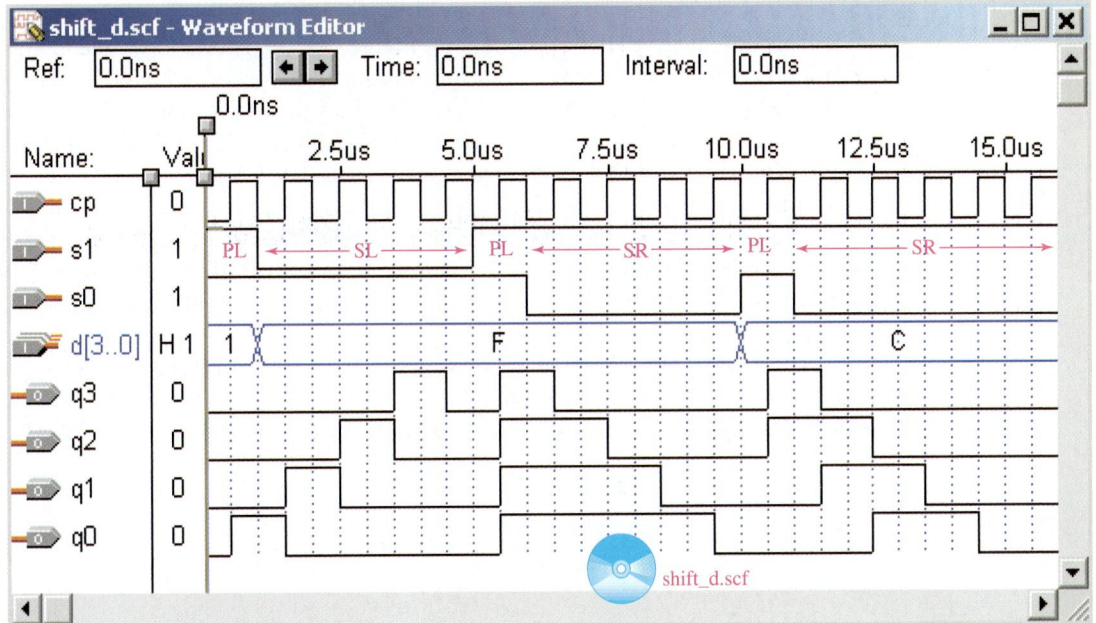

Figure 13–37   Simulation of the 74194 shift register in Figure 13–36.

13–12 Using VHDL Components and Instantiations

A **structural** approach to writing VHDL programs employs the use of multiple **components** connected together to form a complete program. The components are predefined VHDL program modules that can be used repeatedly like subroutines in a computer language. For example, the key component in a shift register is a *D* flip-flop. Using the structural approach, the *D* flip-flop must first be defined at the beginning of the VHDL program or it may be previously defined and stored in a library. Then in the main body of the VHDL program, the component will be used repeatedly for each occurrence of a *D* flip-flop in the shift register. To use the previously defined component, it must be declared in the beginning of the architecture, and then each instance of the component is defined using the **PORT MAP** keyword to describe how it is connected in the circuit. This is known as the component **instantiation.**

Figure 13–38 shows a block diagram of how the *serial-in* and *cp* inputs are connected to the four *dflipflop* components, which then provide the 4-bit parallel output at *q3, q2, q1,* and *q0.* Notice that the *dflipflop* component uses the spelling *d, clk,* and *q* for its I/O and the overall shift register (*shiftreg*) uses the spelling *serial_in, cp, q3, q2, q1,* and *q0.* The clock signal, *cp,* is connected to the *clk* inputs of all four *dflipflop* components. The *serial_in* is input to *ff3's d* input. Its *q* output is sent to *q3,* which also becomes the input to *ff2's d* input, and so on for *ff1* and *ff0.*

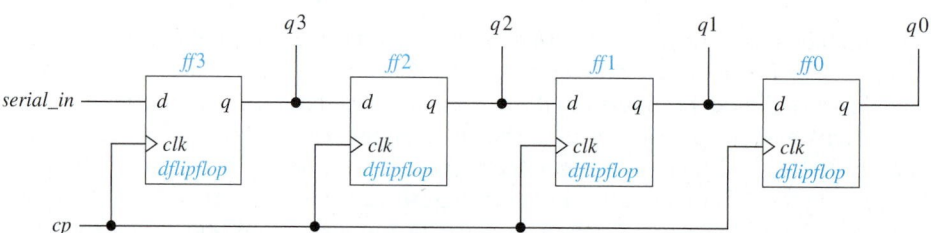

Figure 13–38   Block diagram of a 4-bit serial-in, parallel-out shift register using component instantiations.

The complete VHDL program in Figure 13–39 shows how the component *dflipflop* is defined and later used in four instances to form a 4-bit shift register. The *dflipflop* ENTITY-ARCHITECTURE group defines the *D* flip-flop as a positive edge-triggered component similar to Example 10–9. The *shiftreg* ENTITY-ARCHITECTURE group defines how the 4-bit serial-in parallel-out shift register is connected using four instances of the *dflipflop* component.

The ENTITY of the *shiftreg* uses the PORT keyword to declare the inputs (*serial_in, cp*) and outputs (*q3, q2, q1, q0*) to the overall shift register. The outputs are declared as BUFFER because they are used as inputs and outputs. (i.e. *q3* is an output, but it is also used as an input to *d* of *ff2.*)

The ARCHITECTURE of the *shiftreg* declares the use of COMPONENT *dflipflop* and then uses four instances of the *dflipflop* labeled *ff3, ff2, ff1,* and *ff0.* The keyword PORT MAP is followed by a description of how the component is connected. The symbol (=>) is used to signify an internal connection wire, not the direction of data flow. For example, *ff3* is an instance of the component *dflipflop* whose *d* input is connected to *serial_in* and whose clock (*clk*) is connected to the common clock (*cp*) and whose *q* output is connected to the parallel output *q3*. The remaining PORT MAP connections required to fulfill the design of Figure 13–38 are provided for *ff2*, *ff1*, and *ff0* as shown.

Figure 13–40 shows a simulation of the serial-in parallel-out shift register. It illustrates that at each positive edge of *cp*, *q3* receives the level of *serial_in*, *q2* receives the level of *q3*, and so on.

```
shiftreg.vhd - Text Editor                                    _ □ ×
    LIBRARY ieee;               ------------------------------------
    USE ieee.std_logic_1164.ALL; --Shift register defined using    --
                                 --Dflipflop component instantiations --
    ENTITY dflipflop IS          ------------------------------------
        PORT(d, clk           : IN     std_logic;
               q              : OUT    std_logic);
    END dflipflop ;

    ARCHITECTURE arc OF dflipflop IS
    BEGIN
        PROCESS (clk)
        BEGIN
            IF (clk'EVENT AND clk= '1') THEN q<=d;
            END IF;
        END PROCESS;
    END arc;

    LIBRARY ieee;
    USE ieee.std_logic_1164.ALL;

    ENTITY shiftreg IS
        PORT(serial_in, cp  : IN       STD_LOGIC;
               q3,q2,q1,q0   : BUFFER   STD_LOGIC);
    END shiftreg ;

    ARCHITECTURE arc of shiftreg IS
        COMPONENT dflipflop
            PORT (d, clk          : IN  STD_LOGIC;
                    q             : OUT STD_LOGIC);
        END COMPONENT;
    BEGIN
        ff3: dflipflop PORT MAP (d=>serial_in, clk=>cp, q=>q3);
        ff2: dflipflop PORT MAP (d=>q3,        clk=>cp, q=>q2);
        ff1: dflipflop PORT MAP (d=>q2,        clk=>cp, q=>q1);
        ff0: dflipflop PORT MAP (d=>q1,        clk=>cp, q=>q0);
    END arc;
Line   1    Col   1    INS ◄
```

Annotations on figure:
- D flip-flop from Example 10-9 (points to the dflipflop entity/architecture)
- External I/O for shiftreg (points to PORT declaration)
- Declare the use of the component dflipflop from above (points to COMPONENT declaration)
- I/O for each component dflipflop (points to PORT in COMPONENT)
- This defines the internal and external connections
- Four instantiations (points to ff3, ff2, ff1, ff0)
- Keyword (points to PORT MAP)
- component name (points to dflipflop)
- instantiation label (points to ff3)

Figure 13–39 VHDL program listing for the 4-bit serial-in, parallel-out shift register using component instantiations.

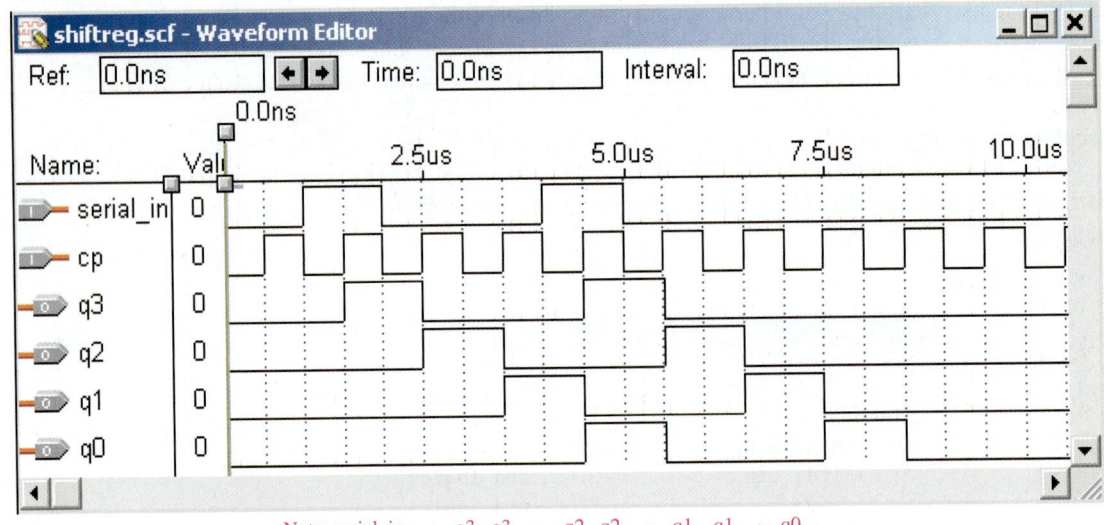

Note: serial_in ⟶ q3 , q3 ⟶ q2 , q2 ⟶ q1 , q1 ⟶ q0

Figure 13–40 Simulation of the 4-bit shift register described in Figure 13–39.

EXAMPLE 13-8

Use the structured programming approach to design a 4-bit synchronous counter using component instantiations as shown in Figure 13–41.

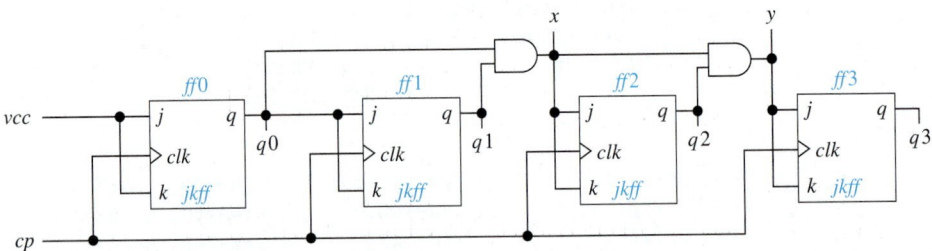

Figure 13–41 Block diagram of a 4-bit synchronous counter using component instantiations.

Solution: The VHDL solution for the 4-bit synchronous counter using component instantiations is given in Figure 13–42. You will notice that the *J-K* flip-flop component (*jk_ff*) was taken from the program written for Example 10–14. The ARCHITECTURE of *sync_count* uses four instances of *jk_ff*. An internal SIGNAL was required for defining the signals *vcc* (which is set to '1' for input to the first *jk_ff*) and *x* and *y* (which are the outputs of internal AND gates.)

Figure 13–43 shows the simulation of the circuit counting from 0000 up to 1111 on the *q*-outputs.

```
 sync_count.vhd - Text Editor                                        _ □ X
    LIBRARY ieee;                      --------------------------------   ▲
    USE ieee.std_logic_1164.ALL; -- Synchronous counter using    --
                                 -- jk_ff component instantiations --
    ENTITY jk_ff IS                    --------------------------------
        PORT(j, k, clk      : IN      std_logic;
             q              : BUFFER std_logic);
    END jk_ff ;


    ARCHITECTURE arc OF jk_ff IS
    SIGNAL jk : std_logic_vector (1 DOWNTO 0);
    BEGIN
        jk<=j&k;
        PROCESS (clk)
        BEGIN
            IF (clk'EVENT AND clk= '1') THEN
                CASE jk IS
                    WHEN "00" => q <= q;
                    WHEN "01" => q <= '0';
                    WHEN "10" => q <= '1';
                    WHEN "11" => q <= NOT q;
                    WHEN OTHERS => q <= q;
                END CASE;
            END IF;
        END PROCESS;
    END arc;


    LIBRARY ieee;
    USE ieee.std_logic_1164.ALL;


    ENTITY sync_count IS
        PORT(cp              : IN          STD_LOGIC;
             q3,q2,q1,q0     : BUFFER      STD_LOGIC);
    END sync_count ;


    ARCHITECTURE arc OF sync_count IS
    SIGNAL  vcc, x, y   :STD_LOGIC;
        COMPONENT jk_ff
            PORT (j,k, clk  : IN      STD_LOGIC;
                  q         : OUT     STD_LOGIC);
        END COMPONENT;
    BEGIN
        vcc<='1';
        ff0: jk_ff PORT MAP (j=>vcc, k=>vcc, clk=>cp, q=>q0);
        ff1: jk_ff PORT MAP (j=>q0,  k=>q0,  clk=>cp, q=>q1);
        x<=q0 AND q1;
        ff2: jk_ff PORT MAP (j=>x,   k=>x,   clk=>cp, q=>q2);
        y<=x AND q2;
        ff3: jk_ff PORT MAP (j=>y,   k=>y,   clk=>cp, q=>q3);
    END arc;
 Line   1    Col   1      INS ◄                                       ►
```

J-K flipflop from Example 10-14

External I/O for sync-count

Signal for internal interconnection

Declare the use of the component jk_ff from above

I/O for each component jk_ff

use 1's for jk of ff0

Four instantiations

use x for jk of ff2

use y for jk of ff3

Figure 13–42 VHDL solution for the 4-bit synchronous counter using component instantiations for Example 13–8.

629

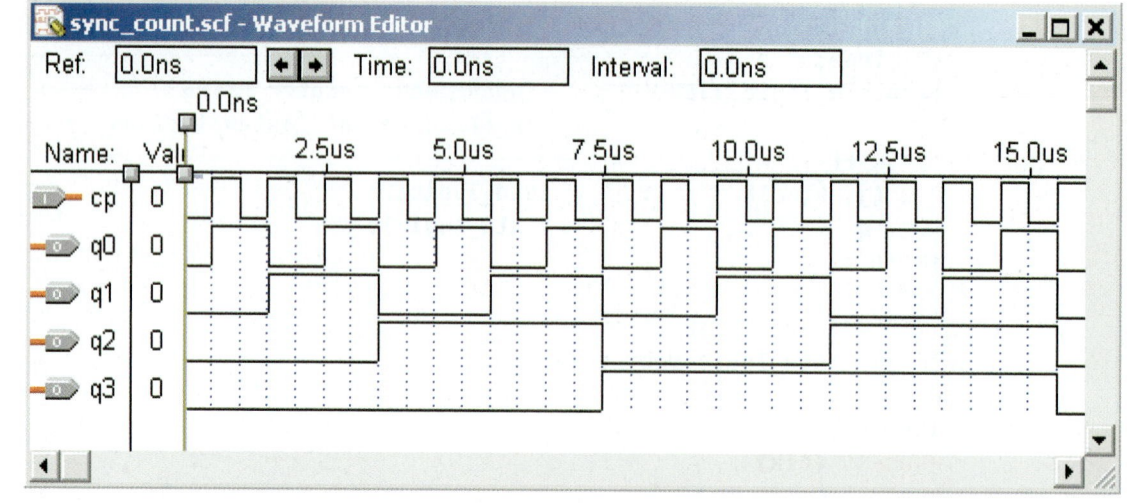

Figure 13–43   Simulation of the 4-bit synchronous counter of Example 13–8.

Summary

In this chapter, we have learned that

1. Shift registers are used for serial-to-parallel and parallel-to-serial conversions.

2. One common form of digital communication is for a sending computer to convert its data from parallel to serial and then transmit over a telephone line to a receiving computer, which converts back from serial to parallel.

3. Simple shift registers can be constructed by connecting the Q-outputs of one J-K flip-flop into the J-K inputs of the next flip-flop. Several flip-flops can be cascaded together this way, driven by a common clock, to form multibit shift registers.

4. The ring and Johnson shift counters are two specialized shift registers used to create sequential control waveforms.

5. Several multibit shift register ICs are available for the designer to choose from. They generally have four or eight internal flip-flops and are designed to shift either left or right and perform either serial-to-parallel or parallel-to-serial conversions.

6. The 74194 is called a universal 4-bit shift register because it can shift in either direction and can receive and convert to either format.

7. Three-state outputs are used on ICs that must have their outputs go to a common point. They are capable of the normal HIGH/LOW levels but can also output a float (or high-impedance) state.

8. The rotation of a stepper motor is made by taking small angular steps. This is controlled by sequential digital strings often generated by a recirculating shift register.

9. Three-state buffers, latches, and transceivers are an integral part of microprocessor interface circuitry. They allow the microprocessor system to have external control of 8-bit groups of data. The *buffer* can be used to allow multiple input devices to feed a common point or to provide high output current to a connected load. The *latch* can be used to remember momentary data from the microprocessor that needs to be held for other devices in the system. The *transceiver* provides bidirectional (input or output) control of interface circuitry.

10. Instantiations of VHDL component modules can be linked together to solve larger system design applications.

Glossary

Bidirectional: Allowing data to flow in either direction.

Buffer: A logic device connected between two digital circuits, providing isolation, high sink, and source current, and usually, three-state control.

Clock Enable: A separate input pin included on some ICs and used to enable or disable the clock input signal.

Closed-Loop Feedback: A system that sends information about an output device back to the device that is driving the output device to keep track of the particular activity.

Components: In VHDL it is a module or sub-program that can be declared as a part of a larger design entity and used repeatedly.

Data Bit: A single binary representation (0 or 1) of digital information.

Data Bus: A group of 8, 16, or 32 lines or electrical conductors, usually connected to a microprocessor and shared by a number of other devices connected to it.

Data Conversion: Transformation of digital information from one format to another (e.g., serial-to-parallel conversion).

Data Transmission: The movement of digital information from one location to another.

Digital Sequencer: A system (like a shift counter) that can produce a specific series of digital waveforms to drive another device in a specific sequence.

Ferromagnetic: A material in which magnetic flux lines can easily pass (high permeability).

Float: A digital output level that is neither HIGH nor LOW but, instead, is in a *high-impedance state.* In this state, the output acts like a high impedance with respect to ground and will float to any voltage level that happens to be connected to it.

Flux Lines: The north-to-south magnetic field set up by magnets is made up of flux lines.

Handshaking: The communication between a data sending device and receiving device that is necessary to determine the status of the transmitted data.

High-Impedance State: *See* float.

Instantiation: An implementation of a VHDL program module that is defined by specifying the inputs and outputs of a previously defined component.

Mode Control: Input pins available on some ICs used to control the operating functions of that IC.

Octal: When referring to an IC, octal means that a single package contains *eight* logic devices.

Output Enable: An input pin on an IC that can be used to enable or disable the outputs. When disabled, the outputs are in the float condition.

Parallel Enable: An IC input pin used to enable or disable a synchronous parallel load of data bits.

Pole Pair: Two opposing magnetic poles situated opposite each other in a motor housing and energized concurrently.

PORT MAP: A keyword in VHDL that is used to define the internal connections between inputs, outputs, and internal signals.

Recirculating: In a shift register, instead of letting the shifting data bits drop out of the end of the register, a recirculating connection can be made to pass the bits back into the front end of the register.

Register: Two or more flip-flops (or storage units) connected as a group and operated simultaneously.

Rotor: The rotating part of the stepper motor.

Shift Counter: A special-purpose shift register with modifications to its connections and preloaded with a specific value to enable it to output a special sequence of digital waveforms. It does not count in true binary but, instead, is used for special sequential waveform generation.

Shift Register: A storage device containing two or more data bits and capable of moving the data to the left or right and performing conversions between serial and parallel.

Stator Coil: A stationary coil, mounted on the inside of the motor housing.

Step Angle: The number of degrees that a stepper motor rotates for each change in the digital input signal (usually 15° or 7.5°).

Stepper Motor: A motor whose rotation is made in steps that are controlled by a digital input signal.

Strobe: A connection used in digital circuits to enable or disable a particular function.

Structural: A VHDL design method that employs the technique of linking together instantiations of VHDL component modules to solve larger system design applications.

Three-State Output: A feature on some ICs that allows you to connect several outputs to a common point. When one of the outputs is HIGH or LOW, all others will be in the float condition (the three output levels are HIGH, LOW, and float).

Transceiver: A data transmission device that is bidirectional, allowing data to flow through it in either direction.

Transparent Latch: An asynchronous device whose outputs hold onto the most recent digital state of the inputs. The outputs immediately follow the state of the inputs (transparent) while the trigger input is active and then latch onto that information when the trigger is removed.

Sections 13–1 Through 13–4

13–1. In Figure P13–1, will the data bits be shifted right or left with each clock pulse? Will they be shifted on the positive or negative clock edge?

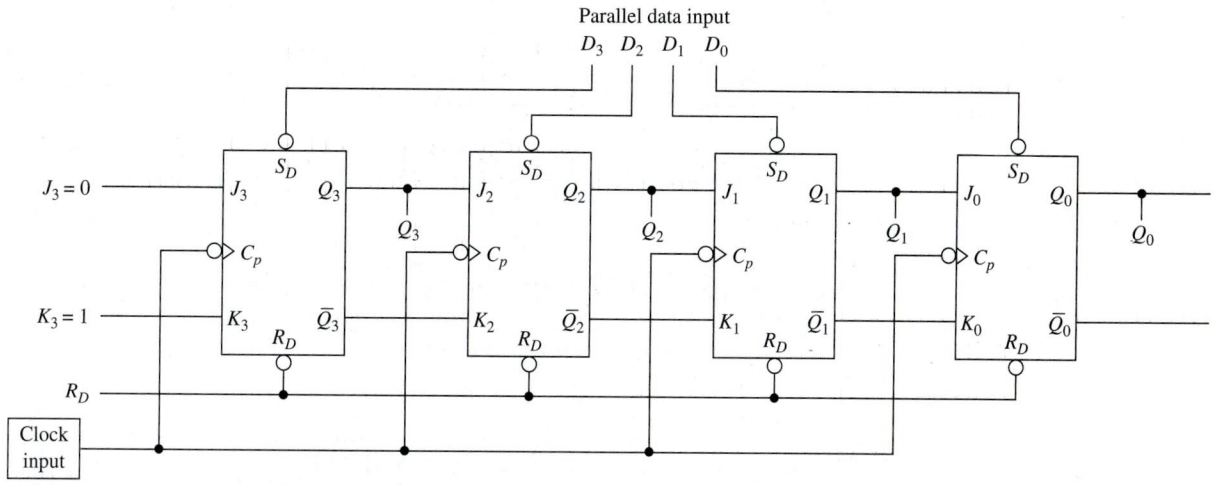

Parallel data input
D_3 D_2 D_1 D_0

Figure P13–1

13–2. If the register of Figure P13–1 is initially parallel loaded with $D_3 = 0$, $D_2 = 1$, $D_1 = 0$, and $D_0 = 0$, what will the output at Q_3 to Q_0 be after two clock pulses? After four clock pulses?

13–3. Repeat Problem 13–2 for $J_3 = 1$, $K_3 = 0$.

13–4. Change Figure P13–1 to a recirculating shift register by connecting Q_0 back to J_3 and $\overline{Q_0}$ back to K_3. If the register is initially loaded with $D_3 - D_0 = 1001$, what is the output at Q_3 to Q_0:

(a) After two clock pulses?

(b) After four clock pulses?

13–5. Outline the steps that you would take to parallel load the binary equivalent of a hex B into the register of Figure P13–1.

13–6. To use Figure P13–1 as a parallel-to-serial converter, where are the data input line(s) and data output line(s)?

13–7. Repeat Problem 13–6 for a serial-to-parallel converter.

Section 13–5

13–8. What changes have to be made to the circuit of Figure P13–1 to make it a Johnson shift counter?

13–9. How many flip-flops are required to produce the waveform shown in Figure P13–9 at the Q_0 output of a ring shift counter?

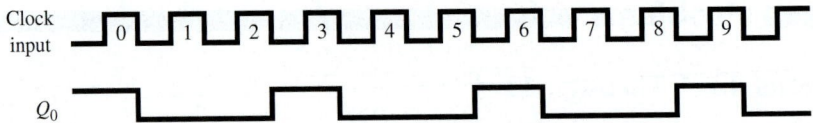

Clock input

Q_0

Figure P13–9

13–10. Repeat Problem 13–9 for the waveforms shown in Figure P13–10.

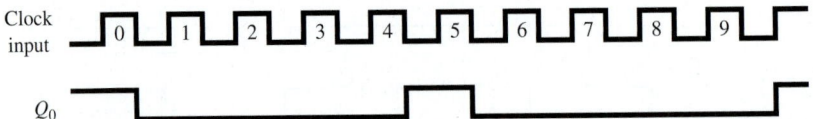

Clock input

Q_0

Figure P13–10

13–11. Which flip-flop(s) of a 4-bit ring shift counter must be initially Set to produce the waveform shown in Figure P13–11 at Q_0?

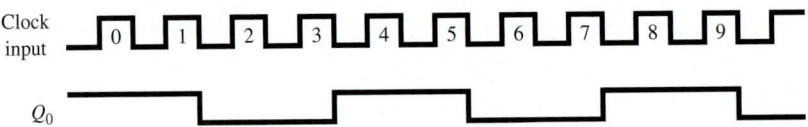

Clock input

Q_0

Figure P13–11

C **13–12.** Sketch the waveforms at Q_2 for the first seven clock pulses generated by the circuit shown in Figure P13–12.

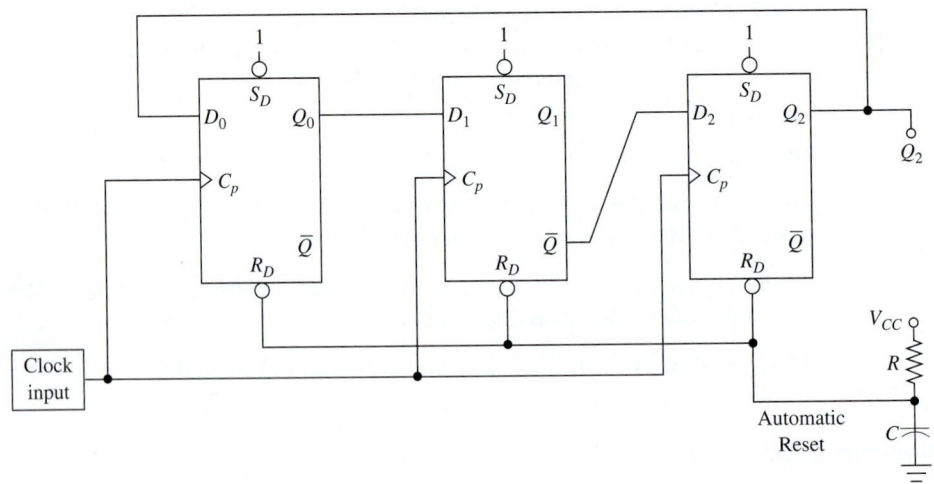

Figure P13–12

C **13–13.** In Figure P13–12, connect the automatic Reset line to the three $\overline{S_D}$ inputs instead of the three $\overline{R_D}$ inputs, and sketch the waveforms at Q_2 for the first seven clock pulses.

CHAPTER 13 I SHIFT REGISTERS

C   **13–14.** Sketch the waveforms at C_p, Q_0, Q_1, and Q_2 for seven clock pulses for the ring shift counter shown in Figure P13–14.

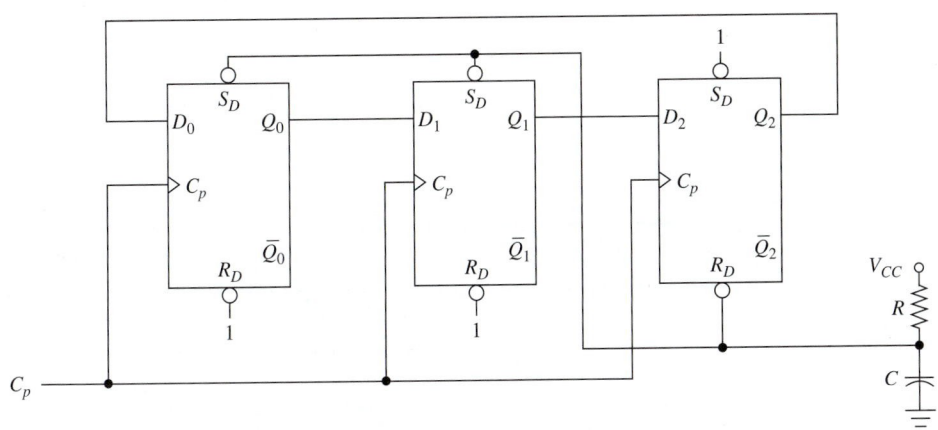

Figure P13–14

13–15. Using the Johnson shift counter output waveforms in Figure 13–6, add some logic gates to produce the waveforms at X, Y, and Z shown in Figure P13–15.

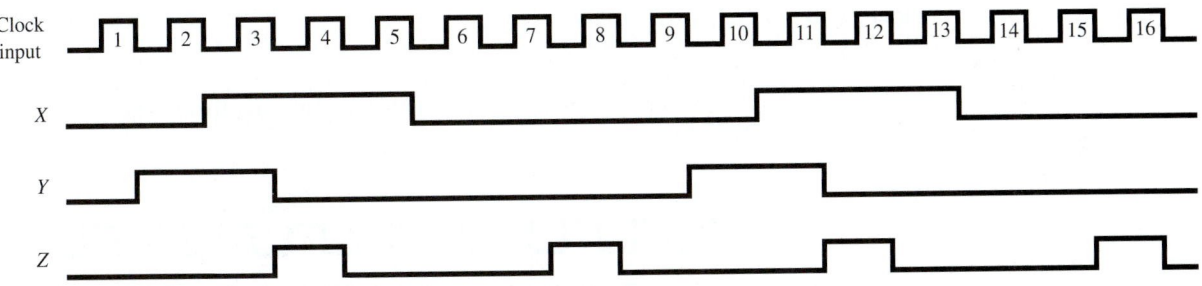

Figure P13–15

D   **13–16.** Redesign the Johnson shift counter of Figure 13–6(a) using 7474 D flip-flops in place of the J-K flip-flops.

Sections 13–6 and 13–7

D   **13–17.** What modification could be made to the circuit in Figure 13–20 to cause the yellow light to flash all day on Sundays? (Assume that someone will throw a switch at the beginning and end of each Sunday.)

13–18. Sketch the output waveforms at Q_0 to Q_3 for the 74194 circuit shown in Figure P13–18. Also, list the operating mode at each positive clock edge.

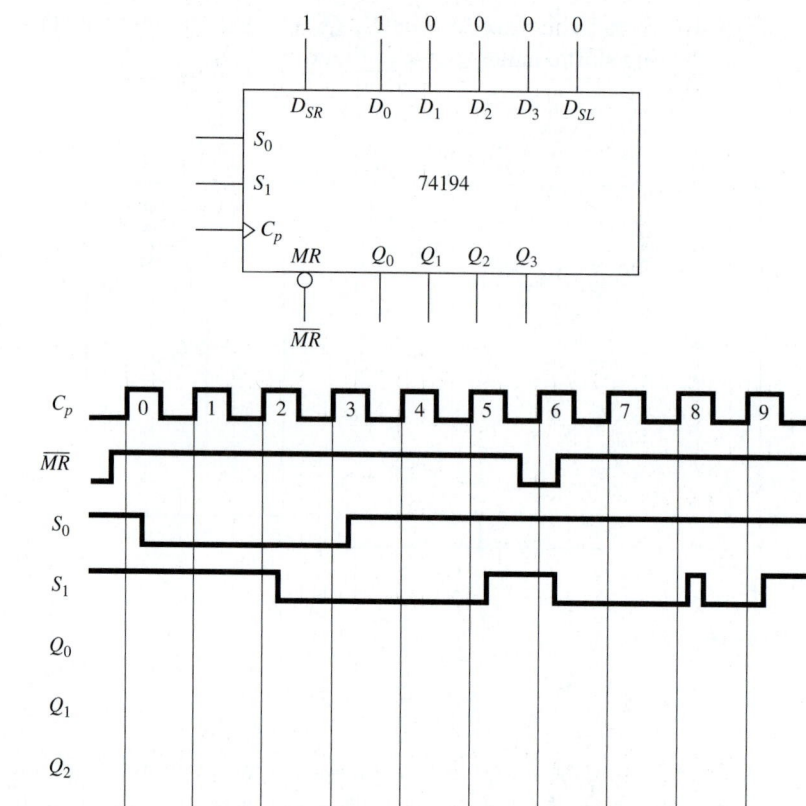

Figure P13–18

13–19. Repeat Problem 13–18 for the input waveforms shown in Figure P13–19.

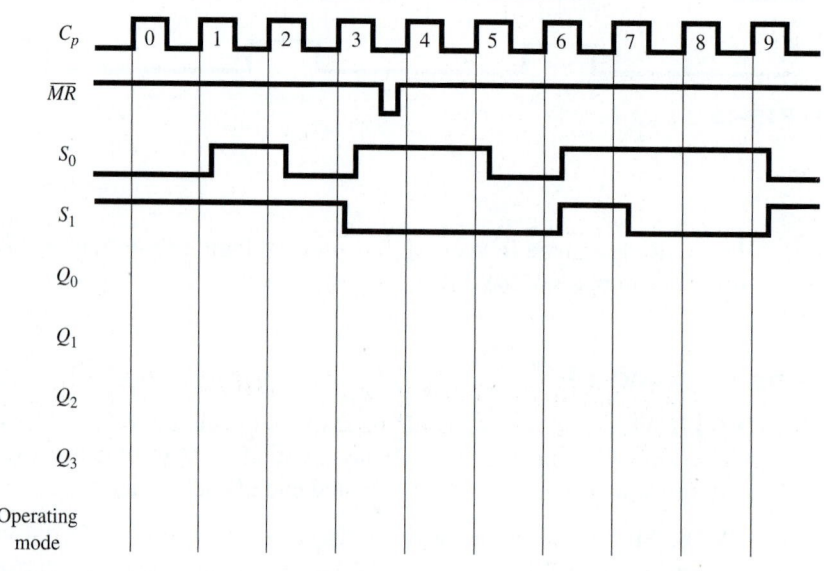

Figure P13–19

13–20. Sketch the output waveforms at Q_0 to Q_3 for the 74194 circuit shown in Figure P13–20. Also, list the operating mode at each positive clock edge.

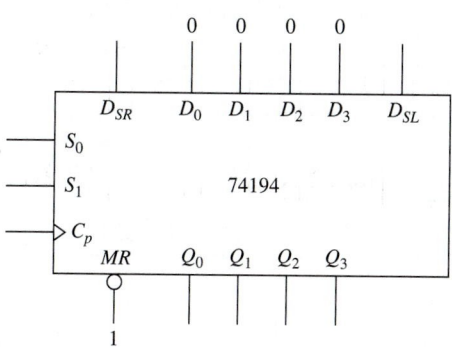

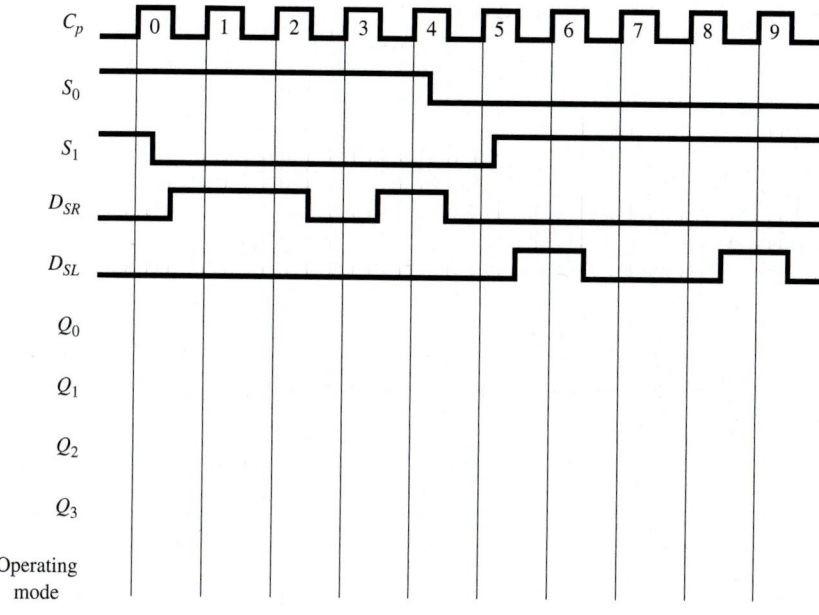

Figure P13–20

13–21. Repeat Problem 13–20 for the waveforms shown in Figure P13–21.

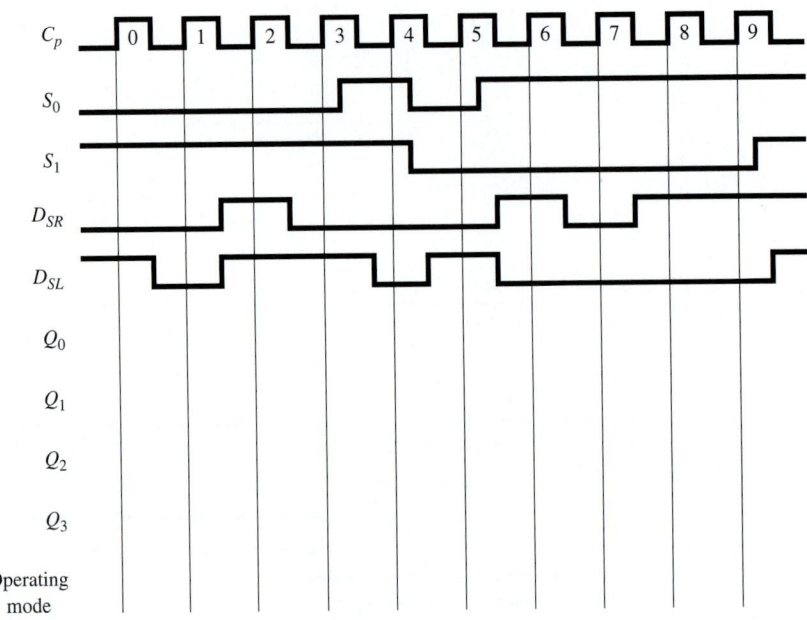

Figure P13–21

13–22. Draw the timing waveforms (similar to Figure 13–12) for a 74164 used to convert the serial binary number 10010110 into parallel.

13–23. Draw the circuit connections and timing waveforms for a 74165 used to convert the parallel binary number 1001 0110 into serial, MSB first.

13–24. Using your TTL data manual (or a manufacturer's Web site), describe the differences between the 74195 and the 74395A.

13–25. Using your TTL data manual (or a manufacturer's Web site), describe the differences between the 74164 and the 74165.

13–26. Describe how the procedure for parallel loading the 74165 differs from parallel loading a 74166.

C D   **13–27.** Design a system that can be used to convert an 8-bit serial number LSB first into an 8-bit serial number MSB first. Show the timing waveforms for 16 clock pulses and any control pulses that may be required for the binary number 10110100.

Section 13–8

13–28. How many clock pulses are required at C_p to cause the stepper motor to make one revolution in Figure 13–27?

C   **13–29.** Sketch the waveforms at C_p, S_0, S_1, Q_0, Q_1, Q_2, and Q_3 for six steps of the motor in Figure 13–27.

C   **13–30.** What must the clock frequency be in Figure 13–27 to make the stepper motor revolve at 600 rpm (rotations per minute)?

13–31. Describe the difference between a buffer and a latch and between a buffer and a transceiver.

13–32. Why is it important to use devices with three-state outputs when interfacing to a microprocessor data bus?

Schematic Interpretation Problems

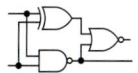

See Appendix G for the schematic diagrams.

S **13–33.** Identify the following ICs on the 4096/4196 schematic (sheets 1 and 2):

(a) The three-state octal buffers

(b) The three-state octal *D* flip-flops

(c) The three-state octal transceivers

(d) The three-state octal latches

S C **13–34.** Describe the operation of U6 in the 4096/4196 schematic. Use the names of the input/output labels provided on the IC for your discussion.

S **13–35.** Refer to sheet 2 of the 4096/4196 schematic. Describe the sequence of operations that must take place to load the 8-bit data string labeled IA0–IA7 and the 8-bit data string labeled ID0–ID7. Include reference to U30, U32, U23, U13:A, U1:F, and U33.

Electronics Workbench®/MultiSIM® Exercises

E13–1. Load the circuit file for **Section 13–2a.** This is a 4-bit shift register made from four JK flip-flops similar to Section 13–2. Press R (Rd′) to Reset all flip-flops. Press S (Sd′) to Set the first flip-flop.

(a) How many times must you now press C (C_p') to shift the ON bit to the rightmost flip-flop? Try it.

(b) What steps would you follow to shift an ON bit into all four flip-flops? Try it.

E13–2. Load the circuit file for **Section 13–2b.** This is a 4-bit shift register made from four JK flip-flops similar to Section 13–2. Make the necessary connections to form a recirculating shift register. Reset all FFs, then load the number 1000 (Q^3 light ON).

(a) How many times must you press C to rotate the ON bit to Q^0? Try it.

(b) Reset and turn Q^3 ON. Which light will be ON if you now press C six times? Try it.

E13–3. Load the circuit file for **Section 13–5a.** This is a Ring Shift Counter similar to Figure 13–5. Measure the time width of each positive output pulse.

(a) T = _____?

(b) On a piece of paper sketch, what you think the outputs would look like if the Cp′ connection to the last flip-flop was broken. Try it to see if you were right.

D

E13–4. Load the circuit file for **Section 13–5b.** This is a Ring Shift Counter similar to Figure 13–5.

(a) What modifications must be made to it to form a Johnson Shift Counter?

(b) Observe the output waveforms on the Logic Analyzer. Measure the positive pulse width of one of the output waveforms. T = _____?

C D

E13–5. Load the circuit file for **Section 13–6a.** The 74164 IC is an 8-bit, serial-in, parallel-out shift register. List the steps that you must perform to serially load the ASCII code for the letter M into this shift register. Try it, and demonstrate your results to your instructor.

C D

E13–6. Load the circuit file for **Section 13–6b.** Make the necessary connections to the 74194 universal shift register to be able to parallel load a hex D (1101), and then shift it to the right four positions. (Outputs would then show 0000.) Demonstrate it to your instructor.

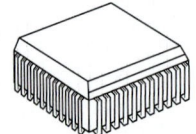

CPLD Problems

C13–1. The VHDL program in Figure 13–7 is the implementation of a serial-in, shift-right shift register.

(a) Load the files *shift_a.vhd* and *shift_a.scf* from the textbook CD. Save these files with the new name *prob_c13_1.vhd* and *prob_c13_1.scf.*

(b) Compile and simulate this program. (Remember, the Entity name must be changed to the new name in all three locations before attempting to compile.)

(c) Convert the program to a shift-left shift register. (The serial data enters *q0* instead of *q3.*)

(d) Run the simulation demonstrating the shifting sequence *q0*-to-*q1*-to-*q2*-, etc.

(e) Download your design of part (c) to a CPLD IC. Discuss your observations of the *q* output LEDs with your instructor as you demonstrate the shift-left operation. Use a debounced switch for *n_cp.* (The UP-1 board does not have debounced switches so you must build an off-board switch debouncer similar to the one explained in Figure 11–47.)

C13–2. The VHDL program in Figure 13–9 is the implementation of a parallel-load, shift-right shift register.

(a) Load the files *shift_b.vhd* and *shift_b.scf* from the textbook CD. Save these files with the new name *prob_c13_2.vhd* and *prob_c13_2.scf.*

(b) Compile and simulate this program. (Remember, the Entity name must be changed to the new name in all three locations before attempting to compile.)

(c) Convert the program to a recirculating shift register. (The data leaving *q0* is recirculated back to *q3.*)

(d) Run the simulation demonstrating the shifting sequence *q3*-to-*q2*-to-*q1*-to-*q0*-to-*q3*, etc.

(e) Download your design of part (c) to a CPLD IC. Discuss your observations of the *q* output LEDs with your instructor as you demonstrate the recirculating shift-right operation. Use a debounced switch for *n_cp*. (The UP-1 board does not have debounced switches so you must build an off-board switch debouncer similar to the one explained in Figure 11–47.)

C13–3. Redo problem C13-2, adding recirculation and Left/Right (*lr*) control to the shifting sequence. When *lr* = '0', shift left, when *lr* = '1' shift right. The simulation and CPLD demonstration should show recirculating data bits with left/right direction control.

C13–4. Redo problem C13-2, adding recirculation and clock strobe control to the shifting sequence. The clock strobe (*strobe*) operates as shown in Figure 13–4 (with *strobe* = '0' the clock is disabled.). The simulation and CPLD demonstration should show recirculating data bits with clock strobe control.

C13–5. The LPM shift register in Figure 13–34 is set up as a 4-bit serial or parallel-in shift-right shift register. Build new *gdf* and *scf* files called *prob_c13_5.gdf* and *prob_c13_5.scf* with the LPM reconfigured as shift-left. The simulation and CPLD demonstration should show an initial HIGH on the *shiftin* input, then several shift-left operations, followed by a parallel load and several shift-left operations.

C13–6. The LPM shift register in Figure 13–34 is set up as a 4-bit serial or parallel-in shift-right shift register. Build new *gdf* and *scf* files called *prob_c13_6.gdf* and *prob_c13_6.scf* with the LPM reconfigured as a recirculating shift-left with control inputs for the ports labeled *aclr, aset,* and *enable*. (These ports are described in the help screen of the **Edit Ports/ Parameters** menu.) The simulation and CPLD demonstration should exercise all of the inputs to show an asynchronous parallel load (*aset*), recirculating bits, asynchronous clear (*aclr*), and clock enable/disable. (The *cp* input has to be debounced, but the others do not.)

C13–7. The 74194 macrofunction in Figure 13–36 is set up as a parallel-load shift-right, shift-left register. Build new *gdf* and *scf* files called *prob_c13_7.gdf* and *prob_c13_7.scf* with the 74194 reconfigured as a recirculating shift-left, shift-right register with parallel-load and asynchronous clear. The simulation and CPLD demonstration should exercise all of the inputs to show a parallel-load, shift right, shift left, hold, recirculation of data bits, and asynchronous clear. (The *cp* input has to be debounced, but the others do not.)

C13–8. Figure 13–39 is the VHDL implementation of a 4-bit shift-right shift register.

(a) Load the files *shiftreg.vhd* and *shiftreg.scf* from the textbook CD. Save these files with the new name *prob_c13_8.vhd* and *prob_c13_8.scf.*

(b) Compile and simulate this program. (Remember, the Entity name must be changed to the new name in all three locations before attempting to compile.)

(c) Convert this program to an 8-bit shift-right shift register.

(d) Modify the simulation to demonstrate the 8-bit shifting operation.

(e) Download your design of part (c) to a CPLD IC. Discuss your observations of the *q* output LEDs with your instructor as you demonstrate the shifting operation. Use a debounced switch for *cp*. (The UP-1 board does not have debounced switches so you must build an off-board switch debouncer similar to the one explained in Figure 11–47.)

C13–9. Figure 13–39 is the VHDL implementation of a 4-bit shift-right shift register.

(a) Load the files *shiftreg.vhd* and *shiftreg.scf* from the textbook CD. Save these files with the new name *prob_c13_9.vhd* and *prob_c13_9.scf.*

(b) Compile and simulate this program. (Remember, the Entity name must be changed to the new name in all three locations before attempting to compile.)

(c) Convert this program to a 4-bit Johnson shift counter similar to Figure 13–6 but using *D* flip flops instead of *J-K* flip flops.

(d) Modify the simulation to demonstrate the shifting operation.

(e) Download your design of part (c) to a CPLD IC. Discuss your observations of the *q* output LEDs with your instructor as you demonstrate the shifting operation. Use a debounced switch for *cp*. (The UP-1 board does not have debounced switches so you must build an off-board switch debouncer similar to the one explained in Figure 11–47.)

C13–10. Figure 13–42 is the VHDL implementation of a 4-bit synchronous counter.

(a) Load the files *sync_count.vhd* and *sync_count.scf* from the textbook CD. Save these files with the new name *prob_c13_10.vhd* and *prob_c13_10.scf.*

(b) Compile and simulate this program. (Remember, the Entity name must be changed to the new name in all three locations before attempting to compile.)

(c) Convert this program to a 4-bit MOD-10 counter.

(d) Modify the simulation to demonstrate the MOD-10 counting operation.

(e) Download your design of part (c) to a CPLD IC. Discuss your observations of the *q* output LEDs with your instructor as you demonstrate the counting operation. Use a debounced switch for *cp*. (The UP-1 board does not have debounced switches so you must build an off-board switch debouncer similar to the one explained in Figure 11–47.)

C13–11. Figure 13–42 is the VHDL implementation of a 4-bit synchronous counter.

(a) Load the files *sync_count.vhd* and *sync_count.scf* from the textbook CD. Save these files with the new name *prob_c13_11.vhd* and *prob_c13_11.scf.*

(b) Compile and simulate this program. (Remember, the Entity name must be changed to the new name in all three locations before attempting to compile.)

(c) Convert this program to a 5-bit counter.

(d) Modify the simulation to demonstrate the MOD-32 counting operation.

(e) Download your design of part (c) to a CPLD IC. Discuss your observations of the *q* output LEDs with your instructor as you demonstrate the counting operation. Use a debounced switch for *cp*. (The UP-1 board does not have debounced switches so you must build an off-board switch debouncer similar to the one explained in Figure 11–47.)

13–1. True

13–2. The output of a flip-flop connected to the input of the next flip-flop (Q to J, $\overline{Q}$ to K)

13–3. By using the active-LOW asynchronous set $(\overline{S_D})$

13–4. $Q_3 = 0$, $Q_2 = 0$, $Q_1 = 0$, $Q_0 = 0$; $Q_3 = 1$, $Q_2 = 1$, $Q_1 = 0$, $Q_0 = 0$

13–5. The data would continue shifting out the register and would be lost.

13–6. 1000,0000

13–7. HIGH

13–8. It's an active-LOW clock enable for starting/stopping the clock.

13–9. Because the data can be input or output, serial or parallel, shifted left or right, held and reset

13–10. $\overline{MR} = 1$, $S_1 = 1$, $S_0 = 1$, $D_0 = 1$, $D_1 = 1$, $D_2 = 0$, $D_3 = 1$. Data are loaded on the rising edge of C_p.

13–11. D_{SL}, Q_0, 0 1

13–12. Parallel enable (*PE*) enables the data to be loaded synchronously on the negative clock edge, and parallel load (*PL*) loads the data asynchronously.

13–13. HIGH, float

13–14. HIGH, enables

13–15. RC circuit and the Schmitt trigger

13–16. True

13–17. Instead of recycling, the parallel-loaded information would be lost because D_{SL} is ignored.

13–18. It is used as a strobe to enable the clock.

13–19. Clockwise. Do a shift-left in the 74194.

13–20. 8, $\overline{OE_a}$ $\overline{OE_b}$

13–21. Synchronous, asynchronous

13–22. True

14

Multivibrators and the 555 Timer

OUTLINE

OBJECTIVES

Upon completion of this chapter, you should be able to:

- Calculate capacitor charging and discharging rates in series RC timing circuits.
- Sketch the waveforms and calculate voltage and time values for astable and monostable multivibrators.
- Connect IC monostable multivibrators to output a waveform with a specific pulse width.
- Explain the operation of the internal components of the 555 IC timer.
- Connect a 555 IC timer as an astable multivibrator and as a monostable multivibrator.
- Discuss the operation and application of crystal oscillator circuits.

INTRODUCTION

We have seen that timing is very important in digital electronics. Clock oscillators, used to drive counters and shift registers, must be designed to oscillate at a specific frequency. Specially designed **pulse-stretching** and time-delay circuits are also required to produce specific pulse widths and delay periods.

14–1 Multivibrators

Multivibrator circuits have been around for years, designed from various technologies, to fulfill electronic circuit timing requirements. A **multivibrator** is a circuit that changes between the two digital levels on a continuous, free-running basis or on demand from some external trigger source. Basically, there are three types of multivibrators: bistable, astable, and monostable.

The *bistable* multivibrator is triggered into one of the two digital states by an external source and stays in that state until it is triggered into the opposite state. The *S-R* flip-flop is a bistable multivibrator; it is in either the Set or Reset state.

The *astable* multivibrator is a free-running oscillator that alternates between the two digital levels at a specific frequency and duty cycle.

The *monostable* multivibrator, also known as a *one shot,* provides a single output pulse of a specific time length when it is triggered from an external source.

The bistable multivibrator (*S-R* flip-flop) was discussed in detail in Chapter 10. The astable and monostable multivibrators discussed in this chapter can be built from basic logic gates or from special ICs designed specifically for timing applications. In either case, the charging and discharging rate of a capacitor is used to provide the specific time durations required for the circuits to operate.

14–2 Capacitor Charge and Discharge Rates

Because the capacitor is so critical in determining the time durations, let's briefly discuss the capacitor charge and discharge formulas. We will use Figure 14–1 to determine the voltages on the capacitor at various periods of time after the switch is closed. In Figure 14–1, with the switch in position 1, conventional current will flow clockwise from the *E* source through the *RC* circuit. The capacitor will charge at an exponential rate toward the value of the *E* source. The rate that the capacitor charges is dependent on the product of *R* times *C*:

$$\Delta v = E(1 - e^{-t/RC}) \qquad\qquad 14\text{–}1$$

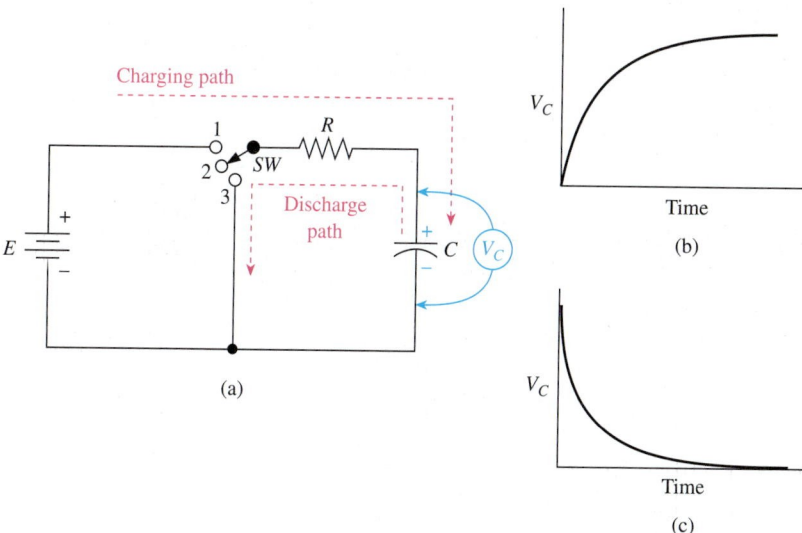

Figure 14–1 Basic *RC* charging/discharging circuit: (a) *RC* circuit; (b) charging curve; (c) discharging curve.

where Δv = change in capacitor voltage over a period of time t
        E = voltage difference between the initial voltage on the capacitor and the total voltage that it is trying to reach
        e = base of the natural log (2.718)
        t = time that the capacitor is allowed to charge
        R = resistance, ohms
        C = capacitance, farads

In some cases, the capacitor is initially discharged, and Δv is equal to the final voltage on the capacitor. But, with astable multivibrator circuits, the capacitor usually is not fully discharged, and Δv is equal to the final voltage minus the starting voltage. If you think of the y axis in the graph of Figure 14–1(b) as a distance that the capacitor voltage is traveling through, the variables in Equation 14–1 take on new meaning, as follows:

$$\Delta v \equiv \text{distance that the capacitor voltage travels}$$

$$E \equiv \text{total distance that the capacitor voltage is trying to travel}$$

Using these new definitions, Equation 14–1 can be used whether the capacitor is charging *or* discharging (a discharging capacitor can be thought of as *charging to a lower voltage*).

When the switch in Figure 14–1 is thrown to position 3, the capacitor discharges counterclockwise through the RC circuit. The values for the variables in Equation 14–1 are determined the same way as they were for the charging condition, except that the voltage on the capacitor is decreasing **exponentially,** as shown in Figure 14–1(c).

Transposing the Capacitor Charging Formula to Solve for t

Often in the design of timing circuits, it is necessary to solve for t,[*] given Δv, E, R, and C. To make life easy for ourselves, let's develop a new equation by rearranging Equation 14–1 to solve for t instead of Δv.

$$\Delta v = E(1 - e^{-t/RC})$$

$$\frac{\Delta v}{E} = 1 - e^{-t/RC} \qquad \text{Divide both sides by } E.$$

$$\frac{\Delta v}{E} - 1 = -e^{-t/RC} \qquad \text{Subtract 1 from both sides.}$$

$$1 - \frac{\Delta v}{E} = e^{-t/RC} \qquad \text{Multiply both sides by } (-1).$$

$$\frac{1}{1 - \Delta v/E} = \frac{1}{e^{-t/RC}} \qquad \text{Take the reciprocal of both sides.}$$

$$\frac{1}{1 - \Delta v/E} = e^{t/RC} \qquad \frac{1}{e^{-x}} = e^{x}$$

$$\ln\left(\frac{1}{1 - \Delta v/E}\right) = \ln e^{t/RC} \qquad \text{Take the natural logarithm of both sides.}$$

$$\ln\left(\frac{1}{1 - \Delta v/E}\right) = \frac{t}{RC} \qquad \ln e^{x} = x$$

$$t = RC \ln\left(\frac{1}{1 - \Delta v/E}\right) \qquad \qquad \textbf{14–2}$$

---

[*]A practical assumption often used by technicians is that a capacitor is 99% charged after a time equal to $5 \times R \times C$.

The following examples illustrate the use of Equations 14–1 and 14–2 for solving capacitor timing problems.

EXAMPLE 14–1

The capacitor in Figure 14–2 is initially discharged. Determine the voltage on the capacitor 0.5 ms after the switch is moved from position 2 to position 1.

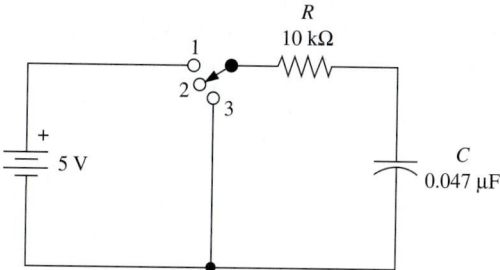

Figure 14–2 Circuit for Examples 14–1 through 14–4.

Solution: E, the total distance that the capacitor voltage is trying to charge to, is 5 V. Using Equation 14–1 yields

$$\Delta v = E(1 - e^{-t/RC})$$
$$= 5.0(1 - e^{-0.5\ \text{ms}/(10\ \text{k}\Omega \times 0.047\ \mu\text{F})})$$
$$= 5.0(1 - e^{-1.06})$$
$$= 5.0(1 - 0.345)$$
$$= 5.0(0.655)$$
$$= 3.27\ \text{V} \qquad \textit{Answer}$$

Thus, the distance the capacitor voltage traveled in 0.5 ms is 3.27 V. Because it started at 0 V, $V_{\text{cap}} = 3.27$ V.

Team Discussion

Would the capacitor voltage be more than or less than 3.27 V if the resistor is doubled? If the capacitor is doubled?

EXAMPLE 14–2

The capacitor in Figure 14–2 is initially discharged. How long after the switch is moved from position 2 to position 1 will it take for the capacitor to reach 3 V?

Solution: Δv, the distance that the capacitor voltage travels through, is 3 V. E, the total distance that the capacitor voltage is trying to travel, is 5 V. Using Equation 14–2, we obtain

$$t = RC \ln\left(\frac{1}{1 - \Delta v/E}\right)$$
$$= (10\ \text{k}\Omega)(0.047\ \mu\text{F}) \ln\left(\frac{1}{1 - 3/5}\right)$$
$$= 0.00047 \ln\left(\frac{1}{0.4}\right)$$
$$= 0.00047 \ln(2.5)$$
$$= 0.00047(0.916)$$
$$= 0.0431\ \text{ms} \qquad \textit{Answer}$$

EXAMPLE 14-3

For this example, let's assume that the capacitor in Figure 14–2 is initially charged to 1 V. How long after the switch is thrown from position 2 to position 1 will it take for the capacitor to reach 3 V?

Solution: Δv, the distance through which the capacitor voltage travels, is 2 V (3 V − 1 V). E, the total distance that the capacitor voltage is trying to travel, is 4 V (5 V − 1 V). Using Equation 14–2 gives us

$$t = RC \ln\left(\frac{1}{1 - \Delta v/E}\right)$$

$$= (10\ k\Omega)(0.047\ \mu F) \ln\left(\frac{1}{1 - 2/4}\right)$$

$$= 0.326\ ms \quad \textit{Answer}$$

The graph of the capacitor voltage is shown in Figure 14–3.

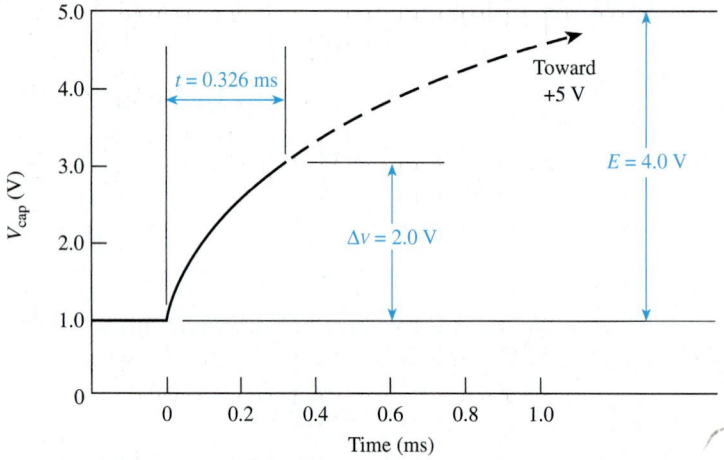

Figure 14-3 Graphic illustration of the capacitor voltage for Example 14–3.

EXAMPLE 14-4

The capacitor in Figure 14–2 is initially charged to 4.2 V. How long after the switch is thrown from position 2 to position 3 will it take to drop to 1.5 V?

Solution: Equation 14–2 can be used to solve for t by thinking of the capacitor as *charging to a lower voltage*. Δv, the distance that the capacitor voltage travels through, is 2.7 V (4.2 V − 1.5 V). E, the total distance that the capacitor voltage is trying to travel, is 4.2 V (4.2 V − 0 V). Using Equation 14–2 yields

$$t = RC \ln\left(\frac{1}{1 - \Delta v/E}\right)$$

$$= (10\ k\Omega)(0.047\ \mu F) \ln\left(\frac{1}{1 - 2.7\ V/4.2\ V}\right)$$

$$= 0.484\ ms \quad \textit{Answer}$$

The graph of the capacitor voltage is shown in Figure 14–4.

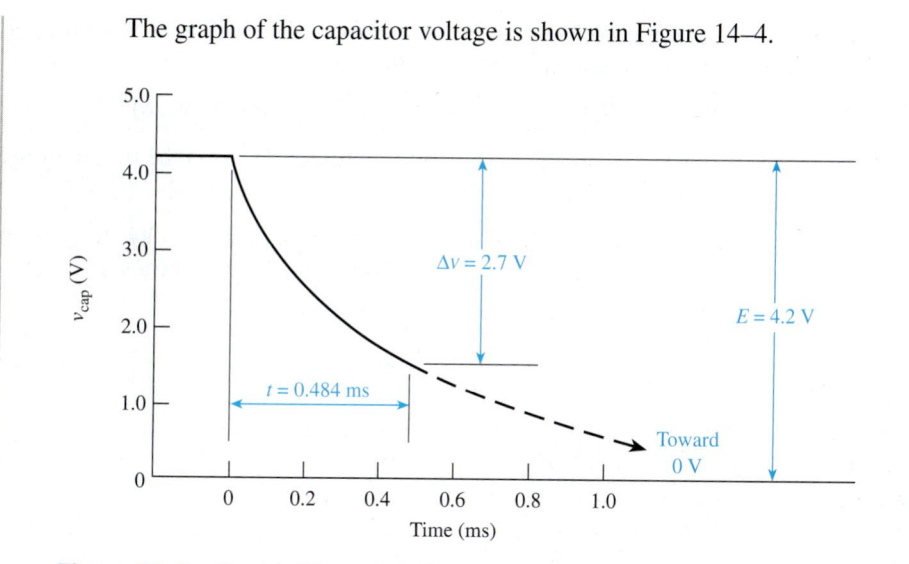

Figure 14–4   Graphic illustration of the capacitor voltage for Example 14–4.

Review Questions

14–1. The *astable* multivibrator, also known as a *one shot,* produces a single output pulse after it is triggered. True or false?

14–2. The voltage on a charging capacitor will increase _____ (faster, slower) if its series resistor is increased.

14–3. A 1-μF capacitor with a 10-kΩ series resistor will have the same charging rate as a 10-μF capacitor with a 1-kΩ series resistor. True or false?

14–3 Astable Multivibrators

The predictable charging rate of capacitors discussed in the previous section will now be used in the design of oscillator and timing circuits. A very simple astable multivibrator (free-running oscillator) can be built from a single Schmitt trigger inverter and an *RC* circuit, as shown in Figure 14–5. The oscillator of Figure 14–5 operates as follows:

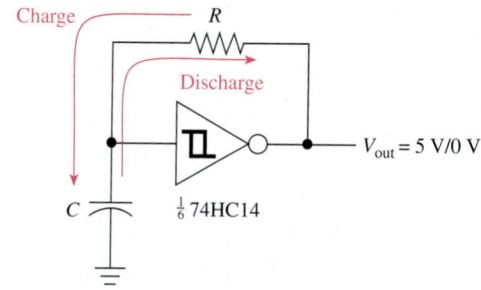

Figure 14–5   Schmitt trigger astable multivibrator.

Team Discussion

How would the operation of this circuit change if a 7414 were used in place of the 74HC14? (Consider I_I's and V_o's.)

1. When the IC supply power is first turned on, V_{cap} is 0 V, so V_{out} will be HIGH ($\approx$5.0 V for high-speed CMOS).

2. The capacitor will start charging toward the 5 V at V_{out}.

3. When V_{cap} reaches the positive-going threshold (V_{T+}) of the Schmitt trigger, the output of the Schmitt will change to a LOW (≈ 0 V).

4. Now, with $V_{out} \approx 0$ V, the capacitor will start discharging toward 0 V.

5. When V_{cap} drops below the negative-going threshold (V_{T-}) the output of the Schmitt will change back to a HIGH.

6. The cycle repeats now, with the capacitor charging back up to V_{T+} then down to V_{T-} then up to V_{T+} and so on. (The waveform at V_{out} will be a square wave oscillating between V_{OH} and V_{OL}, as shown in Figure 14–6.)

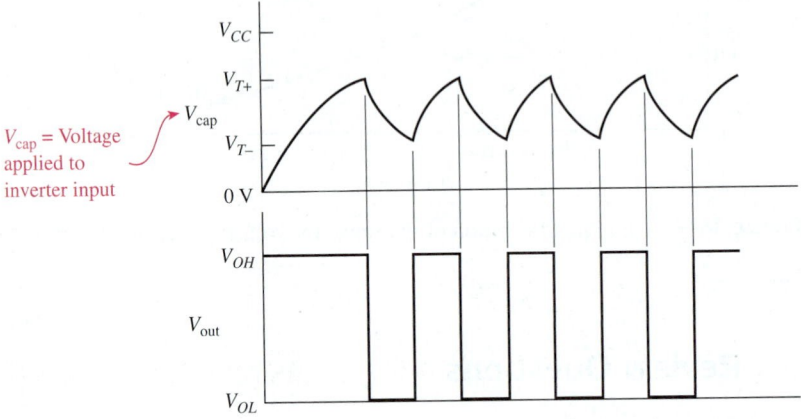

V_{cap} = Voltage applied to inverter input

Figure 14–6   Waveforms from the oscillator circuit of Figure 14–5.

EXAMPLE 14–5

(a) Sketch and label the waveforms for the Schmitt RC oscillator of Figure 14–5, given the following specifications for a 74HC14 high-speed CMOS Schmitt inverter ($V_{CC} = 5.0$ V):

$$V_{OH} = 5.0 \text{ V}, \qquad V_{OL} = 0.0 \text{ V}$$
$$V_{T+} = 2.75 \text{ V}, \qquad V_{T-} = 1.67 \text{ V}$$

(b) Calculate the time HIGH (t_{HI}), time LOW (t_{LO}), duty cycle, and frequency if $R = 10$ kΩ and $C = 0.022$ μF.

Solution: **(a)** The waveforms for the oscillator are shown in Figure 14–7.

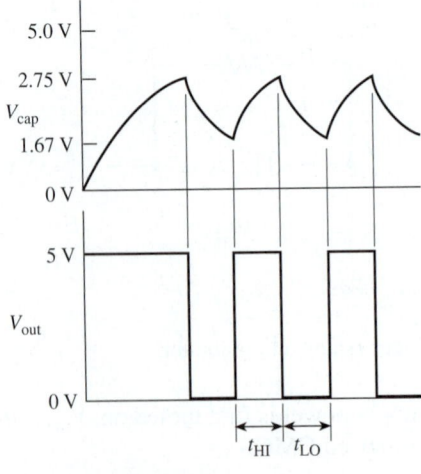

Figure 14–7   Solution to Example 14–5.

CHAPTER 14 | MULTIVIBRATORS AND THE 555 TIMER

(b)   To solve for t_{HI}:

$$\Delta V = V_{T^+} - V_{T^-}$$
$$\Delta V = 2.75 - 1.67 = 1.08 \text{ V}$$
$$E = 5.00 - 1.67 = 3.33 \text{ V}$$

$$t_{HI} = RC \ln\left(\frac{1}{1 - \Delta v/E}\right)$$

$$= (10 \text{ k}\Omega)(0.022 \text{ } \mu\text{F}) \ln\left(\frac{1}{1 - 1.08 \text{ V}/3.33 \text{ V}}\right)$$

$$= 86.2 \text{ } \mu s$$

To solve for t_{LO}:

$$\Delta v = 2.75 - 1.67 = 1.08 \text{ V}$$
$$E = 2.75 - 0 = 2.75 \text{ V}$$

$$t_{LO} = RC \ln\left(\frac{1}{1 - \Delta v/E}\right)$$

$$= (10 \text{ k}\Omega)(0.022 \text{ } \mu\text{F}) \ln\left(\frac{1}{1 - 1.08 \text{ V}/2.75 \text{ V}}\right)$$

$$= 110 \text{ } \mu s$$

To solve for duty cycle: **Duty cycle** is a ratio of the length of time a square wave is HIGH to the total period:

$$D = \frac{t_{HI}}{t_{HI} + t_{LO}}$$

$$= \frac{86.2 \text{ } \mu s}{86.2 \text{ } \mu s + 110 \text{ } \mu s}$$

$$= 0.439 = 43.9\%$$

To solve for frequency:

$$f = \frac{1}{t_{HI} + t_{LO}}$$

$$= \frac{1}{86.2 \text{ } \mu s + 110 \text{ } \mu s}$$

$$= 5.10 \text{ kHz}$$

Review Questions

14–4. The capacitor voltage levels in a Schmitt trigger astable multivibrator are limited by _____ and the output voltage is limited by _____.

14–5. One way to increase the frequency of a Schmitt trigger astable multivibrator is to _____ (increase, decrease) the resistor.

14–4   Monostable Multivibrators

The block diagram and I/O waveforms for a monostable multivibrator (commonly called a one shot) are shown in Figure 14–8. The one shot has one *stable state,* which is Q = LOW and $\overline{Q}$ = HIGH. The outputs switch to their opposite state for a length of time t_w only when a trigger is applied to the $\overline{A}$ input. $\overline{A}$ is a negative edge trigger in this case (other one shots use a positive edge trigger or both). The I/O waveforms in Figure 14–8 show the effect that $\overline{A}$ has on the Q output. Q is LOW until the HIGH-to-LOW edge of $\overline{A}$ causes Q to go HIGH for the length of time t_w. The output pulse width (t_w) is determined by the discharge rate of a capacitor in an RC circuit.

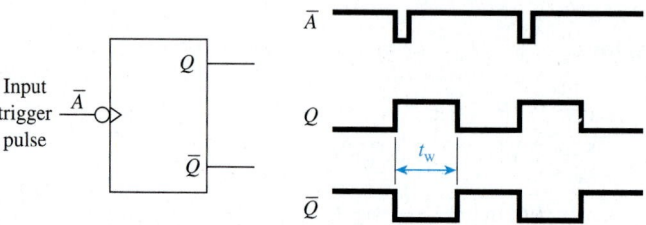

Figure 14–8   Block diagram and input/output waveforms for a monostable multivibrator.

A simple monostable multivibrator can be built from NAND gates and an RC circuit as shown in Figure 14–9. The operation of Figure 14–9 is as follows:

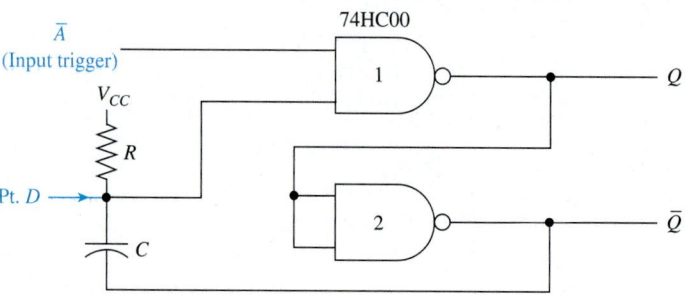

Figure 14–9   Two-gate monostable multivibrator.

1. When power is first applied, make the following assumptions: $\overline{A}$ is HIGH, Q is LOW, $\overline{Q}$ is HIGH, and C is discharged. Therefore, point D is HIGH.

2. When a negative-going pulse is applied at $\overline{A}$, Q is forced HIGH, which forces $\overline{Q}$ LOW.

3. Because the capacitor voltage cannot change instantaneously, point D will drop to 0 V.

4. The 0 V at point D will hold one input to gate 1 LOW, even if the $\overline{A}$ trigger goes back HIGH. Therefore, Q stays HIGH, and $\overline{Q}$ stays LOW.

5. Meanwhile, the capacitor is charging toward V_{CC}. When the capacitor voltage at point D reaches the HIGH-level input voltage rating (V_{IH}) of gate 1, Q will switch to a LOW, making $\overline{Q}$ HIGH.

6. The circuit is back in its stable state, awaiting another trigger signal from $\overline{A}$. The capacitor will discharge back to $\approx$ 0 V ($\approx V_{CC}$ on each side).

The waveforms in Figure 14–10 show the I/O characteristics of the circuit and will enable us to develop an equation to determine t_w. In the stable state ($\overline{Q}$ = HIGH), the voltage at point D will sit at V_{CC} ① because the capacitor is discharged (it has V_{CC} on both sides of it). When $\overline{Q}$ goes LOW due to an input trigger at $\overline{A}$, point D will follow $\overline{Q}$ LOW ② because the capacitor is still discharged. Now, the capacitor will start charging toward V_{CC}. When point D reaches V_{IH} ③, $\overline{Q}$ will switch back HIGH. The capacitor still has V_{IH} volts across it: $\rightarrow V_{IH} \vdash$. The capacitor voltage is added to the HIGH-level output of $\overline{Q}$, which causes point D to shoot up to $V_{CC} + V_{IH}$ ④. As the capacitor voltage discharges back to 0 V, point D drops back to the V_{CC} level ⑤ and awaits the next trigger.

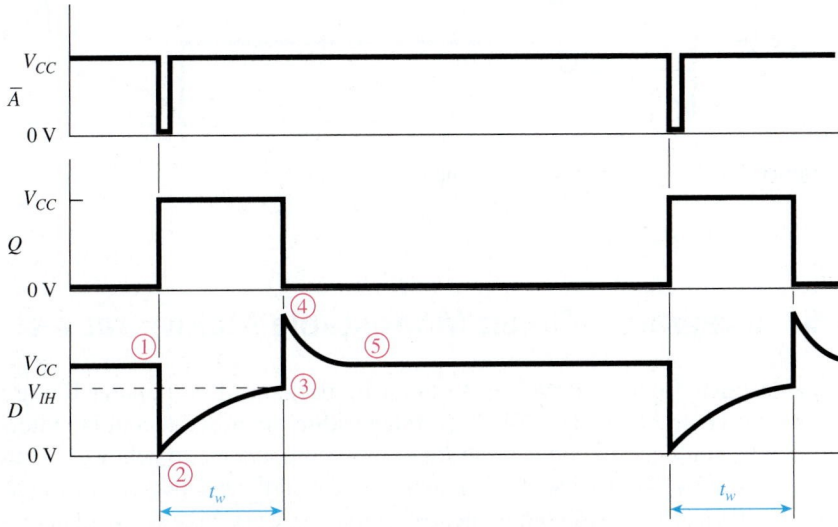

Figure 14–10 Input/output waveforms for the circuit of Figure 14–9.

━━━━ **EXAMPLE 14–6** ━━━━

(a) Sketch and label the waveforms for the monostable multivibrator of Figure 14–9, given the input waveform at $\overline{A}$ and the following specifications for a 74HC00 high-speed CMOS NAND gate (V_{CC} = 5.0 V):

$$V_{OH} = 5.0 \text{ V}, \qquad V_{OL} = 0.0 \text{ V}$$
$$V_{IH} = 3.5 \text{ V}, \qquad V_{IL} = 1.0 \text{ V}$$

(b) Calculate the output pulse width (t_w) for R = 4.7 kΩ and $C = 0.0047 \ \mu F$.

Solution:

(a) The waveforms for the multivibrator are shown in Figure 14–11.

(b) To solve for t_w using Equation 14–2:

$$\Delta v = 3.5 \text{ V} - 0 \text{ V} = 3.5 \text{ V}$$
$$E = 5.0 \text{ V} - 0 \text{ V} = 5.0 \text{ V}$$

$$t_w = RC \ln\left(\frac{1}{1 - \Delta v/E}\right)$$

$$= (4.7 \text{ k}\Omega)(0.0047 \ \mu F) \ln\left(\frac{1}{1 - 3.5 \text{ V}/5.0 \text{ V}}\right)$$

$$= 26.6 \ \mu s \qquad \textit{Answer}$$

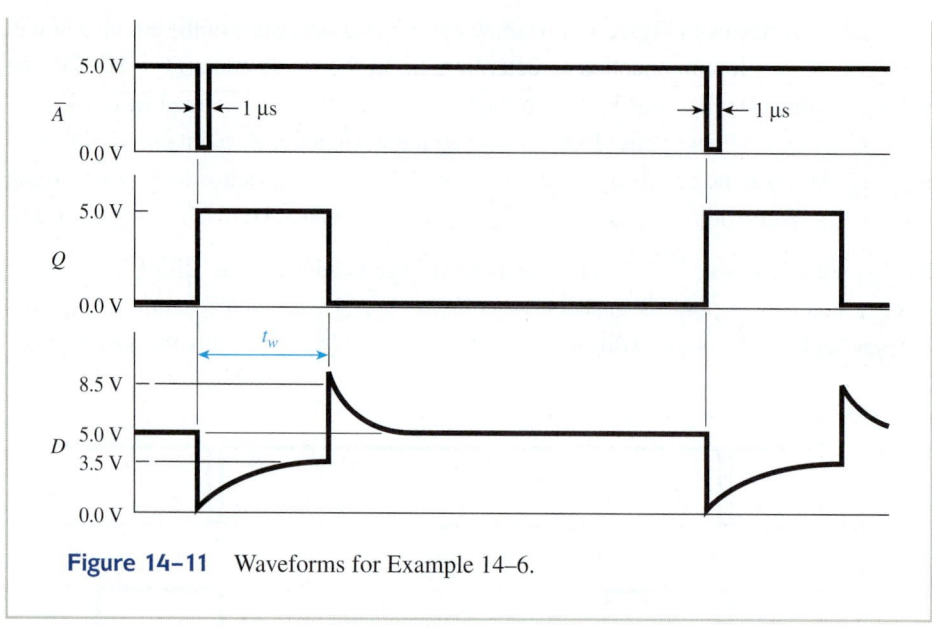

Figure 14–11 Waveforms for Example 14–6.

14–5 Integrated-Circuit Monostable Multivibrators

Monostable multivibrators are available in an IC package. Two popular ICs are the 74121 (nonretriggerable) and the 74123 (**retriggerable**) monostable multivibrators. To use these ICs, you need to connect the RC timing components to achieve the proper pulse width. The 74121 provides for two active-LOW and one active-HIGH trigger inputs $(\bar{A}_1, \bar{A}_2, B)$ and true and complemented outputs $(Q, \bar{Q})$. Figure 14–12 shows the 74121 block diagram and function table that we can use to figure out its operation.

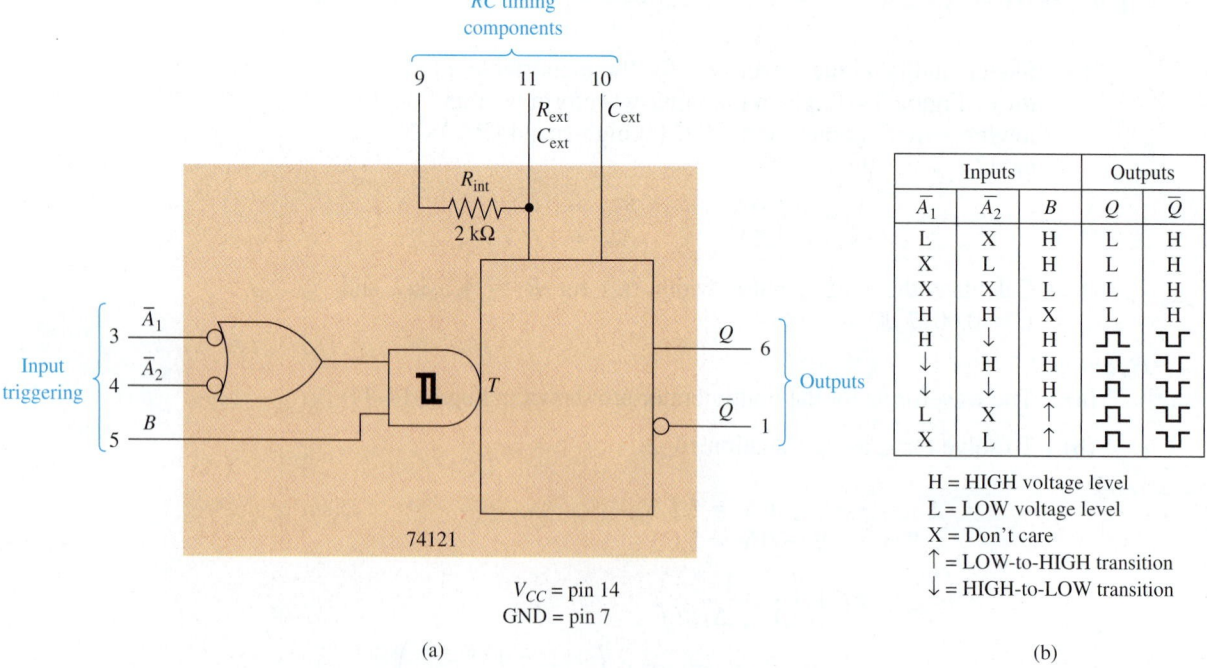

Figure 14–12 The 74121 monostable multivibrator one shot: (a) block diagram; (b) function table.

To trigger the multivibrator at point T in Figure 14–12, the inputs to the Schmitt AND gate must both be HIGH. To do that, you need B with ($\overline{A_1}$ or $\overline{A_2}$). Holding $\overline{A_1}$ or $\overline{A_2}$ LOW and bringing the input trigger in on B is useful if the trigger signal is slow rising or has noise on it, because the Schmitt input will provide a definite trigger point.

The RC timing components are set up on pins 9, 10, and 11. If you can use the 2-kΩ *internal* resistor, just connect pin 9 to V_{CC} and put a timing capacitor between pins 10 and 11. An *external* timing resistor can be used instead by placing it between pin 11 and V_{CC} and putting the timing capacitor between pins 10 and 11. If the external timing resistor is used, pin 9 must be left open. The allowable range of R_{ext} is 1.4 to 40 kΩ, and C_{ext} is 0 to 1000 μF. If an electrolytic capacitor is used, its positive side must be connected to pin 11.

The formula that the IC manufacturer gives for determining the output pulse width is

$$t_w = R_{ext}C_{ext} \ln 2 \qquad \textbf{14–3}$$

(Substitute 2 kΩ for R_{ext} if the internal timing resistor is used.)

For example, if the external timing RC components are 10 kΩ and 0.047 μF, then t_w will equal 10 kΩ $\times$ 0.047 μF $\times$ ln 2, which works out to be 326 μs. Using the maximum allowed values of R_{ext} C_{ext}, the maximum pulse width is almost 28 s (40 kΩ $\times$ 1000 μF $\times$ ln 2).

The function table in Figure 14–12 shows that the Q output is LOW and the $\overline{Q}$ output is HIGH as long as the $\overline{A_1}$, $\overline{A_2}$, B inputs do not provide a HIGH–HIGH to the Schmitt-AND inputs. But, by holding B HIGH and applying a HIGH-to-LOW edge to $\overline{A_1}$ or $\overline{A_2}$, the outputs will produce a pulse. Also, the function table shows, in its last two entries, that a LOW-to-HIGH edge at input B will produce an output pulse as long as either $\overline{A_1}$ or $\overline{A_2}$ is held LOW.

The following examples illustrate the use of the 74121 for one-shot operation.

Common Misconception

Students sometimes mistakenly sketch the output at Q as a negative-going pulse.

═══ **EXAMPLE 14–7** ═══

Design a circuit using a 74121 to convert a 50-kHz, 80% duty cycle square wave to a 50-kHz, 50% duty cycle square wave. (In other words, stretch the negative-going pulse to cover 50% of the total period.)

Solution: First, let's draw the original square wave [Figure 14–13(a)] to see what we have to work with ($t = 1/50$ kHz $= 20$ μs, $t_{HI} = 80\% \times 20$ μs $= 16$ μs).

Now, we want to stretch the 4-μs negative pulse out to 10 μs to make the duty cycle 50%. If we use the HIGH-to-LOW edge on the negative pulse to trigger the $\overline{A_1}$ input to a 74121 and set the output pulse width (t_w) to 10 μs, we should have the solution. The output will be taken from $\overline{Q}$ because it provides a negative pulse when triggered.

Using the formula given in the IC specifications, we can calculate an appropriate R_{ext}, C_{ext} to yield 10 μs.

$$t_w = R_{ext}C_{ext} \ln (2)$$
$$10\ \mu s = R_{ext}C_{ext} (0.693)$$
$$R_{ext}C_{ext} = 14.4\ \mu s$$

Pick $C_{ext} = 0.001$ μF (1000 pF); then,

$$R_{ext} = \frac{14.4\ \mu s}{0.001\ \mu F}$$

$$= 14.4\ k\Omega\ \text{(Use a 10-k}\Omega\text{ fixed resistor with a 5-k}\Omega\text{ potentiometer.}$$

The value 0.001 μF is a good choice for C_{ext} because it is much larger than any stray capacitance that might be encountered in a typical circuit. Values of capacitance less than 100 pF (0.0001 μF) may be unsuitable because 50 pF of stray capacitance between traces is not uncommon in a printed-circuit board. Also, resistances in the kilohm range are a good choice because they are big enough to limit current flow, but not so big as to be susceptible to electrostatic noise. The final circuit design and waveforms are given in Figure 14–13(b) and (c).

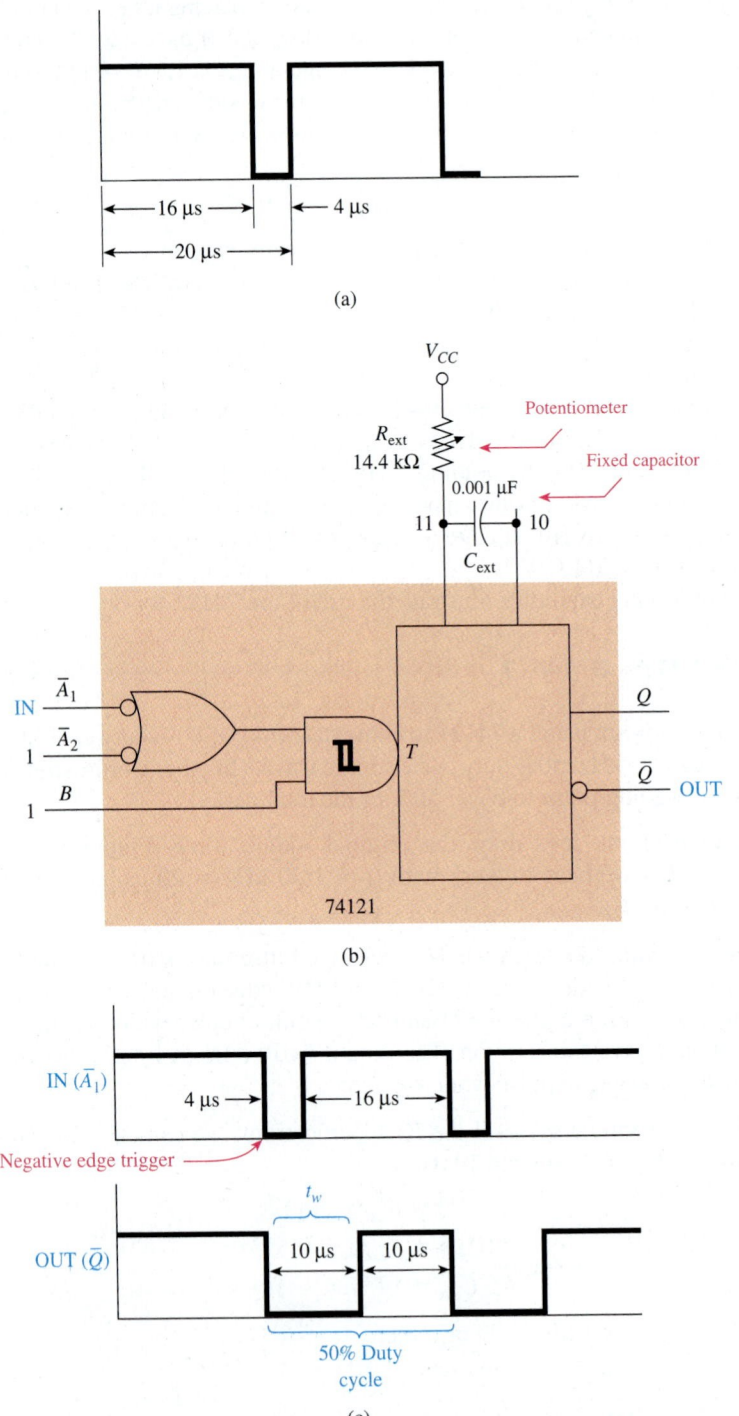

Figure 14–13 (a) Original square wave for Example 14–7; (b) monostable multivibrator circuit connections; (c) input/output waveforms.

EXAMPLE 14–8

In microprocessor systems, most control signals are active-LOW, and often one shots are required to introduce delays for certain devices to wait for other, slower devices to respond. For example, to read from a memory device, a line called $\overline{\text{READ}}$ goes LOW to enable the memory device. Most systems have to introduce a delay after the memory device is enabled (to allow for internal propagation delays) before the microprocessor actually reads the data. Design a system using two 74121s to output a 200-ns LOW pulse (called Data-Ready) 500 ns after the $\overline{\text{READ}}$ line goes LOW.

Solution: The first 74121 will be used to produce the 500-ns delay pulse as soon as the $\overline{\text{READ}}$ line goes LOW (see Figure 14–14). The second 74121 will be triggered by the end of the 500-ns delay pulse and will output its own 200-ns LOW pulse for the Data-Ready line. (The 74121s are edge triggered, so they will trigger only on a HIGH-to-LOW or LOW-to-HIGH *edge*.)

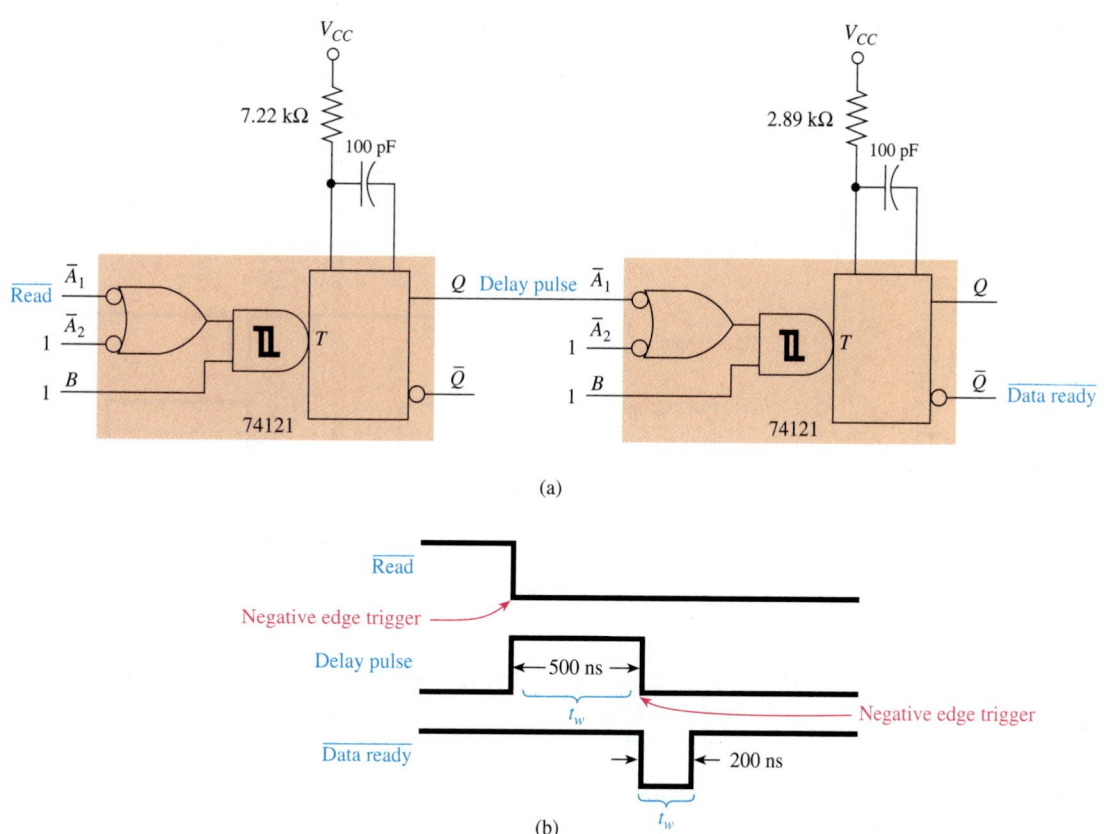

(a)

(b)

Figure 14–14 Solution to Example 14–8: (a) circuit connections for creating a delayed pulse; (b) input/output waveforms.

For $t_w = 500$ ns (output for first 74121):

$$t_w = R_{\text{ext}}C_{\text{ext}} \ln (2)$$
$$500 \text{ ns} = R_{\text{ext}}C_{\text{ext}} (0.693)$$
$$R_{\text{ext}}C_{\text{ext}} = 0.722 \ \mu\text{s}$$

Pick $C_{\text{ext}} = 100$ pF; then,

$$R_{ext} = \frac{07.22\ \mu s}{0.0001\ \mu F} = 7.22\ k\Omega$$

For $t_w = 200$ ns (output for second 74121):

$$t_w = R_{ext}C_{ext} \ln(2)$$
$$200\ \text{ns} = R_{ext}C_{ext}\ (0.693)$$
$$R_{ext}C_{ext} = 0.289\ \mu s$$

Pick $C_{ext} = 100$ pF; then,

$$R_{ext} = 2.89\ k\Omega$$

14–6 Retriggerable Monostable Multivibrators

Have you wondered what might happen if a second input trigger came in before the end of the multivibrator's timing cycle? With the 74121 (which is nonretriggerable), any triggers that come in before the end of the timing cycle are ignored.

Retriggerable monostable multivibrators (such as the 74123) are available, which will start a new timing cycle each time a new trigger is applied. Figure 14–15 illustrates the differences between the retriggerable and nonretriggerable types, assuming that $t_w = 500$ ns and a negative edge-triggered input.

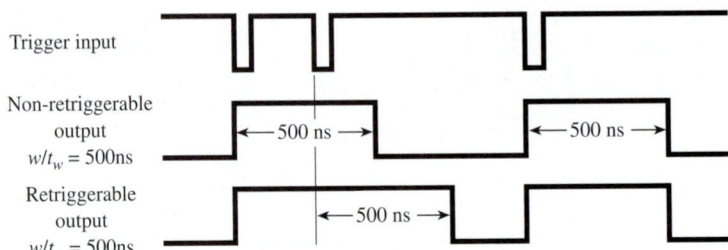

Figure 14–15   Comparison of retriggerable and nonretriggerable one-shot outputs.

As you can see in Figure 14–15, the retriggerable device starts its timing cycle all over again when the second (or subsequent) input trigger is applied. The nonretriggerable device ignores any additional triggers until it has completed its 500-ns timing pulse.

The logic symbol and function table for the 74123 retriggerable monostable multivibrator are given in Figure 14–16.

Besides being retriggerable, some of the other important differences of the 74123 are as follows:

1. It is a *dual* multivibrator (two multivibrators in a single IC package).

2. It has an active-LOW Reset (R_D), which terminates all timing functions by forcing Q LOW, $\overline{Q}$ HIGH.

3. It has no internal timing resistor.

4. It uses a different method for determining the output pulse width, as explained next.

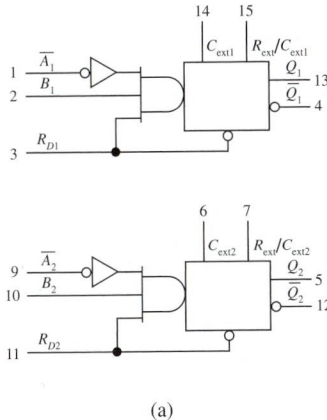

Input			Output	
R_D	$\overline{A}$	B	Q	$\overline{Q}$
L	X	X	L	H
X	H	X	L	H
X	X	L	L	H
H	L	↑	⊓	⊔
H	↓	H	⊓	⊔
↑	L	H	⊓	⊔

H = HIGH voltage level
L = LOW voltage level
X = Don't care
↑ = LOW-to-HIGH transition
↓ = HIGH-to-LOW transition
⊓ = One HIGH-level pulse
⊔ = One LOW-level pulse

(b)

Figure 14–16 The 74123 retriggerable monostable multivibrator: (a) logic symbol; (b) function table.

Output Pulse Width of the 74123

If $C_{ext} > 1000$ pF, the output pulse width is determined by the formula

$$t_w = 0.28 R_{ext} C_{ext}\left(1 + \frac{700}{R_{ext}}\right) \qquad\qquad \textbf{14–4}$$

If $C_{ext} \leq 1000$ pF, the timing chart shown in Figure 14–17 must be used to find t_w.

For example, let's say that we need an output pulse width of 200 ns. Using the chart in Figure 14–17, one choice for the timing components would be $R_{ext} = 10$ kΩ, $C_{ext} = 30$ pF, or a better choice might be $R_{ext} = 5$ kΩ, $C_{ext} = 90$ pF. (With such small capacitances, required for nanosecond delays, we must be careful to minimize stray capacitance by using proper printed-circuit layout and component placement techniques.)

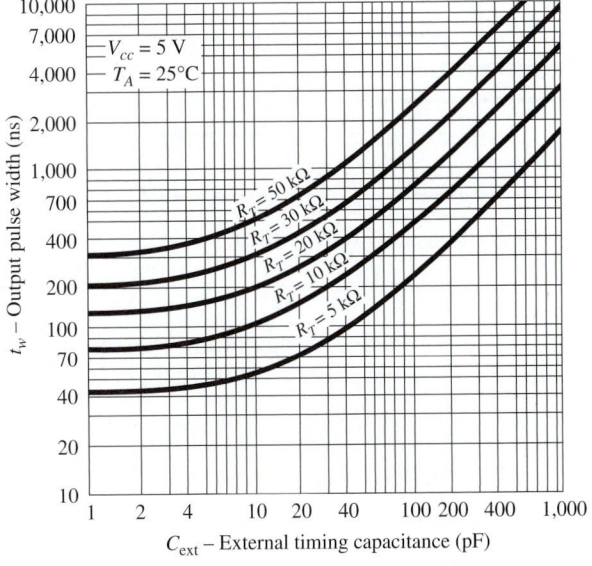

Figure 14–17 The 74123 timing chart for determining t_w when $C_{ext} \leq 1000$ pF.

EXAMPLE 14–9

To the multivibrator circuit of Figure 14–18, apply the trigger input waveforms shown in Figure 14–19. Determine the output at Q_1.

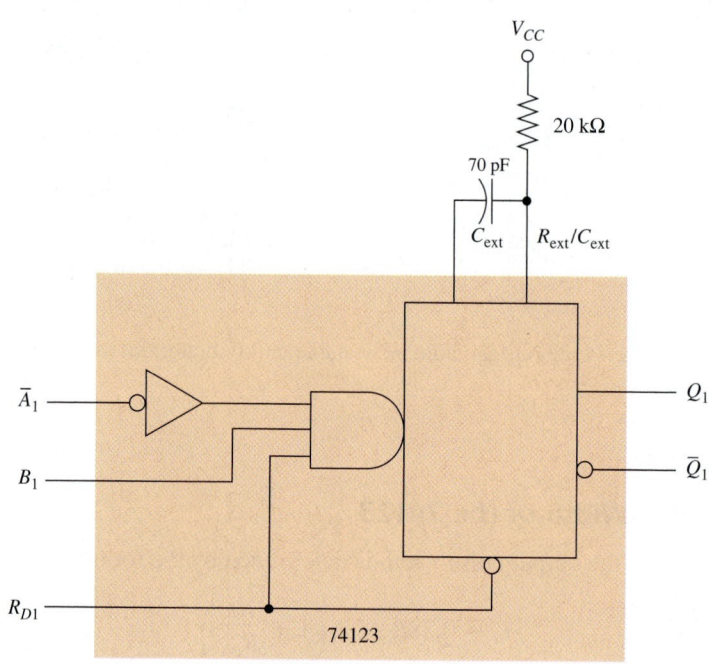

Figure 14–18   The 74123 circuit.

Team Discussion

The R_{D1} input serves three functions between 3.8 and 4.6. What are they?

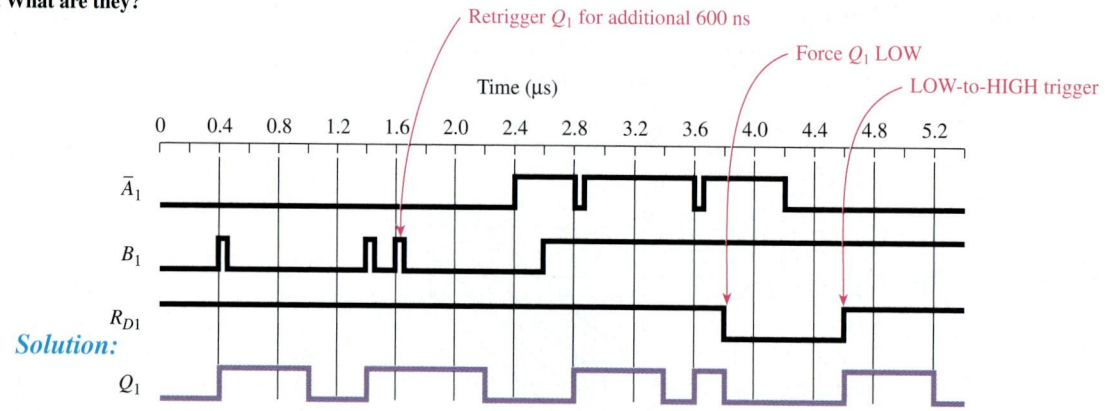

Figure 14–19   Waveforms for Example 14–9.

Solution:

Explanation: t_w is determined from the timing chart for the 74123 (Figure 14–17). With $C_{ext} = 70$ pF and $R_{ext} = 20$ kΩ, $t_w = 0.6$ μs (600 ns).

0 to 0.4 μs	Q is in its stable state (LOW); no trigger has been applied.
0.4 to 1.0 μs	Q is triggered HIGH for 0.6 μs by the pulse on B_1.
1.0 to 1.4 μs	Q returns to its stable state.
1.4 to 2.2 μs	Q is triggered HIGH by B_1 at 1.4 μs; then, Q is retriggered at 1.6 μs.

2.2 to 2.8 μs	Q returns LOW; the conditions on $A_1B_1R_{D1}$ are not right to create a trigger.
2.8 to 3.4 μs	Q is triggered HIGH by the LOW pulse on $\overline{A}_1$ while B_1 = HIGH and R_{D1} = HIGH.
3.4 to 3.6 μs	Q returns LOW.
3.6 to 3.8 μs	Q is triggered HIGH by $\overline{A}_1$, but the output pulse is terminated by a LOW on R_{D1}.
3.8 to 4.6 μs	Q is held LOW by R_{D1} no matter what the other inputs are doing.
4.6 to 5.2 μs	Q is triggered HIGH by the LOW-to-HIGH edge of R_{D1} while $\overline{A}_1$ = LOW, B_1 = HIGH.
5.2 to 5.4 μs	Q returns LOW.

Review Questions

14–6. The output of a monostable multivibrator has a predictable pulse width based on the width of the input trigger pulse. True or false?

14–7. To trigger a 74121 one-shot IC, A_1 _____ (and, or) A_2 must be made _____ (LOW, HIGH) _____ (and, or) B must be made _____ (HIGH, LOW).

14–8. Which of the three trigger inputs of the 74121 would you use to trigger from a falling edge of a pulse? What would you do with the other two inputs?

14–9. When the 74121 receives a trigger, the Q output goes _____ (HIGH, LOW) for a time duration t_w.

14–10. The 74123 one-shot IC is *retriggerable,* whereas the 74121 is not. What does this statement mean?

14–7 Astable Operation of the 555 IC Timer

The 555 is a very popular, general-purpose timer IC. It can be connected as a one shot or an astable oscillator as well as used for a multitude of custom designs. Figure 14–20 shows a block diagram of the chip with its internal components and the external components that are required to set it up as an astable oscillator.

The 555 got its name from the three 5-kΩ resistors. They are set up as a voltage divider from V_{CC} to ground. The top of the lower 5 kΩ is at $\frac{1}{3}V_{CC}$, and the top of the middle 5 k is at $\frac{2}{3}V_{CC}$. For example, if V_{CC} is 6 V, each resistor will drop 2 V.

The triangle-shaped symbols represent **comparators.** A comparator simply outputs a HIGH or LOW based on a comparison of the analog voltage levels at its input. If the + input is *more positive* than the − input, it outputs a HIGH. If the + input is *less positive* than the − input, it outputs a LOW.

The *S-R flip-flop* is driven by the two comparators. It has an active-LOW Reset, and its output is taken from the $\overline{Q}$.

The *discharge transistor* is an NPN, which is used to short pins 7 to 1 when $\overline{Q}$ is HIGH.

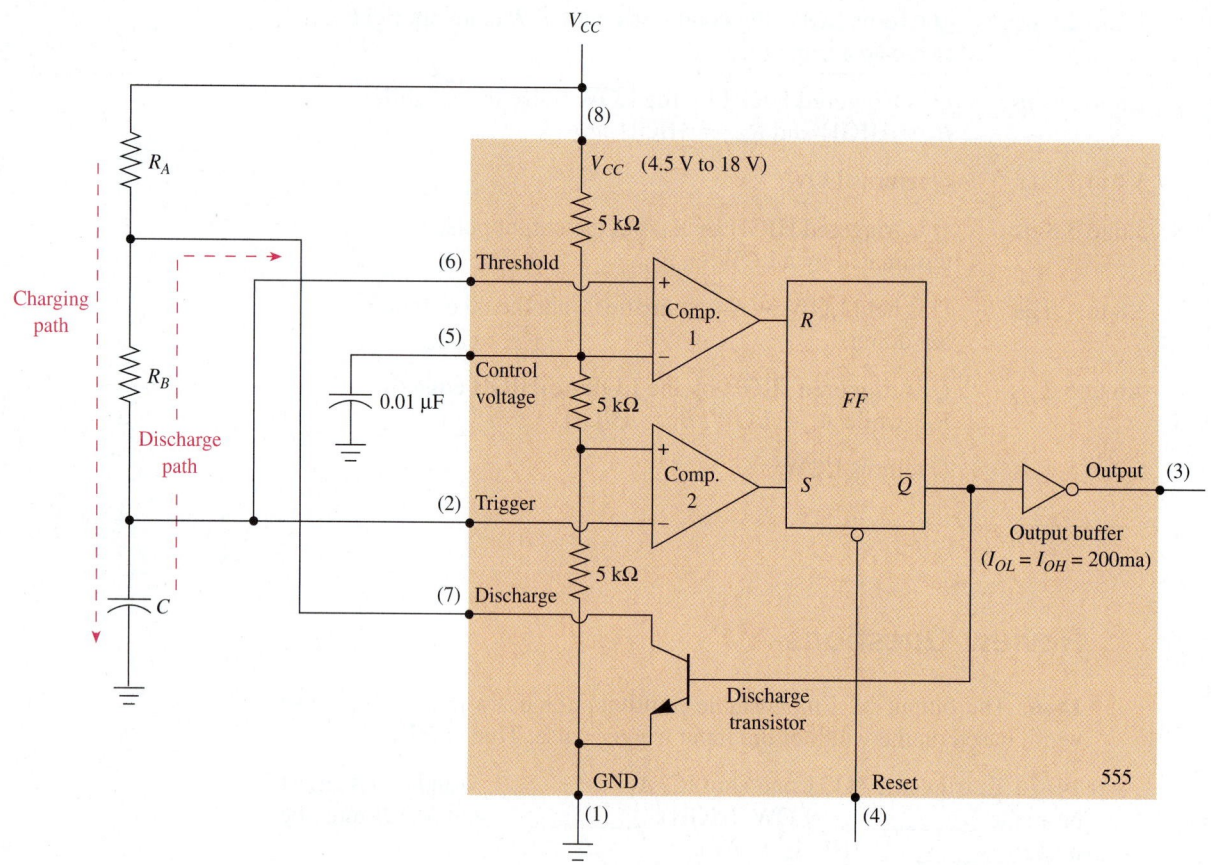

Figure 14–20 Simplified block diagram of a 555 timer with the external timer components to form an astable multivibrator.

The operation and function of the 555 pins are as follows:

Pin 1 (ground): System ground.

Pin 2 (trigger): Input to the lower comparator, which is used to Set the flip-flop. When the voltage at pin 2 crosses from above to below $\frac{1}{3}V_{CC}$, the comparator switches to a HIGH, setting the flip-flop.

Pin 3 (output): The output of the 555 is driven by an inverting buffer capable of sinking or sourcing 200 mA. The output voltage levels are dependent on the output current but are approximately $V_{OH} = V_{CC} - 1.5$ V and $V_{OL} = 0.1$ V.

Pin 4 (Reset): Active-LOW Reset, which forces $\overline{Q}$ HIGH and pin 3 (output) LOW.

Pin 5 (control): Used to override the $\frac{2}{3}V_{CC}$ level, if required. Usually, it is connected to a grounded 0.01-μF capacitor to bypass noise on the V_{CC} line.

Pin 6 (threshold): Input to the upper comparator, which is used to Reset the flip-flop. When the voltage at pin 6 crosses from below to above $\frac{2}{3}V_{CC}$, the comparator switches to a HIGH, resetting the flip-flop.

Pin 7 (discharge): Connected to the open collector of the *NPN* transistor. It is used to short pin 7 to ground when $\overline{Q}$ is HIGH (pin 3 LOW), which will discharge the external capacitor.

Pin 8 (V_{CC}): Supply voltage. V_{CC} can range from 4.5 to 18 V.

The operation of the 555 connected in the astable mode shown in Figure 14–20 is explained as follows:

1. When power is first turned on, the capacitor is discharged, which places 0 V at pin 2, forcing the lower comparator HIGH. This sets the flip-flop ($\overline{Q}$ = LOW, output = HIGH).

2. With the output HIGH ($\overline{Q}$ LOW), the discharge transistor is open, which allows the capacitor to charge toward V_{CC} via $R_A + R_B$.

3. When the capacitor voltage exceeds $\frac{1}{3}V_{CC}$, the lower comparator goes LOW, which has no effect on the *S-R* flip-flop, but when the capacitor voltage exceeds $\frac{2}{3}V_{CC}$, the upper comparator goes HIGH, resetting the flip-flop and forcing $\overline{Q}$ HIGH and the output LOW.

4. With $\overline{Q}$ HIGH, the transistor shorts pin 7 to ground, which discharges the capacitor via R_B.

5. When the capacitor voltage drops below $\frac{1}{3}V_{CC}$, the lower comparator goes back HIGH again, setting the flip-flop and making $\overline{Q}$ LOW, output HIGH.

6. Now, with $\overline{Q}$ LOW, the transistor opens again, allowing the capacitor to start charging up again.

7. The cycle repeats, with the capacitor charging up to $\frac{2}{3}V_{CC}$ and then discharging down to $\frac{1}{3}V_{CC}$, continuously. While the capacitor is charging, the output is HIGH, and when the capacitor is discharging, the output is LOW.

The theoretical waveforms depicting the operation of the 555 as an astable oscillator are shown in Figure 14–21(a). Figure 14–21(b) shows the actual circuit and oscilloscope waveforms at V_C and V_{out}.

The formulas for the time durations t_{LO} and t_{HI} can be derived using the theory presented in Section 14–2 and Equation 14–2. The time duration t_{LO} is determined by realizing that the capacitor voltage (Δv) travels a distance of $\frac{1}{3}V_{CC}$ (from $\frac{2}{3}V_{CC}$ to $\frac{1}{3}V_{CC}$) and that the total path it is trying to travel (E) is equal to $\frac{2}{3}V_{CC}$ (from $\frac{2}{3}V_{CC}$ to 0 V). The path of the discharge current is through R_B and C, so the time constant, τ (tau), is $R_B \times C$. Therefore, the equation for t_{LO} is derived as follows:

$$t = RC \ln\left(\frac{1}{1 - \Delta v/E}\right) \qquad \textbf{14–2}$$

$$t_{LO} = R_B C \ln\left(\frac{1}{1 - \frac{1}{3}V_{CC}/\frac{2}{3}V_{CC}}\right)$$

$$t_{LO} = R_B C \ln\left(\frac{1}{1 - 0.5}\right)$$

$$t_{LO} = R_B C \ln(2)$$

$$t_{LO} = 0.693 R_B C \qquad \textbf{14–5}$$

To derive the equation for t_{HI}:

Δv = distance the capacitor voltage travels = $\frac{1}{3}V_{CC}$ (i.e., $\frac{2}{3}V_{CC} - \frac{1}{3}V_{CC}$)

E = total distance that the capacitor voltage is trying to travels = $\frac{2}{3}V_{CC}$ (i.e., $V_{CC} - \frac{1}{3}V_{CC}$)

τ = **time constant,** or the path that the charging current flows through = $(R_A + R_B) \times C$

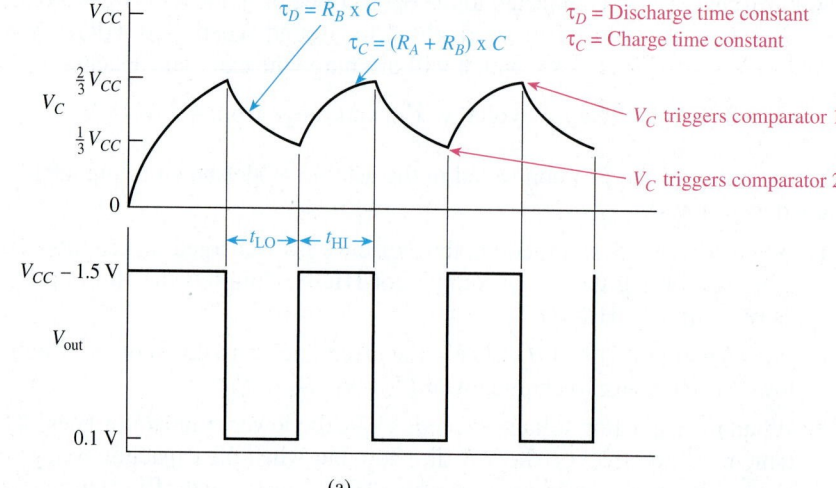

Team Discussion

Why can't the duty cycle of V_{out} ever be less than 50% in this configuration?

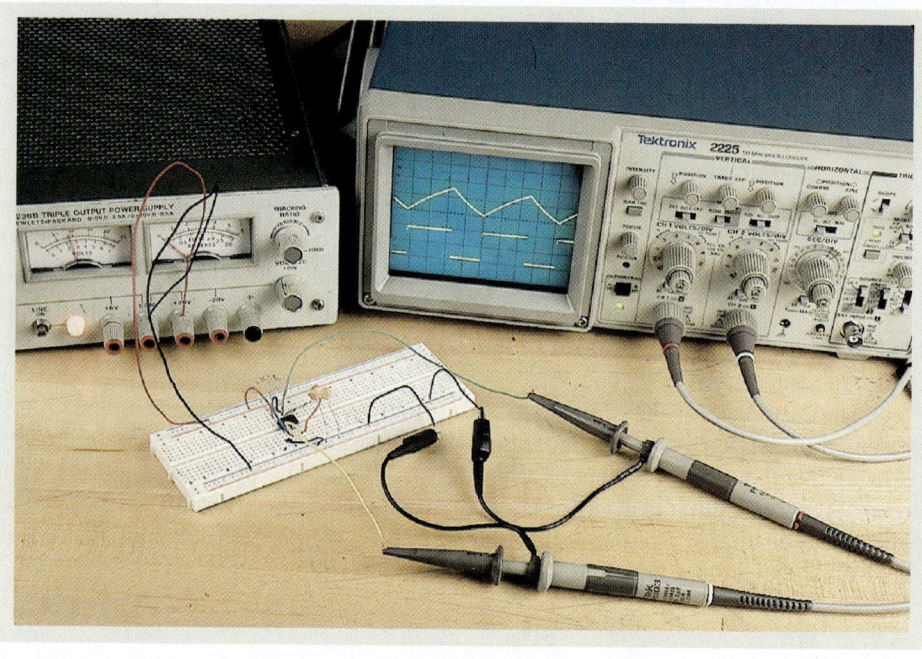

(b)

Figure 14–21 The 555 astable multivibrator: (a) theoretical V_C and V_{out} versus time waveforms; (b) actual breadboarded circuit, power supply, and oscilloscope displaying the measured V_C and V_{out} waveforms.

$$t_{HI} = (R_A + R_B)C \ln\left(\frac{1}{1 - \frac{1}{3}V_{CC}/\frac{2}{3}V_{CC}}\right)$$

$$t_{HI} = (R_A + R_B)C \ln\left(\frac{1}{1 - 0.5}\right)$$

$$t_{HI} = (R_A + R_B)C \ln(2)$$

$$t_{HI} = 0.693(R_A + R_B)C \qquad 14\text{–}6$$

EXAMPLE 14–10

Determine t_{HI}, t_{LO}, duty cycle, and frequency for the 555 astable multivibrator circuit of Figure 14–22.

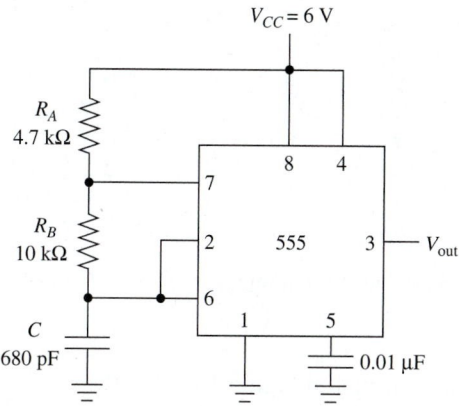

Figure 14–22 The 555 astable connections for Example 14–10.

Team Discussion

If you want to increase the frequency of this oscillator without affecting the duty cycle, do both resistors need adjusting?

Solution:

$$t_{LO} = 0.693 R_B C$$
$$= 0.693(10\ k\Omega)680\ pF$$
$$= 4.71\ \mu s$$
$$t_{HI} = 0.693(R_A + R_B)C$$
$$= 0.693(4.7\ k\Omega + 10\ k\Omega)680\ pF$$
$$= 6.93\ \mu s$$
$$\text{duty cycle} = \frac{t_{HI}}{t_{HI} + t_{LO}}$$
$$= \frac{6.93\ \mu s}{6.93\ \mu s + 4.71\ \mu s}$$
$$= 59.5\%$$
$$\text{frequency} = \frac{1}{t_{HI} + t_{LO}}$$
$$= \frac{1}{6.93\ \mu s + 4.71\ \mu s}$$
$$= 85.9\ kHz$$

50% Duty Cycle Astable Oscillator

By studying Figure 14–21, you should realize that to get a 50% duty cycle t_{LO} must equal t_{HI}. You should also realize that for this to occur, the capacitor charging time constant (τ) must equal the discharging time constant. But, with the astable circuits that we have seen so far, this can never be true because the resistance in one case is just R_B but in the other case is R_B *plus* R_A. You cannot just make R_A equal to 0 Ω because that would put V_{CC} directly on pin 7.

However, if we make $R_A = R_B$ and short R_B with a diode during the capacitor charging cycle, we can achieve a 50% duty cycle. The circuit for a 50% duty cycle is shown in Figure 14–23.

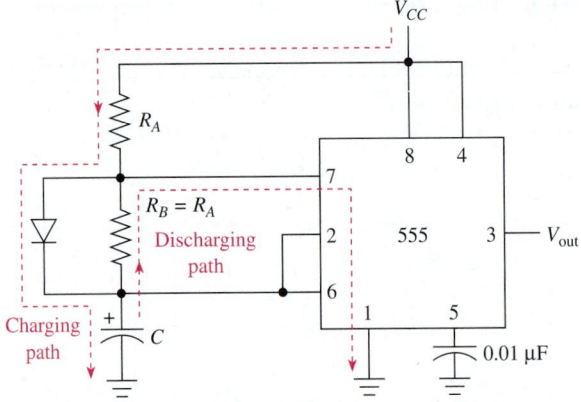

Figure 14–23 The 555 astable multivibrator set up for a 50% duty cycle.

The charging time constant in Figure 14–23 is $R_A \times C$, and the discharging time constant is $R_B \times C$. The formulas, therefore, become

$$t_{HI} = 0.693 R_A C \qquad \text{14–7}$$

$$t_{LO} = 0.693 R_B C \qquad \text{14–8}$$

If R_B is set equal to R_A, t_{HI} will equal t_{LO}, and the duty cycle will be 50%. Also, duty cycles of *less than* 50% can be achieved by using the diode and making R_A less than R_B.

14–8 Monostable Operation of the 555 IC Timer

Another common use for the 555 is as a monostable multivibrator, as shown in Figure 14–24. The one shot of Figure 14–24 operates as follows:

1. Initially (before the trigger is applied), V_{out} is LOW, shorting pin 7 to ground and discharging C.

2. Pin 2 is normally held HIGH by the 10-kΩ pull-up resistor. To trigger the one shot, a negative-going pulse (less than $\frac{1}{3} V_{CC}$) is applied to pin 2.

3. The trigger forces the lower comparator HIGH (see Figure 14–20), which sets the flip-flop, making V_{out} HIGH and opening the discharge transistor (pin 7).

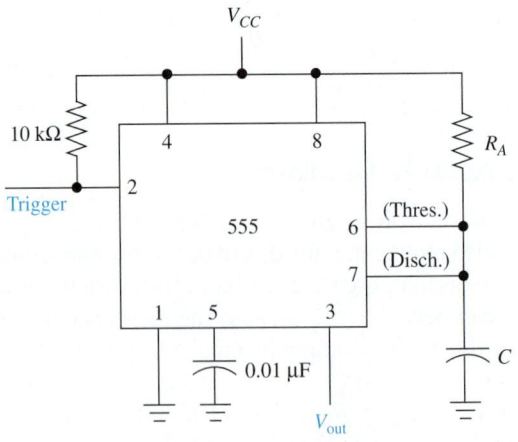

Figure 14–24 The 555 connections for one-shot operation.

CHAPTER 14 | MULTIVIBRATORS AND THE 555 TIMER

4. Now the capacitor is free to charge from 0 V up toward V_{CC} via R_A.

5. When V_C crosses the threshold of $\frac{2}{3}V_{CC}$, the upper comparator goes HIGH, resetting the flip-flop, making V_{out} LOW, and shorting the discharge transistor.

6. The capacitor discharges rapidly to 0 V, and the one shot is held in its stable state (V_{out} = LOW) until another trigger is applied.

The waveforms that are generated for the one-shot operation are shown in Figure 14–25.

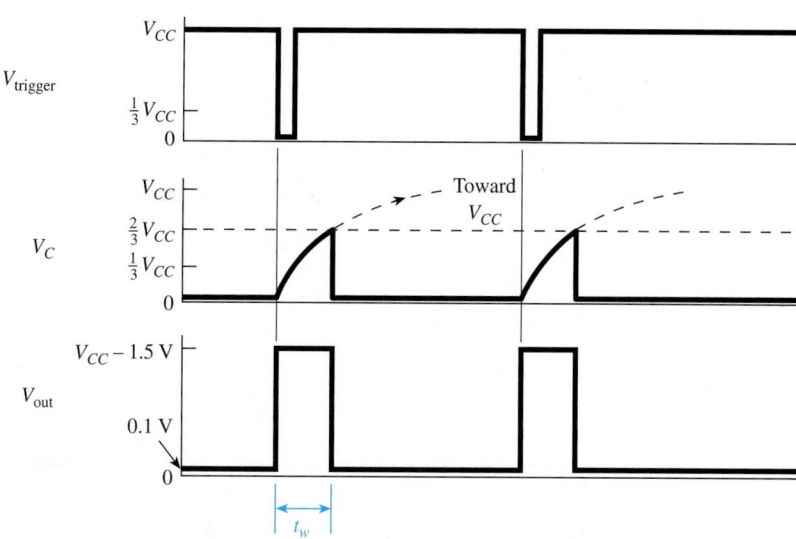

Figure 14–25 Waveforms for the one-shot circuit of Figure 14–24.

To derive the equation for t_w of the one shot, we start with the same capacitor-charging formula (Equation 14–2):

$$t = RC \ln\left(\frac{1}{1 - \Delta v/E}\right) \qquad \text{14–2}$$

where $R = R_A$
Δv = distance that the capacitor voltage travels = $\frac{2}{3}V_{CC}$ (from 0 V up to $\frac{2}{3}V_{CC}$)
E = distance that the capacitor voltage is trying to travel = V_{CC} (from 0 V up to V_{CC})

Substitution yields

$$t_w = R_A C \ln\left(\frac{1}{1 - \frac{2}{3}V_{CC}/V_{CC}}\right)$$

$$= R_A C \ln\left(\frac{1}{1 - 0.667}\right)$$

$$= R_A C \ln(3)$$

$$= 1.10 R_A C \qquad \text{14–9}$$

EXAMPLE 14–11

Design a circuit using a 555 one shot that will stretch a 1-μs negative-going pulse that occurs every 60 μs into a 10-μs negative-going pulse.

Solution: To set the output pulse width to 10 μs:

$$t_w = 1.10 R_A C$$
$$10 \ \mu s = 1.10 R_A C$$
$$R_A C = 9.09 \ \mu s$$

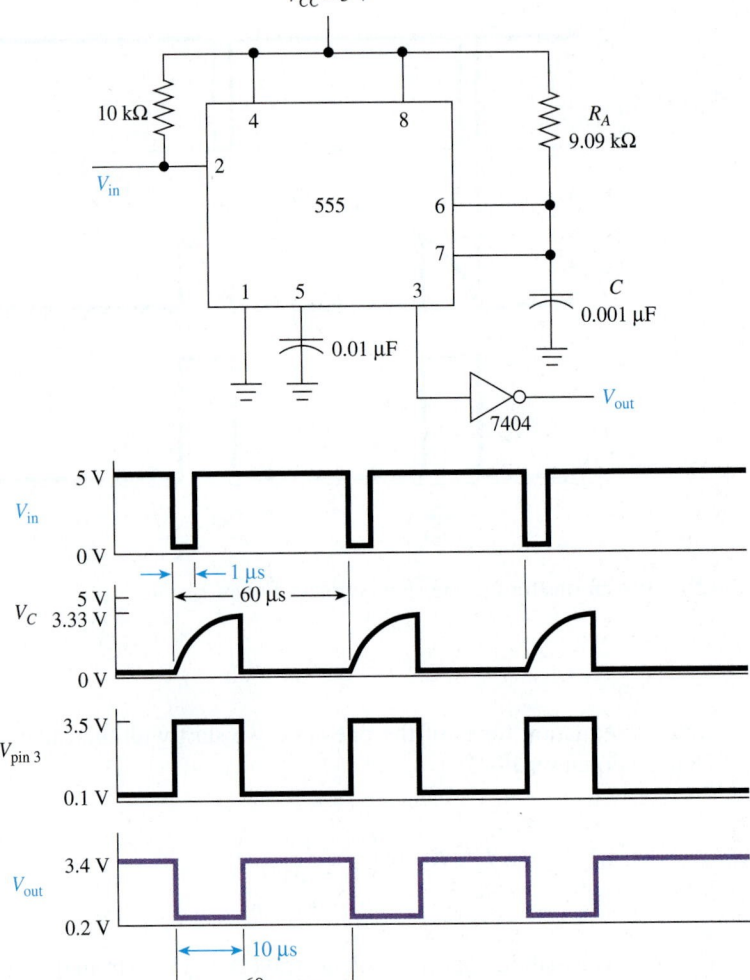

Figure 14–26 Solution to Example 14–11.

Pick $C = 0.001 \ \mu F$; then,

$$R_A = 9.09 \ k\Omega$$

Also, because the 555 outputs a positive-going pulse, an inverter must be added to change it to a negative-going pulse. (7404 inverter: $V_{OH} = 3.4$ V, $V_{OL} = 0.2$ V). The final circuit design and waveforms are shown in Figure 14–26.

14–9   Crystal Oscillators

None of the *RC* oscillators or one shots presented in the previous sections are extremely stable. In fact, the standard procedure for building those timing circuits is to prototype them based on the *R* and *C* values calculated using the formulas and then "tweak" (or make adjustments to) the resistor values while observing the time period on an oscilloscope. Normally, standard values are chosen for the capacitors, and potentiometers are used for the resistors.

However, even after a careful calibration of the time period, changes in the components and IC occur as the devices age and as the ambient temperature varies. To partially overcome this problem, some manufacturers will allow their circuits to **burn in,** or age for several weeks, before the final calibration and shipment.

Instead of using *RC* components, another timing component is available to the design engineer when extremely critical timing is required. This highly stable and accurate timing component is the *quartz* **crystal** (see Figure 14–27). A piece of quartz crystal is cut to a specific size and shape to vibrate at a specific frequency, similar to an *RLC* resonant circuit. Its frequency is typically in the range 10 kHz to 10 MHz. Accuracy of more than *five significant digits* can be easily achieved using this method.

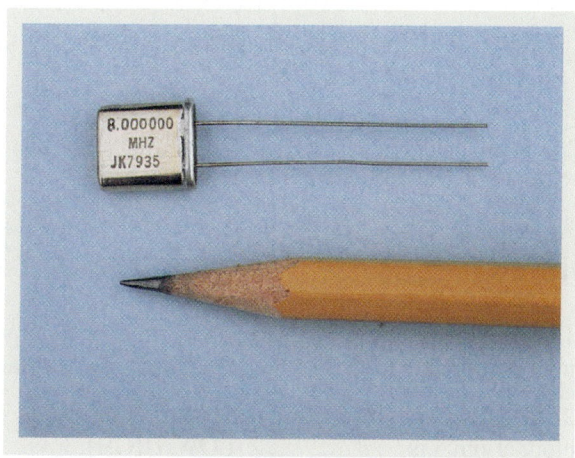

Figure 14–27   Photograph of an 8.000000-MHz quartz crystal.

Crystal **oscillators** are available as an IC package or can be built using an external quartz crystal in circuits such as those shown in Figure 14–28. The circuits shown in the figure will oscillate at a frequency dictated by the crystal chosen. In Figure 14–28(a), the 100-kΩ pot may need adjustment to start oscillation.

The 74S124 TTL chip in Figure 14–28(b) is a **voltage-controlled oscillator,** set up to generate a specific frequency at V_{out}. By changing the crystal to a capacitor, the output frequency will vary, depending on the voltage level at the frequency control (pin 1) and frequency range (pin 14) inputs. Using specifications presented in the manufacturer's data manual, you can determine the output frequency of the VCO based on the voltage level applied to its inputs.

The oscillator in Figure 14–28(c) uses a 2-MHz crystal and a single HCT-Schmitt to create the highly accurate waveform shown.

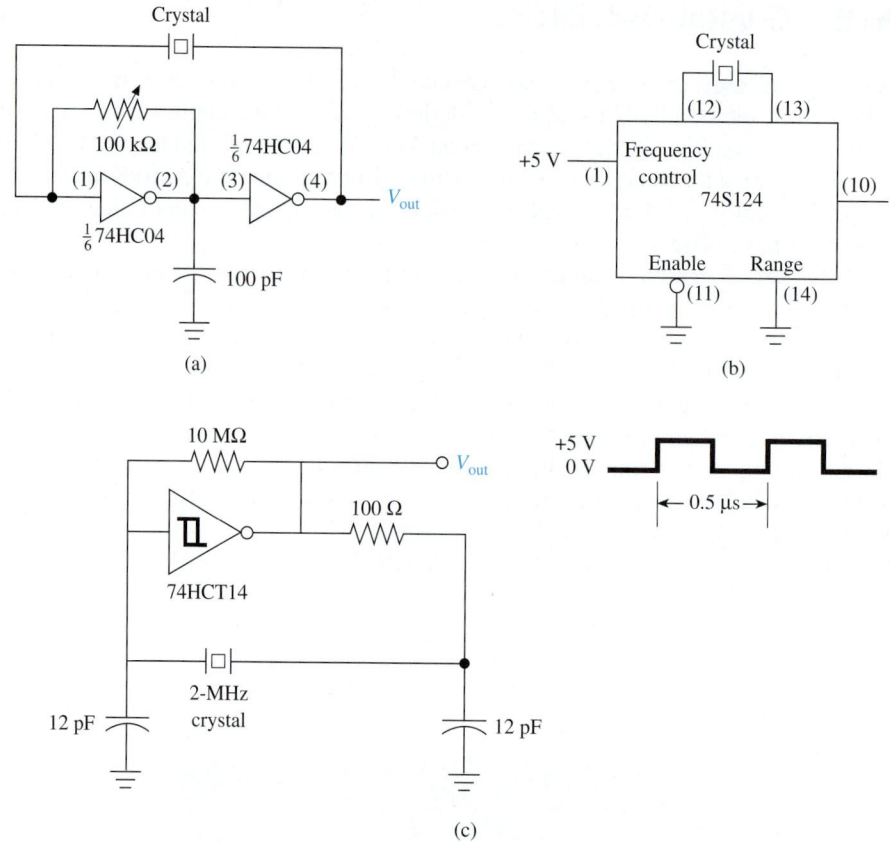

Figure 14–28 Crystal oscillator circuits: (a) high-speed CMOS oscillator; (b) Schottky-TTL oscillator; (c) Schmitt-inverter oscillator.

Review Questions

14–11. The comparators inside the 555 IC timer will output a LOW if their $(-)$ input is more positive than their $(+)$ input. True or false?

14–12. The discharge transistor inside the 555 shorts pin 7 to ground when the output at pin 3 is _____ (LOW, HIGH).

14–13. When pin 6 (Threshold) of the 555 IC exceeds _____ ($\frac{1}{3}V_{CC}$, $\frac{2}{3}V_{CC}$), the flip-flop is _____ (Reset, Set), making the output at pin 3 _____ (LOW, HIGH).

14–14. The 555 is connected as an astable multivibrator in Figure 14–20. V_{out} is HIGH while the capacitor charges through resistor(s) _____, and V_{out} is LOW while the capacitor discharges through resistor(s) _____.

14–15. The 555 astable multivibrator in Figure 14–20 will always have a duty cycle _____ (greater than, less than) 50% because _____.

14–16. When using the 555 as a monostable (one-shot) multivibrator, a _____ (LOW, HIGH) trigger is applied to pin 2, which forces V_{out} _____ (LOW, HIGH) and initiates the capacitor to start _____ (charging, discharging).

14–17. What advantage does a quartz crystal have over an RC circuit when used in timing applications?

Summary

In this chapter, we have learned that

1. Multivibrator circuits are used to produce free-running clock oscillator waveforms or to produce a timed digital level change triggered by an external source.

2. Capacitor voltage charging and discharging rates are the most common way to produce predictable time duration for oscillator and timing operations.

3. An astable multivibrator is a free-running oscillator whose output oscillates between two voltage levels at a rate determined by an attached RC circuit.

4. A monostable multivibrator is used to produce an output pulse that starts when the circuit receives an input trigger and lasts for a length of time dictated by the attached RC circuit.

5. The 74121 is an IC monostable multivibrator with two active-LOW and one active-HIGH input trigger sources and an active-HIGH and an active-LOW pulse output terminal.

6. Retriggerable monostable multivibrators allow multiple input triggers to be acknowledged even if the output pulse from the previous trigger has not expired.

7. The 555 IC is a general-purpose timer that can be used to make astable and monostable multivibrators and to perform any number of other timing functions.

8. Crystal oscillators are much more accurate and stable than RC timing circuits. They are used most often for microprocessor and digital communication timing.

Glossary

Burn In: A step near the end of the production process in which a manufacturer exercises the functions of an electronic circuit and ages the components before the final calibration step.

Comparator: As used in a 555 timer, it compares the analog voltage level at its two inputs and outputs, a HIGH or a LOW, depending on which input was higher. (If the voltage level on the $+$ input is higher than the voltage level on the $-$ input, the output is HIGH; otherwise, it is LOW.)

Crystal: A material, usually made from quartz, that can be cut and shaped to oscillate at a very specific frequency. It is used in highly accurate clock and timing circuits.

Duty Cycle: A ratio of the lengths of time that a digital signal is HIGH versus its total period:

$$\text{duty cycle} = \frac{t_{HI}}{t_{HI} + t_{LO}}$$

Exponential Charge/Discharge: An exponential rate of charge or discharge is nonlinear, meaning that the rate of change of capacitor voltage is greater in the beginning and then slows down toward the end.

Multivibrator: An electronic circuit or IC used in digital electronics to generate HIGH and LOW logic states. The *bistable* multivibrator is an *S-R* flip-flop triggered into its HIGH or LOW state. The *astable* multivibrator is a free-running oscillator that continuously alternates between its HIGH and LOW states. The *monostable* multivibrator is a one shot that, when triggered, outputs a single pulse of a specific time duration.

Oscillator: An electronic circuit whose output continuously alternates between HIGH and LOW states at a specific frequency.

Pulse Stretching: Increasing the time duration of a pulse width.

Retriggerable: A device that is capable of reacting to a second or subsequent trigger before the action initiated by the first trigger is complete.

Time Constant (tau, τ): τ is equal to the product of resistance times capacitance and is used to determine the *rate* of charge or discharge in a series *RC* circuit. (1τ is equal to the number of seconds that it takes for a capacitor's voltage to reach 63% of its final value.)

Voltage-Controlled Oscillator (VCO): An oscillator whose output frequency is dependent on the analog voltage level at its input.

▬▬ Problems ▬▬

Sections 14–1 and 14–2

14–1. Which type of multivibrator is also known as a:

(a) One shot?

(b) *S-R* flip-flop?

(c) Free-running oscillator?

14–2.

(a) For the *RC* circuit of Figure P14–2, determine the voltage on the capacitor 50 μs after the switch is moved from position 2 to position 1. (Assume that $V_C = 0$ V initially.)

(b) Repeat part (a) for 100 μs.

(c) Repeat part (a) for 150 μs.

(d) Sketch and label a graph of capacitor voltage versus time for the values that you found in parts (a), (b), and (c).

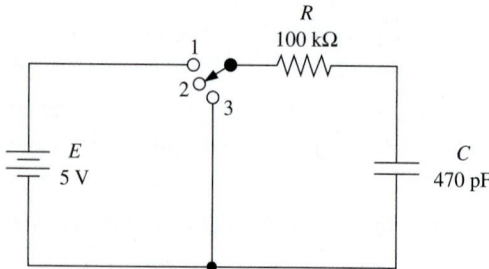

Figure P14–2

14–3. The capacitor in Figure P14–2 is initially discharged. How long after the switch is moved from position 2 to position 1 will it take for the capacitor to reach 4 V?

14–4. Assume that the capacitor in Figure P14–2 is initially charged to 2 V. How long after the switch is moved from position 2 to position 1 will it take for the capacitor voltage to reach 4 V?

14–5. Assume that the capacitor in Figure P14–2 is initially charged to 4 V. How long after the switch is moved from position 2 to position 3 will it take for the voltage to drop to 2 V?

14–6. If you were successful at solving for the time in Problems 14–4 and 14–5, you will notice that it takes longer for the capacitor voltage to go from 2 to 4 V than it did to go from 4 to 2 V. Why is that true?

Section 14–3

14–7. Why is a Schmitt trigger inverter used for the astable multivibrator circuit of Figure 14–5 instead of a regular inverter like a 74HC04?

14–8. In a Schmitt trigger astable multivibrator, if the hysteresis voltage (V_{T^+} minus V_{T^-}) decreases due to a temperature change, what happens to:

(a) The output frequency?

(b) The output voltage?

14–9. Specifications for the 74HC14 Schmitt inverter when powered from a 6-V supply are as follows: $V_{OH} = 6.0$ V, $V_{OL} = 0.0$ V, $V_{T^+} = 3.3$ V, and $V_{T^-} = 2.0$ V.

(a) Sketch and label the waveforms for V_{cap} and V_{out} in the astable multivibrator circuit of Figure 14–5. (Use $R = 68$ kΩ and $C = 0.0047$ μF.)

(b) Calculate t_{HI}, t_{LO}, duty cycle, and frequency.

Sections 14–4 and 14–5

D   **14–10.** Design a monostable multivibrator using two 74HC00 NAND gates similar to Figure 14–9. Determine the values for R and C such that a negative-going, 2-μs input trigger will create a 50-μs positive-going output pulse.

D   **14–11.** Make the external connections to a 74121 monostable multivibrator to convert a 100-kHz, 30% duty cycle square wave to a 100-kHz, 50% duty cycle square wave.

C D   **14–12.** Use two 74121s as a delay line to reproduce the waveforms shown in Figure P14–12. (The output pulse will look just like the input pulse but delayed by 30 μs.)

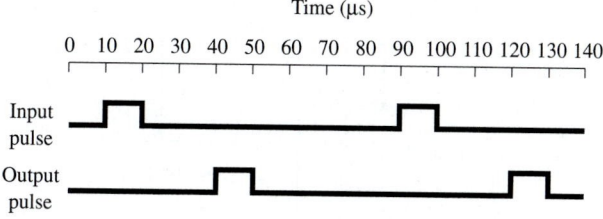

Figure P14–12

Section 14–6

C D **14–13.** The NPRO microprocessor control line is supposed to issue a 10-μs LOW pulse every 150 μs as long as a certain process is running smoothly. Design a "missing pulse detector" using a 74123 that will normally output a HIGH but will output a LOW if a single pulse on the NPRO line is skipped. (*Hint:* The 74123 will be retriggered by NPRO every 150 μs.)

14–14. Using the timing chart in Figure 14–17 for a 74123, determine a good value for R_{ext} and C_{ext} to give an output pulse width of 400 ns.

14–15. Sketch the output waveforms at Q_1 and Q_2 of the multivibrators given in Figure P14–15 if the $\overline{A}_1$ and B waveforms are as shown.

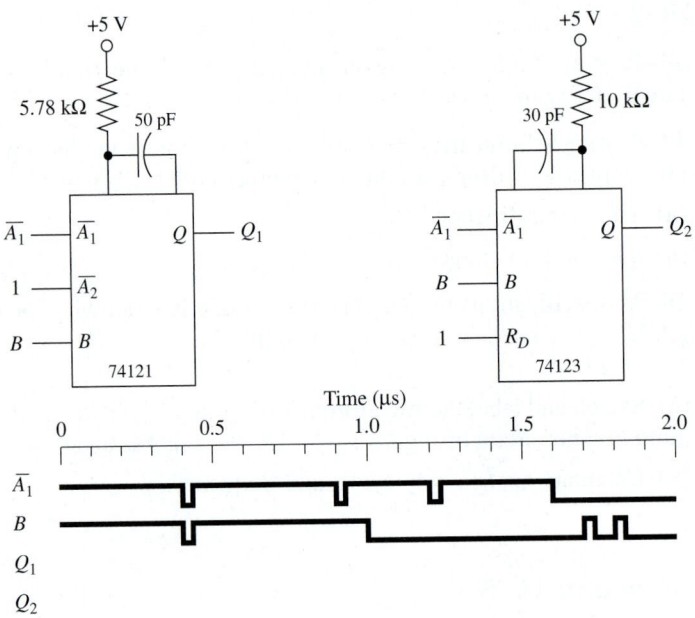

Figure P14–15

Sections 14–7 and 14–8

14–16. Sketch and label the waveforms at V_{out} for the 555 circuit of Figure P14–16 with the potentiometer set at 0 Ω.

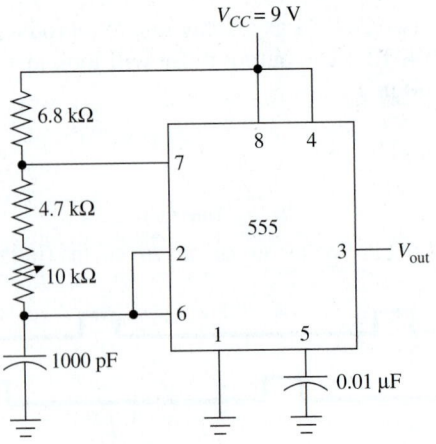

Figure P14–16

674 CHAPTER 14 | MULTIVIBRATORS AND THE 555 TIMER

C   **14–17.** Determine the maximum and minimum frequency and the maximum and minimum duty cycle that can be achieved by adjusting the potentiometer in Figure P14–16.

C   **14–18.** Derive formulas for duty cycle and frequency in terms of R_A, R_B, and C for a 555 astable multivibrator. (Test your formulas by re-solving Problem 14–17.)

C D   **14–19.** Using a 555, design an astable multivibrator that will oscillate at 50 kHz, 60% duty cycle. (So that we all get the same answer, let's pick $C = 0.0022 \ \mu F$.)

C D   **14–20.** Design a circuit that will produce a 100-kHz square wave using:

(a) A 74HC14   (c) A 74123

(b) Two 74121s   (d) A 555

14–21. Sketch and label the waveforms at $V_{trigger}$, V_{cap}, and V_{out} for the 555 one-shot circuit of Figure 14–24. Assume that $V_{trigger}$ is a 5-μs negative-going pulse that occurs every 100 μs and $V_{CC} = 5$ V, $R_A = 47$ kΩ, and $C = 1000$ pF.

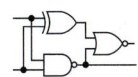

▬▬ Schematic Interpretation Problems ▬▬

See Appendix G for the schematic diagrams.

S D   **14–22.** The 68HC11 microcontroller in the HC11D0 Master Board schematic provides a clock output signal at the pin labeled E. This clock signal is used as the input to the LCD controller, M1 (grid location E-7). The frequency of this signal is 9.8304 MHz, as dictated by the crystal on the 68HC11. To experiment with different clock speeds on the LCD controller, you want to design a variable frequency oscillator that can scan the frequency range of 100 kHz to 1 MHz. Design this oscillator using a 555 that will output its signal to pins 6 and 10 of the LCD controller.

C S D   **14–23.** Design a "missing pulse detector" similar to the one designed in Problem 14–13. It will be used to monitor the DAV input line in the 4096/4196 schematic. Assume that the DAV line is supposed to provide a 2-μs HIGH pulse every 100 μs. Monitor the DAV line with a 74123 monostable multivibrator. Have the 74123 output a HIGH to port 1, bit 7 (P1.7) of the 8031 microcontroller if a missing pulse is detected.

▬▬ Electronics Workbench®/MultiSIM® Exercises ▬▬

E14–1. Load the circuit file for **Section 14–5a.** This monostable multivibrator (one-shot) is used to create a positive output pulse having a width equal to 0.693 × Rext × Cext.

(a) Use your calculator to determine the pulse width of Vout. Use the oscilloscope to measure the actual output pulse width. Tcalc. = _____ Tmeas. = _____.

(b) Measure the duty cycle (DC) of Vtrigger and Vout. DC Vtrig = _____ DC Vout = _____.

(c) Calculate a new value for Rext to make the duty cycle of Vout equal to 50%. Rext = _____. Change Rext in the circuit to your calculated value and measure the new duty cycle to see if you were accurate.

C

E14–2. Load the circuit file for **Section 14–5b.** Use a calculator to predict the pulse width of Vout. Notice that this one shot uses the active-LOW trigger input and provides an active-LOW output pulse. On a piece of graph paper, sketch the expected waveforms at Vtrigger and Vout. (Show all times and voltage levels.)

Connect the oscilloscope to measure the waveforms. What is the measured LOW pulse width of Vtrigger and Vout? Are they close to what you predicted?

C D

E14–3. Load the circuit file for **Section 14–5c.** Design a circuit using a monostable multivibrator that will convert a 10-kHz, 70% duty cycle square wave into a 50% duty cycle square wave. Demonstrate your waveforms to your instructor.

E14–4. Load the circuit file for **Section 14–8a.** This is an exact duplicate of Example 14–10 in the textbook. The calculated Thi = 6.93uS and Tlo = 4.71uS. Use the oscilloscope to compare calculated to measured values.

(a) Measured Thi = _____ Tlo = _____.

(b) Changing Vcc to 9V makes Thi and Tlo increase, decrease, or remain the same?

(c) Changing Ra to 10k makes Tlo increase, decrease, or remain the same?

(d) Changing C to 330pF makes the frequency increase, decrease, or remain the same?

E14–5. Load the circuit file for **Section 14–8b.** On a piece of paper, calculate and sketch the new waveforms for Vc and Vout if Rb is changed to 20k-ohms. Create a data table of calculated versus measured time values for Vout.

C D

E14–6. Load the circuit file for **Section 14–8c.** Design a 555 astable multivibrator that produces a 50-kHz, 60% duty cycle square wave at Vout. Demonstrate your design and waveforms to your instructor.

Answers to Review Questions

14–1. False

14–2. Slower

14–3. True

14–4. V_{T^+} and V_{T^-}, V_{OH} and V_{OL}

14–5. Decrease

14–6. False

14–7. Or, LOW, and, HIGH

14–8. $\overline{A}_1$ or $\overline{A}_2$, HIGH–HIGH

14–9. HIGH

14–10. It means that input triggers are ignored during the timing cycle of a 74121, but not for the 74123. A new timing cycle is started each time a trigger is applied to a 74123.

14–11. True

14–12. LOW

14–13. $\frac{2}{3}V_{CC}$, Reset, LOW

14–14. $R_A + R_B$, R_B

14–15. Greater than; the charging path consists of $R_A + R_B$, whereas the discharging path consists of only R_B.

14–16. LOW, HIGH, charging

14–17. It has greater stability and accuracy than an RC circuit.

Interfacing to the Analog World

OUTLINE

OBJECTIVES

Upon completion of this chapter, you should be able to:

- Perform the basic calculations involved in the analysis of operational amplifier circuits.
- Explain the operation of binary-weighted and $R/2R$ digital-to-analog converters.
- Make the external connections to a digital-to-analog IC to convert a numeric binary string into a proportional analog voltage.
- Discuss the meaning of the specifications for converter ICs as given in a manufacturer's data manual.
- Explain the operation of parallel-encoded, counter-ramp, and successive-approximation analog-to-digital converters.
- Make the external connections to an analog-to-digital converter IC to convert an analog voltage to a corresponding binary string.
- Discuss the operation of a typical data acquisition system.

INTRODUCTION

Most physical quantities that we deal with in this world are *analog* in nature. For example, temperature, pressure, and speed are not simply 1's and 0's but, instead, take on an infinite number of possible values. To be understood by a digital system, these values must be converted into a binary string representing their value; thus, we have the need for *analog-to-digital* conversion. Also, it is important when we need to use a computer to control analog devices to be able to convert from *digital to analog*.

Devices that convert physical quantities into electrical quantities are called **transducers.** Transducers are readily available to convert such quantities as temperature, pressure, velocity, position, and direction into a proportional analog voltage or current. For example, a common transducer for measuring temperature is a thermistor. A **thermistor** is simply a temperature-sensitive resistor. As its temperature changes, so does its resistance. If we send a constant current through the thermistor and then measure the voltage across it, we can determine its resistance *and* temperature.

Team Discussion

What are some scientific and commercial applications for analog-to-digital conversion and digital-to-analog conversion?

15–1   Digital and Analog Representations

For *analog-to-digital* (**A/D**) or *digital-to-analog* (**D/A**) **converters** to be useful, there has to be a meaningful representation of the analog quantity as a digital representation and the digital quantity as an analog representation. If we choose a convenient range of analog levels such as 0 to 15 V, we could easily represent each 1-V step as a unique digital code, as shown in Figure 15–1.

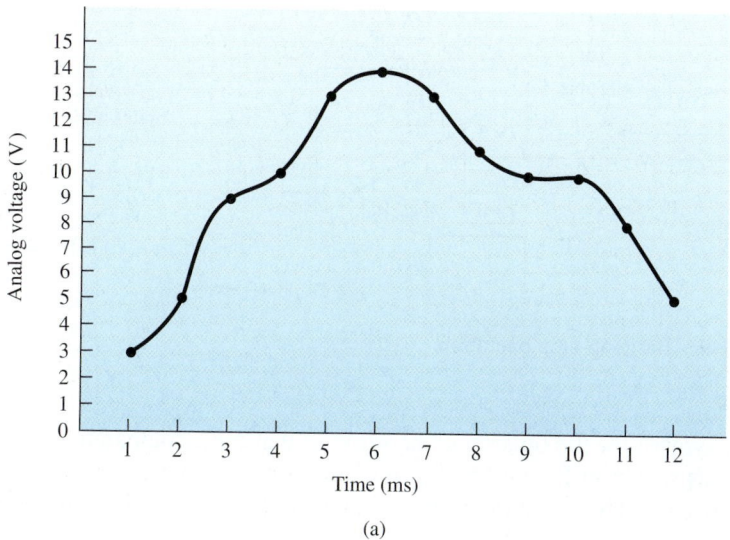

(a)

Time (ms)	Representation	
	Analog	Digital
1	3	0011
2	5	0101
3	9	1001
4	10	1010
5	13	1101
6	14	1110
7	13	1101
8	11	1011
9	10	1010
10	10	1010
11	8	1000
12	5	0101

(b)

Figure 15–1   Analog and digital representations: (a) voltage versus time; (b) representations at 1-ms intervals.

Figure 15–1 shows that for each analog voltage, we can determine an equivalent digital representation. Using four binary positions gives us 4-bit **resolution,** which allows us to develop 16 different representations, with the increment between each being 1 part in 16. If we need to represent more than just 16 different analog levels, we would have to use a digital code with more than four binary positions. For example, a D/A converter with 8-bit resolution will provide increments of 1 part in 256, which provides much more precise representations.

15–2 Operational Amplifier Basics

Most A/D and D/A circuits require the use of an **op amp** for signal conditioning. Three characteristics of op amps make them an almost *ideal amplifier:* (1) very high input impedance, (2) very high voltage gain, and (3) very low output impedance. In this section, we gain a basic understanding of how an op amp works, and in future sections, we see how it is used in the conversion process. A basic op-amp circuit is shown in Figure 15–2.

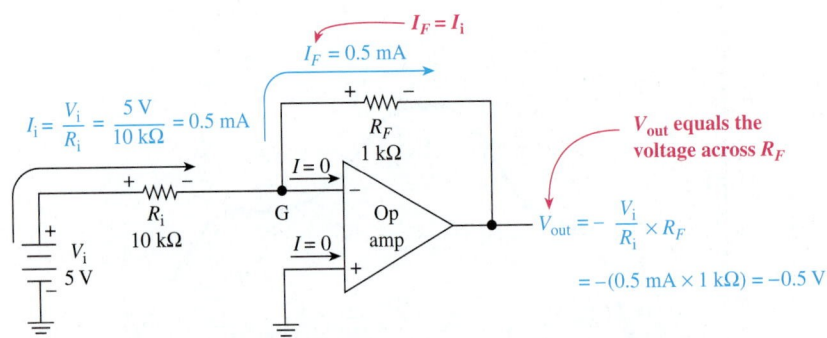

Figure 15–2 Basic op-amp operation.

The symbol for the op amp is the same as that for a comparator, but when it is connected as shown in Figure 15–2, it provides a much different function. The basic theory involved in the operation of the op-amp circuit in Figure 15–2 is as follows:

1. The impedance looking into the (+) and (−) input terminals is assumed to be infinite; therefore, I_{in} (+), (−) = 0 A.

2. Point G is assumed to be at the same potential as the (+) input; therefore, point G is at 0 V, called **virtual ground.** (Virtual means "in effect" but not actual. It is at 0 V, but it cannot sink current.)

3. With point G at 0 V, there will be 5 V across the 10-kΩ resistor, causing 0.5 mA to flow.

4. The 0.5 mA cannot flow into the op amp; therefore, it flows up through the 1-kΩ resistor.

5. Because point G is at virtual ground and V_{out} is measured with respect to ground, V_{out} is equal to the voltage across the 1-kΩ resistor, which is −0.5 V.

EXAMPLE 15–1

Find V_{out} in Figure 15–3.

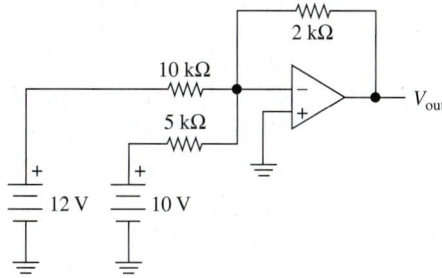

Figure 15–3 Op-amp circuit for Example 15–1.

Solution:

$$I_{10\ k\Omega} = \frac{12\ V}{10\ k\Omega} = 1.2\ mA$$

$$I_{5\ k\Omega} = \frac{10\ V}{5\ k\Omega} = 2\ mA$$

$$I_{2\ k\Omega} = 1.2\ mA + 2\ mA = 3.2\ mA$$

$$V_{out} = -(3.2\ mA \times 2\ k\Omega) = 6.4\ V$$

Review Questions

15–1. Transducers are devices that convert physical quantities like pressure and temperature into electrical quantities. True or false?

15–2. An 8-bit A/D converter is capable of producing how many unique digital output codes?

15–3. The input impedance to an operational amplifier is assumed to be _____. The voltage difference between the (+) input and (−) input is approximately _____ volts.

15–3 Binary-Weighted D/A Converters

A basic D/A converter can be built by expanding on the information presented in Section 15–2. Example 15–1 showed us that the 2-kΩ resistor receives the *sum* of the currents heading toward the op amp from the two input resistors. If we scale the input resistors with a **binary weighting** factor, each input can be made to provide a binary-weighted amount of current, and the output voltage will represent a sum of all the binary-weighted input currents, as shown in Figure 15–4.

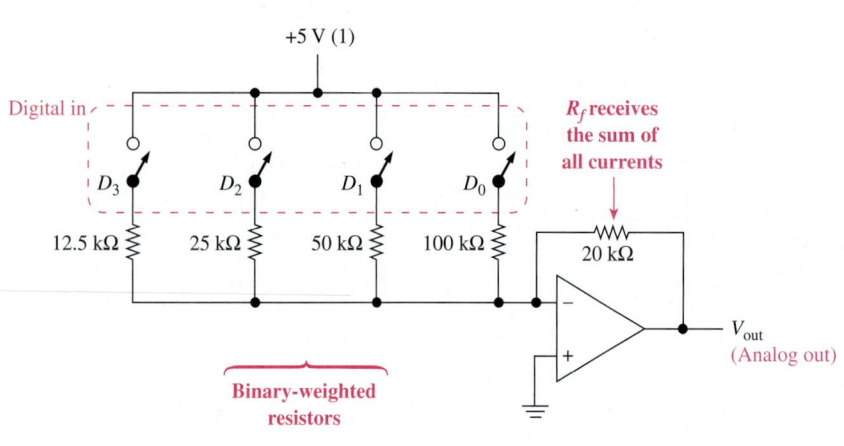

	Digital			Analog
D_3	D_2	D_1	D_0	V_{out} (−V)
0	0	0	0	0
0	0	0	1	1
0	0	1	0	2
0	0	1	1	3
0	1	0	0	4
0	1	0	1	5
0	1	1	0	6
0	1	1	1	7
1	0	0	0	8
1	0	0	1	9
1	0	1	0	10
1	0	1	1	11
1	1	0	0	12
1	1	0	1	13
1	1	1	0	14
1	1	1	1	15

Figure 15–4 Binary-weighted D/A converter.

In Figure 15–4, the 20-kΩ resistor *sums* the currents that are provided by closing any of switches D_0 to D_3. The resistors are scaled in such a way as to provide a binary-weighted amount of current to be summed by the 20-k resistor. Closing D_0 causes 50 μA to flow through the 20 kΩ, creating −1.0 V at V_{out}. Closing each successive switch creates *double* the amount of current of the previous switch. Work through several of the switch combinations presented in Figure 15–4 to prove its operation.

If we were to expand Figure 15–4 to an 8-bit D/A converter, the resistor for D_4 would be one-half of 12.5 kΩ, which is 6.25 kΩ. Each successive resistor is one-half of the previous one. Using this procedure, the resistor for D_7 would be 0.78125 kΩ!

Coming up with accurate resistances over such a large range of values is very difficult. This limits the practical use of this type of D/A converter for any more than 4-bit conversions.

EXAMPLE 15–2

Determine the voltage at V_{out} in Figure 15–4 if the binary equivalent of 10_{10} is input on switches D_3 to D_0.

Solution: $10_{10} = 1010_2$ (switches D_3 and D_1 are closed)

$$I_3 = \frac{5\ V}{12.5\ k\Omega} = 0.4\ mA$$

$$I_1 = \frac{5\ V}{50\ k\Omega} = 0.1\ mA$$

$$V_{out} = -[(0.4\ mA + 0.1\ mA) \times 20\ k\Omega] = -10\ V$$

15–4 *R/2R* Ladder D/A Converters

The method for D/A conversion that is most often used in D/A converters is known as the *R/2R* ladder circuit. In this circuit, only two resistor values are required, which lends itself nicely to the fabrication of ICs with a resolution of 8, 10, or 12 bits, and higher. Figure 15–5 shows a 4-bit D/A *R/2R* converter. To form converters with higher resolu-

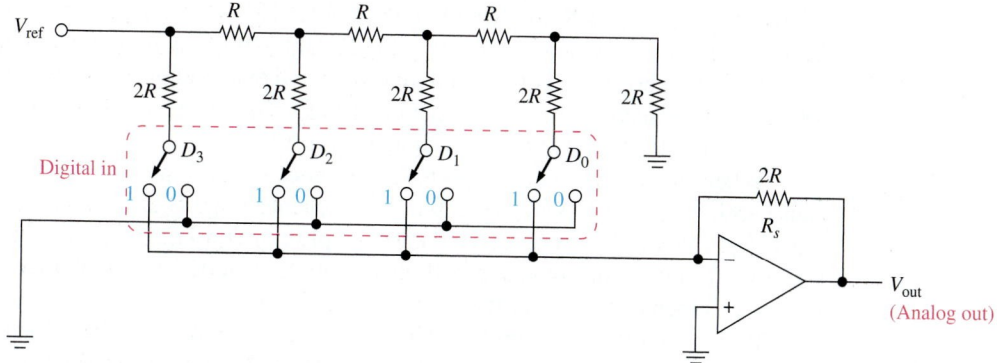

Figure 15–5   The R/2R ladder D/A converter.

tion, all that needs to be done is to add more *R/2R* resistors and switches to the left of D_3. Commercially available D/A converters with resolutions of 8, 10, 12, 14, and 16 bits are commonly made this way.

In Figure 15–5, the 4-bit digital information to be converted to analog is entered on the D_0 to D_3 switches. (In an actual IC, those switches would be transistor switches.) The arrangement of the circuit is such that as the switches are moved to the 1 position, they cause a current to flow through the summing resistor, R_s, that is proportional to their binary equivalent value. (Each successive switch is worth double the previous one.)

This circuit, which is designed in the shape of a ladder, is an ingenious way to form a binary-weighted current-division circuit. Refer to Figure 15–6 to see how we arrive at the current levels and value of V_{out}. First, keep in mind the op-amp rules presented in Section 15–2. In particular, remember that the $(-)$ input to the op amp is at virtual ground, and any current that reaches this point will continue to flow past it, up through the summing resistor (R_s). With this knowledge, we can determine that the resistance of, and the current through, each rung of the ladder is unaffected by the position of any of the data switches (D_3 to D_0). This is because: (1) with a data switch in the 0 position, the bottom of the corresponding 20-kΩ resistor is connected to ground; and (2) when it is in the 1 position, it is connected to virtual ground, which acts the same as ground.

To calculate the current contributed by each rung in Figure 15–6, we must first calculate the total current leaving V_{ref}. By collapsing the circuit from the right, we have

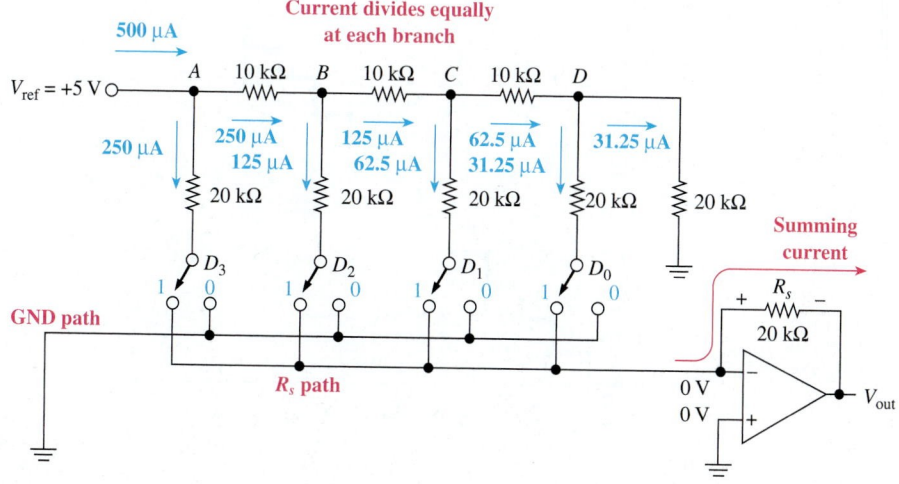

Figure 15–6   Current division in each rung of an R/2R ladder D/A converter.

20 kΩ in parallel with 20 kΩ, which gives 10 kΩ. This 10 kΩ is in series with the 10-kΩ resistor between C and D, making 20 kΩ. This 20 kΩ is now in parallel with the 20-kΩ resistor above D_1. This procedure is repeated over and over until we arrive at the total resistance seen by V_{ref} being equal to 10 kΩ. Therefore, the total current leaving V_{ref} is 5 V divided by 10 kΩ, which equals 500 μA.

When that current reaches point A, it divides equally into the two branches, because each branch is 20 kΩ. The 250 μA that reaches point B also splits equally, sending 125 μA down the D_2 rung. This current splitting continues for each rung. Notice that each resulting current is one-half the current to its left, forming a binary-weighted ratio that is then available for the op-amp summing resistor.

If D_3 is in the 1 position, the 250 μA is routed through the 20-kΩ summing resistor (R_s), creating -5.0 V at V_{out}. (If it is in the 0 position, the current is routed directly to ground and does not contribute to V_{out}.) The portion of V_{out} contributed by the current through each data switch can be summarized as follows:

$$(D_3)\; V_{out} = -250\,\mu A \times 20\,k\Omega\; = -5.000\;V$$

$$(D_2)\; V_{out} = -125\,\mu A \times 20\,k\Omega\; = -2.500\;V$$

$$(D_1)\; V_{out} = -62.5\,\mu A \times 20\,k\Omega\; = -1.250\;V$$

$$(D_0)\; V_{out} = -31.25\,\mu A \times 20\,k\Omega = \underline{-0.625\;V}$$

$$\text{Total } (D_3 \text{ to } D_0 = 1111)\quad V_{out} = -9.375\;V$$

A 4-bit D/A converter such as this can have 16 different combinations of D_3 to D_0. The output voltage for any combination of binary input (B_{in}) of a 4-bit D/A converter can be determined by the following equation:

$$V_{out} = -\left(V_{ref} \times \frac{B_{in}}{8}\right) \qquad \textbf{15--1}$$

The analog output voltages for our 4-bit converter are given in Figure 15–7.

D_3	D_2	D_1	D_0	V_{out} (V)
0	0	0	0	0.000
0	0	0	1	–0.625
0	0	1	0	–1.250
0	0	1	1	–1.875
0	1	0	0	–2.500
0	1	0	1	–3.125
0	1	1	0	–3.750
0	1	1	1	–4.375
1	0	0	0	–5.000
1	0	0	1	–5.625
1	0	1	0	–6.250
1	0	1	1	–6.875
1	1	0	0	–7.500
1	1	0	1	–8.125
1	1	1	0	–8.750
1	1	1	1	–9.375

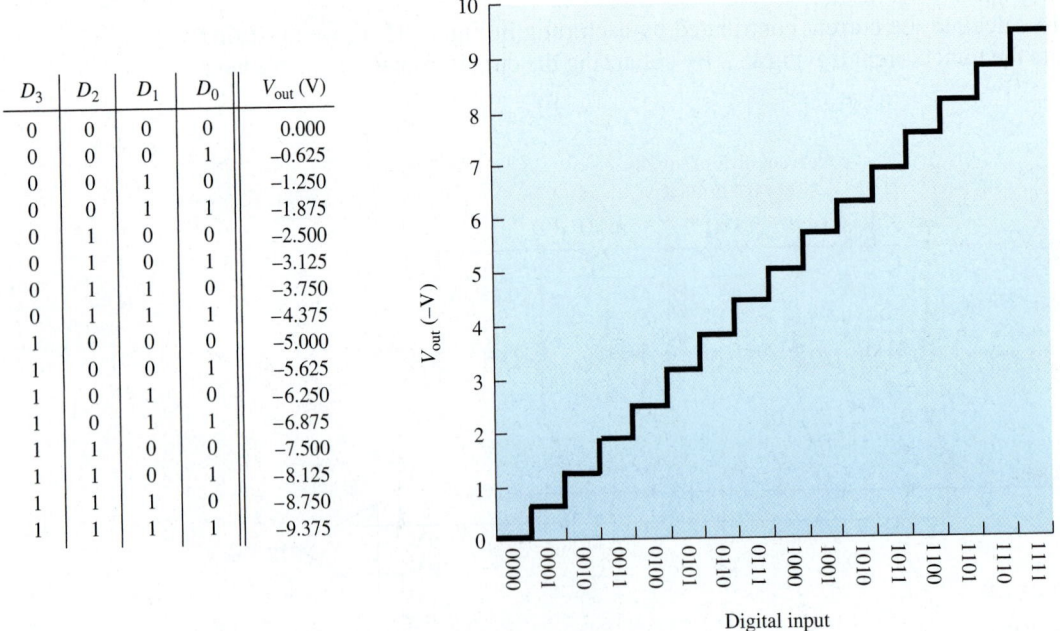

Figure 15–7 Analog output versus digital input for Figure 15–6.

15–4. If the first three resistors in a binary-weighted D/A converter are 30, 60, and 120 kΩ, the fourth resistor, used for the D_0 input, must be _____ ohms.

15–5. Why is it difficult to build an accurate *8-bit* binary-weighted D/A converter?

15–6. To build an 8-bit *R/2R* ladder D/A converter, you would need at least eight different resistor sizes. True or false?

15–7. If V_{ref} is changed to 6 V in the *R/2R* converter of Figure 15–6, the maximum value at V_{out} will be 11.25 V. True or false?

15–5 Integrated-Circuit D/A Converters

One very popular and inexpensive 8-bit D/A converter (DAC) is the DAC0808 and its equivalent, the MC1408. A block diagram, pin configuration, and typical application are shown in Figure 15–8. The circuit in Figure 15–8(c) is set up to accept an 8-bit digital input and provide a 0- to +10-V analog output. A reference current (I_{ref}) is required for the D/A and is provided by the 10-V, 5-kΩ combination shown. The negative reference (pin 15) is then tied to ground via an equal-size (5-kΩ) resistor.

The 2-mA reference current dictates the full-scale output current (I_{out}) to also be approximately 2 mA. To calculate the *actual* output current, use the formula

$$I_{out} = I_{ref} \times \left(\frac{A_1}{2} + \frac{A_2}{4} + \ldots + \frac{A_8}{256} \right) \qquad \textbf{15–2}$$

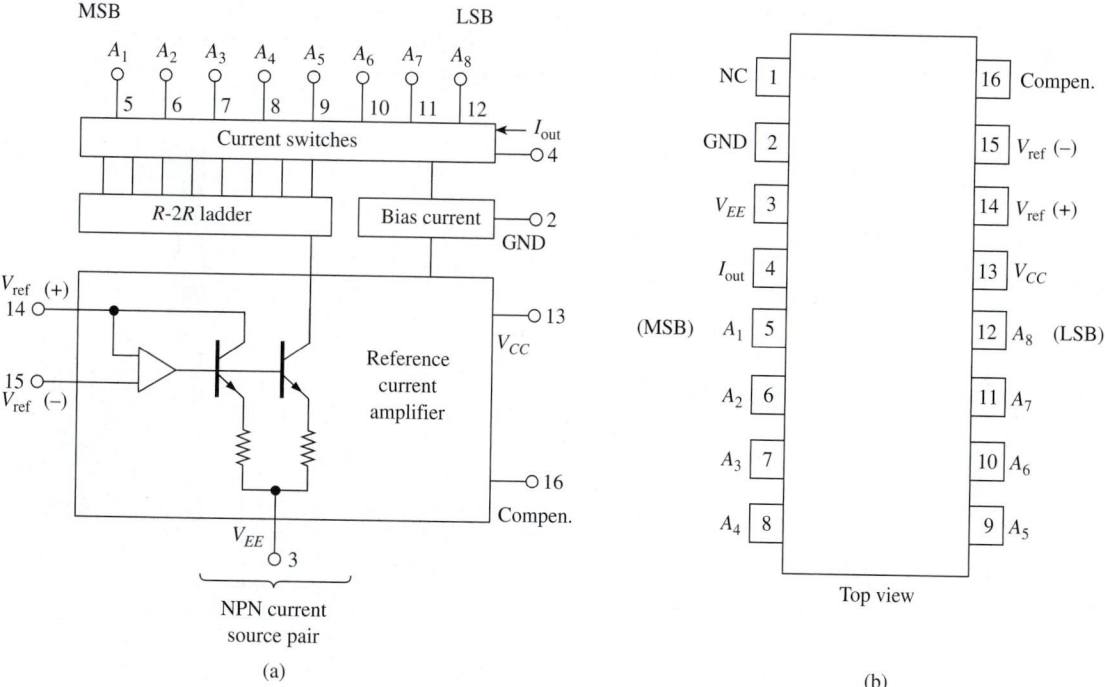

Figure 15–8 The MC 1408 D/A converter: (a) block diagram; (b) pin configuration

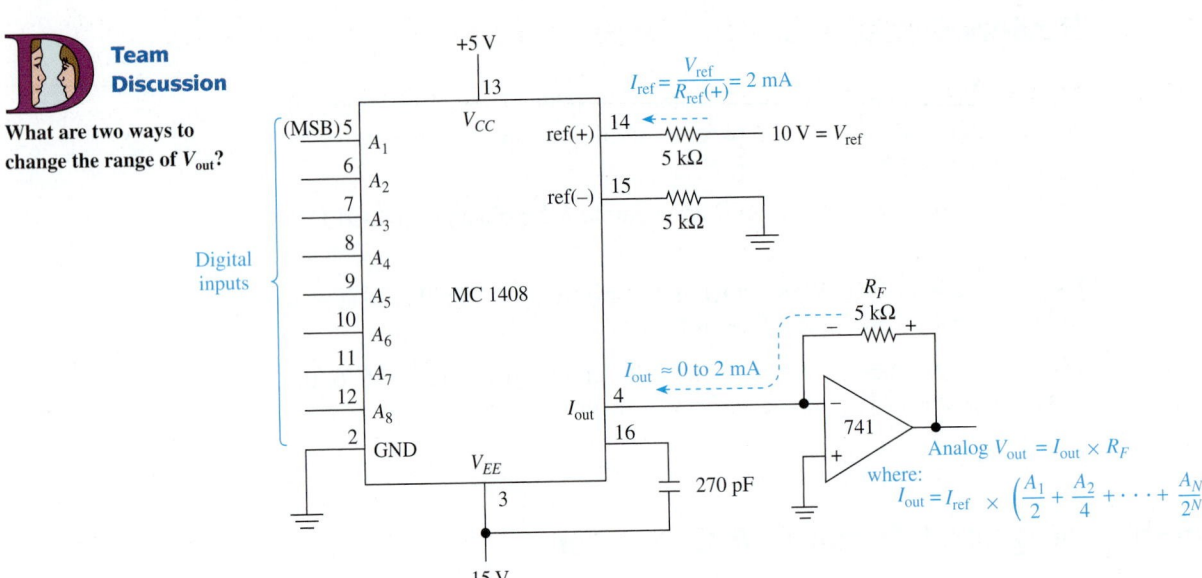

Team Discussion

What are two ways to change the range of V_{out}?

Figure 15–8   *(Continued)* (c) typical application.

[For example, with all inputs (A_1 to A_8) HIGH, $I_{out} = I_{ref} \times \left(\dfrac{255}{256}\right)$]. To convert an output current to an output voltage, a series resistor could be connected from pin 4 to ground and the output taken across the resistor. This method is simple, but it may cause inaccuracies as various-size loads are connected to it.

A more accurate method uses an op amp such as the 741 shown in Figure 15–8(c). The output current flows through R_F, which develops an output voltage equal to $I_{out} \times R_F$. The range of output voltage can be changed by changing R_F and is limited only by the specifications of the op amp used.

Team Discussion

How could the D/A converter be driven to create a triangle waveform? A square wave?

To test the circuit, an oscillator and an 8-bit counter can be used to drive the digital inputs, and the analog output can be observed on an oscilloscope, as shown in Figure 15–9. In Figure 15–9, as the counters count from 0000 0000 up to 1111 1111, the analog output will go from 0 V up to almost +10 V in 256 values. The time per step will be equal to the reciprocal of the input clock frequency.

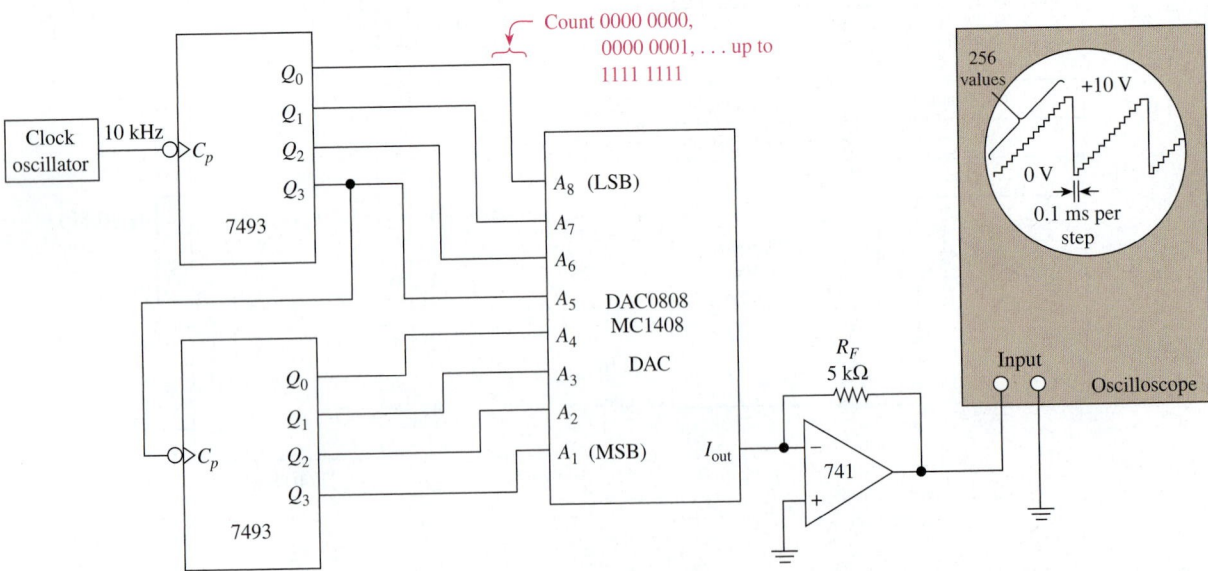

Figure 15–9   Using an 8-bit counter to test the 256-step output of a DAC.

EXAMPLE 15–3

Determine I_{out} and V_{out} in Figure 15–8(c) if the following binary strings are input at A_1 to A_8: (a) 1111 1111; (b) 1001 1011.

Solution:

(a) Using Equation 15–2,

$$I_{out} = I_{ref} \times \left(\frac{A_1}{2} + \frac{A_2}{4} + \ldots + \frac{A_8}{256} \right)$$

$$= 2 \text{ mA} \times \left(\frac{1}{2} + \frac{1}{4} + \frac{1}{8} + \frac{1}{16} + \frac{1}{32} + \frac{1}{64} + \frac{1}{128} + \frac{1}{256} \right)$$

$$= 2 \text{ mA} \times \left(\frac{255}{256} \right)$$

$$= 1.99 \text{ mA}$$

and $V_{out} = I_{out} \times R_F$

$$= 1.99 \text{ mA} \times 5 \text{ k}\Omega$$

$$= 9.96 \text{ V}$$

(b) Using Equation 15–2,

$$I_{out} = I_{ref} \times \left(\frac{A_1}{2} + \frac{A_2}{4} + \ldots + \frac{A_8}{256} \right)$$

$$= 2 \text{ mA} \times \left(\frac{1}{2} + \frac{1}{16} + \frac{1}{32} + \frac{1}{128} + \frac{1}{256} \right)$$

$$= 2 \text{ mA} \times \left(\frac{155}{256} \right)$$

$$= 1.21 \text{ mA}$$

and $\qquad V_{out} = I_{out} \times R_F$

$$= 1.21 \text{ mA} \times 5 \text{ k}\Omega$$

$$= 6.05$$

There are two ways to arrive at the fraction 155/256 in the above solution. One way is to find a common denominator for all the fractions and then add them together, as we did. The other way is to convert the binary input 1001 1011 into decimal, which equals 155, and then divide that by 256. The new equation becomes

$$I_{out} = I_{ref} \times \left(\frac{\text{Binary input}_{10}}{256_{10}} \right)$$

15–6 Integrated-Circuit Data Converter Specifications

Besides resolution, several other specifications are important in the selection of D/A and analog-to-digital (A/D) converters (DACs and ADCs). It is important that the specifications and their definitions given in the manufacturer's data book be studied and understood before selecting a particular DAC or ADC. Figure 15–10 lists some of the more important specifications of data conversion ICs.

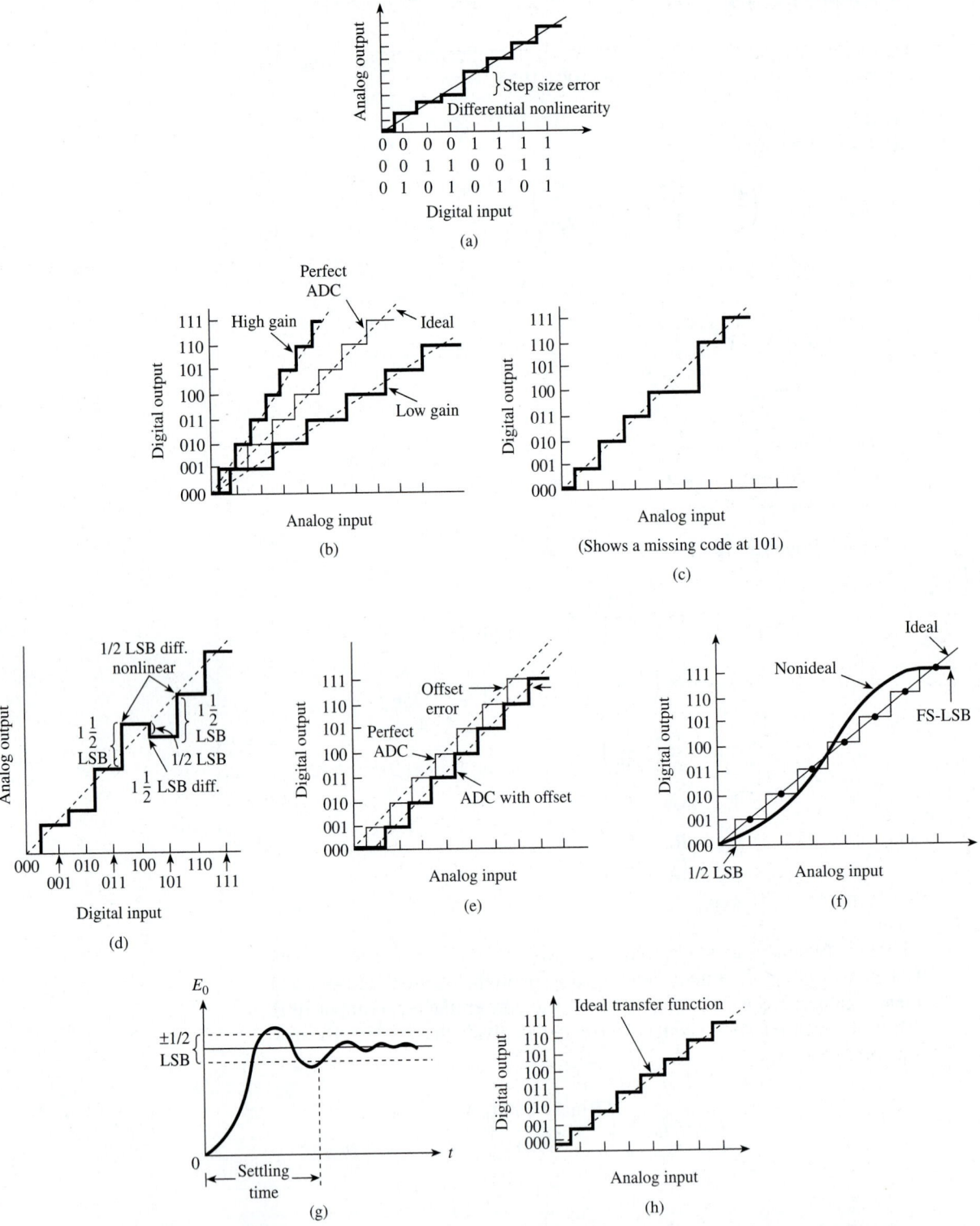

Figure 15–10 DAC and ADC specification definition: (a) differential **nonlinearity;**
(b) gain error; (c) missing codes; (d) **nonmonotonic** (must be $>\pm\frac{1}{2}$ LSB nonlinear);
(e) offset error; (f) relative accuracy; (g) settling time; (h) 3-bit ADC transfer characteristic.

Review Questions

15–8. The analog output of the MC1408 DAC IC is represented by current or voltage?

15–9. Which digital input to the MC1408 DAC IC has the most significant effect on the analog output: A_1, or A_8?

15–10. The *resolution* of a DAC or ADC specifies the _____.

15–11. A DAC is *nonmonotonic* if its analog output *drops* after a 1-bit increase in digital input. True or false?

15–12. Which error affects the rate of change, or slope, of the ideal transfer function of an ADC, the gain error or the offset error?

15–7 Parallel-Encoded A/D Converters

The process of taking an analog voltage and converting it to a digital signal can be done in several ways. One simple way that is easy to visualize is by means of parallel encoding (also known as *simultaneous, multiple comparator,* or *flash* converting). In this method, several comparators are set up, each at a different voltage reference level with their outputs driving a priority encoder, as shown in Figure 15–11. The voltage-divider network in Figure 15–11 is designed to drop 1 V across each resistor. This sets up a voltage reference at each comparator input in 1-V steps.

When V_{in} is 0 V, the (+) input on all seven comparators will be higher than the (−) input, so they will all output a HIGH. In this case, $\overline{I_0}$ is the only active-LOW input that is enabled, so the 74148 will output an active-LOW binary 0 (111).

When V_{in} exceeds 1.0 V, comparator 1 will output a LOW. Now $\overline{I_0}$ and $\overline{I_1}$ are both enabled, but because it is a *priority* encoder, the output will be a binary 1 (110). As V_{in} increases further, each successive comparator outputs a LOW. The highest input that receives a LOW is encoded into its binary equivalent output.

The A/D converter in Figure 15–11 is set up to convert analog voltages in the range from 0 to 7 V. The range can be scaled higher or lower, depending on the input voltage levels that are expected. The resolution of this converter is only 3 bits, so it can only distinguish among eight different analog input levels. To expand to 4-bit resolution, eight more comparators are required to differentiate the 16 different voltage levels. To expand to 8-bit resolution, 256 comparators would be required! As you can see, circuit complexity becomes a real problem when using parallel encoding for high resolution conversion. However, a big advantage of using parallel encoding is its high speed. The conversion speed is limited only by the propagation delays of the comparators and encoder (less than 20 ns total).

15–8 Counter-Ramp A/D Converters

Helpful Hint

This circuit is made completely of devices that have already been discussed, which makes it a good circuit to illustrate timing and interface between multiple ICs.

The counter-ramp method of A/D conversion (ADC) uses a counter in conjunction with a D/A converter (DAC) to determine a digital output that is equivalent to the unknown analog input voltage. In Figure 15–12, depressing the start conversion push button clears the counter outputs to 0, which sets the DAC output to 0 V. The (−) input to the comparator is now 0 V, which is less than the positive analog input voltage at the (+) input. Therefore, the comparator outputs a HIGH, which enables the AND gate, allowing the counter to start counting. As the counter's binary output increases, so does the DAC output voltage in the form of a staircase.

**Helpful
Hint**

This is a perfect example
of the *priority* feature of
the '148.

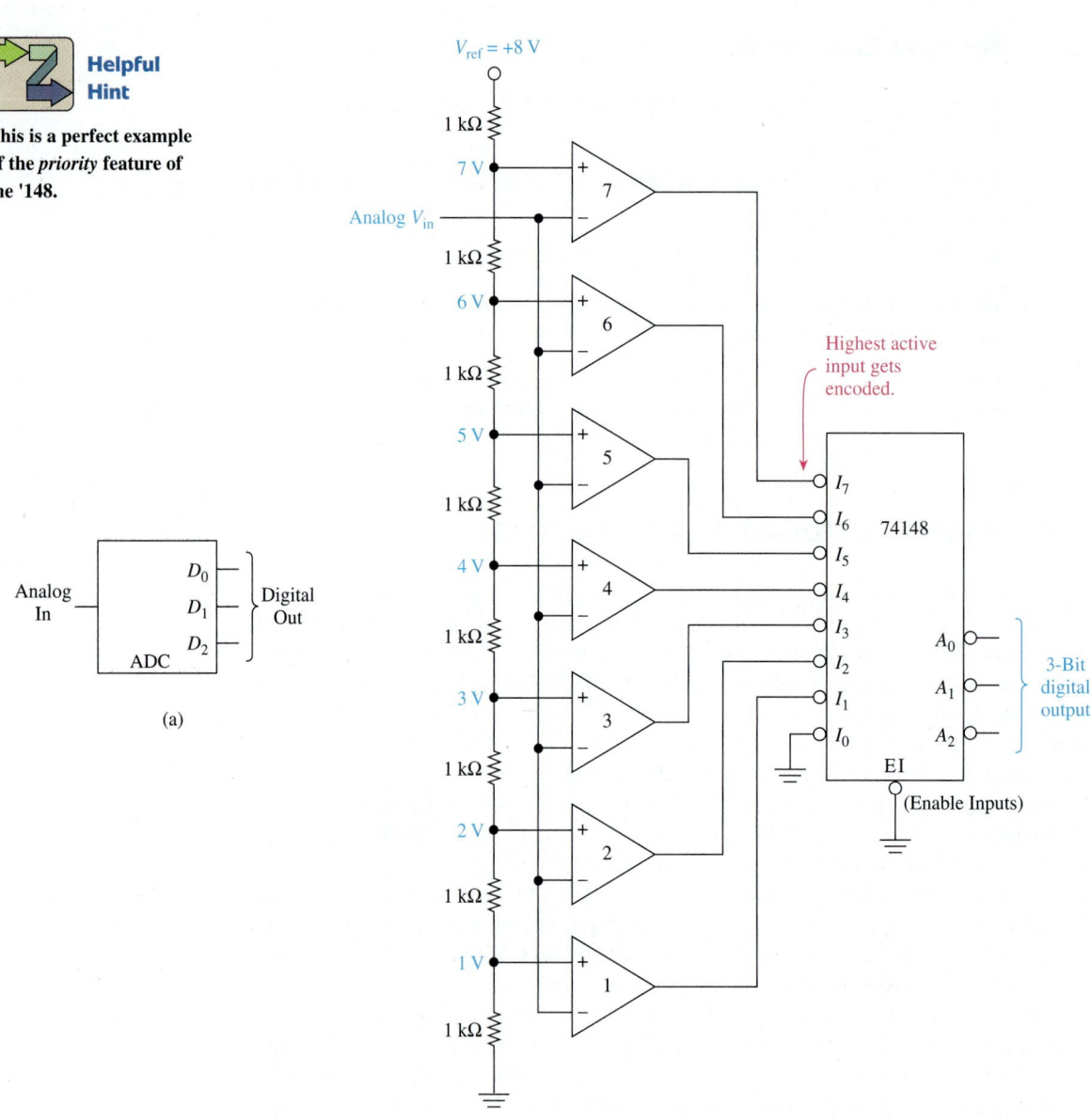

Figure 15-11   Three-bit parallel-encoded ADC: (a) block diagram; (b) circuit diagram.

When the staircase voltage reaches and then exceeds the analog input voltage, the comparator output goes LOW, disabling the clock and stopping the counter. The counter output at that point is equal to the binary number that caused the DAC to output a voltage slightly greater than the analog input voltage. Thus, we have the binary equivalent of the analog voltage!

The HIGH-to-LOW transition of the comparator is also used to trigger the D flip-flop to latch on to the binary number at that instant. To perform another conversion, the start push button is depressed again, and the process repeats. The result from the previous conversion remains in the D flip-flop until the next end-of-conversion HIGH-to-LOW edge comes along.

To change the circuit to perform **continuous conversions**, the end-of-conversion line could be tied back to the clear input of the counter. A short delay needs to be inserted into this new line, however, to allow the D flip-flop to read the binary number

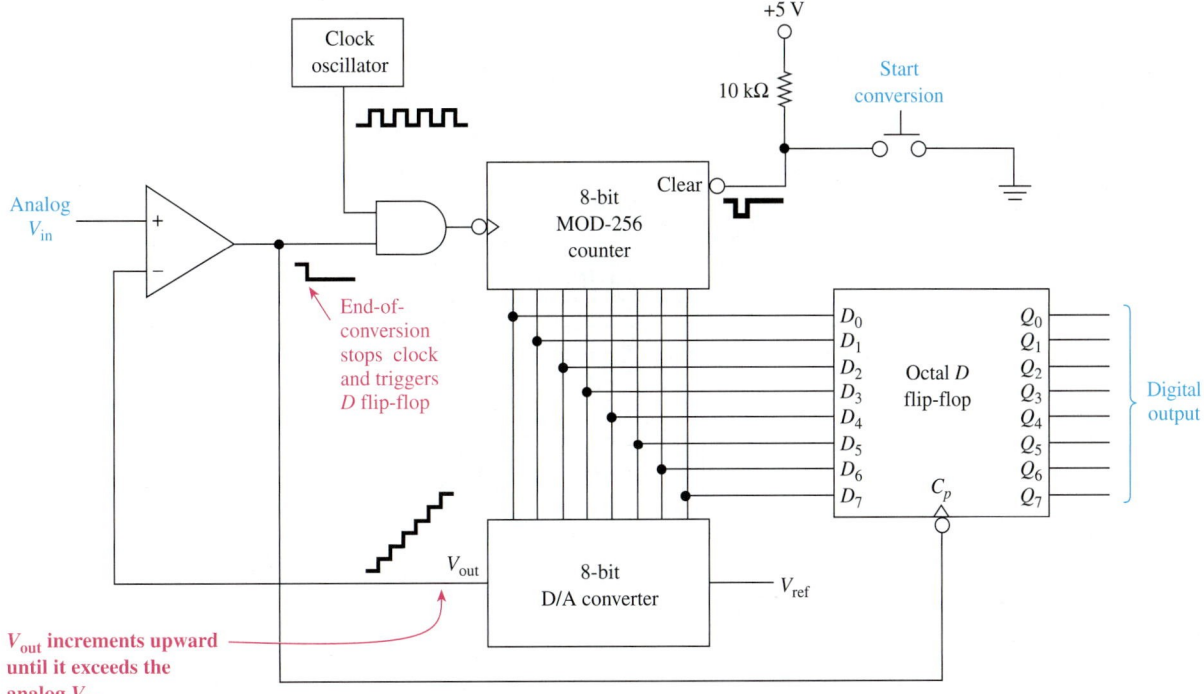

Figure 15–12   Counter-ramp A/D converter.

before the counter is Reset. Two inverters placed end to end in the line will produce a sufficient delay.

The main *disadvantage* of the counter-ramp method of conversion is its slow conversion speed. The worst-case maximum **conversion time** will occur when the counter has to count all 255 steps before the DAC output voltage matches the analog input voltage. ($t_{max} = 256 \times (1/f_{cp})$)

Review Questions

15–13. In the parallel-encoded ADC of Figure 15–11, *only one* of the comparators will output a LOW for each analog input value. True or false?

15–14. One difficulty in building a high-resolution, 10-bit, parallel-encoded ADC is that it would take _____ comparators to complete the design.

15–15. The counter-ramp ADC of Figure 15–12 signifies an "end of conversion" when the $(-)$ input voltage to the comparator drops below the $(+)$ input voltage. True or false?

15–16. The digital output of the octal *D* flip-flop in Figure 15–12 is continuously changing at the same rate as the MOD-256 counter. True or false?

15–9 Successive-Approximation A/D Conversion

Other methods of A/D conversion employ *up/down-counters* and *integrating slope converters* to track the analog input, but the method used in most modern IC ADCs is called **successive approximation.** This converter circuit is similar to the counter-ramp ADC circuit, except that the method of narrowing in on the unknown analog input voltage is

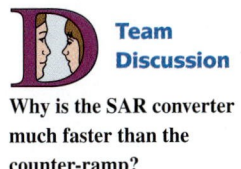

Team Discussion

Why is the SAR converter much faster than the counter-ramp?

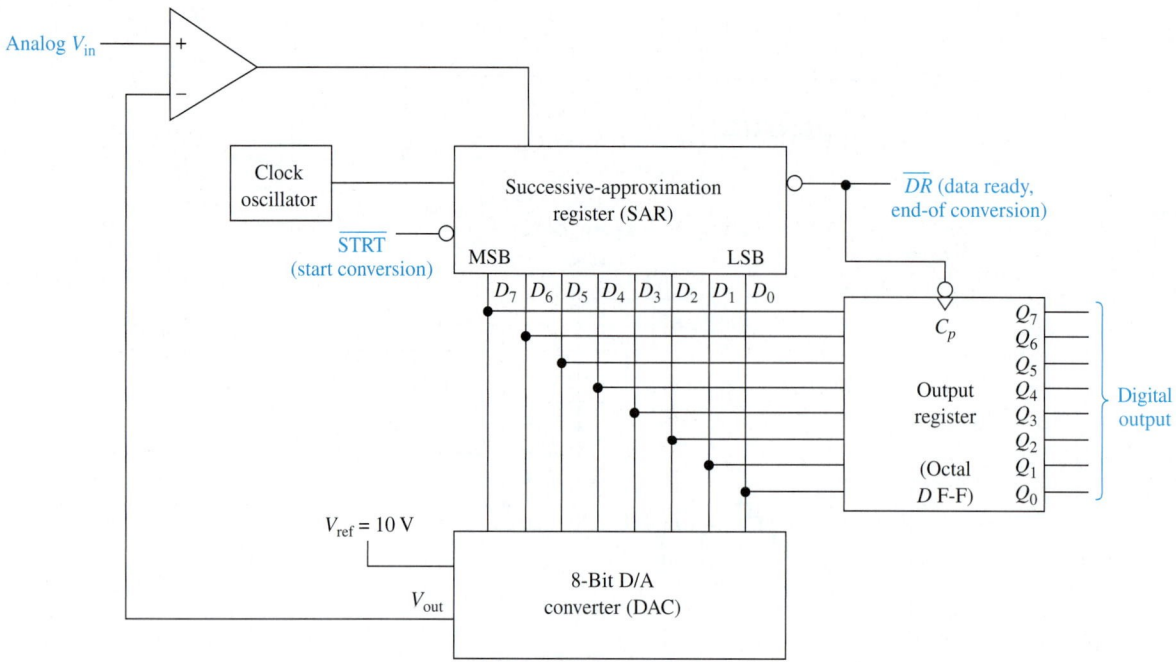

Figure 15–13 Simplified SAR A/D converter.

much improved. Instead of counting up from 0 and comparing the DAC output each step of the way, a successive-approximation register (SAR) is used in place of the counter (see Figure 15–13).

In Figure 15–13, the conversion is started by dropping the $\overline{STRT}$ line LOW. Then the SAR first tries a HIGH on the MSB (D_7) line to the DAC. (Remember, D_7 will cause the DAC to output half of its full-scale output.) If the DAC output is then *higher* than the unknown analog input voltage, the SAR returns the MSB LOW. If the DAC output was still *lower* than the unknown analog input voltage, the SAR leaves the MSB HIGH.

Now, the next lower bit (D_6) is tried. If a HIGH on D_6 causes the DAC output to be higher than the analog V_{in}, it is returned LOW. If not, it is left HIGH. The process continues until all 8 bits, down to the LSB, have been tried. At the end of this eight-step conversion process, the SAR contains a valid 8-bit binary output code that represents the unknown analog input. The $\overline{DR}$ output now goes LOW, indicating that the *conversion is complete* and the data are ready. The HIGH-to-LOW edge on $\overline{DR}$ clocks the D_0 to D_7 data into the octal D flip-flop to make the digital output results available at the Q_0 to Q_7 lines.

The main advantage of the SAR ADC method is its high speed. The ADC in Figure 15–13 takes only eight clock periods to complete a conversion, which is a vast improvement over the counter-ramp method. [$t_{max} = 8 \times (1/f_{cp})$] Notice with the SAR, the final binary result is always slightly *less than* the equivalent analog input, where as with the counter-ramp, it is slightly *more.*

EXAMPLE 15–4

Show the timing waveforms that would occur in the successive approximation ADC of Figure 15–13 when converting the analog voltage of 6.84 V to
8-bit binary, assuming that the full-scale input voltage to the DAC is 10 V
(V_{ref} = 10 V).

**Team
Discussion**

**What is the highest
percent error that you
could have with the 8-bit
ADC in Example 15–4?**

Solution: Each successive bit, starting with the MSB, will cause the DAC
part of the system to output a voltage to be compared. If the full-scale
output is 10 V, D_7 will be worth 5 V, D_6 will be worth 2.5 V, D_5 will be
worth 1.25 V, and *so on*, as shown in Table 15–1.

Now, when $\overline{STRT}$ goes LOW, successive bits starting with D_7 will be
tried, creating the waveforms shown in Figure 15–14. The HIGH-to-LOW
edge on $\overline{DR}$ clocks the final binary number 1010 1111 into the D flip-flop
and Q_0 to Q_7 outputs.

Now the Q_0 to Q_7 lines contain the 8-bit binary representation of the
analog number 6.8359375, which is an error of only 0.0594% from the tar-
get number of 6.84:

$$\% \text{ error} = \frac{\text{actual voltage} - \text{final DAC output}}{\text{actual voltage}} \times 100\% \qquad \textbf{15–3}$$

TABLE 15–1	Voltage-Level Contributions by Each Successive Approximation Register Bit
DAC Input	**DAC V_{out}**
D_7	5.0000
D_6	2.5000
D_5	1.2500
D_4	0.6250
D_3	0.3125
D_2	0.15625
D_1	0.078125
D_0	0.0390625

To watch the conversion in progress, an eight-channel oscilloscope or
logic analyzer can be connected to the D_0 to D_7 outputs of the SAR.

For *continuous conversions,* the $\overline{DR}$ line can be connected back to the
$\overline{STRT}$ line. That way, as soon as the conversion is complete, the HIGH-to-
LOW on $\overline{DR}$ will issue another start conversion ($\overline{STRT}$), which forces the
data ready ($\overline{DR}$) line back HIGH for eight clock periods while the new conversion is being made. The latched Q_0 to Q_7 digital outputs will always display the results of the *previous conversion.*

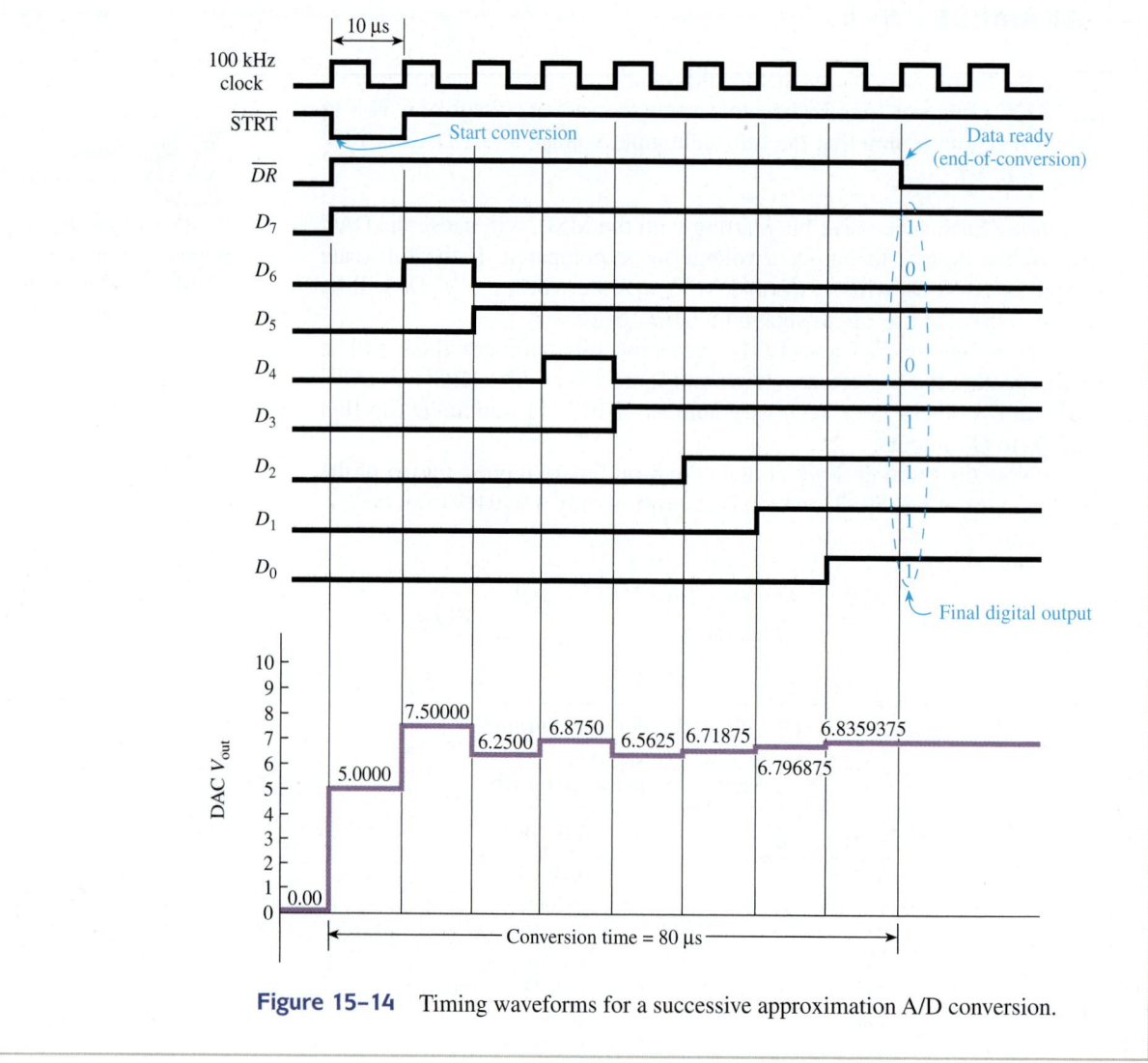

Figure 15–14 Timing waveforms for a successive approximation A/D conversion.

15–10 Integrated-Circuit A/D Converters

Examples of two basic ADCs are the NE5034 and the ADC0804.

The NE5034

The block diagram and pin configuration for the NE5034 are given in Figure 15–15. Operation of the NE5034 is almost identical to that of the SAR ADC presented in Section 15–9. One difference is that the NE5034 uses a three-state output buffer instead of a D flip-flop. With three-state outputs, when $\overline{OE}$ (Output Enable) is LOW, the DB_7 to DB_0 outputs display the continuous status of the eight SAR lines, and when the $\overline{OE}$ line goes HIGH, the DB_7 to DB_0 outputs return to a float or high-impedance state. This way, if the ADC outputs go to a common *data* **bus** shared by other devices, when DB_7 to DB_0 float, one of the other devices can output information to the data bus without interference.

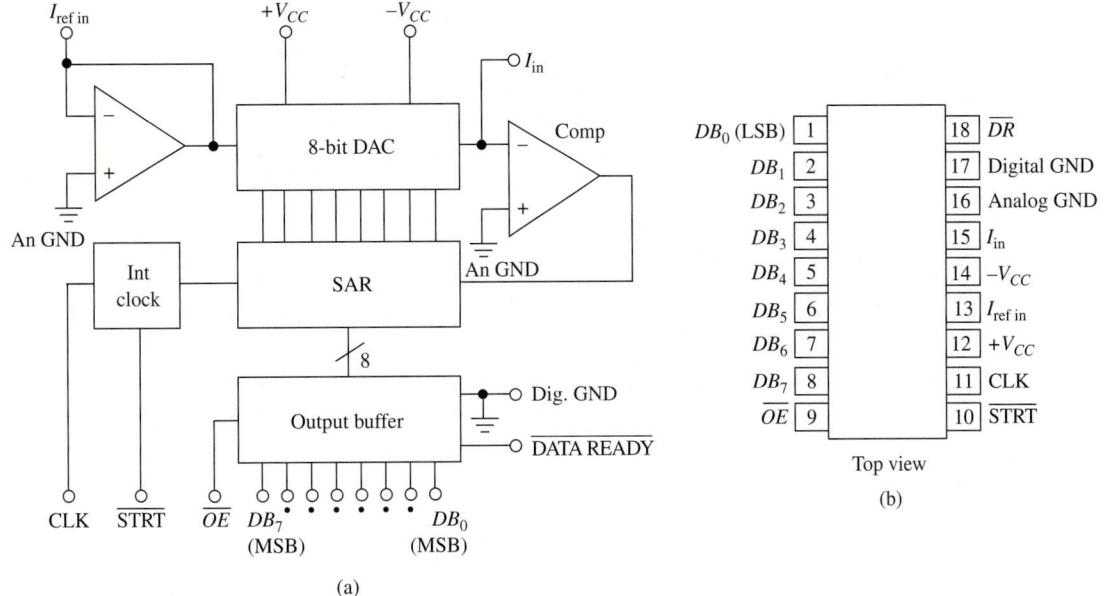

Figure 15–15   The NE5034 A/D converter: (a) block diagram; (b) pin configuration.

The NE5034 can provide conversion speeds as high as one per 17 μs, and its three-state outputs make it compatible with bus-oriented microprocessor systems. It also has its own internal clock for providing timing pulses. The frequency is determined by an external capacitor placed between pins 11 and 17. Figure 15–16 shows the frequency and conversion time that can be achieved using the internal clock.

The ADC0804

The pin configuration and block diagram for the ADC0804 are given in Figure 15–17. The ADC0804 uses the successive-approximation method to convert an analog input to an 8-bit binary code. Two analog inputs are provided to allow differential measurements [analog $V_{in} = V_{in(+)} - V_{in(-)}$]. It has an internal clock that generates its own timing pulses at a frequency equal to $f = 1/(1.1RC)$ (Figure 15–18 shows the connections for the external R and C). It uses output D latches that are three-stated to facilitate easy bus interfacing.

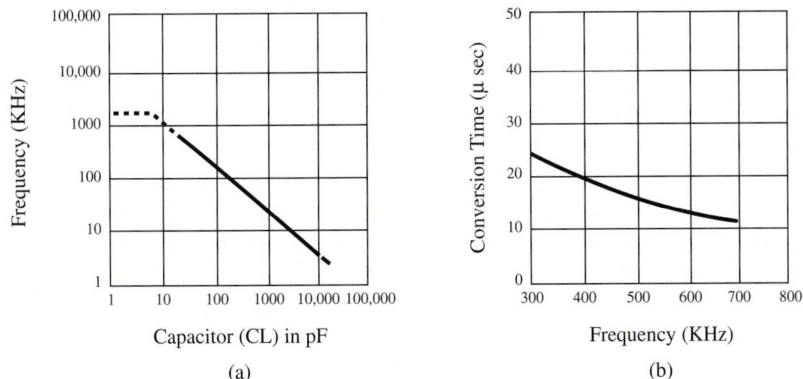

Figure 15–16   The NE5034 internal clock characteristics: (a) internal clock frequency versus external capacitor (CL); (b) conversion time versus clock frequency.

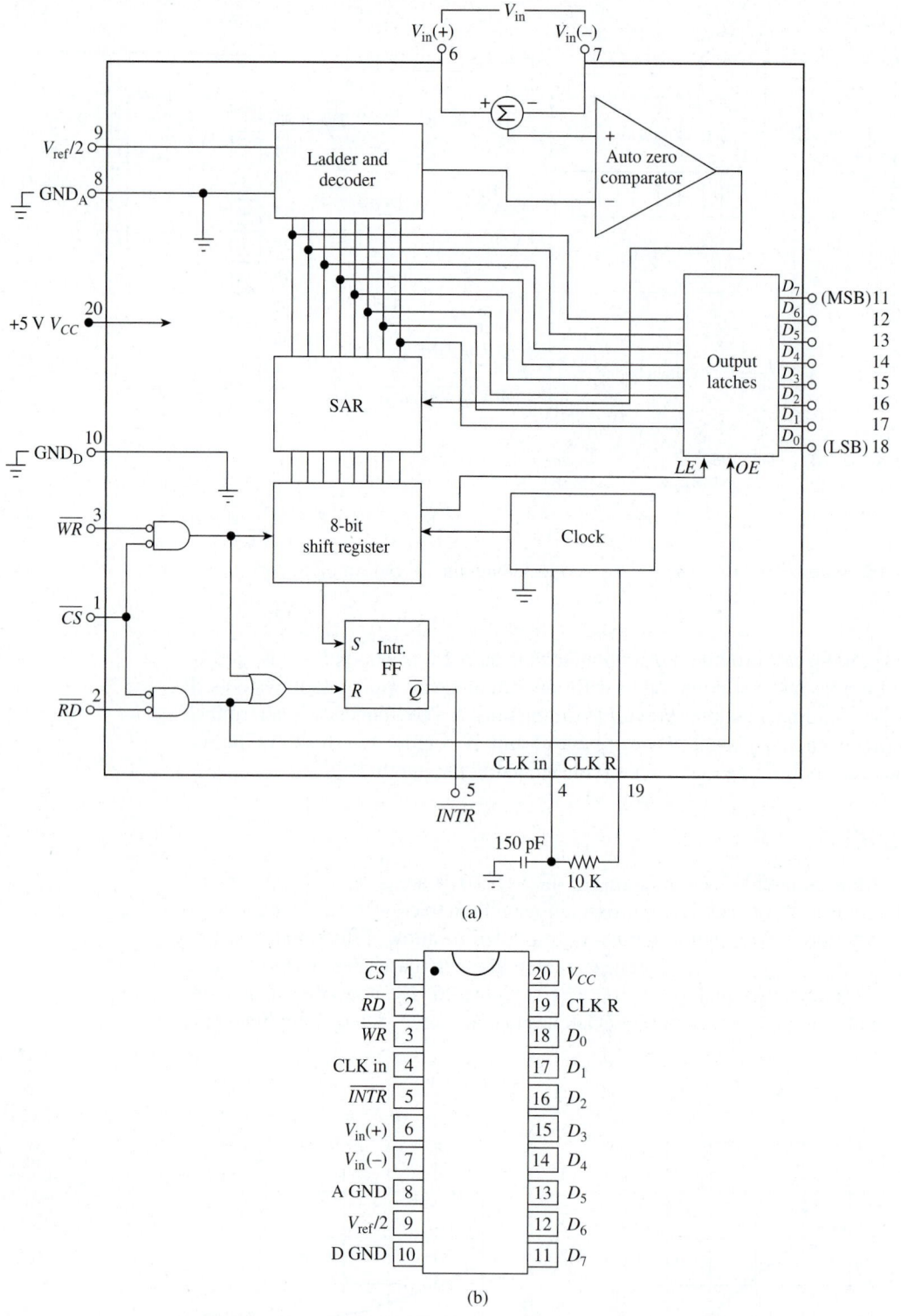

(a)

CHAPTER 15 | INTERFACING TO THE ANALOG WORLD

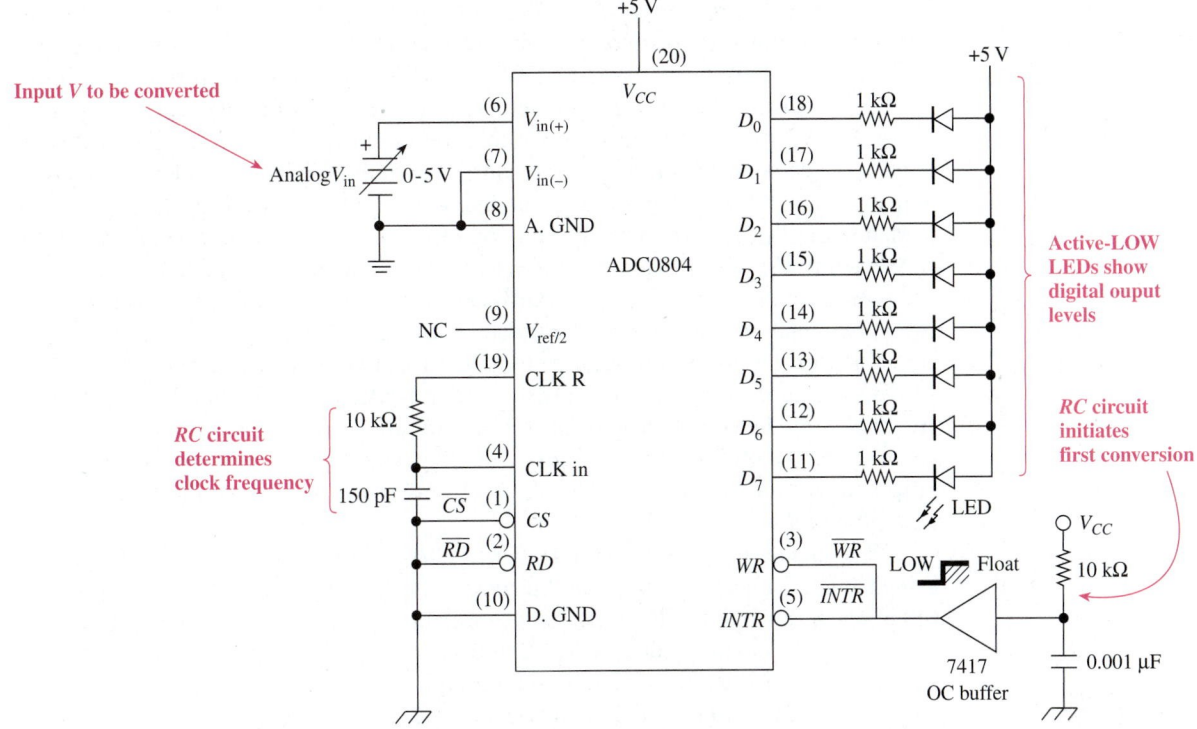

Figure 15–18 Connections for continuous conversions using the ADC0804.

The convention for naming the ADC0804 pins follows that used by microprocessors to ease interfacing. Basically, the operation of the ADC0804 is similar to that of the NE5034. The ADC0804 pins are defined as follows:

$\overline{CS}$—active-LOW *Chip Select*

$\overline{RD}$—active-LOW *Output Enable*

$\overline{WR}$—active-LOW *Start Conversion*

CLK IN—external clock input or capacitor connection point for the internal clock

$\overline{INTR}$—active-LOW *End-of-Conversion* (Data Ready)

$V_{in(+)}$, $V_{in(-)}$—differential analog inputs (ground one pin for single-ended measurements)

A. GND—analog ground

$V_{ref}/2$—optional **reference voltage** (used to override the reference voltage assumed at V_{CC})

D. GND—digital ground

V_{CC}—5-V power supply and assumed reference voltage

CLK R—resistor connection for the internal clock

D_0 to D_7—digital outputs

To set the ADC0804 up for continuous A/D conversions, the connections shown in Figure 15–18 should be made. The external *RC* will set up a clock frequency of

$$f = \frac{1}{1.1RC} = \frac{1}{1.1(10 \text{ k}\Omega)150 \text{ pF}} = 606 \text{ kHz}$$

The connection from $\overline{INTR}$ to $\overline{WR}$ will cause the ADC to start a new conversion each time the $\overline{INTR}$ (end-of-conversion) line goes LOW. The RC circuit with the 7417 open-collector buffer will issue a LOW-to-float pulse at power-up to ensure initial startup. An *open-collector* output gate is required instead of a totem-pole output because the $\overline{INTR}$ is forced LOW by the internal circuitry of the ADC0804 at the end of each conversion. This LOW would conflict with the HIGH output level if a totem-pole output were used. The $\overline{CS}$ is grounded to enable the ADC chip. $\overline{RD}$ is grounded to enable the D_0 to D_7 outputs. The analog input voltage is positive, 0 to 5 V, so it is connected to $V_{in(+)}$. If it were negative, $V_{in(+)}$ would be grounded, and the input voltage would be connected to $V_{in(-)}$. **Differential measurements** (the difference between two analog voltages) can be made by using both $V_{in(+)}$ and $V_{in(-)}$. The LEDs connected to the digital output will monitor the operation of the ADC outputs. An LED ON indicates a LOW, and an LED OFF indicates a HIGH. (In other words, they are displaying the *complement* of the binary output.) To test the circuit operation, you could watch the OFF LEDs count up in binary from 0 to 255 as the analog input voltage is slowly increased from 0 to +5 V. (In other words, you could watch the *ON* LEDs count *down*.)

The analog input voltage range can be changed to values other than 0 to 5 V by using the $V_{ref}/2$ input. This provides the means of encoding small analog voltages to the full 8 bits of resolution. The $V_{ref}/2$ pin is normally not connected, and it sits at 2.500 V ($V_{CC}/2$). By connecting 2.00 V to $V_{ref}/2$, the analog input voltage range is changed to 0 through 4 V; 1.5 V would change it to 0 through 3.0 V; and so on. However, the accuracy of the ADC suffers as the input voltage range is decreased.

Because the analog input voltage is directly proportional to the digital output, the following ratio can be used as an equation to solve for the digital output value:

$$\frac{A_{in}}{V_{ref}} = \frac{D_{out}}{256} \qquad\qquad \text{15–4}$$

or

$$D_{out} = \frac{A_{in}}{V_{ref}} \times 256 \qquad\qquad \text{15–5}$$

where A_{in} = analog input voltage
   V_{ref} = reference voltage ($V_{pin\,20}$ or $V_{pin\,9} \times 2$)
   D_{out} = digital output (converted to base 10)
   256 = total number of digital output steps (0 to 255, inclusive)

One final point on the ADC0804. An analog ground *and* a digital ground are both provided to enhance the accuracy of the system. The V_{CC}-to-digital ground lines are inherently noisy due to the switching transients of the digital signals. Using separate analog and digital grounds is not mandatory, but when used, they ensure that the analog voltage comparator will not switch falsely due to digital noise and jitter.

EXAMPLE 15–5

Change V_{CC} (pin 20) in Figure 15–18 to 5.12 V. Determine which LEDs will be on for the following analog input voltages:

(a) 5.100 V; (b) 2.26 V.

Solution:

(a)
$$D_{out} = \frac{A_{in}}{V_{ref}} \times 256$$
$$= \frac{5.100 \text{ V}}{5.120 \text{ V}} \times 256$$
$$= 255_{10}$$
$$= 1111\ 1111_2$$

Because the LEDs are active-LOW, none of them will be ON.

(b)
$$D_{out} = \frac{A_{in}}{V_{ref}} \times 256$$
$$= \frac{2.26 \text{ V}}{5.12 \text{ V}} \times 256$$
$$= 113_{10}$$
$$= 0111\ 0001_2$$

The following LEDs will be ON: LED_7, LED_3, LED_2, LED_1.

Review Questions

15–17. The SAR ADC in Figure 15–13 starts making a conversion when _____ goes LOW and signifies that the conversion is complete when _____ goes LOW.

15–18. The SAR method of A/D conversion is faster than the counter/ramp method because the SAR clock oscillator operates at a higher speed. True or false?

15–19. Decide if the following pins on the ADC0804 are for input or output signals.

(a) $\overline{CS}$ (c) $\overline{WR}$

(b) $\overline{RD}$ (d) $\overline{INTR}$

15–20. List the order in which the signals listed in Question 15–19 become active to perform an A/D conversion.

15–11 Data Acquisition System Application

The computerized acquisition of analog quantities is becoming more important than ever in today's automated world. Computer systems are capable of scanning several analog inputs on a particular schedule and sequence to monitor critical quantities and acquire data for future recall. A typical eight-channel computerized **data acquisition system (DAS)** is shown in Figure 15–19.

The entire system in Figure 15–19 communicates via two common buses, the *data bus* and the *control bus*. The data bus is simply a common set of eight electrical conductors shared by as many devices as necessary to send and receive 8 bits of parallel data to and from anywhere in the system. In this case, there are three devices on the

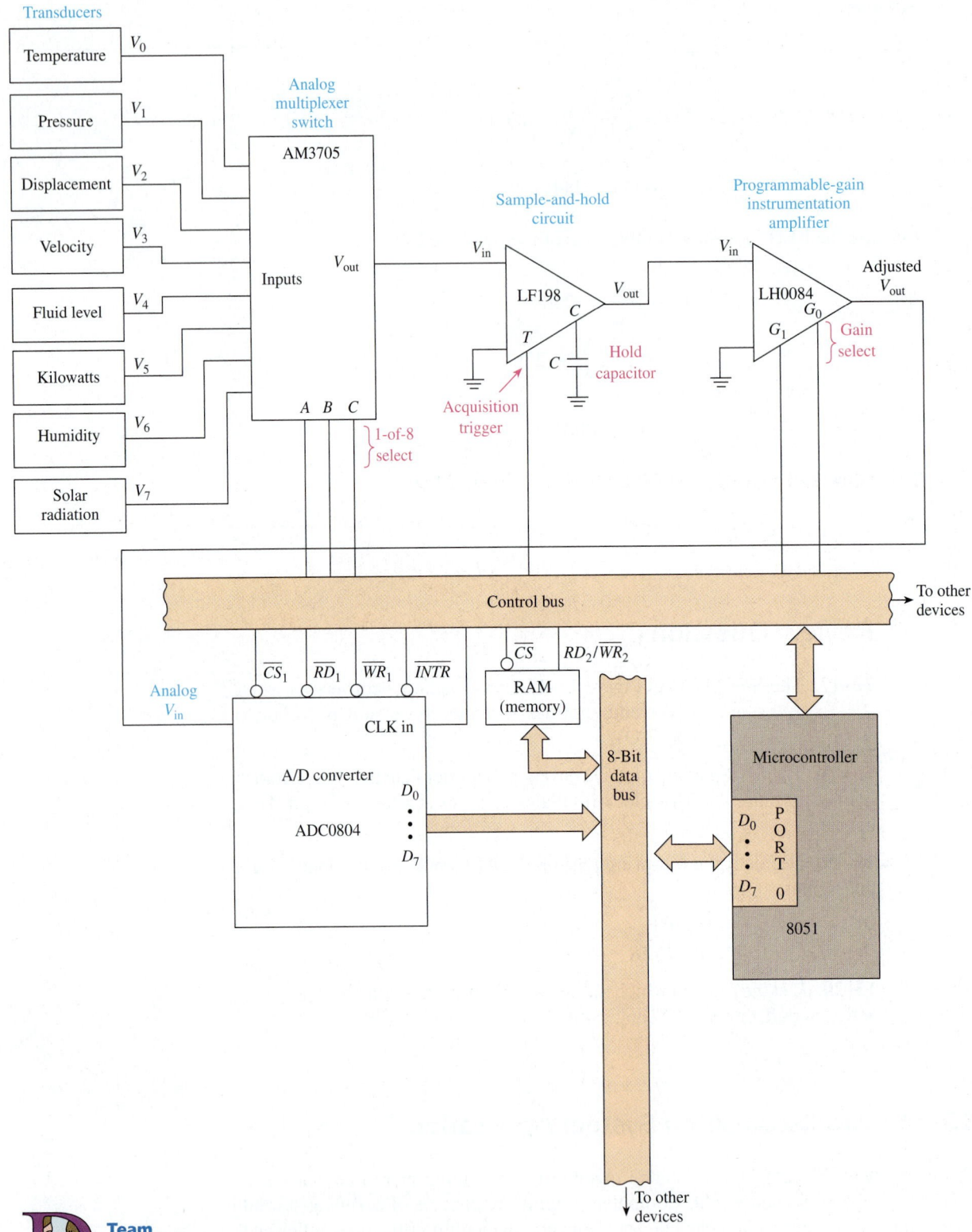

Figure 15–19 Data acquisition system.

Team Discussion

Discuss the flow of a signal as it travels from the temperature transducer through each of the ICs to the microcontroller.

data bus: the ADC, the microprocessor (or microcontroller), and **memory.** The control bus passes control signals to and from the various devices for such things as chip select $(\overline{CS})$, output enable $(\overline{RD})$, system clock, triggers, and selects.

Each of the eight transducers is set up to output a voltage that is proportional to the analog quantity being measured. The task of the microprocessor is to scan all the quantities at some precise interval and store the digital results in memory for future use. Because the analog values vary so slowly, scanning (reading) the variables at 1-s intervals is usually fast enough to get a very accurate picture of their levels. This can easily be accomplished by **microprocessors** with clock rates in the microsecond range.

To do this, the microprocessor must enable and send the proper control signals to each of the devices, in order, starting with the multiplexer and ending with the ADC. This is called **handshaking,** or *polling,* and is all done with software statements. If you are fortunate enough to take a course in microprocessor programming, you will learn how to perform some of these tasks.

All the hardware **interfacing** and handshaking that takes place between the microprocessor and the transducers can be explained by taking a closer look at each of the devices in the system.

Analog Multiplexer Switch (AM3705)

The multiplexer reduces circuit complexity and eliminates duplication of circuitry by allowing each of the eight transducer outputs to take turns traveling through the other devices. The microprocessor selects each of the transducers at the appropriate time by setting up the appropriate binary select code on the *A, B, C* inputs via the control bus. This allows the selected transducer signal to pass through to the next device.

Sample-and-Hold Circuit (LF198)

Because analog quantities can be constantly varying, it is important to be able to select a precise time to take the measurement. The **sample-and-hold** circuit, with its external *Hold capacitor,* allows the system to take (Sample) *and Hold* an analog value at the precise instant that the microprocessor issues the *acquisition trigger.*

Programmable-Gain Instrumentation Amplifier (LH0084)

Each of the eight transducers has different full-scale output ratings. For instance, the temperature transducer may output in the range from 0 to 5 V, whereas the pressure transducer may only output 0 to 500 mV. The LH0084 is a **programmable-gain amplifier** capable of being programmed, via the gain select inputs, for gains of 1, 2, 5, or 10. When it is time to read the pressure transducer, the microprocessor will program the gain for 10 so that the range will be 0 to 5 V, to match that of the other transducers. This way, the ADC can always operate in its most accurate range, 0 to 5 V.

Analog-to-Digital Converter (ADC0804)

The ADC receives the adjusted analog voltage and converts it to an equivalent 8-bit binary string. To do this, the microprocessor issues chip select $(\overline{CS_1})$ and start conversion $(\overline{WR_1})$ pulses. When the end-of-conversion $(\overline{INTR})$ line goes LOW, the microprocessor issues an output enable $(\overline{RD_1})$ to read the data (D_0 and D_7) that pass, via the data bus, into the microprocessor and then into the random-access memory (RAM) chip (more on memory in Chapter 16).

This cycle repeats for all eight transducers whenever the microprocessor determines that it is time for the next scan. Other software routines executed by the microprocessor will act on the data that have been gathered. Some possible responses to the measured results might be to sound an alarm, speed up a fan, reduce energy consumption, increase a fluid level, or simply produce a tabular report of the measured quantities.

15–12 Transducers and Signal Conditioning

Hundreds of *transducers* are available today that convert physical quantities such as heat, light, or force into electrical quantities (or vice versa). The *electrical quantities* (or signal levels) must then be *conditioned* (or modified) before they can be interpreted by a digital computer.

Signal conditioning is required because transducers each output different ranges and types of electrical signals. For example, transducers can produce output voltages or output currents or act like variable resistances. A transducer may have a nonlinear response to input quantities, may be inversely proportional, and may output signals in the microvolt range.

A transducer's response specifications are given by the manufacturer and must be studied carefully to determine the appropriate analog signal-conditioning circuitry required to interface it to an A/D converter. After the information is read into a digital computer, software instructions convert the binary input into a meaningful output that can be used for further processing. Let's take a closer look at three commonly used transducers: a thermistor, an IC temperature sensor, and a strain gage.

Thermistors

A thermistor is an electronic component whose resistance is highly dependent on temperature. Its resistance changes by several percent with each degree change in temperature. It is a very sensitive temperature-measuring device. One problem, however, is that its response is nonlinear, meaning that 1° step changes in temperature will not create equal step changes in resistance. This fact is illustrated in the characteristic curve of the 10-kΩ (at 25°C) thermistor shown in Figure 15–20.

From the characteristic curve, you can see that not only is this thermistor nonlinear, but it also has a negative temperature coefficient (i.e., its resistance decreases with increasing temperatures).

To use a thermistor with an ADC like the ADC0804, we need to convert the thermistor resistance to a voltage in the range of 0 to 5 V. One way to accomplish this task is with the circuit shown in Figure 15–21.

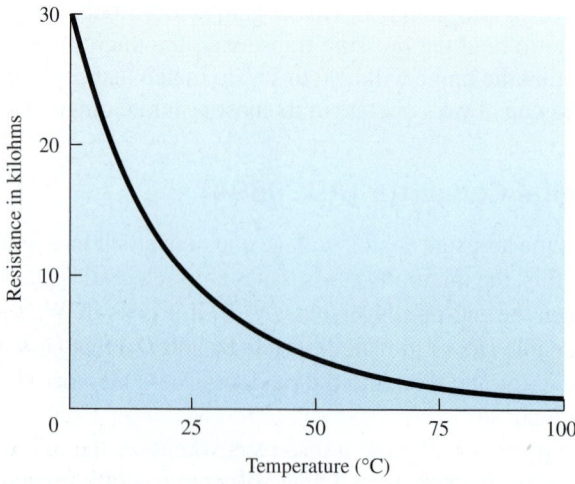

Figure 15–20 Thermistor characteristic curve of resistance versus temperature.

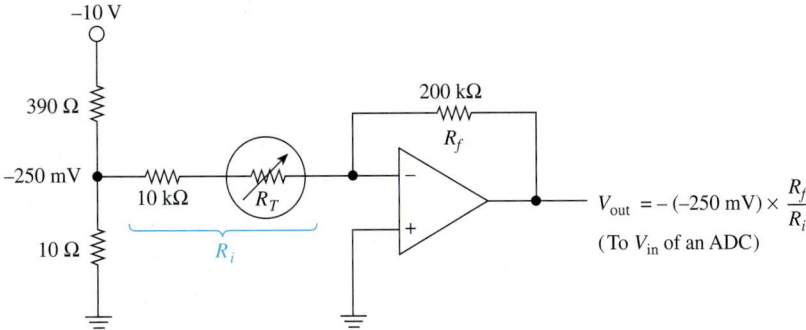

Figure 15–21 Circuit used to convert thermistor ohms to a dc voltage.

This circuit operates similarly to the op-amp circuit explained in Section 15–2. The output of the circuit is found using the formula

$$V_{out} = -V_{in} \times \frac{R_f}{R_i}$$

where V_{in} is a fixed reference voltage of -250 mV and R_i is the sum of the thermistor's resistance, R_T, plus 10 kΩ. Using specific values for R_T found in the manufacturer's data manual, we can create Table 15–2, which shows V_{out} as a function of temperature.

TABLE 15–2	**Tabulation of Output Voltage Levels for a Temperature Range of 0° to 100° C in Figure 15–21**		
Temperature (in °C)	**R_T (in kΩ)**	**R_i (in kΩ)**	**V_{out} (in V)**
0	29.490	39.490	1.27
25	10.000	20.000	2.50
50	3.893	13.893	3.60
75	1.700	11.700	4.27
100	0.817	10.817	4.62

The output voltage, V_{out}, is fed into an ADC that converts it into an 8-bit binary number. The binary number is then read by a microprocessor that converts it into the corresponding degrees Celsius using software program instructions.

Linear IC Temperature Sensors

The computer software required to convert the output voltages of the previous thermistor circuit is fairly complicated because of the nonlinear characteristics of the device. *Linear temperature sensors* were developed to simplify the procedure. One such device is the LM35 IC temperature sensor. It is fabricated in a three-terminal transistor package and is designed to output 10 mV for each degree Celsius above zero. (Another temperature sensor, the LM34, is calibrated in degrees Fahrenheit.) For example, at 25° C the sensor outputs 250 mV, at 50°C it outputs 500 mV, and so on, in linear steps for its entire range. Figure 15–22 shows how we can interface the LM35 to an ADC and microprocessor.

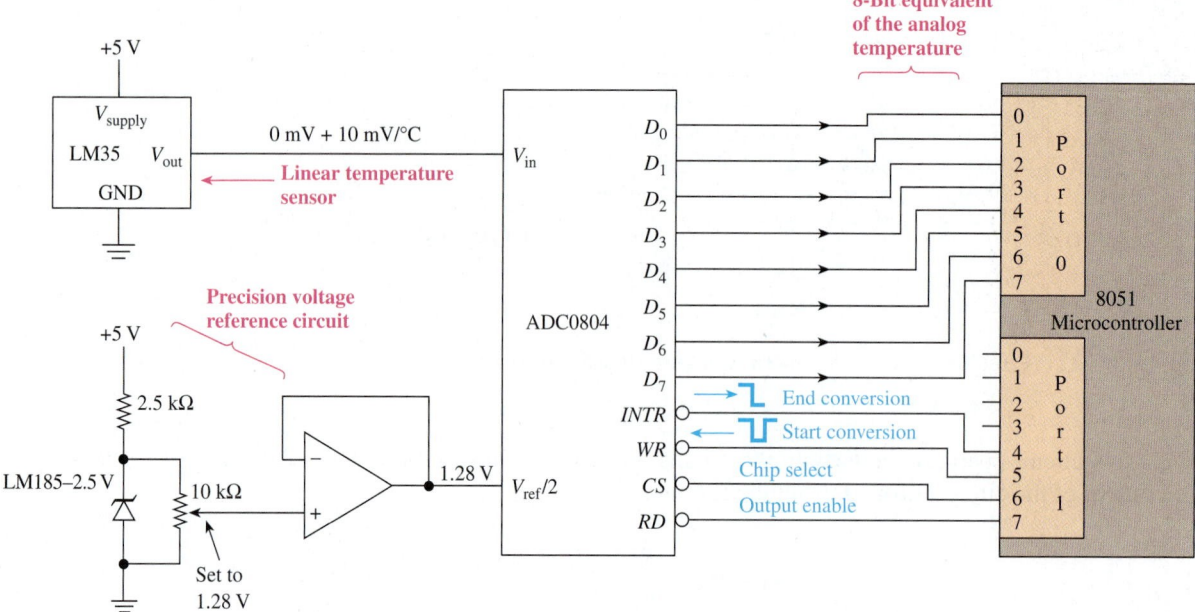

Figure 15–22   Interfacing the LM35 linear temperature sensor to an ADC and a microcontroller.

Team Discussion

Discuss the order of the control signals between the microcontroller and the ADC.

The 1.28-V reference level used in Figure 15–22 is the key to keeping the conversion software programming simple. With 1.28 V at $V_{ref}/2$, the maximum full-scale analog V_{in} is defined as 2.56 V (2560 mV). This value corresponds one-for-one with the 256 binary output steps provided by an 8-bit ADC. A 1° rise in temperature increases V_{in} by 10 mV, which increases the binary output by 1. Therefore, if $V_{in} = 0$ V, D_7 to D_0 = 0000 0000; if $V_{in} = 2.55$ V, D_7 to D_0 = 1111 1111; and if $V_{in} = 1.00$ V, D_7 to D_0 = the binary equivalent of 100, which is 0110 0100. Table 15–3 lists some representative values of temperature versus binary output.

TABLE 15–3	Tabulation of Temperature Versus Binary Output for a Linear Temperature Sensor and an ADC Set Up for 2560 mV Full Scale

Temperature (in °C)	V_{in} (in mV)	Binary Output (D_7 to D_0)
0	0	0000 0000
1	10	0000 0001
2	20	0000 0010
25	250	0001 1001
50	500	0011 0010
75	750	0100 1011
100	1000	0110 0100

In Figure 15–22, the LM185 is a 2.5-V precision voltage reference diode. This diode maintains a steady 2.5 V across the 10-kΩ potentiometer even if the 5-V power supply line fluctuates. The 10-kΩ potentiometer must be set to output exactly 1.280 V. The op amp is used as a unity-gain buffer between the potentiometer and ADC and will maintain a steady 1.280 V for the $V_{ref}/2$ pin.

The Strain Gage

The strain gage is a device whose resistance changes when it is stretched. The gage is stretched, or elongated, when it is "strained" by a physical force. This property makes it useful for measuring weight, pressure, flow, and acceleration.

Several types of strain gages exist, the most common being the foil type illustrated in Figure 15–23. The gage is simply a thin electrical conductor that is looped back and forth and bonded securely to the piece of material to be strained (see Figure 15–24). Applying a force to the metal beam bends the beam slightly, which stretches the strain gage in the direction of its sensitivity axis.

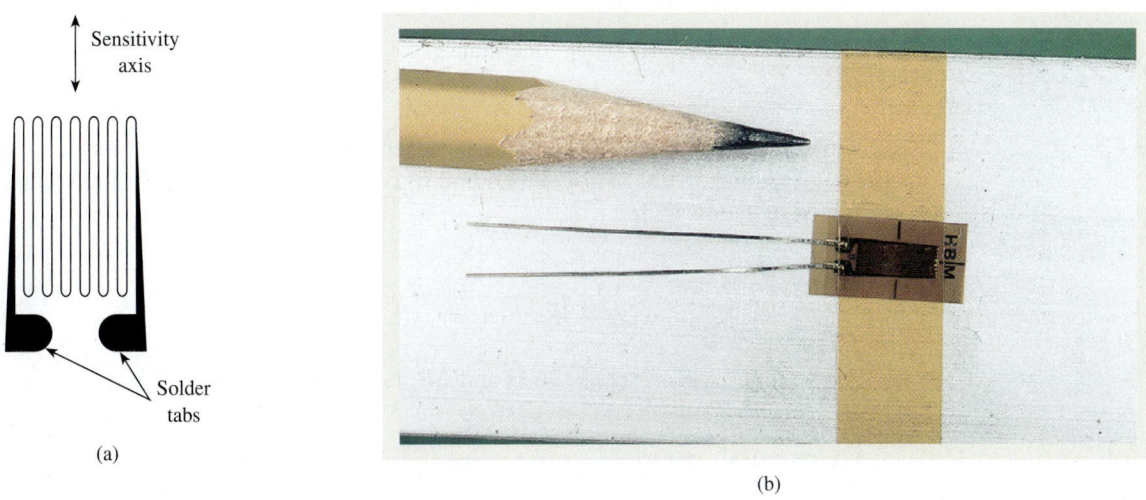

(a)

(b)

Figure 15–23 A foil-type strain gage: (a) sketch; (b) photograph.

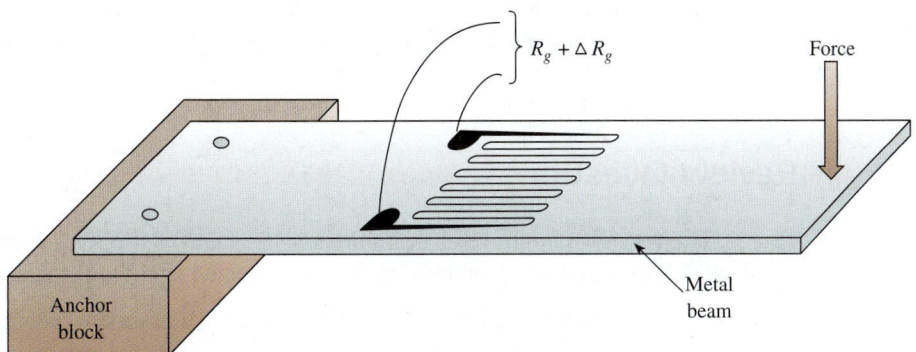

Figure 15–24 Using a strain gage to measure force.

As the strain gage conductor is stretched, its cross-sectional area decreases and its length increases, thus increasing the resistance measured at the solder tabs. The change in resistance is linear with respect to changes in the length of the strain gage. However, the change in resistance is very slight, usually milliohms, and it must be converted to a voltage and amplified before it is input to an ADC. Figure 15–25 shows the signal conditioning circuitry for a 120-Ω strain gage [R_g (unstrained) = 120 Ω]. The instrumentation amplifier is a special-purpose IC used to amplify small-signal differential voltages such as those at points A and B in the bridge circuit in Figure 15–25.

With the metal beam unstrained (no force applied), the 120-Ω potentiometer is adjusted so that V_{out} equals 0 V (called the "null adjustment"). Next, force is applied to the metal beam, elongating the strain gage and increasing its resistance slightly from

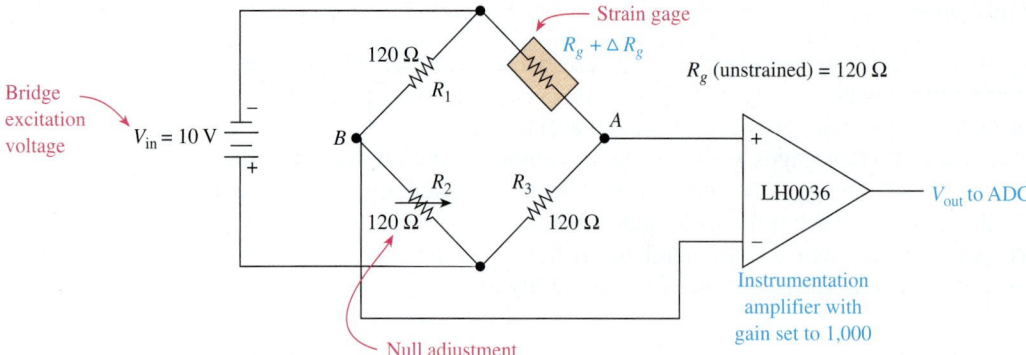

Figure 15–25   Signal conditioning for a strain gage.

its unstrained value of 120 Ω. This increase causes the bridge circuit to become unbalanced, thus creating a voltage at points A and B. From basic circuit theory, the voltage is calculated by the formula

$$V_{AB} = V_{in} \left[\frac{R_3}{R_3 + (R_g + \Delta R_g)} - \frac{R_2}{R_1 + R_2} \right]$$ **15–6**

For example, if R_2 is set at 120 Ω and ΔR_g is 150 mΩ, the voltage is

$$V_{AB} = -10 \left[\frac{120}{120 + (120.150)} - \frac{120}{120 + 120} \right] = 3.12 \text{ mV}$$

Because the instrumentation amplifier gain is set at 1000, the output voltage sent to the ADC will be 3.12 V.

Through experimentation with several known weights, the relationship of force versus V_{out} can be established and programmed into the microprocessor reading the ADC output.

Review Questions

15–21. The AM7305 multiplexer in Figure 15–19 is used to apply the appropriate voltage gain to each of the transducers connected to it. True or false?

15–22. The 8-bit data out from the ADC in Figure 15–19 passes to the microprocessor via the data bus. True or false?

15–23. Signal conditioning is required to make transducer output levels compatible with ADC input requirements. True or false?

15–24. What is the advantage of using the LM35 linear temperature sensor over a thermistor for measuring temperature?

Summary

In this chapter, we have learned that

1. Any analog quantity can be represented by a binary number. Longer binary numbers provide higher resolution, which gives a more accurate representation of the analog quantity.

2. Operational amplifiers are important building blocks in analog-to-digital (A/D) and digital-to-analog (D/A) converters. They provide a means for summing currents at the input and converting a current to a voltage at the output of converter circuits.

3. The binary-weighted D/A converter is the simplest to construct, but it has practical limitations in resolution (number of input bits).

4. The $R/2R$ ladder D/A converter uses only two different resistor values, no matter how many binary input bits are included. This allows for very high resolution and ease of fabrication in IC form.

5. The DAC0808 (or MC1408) IC is an 8-bit D/A converter that uses the $R/2R$ ladder method of conversion. It accepts 8 binary input bits and outputs an equivalent analog current. Having 8 input bits means that it can resolve up to 256 unique binary values into equivalent analog values.

6. Applying an 8-bit counter to the input of an 8-bit D/A converter will produce a 256-step sawtooth waveform at its output.

7. The simplest way to build an analog-to-digital (A/D) converter is to use the parallel encoding method. The disadvantage is that it is practical only for low-resolution applications.

8. The counter-ramp A/D converter employs a counter, a D/A converter, and a comparator to make its conversion. The counter counts from zero up to a value that causes the D/A output to exceed the analog input value slightly. That binary count is then output as the equivalent to the analog input.

9. The method of A/D conversion used most often is called successive approximation. In this method, successive bits are tested to see if they contribute an equivalent analog value that is greater than the analog input to be converted. If they do, they are returned to zero. After all bits are tested, the ones that are left ON are used as the final digital equivalent to the analog input.

10. The NE5034 and the ADC0804 are examples of A/D converter ICs. To make a conversion, the *start-conversion* pin is made LOW. When the conversion is completed the *end-of-conversion* pin goes LOW. Then to read the digital output, the *output enable* pin is made LOW.

11. Data acquisition systems are used to read several different analog inputs, respond to the values read, store the results, and generate reports on the information gathered.

12. Transducers are devices that convert physical quantities such as heat, light, or force into electrical quantities. Those electrical quantities must then be conditioned (or modified) before they can be interpreted by a digital computer.

Glossary

A/D Converter (ADC): Analog-to-digital converter.

Binary Weighting: Each binary position in a string is worth double the amount of the bit to its right. By choosing resistors in that same proportion, binary-weighted current levels will flow.

Bus: A common set of electrical conductors shared by several devices and ICs.

Continuous Conversions: An ADC that is connected to repeatedly perform analog-to-digital conversions by using the end-of-conversion signal to trigger the start-conversion input.

Conversion Time: The length of time between the start of conversion and end of conversion of an ADC.

D/A Converter (DAC): Digital-to-analog converter.

Data Acquisition: A term generally used to refer to computer-controlled acquisition and conversion of analog values.

Differential Measurement: The measurement of the difference between two values.

Handshaking: Devices and ICs that are interfaced together must follow a specific protocol, or sequence of control operations, to be understood by each other.

Interfacing: The device control and interconnection schemes required for electronic devices and ICs to communicate with each other.

Memory: A storage device capable of holding data that can be read by some other device.

Microprocessor: A large-scale IC capable of performing several functions, including the interpretation and execution of programmed software instructions.

Nonlinearity: Nonlinearity error describes how far the actual transfer function of an ADC or DAC varies from the ideal straight line drawn from zero up to the full-scale values.

Nonmonotonic: A nonmonotonic DAC is one in which, for every increase in the input digital code, the output level does not either remain the same or increase.

Op Amp: An amplifier that exhibits almost ideal features (i.e., infinite input impedance, infinite gain, and zero output impedance).

Programmable-Gain Amplifier: An amplifier that has a variable voltage gain that is set by inputting the appropriate digital levels at the gain select inputs.

Reference Voltage: In DAC and ADC circuits, a reference voltage or current is provided to the circuit to set the relative scale of the input and output values.

Resolution: The number of bits in an ADC or DAC. The higher the number, the closer the final representation can be to the actual input quantity.

Sample and Hold: A procedure of taking a reading of a varying analog value at a precise instant and holding that reading.

Successive Approximation: A method of arriving at a digital equivalent of an analog value by successively trying each of the individual digital bits, starting with the MSB.

Thermistor: An electronic component whose resistance changes with a change in temperature.

Transducer: A device that converts a physical quantity, such as heat or light, into an electrical quantity such as amperes or volts.

Virtual Ground: In certain op-amp circuit configurations, with one input at actual ground potential, the other input will be held at a 0-V potential but will not be able to sink or source current.

Sections 15–1 and 15–2

15–1. Describe the function of a transducer.

15–2. How many different digital representations are allowed with:

(a) A 4-bit converter?   **(c)** An 8-bit converter?

(b) A 6-bit converter?   **(d)** A 12-bit converter?

15–3. List three characteristics of op amps that make them almost ideal amplifiers.

15–4. Determine V_{out} for the op-amp circuits of Figure P15–4.

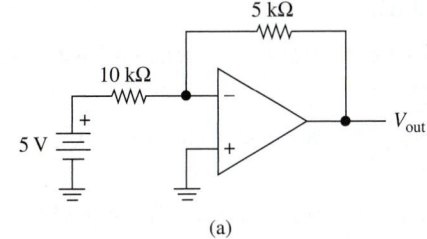

(a)

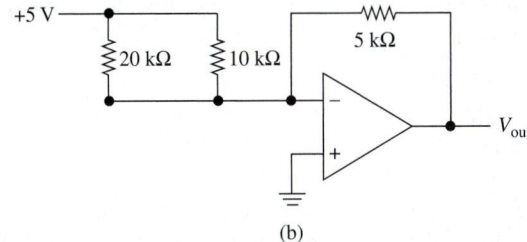

(b)

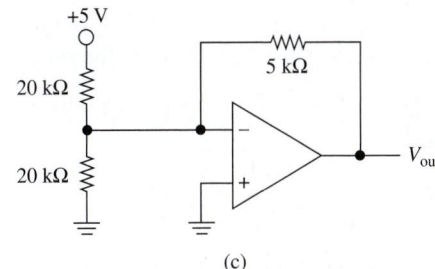

(c)

Figure P15–4

15–5. The virtual ground concept simplifies the analysis of op-amp circuits by allowing us to assume what?

Sections 15–3 and 15–4

15–6. Calculate the current through each switch and the resultant V_{out} in Figure 15–4 if the number 12_{10} is input.

D   **15–7.**

(a) Change the resistor that is connected to the D_3 switch in Figure 15–4 to 10 kΩ. What values must be used for the other three resistors to ensure the correct binary weighting factors?

(b) Reconstruct the data table in Figure 15–4 with new values for V_{out} using the resistor values found in part (a).

15–8. What effect would doubling the 20-kΩ resistor have on the values for V_{out} in Figure 15–4?

15–9. What effect would changing the reference voltage (V_{ref}) in Figure 15–6 from +5 to −5 V have on V_{out}?

15–10. Change V_{ref} in Figure 15–6 to +2 V, and calculate V_{out} for $D_0 = 0$, $D_1 = 0, D_2 = 0, D_3 = 1$.

15–11. Reconstruct the data table in Figure 15–7 for a V_{ref} of +2 V instead of +5 V.

Sections 15–5 and 15–6

15–12. Does the MC1408 DAC use a binary-weighted or an $R/2R$ method of conversion?

15–13. What is the purpose of the op amp in the DAC application circuit of Figure 15–8(c)?

15–14. What is the resolution of the DAC0808/MC1408 DAC shown in Figure 15–8?

C

15–15. Calculate V_{out} in Figure 15–8(c) for the following input values:

(a) $0100\ 0000_2$

(b) $0011\ 0110_2$

(c) 32_{10}

(d) 30_{10}

D

15–16. Sketch a partial transfer function of analog output versus digital input in Figure 15–8(c) for digital input values of 0000 0000 through 0000 0111.

15–17. How could the reference current (I_{ref}) in Figure 15–8 be changed to 1.5 mA? What effect would that have on the range of I_{out} and V_{out}?

15–18. In Figure 15–8(c), if V_{ref} is changed to 5 V, find V_{out} full-scale (A_1 to A_8 = HIGH).

Sections 15–7 and 15–8

15–19. Draw a graph of the transfer function (digital output versus analog input) for the parallel-encoded ADC of Figure 15–11.

15–20. What is one advantage and one disadvantage of using the multiple-comparator parallel encoding method of A/D conversion?

15–21. Refer to the counter-ramp ADC of Figure 15–12.

(a) What is the level at the DAC output the *instant after* the start conversion push button is pressed?

(b) What is the relationship between the $V(+)$ and $V(-)$ comparator inputs the *instant before* the HIGH-to-LOW edge of end of conversion?

15–22. In Figure 15–12, what is the worst-case (longest) conversion time that might be encountered if the clock frequency is 100 kHz?

Sections 15–9 and 15–10

15–23. Determine the conversion time for an 8-bit ADC that uses a successive-approximation circuit similar to Figure 15–13 if its clock frequency is 50 kHz.

15–24. What connections could be made in the ADC shown in Figure 15–13 to enable it to make continuous conversions?

C **15–25.** Use the SAR ADC of Figure 15–13 to convert the analog voltage of 7.28 to 8-bit binary. If $V_{ref} = 10$ V, determine the final binary answer and the percent error.

15–26. Why is the three-state buffer at the output of the NE5034 ADC an important feature?

15–27. Referring to the block diagram of the ADC0804 (Figure 15–17), which inputs are used to enable the three-state output latches? Are they active-LOW or active-HIGH inputs?

15–28. What type of application might require the use of the differential inputs $[V_{in(+)}, V_{in(-)}]$ on the ADC0804?

15–29. Refer to Figures 15–17 and 15–18.

(a) How would the operation change if $\overline{RD}$ were connected to +5 V instead of ground?

(b) How would the operation change if $\overline{CS}$ were connected to +5 V instead of ground?

(c) What is the purpose of the 10-kΩ–0.001-μF RC circuit?

(d) What is the maximum range of the analog V_{in} if $V_{ref}/2$ is changed to 0.5 V?

C **15–30.** Change V_{CC} to 5.12 V in Figure 15–18, and determine which LEDs will be ON for the following values of V_{in}:

(a) 3.6 V

(b) 1.86 V

Sections 15–11 and 15–12

15–31. Briefly describe the flow of the signal from the temperature transducer as it travels through the circuit to the RAM memory in the data acquisition system of Figure 15–19.

C D **15–32.** Table 15–2 gives the values for the output of the thermistor circuit of Figure 15–21 for various temperatures. Redesign the circuit by changing R_F so that V_{out} equals 5 V at 25°C.

C **15–33.** The output voltage of Figure 15–21 is exactly 2.5 V at 25°C. Calculate the range of output voltage if the 10 kΩ thermistor has a tolerance of 5%.

15–34. What binary value will the 8051 microcontroller read in Figure 15–22 if the temperature is 60°C?

C **15–35.** Through experimentation, it is determined that the strain gage in Figures 15–24 and 15–25 changes in resistance by 20 mΩ for each kilogram that is added to the metal beam. Determine how many kilograms of force are on the metal beam if V_{out} is 4 V in Figure 15–25.

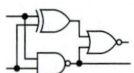

Schematic Interpretation Problems

See Appendix G for the schematic diagrams.

S C D 15–36. Design a circuit interface that will provide analog output capability to the 4096/4196 control card. Assume that software will be written by a programmer to output the appropriate digital strings to port 1 (P1.7–P1.0) of the 8031 microcontroller. Devise an analog output circuit using an MC1408 DAC with a 741 op amp to output analog voltages in the range of 0 to 5 V.

S C D 15–37. Design a circuit interface that will provide analog input capability to the 4096/4196 control card. The design must be capable of inputting the 8-bit digital results from *two* ADC0804 converters into port 1 (P1.7–P1.0) of the 8031 microcontroller. Assume that a single-bit control signal will be output on port 2, bit 0 (P2.0) to tell which ADC results are to be transmitted. (Assume 1 = ADC 1 and 0 = ADC 2). Set up the ADCs for continuous conversions and 0- to 5-V analog input level.

Electronics Workbench®/MultiSIM® Exercises

E15–1. Load the circuit file for **Section 15–3a.** This is a binary-weighted D/A converter like that shown in Section 15–3.

(a) Close each switch individually to determine how many volts each switch contributes to Vout. List those four values in a data table.

(b) The current through If is always equal to Isum (T or F)?

(c) Which switches need to be closed to yield the following outputs: −8V, −4V, −6V, −11V?

(d) How many different digital numbers can be represented with the four input switches?

(e) To make a 5-bit D/A you would need to add another switch and resistor to the left of D3. What size would that resistor be?

C

E15–2. Load the circuit file for **Section 15–5a.** This DAC outputs a voltage proportional to its 8-bit binary input. The +Vref input is set to 5V. This in turn sets the maximum output at 5V. Because there are 256 possible binary input steps, then the volts per step will be 5V/256 = 19.53 mV. The word generator is used to count from 0 to F repeatedly. (The four high-order inputs are not used in this case.)

(a) On a piece of paper, sketch and label the analog output value.

(b) Change the "Final" address in the Word Generator to FF so that all 256 steps will be output. Sketch and label the resultant waveform.

E15–3. Load the circuit file for **Section 15–10a.** This ADC is used to convert the Analog Vin into a binary number at the output of the ADC. This binary output is displayed in hex on the two seven-segment displays. The Analog Vin is varied by changing the potentiometer setting (R and Shift-R).

(a) Create a data table of Analog Vin versus hex output for 11 steps in Vin from 0 to 5V.

(b) Convert all of your hex answers to decimal. Do they appear to be taking approximately equal steps?

C   **E15–4.** Load the circuit file for **Section 15–10b.** The object of this design is to digitize several points along a slow-moving sine wave. The analog sine wave that is input to the ADC is set up to oscillate between the levels 0 to 5V. The ADC converts the analog signal into its digital equivalent at about 30 data points. As you drag the #1 cursor the hex numbers change. (First, they increase as the sine wave increases, and then they decrease as the sine wave's amplitude decreases.)

    (a) List on a piece of paper the hex digits that you read as you drag the #1 cursor.

    (b) Identify the hex numbers that represent the midpoint, peak, and valley of the sine wave.

Answers to Review Questions

15–1. True

15–2. 256

15–3. Infinite, 0

15–4. 240 kΩ

15–5. Because finding accurate resistances over such a large range of values would be very difficult

15–6. False

15–7. True

15–8. Current

15–9. A_1

15–10. Number of bits at the input or output

15–11. True

15–12. Gain error

15–13. False

15–14. 1023

15–15. False

15–16. False

15–17. $\overline{STRT}$, $\overline{DR}$

15–18. False

15–19. (a) Input (b) input (c) input (d) output

15–20. $\overline{CS}$, $\overline{WR}$, $\overline{INTR}$, $\overline{RD}$

15–21. False

15–22. True

15–23. True

15–24. It is a linear device, whereas the thermistor is a nonlinear device.

16

Semiconductor, Magnetic, and Optical Memory

OUTLINE

OBJECTIVES

Upon completion of this chapter, you should be able to:

- Explain the basic concepts involved in memory addressing and data storage.
- Interpret the specific timing requirements given in a manufacturer's data manual for reading or writing to a memory IC.
- Discuss the operation and application for the various types of semiconductor memory ICs.
- Design circuitry to facilitate memory expansion.
- Explain the refresh procedure for dynamic RAMs.
- Explain the differences between the various types of magnetic and optical storage.

INTRODUCTION

In digital systems, memory circuits provide the means of storing information (data) on a temporary or permanent basis for future recall. The storage medium can be either a semiconductor IC, a magnetic device such as magnetic tape or disk, or optical storage such as CD or DVD. **Magnetic** and **optical memory** generally are capable of storing larger quantities of data than semiconductor memories, but the access time (time it takes to locate and then read or write data) is usually much more. With magnetic and optical disks, it takes time to physically move the read/write mechanism to the exact location to be written to or read form.

With **semiconductor memory** ICs, electrical signals are used to identify a particular memory location within the IC, and data can be stored in or read from that location in a matter of nanoseconds.

The technology used in the fabrication of memory ICs can be based on either bipolar or MOS transistors. In general, bipolar memories are faster than MOS memories, but MOS can be integrated more densely, providing much more memory locations in the same amount of area.

16–1 Memory Concepts

Let's say that you have an application where you must store the digital states of eight binary switches once every hour for 16 hours. This would require 16 *memory locations,* each having a *unique 4-bit* **memory address** (0000 to 1111) and each being capable of containing 8 bits of data as the **memory contents.** A group of 8 bits is also known as 1 **byte,** so what we would have is a 16-byte memory, as shown in Figure 16–1.

To set up this memory system using actual ICs, we could use 16 8-bit flip-flop registers to contain the 16 bytes of data. To identify the correct address, a 4-line-to-16-line decoder can be used to decode the 4-bit address location into an active-LOW chip select to select the appropriate (1-of-16) data register for input/output. Figure 16–2 shows the circuit used to implement this memory application.

The 74LS374s are octal (eight) D flip-flops with three-state outputs. To store data in them, 8 bits of data are put on the D_0 to D_7 data inputs via the data bus. Then, a LOW-to-HIGH edge on the C_p clock input will cause the data at D_0 to D_7 to be latched into each flip-flop. The value stored in the D flip-flops is observed at the Q_0 to Q_7 outputs by making the Output Enable $(\overline{OE})$ pin LOW.

To select the appropriate (1-of-16) memory location, a 4-bit address is input to the 74LS154 (4-line-to-16-line decoder), which outputs a LOW pulse on one of the output lines when the $\overline{WRITE}$ enable input is pulsed LOW.

As you can see, the timing for setting up the address bus and data bus, and pulsing the $\overline{WRITE}$ line is critical. Timing diagrams are necessary for understanding the operation of memory ICs, especially when you are using larger-scale memory ICs. The timing diagram for our 16-byte memory design of Figure 16–2 is given in Figure 16–3.

Figure 16–3 begins to show us some of the standard ways that manufacturers illustrate timing parameters for bus-driven devices. Rather than showing all four address lines and all eight data lines, they group them together and use an X (crossover) to show where any or all of the lines are allowed to change digital levels.

In Figure 16–3 the address and data lines must be set up some time (t_s) before the LOW-to-HIGH edge of $\overline{WRITE}$. In other words, the address and data lines must *be valid* (be at the appropriate levels) some period of time (t_s) *before* the LOW-to-HIGH edge of $\overline{WRITE}$ in order for the 74LS374 D flip-flop to interpret the input correctly.

When the $\overline{WRITE}$ line is pulsed, the 74LS154 decoder outputs a LOW pulse on one of its 16 outputs, which clocks the appropriate memory location to receive data from the data bus. After the propagation delay (t_p), the data output at Q_0 to Q_7 will be the new data just entered into the D flip-flop. The t_p will include the propagation delay of the decoder *and* the C_p-to-Q of the D flip-flop.

Figure 16–1 Layout for 16 8-bit memory locations.

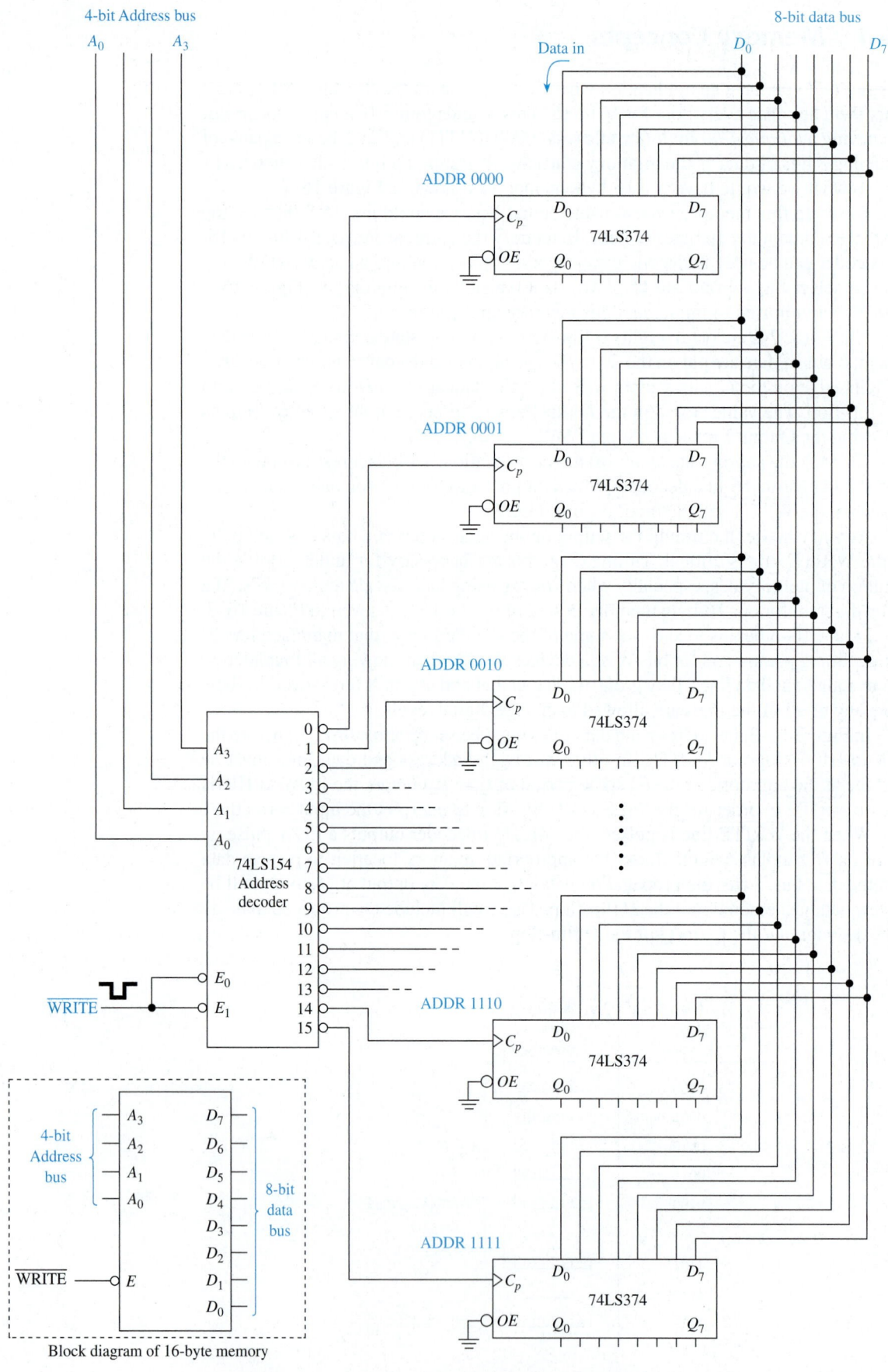

Figure 16–2 Writing to a 16-byte memory constructed from 16 octal *D* flip-flops and a 1-of-16 decoder. (Data travel down the data bus to the addressed *D* flip-flop.)

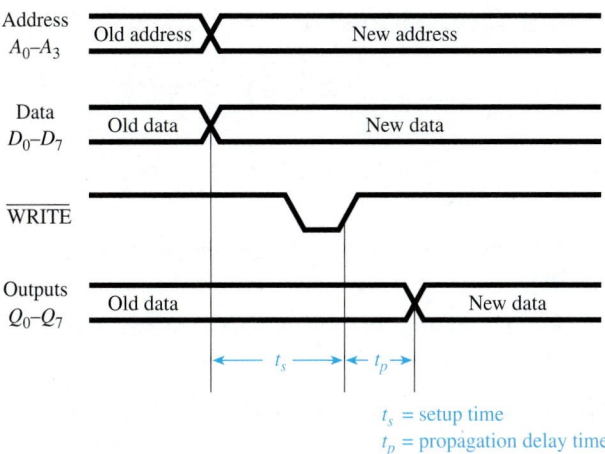

t_s = setup time
t_p = propagation delay time

Figure 16–3 Timing requirements for writing data to the 16-byte memory circuit of Figure 16–2.

Team Discussion

What is the order of operation between the data bus, the address bus, and the write pulse?

In Figure 16–2 all the three-state outputs are continuously enabled so that their Q outputs are always active. To connect the Q_0 to Q_7 outputs of all 16 memory locations back to the data bus, the $\overline{OE}$ enables would have to be individually selected at the appropriate time to avoid a conflict on the data bus, called **bus contention.** Bus contention occurs when two or more devices are trying to send their own digital levels to the shared data bus at the same time. To individually select each group of Q outputs in Figure 16–2, the grounds on the $\overline{OE}$ enables would be removed and, instead, be connected to the output of another 74LS154 1-of-16 decoder, as shown in Figure 16–4.

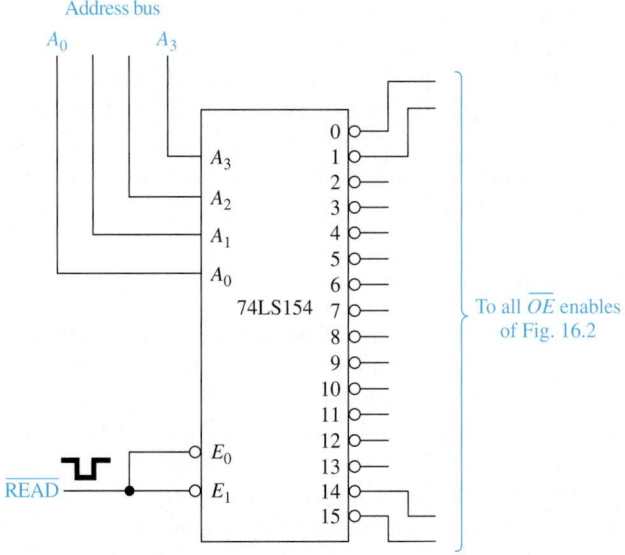

Figure 16–4 Using another decoder to individually select memory locations for Read operations.

Team Discussion

Discuss the entire sequence of events and timing required to store 1 byte of data at each memory location in Figure 16–2. Repeat for the retrieval of each byte of data using the decoder in Figure 16–4.

In Figures 16–2, 16–3, and 16–4, we have designed a small 16-byte (16 × 8) random-access memory (**RAM**). Commercially available RAM ICs combine all the decoding and storage elements in a single package, as seen in the next section.

Review Questions

16–1. The _____ bus is used to specify the location of the data stored in a memory circuit.

16–2. Once a memory location is selected, data travel via the _____ bus.

16–3. In Figure 16–2, how is the correct octal D flip-flop chosen to receive data?

16–4. What is the significance of the X (crossover) on the address and data waveforms in Figure 16–3?

16–5. Why would a second 74LS154 decoder be required in Figure 16-2 to read the data from the D flip-flops if all Q outputs were connected back to the data bus? (*Hint:* See Figure 16–4.)

16–2   Static RAMs

Inside Your PC

A PC basically has two kinds of semiconductor memory, SRAM and DRAM. The SRAM is faster but more expensive per bit. It is used for the cache memory, which is a temporary holding area for frequently used data items from the much slower DRAM memory. A typical cache size is 512kB (kilo-Bytes) having access times of less than 2 ns. DRAM, on the other hand, is much larger because it is used for all of the applications and graphics that are active at the time. A typical DRAM memory size is 96 or 128MB (Mega-Bytes) and will have access times more than 10 ns. Computer DRAM is usually referred to simply as RAM.

Large-scale *random-access memory* (RAM), also known as *read/write memory,* is used for temporary storage of data and program instructions in microprocessor-based systems. The term *random access* means that the user can access (read or write) data at any location within the entire memory device randomly without having to sequentially read through several data values until positioned at the desired memory location. [An example of a sequential (nonrandom) memory device is magnetic tape. A CD player or hard-disk drive has random access capability.]

A better term for RAM is *read/write memory* (RWM) because all semiconductor and disk memories have random access. RWM is more specific because it tells us that data can be *read or written* to any memory location, but RAM is the industry-standard term.

RAM is classified as either static or dynamic. *Static* RAMs (SRAMs) use flip-flops as basic storage elements, whereas *dynamic* RAMs (DRAMs) use internal capacitors as basic storage elements. Additional *refresh* circuitry is needed to maintain the charge on the internal capacitors of a dynamic RAM, which makes it more difficult to use. Dynamic RAMs can be packed very densely, however, yielding much more storage capacity per unit area than a static RAM. The cost per bit of dynamic RAM is also much less than that of the static RAM.

The 2147H Static MOS RAM

The 2147H is a static RAM that uses MOS technology. The 2147H is set up with 4096 (abbreviated 4K, where 1K = 1024) memory locations, with each location containing 1 bit of data. This configuration is called 4096 × 1. This is very small by today's standard, but studying its operation is a good start for understanding the larger SRAMs.

To develop a unique address for each of the 4096 locations, 12 address lines must be input (2^{12} = 4096). The storage locations are set up as a 64 × 64 array with A_0 to A_5 identifying the row and A_6 to A_{11} identifying the column to pinpoint the specific location to be used. The data sheet for the 2147H is given in Figure 16–5(a). This figure shows the row and column circuitry used to pinpoint the **memory cell** within the 64 × 64 array. The box labeled "Row Select" is actually a 6-to-64 decoder for identifying the appropriate 1-of-64 row. The box labeled "Column Select" is also a 6-to-64 decoder for identifying the appropriate 1-of-64 column.

Figure 16–5(b) can be used to show how individual memory cells within the 2147H memory array are pinpointed. The *Row Select Decoder* uses the lower address lines (A_5 to A_0) to develop a single active-HIGH line based on their binary value. For example, if A_5 to $A_0 = 000010$, then ROW 2 will be HIGH. Then, to pinpoint the exact cell, the *Column Select Decoder* uses the high-order address lines (A_{11} to A_6) to determine which of the 64 possible columns should be active-HIGH.

For example, if the complete 12-bit address (A_{11} to A_0) is 111111 000010, we will be accessing the memory cell that is in row 2, column 63, as illustrated in Figure 16–5(b).

Once the location is selected, the AND gates at the bottom of the block diagram allow the data bit to either pass into (D_{in}) or come out of (D_{out}) the memory location selected. Each memory location, or cell, is actually a configuration of transistors that functions like a flip-flop that can be Set (1) or Reset (0).

During *write operations*, for D_{in} to pass through its three-state buffer, the Chip Select ($\overline{CS}$) must be LOW *and* the Write Enable ($\overline{WE}$) must also be LOW. During *read*

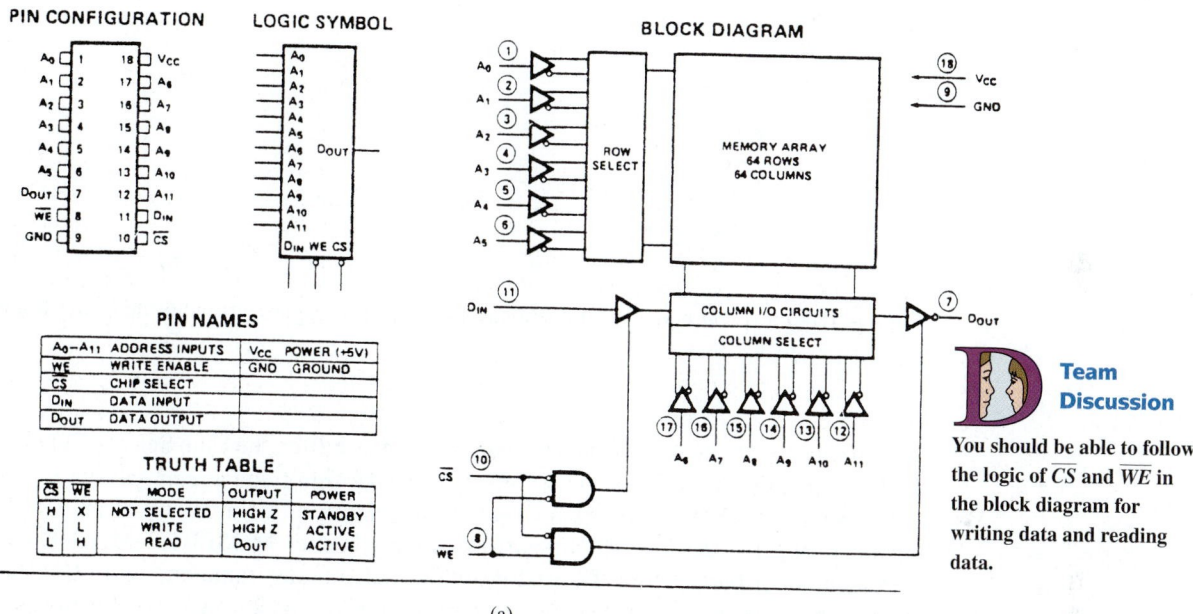

You should be able to follow the logic of $\overline{CS}$ and $\overline{WE}$ in the block diagram for writing data and reading data.

(a)

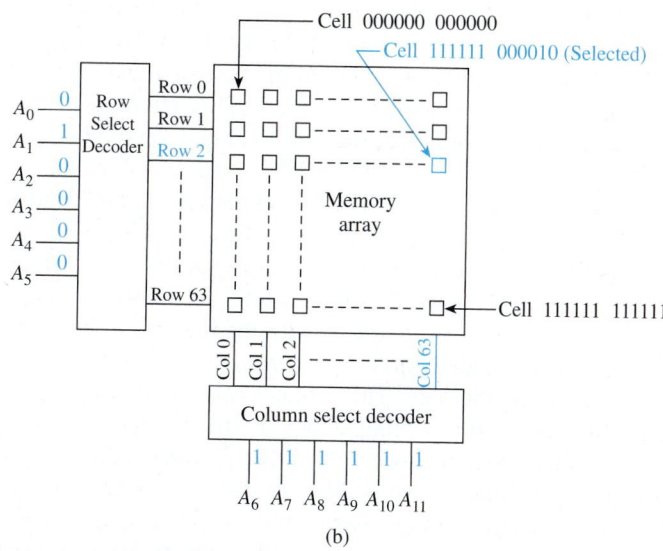

(b)

Figure 16–5 The 2147H 4K × 1 static RAM: (a) Data sheet; (b) Address row and column decoders select memory cell 111111 000010 in the 2147H memory array;

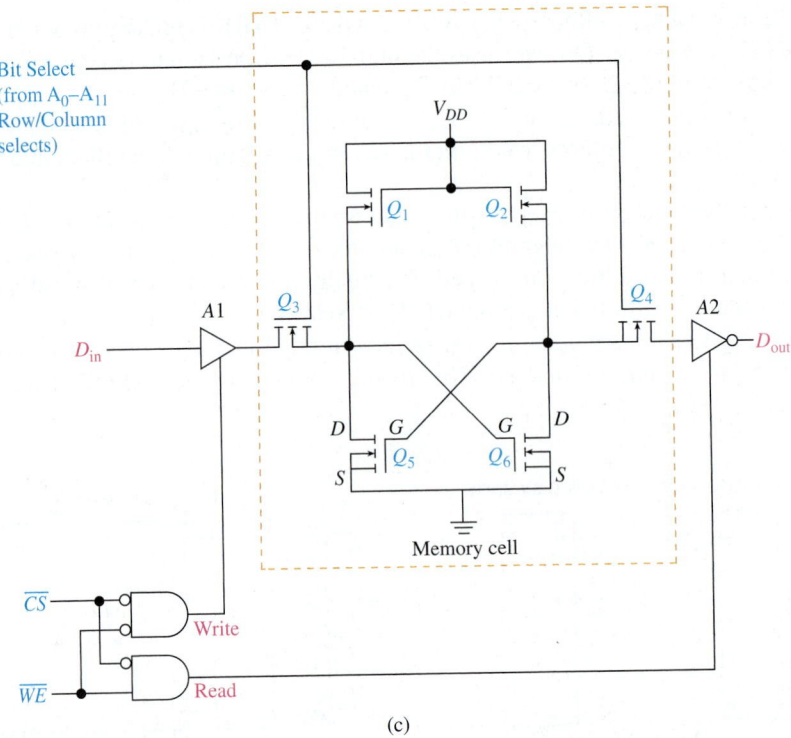

Bit Select
(from A_0–A_{11}
Row/Column
selects)

V_{DD}

Q_1 Q_2

A1 Q_3 Q_4 A2

D_{in} D_{out}

D   G   G   D
Q_5 Q_6
S   S

Memory cell

(c)

$\overline{CS}$
Write

$\overline{WE}$
Read

Figure 16–5   *(Continued)* (c) Functional diagram of a single cell within the static RAM memory array.

operations, for D_{out} to receive data from its three-state buffer, the Chip Select $(\overline{CS})$ must be LOW *and* the Write Enable $(\overline{WE})$ must be HIGH, signifying a *read operation.*

Figure 16–5(c) shows the internal configuration of a single cell of an SRAM memory like the 2147H. All transistors are N-channel MOSFETs (Metal Oxide Semiconductor Field-Effect Transistor), which, as you may remember, turn ON (become a short) when a positive voltage (a logic 1) is placed on it's gate. (See Section 9–5 for a review of MOSFETs). A_1 and A_2 are three-state buffers used to enable a data bit to be *written into* the memory cell if $\overline{WE} = 0$ or *read from* the memory cell if $\overline{WE} = 1$. To select this particular memory cell, the A_0 to A_{11} address lines on the 2147H are decoded into a single row and column in such a way to place a HIGH on the internal memory array line labeled *bit select.* This HIGH turns on Q_3 and Q_4, which allow the data to enter (or leave) the cross-connection circuit (Q_5 and Q_6) that holds the data bit.

To store a 1 into the memory cell, D_{in} is made HIGH, and $\overline{CS}$ and $\overline{WE}$ are made LOW. The HIGH will pass through Q_3 to the gate of Q_6, turning it ON. An ON transistor essentially acts like a short, placing 0 V (0) at the drain of Q_6, which also places 0 V at the gate of Q_5, turning it off. Because Q_5 is OFF, its drain voltage will be close to V_{DD} (1). Q_1 and Q_2 are used to provide the bias for the memory transistors, Q_5 and Q_6.

Now that the data are loaded into the cell, this *bit select* line can go LOW, allowing the 2147H to access another cell. This isolates this memory cell from the outside world by turning Q_3 and Q_4 OFF. As long as V_{DD} is still applied, the 1 that was loaded into the cell will remain there in the form of an ON Q_6, which holds Q_5 in an OFF state, which in turn holds Q_6 in an ON state similar to the cross-NAND *S-R* flip-flop. The need to maintain a V_{DD} supply voltage makes this a **volatile** memory. If V_{DD} is turned off, no one could predict what state Q_5 and Q_6 will return to when power is reapplied. (Now you know why you lost the almost-completed term paper that you were working on when you inadvertently hit the power switch on your PC.)

To read data out of the memory cell, $\overline{CS}$ is made LOW, and $\overline{WE}$ is made HIGH, enabling the A_2 inverting buffer. Then, when this particular cell is selected by a HIGH on *bit select*, the value on Q_6 will be inverted and passed out to D_{out} as the original level that was input at D_{in}.

The timing waveforms for the Read and Write cycles are given in Figure 16–6.

Read Operation: The circuit connections and waveforms for reading data from a location in a 2147H are given in Figure 16–6(b). The 12 address lines are brought in from the address bus for address selection. The $\overline{WE}$ input is held HIGH to enable the Read operation.

Referring to the timing diagram, when the new address is entered in the A_0 to A_{11} inputs and the $\overline{CS}$ line goes LOW, it takes a short period of time, called the *access time*, before the data output is valid. The access time is the length of time from the beginning of the read cycle to the end of t_{ACS}, or t_{AA}, whichever ends last. Before the $\overline{CS}$ is

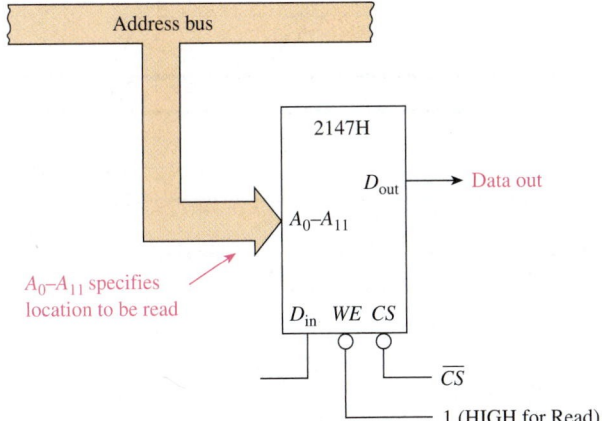

Team Discussion

Discuss the differences in the timing waveforms of D_{out} and D_{in} in Figure 16–6.

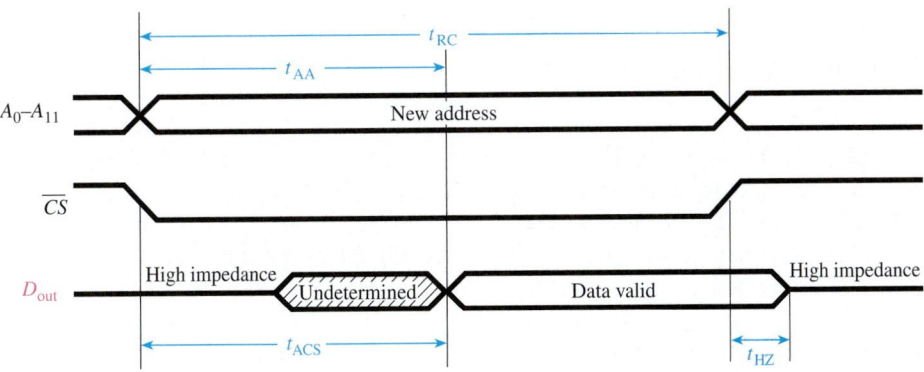

Symbol	Parameter	Min.	Max.	Unit
t_{RC}	Read cycle time	35		ns
t_{AA}	Address access time		35	ns
t_{ACS}	Chip select access time		35	ns
t_{HZ}	Chip deselection to high-Z out	0	30	ns

(a)

Figure 16–6 The 2147H static RAM timing waveforms: (a) Read cycle;

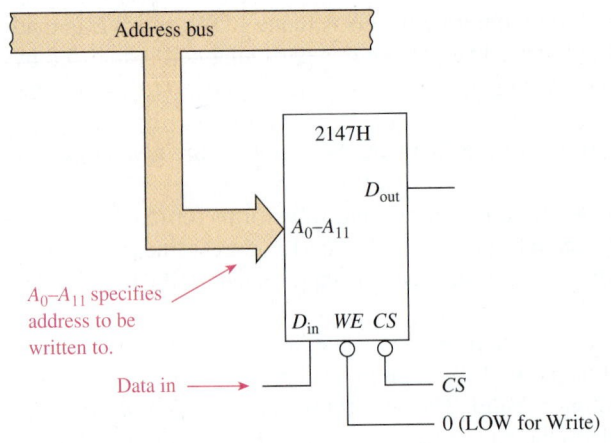

Address bus

2147H

D_{out}

A_0–A_{11}

A_0–A_{11} specifies address to be written to.

D_{in} WE CS

Data in →

$\overline{CS}$

0 (LOW for Write)

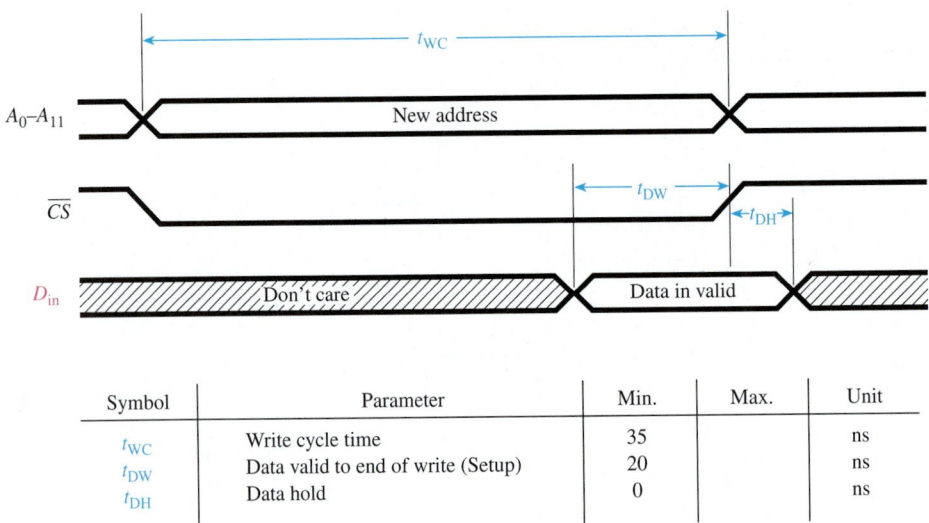

Symbol	Parameter	Min.	Max.	Unit
t_{WC}	Write cycle time	35		ns
t_{DW}	Data valid to end of write (Setup)	20		ns
t_{DH}	Data hold	0		ns

(b)

Figure 16–6 *(Continued)* (b) Write cycle.

brought LOW, D_{out} is in a high-impedance (float) state. The $\overline{CS}$ *and* A_0 to A_{11} inputs must both be held stable for a minimum length of time, t_{RC}, before another Read cycle can be initiated.

After $\overline{CS}$ goes back HIGH, the data out is still valid for a short period of time, t_{HZ}, before returning to its high-impedance state.

Write Operation: A similar set of waveforms is used for the write operation [Figure 16–6(b)]. In this case, the D_{in} is written into memory while the $\overline{CS}$ and $\overline{WE}$ are both LOW. The D_{in} must be set up for a length of time *before* either $\overline{CS}$ or $\overline{WE}$ go back HIGH (t_{DW}), and it must also be held for a length of time *after* either $\overline{CS}$ or $\overline{WE}$ go back HIGH (t_{DH}).

Memory Expansion: Because the contents of each memory location in the 2147H is only 1 bit, to be used in an 8-bit computer system, eight 2147Hs must be set up in such a way that, when an address is specified, 8 bits of data will be read or written. With eight 2147s, we have a 4096 × 8 (4K × 8) memory system, as shown in Figure 16–7(a).

The address selection for each 2147H in Figure 16–7(a) is identical because they are all connected to the same address bus lines. This way, when reading or writing from

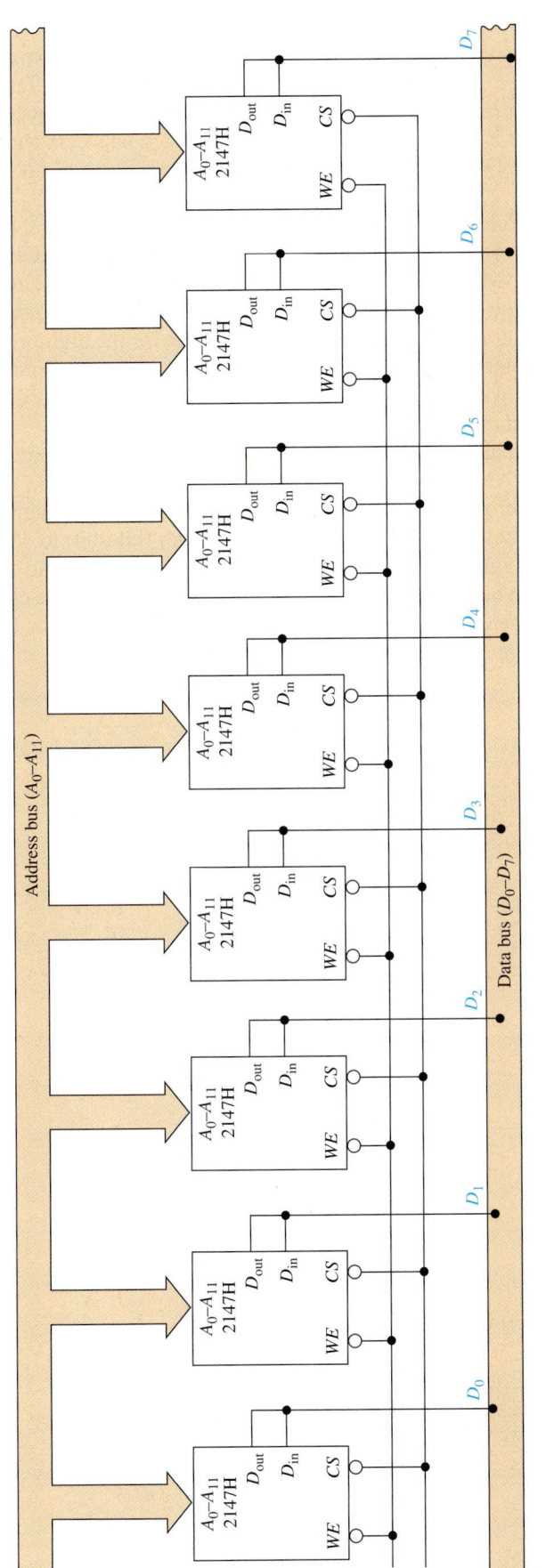

Figure 16–7 RAM memory expansion: (a) eight 2147Hs configured as a 4K × 8; (b) thirty-two 2114s configured as two banks of 8K × 8.

(a)

(b)

723

a specific address, 8 bits, each at the same address, will be sent to or received from the data bus simultaneously. The $\overline{WE}$ input determines which internal three-state buffer is enabled, connecting *either* D_{in} or D_{out} to the data bus. The $\overline{WE}$ input is sometimes labeled READ/$\overline{WRITE}$, meaning that it is HIGH for a Read operation, which puts data out to the data bus via D_{out}, and it is LOW for a Write operation, which writes data into the memory via D_{in}.

Several other configurations of RAM memory are available. For example, the 2114 is configured as a 1024 × 4-bit (1K × 4) RAM instead of the 4096 × 1 used by the 2147H. A 1024 × 4-bit RAM will input/output 4 bits at a time for each address specified. This way, interfacing to an 8-bit data bus is simplified by having to use only two 2114s, one for the low-order data bits (D_0 to D_3) and the other for the high-order data bits (D_4 to D_7). For example, the photograph in Figure 16–7(b) shows a memory expansion printed-circuit board with thirty-two 2114 RAM ICs. They are set up as two memory banks, each configured as 8K × 8.

The 2147H's extremely small bit size and complexity make it a suitable choice to learn the fundaments of static RAM circuitry, but it will never find its way into new designs. Much faster, denser, lower-power SRAMs are available for the digital designer. For example, modern SRAMs have bit densities ranging from 64kb (kilo-bit) to 4Mb and access speeds as fast as 7.5 ns. Power demand is as low as 3.3 μW when in the standby mode. The memory can be placed in the standby mode by deselecting the chip by making $\overline{CS}$ = HIGH. Examples for a few SRAM memory configurations for Hitachi Semiconductor SRAMs are shown in Figure 16–8 and Table 16–1.

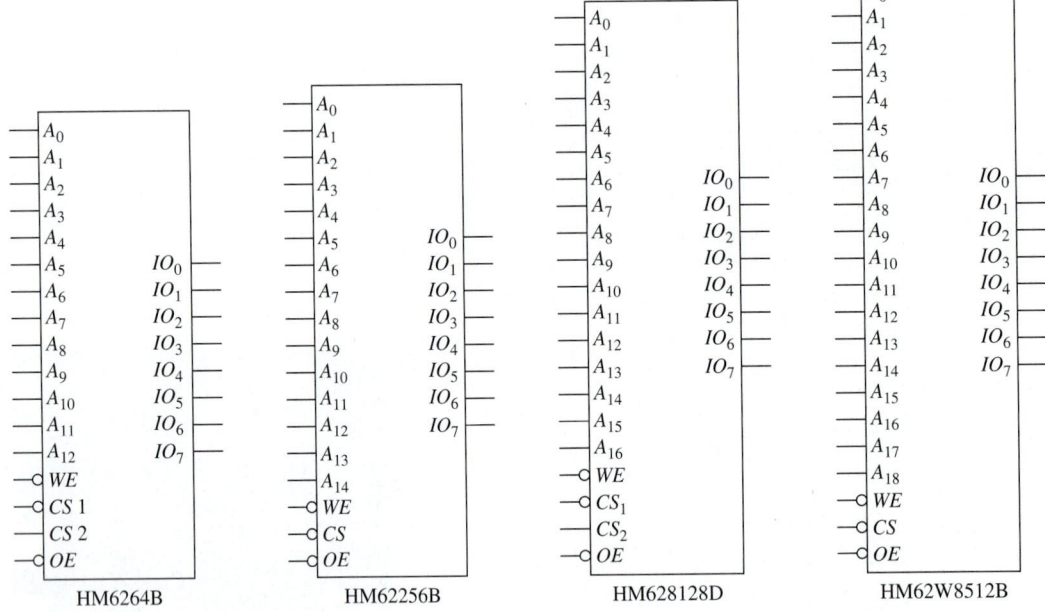

Figure 16–8 Logic symbols for several Hitachi low-power SRAMs.

TABLE 16–1	Sample Hitachi Low-Power Static RAMs	
Part No.	Density	Configuration
HM6264B	64kb	8k × 8
HM62256B	256kb	32k × 8
HM628128D	1Mb	128k × 8
HM62W8512B	4Mb	512k × 8

The HM6264B has a density of 64kb configured as 8k $\times$ 8. This is relatively small, but if your application only calls for that much memory, you can save a lot of space on your circuit board because it is only a 28-pin IC. It is available as a through-hole DIP or as a surface-mount SOP28 package.

On the other end of the spectrum, the HM62W8512B is a 4Mb SRAM configured as 512k $\times$ 8. To access 512k memory locations, we need 19 address lines (A_0 to A_{18}), as shown in Figure 16–8 ($2^{19} = 512$k). As you can see in the logic symbol, it has 8 I/O lines. It is written to by making $\overline{WE}$ = LOW and read from by making $\overline{WE}$ = HIGH. A LOW $\overline{CS}$ performs a chip select to access memory, and a HIGH puts it in standby mode, reducing its power consumption to 3.3 μW (typical). Several of the higher-end SRAMs such as this are made to operate with $V_{CC} = 3.3$ V, making it compatible with the low-voltage families such as LV-TTL. It is available in 32-pin surface-mount package styles only. A list of the Internet sites for Hitachi and other memory manufacturers is given in Appendix A.

Review Questions

16–6. The 2114 memory IC is a 1K $\times$ 4 static RAM, which means that it has _____ memory locations with _____ data bits at each location.

16–7. To perform a *read operation* with a 2147H RAM, $\overline{CS}$ must be _____ (LOW, HIGH), and $\overline{WE}$ must be _____ (LOW, HIGH).

16–8. In the 2147H memory array, address lines _____ through _____ are used to select the *row,* and _____ through _____ are used to select the *column.*

16–9. In Figure 16–5(b), what is the row and column selected if the address (A_{11} to A_0) is 001001000111?

16–10. Which transistor(s) in Figure 16–5(c) store the memory bit?

16–11. According to the Read cycle timing waveforms for the 2147H, how long must you wait after a chip select ($\overline{CS}$) before valid data are available at D_{out}?

16–3 Dynamic RAMs

Although **dynamic** RAMs (DRAMs) require more support circuitry and are more difficult to use than static RAMs (SRAMs), they are less expensive per bit and have a much higher density, minimizing circuit-board area. Most applications requiring large amounts of read/write memory will use DRAMs instead of static. The main memory in PCs use 64, 96, 128, or more megabytes of DRAM.

The most important difference in a DRAM memory is that its storage technique is to *place a charge on a capacitor* inside each memory location instead of using the cross-transistor configuration of SRAMs. Using this method *greatly* reduces the size per bit. However, the time it takes to read or write a bit is greater for the DRAM because it has to change the charge on the capacitor, which takes longer than changing the state of the MOS transistors in the SRAM.

Figure 16–9 shows a simplified schematic of a single DRAM memory cell. The storage capacitor is isolated from the D_{in} or D_{out} line until the MOSFET control transistor is turned ON momentarily with a write or read pulse at its gate. To write a 1 to the capacitor, D_{in} is set to 1 ($+V_{CC}$), and the DRAM control circuitry provides a pulse

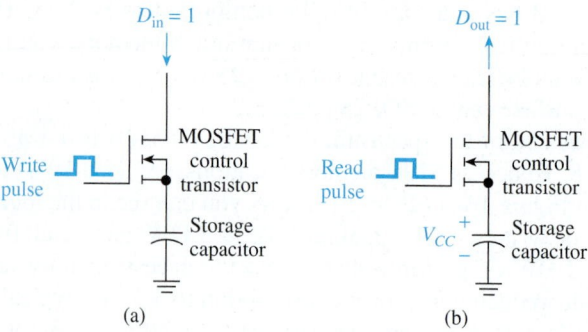

Figure 16–9 Simplified DRAM memory cell: (a) writing a 1 to the storage capacitor; (b) reading a 1 from the storage capacitor.

to the gate of the selected memory cell. This pulse turns the transistor ON, shorting the drain to source and placing $+V_{CC}$ on the capacitor. (Depending the DRAM family, $+V_{CC}$ will be $+5$ or $+3.3$ V, and in some cases even lower.) To read the data at the cell location, the DRAM control circuitry redirects the drain of the MOSFET to the D_{out} line and issues a read pulse on its gate. This shorts the transistor, connecting the capacitor directly to D_{out}.

As you know, capacitors cannot hold their charge forever, even when the MOSFET control transistor is an open. Therefore, the capacitor has to be refreshed on a regular basis, called the *refresh period*. The refresh period for the earlier DRAMs like the 2118 was 2 ms, but newer technology has extended the period to 64 ms. Figure 16–10 shows the voltage on a storage capacitor that is initially 0 V, then loaded with a 1 ($+V_{CC}$). The voltage, $+V_{CC}$, will immediately start dropping and require refreshing before its level drops below a recognizable 1-level.

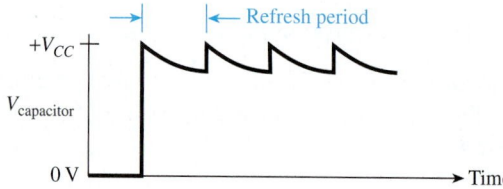

Figure 16–10 Refreshing the voltage on the DRAM storage capacitor.

An example of a 16K $\times$ 1-bit DRAM is the Intel 2118, whose data sheet is shown in Figure 16–11(a).

To uniquely address 16,384 locations, 14 address lines are required ($2^{14} = $ 16,384). However, Figure 16–11(a) shows only seven address lines (A_0 to A_6). This is because with larger memories such as this, to keep the IC pin count to a minimum, the address lines are *multiplexed* into two groups of seven. An external 14-line-to-7-line multiplexer is required in conjunction with the control signals, $\overline{RAS}$ and $\overline{CAS}$, to access a complete 14-line address.

The controlling device must put the valid 7-bit address of the desired memory array *row* on the A_0 to A_6 inputs and then send the Row Address Strobe ($\overline{RAS}$) LOW. Next, the controlling device must put the valid 7-bit address of the desired memory array *column* on the *same* A_0 to A_6 inputs and then send the Column Address Strobe ($\overline{CAS}$) LOW. Each of these 7-bit addresses is latched and will pinpoint the desired 1-bit memory location by its row–column coordinates.

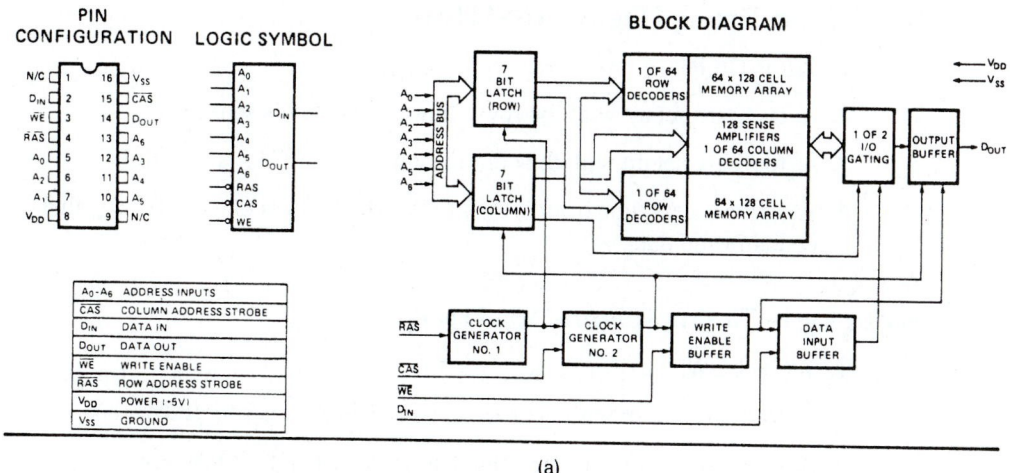

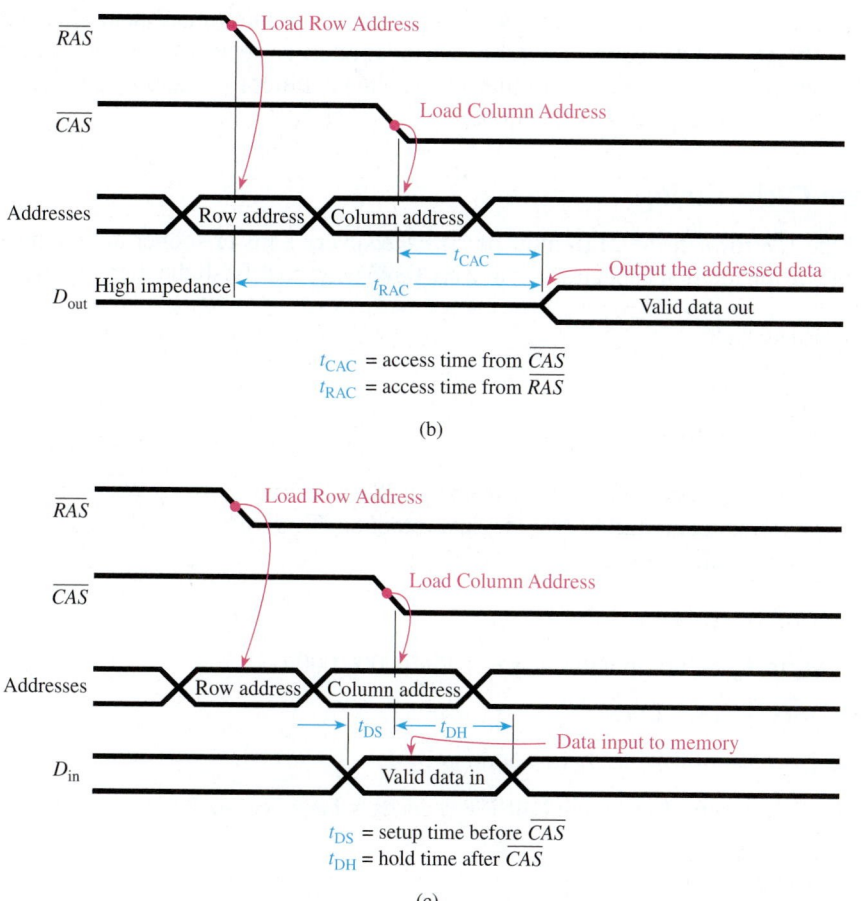

Figure 16–11 (a) The 2118 16K $\times$ 1 dynamic RAM; (b) dynamic RAM Read cycle timing ($\overline{WE}$ = HIGH); (c) dynamic RAM Write cycle timing ($\overline{WE}$ = LOW).

Once the memory location is identified, the $\overline{WE}$ input is used to direct either a Read or Write cycle similar to the static RAM operation covered in Section 16–2. When $\overline{WE}$ is LOW, data are written to the RAM via D_{in}; when $\overline{WE}$ is HIGH, data are read from the RAM via D_{out}.

Read Cycle Timing [Figure 16–11(b)]

1. $\overline{WE}$ is HIGH.

2. A_0 to A_6 are set up with the row address, and $\overline{RAS}$ is sent LOW.

3. A_0 to A_6 are set up with the column address, and $\overline{CAS}$ is sent LOW.

4. After the access time from $\overline{RAS}$ or $\overline{CAS}$ (whichever is longer), the D_{out} line will contain valid data.

Write Cycle Timing [Figure 16–11(c)]

1. $\overline{WE}$ is LOW.

2. A_0 to A_6 are set up with the row address, and $\overline{RAS}$ is sent LOW.

3. A_0 to A_6 are set up with the column address, and $\overline{CAS}$ is sent low.

4. At the HIGH-to-LOW edge of $\overline{CAS}$, the level at D_{in} is stored at the specified row–column memory address. D_{in} must be set up before and held after the HIGH-to-LOW edge of $\overline{CAS}$ to be interpreted correctly. (There are other setup, hold, and delay times that are not shown. Refer to a memory data book for more complete specifications.)

Refresh Cycle Timing

Each of the 128 rows of the 2118 must be *refreshed* every 2 ms or sooner to replenish the charge on the internal capacitors. There are three ways to refresh the memory cells:

1. Read cycle

2. Write cycle

3. $\overline{RAS}$-only cycle

Unless you are reading or writing from all 128 rows every 2 ms, the $\overline{RAS}$-only cycle is the preferred technique to provide data retention. To perform a $\overline{RAS}$-only cycle, the following procedure is used:

1. $\overline{CAS}$ is HIGH.

2. A_0 to A_6 are set up with the row address 000 0000.

3. $\overline{RAS}$ is pulsed LOW.

4. Increment the A_0 to A_6 row address by 1.

5. Repeat steps 3 and 4 until all 128 rows have been accessed.

Dynamic RAM Controllers

It seems like a lot of work demultiplexing the addresses and refreshing the memory cells, doesn't it? Well, most manufacturers of DRAMs have developed controller ICs to simplify the task. Some of the newer dynamic RAMs have refresh and error detection/correction circuitry built right in, which makes the DRAM look **static** to the user.

A popular controller IC to interface to the 2118 is the Intel 3242 address multiplexer and refresh counter for 16K dynamic RAMs. Figure 16–12 shows how this controller IC is used in conjunction with four 2118 DRAMs.

The 3242 in Figure 16–12 is used to multiplex the 14 input address A_0 to A_{13} to seven active-LOW output addresses $\overline{Q_0}$ to $\overline{Q_6}$. When the Row Enable input is HIGH, A_0 to A_6 are output inverted to $\overline{Q_0}$ to $\overline{Q_6}$ as the row addresses. When the Row Enable input is

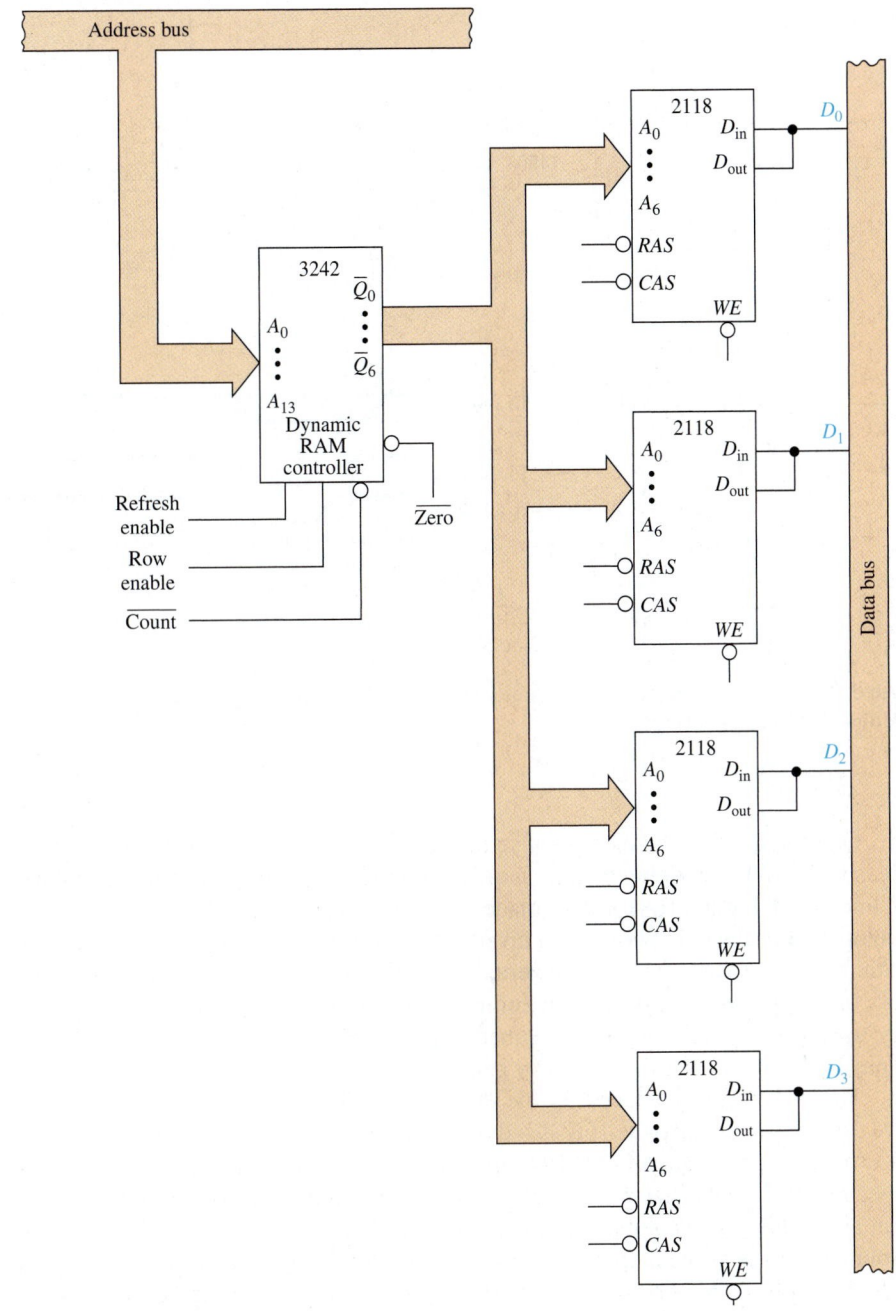

Figure 16–12 Using a 3242 address multiplexer and refresh counter in a 16K × 4 dynamic RAM memory system.

LOW, A_7 to A_{13} are output inverted to $\overline{Q}_0$ to $\overline{Q}_6$ as the column address. Of course, the timing of the $\overline{RAS}$ and $\overline{CAS}$ on the 2118s must be synchronized with the Row Enable signal.

To provide a "burst" refresh to all 128 rows of the 2118s, the Refresh Enable input in Figure 16–12 is made HIGH. This causes the $\overline{Q}_0$ to $\overline{Q}_6$ outputs to count from 0 to 127 at a rate determined by the count input clock signal. When the first 6 significant bits of the counter sequence to all zeros, the zero output goes LOW, signifying the completion of the first 64 refresh cycles.

Modern DRAMs have capacities in the megabyte range and, thus, need controllers with additional address lines. National semiconductor DM8420 series provides control for 4, 16, or 64 MB of DRAM.

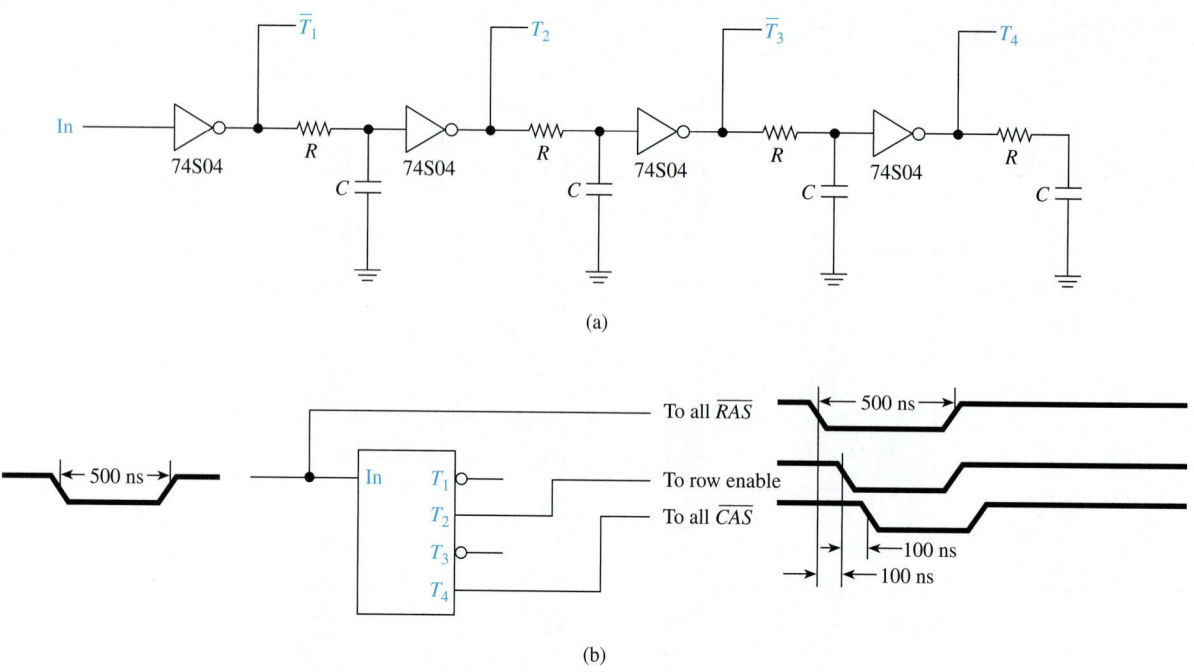

Figure 16–13   Four-tap, 50-ns delay line used for DRAM timing: (a) logic diagram; (b) logic symbol and timing.

Inside Your PC

Modern PCs use 64MB, 128MB, or more of DRAM. To physically mount so much memory, manufacturers solder several memory ICs on a single board, which is then inserted into a special socket on the motherboard. These memory boards are called a SIMMs (Single-Inline Memory Module), DIMMs (Dual-Inline Memory Module), or RIMMs (Rambus-Inline Memory Module). As memory requirements increase, these memory modules can easily be upgraded by the user without any special tools. (For SIMM, DIMM, and RIMM availability, visit the DRAM manufacturers' Web sites listed in Appendix A.)

One commonly used method of setting up the timing for $\overline{RAS}$, $\overline{CAS}$, and Row Enable is with a multitap **delay line,** as shown in Figure 16–13. Basically, the four-tap delay line IC of Figure 16–13(a) is made up of four inverters with precision RCs to develop a 50-ns delay between each inverter. The pulses out of each tap have the same width, but each successive tap is inverted and delayed by 50 ns. [In Figure 16–13(b), every other tap was used to arrive at noninverted, 100-ns delay pulses.] Delay lines are very useful for circuits requiring sequencing, as DRAM memory systems do. (See Figure 11-16 for a 5 ns multitap delay gate.)

The waveforms produced by the delay line of Figure 16–13(b) can be used to drive the control inputs to the 16K × 4 dynamic RAM memory system of Figure 16–12 (a LOW $\overline{RAS}$ pulse, then a LOW Row Enable pulse, then a LOW $\overline{CAS}$ pulse). Careful inspection of the data sheets for the 3242 and 2118 is required to determine the maximum and minimum allowable values for pulse widths and delay times. To design the absolute fastest possible memory circuit, all the times would be kept at their minimum value. But, it is a good practice to design in a 10% to 20% margin to be safe.

Because of their extremely high circuit density, DRAMs are the choice for a PC's large RAM requirement. Typical DRAM capacities for PCs are 64, 96, and 128 MB and higher. Table 16–2 lists several popular high-density dynamic RAM ICs available from Hitachi Semiconductor.

TABLE 16–2	Sample Hitachi Low-Power DRAMs		
Part No.	**Density**	**Configuration**	**Voltage**
HM5165805	64Mb	8M × 8	3.3 V
HM5113805	128Mb	16M × 8	3.3 V
HM5225805	256Mb	32M × 8	3.3 V
HM5251805	512Mb	64M × 8	3.3 V

Review Questions

16–12. Why are address lines on larger memory ICs multiplexed?

16–13. What is the primary storage medium in a DRAM memory cell?

16–14. What is meant by *refreshing* a DRAM?

16–15. What is a typical refresh period for a DRAM (60 s, 1 s, 60 ms)?

16–4 Read-Only Memories

Read-Only Memories (**ROMs**) are memory ICs used to store data on a permanent basis. They are capable of random access and are *nonvolatile,* meaning that they do not lose their memory contents when power is removed. This makes them very useful for the storage of computer operating systems, software language compilers, table look-ups, specialized code conversion routines, and programs for dedicated microprocessor applications.

ROMs are generally used for read-only operations and are not written to after they are initially programmed. However, there is an erasable variety of ROM called an **EPROM** (erasable-programmable-read-only memory) that is very useful because it can be erased and then reprogrammed if desired.

To use a ROM, the user simply specifies the correct address to be read and then enables the chip select ($\overline{CS}$). The data contents at that address (usually 8 bits) will then appear at the outputs of the ROM (some ROM outputs will be three stated, so you will have to enable the output with a LOW on $\overline{OE}$).

Mask ROMs

Manufacturers will make a custom **mask** ROM for users who are absolutely sure of the desired contents of the ROM and have a need for at least 1000 chips. To fabricate a custom IC like the mask ROM, the manufacturer charges a one-time fee of more than $1000 for the design of a unique mask that is required in the fabrication of the IC. After that, each identical ROM that is produced is very inexpensive. In basic terms, a mask is a cover placed over the silicon chip during fabrication that determines the permanent logic state to be formed at each memory location. Of course, before the mass production of a quantity of mask ROMs, the user should have thoroughly tested the program or data that will be used as the model for the mask. Most desktop computers use mask ROMs to contain their operating system and for executing procedures that do not change, such as decoding the keyboard and the generation of characters for the display. Figure 16–14 shows how a 1 or 0 is derived when the IC manufacturer alters the source connection of a single MOSFET.

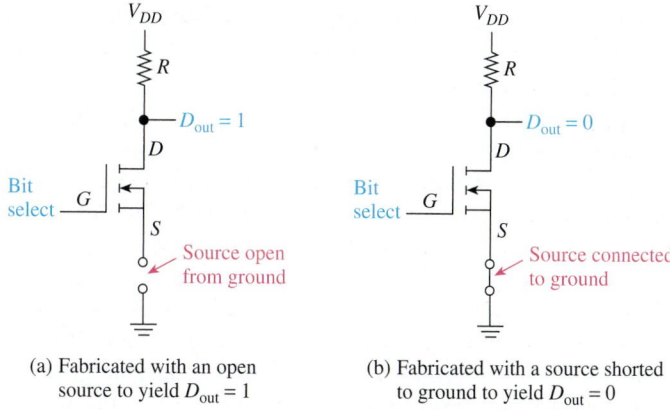

Figure 16–14 Simplified schematic of a single mask-ROM memory cell.

Fusible-Link PROMs

To avoid the high one-time cost of producing a custom mask, IC manufacturers provide user-programmable ROMs (**PROMs**). They are available in standard configurations such as 4K × 4, 4K × 8, 8K × 4, and so on.

Initially, every memory cell has a **fusible link,** keeping its output at 0 (see Figure 16–15). A 0 is changed to a 1 by sending a high-enough current through the fuse to permanently open it, making the output of that cell a 1. The programming procedure involves addressing each memory location, in turn, and placing the 4- or 8-bit data to be programmed at the PROM outputs and then applying a programming pulse (either a HIGH voltage or a constant current to the programming pin). Details for programming are given in the next section.

Once the fusible link is burned open, the data are permanently stored in the PROM and can be read over and over again just by accessing the correct memory address. The process of programming such a large number of locations is best done by a PROM programmer or microprocessor development system (MDS). These systems can copy a good PROM or the data can be input via a computer keyboard or from a magnetic disk.

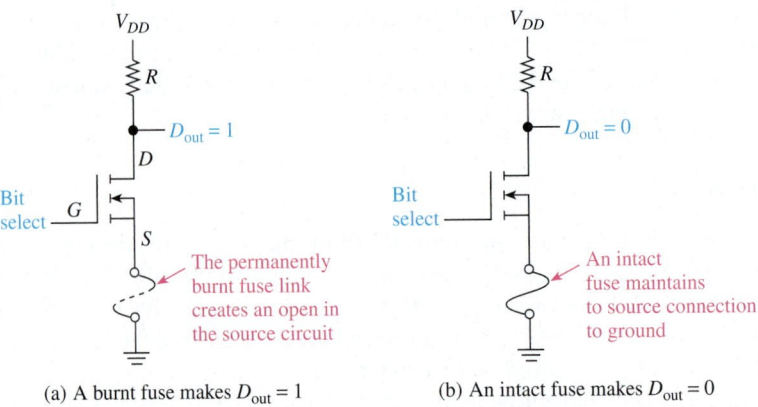

(a) A burnt fuse makes $D_{out} = 1$ (b) An intact fuse makes $D_{out} = 0$

Figure 16–15   Simplified schematic of a single fusible-link PROM memory cell.

EPROMs, EEPROMs, and Flash Memory

When using mask ROMs or PROMs, if you need to make a change in the memory contents or if you make a mistake in the initial programming, you are out of luck! One solution to that problem is to use an erasable PROM (EPROM). These PROMs are erased by exposing an open "window" (see Figure 16–16) in the IC to an ultraviolet (UV) light source for a specified length of time.

Two other types of EPROMs are the Electrically-Erasable PROM (**EEPROM**) and **Flash memory.** The EEPROM provides nonvolatile memory but can be programmed and erased while still in the circuit. Like the EEPROM, Flash memory is also electrically erasable. It derives its name from the fact that its internal processing capability provides for much faster access times and reduced system overhead during the erasure process.

Any of these three memories can be erased and reprogrammed thousands of times reliably. There is, however, one type of EPROM having a part number suffix-*OTP,* which stands for *One-Time-Programmable.* Because it is only programmed once, it is used like a PROM but has pin compatibility and I/O specs the same as its EPROM part number.

Figure 16–16 A 2716 EPROM IC showing the window that allows UV light to strike the memory cells for erasure.

The UV-erasable EPROM has the slowest erasure time. It takes several minutes of intense UV radiation, and you have no choice but to erase the entire chip. The EEPROMs and Flash memory are more flexible, allowing for individual bits or bytes to be erased in less than a millisecond. Flash memory even goes one step further, allowing you to erase entire blocks or the entire chip in the same time that it takes an EEPROM to erase 1 byte.

The UV-erasable EPROMs are the least expensive and are generally used during the initial design and debug stages of new product implementation. Designers will program their initial versions of software and data on an EPROM to test their ideas. Once a series of tests and retests have been passed, they will order the mask ROM version or implement the design with a CPLD or EEPROM-type device.

Commercially available EPROM programmers usually connect to a PC. A blank EPROM is plugged into a socket on the programmer, and then a command is given to the PC to download the binary program or data bytes that the designer wants put on the EPROM. An opaque cover is then placed over the exposed window to avoid inadvertent erasure, which can be caused by sunlight or fluorescent lighting. To erase the EPROM, the opaque cover is removed, and the IC is placed in a UV eraser device, which bombards the open EPROM window with UV radiation for a specified length of time (usually several minutes).

Because of their in-circuit reprogrammability, EEPROMs are becoming more popular than EPROMs. EEPROMs are often sold as a serial I/O memory devices. This significantly cuts down on its pin count and chip size. Its reduced size makes it a perfect solution for handheld devices like TV remote controls to remember favorite user settings and in cell phones for last-number radial and speed dial.

Flash Memory: Flash memory has become the most popular nonvolatile memory solution for many of the new electronic devices recently introduced. Digital cameras and personal digital assistants (PDAs) use Flash cards as their medium to store data. Also, PCs store operating system firmware, and printers store fonts on Flash memory.

The most common device used for individual memory cells for each of these three nonvolatile memories is the **floating-gate MOSFET,** as shown in Figure 16–17. The floating gate is the actual storage element for these devices. It is insulated from the three transistor connections by a dielectric layer on one side and by a thin oxide layer on the other. By placing a high-enough voltage at the control gate, an electric-field effect is created across the floating gate, which forces it to gain an excess of electrons. Manufacturers guarantee that this electron charge will remain on the floating gate for more than 10 years unless drained off by electrically erasing the cell.

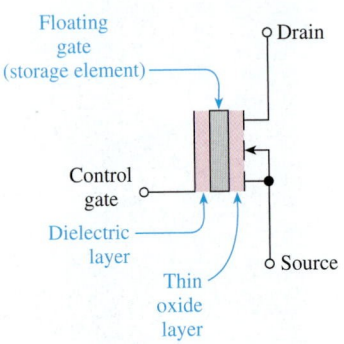

Figure 16–17   Floating-gate MOSFET.

So, basically we need to perform three operations on this memory cell: write, erase, and read. Figure 16–18(a) shows how a 1 is *written* to a cell. Internal chip circuitry places a high voltage (usually 12 V), called V_{PP}, on the control gate and connects the transistor source to ground. This creates an extremely high electric field across the floating gate. The floating gate responds by absorbing electrons that jump across the thin oxide layer from the transistor source. This continues for a few nanoseconds, until a sufficient charge is obtained. When the V_{PP} voltage is removed, the charge remains trapped on the floating gate because it is insulated by the dielectric and oxide layers.

To *erase* the memory cell of an EEPROM or flash memory, the electrons must be drained off the floating gate, as shown in Figure 16–18(b). To do this, the internal control circuitry reverses the connections to the control gate and source, reversing the write operation and leaving the floating gate empty of excess electrons. (A UV-erasable EPROM has the excess electrons removed when it absorbs UV light energy through its open window.)

To *read* a cell, the address lines connected to the memory IC are decoded into a specific row and column to select a particular memory cell within the memory array, as was done with RAM ICs. This causes the bit-select line to place V_{CC} on the control gate of the selected memory cell, as shown in Figure 16–18(c) and (d). If the floating gate has no electron charge on it [Figure 16–18(c)], the V_{CC} voltage (usually 5 or 3.3 V) is sufficient to turn the transistor ON, but not high enough to add an electron charge to the floating gate. The ON transistor shorts the drain to source, which places 0 V at D_{out}. If the floating gate has a charge on it [Figure 16–18(d)], the V_{CC} voltage will not be high enough to overcome the threshold required to turn the transistor ON. An OFF transistor acts like an open, keeping the current flow to zero and making the drain voltage equal to V_{CC}, so $D_{out} = 1$.

Very specific *timing requirements* must be met when reading and programming EPROM ICs. A sample of the interface circuitry and timing for a small EPROM, the 2716, is given in Figure 16–19.

The 2716 EPROM: The data sheet for the 2716 EPROM is given in Appendix B. Referring to the data sheet, notice that the 2716 has 16K bits of memory, organized as

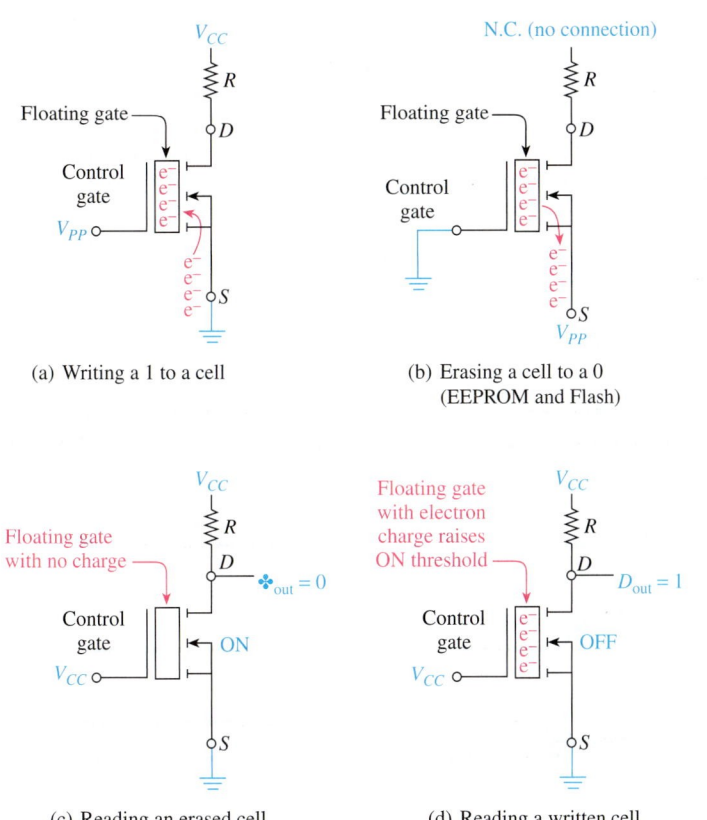

(a) Writing a 1 to a cell

(b) Erasing a cell to a 0
(EEPROM and Flash)

(c) Reading an erased cell

(d) Reading a written cell

Figure 16–18 Simplified diagram of the memory cell used for EPROMs, EEPROMs, and Flash memories.

2K $\times$ 8.2K locations that require 11 address inputs (2^{11} = 2048), which are labeled A_0 to A_{10}.

To read a byte (8 bits) of data from the chip, the 11 address lines are set up, and then $\overline{CE}$ and $\overline{OE}$ are brought LOW to enable the chip and to enable the output. The timing waveforms for the chip show that the data outputs (O_0 to O_7) become valid after a time delay for setting up the addresses (t_{ACC}), or enabling the chip (t_{CE}), or enabling the output (t_{OE}), whichever is completed last. Figure 16–19(a) shows the circuit connections and waveforms for reading the 2716 EPROM.

In Figure 16–19(a), the X in the address waveform signifies the point where the address lines must change (1 to 0 or 0 to 1), if they are going to change. The $\overline{CE}$/PGM line is LOW for Chip Enable and HIGH for programming mode. Outputs O_0 to O_7 are in the high-impedance state (float) until $\overline{OE}$ goes LOW. The outputs are then undetermined until the delay time t_{OE} has expired, at which time they become the valid levels from the addressed memory contents.

Programming the 2716: Initially, and after an erasure, all bits in the 2716 are 1's. To program the 2716, the following procedure is used:

1. Set V_{PP} to 25 V and $\overline{OE}$ = HIGH (5 V).

2. Set up the address of the byte location to be programmed.

3. Set up the 8-bit data to be programmed on the O_0 to O_7 outputs.

4. Apply a 50-ms positive TTL pulse to the $\overline{CE}$/PGM input.

5. Repeat steps 2, 3, and 4 until all the desired locations have been programmed.

16–16. What is meant by the term *volatile?*

16–17. Describe a situation where you would want to convert your EPROM memory design over to the mask ROMs.

16–18. Describe how a memory cell in each of the following memories is made to output a 1:

(a) mask ROM

(b) PROM

(c) EPROM

(d) EEPROM

(e) Flash memory

16–19. Describe how you erase a memory cell in each of the following memories:

(a) EPROM

(b) EEPROM

(c) Flash memory

16–20. According to the 2716 EPROM Read cycle timing waveforms, you must wait _____ nanoseconds after $\overline{OE}$ goes LOW before the data at O_0 to O_7 are valid (assuming $\overline{CE}$ has already been LOW for at least _____ nanosecond).

16–21. The time for the outputs to return to a float state for the 2716 is $t_{DF} = 100$ ns. Under what circumstances would that time be important to know?

16–5 Memory Expansion and Address Decoding Applications

When more than one memory IC is used in a circuit, a decoding technique (called **address decoding**) must be used to identify *which IC* is to be read or written to. Most 8-bit microprocessors use 16 separate address lines to identify unique addresses within the computer system. Some of those 16 lines will be used to identify the chip to be accessed, whereas the others pinpoint the exact memory location. For instance, the 2732 is a $4K \times 8$ EPROM that requires 12 of these address lines (A_0 to A_{11}) just to locate specific contents within its memory. This leaves four address lines (A_{12} to A_{15}) free for chip address decoding. A_{12} to A_{15} can be used to identify which IC within the system is to be accessed.

With 16 total address lines, there will be 64K, or 65,536 ($2^{16} = 65,536$), unique address locations. One 2732 will use up 4K of those. To design a large EPROM memory system, let's say, 16K bytes, four 2732s would be required. The address decoding scheme shown in Figure 16–20 could be used to set up the four EPROMS consecutively in the first 16K addresses of a computer system.

The four EPROMs in Figure 16–20 are set up in consecutive memory locations between 0 to 16K and are individually enabled by the 74LS138 address decoder. The $4K \times 8$ EPROMs each require 12 address lines for internal memory selection, leaving the four HIGH-order address lines (A_{12} to A_{15}) free for chip selection by the 74LS138.

To read from the EPROMs, the microprocessor first sets up on the address bus the unique 16-bit address that it wants to read from. Then it issues a LOW level on its $\overline{RD}$ output. This satisfies the three enable inputs for the 74LS138, which then uses A_{12}, A_{13},

Figure 16–19(b) shows the circuit connections and waveforms for programming a 2716.

Table 16–3 shows a representation of some EPROMs, EEPROMs, and Flash memory available today. In general, EEPROMs have the smallest bit size, and the Flash memory has the largest. Flash memory is the fastest-growing family of this type, finding applications in many PCs and other consumer electronic devices. Flash memory is now packaged in modules (Flash cards) that can be handled by the consumer as they insert it into a slot on their PC or handheld device. Flash cards can have densities exceeding 1 gigabit.

Table 16–4 summarizes all of the semiconductor memories.

TABLE 16–3	Representative EPROMs, EEPROMs, and Flash Memory	
Part Number	**Organization**	**Description**
27C16	2k × 8	16k UV-Erasable CMOS EPROM
27C256	32k × 8	256k UV-Erasable CMOS EPROM
27C040	512k × 8	4 Meg UV-Erasable CMOS EPROM
28C64	8k × 8	64k CMOS EEPROM
24C64	8k × 8	64k Serial CMOS EEPROM
24C256	32k × 8	256k Serial CMOS EEPROM
29F080	1M × 8	8Meg CMOS Flash Memory
29F032	4M × 8	32Meg CMOS Flash Memory
29W25611	32M × 8	256Meg CMOS Flash memory

TABLE 16–4	Summary of Semiconductor Memory				
Memory Type	**Basic Cell Structure**	**Volatile**	**In-Circuit Re-Writeable**	**Density**	**Comments**
SRAM	Cross-connected MOSFETs	Yes	Yes	Low	Fastest speed makes them well suited for PC cache memory.
DRAM	MOSFET with capacitor	Yes	Yes	High	Slower than SRAM, but extremely high density. Used for PC main RAM memory.
Mask ROM	MOSFET	No	No	High	High initial cost, but useful for high production runs such as PC BIOS.
PROM	MOSFET with fused link	No	No	High	Program once. Used for small-volume or trial circuits and data tables.
EPROM	Floating-gate MOSFET	No	No	High	Slow erasure time, but very high density and fast access time. Used for initial troubleshooting and design.
EEPROM	Floating-gate MOSFET	No	Yes	Low	Small size and ability to rewrite make it useful for handheld portable electronic devices.
Flash	Floating-gate MOSFET	No	Yes	High	Extremely high density and ability to rewrite facilitate portable mega-byte memory like the Flash card.

16–16. What is meant by the term *volatile?*

16–17. Describe a situation where you would want to convert your EPROM memory design over to the mask ROMs.

16–18. Describe how a memory cell in each of the following memories is made to output a 1:

(a) mask ROM

(b) PROM

(c) EPROM

(d) EEPROM

(e) Flash memory

16–19. Describe how you erase a memory cell in each of the following memories:

(a) EPROM

(b) EEPROM

(c) Flash memory

16–20. According to the 2716 EPROM Read cycle timing waveforms, you must wait _____ nanoseconds after $\overline{OE}$ goes LOW before the data at O_0 to O_7 are valid (assuming $\overline{CE}$ has already been LOW for at least _____ nanosecond).

16–21. The time for the outputs to return to a float state for the 2716 is $t_{DF} = 100$ ns. Under what circumstances would that time be important to know?

16–5 Memory Expansion and Address Decoding Applications

When more than one memory IC is used in a circuit, a decoding technique (called **address decoding**) must be used to identify *which IC* is to be read or written to. Most 8-bit microprocessors use 16 separate address lines to identify unique addresses within the computer system. Some of those 16 lines will be used to identify the chip to be accessed, whereas the others pinpoint the exact memory location. For instance, the 2732 is a 4K × 8 EPROM that requires 12 of these address lines (A_0 to A_{11}) just to locate specific contents within its memory. This leaves four address lines (A_{12} to A_{15}) free for chip address decoding. A_{12} to A_{15} can be used to identify which IC within the system is to be accessed.

With 16 total address lines, there will be 64K, or 65,536 ($2^{16} = 65,536$), unique address locations. One 2732 will use up 4K of those. To design a large EPROM memory system, let's say, 16K bytes, four 2732s would be required. The address decoding scheme shown in Figure 16–20 could be used to set up the four EPROMS consecutively in the first 16K addresses of a computer system.

The four EPROMs in Figure 16–20 are set up in consecutive memory locations between 0 to 16K and are individually enabled by the 74LS138 address decoder. The 4K × 8 EPROMs each require 12 address lines for internal memory selection, leaving the four HIGH-order address lines (A_{12} to A_{15}) free for chip selection by the 74LS138.

To read from the EPROMs, the microprocessor first sets up on the address bus the unique 16-bit address that it wants to read from. Then it issues a LOW level on its $\overline{RD}$ output. This satisfies the three enable inputs for the 74LS138, which then uses A_{12}, A_{13},

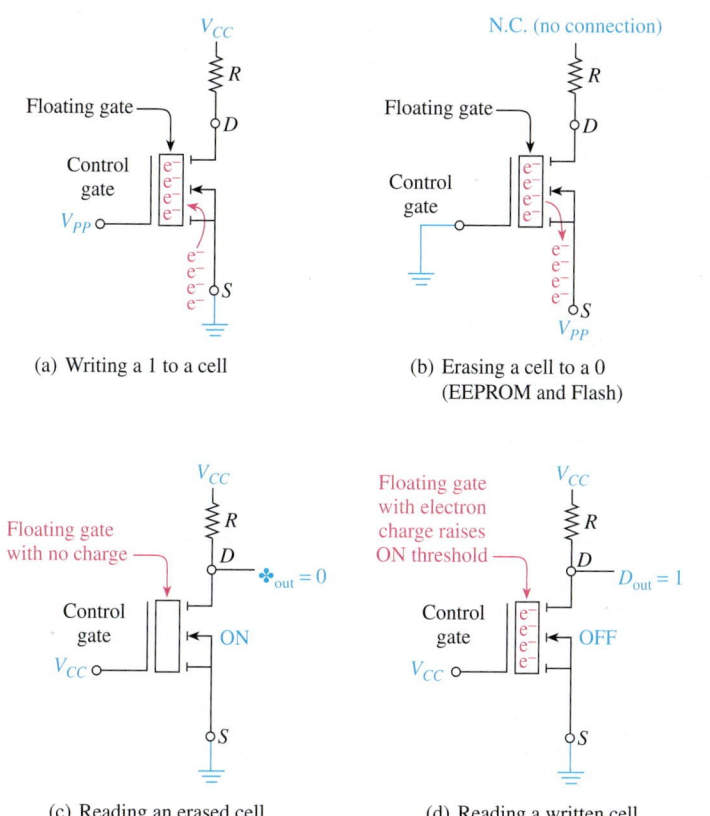

(a) Writing a 1 to a cell

(b) Erasing a cell to a 0
(EEPROM and Flash)

(c) Reading an erased cell

(d) Reading a written cell

Figure 16–18   Simplified diagram of the memory cell used for EPROMs, EEPROMs, and Flash memories.

2K $\times$ 8.2K locations that require 11 address inputs ($2^{11} = 2048$), which are labeled A_0 to A_{10}.

   To read a byte (8 bits) of data from the chip, the 11 address lines are set up, and then $\overline{CE}$ and $\overline{OE}$ are brought LOW to enable the chip and to enable the output. The timing waveforms for the chip show that the data outputs (O_0 to O_7) become valid after a time delay for setting up the addresses (t_{ACC}), or enabling the chip (t_{CE}), or enabling the output (t_{OE}), whichever is completed last. Figure 16–19(a) shows the circuit connections and waveforms for reading the 2716 EPROM.

   In Figure 16–19(a), the X in the address waveform signifies the point where the address lines must change (1 to 0 or 0 to 1), if they are going to change. The $\overline{CE}/PGM$ line is LOW for Chip Enable and HIGH for programming mode. Outputs O_0 to O_7 are in the high-impedance state (float) until $\overline{OE}$ goes LOW. The outputs are then undetermined until the delay time t_{OE} has expired, at which time they become the valid levels from the addressed memory contents.

Programming the 2716:   Initially, and after an erasure, all bits in the 2716 are 1's. To program the 2716, the following procedure is used:

1. Set V_{PP} to 25 V and $\overline{OE}$ = HIGH (5 V).

2. Set up the address of the byte location to be programmed.

3. Set up the 8-bit data to be programmed on the O_0 to O_7 outputs.

4. Apply a 50-ms positive TTL pulse to the $\overline{CE}/PGM$ input.

5. Repeat steps 2, 3, and 4 until all the desired locations have been programmed.

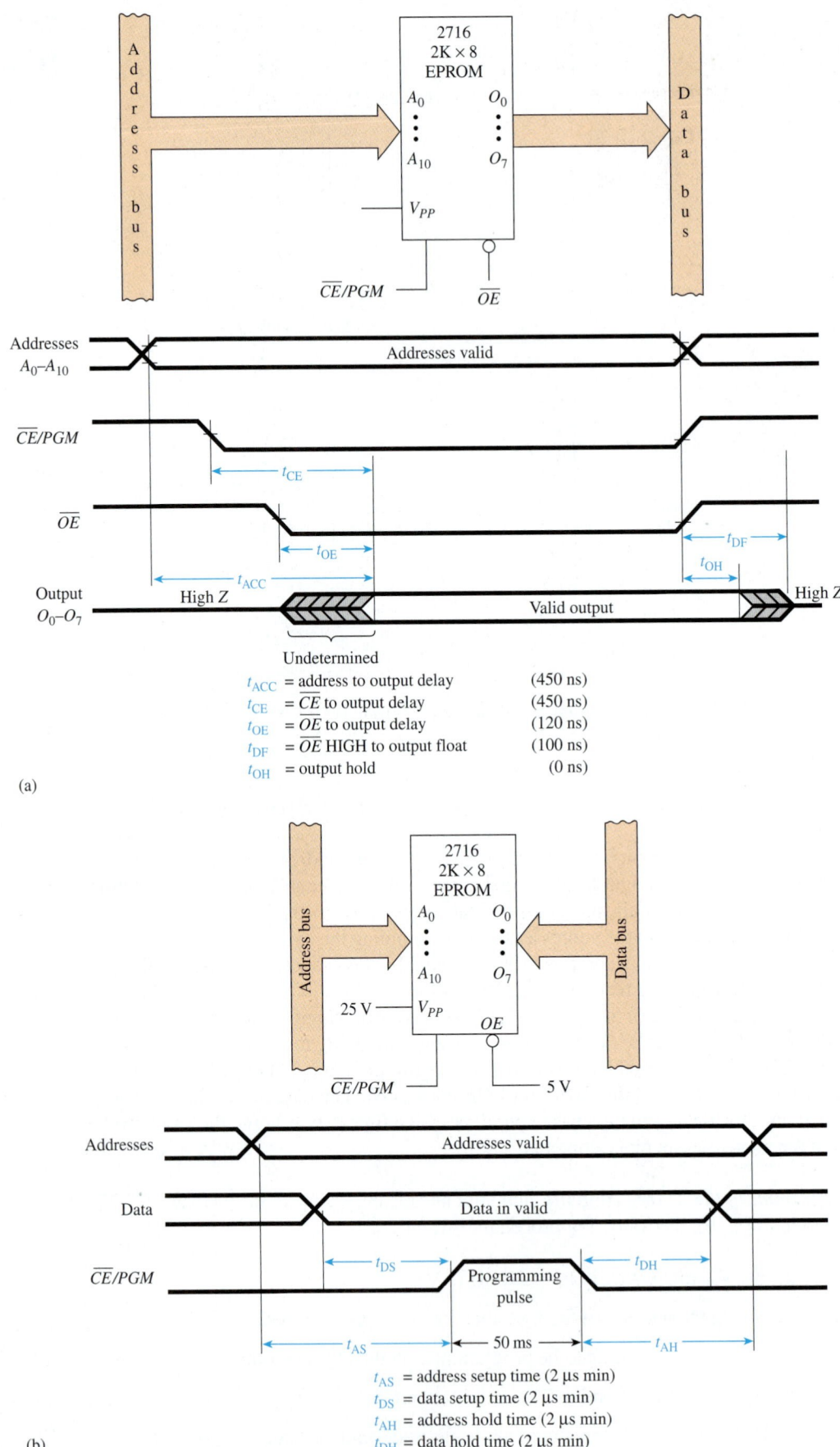

t_{ACC} = address to output delay   (450 ns)
t_{CE} = $\overline{CE}$ to output delay   (450 ns)
t_{OE} = $\overline{OE}$ to output delay   (120 ns)
t_{DF} = $\overline{OE}$ HIGH to output float   (100 ns)
t_{OH} = output hold   (0 ns)

(a)

t_{AS} = address setup time (2 μs min)
t_{DS} = data setup time (2 μs min)
t_{AH} = address hold time (2 μs min)
t_{DH} = data hold time (2 μs min)

(b)

Figure 16–19   The 2716 EPROM: (a) Read cycle; (b) Program cycle.

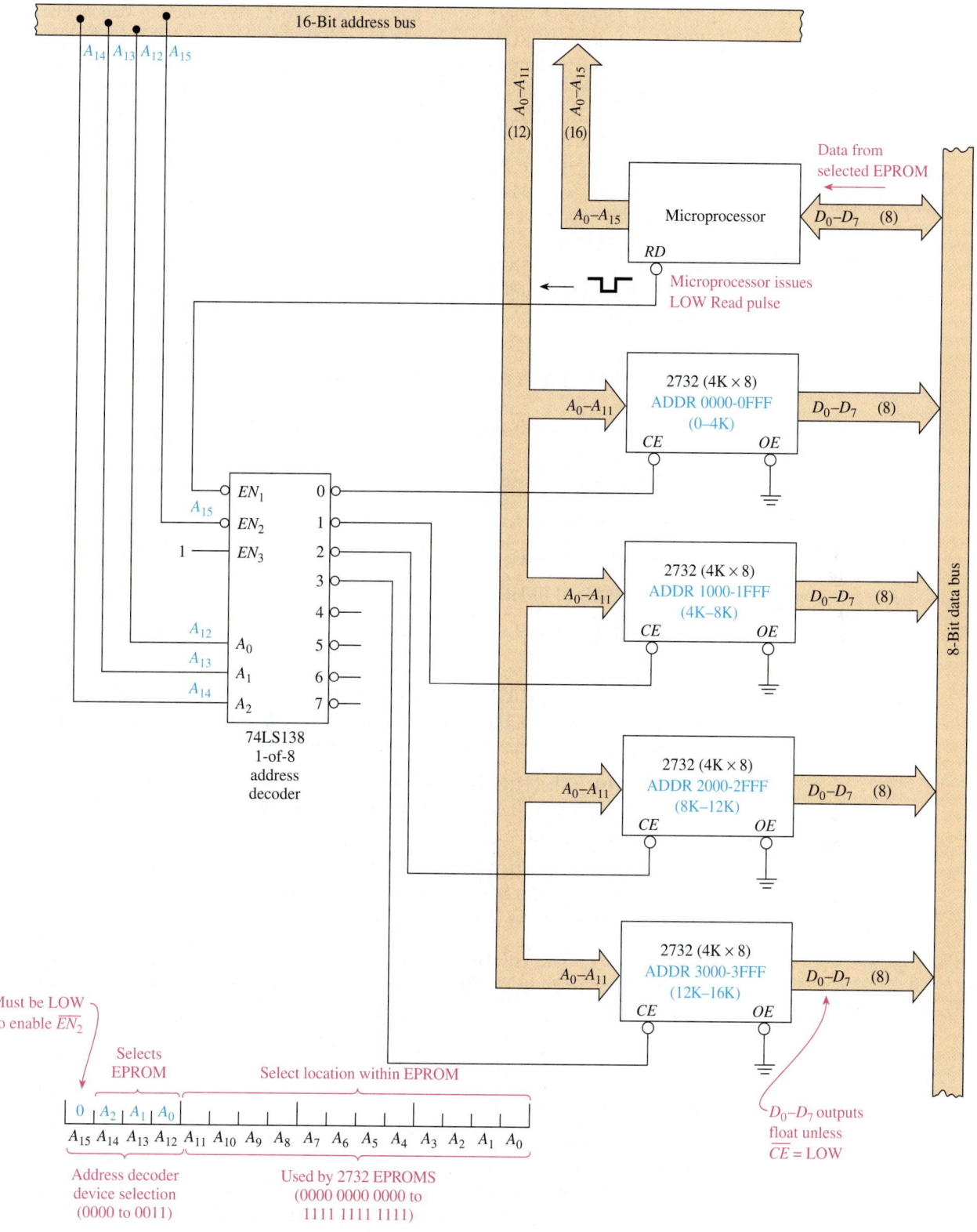

Figure 16–20 Address decoding scheme for a 16K-byte EPROM memory system.

and A_{14} to determine which of its outputs is to go LOW, selecting one of the four EPROMs. Once an EPROM has been selected, it outputs its addressed 8-bit contents to the data bus. The outputs of the other EPROMs will float because their $\overline{CE}$'s are HIGH. The microprocessor gives all the chips time to respond and then reads the data that it requested from the data bus.

The address decoding scheme shown in Figure 16–20 is a very common technique used for mapping out the memory allocations in microprocessor-based systems (called *memory mapping*). RAM (or RWM) is added to the memory system the same way.

For example, if we wanted to add four 4K × 8 RAMs, their chip enables would be connected to the 4–5–6–7 outputs of the 74LS138, and they would occupy locations 4XXX, 5XXX, 6XXX, and 7XXX. Then, when the microprocessor issues a Read or Write command for, let's say, address 4007H (4007 Hex), the first RAM would be accessed.

Team Discussion

This is a very popular scheme for address decoding (Figure 16–20). To be sure that you thoroughly understand it, determine the new addresses if A_{15} is moved to the EN3 input.

EXAMPLE 16–1

Determine which EPROM and which EPROM address are accessed when the microprocessor of Figure 16–20 issues a Read command for the following hex addresses:

(a) READ 0007H;

(b) READ 26C4H;

(c) READ 3FFFH;

(d) READ 5007H.

Solution:

    (a)   The HIGH-order hex digit (0) will select the first EPROM. Address 007 (0000 0000 0111) in the first EPROM will be accessed. (Address 007H is actually the *eighth* location in that EPROM.)

    (b)   The HIGH-order hex digit (2) will select the third EPROM ($A_{15} = 0, A_{14} = 0, A_{13} = 1, A_{12} = 0$). Address 6C4H in the third EPROM will be accessed.

    (c)   The HIGH-order hex digit (3) will select the fourth EPROM ($A_{15} = 0, A_{14} = 0, A_{13} = 1, A_{12} = 1$). Address FFFH (the last location) in the fourth EPROM will be accessed.

    (d)   The HIGH-order hex digit (5) will cause the output 5 of the 74LS138 to go LOW. Because no EPROM is connected to it, nothing will be read.

Expansion to 64K

The memory system of Figure 16–20 can be expanded to 64K bytes by utilizing two 74LS138 decoders, as shown in Figure 16–21. Address lines A_0 to A_{11} are not shown in Figure 16–21, but they would go to each 2732 EPROM, just as they did in Figure 16–20. The HIGH-order addresses (A_{12} to A_{15}) are used to select the individual EPROMs. When A_{15} is LOW, the upper decoder in Figure 16–21 is enabled, and EPROMs 1 to 8 can be selected. When A_{15} is HIGH, the lower decoder is enabled, and EPROMs 9 to 16 can be selected. Using the circuit in Figure 16–21 will allow us to map in sixteen 4K × 8 EPROMs, which will *completely* fill the memory map in a 16-bit address system. Actually, this would not be practical because some room must be set aside for RAM and I/O devices.

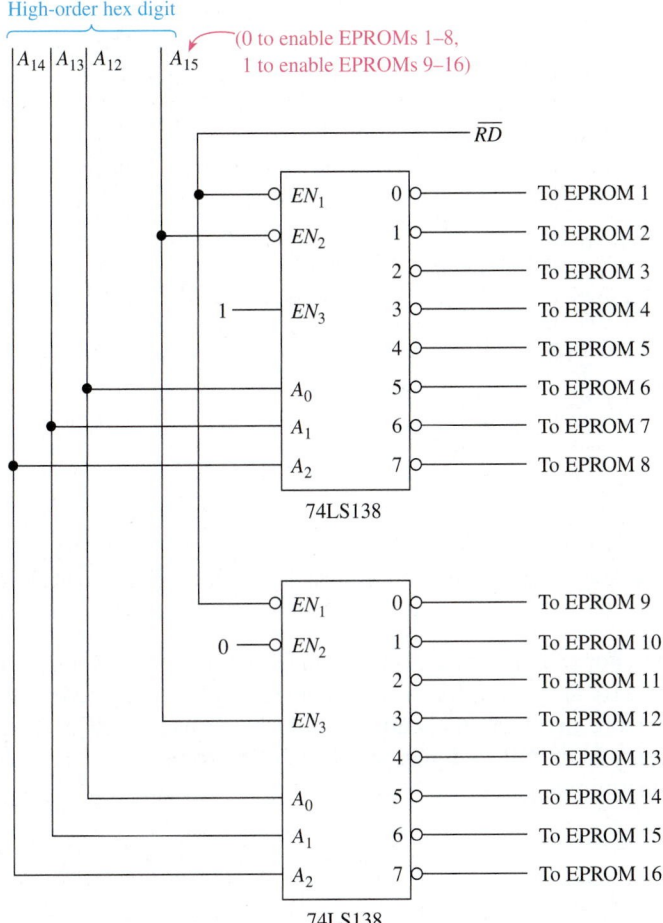

Helpful Hint

This decoding scheme can also be used to access RAM ICs. In that case the $\overline{RD}$ line would need to be ORed with $\overline{WR}$.

Figure 16–21 Expanding the memory of Figure 16–20 to 64K bytes.

One final point on memory and bus operation: microprocessors and MOS memory ICs are generally designed to drive only a single TTL load. Therefore, when several inputs are being fed from the same bus, an MOS device driving the bus must be buffered. An octal **buffer** IC such as the 74244 connected between an MOS IC output and the data bus will provide the current capability to drive a heavily loaded data bus. Bidirectional bus drivers (or transceivers) such as the 74LS640 provide buffering in both directions for use by read/write memories (RAM or RWM).

APPLICATION 16–1

A PROM Look-Up Table

Besides being used strictly for memory, ROMs, PROMs, and EPROMS can also be programmed to provide special-purpose functions. One common use is as a **look-up table.** A simple example is to use a PROM as a 4-bit binary–to–Gray code converter, as shown in Figure 16–22.

The PROM chosen for Figure 16–22 must have 16 memory locations, each location containing a 4-bit Gray code. The 4-bit binary string to be converted is used as the address inputs to the PROM. The PROM must be programmed such that each memory contains the equivalent Gray code to

Team Discussion

How could a PROM or EPROM be used as an ASCII look-up table?

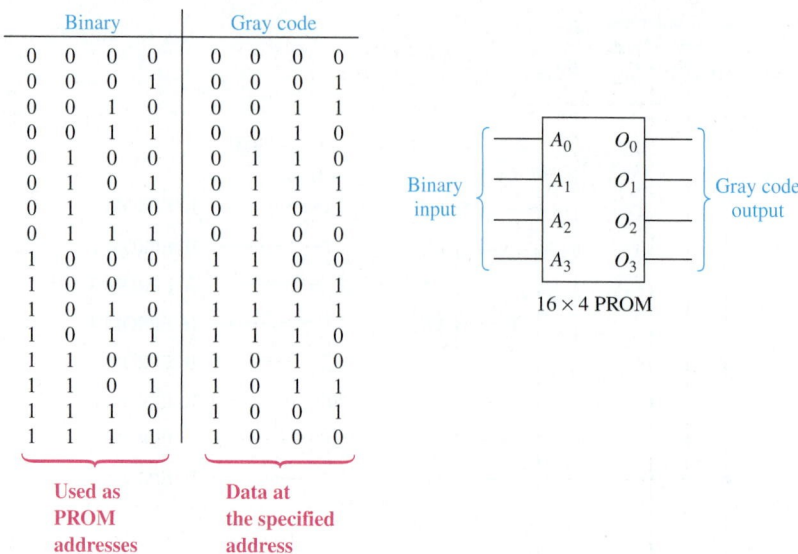

Binary				Gray code			
0	0	0	0	0	0	0	0
0	0	0	1	0	0	0	1
0	0	1	0	0	0	1	1
0	0	1	1	0	0	1	0
0	1	0	0	0	1	1	0
0	1	0	1	0	1	1	1
0	1	1	0	0	1	0	1
0	1	1	1	0	1	0	0
1	0	0	0	1	1	0	0
1	0	0	1	1	1	0	1
1	0	1	0	1	1	1	1
1	0	1	1	1	1	1	0
1	1	0	0	1	0	1	0
1	1	0	1	1	0	1	1
1	1	1	0	1	0	0	1
1	1	1	1	1	0	0	0

Used as PROM addresses ⎵   Data at the specified address ⎵

Figure 16–22 Using a PROM look-up table to convert binary to Gray code.

be output. For example, address location 0010 will contain 0011, 0100 will contain 0110, and so on, for the complete binary–to–Gray code data table.

A more practical application would be to use a PROM to convert 7-bit binary to two BCD digits, which is a very complicated procedure using ordinary logic gates.

APPLICATION 16–2

A Digital LCD Thermometer

Another application, one that covers several topics from within this text, is a digital Celsius thermometer. In this application, using a PROM look-up table simplifies the task of converting meaningless digital strings into decimal digits. Figure 16–23 shows a block diagram of a two-digit Celsius thermometer.

For the circuit of Figure 16–23 to work, a binary-to-two-digit BCD look-up table has to be programmed into the PROM. Because a standard thermistor is a nonlinear device, as the temperature varies the binary output of the ADC will not change in proportional steps. Programming the PROM with the appropriate codes can compensate for that and can also ensure that the output being fed to the two decoders is in the form of two BCD codes, each within the range 0 to 9. The appropriate codes for the PROM contents are best determined through experimentation. For example, if at 30°C the output of the ADC is 0100 1100 ($4C_{16}$), then location $4C_{16}$ of the PROM should be programmed with 0011 0000 (30_{BCD}).

The 74HC4543 will convert its BCD input into a seven-segment code for the liquid-crystal displays (**LCDs**). Liquid-crystal displays consume significantly less power than LED displays but require a separate square-wave oscillator to drive their backplane. As shown in Figure 16–23, a 40-Hz oscillator is connected to the phase input (PH) of each decoder and the backplane (BP) of each LCD.

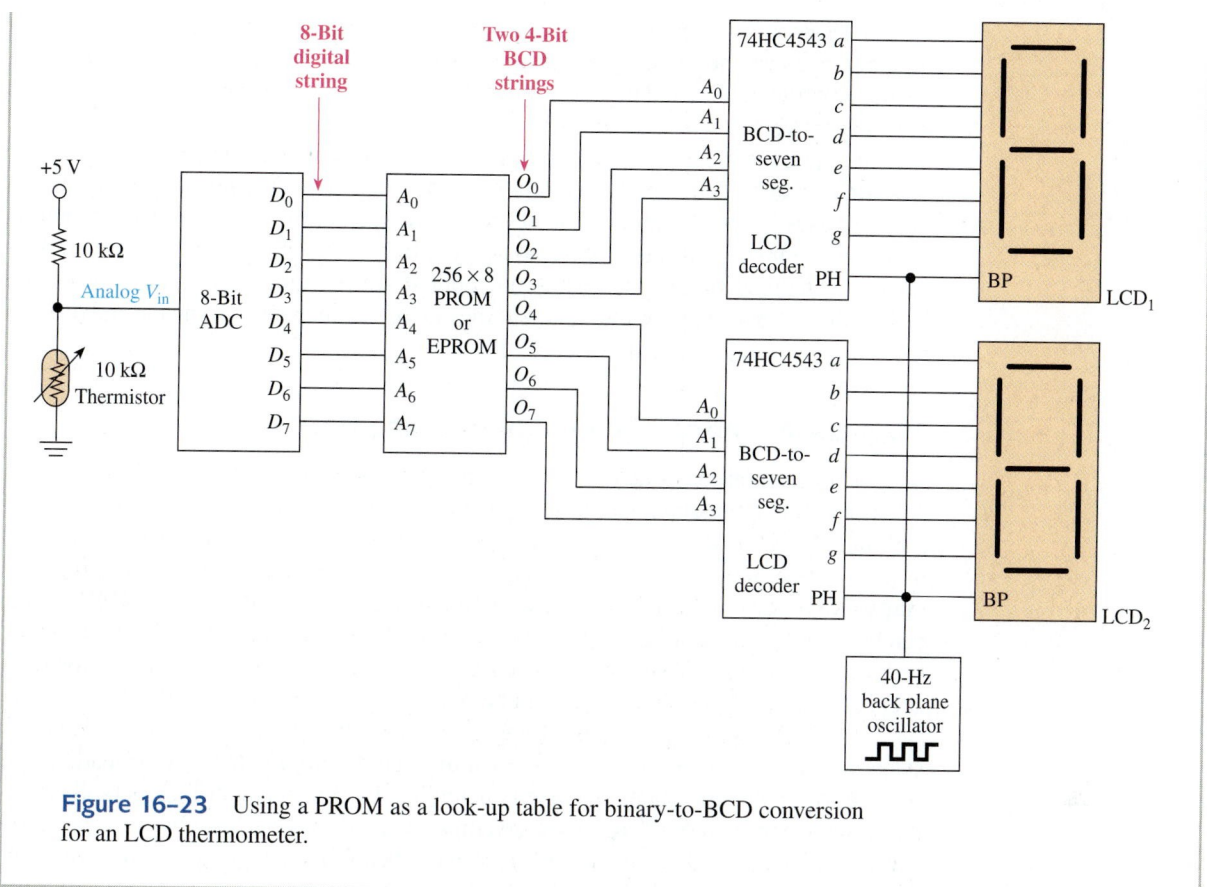

Figure 16–23 Using a PROM as a look-up table for binary-to-BCD conversion for an LCD thermometer.

Review Questions

16–22. Determine which EPROM and which EPROM address are accessed when the microprocessor of Figure 16–20 issues a Read command for the following hexadecimal addresses:

    (a) READ 2002H

    (b) READ 0AF7H

16–23. In Figure 16–20, connect A_{15} to EN_3 and ground EN_2. What is the new range of addresses that will access the second EPROM that was formerly accessed at 1000H through 1FFFH?

16–24. In Figure 16–23, assume that the thermistor resistance is 10 kΩ at 25°C, thus making D_7 to D_0 equal to 1000 0000 (one-half of full scale). Determine what data value should be programmed into the EPROM at address 1000 0000.

16–6 Magnetic and Optical Storage

The types of memory storage discussed so far have all been based on semiconductor material. They have relied on turning transistors ON or OFF or storing an electrical charge on a capacitor or on a floating gate of a transistor.

The most common magnetic and optical memories are more electromechanical in nature because the memory material containing the 1's and 0's physically spins beneath a read/write head. They are nonvolatile, providing permanent storage and retrieval even when power is removed.

In the case of magnetic memory, the 1 or 0 is represented on the magnetic medium as a tiny north–south or south–north polarity magnet. Optical memory differs in that it uses a laser to optically read the data stored on the medium. An indentation area (called a *pit*) represents 1, and a non-pit (called *land*) represents 0. Because of their electromechanical nature, magnetic and optical storage units are much slower and bulkier than semiconductor memory, but they are also much less expensive and provide much higher storage capacity.

Magnetic Memory: The Floppy Disk and Hard Disk

The most common removable magnetic storage for a PC is the 3.5-inch *floppy disk*. It consists of a magnetizable medium that spins inside a rigid plastic jacket at a speed of 300 rpm (revolutions per minute). The disk drive that holds the floppy disk (or diskette) has two read/write heads, one for each side of the diskette. They are used to record a magnetic charge to, or read a magnetic charge from, the diskette. Up to 1.44MB of data can be stored on the double-sided diskette. The beauty of the diskette is that it can then be removed from its drive unit and used at a different location or filed away for safe-keeping. Additional diskettes can then be purchased for less than one dollar.

The highest-speed magnetic storage is achieved using a *hard-disk drive*. This disk system is not considered a removable medium like the floppy disk but, instead, uses a series of rigid platters mounted in a sealed unit inside a PC. Its recording method is similar to the floppy, except that it uses several two-sided platters (usually three or more) to hold the magnetic data and has one read/write head for each platter surface. Because the platters are rigid and mounted in a permanent housing, their tolerances are more precisely defined, allowing for tighter spacing between bits. This also enables the disks to spin at much higher speeds (thousands of rpm). Storage capacity of hard drives are in the gigabyte range.

The method of storing 1's and 0's magnetically is shown in Figure 16–24. Each disk surface is made of an iron-oxide layer capable of becoming magnetized. This surface is mounted on a substrate (foundation), which is usually a flexible mylar for the floppy disk and aluminum for the hard-drive platter. Initially, the magnetic layer consists of totally nonaligned particles representing no particular magnetic direction. By passing magnetic flux lines through the material, the particles align themselves in a specific north–south or south–north direction.

As the disk surface revolves, the read/write head is moved laterally to the precise track (ring) and bit position as dictated by the PC operating system software. If the software is *writing* data to the disk, the control circuitry places the appropriate polarity on the electromagnet in the read/write head. The example in Figure 16–24(b) shows how to write a 1 onto the disk surface. (Four previously written bits are also shown.) With the + to − polarity shown, magnetic flux lines will flow clockwise through the electromagnet core. As the flux lines pass through the magnetic surface of the disk, they force the particles to align in a specific direction, leaving behind a north–south magnetic charge (south–north as you look at it.) To store a 0, the + to − polarity is reversed, which reverses the flux lines, which in turn reverses the direction of the stored magnetic charge. To *read* the data, the read/write head is used as a magnetic sensor, reading the magnetic polarity as it passes beneath it.

A hard drive achieves higher bit capacity and data access speed because it uses rigid disk platters revolving at a much higher speed within a precision, sealed unit. Typical rotation speed is 3600 rpm. (Some newer drives now exceed 10,000 rpm.) Because the hard drive operates in such a controlled environment, the bits can be packed closer together. Figure 16–25 shows how the bits are laid out on a disk surface. The concen-

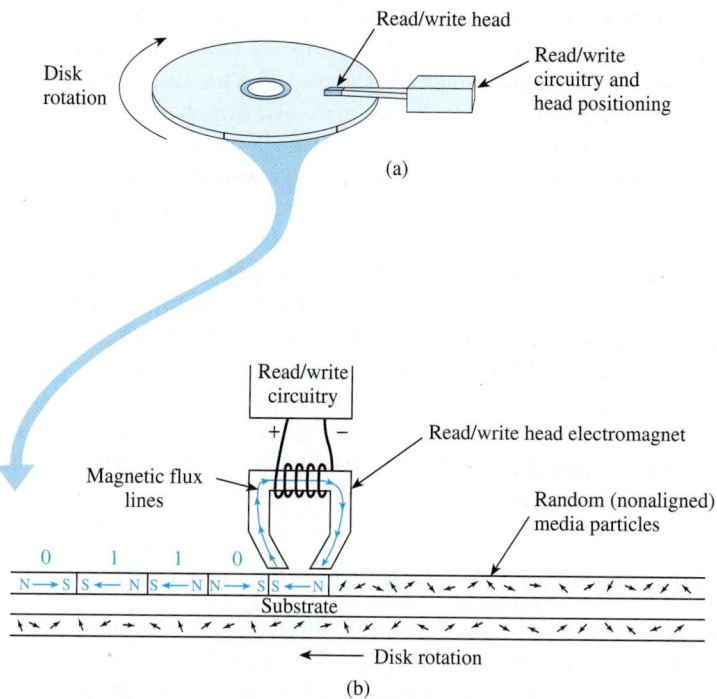

Read/write head

Read/write
circuitry and
head positioning

Disk
rotation

(a)

Read/write
circuitry

+ −

Read/write head electromagnet

Magnetic flux
lines

Random (nonaligned)
media particles

0   1   1   0

N→S S← N S←N N→ S S←N

Substrate

Disk rotation

(b)

Figure 16–24 Magnetic memory: (a) a floppy disk or hard-drive platter with one of its read/write heads; (b) a cutaway view of Figure 16–24(a), showing how to store a south-north magnetic charge.

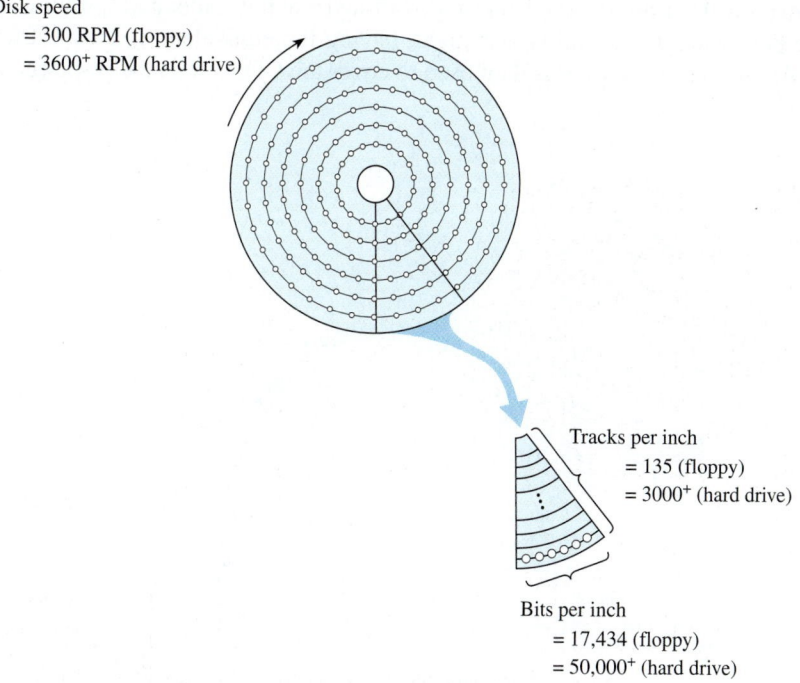

Disk speed
= 300 RPM (floppy)
= 3600$^+$ RPM (hard drive)

Tracks per inch
= 135 (floppy)
= 3000$^+$ (hard drive)

Bits per inch
= 17,434 (floppy)
= 50,000$^+$ (hard drive)

Figure 16–25 Bit density of a floppy disk and a hard drive.

tric circles on the disk are called *tracks*. (In the case of a hard drive, the term *cylinder* is used to describe the cylindrical shape that appears from the series of tracks having the same diameter on the entire stack of platters.) The tracks and the bits on the hard-disk drive are packed much closer together than on a floppy diskette. Newer drives exceed 20,000 tracks per inch and 300 kilobits per inch on each track.

One of the most important specifications on magnetic media is the *data transfer rate*. A floppy can transfer data at 45 kB/s, whereas a hard drive can exceed 30 MB/s. The slow speed of the floppy becomes annoying when you have to wait more than half a minute to transfer a large file that only takes a couple of seconds for a hard drive.

Higher-density, removable magnetic devices are now finding their way into common use with PC users. Two of these are the *Zip disk* and the *Jaz cartridge*. They are removable media like the floppy diskette, but they have much higher storage capacity and faster read/write speed. One reason for the high speed is that the Zip disk spins at 3000 rpm. The magnetic images stored on the Zip disk are spaced much more tightly than on a standard floppy disk, giving it a storage capacity of 100MB. The Jaz cartridge employs two rigid platters and has a capacity as high as 2GB.

Optical Memory: The CD, CD-R, CD-RW, and DVD

Music CDs have been around since the late 1970s as digital medium for the storage and playback of analog music. In the mid-1980s, they were adopted for the storage and retrieval of digital computer data. Their data transfer rate is generally not as fast as a hard-disk drive, but manufacturers are constantly improving the speed. Because it is a removable media and capable of holding up to 650MB of data, most new operating systems and applications software are provided on CDs.

Figure 16–26 illustrates the construction of a CD. It is made of an aluminum alloy coating on the bottom of a rigid polycarbonate wafer. Binary data are stored on the CD by a series of indentations (called *pits*) representing 1's and non-pits (called *lands*) representing 0's. Pits are formed in the CD by stamping tiny indentations into the aluminum alloy. Data are recorded on a CD starting from the center and spiraling outward to the perimeter. The spiral is very tight, having the equivalent of 16,000 tracks per inch. A thin, plastic coating is then used to cover the CD, and a label is placed on top.

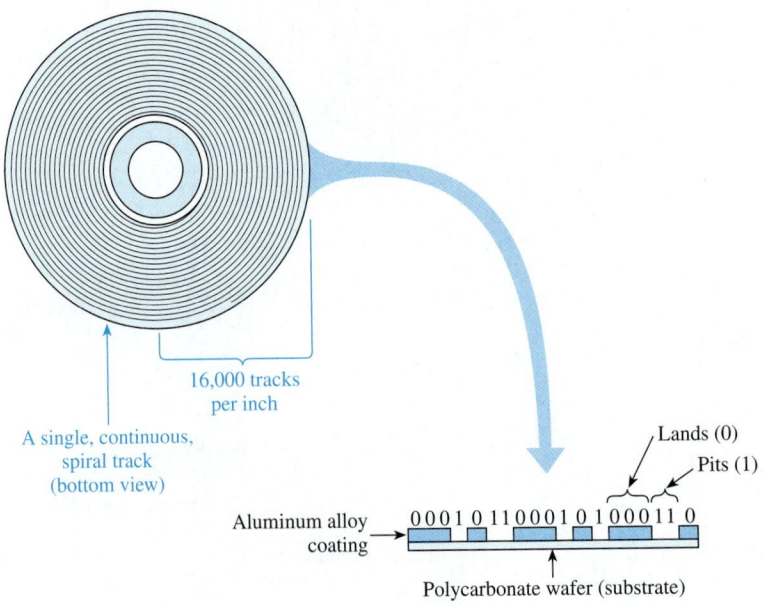

16,000 tracks
per inch

A single, continuous,
spiral track
(bottom view)

Lands (0)

Pits (1)

Aluminum alloy
coating

0 0 0 1 0 1 1 0 0 0 1 0 1 0 0 0 1 1 0

Polycarbonate wafer (substrate)

Figure 16–26 CD construction.

As the CD spins, each bit position is read optically when a laser beam is reflected off the bottom of the CD surface. A light receptor receives the reflected light and distinguishes between light that is strongly reflected (from the land) versus light that is diffused or absent (from the pit).

User-recordable CDs (CD-Rs) are also available. To make CDs recordable, they are manufactured with a photosensitive dye on a reflective gold layer placed on the bottom of the rigid polycarbonate wafer. To form a 1, the CD recorder superheats a tiny spot on the dye/gold layer, changing its composition so that it will not reflect the laser light in that one spot. This is not actually a pit but, instead, is just an area with reduced reflective properties. These CDs are called WORM (Write Once, Read Many) media because once they are full, they cannot be erased or rewritten.

The *re-writable* type of CD is the CD-RW. Its surface is a silver alloy crystalline structure that is also converted into a 1 by superheating the specific bit area. The heat turns that area into its amorphous (nonreflective) state. The beauty of the CD-RW technology is that the 1 can be converted back to look like a 0 by reapplying a lower-level heat to that area, which returns the silver alloy back to its crystalline (reflective) state. Again, these are not actual pits and lands but, instead, are reflective and nonreflective areas.

The newest standard for the CD is called DVD (Digital Versatile Disk). DVDs are not only important in the computer industry for data storage but also in home entertainment, where they are replacing prerecorded VCR tapes, offering much higher quality and greater content. They are constructed similar to the CD but have a much higher data capacity, because the pits and tracks are packed much closer on the platter. DVDs can be single- or double-sided and have data capacities of 4.7GB up to 17GB.

Summary

In this chapter, we have learned that

1. A simple 16-byte memory circuit can be constructed from 16 octal *D* flip-flops and a decoder. This circuit would have 16 memory locations (addresses) selectable by the decoder, with 1 byte (8 bits) of data at each location.

2. Static RAM (random-access memory) ICs are also called read/write memory. They are used for the temporary storage of data and program instructions in microprocessor-based systems.

3. A typical RAM IC is the 2114A. It is organized as $1K \times 4$, which means that it has 1K locations, with 4 bits of data at each location. (1K is actually an abbreviation for 1024.) An example of a higher-density RAM IC is the 6206, which is organized as $32K \times 8$.

4. Dynamic RAMs are less expensive per bit and have a much higher density than static RAMs. Their basic storage element is an internal capacitor at each memory cell. External circuitry is required to refresh the charge on all capacitors every 2 ms or less.

5. Dynamic RAMs generally multiplex their address bus. This means that the high-order address bits share the same pins as the low-order address bits. They are demultiplexed by the RAS and CAS (Row Address Strobe and Column Address Strobe) control signals.

6. Read-only memory (ROM) is used to store data on a permanent basis. It is nonvolatile, which means that it does not lose its memory contents when power is removed.

7. Three common ROMs are (1) the mask ROM, which is programmed once by a masking process by the manufacturer; (2) the fusible-link programmable ROM (PROM), which is programmed once by the user; and (3) the erasable-programmable ROM (EPROM), which is programmable and UV-erasable by the user.

8. Memory expansion in microprocessor systems is accomplished by using octal or hexadecimal decoders as address decoders to select the appropriate memory IC.

9. The electrically erasable PROM (EEPROM) and Flash memory use a floating-gate MOSFET for their primary storage element. A charge on the floating gate represents the stored data.

10. Magnetic storage like the floppy or hard disk use magnetized particles to represent the stored 1 or 0. Individual data bits are read and written using an electromagnetic read/write head.

11. Optical memory like the CD or DVD use a laser beam to reflect light off a rigid platter. This CD or DVD platter will have either a nonreflective pit to represent a 1 or a non-pit (land) to represent a 0.

Glossary

Address Decoding: A scheme used to locate and enable the correct IC in a system with several addressable ICs.

Buffer: An IC placed between two other ICs to boost the load-handling capability of the source IC and to provide electrical isolation.

Bus Contention: Bus contention arises when two or more devices are outputting to a common bus at the same time.

Byte: A group of 8 bits.

CAS: Column address strobe. An active-LOW signal provided when the address lines contain a valid column address.

Delay Line: An integrated circuit that has a single pulse input and provides a sequence of true and complemented output pulses, with each output being delayed from the preceding one by some predetermined time period.

Dynamic: A term used to describe a class of semiconductor memory that uses the charge on an internal capacitor as its basic storage element.

EEPROM: Electrically erasable, programmable read-only memory.

EPROM: UV-erasable, programmable read-only memory.

Flash Memory: An electrically erasable nonvolatile high density type of memory often used in hand held consumer electronic devices.

Floating-Gate MOSFET: A special type of MOSFET, the gate of which can permanently hold an electron charge when a strong electric field is placed across it. This stored charge determines if the transistor is read as a 1 or a 0. It is the basic storage element in most EEPROM and Flash-memory ICs.

Fusible Link: Used in programmable ICs to determine the logic level at that particular location. Initially, all fuses are intact. Programming the IC either blows the fuse to change the logic state or leaves it intact.

LCD: Liquid-crystal display. A multisegmented display similar to LED displays, except that it uses liquid-crystal technology instead of light-emitting diodes.

Look-Up Table: A table of values that is sometimes programmed into an IC to provide a translation between two quantities.

Magnetic Memory: The most common types of magnetic memory are the floppy disk and hard disk used in a PC. Magnetically charged north–south particles are used to represent 1's and 0's. Individual data bits are read and written using an electromagnetic read/write head.

Mask: A material covering the silicon of a masked ROM during the fabrication process. It determines the permanent logic state to be formed at each memory location.

Memory Address: The location of the stored data to be accessed.

Memory Cell: The smallest division of a memory circuit or IC. It contains a single bit of data (1 or 0).

Memory Contents: The binary data quantity stored at a particular memory address.

Optical Memory: The most common types of optical memory are the CD and DVD used by a PC. They use the reflection of a laser beam off a rigid disk platter to represent a 1 or 0. This CD or DVD platter will have either a nonreflective pit to represent a 1 or a non-pit (land) to represent a 0.

PROM: Programmable read-only memory.

RAM: Random-access memory (read/write memory).

RAS: Row address strobe. An active-LOW signal provided when the address lines contain a valid row address.

ROM: Read-only memory.

Semiconductor Memory: Digital ICs used for the storage of large amounts of binary data. The binary data at each memory cell are stored as the state of a flip-flop (RAM), the charge on a capacitor (DRAM), an internal transistor connection (ROM), or the charge on the gate of a floating-gate MOSFET (EEPROM, Flash).

Static: A term used to describe a class of semiconductor memory that uses the state on an internal flip-flop as its basic storage element.

Volatile: ICs that lose their memory contents when power is removed.

Problems

Section 16–1

16–1. Most semiconductor memory is implemented with either bipolar or MOS transistors.

(a) In general, which type of memory technology is faster, bipolar or MOS?

(b) Which is more dense, bipolar or MOS?

16–2. Describe the difference between the columns labeled "address" and "data" in Figure 16–1.

16–3. In Figure 16–2, all $\overline{OE}$'s are grounded. Why isn't there a problem with bus contention?

T　**16–4.** Assume that the 74LS374s in the memory system in Figure 16–2 are loaded with the output of a hex counter from 00H to 0FH (H = Hex). Connecting the circuit of Figure 16–4 to Figure 16–2 allows you to test the memory system. When you read the data from the memory, you find that addresses 0000 to 0111 have the hex numbers 00H to 07H as they are supposed to, but addresses 1000 to 1111 have the same data (00H to 07H). What do you suppose is wrong, and how would you troubleshoot the system?

D　**16–5.** Design and sketch an 8-byte memory system similar to Figure 16–2, using eight 74LS374s and one 74LS138.

Section 16–2

16–6. Briefly describe the difference between SRAMs and DRAMs. What are the advantages and disadvantages of each?

16–7. Use the block diagram for the 2147H in Figure 16–5 to determine what state $\overline{CS}$ and $\overline{WE}$ must be in to enable the three-state buffer connected to D_{in}.

16–8. What is the level of D_{out} on a 2147H when $\overline{CE}$ and $\overline{WE}$ are both LOW?

16–9. How many address lines are required to select a specific memory location within a RAM having:

(a) 1024 locations?　　　(c) 8192 locations?

(b) 4096 locations?

16–10. How many memory *locations* do the following RAM configurations have?

(a) 2048 × 1　　　(d) 1024 × 4

(b) 2K × 4　　　(e) 4K × 8

(c) 8192 × 8　　　(f) 16K × 1

16–11. What is the total number of *bits* that can be stored in the following RAM configurations?

(a) 1K × 8　　　(c) 8K × 8

(b) 4K × 4　　　(d) 16K × 1

D　**16–12.** Design and sketch a 1K × 8 RAM memory system using two 2148Hs.

T　**16–13.** When troubleshooting the memory system in Figure 16–7, you keep reading incorrect values at D_0 to D_7. Using a logic analyzer, you observe the waveforms of Figure P16–13 at A_0 to A_{11} and $\overline{CE}$. What is wrong and how would you correct it?

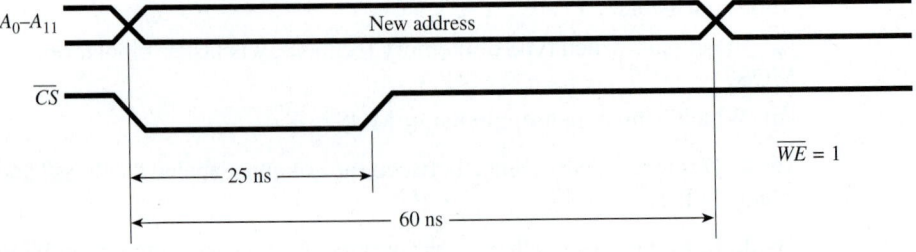

Figure P16–13

D   **16–14.** Use Table 16–1 to determine how many 62256 RAM ICs it would take to design a 1MB memory system for a computer that uses 8-bit data storage.

Section 16–3

16–15. Which lines are multiplexed on DRAMs, and why?

16–16. What is the purpose of $\overline{RAS}$ and $\overline{CAS}$ on DRAMs?

16–17.

(a) Draw the timing diagrams for a Read cycle and a Write cycle of a DRAM similar to Figure 16–11(b) and (c). Assume that $\overline{CAS}$ is delayed from $\overline{RAS}$ by 100 ns. Also assume that $t_{CAC} = 120$ ns (max), $t_{RAC} = 180$ ns (max), $t_{DS} = 40$ ns (min), and $t_{DH} = 30$ ns (min).

(b) How long after the falling edge of $\overline{RAS}$ will the *data out* be valid?

(c) How soon after the falling edge of $\overline{RAS}$ must the *data in* be set up?

16–18. How often does a 2118 DRAM have to be *refreshed,* and why?

16–19. What functions does the 3242 DRAM controller take care of?

Section 16–4

16–20. Are the following memory ICs volatile or nonvolatile?

(a) Mask ROM   **(c)** DRAM

(b) SRAM   **(d)** EPROM

16–21. Which EPROM is electrically erasable, the 2716 or the 2864?

Section 16–5

16–22. Which EPROM and which EPROM address are accessed when the microprocessor of Figure 16–20 issues a read command for the following addresses?

(a) READ 1020H   **(c)** READ 7001H

(b) READ 0ABCH   **(d)** READ 3FFFH?

D   **16–23.** Redesign the connections to the 74LS138 in Figure 16–20 so that the four EPROMs are accessed at addresses 8000H to BFFFH.

T   **16–24.** When testing the EPROM memory system of Figure 16–20, the microprocessor reads valid data from all EPROMs except EPROM2 (1000H–1FFFH). What are two probable causes?

T   **16–25.** When the microprocessor of Figure 16–20 reads data from addresses 8000H to 8FFFH, it finds the same data as that at 0000H to 0FFFH. What is the problem?

D   **16–26.** In Figure 16–20, should the microprocessor software be designed to read data from the EPROMs when it issues the HIGH-to-LOW or LOW-to-HIGH edge on $\overline{RD}$? Why?

C D   **16–27.** Design and sketch an address decoding scheme similar to Figure 16–20 for an 8K × 8 EPROM memory system using 2716 EPROMs. (The 2716 is a 2K × 8 EPROM.)

16–28. What single decoder chip could be used in Figure 16–21 in place of the two 74LS138s?

C D **16–29.** Design a PROM IC to act like a 3-bit *controlled inverter.* Use a 16 × 4 PROM similar to that in Figure 16–22. When A_3 is HIGH, the input at A_0 to A_2 is to be inverted; otherwise, it is not. Build a truth table showing the 16 possible inputs and the resultant output at Q_0 to Q_2 (Q_3 is not used).

16–30. If at 65°C the output of the ADC in Figure 16–23 is 0111 1010 (7AH), determine what location 7AH in the EPROM should contain.

Section 16–6

16–31. Which of the following memories are nonvolatile?

(a) RAM

(b) DRAM

(c) EEPROM

(d) Floppy disk

(e) Hard disk

(f) CD

16–32. How is a single data bit stored in the following types of memory?

(a) Magnetic memory

(b) Optical memory

16–33.

(a) Which magnetic memory system has a higher bit density, floppy disk or hard disk?

(b) Which optical memory system has a higher bit density, CD or DVD?

16–34. How does the construction of a CD-RW differ from that of a CD-R to make it rewritable?

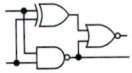

Schematic Interpretation Problems

See Appendix G for the schematic diagrams.

S **16–35.** The 62256 (U10) IC in the 4096/4196 schematic is a MOS static RAM. By looking at the number of address lines and data lines, determine the size and configuration of the RAM.

S **16–36.** Repeat Problem 16–35 for U6 of the HC11D0 schematic. Looking at the connections to the address lines, determine how much of the RAM is actually accessible.

S C **16–37.** The HC11D0 schematic uses two 27C64 EPROMS.

(a) What are their size and configuration?

(b) What are the labels of the control signals used to determine which EPROM is selected?

(c) Place a jumper from pin 2 to pin 3 of jumper J1 (grid location D-6). Determine the range of addresses that make SMN_SL active (active-LOW).

(d) Determine the range of addresses that make MON_SL active (active-LOW).

Electronics Workbench®/MultiSIM® Exercises

E16–1. Load the circuit file for **Section 16–1a.** The 74374 will be used as an 8-bit memory system in this circuit. The Word Generator is set up to output an 8-bit counter from 00H to FFH repeatedly. The instant that "C" is pressed, whatever is output by the Word Generator will be captured by the

CHAPTER 16 I SEMICONDUCTOR, MAGNETIC, AND OPTICAL MEMORY

8-bit D flip-flop memory system. Turn the power switch ON and press "C". The lights that come on will be equivalent to the hex number that appeared in the Word Generator at the instant "C" was pressed.

(a) Why don't the lights change after the Word Generator continues to count up?

(b) What happens if OC' is connected to Vcc instead of ground?

Answers to Review Questions

16–1. Address

16–2. Data

16–3. By inputting the 4-bit address to the 74LS154, which outputs a LOW pulse on one of the output lines when $\overline{WRITE}$ is pulsed LOW

16–4. It shows when any or all of the lines are allowed to change digital levels.

16–5. To avoid a bus conflict

16–6. 1024, 4

16–7. LOW, HIGH

16–8. $A_0 A_5$, $A_6 A_{11}$

16–9. Row 7, column 9

16–10. Q_5 and Q_6

16–11. 35 ns max

16–12. To keep the IC pin count to a minimum. They are demultiplexed by using control signals $\overline{RAS}$ and $\overline{CAS}$.

16–13. A capacitor

16–14. The charge on the internal capacitors in the RAM is replenished.

16–15. 60 ms

16–16. It means that when the power is removed, the memory contents are lost.

16–17. After the EPROM program has been thoroughly tested

16–18. (a) Open source on a MOS transistor.

(b) An open fuse in the source of a MOS transistor.

(c) Having an electron charge on the floating gate of a MOS transistor.

(d) Same as (c).

(e) Same as (c).

16–19. (a) Dissipate the electron charge on the floating gate by exposing it to UV radiation through the open IC window.

(b) Ground the control gate and put a high voltage (V_{PP}) on the source to drain the electron charge off of the floating gate.

(c) Same as (b)

16–20. 120, 330 (450 total)

16–21. When another device has to use the bus

16–22. (a) third EPROM, address 002H,

(b) first EPROM, address AF7H.

16–23. 9000H–9FFFH

16–24. 0010 0101

17

Microprocessor Fundamentals

OBJECTIVES

Upon completion of this chapter, you should be able to:

- Describe the benefits that microprocessor design has over hard-wired IC logic design.
- Discuss the functional blocks of a microprocessor-based system having basic I/O capability.
- Describe the function of the address, data, and control buses.
- Discuss the timing sequence on the three buses required to perform a simple I/O operation.
- Explain the role of software program instructions in a microprocessor-based system.
- Understand the software program used to read data from an input port and write it to an output port.
- Discuss the basic function of each of the internal blocks of the 8085A microprocessor.
- Follow the flow of data as they pass through the internal parts of the 8085A microprocessor.
- Make comparisons between assembly language, machine language, and high-level languages.
- Discuss the fundamental circuitry and timing sequence for external microprocessor I/O.

INTRODUCTION

The design applications studied in the previous chapters have all been based on combinational logic gates and sequential logic ICs. One example is a traffic light controller that goes through the sequence green–yellow–red. To implement the circuit using combinational and sequential logic, we would use some counter ICs for the timing, a shift register for sequencing the lights, and a D flip-flop if we want to interrupt the sequence with a pedestrian cross-walk push button. A complete design solution is easily within the realm of SSI and MSI ICs.

On the other hand, think about the complexity of electronic control of a modern automobile. There are several analog quantities to monitor, such as engine speed, manifold pressure, and coolant temperature, and there are several digital control functions to perform, such as spark plug timing, fuel mixture control, and radiator circulation control. The operation is further complicated by the calculations and decisions that have to be made on a continuing basis. This is definitely an application for a *microprocessor-based system.*

A system designer should consider a microprocessor-based solution whenever an application involves making calculations, making decisions based on external stimulus, and maintaining memory of past events. A microprocessor offers several advantages over the hard-wired SSI/MSI IC approach. First, the microprocessor itself is a general-purpose device. It takes on a unique personality by the software program instructions given by the designer. If you want it to count, you tell it to do so, with software. If you want to shift its output level left, there's an instruction for that. And if you want to add a new quantity to a previous one, there's another instruction for that. Its capacity to perform arithmetic, make comparisons, and update memory makes it a very powerful digital problem solver. Making changes to an application can usually be done by changing a few program instructions, unlike the hard-wired system that may have to be totally redesigned and reconstructed.

New microprocessors are introduced every year to fill the needs of the design engineer. However, the theory behind microprocessor technology remains basically the same. It is a general-purpose digital device that is driven by software instructions and communicates with several external *support chips* to perform the necessary I/O of a specific task. Once you have a general understanding of one of the earlier microprocessors that came on the market, such as the Intel 8080/8085, the Motorola 6800, or the Zilog Z80, it is an easy task to teach yourself the necessary information to upgrade to the new microprocessors as they are introduced. Typically, when a new microprocessor is introduced, it will have a few new software instructions available and have some of the I/O features, previously handled by external support chips, integrated right into the microprocessor chip. Learning the basics on these new microprocessor upgrades is more difficult, however, because some of their advanced features tend to hide the actual operation of the microprocessor and may hinder your complete understanding of the system.

17–1 *Introduction to System Components and Buses

Figure 17–1 shows a **microprocessor** with the necessary **support circuitry** to perform basic input and output functions. We will use this figure to illustrate how the microprocessor acts like a general-purpose device, driven by software, to perform a specific task related to the input data switches and output data LEDs. First, let's discuss the components of the system.

Microprocessor

The heart of the system is an 8-bit microprocessor. It could be any of the popular 8-bit microprocessors such as the Intel 8085, the Motorola 6800, or the Zilog Z80. They are

---

*From *Digital and Microprocessor Fundamentals Fourth Edition,* by William Kleitz, Prentice-Hall, Upper Saddle River, N.J. 2003.

Team Discussion

Discuss the flow of data as the microprocessor reads program instructions from memory, which then tell it to read the input data switches and transfer what it has read out to the LEDs.

Team Discussion

Which port could be connected to a D/A converter? To an A/D converter?

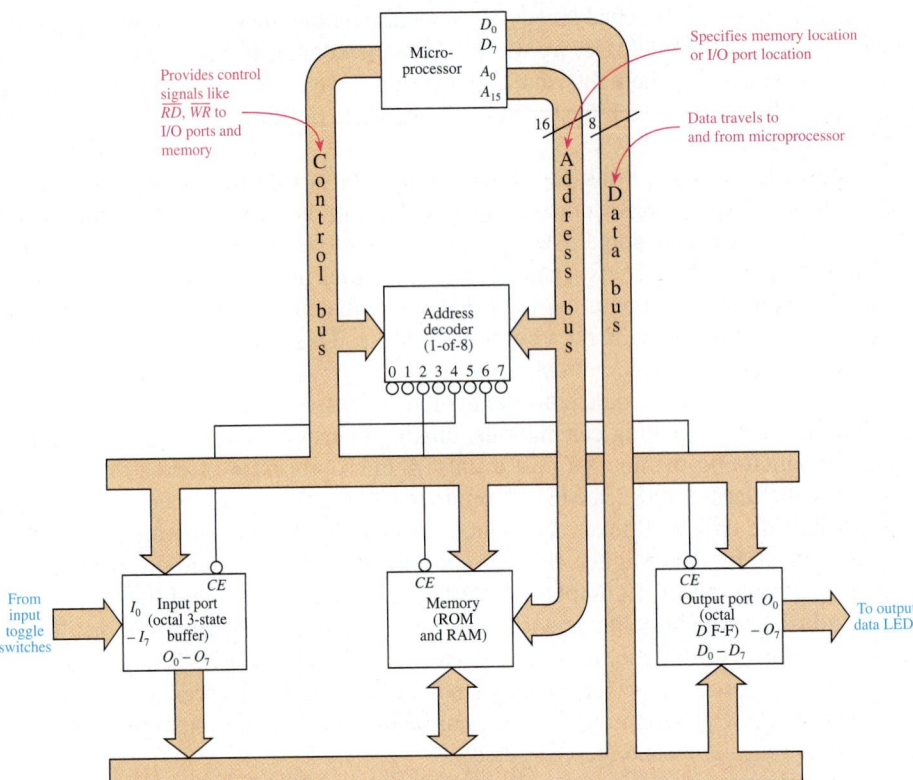

Figure 17–1 Example of a microprocessor-based system used for simple I/O operations.

called 8-bit microprocessors because external and internal data movement is performed on 8 bits at a time. It will read *program instructions* from memory and execute those instructions that drive the three *external buses* with the proper levels and timing to make the connected devices perform specific operations. The buses are simply groups of conductors that are routed throughout the system and tapped into by various devices (or ICs) that need to share the information that is traveling on them.

Address Bus

The **address bus** is 16 bits wide and is generated by the microprocessor to select a particular location or IC to be active. In the case of a selected memory IC, the low-order bits on the address bus select a particular location within the IC (see Section 16–5). Because the address bus is 16 bits wide, it can actually specify 65,536 (2^{16}) different addresses. The input port is one address, the output port is one address, and the memory in a system of this size may be 4K (4096) addresses. This leaves about 60K addresses available for future expansion.

Data Bus

Once the address bus is set up with the particular address that the microprocessor wants to access, the microprocessor then sends or receives 8 bits of data to or from that address via the **bidirectional** (two-way) **data bus.**

Control Bus

The **control bus** is of varying width, depending on the microprocessor being used. It carries control signals that are tapped into by the other ICs to tell what type of operation is being performed. From these signals, the ICs can tell if the operation is a read, a write, an I/O, a memory access, or some other operation.

Address Decoder

The address decoder is usually an octal decoder like the 74LS138 studied in Chapters 8 and 16. Its function is to provide active-LOW Chip Enables ($\overline{CE}$) to the external ICs based on information it receives from the microprocessor via the control and address buses. Because there are multiple ICs on the data bus, the address decoder ensures that only one IC is active at a time to avoid a bus conflict caused by two ICs writing different data to the same bus.

Memory

There will be at least two memory ICs: ROM or EPROM, and a RAM. The ROM will contain the *initialization* instructions, telling the microprocessor what to do when power is first turned on. This includes tasks like reading the keyboard and driving the CRT display. It will also contain several subroutines that can be called by the microprocessor to perform such tasks as time delays or I/O data translation. These instructions, which are permanently stored in ROM, are referred to as the **monitor program** or **operating system.** The RAM part of memory is volatile, meaning that it loses its contents when power is turned off and, therefore, is used only for temporary data storage.

Input Port

The input port provides data to the microprocessor via the data bus. In this case, it is an octal buffer with three-stated outputs. The input to the buffer will be provided by some input device like a keyboard or, as in this case, from eight HIGH–LOW toggle switches. The input port will dump its information to the data bus when it receives a Chip Enable ($\overline{CE}$) from the address decoder and a Read command ($\overline{RD}$) from the control bus.

Output Port

The output port provides a way for the microprocessor to talk to the outside world. It could be sending data to an output device like a printer or, as in this case, it could send data to eight LEDs. An octal *D* flip-flop is used as the interface because, after the microprocessor sends data to it, the flip-flop will latch on to the data, allowing the microprocessor to continue with its other tasks.

To load the *D* flip-flop, the microprocessor must first set up the data bus with the data to be output. Then, it sets up the address of the output port so that the address decoder will issue a LOW $\overline{CE}$ to it. Finally, it issues a pulse on its $\overline{WR}$ (write) line that travels the control bus to the clock input of the *D* flip-flop. When the *D* flip-flop receives the clock trigger pulse, it latches onto the data that are on the data bus at that time and drives the LEDs.

Review Questions

17–1. What are the names of the three buses associated with microprocessors?

17–2. How much of the circuitry shown in Figure 17–1 is contained inside a microcontroller IC?

17–3. Why must the data bus be bidirectional?

17–4. The purpose of the address decoder IC in Figure 17–1 is to enable two or more of the external ICs to be active at the same time to speed up processing. True or false?

17–5. The input port in Figure 17–1 must have *three-state* outputs so that its outputs are floating whenever any other IC is writing data to the data bus. True or false?

17–6. What would be the consequence of using an octal buffer for the output port in Figure 17–1 instead of an octal *D* flip-flop?

17–2 Software Control of Microprocessor Systems

The nice thing about microprocessor-based systems is that, once you have a working prototype, you can put away the soldering iron because all operational changes can then be made with software. The student of electronics has a big advantage when writing microprocessor software because he or she understands the hardware at work as well as the implications that software will have on the hardware. Areas such as address decoding, chip enables, instruction timing, and hardware interfacing become important when programming microprocessors.

As a brief introduction to microprocessor software, let's refer back to Figure 17–1 and learn the statements required to perform some basic I/O operations. To route the data from the input switches to the output LEDs, the data from the input port must first be read into the microprocessor before they can be sent to the output port. The microprocessor has an 8-bit internal register called the **accumulator** that can be used for this purpose.

The software used to drive microprocessor-based systems is called **assembly language.** The Intel 8080/8085 assembly language statement to load the contents of the input port into the accumulator is LDA *addr.* LDA is called a **mnemonic,** an abbreviation of the operation being performed, which in this case is "Load Accumulator." The suffix *addr* will be replaced with a 16-bit address (4 hex digits) specifying the address of the input port.

After the execution of LDA *addr,* the accumulator will contain the digital value that was on the input switches. Now, to write these data to the output port, we use the command STA *addr.* STA is the mnemonic for "Store Accumulator," and *addr* is the 16-bit address where you want the data stored.

Execution of those two statements is all that is necessary to load the value of the switches into the accumulator and then transfer these data to the output LEDs. The microprocessor takes care of the timing on the three buses, and the address decoder takes care of providing chip enables to the appropriate ICs.

If the system is based on Motorola or Zilog technology, the software in this case will be almost the same. Table 17–1 makes a comparison of the three assembly languages.

Helpful Hint

It is beyond the scope of this text to cover an entire software instruction set, but it might be instructive for you to locate a programmer's manual for one of the more popular microprocessors to see an entire list of instructions so that you can get a feel for its power.

TABLE 17–1	Comparison of I/O Software on Three Different Microprocessors		
Operation	Intel 8080/8085	Motorola 6800	Zilog Z80
Load accumulator with contents of location *addr*	LDA *addr*	LDAA *addr*	LD A, (*addr*)
Store accumulator to location *addr*	STA *addr*	STAA *addr*	LD (*addr*), A

17–3   Internal Architecture of a Microprocessor

The design for the Intel 8085A microprocessor was derived from its predecessor, the 8080A. The 8085A is **software** *compatible* with the 8080A, meaning that software programs written for the 8080A can run on the 8085A without modification. The 8085A has a few additional features not available on the 8080A. The 8085A also has a higher level of hardware integration, allowing the designer to develop complete microprocessor-based systems with fewer external support ICs than were required by the 8080A. Studying the internal **architecture** of the 8085A in Figure 17–2 and its pin configuration in Figure 17–3 will give us a better understanding of its operation.

The 8085A is an 8-bit parallel **central processing unit (CPU).** The accumulator discussed in the previous section is connected to an 8-bit *internal data bus.* Six other *general-purpose registers* labeled B, C, D, E, H, and L are also connected to the same bus.

All arithmetic operations take place in the **arithmetic logic unit (ALU).** The accumulator, along with a temporary register, is used as input to all arithmetic operations. The output of the operations is sent to the internal data bus and to five *flag flip-flops* that record the status of the arithmetic operation.

The **instruction register** and **decoder** receive the software instructions from external memory, interpret what is to be done, and then create the necessary timing and control signals required to execute the instruction.

The block diagram also shows **interrupt** control, which provides a way for an external digital signal to interrupt a software program while it is executing. This is accomplished by applying the proper digital signal on one of the interrupt inputs: INTR, RSTx.x, or TRAP. *Serial communication* capabilities are provided via the SID and SOD I/O pins (Serial Input Data, Serial Output Data).

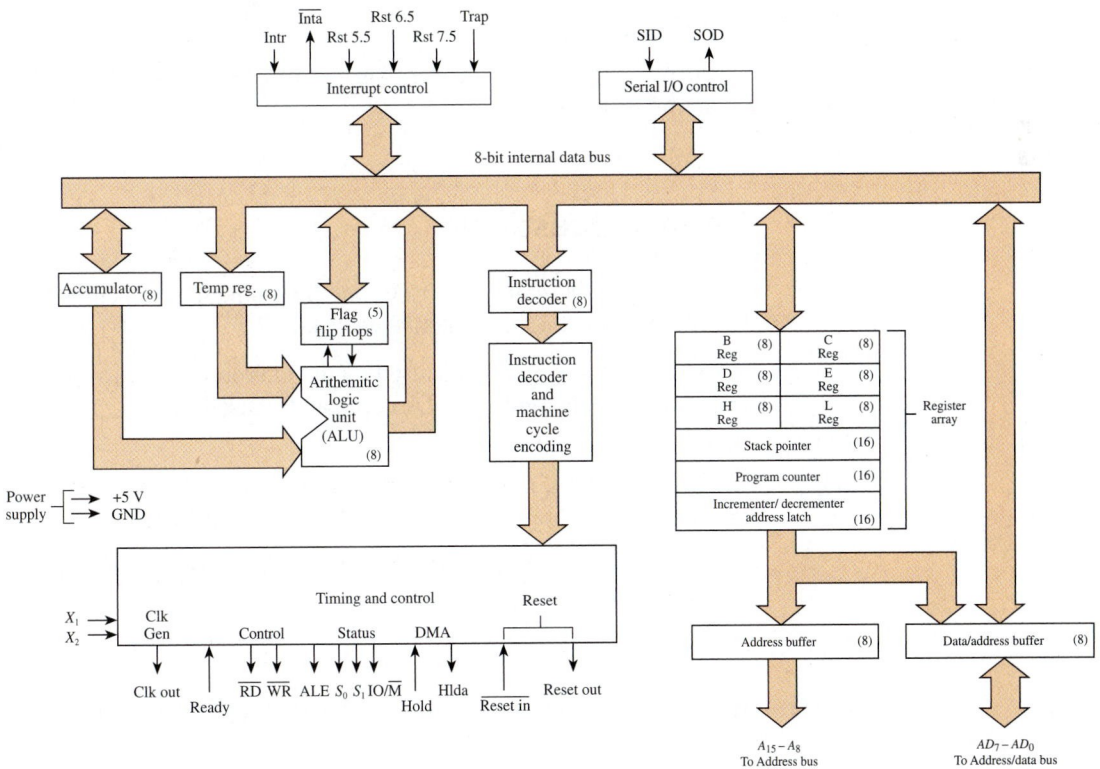

Figure 17–2   The 8085A CPU functional block diagram. (Courtesy of Intel Corporation.)

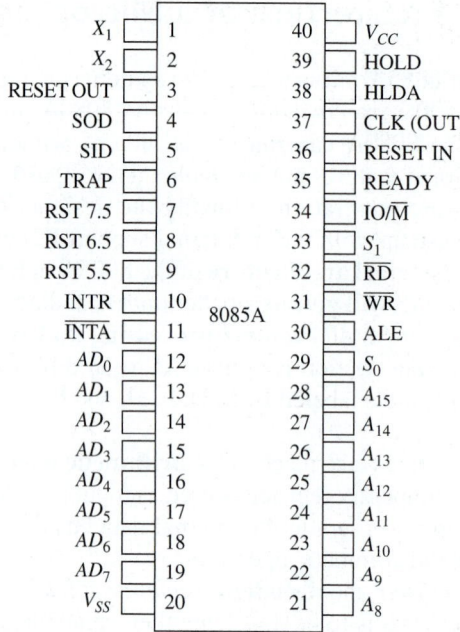

X_1	1		40	V_{CC}
X_2	2		39	HOLD
RESET OUT	3		38	HLDA
SOD	4		37	CLK (OUT)
SID	5		36	RESET IN
TRAP	6		35	READY
RST 7.5	7		34	$IO/\overline{M}$
RST 6.5	8		33	S_1
RST 5.5	9		32	$\overline{RD}$
INTR	10	8085A	31	$\overline{WR}$
$\overline{INTA}$	11		30	ALE
AD_0	12		29	S_0
AD_1	13		28	A_{15}
AD_2	14		27	A_{14}
AD_3	15		26	A_{13}
AD_4	16		25	A_{12}
AD_5	17		24	A_{11}
AD_6	18		23	A_{10}
AD_7	19		22	A_9
V_{SS}	20		21	A_8

Figure 17–3 The 8085A pin configuration. (Courtesy of Intel Corporation.)

The *register array* contains the six general-purpose 8-bit registers and three 16-bit registers. Sixteen-bit registers are required whenever you need to store addresses. The **stack pointer** stores the address of the last entry on the stack. The stack is a data storage area in RAM used by certain microprocessor operations. The **program counter** contains the 16-bit address of the next software instruction to be executed. The third 16-bit register is the *address latch,* which contains the current 16-bit address that is being sent to the address bus.

The six general-purpose 8-bit registers can also be used in pairs (*B–C, D–E, H–L*) to store addresses or 16-bit data.

Review Questions

17–7. The suffix *addr* in the LDA and STA mnemonics is used to specify the address of the accumulator. True or false?

17–8. The LDA command is used by the microprocessor to _____ (read, write) data, and the STA command is used to _____ (read, write) data.

17–9. The 8085A microprocessor has an *internal* accumulator and six *external* registers called B, C, D, E, H, and L. True or false?

17–10. The ALU block inside the 8085A determines the timing and control signals required to execute an instruction. True or false?

17–4 Instruction Execution Within a Microprocessor

Now, referring back to the basic I/O system diagram of Figure 17–1, let's follow the flow of the LDA and STA instructions as they execute in the block diagram of the

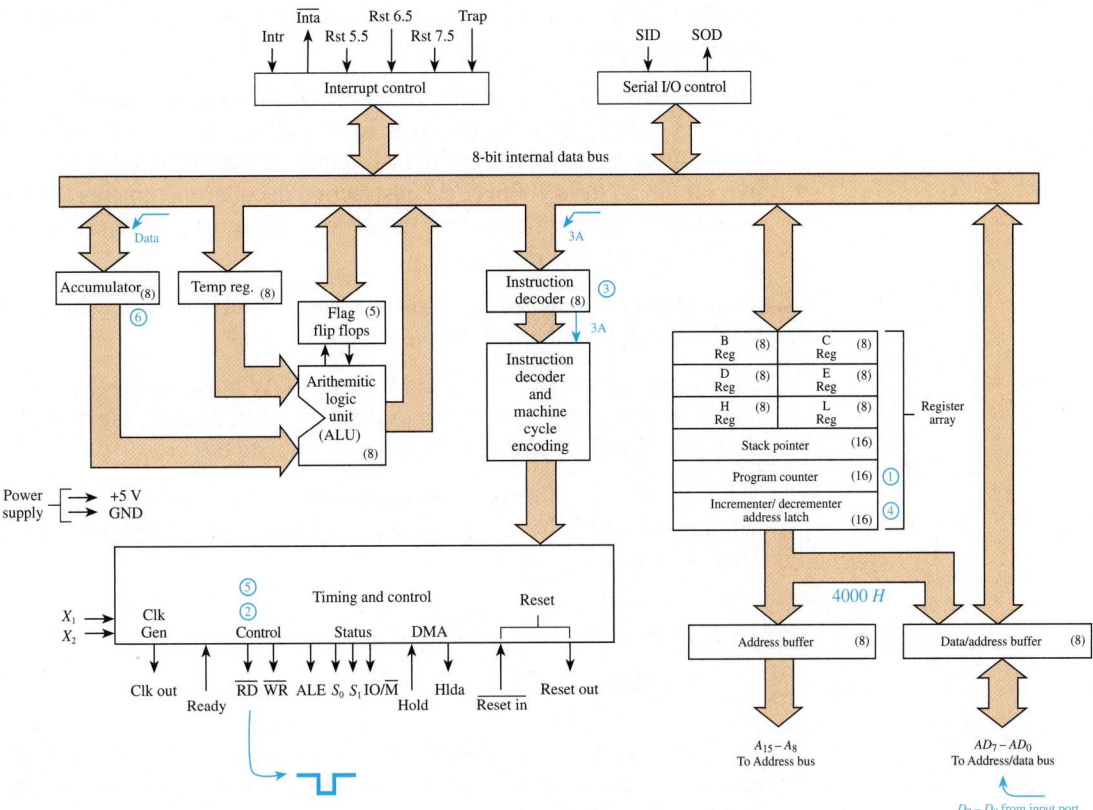

Figure 17–4 Execution of the LDA instruction within the 8085A.

8085A. Figure 17–4 shows the 8085A block diagram with numbers indicating the succession of events that occurs when executing the LDA instruction.

Remember, LDA *addr* and STA *addr* are assembly language instructions, stored in an external memory IC, that tell the 8085A CPU what to do. LDA *addr* tells the CPU to load its accumulator with the data value that is at address *addr*. STA *addr* tells the CPU to store (or send) the 8-bit value that is in the accumulator to the output port at address *addr*.

The mnemonics LDA and STA cannot be understood by the CPU as they are; they have to be *assembled,* or converted, into a binary string called **machine code.** Binary or hexadecimal machine code is what is actually read by the CPU and passed to the instruction register and decoder to be executed. The Intel 8085A Users Manual gives the machine code translation for LDA as $3A_{16}$ (or 3AH) and STA as 32H.

Before studying the flow of execution in Figure 17–4, we need to make a few assumptions. Let's assume that the input port is at address 4000H and the output port is at address 6000H. Let's also assume that the machine code program LDA 4000H, STA 6000H is stored in RAM starting at address 2000H.

Load Accumulator

The sequence of execution of LDA 4000H in Figure 17–4 will be as follows:

1. The program counter will put the address 2000H on the address bus.

2. The timing and control unit will issue a LOW pulse on the $(\overline{RD})$ line. This pulse will travel the control bus to the RAM in Figure 17–1 and will cause the contents at location 2000H to be put onto the external data bus. RAM

Team Discussion

Identify the route that data travel, from an external input to the accumulator and then to an output port.

Common Misconception

Students often confuse the external data bus with the internal data bus.

(2000H) has the machine code 3AH, which will travel across the internal data bus to the instruction register.

3. The instruction register passes the 3AH to the instruction decoder, which determines that 3AH is the code for LDA and that a 16-bit (2-byte) address must follow. Because the entire instruction is 3 bytes (one for the 3AH and two for the address 4000H), the instruction decoder increments the program counter two more times so that the address latch register can read and store bytes 2 and 3 of the instruction.

4. The address latch and address bus now have 4000H on them, which provides the LOW $\overline{CE}$ for the input port in Figure 17–1.

5. The timing and control unit again issues a LOW pulse on the $(\overline{RD})$ line. This pulse will travel the control bus to the input port, causing the data at the input port (4000H) to be put onto the external data bus.

6. That data will travel across the external data bus in Figure 17–1, to the internal data bus in Figure 17–4, to the accumulator, where they are now stored. The instruction is complete.

Store Accumulator

Figure 17–5 shows the flow of execution of the STA 6000H instruction.

1. After the execution of the 3-byte LDA 4000H instruction, the program counter will have 2003H in it. (Instruction LDA 4000H resided in locations 2000H, 2001H, 2002H.)

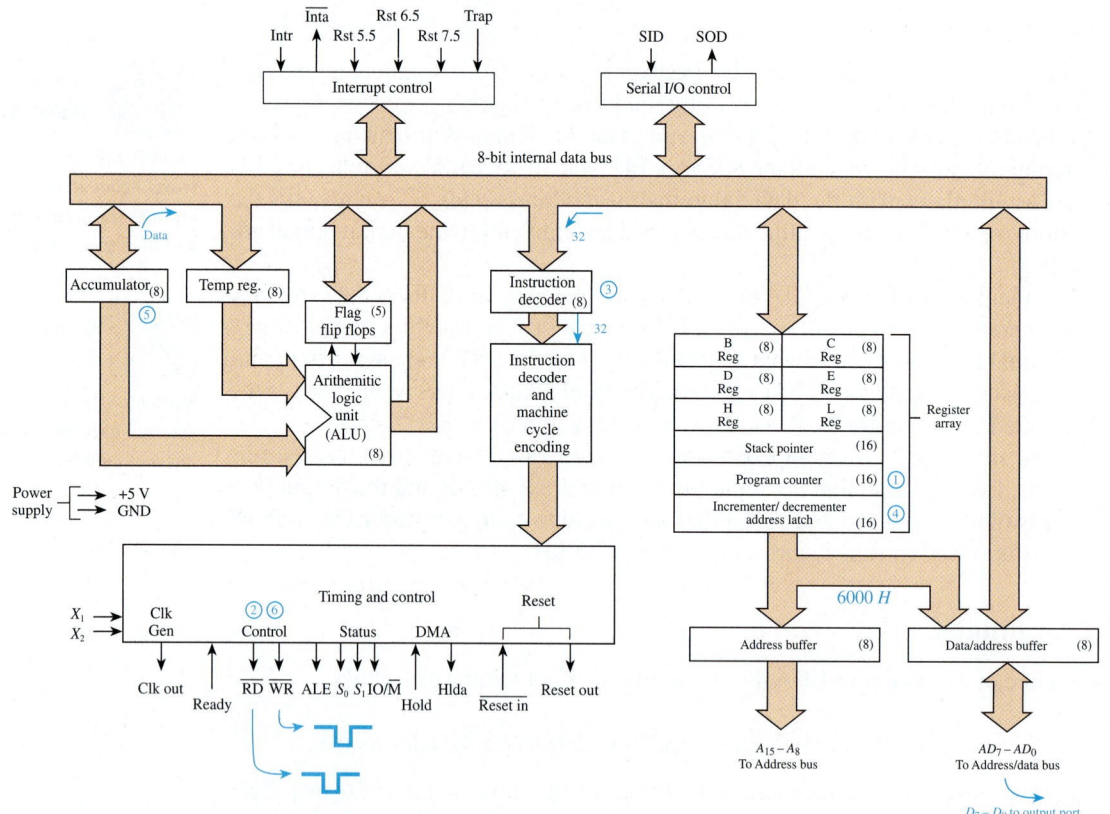

Figure 17–5 Execution of the STA instruction within the 8085A.

CHAPTER 17 I MICROPROCESSOR FUNDAMENTALS

2. The timing and control unit will issue a LOW pulse on the $\overline{RD}$ line. This will cause the contents of RAM location 2003H to be put onto the external data bus. RAM (2003H) has the machine code 32H, which will travel up the internal data bus to the instruction register.

3. The instruction register passes the 32H to the instruction decoder, which determines that 32H is the code for STA and that a 2-byte address must follow. The program counter gets incremented two more times, reading and storing bytes 2 and 3 of the instruction into the address latch.

4. The address latch and address bus now have 6000H on them, which is the address of the output port in Figure 17–1.

5. The instruction decoder now issues the command to place the contents of the accumulator onto the data bus.

6. The timing and control unit issues a LOW pulse on the $\overline{WR}$ line. Because the $\overline{WR}$ line is used as a clock input to the D flip-flop of Figure 17–1, the data from the data bus will be stored and displayed on the LEDs. (The $\overline{WR}$ line from the microprocessor is part of the control bus in Figure 17–1.)

The complete assembly language and machine code program for the preceding I/O example is given in Table 17–2.

TABLE 17–2	Assembly Language and Machine Code Listing for the LDA-STA Program

Memory Location	Assembly Language	Machine Code	
2000H	LDA 4000H	3A	⎫ Three-byte instruction to load accumulator
2001H		00	⎬ with contents from address 4000H
2002H		40	⎭
2003H	STA 6000H	32	⎫ Three-byte instruction to store accumulator
2004H		00	⎬ out to address 6000H
2005H		60	⎭

Review Questions

17–11. What is the difference between assembly language and machine code?

17–12. Why is the instruction LDA 4000H called a 3-byte instruction?

17–13. When executing the instruction LDA 4000H, the microprocessor fetches the machine code from RAM location 4000H. True or false?

17–14. Which instruction, STA or LDA, issues a pulse on the $\overline{WR}$ line? Why?

17–15. How many bytes of RAM does the program in Table 17–2 occupy?

17–5 Hardware Requirements for Basic I/O Programming

A good way to start out in microprocessor programming is to illustrate program execution by communicating to the outside world. In Section 17–4, we read input switches at memory location 4000H using the LDA instruction and wrote their value to output

LEDs at location 6000H using the STA instruction. This was an example of **memory-mapped I/O.** Using this method, the input and output devices were accessed *as if they were memory locations* by specifying their unique 16-bit address (4000H or 6000H).

The other technique used by the 8085A microprocessor for I/O mapping is called *standard I/O* or **I/O-mapped I/O.** *I/O-mapped systems* identify their input and output devices by giving them an 8-bit **port number.** The microprocessor then accesses the I/O ports by using the instructions OUT *port* and IN *port,* where port is 00H to FFH.

Special **hardware** external to the 8085A is required to provide the source for the IN instruction and the destination for the OUT instruction. Figure 17–6 shows a basic hardware configuration, using standard SSI and MSI ICs, that could be built to input data from eight switches and to output data to eight LEDs using I/O-mapped I/O.

Helpful Hint

Review the operation of the 74LS244 octal buffer and the 74LS374 octal *D* flip-flop before studying this figure.

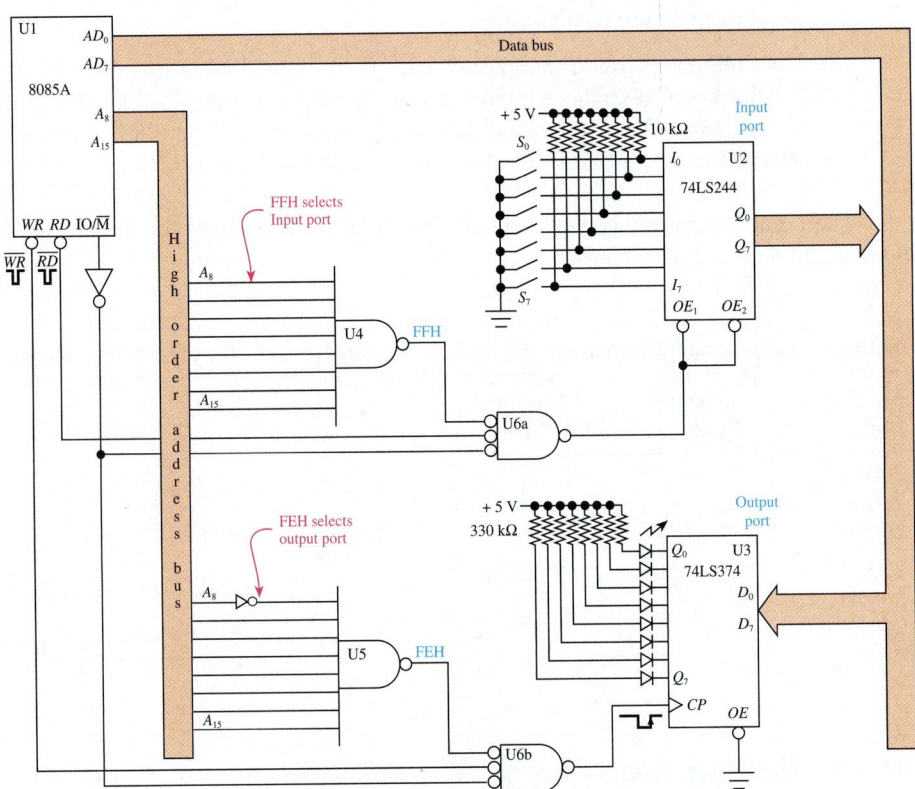

Figure 17–6 Hardware requirements for the IN FFH and OUT FEH instructions.

Figure 17–6 is set up to decode the input switches as port FFH and the output LEDs as port FEH. The $IO/\overline{M}$ line from the microprocessor goes HIGH whenever an IN or OUT instruction is being executed (I/O-mapped I/O). All instructions that access memory and memory-mapped devices will cause the $IO/\overline{M}$ line to go LOW. The $\overline{RD}$ line from the microprocessor will be pulsed LOW when executing the IN instruction, and the $\overline{WR}$ line will be pulsed LOW when executing the OUT instruction.

IN FFH

The 74LS244 (see Section 13–10) is an octal three-state buffer that is set up to pass the binary value of the input switches over to the data bus as soon as $\overline{OE}_1$ and $\overline{OE}_2$ are brought LOW. To get that LOW, U6a, the inverted-input NAND gate (OR gate), must receive three LOWs at its input. We know that the IN instruction will cause the *inverted* $IO/\overline{M}$ line to go LOW and the $\overline{RD}$ line to go LOW. The other input is dependent on the output from the right-input NAND gate (U4). Gate U4 *will* output a LOW because the binary value of the port number (1111 1111) used in the IN instruction is put onto the high-order address bus during the execution of the IN FFH instruction.

All conditions are now met; U6a will output a LOW pulse (the same width as the LOW $\overline{RD}$ pulse), which will enable the outputs of U2 to pass to the data bus. After the microprocessor drops the $\overline{RD}$ line LOW, it waits a short time for external devices (U2 in this case) to respond, and then it reads the data bus and raises the $\overline{RD}$ line back HIGH. The data from the input switches are now stored in the accumulator.

OUT FEH

The 74LS374 (see Section 13–10) is an octal *D* flip-flop set up to sink current to illuminate individual LEDs based on the binary value it receives from the data bus. The outputs at Q_0 to Q_7 will latch onto the binary values at D_0 to D_7 at the LOW-to-HIGH edge of C_p. U5 and U6b are set up similarly to U4 and U6a, except U5's output goes LOW when FEH (1111 1110) is input. Therefore, during the execution of OUT FEH, U6b will output a LOW pulse, the same width as the $\overline{WR}$ pulse issued by the microprocessor.

The setup time of the 74LS374 latch is accounted for by the microprocessor timing specifications. The microprocessor issues a HIGH-to-LOW edge at $\overline{WR}$ that makes its way to C_p. At the same time, the microprocessor also sends the value of the accumulator to the data bus. After a time period greater than the setup time for U3, $\overline{WR}$ goes back HIGH, which applies the LOW-to-HIGH trigger edge for U3, latching the data at Q_0 to Q_7.

To summarize, the instruction IN FFH reads the binary value at port FFH into the accumulator. The instruction OUT FEH writes the binary value in the accumulator out to port FEH. Port selection is taken care of by eight-input NAND gates attached to the high-order address bus and by use of the $\overline{RD}$, $\overline{WR}$, and *IO/M̄* lines.

17–6 Writing Assembly Language and Machine Language Programs

The microprocessor is driven by software instructions to perform specific tasks. The instructions are first written in assembly language using mnemonic abbreviations and then converted to machine language so that they can be interpreted by the microprocessor. The conversion from assembly language to machine language involves translating each mnemonic into the appropriate hexadecimal machine code and storing the codes in specific memory addresses. This can be done by a software package called an **assembler,** provided by the microprocessor manufacturer, or it can be done by the programmer by looking up the codes and memory addresses (called **hand assembly**).

Assembly language is classified as a low-level language because the programmer has to take care of all the most minute details. High-level languages such as Pascal, FORTRAN, C++, and BASIC are much easier to write but are not as memory efficient or as fast as assembly language. All languages, whether Pascal, C++, BASIC, or FORTRAN, get reduced to machine language code before they can be executed by the microprocessor. The conversion from high-level languages to machine code is done by a **compiler.** The compiler makes memory assignments and converts the English-language-type instructions into executable machine code.

On the other hand, assembly language translates *directly* into machine code. This allows the programmer to write the most streamlined, memory-efficient, and fastest programs possible on the specific hardware configuration that is being used.

Assembly language and its corresponding machine code differ from processor to processor. The fundamentals of the different assembly languages are the same, however, and once you have become proficient on one microprocessor, it is easy to pick it up on another.

Let's start off our software training by studying a completed assembly language program and comparing it to the same program written in the **BASIC** computer **language.** BASIC is a high-level language that uses English-language-type commands that are fairly easy to figure out, even by the inexperienced programmer.

Program Definition

Write a program that will function as a down-counter, counting 9 to 0 repeatedly. First draw a **flowchart,** and then write the program statements in the BASIC language, assembly language, and machine language.

Solution

The flowchart in Figure 17–7 is used to show the sequence of program execution, including the branching and looping that takes place.

According to the flowchart, the counter is decremented repeatedly until zero is reached, at which time the counter is reinitialized to 9 and the cycle repeats. The instructions used to implement the program are given in Table 17–3.

BASIC

BASIC uses the variable COUNT to hold the counter value. Line 30 checks the count. If COUNT is equal to zero, then the program goes back to the beginning. Otherwise, it goes back to subtract 1 from COUNT and checks COUNT again.

The 8085A version of the program is first written in assembly language, and then it is either hand assembled into machine language or computer assembled using a personal computer with an assembler software package.

Assembly Language

Assembly language is written using *mnemonics:* MVI, DCR, JZ, and the like. The term mnemonics is defined as "abbreviations used to assist the memory." The first mnemonic, MVI, stands for "Move Immediate." The instruction MVI A,09H will move the data value 09H into register A (register A and the accumulator are the same). The next instruction, DCR A, decrements register A by 1.

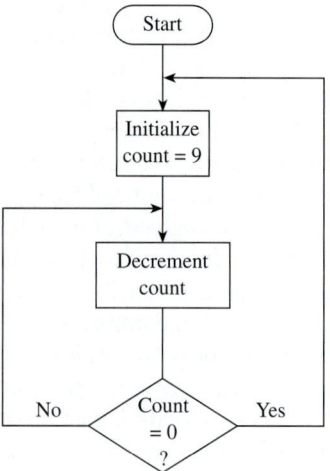

Figure 17–7   Flowchart for Table 17–3.

TABLE 17–3 | Down-Counter Program in Three Languages

BASIC Language		8085A Assembly Language		8085A Machine Language	
Line	Instruction	Label	Instruction	Address	Contents
10	COUNT=9	START:	MVI A,09H	2000	3E (opcode)
				2001	09 (data)
20	COUNT=COUNT-1	LOOP:	DCR A	2002	3D (opcode)
30	IF COUNT=0		JZ START	2003	CA (opcode)
	THEN GO TO 10			2004	00 } (address)
				2005	20 }
40	GO TO 20		JMP LOOP	2006	C3 (opcode)
				2007	02 } (address)
				2008	20 }

The third instruction, JZ START, is called a *conditional jump.* The condition that it is checking for is the *zero* condition. As the *A* register is decremented, if *A* reaches 0, then a flag bit, called the **zero flag,** gets set (a *set* flag is equal to 1). The instruction JZ START is interpreted as "jump to **statement label** START if the zero flag is set."

If the condition is not met (zero flag not set), then control passes to the next instruction, JMP LOOP, which is an *unconditional jump.* This instruction is interpreted as "jump to label LOOP regardless of any condition flags."

At this point you should see how the assembly language program functions exactly like the BASIC language program.

Machine Language

Machine language is the final step in creating an executable program for the microprocessor. In this step, we must determine the actual hexadecimal codes that will be stored in memory to be read by the microprocessor. First, we have to determine what memory locations will be used for our program. This depends on the memory-map assignments made in the system hardware design. We have 64K of addressable memory locations (0000H to FFFFH). We'll make an assumption that the user program area was set up in the hardware design to start at location 2000H. The length of the program memory area depends on the size of the ROM or RAM memory IC being used. A 256 × 8 RAM memory is usually sufficient for introductory programming assignments and is commonly used on educational microprocessor trainers. The machine language program listed in Table 17–3 fills up 9 bytes of memory (2000H to 2008H).

The first step in the hand assembly is to determine the code for MVI A. This is known as the **opcode** (operation code) and is found in an 8085A Assembly Language Reference Chart. The opcode for MVI A is 3E. The programmer will store the binary equivalent for 3E (0011 1110) into memory location 2000H. Instructions for storing your program into memory are given by the manufacturer of the microprocessor trainer that you are using. If you are using an assembler software package, then the machine code that is generated will usually be saved on a computer disk or used to program an EPROM to be placed in a custom microprocessor hardware design.

The machine language instruction MVI A,09H in Table 17–3 requires 2 bytes to complete. The first byte is the opcode, 3E, which identifies the instruction for the microprocessor. The second byte (called the **operand**) is the data value, 09H, which is to be moved into register *A.*

The second instruction, DCR A, is a 1-byte instruction. It requires just its opcode, 3D, which is found in the reference chart.

The opcode for the JZ instruction is CA. It must be followed by the 16-bit (2-byte) address to jump to if the condition (zero) is met. This makes it a 3-byte instruction. Byte

2 of the instruction (location 2004H) is the low-order byte of the address, and byte 3 is the high-order byte of the address to jump to. (Be careful to always enter addresses as low-order first, then high-order.)

The opcode for JMP is C3 and must also be followed by a 16-bit (2-byte) address specifying the location to jump to. Therefore, this is also a 3-byte instruction where byte 2–byte 3 gives a jump address of 2002H.

17–7 Survey of Microprocessors and Manufacturers

Since its introduction in the early 1970s, the microprocessor has had a huge impact on the electronics industry. The first microprocessors had a 4-bit internal data width. In 1974, Intel introduced the first 8-bit microprocessor, the 8008. Within a year it offered an upgrade, the 8080, which served as a point of comparison for all other manufacturers.

The first challenger to the 8080 was the Motorola 6800. Other IC manufacturers (National Semiconductor, Texas Instruments, Zilog, RCA, and Fairchild) soon introduced their own versions of the microprocessor. The race had begun. Since then, 16-, and 32- and 64-bit architectures have been developed and are finding their way into most new high-end applications.

Along the way, manufacturers started integrating whole multichip systems with RAM, ROM, and I/O into a single package called a *microcontroller*. Today, the microcontroller is the most popular choice for embedded control applications such as those found in automobiles, home entertainment systems, and data acquisition and control systems.

Each microprocessor and microcontroller has its own special niche, but throughout the years, the two most important players have been Intel and Motorola. Table 17–4 lists the most popular processors manufactured by these two companies. You can identify the manufacturer by the first two numbers in the part number (68 for Motorola and 80 for Intel). Appendix A lists the Web site addresses of several microprocessor manufacturers.

TABLE 17–4	Popular Intel and Motorola Microprocessors and Microcontrollers		
Part No.	Data Bits	Address Bits	Comments
8085	8	16	Upgrade of the 8080
8051	8	16	Microcontroller with on-chip ROM, RAM, and I/O
6809	8	16	Upgrade of the original 6800
68HC11	8	16	Microcontroller with on-chip ROM, RAM, I/O, and A/D Converter
8088	8	20	8-Bit downgrade of the 8086
8086	16	20	Made popular by its use in IBM PC-compatible computers
80186	16	20	8086 With on-chip support functions
80286	16	24	8086 Upgrade with extended addressing capability; used in IBM AT-compatible computers
8096	16	16	Microcontroller with on-chip ROM, RAM, I/O, and A/D converter
68000	16	23	Made popular by its use in Apple Macintosh II and Unix-based workstations
68010	16	24	Upgrade of the 68000
80386	32	32	32-Bit upgrade of the 80286
80486	32	32	Upgrade of the 80386
68020	32	32	32-Bit upgrade of the 68010
68030	32	32	Upgrade of the 68020
Pentium	64	32	Dual instruction pipelining allows concurrent execution of instructions, decreasing processing times
Pentium II	64	36	Higher speeds and memory capacity
Pentium III	64	36	Upgrade of the Pentium II
Pentium IV	64	36	Upgrade of the Pentium III

CHAPTER 17 | MICROPROCESSOR FUNDAMENTALS

17–16. Programs written in assembly language must be converted to machine code before being executed by a microprocessor. True or false?

17–17. Programs written in a high-level language are more memory efficient than those written in assembly language. True or false?

17–18. What is the *port number* of the input switches and of the output LEDs in Figure 17–6?

17–19. The OUT FEH statement is used to output the switch data to the microprocessor, and the IN FFH statement is used to input data to the LEDs. True or false?

17–20. In Table 17–3, addresses 2000 through 2008 are where the _____ is stored.

17–21. How does the JZ instruction differ from the JMP instruction?

Summary of Instructions

LDA Addr: (Load Accumulator Direct) Load the accumulator with the contents of memory whose address (*addr*) is specified in byte 2–byte 3 of the instruction.

STA Addr: (Store Accumulator Direct) Store the contents of the accumulator to memory whose address (*addr*) is specified in byte 2–byte 3 of the instruction.

IN Port: (Input) Load the accumulator with the contents of the specified port.

OUT Port: (Output) Move the contents of the accumulator to the specified port.

MVI r,data: (Move Immediate) Move into register *r* the data specified in byte 2 of the instruction.

DCR r: (Decrement Register) Decrement the value in register *r* by 1.

JMP Addr: (Jump) Transfer control to address *addr* specified in byte 2–byte 3 of the instruction.

JZ Addr: (Jump If Zero) Transfer control to address *addr* if the zero flag is set.

Summary

In this chapter, we have learned that

1. A system designer should consider using a microprocessor instead of logic circuitry whenever an application involves making calculations, making decisions based on external stimuli, and maintaining memory of past events.

2. A microprocessor is the heart of a computer system. It reads and acts on program instructions given to it by a programmer.

3. A microprocessor system has three buses: address, data, and control.

4. Microprocessors operate on instructions given to them in the form of machine code (1's and 0's). The machine code is generated by a higher-level language like C or assembly language.

5. The Intel 8085A is an 8-bit microprocessor. It has seven internal registers, an 8-bit data bus, an arithmetic/logic unit, and several I/O functions.

6. Program instructions are executed inside the microprocessor by the instruction decoder, which issues the machine cycle timing and initiates I/O operations.

7. The microprocessor provides the appropriate logic levels on the data and address buses and takes care of the timing of all control signals output to the connected interface circuitry.

8. Assembly language instructions are written using mnemonic abbreviations and then converted into machine language so that they can be interpreted by the microprocessor.

9. Higher-level languages like C++ or Pascal are easier to write than assembly language, but they are not as memory efficient or as fast. All languages must be converted into a machine language matching that of the microprocessor before they can be executed.

Glossary

Accumulator: The parallel register in a microprocessor that is the focal point for all arithmetic and logic operations.

Address Bus: A group of conductors that are routed throughout a computer system and used to select a unique memory or I/O location based on their binary value.

Architecture: The layout and design of a system.

Arithmetic Logic Unit (ALU): The part of a microprocessor that performs all the arithmetic and digital logic functions.

Assembler: A software package that is used to convert assembly language into machine language.

Assembly Language: A low-level programming language unique to each microprocessor. It is converted, or assembled, into machine code before it can be executed.

BASIC Language: A high-level computer programming language that uses English-language-type instructions that are converted to executable machine code.

Bidirectional: Systems capable of transferring digital information in two directions.

Central Processing Unit (CPU): The "brains" of a computer system. The term is used interchangeably with "microprocessor."

Compiler: A software package that converts a high-level language program into machine language code.

Control Bus: A group of conductors that is routed throughout a computer system and used to signify special control functions, such as Read, Write, I/O, Memory, and Ready.

Data Bus: A group of conductors that is routed throughout a computer system and contains the binary data used for all arithmetic and I/O operations.

Flowchart: A diagram used by the programmer to map out the looping and conditional branching that a program must make. It becomes the blueprint for the program.

Hand Assembly: The act of converting assembly language instructions into machine language codes by hand, using a reference chart.

Hardware: The ICs and electronic devices that make up a computer system.

Instruction Decoder: The circuitry inside a microprocessor that interprets the machine code and produces the internal control signals required to execute the instruction.

Instruction Register: A parallel register in a microprocessor that receives the machine code and produces the internal control signals required to execute the instruction.

Interrupt: A digital control signal that is input to a microprocessor IC pin that suspends current software execution and performs another predefined task.

I/O-Mapped I/O: A method of input/output that addresses each I/O device as a port selected by a binary (or Hex) port number.

Machine Code: The binary codes that make up a microprocessor's program instructions.

Memory-Mapped I/O: A method of input/output that addresses each I/O device as a memory location selected by a binary (or Hex) address.

Microprocessor: An LSI or VLSI integrated circuit that is the fundamental building block of a digital computer. It is controlled by software programs that allow it to do all digital arithmetic, logic, and I/O operations.

Mnemonic: The abbreviated spellings of instructions used in assembly language.

Monitor Program: The computer software program initiated at power-up that supervises system operating tasks such as reading the keyboard and driving the CRT.

Opcode: Operation code. It is the unique 1-byte code given to identify each instruction to the microprocessor.

Operand: The parameters that follow the assembly language mnemonic to complete the specification of the instruction.

Operating System: *See* Monitor program.

Port Number: A number used to select a particular I/O port.

Program Counter: An internal register that contains the address of the next program instruction to be executed.

Software: Computer program statements that give step-by-step instructions to a computer to solve a problem.

Stack Pointer: An internal register that contains the address of the last entry on the RAM stack.

Statement Label: A meaningful name given to certain assembly language program lines so that they can be referred to from different parts of the program, using statements like JUMP or CALL.

Support Circuitry: The integrated circuits and electronic devices that assist the microprocessor in performing I/O and other external tasks.

Zero Flag: A bit internal to the microprocessor that, when set (1), signifies the last arithmetic or logic operation had a result of zero.

Problems

Section 17–1

17–1. Describe the circumstances that would prompt you to use a microprocessor-based design solution instead of a hard-wired IC logic design.

17–2. In an 8-bit microprocessor system, how many lines are in the data bus? The address bus?

17–3. What is the function of the address bus?

D **17–4.** Use a TTL data manual to find an IC that you could use for the *output port* in Figure 17–1. Draw its logic diagram and external connections.

D **17–5.** Repeat Problem 17–4 for the *input port.*

C D **17–6.** Repeat Problem 17–4 for the *address decoder.* Assume that the input port is at address 4000H, the output port is at address 6000H, and memory is at address 2000H. (*Hint:* Use an address decoding scheme similar to that found in Section 16–5.)

17–7. Why does the input port in Figure 17–1 have to have three-stated outputs?

17–8. What two control signals are applied to the input port in Figure 17–1 to cause it to transfer the switch data to the data bus?

17–9. How many different addresses can be accessed using a 16-bit address bus?

Sections 17–2 and 17–3

17–10. In the assembly language instruction LDA 4000H, what does the LDA signify and what does the 4000H signify?

17–11. Describe what the statement STA 6000H does.

17–12. What are the names of the six internal 8085A general-purpose registers?

17–13. What is the function of the 8085A's instruction register and instruction decoder?

17–14. Why is the program counter register 16 bits instead of 8?

Section 17–4

C **17–15.** During the execution of the LDA 4000 instruction in Figure 17–4, the $\overline{RD}$ line goes LOW four times. Describe the activity initiated by each LOW pulse.

17–16. What action does the LOW $\overline{WR}$ pulse initiate during the STA 6000 instruction in Figure 17–5?

Section 17–5

17–17. Describe one advantage and one disadvantage of writing programs in a high-level language instead of assembly language.

17–18. Are the following instructions used for memory-mapped I/O or for I/O-mapped I/O?

(a) LDA *addr*   (c) IN *port*

(b) STA *addr*   (d) OUT *port*

17–19. What is the digital level on the microprocessor's $IO/\overline{M}$ line for each of the following instructions?

(a) LDA *addr*   (c) IN *port*

(b) STA *addr*   (d) OUT *port*

D

17–20. List the new IN and OUT instructions that would be used to I/O to the switches and LEDs if the following changes to U4 and U5 were made in Figure 17–6.

(a) Add inverters to inputs A_8 and A_9 of U4 and to A_9 and A_{10} of U5.

(b) Add inverters to inputs A_{14} and A_{15} of U4 and to A_{14} and A_{15} of U5.

17–21. U6a and U6b in Figure 17–6 are OR gates. Why are they drawn as inverted-input NAND gates?

17–22. Are the LEDs in Figure 17–6 active-HIGH or active-LOW?

17–23. Is the $\overline{RD}$ line or the $\overline{WR}$ line pulsed LOW by the microprocessor during the:

(a) IN instruction?   **(b)** OUT instruction?

17–24. What three conditions must be met to satisfy the output enables of U2 in Figure 17–6?

17–25. What three conditions must be met to provide a pulse to the C_p input of U3 in Figure 17–6?

17–26. Which internal data register is used for the IN and OUT instructions?

Section 17–6

17–27. Write the assembly language instruction that would initialize the accumulator to 4FH.

17–28. Describe in words what the instruction JZ LOOP does.

C

17–29. Write the machine language code for the following assembly language program. (Start the machine code at address 2010H.)

```
INIT:    MVI A, 04H
X1:      DCR A
         JZ INIT
         JMP X1
```

Schematic Interpretation Problems

See Appendix G for the schematic diagrams.

S

17–30. Locate the 68HC11 microcontroller in the HC11D0 schematic. (A microcontroller is a microprocessor with built-in RAM, ROM, and I/O ports). Pins 31–38 are the low-order address bus (A0–A7) multiplexed (shared) with the data bus (D0–D7). Pins 9–16 are the high-order address bus (A8–A15). The low-order address bus is demultiplexed (selected and latched) from the shared address/data lines by U2 and the AS (Address Strobe) line.

(a) Which ICs are connected to the data bus (DB0–DB7)?

(b) Which ICs are connected to the address bus (AD0–AD15)?

S C

17–31. U9 and U5 in the HC11D0 schematic are used for address decoding. Determine the levels on AD11–AD15 and AD3–AD5 to select (a) the LCD (LCD_SL) and (b) the keyboard (KEY_SL).

S

17–32. Locate the microcontroller in the 4096/4196 schematic.

(a) What is its grid location and part number?

(b) Its low-order address is multiplexed like the 68HC11 in Problem 17–30. What IC and control signal are used to demultiplex the address/data bus (AD0–AD7) into the low-order address bus (A0–A7)?

(c) What IC and control signal are used to demultiplex the address/data bus (AD0–AD7) into the data bus (D0–D7)?

Answers to Review Questions

17–1. Address, data, control

17–2. All of it

17–3. To use the same path for both input and output data

17–4. False

17–5. True

17–6. The output to the LEDs would float when $\overline{WR}$ returns HIGH.

17–7. False

17–8. Read, write

17–9. False

17–10. False

17–11. Assembly language, which is written in short mnemonics, needs to be assembled or converted into a binary string called a machine code, which is read by the CPU.

17–12. 1 byte is used to store 3AH, 2 bytes are used to store 4000H.

17–13. False

17–14. STA, the $\overline{WR}$ is used as the clock input to the D flip-flop.

17–15. 6 bytes

17–16. True

17–17. False

17–18. Input is FFH, output is FEH

17–19. False

17–20. Machine code

17–21. The JZ instruction looks for a zero flag and jumps to the label START if the flag is set. The JMP instruction jumps to the label LOOP, regardless of any flags.

18

The 8051 Microcontroller*

OUTLINE

OBJECTIVES

Upon completion of this chapter you should be able to:

- Describe the advantages that a microcontroller has over a microprocessor for control applications.
- Describe the functional blocks within the 8051 microcontroller.
- Make comparisons between the different microcontrollers available within the 8051 family.
- Write the software instructions to read and write data to and from the I/O ports.
- Describe the use of the alternate functions on the I/O ports.
- Make the distinction between the internal and external data and code memory spaces.
- Use various addressing modes to access internal data memory and the Special Function Registers (SFRs).
- Interface an external EPROM and a RAM to the 8051.
- Use some of the more commonly used instructions of the 8051 instruction set.
- Write simple 8051 I/O programs.
- Use the bit operations of the 8051.
- Understand program solutions to applications such as a keyboard decoder and analog-to-digital converter.

---

*Visit the following sites for microcontroller trainers and downloadable software tools: www.acebus.com, www.bipom.com, www.elexp.com, www.emacinc.com, and www.8052.com

INTRODUCTION

In the previous chapter you should have recognized several common components that are incorporated into most microprocessor-based system applications. These are the microprocessor, RAM, ROM (or EPROM), and parallel I/O ports. Microprocessor manufacturers have recognized this and over the years since the introduction of the 8085A, have been developing a new line of microprocessors specifically designed for control applications. They are called **microcontrollers.**

The microcontroller has the CPU, RAM, ROM, timer/counter, and parallel and serial I/O ports fabricated into a single IC. It is often called "a computer on a chip." The CPU's instruction set is improved for control applications and offers bit-oriented data manipulation, branching, and I/O, as well as multiply and divide instructions.

The microcontroller is most efficiently used in systems that have a fixed program for a dedicated application. A microcontroller is used in a keyboard for a personal computer to scan and decode the keys. It is used in an automobile for sensing and controlling engine operation. Other dedicated applications such as microwave ovens, videocassette recorders, gas pumps, and automated teller machines use the microcontroller IC also.

The previous chapters have provided a good background for understanding the microcontroller. The theory behind microprocessor buses, external memory, and I/O ports is necessary to utilize the features available on a microcontroller. In this chapter we'll be introduced to one of the most widely used microcontrollers available: the 8051. You'll see several similarities between it and the 8085A-based systems previously discussed. You should appreciate the way that you were eased into microprocessor operational theory using the 8085A, but now you will see why the 8051 can be a much better solution to dedicated control applications.

Most control applications require extensive I/O and need to work with individual bits. The 8051 addresses both of these needs by having 32 I/O lines and a CPU instruction set that handles single-bit I/O, bit manipulation, and bit checking.

18–1 The 8051 Family of Microcontrollers

The basic architectural structure of the 8051 is given in Figure 18–1 . The block diagram gives us a good picture of the hardware included in the 8051 IC. For internal memory it has a 4K $\times$ 8 ROM and a 128 $\times$ 8 RAM. It has two 16-bit counter/timers and interrupt control for five interrupt sources. Serial I/O is provided by TXD and RXD (transmit and receive), and it also has four 8-bit parallel I/O ports (P0, P1, P2, P3). There is also an 8052 series of microcontrollers available that has an 8K ROM, 256 RAM, and three counter/timers.

Other versions of the 8051 are the 8751, which has an internal EPROM for program storage in place of the ROM, and the 8031, which has *no* internal ROM, but instead accesses an external ROM or EPROM for program instructions. Table 18–1 summarizes the 8051 family of microcontrollers. Note that the 8052/8752/8032 series has an extra 4K of program space (except the 8032), double the RAM area, an extra timer/counter, and one additional interrupt source. All parts use the same CPU instruction set. The ROMless versions (8031 and 8032) are the least expensive parts but require an external ROM or EPROM like a 2732 or 2764 for program storage.

18–2 8051 Architecture

In order to squeeze so many functions on a single chip, the designers had to develop an architecture that uses the same internal address spaces and external pins for more than one function. This technique is similar to that used by the 8085A for the multiplexed AD_0–AD_7 lines.

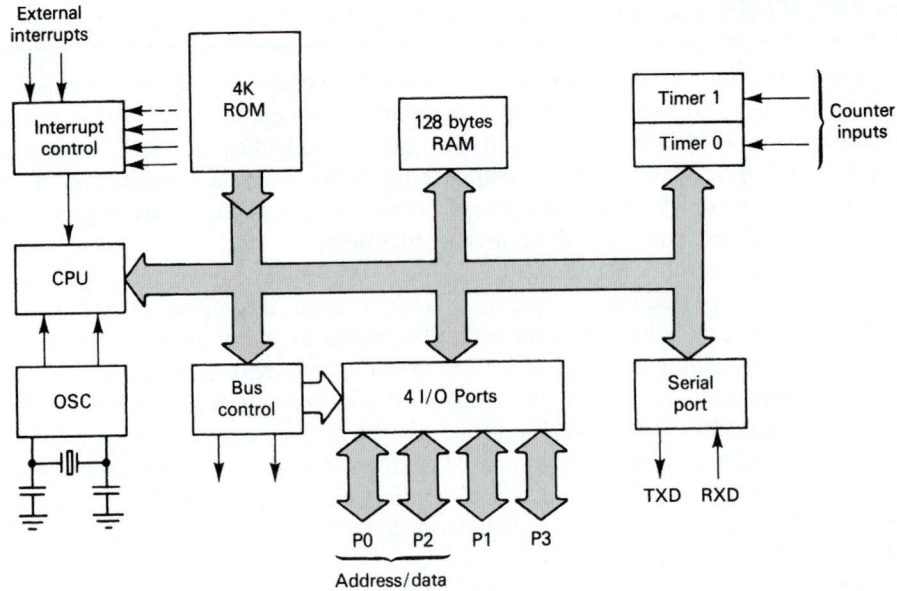

Figure 18–1 Block diagram of the 8051 microcontroller.

Device number	Internal memory		Timers/event counters	Interrupt sources
	Program	**Data**		
8051	4K × 8 ROM	128 × 8 RAM	2 × 16-bit	5
8751H	4K × 8 EPROM	128 × 8 RAM	2 × 16-bit	5
8031	None	128 × 8 RAM	2 × 16-bit	5
8052AH	8K × 8 ROM	256 × 8 RAM	3 × 16-bit	6
8752BH	8K × 8 EPROM	256 × 8 RAM	3 × 16-bit	6
8032AH	None	256 × 8 RAM	3 × 16-bit	6

TABLE 18–1 | **The 8051 Family of Microcontrollers**

The 8051 is a 40-pin IC. Thirty-two pins are needed for the four I/O ports. In order to provide for the other microcontroller control signals, most of those pins have alternate functions, which will be described in this section. Also in this section, we'll see how the 8051 handles the overlapping address spaces used by internal memory, external memory, and the special function registers.

The pin configuration for the 8051 is given in Figure 18–2.

Port 0

Port 0 is dual purpose, serving as either an 8-bit bidirectional I/O port (P0.0–P0.7) or the low-order multiplexed address/data bus (AD_0–AD_7). As an I/O port, it can sink up to 8 LS TTL loads in the LOW condition and is a float for the HIGH condition (I_{OL} = 3.2 mA). The alternate port designations, AD_0–AD_7, are used to access external memory. They are activated automatically whenever reference is made to external memory. The AD lines are demultiplexed into A_0–A_7 and D_0–D_7 by using the ALE signal, the same way that it was done with the 8085A.

Port 1

Port 1 is an 8-bit bidirectional I/O port that can sink or source up to 4 LS TTL loads. (I_{OL} = 1.6 mA, I_{OH} = −80 μA.)

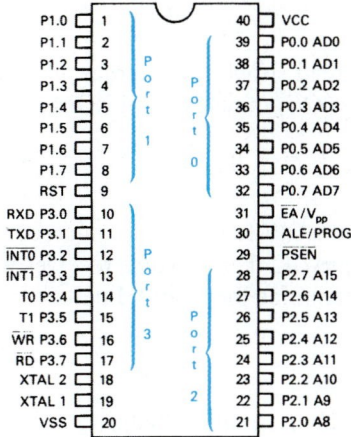

Figure 18–2   The 8051 pin configuration.

Port 2

Port 2 is dual purpose, serving as either an 8-bit **bidirectional** I/O port (P2.0–P2.7), or as the high-order address bus (A_8–A_{15}) for access to external memory. As an I/O port it can sink or source up to 4 LS TTL loads. The port becomes active as the high-order address bus whenever reference to external memory is made.

Port 3

Port 3 is dual purpose, serving as an 8-bit bidirectional I/O port that can sink or source up to 4 LS TTL loads or as special-purpose I/O to provide the functions listed in Table 18–2.

TABLE 18–2	Alternative Functions of Port 3
Port pin	**Alternative function**
P3.0	RXD (serial input port)
P3.1	TXD (serial output port)
P3.2	$\overline{INT0}$ (external interrupt 0)
P3.3	$\overline{INT1}$ (external interrupt 1)
P3.4	T0 (Timer 0 external input)
P3.5	T1 (Timer 1 external input)
P3.6	$\overline{WR}$ (external data memory write strobe)
P3.7	$\overline{RD}$ (external data memory read strobe)

RST

Reset input. A HIGH on this pin resets the microcontroller.

ALE/$\overline{PROG}$

Address Latch Enable output pulse for latching the low-order byte of the address during accesses to external memory. This pin is also the program pulse input ($\overline{PROG}$) during programming of the EPROM parts.

PSEN

Program Store Enable is a read strobe for external program memory. It will be connected to the Output Enable ($\overline{OE}$) of an external ROM or EPROM.

EA/VPP

External Access ($\overline{EA}$) is tied LOW to enable the microcontroller to fetch its program code from an external memory IC. This pin also receives the 21-V programming supply voltage (VPP) for programming the EPROM parts.

XTAL1,XTAL2

Connections for a crystal or an external oscillator.

Address Spaces

The address spaces of the 8051 are divided into four distinct areas: internal data memory, external data memory, internal code memory, and external code memory (see Figure 18–3).

The 8051 allows for up to 64K of external data memory (RAM) and 64K of external code memory (ROM/EPROM). The only disadvantage of external memory, besides the additional circuitry, is that ports 0 and 2 get tied up for the address and data bus. The actual hardware interfacing for external memory will be given later in this chapter.

When the 8051 is first reset, the program counter starts at 0000H. This points to the first program instruction in the *internal code memory* unless $\overline{EA}$ (**External Access**) is tied LOW. If $\overline{EA}$ is tied LOW, the CPU issues a LOW on **PSEN (Program Store En-**

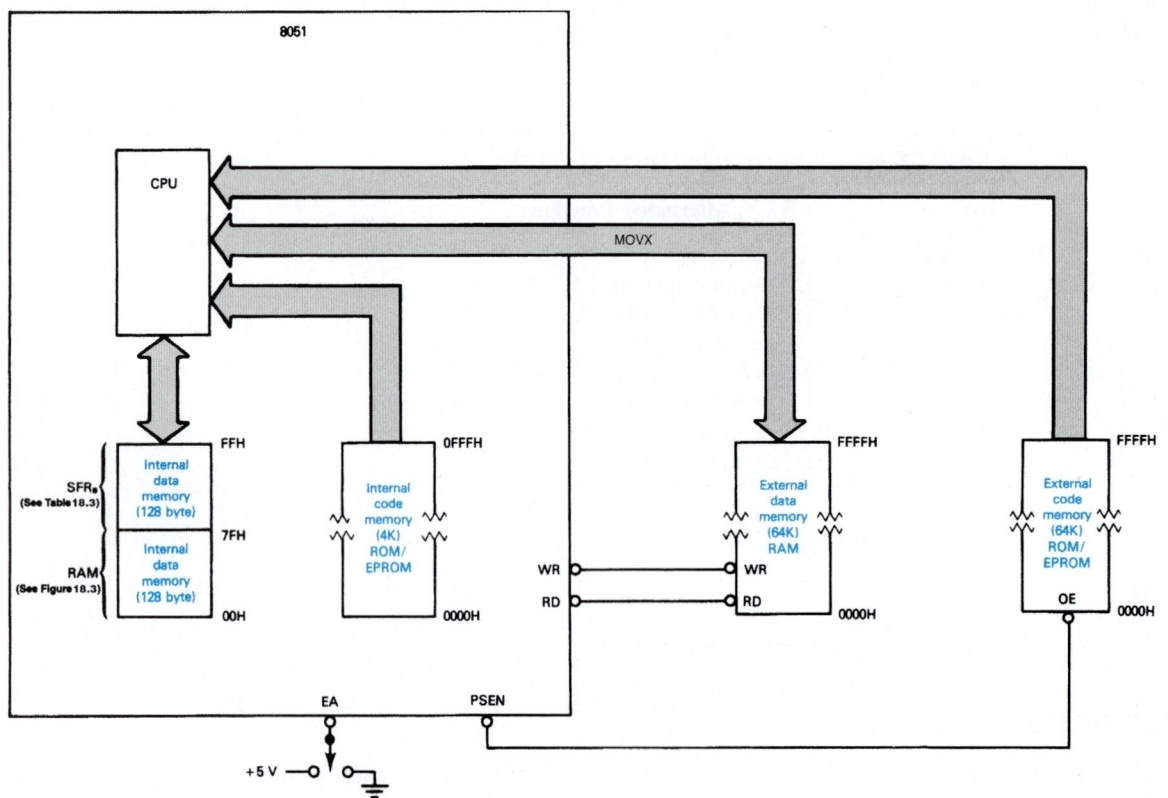

Figure 18–3 8051 address spaces.

able), enabling the *external code memory* ROM/EPROM instead. $\overline{EA}$ *must* be tied LOW if using the ROMless (8031) version.

With $\overline{EA}$ tied HIGH, the first 4K of program instruction fetches are made from the internal code memory. Any fetches from code memory above 4K (1000H to FFFFH) will automatically be made from the external code memory.

If the application has a need for large amounts of data memory, then *external data memory* (RAM) can be used. I/O to external RAM can only be made by using one of the MOVX instructions. When executing a MOVX instruction, the 8051 knows that you are referring to external RAM and issues the appropriate $\overline{WR}$ or $\overline{RD}$ signal.

If 128 bytes (256 for the 8052) of RAM storage is enough, then the *internal data memory* is better to use because it has a much faster response time and a broad spectrum of instructions to access it. Figure 18–3 shows two blocks of internal data memory. The first is the 128-byte RAM at address 00H to 7FH. The second is made up of the **Special Function Registers (SFRs)** at addresses 80H to FFH. Note that *all addresses are only 1 byte wide,* which allows for more efficient use of code space and faster access time. The use of the address space in the 128-byte RAM is detailed in Figure 18–4.

The first 32 locations are set aside for four banks of data registers (called **register banks**), giving the programmer the use of 32 separate registers. The registers within each bank are labeled R0 through R7. The bank that is presently in use is defined by the setting of the two "bank select bits" in the PSW (Program Status Word).

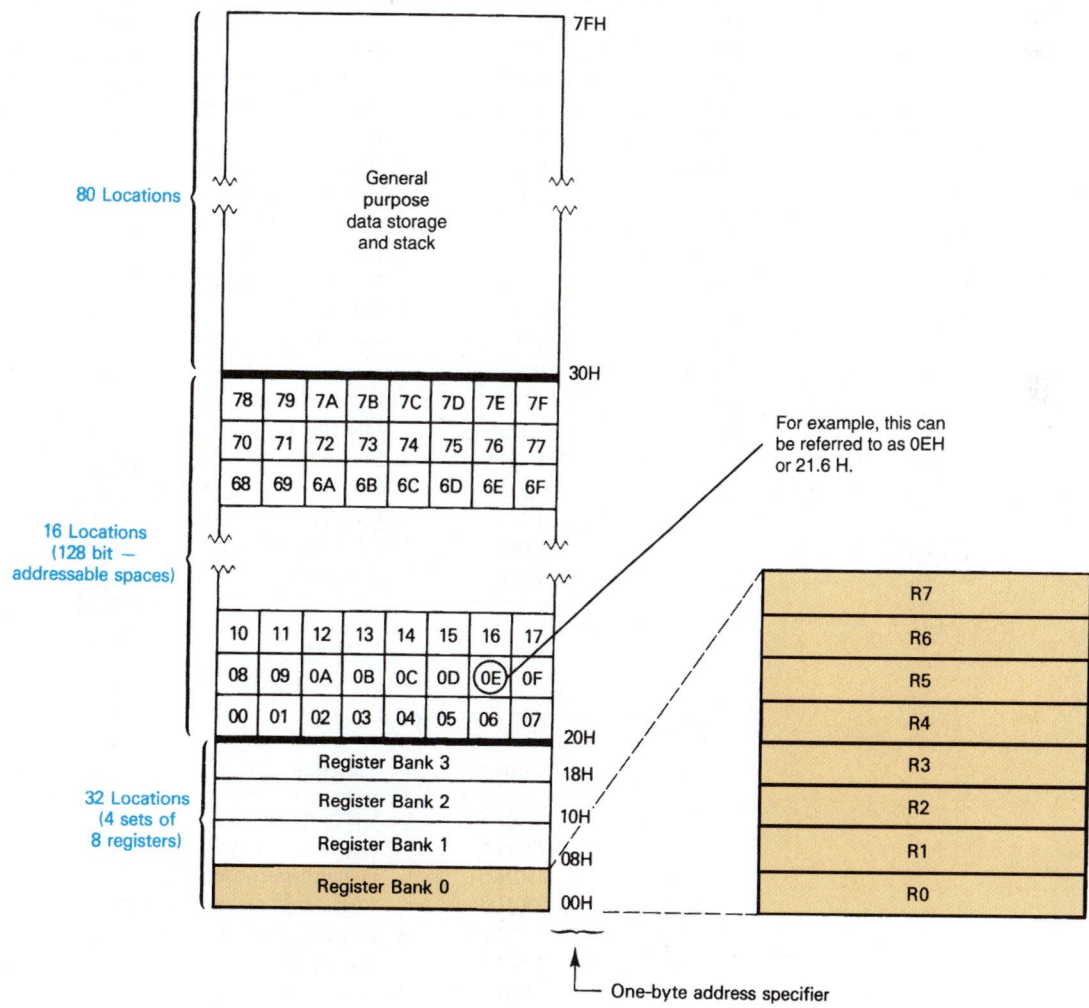

Figure 18–4   Address space in the 128-byte internal data memory RAM.

RAM addresses 20H through 2FH are designated as **bit-addressable** memory locations. This meets the needs of control applications for a large number of ON/OFF bit flags, by providing 128 uniquely addressable bit locations.

The last 80 locations are set aside for general-purpose data storage and stack operations.

The Special Function Registers are maintained in the next 128 addresses in the internal data memory of the 8051 (address 80H to FFH). It contains registers that are required for software instruction execution as well as those used to service the special hardware features built into the 8051. Table 18–3 lists the SFRs and their addresses.

The address of an SFR is also only 1-byte wide. However, instead of specifying the 1-byte hex address, you also have the option of addressing any SFR location by simply specifying its register name, like P0 or SP. Several of the SFRs are bit addressable. For example, to address bit 3 of the accumulator, you would use the address E0H.3 (or you could use ACC.3). The PSW is also bit addressable, having the bit addresses listed in Table 18–4.

For example, to address the parity bit in the PSW, you would use the address D0H.0 (or you could use PSW.0).

TABLE 18–3	The Special Function Registers (SFRs) [Internal Data Memory (80H to FFH)]	
Register	**Address**	**Function**
P0	80H[a]	Port 0
SP	81H	Stack pointer
DPL	82H	Data pointer (Low)
DPH	83H	Data pointer (High)
TCON	88H[a]	Timer register
TMOD	89H	Timer mode register
TL0	8AH	Timer 0 low byte
TL1	8BH	Timer 1 low byte
TH0	8CH	Timer 0 high byte
TH1	8DH	Timer 1 high byte
P1	90H[a]	Port 1
SCON	98H[a]	Serial port control register
SBUF	99H	Serial port data buffer
P2	A0H[a]	Port 2
IE	A8H[a]	Interrupt enable register
P3	B0H[a]	Port 3
IP	B8H[a]	Interrupt priority register
PSW	D0H[a]	Program status word
ACC	E0H[a]	Accumulator (direct address)
B	F0H[a]	B register

[a]Bit-addressable register.

TABLE 18–4	The PSW Bit Addresses	
Symbol	**Address**	**Function**
CY	D0H.7	Carry flag
AC	D0H.6	Auxiliary carry flag
F0	D0H.5	General-purpose flag
RS1	D0H.4	Register bank select (MSB)
RS0	D0H.3	Register bank select (LSB)
OV	D0H.2	Overflow flag
—	D0H.1	User-definable flag
P	D0H.0	Parity flag

18–3 Interfacing to External Memory

Up to 64K of code memory (ROM/EPROM) and 64K of data memory (RAM) can be added to any of the 8051 family members. If you are using the 8031 (ROMless) part, then you *have* to use external code memory for storing your program instructions. As mentioned earlier, the alternate function of port 2 is to provide the high-order address byte (A_8–A_{15}), and the alternate function of port 0 is to provide the multiplexed low-order address/data byte (AD_0–AD_7).

If you are interfacing to a general-purpose EPROM like the 2732, then the ALE signal provided by the 8031 is used to demultiplex the AD_0–AD_7 lines via an address latch, as shown in Figure 18–5.

As you can see, two of the I/O ports are used up in order to provide the address and data buses. The AD_0–AD_7 lines that are output on port 0 are demultiplexed by the ALE signal and the address latch, the same way as they were in the 8085A circuits studied earlier. The $\overline{PSEN}$ signal is asserted at the end of each instruction fetch cycle to enable the EPROM outputs to put the addressed code byte on the data bus to be read by port 0. The $\overline{EA}$ line is tied LOW so that the 8031 knows to fetch all program code from external memory.

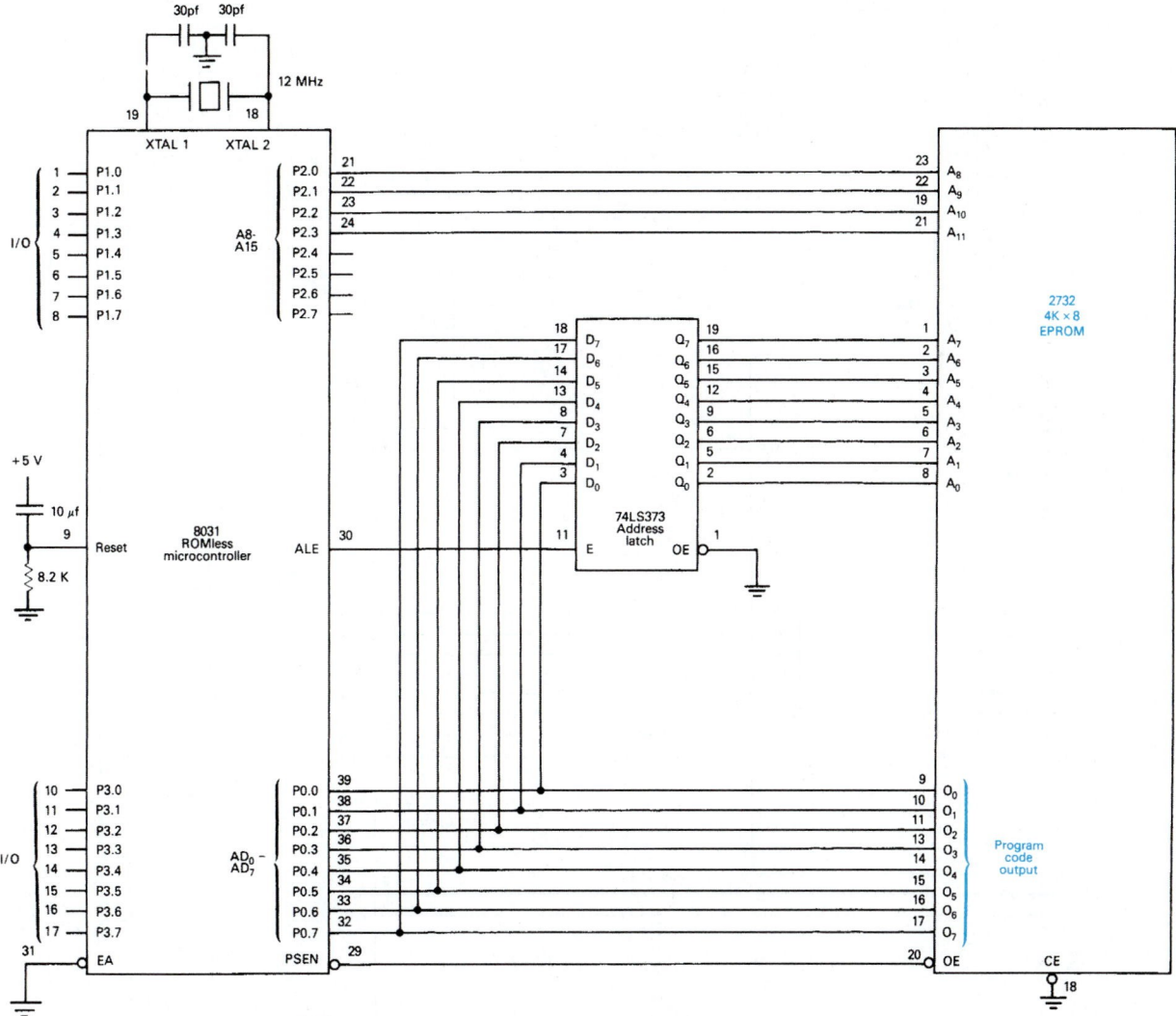

Figure 18–5 Interfacing a 2732 EPROM to the ROMless 8031 microcontroller.

If extra *data memory* is needed, then the interface circuit given in Figure 18–6 can be used. The 8155 RAM accepts the AD_0–AD_7 lines directly and uses the ALE signal to internally demultiplex, eliminating the need for an address latch IC. We have to use 11 of the 8051 pins to interface to the 8155, but the 8155 provides an additional 22 I/O lines, giving us a net gain of 11.

The addresses of the external RAM locations are 0000H to 00FFH, overlapping the addresses of the internal data memory. There is not a conflict, however, because all I/O to the external data memory is made using the MOVX instruction. The MOVX instruction ignores internal memory and instead, activates the appropriate control signal, $\overline{RD}$ or $\overline{WR}$ via port 3. The LOW $\overline{RD}$ or $\overline{WR}$ signal allows the 8155 to send or receive data to or from port 0 of the 8051.

The I/O ports on the 8155 are accessed by making the IO/$\overline{M}$ line HIGH. This is done by using memory-mapped I/O and specifying an address whose bit A_{15} is HIGH (8000H or higher).

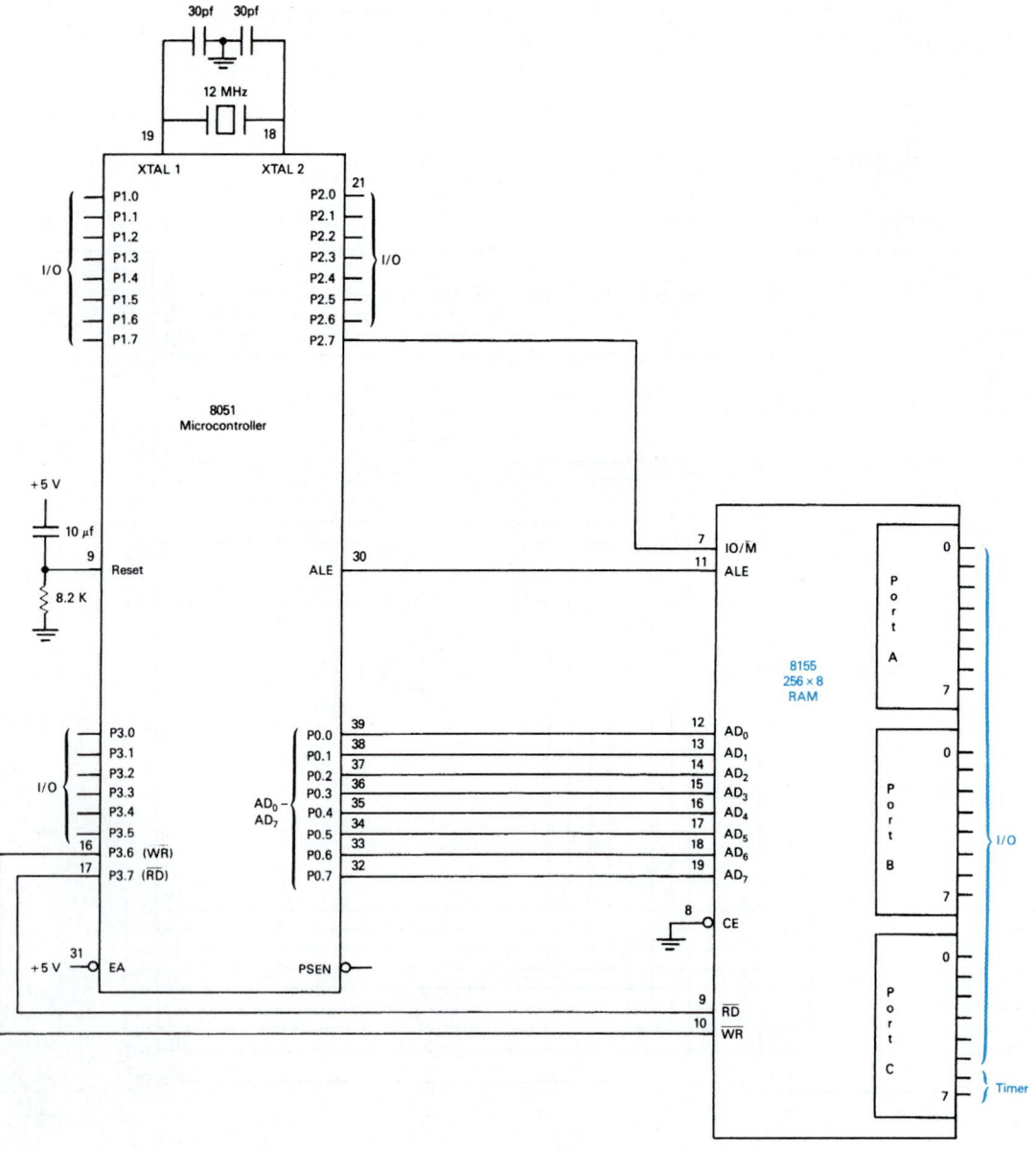

Figure 18–6 Interfacing an 8155 RAM/IO/Timer to the 8051 microcontroller.

CHAPTER 18 I THE 8051 MICROCONTROLLER

18-4 The 8051 Instruction Set

All the members of the 8051 family use the same instruction set. (The complete instruction set is given in the 8051 Instruction Set Summary in Appendix H.) Several new instructions in the 8051 make it especially well suited for control applications. The discussion that follows assumes that you are using a commercial **assembler** software package. Hand assembly of the 8051 instructions into executable machine code is very difficult and misses out on several of the very useful features available to the ASM51 programmer.

Addressing Modes

The instruction set provides several different means to address data memory locations. We'll use the MOV instruction to illustrate several common addressing modes. For example, to move data into the accumulator, any of the following instructions could be used:

MOV A,Rn: *Register addressing,* the contents of register *Rn* (where $n = 0-7$) is moved to the accumulator.

MOV A,@Ri: *Indirect addressing,* the contents of memory whose address is in *Ri* (where $i = 0$ or 1) is moved to the accumulator. (*Note:* Only registers R_0 and R_1 can be used to hold addresses for the indirect-addressing instructions.)

MOV A,20H: *Direct addressing,* move the contents of RAM location 20H to the accumulator. I/O ports can also be accessed as a direct address as shown in the following instruction.

MOV A,P3: *Direct addressing,* move the contents of port 3 to the accumulator. Direct addressing allows you to specify the address by giving its actual hex address (e.g., B0H) or by giving its abbreviated name (e.g., P3), which is listed in the SFR table (Table 18-3).

MOV A,#64H: *Immediate constant,* move the number 64H into the accumulator.

In each of the previous instructions, the *destination* of the move was the accumulator. The destination in any of those instructions could also have been a register, a direct address location, or an indirect address location.

EXAMPLE 18-1

Write the 8051 instruction to perform each of the following operations.

(a) Move the contents of the accumulator to register 5.

(b) Move the contents of RAM memory location 42H to port 1.

(c) Move the value at port 2 to register 3.

(d) Send FFH to port 0.

(e) Send the contents of RAM memory, whose address is in register 1, to port 3.

Solution:
(a) MOV R5,A

(b) MOV P1,42H

(c) MOV R3,P2

(d) MOV P0,#0FFH (Note the addition of the leading 0 to the FFH. This is because the ASM51 assembler requires that the first position of any hexadecimal number must be a numeric digit from 0 to 9.)

(e) MOV P3,@R1

Program Branching Instructions

8051 assembly language provides several different ways to branch (jump) to various program segments within the 64K of code memory area. Below are some of the most useful jump instructions.

JMP *label (Unconditional jump):* Program control passes to location *label*. The JMP instruction is converted by the ASM51 assembler into an absolute jump, AJMP; a short jump, SJMP; or a long jump, LJMP, depending on the destination pointed to by *label*.

JZ *label (Jump if accumulator zero):* Program control passes to location *label* if the accumulator equals zero.

JNZ *label (Jump if accumulator not zero):* Program control passes to location *label* if the accumulator is not equal to zero.

JB *bit,label (Jump if bit set):* Program control passes to location *label* if the specified bit, *bit,* is set.

JNB *bit,label (Jump if bit not set):* Program control passes to location *label* if the specified bit, *bit,* is not set.

DJNZ *Rn,label (Decrement register and jump if not zero):* Program control passes to location *label* if, after decrementing register *Rn,* the register is not equal to zero ($n = 0$–7).

CJNE *Rn,#data,label (Compare immediate data to register and jump if not equal):* Program control passes to location *label* if register *Rn* is not equal to the immediate data #*data*. The compare can also be made to the accumulator by specifying *A* instead of *Rn*.

CALL *label (Call subroutine):* Program control passes to location *label*. The return address is stored on the stack. The CALL instruction is converted by the ASM51 assembler into an absolute call, ACALL, or a long call, LCALL, depending on the destination pointed to by *label*.

RET *(Return):* Program control is returned back to the instruction following the CALL instruction.

EXAMPLE 18–2

Write a program that continuously reads a byte from port 1 and writes it to port 0 until the byte read equals zero.

Solution:

```
READ:       MOV A,P1          ; A ← P1
            MOV P0,A          ; P0 ← A
            JNZ READ          ; Repeat until A = 0
            NOP               ; Remainder of program
            etc.
```

EXAMPLE 18-3

Repeat Example 18–2, except stop the looping when the number 77H is read.

Solution:

```
READ:      MOV A,P1          ; A ← P1
           MOV P0,A          ; P0 ← A
           CJNE A,#77H,READ  ; Repeat until A = 77H
           NOP               ; Remainder of program
           etc.
```

EXAMPLE 18-4

Repeat Example 18–2, except stop the looping when bit 3 of port 2 is set.

Solution:

```
READ:      MOV A,P1      ; A ← P1
           MOV P0,A      ; P0 ← A
           JNB P2.3,READ ; Repeat until bit 3 of port 2 is set
           NOP           ; Remainder of program
           etc.          ;
```

EXAMPLE 18-5

Write a program that will produce an output at port 0 that counts down from 80H to 00H.

Solution:

```
           MOV R0,#80H     ; R0 ← 80H
COUNT:     MOV P0,R0       ; P0 ← R0
           DJNZ R0,COUNT   ; Decrement R0, jump to COUNT if not 0
           NOP             ; Remainder of program
           etc.            ;
```

Logical and Bit Operations

The 8051 instruction set provides the basic logical operations (OR, AND, Ex-OR, and NOT), rotates (left or right, with or without carry), and bit operations (Clear, Set, and complement). The following list shows some of the most commonly used of those operations.

ANL A,*Rn* (*AND register to accumulator*): Logically AND register *Rn,* bit by bit, with the accumulator and store the result in the accumulator.

ANL A,#*data* (*AND data byte to accumulator*): Logically AND data byte #*data,* bit by bit, with the accumulator and store the result in the accumulator.

The descriptions of logical ORs and exclusive-ORs are similar to the AND, and use the following instructions:

ORL A,*Rn* *(OR register to accumulator)*

ORL A,#*data* *(OR data byte to accumulator)*

XRL A,*Rn* *(Ex-OR register to accumulator)*

XRL A,#*data* *(Ex-OR data byte to accumulator)*

The most commonly used instructions to operate on individual bits are as follows:

CLR *bit* *(Clear bit):* Clear (reset to 0) the value of the bit located at address *bit*.

SETB *bit* *(Set bit):* Set (set to 1) the value of the bit located at address *bit*.

CPL *bit* *(Complement bit):* Complement the value of the bit located at address *bit*.

The rotate commands available to the 8051 programmer are as follows:

RL A *(Rotate accumulator left):* Rotate the 8 bits of the accumulator one position to the left.

RLC A *(Rotate accumulator left through carry):* Rotate the 9 bits of the accumulator including carry one position to the left.

RR A *(Rotate accumulator right):* Rotate the 8 bits of the accumulator one position to the right.

RRC A *(Rotate accumulator right through carry):* Rotate the 9 bits of the accumulator including carry one position to the right.

The following examples illustrate the use of the logical and bit operations.

EXAMPLE 18–6

Determine the contents of the accumulator after the execution of the following program segments.

```
(a) MOV A,#3CH          ; A ← 0011 1100
    MOV R4,#66H         ; R4 ← 0110 0110
    ANL A,R4            ; A ← A AND R4
```

Solution:

$$A = 0011\ 1100$$
$$R4 = 0110\ 0110$$
$$A\ AND\ R4 = 0010\ 0100 = 24H \quad \textit{Answer}$$

```
(b) MOV A,#3FH          ; A ← 0011 1111
    XRL A,#7CH          ; A ← EX-OR 7CH
```

Solution:

$$A = 0011\ 1111$$
$$7CH = 0111\ 1100$$
$$A\ EX\text{-}OR\ 7CH = 0100\ 0011 = 43H \quad \textit{Answer}$$

(c) MOV A,#0A3H   ; A ← 1010 0011
    RR A   ; Rotate right

Solution:

        A = 1010 0011
        Rotate right
        A = 1101 0001   *Answer*

(d) MOV A,#0C3H   ; A ← 1100 0011
    RLC A   ; Rotate left through carry

Solution:

        Assume carry = 0 initially
        A = 1100 0011
        Rotate left through carry
        A = 1000 0110, carry = 1   *Answer*

EXAMPLE 18–7

Use the SETB, CLR, and CPL instructions to do the following operations:

(a) Clear bit 7 of the accumulator.

(b) Output a 1 on bit 0 of port 3.

(c) Complement the parity flag (bit 0 of the PSW).

Solution:
(a) CLR ACC.7 (*Note:* According to Table 18–3, the symbol for the direct address of the accumulator is ACC.)

(b) SETB P3.0

(c) CPL PSW.0

EXAMPLE 18–8

Describe the activity at the output of port 0 during the execution of the following program segment:

```
            MOV R7,#0AH
            MOV P0,#00H
LOOP:       CPL P0.7
            DJNZ R7,LOOP
```

Solution: Register 7 is used as a loop counter with the initial value of 10 (0AH). Each time that the complement instruction (CPL) is executed, bit 7 will toggle to its opposite state. Toggling bit 7 ten times will create a waveform with five positive pulses.

Arithmetic Operations

The 8051 is capable of all the basic arithmetic functions: addition, subtraction, incrementing, decrementing, multiplication, and division. The following list outlines the most commonly used forms of the arithmetic instructions.

ADD A, Rn *(Add register to accumulator):* Add the contents of register *Rn* (where $n = 0-7$) to the accumulator and place the result in the accumulator.

ADD A, #data *(Add immediate data to accumulator):* Add the value of #*data* to the accumulator and place the result in the accumulator.

SUBB A, Rn *(Subtract register from accumulator with borrow):* Subtract the contents of register *Rn* and borrow (carry flag) from the accumulator and place the result in the accumulator.

SUBB A, #data *(Subtract immediate data from accumulator with borrow):* Subtract the value of #*data* and borrow (carry flag) from the accumulator and place the result in the accumulator.

INC A *(Increment accumulator)*

INC Rn *(Increment register)*

DEC A *(Decrement accumulator)*

DEC Rn *(Decrement register)*

MUL AB *(Multiply A times B):* Multiply the value in the accumulator times the value in the *B* register. The low-order byte of the 16-bit product is left in the accumulator and the high-order byte is placed in register *B*.

DIV AB *(Divide A by B):* Divide the value in the accumulator by the value in the *B* register. The accumulator receives the quotient and register *B* receives the remainder.

DA A *(Decimal adjust accumulator):* Adjust the value in the accumulator, resulting from an addition, into two BCD digits.

The following examples illustrate the use of arithmetic instructions.

EXAMPLE 18–9

Add the value being input at port 1 to the value at port 2 and send the result to port 3.

Solution:

```
MOV R0,P1      ; R0 ← P1
MOV A,P2       ; A ← P2
ADD A,R0       ; A ← A + R0
MOV P3,A       ; P3 ← A
```

EXAMPLE 18–10

Multiply the value being input at port 0 times the value at port 1 and send the result to ports 3 and 2 (high order, low order).

```
MOV A,P0      ; A ← P0
MOV B,P1      ; B ← P1
MUL AB        ; A × B
MOV P2,A      ; P2 ← A (low order)
MOV P3,B      ; P3 ← B (high order)
```

18–5   8051 Applications

Having bit-handling instructions and built-in I/O makes the 8051 a good choice for data acquisition and control applications. In this section we'll look at a few applications that illustrate how we can utilize these new features to simplify our program solutions.

Instruction Timing

The 8051 circuitry that we looked at earlier was driven with a 12-MHz crystal. The 12 MHz is a convenient choice because each instruction machine cycle takes 12 clock periods to complete. This means that one machine cycle takes 1 μs. All 8051 instructions are completed in one or two machine cycles (12 or 24 oscillator periods), except MUL and DIV, which take four. The oscillator periods for each instruction are given in the 8051 Instruction Set Summary Appendix. For example, the MOV A,Rn instruction requires 12 oscillator periods to complete, which will take 1 μs [12 × (1/12 MHz)].

Time Delay

Knowing the time duration of each instruction, we can write accurate time delays for our applications programs. Writing counter/loop programs is much easier now, with the introduction of the DJNZ and the CJNE instructions. Table 18–5 and Figure 18–7 show a time-delay program that we could call as a subroutine from an applications program.

TABLE 18–5	Time-Delay Subroutine
DELAY:	MOV R1,#0F3H ; Outer-loop counter (F3H = 243)
	MOV R0,#00H ; Inner-loop counter
LOOP:	DJNZ R0,LOOP ; Loop 256 times
	DJNZ R1,LOOP ; Loop 243 times
	RET ; Return

The DJNZ instruction takes 24 oscillator periods, or 2 μs. By initializing register 0 at 00H, the first time DJNZ R0,LOOP is executed, R0 will become FFH. The program will loop within that same instruction 255 more times until R0 equals 0. At that point, control drops down to the next DJNZ, which functions as an outer loop as shown in the flowchart.

The R0 (inner) loop will take 512 μs to complete. The R1 (outer) loop executes the inner loop 243 times for a total time delay of approximately one-eighth second. A third loop, using R2, could be added to increase the time by another factor of 256.

TABLE 18–6 | Centigrade Thermometer Software

Addr	Hex	Label	ASM51		Comments
0000			ORG	0000H	; START OF EPROM
0000	802E		SJMP	START	; SKIP OVER 8051 RESTART
					; LINK AREA
0030			ORG	0030H	; START OF PGM
0030	1138	START:	ACALL	ADC	; PERFORM AN A-TO-D
					; CONVERSION
0032	1159		ACALL	DAC	; DISPLAY RESULT AND
					; PERFORM D-TO-A
					; CONVERSION
0034	114C		ACALL	DELAY	; DELAY 4 SECOND
0036	80F8		SJMP	START	; REPEAT
					;
0038	75B0FF	ADC:	MOV	P3,#0FFH	; FLOAT THE BUS
003B	7590FF		MOV	P1,#0FFH	; FLOAT THE BUS
003E	C2B7	SC:	CLR	P3.7	; DROP SC LINE LOW
0040	D2B7		SETB	P3.7	; THEN RAISE HIGH
0042	20B6FD	WAIT:	JB	P3.6, WAIT	; WAIT TILL EOC GOES LOW
0045	C2B6		CLR	P3.6	; CLEAR P3.6 WHICH WAS
					; USED FOR EOC
0047	C2B7		CLR	P3.7	; CLEAR P3.7 WHICH WAS
					; USED FOR SC
0049	A8B0		MOV	R0,P3	; STORE FINAL ADC RESULT
					; INTO R0
004B	22		RET		
					;
004C	7F20	DELAY:	MOV	R7,#20H	; OUTERMOST LOOP (32 times)
004E	7D00		MOV	R5,#00H	; INNERMOST LOOP (256 times)
0050	7EF3	LOOP2:	MOV	R6,#0F3H	; MIDDLE LOOP (243 times)
0052	DDFE	LOOP1:	DJNZ	R5,LOOP1	;
0054	DEFC		DJNZ	R6,LOOP1	;
0056	DFF8		DJNZ	R7,LOOP2	;
0058	22		RET		
					; CONVERT HEX TO BCD
					; AND DISPLAY ON DAC
					; MODULE
0059	7400	DAC:	MOV	A,#00H	; ZERO OUT ACCUMULATOR
005B	2401	LOOP3:	ADD	A,#01H	; Acc WILL HAVE VALID BCD,
					; R0=ADC HEX
					; RESULT
005D	D4		DA	A	; DECIMAL ADJUST Acc IF
					; INVALID BCD
005E	D8FB		DJNZ	R0,LOOP3	; DECR R0 WHILE INCR AND
					; ADJUSTING Acc TILL
					; R0=0
0060	F590		MOV	P1,A	; MOV CORRECTED BCD TO
					; DAC DISPLAY
0062	22		RET		;
					;
0000			END		

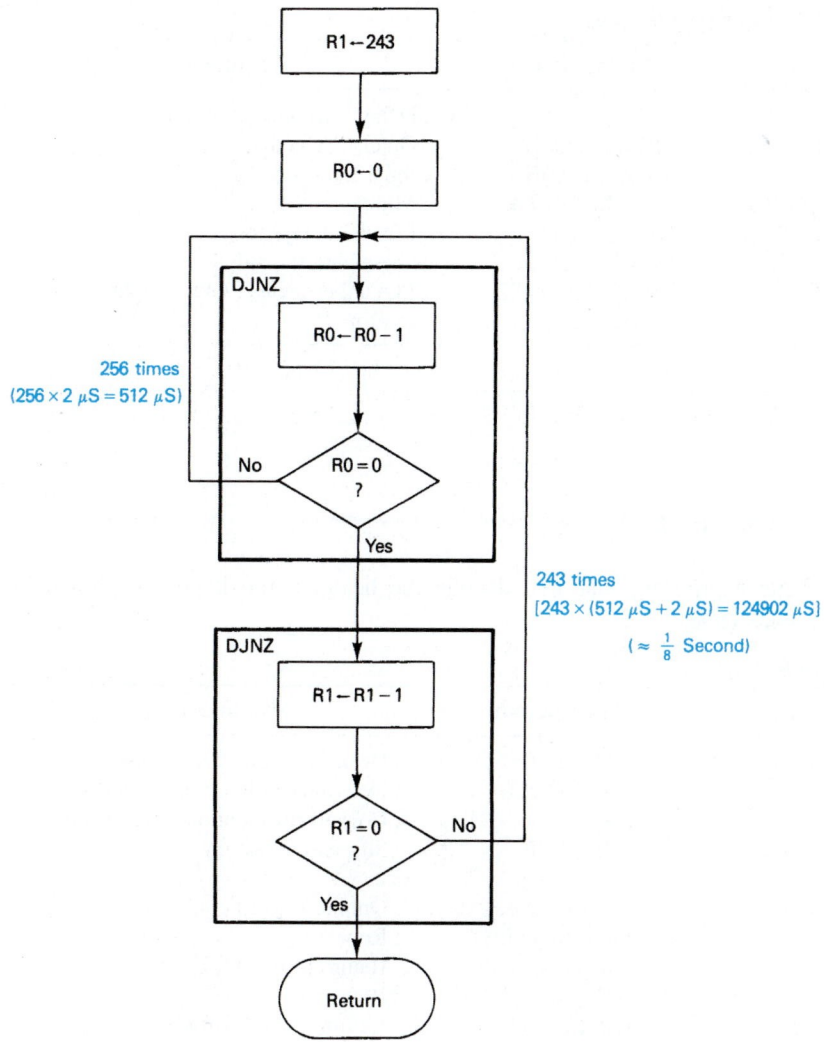

Figure 18–7 Flowchart for the time-delay subroutine of Table 18–5.

EXAMPLE 18–11

Assume that there are input switches connected to port 0 and output LEDs connected to port 1 of an 8051. Write a program that will flash the LEDs ON one second, OFF one second, the number of times indicated on the input switches.

Solution:

Label	Instruction	Comments
READ:	MOV A,P0	; Keep reading port 0 switches
	JZ READ	; into A until A ≠ 0
	MOV R0,A	; Transfer A to register 0
ON:	MOV P1,#0FFH	; Turn ON port 1 LEDs
	CALL DELAY	; Delay 1 second
OFF:	MOV P1,#00H	; Turn OFF port 1 LEDs
	CALL DELAY	; Delay 1 second
	DJNZ R0,ON	; Loop back number of times on switches
STOP:	JMP STOP	; Suspend operation
		;

Label	Instruction	Comments
		; Delay 1 second subroutine
DELAY:	MOV R7,#08H	; Outer loop counter
	MOV R5,#00H	; Inner loop counter
LOOP2:	MOV R6,#0F3H	; Middle loop counter
LOOP1:	DJNZ R5,LOOP1	; LOOP1 delays for
	DJNZ R6,LOOP1	; one-eighth second
	DJNZ R7,LOOP2	; LOOP2 executes LOOP1 eight times
	RET	; Return

EXAMPLE 18–12

Write a program that will decode the hexadecimal keyboard shown in Figure 18–8.

Solution:

Label	Instruction	Comments
KEYSCAN:	CALL ROWRD	; Determine row of key pressed
	CALL COLRD	; Determine column of key pressed
	CALL CONVRT	; Convert row/column to key value
STOP:	JMP STOP	; Suspend operation
		;
ROWRD:	MOV P0,#0FH	; Output 0s to all columns
	MOV R0,#00H	; Row = 0
	JNB P0.0,RET1	; Return if row 0 is LOW
	MOV R0,#01H	; Row = 1
	JNB P0.1,RET1	; Return if row 1 is LOW
	MOV R0,#02H	; Row = 2
	JNB P0.2,RET1	; Return if row 2 is LOW
	MOV R0,#03H	; Row = 3
	JNB P0.3,RET1	; Return if row 3 is LOW
	JMP ROWRD	; Else keep reading
RET1:	RET	; Return
		;
COLRD:	MOV P0,#0F0H	; Output 0s to all rows
	MOV R1,#00H	; Column = 0
	JNB P0.4,RET2	; Return if column 0 is LOW
	MOV R1,#01H	; Column = 1
	JNB P0.5,RET2	; Return if column 1 is LOW
	MOV R1,#02H	; Column = 2
	JNB P0.6,RET2	; Return if column 2 is LOW
	MOV R1,#03H	; Column = 3
	JNB P0.7,RET2	; Return if column 3 is LOW
	JMP COLRD	; Else keep reading
RET2:	RET	; Return
		;
CONVRT:	MOV B,#04H	; B = Multiplication factor
	MOV A,R0	; Move row number to A
	MUL AB	; A = row × 4
	ADD A,R1	; A = row × 4 + column
	RET	; A now contains value of key pressed

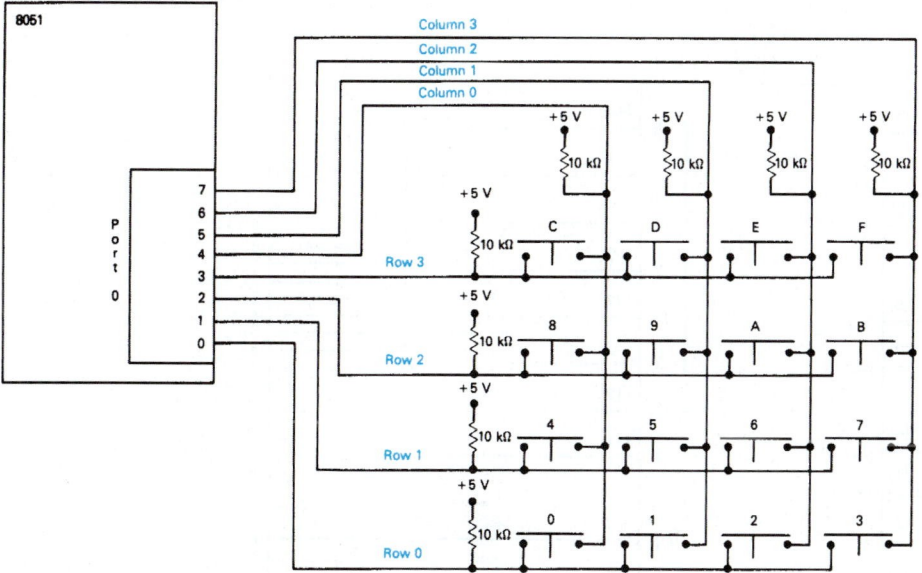

Figure 18–8 Keyboard interface to an 8051.

Explanation: Explanation:The keyboard is wired as a 4 × 4 row-column matrix. The low-order nibble of port 0 is connected to the rows, and the high order is connected to the columns. The rows *and* columns are held HIGH with the 10-kΩ pull-up resistors. Since the I/O ports of the 8051 can be both read *and* written to (bidirectional), we can use a different technique to scan a keyboard than we would use with the 8085A. To determine the row of a depressed key, we drive all of the columns LOW and all of the rows HIGH by executing the instruction MOV P0,#0FH. The HIGH on the rows is actually a *float* state, allowing the rows to be read. The next series of instructions in the ROWRD subroutine read each row to determine the row number that is LOW, if any. If, for example, key 5 is depressed, then row 1 will be LOW and the other three rows will be HIGH.

Now that we know the row, we must next determine the column. The instruction MOV P0,#0F0H will drive the rows LOW and float the columns. If the number 5 key is still depressed, column 1 will be LOW.

The last subroutine converts the row–column combination to the numeric value of the key pressed. Rows 0, 1, 2, and 3 have weighting factors of 0, 4, 8, and 12, respectively, and columns 0, 1, 2, and 3 have weighting factors of 0, 1, 2, and 3, respectively. Knowing that, the CONVRT subroutine uses the following formula to determine the numeric value of the key pressed:

$$\text{Key pressed} = \text{row} \times 4 + \text{column}$$

EXAMPLE 18–13

Figure 18–9 shows how an ADC0801 is interfaced to an 8051. Write a program that takes care of the handshaking requirements for $\overline{SC}$ and $\overline{EOC}$ to complete an analog-to-digital conversion.

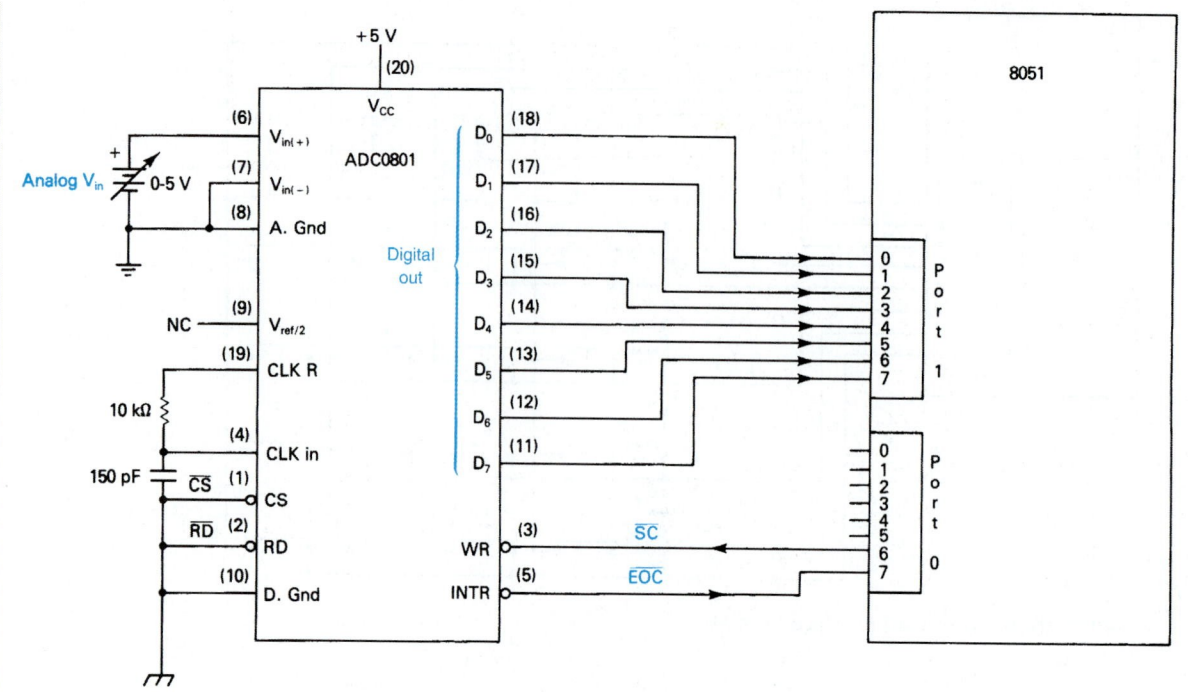

Figure 18–9   Interfacing an analog-to-digital converter to an 8051.

Solution:

Label	Instruction	Comments
FLOAT:	MOV P1,#0FFH	; Write 1's to port 1
	MOV P0,#0FFH	; Write 1's to port 0
SC:	CLR P0.6	; Output LOW-then-HIGH on $\overline{SC}$
	SETB P0.6	;
WAIT:	JB P0.7, WAIT	; Wait here until $\overline{EOC}$ goes LOW
DONE:	MOV R0,P1	; Transfer ADC result to register 0

Explanation: The I/O ports are in the float condition after the initial reset of the 8051. They must be floating in order to be used as inputs. The first two instructions in this program write 1's to ports 1 and 0, which make them float just in case they had 0's in them from a previous step in the program. To start the conversion process, bit 6 of port 0 ($\overline{SC}$) is pulsed LOW then HIGH using the clear (CLR) and set bit (SETB) instructions. The jump-if-bit-set instruction (JB) then monitors bit 7 of port 0 ($\overline{EOC}$) and remains in a WAIT loop until it goes LOW. Finally, at the end of conversion, port 1 is read into register 0, which is used to hold the ADC result.

18–6   Data Acquisition and Control System Application

The object of this section is to implement the 8051-based data acquisition and control system shown in Figure 18–10. Three modules will be discussed using several external transducers and instruments:

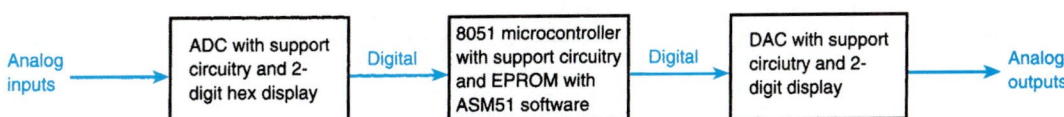

Figure 18–10 Modular 8051-based data acquisition and control application.

1. ADC module with hex display of the output.

2. DAC module with hex display of the input.

3. 8051 microcontroller module with connection points for (1) and (2).

Assembly-language software will be written and stored on an EPROM connected to the 8051. This software will exercise the ADC/DAC modules and perform other I/O functions.

One of the outcomes of these designs is for their use in colleges for teaching the concepts of data acquisition and control, so careful attention will be paid to the cost, durability, and ease of reproduction of each module.

Hardware

The 8051 Microcontroller Module The 8051 microcontroller module is shown in Figure 18–11. The heart of this module is the 8051 microcontroller. Also on the module are input and output buffers, an address latch, and an EPROM interface.

In Figure 18–11, the microcontroller used is the 8031, which is a subset of the 8051. The 8031 differs in that it is the ROMless part. Instead of using the 8051, which has an internal mask-programmable ROM, we will use the 8031 and interface an EPROM to it to supply our program statements.

The 8031 will be run with a 12 MHz crystal. This was chosen because most of the instructions take 12 clock periods to execute. This way, accurate timing can be developed based on one microsecond instruction execution time.

The EPROM is interfaced to the 8031 via port 0 and port 2. The 12-bit address required to access the EPROM comes from the 8 bits provided from port 0 and the 4 bits provided from port 2. The data output from the EPROM is sent back into the microcontroller via port 0. Since port 0 has a dual purpose (i.e., address and data), a control signal from the ALE line is used to tell the address latch when port 0 has valid addresses and when it has data. Also, the $\overline{EA}$ signal is grounded, which signifies that all instructions will be fetched from an external ROM. The PSEN line goes LOW when the microcontroller is ready to read data from the EPROM. This LOW control signal is attached to the $\overline{OE}$ line of the 2732 EPROM, which changes the outputs at D0 through D7 from their float condition to an active condition. At that point the microcontroller reads the data.

Hardware reset is provided at pin 9 via the R-C circuit of R1–C3. At power-up the capacitor is initially discharged and the large inrush current through R1 provides a HIGH-to-LOW level for PIN 9 to be reset. Resetting can also be accomplished with the push-button, which shorts the 5-V supply directly to pin 9 any time you want to have a user reset.

Two eight-bit I/Os are provided via port 1 and port 3. In this circuit, port 1 is used as an output port capable of sinking or sourcing up to four LSTTL loads. This is equivalent to a sink current of approximately 1.6 mA, which is normally not enough to drive loads such as LED indicators. Because of this, the 74LS244 (U1) is used as an output buffer. The 74LS244 can sink 24 mA and source 15 mA. The outputs of this buffer are always enabled by grounding pins 1 and 19. This way, whatever appears at port 1 will immediately be available to the LEDs. The LEDs are connected as active-LOW by providing a series 270 ohm resistor to the + 5-V supply. When any of the

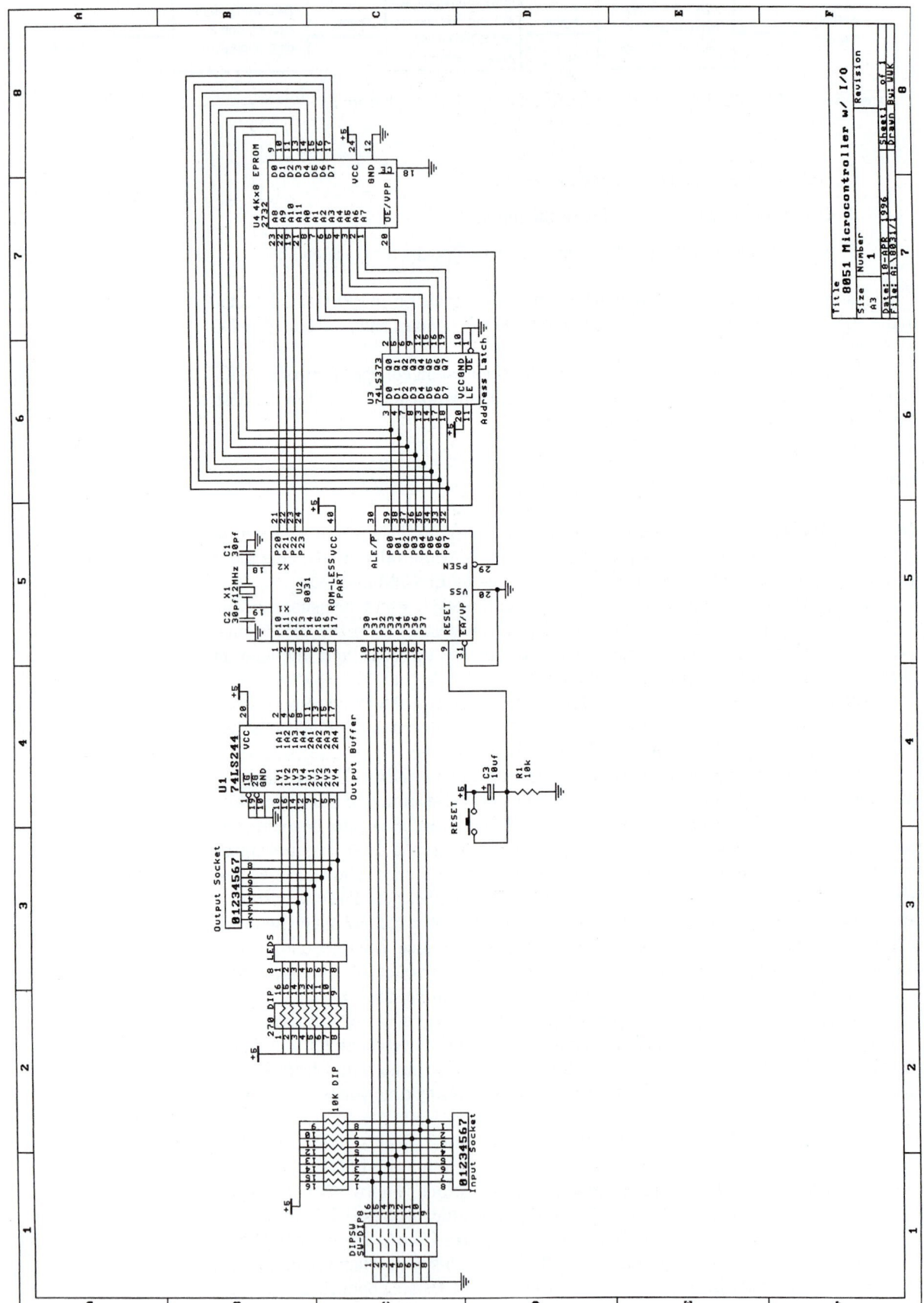

Figure 18–11 8051 Microcontroller with input/output.

outputs of U1 go LOW, the corresponding LED will become active. The output socket is also provided on this module so that other devices can be driven from this port at the same time that the LEDs are being activated.

In this circuit, port 3 is used for input. The 8-bit DIP switch is connected to port 3 in conjunction with the 10 k DIP pack. Therefore, each of the bits of port 3 is normally HIGH via a 10 k pullup. They are made LOW any time one of the DIP switches is closed. An input socket is also provided at this port so that external digital inputs can be input to port 3. To use the input socket, all eight of the DIP switches must be in their open position so that the devices driving the input socket can drive the appropriate line LOW or HIGH without a conflict with the LOW from the DIP switch.

The 74LS373 is an octal transparent latch used as an address demultiplexer. Data entered at the D inputs are transferred into the Q outputs whenever the LE input is HIGH. The data at the D inputs are latched when the LE goes LOW. This address latch is used to demultiplex port 0's address/data information. When port 0 has valid address information, the ALE line is pulsed HIGH. This causes the address latch to grab hold of the information on port 0 and pass it through to the Qs and latch onto it. After the addresses from port 0 are latched, the port is then placed in the float condition so that it can be used to receive the data from the EPROM as the PSEN line is pulsed LOW. The LOW pulse on the PSEN line activates the active-LOW $\overline{OE}$ on the EPROM, which causes its outputs to become active. This data is routed back into port 0 of the microcontroller where it is used as program instructions.

The ADC Interface Module   The heart of the ADC interface module (Figure 18–12) is the ADC0804 IC. This analog-to-digital converter provides 8-bit output resolution and can be directly interfaced to a microcontroller. The object of this interface module is to convert analog quantities brought into $V_{in(+)}$ into 8-bit digital outputs. The digital outputs will be routed to an I/O connector, which can be interfaced to other devices or the microcontroller module. It will also be connected to the 2-digit hex display so that you can constantly monitor the 8-bit values that were converted.

The clock oscillator for the ADC is provided via R1–C1. The frequency of oscillation can be calculated by taking the reciprocal of ($1.1 \times 10\ k\Omega \times 150$ picofarads), which equals approximately 606 KHz. Since this is a successive approximation ADC we can assume that the conversion will take place in 10 to 12 clock periods, which provides a conversion time of about 20 ms.

The chip select (CS) pin on U1 is connected to ground so that the chip is always activated. The RD pin is an active-LOW output enable. This is connected to ground also, so that the outputs are always active. To start a conversion, the $\overline{WR}$ line, which is also known as the start-conversion ($\overline{SC}$), is pulsed LOW-then-HIGH. After the SAR has completed its conversion, it issues a LOW pulse on the INTR line [also known as end-of-conversion ($\overline{EOC}$)], which will be read by the microcontroller to determine if the conversion is complete.

The $V_{ref/2}$ pin is connected to a reference voltage that overrides the normal reference of 5 V provided by V_{cc}. By setting the $V_{ref/2}$ at 1.28 volts, the actual V_{ref} will be 2.56 V. This way, with a V_{ref} of 2.56 volts, and an ADC that has a maximum number of steps of 256, the change in the output will be 1 bit change or 1 binary step for every 10 mV of input change. This provides a convenient conversion for the linear temperature sensor (T1) because T1 provides a 10 mV change for every degree Celsius. It also makes the calculations for the linear phototransistor sensor much simpler.

The DAC Interface Module   The heart of the DAC interface module (Figure 18–13) is the MC1408 (U1). This is an 8-bit digital-to-analog converter. The 8 bits to be converted are brought into the A8–A1 digital input side with A1 receiving the MSB. The DAC provides an analog output current that is proportional to that digital input string. The amount of full scale output current is dictated by the R1–R2 reference resistors. With a 15 V supply and a 10-kΩ resistor, the maximum output current will be 1.5 mA.

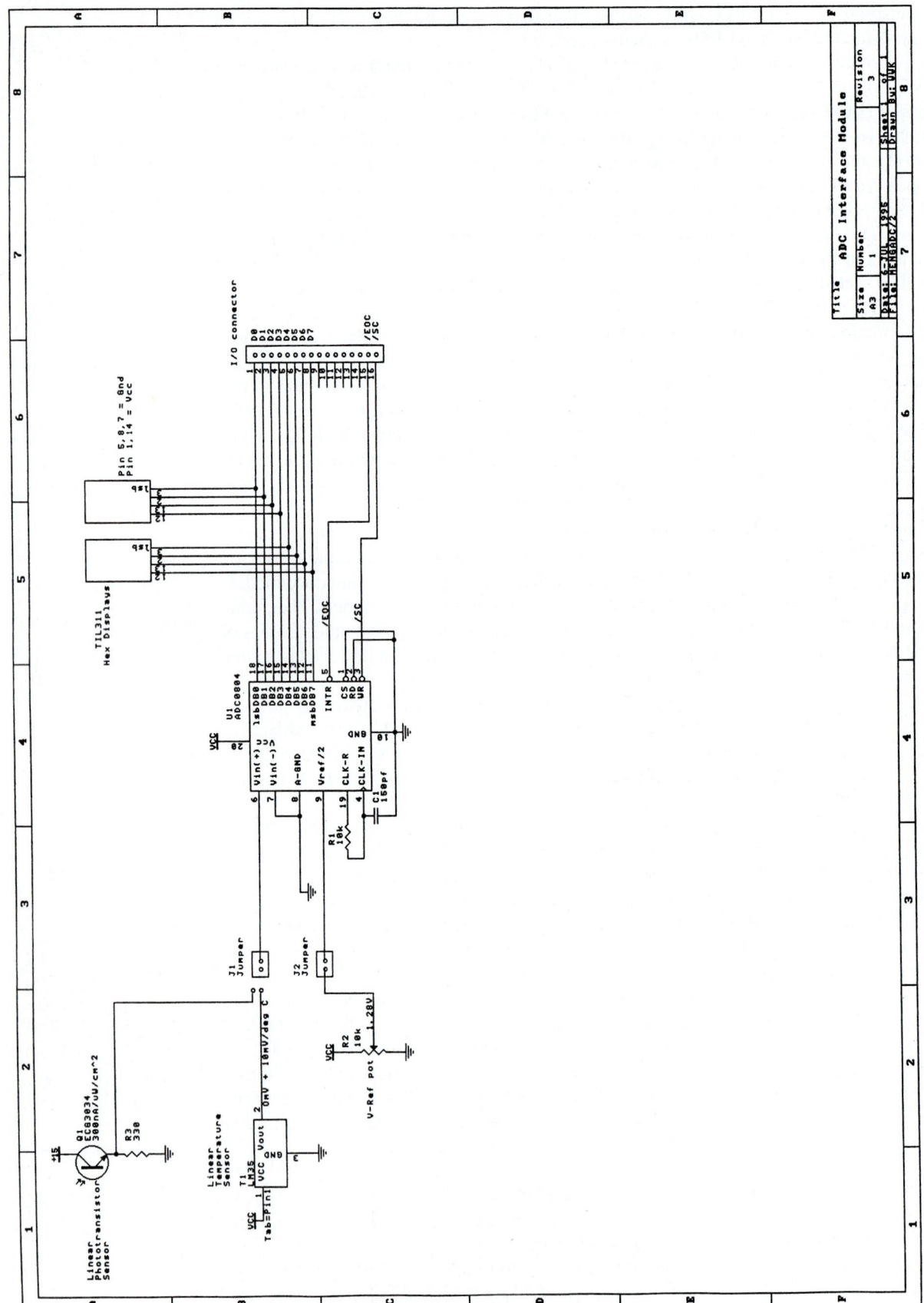

Figure 18–12 The ADC 0804 interface module.

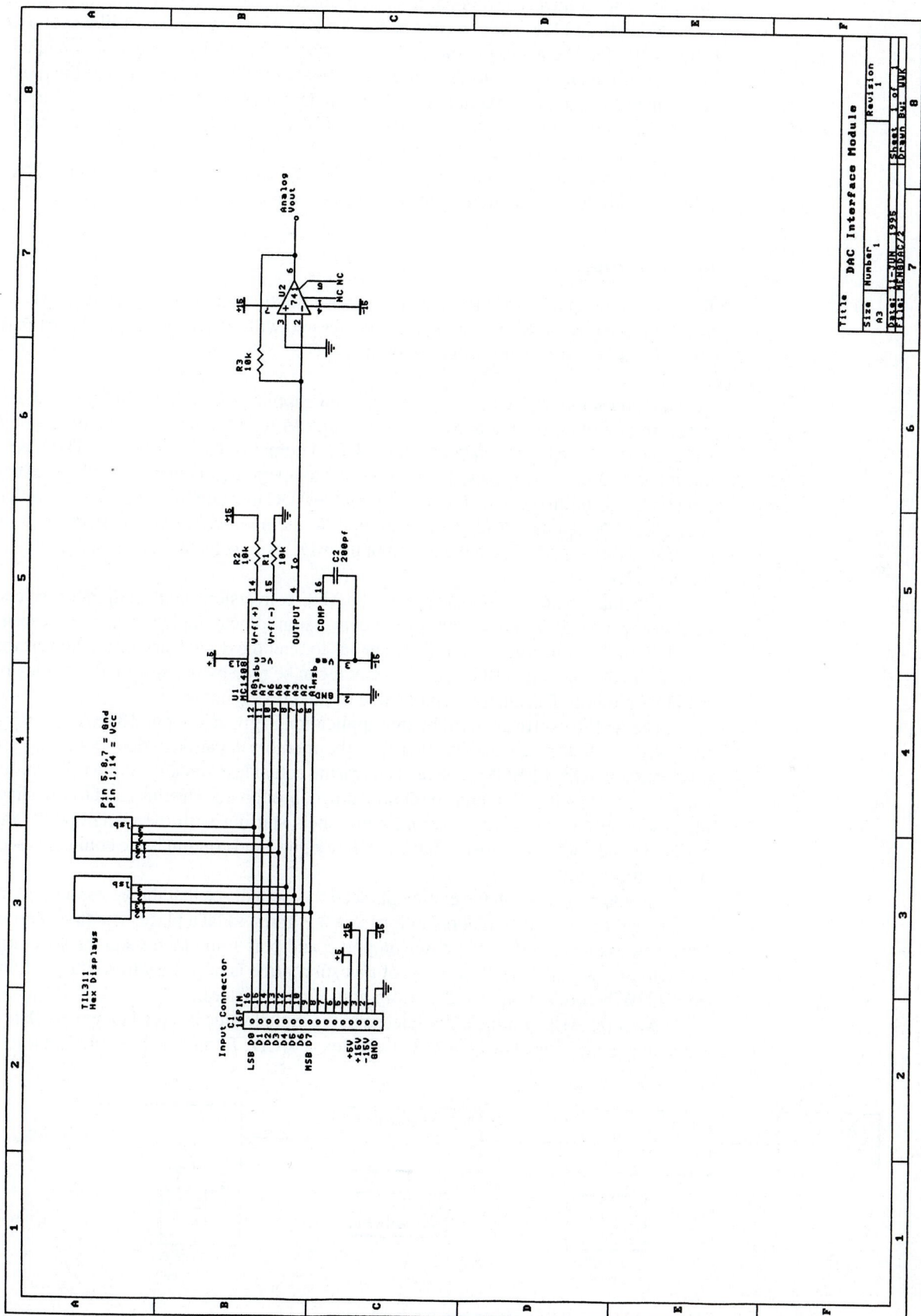

Figure 18–13 The MC1408 DAC interface module.

The analog output current must be converted into a usable voltage. To perform this current-to-voltage conversion, a 741 op amp is used. The current at I_o is drawn through the 10-kΩ feedback resistor of U2. This way, the analog V_{out} will be equal to I_o times 10 kΩ. The analog output voltage can be predicted by using a ratio method. The ratio of the analog V_{out} to the binary input is proportional to the ratio of the maximum analog V_{out} to the maximum binary input (15 V maximum divided by 256 maximum). (Since the supply on the op amp is $\pm$ 15 V, we will never be able to achieve 15 V at V_{out} but instead will be limited to about 13.5.)

A two-digit hexadecimal display is connected directly to the input to the DAC chip. This provides a way to monitor the binary input at any time.

Applications

Three separate applications are used to exercise the three modules. Each of these applications are controlled by the microcontroller and will read analog input values with the ADC and output analog values with the DAC.

Celsius Thermometer   The celsius thermometer application is shown in Figure 18–14. Software is written for the 8051 microcontroller to read the temperature input to the ADC and display a binary coded decimal (BCD) output of the temperature. The LM35 linear temperature sensor is used to measure the temperature. It outputs 10 mV for every degree C. This millivolt value is converted by the ADC into an 8-bit binary string, which is displayed on the hex display and also input to the microcontroller. Since the V_{ref} on the ADC is set at 2.56 V, then the output of the ADC is an actual hexadecimal display of the degrees C.

The microcontroller is used to start the ADC conversion and then monitor the end of conversion ($\overline{EOC}$) line to wait for the conversion to be complete. Once the conversion is complete, the microcontroller then has to read the digital string and convert the hexadecimal value into a BCD value so that it can be read by the user on the display of the DAC module. The analog output value of the DAC is not used.

The ASM51 software used by this application is given in Table 18–6. This listing basically has five columns. The first lists the memory locations 0000–0062 that are programmed in the EPROM. The next shows the actual hex contents of the memory locations, starting with 802E and continuing down through 22. The third and fourth show the assembly language programming with labels starting with the ORG 0008H and ending with the RET statement. In the extreme right-hand column is the comments section for the program.

As you read through the comments section you will see that three subroutines are used to call the three different operations that are going to take place. The first subroutine, ADC, is used to float the buses on port 3 and port 1 and then issue the start conversion ($\overline{SC}$) signal. The "wait" part of that subroutine is necessary to wait until $\overline{EOC}$ goes LOW before reading the data from port 3 into register 0.

After the ADC routine is complete and the digital value is stored in register 0, that value in register 0 must be converted from hex to BCD. To do this, a counter routine is

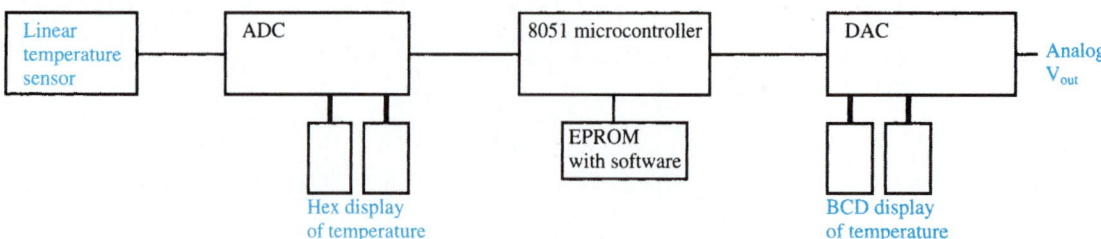

Figure 18–14   Celsius thermometer.

used to count up in BCD until the hex value is reached. At that point, the accumulator will have in it the correct BCD value that was achieved by using the decimal adjust instruction (DA A). After the DAC displays the BCD result, a delay is used for approximately 4 seconds so that there will be a 4-second delay before the temperature is refreshed.

Temperature-Dependent PWM Speed Control   This application is shown in Figure 18–15. It uses most of the same hardware as the Celsius thermometer. The object of this application is to monitor the temperature and produce a pulse-width-modulated (PWM) square wave whose duty cycle is proportional to the temperature. This PWM square wave can be used to drive a DC motor. The effective DC value will increase as the duty cycle increases. Since the duty cycle increases with increasing temperature, the speed of the motor will also increase with increasing temperature.

The ASM51 software used to drive this application is given in Table 18–7. Basically, this software has two subroutines that will be called. The first one is the ADC subroutine. This subroutine starts the conversion, waits until the conversion is complete, and moves the ADC result into register 0. The next subroutine, the PWM, is used to produce the PWM square wave. To produce a PWM square wave, the output of the DAC will first be held LOW (0 V) for a delay multiple equivalent to 10 hexadecimal. Then the DAC will be sent HIGH, which in this case is one-half the full scale output (80 hex). This HIGH is held for a delay multiple equivalent to the temperature value received from the ADC conversion. This way, as the ADC result increases with increasing temperature, the duty cycle of the PWM will increase, thus increasing the effective DC value sent to the motor.

Integrating Solar Radiometer   The block diagram for the integrating solar radiometer is given in Figure 18–16. The object of this application is to measure sunlight intensity and produce a real-time display of the irradiation from the sunlight and then accumulate the irradiation, the same as taking the integral, and display the integrated output as an analog voltage and also on a hex display. To measure the irradiation, an ECG 3034 phototransistor is used. It produces 300 nanoamps per microwatt per centimeter squared. The current is changed to a voltage by the 330 ohm resistor. This analog voltage is converted with the ADC and is then integrated using an ASM51 software program. The integrated output of the ADC will be proportional to the amount of sunlight that struck the measured surface over a certain length of time. This length of time could be one hour, one day, one week, or whatever time is desired.

The software used to execute this application is given in Table 18–8. The first subroutine is the ADC subroutine, which is used to convert the irradiation into a digital value. After the microcontroller has this digital value, the DAC routine performs the integration simply by accumulating the ADC result over each period of time. The DAC provides both the hexadecimal equivalent of the integrated irradiation as well as an analog output voltage that is proportional to the integrated irradiation.

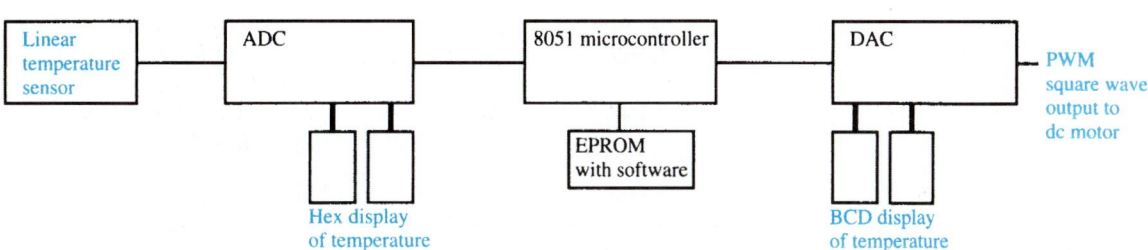

Figure 18–15   Temperature-dependent PWM speed control.

TABLE 18–7 Temperature-Dependent PWM Speed Control Software

Addr	Hex	Label	ASM51		Comments
0000			ORG	0000H	; START OF EPROM
0000	8030		SJMP	START	; SKIP OVER 8051 RESTART ; LINK AREA
0030			ORG	0030H	; START OF PGM
0030	7F01	START:	MOV	R7,#01H	; TO ENABLE ADC READ ON ; 1st LOOP
0032	1138	LOOP:	ACALL	ADC	; PERFORM AN A-TO-D ; CONVERSION
0034	114C		ACALL	PWM	; DRIVE DC MOTOR WITH ; PWM SIGNAL
0036	80FA		SJMP	LOOP	; REPEAT
0038	DF11	ADC:	DJNZ	R7,ERET	; TAKE AN ADC READ EVERY ; 256th TIME FOR ; STABILITY
003A	75B0FF		MOV	P3,#0FFH	; FLOAT THE BUS
003D	C2B7	SC:	CLR	P3.7	; DROP SC LINE LOW
003F	D2B7		SETB	P3.7	; THEN RAISE HIGH ; (PROVIDES POSITIVE EDGE)
0041	20B6FD	WAIT:	JB	P3.6,WAIT	; WAIT TILL EOC GOES LOW
0044	E5B0		MOV	A,P3	; STORE ADC RESULT INTO ; Acc
0046	547F		ANL	A,#01111111B	; CLEAR BIT 7 WHICH WAS ; USED FOR SC
0048	F8		MOV	R0,A	; MOVE ADC RESULT TO ; REG 0
0049	AF40		MOV	R7,40H	; RESET REG 7 to 64 DECIMAL ;
004B	22	ERET:	RET		 ;
004C	759000	PWM:	MOV	P1,#00H	; OUTPUT A LOW
004F	743F		MOV	A,#3FH	; 3FH IS HIGHEST ASSUMED ; ADC OUTPUT VALUE
0051	98		SUBB	A,R0	; SUBTRACT THE ADC ; VALUE FROM 3FH
0052	F9		MOV	R1,A	; R1 (DELAY VALUE) FOR ; LOW OUT (DEC W/INC TEMP)
0053	115D		ACALL	DELAY	; CALL DELAY TO HOLD LOW
0055	7590FF		MOV	P1,#0FFH	; OUTPUT FULL SCALE ; OUTPUT FOR A HIGH
0058	A900		MOV	R1,R0	; USE TEMPERATURE AS ; DELAY MULTIPLIER ; FOR HIGH
005A	115D		ACALL	DELAY	; CALL DELAY TO HOLD ; HIGH
005C	22		RET		; RETURN ;
005D	7D80	DELAY:	MOV	R5,#80H	; START COUNT OF R5
005F	DDFE	LOOP1:	DJNZ	R5,LOOP1	; R5=INNERLOOP COUNT ; DOWN TO ZERO
0061	D9FA		DJNZ	R1,DELAY	; R1=MULTIPLIER FROM ; PWM ROUTINE
0063	22		RET		
0000			END		

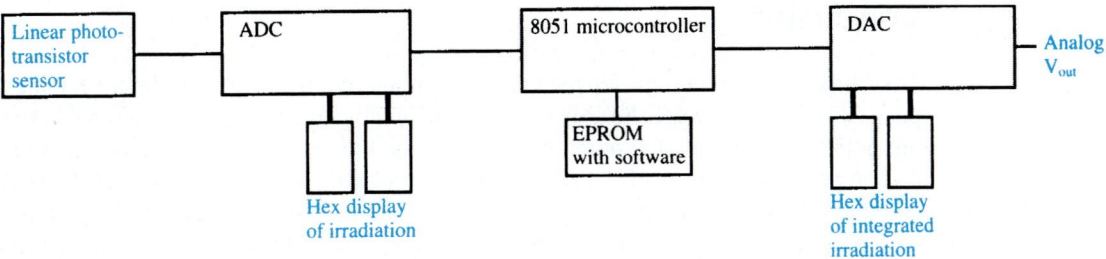

Figure 18–16 Integrating solar radiometer.

TABLE 18–8	Integrating Solar Radiometer Software				

Addr	Hex	Label	ASM51		Comments
0000		ORG	0000H		; START OF EPROM
0000	8030		SJMP	START	; SKIP OVER 8051 RESTART
					; LINK AREA
0030			ORG	0030H	; START OF PGM
0030	7400	START:	MOV	A,#00H	; RESET ACCUMULATOR
0032	113A	LOOP:	ACALL	ADC	; PERFORM AN A-TO-D
					; CONVERSION
0034	1158		ACALL	DAC	; DISPLAY RESULT AND
					; PERFORM D-TO-A
					; CONVERSION
0036	114B		ACALL	DELAY	; DELAY 4 SECOND
0038	80F8		SJMP	LOOP	; REPEAT
003A	75B0FF	ADC:	MOV	P3,#0FFH	; FLOAT THE BUS
003D	C2B7	SC:	CLR	P3.7	; DROP SC LINE LOW
003F	D2B7		SETB	P3.7	; THEN RAISE HIGH
0041	20B6FD	WAIT:	JB	P3.6,WAIT	; WAIT TILL EOC GOES LOW
0044	C2B6		CLR	P3.6	; CLEAR P3.6 WHICH WAS
					; USED FOR EOC
0046	C2B7		CLR	P3.7	; CLEAR P3.7 WHICH WAS
					; USED FOR SC
0048	A8B0		MOV	R0,P3	; STORE FINAL ADC RESULT
					; INTO R0
004A	22		RET		
					;**DELAY 4 SECONDS**
004B	7F20	DELAY:	MOV	R7,#20H	; OUTERMOST LOOP (32X)
004D	7D00		MOV	R5,#00H	; INNERMOST LOOP (256X)
004F	7EF3	LOOP2:	MOV	R6,#0F3H	; MIDDLE LOOP (243X)
0051	DDFE	LOOP1:	DJNZ	R5,LOOP1	;
0053	DEFC		DJNZ	R6,LOOP1	;
0055	DFF8		DJNZ	R7,LOOP2	;
0057	22		RET		
					; INTEGRATE BY ACCUMU-
					; LATING ADC RESULT
0058	8840	DAC:	MOV	40H,R0	; MOVE ADC RESULT TO
					; RAM LOCATION 40H
005A	53403F		ANL	40H,#00111111B	; MASK OFF UNWANTED
					; DATA
005D	2540		ADD	A,40H	; ACCUMULATE NEW ADC
					; VALUE TO PREVIOUS
					; TOTAL
005F	F590		MOV	P1,A	; MOVE TO DAC DISPLAY
0061	22		RET		;
					;
0000			END		

Conclusion

This three-module method for data acquisition and control proves to be a very inexpensive and simple way to be able to input both analog and digital quantities and to output both digital and analog quantities. Having the hex displays connected to the input and output modules is an effective way of monitoring the real-time activity of the DAC and the ADC. The entire three-module data acquisition set up is very portable in that it only requires a bipolar power supply to operate it. The software used to exercise these modules is very straightforward and is effectively written using the ASM51 assembler on a personal computer and downloaded to an EPROM using a standard EPROM programming software package. Once you have the programmed EPROM, all that is required for a stand-alone data acquisition system is a power supply.

Summary

In this chapter we have learned that

1. The 8051 microcontroller is different from a microprocessor because it has the CPU, ROM, RAM, timer/counter, and parallel and serial ports fabricated into a single IC.

2. Thirty-two of the 40 pins of the 8051 are used for the four 8-bit parallel I/O ports. Three of the ports share their function with the address, data, and control buses.

3. The address spaces of the 8051 are divided into four distinct areas: internal data memory, external data memory, internal code memory, and external code memory. The internal data memory is further divided into user RAM and Special Function Registers (SFRs).

4. To interface to an external EPROM like the 2732, an octal D-latch is required to demultiplex the address/data bus, which is shared with port 0. The External Access $(\overline{EA})$ pin is tied LOW and the $(\overline{PSEN})$ output is used to enable the output of the EPROM.

5. Extra data memory and I/O ports can be interfaced by using the 8155 IC. The 8155 demultiplexes the address/data bus internally so an octal D-latch is not required.

6. The MOV instruction is very powerful, providing the ability to move data almost anywhere internal or external to the microcontroller and to the I/O ports.

7. Program branching is accomplished by use of many different conditional and unconditional jumps and calls.

8. The 8051 instruction set provides the ability to work with individual bits, which makes it very efficient for on/off control operations. Instructions are available for all the logic functions, rotates, and bit manipulations.

9. Instructions are provided for all the basic arithmetic instructions: addition, subtraction, multiplication, division, incrementing, and decrementing.

10. Each instruction machine cycle takes 12 clock periods to complete. This means that if a 12 MHz crystal is used, each machine cycle takes 1 microsecond to complete. One complete instruction takes 1, 2, or 4 machine cycles.

11. A 4-by-4 matrix keyboard can be scanned using bit operations on a single I/O port.

12. Interfacing an 8-bit analog-to-digital converter to an 8051 is accomplished with one port and two bits on a second port. The *start-conversion* LOW pulse is issued with bit-setting instructions and the *end-of-conversion* signal is monitored with bit-checking instructions.

▬ Glossary ▬

Assembler: A software package used to convert assembly language programs into executable machine code.

Bidirectional: An I/O port that can be used for both input and output.

Bit Addressable: Memory spaces in the internal RAM and SFR areas that allow for the addressing of individual bits.

External Access ($\overline{EA}$): An active-LOW input that when forced LOW, tells the processor that all program memory is external

Microcontroller: A computer on a chip. It contains a CPU, ROM, RAM, counter/timers, and I/O ports. It is especially well suited for control applications.

Program Store Enable ($\overline{PSEN}$): An|hyphen|active|hyphen|Low output control signal issued by the microcontroller as the read strobe for external program memory.

Register Bank: A group of eight registers. The 8051 has four sets of banks that are selected by writing to the two bank-select bits in the PSW.

Special Function Register (SFR): A data register that has a dedicated space within the microcontroller. It is used to maintain the data required for microcontroller operations.

▬ Problems ▬

18–1. Why is a microcontroller sometimes referred to as a "computer on a chip"?

18–2. Describe the difference between the 8031, 8051, and the 8751.

18–3. What additional features does the 8052 have over the 8051?

18–4. Port 2 has a dual purpose. One is as a bidirectional I/O port. What is its other purpose?

18–5. If you are using the internal ROM in the 8051 for your program code memory, should the $\overline{EA}$ be tied HIGH or LOW?

18–6. What is the maximum size EPROM and RAM that can be interfaced to the 8051?

18–7. What is the address range of the Special Function Registers (SFRs)?

18–8. What are the SFR addresses of the four I/O ports?

18–9. The 8051 has four register banks with eight registers in each bank. How does the CPU know which bank is currently in use?

18–10. Why is there a need for the 74LS373 address latch when interfacing the 2732 EPROM in Figure 18–5 but not when interfacing the 8155 RAM in Figure 18–6?

18–11. Determine the value of the accumulator after the execution of instruction A:, B:, C:, and D:

```
            MOV 40H,#88H
            MOV R0,#40H
A:          MOV A,R0              ; A = _____
B:          MOV A,@R0             ; A = _____
C:          MOV A,40H             ; A = _____
D:          MOV A,#40H            ; A = _____
```

18–12. Repeat Problem 18–11 for the following instructions:

```
            MOV 50H,#33H
            MOV 40H,#22H
            MOV R0,#30H
            MOV 30H,50H
            MOV R1,#40H
A:          MOV A,40H             ; A = _____
B:          MOV A,@R0             ; A = _____
C:          MOV A,R1              ; A = _____
D:          MOV A,30H             ; A = _____
```

18–13. Write the instructions to perform each of the following operations:

(a) Output a C7H to port 3.

(b) Load port 1 into register 7.

(c) Load the accumulator with the number 55H.

(d) Send the contents of RAM memory, whose address is in register 0, to the accumulator.

(e) Output the contents of register 1 to port 0.

18–14. Write a program that continuously reads port 0 until the byte read equals A7H. At that time, turn on the output LEDs connected at port 1.

18–15. Write a program that will produce an output at port 1 that counts up from 20H to 90H repeatedly.

18–16. Determine the value of the accumulator after the execution of instruction A:, B:, and C:

```
            MOV A,#00H
            MOV R0,#36H
A:          XRL A,R0              ; A = _____
B:          ORL A,#71H            ; A = _____
C:          ANL A,#0F6H           ; A = _____
```

18–17. Repeat Problem 18–16 for the following program:

```
            MOV A,#77H
A:          CLR ACC.1             ; A = _____
B:          SETB ACC.7            ; A = _____
C:          RL A                  ; A = _____
```

18–18. Modify one instruction in Example 18–8 so that it will output eight pulses at port 0 instead of five.

18–19. Write a program that adds 05H to the number read at port 0 and outputs the result to port 1.

18–20. What is the length of time of the following time-delay subroutine? (Assume a 12-MHz crystal is being used.)

DELAY:	MOV R1,#00H
	MOV R0,#00H
LOOP:	DJNZ R0,LOOP
	DJNZ R1,LOOP
	RET

Schematic Interpretation Problems

18–21. Find Port 2 (P2.7–P2.0) of U8 in the 4096/4196 schematic. This port outputs the high-order address bits for the system ($A8$–$A15$). On a separate piece of paper, draw a binary comparator that compares the four bits $A8$–$A11$ to the four bits $A12$–$A15$. The HIGH output for an equal comparison is to be input to P3.4 (pin 14).

18–22. Locate the microcontroller in the 4096/4196 schematic.

(a) What is its grid location and part number?

(b) Its low-order address is multiplexed. What IC and control signal is used to demultiplex the address/data bus ($AD0$–$AD7$) into the low-order address bus ($A0$–$A7$)?

(c) What IC and control signal is used to demultiplex the address/data bus ($AD0$–$AD7$) into the data bus ($D0$–$D7$)?

WWW Sites

Useful WWW sites for this book:

www.acebus.com (AceBus) Downloadable 8051 microcontroller editor, assembler, and simulator

www.ahinc.com/scsi.htm (Advanced Horizons, Inc.) SCSI, CD, and DVD definitions and standards

www.altera.com (Altera Corporation) CPLDs and FPGAs

www.amanb.net/8085.html (ABCreations) 8085 simulator

www.amd.com (Advanced Micro Devices, Inc.) Microprocessors, EPROMs, and Flash memory

www.analog.com (Analog Devices, Inc.) Analog ICs, ADCs, and DACs

www.bipom.com (BiPom Electronics, Inc.) Downloadable 8051 microcontroller editor, assembler, and simulator

www.chips.ibm.com (IBM Microelectronics Corporation) Microprocessors, SRAMs, DRAMs, and custom logic

www.cyrix.com (Cyrix Corporation) Microprocessors

www.digikey.com Sales catalog of electronic products and components

www.elexp.com (Electronics Express, Inc.) Sales of electronic products, CPLD programmer boards, microprocessor trainers and components

www.emacinc.com (EMAC, Inc.) Microprocessor and microcontroller trainers and single-board computers

www.fairchildsemi.com (Fairchild Semiconductor, Inc.) Digital Logic ICs, Analog ICs, discrete semiconductors, EEPROMs, and microcontrollers

www.fujitsumicro.com (Fujitsu, Inc.) Microcontrollers, DRAMs, and Flash memory

www.hitachi.com (Hitachi, LTD.) Microprocessors, DRAMs, SRAMs, Flash memories, EEPROMs, digital logic ICs, analog ICs, and discrete semiconductors www.idt.com (Integrated Device Technology, Inc.) SRAMs and digital logic ICs

www.intel.com (Intel Corporation) Microprocessors, microcontrollers, and Flash memory

www.interactiv.com (Interactive Image Technology, Inc.) MultiSIM software

www.intersil.com (Intersil Corporation) Analog ICs, ADCs, DACs, SRAMs, microprocessors, microcontrollers, and discrete semiconductors

www.jameco.com Sales catalog of electronic products and components

www.jdr.com Sales catalog of electronic products and components

www.latticesemi.com (Lattice Semiconductor Corporation) CPLDs

www.linear-tech.com (Linear Technology Corporation) ADCs and DACs

www.microchip.com (Microchip Technology, Inc.) EEPROMs and microcontrollers

www.micronsemi.com (Micron Semiconductor Products, Inc.) DRAMs, SRAMs, and Flash memory

www.mot-sps.com (Motorola Semiconductors, Inc.) Analog ICs, digital logic ICs, microcontrollers, microprocessors, SRAMs, and discrete semiconductors

www.mp3-tech.org (MP3 Tech) MP3 definitions and standards

www.national.com (National Semiconductor Corp.) Analog ICs, digital logic ICs, and microcontrollers

www.onsemi.com (ON Semiconductor, Inc.) Analog ICs, digital logic ICs, and discrete semiconductors

www.semiconductors.philips.com (Philips Semiconductors, Inc.) Analog ICs, digital logic ICs, microcontrollers, and discrete semiconductors

www.sharpmeg.com (Sharp Microelectronics Corporation) Microprocessors, microcontrollers, SRAMs, and Flash memory

www.ti.com (Texas Instruments, Inc.) Analog ICs, digital logic ICs, and microcontrollers

www.toshiba.com (Toshiba America, Inc.) Analog ICs, digital logic ICs, microcontrollers, DRAMs, SRAMs, and Flash memory

www.usb.org (Universal Serial Bus) USB definitions and standards

www.webopedia.com Glossary of PC and internet terms with links to other sites

www.xess.com (XESS Corporation) CPLD programmer boards

www.xilinx.com (Xilinx, Inc.) CPLDs and FPGAs

www.zilog.com (Zilog Inc.) Microprocessors and microcontrollers

Appendix B

Manufacturers' Data Sheets*

Datasheet IC Numbers	Function
74HC00	High-speed CMOS NAND
74LV00	Low-voltage NAND
74ABT244	BiCMOS Octal Buffer
7400	Standard TTL NAND
7414	Inverter Schmitt Trigger
74121	Monostable Multivibrator
2716	EPROM (2K $\times$ 8)
ADC0801	Analog-to-Digital Converter
MC1508/1408	Digital-to-Analog Converter
μA741	Operational Amplifier
NE555	Timer
MAX7000	Altera CPLD Family
XC95108	Xilinx CPLD

---

*Courtesy of Signetics (Philips Semiconductors) Corporation, Intel Corporation, Texas Instruments, Inc., Xilinx, Inc., and Altera Corporation.

DESCRIPTION

The 54/74HC00 and 54/74HCT00 are high-speed Si-gate CMOS devices and are pin compatible with low power Schottky TTL (LSTTL). They are specified in compliance with JEDEC standard no. 7.

The 54/74HC00 and 54/74HCT00 provide the positive 2-input NAND function.

SYMBOL AND PARAMETER		CONDITIONS	TYPICAL		UNIT
			HC	HCT	
t_{PHL} t_{PLH}	Propagation delay nA, nB to nY	$C_L = 15pF$	8	8	ns
C_I	Input capacitance		3.5	3.5	pF
C_{PD}^1	Power dissipation capacitance per gate	See note 2	22	22	pF

NOTES:
1. C_{PD} is used to determine the dynamic power consumption -
   $$P_D = C_{PD} \cdot V_{CC}^2 \cdot f_i + \Sigma C_L \cdot V_{CC}^2 \cdot f_o \text{ where:}$$
   f_i = input frequency; f_o = output frequency
   C_L = output load capacitance; V_{CC} = supply voltage
2. For HC, condition is V_I = GND to V_{CC}
   For HCT, condtion is V_I = GND to $V_{CC} - 1.5V$

ORDERING CODE

PACKAGES	COMMERCIAL RANGES $T_A = -40°C$ to $+85°C$	MILITARY RANGES $T_A = -55°C$ to $+125°C$
Plastic DIP	N74HC00N, N74HCT00N	
Ceramic DIP		
Plastic SO	N74HC00D, N74HCT00D	

PIN DESCRIPTION

PIN NO.	SYMBOL	NAME AND FUNCTION
1,4,9,12	1A to 4A	Data inputs
2,5,10,13	1B to 4B	Data inputs
3,6,8,11	1Y to 4Y	Data outputs
7	GND	Ground (OV)
14	V_{CC}	Positive supply voltage

PIN CONFIGURATION

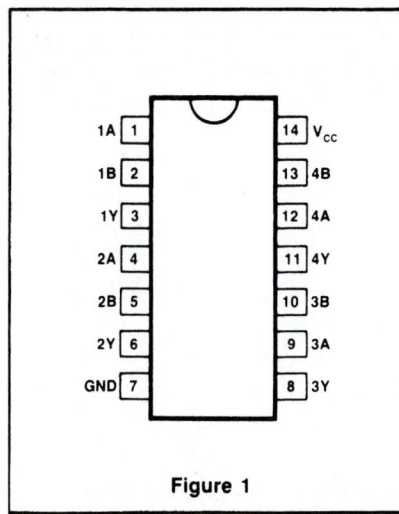

Figure 1

LOGIC SYMBOL

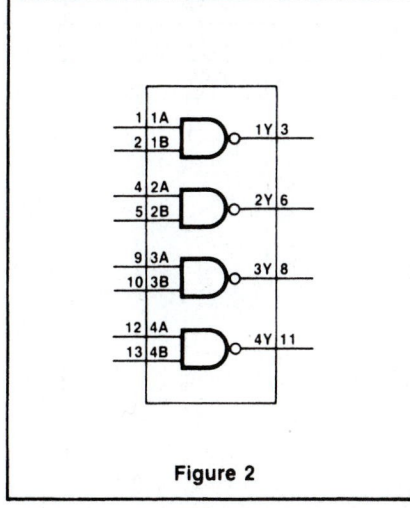

Figure 2

LOGIC SYMBOL (IEEE/IEC)

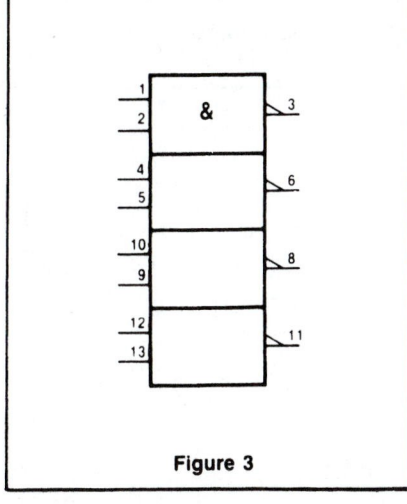

Figure 3

ABSOLUTE MAXIMUM RATINGS

Limiting values in accordance with the Absolute Maximum System (IEC134)

SYMBOL AND PARAMETER		RATING	UNIT
V_{CC}	Supply voltage		-0.5 to $+7.0$	V
$\pm I_{IK}$	Input diode current $V_I \le -0.5V$ or $V_I \ge V_{CC}+0.5V$		20	mA
$+I_{DK}$	Output diode current $V_O \le -0.5V$ or $V_O \ge V_{CC}+0.5V$		20	mA
$\pm I_{OH}$, $\pm I_{OL}$	Output source or sink current $-0.5V \le V_O \le V_{CC}+0.5V$	Standard output	25	mA
$\pm I_{CC}$, $\pm I_{GND}$	V_{CC} or GND current	Standard outputs	50	mA
T_{stg}	Storage temperature range		-65 to $+150$	°C
P_{tot}	Power dissipation per package Plastic and Ceramic (Cerdip) DIL (above $+60°C$; derate linearly with 8mW/K)	Standard temp -40 to $+85°C$	500	mW
	Power dissipation per package Plastic minipack (SO) (above $+70°C$; derate linearly with 5mW/K)	Standard temp -40 to $+85°C$	200	mW
	Power dissipation per package Ceramic (Cerdip) DIL (above $+100°C$; derate linearly with 8mW/K)	Extended temp -55 to $+125°C$	500	mW

Voltages are referenced to GND (ground = OV)

RECOMMENDED OPERATION CONDITIONS

SYMBOL AND PARAMETER			54/74 HC			54/74 HCT			UNIT
			Min	Typ	Max	Min	Typ	Max	
V_{CC}	Supply voltage		2.0	5.0	6.0	4.5	5.0	5.5	V
V_I	Input voltage range		0		V_{CC}	0		V_{CC}	V
V_O	Output voltage range		0		V_{CC}	0		V_{CC}	V
T_{amb}	Operating ambient temperature range	54	-55		$+125$	-55		$+125$	°C
		74	-40		$+85$	-40		$+85$	
t_r t_f	Input rise and fall times	$V_{CC}=2.0V$			1000				ns
		$V_{CC}=4.5V$		6.0	500		6.0	500	
		$V_{CC}=6.0V$			400				

DC ELECTRICAL CHARACTERISTICS: 54/74HC

SYMBOL AND PARAMETER			T_{amb} (°C)						UNIT	TEST CONDITIONS[1]		
			54/74HC +25			74HC −40 to 85		54HC −55 to 125					
			Min	Typ	Max	Min	Max	Min	Max		V_{CC}	V_{IN}	OTHER
V_{IH}	HIGH-level input voltage		1.5 3.15 4.2			1.5 3.15 4.2		1.5 3.15 4.2		V	2V 4.5V 6V		
V_{IL}	LOW-level input voltage				0.3 0.9 1.2		0.3 0.9 1.2		0.3 0.9 1.2	V	2V 4.5V 6V		
V_{OH}	HIGH-level output voltage		1.9 4.4 5.9			1.9 4.4 5.9		1.9 4.4 5.9		V	2V 4.5V 6V	V_{IH} or V_{IL} V_{IH} or V_{IL} V_{IH} or V_{IL}	$-I_O = 20\mu A$ $-I_O = 20\mu A$ $-I_O = 20\mu A$
			3.98 5.48			3.84 5.84		3.7 5.2		V	4.5V 6V	V_{IH} or V_{IL} V_{IH} or V_{IL}	$-I_O = 4mA$ $-I_O = 5.2mA$
V_{OL}	LOW-level output voltage				0.1 0.1 0.1		0.1 0.1 0.1		0.1 0.1 0.1	V	2V 4.5V 6V	V_{IH} or V_{IL} V_{IH} or V_{IL} V_{IH} or V_{IL}	$I_O = 20\mu A$ $I_O = 20\mu A$ $I_O = 20\mu A$
					0.26 0.26		0.33 0.33		0.4 0.4	V	4.5V 6V	V_{IH} or V_{IL} V_{IH} or V_{IL}	$I_O = 4mA$ $I_O = 5.2mA$
$\pm I_I$	Input leakage current				0.1		1.0		1.0	μA	6V	V_{CC} or GND	
I_{CC}	Quiescent supply current	SSI			2.0		20.0		40.0	μA	6V	V_{CC} or GND	$I_O = 0$

NOTE:
1. Voltages are referenced to GND (ground = OV).

AC ELECTRICAL CHARACTERISTICS: 54/74HC
GND = OV; $t_r = t_f = 6ns$; $C_L = 50pF$

SYMBOL AND PARAMETER		T_{amb}(°C)							UNIT	TEST CONDITIONS	
		54/74HC +25			74HC −40 to +85		54HC −55 to +125				
		Min	Typ	Max	Min	Max	Min	Max		V_{CC}	Figure
t_{PHL} t_{PLH}	Propagation delay nA, nB, to nY			100 20 17		125 25 21		150 30 26	ns	2V 4.5V 6V	4
t_{THL} t_{TLH}	Output transition time			75 15 13		95 19 16		112 22 19	ns	2V 4.5V 6V	4

DC ELECTRICAL CHARACTERISTICS: 54/74HCT

SYMBOL AND PARAMETER			T_{amb} (°C)						UNIT	TEST CONDITIONS[1]		
			54/74HCT +25			74HCT −40 to 85		54HCT −55 to 125					
			Min	Typ	Max	Min	Max	Min	Max		V_{CC}	V_{IN}	OTHER
V_{IH}	HIGH-level input voltage		2.0			2.0		2.0		V	4.5 to 5.5V		
V_{IL}	LOW-level input voltage				0.8		0.8		0.8	V	4.5 to 5.5V		
V_{OH}	HIGH-level output voltage		4.4			4.4		4.4		V	4.5	V_{IH} or V_{IL}	$-I_O = 20\mu A$
			3.98			3.84		3.7		V	4.5	V_{IH} or V_{IL}	$-I_O = 4mA$
V_{OL}	LOW-level output voltage				0.1		0.1		0.1	V	4.5	V_{IH} or V_{IL}	$I_O = 20\mu A$
					0.26		0.33		0.4	V	4.5	V_{IH} or V_{IL}	$I_O = 4mA$
$\pm I_I$	Input leakage current				0.1		0.1		0.1	μA	5.5V	V_{IH} or V_{IL}	
I_{CC}	Quiescent supply current	SSI			2.0		20.0		40.0	μA	5.5V	V_{CC} or GND	$I_O = 0$
I_C	Supply current									μA	5.5V	2.4V or 0.5V	$I_O = 0$[2]

NOTES:
1. Voltages are referenced to GND (ground = OV).
2. Per input-pin, other inputs at V_{CC} or GND.

AC ELECTRICAL CHARACTERISTICS: 54/74HCT
GND = OV; t_r = t_f = 6ns; C_L = 50pF

SYMBOL AND PARAMETER			T_{amb}(°C)						UNIT	TEST CONDITIONS	
			54/74HCT +25			74HCT −40 to +85		54HCT −55 to +125				
		Min	Typ	Max	Min	Max	Min	Max		V_{CC}	Figure
t_{PHL} t_{PLH}	Propagation delay nA, nB to nY			20		25		30	ns	4.5V	4
t_{THL} t_{TLH}	Output transition time			15		19		22	ns	4.5V	4

AC WAVEFORM

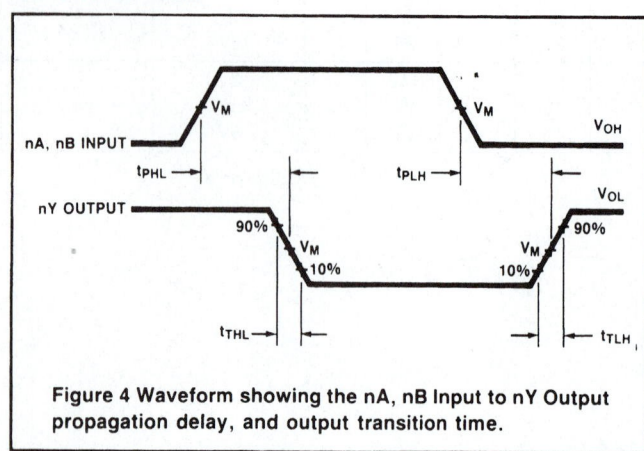

Figure 4 Waveform showing the nA, nB Input to nY Output propagation delay, and output transition time.

NOTE:
HC: V_I = GND to V_{CC}
    V_M = ½ V_{CC}
HCT: V_I = GND to 3.0V
    V_M = 1.3V

FEBRUARY 1993

- **Space-Saving Package Option:** Shrink Small-Outline Package (DB) Features EIAJ 0.65-mm Lead Pitch
- *EPIC* ™ (Enhanced-Performance Implanted CMOS) 2-μm Process
- Typical V_{OLP} (Output Ground Bounce) < 0.8 V at V_{CC} = 3.3 V, T_A = 25°C
- Typical V_{OHV} (Output V_{OH} Undershoot) > 2 V at V_{CC} = 3.3 V, T_A = 25°C
- ESD Protection Exceeds 2000 V Per MIL-STD-883C, Method 3015; Exceeds 200 V Using Machine Model (C = 200 pF, R = 0)
- Latch-Up Performance Exceeds 250 mA Per JEDEC Standard JESD-17
- Package Options Include Plastic Small-Outline and Thin Shrink Small-Outline Packages

D, DB, OR PW PACKAGE
(TOP VIEW)

1A	1	14	V_{CC}
1B	2	13	4B
1Y	3	12	4A
2A	4	11	4Y
2B	5	10	3B
2Y	6	9	3A
GND	7	8	3Y

PRODUCT PREVIEW

description

This quadruple 2-input positive-NAND gate is designed for 2.7-V to 3.6-V V_{CC} operation.

The SN74LV00 performs the Boolean functions $Y = \overline{A \cdot B}$ or $Y = \overline{A} + \overline{B}$ in positive logic.

The SN74LV00 is packaged in TI's shrink small-outline package (DB), which provides the same I/O pin count and functionality of standard small-outline packages in less than half the printed-circuit-board area.

The SN74LV00 is characterized for operation from −40°C to 85°C.

FUNCTION TABLE
(each gate)

INPUTS		OUTPUT
A	B	Y
H	H	L
L	X	H
X	L	H

EPIC is a trademark of Texas Instruments Incorporated.

TEXAS INSTRUMENTS
POST OFFICE BOX 655303 ● DALLAS, TEXAS 75265

SN74LV00
QUADRUPLE 2-INPUT POSITIVE-NAND GATE

FEBRUARY 1993

logic symbol†

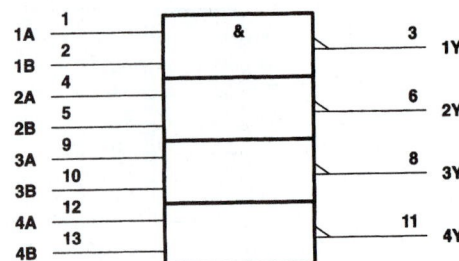

† This symbol is in accordance with ANSI/IEEE Std 91-1984 and
IEC Publication 617-12.

logic diagram, each gate (positive logic)

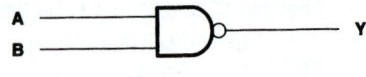

absolute maximum ratings over operating free-air temperature range (unless otherwise noted)‡

Supply voltage range, V_{CC} .... −0.5 V to 4.6 V
Input voltage range, V_I (see Note 1) .... −0.5 V to V_{CC} + 0.5 V
Output voltage range, V_O (see Notes 1 and 2) .... −0.5 V to V_{CC} + 0.5 V
Input clamp current, I_{IK} (V_I < 0 or V_I > V_{CC}) .... ±20 mA
Output clamp current, I_{OK} (V_O < 0 or V_O > V_{CC}) .... ±50 mA
Continuous output current, I_O (V_O = 0 to V_{CC}) .... ±25 mA
Continuous current through V_{CC} or GND pins .... ±50 mA
Maximum power dissipation at T_A = 55°C (in still air): D package .... 0.7 W
                                              DB package .... 0.4 W
                                              PW package .... 0.4 W
Storage temperature range .... −65°C to 150°C

‡ Stresses beyond those listed under "absolute maximum ratings" may cause permanent damage to the device. These are stress ratings only and
functional operation of the device at these or any other conditions beyond those indicated under "recommended operating conditions" is not
implied. Exposure to absolute-maximum-rated conditions for extended periods may affect device reliability.
NOTES: 1. The input and output voltage ratings may be exceeded if the input and output clamp-current ratings are observed.
          2. This value is limited to 4.6 V maximum.

recommended operating conditions (see Note 3)

			MIN	NOM	MAX	UNIT
V_{CC}	Supply voltage		2.7	3.3	3.6	V
V_{IH}	High-level input voltage	V_{CC} = 2.7 V to 3.6 V	2			V
V_{IL}	Low-level input voltage	V_{CC} = 2.7 V to 3.6 V			0.8	V
V_I	Input voltage		0		V_{CC}	V
V_O	Output voltage		0		V_{CC}	V
I_{OH}	High-level output current				−6	mA
I_{OL}	Low-level output current				6	mA
$\Delta t/\Delta v$	Input transition rise or fall rate		0		100	ns/V
T_A	Operating free-air temperature		−40		85	°C

NOTE 3: Unused or floating inputs must be held high or low.

PRODUCT PREVIEW

POST OFFICE BOX 655303 ● DALLAS, TEXAS 75265

electrical characteristics over recommended operating free-air temperature range (unless otherwise noted)

PARAMETER	TEST CONDITIONS		V_{CC}†	MIN	TYP	MAX	UNIT
V_{IK}	$I_I = -18$ mA		2.7 V			-1.5	V
V_{OH}	$I_{OH} = -100$ μA		MIN to MAX	$V_{CC}-0.2$			V
	$I_{OH} = -6$ mA		3 V	2.4			
V_{OL}	$I_{OL} = 100$ μA		MIN to MAX			0.2	V
	$I_{OL} = 6$ mA		3 V			0.4	
I_I	$V_I = V_{CC}$ or GND		3.6 V			±1	μA
I_{OZ}	$V_O = V_{CC}$ or GND		3.6 V			±5	μA
I_{CC}	$V_I = V_{CC}$ or GND,	$I_O = 0$	3.6 V			20	μA
ΔI_{CC}	$V_{CC} = 3$ V to 3.6 V, Other inputs at V_{CC} or GND	One input at $V_{CC} - 0.6$ V,				500	μA
C_I	$V_I = V_{CC}$ or GND		3.3 V		TBD		pF
C_o	$V_O = V_{CC}$ or GND		3.3 V		TBD		pF

† For conditions shown as MIN or MAX, use the appropriate values under recommended operating conditions.

PRODUCT PREVIEW

TEXAS INSTRUMENTS
POST OFFICE BOX 655303 ● DALLAS, TEXAS 75265

Philips Components—Signetics

Document No.	
ECN No.	853-1444 00227
Date of Issue	August 20, 1990
Status	Product Specification
Advanced BiCMOS Products	

74ABT244
Octal buffer/line driver
(3-State)

FEATURES

- Octal bus interface
- 3-State buffers
- Output capability: +64 mA/-32mA
- Latch-up protection exceeds 500mA per Jedec JC40.2 Std 17
- ESD protection exceeds 2000 V per MIL STD 883C Method 3015.6 and 200 V per Machine Model

DESCRIPTION

The 74ABT244 high-performance BiCMOS device combines low static and dynamic power dissipation with high speed and high output drive.

The 74ABT244 device is an octal buffer that is ideal for driving bus lines or buffer memory address registers. The device features two Output Enables ($1\overline{OE}$, $2\overline{OE}$), each controlling four of the 3-State outputs.

QUICK REFERENCE DATA

SYMBOL	PARAMETER	CONDITIONS $T_{amb} = 25°C$; GND = 0V	TYPICAL	UNIT
t_{PLH} t_{PHL}	Propagation delay A_n to Y_n	$C_L = 50pF$; $V_{CC} = 5V$	2.9	ns
C_{IN}	Input capacitance	$V_I = 0V$ or V_{CC}	4	pF
C_{OUT}	Output capacitance	$V_I = 0V$ or V_{CC}	7	pF
I_{CCZ}	Total supply current	Outputs Disabled; $V_{CC} = 5.5V$	500	nA

ORDERING INFORMATION

PACKAGES	TEMPERATURE RANGE	ORDER CODE
20-Pin Plastic DIP	-40°C to +85°C	74ABT244N
20-Pin Plastic SOL	-40°C to +85°C	74ABT244D

PIN CONFIGURATION

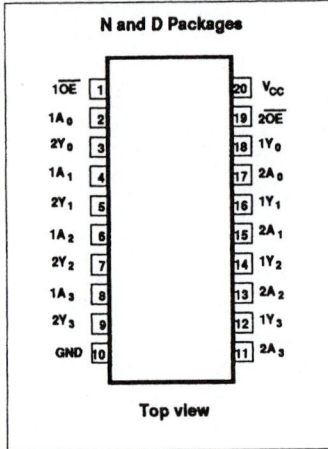

N and D Packages

Top view

LOGIC SYMBOL

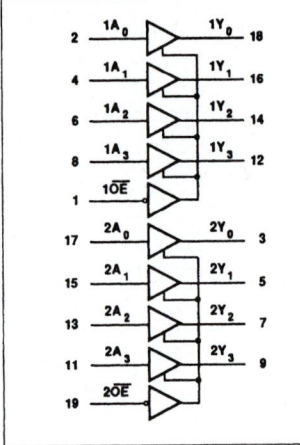

LOGIC SYMBOL (IEEE/IEC)

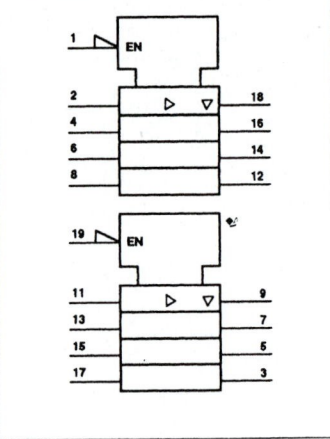

Octal buffer/line driver (3-State)

74ABT244

RECOMMENDED OPERATING CONDITIONS

SYMBOL	PARAMETER	LIMITS		UNIT
		Min	Max	
V_{CC}	DC supply voltage	4.5	5.5	V
V_I	Input voltage	0	V_{CC}	V
V_{IH}	High-level input voltage	2.0		V
V_{IL}	Input voltage		0.8	V
I_{OH}	High level output current		-32	mA
I_{OL}	Low level output current		64	mA
$\Delta t/\Delta v$	Input transition rise or fall rate	0	5	ns/V
T_{amb}	Operating free-air temperature range	-40	+85	°C

ABSOLUTE MAXIMUM RATINGS[1]

SYMBOL	PARAMETER	CONDITIONS	RATING	UNIT
V_{CC}	DC supply voltage		-0.5 to +7.0	V
I_{IK}	DC input diode current	$V_I < 0$	-18	mA
V_I	DC input voltage[2]		-1.2 to +7.0	V
I_{OK}	DC output diode current	$V_O < 0$	-50	mA
V_O	DC output voltage[2]	output in Off or High state	-0.5 to +5.5	V
I_O	DC output current	output in Low state	128	mA
T_{stg}	Storage temperature range		-65 to 150	°C

NOTES:
1. Stresses beyond those listed may cause permanent damage to the device. These are stress ratings only and functional operation of the device at these or any other conditions beyond those indicated under "recommended operating conditions" is not implied. Exposure to absolute-maximum-rated conditions for extended periods may affect device reliability.
2. The input and output voltage ratings may be exceeded if the input and output current ratings are observed.

FUNCTION TABLE

INPUTS				OUTPUT	
$1\overline{OE}$	$1A_n$	$2\overline{OE}$	$2A_n$	$1Y_n$	$2Y_n$
L	L	L	L	L	L
L	H	L	H	H	H
H	X	H	X	Z	Z

PIN DESCRIPTION

PIN NUMBER	SYMBOL	FUNCTION
2, 4, 6, 8	$1A_0$ - $1A_3$	Data inputs
17, 15, 13, 11	$2A_0$ - $2A_3$	Data inputs
18, 16, 14, 12	$1Y_0$ - $1Y_3$	Data outputs
3, 5, 7, 9	$2Y_0$ - $2Y_3$	Data outputs
1, 19	$1\overline{OE}$, $2\overline{OE}$	Output enables
10	GND	Ground (0V)
20	V_{CC}	Positive supply voltage

Octal buffer/line driver (3-State)

74ABT244

DC ELECTRICAL CHARACTERISTICS

SYMBOL	PARAMETER	TEST CONDITIONS	LIMITS					UNIT
			$T_{amb} = +25°C$			$T_{amb} = -40°C$ to +85°C		
			Min	Typ	Max	Min	Max	
V_{IK}	Input clamp voltage	$V_{CC} = 4.5V; I_{IK} = -18mA$			-1.2		-1.2	V
V_{OH}	High-level output voltage	$V_{CC} = 4.5V; I_{OH} = -3mA; V_I = V_{IL}$ or V_{IH}	2.5			2.5		V
		$V_{CC} = 5.0V; I_{OH} = -3mA; V_I = V_{IL}$ or V_{IH}	3.0			3.0		
		$V_{CC} = 4.5V; I_{OH} = -32mA; V_I = V_{IL}$ or V_{IH}	2.0	2.4		2.0		
V_{OL}	Low-level output voltage	$V_{CC} = 4.5V; I_{OL} = 64mA; V_I = V_{IL}$ or V_{IH}		0.42	0.55		0.55	V
I_I	Input leakage current	$V_{CC} = 5.5V; V_I = $ GND or 5.5V		±0.01	±1.0		±1.0	μA
I_{OZH}	3-State output High current	$V_{CC} = 5.5V; V_O = 2.7V; V_I = V_{IL}$ or V_{IH}		5.0	50		50	μA
I_{OZL}	3-State output Low current	$V_{CC} = 5.5V; V_O = 0.5V; V_I = V_{IL}$ or V_{IH}		-5.0	-50		-50	μA
I_O	Short-circuit output current[1]	$V_{CC} = 5.5V; V_O = 2.5V$	-50	-100	-180	-50	-180	mA
I_{CCH}	Quiescent supply current	$V_{CC} = 5.5V;$ Outputs High; $V_I = $ GND or V_{CC}		0.5	50		50	μA
I_{CCL}		$V_{CC} = 5.5V;$ Outputs Low; $V_I = $ GND or V_{CC}		24	30		30	mA
I_{CCZ}		$V_{CC} = 5.5V;$ Outputs 3-State; $V_I = $ GND or V_{CC}		0.5	50		50	μA
ΔI_{CC}	Additional supply current per input pin[2]	Outputs enabled, one input at 3.4V, other inputs at V_{CC} or GND; $V_{CC} = 5.5V$		0.5	1.5		1.5	mA
		Outputs 3-State, one data input at 3.4V, other inputs at V_{CC} or GND; $V_{CC} = 5.5V$		0.5	50		50	μA
		Outputs 3-State, one enable input at 3.4V, other inputs at V_{CC} or GND; $V_{CC} = 5.5V$		0.5	1.5		1.5	mA

NOTES:
1. Not more than one output should be tested at a time, and the duration of the test should not exceed one second.
2. This is the increase in supply current for each input at 3.4V.

Octal buffer/line driver (3-State)

74ABT244

AC CHARACTERISTICS

GND = 0V; t_R = t_F = 2.5ns; C_L = 50pF, R_L = 500Ω

SYMBOL	PARAMETER	WAVEFORM	LIMITS					UNIT
			T_{amb} = +25°C V_{CC} = +5.0V			T_{amb} = -40°C to +85°C V_{CC} = +5.0V ±0.5V		
			Min	Typ	Max	Min	Max	
t_{PLH} t_{PHL}	Propagation delay A_n to Y_n	1	1.0 1.0	2.6 2.9	4.1 4.2	1.0 1.0	4.6 4.6	ns
t_{PZH} t_{PZL}	Output enable time to High and Low level	2	1.1 2.1	3.1 4.1	4.6 5.6	1.1 2.1	5.1 6.1	ns
t_{PHZ} t_{PLZ}	Output disable time from High and Low level	2	2.1 1.7	4.1 3.7	5.6 5.2	2.1 1.7	6.6 5.7	ns

AC WAVEFORMS

(V_M = 1.5V, V_{IN} = GND to 3.0V)

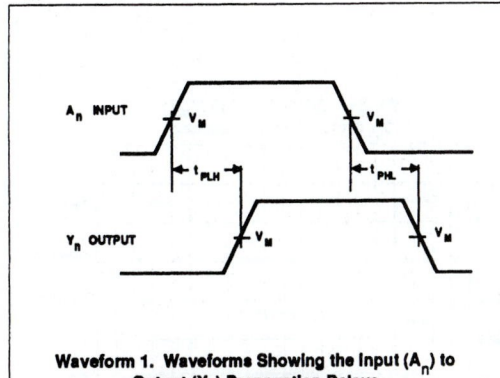

Waveform 1. Waveforms Showing the Input (A_n) to Output (Y_n) Propagation Delays

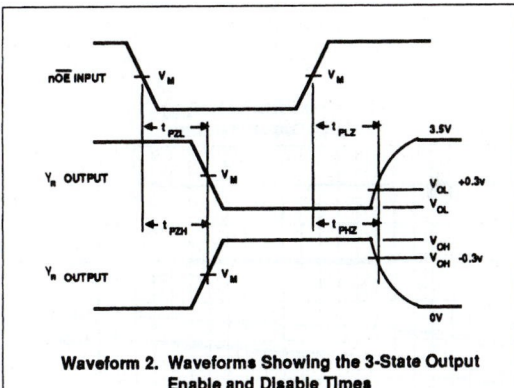

Waveform 2. Waveforms Showing the 3-State Output Enable and Disable Times

TEST CIRCUIT AND WAVEFORMS

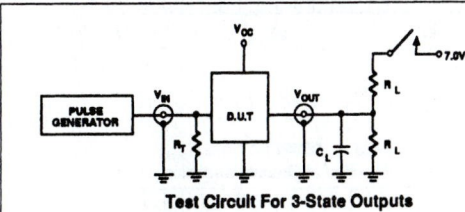

Test Circuit For 3-State Outputs

SWITCH POSITION

TEST	SWITCH
t_{PLZ}	closed
t_{PZL}	closed
All other	open

DEFINITIONS

R_L = Load resistor; see AC CHARACTERISTICS for value.

C_L = Load capacitance includes jig and probe capacitance; see AC CHARACTERISTICS for value.

R_T = Termination resistance should be equal to Z_{OUT} of pulse generators.

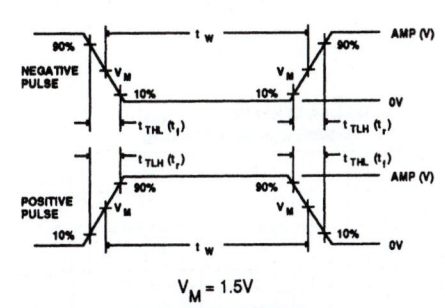

V_M = 1.5V
Input Pulse Definition

FAMILY	INPUT PULSE REQUIREMENTS				
	Amplitude	Rep. Rate	t_W	t_R	t_F
74ABT	3.0V	1MHz	500ns	2.5ns	2.5ns

Octal buffer/line driver (3-State)

74ABT244

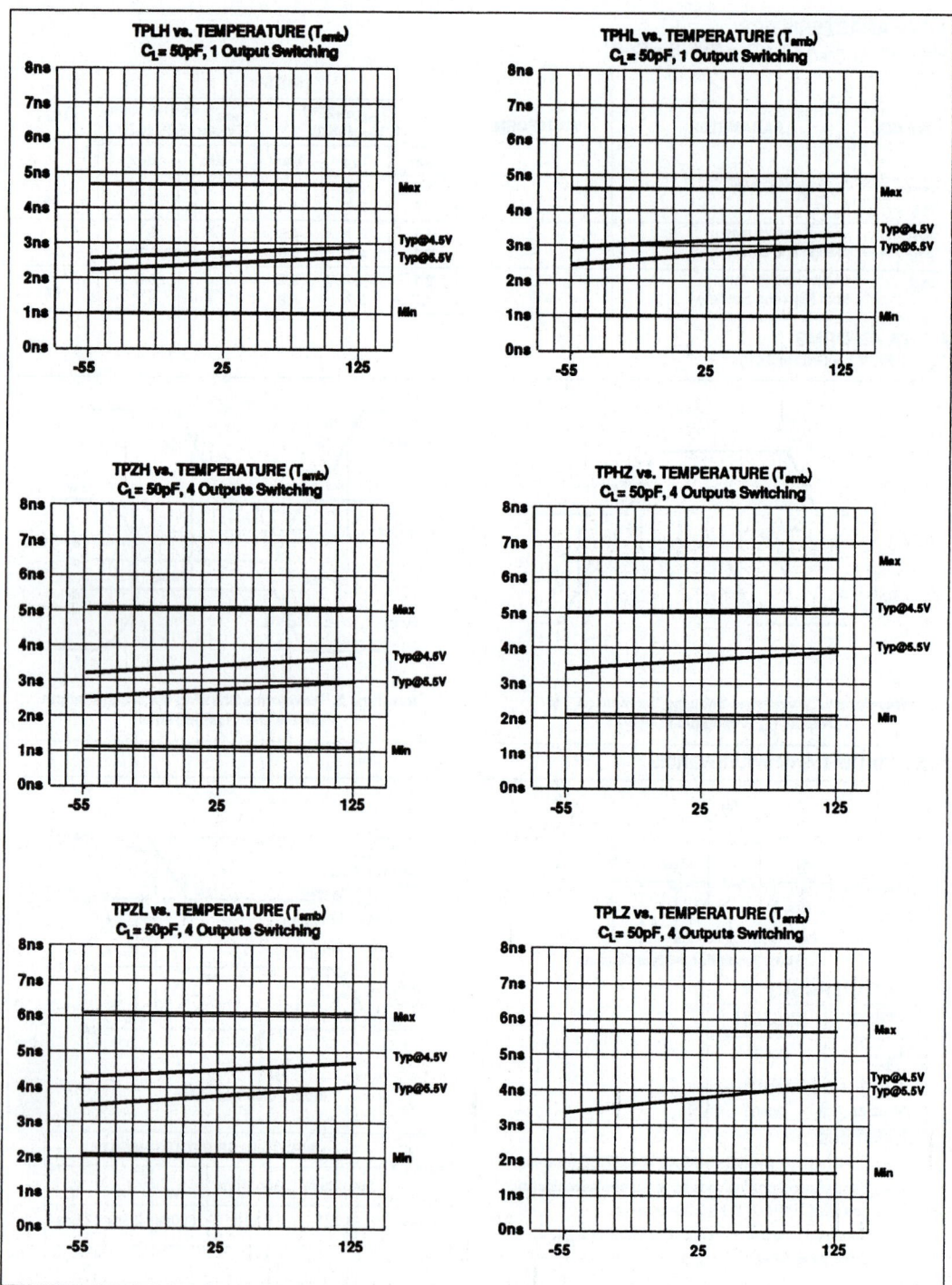

Octal buffer/line driver (3-State)

74ABT244

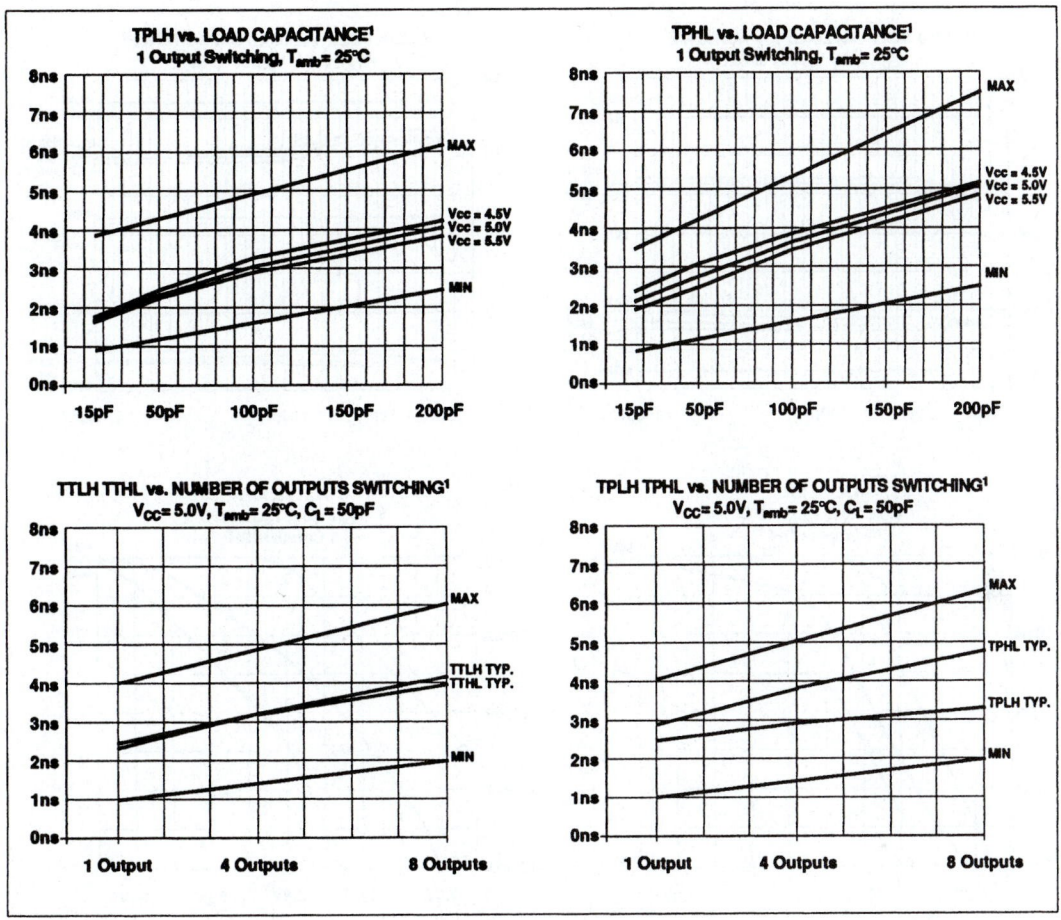

NOTES:
1. MIN and MAX lines are design characteristics and are not necessarily guaranteed by test.

Octal buffer/line driver (3-State)

74ABT244

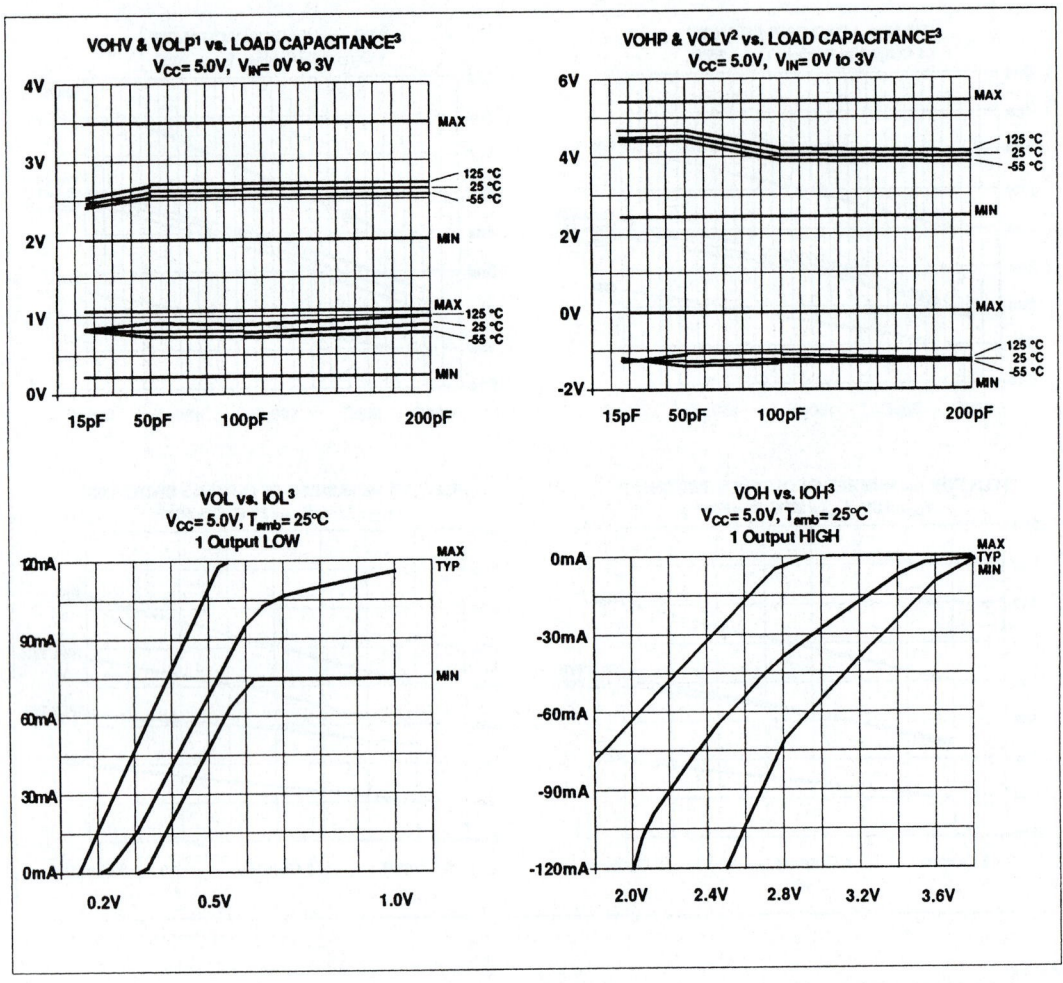

NOTES:
1. VOHV is defined as the minimum (valley) voltage induced on a quiescent high-level output during switching of other outputs. VOLP is defined as the maximum (peak) voltage induced on a quiescent low-level output during switching of other outputs.
2. VOHP is defined as the maximum (peak) voltage induced on a quiescent high-level output during switching of other outputs. VOLV is defined as the minimum (valley) voltage induced on a quiescent low-level output during switching of other outputs.
3. MIN and MAX lines are design and process characteristics. They are not necessarily guaranteed by test.

DC ELECTRICAL CHARACTERISTICS (Over recommended operating free-air temperature range unless otherwise noted.)

PARAMETER	TEST CONDITIONS		54/7400 Min	Typ²	Max	54/74LS00 Min	Typ²	Max	54/74S00 Min	Typ²	Max	UNIT
V_{OH} HIGH-level output voltage	V_{CC}=MIN, V_{IH}=MIN, V_{IL}=MAX, I_{OH}=MAX	Mil	2.4	3.4		2.5	3.4		2.7	3.4		V
		Com'l	2.4	3.4							0.5⁴	V
V_{OL} LOW-level output voltage	V_{CC}=MIN, V_{IH}=MIN	I_{OL}=MAX		0.2	0.4		0.25	0.4		0.25	0.5	V
		I_{OL}=4mA 74LS		0.2	0.4		0.35	0.5				V
V_{IK} Input clamp voltage	V_{CC}=MIN, I_I=I_{IK}				−1.5			−1.5			−1.2	V
I_I Input current at maximum input voltage	V_{CC}=MAX	V_I=5.5V			1.0						1.0	mA
		V_I=7.0V						0.1				mA
I_{IH} HIGH-level input current	V_{CC}=MAX	V_I=2.4V			40							µA
		V_I=2.7V						20			50	µA
I_{IL} LOW-level input current	V_{CC}=MAX	V_I=0.4V			−1.6			−0.4				mA
		V_I=0.5V									−2.0	mA
I_{OS} Short-circuit output current³	V_{CC}=MAX	Mil	−20		−55	−20		−100	−40		−100	mA
		Com'l	−18		−55	−20		−100	−40		−100	mA
I_{CC} Supply current (total)	V_{CC}=MAX	I_{CCH} Outputs HIGH		4	8		0.8	1.6		10	16	mA
		I_{CCL} Outputs LOW		12	22		2.4	4.4		20	36	mA

NOTES
1. For conditions shown as MIN or MAX, use the appropriate value specified under recommended operating conditions for the applicable type.
2. All typical values are at V_{CC}=5V, T_A=25°C.
3. I_{OS} is tested with V_{OUT}=0V and V_{CC}=V_{CC} MAX + 0.5V. Not more than one output should be shorted at a time and duration of the short circuit should not exceed one second.
4. V_{OL}=+0.45V MAX for 54S at T_A=+125°C only

AC WAVEFORM

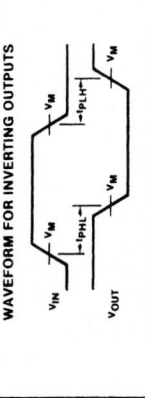

WAVEFORM FOR INVERTING OUTPUTS

V_M=1.3V for 54LS/74LS, V_M=1.5V for all other TTL families

Waveform 1

AC CHARACTERISTICS T_A=25°C, V_{CC}=5.0V

PARAMETER	TEST CONDITIONS	5474 C_L=15pF, R_L=400Ω Min	Max	5474LS C_L=15pF, R_L=2kΩ Min	Max	5474S C_L=15pF, R_L=280Ω Min	Max	UNIT
t_{PLH}	Propagation delay	Waveform 1	22		15		4.5	ns
t_{PHL}			15		15		5.0	ns

Quad Two-Input NAND Gate

TYPE	TYPICAL PROPAGATION DELAY	TYPICAL SUPPLY CURRENT (Total)
7400	9ns	8mA
74LS00	9.5ns	1.6mA
74S00	3ns	15mA

ORDERING CODE

PACKAGES	COMMERCIAL RANGES V_{CC}=5V ±5%; T_A=0°C to +70°C	MILITARY RANGES V_{CC}=5V ±10%; T_A=−55°C to +125°C
Plastic DIP	N7400N • N74LS00N N74S00N	
Plastic SO	N74LS00D N74S00D	
Ceramic DIP	S5400F • S54S00F	S54LS00F
Flatpack	S5400W	S54S00W S54LS00W
LLCC		S54LS00G

INPUT AND OUTPUT LOADING AND FAN-OUT TABLE

PINS	DESCRIPTION	5474	5474S	5474LS
A, B	Inputs	1ul	1Sul	1LSul
Y	Output	10ul	10Sul	10LSul

NOTE
Where a 54/74 unit load (ul) is understood to be 40µA I_{IH} and −1.6mA I_{IL}, a 54/74S unit load (Sul) is 50µA I_{IH} and −2.0mA I_{IL}, and 54/74LS unit load (LSul) is 20µA I_{IH} and −0.4mA I_{IL}.

FUNCTION TABLE

INPUTS A	B	OUTPUT Y
L	L	H
L	H	H
H	L	H
H	H	L

H = HIGH voltage level
L = LOW voltage level

LOGIC SYMBOL

LOGIC SYMBOL (IEEE/IEC)

PIN CONFIGURATION

GATES

54/7400, LS00, S00

ABSOLUTE MAXIMUM RATINGS (Over operating free-air temperature range unless otherwise noted.)

	PARAMETER	54	54LS	54S	74	74LS	74S	UNIT
V_{CC}	Supply voltage	7.0	7.0	7.0	7.0	7.0	7.0	V
V_{IN}	Input voltage	-0.5 to +5.5	-0.5 to +7.0	-0.5 to +5.5	-0.5 to +5.5	-0.5 to +7.0	-0.5 to +5.5	V
I_{IN}	Input current	-30 to +5	-30 to +1	-30 to +5	-30 to +5	-30 to +1	-30 to +5	mA
V_{OUT}	Voltage applied to output in HIGH output state	-0.5 to +V_{CC}	-0.5 to +V_{CC}	-0.5 to +V_{CC}	-0.5 to +V_{CC}	-0.5 to +V_{CC}	-0.5 to +V_{CC}	V
T_A	Operating free-air temperature range	-55 to +125			0 to 70			°C

RECOMMENDED OPERATING CONDITIONS

	PARAMETER		54/74 Min	Nom	Max	54/74LS Min	Nom	Max	54/74S Min	Nom	Max	UNIT
V_{CC}	Supply voltage	Mil	4.5	5.0	5.5	4.5	5.0	5.5	4.5	5.0	5.5	V
		Com'l	4.75	5.0	5.25	4.75	5.0	5.25	4.75	5.0	5.25	V
V_{IH}	HIGH-level input voltage		2.0			2.0			2.0			V
V_{IL}	LOW-level input voltage	Mil			+0.8			+0.7			+0.8	V
		Com'l			+0.8			+0.8			+0.8	V
I_{IK}	Input clamp current				-12			-18			-18	mA
I_{OH}	HIGH-level output current	Mil			-400			-400			-1000	µA
		Com'l			-400			-400			-1000	µA
I_{OL}	LOW-level output current	Mil			16			8			20	mA
		Com'l			16			8			20	mA
T_A	Operating free-air temperature	Mil	-55		+125	-55		+125	-55		+125	°C
		Com'l	0		70	0		70	0		70	°C

NOTE
V_{IL} = +0.7V MAX for 54S at T_A = +125°C only

TEST CIRCUITS AND WAVEFORMS

TEST CIRCUIT FOR 54/74 TOTEM-POLE OUTPUTS

INPUT PULSE DEFINITIONS

NEGATIVE PULSE
POSITIVE PULSE

V_M = 1.3V for 54LS/74LS, V_M = 1.5V for all other TTL families

INPUT PULSE REQUIREMENTS

FAMILY	Amplitude	Rep. Rate	Pulse Width	t_{TLH}	t_{THL}
54/74	3.0V	1MHz	500ns	7ns	7ns
54LS/74LS	3.0V	1MHz	500ns	15ns	6ns
54S/74S	3.0V	1MHz	500ns	2.5ns	2.5ns

DEFINITIONS

R_L = Load resistor to V_{CC}; see AC CHARACTERISTICS for value
C_L = Load capacitance includes jig and probe capacitance; see AC CHARACTERIS-TICS for value
R_T = Termination resistance should be equal to Z_{OUT} of Pulse Generators
D = Diodes are 1N916, 1N3064, or equivalent
t_{TLH}, t_{THL} Values should be less than or equal to the table entries.

54/7414, LS14

SCHMITT TRIGGERS

Hex Inverter Schmitt Trigger

TYPE	TYPICAL PROPAGATION DELAY	TYPICAL SUPPLY CURRENT (Total)
7414	15ns	31mA
74LS14	15ns	10mA

ORDERING CODE

PACKAGES	COMMERCIAL RANGES V_{CC}=5V ±5%; T_A=0°C to +70°C		MILITARY RANGES V_{CC}=5V ±10%; T_A= -55°C to +125°C	
Plastic DIP	N7414N	N74LS14N		
Plastic SO		N74LS14D		
Ceramic DIP			S5414F	S54LS14F
Flatpack			S5414W	S54LS14W

DESCRIPTION

The '14 contains six logic inverters which accept standard TTL input signals and provide standard TTL output levels. They are capable of transforming slowly changing input signals into sharply defined, jitter-free output signals. In addition, they have greater noise margin than conventional inverters.

Each circuit contains a Schmitt trigger followed by a Darlington level shifter and a phase splitter driving a TTL totem-pole output. The Schmitt trigger uses positive feedback to effectively speed-up slow input transition, and provide different input threshold voltages for positive and negative-going transitions. This hysteresis between the positive-going and negative-going input thresholds (typically 800mV) is determined internally by resistor ratios and is essentially insensitive to temperature and supply voltage variations.

FUNCTION TABLE

INPUT	OUTPUT
A	Y
0	1
1	0

INPUT AND OUTPUT LOADING AND FAN-OUT TABLE

PINS	DESCRIPTION	54/74	54/74LS
A	Inputs	1ul	1LSul
Y	Output	10ul	10LSul

NOTE
Where a 54/74 unit load (ul) is understood to be 40µA I_{IH} and -1.6mA I_{IL}, and a 54/74LS unit load (LSul) is 20µA I_{IH} and -0.4mA I_{IL}.

PIN CONFIGURATION

LOGIC SYMBOL

LOGIC SYMBOL (IEEE/IEC)

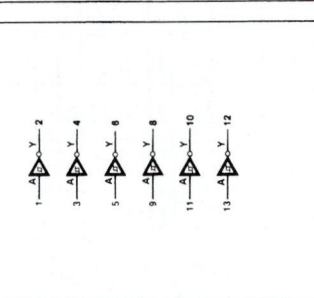

DC ELECTRICAL CHARACTERISTICS (Over recommended operating free-air temperature range unless otherwise noted.)

PARAMETER		TEST CONDITIONS[1]	54/7414 Min	54/7414 Typ[2]	54/7414 Max	54/74LS14 Min	54/74LS14 Typ[2]	54/74LS14 Max	UNIT
V_{T+} Positive-going threshold		$V_{CC} = 5.0V$	1.5	1.7	2.0	1.4	1.6	1.9	V
V_{T-} Negative-going threshold		$V_{CC} = 5.0V$	0.6	0.9	1.1	0.5	0.8	1.0	V
ΔV_T Hysteresis $(V_{T+} - V_{T-})$		$V_{CC} = 5.0V$	0.4	0.8		0.4	0.8		V
V_{OH} HIGH-level output voltage	Mil	$V_{CC} = MIN, V_I = V_{T-MIN}, I_{OH} = MAX$	2.4	3.4		2.5	3.4		V
	Com'l		2.4	3.4		2.7	3.4		V
V_{OL} LOW-level output voltage	Mil	$V_{CC} = MIN, V_I = V_{T+MAX}$, $I_{OL} = MAX$		0.2	0.4		0.25	0.4	V
	Com'l			0.2	0.4		0.35	0.5	V
	74LS	$I_{OL} = 4mA$					0.25	0.4	V
V_{IK} Input clamp voltage		$V_{CC} = MIN, I_I = I_{IK}$			-1.5			-1.5	V
I_{T+} Input current at positive-going threshold		$V_{CC} = 5.0V, V_I = V_{T+}$		-0.43			-0.14		mA
I_{T-} Input current at negative-going threshold		$V_{CC} = 5.0V, V_I = V_{T-}$		-0.56			-0.18		mA
I_I Input current at maximum input voltage		$V_{CC} = MAX$, $V_I = 5.5V$ / $V_I = 7.0V$			1.0			0.1	mA
I_{IH} HIGH-level input current		$V_{CC} = MAX$, $V_I = 2.4V$ / $V_I = 2.7V$			40			20	μA
I_{IL} LOW-level input current		$V_{CC} = MAX, V_I = 0.4V$			-1.2			-0.4	mA
I_{OS} Short-circuit output current[3]	Mil	$V_{CC} = MAX$	-20		-55	-20		-100	mA
	Com'l		-18		-55	-20		-100	mA
I_{CC} Supply current (total)	I_{CCH} Outputs HIGH	$V_{CC} = MAX$		22	36		8.6	16	mA
	I_{CCL} Outputs LOW			39	60		12	21	mA

AC WAVEFORMS

Waveform 1

$V_{ref(H)} = 1.7V$ for '14 $V_{ref(H)} = 1.6V$ for LS14
$V_{ref(L)} = 0.9V$ for '14 $V_{ref(L)} = 0.8V$ for LS14
$V_M = 1.5V$ for 54/74 $V_M = 1.3V$ for 54LS/74LS

Labels: $V_{ref(H)}$, $V_{ref(L)}$, V_M, t_{PLH}, t_{PHL}

NOTES
1. For conditions shown as MIN or MAX, use the appropriate value specified under recommended operating conditions for the applicable type.
2. All typical values are at $V_{CC} = 5V$, $T_A = 25°C$.
3. I_{OS} is tested with $V_{OUT} = +0.5V$ and $V_{CC} = V_{CC}$ MAX + 0.5V. Not more than one output should be shorted at a time and duration of the short circuit should not exceed one second.

AC CHARACTERISTICS $T_A = 25°C$, $V_{CC} = 5.0V$

PARAMETER	TEST CONDITIONS	54/74 ($C_L = 15pF$, $R_L = 400\Omega$) Min	54/74 Max	54LS/74LS ($C_L = 15pF$, $R_L = 2k\Omega$) Min	54LS/74LS Max	UNIT
t_{PLH} / t_{PHL} Propagation delay	Waveform 1		22 / 22		22 / 22	ns

---

SCHMITT TRIGGERS   54/7414, LS14

ABSOLUTE MAXIMUM RATINGS (Over operating free-air temperature range unless otherwise noted.)

PARAMETER	54	74	54LS	74LS	UNIT
V_{CC} Supply voltage	7.0	7.0	7.0	7.0	V
V_{IN} Input voltage	-0.5 to +5.5	-0.5 to +5.5	-0.5 to +7.0	-0.5 to +7.0	V
I_{IN} Input current	-30 to +5	-30 to +5	-30 to +1	-30 to +1	mA
V_{OUT} Voltage applied to output in HIGH output state	-0.5 to +V_{CC}	-0.5 to +V_{CC}	-0.5 to +V_{CC}	-0.5 to +V_{CC}	V
T_A Operating free-air temperature range	-55 to +125	0 to 70			°C

RECOMMENDED OPERATING CONDITIONS

PARAMETER		54/74 Min	54/74 Nom	54/74 Max	54/74LS Min	54/74LS Nom	54/74LS Max	UNIT
V_{CC} Supply voltage	Mil	4.5	5.0	5.5	4.5	5.0	5.5	V
	Com'l	4.75	5.0	5.25	4.75	5.0	5.25	V
I_{IK} Input clamp current				-12			-18	mA
I_{OH} HIGH-level output current				-800			-400	μA
I_{OL} LOW-level output current	Mil			16			4	mA
	Com'l			16			8	mA
T_A Operating free-air temperature	Mil	-55		+125	-55		+125	°C
	Com'l	0		70	0		70	°C

TEST CIRCUITS AND WAVEFORMS

TEST CIRCUIT FOR 54/74 TOTEM-POLE OUTPUTS

INPUT PULSE DEFINITIONS

NEGATIVE PULSE / POSITIVE PULSE

$V_M = 1.3V$ for 54LS/74LS, $V_M = 1.5V$ for all other TTL families

FAMILY	Amplitude	Rep. Rate	Pulse Width	t_{TLH}	t_{THL}
54/74	3.0V	1MHz	500ns	7ns	7ns
54LS/74LS	3.0V	1MHz	500ns	15ns	6ns
54S/74S	3.0V	1MHz	500ns	2.5ns	2.5ns

DEFINITIONS
R_L = Load resistor to V_{CC}; see AC CHARACTERISTICS for value.
C_L = Load capacitance includes jig and probe capacitance; see AC CHARACTERISTICS for value.
R_T = Termination resistance should be equal to Z_{OUT} of Pulse Generators.
D = Diodes are 1N916, 1N3064, or equivalent.
t_{TLH}, t_{THL} Values should be less than or equal to the table entries.

MULTIVIBRATOR

Monostable Multivibrator

- Very good pulse width stability
- Virtually immune to temperature and voltage variations
- Schmitt trigger input for slow input transitions
- Internal timing resistor provided

DESCRIPTION

These multivibrators feature dual active LOW going edge inputs and a single active HIGH going edge input which can be used as an active HIGH enable input. Complementary output pulses are provided.

Pulse triggering occurs at a particular voltage level and is not directly related to the transition time of the input pulse. Schmitt-trigger input circuitry (TTL hysteresis) for the B input allows jitter-free triggering from inputs with transition rates as slow as 1 volt/second, providing the circuit with an excellent noise immunity of typically 1.2 volts. A high immunity to V_{CC} noise of typically 1.5 volts is also provided by internal latching circuitry. Once fired, the outputs are independent of further transitions of the inputs and are a function only of the timing components. Input pulses may be of any duration relative to the output pulse. Output pulse length may be varied from 20 nanoseconds to 28 seconds by choosing appropriate timing components. With no external timing components (i.e. R_{int} connected to V_{CC}, C_{ext} and R_{ext}/C_{ext} open), an output pulse of typically 30 or 35 nanoseconds is achieved which may be used as a dc triggered reset signal. Output rise and fall times are TTL compatible and independent of pulse length.

Pulse width stability is achieved through internal compensation and is virtually independent of V_{CC} and temperature. In most applications, pulse stability will only be limited by the accuracy of external timing components.

Jitter-free operation is maintained over the full temperature and V_{CC} ranges for more than six decades of timing capacitance (10pF to 10μF) and more than one decade of timing resistance (2kΩ to 30kΩ)

for the 54121 and 2kΩ to 40kΩ for the 74121). Throughout these ranges, pulse width is defined by the relationship: (see Figure 1)

$$t_W(out) = C_{ext} R_{ext} \ln 2$$
$$t_W(out) = 0.7 C_{ext} R_{ext}$$

In circuits where pulse cutoff is not critical, timing capacitance up to 1000μF and timing resistance as low as 1.4kΩ may be used.

TYPE	TYPICAL PROPAGATION DELAY	TYPICAL SUPPLY CURRENT (Total)
74121	43ns	18mA

ORDERING CODE

PACKAGES	COMMERCIAL RANGES $V_{CC} = 5V \pm 5\%$; $T_A = 0°C$ to $+70°C$	MILITARY RANGES $V_{CC} = 5V \pm 10\%$; $T_A = -55°C$ to $+125°C$
Plastic DIP	N74121N	
Plastic SO		
Ceramic DIP	N74121D	
Flatpack		S54121F
		S54121W

FUNCTION TABLE

INPUTS			OUTPUTS	
$\overline{A}_1$	$\overline{A}_2$	B	Q	$\overline{Q}$
L	X	H	L	H
X	L	H	L	H
X	X	L	L	H
H	↓	H	⊓	⊔
↓	H	H	⊓	⊔
↓	↓	H	⊓	⊔
L	X	↑	⊓	⊔
X	L	↑	⊓	⊔

H = HIGH voltage level
L = LOW voltage level
X = Don't care
↑ = LOW-to-HIGH transition
↓ = HIGH-to-LOW transition

INPUT AND OUTPUT LOADING AND FAN-OUT TABLE

PINS	DESCRIPTION	54/74
$\overline{A}_1$, $\overline{A}_2$	Inputs	1ul
B	Input	2ul
Q, $\overline{Q}$	Outputs	10ul

NOTE
A 54/74 unit load (ul) is understood to be 40μA I_{IH} and −1.6mA I_{IL}.

PIN CONFIGURATION

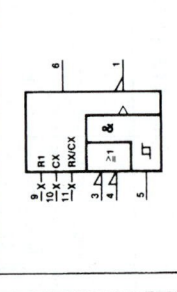

LOGIC SYMBOL

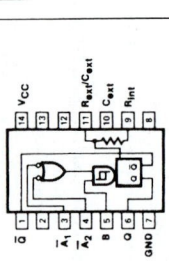

LOGIC SYMBOL (IEEE/IEC)

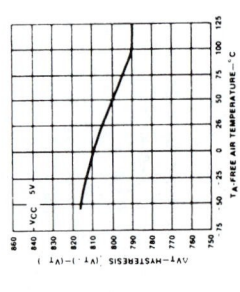

SCHMITT TRIGGERS

TYPICAL CHARACTERISTICS

(54/74, 54LS/74LS)
VIN vs VOUT
TRANSFER FUNCTION

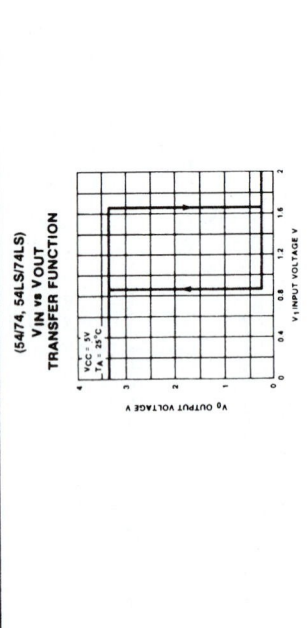

(54LS/74LS)
THRESHOLD VOLTAGE AND HYSTERESIS vs POWER SUPPLY VOLTAGE

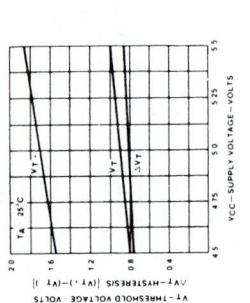

(54LS/74LS)
THRESHOLD VOLTAGE AND HYSTERESIS vs AMBIENT TEMPERATURE

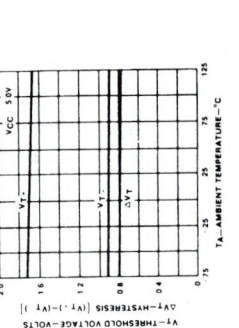

(54/74)
THRESHOLD VOLTAGE AND HYSTERESIS vs POWER SUPPLY VOLTAGE

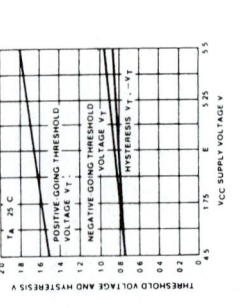

(54/74)
HYSTERESIS vs TEMPERATURE

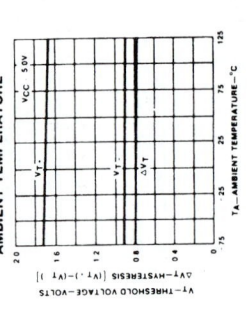

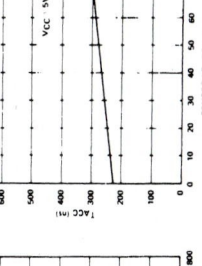

intel®

2716*
16K (2K × 8) UV ERASABLE PROM

- **Fast Access Time**
  - 350 ns Max. 2716-1
  - 390 ns Max. 2716-2
  - 450 ns Max. 2716
  - 490 ns Max. 2716-5
  - 650 ns Max. 2716-6

- **Single + 5V Power Supply**

- **Low Power Dissipation**
  - 525 mW Max. Active Power
  - 132 mW Max. Standby Power

- **Pin Compatible to Intel® 2732 EPROM**

- **Simple Programming Requirements**
  - Single Location Programming
  - Programs with One 50 ms Pulse

- **Inputs and Outputs TTL Compatible during Read and Program**

- **Completely Static**

The Intel® 2716 is a 16,384 bit ultraviolet erasable and electrically programmable read-only memory (EPROM). The 2716 operates from a single 5-volt power supply, has a static standby mode, and features fast single address location programming. It makes designing with EPROMs faster, easier and more economical.

The 2716, with its single 5-volt supply and with an access time up to 350 ns, is ideal for use with the newer high performance +5V microprocessors such as Intel's 8085 and 8086. A selected 2716-5 and 2716-6 is available for slower speed applications. The 2716 is also the first EPROM with a static standby mode which reduces the power dissipation without increasing access time. The maximum active power dissipation is 525 mW while the maximum standby power dissipation is only 132 mW, a 75% savings.

The 2716 has the simplest and fastest method yet devised for programming EPROMs — single pulse TTL level programming. No need for high voltage pulsing because all programming controls are handled by TTL signals. Program any location at any time—either individually, sequentially or at random, with the 2716's single address location programming. Total programming time for all 16,384 bits is only 100 seconds.

PIN CONFIGURATION

2716 / 2732†

†Refer to 2732 data sheet for specifications

PIN NAMES

A0-A10	ADDRESSES
CE/PGM	CHIP ENABLE/PROGRAM
OE	OUTPUT ENABLE
O0-O7	OUTPUTS

MODE SELECTION

PINS / MODE	CE/PGM (18)	OE (20)	Vpp (21)	Vcc (24)	OUTPUTS (9-11, 13-17)
Read	VIL	VIL	+5	+5	Dout
Standby	VIH	Don't Care	+5	+5	High Z
Program	Pulsed VIL to VIH	VIH	+25	+5	Din
Program Verify	VIL	VIL	+25	+5	Dout
Program Inhibit	VIL	VIH	+25	+5	High Z

BLOCK DIAGRAM

2716

PROGRAMMING
The programming specifications are described in the Data Catalog PROM/ROM Programming Instructions Section.

Absolute Maximum Ratings*

Temperature Under Bias	-10°C to +80°C
Storage Temperature	-65°C to +125°C
All Input or Output Voltages with Respect to Ground	+6V to -0.3V
Vpp Supply Voltage with Respect to Ground During Program	+26.5V to -0.3V

DC and AC Operating Conditions During Read

	2716	2716-1	2716-2	2716-5	2716-6
Temperature Range	0°C – 70°C	0°C – 70°C	0°C – 70°C	0°C – 70°C	0°C – 70°C
Vcc Power Supply[1,2]	5V ±5%	5V ±10%	5V ±5%	5V ±5%	5V ±5%
Vpp Power Supply[2]	Vcc	Vcc	Vcc	Vcc	Vcc

READ OPERATION
D.C. and Operating Characteristics

Symbol	Parameter	Limits Min.	Typ.[3]	Max.	Unit	Conditions
I_{LI}	Input Load Current			10	μA	$V_{IN} = 5.25V$
I_{LO}	Output Leakage Current			10	μA	$V_{OUT} = 5.25V$
I_{PP1}[2]	Vpp Current			5	mA	$V_{PP} = 5.25V$
I_{CC1}[2]	Vcc Current (Standby)		10	25	mA	$\overline{CE} = V_{IH}, \overline{OE} = V_{IL}$
I_{CC2}[2]	Vcc Current (Active)		57	100	mA	$\overline{OE} = \overline{CE} = V_{IL}$
V_{IL}	Input Low Voltage	-0.1		0.8	V	
V_{IH}	Input High Voltage	2.0		$V_{CC}+1$	V	
V_{OL}	Output Low Voltage			0.45	V	$I_{OL} = 2.1$ mA
V_{OH}	Output High Voltage	2.4			V	$I_{OH} = -400$ μA

NOTES:
1. Vcc must be applied simultaneously or before Vpp and removed simultaneously or after Vpp.
2. Vpp may be connected directly to Vcc except during programming. The supply current would then be the sum of Icc and Ipp1.
3. Typical values are for TA = 25°C and nominal supply voltages.
4. This parameter is only sampled and is not 100% tested.

Typical Characteristics

Icc CURRENT vs. TEMPERATURE

ACCESS TIME vs. CAPACITANCE

ACCESS TIME vs. TEMPERATURE

ERASURE CHARACTERISTICS

The erasure characteristics of the 2716 are such that erasure begins to occur when exposed to light with wavelengths shorter than approximately 4000 Angstroms (Å). It should be noted that sunlight and certain types of fluorescent lamps have wavelengths in the 3000—4000Å range. Data show that constant exposure to room level fluorescent lighting could erase the typical 2716 in approximately 3 years, while it would take approximatley 1 week to cause erasure when exposed to direct sunlight. If the 2716 is to be exposed to these types of lighting conditions for extended periods of time, opaque labels are available from Intel which should be placed over the 2716 window to prevent unintentional erasure.

The recommended erasure procedure (see Data Catalog PROM/ROM Programming Instruction Section) for the 2716 is exposure to shortwave ultraviolet light which has a wavelength of 2537 Angstroms (Å). The integrated dose (i.e., UV intensity X exposure time) for erasure should be a minimum of 15 W-sec/cm². The erasure time with this dosage is approximately 15 to 20 minutes using an ultraviolet lamp with a 12000 μW/cm² power rating. The 2716 should be placed within 1 inch of the lamp tubes during erasure. Some lamps have a filter on their tubes which should be removed before erasure.

DEVICE OPERATION

The five modes of operation of the 2716 are listed in Table I. It should be noted that all inputs for the five modes are at TTL levels. The power supplies required are a +5V Vcc and a Vpp. The Vpp power supply must be at 25V during the three programming modes, and must be at 5V in the other two modes.

READ MODE

The 2716 has two control functions, both of which must be logically satisfied in order to obtain data at the outputs. Chip Enable (CE) is the power control and should be used for device selection. Output Enable (OE) is the output control and should be used to gate data to the output pins, independent of device selection. Assuming that addresses are stable, address access time (t_{ACC}) is equal to the delay from CE to output (t_{CE}). Data is available at the outputs 120 ns (t_{OE}) after the falling edge of OE, assuming that CE has been low and addresses have been stable for at least $t_{ACC} - t_{OE}$.

STANDBY MODE

The 2716 has a standby mode which reduces the active power dissipation by 75%, from 525 mW to 132 mW. The 2716 is placed in the standby mode by applying a TTL high signal to the CE input. When in standby mode, the outputs are in a high impedence state, independent of the OE input.

OUTPUT OR-TIEING

Because 2716's are usually used in larger memory arrays, Intel has provided a 2 line control function that accomodates this use of multiple memory connections. The two line control function allows for:

a) the lowest possible memory power dissipation, and
b) complete assurance that output bus contention will not occur.

To most efficiently use these two control lines, it is recommended that CE (pin 18) be decoded and used as the primary device selecting function, while OE (pin 20) be made a common connection to all devices in the array and connected to the READ line from the system control bus.

This assures that all deselected memory devices are in their low power standby mode and that the output pins are only active when data is desired from a particular memory device.

PROGRAMMING (See Programming Instruction Section for Waveforms.)

Initially, and after each erasure, all bits of the 2716 are in the "1" state. Data is introduced by selectively programming "0's" into the desired bit locations. Although only "0's" will be programmed, both "1's" and "0's" can be presented in the data word. The only way to change a "0" to a "1" is by ultraviolet light erasure.

The 2716 is in the programming mode when the Vpp power supply is at 25V and OE and CE is at V_{IH}. The data to be programmed is applied 8 bits in parallel to the data output pins. The levels required for the address and data inputs are TTL.

When the address and data are stable, a 50 msec, active high, TTL program pulse is applied to the CE/PGM input. A program pulse must be applied at each address location to be programmed. You can program any location at any time – either individually, sequentially, or at random. The program pulse has a maximum width of 55 msec. The 2716 must not be programmed with a DC signal applied to the CE/PGM input.

Programming of multiple 2716s in parallel with the same data can be easily accomplished due to the simplicity of the programming requirements. Like inputs of the paralleled 2716s may be connected together when they are programmed with the same data. A high level TTL pulse applied to the CE/PGM input programs the paralleled 2716s.

PROGRAM INHIBIT

Programming of multiple 2716s in parallel with different data is also easily accomplished. Except for CE/PGM, all like inputs (including OE) of the parallel 2716s may be common. A TTL level program pulse applied to a 2716's CE/PGM input with Vpp at 25V will program that 2716. A low level CE/PGM input inhibits the other 2716 from being programmed.

PROGRAM VERIFY

A verify should be performed on the programmed bits to determine that they were correctly programmed. The verify may be performed with Vpp at 25V. Except during programming and program verify, Vpp must be at 5V.

A.C. Characteristics

Symbol	Parameter	Limits (ns) 2716 Min	2716 Max	2716-1 Min	2716-1 Max	2716-2 Min	2716-2 Max	2716-5 Min	2716-5 Max	2716-6 Min	2716-6 Max	Test Conditions
t_{ACC}	Address to Output Delay		450		350		390		450		450	CE=OE=V_{IL}
t_{CE}	CE to Output Delay		450		350		390		490		650	OE=V_{IL}
t_{OE}	Output Enable to Output Delay		120		120		120		160		200	CE=V_{IL}
t_{DF}	Output Enable High to Output Float	0	100	0	100	0	100	0	100	0	100	CE=V_{IL}
t_{OH}	Output Hold from Addresses, CE or OE Whichever Occurred First	0		0		0		0		0		CE=OE=V_{IL}

Capacitance [4] $T_A = 25°C$, $f = 1$ MHz

Symbol	Parameter	Typ.	Max.	Unit	Conditions
C_{IN}	Input Capacitance	4	6	pF	V_{IN}=0V
C_{OUT}	Output Capacitance	8	12	pF	V_{OUT}=0V

A.C. Test Conditions:

Output Load: 1 TTL gate and $C_L = 100$ pF
Input Rise and Fall Times: ≤20 ns
Input Pulse Levels: 0.8V to 2.2V
Timing Measurement Reference Level:
  Inputs   1V and 2V
  Outputs 0.8V and 2V

A.C. Waveforms [1]

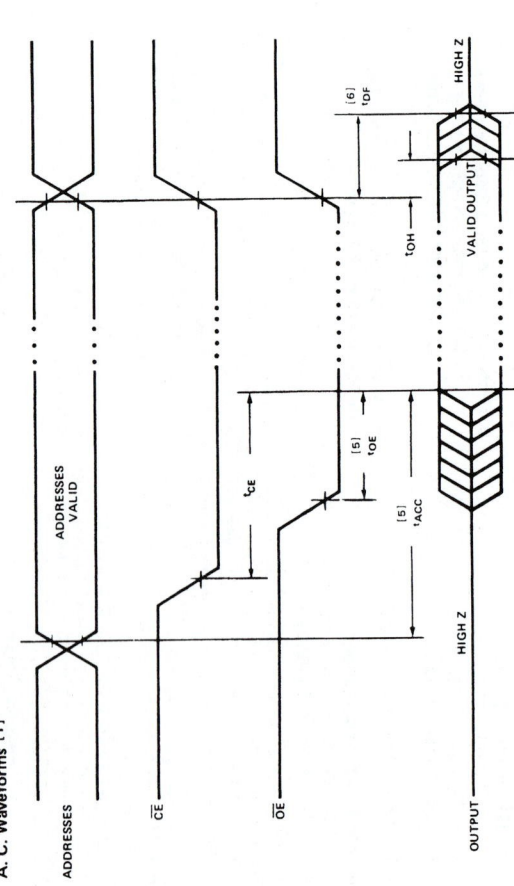

TABLE I MODE SELECTION

PINS / MODE	CE/PGM (18)	OE (20)	Vpp (21)	Vcc (24)	OUTPUTS (9-11,13-17)
Read	V_{IL}	V_{IL}	+5	+5	D_{OUT}
Standby	V_{IH}	Don't Care	+5	+5	High Z
Program	Pulsed V_{IL} to V_{IH}	V_{IH}	+25	+5	D_{IN}
Program Verify	V_{IL}	V_{IL}	+25	+5	D_{OUT}
Program Inhibit	V_{IL}	V_{IH}	+25	+5	High Z

NOTE:
1. VCC must be applied simultaneously or before Vpp and removed simultaneously or after Vpp.
2. Vpp may be connected directly to VCC except during programming. The supply current would then be the sum of I_{CC} and I_{PP1}.
3. Typical values are for $T_A = 25°C$ and nominal supply voltages.
4. This parameter is only sampled and is not 100% tested.
5. This parameter is only sampled and is not 100% tested.
6. OE may be delayed up to $t_{ACC} - t_{OE}$ after the falling edge of CE without impact on t_{ACC}.
7. t_{DF} is specified from OE or CE, whichever occurs first.

CMOS 8-BIT A/D CONVERTERS

ADC0801/2/3/4/5-1

Preliminary

DESCRIPTION

The ADC0801 family is a series of five CMOS 8-bit successive approximation A/D converters using a resistive ladder and capacitive array together with an auto-zero comparator. These converters are designed to operate with microprocessor controlled buses using a minimum of external circuitry. The three-state output data lines can be connected directly to the data bus.

The differential analog voltage input allows for increased common-mode rejection and provides a means to adjust the zero scale offset. Additionally, the voltage reference input provides a means of encoding small analog voltages to the full 8 bits of resolution.

FEATURES

- Compatible with most microprocessors
- Differential inputs
- Three-state outputs
- Logic levels TTL and MOS compatible
- Can be used with internal or external clock
- Analog input range 0V to V_{CC}
- Single 5V supply
- Guaranteed specification with 1MHz clock

APPLICATIONS

- Transducer to microprocessor interface
- Digital thermometer
- Digitally-controlled thermostat
- Microprocessor-based monitoring and control systems

PIN CONFIGURATION

F,N PACKAGE

TOP VIEW

ORDER NUMBERS
ADC0801/02-1 F
ADC0801/02/03-1 LCF
ADC0801/02/03/04/05-1 LCN
ADC0804-1 CN

ABSOLUTE MAXIMUM RATINGS

SYMBOL & PARAMETER		RATING	UNIT
V_{CC}	Supply Voltage	6.5	V
	Logic Control Input Voltages	-0.3 to $+16$	V
	All Other Input Voltages	-0.3 to $(V_{CC} - 0.3)$	V
T_A	Operating Temperature Range		
	ADC0801/02-1 F	-55 to $+125$	°C
	ADC0801/02/03-1 LCF	-40 to $+85$	°C
	ADC0801/02/03/04/05-1 LCN	-40 to $+85$	°C
	ADC0804-1 CN	0 to $+70$	°C
T_{STG}	Storage Temperature	-65 to $+150$	°C
T_{SOLD}	Lead Soldering Temperature (10 seconds)	300	°C
P_D	Package Power Dissipation at $T_A = 25$°C	875	mW

CMOS 8-BIT A/D CONVERTERS

ADC0801/2/3/4/5-1

Preliminary

BLOCK DIAGRAMS

CMOS 8-BIT A/D CONVERTERS

Preliminary

AC ELECTRICAL CHARACTERISTICS

SYMBOL & PARAMETER	TO	FROM	TEST CONDITIONS	ADC0801/2/3/4/5			UNIT
				Min	Typ	Max	
Conversion Time			f_{CLK} = 1MHz[1]	66		73	μs
t_{CLK} Clock Frequency			See Note 1.	0.1	1.0	3.0	MHz
Clock Duty Cycle			See Note 1.	40		60	%
CR Free-Running Conversion Rate			$\overline{CS}$ = 0, t_{CLK} = 1MHz $\overline{INTR}$ Tied To $\overline{WR}$			13690	conv s
$t_{W(WR)L}$ Start Pulse Width			$\overline{CS}$ = 0	30			ns
t_{ACC} Access Time	Output	$\overline{RD}$	$\overline{CS}$ = 0, C_L = 100 pF		75	100	ns
t_{1H}, t_{0H} Three-State Control	Output	$\overline{RD}$	CL = 10 pF, RL = 10K See Three-State Test Circuit		70	100	ns
t_{WI}, t_{RI} $\overline{INTR}$ Delay	$\overline{INTR}$	$\overline{WD}$ or $\overline{RD}$			100	150	ns
C_{IN} Logic Input = Capacitance					5	7.5	pF
C_{OUT} Three-State Output Capacitance					5	7.5	pF

NOTE:
1. Accuracy is guaranteed at f_{CLK} = 1MHz. Accuracy may degrade at higher clock frequencies.

CMOS 8-BIT A/D CONVERTERS

Preliminary

DC ELECTRICAL CHARACTERISTICS V_{CC} = 5.0V, f_{CLK} = 1MHz, $T_{MIN} \leq T_A \leq T_{MAX}$, unless otherwise specified.

SYMBOL & PARAMETER	TEST CONDITIONS	ADC0801/2/3/4/5			UNIT
		Min	Typ	Max	
ADC0801 Relative Accuracy Error (Adjusted)	Full Scale Adjusted			0.25	LSB
ADC0802 Relative Accuracy Error (Unadjusted)	$\frac{V_{REF}}{2}$ = 2.500 V_{DC}			0.50	LSB
ADC0803 Relative Accuracy Error (Adjusted)	Full Scale Adjusted			0.50	LSB
ADC0804 Relative Accuracy Error (Unadjusted)	$\frac{V_{REF}}{2}$ = 2.500 V_{DC}			1	LSB
ADC0805 Relative Accuracy Error (Unadjusted)	$\frac{V_{REF}}{2}$ = has no connection			1	LSB
$\frac{V_{REF}}{2}$ Input Resistance		400	640		Ω
Analog Input Voltage Range	Over Analog Input Voltage Range	-0.05		V_{CC} +0.05	V
DC Common Mode Error			1/16	1/8	LSB
Power Supply Sensitivity	V_{CC} = 5V ±10%[1]				
CONTROL INPUTS					
V_{IH} Logical "1" Input Voltage	V_{CC} = 5.25V_{DC}	2.0		15	V_{DC}
V_{IL} Logical "0" Input Voltage	V_{CC} = 4.75V_{DC}			0.8	V_{DC}
I_{IH} Logical "1" Input Current	V_{IN} = 5V_{DC}		0.005	1	μA_{DC}
I_{IL} Logical "0" Input Current	V_{IN} = 0V_{DC}	-1	-0.005		μA_{DC}
CLOCK IN AND CLOCK R					
V_{T+} Clk In Positive-Going Threshold Voltage		2.7	3.1	3.5	V_{DC}
V_{T-} Clk In Negative-Going Threshold Voltage		1.5	1.8	2.1	V_{DC}
V_H Clk In Hysteresis (V_{T+}) - (V_{T-})		0.6	1.3	2.0	V_{DC}
V_{OL} Logical "0" Clk R Output Voltage	I_{OL} = 360μA, V_{CC} = 4.75 V_{DC}			0.4	V_{DC}
V_{OH} Logical "1" Clk R Output Voltage	I_{OH} = -360μA, V_{CC} = 4.75 V_{DC}	2.4			V_{DC}
DATA OUTPUT AND $\overline{INTR}$					
V_{OL} Logical "0" Output Voltage					
Data Outputs	I_{OL} = 1.6mA, V_{CC} = 4.75 V_{DC}			0.4	V_{DC}
$\overline{INTR}$ Outputs	I_{OL} = 1.0mA, V_{CC} = 4.75 V_{DC}			0.4	V_{DC}
V_{OH} Logical "1" Output Voltage	I_{OH} = -360μA, V_{CC} = 4.75 V_{DC}	2.4			V_{DC}
	I_{OH} = -10μA, V_{CC} = 4.75 V_{DC}	4.5			V_{DC}
I_{OZL} 3-State Output Leakage	V_{OUT} = 0V_{DC}, $\overline{CS}$ = Logical "1"	-3			μA_{DC}
I_{OZH} 3-State Output Leakage	V_{OUT} = 5V_{DC}, $\overline{CS}$ = Logical "1"			3	μA_{DC}
I_{SC} + Output Short Circuit Current	V_{OUT} = 0V, T_A = 25°C	4.5	6		mA_{DC}
I_{SC} - Output Short Circuit Current	V_{OUT} = V_{CC}, T_A = 25°C	9.0	16		mA_{DC}
I_{CC} Power Supply Current	t_{CLK}=1MHz, V_{REF} 2 = Open $\overline{CS}$ = Logical "1", T_A = 25°C		3.0	3.5	mA

NOTE:
1. Analog inputs must remain within the range: -0.05 ≤ V_{IN} ≤ V_{CC} + 0.05V.

Preliminary

TIMING DIAGRAMS (All timing is measured from the 50% voltage points)

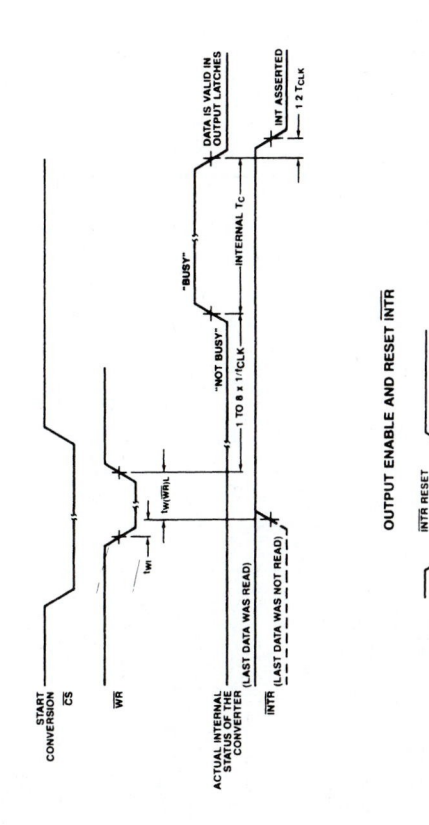

OUTPUT ENABLE AND RESET INTR

Note: Read strobe must occur 8 clock periods (8 t$_{CLK}$) after assertion of interrupt to guarantee reset of INTR

Preliminary

FUNCTIONAL DESCRIPTION

The ADC0801 through ADC0805 series of A/D converters are successive approximation A/D converters with 8-bit resolution and no missing codes. The most significant bit is tested first and after 64 clock cycles a digital 8-bit binary word is transferred to an output latch and the INTR pin goes low, indicating that conversion is complete. A conversion in progress can be interrupted by issuing another start command. The device may be operated in a continuous conversion mode by connecting the INTR and WR pins together and holding the CS pin low. To insure start-up when connected this way, an external WR pulse is required at power-up.

As the WR input goes low, when CS is low, the SAR is cleared and remains so as long as these two inputs are low. Conversion begins between 1 and 8 clock periods after at least one of these inputs goes high. As the conversion begins, the INTR line goes high. Note that the INTR line will remain low until 1 to 8 clock cycles after either the WR or CS input (or both) goes high.

When the CS and RD inputs are both brought low to read the data, the INTR line will go low and the three-state output latches are enabled.

The digital control lines (CS, RD, and WR) operate with standard TTL levels and have been renamed when compared with standard A/D Start and Output Enable labels. For non-microprocessor based applications, the CS pin can be grounded, the WR pin can be interpreted as a START pulse pin, and the RD pin performs the OE (Output Enable) function.

The V$_{IN}$(−) input can be used to subtract a fixed voltage from the input voltage. Because there is a time interval between sampling the V$_{IN}$(+) and the V(−) inputs, it is important that these inputs remain constant, during the entire conversion cycle.

THREE-STATE TEST CIRCUITS AND WAVEFORMS

t$_{1H}$

t$_{1H}$, C$_L$ = 10 pF

t$_r$ = 20 ns

t$_{0H}$

t$_{0H}$, C$_L$ = 10 pF

t$_r$ = 20 ns

DC ELECTRICAL CHARACTERISTICS[1]

Pin 3 must be 3V more negative than the potential to which R15 is returned

$V_{CC} = +5.0$ Vdc, $V_{EE} = -15$ Vdc, $V_{ref}/R_{14} = 2.0$ mA
unless otherwise specified.
MC1508: $T_A = -55°C$ to $125°C$. MC1408: $T_A = 0°C$ to $75°C$
unless otherwise noted.

PARAMETER		TEST CONDITIONS	MC1508-8			MC1408-8			MC1408-7			UNIT
			Min	Typ	Max	Min	Typ	Max	Min	Typ	Max	
E_r	Relative accuracy	Error relative to full scale I_O, Figure 3			±0.19			±0.19			±0.39	%
t_s	Setting time[1]	To within 1/2 LSB includes t'PLH, $T_A = +25°C$, Figure 4		70			70			70		ns
	Propagation delay time											ns
t_{PLH}	Low-to-high	Figure 4		35	100		35	100		35	100	
t_{PHL}	High-to-low											
TCI_O	Output full scale current drift			-20			-20			-20		PPM/°C
	Digital input logic level (MSB)											Vdc
V_{IH}	High	Figure 5	2.0			2.0			2.0			
V_{IL}	Low	Figure 5			0.8			0.8			0.8	
	Digital input current (MSB)											mA
I_H	High	$V_{IH} = 5.0V$		0	0.04		0	0.04		0	0.04	
I_L	Low	$V_{IL} = 0.8V$		-0.4	-0.8		-0.4	-0.8		-0.4	-0.8	
I_{15}	Reference input bias current	Pin 15, Figure 5		-1.0	-5.0		-1.0	-5.0		-1.0	-5.0	μA
I_{OR}	Output current range	Figure 5										mA
		$V_{EE} = -5.0V$	0	2.0	2.1	0	2.0	2.1	0	2.0	2.1	
		$V_{EE} = -7.0V$ to $-15V$	0	2.0	4.2	0	2.0	4.2	0	2.0	4.2	
I_O	Output current	Figure 5 $V_{ref} = 2.000V$, R14 = 1000Ω	1.9	1.99	2.1	1.9	1.99	2.1	1.9	1.99	2.1	mA
$I_{O(min)}$	Off-state	All bits low		0	4.0		0	4.0		0	4.0	μA
V_O	Output voltage compliance	$E_r \leq 0.19\%$ at $T_A = +25°C$, Figure 5 $V_{EE} = -5V$		-0.6, +10	-0.55, +0.5		-0.6, +10	-0.55, +0.5		-0.6, +10	-0.55, +0.5	Vdc
		V_{EE} below $-10V$		-5.5, +10	-5.0, +0.5		-5.5, +10	-5.0, +0.5		-5.5, +10	-5.0, +0.5	
SR_{Iref}	Reference current slew rate	Figure 6		8.0			8.0			8.0		mA/μs
$PSRR_{(-)}$	Output current power supply sensitivity	$I_{ref} = 1mA$		0.5	2.7		0.5	2.7		0.5	2.7	μA/V
	Power supply current											mA
I_{CC}	Positive	All bits low.		+2.5	+22		+2.5	+22		+2.5	+22	
I_{EE}	Negative	Figure 5		-6.5	-13		-6.5	-13		-6.5	-13	
	Power supply voltage range											Vdc
V_{CCR}	Positive	$T_A = +25°C$,	+4.5	+5.0	+5.5	+4.5	+5.0	+5.5	+4.5	+5.0	+5.5	
V_{EER}	Negative	Figure 5	-4.5	-15	-16.5	-4.5	-15	-16.5	-4.5	-15	-16.5	
P_D	Power dissipation	All bits low. Figure 5 $V_{EE} = -5.0V$dc $V_{EE} = -15V$dc		34 110	170 305		34 110	170 305		34 110	170 305	mW

NOTES:
1. All bits switched

DESCRIPTION

The MC1508/MC1408 series of 8-bit monolithic digital-to-analog converters provide high speed performance with low cost. They are designed for use where the output current is a linear product of an 8-bit digital word and an analog reference voltage.

FEATURES

- Fast settling time—70ns (typ)
- Relative accuracy ±0.19% (max error)
- Non-inverting digital inputs are TTL and CMOS compatible
- High speed multiplying rate 4.0mA/μs (input slew)
- Output voltage swing +5V to −5.0V
- Standard supply voltages +5.0V and −5.0V to −15V
- Military qualifications pending

APPLICATIONS

- Tracking A-to-D converters
- 2½-digit panel meters and DVM's
- Waveform synthesis
- Sample and hold
- Peak detector
- Programmable gain and attenuation
- CRT character generation
- Audio digitizing and decoding
- Programmable power supplies
- Analog-digital multiplication
- Digital-digital multiplication
- Analog-digital division
- Digital addition and subtraction
- Speech compression and expansion
- Stepping motor drive
- Modems
- Servo motor and pen drivers

CIRCUIT DESCRIPTION

The MC1508/MC1408 consists of a reference current amplifier, an R-2R ladder, and 8 high speed current switches. For many applications, only a reference resistor and reference voltage need be added.

The switches are non-inverting in operation; therefore, a high state on the input turns on the specified output current component.

The switch uses current steering for high speed, and a termination amplifier consisting of an active load gain stage with unity gain feedback. The termination amplifier holds the parasitic capacitance of the ladder at a constant voltage during switching, and provides a low impedance termination of equal voltage for all legs of the ladder.

The R-2R ladder divides the reference amplifier current into binarily-related components, which are fed to the switches. Note that there is always a remainder current which is equal to the least significant bit. This current is shunted to ground, and the maximum output current is 255/256 of the reference amplifier current, or 1.992mA for a 2.0mA reference amplifier current if the NPN current source pair is perfectly matched.

PIN CONFIGURATION

F,N PACKAGE

TOP VIEW

ORDER NUMBERS
MC1508-8F   MC1408-7N
MC1408-8F

D³ PACKAGE

TOP VIEW

ORDER NUMBER
MC1408-8D

NOTES:
1. SOL Released in Large SO package only.
2. SOL and non-standard pinout.
3. SO and non-standard pinouts.

BLOCK DIAGRAM

ABSOLUTE MAXIMUM RATINGS $T_A = +25°C$ unless otherwise specified

PARAMETER		RATING	UNIT
	Power Supply Voltage		
V_{CC}	Positive	+5.5	V
V_{EE}	Negative	−16.5	V
V_5, V_{12}	Digital Input Voltage	0 to V_{CC}	V
V_O	Applied Output Voltage	−5.2 to +18	V
I_{14}	Reference Current	5.0	mA
V_{14}, V_{15}	Reference Amplifier Inputs	V_{EE} to V_{CC}	
P_D	Power Dissipation (Package Limitation)		
	Ceramic Package	1000	mW
	Plastic Package	800	mW
	Lead Soldering Temperature (60 sec)	300	°C
T_A	Operating Temperature Range		
	MC1508	−55 to +125	°C
	MC1408	0 to +75	°C
T_{STG}	Storage Temperature Range	−65 to +150	°C

TYPICAL PERFORMANCE CHARACTERISTICS

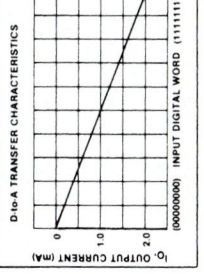

D-to-A TRANSFER CHARACTERISTICS

I₀, OUTPUT CURRENT (mA) — (00000000) INPUT DIGITAL WORD (11111111)

FUNCTIONAL DESCRIPTION

Reference Amplifier Drive and Compensation

The reference amplifier input current must always flow into pin 14 regardless of the setup method or reference voltage polarity.

Connections for a positive reference voltage are shown in Figure 1. The reference voltage source supplies the full reference current For bipolar reference signals, as in the multiplying mode. R_{15} can be tied to either V_{EE} or ground, but using V_{EE} increases negative supply rejection. (Fluctuations in the negative supply have more effect on accuracy than do any changes in the positive supply).

A negative reference voltage may be used if R_{14} is grounded and the reference voltage is applied to R_{15}, as shown in Figure 2. A high input impedance is the main advantage of this method. The negative reference voltage must be at least 3.0V above the V_{EE} supply. Bipolar input signals may be handled by connecting R_{14} to a positive reference voltage equal to the peak positive input level at pin 15.

Capacitive bypass to ground is recommended when a DC reference voltage is used. The 5.0V logic supply is not recommended as a reference voltage, but if a

well regulated 5.0V supply which drives logic is to be used as the reference, R_{14} should be formed of two series resistors and the junction of the two resistors bypassed with $0.1\mu F$ to ground. For reference voltages greater than 5.0V, a clamp diode is recommended between pin 14 and ground

If pin 14 is driven by a high impedance such as a transistor current source, none of the above compensation methods apply and the amplifier must be heavily compensated, decreasing the overall bandwidth.

Output Voltage Range

The voltage at pin 4 must always be at least 4.5 volts more positive than the voltage of the negative supply (pin 3) when the reference current is 2mA or less, and at least 8 volts more positive than the negative supply when the reference current is between 2mA and 4mA. This is necessary to avoid saturation of the output transistors, which would cause serious degradation of accuracy.

Signetics' MC1508/MC1408 does not need a range control because the design extends the compliance range down to 4.5 volts (or 8 volts—see above) above the negative supply voltage without significant degradation of accuracy. Signetics' MC1508/MC1408 can be used in sockets designed for other manufacturers' MC1508/ MC1408 without circuit modification.

Output Current Range

Any time the full scale current exceeds 2mA, the negative supply must be at least 8 volts more negative than the output voltage. This is due to the increased internal voltage drops between the negative supply and the outputs with higher reference currents.

Accuracy

Absolute accuracy is the measure of each output current level with respect to its intended value, and is dependent upon relative accuracy, full scale accuracy and full scale current drift. Relative accuracy is the measure of each output current level as a fraction of the full scale current after zero scale current has been nulled out. The relative accuracy of the MC1508/MC1408 is essentially constant over the operating temperature range because of the excellent temperature tracking of the monolithic resistor ladder. The reference current may drift with temperature, causing a change in the absolute accuracy of output current; however, the MC1508/MC1408 has a very low full scale current drift over the operating temperature range.

The MC1508/MC1408 series is guaranteed accurate to withiu ±1/2 LSB at +25°C at a full scale output current of 1.99mA. The relative accuracy test circuit is shown in Figure 3. The 12-bit converter is calibrated to a full scale output current of 1.99219mA; then the MC1508/MC1408's full scale current is trimmed to the same value with R_{14} so that a zero value appears at the error amplifier output. The counter is activated and the error band may be displayed on the oscilloscope, detected by comparators, or stored in a peak detector.

Two 8-bit D-to-A converters may not be used to construct a 16-bit accurate D-to-A converter. Sixteen-bit accuracy implies a total of ±1/2 part in 65,536, or ±0.000076%, which is much more accurate than the ±0.19°% specification of the MC1508/ MC1408.

Monotonicity

A monotonic converter is one which always provides an analog output greater than or equal to the preceding value for a corresponding increment in the digital input code. The MC1508/MC1408 is monotonic for all values of reference current above 0.5mA. The recommended range for operation is a DC reference current between 0.5mA and 4.0mA.

Settling Time

The worst case switching condition occurs when all bits are switched on, which corresponds to a low-to-high transition for all input bits. This time is typically 70ns for settling to within 1/2 LSB for 8-bit accuracy. This time applies when RL <500 ohms and C_O <25pF. The slowest single switch is the least significant bit, which typically turns on and settles in 65ns. In applications where the D-to-A converter functions in a positive going ramp mode, the worst case condition does not occur and settling times less than 70ns may be realized.

Extra care must be taken in board layout since this usually is the dominant factor in satisfactory test results when measuring settling time. Short leads, 100μF supply bypassing for low frequencies, minimum scope lead length, good ground planes, and avoidance of ground loops are all mandatory.

TEST CIRCUITS

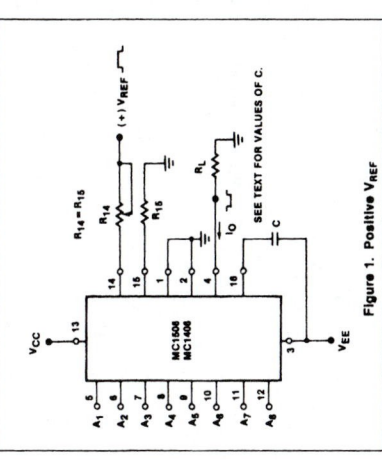

Figure 1. Positive V_{REF}

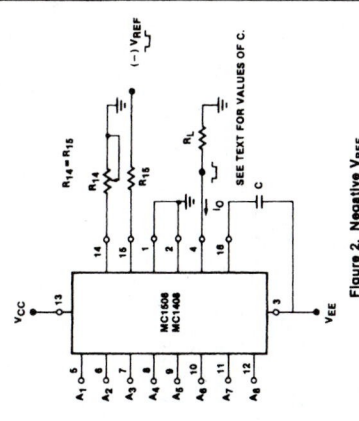

Figure 2. Negative V_{REF}

TEST CIRCUITS (Cont'd)

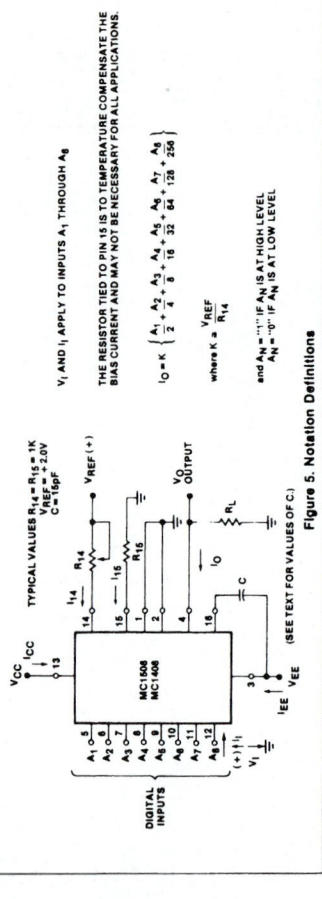

TYPICAL VALUES $R_{14} = R_{15} = 1K$
$V_{REF} = +2.0V$
$C = 15pF$

V_I AND I_I APPLY TO INPUTS A_1 THROUGH A_8.

THE RESISTOR TIED TO PIN 15 IS TO TEMPERATURE COMPENSATE THE BIAS CURRENT AND MAY NOT BE NECESSARY FOR ALL APPLICATIONS.

$$I_O = K \left\{ \frac{A_1}{2} + \frac{A_2}{4} + \frac{A_3}{8} + \frac{A_4}{16} + \frac{A_5}{32} + \frac{A_6}{64} + \frac{A_7}{128} + \frac{A_8}{256} \right\}$$

where $K = \dfrac{V_{REF}}{R_{14}}$

and A_N = "1" IF A_N IS AT HIGH LEVEL
A_N = "0" IF A_N IS AT LOW LEVEL

Figure 5. Notation Definitions

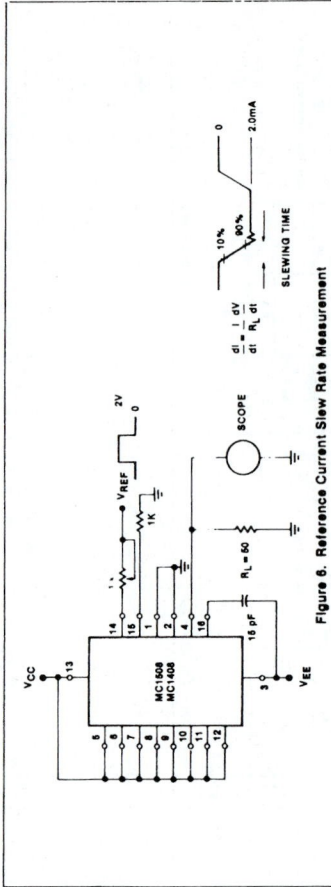

Figure 6. Reference Current Slew Rate Measurement

TEST CIRCUITS (Cont'd)

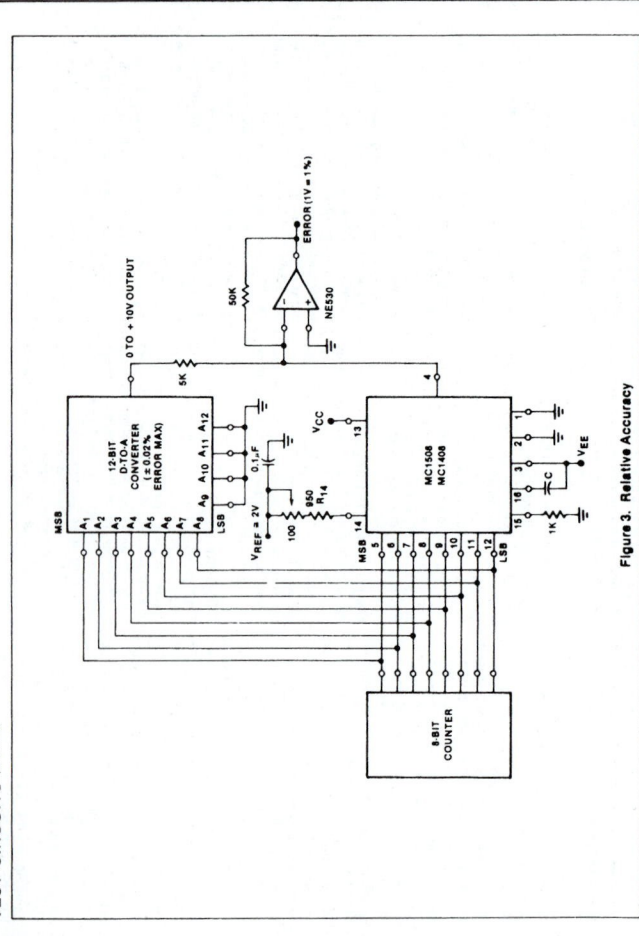

Figure 3. Relative Accuracy

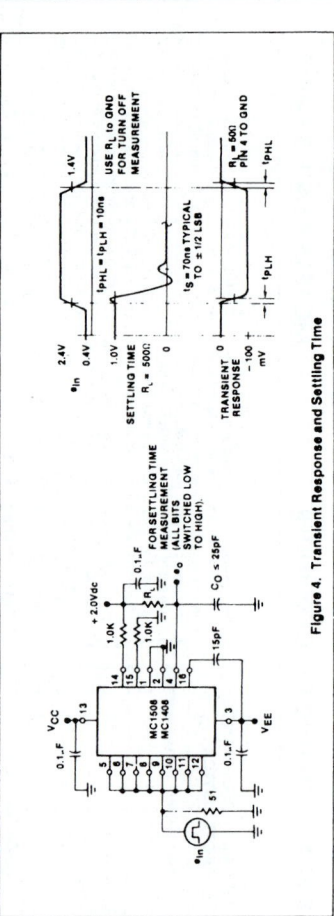

Figure 4. Transient Response and Settling Time

Signetics

µA741/µA741C/SA741C
General Purpose Operational Amplifier

Product Specification

Linear Products

DESCRIPTION
The µA741 is a high performance operational amplifier with high open-loop gain, internal compensation, high common mode range and exceptional temperature stability. The µA741 is short-circuit-protected and allows for nulling of offset voltage.

FEATURES
- Internal frequency compensation
- Short circuit protection
- Excellent temperature stability
- High input voltage range

PIN CONFIGURATION

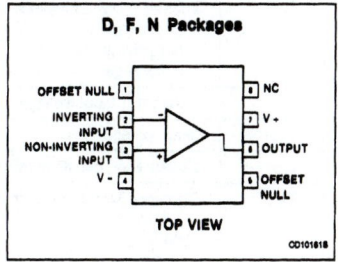

D, F, N Packages

```
OFFSET NULL    1        8  NC
INVERTING      2        7  V +
INPUT
NON-INVERTING  3        6  OUTPUT
INPUT
V -            4        5  OFFSET
                           NULL
```

TOP VIEW

ORDERING INFORMATION

DESCRIPTION	TEMPERATURE RANGE	ORDER CODE
8-Pin Plastic DIP	−55°C to +125°C	µA741N
8-Pin Plastic DIP	0 to +70°C	µA741CN
8-Pin Plastic DIP	−40°C to +85°C	SA741CN
8-Pin Cerdip	−55°C to +125°C	µA741F
8-Pin Cerdip	0 to +70°C	µA741CF
8-Pin SO	0 to +70°C	µA741CD

EQUIVALENT SCHEMATIC

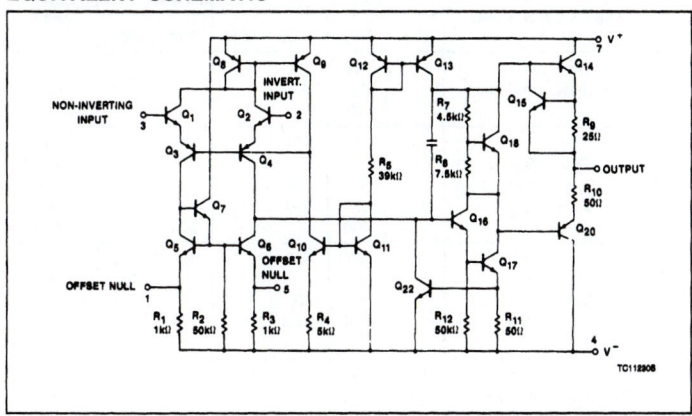

General Purpose Operational Amplifier

μA741/μA741C/SA741C

ABSOLUTE MAXIMUM RATINGS

SYMBOL	PARAMETER	RATING	UNIT
V_S	Supply voltage μA741C μA741	±18 ±22	V V
P_D	Internal power dissipation D package N package F package	500 1000 1000	mW mW mW
V_{IN}	Differential input voltage	±30	V
V_{IN}	Input voltage[1]	±15	V
I_{SC}	Output short-circuit duration	Continuous	
T_A	Operating temperature range μA741C SA741C μA741	0 to +70 −40 to +85 −55 to +125	°C °C °C
T_{STG}	Storage temperature range	−65 to +150	°C
T_{SOLD}	Lead soldering temperature (10sec max)	300	°C

NOTE:
1. For supply voltages less than ±15V, the absolute maximum input voltage is equal to the supply voltage.

DC ELECTRICAL CHARACTERISTICS (μA741, μA741C) T_A = 25°C, V_S = ±15V, unless otherwise specified.

SYMBOL	PARAMETER	TEST CONDITIONS	μA741 Min	μA741 Typ	μA741 Max	μA741C Min	μA741C Typ	μA741C Max	UNIT
V_{OS}	Offset voltage	R_S = 10kΩ R_S = 10kΩ, over temp.		1.0 1.0	5.0 6.0		2.0	6.0 7.5	mV mV
$\Delta V_{OS}/\Delta T$				10			10		μV/°C
I_{OS}	Offset current	 Over temp. T_A = +125°C T_A = −55°C		20 7.0 20	200 200 500		20	200 300	nA nA nA nA
$\Delta I_{OS}/\Delta T$				200			200		pA/°C
I_{BIAS}	Input bias current	 Over temp. T_A = +125°C T_A = −55°C		80 30 300	500 500 1500		80	500 800	nA nA nA nA
$\Delta I_B/\Delta T$				1			1		nA/°C
V_{OUT}	Output voltage swing	R_L = 10kΩ R_L = 2kΩ, over temp.	±12 ±10	±14 ±13		±12 ±10	±14 ±13		V V
A_{VOL}	Large-signal voltage gain	R_L = 2kΩ, V_O = ±10V R_L = 2kΩ, V_O = ±10V, over temp.	50 25	200		20 15	200		V/mV V/mV
	Offset voltage adjustment range			±30			±30		mV
PSRR	Supply voltage rejection ratio	R_S ≤ 10kΩ R_S ≤ 10kΩ, over temp.		10	150		10	150	μV/V μV/V
CMRR	Common-mode rejection ratio	Over temp.	70	90					dB dB
I_{CC}	Supply current	T_A = +125°C T_A = −55°C		1.4 1.5 2.0	2.8 2.5 3.3		1.4	2.8	mA mA mA

General Purpose Operational Amplifier $\quad\quad$ μA741/μA741C/SA741C

DC ELECTRICAL CHARACTERISTICS (Continued) (μA741, μA741C) $T_A = 25°C$, $V_S = ±15V$, unless otherwise specified.

SYMBOL	PARAMETER	TEST CONDITIONS	μA741			μA741C			UNIT
			Min	Typ	Max	Min	Typ	Max	
V_{IN} R_{IN}	Input voltage range Input resistance	(μA741, over temp.)	±12 0.3	±13 2.0		±12 0.3	±13 2.0		V MΩ
P_D	Power consumption	$T_A = +125°C$ $T_A = -55°C$		50 45 45	80 75 100		50	85	mW mW mW
R_{OUT}	Output resistance			75			75		Ω
I_{SC}	Output short-circuit current		10	25	60	10	25	60	mA

DC ELECTRICAL CHARACTERISTICS (SA741C) $T_A = 25°C$, $V_S = ±15V$, unless otherwise specified.

SYMBOL	PARAMETER	TEST CONDITIONS	SA741C			UNIT
			Min	Typ	Max	
V_{OS} $\Delta V_{OS}/\Delta T$	Offset voltage	$R_S = 10kΩ$ $R_S = 10kΩ$, over temp.		2.0 10	6.0 7.5	mV mV μV/°C
I_{OS} $\Delta I_{OS}/\Delta T$	Offset current	Over temp.		20 200	200 500	nA nA pA/°C
I_{BIAS} $\Delta I_B/\Delta T$	Input bias current	Over temp.		80 1	500 1500	nA nA nA/°C
V_{OUT}	Output voltage swing	$R_L = 10kΩ$ $R_L = 2kΩ$, over temp.	±12 ±10	±14 ±13		V V
A_{VOL}	Large-signal voltage gain	$R_L = 2kΩ$, $V_O = ±10V$ $R_L = 2kΩ$, $V_O = ±10V$, over temp.	20 15	200		V/mV V/mV
	Offset voltage adjustment range			±30		mV
PSRR	Supply voltage rejection ratio	$R_S \leq 10kΩ$		10	150	μV/V
V_{IN}	Input voltage range	(μA741, over temp.)	±12	±13		V
R_{IN}	Input resistance		0.3	2.0		MΩ
P_d	Power consumption			50	85	mW
R_{OUT}	Output resistance			75		Ω
I_{SC}	Output short-circuit current			25		mA

AC ELECTRICAL CHARACTERISTICS $T_A = 25°C$, $V_S = ±15V$, unless otherwise specified.

SYMBOL	PARAMETER	TEST CONDITIONS	μA741, μA741C			UNIT
			Min	Typ	Max	
C_{IN}	Parallel input capacitance	Open-loop, f = 20Hz		1.4		pF
	Unity gain crossover frequency	Open-loop		1.0		MHz
t_R SR	Transient response unity gain Rise time Overshoot Slew rate	$V_{IN} = 20mV$, $R_L = 2kΩ$, $C_L \leq 100pF$ $C \leq 100pF$, $R_L \geq 2kΩ$, $V_{IN} = ±10V$		0.3 5.0 0.5		μs % V/μs

General Purpose Operational Amplifier

μA741/μA741C/SA741C

TYPICAL PERFORMANCE CHARACTERISTICS

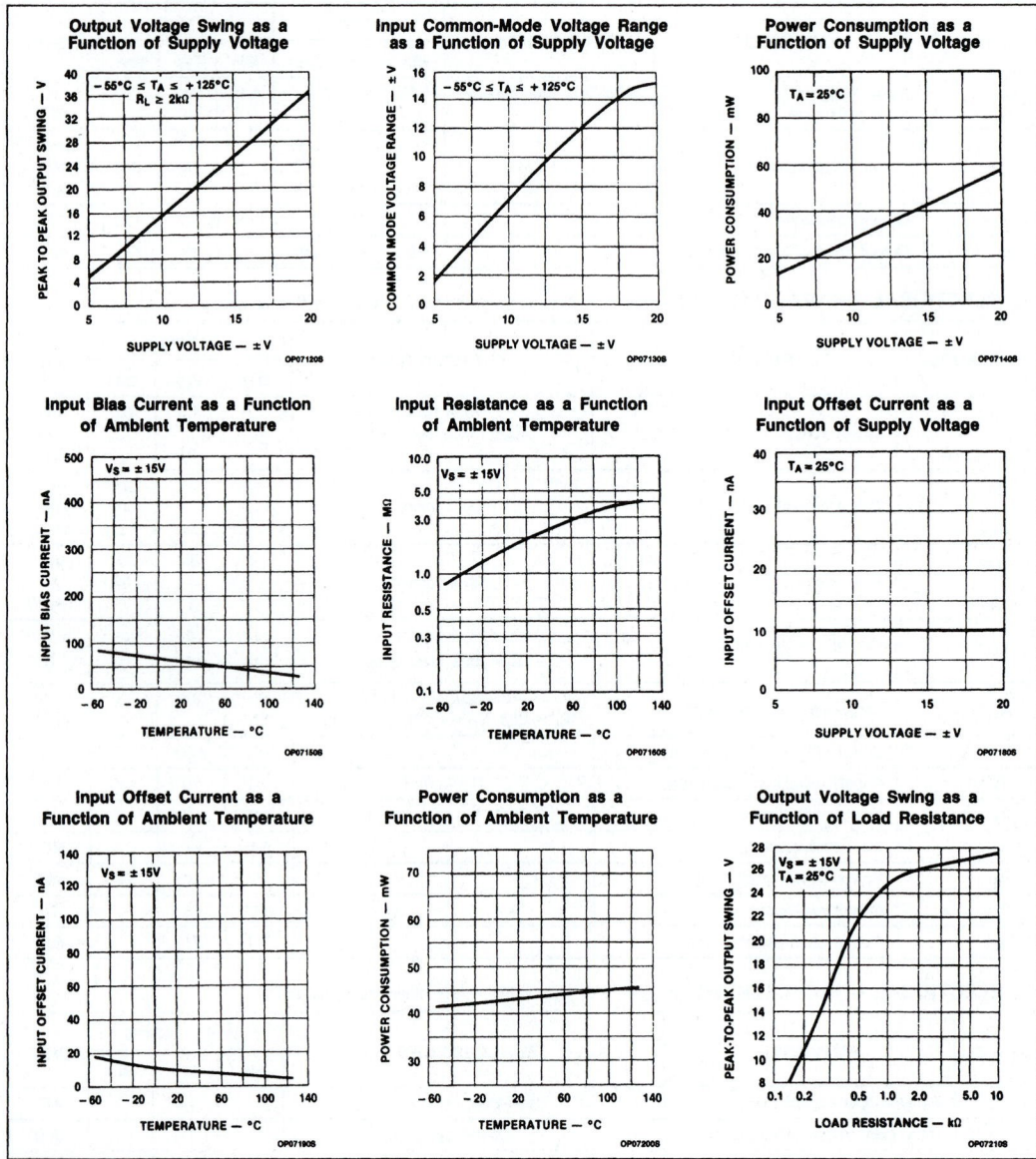

General Purpose Operational Amplifier

μA741/μA741C/SA741C

TYPICAL PERFORMANCE CHARACTERISTICS (Continued)

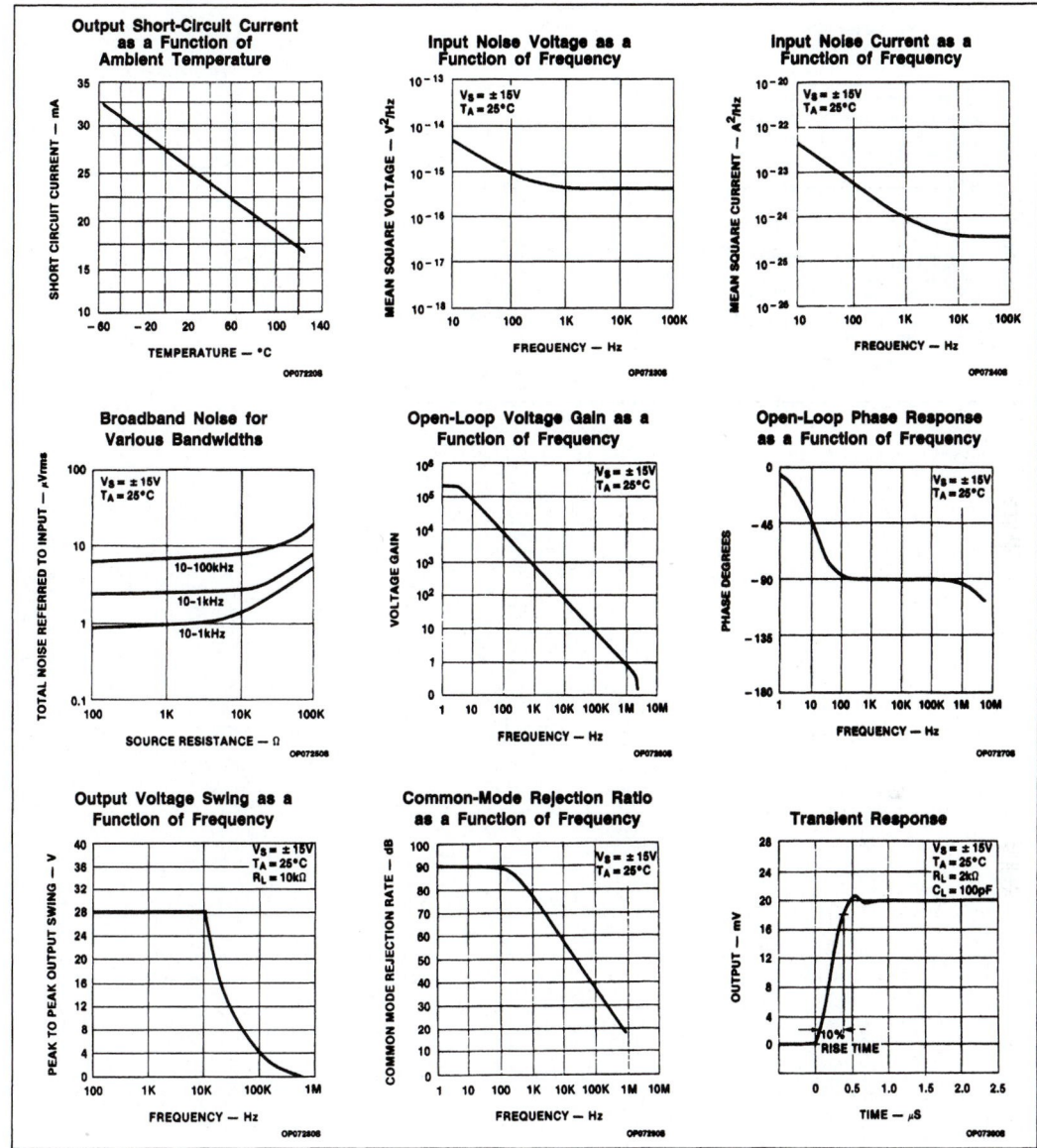

General Purpose Operational Amplifier μA741/μA741C/SA741C

TYPICAL PERFORMANCE CHARACTERISTICS (Continued)

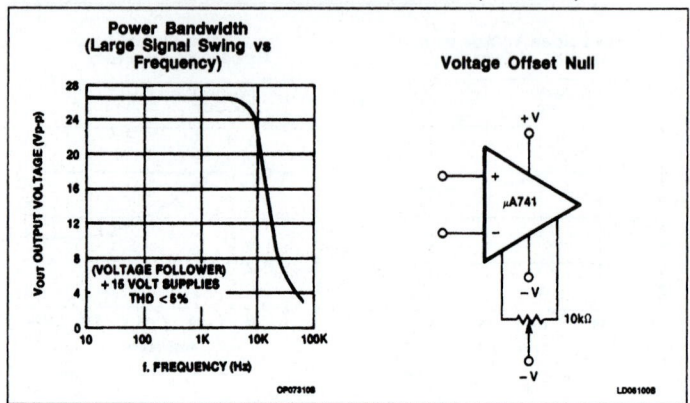

Signetics

NE/SE555/SE555C
Timer

Product Specification

Linear Products

DESCRIPTION

The 555 monolithic timing circuit is a highly stable controller capable of producing accurate time delays, or oscillation. In the time delay mode of operation, the time is precisely controlled by one external resistor and capacitor. For a stable operation as an oscillator, the free running frequency and the duty cycle are both accurately controlled with two external resistors and one capacitor. The circuit may be triggered and reset on falling waveforms, and the output structure can source or sink up to 200mA.

FEATURES

• **Turn-off time less than 2µs**
• **Max. operating frequency greater than 500kHz**
• **Timing from microseconds to hours**
• **Operates in both astable and monostable modes**
• **High output current**
• **Adjustable duty cycle**
• **TTL compatible**
• **Temperature stability of 0.005% per °C**

APPLICATIONS

• **Precision timing**
• **Pulse generation**
• **Sequential timing**
• **Time delay generation**
• **Pulse width modulation**
• **Pulse position modulation**
• **Missing pulse detector**

PIN CONFIGURATIONS

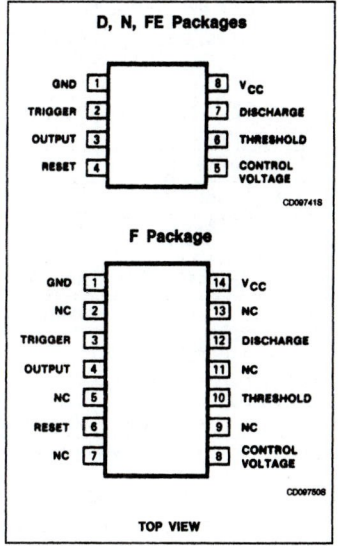

EQUIVALENT SCHEMATIC

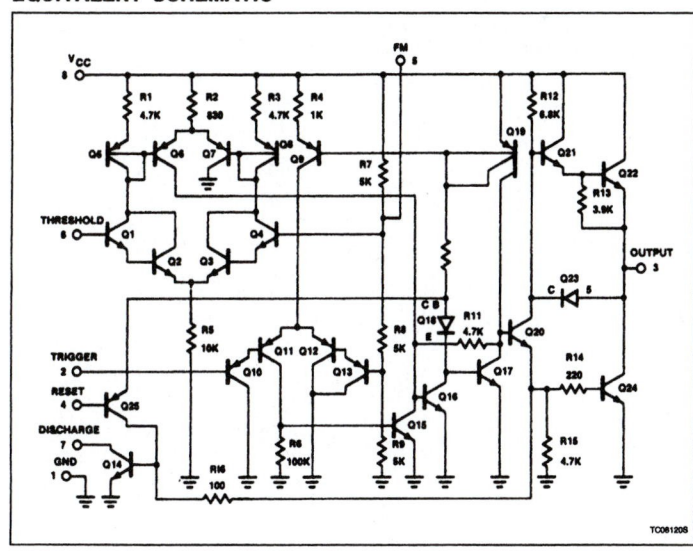

Timer

ORDERING INFORMATION

DESCRIPTION	TEMPERATURE RANGE	ORDER CODE
8-Pin Hermetic Cerdip	0 to +70°C	NE555FE
8-Pin Plastic SO	0 to +70°C	NE555D
8-Pin Plastic DIP	0 to +70°C	NE555N
8-Pin Hermetic Cerdip	−55°C to +125°C	SE555CFE
8-Pin Plastic DIP	−55°C to +125°C	SE555CN
14-Pin Plastic DIP	−55°C to +125°C	SE555N
8-Pin Hermetic Cerdip	−55°C to +125°C	SE555FE
14-Pin Ceramic DIP	0 to +70°C	NE555F
14-Pin Ceramic DIP	−55°C to +125°C	SE555F
14-Pin Ceramic DIP	−55°C to +125°C	SE555CF

BLOCK DIAGRAM

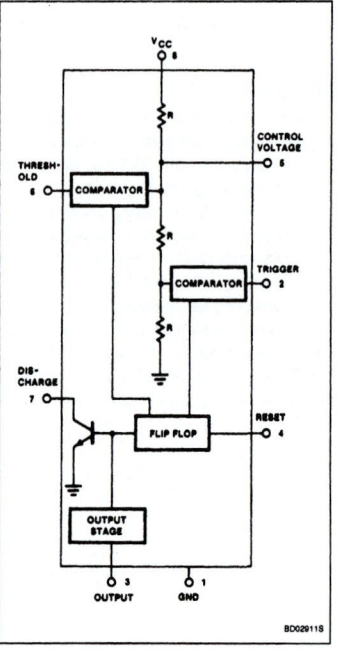

ABSOLUTE MAXIMUM RATINGS

SYMBOL	PARAMETER	RATING	UNIT
V_{CC}	Supply voltage SE555 NE555, SE555C	+18 +16	V V
P_D	Maximum allowable power dissipation[1]	600	mW
T_A	Operating ambient temperature range NE555 SE555, SE555C	0 to +70 −55 to +125	°C °C
T_{STG}	Storage temperature range	−65 to +150	°C
T_{SOLD}	Lead soldering temperature (10sec max)	+300	°C

NOTE:

1. The junction temperature must be kept below 125°C for the D package and below 150°C for the FE, N and F packages. At ambient temperatures above 25°C, where this limit would be derated by the following factors:

    D package 160 °C/W
    FE package 150 °C/W
    N package 100 °C/W
    F package 105 °C/W

Timer

NE/SE555/SE555C

DC AND AC ELECTRICAL CHARACTERISTICS $T_A = 25°C$, $V_{CC} = +5V$ to $+15$ unless otherwise specified.

SYMBOL	PARAMETER	TEST CONDITIONS	SE555			NE555/SE555C			UNIT
			Min	Typ	Max	Min	Typ	Max	
V_{CC}	Supply voltage		4.5		18	4.5		16	V
I_{CC}	Supply current (low state)[1]	$V_{CC} = 5V$, $R_L = \infty$ $V_{CC} = 15V$, $R_L = \infty$		3 10	5 12		3 10	6 15	mA mA
t_M $\Delta t_M/\Delta T$ $\Delta t_M/\Delta V_S$	Timing error (monostable) Initial accuracy[2] Drift with temperature Drift with supply voltage	$R_A = 2k\Omega$ to $100k\Omega$ $C = 0.1\mu F$		0.5 30 0.05	2.0 100 0.2		1.0 50 0.1	3.0 150 0.5	% ppm/°C %/V
t_A $\Delta t_A/\Delta T$ $\Delta t_A/\Delta V_S$	Timing error (astable) Initial accuracy[2] Drift with temperature Drift with supply voltage	R_A, $R_B = 1k\Omega$ to $100k\Omega$ $C = 0.1\mu F$ $V_{CC} = 15V$		4 0.15	6 500 0.6		5 0.3	13 500 1	% ppm/°C %/V
V_C	Control voltage level	$V_{CC} = 15V$ $V_{CC} = 5V$	9.6 2.9	10.0 3.33	10.4 3.8	9.0 2.6	10.0 3.33	11.0 4.0	V V
V_{TH}	Threshold voltage	$V_{CC} = 15V$ $V_{CC} = 5V$	9.4 2.7	10.0 3.33	10.6 4.0	8.8 2.4	10.0 3.33	11.2 4.2	V V
I_{TH}	Threshold current[3]			0.1	0.25		0.1	0.25	μA
V_{TRIG}	Trigger voltage	$V_{CC} = 15V$ $V_{CC} = 5V$	4.8 1.45	5.0 1.67	5.2 1.9	4.5 1.1	5.0 1.67	5.6 2.2	V V
I_{TRIG}	Trigger current	$V_{TRIG} = 0V$		0.5	0.9		0.5	2.0	μA
V_{RESET}	Reset voltage[4]		0.3		1.0	0.3		1.0	V
I_{RESET}	Reset current Reset current	$V_{RESET} = 0V$		0.1 0.4	0.4 1.0		0.1 0.4	0.4 1.5	mA mA
V_{OL}	Output voltage (low)	$V_{CC} = 15V$ $I_{SINK} = 10mA$ $I_{SINK} = 50mA$ $I_{SINK} = 100mA$ $I_{SINK} = 200mA$ $V_{CC} = 5V$ $I_{SINK} = 8mA$ $I_{SINK} = 5mA$		0.1 0.4 2.0 2.5 0.1 0.05	0.15 0.5 2.2 0.25 0.2		0.1 0.4 2.0 2.5 0.3 0.25	0.25 0.75 2.5 0.4 0.35	V V V V V V
V_{OH}	Output voltage (high)	$V_{CC} = 15V$ $I_{SOURCE} = 200mA$ $I_{SOURCE} = 100mA$ $V_{CC} = 5V$ $I_{SOURCE} = 100mA$	 13.0 3.0	12.5 13.3 3.3		 12.75 2.75	12.5 13.3 3.3		V V V
t_{OFF}	Turn-off time[5]	$V_{RESET} = V_{CC}$		0.5	2.0		0.5	2.0	μs
t_R	Rise time of output			100	200		100	300	ns
t_F	Fall time of output			100	200		100	300	ns
	Discharge leakage current			20	100		20	100	ns

NOTES:
1. Supply current when output high typically 1mA less.
2. Tested at $V_{CC} = 5V$ and $V_{CC} = 15V$.
3. This will determine the max value of $R_A + R_B$, for 15V operation, the max total $R = 10M\Omega$, and for 5V operation, the max. total $R = 3.4M\Omega$.
4. Specified with trigger input high.
5. Time measured from a positive going input pulse from 0 to $0.8 \times V_{CC}$ into the threshold to the drop from high to low of the output. Trigger is tied to threshold.

Timer

TYPICAL PERFORMANCE CHARACTERISTICS

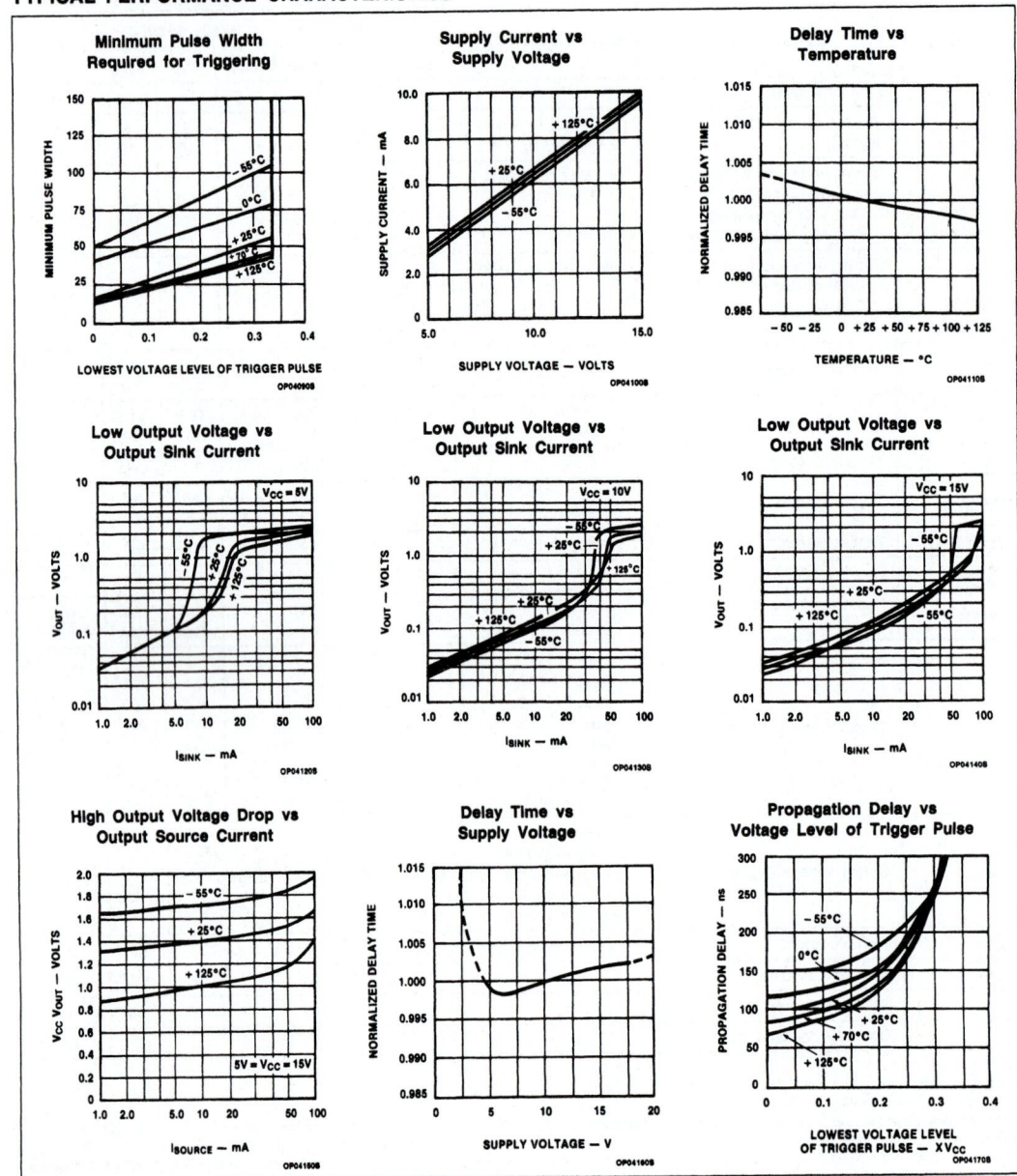

Timer

TYPICAL APPLICATIONS

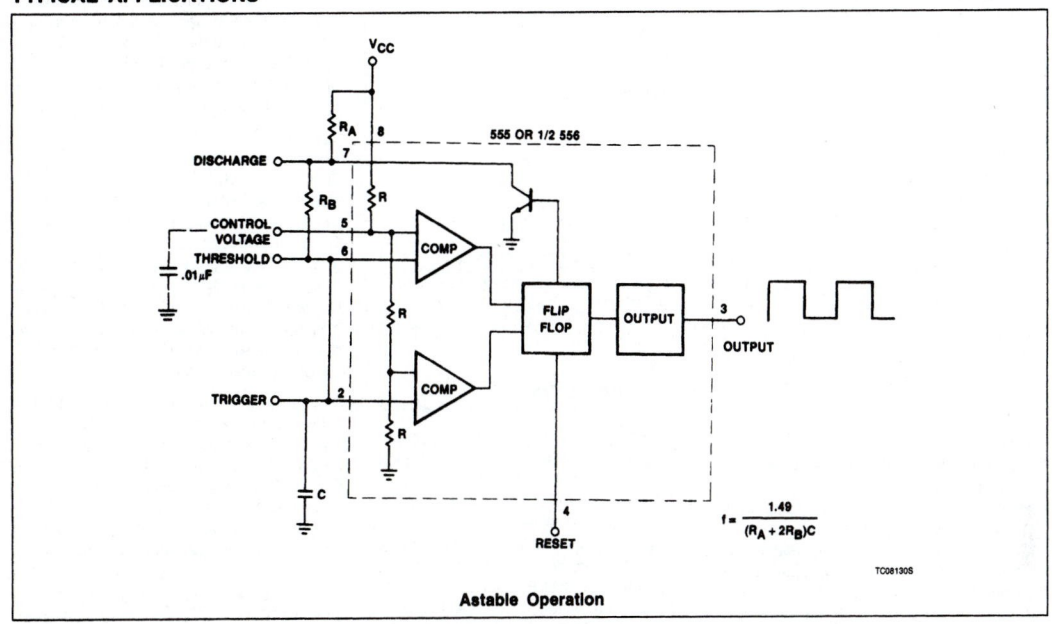

Astable Operation

$$f = \frac{1.49}{(R_A + 2R_B)C}$$

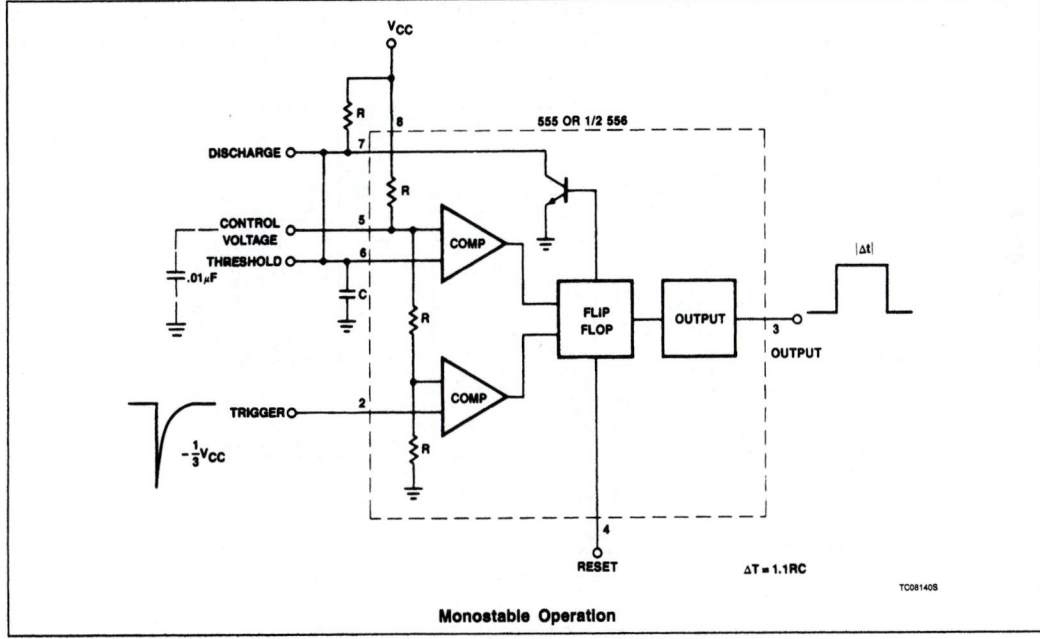

Monostable Operation

$$\Delta T = 1.1RC$$

Timer

TYPICAL APPLICATIONS

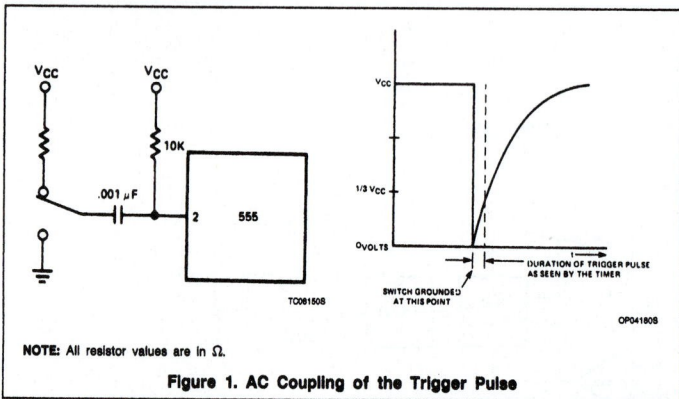

NOTE: All resistor values are in Ω.

Figure 1. AC Coupling of the Trigger Pulse

Trigger Pulse Width Requirements and Time Delays

Due to the nature of the trigger circuitry, the timer will trigger on the negative going edge of the input pulse. For the device to time out properly, it is necessary that the trigger voltage level be returned to some voltage greater than one third of the supply before the time out period. This can be achieved by making either the trigger pulse sufficiently short or by AC coupling into the trigger. By AC coupling the trigger, see Figure 1, a short negative going pulse is achieved when the trigger signal goes to ground. AC coupling is most frequently used in conjunction with a switch or a signal that goes to ground which initiates the timing cycle. Should the trigger be held low, without AC coupling, for a longer duration than the timing cycle the output will remain in a high state for the duration of the low trigger signal, without regard to the threshold comparator state. This is due to the predominance of Q_{15} on the base of Q_{16}, controlling the state of the bistable flip-flop. When the trigger signal then returns to a high level, the output will fall immediately. Thus, the output signal will follow the trigger signal in this case.

Another consideration is the "turn-off time". This is the measurement of the amount of time required after the threshold reaches 2/3 V_{CC} to turn the output low. To explain further, Q_1 at the threshold input turns on after reaching 2/3 V_{CC}, which then turns on Q_5, which turns on Q_6. Current from Q_6 turns on Q_{16} which turns Q_{17} off. This allows current from Q_{19} to turn on Q_{20} and Q_{24} to given an output low. These steps cause the 2µs max. delay as stated in the data sheet.

Also, a delay comparable to the turn-off time is the trigger release time. When the trigger is low, Q_{10} is on and turns on Q_{11} which turns on Q_{15}. Q_{15} turns off Q_{16} and allows Q_{17} to turn on. This turns off current to Q_{20} and Q_{24}, which results in output high. When the trigger is released, Q_{10} and Q_{11} shut off, Q_{15} turns off, Q_{16} turns on and the circuit then follows the same path and time delay explained as "turn off time". This trigger release time is very important in designing the trigger pulse width so as not to interfere with the output signal as explained previously.

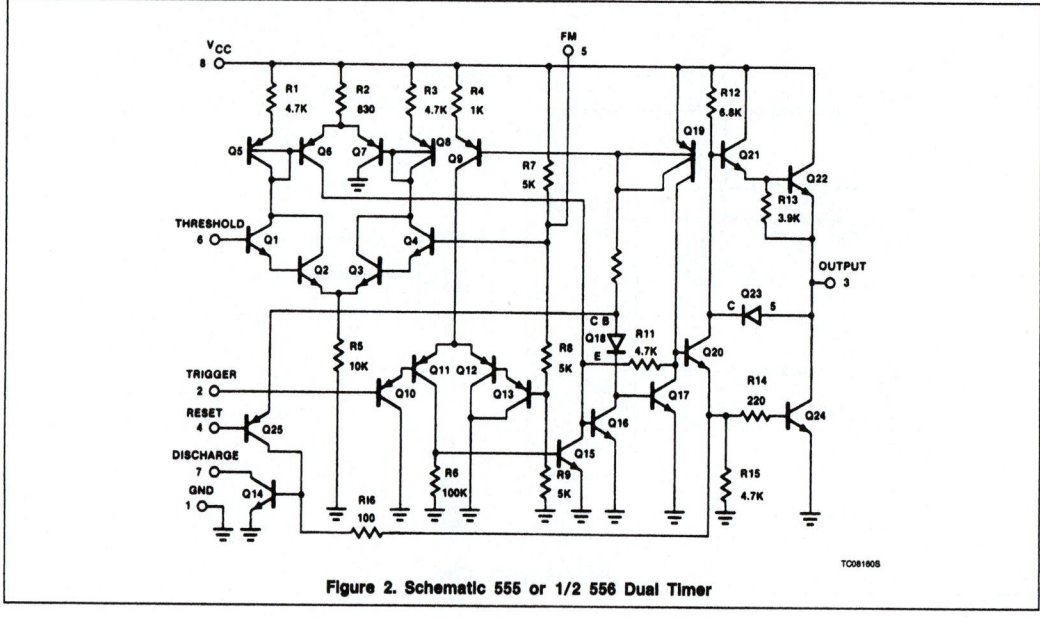

Figure 2. Schematic 555 or 1/2 556 Dual Timer

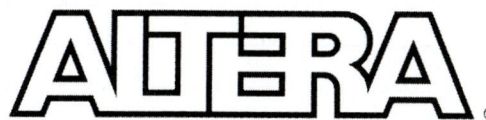

Includes
MAX 7000E &
MAX 7000S

MAX 7000

Programmable Logic Device Family

August 2000, ver. 6.02

Data Sheet

Features...

- High-performance, EEPROM-based programmable logic devices (PLDs) based on second-generation Multiple Array MatriX (MAX®) architecture
- 5.0-V in-system programmability (ISP) through the built-in IEEE Std. 1149.1 Joint Test Action Group (JTAG) interface available in MAX 7000S devices
- Includes 5.0-V MAX 7000 devices and 5.0-V ISP-based MAX 7000S devices
- Built-in JTAG boundary-scan test (BST) circuitry in MAX 7000S devices with 128 or more macrocells
- Complete EPLD family with logic densities ranging from 600 to 5,000 usable gates (see Tables 1 and 2)
- 5-ns pin-to-pin logic delays with up to 175.4-MHz counter frequencies (including interconnect)
- Peripheral component interconnect (PCI)-compliant devices available

For information on in-system programmable 3.3-V MAX 7000A or 2.5-V MAX 7000B devices, see the *MAX 7000A Programmable Logic Device Family Data Sheet* or the *MAX 7000B Programmable Logic Devices Advance Information Brief*.

Table 1. MAX 7000 Device Features

Feature	EPM7032	EPM7064	EPM7096	EPM7128E	EPM7160E	EPM7192E	EPM7256E
Usable gates	600	1,250	1,800	2,500	3,200	3,750	5,000
Macrocells	32	64	96	128	160	192	256
Logic array blocks	2	4	6	8	10	12	16
Maximum user I/O pins	36	68	76	100	104	124	164
t_{PD} (ns)	6	6	7.5	7.5	10	12	12
t_{SU} (ns)	5	5	6	6	7	7	7
t_{FSU} (ns)	2.5	2.5	3	3	3	3	3
t_{CO1} (ns)	4	4	4.5	4.5	5	6	6
f_{CNT} (MHz)	151.5	151.5	125.0	125.0	100.0	90.9	90.9

Table 2. MAX 7000S Device Features

Feature	EPM7032S	EPM7064S	EPM7128S	EPM7160S	EPM7192S	EPM7256S
Usable gates	600	1,250	2,500	3,200	3,750	5,000
Macrocells	32	64	128	160	192	256
Logic array blocks	2	4	8	10	12	16
Maximum user I/O pins	36	68	100	104	124	164
t_{PD} (ns)	5	5	6	6	7.5	7.5
t_{SU} (ns)	2.9	2.9	3.4	3.4	4.1	3.9
t_{FSU} (ns)	2.5	2.5	2.5	2.5	3	3
t_{CO1} (ns)	3.2	3.2	4	3.9	4.7	4.7
f_{CNT} (MHz)	175.4	175.4	147.1	149.3	125.0	128.2

...and More Features

- Open-drain output option in MAX 7000S devices
- Programmable macrocell flipflops with individual clear, preset, clock, and clock enable controls
- Programmable power-saving mode for a reduction of over 50% in each macrocell
- Configurable expander product-term distribution, allowing up to 32 product terms per macrocell
- 44 to 208 pins available in plastic J-lead chip carrier (PLCC), ceramic pin-grid array (PGA), plastic quad flat pack (PQFP), power quad flat pack (RQFP), and 1.0-mm thin quad flat pack (TQFP) packages
- Programmable security bit for protection of proprietary designs
- 3.3-V or 5.0-V operation
  - MultiVolt™ I/O interface operation, allowing devices to interface with 3.3-V or 5.0-V devices (MultiVolt I/O operation is not available in 44-pin packages)
  - Pin compatible with low-voltage MAX 7000A and MAX 7000B devices
- Enhanced features available in MAX 7000E and MAX 7000S devices
  - Six pin- or logic-driven output enable signals
  - Two global clock signals with optional inversion
  - Enhanced interconnect resources for improved routability
  - Fast input setup times provided by a dedicated path from I/O pin to macrocell registers
  - Programmable output slew-rate control
- Software design support and automatic place-and-route provided by Altera's MAX+PLUS® II development system for Windows-based PCs and Sun SPARCstation, HP 9000 Series 700/800, and IBM RISC System/6000 workstations, and the Quartus™ development system for Windows-based PCs and Sun SPARCstation and HP 9000 Series 700 workstations

- Additional design entry and simulation support provided by EDIF 2 0 0 and 3 0 0 netlist files, library of parameterized modules (LPM), Verilog HDL, VHDL, and other interfaces to popular EDA tools from manufacturers such as Cadence, Exemplar Logic, Mentor Graphics, OrCAD, Synopsys, and VeriBest
- Programming support
  - Altera's Master Programming Unit (MPU) and programming hardware from third-party manufacturers program all MAX 7000 devices
  - The BitBlaster™ serial download cable, ByteBlaster™ parallel port download cable, ByteBlasterMV™ parallel port download cable, and MasterBlaster™ serial/universal serial bus (USB) download cable program MAX 7000S devices

General Description

The MAX 7000 family of high-density, high-performance PLDs is based on Altera's second-generation MAX architecture. Fabricated with advanced CMOS technology, the EEPROM-based MAX 7000 family provides 600 to 5,000 usable gates, ISP, pin-to-pin delays as fast as 5 ns, and counter speeds of up to 175.4 MHz. MAX 7000S devices in the -5, -6, -7, and -10 speed grades as well as MAX 7000 and MAX 7000E devices in -5, -6, -7, -10P, and -12P speed grades comply with the PCI Special Interest Group (PCI SIG) *PCI Local Bus Specification, Revision 2.2*. See Table 3 for available speed grades.

Table 3. MAX 7000 Speed Grades

Device	Speed Grade									
	-5	-6	-7	-10P	-10	-12P	-12	-15	-15T	-20
EPM7032		✓	✓		✓		✓	✓	✓	
EPM7032S	✓	✓	✓		✓					
EPM7064		✓	✓		✓		✓	✓		
EPM7064S	✓	✓	✓		✓					
EPM7096		✓			✓		✓	✓		
EPM7128E		✓	✓	✓	✓		✓	✓		✓
EPM7128S	✓	✓	✓		✓			✓		
EPM7160E			✓	✓	✓		✓	✓		✓
EPM7160S	✓	✓			✓			✓		
EPM7192E						✓	✓	✓		✓
EPM7192S		✓			✓			✓		
EPM7256E						✓	✓	✓		✓
EPM7256S			✓		✓			✓		

The MAX 7000E devices—including the EPM7128E, EPM7160E, EPM7192E, and EPM7256E devices—have several enhanced features: additional global clocking, additional output enable controls, enhanced interconnect resources, fast input registers, and a programmable slew rate.

In-system programmable MAX 7000 devices—called MAX 7000S devices—include the EPM7032S, EPM7064S, EPM7128S, EPM7160S, EPM7192S, and EPM7256S devices. MAX 7000S devices have the enhanced features of MAX 7000E devices as well as JTAG BST circuitry in devices with 128 or more macrocells, ISP, and an open-drain output option. See Table 4.

Table 4. MAX 7000 Device Features			
Feature	EPM7032 EPM7064 EPM7096	All MAX 7000E Devices	All MAX 7000S Devices
ISP via JTAG interface			✓
JTAG BST circuitry			✓ *(1)*
Open-drain output option			✓
Fast input registers		✓	✓
Six global output enables		✓	✓
Two global clocks		✓	✓
Slew-rate control		✓	✓
MultiVolt interface *(2)*	✓	✓	✓
Programmable register	✓	✓	✓
Parallel expanders	✓	✓	✓
Shared expanders	✓	✓	✓
Power-saving mode	✓	✓	✓
Security bit	✓	✓	✓
PCI-compliant devices available	✓	✓	✓

Notes:
(1)   Available in EPM7128S, EPM7160S, EPM7192S, and EPM7256S devices only.
(2)   The MultiVolt I/O interface is not available in 44-pin packages.

The MAX 7000 architecture supports 100% TTL emulation and high-density integration of SSI, MSI, and LSI logic functions. It easily integrates multiple devices ranging from PALs, GALs, and 22V10s to MACH and pLSI devices. MAX 7000 devices are available in a wide range of packages, including PLCC, PGA, PQFP, RQFP, and TQFP packages. See Table 5.

Table 5. MAX 7000 Maximum User I/O Pins   *Note (1)*

Device	44-Pin PLCC	44-Pin PQFP	44-Pin TQFP	68-Pin PLCC	84-Pin PLCC	100-Pin PQFP	100-Pin TQFP	160-Pin PQFP	160-Pin PGA	192-Pin PGA	208-Pin PQFP	208-Pin RQFP
EPM7032	36	36	36									
EPM7032S	36		36									
EPM7064	36		36	52	68	68						
EPM7064S	36		36		68		68					
EPM7096				52	64	76						
EPM7128E					68	84		100				
EPM7128S					68	84	84 *(2)*	100				
EPM7160E					64	84		104				
EPM7160S					64		84 *(2)*	104				
EPM7192E								124	124			
EPM7192S								124				
EPM7256E								132 *(2)*		164		164
EPM7256S											164 *(2)*	164

Notes:
(1) When the JTAG interface in MAX 7000S devices is used, four I/O pins become JTAG pins.
(2) Perform a complete thermal analysis before committing a design to this device package. See the *Operating Requirements for Altera Devices Data Sheet* for more information.

MAX 7000 devices use CMOS EEPROM cells to implement logic functions. The user-configurable MAX 7000 architecture accommodates a variety of independent combinatorial and sequential logic functions. The devices can be reprogrammed for quick and efficient iterations during design development and debug cycles, and can be programmed and erased up to 100 times.

MAX 7000 devices contain from 32 to 256 macrocells that are combined into groups of 16 macrocells, called logic array blocks (LABs). Each macrocell has a programmable-AND/fixed-OR array and a configurable register with independently programmable clock, clock enable, clear, and preset functions. To build complex logic functions, each macrocell can be supplemented with both shareable expander product terms and high-speed parallel expander product terms to provide up to 32 product terms per macrocell.

The MAX 7000 family provides programmable speed/power optimization. Speed-critical portions of a design can run at high speed/full power, while the remaining portions run at reduced speed/low power. This speed/power optimization feature enables the designer to configure one or more macrocells to operate at 50% or lower power while adding only a nominal timing delay. MAX 7000E and MAX 7000S devices also provide an option that reduces the slew rate of the output buffers, minimizing noise transients when non-speed-critical signals are switching. The output drivers of all MAX 7000 devices (except 44-pin devices) can be set for either 3.3-V or 5.0-V operation, allowing MAX 7000 devices to be used in mixed-voltage systems.

The MAX 7000 family is supported by the Quartus and MAX+PLUS II development systems, a single, integrated package that allows schematic, text—including VHDL, Verilog HDL, and the Altera Hardware Description Language (AHDL)—and waveform design entry, compilation and logic synthesis, simulation and timing analysis, and device programming. The Quartus and MAX+PLUS II software provides EDIF 2 0 0 and 3 0 0, LPM, VHDL, Verilog HDL, and other interfaces for additional design entry and simulation support from other industry-standard PC- and UNIX-workstation-based EDA tools. The MAX+PLUS II software runs on Windows-based PCs, as well as Sun SPARCstation, HP 9000 Series 700/800, and IBM RISC System/6000 workstations. The Quartus software runs on Windows-based PCs, as well as Sun SPARCstation and HP 9000 Series 700 workstations.

For more information on development tools, go to the *MAX+PLUS II Programmable Logic Development System & Software Data Sheet.*

Functional Description

The MAX 7000 architecture includes the following elements:

- Logic array blocks
- Macrocells
- Expander product terms (shareable and parallel)
- Programmable interconnect array
- I/O control blocks

The MAX 7000 architecture includes four dedicated inputs that can be used as general-purpose inputs or as high-speed, global control signals (clock, clear, and two output enable signals) for each macrocell and I/O pin. Figure 1 shows the architecture of EPM7032, EPM7064, and EPM7096 devices.

Figure 1. EPM7032, EPM7064 & EPM7096 Device Block Diagram

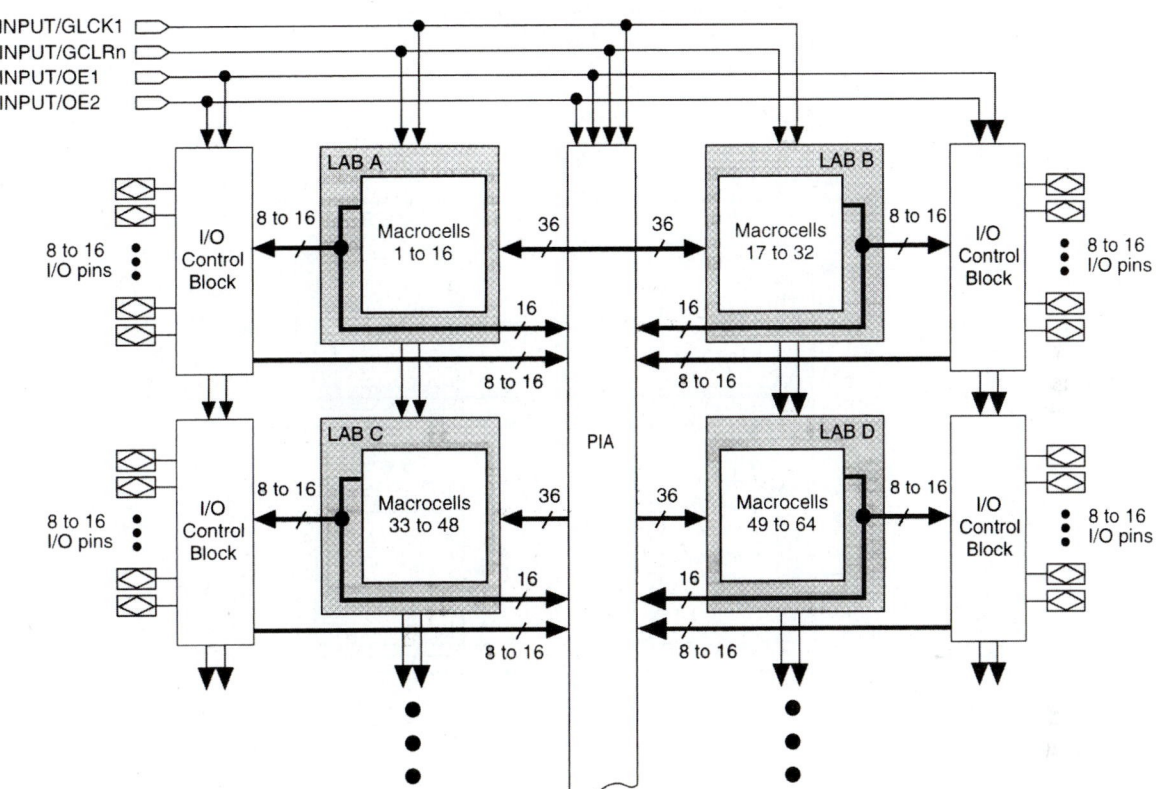

Figure 2 shows the architecture of MAX 7000E and MAX 7000S devices.

Figure 2. MAX 7000E & MAX 7000S Device Block Diagram

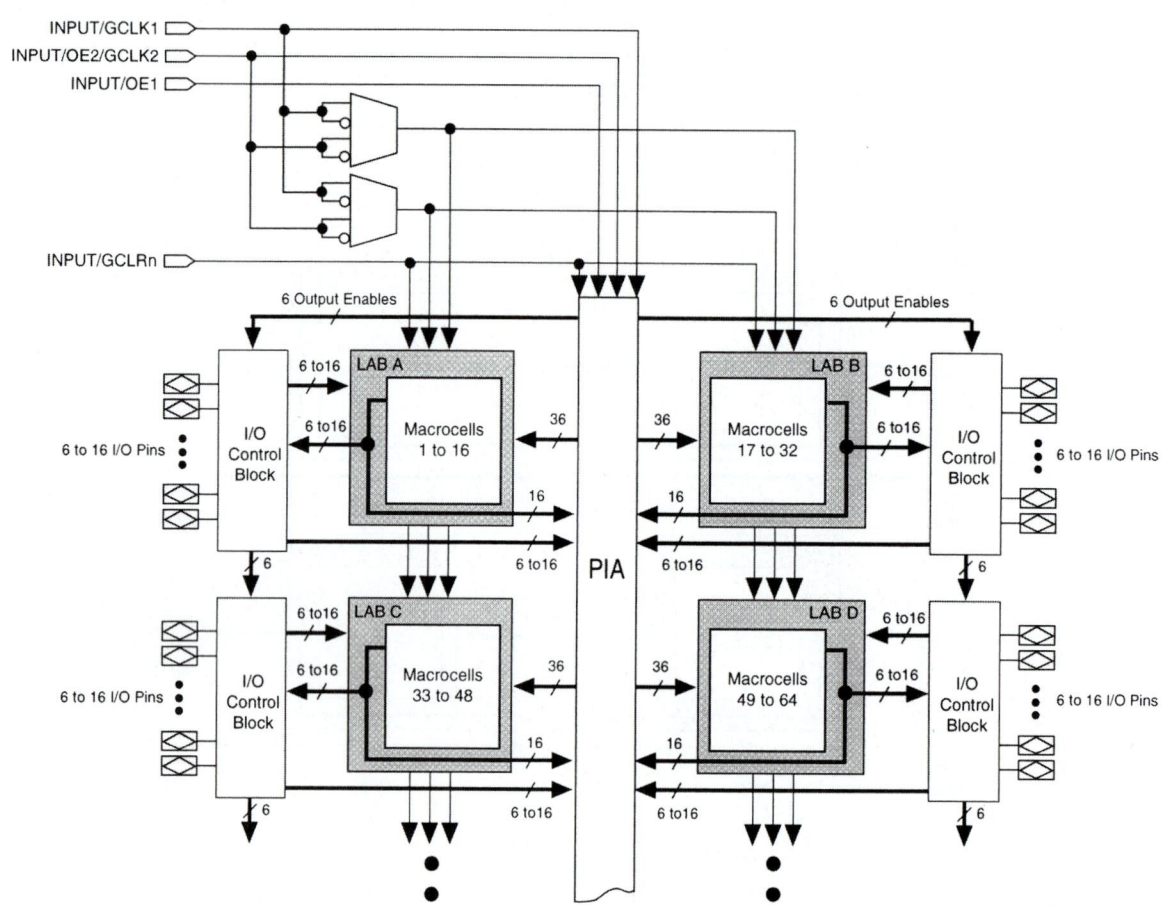

Logic Array Blocks

The MAX 7000 device architecture is based on the linking of high-performance, flexible, logic array modules called logic array blocks (LABs). LABs consist of 16-macrocell arrays, as shown in Figures 1 and 2. Multiple LABs are linked together via the programmable interconnect array (PIA), a global bus that is fed by all dedicated inputs, I/O pins, and macrocells.

Each LAB is fed by the following signals:

- 36 signals from the PIA that are used for general logic inputs
- Global controls that are used for secondary register functions
- Direct input paths from I/O pins to the registers that are used for fast setup times for MAX 7000E and MAX 7000S devices

Macrocells

The MAX 7000 macrocell can be individually configured for either sequential or combinatorial logic operation. The macrocell consists of three functional blocks: the logic array, the product-term select matrix, and the programmable register. The macrocell of EPM7032, EPM7064, and EPM7096 devices is shown in Figure 3.

Figure 3. EPM7032, EPM7064 & EPM7096 Device Macrocell

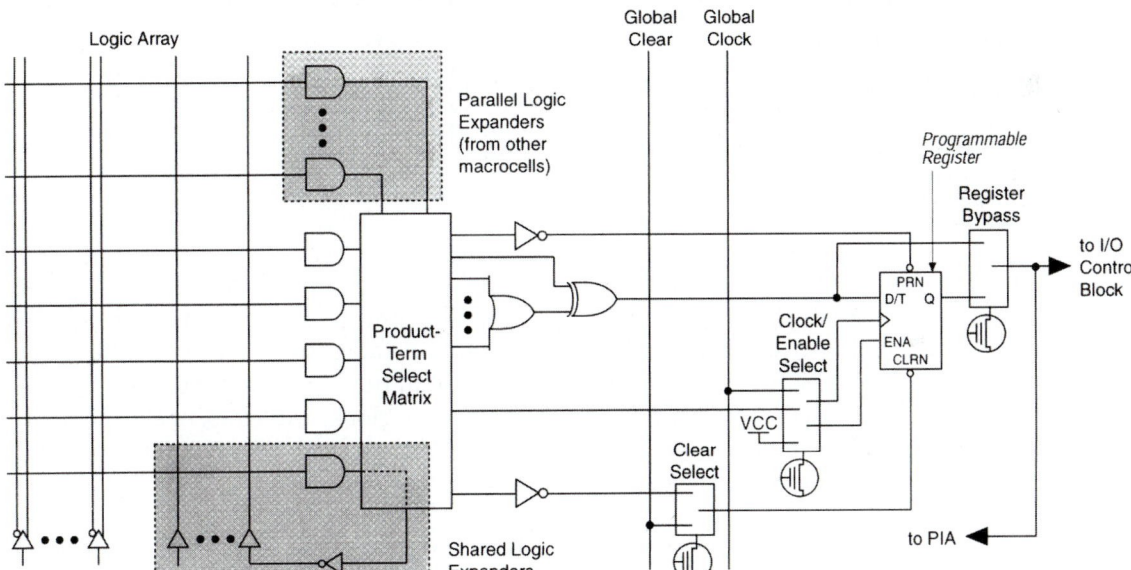

The macrocell of MAX 7000E and MAX 7000S devices is shown in Figure 4.

Figure 4. MAX 7000E & MAX 7000S Device Macrocell

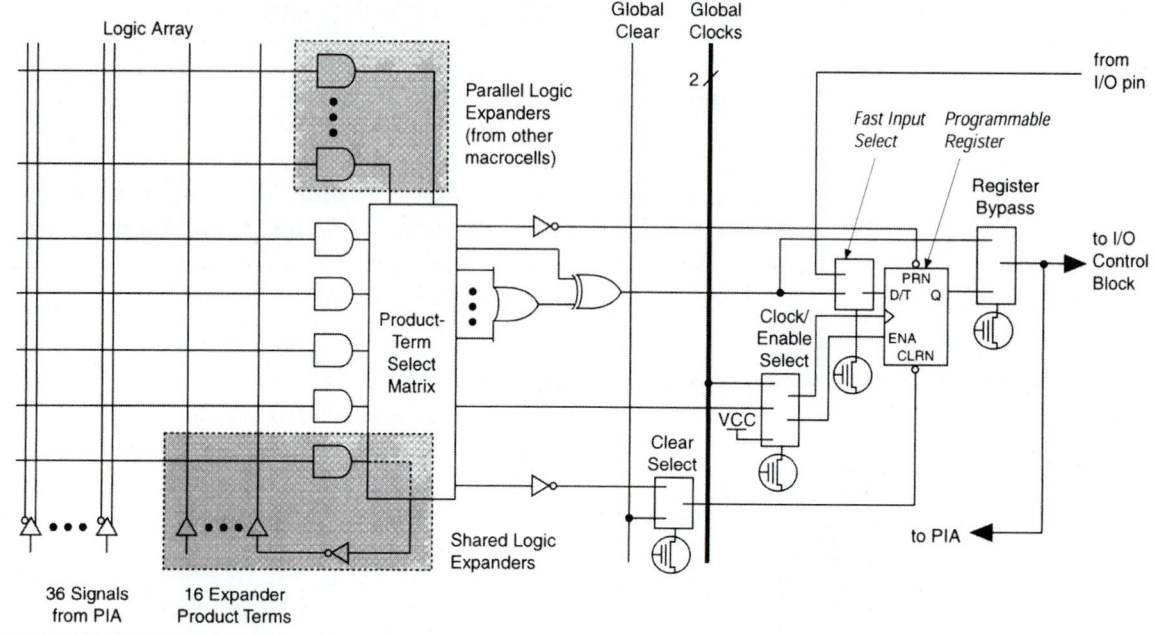

Combinatorial logic is implemented in the logic array, which provides five product terms per macrocell. The product-term select matrix allocates these product terms for use as either primary logic inputs (to the OR and XOR gates) to implement combinatorial functions, or as secondary inputs to the macrocell's register clear, preset, clock, and clock enable control functions. Two kinds of expander product terms ("expanders") are available to supplement macrocell logic resources:

■ Shareable expanders, which are inverted product terms that are fed back into the logic array
■ Parallel expanders, which are product terms borrowed from adjacent macrocells

The Quartus and MAX+PLUS II software automatically optimizes product-term allocation according to the logic requirements of the design.

For registered functions, each macrocell flipflop can be individually programmed to implement D, T, JK, or SR operation with programmable clock control. The flipflop can be bypassed for combinatorial operation. During design entry, the designer specifies the desired flipflop type; the Quartus and MAX+PLUS II software then selects the most efficient flipflop operation for each registered function to optimize resource utilization.

Operating Conditions

Tables 10 through 15 provide information about absolute maximum ratings, recommended operating conditions, operating conditions, and capacitance for 5.0-V MAX 7000 devices.

Table 10. MAX 7000 5.0-V Device Absolute Maximum Ratings   *Note (1)*

Symbol	Parameter	Conditions	Min	Max	Unit
V_{CC}	Supply voltage	With respect to ground *(2)*	−2.0	7.0	V
V_I	DC input voltage		−2.0	7.0	V
I_{OUT}	DC output current, per pin		−25	25	mA
T_{STG}	Storage temperature	No bias	−65	150	° C
T_{AMB}	Ambient temperature	Under bias	−65	135	° C
T_J	Junction temperature	Ceramic packages, under bias		150	° C
		PQFP and RQFP packages, under bias		135	° C

Table 11. MAX 7000 5.0-V Device Recommended Operating Conditions

Symbol	Parameter	Conditions	Min	Max	Unit
V_{CCINT}	Supply voltage for internal logic and input buffers	*(3), (4)*	4.75 (4.50)	5.25 (5.50)	V
V_{CCIO}	Supply voltage for output drivers, 5.0-V operation	*(3), (4)*	4.75 (4.50)	5.25 (5.50)	V
	Supply voltage for output drivers, 3.3-V operation	*(3), (4), (5)*	3.00 (3.00)	3.60 (3.60)	V
V_{CCISP}	Supply voltage during ISP	*(6)*	4.75	5.25	V
V_I	Input voltage		−0.5 *(7)*	$V_{CCINT} + 0.5$	V
V_O	Output voltage		0	V_{CCIO}	V
T_A	Ambient temperature	For commercial use	0	70	° C
		For industrial use	−40	85	° C
T_J	Junction temperature	For commercial use	0	90	° C
		For industrial use	−40	105	° C
t_R	Input rise time			40	ns
t_F	Input fall time			40	ns

DC Characteristics Over Recommended Operating Conditions

Symbol	Parameter	Test Conditions	Min	Max	Units
V_{OH}	Output high voltage for 5 V operation	I_{OH} = -4.0 mA V_{CC} = Min	2.4		V
	Output high voltage for 3.3 V operation	I_{OH} = -3.2 mA V_{CC} = Min	2.4		V
V_{OL}	Output low voltage for 5 V operation	I_{OL} = 24 mA V_{CC} = Min		0.5	V
	Output low voltage for 3.3 V operation	I_{OL} = 10 mA V_{CC} = Min		0.4	V
I_{IL}	Input leakage current	V_{CC} = Max V_{IN} = GND or V_{CC}		±10.0	μA
I_{IH}	I/O high-Z leakage current	V_{CC} = Max V_{IN} = GND or V_{CC}		±10.0	μA
C_{IN}	I/O capacitance	V_{IN} = GND f = 1.0 MHz		10.0	pF
I_{CC}	Operating Supply Current (low power mode, active)	V_I = GND, No load f = 1.0 MHz	100 (Typ)		ma

AC Characteristics

Symbol	Parameter	XC95108-7		XC95108-10		XC95108-15		XC95108-20		Units
		Min	Max	Min	Max	Min	Max	Min	Max	
t_{PD}	I/O to output valid		7.5		10.0		15.0		20.0	ns
t_{SU}	I/O setup time before GCK	4.5		6.0		8.0		10.0		ns
t_H	I/O hold time after GCK	0.0		0.0		0.0		0.0		ns
t_{CO}	GCK to output valid		4.5		6.0		8.0		10.0	ns
f_{CNT}[1]	16-bit counter frequency	125.0		111.1		95.2		83.3		MHz
f_{SYSTEM}[2]	Multiple FB internal operating frequency	83.3		66.7		55.6		50.0		MHz
t_{PSU}	I/O setup time before p-term clock input	0.5		2.0		4.0		4.0		ns
t_{PH}	I/O hold time after p-term clock input	4.0		4.0		4.0		6.0		ns
t_{PCO}	P-term clock to output valid		8.5		10.0		12.0		16.0	ns
t_{OE}	GTS to output valid		5.5		6.0		11.0		16.0	ns
t_{OD}	GTS to output disable		5.5		6.0		11.0		16.0	ns
t_{POE}	Product term OE to output enabled		9.5		10.0		14.0		18.0	ns
t_{POD}	Product term OE to output disabled		9.5		10.0		14.0		18.0	ns
t_{WLH}	GCK pulse width (High or Low)	4.0		4.5		5.5		5.5		ns

Note: 1. f_{CNT} is the fastest 16-bit counter frequency available, using the local feedback when applicable.
f_{CNT} is also the Export Control Maximum flip-flop toggle rate, f_{TOG}.
2. f_{SYSTEM} is the internal operating frequency for general purpose system designs spanning multiple FBs.

Absolute Maximum Ratings

Symbol	Parameter	Value	Units
V_{CC}	Supply voltage relative to GND	-0.5 to 7.0	V
V_{IN}	DC input voltage relative to GND	-0.5 to V_{CC} + 0.5	V
V_{TS}	Voltage applied to 3-state output with respect to GND	-0.5 to V_{CC} + 0.5	V
T_{STG}	Storage temperature	-65 to +150	°C
T_{SOL}	Max soldering temperature (10 s @ 1/16 in = 1.5 mm)	+260	°C

Warning: Stresses beyond those listed under Absolute Maximum Ratings may cause permanent damage to the device. These are stress ratings only, and functional operation of the device at these or any other conditions beyond those listed under Recommended Operating Conditions is not implied. Exposure to Absolute Maximum Rating conditions for extended periods of time may affect device reliability.

Recommended Operation Conditions[1]

Symbol	Parameter	Min	Max	Units
V_{CCINT}	Supply voltage for internal logic and input buffer	4.75 (4.5)	5.25 (5.5)	V
V_{CCIO}	Supply voltage for output drivers for 5 V operation	4.75 (4.5)	5.25 (5.5)	V
	Supply voltage for output drivers for 3.3 V operation	3.0	3.6	V
V_{IL}	Low-level input voltage	0	0.80	V
V_{IH}	High-level input voltage	2.0	V_{CCINT} +0.5	V
V_O	Output voltage	0	V_{CCIO}	V

Note: 1. Numbers in parenthesis are for industrial-temperature range versions.

Endurance Characteristics

Symbol	Parameter	Min	Max	Units
t_{DR}	Data Retention	20	-	Years
N_{PE}	Program/Erase Cycles	10,000	-	Cycles

DC Characteristics Over Recommended Operating Conditions

Symbol	Parameter	Test Conditions	Min	Max	Units
V_{OH}	Output high voltage for 5 V operation	I_{OH} = -4.0 mA V_{CC} = Min	2.4		V
	Output high voltage for 3.3 V operation	I_{OH} = -3.2 mA V_{CC} = Min	2.4		V
V_{OL}	Output low voltage for 5 V operation	I_{OL} = 24 mA V_{CC} = Min		0.5	V
	Output low voltage for 3.3 V operation	I_{OL} = 10 mA V_{CC} = Min		0.4	V
I_{IL}	Input leakage current	V_{CC} = Max V_{IN} = GND or V_{CC}		±10.0	µA
I_{IH}	I/O high-Z leakage current	V_{CC} = Max V_{IN} = GND or V_{CC}		±10.0	µA
C_{IN}	I/O capacitance	V_{IN} = GND f = 1.0 MHz		10.0	pF
I_{CC}	Operating Supply Current (low power mode, active)	V_I = GND, No load f = 1.0 MHz	100 (Typ)		ma

AC Characteristics

Symbol	Parameter	XC95108-7		XC95108-10		XC95108-15		XC95108-20		Units
		Min	Max	Min	Max	Min	Max	Min	Max	
t_{PD}	I/O to output valid		7.5		10.0		15.0		20.0	ns
t_{SU}	I/O setup time before GCK	4.5		6.0		8.0		10.0		ns
t_H	I/O hold time after GCK	0.0		0.0		0.0		0.0		ns
t_{CO}	GCK to output valid		4.5		6.0		8.0		10.0	ns
f_{CNT}[1]	16-bit counter frequency	125.0		111.1		95.2		83.3		MHz
f_{SYSTEM}[2]	Multiple FB internal operating frequency	83.3		66.7		55.6		50.0		MHz
t_{PSU}	I/O setup time before p-term clock input	0.5		2.0		4.0		4.0		ns
t_{PH}	I/O hold time after p-term clock input	4.0		4.0		4.0		6.0		ns
t_{PCO}	P-term clock to output valid		8.5		10.0		12.0		16.0	ns
t_{OE}	GTS to output valid		5.5		6.0		11.0		16.0	ns
t_{OD}	GTS to output disable		5.5		6.0		11.0		16.0	ns
t_{POE}	Product term OE to output enabled		9.5		10.0		14.0		18.0	ns
t_{POD}	Product term OE to output disabled		9.5		10.0		14.0		18.0	ns
t_{WLH}	GCK pulse width (High or Low)	4.0		4.5		5.5		5.5		ns

Note: 1. f_{CNT} is the fastest 16-bit counter frequency available, using the local feedback when applicable.
f_{CNT} is also the Export Control Maximum flip-flop toggle rate, f_{TOG}.
2. f_{SYSTEM} is the internal operating frequency for general purpose system designs spanning multiple FBs.

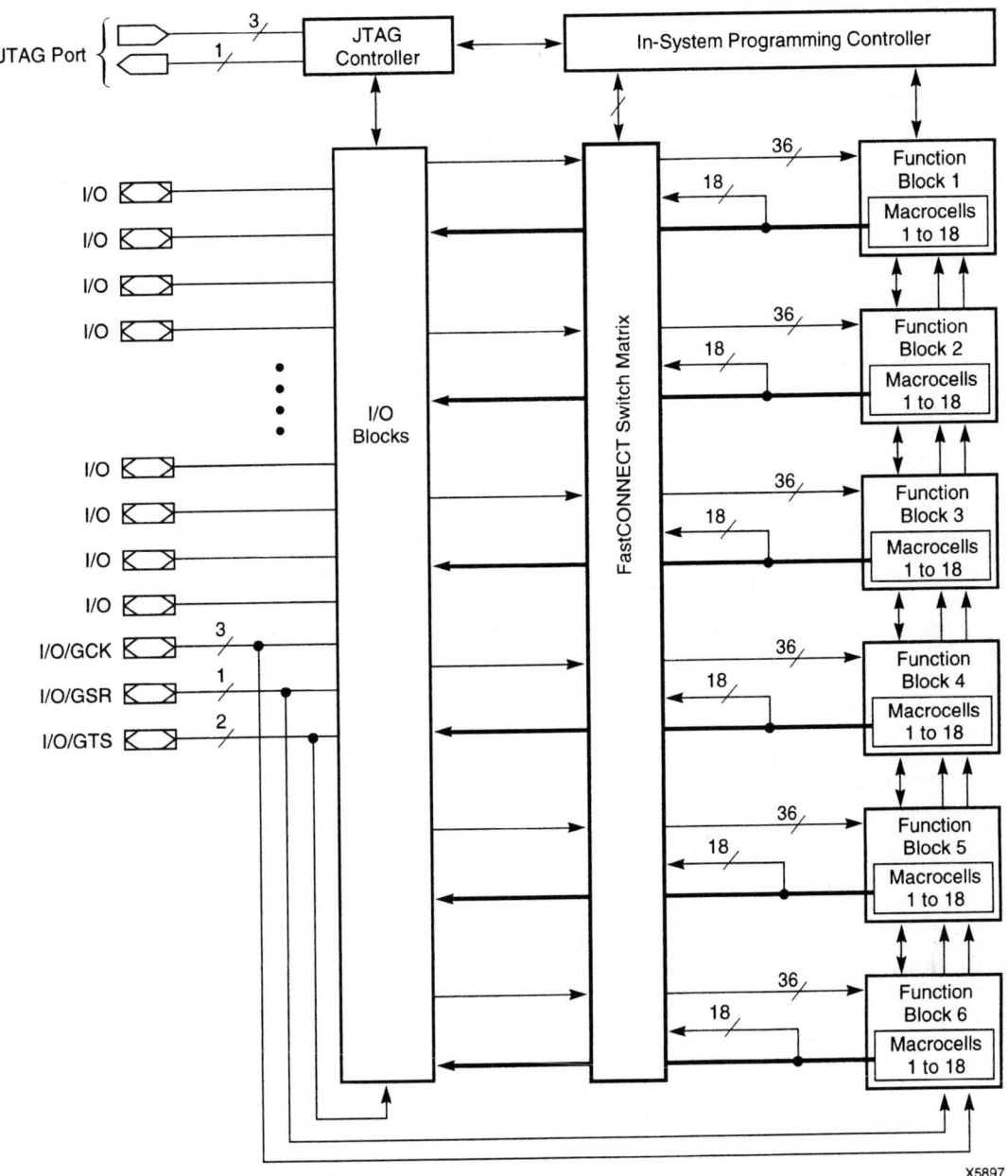

Figure 2: XC95108 Architecture

Note: Function Block outputs (indicated by the bold line) drive the I/O Blocks directly

XC95108 In-System Programmable CPLD

December 4, 1998 (Version 3.0)

Product Specification

Features

- 7.5 ns pin-to-pin logic delays on all pins
- f_{CNT} to 125 MHz
- 108 macrocells with 2400 usable gates
- Up to 108 user I/O pins
- 5 V in-system programmable (ISP)
  - Endurance of 10,000 program/erase cycles
  - Program/erase over full commercial voltage and temperature range
- Enhanced pin-locking architecture
- Flexible 36V18 Function Block
  - 90 product terms drive any or all of 18 macrocells within Function Block
  - Global and product term clocks, output enables, set and reset signals
- Extensive IEEE Std 1149.1 boundary-scan (JTAG) support
- Programmable power reduction mode in each macrocell
- Slew rate control on individual outputs
- User programmable ground pin capability
- Extended pattern security features for design protection
- High-drive 24 mA outputs
- 3.3 V or 5 V I/O capability
- Advanced CMOS 5V FastFLASH technology
- Supports parallel programming of more than one XC9500 concurrently
- Available in 84-pin PLCC, 100-pin PQFP, 100-pin TQFP and 160-pin PQFP packages

Description

The XC95108 is a high-performance CPLD providing advanced in-system programming and test capabilities for general purpose logic integration. It is comprised of six 36V18 Function Blocks, providing 2,400 usable gates with propagation delays of 7.5 ns. See Figure 2 for the architecture overview.

Power Management

Power dissipation can be reduced in the XC95108 by configuring macrocells to standard or low-power modes of operation. Unused macrocells are turned off to minimize power dissipation.

Operating current for each design can be approximated for specific operating conditions using the following equation:

I_{CC} (mA) =

MC_{HP} (1.7) + MC_{LP} (0.9) + MC (0.006 mA/MHz) f

Where:

MC_{HP} = Macrocells in high-performance mode

MC_{LP} = Macrocells in low-power mode

MC = Total number of macrocells used

f = Clock frequency (MHz)

Figure 1 shows a typical calculation for the XC95108 device.

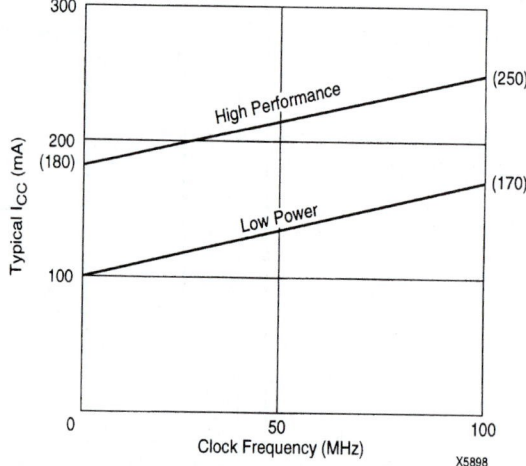

Figure 1: Typical I_{CC} vs. Frequency for XC95108

Operating Conditions

Tables 10 through 15 provide information about absolute maximum ratings, recommended operating conditions, operating conditions, and capacitance for 5.0-V MAX 7000 devices.

Table 10. MAX 7000 5.0-V Device Absolute Maximum Ratings   *Note (1)*

Symbol	Parameter	Conditions	Min	Max	Unit
V_{CC}	Supply voltage	With respect to ground *(2)*	−2.0	7.0	V
V_I	DC input voltage		−2.0	7.0	V
I_{OUT}	DC output current, per pin		−25	25	mA
T_{STG}	Storage temperature	No bias	−65	150	° C
T_{AMB}	Ambient temperature	Under bias	−65	135	° C
T_J	Junction temperature	Ceramic packages, under bias		150	° C
		PQFP and RQFP packages, under bias		135	° C

Table 11. MAX 7000 5.0-V Device Recommended Operating Conditions

Symbol	Parameter	Conditions	Min	Max	Unit
V_{CCINT}	Supply voltage for internal logic and input buffers	*(3), (4)*	4.75 (4.50)	5.25 (5.50)	V
V_{CCIO}	Supply voltage for output drivers, 5.0-V operation	*(3), (4)*	4.75 (4.50)	5.25 (5.50)	V
	Supply voltage for output drivers, 3.3-V operation	*(3), (4), (5)*	3.00 (3.00)	3.60 (3.60)	V
V_{CCISP}	Supply voltage during ISP	*(6)*	4.75	5.25	V
V_I	Input voltage		−0.5 *(7)*	$V_{CCINT} + 0.5$	V
V_O	Output voltage		0	V_{CCIO}	V
T_A	Ambient temperature	For commercial use	0	70	° C
		For industrial use	−40	85	° C
T_J	Junction temperature	For commercial use	0	90	° C
		For industrial use	−40	105	° C
t_R	Input rise time			40	ns
t_F	Input fall time			40	ns

Table 12. MAX 7000 5.0-V Device DC Operating Conditions   Note (8)

Symbol	Parameter	Conditions	Min	Max	Unit
V_{IH}	High-level input voltage		2.0	V_{CCINT} + 0.5	V
V_{IL}	Low-level input voltage		−0.5 *(7)*	0.8	V
V_{OH}	5.0-V high-level TTL output voltage	I_{OH} = −4 mA DC, V_{CCIO} = 4.75 V *(9)*	2.4		V
	3.3-V high-level TTL output voltage	I_{OH} = −4 mA DC, V_{CCIO} = 3.00 V *(9)*	2.4		V
	3.3-V high-level CMOS output voltage	I_{OH} = −0.1 mA DC, V_{CCIO} = 3.0 V *(9)*	V_{CCIO} − 0.2		V
V_{OL}	5.0-V low-level TTL output voltage	I_{OL} = 12 mA DC, V_{CCIO} = 4.75 V *(10)*		0.45	V
	3.3-V low-level TTL output voltage	I_{OL} = 12 mA DC, V_{CCIO} = 3.00 V *(10)*		0.45	V
	3.3-V low-level CMOS output voltage	I_{OL} = 0.1 mA DC, V_{CCIO} = 3.0 V *(10)*		0.2	V
I_I	Leakage current of dedicated input pins	V_I = V_{CC} or ground	−10	10	µA
I_{OZ}	I/O pin tri-state output off-state current	V_O = V_{CC} or ground *(11)*	−40	40	µA

Table 13. MAX 7000 5.0-V Device Capacitance: EPM7032, EPM7064 & EPM7096 Devices   Note (12)

Symbol	Parameter	Conditions	Min	Max	Unit
C_{IN}	Input pin capacitance	V_{IN} = 0 V, f = 1.0 MHz		12	pF
$C_{I/O}$	I/O pin capacitance	V_{OUT} = 0 V, f = 1.0 MHz		12	pF

Table 14. MAX 7000 5.0-V Device Capacitance: MAX 7000E Devices   Note (12)

Symbol	Parameter	Conditions	Min	Max	Unit
C_{IN}	Input pin capacitance	V_{IN} = 0 V, f = 1.0 MHz		15	pF
$C_{I/O}$	I/O pin capacitance	V_{OUT} = 0 V, f = 1.0 MHz		15	pF

Table 15. MAX 7000 5.0-V Device Capacitance: MAX 7000S Devices   Note (12)

Symbol	Parameter	Conditions	Min	Max	Unit
C_{IN}	Dedicated input pin capacitance	V_{IN} = 0 V, f = 1.0 MHz		10	pF
$C_{I/O}$	I/O pin capacitance	V_{OUT} = 0 V, f = 1.0 MHz		10	pF

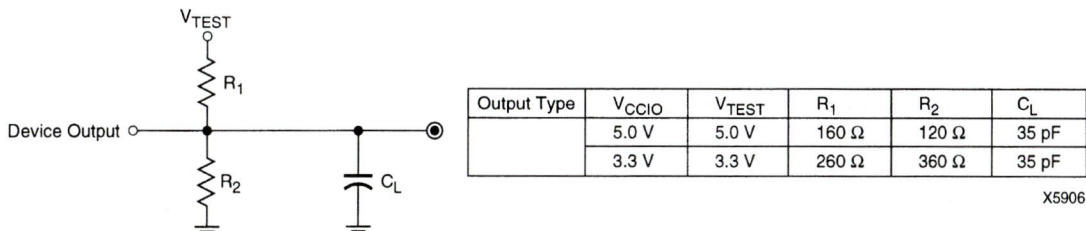

Figure 3: AC Load Circuit

Output Type	V_{CCIO}	V_{TEST}	R₁	R₂	C_L
	5.0 V	5.0 V	160 Ω	120 Ω	35 pF
	3.3 V	3.3 V	260 Ω	360 Ω	35 pF

X5906

Internal Timing Parameters

Symbol	Parameter	XC95108-7		XC95108-10		XC95108-15		XC95108-20		Units
		Min	Max	Min	Max	Min	Max	Min	Max	
Buffer Delays										
t_{IN}	Input buffer delay		2.5		3.5		4.5		6.5	ns
t_{GCK}	GCK buffer delay		1.5		2.5		3.0		3.0	ns
t_{GSR}	GSR buffer delay		4.5		6.0		7.5		9.5	ns
t_{GTS}	GTS buffer delay		5.5		6.0		11.0		16.0	ns
t_{OUT}	Output buffer delay		2.5		3.0		4.5		6.5	ns
t_{EN}	Output buffer enable/disable delay		0.0		0.0		0.0		0.0	ns
Product Term Control Delays										
t_{PTCK}	Product term clock delay		3.0		3.0		2.5		2.5	ns
t_{PTSR}	Product term set/reset delay		2.0		2.5		3.0		3.0	ns
t_{PTTS}	Product term 3-state delay		4.5		3.5		5.0		5.0	ns
Internal Register and Combinatorial delays										
t_{PDI}	Combinatorial logic propagation delay		0.5		1.0		3.0		4.0	ns
t_{SUI}	Register setup time	1.5		2.5		3.5		3.5		ns
t_{HI}	Register hold time	3.0		3.5		4.5		6.5		ns
t_{COI}	Register clock to output valid time		0.5		0.5		0.5		0.5	ns
t_{AOI}	Register async. S/R to output delay		6.5		7.0		8.0		8.0	ns
t_{RAI}	Register async. S/R recovery before clock	7.5		10.0		10.0		10.0		ns
t_{LOGI}	Internal logic delay		2.0		2.5		3.0		3.0	ns
t_{LOGILP}	Internal low power logic delay		10.0		11.0		11.5		11.5	ns
Feedback Delays										
t_F	FastCONNECT matrix feedback delay		8.0		9.5		11.0		13.0	ns
t_{LF}	Function Block local feeback delay		4.0		3.5		3.5		5.0	ns
Time Adders										
t_{PTA}[3]	Incremental Product Term Allocator delay		1.0		1.0		1.0		1.5	ns
t_{SLEW}	Slew-rate limited delay		4.0		4.5		5.0		5.5	ns

Note: 3. t_{PTA} is multiplied by the span of the function as defined in the family data sheet.

XC95108 In-System Programmable CPLD

XC95108 I/O Pins

Function Block	Macrocell	PC84	PQ100	TQ100	PQ160	BScan Order	Notes	Function Block	Macrocell	PC84	PQ100	TQ100	PQ160	BScan Order	Notes
1	1	–	–	–	25	321		3	1	–	–	–	45	213	
1	2	1	15	13	21	318		3	2	14	31	29	47	210	
1	3	2	16	14	22	315		3	3	15	32	30	49	207	
1	4	–	21	19	29	312		3	4	–	36	34	57	204	
1	5	3	17	15	23	309		3	5	17	34	32	54	201	
1	6	4	18	16	24	306		3	6	18	35	33	56	198	
1	7	–	–	–	27	303		3	7	–	–	–	50	195	
1	8	5	19	17	26	300		3	8	19	37	35	58	192	
1	9	6	20	18	28	297		3	9	20	38	36	59	189	
1	10	–	26	24	36	294		3	10	–	45	43	69	186	
1	11	7	22	20	30	291		3	11	21	39	37	60	183	
1	12	9	24	22	33	288	[1]	3	12	23	41	39	62	180	
1	13	–	–	–	34	285		3	13	–	–	–	52	177	
1	14	10	25	23	35	282	[1]	3	14	24	42	40	63	174	
1	15	11	27	25	37	279		3	15	25	43	41	64	171	
1	16	12	29	27	42	276	[1]	3	16	26	44	42	68	168	
1	17	13	30	28	44	273		3	17	31	51	49	77	165	
1	18	–	–	–	43	270		3	18	–	–	–	74	162	
2	1	–	–	–	158	267		4	1	–	–	–	123	159	
2	2	71	98	96	154	264		4	2	57	83	81	134	156	
2	3	72	99	97	156	261		4	3	58	84	82	135	153	
2	4	–	4	2	4	258		4	4	–	82	80	133	150	
2	5	74	1	99	159	255	[1]	4	5	61	87	85	138	147	
2	6	75	3	1	2	252		4	6	62	88	86	139	144	
2	7	–	–	–	9	249		4	7	–	–	–	128	141	
2	8	76	5	3	6	246	[1]	4	8	63	89	87	140	138	
2	9	77	6	4	8	243	[1]	4	9	65	91	89	142	135	
2	10	–	9	7	12	240		4	10	–	–	–	147	132	
2	11	79	8	6	11	237		4	11	66	92	90	143	129	
2	12	80	10	8	13	234		4	12	67	93	91	144	126	
2	13	–	–	–	14	231		4	13	–	–	–	153	123	
2	14	81	11	9	15	228		4	14	68	95	93	146	120	
2	15	82	12	10	17	225		4	15	69	96	94	148	117	
2	16	83	13	11	18	222		4	16	–	94	92	145	114	
2	17	84	14	12	19	219		4	17	70	97	95	152	111	
2	18	–	–	–	16	216		4	18	–	–	–	155	108	

Notes: [1] Global control pin

XC95108 I/O Pins (continued)

Function Block	Macrocell	PC84	PQ100	TQ100	PQ160	BScan Order	Notes	Function Block	Macrocell	PC84	PQ100	TQ100	PQ160	BScan Order	Notes
5	1	–	–	–	76	105		6	1	–	–	–	91	51	
5	2	32	52	50	79	102		6	2	45	67	65	103	48	
5	3	33	54	52	82	99		6	3	46	68	66	104	45	
5	4	–	48	46	72	96		6	4	–	75	73	116	42	
5	5	34	55	53	86	93		6	5	47	69	67	106	39	
5	6	35	56	54	88	90		6	6	48	70	68	108	36	
5	7	–	–	–	78	87		6	7	–	–	–	105	33	
5	8	36	57	55	90	84		6	8	50	72	70	111	30	
5	9	37	58	56	92	81		6	9	51	73	71	113	27	
5	10	–	–	–	84	78		6	10	–	–	–	107	24	
5	11	39	60	58	95	75		6	11	52	74	72	115	21	
5	12	40	62	60	97	72		6	12	53	76	74	117	18	
5	13	–	–	–	87	69		6	13	–	–	–	112	15	
5	14	41	63	61	98	66		6	14	54	78	76	122	12	
5	15	43	65	63	101	63		6	15	55	79	77	124	9	
5	16	–	61	59	96	60		6	16	–	81	79	129	6	
5	17	44	66	64	102	57		6	17	56	80	78	126	3	
5	18	–	–	–	89	54		6	18	–	–	–	114	0	

XC95108 Global, JTAG and Power Pins

Pin Type	PC84	PQ100	TQ100	PQ160
I/O/GCK1	9	24	22	33
I/O/GCK2	10	25	23	35
I/O/GCK3	12	29	27	42
I/O/GTS1	76	5	3	6
I/O/GTS2	77	6	4	8
I/O/GSR	74	1	99	159
TCK	30	50	48	75
TDI	28	47	45	71
TDO	59	85	83	136
TMS	29	49	47	73
V_{CCINT} 5 V	38,73,78	7,59,100	5,57,98	10,46,94,157
V_{CCIO} 3.3 V/5 V	22,64	28,40,53,90	26,38,51,88	1,41,61,81,121,141
GND	8,16,27,42,49,60	2,23,33,46,64,71,77,86	100,21,31,44,62,69,75,84	20,31,40,51,70,80,99
GND	–	–	–	100,110,120,127,137
GND	–	–	–	160
No connects	–	–	–	3,5,7,32,38,39,48,53,55,65,66,67,83,85,93,109,118,119,125,130,131,132,149,150,151

Ordering Information

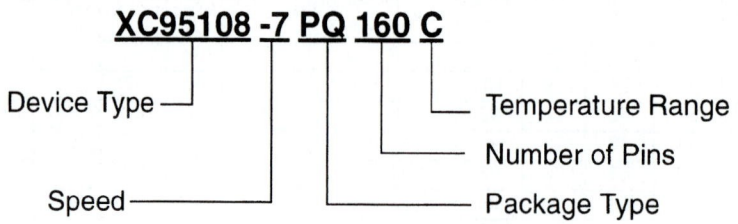

Speed Options

- 20 20 ns pin-to-pin delay
-15 15 ns pin-to-pin delay
-10 10 ns pin-to-pin delay
-7   7 ns pin-to-pin delay

Packaging Options

PC84   84-Pin Plastic Leaded Chip Carrier (PLCC)
PQ100 100-Pin Plastic Quad Flat Pack (PQFP)
TQ100 100-Pin Very Thin Quad Flat Pack (TQFP)
PQ160 160-Pin Plastic Quad Flat Pack (PQFP)

Temperature Options

C   Commercial 0°C to +70°C
I   Industrial   −40°C to +85°C

Component Availability

Pins		84	100		160
Type		Plastic PLCC	Plastic PQFP	Plastic TQFP	Plastic PQFP
Code		PC84	PQ100	TQ100	PQ160
XC95108	−20	C(I)	C(I)	C(I)	C(I)
	−15	C(I)	C(I)	C(I)	C(I)
	−10	C(I)	C(I)	C(I)	C(I)
	−7	C(I)	C(I)	C(I)	C(I)

C = Commercial = 0° to +70°C I = Industrial = −40° to +85°C

Revision Control

Date	Revision
12/04/98	Update AC Characteristics and Internal Parameters

8   December 4, 1998 (Version 3.0)

Appendix C

Explanation of the IEEE/IEC Standard for Logic Symbols (Dependency Notation)*

The IEEE/IEC standard for logic symbols introduces a method of determining the complete logical operation of a given device just by interpreting the notations on the symbol for the device. At the heart of the standard is *dependency notation,* which provides a means of denoting the relationship between inputs and outputs without actually showing all of the internal elements and interconnections involved. The information that follows briefly explains the standards publication IEEE Std. 91–1984 and is intended to help in the understanding of these new symbols.

A complete explanation of the logic symbols is available from Texas Instruments, Inc., in Publication SDYZ001, "Overview of IEEE Std 91–1984." They also have a publication (SZZZ003) entitled "Using Functional Logic Symbols," which explains several applications of the symbols.

Explanation of Logic Symbols

F. A. Mann
Texas Instruments Incorporated

Contents

1.0   Introduction

2.0   Symbol Composition

3.0   Qualifying Symbols

---

*Courtesy of Texas Instruments, Inc.

IEEE standards may be purchased from:

Institute of Electrical and Electronics Engineers, Inc.
IEEE Standards Office
345 East 47th Street
New York, N.Y. 10017

International Electrotechnical Commission (IEC) publications may be purchased from:

American National Standards Institute, Inc.
1430 Broadway
New York, N.Y. 10018

1.0 INTRODUCTION

The International Electrotechnical Commission (IEC) has been developing a very powerful symbolic language that can show the relationship of each input of a digital logic circuit to each output without showing explicitly the internal logic. At the heart of the system is dependency notation, which will be explained in Section 4.

The system was introduced in the USA in a rudimentary form in IEEE/ANSI Standard Y32.14-1973. Lacking at that time a complete development of dependency notation, it offered little more than a substitution of rectangular shapes for the familiar distinctive shapes for representing the basic functions of AND, OR, negation, etc. This is no longer the case.

Internationally, Working Group 2 of IEC Technical Committee TC-3 has prepared a new document (Publication 617-12) that consolidates the original work started in the mid 1960's and published in 1972 (Publication 117-15) and the amendments and supplements that have followed. Similarly for the USA, IEEE Committee SCC 11.9 has revised the publication IEEE Std 91/ANSI Y32.14. Now numbered simply IEEE Std 91-1984, the IEEE standard contains all of the IEC work that has been approved, and also a small amount of material still under international consideration. Texas Instruments is participating in the work of both organizations and this document introduces new logic symbols in accordance with the new standards. When changes are made as the standards develop, future editions will take those changes into account.

The following explanation of the new symbolic language is necessarily brief and greatly condensed from what the standards publications will contain. This is not intended to be sufficient for those people who will be developing symbols for new devices. It is primarily intended to make possible the understanding of the symbols used in various data books and the comparison of the symbols with logic diagrams, functional block diagrams, and/or function tables will further help that understanding.

2.0 SYMBOL COMPOSITION

A symbol comprises an outline or a combination of outlines together with one or more qualifying symbols. The shape of the symbols is not significant. As shown in Figure 1, general qualifying symbols are used to tell exactly what logical operation is performed by the elements. Table I shows general qualifying symbols defined in the new standards. Input lines are placed on the left and output lines are placed on the right. When an exception is made to that convention, the direction of signal flow is indicated by an arrow as shown in Figure 11.

All outputs of a single, unsubdivided element always have identical internal logic states determined by the function of the element except when otherwise indicated by an associated qualifying symbol or label inside the element.

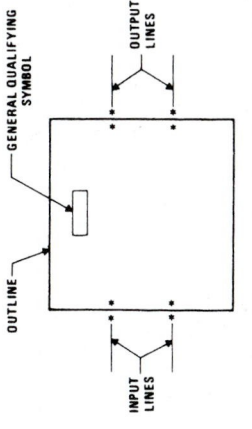

*Possible positions for qualifying symbols relating to inputs and outputs

Figure 1. Symbol Composition

The outlines of elements may be abutted or embedded in which case the following conventions apply. There is no logic connection between the elements when the line common to their outlines is in the direction of signal flow. There is at least one logic connection between the elements when the line common to their outlines is perpendicular to the direction of signal flow. The number of logic connections between elements will be clarified by the use of qualifying symbols and this is discussed further under that topic. If no indications are shown on either side of the common line, it is assumed there is only one connection.

When a circuit has one or more inputs that are common to more than one element of the circuit, the common-control block may be used. This is the only distinctively shaped outline used in the IEC system. Figure 2 shows that unless otherwise qualified by dependency notation, an input to the common-control block is an input to each of the elements below the common-control block.

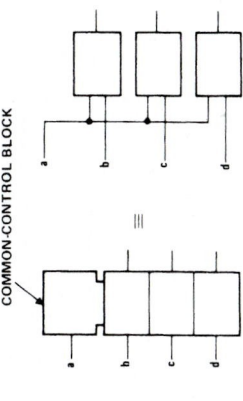

Figure 2. Common-Control Block

873

A common output depending on all elements of the array can be shown as the output of a common-output element. Its distinctive visual feature is the double line at its top. In addition the common-output element may have other inputs as shown in Figure 3. The function of the common-output element must be shown by use of a general qualifying symbol.

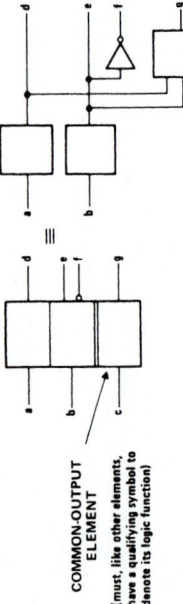

COMMON-OUTPUT ELEMENT

(must, like other elements, have a qualifying symbol to denote its logic function)

Figure 3. Common-Output Element

3.0 QUALIFYING SYMBOLS

3.1 General Qualifying Symbols

Table I shows general qualifying symbols defined by IEEE Standard 91. These characters are placed near the top center or the geometric center of a symbol or symbol element to define the basic function of the device represented by the symbol or of the element.

3.2 Qualifying Symbols for Inputs and Outputs

Qualifying symbols for inputs and outputs are shown in Table II and will be familiar to most users with the possible exception of the logic polarity and analog signal indicators. The older logic negation indicator means that the external 0 state produces the internal 1 state. The internal 1 state means the active state. Logic negation may be used in pure logic diagrams; in order to tie the external 1 and 0 logic states to the levels H (high) and L (low), a statement of whether positive logic (1 = H, 0 = L) or negative logic (1 = L, 0 = H) is being used is required or must be assumed. Logic polarity indicators eliminate the need for calling out the logic convention and are used in various data books in the symbology for actual devices. The presence of the triangular polarity indicator indicates that the L logic level will produce the internal 1 state (the active state) or that, in the case of an output, the internal 1 state will produce the external L level. Note how the active direction of transition for a dynamic input is indicated in positive logic, negative logic, and with polarity indication.

The internal connections between logic elements abutted together in a symbol may be indicated by the symbols shown in Table II. Each logic connection may be shown by the presence of qualifying symbols at one or both sides of the common line and if confusion can arise about the numbers of connections, use can be made of one of the internal connection symbols.

Table I. General Qualifying Symbols

SYMBOL	DESCRIPTION	CMOS EXAMPLE	TTL EXAMPLE
&	AND gate or function.	'HC00	SN7400
≥1	OR gate or function. The symbol was chosen to indicate that at least one active input is needed to activate the output.	'HC02	SN7402
=1	Exclusive OR. One and only one input must be active to activate the output.	'HC86	SN7486
=	Logic identity. All inputs must stand at the same state.	'HC86	SN74180
2k	An even number of inputs must be active.	'HC280	SN74180
2k + 1	An odd number of inputs must be active.	'HC86	SN74ALS86
1	The one input must be active.	'HC04	SN7404
△ or ▽	A buffer or element with more than usual output capability (symbol is oriented in the direction of signal flow).	'HC240	SN74S436
⊓	Schmitt trigger; element with hysteresis.	'HC132	SN74LS18
X:Y	Coder, code converter (DEC:BCD, BIN/OUT, BIN/7-SEG, etc.).	'HC42	SN74LS347
MUX	Multiplexer/data selector	'HC151	SN74150
DMUX or DX	Demultiplexer.	'HC138	SN74138
Σ	Adder.	'HC283	SN74LS385
P−Q	Subtracter.		SN74LS385
CPG	Look-ahead carry generator	'HC182	SN74182
π	Multiplier.		SN74LS384
COMP	Magnitude comparator.	'HC85	SN74LS682
ALU	Arithmetic logic unit.	'HC181	SN74LS381
⊓	Retriggerable monostable.	'HC123	SN74LS422
⊓	Nonretriggerable monostable (one-shot)	'HC221	SN74121
G	Astable element. Showing waveform is optional.		SN74LS320
!G	Synchronously starting astable.		SN74LS624
G!	Astable element that stops with a completed pulse.		
SRGm	Shift register. m = number of bits.	'HC164	SN74LS595
CTRm	Counter. m = number of bits; cycle length = 2^m.	'HC590	SN54LS590
CTR DIVm	Counter with cycle length = m.	'HC160	SN74LS668
RCTRm	Asynchronous (ripple-carry) counter; cycle length = 2^m.	'HC4020	
ROM	Read-only memory.		SN74187
RAM	Random-access read/write memory.	'HC189	SN74170
FIFO	First-in, first-out memory.		SN74LS222
1=0	Element powers up cleared to 0 state.		SN74AS877
1=1	Element powers up set to 1 state.	'HC7022	SN74AS877
Φ	Highly complex function; "gray box" symbol with limited detail shown under special rules.		SN74LS608

*Not all of the general qualifying symbols have been used in TI's CMOS and TTL data books, but they are included here for the sake of completeness.

Table II. Qualifying Symbols for Inputs and Outputs

Logic negation at input. External 0 produces internal 1.

Logic negation at output. Internal 1 produces external 0.

Active-low input. Equivalent to ─o⊣ in positive logic.

Active-low output. Equivalent to ⊳o── in positive logic.

Active-low input in the case of right-to-left signal flow.

Active-low output in the case of right-to-left signal flow.

Signal flow from right to left. If not otherwise indicated, signal flow is from left to right.

Bidirectional signal flow.

	POSITIVE LOGIC	NEGATIVE LOGIC	POLARITY INDICATION
Dynamic inputs active on indicated transition	1 / 0 / not used	0 / not used / 1	not used / H / L

Nonlogic connection. A label inside the symbol will usually define the nature of this pin.

Input for analog signals (on a digital symbol) (see Figure 14).

Input for digital signals (on an analog symbol) (see Figure 14).

Internal connection. 1 state on left produces 1 state on right.

Negated internal connection. 1 state on left produces 0 state on right.

Dynamic internal connection. Transition from 0 to 1 on left produces transitory 1 state on right.

Internal input (virtual input). It always stands at its internal 1 state unless affected by an overriding dependency relationship.

Internal output (virtual output). Its effect on an internal input to which it is connected is indicated by dependency notation.

The internal (virtual) input is an input originating somewhere else in the circuit and is not connected directly to a terminal. The internal (virtual) output is likewise not connected directly to a terminal. The application of internal inputs and outputs requires an understanding of dependency notation, which is explained in Section 4.

Table III. Symbols Inside the Outline

Postponed output (of a pulse-triggered flip-flop). The output changes when input initiating change (e.g., a C input) returns to its initial external state or level. See § 5.

Bi-threshold input (input with hysteresis).

N-P-N open-collector or similar output that can supply a relatively low-impedance L level when not turned off. Requires external pull-up. Capable of positive-logic wired-AND connection.

Passive-pull-up output is similar to N-P-N open-collector output but is supplemented with a built-in passive pull-up.

N-P-N open-emitter or similar output that can supply a relatively low-impedance H level when not turned off. Requires external pull-down. Capable of positive-logic wired-OR connection.

Passive-pull-down output is similar to N-P-N open-emitter output but is supplemented with a built-in passive pull-down.

3-state output.

Output with more than usual output capability (symbol is oriented in the direction of signal flow).

Enable input
When at its internal 1-state, all outputs are enabled.
When at its internal 0-state, open-collector and open-emitter outputs are off, three-state outputs are in the high-impedance state, and all other outputs (e.g., totem-poles) are at the internal 0-state.

J, K, R, S, T   Usual meanings associated with flip-flops (e.g., R = reset, T = toggle).

D   Data input to a storage element equivalent to: ┌S ┐R

Shift right (left) inputs, m = 1, 2, 3, etc. If m = 1, it is usually not shown.

Counting up (down) inputs, m = 1, 2, 3, etc. If m = 1, it is usually not shown.

Binary grouping. m is highest power of 2.

CT = 15   The contents-setting input, when active, causes the content of a register to take on the indicated value.

CT = 9   The content output is active if the content of the register is as indicated.

Input line grouping . . . indicates two or more terminals used to implement a single logic input.
e.g., The paired expander inputs of SN7450.

"1"   Fixed-state output always stands at its internal 1 state. For example, see SN74185,

Answers to Odd-Numbered Problems

Chapter 1

1–1. (a) 6_{10} (b) 11_{10} (c) 9_{10} (d) 7_{10}
(e) 12_{10} (f) 75_{10} (g) 55_{10} (h) 181_{10}
(i) 167_{10} (j) 118_{10}

1–3. (a) 31_8 (b) 35_8 (c) 134_8 (d) 131_8
(e) 155_8

1–5. (a) 23_{10} (b) 31_{10} (c) 12_{10} (d) 58_{10}
(e) 41_{10}

1–7. (a) $B9_{16}$ (b) DC_{16} (c) 74_{16} (d) FB_{16}
(e) $C6_{16}$

1–9. (a) 134_{10} (b) 244_{10} (c) 146_{10}
(d) 171_{10} (e) 965_{10}

1–11. (a) 98_{10} (b) 69_{10} (c) 74_{10} (d) 36_{10}
(e) 81_{10}

1–13. (a) 010 0101
(b) 0100100 0110001 0110100
(c) 1001110 0101101 0110110
(d) 1000011 1010000 1010101
(e) 1010000 1100111

1–15. (a) Tank A, temperature high; tank C, pressure high
(b) Tank D, temperature and pressure high
(c) Tanks B and D, pressure high
(d) Tanks B and C, temperature high
(e) Tank C, temperature and pressure high

1–17. (a) sku43 (b) $534B553433_{16}$

1–19. (a) 16-MAR 1995 Revision A

E1-1. (a) 0000 0101 (b) 11 (c) 0E (d) 27

Chapter 2

2–1. (a) $0.5\ \mu s$ (b) $2\ \mu s$ (c) $0.234\ \mu s$
(d) 58.8 ns (e) 500 kHz (f) 10 kHz
(g) 1.33 kHz (h) 0.667 MHz

2–3. (a) $2.16\ \mu s$ (b) LOW

2–5.

2–7. $V_1 = 0\ V$ $V_5 = 4.3\ V$
$V_2 = 4.3\ V$ $V_6 = 5.0\ V$
$V_3 = 4.3\ V$ $V_7 = 0\ V$
$V_4 = 0\ V$

2–9. That diode will conduct raising V_7 to 4.3 V ("OR").

2–11.

2–13. $V_{out} = 4.998\ V$

2–15. Because, when the transistor is turned on (saturated), the collector current will be excessive ($I_C = 5\ V/R_C$).

2–17. The totem-pole output replaces R_C with a transistor that acts like a variable resistor. The transistor prevents excessive collector current when it is cut off and provides a high-level output when turned on.

2–19. **(a)** 8.0 MHZ **(b)** 125 nS

2–21. P3 parallel, P2 serial

2–23. A HIGH on pin 2 will turn Q1 on, making RESET_B approximately zero.

E2-1. **(a)** Let **(b)** 24

E2-3. **(a)** Cp = 5 V/0 V, Vout3 = 0 V/5 V, inverse of each other
(b) Cp = 5 V/0 V, Vout3 = 0 V/8 V
(c) Cp and Vout3 are in phase.

E2-5. **(a)** V1 = 4.3 V, V2 = 0 V, V3 = 4.3 V, V4 = 0.7 V **(b)** V1 = 0 V, V2 = 4.3 V, V3 = 0 V, V4 = 5.0 V (Both diodes are reverse biased.)

Chapter 3

3–1.

A	B	C	X
0	0	0	0
0	0	1	0
0	1	0	0
0	1	1	0
1	0	0	0
1	0	1	0
1	1	0	0
1	1	1	1

3–3. 256

3–5. The output is HIGH whenever any input is HIGH; otherwise, the output is LOW.

3–7.

3–9.

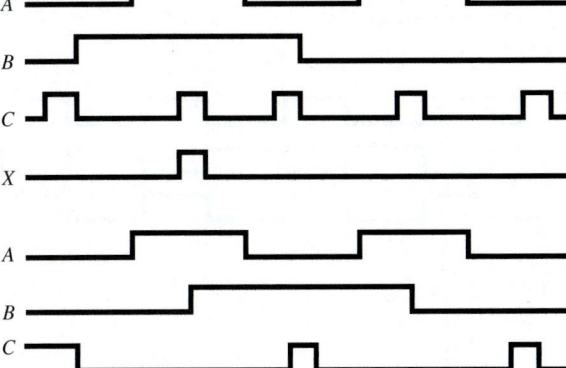

3–11.

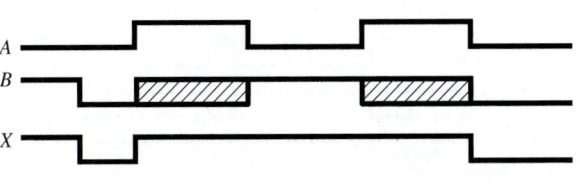

3–13.

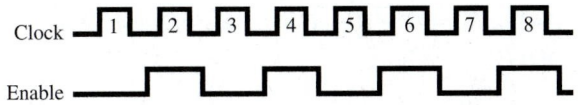

3–15.

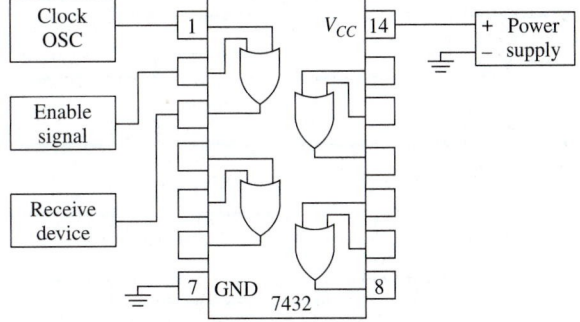

3–17. Two

3–19. To provide pulses to a digital circuit for troubleshooting purposes.

3–21. HIGH, to enable the output to change with pulser (if gate is good).

3–23. Pin 2 should be ON; the Enable switch is bad, or bad Enable connection.

3–25. $X = \overline{A}, X = 0$

3–27.

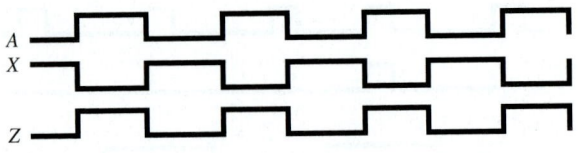

3–29.

A	B	X		C	D	Y
0	0	1		0	0	1
0	1	1		0	1	1
1	0	1		1	0	1
1	1	0		1	1	0

3–31.

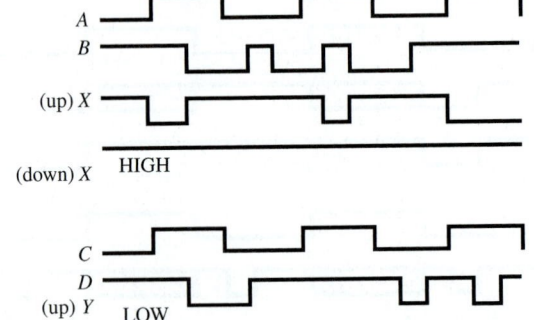

3–33. $X = \overline{A + B + C}$ $Y = \overline{D + E + F}$

3–35.

3–37.

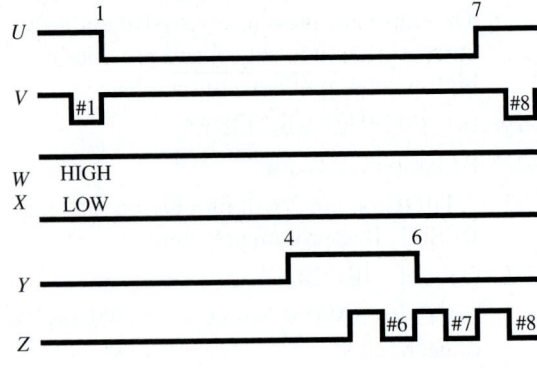

3–39. $U = C_p, A, B$ $W = B, C$
    $V = \overline{C}, \overline{D}$ $X = C_p, C, D$

3–41.

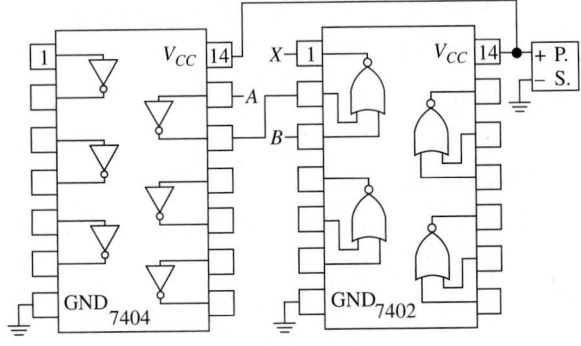

3–43. HIGH; to see inverted output pulses (otherwise, output would always be HIGH).

3–45. There is no problem.

3–47. With all inputs HIGH, pin 8 should be LOW. Next, try making each of the 8 inputs LOW, one at a time, while checking for a HIGH at pin 8.

3–49. AND-74HC08; U3:A = location C2, U3:B = location D2 OR-74HC32; location B7

3–51. Pin 20 = LOW (GND), pin 40 HIGH (+5)

3–53. Place probe-A on the input of the inverter (WATCHDOG_CLK). Using the same scope settings, place probe-B on the output of U4:A. Probe-B should display the complement of probe-A.

3–55. OE_B

E3-1. (a) X = 1, Y = 1
    (b) X = 0, Y = 0
    (c)

A	B	X		A	B	Y
0	0	0		0	0	0
0	1	0		0	1	1
1	0	0		1	0	1
1	1	1		1	1	1

E3-3. (a) Up (b) Down

E3-5. (a) Vcc (b) Logic pulser
(c) Logic probe (d) Ground (e) Vcc

E3-7. (a) X = 0, Y = 0 (b) X = 1, Y = 1

E3-9. (a) Yes (b) X = AB (c) 6mS

E3-11. (a) NAND (b) NOR

E3-13. (a) X = C', D', Cp (b) Y = BD'

Chapter 4

4–1. The 7400-series uses hard-wired logic. The designer must use a different IC for each logic function. Programmable logic contains thousands of logic gates that can be custom-configured by the designer to perform any logic desired.

4–3. Hardware Description Language

4–5. (a) 3 (b) 5

4–7. They receive programming information from a PC and program the on-board CPLD, which can then be tested with actual I/O signals.

4–9. The PLA provides programmable OR gates for combining the product terms.

4–11. So it will not lose its programmed logic design when power is removed.

4–13. The look-up table method

4–15. They must be reprogrammed

4–17. It translates the information from the design entry stage into a binary file that is later used to program the CPLD.

4–19. Text

4–21. ENTITY and3 IS
PORT(
  A, B, C : IN bit;
        X : OUT bit);
END and3;

4–23. (a)

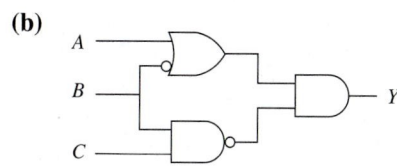

(b)

(c)

Chapter 5

5–1. (a) $W = (A + B)(C + D)$
$X = AB + BC$
$Y = (AB + B)C$
$Z = (AB + B + (B + C))D$

5–3.

A	B	C	D	M	N	Q	R	S		A	B	C	P
0	0	0	0	0	0	0	0	0		0	0	0	0
0	0	0	1	1	0	0	1	1		0	0	1	0
0	0	1	0	1	0	0	0	0		0	1	0	0
0	0	1	1	1	1	0	1	1		0	1	1	1
0	1	0	0	0	0	0	0	0		1	0	0	0
0	1	0	1	1	1	0	1	1		1	0	1	1
0	1	1	0	1	0	0	1	1		1	1	0	0
0	1	1	1	1	1	1	1	1		1	1	1	1
1	0	0	0	0	0	0	0	0					
1	0	0	1	1	1	0	1	1					
1	0	1	0	1	0	0	0	1					
1	0	1	1	1	1	0	1	1					
1	1	0	0	1	0	0	0	1					
1	1	0	1	1	1	0	1	1					
1	1	1	0	1	0	0	1	1					
1	1	1	1	1	1	1	1	1					

5–5. (a) Commutative law (b) Associative law
(c) Distributive law

5–7. $W = (A + B)BC$
$W = BC$

$X = (A + B)(B + C)$
$X = B + AC$

$Y = A + (A + B)BC$
$Y = A + BC$

$Z = AB + B + BC$
$Z = B$

5–9.

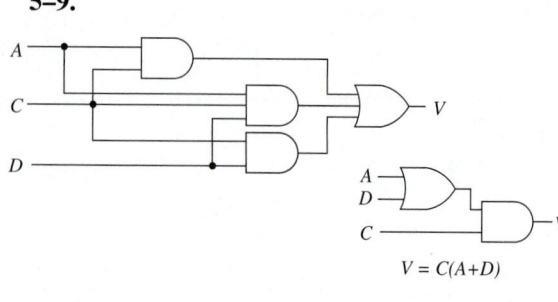

$V = C(A+D)$

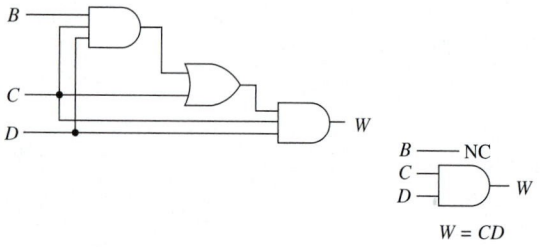

$W = CD$

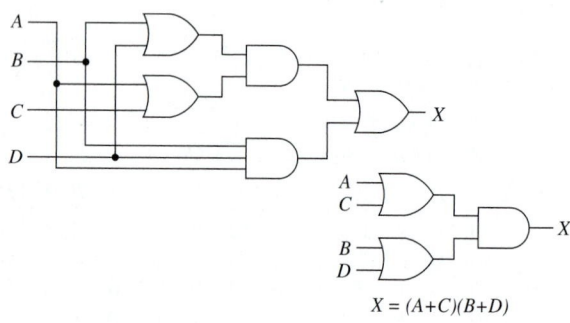

$X = (A+C)(B+D)$

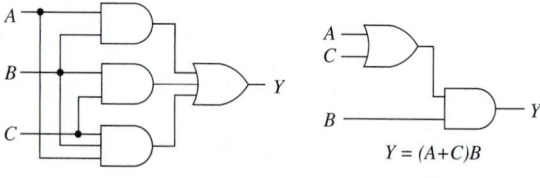

$Y = (A+C)B$

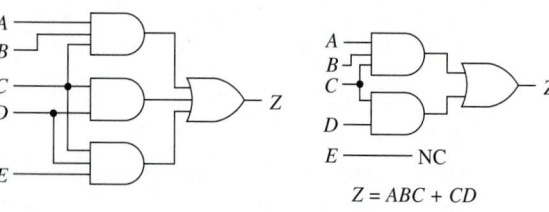

$Z = ABC + CD$

5–11. $X = (A + B)(D + C)$

5–13. Break the long bar and change the AND to an OR, or the OR to an AND.

5–15. Y and Z are both ORs

5–17.

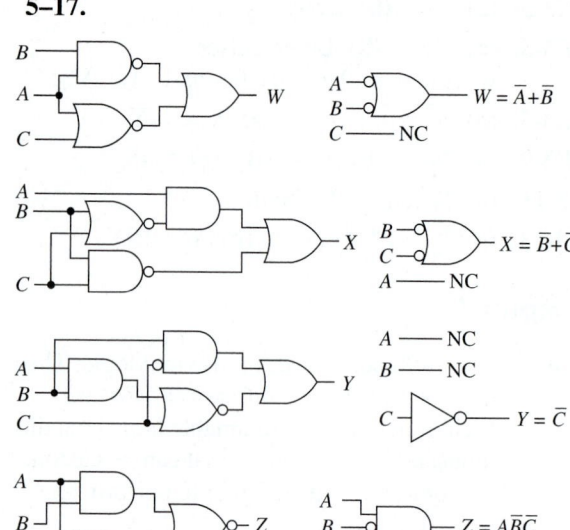

$W = \bar{A} + \bar{B}$

$X = \bar{B} + \bar{C}$

$Y = \bar{C}$

$Z = A\bar{B}\bar{C}$

5–19.

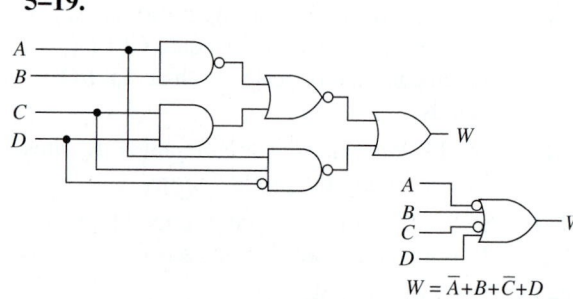

$W = \bar{A} + B + \bar{C} + D$

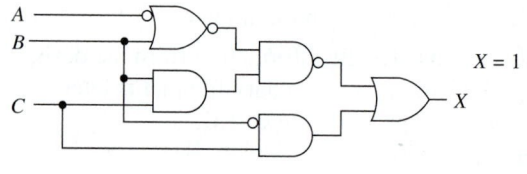

$X = 1$

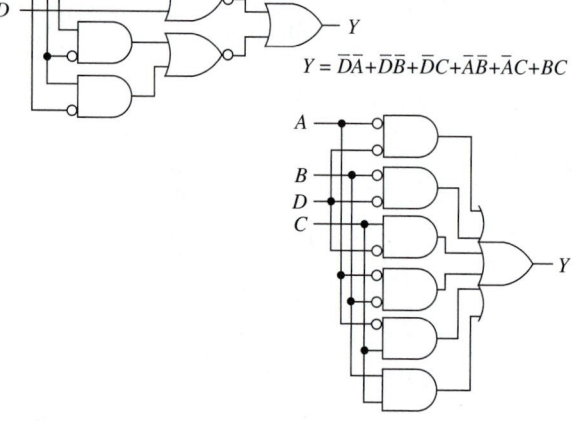

$Y = \bar{D}\bar{A} + \bar{D}\bar{B} + \bar{D}C + \bar{A}\bar{B} + \bar{A}C + BC$

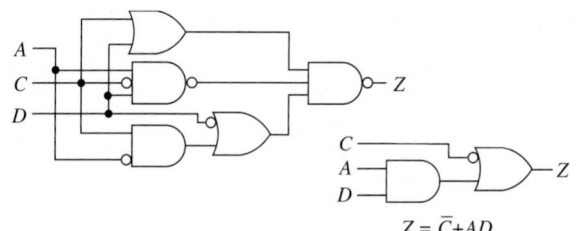

$Z = \bar{C} + AD$

5–21.

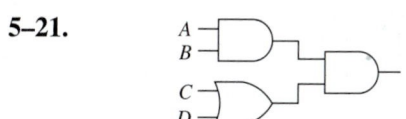

5–23.

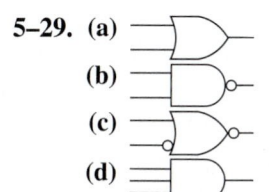

5–25.

$(2^3)\ A$
$(2^2)\ B$ ——— $ABCD>11$
$(2^1)\ C$ ——— NC
$(2^0)\ D$ ——— NC

5–27.

A	B	C	W	X
0	0	0	0	1
0	0	1	1	1
0	1	0	1	1
0	1	1	1	0
1	0	0	1	0
1	0	1	1	1
1	1	0	0	1
1	1	1	0	0

A	B	C	D	Y	Z
0	0	0	0	1	1
0	0	0	1	1	1
0	0	1	0	1	1
0	0	1	1	0	1
0	1	0	0	0	0
0	1	0	1	1	0
0	1	1	0	1	1
0	1	1	1	1	1
1	0	0	0	0	1
1	0	0	1	1	1
1	0	1	0	0	1
1	0	1	1	0	0
1	1	0	0	0	0
1	1	0	1	1	0
1	1	1	0	0	1
1	1	1	1	1	0

5–29. (a)

(b)

(c)

(d)

5–31. (a)

A —— $X = \bar{A}$

(b)

A —— $X = \bar{A}$

5–33.

(a)

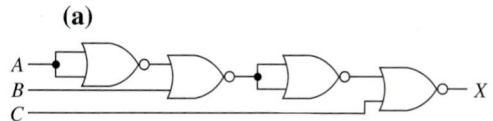

(b)

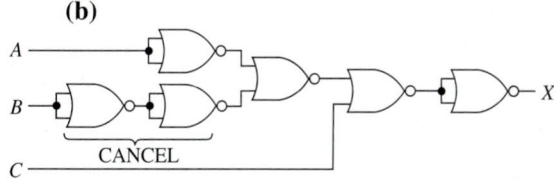

CANCEL

(c)

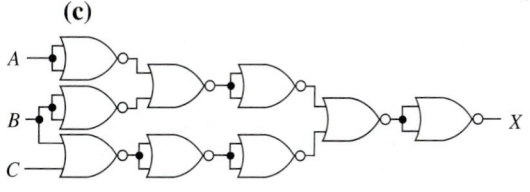

5–35. (u) SOP   **(x)** SOP
(v) POS   **(y)** POS
(w) POS   **(z)** POS, SOP

5–37. $X = \bar{A} + B\bar{C}$
$Y = B + \bar{A}C$
$Z = A\bar{C} + AB + \bar{A}\,\bar{B}C$

5–39. (a) $X = \bar{C}D + AC + B$
(b) $Y = 1$

5–41.

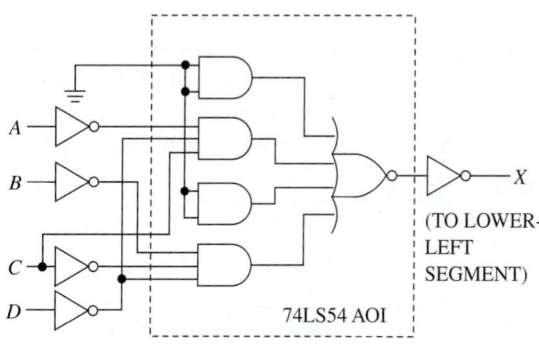

(TO LOWER-LEFT SEGMENT)

74LS54 AOI

$X = \bar{A}C\bar{D} + \bar{B}C\bar{D}$ where A = MSB

5–43. The IC checks out OK. The problem is that pin 9 should be connected to pin 10 (not 9 to GND).

5–45. $\overline{\text{WATCHDOG\_EN} \cdot \text{Qa}}$

5–47. (a) pin 6 = $\overline{\overline{\text{P1.0}} + \overline{\text{A15}}}$
(b) AND
(c) quad 2 input AND
(d) $\overline{\text{RD}}$ is LOW or $\overline{\text{WR}}$ is LOW

E5-1. (a) $B = KD + HD$
(b) $B = D(K + H)$

E5-3. (a) 5
(b) $X = BC + A$

E5-5. (a) 2

    (b) X = BC

    (c)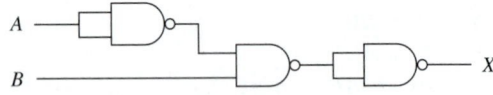

E5-7. X = AB′C′ + A′BC′ + AB′C

E5-9. (a) 2

    (b) X = B′C′

E5-11. (a) 6

    (b) X = B′ + C′

    (c)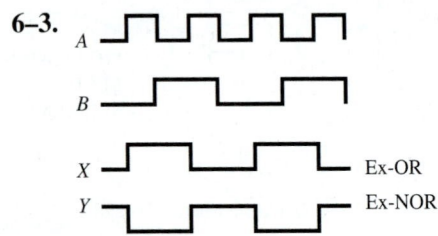

E5-13. (a) AND

    (b) OR

E5-15.

A(2^3) ─┐
B(2^2) ─┘)── X

E5-17. (a) 3

    (b) Gate 2

    (c)

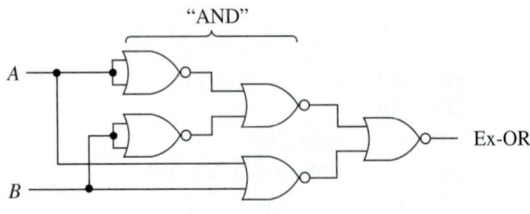

Chapter 6

6–1. (a) Exclusive-OR produces a HIGH output for one or the other input HIGH, but not both.

    (b) Exclusive-NOR produces a HIGH output for both inputs HIGH or both inputs LOW.

6–3.

A ─┐┌─┐┌─┐┌─┐┌─┐┌─

B ──┐┌──┐┌──┐┌──

X ─┐┌──┐┌──┐┌── Ex-OR

Y ─┘└──┘└──┘└─ Ex-NOR

6–5.

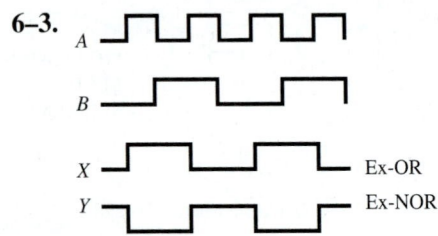

"AND"

A ... Ex-OR

B

6–7. $X = \overline{(AB + \overline{A}\,\overline{B})} + \overline{AB} = A\overline{B}$

     $Y = \overline{\overline{AB} + A\overline{B} \cdot AB} = 1$

6–9. A7 = 1010 0111 0

    4C = 0100 1100 0

    79 = 0111 1001 0

    F3 = 1111 0011 1

    00 = 0000 0000 1

    FF = 1111 1111 1

6–11.

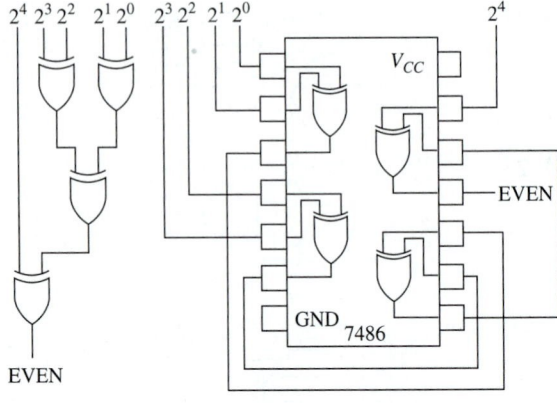

6–13.

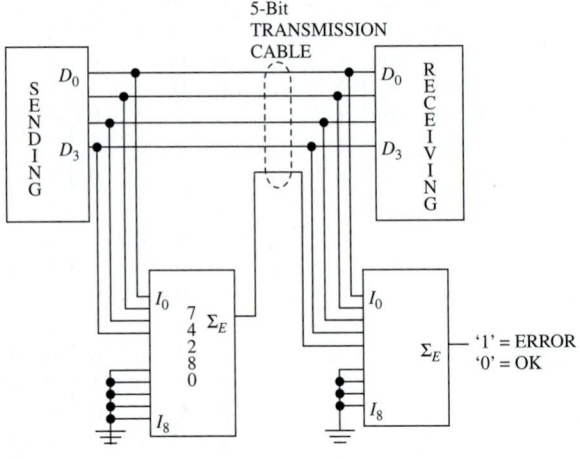

6–15. Yes; LOW

6–17.

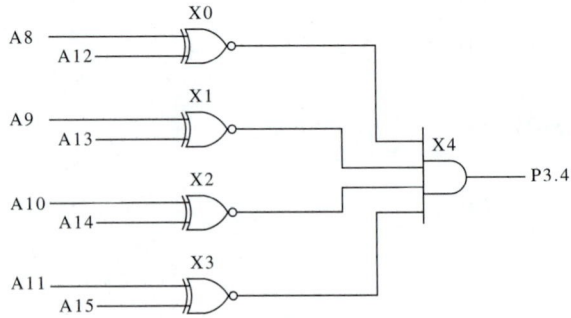

E6-1. (a) $X = 0, Y = 1$
(b) $X = 0, Y = 1$
(c)

A	B	X		A	B	Y
0	0	0		0	0	1
0	1	1		0	1	0
1	0	1		1	0	0
1	1	0		1	1	1

E6-3. Password for *Circuit Restrictions Hide Faults* is: **WK5E**
(a) Yes, because it is an Ex-OR.
(b) Gate 2

E6-5. $X = (A'B + AB')BC = A'BC$

E6-7. Because they both have an odd number of ones

Chapter 7

7–1. (a) 1001 (b) 1111 (c) 1 1100
(d) 100 0010 (e) 1100 1000
(f) 10010 0010 (g) 10100 1111
(h) 10110 0000

7–3. (a) 1 0101 (b) 10 1010 (c) 11 1100
(d) 1 0001 0001 (e) 1 1110 1100 0011
(f) 111 0111 0001 (g) 1 1001 0011
(h) 111 1110 1000 0001

7–5.

+15	0000 1111		−1	1111 1111
+14	0000 1110		−2	1111 1110
+13	0000 1101		−3	1111 1101
+12	0000 1100		−4	1111 1100
+11	0000 1011		−5	1111 1011
+10	0000 1010		−6	1111 1010
+9	0000 1001		−7	1111 1001
+8	0000 1000		−8	1111 1000
+7	0000 0111		−9	1111 0111
+6	0000 0110		−10	1111 0110
+5	0000 0101		−11	1111 0101
+4	0000 0100		−12	1111 0100
+3	0000 0011		−13	1111 0011
+2	0000 0010		−14	1111 0010
+1	0000 0001		−15	1111 0001
0	0000 0000			

7–7. (a) 0001 0110 = +22
(b) 0000 1111 = +15
(c) 0101 1100 = +92
(d) 1000 0110 = −122
(e) 1110 1110 = −18
(f) 1000 0001 = −127
(g) 0111 1111 = +127
(h) 1111 1111 = −1

7–9. (a) 0000 1100 (b) 0000 0110
(c) 0011 0010 (d) 0000 1110
(e) 0000 1010 (f) 0011 1011
(g) 1111 0100 (h) 1010 1100

7–11. (a) E (b) D (c) 21 (d) CA (e) 10C
(f) 162 (g) AB45 (h) A000

7–13. 2CF0H

7–15. b, c, e

7–17. For the LSB addition of two binary numbers.

7–19. Σ_o = OK
C_o = NO

7–21.

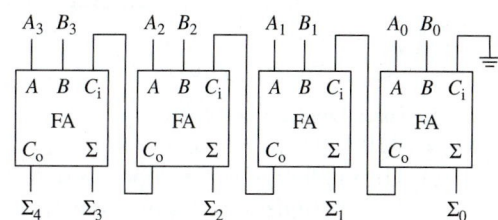

7–23.

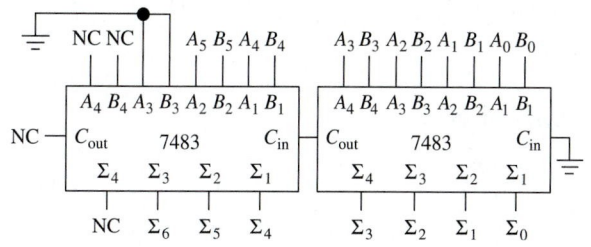

7–25.

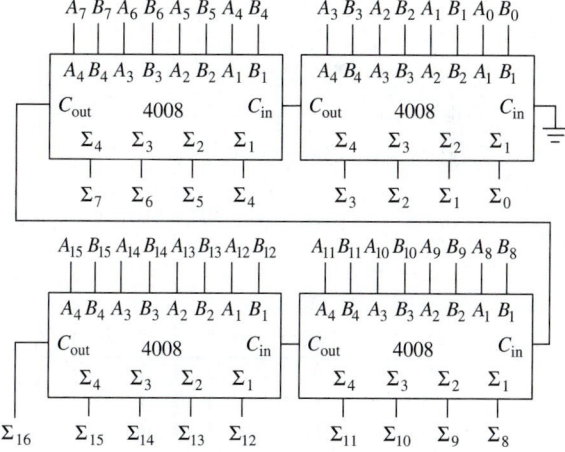

7–27. The Ex-OR gate third from right is bad.
Also, the full-adder fourth from right is bad.

7–29. $S_3 − S_0 = 0100$
(a) 1110 (b) 0001

7–31.

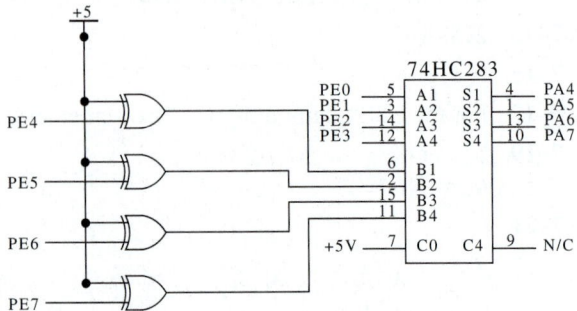

E7-1. (a) Three inputs, two outputs
  (b) Full adder TT
  (c) The number of HIGH inputs is odd. The number of HIGH inputs is two or more.

E7-3. (a) 0011 1011
  (b) 0111 0000

E7-5. (a) 1 0011 (1 = ON) (b) 1 0111 (1 = ON)

Chapter 8

8–1.

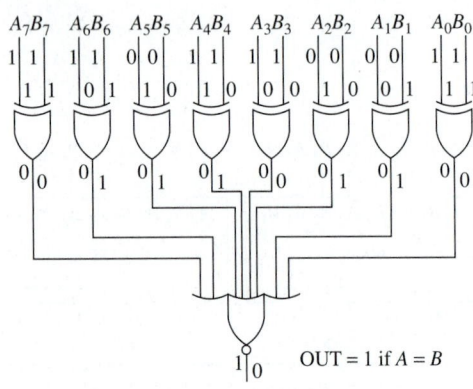

OUT = 1 if A = B

8–3. See #1 (lower numbers)

8–5.

2^3	2^2	2^1	2^0	$\overline{0}$	$\overline{1}$	$\overline{2}$	$\overline{3}$	$\overline{4}$	$\overline{5}$	$\overline{6}$	$\overline{7}$	$\overline{8}$	$\overline{9}$
0	0	0	0	0	1	1	1	1	1	1	1	1	1
0	0	0	1	1	0	1	1	1	1	1	1	1	1
0	0	1	0	1	1	0	1	1	1	1	1	1	1
0	0	1	1	1	1	1	0	1	1	1	1	1	1
0	1	0	0	1	1	1	1	0	1	1	1	1	1
0	1	0	1	1	1	1	1	1	0	1	1	1	1
0	1	1	0	1	1	1	1	1	1	0	1	1	1
0	1	1	1	1	1	1	1	1	1	1	0	1	1
1	0	0	0	1	1	1	1	1	1	1	1	0	1
1	0	0	1	1	1	1	1	1	1	1	1	1	0

8–7. That input is a "don't care" and will have no effect on the output for that particular table entry.

8–9.

Time Interval	Low Output Pulse At:
t_0–t_1	None (E_3 disabled)
t_1–t_2	None (E_3 disabled)
t_2–t_3	$\overline{5}$
t_3–t_4	$\overline{4}$
t_4–t_5	$\overline{3}$
t_5–t_6	$\overline{2}$
t_6–t_7	$\overline{1}$
t_7–t_8	$\overline{0}$
t_8–t_9	$\overline{7}$
t_9–t_{10}	$\overline{6}$
t_{10}–t_{11}	$\overline{5}$
t_{11}–t_{12}	None (E_3 disabled)
t_{12}–t_{13}	None (E_3 disabled)

8–11. All HIGH

8–13. The higher number

8–15.

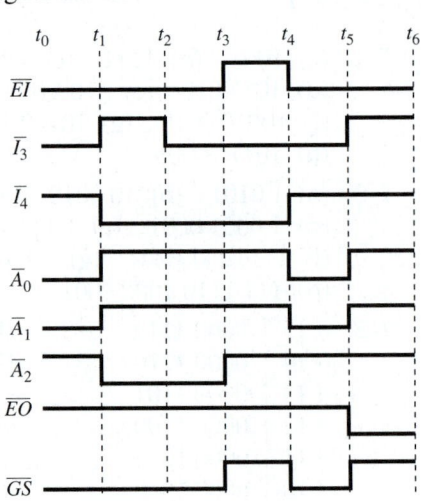

8–17.

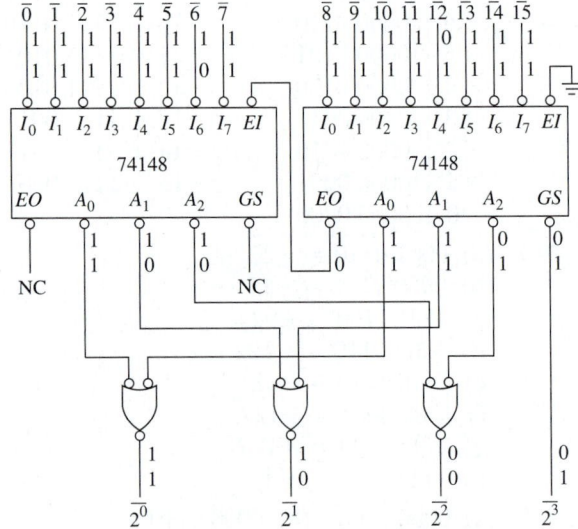

8–19. (a) 100000_2 (b) 101110_2
  (c) 110111_2 (d) 1000100_2

8–21. See #20 (lower numbers).

8–23. (a) $1010_2 = 1111$ (b) $1111_2 = 1000$
(c) $0011_2 = 0010$ (d) $0001_2 = 0001$

8–25.

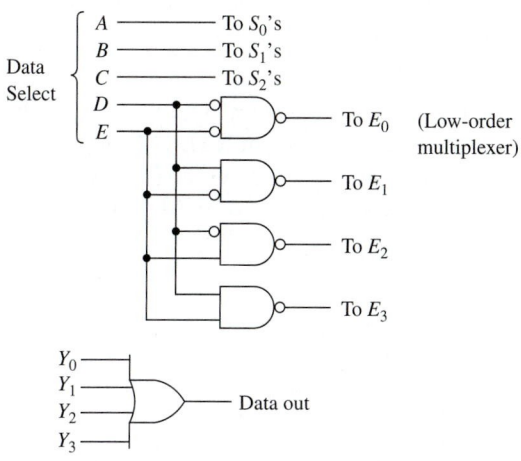

8–27.

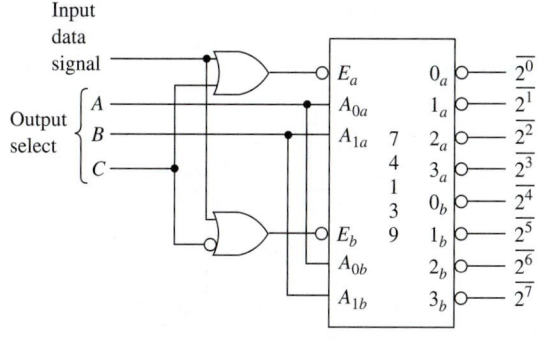

8–29. (a) The 74150 is not working. The data select is set for input D_7, which is 0. Therefore, $\overline{Y}$ should be 1 but it is not.
(b) The 74151 is OK. The data select is set for input I_0, which is 1. Y should equal 1 and $\overline{Y}$ should equal 0, which they do.
(c) The 74139 has two bad decoders. Decoder A is enabled and should output 1011 but does not. Decoder B is disabled and should output 1111 but does not.
(d) The 74154 is OK. The chip is disabled, so all outputs should be HIGH, which they are.

8–31. (a) Write to memory bank 5, location 3H.
(b) Read from memory bank 6, location C7H.

8–33. (a) 05H (b) 07H (c) 82H
(d) Impossible

8–35. $\overline{0} = 0, \overline{1} = 1, \overline{2} = 1, \overline{3} = 1$

8–37. The D input of U1:B

8–39.

	LCD_SL	KEY_SL
AD3	0	1
AD4	0	0
AD5	0	0
AD11	1	1
AD12	1	1
AD13	0	0
AD14	0	0
AD15	0	0
AS	0	0

8–41.

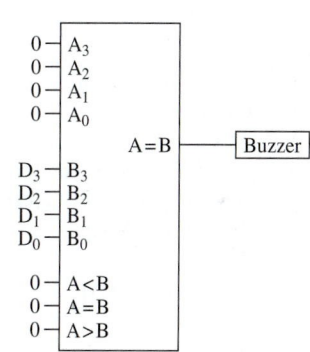

E8-1. (a) A0 to A3 = B0 to B3
(b) NOR

E8-3.

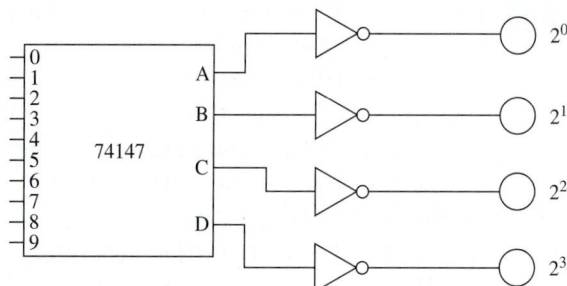

E8-5. (a) All HIGH
(b) All outputs HIGH when G1 = LOW
(c) The 8 (1000) is used to make the top Logic Analyzer trace HIGH.
(d) Put an 8 in the MS hex digit whenever a 7 is in the LS digit.

E8-7.

E8-9. **(a)** 00
(b) 11

E8-11. Password for *Circuit Restrictions Hide Faults* is: **WK5E**
(a) Switch S1 has no effect on flashing speed.
(b) Vcc at S1 is open.

Chapter 9

9–1. D_1 and D_2 provide some protection against negative input voltages.

9–3. From Figure 9-2(a), there is ~0.2 V dropped across the 1.6 k, 0.7 V across V_{BE_3} and 0.7 V across D_3, leaving ~3.4 V at the output terminal.

9–5. Negative sign signifies current *leaving* the input or output of the gate.

9–7a. [STD]
(a) $V_a = 3.4$ V (typ)
$I_a = 120 \, \mu A$
(b) $V_a = 4.6$ V
@V_{in} = LOW, $I_a = 20 \, \mu A$ (typ)
@V_{in} = HIGH, $I_a = 340 \, \mu A$ (typ)
(c) $V_a = V_{OL} = 0.2$ V (typ)
$I_a = 3.2$ mA
(d) $V_a = V_{OH} = 3.4$ V (typ)
$I_a = 380 \, \mu A$

9–7b. [LS]
(a) $V_a = V_{OH} = 3.4$ V (typ)
$I_a = 60 \, \mu A$
(b) $V_a = 4.8$ V
@V_{in} = LOW, $I_a = 35 \, \mu A$ (typ)
@V_{in} = HIGH, $I_a = 340 \, \mu A$ (typ)
(c) $V_a = V_{OL} = 0.35$ V (typ)
$I_a = 0.8$ mA
(d) $V_a = V_{OH} = 3.4$ V (typ)
$I_a = 360 \, \mu A$

9–9.

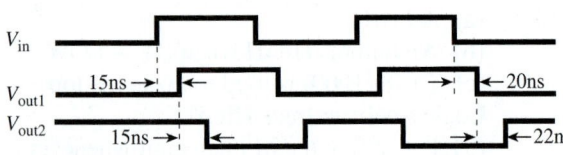

9–11. 7400: $P_D = 75$ mW (max)
74LS00: $P_D = 15$ mW (max)

9–13. **(a)** 7400: HIGH state (min levels) = 0.4 V
LOW state (max levels) = 0.4 V
74LS00: HIGH state (min levels) = 0.7 V
LOW state (max levels) = 0.3 V

(b) The 74LS00 has a wider margin for the HIGH state. The 7400 has a wider margin for the LOW state.

9–15. The open-collector FLOAT level is made a HIGH level by using a pull-up resistor.

9–17. The 7400 series is faster than the 4000B series but dissipates more power.

9–19. Because MOS ICs are prone to electrostatic burnout.

9–21. Where speed is most important, ECL is faster but uses more power.

9–23. The 74HC family

9–25. Interfacing (c) and (e) will require a pull-up resistor to "pull up" the TTL HIGH-level output to meet the minimum HIGH-level input specifications of the CMOS gates.

9–27. **(a)** 10   **(b)** 400

9–29.

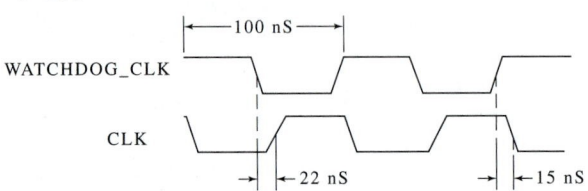

E9-1. **(a)** HIGH   **(b)** ON, OFF   **(c)** OFF, ON
E9-3. **(a)** 12nS, 12nS   **(b)** 6.3nS, 11nS

Chapter 10

10–1.

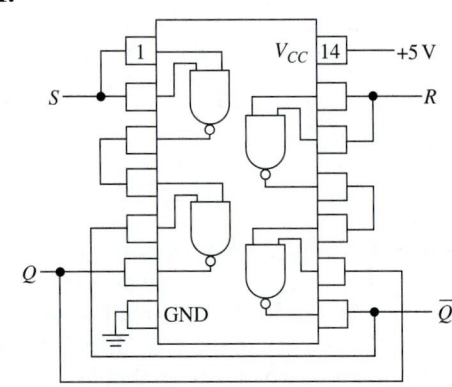

10–3.

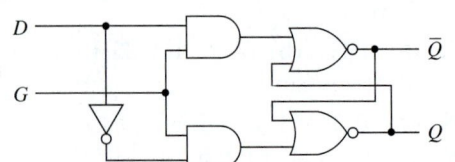

10–5.

10–7.

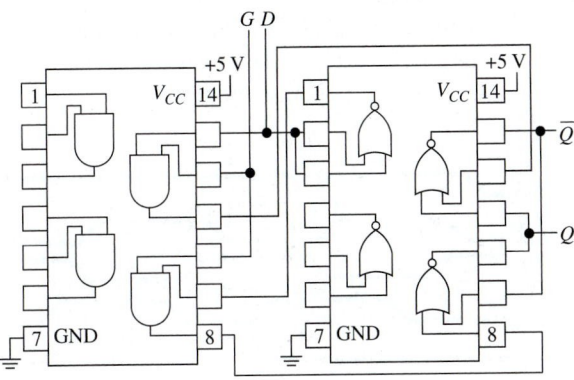

10–9.

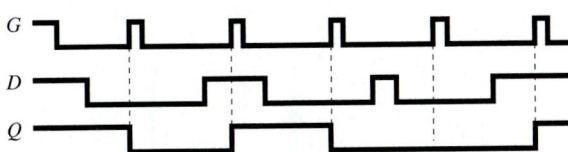

10–11.

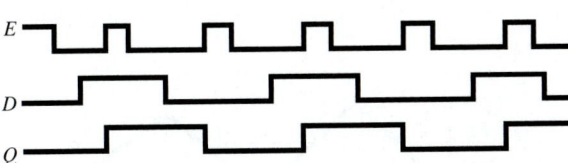

10–13. HIGH, LOW

10–15.

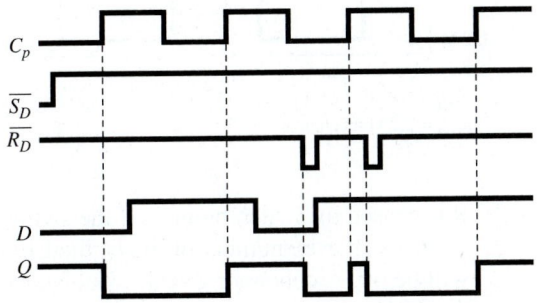

10–17. The 7474 is edge-triggered; the 7475 is pulse-triggered. The 7474 has asynchronous inputs at $\overline{S_D}$ and $\overline{R_D}$.

10–19. The triangle indicates that it is an edge-triggered device as opposed to being pulse-triggered.

10–21.

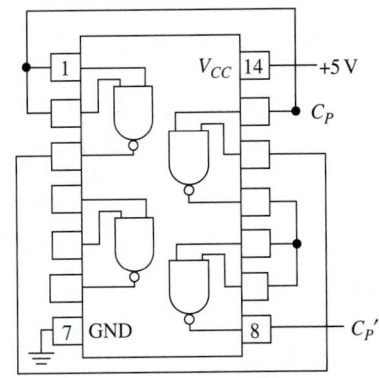

10–23. The toggle mode

10–25. The 7476 accepts J and K data during the entire positive level of CP, whereas the 74LS76 only looks at J and K at the negative edge of CP.

10–27.

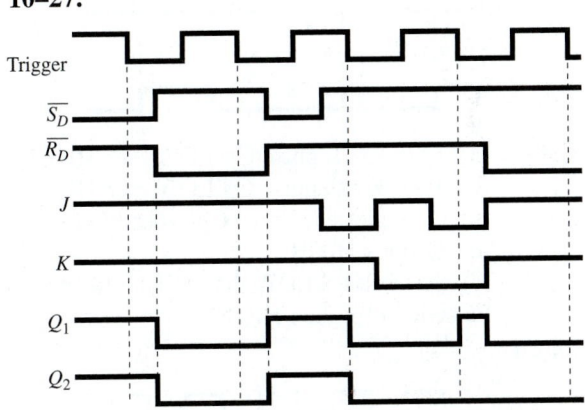

10–29.

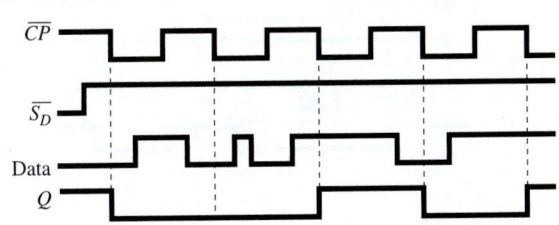

10–31.

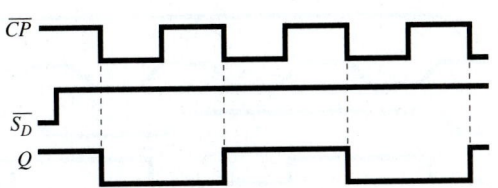

10–33. The '373 is a transparent latch. If the timing pulses are connected to E, the BCD will pass through to Q while E is HIGH and latch on to the data when E goes LOW. As long as the positive timing pulses are very narrow (< 10 ms), the display will not flicker. The '273 is the preferred device, however, because it is edge triggered.

10–35. **(a)** HIGH **(b)** no **(c)** Qa must go HIGH while WATCHDOG_EN is HIGH. (Qa will go HIGH after Qb of U1:B goes HIGH.)

10–37. WATCHDOG_SEL is pulsed

E10-1. **(a)** 1, 0 **(b)** 0, 1 **(c)** 0, 0

E10-3. **(a)** Top inverter **(b)** Vcc **(c)** none **(d)** R switch

E10-5.

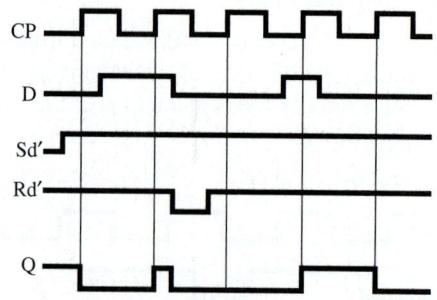

E10-7. **(a)** J=0, K=1, pulse Cp′LOW then HIGH
(b) J=1, K=0, pulse Cp′LOW then HIGH
(c) Q toggles each time Cp′ goes LOW
(d) Q stuck HIGH
(e) Set: Pulse S LOW then HIGH; Reset: Pulse R LOW then HIGH

E10-9.

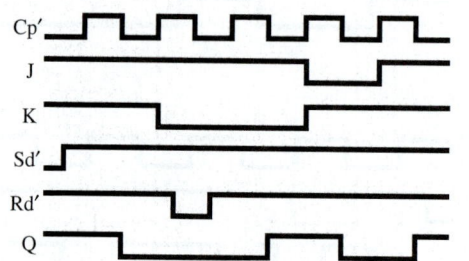

Chapter 11

11–1.

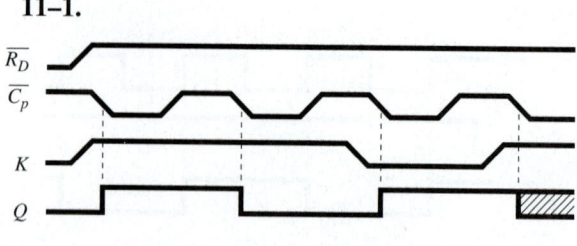

11–3. t_a = 20 ns t_d = 30 ns
t_b = 30 ns t_e = 20 ns
t_c = 20 ns t_f = 30 ns

11–5. Proper circuit operation depends on t_p of the 7432 being ≥10 ns. The worst-case t_p is specified as 15 ns, but the actual t_p may be less. If it's actually less than 10 ns, the circuit won't operate properly.

11–7.

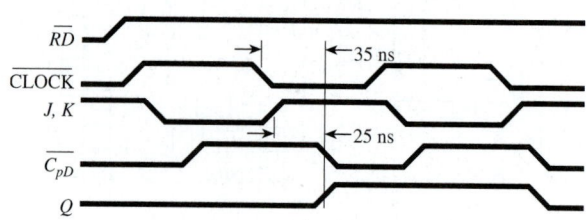

11–9.

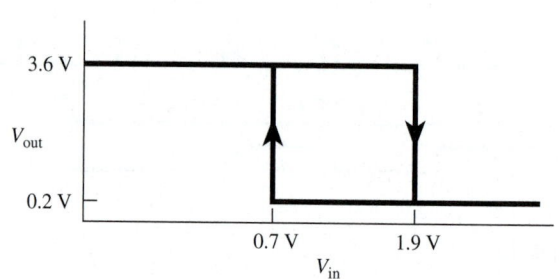

11–11.

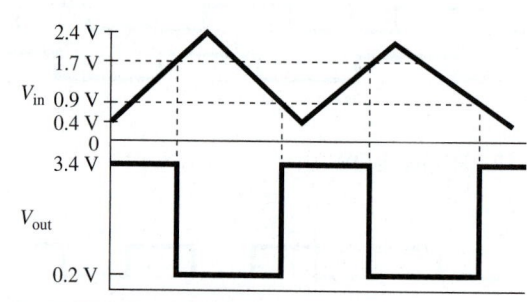

t_{HI} = 1.8 volts change
t_{LOW} = 2.2 volts change
DC = 45%

11–13. It is caused by switch bounce. If the switch bounces an even number of times, the LED will be off. A debounce circuit would correct the problem.

11–15. Check the output of the 7805 for +5 V dc. Check the fuse. If the fuse is OK, you should check for approximately 12.6 V ac at the transformer secondary and 20 V dc at the +/− output of the diode bridge and the input to the 7805.

11–17. Light: V_A = 0.0495 V
Dark: V_A = 4.55 V

11–19. The 7474 can sink 16 mA. Connect the positive lead of the buzzer to +5 V and the negative lead to $\overline{Q}$. When Q is HIGH, the buzzer is energized via the LOW $\overline{Q}$ output.

11–21. $I_{coil} = 240$ mA

11–23.

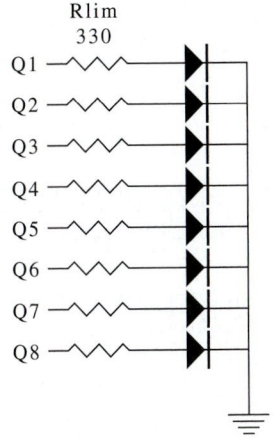

11–25. The switches of U12 and the pull-up resistors are used to place either a HIGH or LOW on the MODA and MODB lines. To place a HIGH on one of these lines, the corresponding switch must be open. A closed switch pulls the line to ground.

E11-1.

Vt+=1.6V, Vt–=1.3V

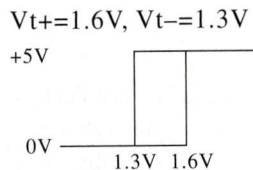

E11-3.

(a) Vt+=1.3V, Vt–=0.83V, Voh=5V, Vol=0V

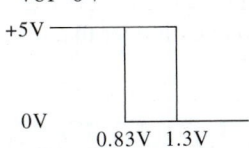

(b) Yes

Chapter 12

12–1. Sequential circuits follow a predetermined sequence of digital states triggered by a timing pulse or clock. Combination logic circuits operate almost instantaneously based on the levels placed at its inputs.

12–3.

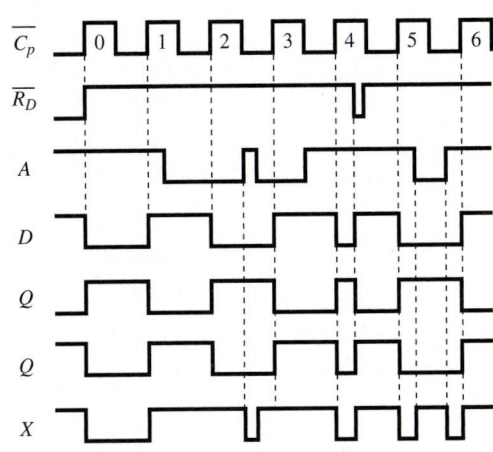

12–5.

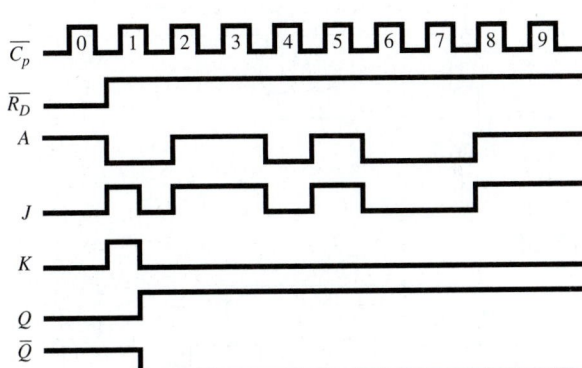

12–7. (a) 3   (b) 3   (c) 1   (d) 5   (e) 6   (f) 4

12–9.

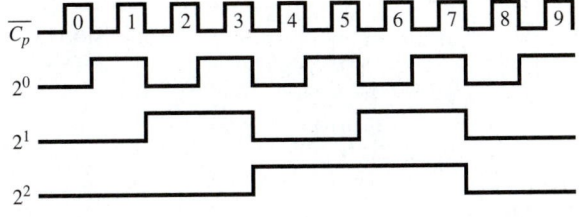

12–11. (a) 3   (b) 15   (c) 127   (d) 1

12–13. (a) 2   (b) 4   (c) 4   (d) 5

12–15.

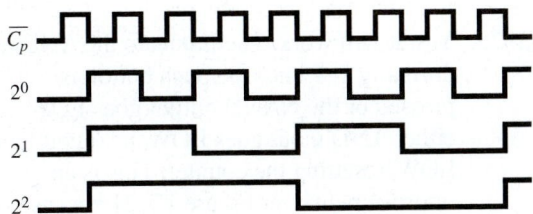

12–17.

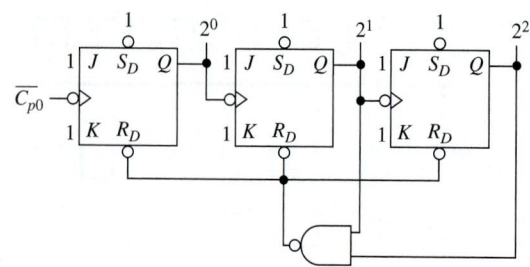

12–19.

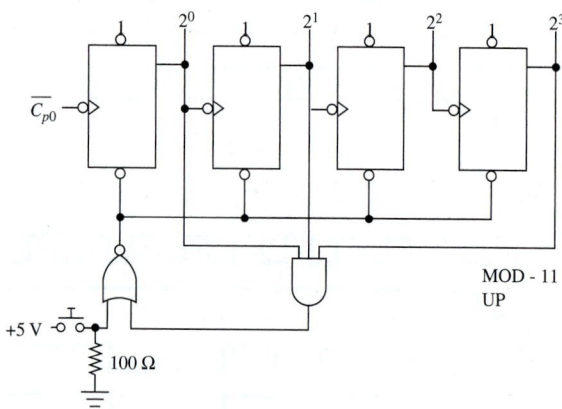

MOD - 11
UP

12–21.

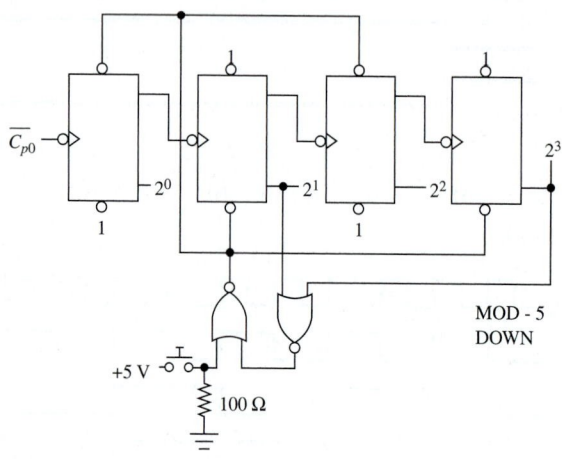

MOD - 5
DOWN

12–23.

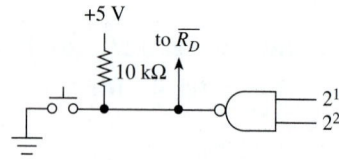

12–25. Yes, it will work. The inputs to the AND are normally 1–1 until the push button is pressed or the NAND output goes LOW. If either AND input goes LOW, its output goes LOW, resetting the counter. This is an improvement over Figure 12–21 because the

10 kΩ draws less current than the 100-Ω resistor when the push button is depressed.

12–27.

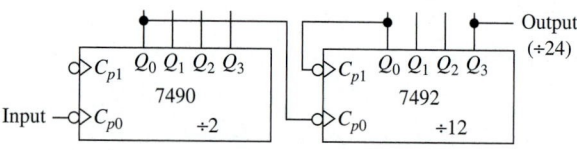

12–29.

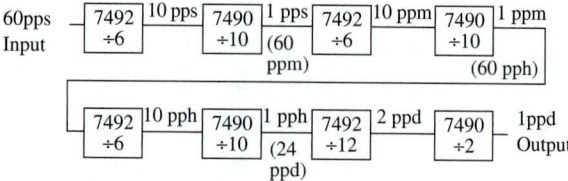

12–31.

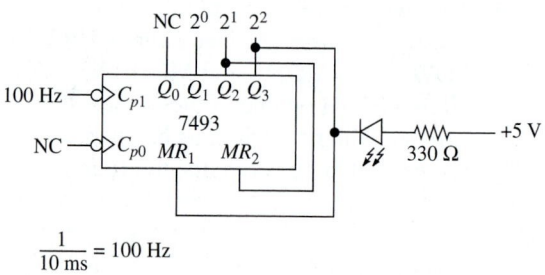

$$\frac{1}{10 \text{ ms}} = 100 \text{ Hz}$$

12–33. Connect a RESET push button across the 0.001-μF capacitor. When momentarily pressed, it will RESET the circuit to its initial condition.

12–35. With all the current going through the same resistor, as more segments are turned ON, the voltage that reaches the segments is reduced, making them dimmer. The displayed 8 is much dimmer than the 1.

12–37.

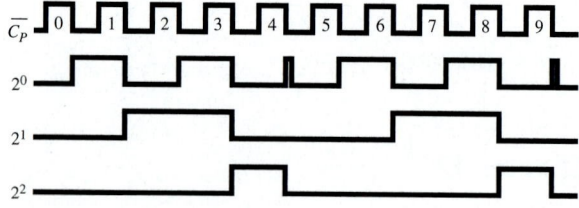

12–39.

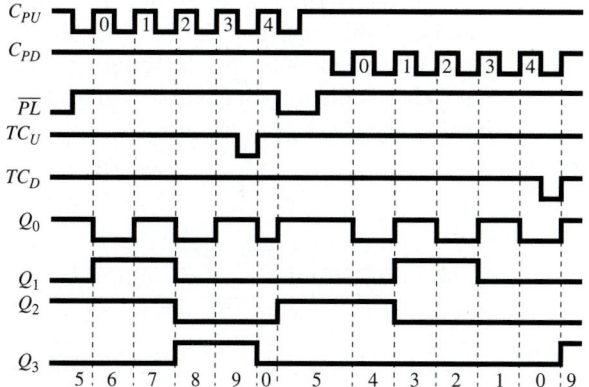

12–41.

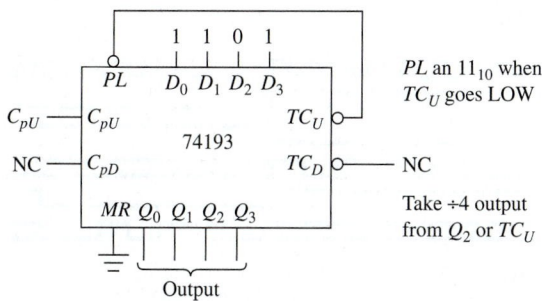

PL an 11_{10} when TC_U goes LOW

Take ÷4 output from Q_2 or TC_U

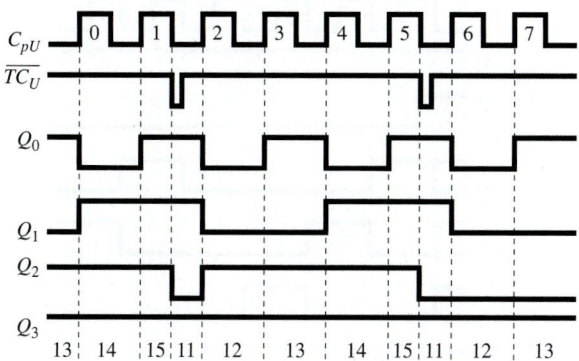

12–43. **(a)** HIGH-order U10, LOW-order U9
(b) no **(c)** by a LOW at the output of U3:B

12–45.

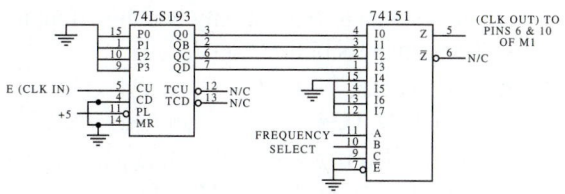

E12-1. **(a)** 2^0 **(b)** 0123456789ABCDEF

E12-3. Password for *Circuit Restrictions Hide Faults* is: **WK5E**
Doesn't count past 7. Bad MS flip-flop

E12-5. **(a)** Mod-10 **(b)** Connect NAND inputs to 2^3 and 2^2 **(c)** 2^2

E12-7. **(a)** Build a circuit similar to Figure 12–41.
(b) Use a third 7490 as div-by-10 with 10 pps driving its Cp0′. (The output of that 7490 is 1 pps.) Make the input to your 00 to 99 counter switch between those two clock speeds.
(c) Place an AND gate before the first Cp′ input. Disable it with a NAND gate that outputs a LOW when 23 is reached.

E12-9.

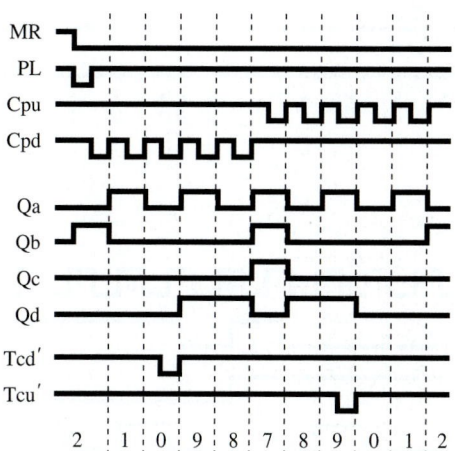

Chapter 13

13–1. Right, negative

13–3. 1110, 1111

13–5. Apply a LOW pulse to $\overline{R_D}$ to RESET all Q outputs to zero. Next, apply a LOW pulse to the active-LOW D_3, D_1, D_0 inputs.

13–7. J_3, K_3 are the data input lines. Q_3, Q_2, Q_1, Q_0 are the data output lines. (D_3, D_2, D_1, D_0 are held HIGH.)

13–9. Three

13–11. The Q_0 flip-flop and the Q_3 flip-flop

13–13.

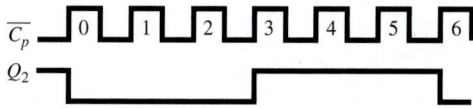

13–15.

13–17. Put a switch in series with the phototransistor's collector. With the switch open, the input to inverter 1 will be HIGH, simulating nighttime conditions, causing the yellow light to flash, day or night.

13–19.

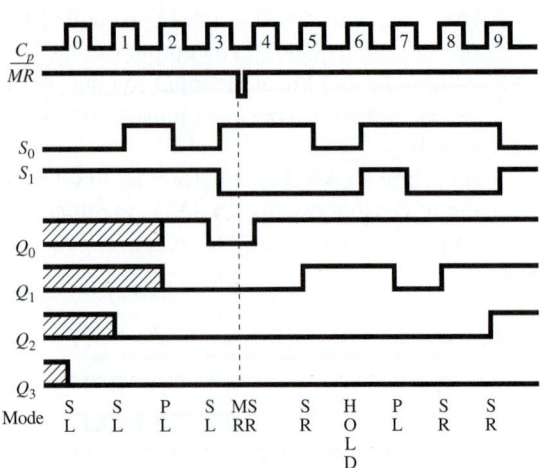

13–21.

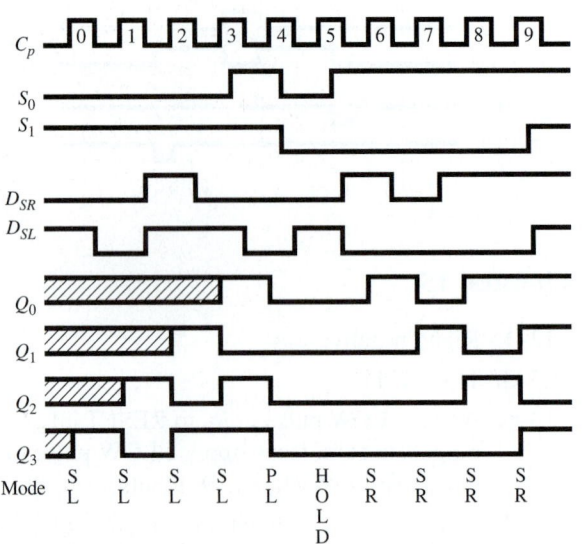

13–23.

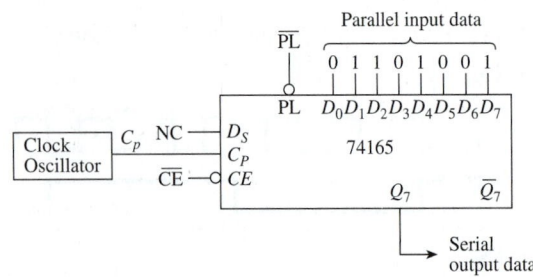

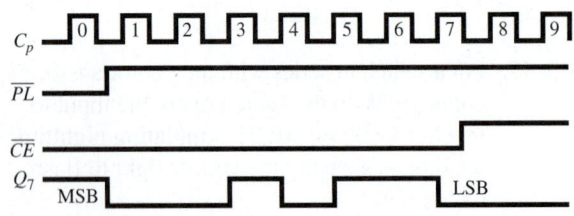

13–25. The 74164 is a serial-in, parallel-out, whereas the 74165 is a serial- or parallel-in, serial-out, shift register. The 74165 provides a clock-enable input, $\overline{CE}$. The serial input to the 74164 is the logical AND of two data inputs $(D_{sa} \cdot D_{sb})$.

13–27.

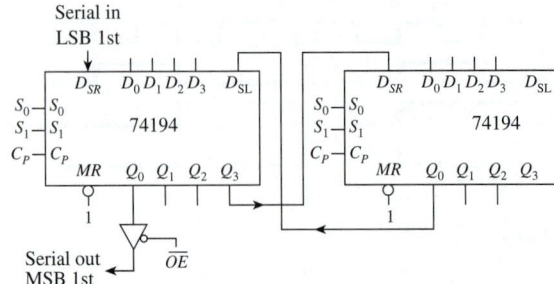

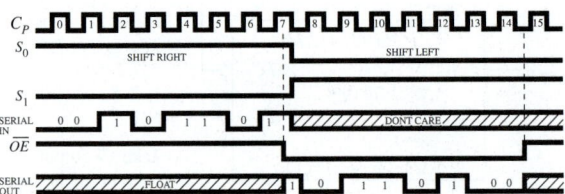

13–29.

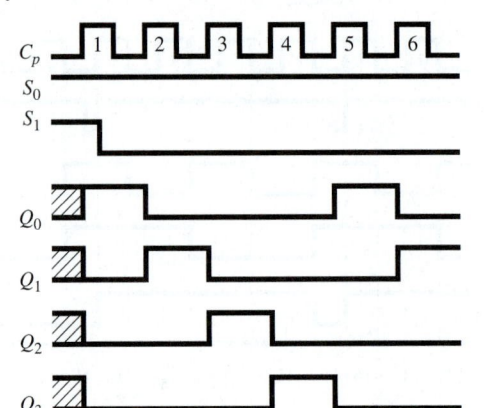

13–31. A buffer is a transparent device that connects two digital circuits; a latch is a storage device that can hold data. A buffer allows data to flow in only one direction; a transceiver is bidirectional.

13–33. (a) U4, U12, (b) U5, U30, U31, U32, U36, U37, U38 (c) U3, U6 (d) U11, U33

13–35. To load IAO-IA7 of U30, valid data must be placed on BD0-BD7 by U33, then a LOW to HIGH pulse must be applied to the CLK input of U30. When pin 3 of U13:A is LOW, a HIGH will appear at LE of U33. This allows the valid data at D0-D7 to pass through to BD0-BD7. Next, pin 3 of U13:A goes HIGH, making LE LOW, which causes the data outputs of U33 to remain latched.

The HIGH at pin 3 of U13:A also enables the decoder U23. Just before pin 3 went HIGH, A0-A1-A2 were set to 1-0-1 to provide an active output at /IOADDR. This is the clock input for U30, which passes the valid data at BD0-BD7 of U30 out to IA0-IA7. To load ID0-ID7 of U32, the same process is followed, except A0-A1-A2 are made 0-1-1 before pin 3 of U13:A is made HIGH.

E13-1. (a) 3 (b) R, S, C, S, C, S, C, S

E13-3. (a) 2mS (b)

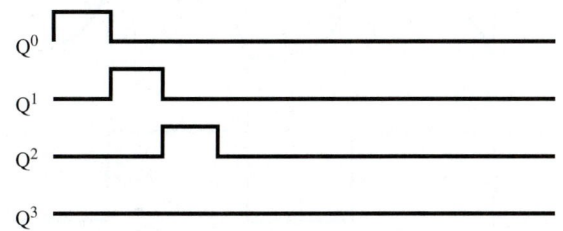

E13-5. M = 0100 1101, Reset 74164. Then Set S = 1, press C; Set S = 0, press C; Set S = 1, press C; Set S = 1, press C; Set S = 0, press C; Set S = 0, press C; Set S = 1, press C; Set S = 0, press C.

Chapter 14

14–1. (a) Monostable (b) Bistable
(c) Astable

14–3. 75.6 μs

14–5. 32.6 μs

14–7. Because a Schmitt device has two distinct switching thresholds, V_{T+} and V_{T-}; a regular inverter does not. The capacitor voltage charges and discharges between those two levels.

14–9.

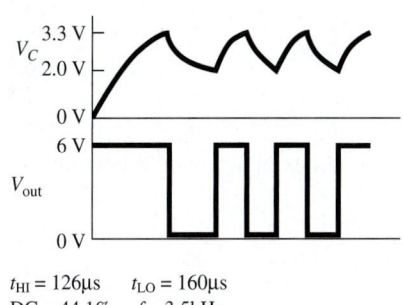

$t_{HI} = 126\mu s$ $t_{LO} = 160\mu s$
DC = 44.1% $f = 3.5$kHz

14–11.

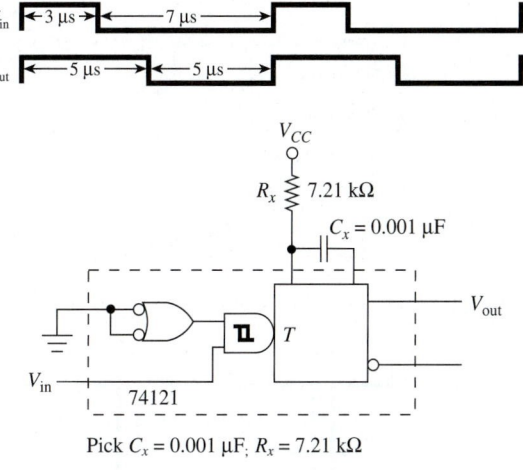

Pick $C_x = 0.001\ \mu F$, $R_x = 7.21\ k\Omega$

14–13. Choose an output pulse width that is longer than 150 μs, let's say, 170 μs. That way, if a pulse is missing after 170 μs (the O.S. is not retriggered), the output will go LOW.

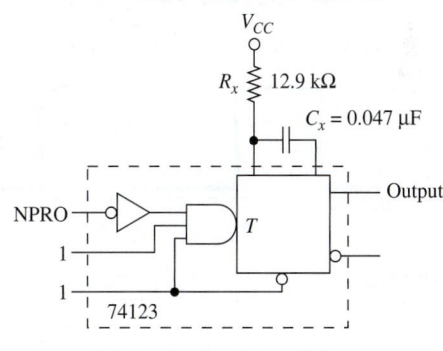

Pick $C_x = 0.047\ \mu F$, $R_x = 12.9\ k\Omega$

14–15.

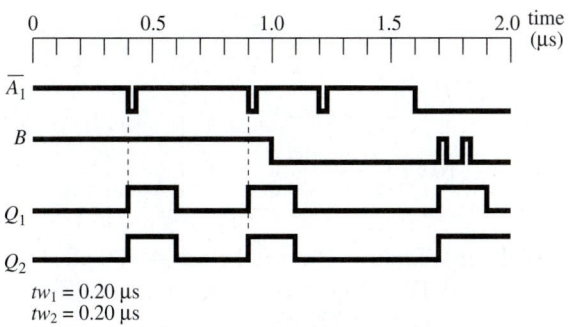

$tw_1 = 0.20\ \mu s$
$tw_2 = 0.20\ \mu s$

14–17.   @ 0Ω, $f = 89.0$ kHz
DC = 71.0%
@ 10 kΩ, $f = 39.8$ kHz
DC = 59.4%

14–19.

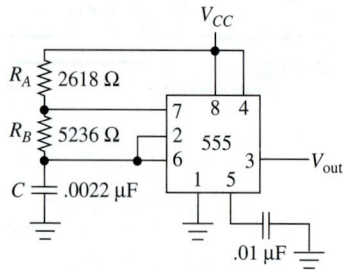

14–21.

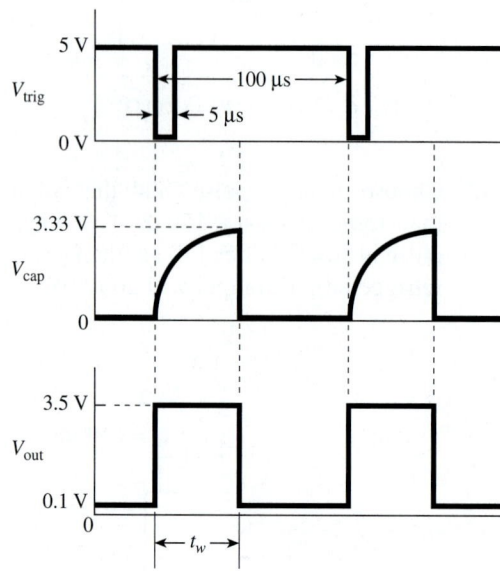

$t_w = 51.7\ \mu s$

14–23.

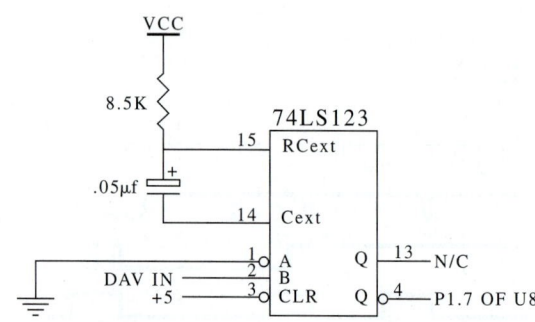

E14-1. (a) Tcalc = 693 μS, Tmeas = 700 μS
(b) DC Vtrig = 20%, DC Vout = 70%
(c) Rext = 722 ohms

E14-3. Ttrigger: Ttot = 1/10 khz = 100 μS,
Thi = 70 μS
Tout: 50% × 100 μS = 50 μS,
50 μS = .693 Rext × Cext,
Pick Cext = .01 μf, Rext = 7220 ohms.
Use those components in a circuit similar to
file *sec 14-5b*.

E14-5.

	Vout (calc.)	Vout (meas.)
th	11.64 µs	11.8 µs
tl	9.42 µs	9.6 µs

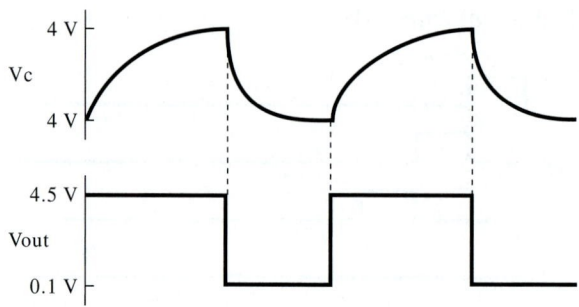

Chapter 15

15–1. Converts physical quantities into electrical quantities.

15–3. Very high input impedance, very high voltage gain, and very low output impedance.

15–5. The (−) input is at 0-V potential.

15–7. (a) 20 kΩ, 40 kΩ, 80 k

(b)

D_3	D_2	D_1	D_0	V_{out}
0	0	0	0	0.00
0	0	0	1	−1.25
0	0	1	0	−2.50
0	0	1	1	−3.75
0	1	0	0	−5.00
0	1	0	1	−6.25
0	1	1	0	−7.50
0	1	1	1	−8.75
1	0	0	0	−10.00
1	0	0	1	−11.25
1	0	1	0	−12.50
1	0	1	1	−13.75
1	1	0	0	−15.00
1	1	0	1	−16.25
1	1	1	0	−17.50
1	1	1	1	−18.75

15–9. All values of V_{out} would become positive.

15–11.

D_3	D_2	D_1	D_0	V_{out}	D_3	D_2	D_1	D_0	V_{out}
0	0	0	0	0.00	1	0	0	0	-2.00
0	0	0	1	-0.25	1	0	0	1	-2.25
0	0	1	0	-0.50	1	0	1	0	-2.50
0	0	1	1	-0.75	1	0	1	1	-2.75
0	1	0	0	-1.00	1	1	0	0	-3.00
0	1	0	1	-1.25	1	1	0	1	-3.25
0	1	1	0	-1.50	1	1	1	0	-3.50
0	1	1	1	-1.75	1	1	1	1	-3.75

15–13. To convert the analog output current (I_{out}) of the MC1408 to a voltage

15–15. (a) 2.5 V (b) 2.11 V (c) 1.25 V
(d) 1.17 V

15–17. By making $V_{REF} = 7.5$ V. The range of I_{out} would then be 0 to 1.5 mA. The range of V_{out} would be 0 to 7.47 V

15–19.

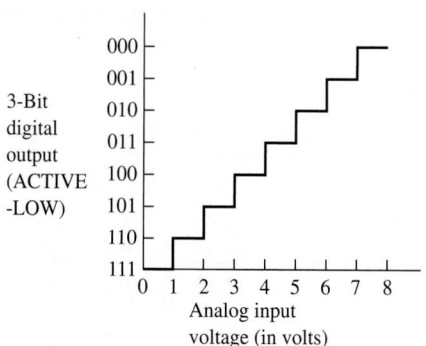

15–21. (a) 0 V (b) They are equal

15–23. $t_{tot} = 8 \times \frac{1}{50 \text{ kHz}} = 0.16$ ms

15–25. % error $= -0.197\%$

15–27. $\overline{CS}$ (chip select) and $\overline{RD}$ (READ); active-LOW

15–29. (a) The three-state output latches (D_0 to D_7) would be in the float condition. (b) The outputs would float and $\overline{WR}$ (start conversion) would be disabled. (c) It issues a LOW at power-up to start the first conversion. (d) 1.0 V

15–31. The temperature transducer is selected by setting up the appropriate code on the *ABC* multiplexer select inputs. The voltage level passes to the LF 198, which takes a sample at some precise time and holds the level on the hold capacitor. The LH0084 adjusts the voltage to an appropriate level to pass into the ADC. The microprocessor issues $\overline{CS_1}$, $\overline{WR_1}$, then waits for $\overline{INTR}$ to go LOW. It then issues $\overline{CS_1}$, $\overline{RD_1}$ to transfer the

converted data to the data bus, then $\overline{CS_2}$, $\overline{WR_2}$, to transfer the data to RAM.

15–33. 2.439 to 2.564 V

15–35. 9.60 kg

15–37.

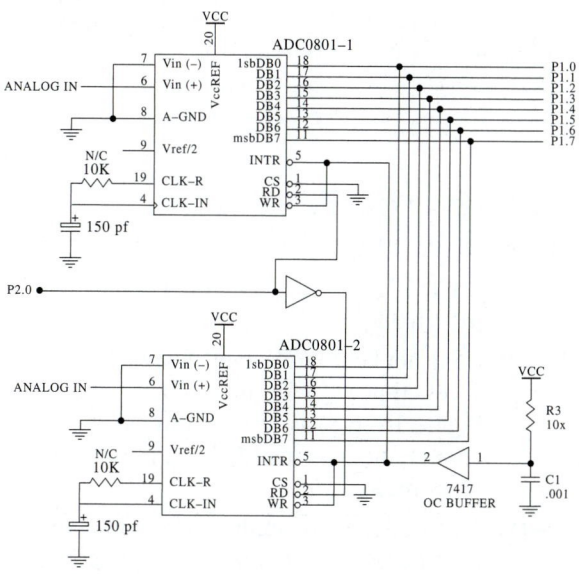

E15-1. (a) D0 = -1 V, D1 = -2 V, D2 = -4 V, D3 = -8 V
(b) True
(c) D3, D2, (D2 and D1), (D3 and D1 and D0)
(d) 16 (e) 6.25 k-ohms

E15-3. (a) Output

Vin	Hex	Dec
0.0	00	0
0.5	19	25
1.0	33	51
1.5	4C	76
2.0	66	102
2.5	7F	127
3.0	99	153
3.5	B2	178
4.0	CC	204
4.5	E5	229
5.0	FF	256

(b) Yes

Chapter 16

16–1. Bipolar, faster; MOS, more dense.

16–3. Data are being written into memory *from* the data bus. Bus contention occurs only when two or more devices are writing *to* the data bus at the same time.

16–5.

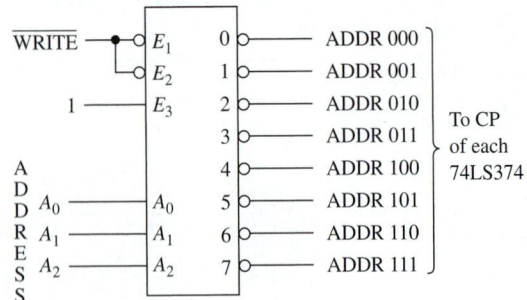

16–7. $\overline{CS}$ = LOW, $\overline{WE}$ = LOW

16–9. 10, 12, 13

16–11. (a) 8192 (b) 16,384 (c) 65,536
(d) 16,384

16–13. $\overline{CS}$ is not held LOW long enough. The access time (t_{acs}) for the 2147H is given in Figure 16–6(a) as 35 ns minimum. To correct, increase the LOW $\overline{CS}$ pulse to 35 ns or more.

16–15. The high- and low-order address lines are multiplexed to minimize the IC pin count.

16–17. (a)

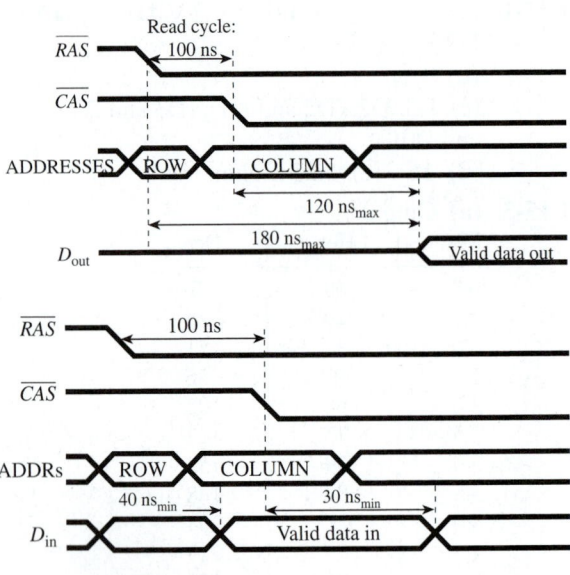

(b) 220 ns max
(c) 60 ns max

16–19. Address line multiplexing and memory refresh cycling

16–21. 2864

16–23.

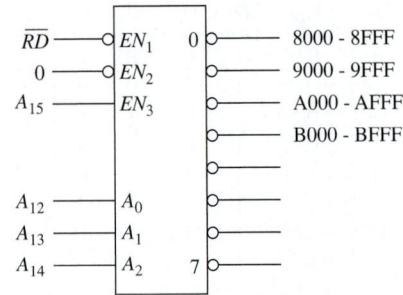

16–25. Address line A_{15} is stuck LOW.

16–27. The circuit design will be similar except the 2716 address inputs are A_0 to A_{10}. The new 74LS138 connections will be as follows:

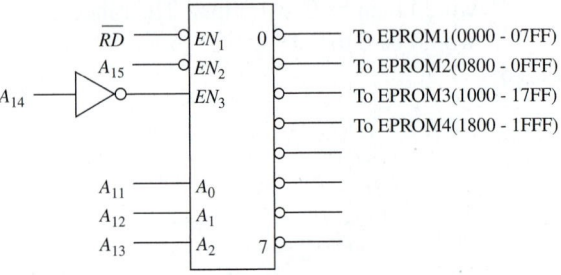

16–29.

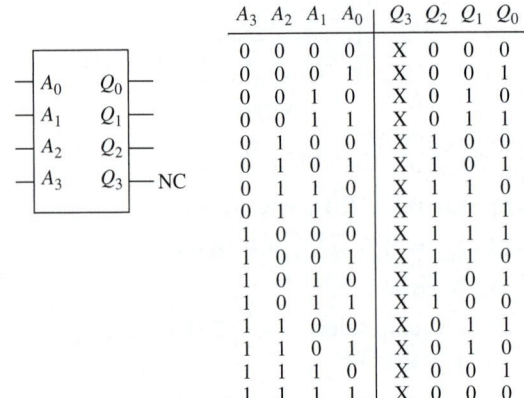

A_3	A_2	A_1	A_0	Q_3	Q_2	Q_1	Q_0
0	0	0	0	X	0	0	0
0	0	0	1	X	0	0	1
0	0	1	0	X	0	1	0
0	0	1	1	X	0	1	1
0	1	0	0	X	1	0	0
0	1	0	1	X	1	0	1
0	1	1	0	X	1	1	0
0	1	1	1	X	1	1	1
1	0	0	0	X	1	1	1
1	0	0	1	X	1	1	0
1	0	1	0	X	1	0	1
1	0	1	1	X	1	0	0
1	1	0	0	X	0	1	1
1	1	0	1	X	0	1	0
1	1	1	0	X	0	0	1
1	1	1	1	X	0	0	0

16–31. c, d, e, f

16–33. (a) Hard disk (b) DVD

16–35. 32k $\times$ 8

16–37. (a) 8k $\times$ 8 (b) SMN_SEL, MON_SEL
(c) A000H-BFFFH (d) E000H-FFFFH

E16-1. (a) The D flip-flop only captures data at the instant C is pressed. (b) Outputs are all stuck LOW.

Chapter 17

17–1. A microprocessor-based system would be used whenever calculations are to be made, decisions based on inputs are to be made, a memory of events is needed, or a modifiable system is needed.

17–3. The address bus is used to select a particular location or device within the system.

17–5.

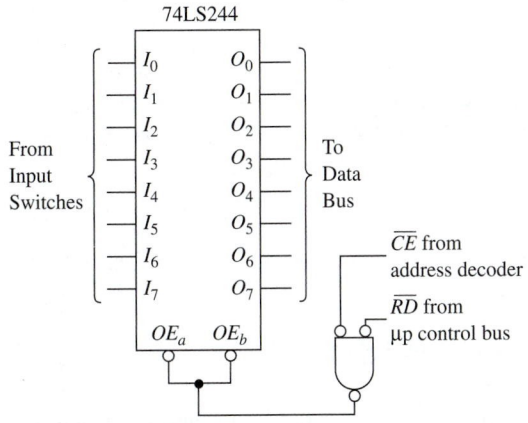

17–7. The input port has three-stated outputs so that it can be disabled when it is not being read.

17–9. 2^{16} (65,536)

17–11. It stores the contents of the accumulator out to address 6000H.

17–13. Instruction decoder and register: register and circuitry inside the microprocessor that receives the machine language code and produces the internal control signals required to execute the instruction.

17–15. (1) Pulse: read memory location 2000 (LDA); (2) and (3) Pulse: read address bytes at 2001, 2002 (4000H); (4) Pulse: read data at address 4000H

17–17. A high-level language (FORTRAN, BASIC, etc.) has the advantage of being easier to write and understand. Its disadvantage is that the programs are not memory efficient.

17–19. (a) LOW (b) LOW (c) HIGH
(d) HIGH

17–21. U6a and U6b are drawn as inverted-input NAND gates to make the logical flow of the schematic easier to understand.

17–23. (a) IN instruction, $\overline{RD}$ is pulsed LOW.
(b) OUT instruction, $\overline{WR}$ is pulsed LOW.

17–25. (1) $A8$ to $A15$ = FEH
(2) $IO/\overline{M}$ = HIGH
(3) $\overline{WR}$ is pulsed LOW/HIGH

17–27. MVI A, 4FH

17–29.

2010	3E
2011	04
2012	3D
2013	CA
2014	10
2015	20
2016	C3
2017	12
2018	20

17–31.

	LCD_SL	KEL_SL
AD3	0	1
AD4	0	0
AD5	0	0
AD11	1	1
AD12	1	1
AD13	0	0
AD14	0	0
AD15	0	0

E17-1. (a) LOW

(b) Press C

(c) Put 7C on the switches, Make OE′ = 0 (Data appears on the data bus), Press C (data appears at output port if OE′ (G) is LOW.)

Chapter 18

18–1. Because it has RAM, ROM, I/O Ports, and a Timer/Counter on it.

18–3. Extra 4 K ROM, 128 bytes RAM, 16 bit Timer/Counter, 1 interrupt.

18–5. High

18–7. 80H-FFH

18–9. By reading the two bank-select bits (D0H.3, D0H.4) in the PSW.

18–11. (a) A = 40H (b) A = 88H
(c) A = 88H (d) A = 40H

18–13. (a) MOV P3,#0C7H (b) MOV R7,P1
(c) MOV A,#55H (d) MOV A,@R0
(e) MOV P0,R1

18–15. START: MOV A,#20H
LOOP: MOV P1,A
            INC A
            CJNE A,#91H,LOOP
            JMP START

18–17. (a) A = 75H (b) A = F5H
(c) A = EBH

18–19. MOV A,90
ADD A,#05H
MOV P1,A

VHDL Language Reference

Library-Entity-Architecture Model

Figure E–1 is a sample VHDL program showing the syntax rules and language format of a basic program.

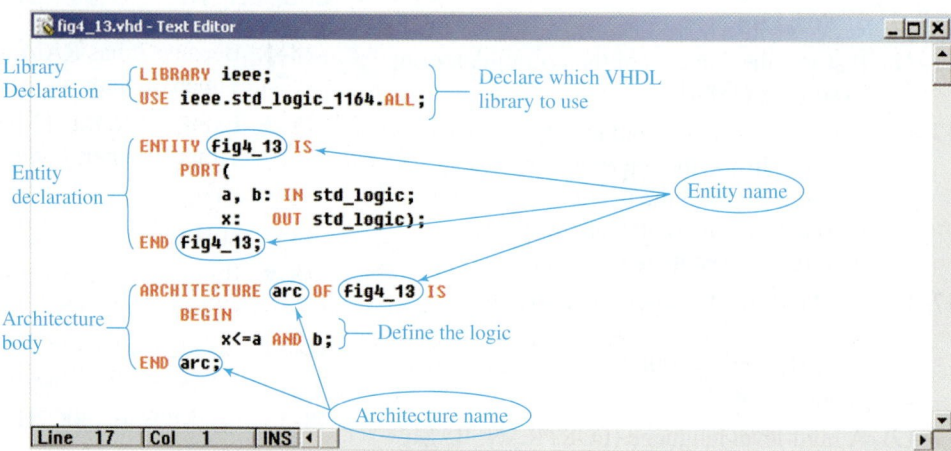

Figure E–1 VHDL program model showing the syntax for Library, Entity, and Architecture declarations.

Program, Entity, and Architecture Naming Conventions

Valid VHDL names can consist of letters, numbers, and the underscore (_) character [hyphens (-) are not allowed]. The name must start with a letter and cannot contain spaces.

Valid names	Invalid Names	
fig4_13	fig4-13	(cannot use hyphen)
boolean3	boolean 3	(cannot use space)
counter_5_a	counter_5.a	(cannot use period)

VHDL Program Comments

Comments can be placed within a VHDL program to document a particular statement or group of statements. Any words preceded by a double hyphen (--) are considered to be a comment for documentation purposes and will be ignored by the VHDL compiler.

Sample Comments

-- Boolean solution to example 5-3

-- Problem C5-3 --

General VHDL Rules

VHDL logical operators have no order of precedence so the order must be explicitly defined using parentheses. For example, the Boolean equation $x = a + bc$ must be written in VHDL as: **x <= a OR (b AND c);** (The symbol <= is used to assign the result of the operation on the right-hand side to the left-hand side.) Also note that VHDL is *not* case-sensitive so the letters *a, b,* and *c* could have been capitalized, but the convention used by most VHDL programmers is to only capitalize the reserved keywords like AND, OR, PORT, ENTITY, and so on. (See Figure E–1.)

Logical Operators

The logic operators are used to perform Boolean bit-wise operations on individual bits or arrays of bits. (Example: x <= a AND b;)

Operator	Description
AND	And
OR	Or
NAND	Not-And
NOR	Not-Or
XOR	Exclusive-Or
XNOR	Exclusive-Nor

Relational Operators

The relational operators are used to test the relative values of two scalar types. The result is a Boolean *true* or *false* value. [Example: **IF a < b THEN result <= "001".** This is interpreted as "if the logic (*a* less than *b*) is *true,* then the bit string "001" is moved to the variable *result.*"]

Operator	Description
=	Equality
/=	Inequality
<	Less than
<=	Less than or equal
>	Greater than
>=	Greater than or equal

Arithmetic Operators

The arithmetic operators are used to perform mathematical operations. (Example: sum_string<=astring+bstring+cin;) The following is a partial list of the arithmetic operators.

Operator	Description
+	Addition
−	Subtraction
&	Concatenation
*	Multiplication
/	Division
**	Exponentiation
mod	Modulus
rem	Remainder
abs	Absolute value
sll	Shift left
srl	Shift right

Data Types

The data type defines the type of value that can be used with the specified input, output, or internal signal. The most common data types used are as follows:

Type	Values
bit	'0', '1'
std_logic	'0', '1', 'U', 'X', 'Z', 'W', 'L', 'H', '-'
integer	Integer values
bit_vector	Multiple instances of '0', '1"
std_vector	Multiple instances of '0', '1', 'U', 'X', 'Z', 'W', 'L', 'H', '-'

std_logic Data Type Values

The values that the *std_logic* data type can have are defined as follows:

Value	Description
'0'	Logic 0
'1'	Logic 1
'U'	Uninitialized
'X'	Unknown
'Z'	High impedance
'W'	Weak unknown
'L'	Weak 0
'H'	Weak 1
'-'	Don't care

VHDL by Example

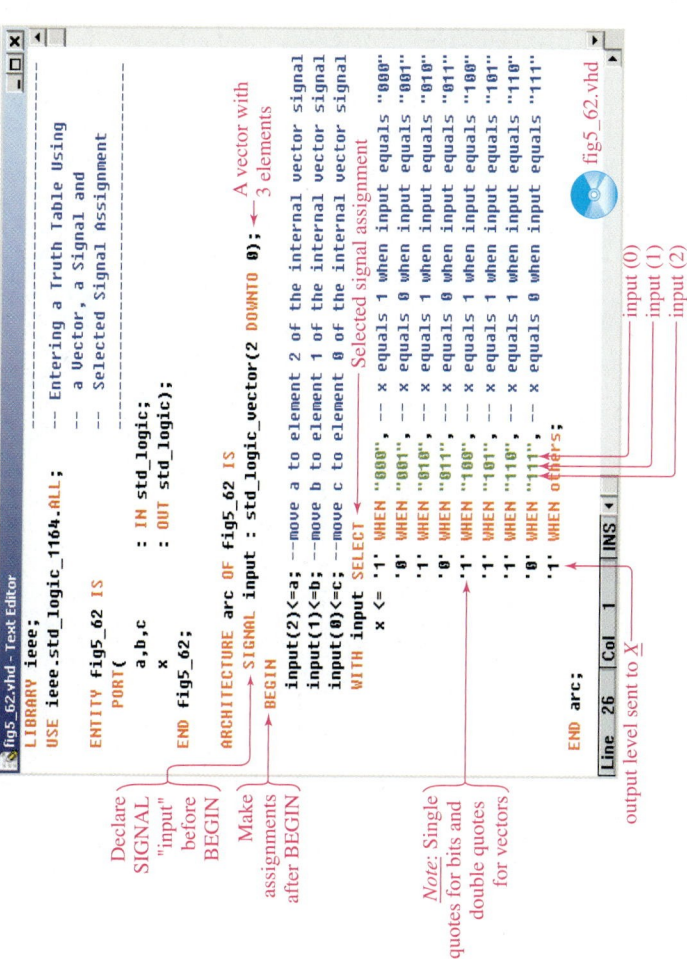

fig4_13.vhd – Text Editor

```
LIBRARY ieee;
USE ieee.std_logic_1164.ALL;      -- Declare which VHDL
                                  --   library to use

ENTITY fig4_13 IS
  PORT(
    a, b: IN std_logic;
    x:   OUT std_logic);
END fig4_13;

ARCHITECTURE arc OF fig4_13 IS
  BEGIN
    x<=a AND b;
END arc;
```
Line 17 Col 1 INS

Library Declaration
Entity name
Entity declaration
Define the logic
Architecture name
Architecture body

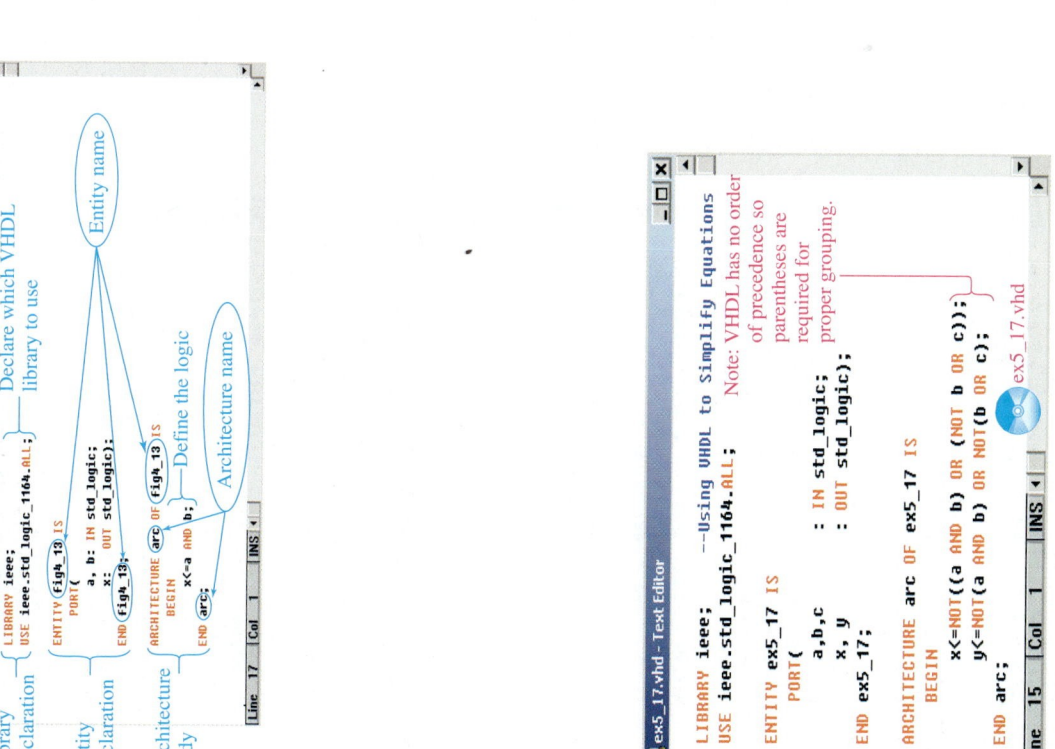

fig5_62.vhd – Text Editor

```
LIBRARY ieee;
USE ieee.std_logic_1164.ALL;      -- Entering a Truth Table Using
                                  -- a Vector, a Signal and
                                  -- Selected Signal Assignment

ENTITY fig5_62 IS
  PORT(
    a,b,c     : IN std_logic;
    x         : OUT std_logic);
END fig5_62;

ARCHITECTURE arc OF fig5_62 IS
  SIGNAL input : std_logic_vector(2 DOWNTO 0);  -- A vector with 3 elements
  BEGIN
    input(2)<=a;   -- move a to element 2 of the internal vector signal
    input(1)<=b;   -- move b to element 1 of the internal vector signal
    input(0)<=c;   -- move c to element 0 of the internal vector signal

    WITH input SELECT                   -- Selected signal assignment
      x <=  '1' WHEN "000",   -- x equals 1 when input equals "000"
            '0' WHEN "001",   -- x equals 0 when input equals "001"
            '1' WHEN "010",   -- x equals 1 when input equals "010"
            '0' WHEN "011",   -- x equals 0 when input equals "011"
            '1' WHEN "100",   -- x equals 1 when input equals "100"
            '1' WHEN "101",   -- x equals 1 when input equals "101"
            '1' WHEN "110",   -- x equals 1 when input equals "110"
            '1' WHEN others;  -- x equals 1 when input equals "111"

END arc;
```
Line 26 Col 1 INS

Declare SIGNAL "input" before BEGIN
Make assignments after BEGIN
Note: Single quotes for bits and double quotes for vectors
output level sent to X
input (0)
input (1)
input (2)

fig5_62.vhd

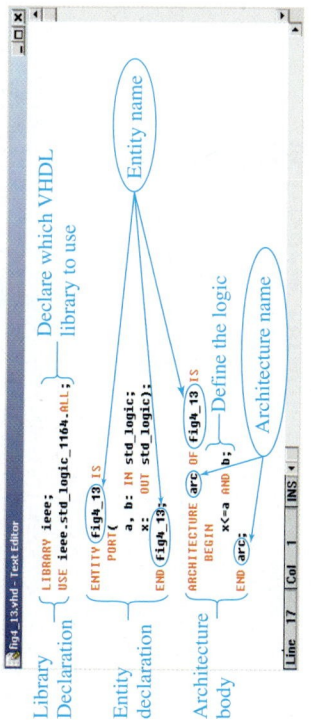

ex5_17.vhd – Text Editor

```
LIBRARY ieee;                 --Using VHDL to Simplify Equations
USE ieee.std_logic_1164.ALL;          Note: VHDL has no order
                                      of precedence so
                                      parentheses are
                                      required for
                                      proper grouping.
ENTITY ex5_17 IS
  PORT(
    a,b,c     : IN std_logic;
    x, y      : OUT std_logic);
END ex5_17;

ARCHITECTURE arc OF ex5_17 IS
  BEGIN
    x<=NOT((a AND b) OR (NOT b OR c));
    y<=NOT(a AND b) OR NOT(b OR c);
END arc;
```
Line 15 Col 1 INS

ex5_17.vhd

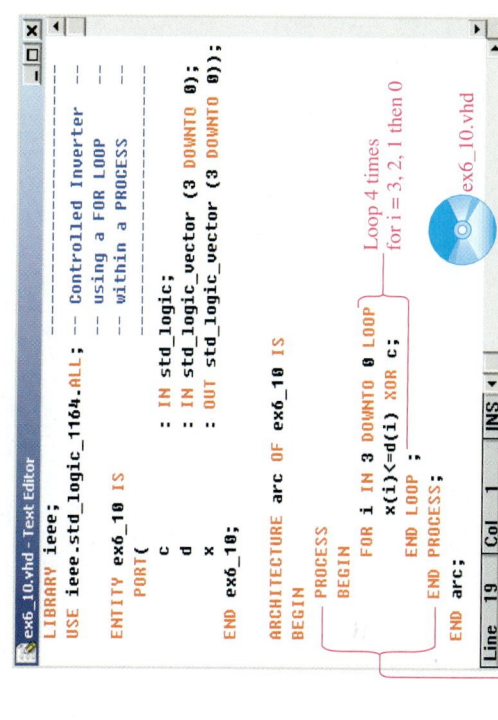

ex6_10.vhd – Text Editor

```
LIBRARY ieee;
USE ieee.std_logic_1164.ALL;      -- Controlled Inverter
                                  --   using a FOR LOOP
                                  --   within a PROCESS

ENTITY ex6_10 IS
  PORT(
    c    : IN std_logic;
    d    : IN std_logic_vector (3 DOWNTO 0);
    x    : OUT std_logic_vector (3 DOWNTO 0));
END ex6_10;

ARCHITECTURE arc OF ex6_10 IS
  BEGIN
    PROCESS
    BEGIN
        FOR i IN 3 DOWNTO 0 LOOP
          x(i)<=d(i) XOR c;
        END LOOP ;
    END PROCESS;
END arc;
```
Line 19 Col 1 INS

Loop 4 times for i = 3, 2, 1 then 0
Sequential process

ex6_10.vhd

compare_8b.vhd - Text Editor

```vhdl
LIBRARY ieee;                          -- 8-bit Comparator using
USE ieee.std_logic_1164.ALL ;          -- IF-THEN-ELSE and ELSIF

ENTITY compare_8b IS
    PORT(a, b         : IN  std_logic_vector(7 DOWNTO 0);
         agb, aeb, alb : OUT std_logic);
END compare_8b;

ARCHITECTURE arc OF compare_8b IS
    SIGNAL result     : std_logic_vector(2 DOWNTO 0);     ← 3-bit internal
BEGIN                                                        SIGNAL result
    PROCESS (a,b)        ← Sensitivity list
    BEGIN
        IF    a<b THEN
            result  <=  "001";    → alb
        ELSIF a=b THEN
            result  <=  "010";    → aeb
        ELSIF a>b THEN
            result  <=  "100";    → agb
        ELSE
            result  <=  "000";
        END IF;
        agb  <=  result(2);     ┐
        aeb  <=  result(1);     ├ Assign vector elements
        alb  <=  result(0);     ┘ to outputs
    END PROCESS;
END arc;
```
compare_8b.vhd

Line 27 | Col 1 | INS

---

decoder_b.vhd - Text Editor

```vhdl
LIBRARY ieee;                          -- Octal decoder using vectors
USE ieee.std_logic_1164.ALL;           -- and
                                       -- selected signal assignment

ENTITY decoder_b IS
    PORT(a : IN  STD_LOGIC_VECTOR (2 downto 0);
         y : OUT STD_LOGIC_VECTOR (7 downto 0));
END decoder_b ;

ARCHITECTURE arc OF decoder_b IS                y7    y0
BEGIN                                       a2  a0
    WITH a SELECT
        y<= "00000001"  WHEN "000",
            "00000010"  WHEN "001",
            "00000100"  WHEN "010",
            "00001000"  WHEN "011",
            "00010000"  WHEN "100",
            "00100000"  WHEN "101",
            "01000000"  WHEN "110",
            "10000000"  WHEN "111",
            "00000000"  WHEN others;
END arc;
```
decoder_b.vhd

Line 21 | Col 1 | INS

---

ex7_26.vhd - Text Editor

```vhdl
-- 8-bit binary adder/subtractor using std_logic vectors  --
-- and the WHEN ELSE conditional signal assignment          --

LIBRARY ieee;
USE ieee.std_logic 1164.ALL;
USE ieee.std_logic_signed.ALL;     ← Required for vector arithmetic

ENTITY ex7_26 IS
    PORT
    (
        add_sub   : IN   std_logic;  ← add or subtract control
        astring   : IN   std_logic_vector(7 DOWNTO 0);   ┐
        bstring   : IN   std_logic_vector(7 DOWNTO 0);   ├ 8-bit vectors
        result    : OUT  std_logic_vector(7 DOWNTO 0)    ┘
    );
END ex7_26 ;

ARCHITECTURE arc OF ex7_26 IS
BEGIN
    result<=astring+bstring WHEN add_sub='0'   ┐ WHEN ELSE
          ELSE astring-bstring;                ┘ conditional assignment
END arc;
```
ex7_26.vhd

Line 27 | Col 1 | INS

---

ex7_27.vhd - Text Editor

```vhdl
LIBRARY ieee;                          -- BCD Correction Adder  --
USE ieee.std_logic_1164.ALL;           -- using IF-THEN-ELSE     --
USE ieee.std_logic_unsigned.ALL;

ENTITY ex7_27 IS
    PORT
    (
        astring    : IN   std_logic_vector(3 DOWNTO 0);
        bstring    : IN   std_logic_vector(3 DOWNTO 0);
        bcd_result : OUT  std_logic_vector(7 DOWNTO 0)
    );
END ex7_27;

ARCHITECTURE arc OF ex7_27 IS
SIGNAL bin_result :    std_logic_vector(7 DOWNTO 0);   ← Internal interconnection signal
BEGIN
    bin_result<=astring+bstring;
    PROCESS
    BEGIN
        IF bin_result>"0100"
            THEN bcd_result<=bin_result+"0110";
            ELSE bcd_result<=bin_result;
        END IF;
    END PROCESS;
END arc;
```
ex7_27.vhd

IF-THEN-ELSE sequential statements

Line 25 | Col 2 | INS

902

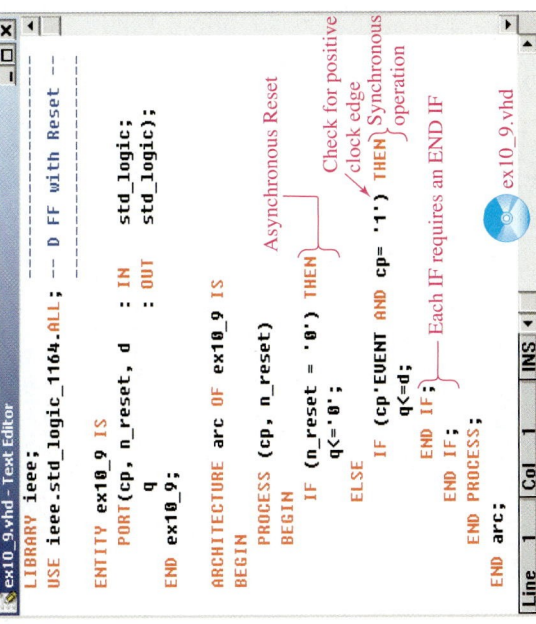

ex10_9.vhd - Text Editor

```vhdl
LIBRARY ieee;
USE ieee.std_logic_1164.ALL;   -- D FF with Reset --

ENTITY ex10_9 IS
    PORT(cp, n_reset, d   : IN    std_logic;
         q                : OUT   std_logic);
END ex10_9;

ARCHITECTURE arc OF ex10_9 IS
BEGIN
    PROCESS (cp, n_reset)
    BEGIN
        IF (n_reset = '0') THEN          -- Asynchronous Reset
            q<='0';
        ELSE
            IF (cp'EVENT AND cp= '1') THEN   -- Check for positive clock edge / Synchronous operation
                q<=d;
            END IF;                      -- Each IF requires an END IF
        END PROCESS;
END arc;
```

ex10_9.vhd

ex10_14.vhd - Text Editor

```vhdl
LIBRARY ieee;
USE ieee.std_logic_1164.ALL;   --   J-K FF   --

ENTITY ex10_14 IS
    PORT(n_cp, j, k   : IN     std_logic;
         q            : BUFFER std_logic);    -- Buffer declares q as output and input
END ex10_14;

ARCHITECTURE arc OF ex10_14 IS
SIGNAL jk : std_logic_vector (1 DOWNTO 0);
BEGIN
    jk<=j&k;                 -- Concatenate j and k into a 2-bit vector
    PROCESS (n_cp, j, k)
    BEGIN
        IF (n_cp'EVENT AND n_cp= '0') THEN   -- Neg edge trigger / Double quote required
            CASE jk IS
                WHEN "00" => q <= q;      --Hold    for vector
                WHEN "01" => q <= '0';    --Reset   quantities
                WHEN "10" => q <= '1';    --Set
                WHEN "11" => q <= NOT q;  --Toggle
                WHEN OTHERS => q <= q;
            END CASE;
        END IF;
    END PROCESS;
END arc;
```

ex10_14.vhd

decoder_d.vhd - Text Editor

```vhdl
LIBRARY ieee;
USE ieee.std_logic_1164.ALL;   -- Octal decoder with enable --
                               -- using IF-THEN-ELSE --
                               -- and CASE statement --

ENTITY decoder_d IS
    PORT(en : IN    std_logic;
         a  : IN    STD_LOGIC_VECTOR (2 downto 0);
         y  : OUT   STD_LOGIC_VECTOR (7 downto 0));
END decoder_d;

ARCHITECTURE arc OF decoder_d IS
BEGIN
    PROCESS (a,en)     -- sensitivity list
    BEGIN
    IF (en='1')        -- Enable must be HIGH to execute the THEN clause
        THEN
            CASE a IS
                WHEN "000" =>y<="00000001";    -- when a = 000, y = 00000001
                WHEN "001" =>y<="00000010";
                WHEN "010" =>y<="00000100";
                WHEN "011" =>y<="00001000";
                WHEN "100" =>y<="00010000";
                WHEN "101" =>y<="00100000";
                WHEN "110" =>y<="01000000";
                WHEN "111" =>y<="10000000";
                WHEN others=>y<="00000000";
            END CASE;
        ELSE y<="00000000";    -- all LOW outputs if en is not HIGH
    END IF;
    END PROCESS;
END arc;
```

decoder_d.vhd

ex8_5.vhd - Text Editor

```vhdl
LIBRARY ieee;
USE ieee.std_logic_1164.ALL;   -- Octal priority encoder --
                               -- using --
                               -- conditional signal assignment --

ENTITY ex8_5 IS
    PORT(i  : IN    std_logic_vector (7 DOWNTO 0);
         a  : OUT   std_logic_vector (2 DOWNTO 0));
END ex8_5 ;

ARCHITECTURE arc OF ex8_5 IS
BEGIN
    a<="111" WHEN i(7)='1' ELSE     -- The first clause that is TRUE has priority, all others are skipped.
       "110" WHEN i(6)='1' ELSE
       "101" WHEN i(5)='1' ELSE
       "100" WHEN i(4)='1' ELSE
       "011" WHEN i(3)='1' ELSE
       "010" WHEN i(2)='1' ELSE
       "001" WHEN i(1)='1' ELSE
       "000" WHEN i(0)='1' ELSE
       "000";
END arc;
```

903

counter_b.vhd - Text Editor

```vhdl
LIBRARY ieee;                                   --4-bit Up/Down-Counter   --
USE ieee.std_logic_1164.ALL;                    -- with Async Reset and Load --
                                                -- and clock enable        --

ENTITY counter_b IS
    PORT(cp, n_rd, n_pl, n_ce, u_d   : IN std_logic;
         pl_data                      : IN integer RANGE 0 TO 15;
         q                            : BUFFER integer RANGE 0 TO 15);
END counter_b;

ARCHITECTURE arc OF counter_b IS
BEGIN
    PROCESS (cp, n_rd, n_pl)
    BEGIN
        IF   (n_rd='0' AND n_pl='1') THEN          -- Asyncronous
             q<= 0;
        ELSIF (n_rd='1' AND n_pl='0') THEN
             q<=pl_data;
        ELSIF (cp'EVENT AND cp='1') THEN           -- Clock enable
             IF (n_ce='0') THEN
                 IF (u-d='0') THEN                 -- Up/down control
                     q<=q+1;
                 ELSE
                     q<=q-1;
                 END IF;
             END IF;
        END IF;
    END PROCESS;
END arc;
```

counter_b.vhd

shiftreg.vhd - Text Editor

```vhdl
LIBRARY ieee;                                    --Shift register defined using  --
USE ieee.std_logic_1164.ALL;                     --Dflipflop component instantiations --

ENTITY dflipflop IS
    PORT(d, clk    : IN    std_logic;
         q          : OUT   std_logic);
END dflipflop ;

ARCHITECTURE arc OF dflipflop IS
BEGIN
    PROCESS (clk)
    BEGIN
        IF (clk'EVENT AND clk= '1') THEN q<=d;
        END IF;
    END PROCESS;
END arc;

LIBRARY ieee;
USE ieee.std_logic_1164.ALL;

ENTITY shiftreg IS
    PORT(serial_in, cp  : IN     STD_LOGIC;
         q3,q2,q1,q0    : BUFFER STD_LOGIC);
END shiftreg ;

ARCHITECTURE arc OF shiftreg IS
    COMPONENT dflipflop
        PORT (d, clk   : IN  STD_LOGIC;
              q        : OUT STD_LOGIC);
    END COMPONENT;  --This defines the internal and external connections
BEGIN
    ff3: dflipflop PORT MAP (d=>serial_in, clk=>cp, q=>q3);
    ff2: dflipflop PORT MAP (d=>q3,        clk=>cp, q=>q2);
    ff1: dflipflop PORT MAP (d=>q2,        clk=>cp, q=>q1);
    ff0: dflipflop PORT MAP (d=>q1,        clk=>cp, q=>q0);
END arc;
```

D flip-flop from Example 10-9

External I/O for shiftreg

I/O for each component dflipflop

Declare the use of the component dflipflop from above

Four instantiations

component name

instantiation label

Keyword

mod16up.vhd - Text Editor

```vhdl
LIBRARY ieee;                     --
USE ieee.std_logic_1164.ALL ;     -- Mod-16 Up-Counter --
                                  --

ENTITY mod16up IS
    PORT(n_cp, n_rd    : IN     std_logic;
         q              : BUFFER integer RANGE 0 TO 15);
END mod16up;

ARCHITECTURE arc OF mod16up IS
BEGIN
    PROCESS (n_cp, n_rd)
    BEGIN
        IF  (n_rd='0') THEN
            q <= 0;
        ELSIF (n_cp'EVENT AND n_cp='0') THEN
            q <=q+1;
        END IF;
    END PROCESS;
END arc;
```

4-bit output

Input and output

Asynchronous Reset has priority

Negative clock edge

Input to equation

Output to CPLD pins

mod16up.vhd

state_gray.vhd - Text Editor

```vhdl
LIBRARY ieee;                                    -- State Machine Gray Code Counter --
USE ieee.std_logic_1164.ALL;

ENTITY state_gray IS
    PORT(clk  : IN    STD_LOGIC;
         q     : OUT   STD_LOGIC_VECTOR(2 downto 0));
END state_gray;

ARCHITECTURE arc OF state_gray IS
    TYPE state_type IS (s0, s1, s2, s3, s4, s5, s6, s7);
    SIGNAL state: state_type;
BEGIN
    PROCESS (clk)
    BEGIN
        IF clk'EVENT AND clk = '1' THEN
            CASE state IS
                WHEN s0 => state <= s1;
                WHEN s1 => state <= s2;
                WHEN s2 => state <= s3;
                WHEN s3 => state <= s4;
                WHEN s4 => state <= s5;
                WHEN s5 => state <= s6;
                WHEN s6 => state <= s7;
                WHEN s7 => state <= s0;
            END CASE;
        END IF;
    END PROCESS;

    WITH state SELECT
        q <= "000"   WHEN s0,
             "001"   WHEN s1,
             "011"   WHEN s2,
             "010"   WHEN s3,
             "110"   WHEN s4,
             "111"   WHEN s5,
             "101"   WHEN s6,
             "100"   WHEN s7;
END arc;
```

Defines a new data type having the allowable values s0-s7.

Increment state to next state.

Assign binary to q based on value of state.

state_gray.vhd

Review of Basic Electricity Principles

Definitions for Figure F–1

$V \equiv$ voltage source that pushes the current (I) through the circuit, like water through a pipe

$I \equiv$ current that flows through the circuit [conventional current flows (+) to (−).]

$R \equiv$ resistance to the flow of current

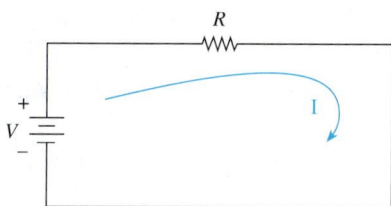

Figure F–1   Series circuit used to illustrate Ohm's Law.

Units

voltage = volts (V), for example, 12 V, 6 mV

current = amperes (A), for example, 2 A, 2.5 mA

resistance = ohms (Ω), for example, 100Ω, 4.7 kΩ

Common Engineering Prefixes

Prefix	Abbreviation	Value
Mega-	M	1,000,000 or 10^6
kilo-	k	1,000 or 10^3
milli-	m	0.001 or 10^{-3}
micro-	μ	0.000001 or 10^{-6}

Ohm's Law

The current (I) in a complete circuit is proportional to the applied voltage (V) and inversely proportional to the resistance (R) of the circuit (see Figure F–2).

Formulas

$$I = \frac{V}{R}$$
$$V = I \times R$$
$$R = \frac{V}{I}$$

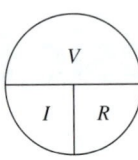

Figure F–2 Ohm's Law circle.

EXAMPLE F–1

(a) Determine the current (I) in the circuit of Figure F–1 if $V = 10$ V and $R = 2\Omega$.

(b) Recalculate the current if $R = 4\Omega$.

(c) Describe what happened to the value of the current when the resistance doubled to 4Ω in (b).

(d) Calculate the voltage required in Figure F–1 to make 2 A flow if $R = 10\Omega$.

(e) What voltage would be required in (d) if you only need one-half that current?

(f) If $V = 12$ V in Figure F–1, what resistance is required to limit the current to 2 A?

(g) To limit the current to 1 A in (f), what resistance would you need?

Solution:

(a) $I = \dfrac{V}{R} = \dfrac{10 \text{ V}}{2\ \Omega} = 5$ A

(b) $I = \dfrac{V}{R} = \dfrac{10 \text{ V}}{4\ \Omega} = 2.5$ A

(c) As the resistance to current flow doubled, the current dropped to one-half.

(d) $V = I \times R = 2\,\text{A} \times 10\Omega = 20$ V

(e) $V = I \times R = 1\,\text{A} \times 10\Omega = 10$ V

Note we only need one-half the voltage to get one-half the current.

(f) $R = \dfrac{V}{I} = \dfrac{12 \text{ V}}{2\ \text{A}} = 6\ \Omega$

(g) $R = \dfrac{V}{I} = \dfrac{12 \text{ V}}{1\ \text{A}} = 12\ \Omega$

Note that, to reduce the current to 1 A, we needed to increase the circuit's resistance.

EXAMPLE F–2

Apply the values listed below to the circuit of Figure F–1 to determine the unknown quantity.

(a) $I = 2$ mA, $\qquad R = 4$ kΩ, $\qquad V = $ _____ ?
(b) $I = 6\,\mu$A, $\qquad R = 200$ kΩ, $\qquad V = $ _____ ?
(c) $I = 24\,\mu$A, $\qquad V = 12$ V, $\qquad R = $ _____ ?
(d) $I = 100$ mA, $\qquad V = 5$ V, $\qquad R = $ _____ ?
(e) $V = 5$ V, $\qquad R = 50$ kΩ, $\qquad I = $ _____ ?
(f) $V = 12$ V, $\qquad R = 600$ Ω, $\qquad I = $ _____ ?

Solution:

(a) $V = I \times R = 2$ mA $\times 4$ kΩ $= (2 \times 10^{-3}$ A$) \times (4 \times 10^{3}\,\Omega) = 8$ V

(b) $V = I \times R = 6\,\mu$A $\times 200$ kΩ $= (6 \times 10^{-6}$ A$) \times (200 \times 10^{3}\,\Omega) = 1.2$ V

(c) $R = \dfrac{V}{I} = \dfrac{12\ \text{V}}{24\,\mu\text{A}} = \dfrac{12\ \text{V}}{24 \times 10^{-6}\ \text{A}} = 500$ kΩ (or 0.5 MΩ)

(d) $R = \dfrac{V}{I} = \dfrac{5\ \text{V}}{100\ \text{mA}} = 50$ Ω

(e) $I = \dfrac{V}{R} = \dfrac{5\ \text{V}}{50\ \text{k}\Omega} = 100\,\mu$A

(f) $I = \dfrac{V}{R} = \dfrac{12\ \text{V}}{600\ \Omega} = 20$ mA

EXAMPLE F–3

A *series circuit* has two or more resistors end to end. The total resistance is equal to the sum of the individual resistances ($R_T = R_1 + R_2$). Also, the sum of the voltage drops across all resistors will equal the total supply voltage ($V_S = V_{R1} + V_{R2}$).

Find the current in the circuit (I), the voltage across R_1 (V_{R1}), and the voltage across R_2 (V_{R2}) in Figure F–3.

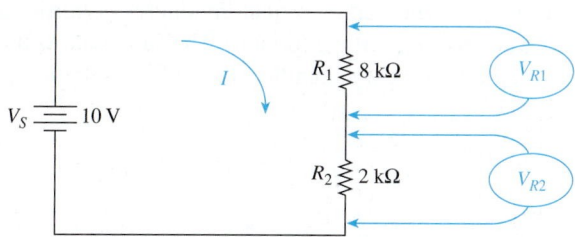

Figure F–3   A series circuit used to derive the voltage divider equation.

Solution:

$$R_T = 8\ \text{k}\Omega + 2\ \text{k}\Omega = 10\ \text{k}\Omega$$

$$I = \frac{10\ \text{V}}{10\ \text{k}\Omega} = 1\ \text{mA}$$

$$V_{R1} = 1\ \text{mA} \times 8\ \text{k}\Omega = 8\ \text{V}$$

$$V_{R2} = 1\ \text{mA} \times 2\ \text{k}\Omega = 2\ \text{V}$$

Check:

$$V_S = 8\,\text{V} + 2\,\text{V} = 10\,\text{V}$$

Notice that the voltage across any resistor in the series circuit is proportional to the size of the resistor. That fact is used in developing the *voltage-divider equation*:

$$V_{R2} = V_S \times \frac{R_2}{R_1 + R_2}$$

$$= 10\,\text{V} \times \frac{2\,\text{k}\Omega}{2\,\text{k}\Omega + 8\,\text{k}\Omega}$$

$$= 2\,\text{V}$$

EXAMPLE F–4

Use the voltage-divider equation to find V_{out} in Figure F–4. (V_{out} is the voltage from the point labeled V_{out} to the ground symbol.)

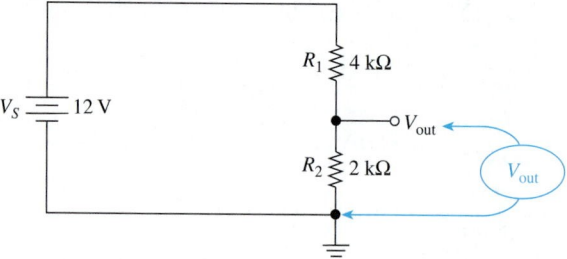

Figure F–4   Circuit used to calculate the output voltage (V_{out}) with respect to ground.

Solution:

$$V_{out} = 12\,\text{V} \times \frac{2\,\text{k}\Omega}{2\,\text{k}\Omega + 4\,\text{k}\Omega}$$

$$= 4\,\text{V}$$

EXAMPLE F–5

A *short circuit* occurs when an electrical conductor is purposely or inadvertently placed across a circuit component. The short causes the current to bypass the shorted component. Calculate V_{out} in Figure F–5.

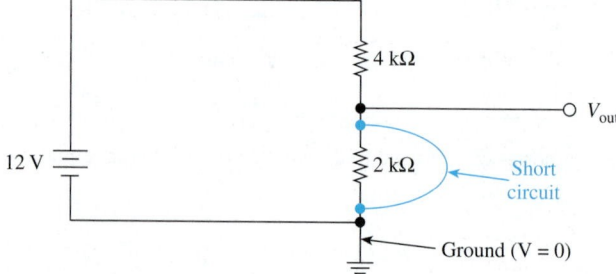

Figure F–5   A short circuit across the 2 kΩ resistor forces the output to zero V.

Solution: V_{out} is connected directly to ground; therefore, $V_{out} = 0$ V. All of the 12-V supply is dropped across the 4-kΩ resistor.

EXAMPLE F-6

Find V_{out} in Figure F-6.

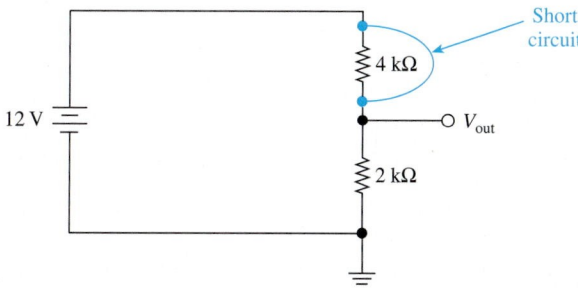

Figure F-6   A short across the top resistor causes the entire 12 volts to reach V_{out}.

Solution: V_{out} is connected directly to the top of the 12-V source battery. Therefore, $V_{out} = 12$ V. The entire 12-V source voltage is now across the 2-kΩ resistor.

EXAMPLE F-7

An *open circuit* is a break in a circuit. This can be done purposely by an electronic switching component, or it could be a circuit fault caused by a bad connector or burnt-out component. This break will cause the current to stop flowing to all components fed from that point, Calculate V_{out} in Figure F-7.

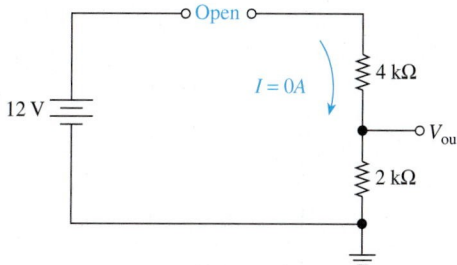

Figure F-7   An open circuit causes current to stop flowing.

Solution: Because $I = 0$ A,

$$V_{2\,k\Omega} = 0 \text{ A} \times 2 \text{ k}\Omega = 0 \text{ V}$$
$$V_{out} = V_{2\,k\Omega} = 0 \text{ V}$$

EXAMPLE F–8

Calculate V_{out} in Figure F–8.

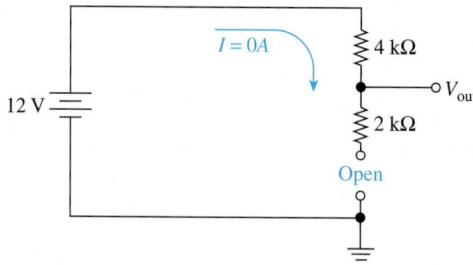

Figure F–8 An open circuit below the measurement point allows the entire supply voltage to reach the output.

Solution: Because $I = 0$ A,

$$V_{drop(4\ k\Omega)} = 0\ A \times 4\ k\Omega = 0\ V$$
$$V_{out} = 12\ V - V_{drop} = 12\ V$$

Note:

This is probably the hardest concept to understand. Another way to explain why the entire supply reaches V_{out} is to assume that an open circuit can be modeled by an extremely large resistance, let's say, 10 MΩ. If you then apply the voltage-divider equation to the circuit with 10 MΩ in place of the open, the calculation will be

$$V_{out} = 12\ V \times \frac{10{,}002{,}000}{10{,}002{,}000 + 4{,}000} = 11.995\ V$$

EXAMPLE F–9

The symbol for a battery is seldom drawn in schematic diagrams. Figure F–9 is an alternative schematic for a series circuit. Solve for V_{out}.

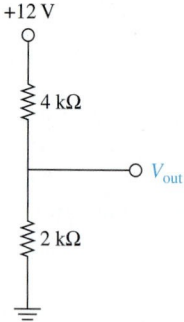

Figure F–9 Series circuit drawn without the battery symbol.

Solution:

$$V_{out} = 12\ V \times \frac{2\ k\Omega}{2\ k\Omega + 4\ k\Omega}$$
$$= 4\ V$$

A relay's contacts or a transistor's collector–emitter can be used to create opens and shorts. Figure F–10(a) uses a relay to short one resistor in a series circuit (relay operation is described in Section 2–6). Sketch the waveform at V_{out} in Figure F–10(a).

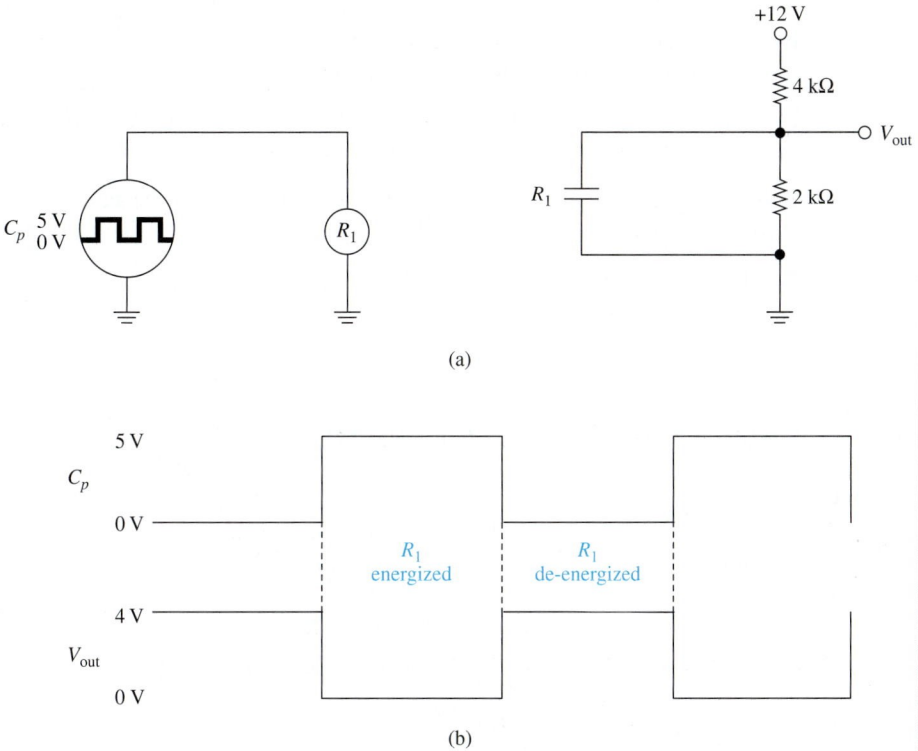

(a)

(b)

Figure F–10   Using a relay to intermittently short the 2kΩ resistor.

Solution: When the R_1 coil energizes, the R_1 contacts close, shorting the 2-kΩ resistor and making $V_{out} = 0$ V. When the coil is deenergized, the contacts are open, and V_{out} is found using the voltage-divider equation.

$$V_{out} = 12 \text{ V} \times \frac{2 \text{ k}\Omega}{2 \text{ k}\Omega + 4 \text{ k}\Omega}$$

$$= 4 \text{ V}$$

The clock oscillator (C_p) and V_{out} waveforms are given in Figure F–10(b).

Review Questions

F–1. What value of voltage will cause 6 mA to flow in Figure F–1 if $R = 2$ kΩ (3 V, 0.333 V, or 12 V)?

F–2. To increase the current in Figure F–1, the resistor value should be _____ (increased, decreased).

F–3. In a series voltage-divider circuit like Figure F–3, the larger resistor will have the larger voltage across it. True or false?

F–4. In Figure F–3, if R_1 is changed to 8 MΩ and R_2 is changed to 2Ω, V_{R2} will be close to _____ (0 V, 10 V).

F–5. If the supply voltage in Figure F–4 is increased to 18 V, V_{out} becomes _____ (6 V, 12 V).

F–6. A short circuit causes current to stop flowing in the part of the circuit not being shorted. True or false?

F–7. The current leaving the battery in Figure F–5 _____ (increases, decreases) if the short circuit is removed.

F–8. The short circuit in Figure F–6 causes V_{out} to become 12 V because the current through the 2-kΩ resistor becomes 0 A. True or false?

F–9. In Figure F–7, the voltage across the 2-kΩ resistor is the same as that across the 4-kΩ resistor. True or false?

F–10. If the 4-kΩ resistor in Figure F–8 is doubled, V_{out} will _____ (increase, decrease, remain the same)?

Answers to Review Questions

F–1. 12 V

F–2. Decreased

F–3. True

F–4. 0 V

F–5. 6 V

F–6. False

F–7. Decreases

F–8. False

F–9. True

F–10. Remain the same

Problems

F–1. Refer to Figure F–1 to solve for the unknown quantities in the following table. (*Example:* For part (A), calculate the resistance if $V = 12$ V and $I = 4$ A.)

	Voltage	Current	Resistance
A	12 V	4 A	
B	8 V	2 A	
C	100 V		20 Ω
D	6 V		2 Ω
E		6 A	2 Ω
F		0.5 A	5 Ω

F–2. Repeat Problem F–1 for the following table.

	Voltage	Current	Resistance
A	12 mV	3 μA	
B	6 V	2 mA	
C	100 mV		20 kΩ
D	8 V		2 MΩ
E		6 mA	3 kΩ
F		0.5 A	5 kΩ

F–3. Refer to Figure F–4 to solve for the unknown quantities in the following table.

	V_S	R_1	R_2	V_{out}
A	18 V	6 kΩ	3 kΩ	
B	18 V	3 kΩ	6 kΩ	
C	12 V	20 kΩ	20 kΩ	
D	6 V	1 kΩ	100 kΩ	

F–4. F–4 Refer to the original Figure F–4 to solve for V_{out} given the following open- and short-circuit conditions.

		V_{out}
A	Short R_1	
B	Short R_2	
C	Open R_1	
D	Open R_2	

Answers to Problems

F–1.
  (a) 3 Ω
  (b) 4 Ω
  (c) 5 A
  (d) 3 A
  (e) 12 V
  (f) 2.5 V

F–2.
  (a) 4 kΩ
  (b) 3 kΩ
  (c) 5 μA
  (d) 4 μA
  (e) 18 V
  (f) 2.5 mV

F–3.
  (a) 6 V
  (b) 12 V
  (c) 6 V
  (d) 5.94 V

F–4.
  (a) 12 V
  (b) 0 V
  (c) 0 V
  (d) 12 V

Appendix G

Schematic Diagrams for Chapter-End Problems

Schematic	Description
4096/4196 Control Card (Sheet 1 of 2)	An 8031 microcontroller-based circuit used to control the front panel of an analog frequency filter.
4096/4196 Control Card (Sheet 2 of 2)	The power supply regulators and I/O buffers for the front panel controller.
HD11D0 Master Board	A 68HC11 microcontroller-based circuit used to interface to a PC and drive a multiline LCD display.
Watchdog Timer	Control Circuit for an AT&T printer used to ensure a safe power-down of the internal printer circuitry.

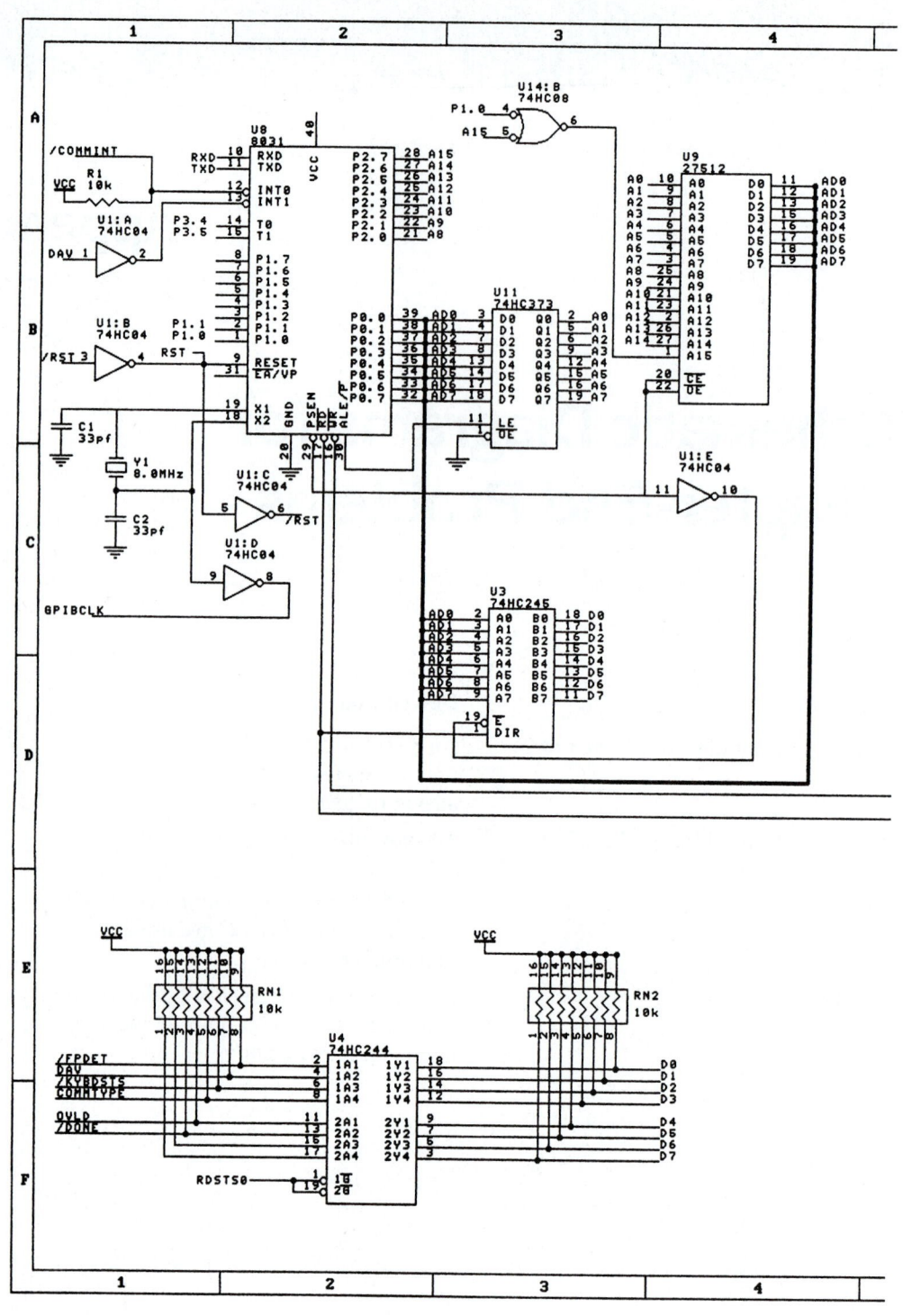

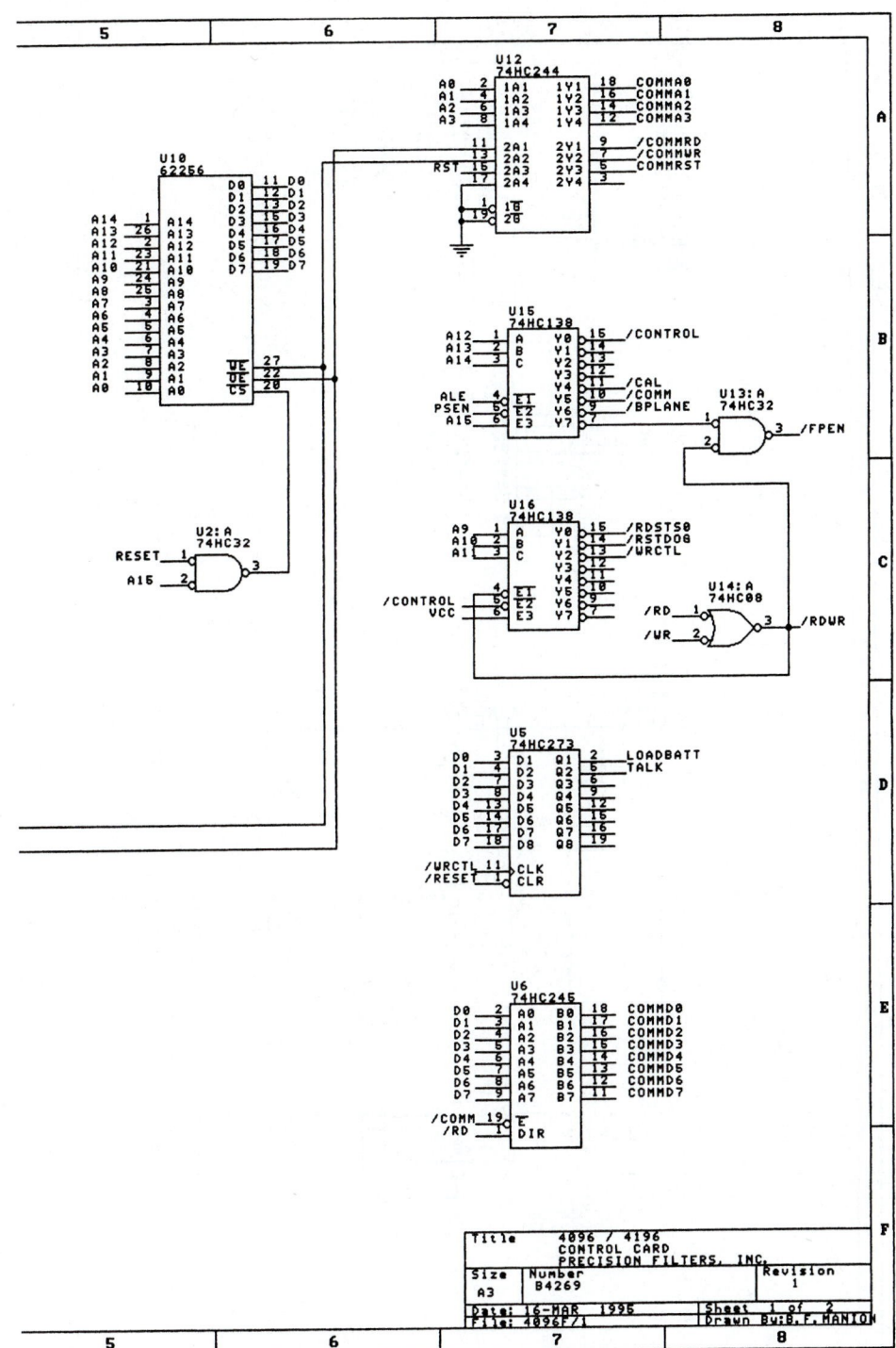

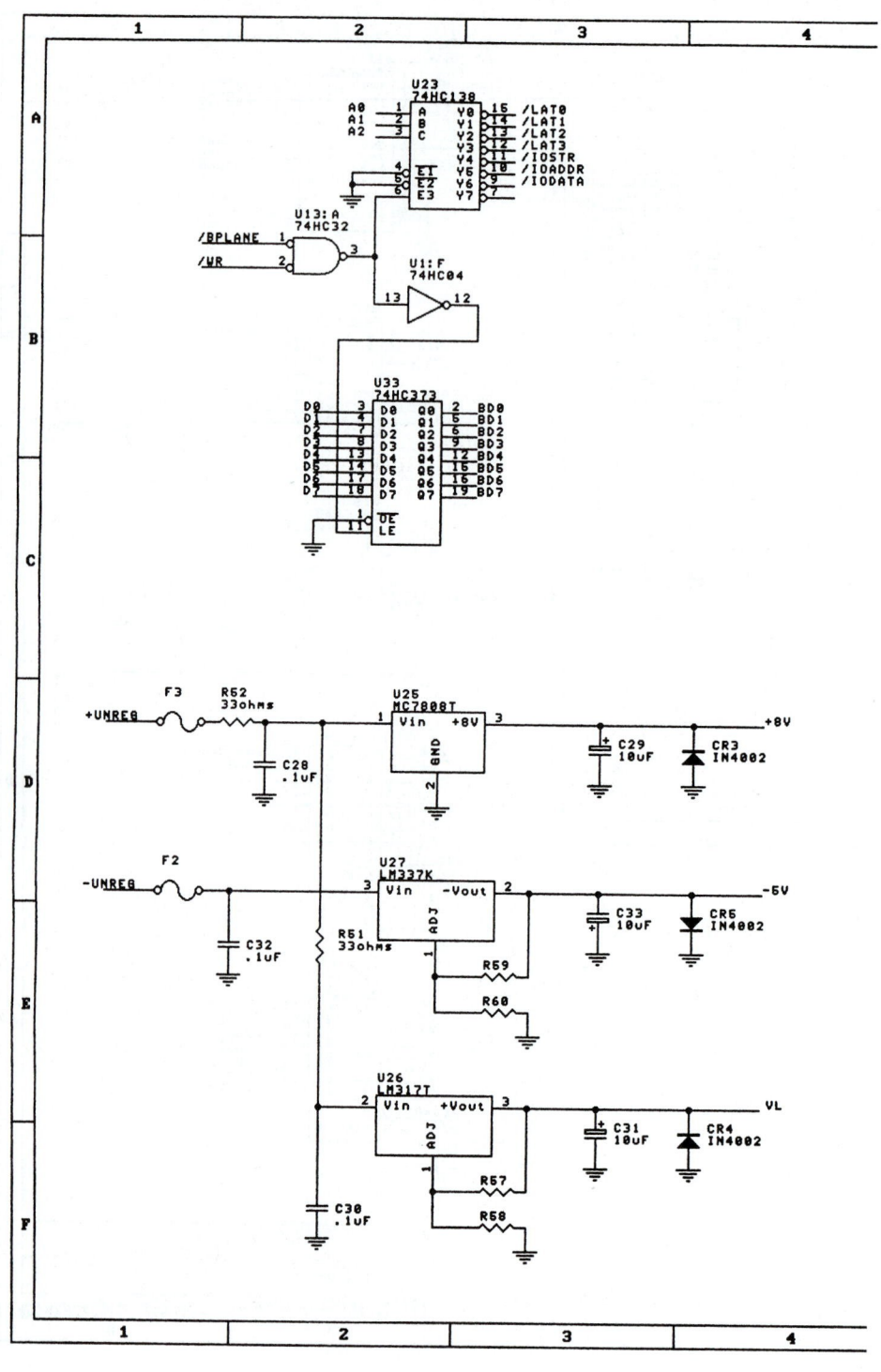

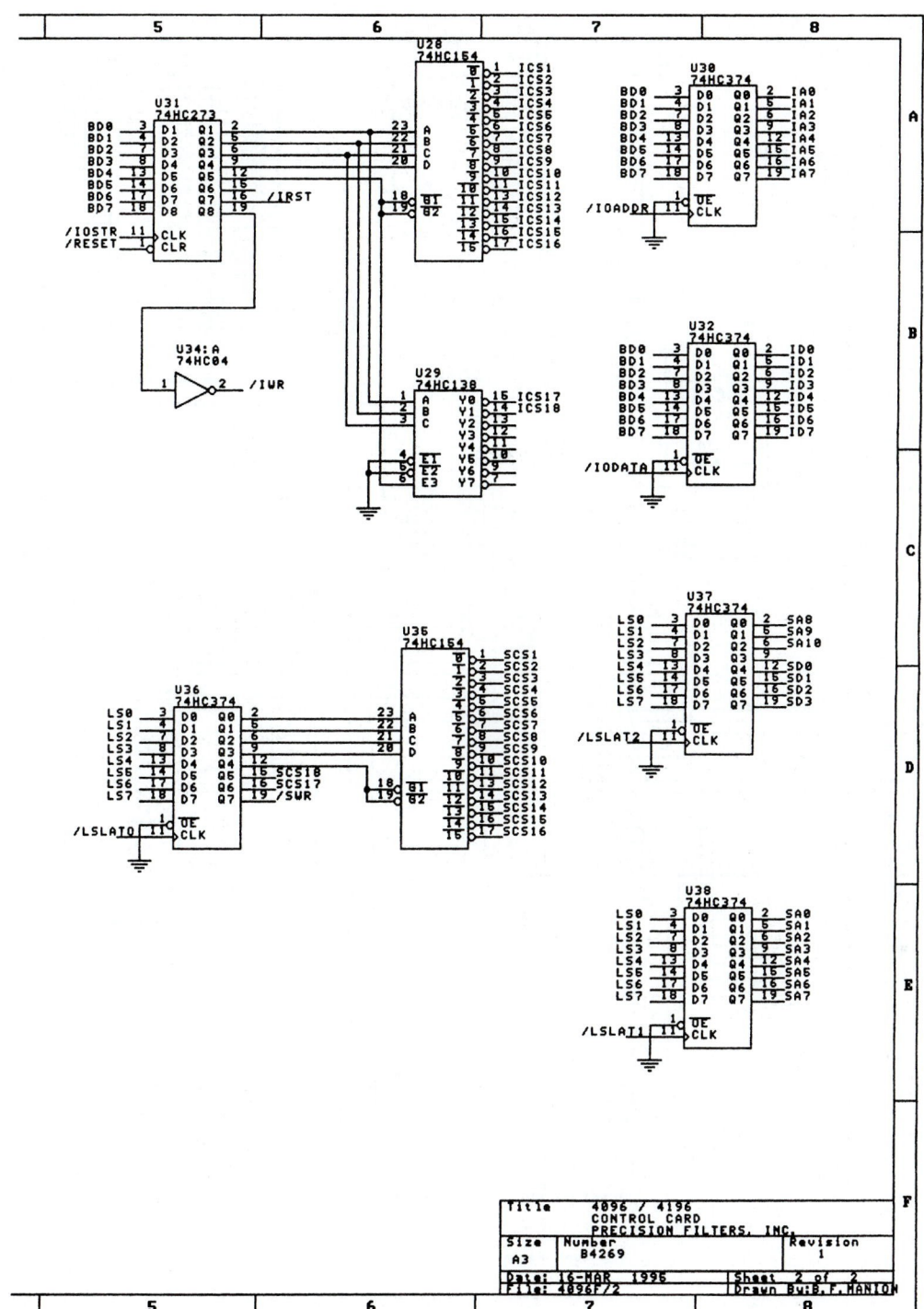

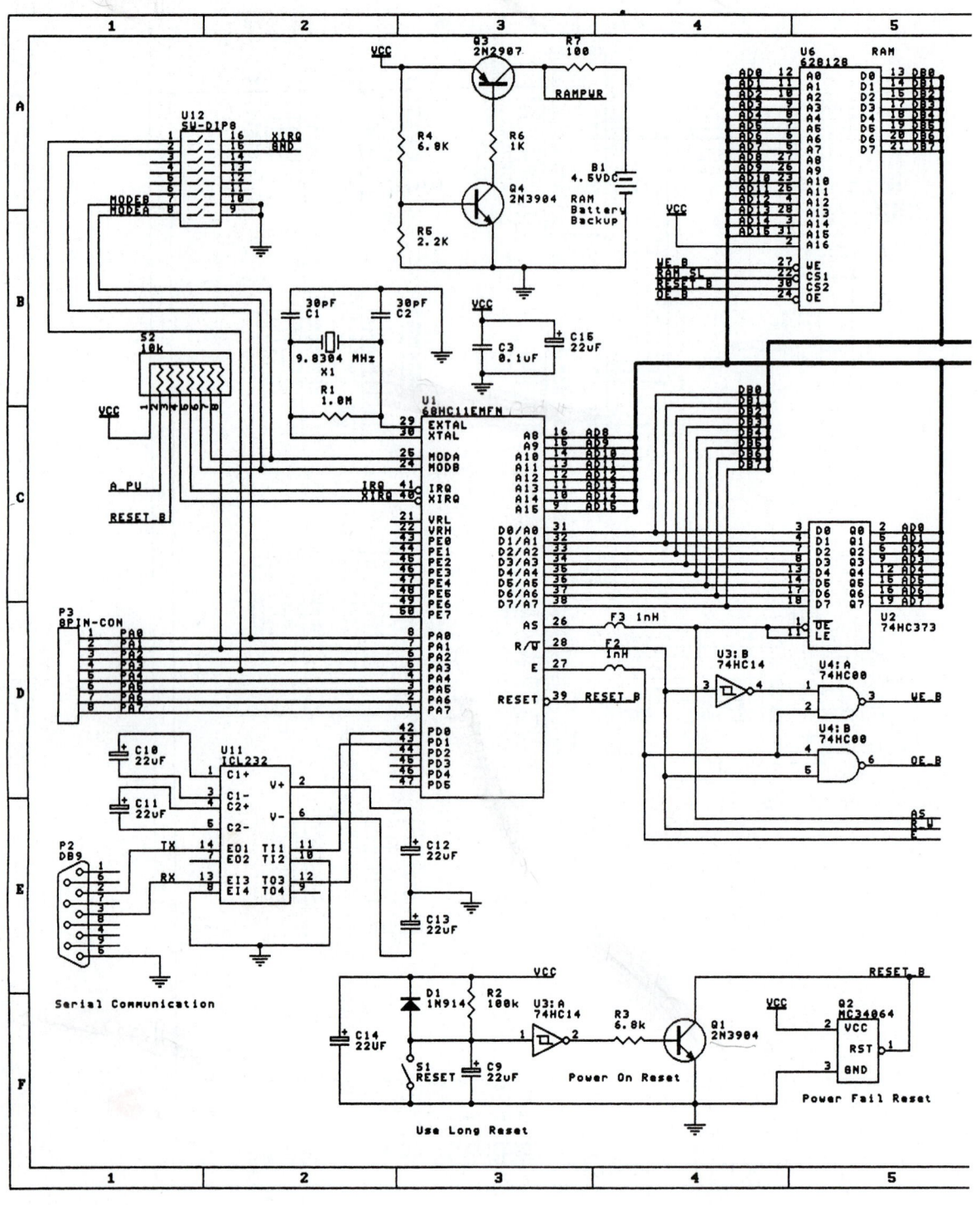

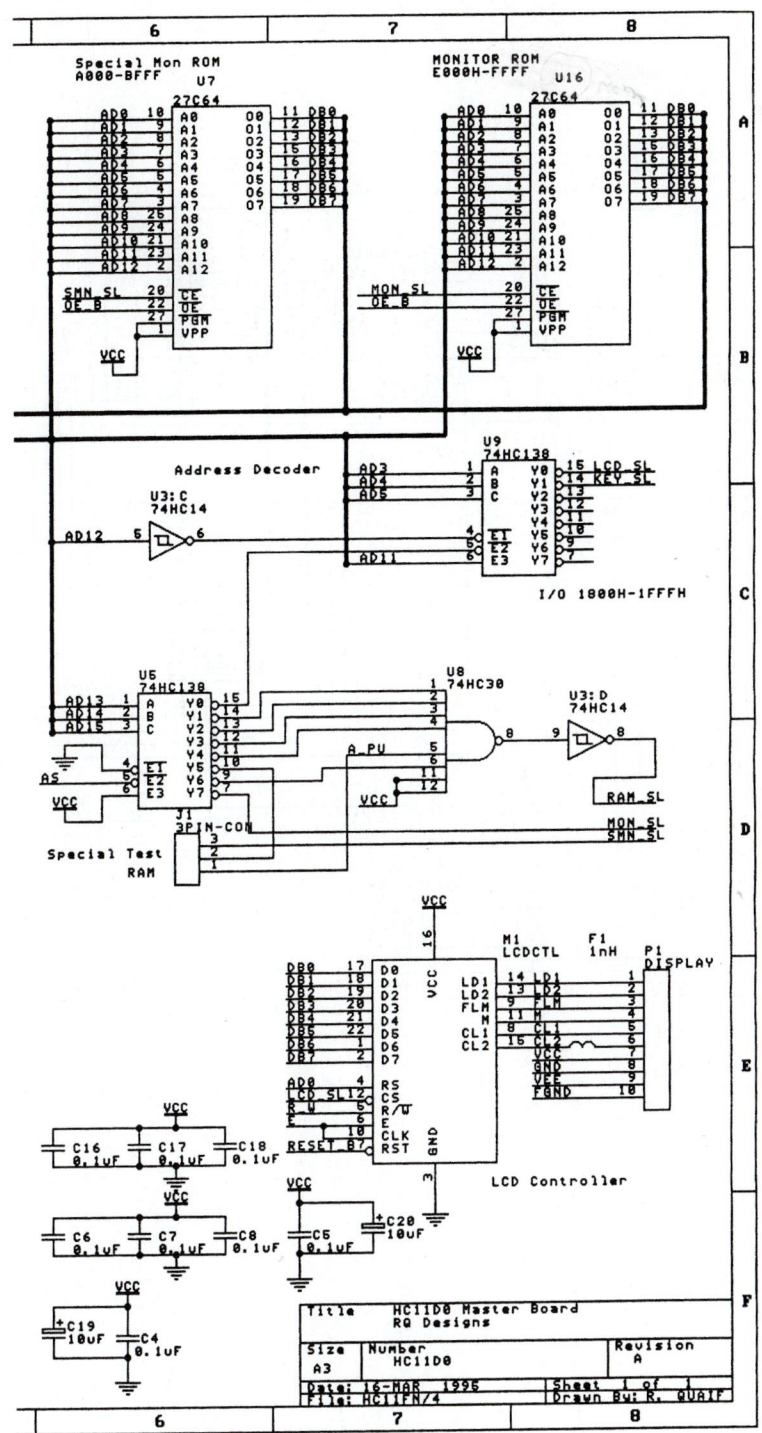

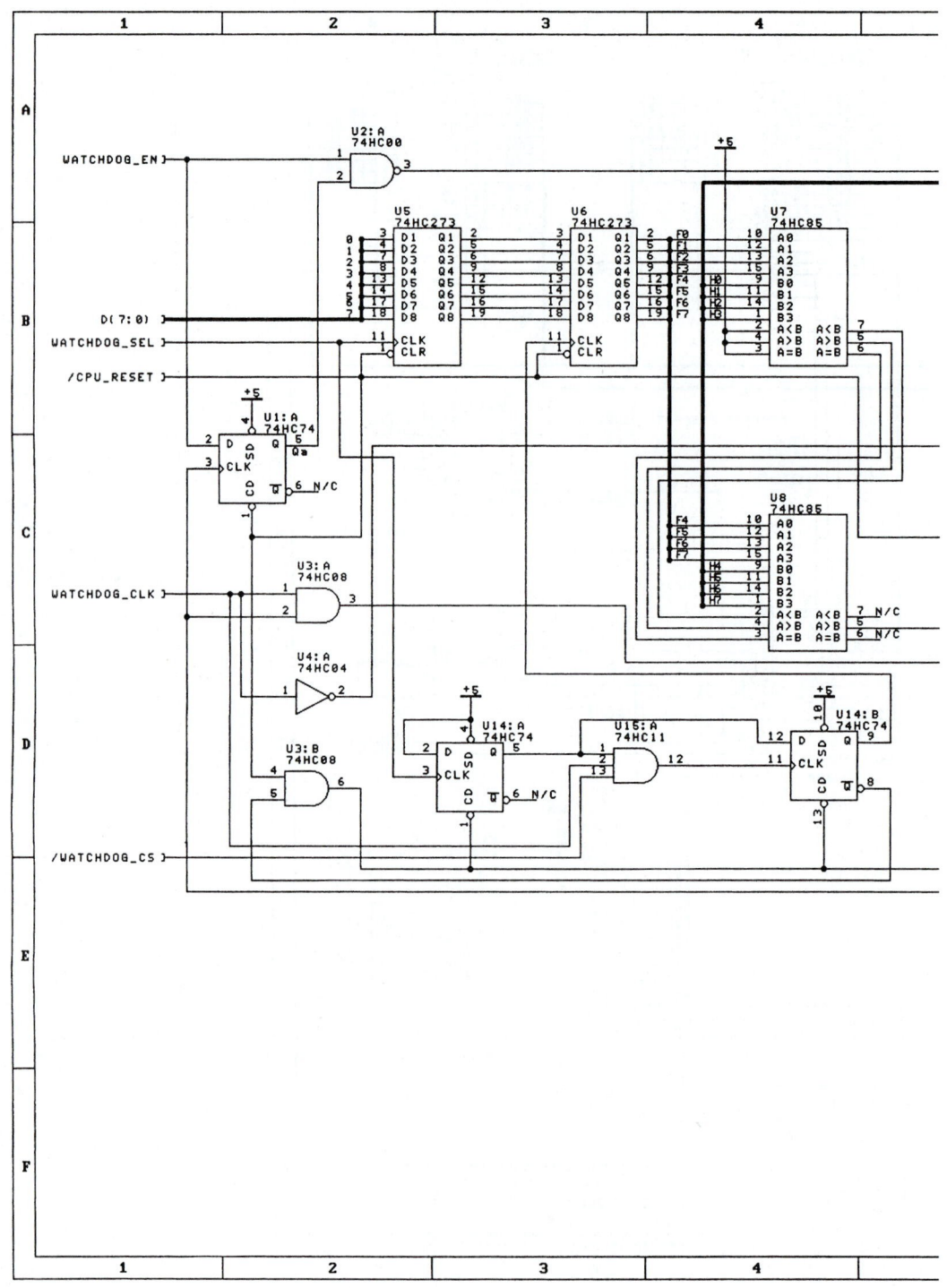

APPENDIX G | SCHEMATIC DIAGRAMS FOR CHAPTER-END PROBLEMS

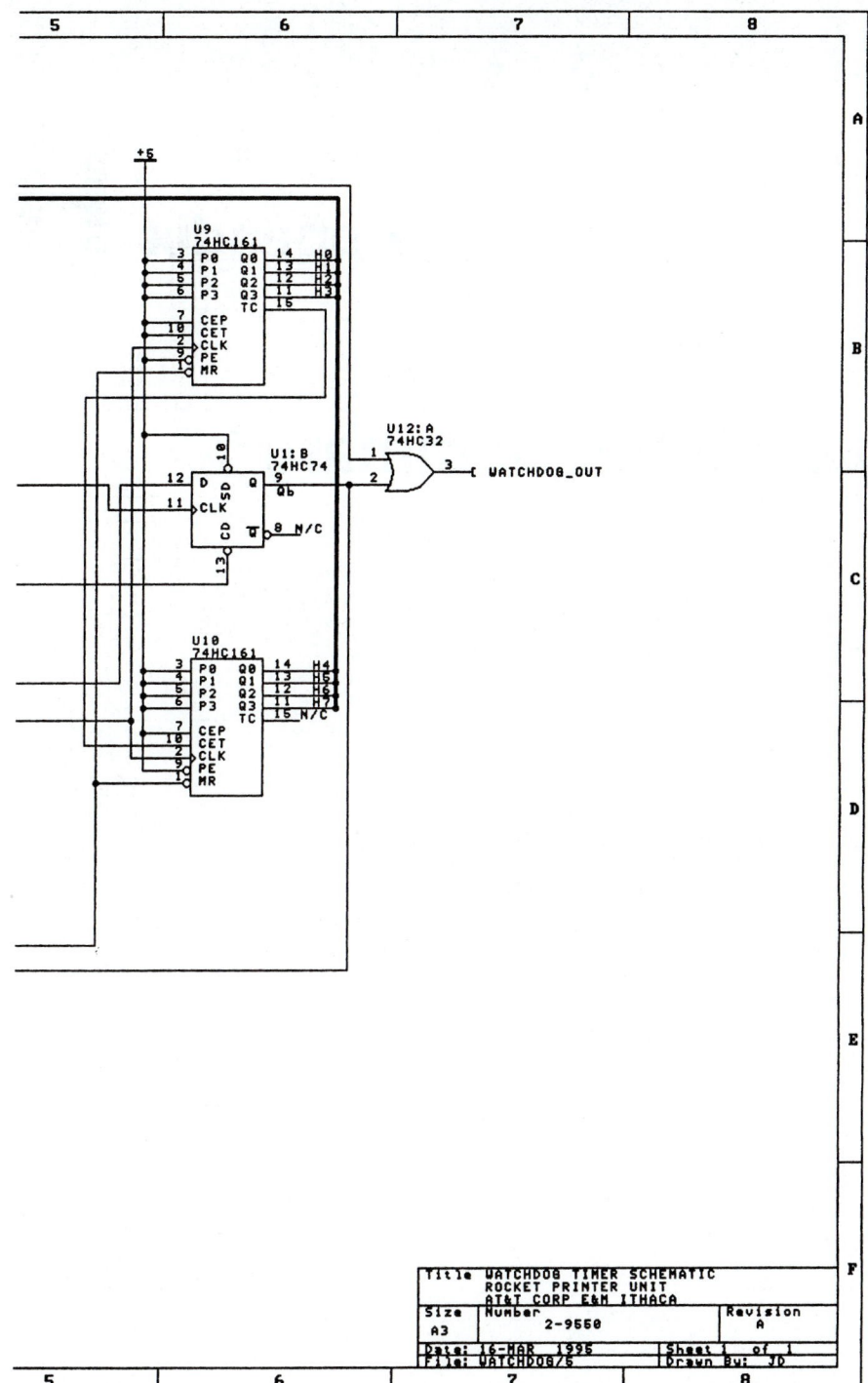

Appendix H

8051 Instruction Set Summary*

*Courtesy of Intel Corporation.

MCS®-51 INSTRUCTION SET

8051 Instruction Set Summary

Interrupt Response Time: Refer to Hardware Description Chapter.

Instructions that Affect Flag Settings(1)

Instruction	Flag			Instruction	Flag		
	C	OV	AC		C	OV	AC
ADD	X	X	X	CLR C	O		
ADDC	X	X	X	CPL C	X		
SUBB	X	X	X	ANL C,bit	X		
MUL	O	X		ANL C,/bit	X		
DIV	O	X		ORL C,bit	X		
DA	X			ORL C,bit	X		
RRC	X			MOV C,bit	X		
RLC	X			CJNE	X		
SETB C	1						

(1)Note that operations on SFR byte address 208 or bit addresses 209-215 (i.e., the PSW or bits in the PSW) will also affect flag settings.

Note on instruction set and addressing modes:

Rn — Register R7–R0 of the currently selected Register Bank.

direct — 8-bit internal data location's address. This could be an Internal Data RAM location (0–127) or a SFR [i.e., I/O port, control register, status register, etc. (128–255)].

@Ri — 8-bit internal data RAM location (0–255) addressed indirectly through register R1 or R0.

#data — 8-bit constant included in instruction.

#data 16 — 16-bit constant included in instruction.

addr 16 — 16-bit destination address. Used by LCALL & LJMP. A branch can be anywhere within the 64K-byte Program Memory address space.

addr 11 — 11-bit destination address. Used by ACALL & AJMP. The branch will be within the same 2K-byte page of program memory as the first byte of the following instruction.

rel — Signed (two's complement) 8-bit offset byte. Used by SJMP and all conditional jumps. Range is −128 to +127 bytes relative to first byte of the following instruction.

bit — Direct Addressed bit in Internal Data RAM or Special Function Register.

* — New operation not provided by 8048AH/8049AH.

Mnemonic		Description	Byte	Oscillator Period
ARITHMETIC OPERATIONS				
ADD	A,Rn	Add register to Accumulator	1	12
ADD	A,direct	Add direct byte to Accumulator	2	12
ADD	A,@Ri	Add indirect RAM to Accumulator	1	12
ADD	A,#data	Add immediate data to Accumulator	2	12
ADDC	A,Rn	Add register to Accumulator with Carry	1	12
ADDC	A,direct	Add direct byte to Accumulator with Carry	2	12
ADDC	A,@Ri	Add indirect RAM to Accumulator with Carry	1	12
ADDC	A,#data	Add immediate data to Acc with Carry	2	12
SUBB	A,Rn	Subtract Register from Acc with borrow	1	12
SUBB	A,direct	Subtract direct byte from Acc with borrow	2	12
SUBB	A,@Ri	Subtract indirect RAM from ACC with borrow	1	12
SUBB	A,#data	Subtract immediate data from Acc with borrow	2	12
INC	A	Increment Accumulator	1	12
INC	Rn	Increment register	1	12
INC	direct	Increment direct byte	2	12
INC	@Ri	Increment direct RAM	1	12
DEC	A	Decrement Accumulator	1	12
DEC	Rn	Decrement Register	1	12
DEC	direct	Decrement direct byte	2	12
DEC	@Ri	Decrement indirect RAM	1	12

All mnemonics copyrighted © Intel Corporation 1980

8051 Instruction Set Summary (Continued)

Mnemonic		Description	Byte	Oscillator Period
ARITHMETIC OPERATIONS (Continued)				
INC	DPTR	Increment Data Pointer	1	24
MUL	AB	Multiply A & B	1	48
DIV	AB	Divide A by B	1	48
DA	A	Decimal Adjust Accumulator	1	12
LOGICAL OPERATIONS				
ANL	A,Rn	AND Register to Accumulator	1	12
ANL	A,direct	AND direct byte to Accumulator	2	12
ANL	A,@Ri	AND indirect RAM to Accumulator	1	12
ANL	A,#data	AND immediate data to Accumulator	2	12
ANL	direct,A	AND Accumulator to direct byte	2	12
ANL	direct,#data	AND immediate data to direct byte	3	24
ORL	A,Rn	OR register to Accumulator	1	12
ORL	A,direct	OR direct byte to Accumulator	2	12
ORL	A,@Ri	OR indirect RAM to Accumulator	1	12
ORL	A,#data	OR immediate data to Accumulator	2	12
ORL	direct,A	OR Accumulator to direct byte	2	12
ORL	direct,#data	OR immediate data to direct byte	3	24
XRL	A,Rn	Exclusive-OR register to Accumulator	1	12
XRL	A,direct	Exclusive-OR direct byte to Accumulator	2	12
XRL	A,@Ri	Exclusive-OR indirect RAM to Accumulator	1	12
XRL	A,#data	Exclusive-OR immediate data to Accumulator	2	12
XRL	direct,A	Exclusive-OR Accumulator to direct byte	2	12
XRL	direct,#data	Exclusive-OR immediate data to direct byte	3	24
CLR	A	Clear Accumulator	1	12
CPL	A	Complement Accumulator	1	12

Mnemonic		Description	Byte	Oscillator Period
LOGICAL OPERATIONS (Continued)				
RL	A	Rotate Accumulator Left	1	12
RLC	A	Rotate Accumulator Left through the Carry	1	12
RR	A	Rotate Accumulator Right	1	12
RRC	A	Rotate Accumulator Right through the Carry	1	12
SWAP	A	Swap nibbles within the Accumulator	1	12
DATA TRANSFER				
MOV	A,Rn	Move register to Accumulator	1	12
MOV	A,direct	Move direct byte to Accumulator	2	12
MOV	A,@Ri	Move indirect RAM to Accumulator	1	12
MOV	A,#data	Move immediate data to Accumulator	2	12
MOV	Rn,A	Move Accumulator to register	1	12
MOV	Rn,direct	Move direct byte to register	2	24
MOV	Rn,#data	Move immediate data to register	2	12
MOV	direct,A	Move Accumulator to direct byte	2	12
MOV	direct,Rn	Move register to direct byte	2	24
MOV	direct,direct	Move direct byte to direct	3	24
MOV	direct,@Ri	Move indirect RAM to direct byte	2	24
MOV	direct,#data	Move immediate data to direct byte	3	24
MOV	@Ri,A	Move Accumulator to indirect RAM	1	12

All mnemonics copyrighted © Intel Corporation 1980

8051 Instruction Set Summary (Continued)

Mnemonic		Description	Byte	Oscillator Period
DATA TRANSFER (Continued)				
MOV	@Ri,direct	Move direct byte to indirect RAM	2	24
MOV	@Ri,#data	Move immediate data to indirect RAM	2	12
MOV	DPTR,#data16	Load Data Pointer with a 16-bit constant	3	24
MOVC	A,@A+DPTR	Move Code byte relative to DPTR to Acc	1	24
MOVC	A,@A+PC	Move Code byte relative to PC to Acc	1	24
MOVX	A,@Ri	Move External RAM (8-bit addr) to Acc	1	24
MOVX	A,@DPTR	Move External RAM (16-bit addr) to Acc	1	24
MOVX	@Ri,A	Move Acc to External RAM (8-bit addr)	1	24
MOVX	@DPTR,A	Move Acc to External RAM (16-bit addr)	1	24
PUSH	direct	Push direct byte onto stack	2	24
POP	direct	Pop direct byte from stack	2	24
XCH	A,Rn	Exchange register with Accumulator	1	12
XCH	A,direct	Exchange direct byte with Accumulator	2	12
XCH	A,@Ri	Exchange indirect RAM with Accumulator	1	12
XCHD	A,@Ri	Exchange low-order Digit indirect RAM with Acc	1	12

Mnemonic		Description	Byte	Oscillator Period
BOOLEAN VARIABLE MANIPULATION				
CLR	C	Clear Carry	1	12
CLR	bit	Clear direct bit	2	12
SETB	C	Set Carry	1	12
SETB	bit	Set direct bit	2	12
CPL	C	Complement Carry	1	12
CPL	bit	Complement direct bit	2	12
ANL	C,bit	AND direct bit to CARRY	2	24
ANL	C,/bit	AND complement of direct bit to Carry	2	24
ORL	C,bit	OR direct bit to Carry	2	24
ORL	C,/bit	OR complement of direct bit to Carry	2	24
MOV	C,bit	Move direct bit to Carry	2	12
MOV	bit,C	Move Carry to direct bit	2	24
JC	rel	Jump if Carry is set	2	24
JNC	rel	Jump if Carry not set	2	24
JB	bit,rel	Jump if direct Bit is set	3	24
JNB	bit,rel	Jump if direct Bit is Not set	3	24
JBC	bit,rel	Jump if direct Bit is set & clear bit	3	24
PROGRAM BRANCHING				
ACALL	addr11	Absolute Subroutine Call	2	24
LCALL	addr16	Long Subroutine Call	3	24
RET		Return from Subroutine	1	24
RETI		Return from interrupt	1	24
AJMP	addr11	Absolute Jump	2	24
LJMP	addr16	Long Jump	3	24
SJMP	rel	Short Jump (relative addr)	2	24

All mnemonics copyrighted © Intel Corporation 1980

8051 Instruction Set Summary (Continued)

Mnemonic		Description	Byte	Oscillator Period
PROGRAM BRANCHING (Continued)				
JMP	@A + DPTR	Jump indirect relative to the DPTR	1	24
JZ	rel	Jump if Accumulator is Zero	2	24
JNZ	rel	Jump if Accumulator is Not Zero	2	24
CJNE	A,direct,rel	Compare direct byte to Acc and Jump if Not Equal	3	24
CJNE	A, # data,rel	Compare immediate to Acc and Jump if Not Equal	3	24

Mnemonic		Description	Byte	Oscillator Period
PROGRAM BRANCHING (Continued)				
CJNE	Rn, # data,rel	Compare immediate to register and Jump if Not Equal	3	24
CJNE	@Ri, # data,rel	Compare immediate to indirect and Jump if Not Equal	3	24
DJNZ	Rn,rel	Decrement register and Jump if Not Zero	2	24
DJNZ	direct,rel	Decrement direct byte and Jump if Not Zero	3	24
NOP		No Operation	1	12

All mnemonics copyrighted © Intel Corporation 1980

Index

NOTE: Page number in **boldface** type indicates end-of-chapter glossary definition for the term.

Supplementary Index of ICs

This is an index of the integrated circuits (ICs) used in this book. The page numbers indicate where the IC is first discussed. Page numbers in **boldface** type indicate pages containing a data sheet for the device.